The Behavior of Systems in the Space Environment

NATO ASI Series

Advanced Science Institutes Series

A Series presenting the results of activities sponsored by the NATO Science Committee, which aims at the dissemination of advanced scientific and technological knowledge, with a view to strengthening links between scientific communities.

The Series is published by an international board of publishers in conjunction with the NATO Scientific Affairs Division

A Life Sciences **B Physics**	Plenum Publishing Corporation London and New York
C Mathematical and Physical Sciences **D Behavioural and Social Sciences** **E Applied Sciences**	Kluwer Academic Publishers Dordrecht, Boston and London
F Computer and Systems Sciences **G Ecological Sciences** **H Cell Biology** **I Global Environmental Change**	Springer-Verlag Berlin, Heidelberg, New York, London, Paris and Tokyo

NATO-PCO-DATA BASE

The electronic index to the NATO ASI Series provides full bibliographical references (with keywords and/or abstracts) to more than 30000 contributions from international scientists published in all sections of the NATO ASI Series.
Access to the NATO-PCO-DATA BASE is possible in two ways:

– via online FILE 128 (NATO-PCO-DATA BASE) hosted by ESRIN, Via Galileo Galilei, I-00044 Frascati, Italy.

– via CD-ROM "NATO-PCO-DATA BASE" with user-friendly retrieval software in English, French and German (© WTV GmbH and DATAWARE Technologies Inc. 1989).

The CD-ROM can be ordered through any member of the Board of Publishers or through NATO-PCO, Overijse, Belgium.

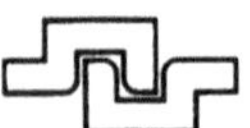

Series E: Applied Sciences - Vol. 245

The Behavior of Systems in the Space Environment

edited by

Robert N. DeWitt

U.S. Department of Energy,
Washington, D.C., U.S.A.

Dwight Duston

Strategic Defense Initiative Organization,
The Pentagon, Washington, D.C., U.S.A.

and

Anthony K. Hyder

University of Notre Dame,
Notre Dame, IN, U.S.A.

Springer-Science+Business Media, B.V.

Proceedings of the NATO Advanced Study Institute on
The Behavior of Systems in the Space Environment
Pitlochry, Scotland
July 7-19, 1991

Library of Congress Cataloging-in-Publication Data

NATO Advanced Study Institute on the Behavior of Systems in the Space Environment (1991 : Pitlochry, Scotland)
The behavior of systems in the space environment / edited by Robert N. DeWitt, Dwight Duston, and Anthony K. Hyder.
p. cm. -- (NATO ASI series. Series E, Applied sciences ; vol. 245)
"Proceedings of the NATO Advanced Study Institute on The Behavior of Systems in the Space Environment, Pitlochry, Scotland, July 7-19, 1991."
"Published in cooperation with NATO Scientific Affairs Division."
Includes index.

1. Space environment--Congresses. 2. Astronautical instruments--Defects--Congresses. 3. Space vehicles--Materials--Deterioration. I. DeWitt, Robert N. II. Duston, Dwight. III. Hyder, A. K. (Anthony K.) IV. North Atlantic Treaty Organization. Scientific Affiars Division. V. Title. VI. Series: NATO ASI series. Series E, Applied sciences ; no. 245.
TL1489.N38 1991
620'.419--dc20 93-26845

DOI 10.1007/978-94-011-2048-7

Printed on acid-free paper

CONTENTS

OVERVIEW OF THE SPACE ENVIRONMENT

ENVIRONMENTAL INTERACTIONS WITH MATERIALS AND COMPONENTS

ENVIRONMENTAL INTERACTIONS WITH SPACE SYSTEMS

APPENDICES

PREFACE

A NATO Advanced Study Institute (ASI) on the Behavior of Systems in the Space Environment was held at the Atholl Palace Hotel, Pitlochry, Perthshire, Scotland, from July 7 through July 19, 1991. This publication is the Proceedings of the Institute.

The NATO Advanced Study Institute Program of the NATO Science Committee is a unique and valuable forum, under whose auspices almost one thousand international tutorial meetings have been held since the inception of the program in 1959. The ASI is intended to be primarily a high-level teaching activity at which a carefully defined subject is presented in a systematic and coherently structured program. The subject is treated in considerable depth by lecturers eminent in their field and of international standing. The subject is presented to other scientists who either will already have specialized in the field or possess an advanced general background. The ASI is aimed at approximately the post-doctoral level.

This ASI emphasized the basic physics of the space environment and the engineering aspects of the environment's interactions with spacecraft. The objective of the ASI was to bring together the latest data characterizing the space environment and the analyses of the interactions of spacecraft systems operating in that environment. The timeliness of this ASI is highlighted by the convergence of two major science and technology endeavors. The first is the emerging perspective of the space environment that has resulted from the vast quantity of new data on space physics that has been obtained recently. These data have provided a revised understanding of the near-earth space environment as well as the interplanetary regions. The second is related to the worldwide renewal of interest in extended space operations for military, commercial, and scientific missions. The ability of the spacecraft engineers to properly design and build spacecraft to accommodate the interactions of their systems with the space environment will pace the future uses of space.

The theme of the Institute was the enhancement of scientific communication and exchange among academic, industrial, and government laboratory groups having a common interest in the behavior of systems in the environment of space. In line with the focus of the Institute, the program was organized into three main sessions: an introduction to and historical perspective of the space environment; the physics of the interactions of materials and components with the space environment; and, lastly, the engineering of systems for operations in the environment.

The first session opened with an historical overview of the exploration of the space environment. Early scientific endeavors to understand the upper atmosphere, the Earth's magnetic field, solar radiation, the ionosphere, and Van Allen Belts were presented by a pioneer in the field. This was followed by an overview of the environment from a systems perspective with an emphasis on the environmental factors affecting system performance and lifetime. Individual lectures were then given to detail and update the data and understanding of the roles of the Sun, magnetosphere, upper atmosphere, ionosphere, cosmic and energetic particle radiation, and micrometeoroid and debris fluxes. The dynamics and cyclic variations of these sources and their features were discussed as well as the attempts to model them.

To understand the fundamentals of the behavior of systems in the space environment, lectures were organized in the second major session on the interaction of the space environment with materials and system components. As the multi-disciplinary character of the interacting environment developed, the lectures began to focus on the total, synergistic effect of the environment on systems: the erosion of spacecraft materials by atomic oxygen, Shuttle experiments including the Long Duration Exposure Facility (LDEF), synergistic effects due to VUV radiation and a detailed description of the local neutral spacecraft environment due to outgasing and plume impingements. Detailed modeling of particle emission by surfaces, low-density wake phenomena and anomalies due to spacecraft orientation, and radiation damage to surfaces and materials were also presented. The session concluded with lectures on particle transport simulation methods useful in the study of shielding effects, spacecraft charging from the space plasma, and several other plasma phenomena affecting spacecraft operations.

The final session included lectures covering specific space systems and missions. The experimental, theoretical, and computational research efforts that comprise the SPEAR (Space Power Experiments Aboard Rockets) program were presented together with a comprehensive review of the latest results from LDEF, a detailed analysis of the Oedipus-A experiment (a rocket-launched double payload connected by an electrically conducting tether, 960 m long), and an analysis of the CRRES mission. Finally, the limitations imposed by the environment on human performance and on future commercial activity were reviewed.

The initial pace of the Institute allowed ample time for informal discussion sessions to be organized and scheduled by the participants and lecturers with the encouragement and assistance of the Directors. As the Institute progressed, the interest and demand for these additional sessions grew and they consumed much of the unscheduled time. A departure from the normal format of an ASI which greatly enhanced this Institute was the opportunity for participants to contribute poster papers in two evening sessions. The posters remained in place during virtually the entire ASI and served as a catalyst for technical interaction among the participants during all of the breaks. Thirteen of these poster papers have been selected by the editors for inclusion in the proceedings and are presented in the appendix.

The Institute was attended by one hundred participants and lecturers representing Canada, Finland, France, Germany, Greece, India, Ireland, the Netherlands, the Republic of China, Turkey, the United Kingdom, and the

United States. A distinguished faculty of lecturers was assembled and the technical program organized with the generous and very capable assistance of an Advisory Committee composed of Dr. Jean-Claude Mandeville (France), Dr. Ranty Liang (USA), Professor J. A.M. McDonnell (UK), Dr. Gordon L. Wrenn (UK), Dr. Carolyn Purvis (USA), and Professor R. C. Tennyson (Canada). The Institute was organized and directed by Dr. Robert N. DeWitt (USA), Dr. Dwight Duston (USA), and Dr. Anthony K. Hyder (USA).

The value to be gained from any ASI depends on the faculty - the lecturers who devote so much of their time and talents to make an Institute successful. As the reader of these proceedings will see, this ASI was particularly honored with an exceptional group of lecturers to whom the organizers and participants offer their deep appreciation.

We are grateful also to a number of organizations for providing the financial assistance that made the Institute possible. Foremost is the NATO Scientific Affairs Division which provided not only important financial support for the Institute, but equally important organizational and moral support. In addition, the following institutions made significant contributions: Auburn University, Naval Surface Warfare Center, Strategic Defense Initiative Organization, Air Force Office of Scientific Research, Air Force Phillips Laboratory, EOARD in London, Industrial Technology Research Institute, Naval Research Laboratory, Jet Propulsion Laboratory, and the NASA Centers for the Commercial Development of Space.

It is a pleasure to acknowledge the work of the Institute administrative and secretarial staff whose tireless efforts were critical to the success of the Institute: Don Parrotte and Jerri LaHaie of the Continuing Education Office of Auburn University, Alice Gehl of the Strategic Defense Initiative Organization, Maria Baxter of the Naval Surface Warfare Center, and Alison Rook of the University of Kent.

We would also like to thank Dougal Spaven, the general manager of the hotel hosting the ASI; his wife, Sally, whose extensive knowledge of Scottish lore made the excursions as informative as the Institute; and all of the staff at the Atholl Palace Hotel for a truly enjoyable and memorable two weeks in the Scottish Highlands. Their warm friendliness made all the attendees feel most welcomed and the superb accommodations, meals, service, and meeting facilities of the Atholl Palace made the venue ideal for an ASI. To the Tenth Duke of Atholl, our thanks for the use of Blair Castle for a magnificent banquet and for an evening that will long remain in the memories of all who were there.

A very special acknowledgment goes to Susie M. Anderson and her staff at the EG&G Washington Analytical Service Center in Dahlgren, Virginia. They undertook the very challenging task of centrally retyping the lecturers' manuscripts and of producing a camera-ready document for Kluwer Academic Publishers. Thank you all for your long hours and hard work.

One of the editors (AKH) would like to acknowledge especially the Space Power Institute and Auburn University for the encouragement they have given before, during, and long after the time of the ASI. Much of the work of organizing the ASI was done while this editor was on the faculty of Auburn and the continued support of the University is greatly appreciated. To Paul Parks, Frank Rose, Ray Askew, and Cal Johnson: thank you!

And, finally, all of our thanks go to the people of Perthshire, Scotland, who certainly displayed, in every way, 'Ceud Mile Failte' (Gaelic for 'A Hundred Thousand Welcomes').

Robert N. DeWitt
Dwight Duston
Washington, DC USA

Anthony K. Hyder
Notre Dame, Indiana USA

December, 1992

NEAR EARTH SPACE, A HISTORICAL PERSPECTIVE

H. FRIEDMAN
U.S. Naval Research Laboratory
Washington, DC

ABSTRACT. Early scientific interest in the upper atmosphere stemmed from observations of the earth's magnetic field and studies of auroras. Marconi's successful transmission of a radio signal across the Atlantic led to the concept of the ionosphere. Further progress was slow until the advent of rockets and satellites to study the terrestrial environment directly at high altitudes. The discoveries of solar x-rays, extreme ultraviolet light and the solar wind, followed by the Van Allen Belts, led to a rapid maturing of contemporary solar-terrestrial and magnetospheric physics.

Introduction

William Gilbert wrote in 1600 – "the terrestrial globe is itself a great magnet" – with north and south poles defining the axis of a simple dipole. Its field was believed to stretch undistorted into the far reaches of outer space. Modern space observations have shown that the dipole field is grossly distorted by the pressure of a solar wind of charged particles in the near space environment to form a magnetosphere, a magnetically confined plasma regime of great complexity and dynamic variability. The interface between the plasma environment entrained by the magnetosphere and the interplanetary medium is called the magnetopause.

Across the interplanetary gap, the solar wind of protons and electrons (density = $10^7/m^3$, temperature = 10^5 K) flows at several hundred km/s toward the Earth and is diverted around the magnetopause by the pressure of the terrestrial magnetic field. A collisionless shock wave forms upstream of the magnetopause, and the solar wind is heated and decelerated at a bow shock. Near the base of the magnetosphere, where the magnetic field enters the upper atmosphere, solar ionizing radiation creates the ionospheric plasma. In the context of modern astrophysics, magnetospheric studies of the terrestrial environment have broad relevance to most of the solar system planets and such exotic objects as neutron stars with magnetic fields trillions of times as strong as Earth's.

R. N. DeWitt et al. (eds.), The Behavior of Systems in the Space Environment, 1–22.

Early Studies of the Ionosphere

The perception of solar terrestrial connections did not develop until late in the 19th century. Earlier, the great physicist Lord Kelvin claimed absolutely conclusive evidence against the supposition "that terrestrial magnetic storms are due to the magnetic action of the Sun: or to any kind of dynamical action taking place within the Sun, or in connection with hurricanes in his atmosphere, or anywhere near the Sun outside," and further, "that the supposed connection between magnetic storms and sunspots is unreal, and the seeming agreement between the periods has been mere coincidence." Kelvin's great prestige set back studies of solar-terrestrial relationships for decades.

Late in the nineteenth Century, Great Britain initiated the operation of a network of magnetic observatories throughout the colonial empire. Colonel Sabine of the British Army noted that by "a most curious coincidence" the magnitude and frequency of magnetic disturbances followed in rhythm with the appearance and disappearance of sunspots. Careful observations also showed that the direction of the compass needle swung in a regular fashion over a diurnal cycle. At times the fluctuations became more intense and rapid, especially in the auroral zones. Such disturbances were called magnetic storms. Earth and Sun were presumed to be separated by empty space and the sporadic appearance of auroras to be caused by direct streams of particles from the sun with no sensible interplanetary medium to intervene.

With later understanding of the interactions of charged particles and magnetic fields, solar energetic particles were thought to be guided from Sun to Earth along magnetic field lines. Arriving at the north and south magnetic poles, they were deposited in two ring shaped ovals within the Arctic and Antarctic zones. By timing the excitation of auroras and magnetic storms relative to the appearance of flares on the Sun, the travel speed was estimated at 500 miles per second over the 93 million miles from Sun to Earth.

Friedrich Gauss, as early as 1839, had interpreted fluctuations in the magnetic compass to indicate the passage of electric currents at high altitude. In 1882, the Scotsman, Balfour Stewart, wrote an article for the Encyclopedia Britannica in which he envisioned a great dynamo in the sky. In essence, Stewart proposed the existence of an ionosphere in which electric currents were generated by tidal movement of ionized air above 100 km driven by solar heating. The vertical movement is two or three km, sufficient to generate great horizontal current sheets of electricity. As the sub-solar point travels westward around the earth, the current sheets remain pinned under the Sun, thus producing a daily pattern of magnetic variation at any particular point on earth. Observations at magnetic observatories near the geomagnetic equator showed very strong variations, indicative of circulating systems of electric currents of opposite symmetry in the northern and southern hemispheres. These currents jointed at the equator to form a strong flow from west to east at 11 A.M. local time. Sydney Chapman coined the name Equatorial Electrojet.

The earliest history of ionospheric studies is linked to the experiments of Guglielmo Marconi. In 1895, when he was only 21, he built a demonstration wireless telegraph on his father's estate near Bologna.

On December 12, 1901, he transmitted a simple Morse Code signal from England to Newfoundland, a distance of 1800 miles, to the amazement of scientists who could not fathom how the waves got around the curvature of the earth. In 1902, Arthur E. Kennelly proposed that radio waves overcame the curvature of the Earth by reflection from an electrically conducting layer at a height of about 50 miles. Oliver Heaviside made a similar proposal almost simultaneously; the ionized layer came to be called the Heaviside layer.

Pioneering research on the ionosphere was carried out in England by Edward Appleton and his student Miles Barnett and in the United States by Edward O. Hulburt and E. Hoyt Taylor at the Naval Research laboratory and Gregory Breit and Merle Tuve at the Carnegie Institution. The British team set out to determine the height of reflection of a continuous wave. Appleton prevailed on the British Broadcasting Company to provide him with a continuously varying wavelength from London at the end of the broadcast day so that he could detect the interference pattern of ground and sky wave at Oxford. With Barnett he observed the elapsed time between emission and reception of the same frequency, as the broadcast frequency was oscillated back and forth. Early experiments used lower frequencies that only probed the lowest portion of the reflecting layer. Later, as Appleton increased the frequency, he found a higher region of reflection. By 1927 he could discriminate three regions which he labelled D, E and F layers in order of increasing height. For these experiments Appleton subsequently received the Nobel Prize.

Skip Distances

Much of the early research by the NRL scientists co-opted the participation of large numbers of radio amateurs ("hams") who used vacuum tubes with power outputs of less than 50 watts to communicate with each other around the world. Transmissions were confined to very short waves - less than 200 meters. Analysis of their records revealed that shortwaves skipped over a "zone of silence" surrounding the transmitter to a distance of 20 or 30 miles and were received out to distances of hundreds of miles.

Hulburt and Taylor showed that the waves were reflected only when the angle of incidence exceeded a minimum value. At smaller angles the waves penetrated the reflecting region and escaped into space. At night, skip distances were greater than during the day and greater in winter than in summer in temperate latitudes. From these simple observations Hulburt calculated the height of reflection and derived a good value for the electron density - about one million electrons per cubic inch at a reflection height of about 100 miles. By 1926, Hulburt and Taylor had enough evidence to publish a remarkably accurate account of the diurnal variation of ionospheric density as it related to solar elevation.

As the work on skip distances with their amateur cohorts progressed, Hulburt and Taylor soon demonstrated that transmissions at 100 watts could carry over several thousand miles; raised to kilowatts, the signals could be detected even after twice circling the earth. Perhaps the most impressive demonstration of the usefulness of their new mode of communication came in connection with Admiral Richard E. Byrd's pioneering flight to the South Pole on November 29, 1929. His plane had very little power available for radio communication. The solution was to match a program of progressive change in wavelength with distance over the course of the flight. For the first 200 miles it was reduced to 45 meters and for the final 380 miles to 34 meters. Thus was radio contact maintained over the full distance of the flight.

The possibility of pulse reflection methods occurred to several groups of investigators during the early twenties, but the difficulties of measuring millisecond intervals between transmission and reception with the string galayanometers of the time discouraged such efforts. Finally, in the winter of 1924-25 Breit and Tuve at the Carnegie joined with Hulburt and Taylor at NRL to beam pulses at 4.2 Mhz from a 10-kw shortwave transmitter to the Carnegie receiver 7 miles away. Breit and Tuve devised a multivibrator keying circuit to pulse at full power for one tenth of a millisecond at 80 cycles per second. The receiver was then moved to NRL and placed close to the transmitter with both looking upward. Precise phase measurements with a range sensitivity of about 20 meters were achieved. This "ionosonde" was the forerunner of the ionospheric probes that came to be used worldwide and were a major component of the program of the International Geophysical Year (IGY). It was incidental to these early experiments that Tuve became exasperated by the repeated interference caused by approaching aircraft. When his complaints reached military ears the seeds of the concept of radar were planted.

The Rocket Era

As ionosonde efforts proliferated, the motivation was primarily to improve the quality of long distance radio communication. Fundamental understanding of Sun-Earth relationships could not have developed significantly without the new capability of in-situ measurements at high altitude that came with post-World War II rocketry. Among the more notable achievements were the direct measurement of atmospheric composition with mass spectrometers and the sensing of the electron density profile of the ionosphere with dual frequency radio transmitters carried on rockets.

Before V-2 rockets were brought to the White Sands Missile Range, no direct measurements were available of solar ionizing radiation nor of the basic parameters of the upper atmosphere-density, temperature and composition. Solar flares certainly were responsible for radio fadeout but there was no physical connection that could be attributed to the bright flash of optical radiation that signaled the flare. Early correlations of cosmic ray intensities with magnetic field variations seemed clear, but the physical connection was mysterious. Strangely, scientists did not conceptualize a magnetosphere even though Carl Stormer began to develop the essentials of a mathematical model as early as 1907. The word magnetosphere was coined only after the discovery of the Van Allen radiation belts in 1958.

Auroral studies had a high priority in IGY planning even though the enormous advantages to come with the use of rockets and satellites were barely appreciated. In the antecedent First and Second Polar Years adventurous scientists trekked to Arctic regions with their optical instruments. For the IGY, auroral morphology was documented by an extensive network of all-sky cameras that photographed the Arctic sky from horizon to horizon. All told, there were 114 cameras in operation in the Arctic and Antarctic. To follow the development of auroral forms hundreds of thousands of photographs were taken at one-minute intervals through the night. Most auroral observations were made by amateurs who fed 18,000 hourly reports on standardized forms to an Auroral Data Center at Cornell University. Contrast that tedious effort with the remarkable images from space with later generation satellites like the Dynamics Explorers.

The V-2 Rockets

Early on there were no significant hints of the importance of solar x-rays in establishing the ionization balance of the lower ionosphere. In 1928, Hulburt proposed that it was solar ultraviolet at extremely short wavelengths that was ionizing the upper atmosphere. By 1930, he had become convinced that there was a connection between solar ultraviolet, sunspots, magnetic storms and the disturbance of radio communications. In 1935, J.H. Dellinger reported on a series of sudden ionospheric disturbances (SID) and urged a collaboration of astronomers and ionospheric scientists to try to understand the physical process. At roughly the same time, Robert H. Goddard was progressing with his development of a liquid fueled rocket to carry instruments to high altitudes. In correspondence with John Fleming at the Carnegie in the early 1930s, Hulburt commented on the suggested theoretical match between atmospheric absorption of soft x-rays and the overhead ionization. He suggested that a good experiment would be to fly a photographic film covered with a thin aluminum foil in one of Goddard's rockets to detect solar x-rays.

World War II diverted the thoughts of most American scientists from upper air research and solar astronomy. At the time, the only serious American effort went into developing the bazooka, a hand held tube that rested on an infantryman's shoulder to launch a military rocket. By 1942, however, NRL undertook a program to develop guided missiles. The JB-2 was a U.S. version of the German V-1 buzz bomb and the Lark was a rocket propelled guided missile for ship to air.

By the end of the war the U.S. military was convinced of the important need for a substantial effort in rocket development. Enter the German V-2, captured by the victorious western allies and transported to the U.S. testing and proving grounds at White Sands, N.M. The V-2 created the opportunity for the new age of rocket high altitude scientific research. The first V-2 launch from White Sands took place on April 16, 1946. Some 60 launches were carried out up to the fall of 1952 when the supply was exhausted.

The V-2 was powered by 10 tons of alcohol and LOX. At takeoff it generated 28 tons of thrust and accelerated to 6 Gs. The most successful flight reached 170 km and lasted 450 seconds, with about 270 seconds above 80 km. During powered flight, the rocket was guided by graphite steering vanes that deflected the exhaust stream of hot gas. Upon landing in streamline flight on the desert floor, the rocket usually created a crater more than 80 feet in diameter, and the remains were almost totally incinerated. Rarely did the V-2 fly like an arrow; as often as not the rockets tumbled and faltered. Some burned up furiously upon ignition. One took off in horizontal flight to land on the edge of Juarez, across the Mexican border.

Early participants in the V-2 program included The Johns Hopkins Applied Physics Laboratory, Harvard, Michigan and Princeton Universities, the U.S. Army Signal Corps, the U.S. Air Force and the U.S. Naval Research Laboratory. The NRL program had several components directed toward studies of the upper atmosphere and ionosphere and the radiation spectrum of the Sun. From the standpoint of understanding solar terrestrial physics, the early observations with broadband photon counters in the ionizing ultraviolet and x-ray regions of the spectrum were the most successfully diagnostic. The picture that emerged from solar rocket astronomy was markedly different from that previously held. The visible Sun resembles a 6000 K black body with a spectral maximum near 5000 Å. At shorter wavelengths, the black body emission would be expected to decrease rapidly and at x-ray wavelengths to be almost inconsequential. Some clue to x-ray emission came from observations of an extended corona at times of total eclipse. To support such a far-reaching bag of gas against the pull of solar gravity required a coronal temperature of the order of a million degrees — an x-ray radiating plasma, but so thin as to have a very small emissivity.

Solar X-rays and Extreme Ultraviolet

In 1949 NRL carried out a definitive experiment on V-2 #49 in which the solar radiation flux was observed by telemetered signals from a set of photon counters over the altitude range from the base of the ionosphere up to 50 km. At the time, the simple Chapman theory of the formation of an ionized layer by the absorption of monochromatic radiation was thought to apply. The new rocket data in a band centered near 8 Å showed that the ionizing radiation was an x-ray continuum that produced an E-region rather than an E-layer. Subsequent measurements traced the merging of E-region with the higher F-region produced by extreme ultraviolet radiation.

The D-region, near 80 km, coincided with the altitude profile of absorption of the resonance line of hydrogen Lyman-alpha at 1216 Å, but D-region formation presented a puzzle. H Lyman alpha, which originates in a thin layer of the solar chromosphere, contains most of the solar energy in the extreme ultraviolet, but it cannot ionize any major component of the atmosphere. Marcel Nicolet found the answer in nitric oxide. A concentration of only one part in 10^8 could be ionized with high enough efficiency by Lyman-alpha to produce the observed D-region.

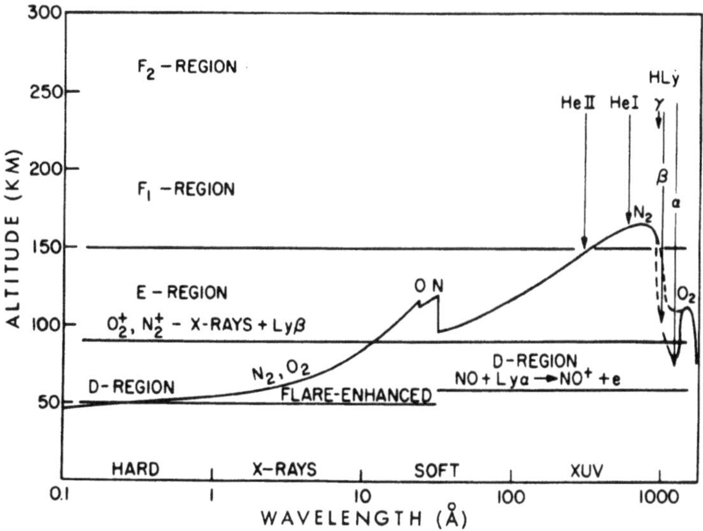

Fig. 1: The depth at which solar radiation penetrating the atmosphere is attenuated to e^{-1} in the wavelength range 0.1 to 1000 Å.

In the broad span of the Schumann region from 1450 Å to 1750 Å, the profile of solar radiation absorption indicated the persistence of molecular oxygen well above the altitudes at which dissociation was thought to be complete. Schumann radiation produced no ionization but played an important role in shaping the ionosphere by dissociating molecular oxygen. Molecular oxygen ions control the rate of neutralization of ionospheric electron density much more rapidly than atomic oxygen ions. The concentration of molecular oxygen provided dissociative recombination at a high rate well into the F-region. In effect, these early results were sufficient to explain the basic outline of the ionosphere under a quiet sun. All that remained to complete the model of solar control of the ionosphere were measurements of the helium resonance lines in the extreme ultraviolet at 304 Å and 584 Å plus groups of spectral lines of lesser combined intensities in neighboring wavelengths that affected the highest F-region.

Following the initial measurements of solar x-rays (1 to 8 Å), broad band photometry of the x-ray spectrum was progressively extended to about 44 Å. The spectral distribution fit well with a thermal source at a temperature of a few million K. Successive measurements at intervals of months to years showed positive correlations of flux variations as much as a factor of 7 for x-rays (8-20 Å) with the concentration of sunspots and with variations in electron density in the ionosphere.

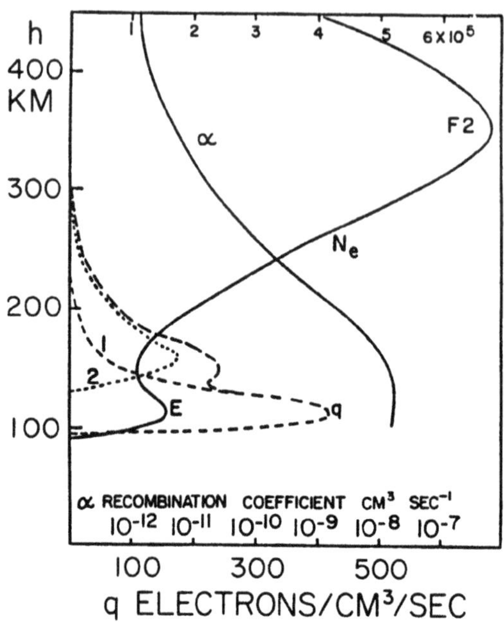

Fig. 2: The dashed curve, 1, is the ionization rate produced by the X-ray spectrum containing 0.1 erg cm^{-2} sec^{-1} between 10 and 100 Å. Curve 2 is the ionization rate produced by the helium resonance lines containing 0.05 erg cm^{-2} sec^{-1}. Rocket Panel densities were used in the computations. The shape of the electron density curve in the F-region is controlled primarily by the effective recombination coefficient, α.

The variation of solar x-ray flux observed over a solar cycle made it clear that the concept of solar emission from a spherically symmetrical corona was altogether inadequate. The corona appeared to be structured in condensations that formed over sunspots and produced enhanced localized x-ray emission. Since sunspots are characterized by powerful magnetic fields, the condensations of the corona were believed to exist in the form of high loops of magnetic lines of force spanning the adjacent spots of opposite polarity. The first x-ray picture to confirm these ideas of coronal condensations was obtained with a pinhole camera in 1960. In more recent years, fine details of magnetic structures connected with x-rays have been revealed with reflecting x-ray telescopes of arc-second resolution.

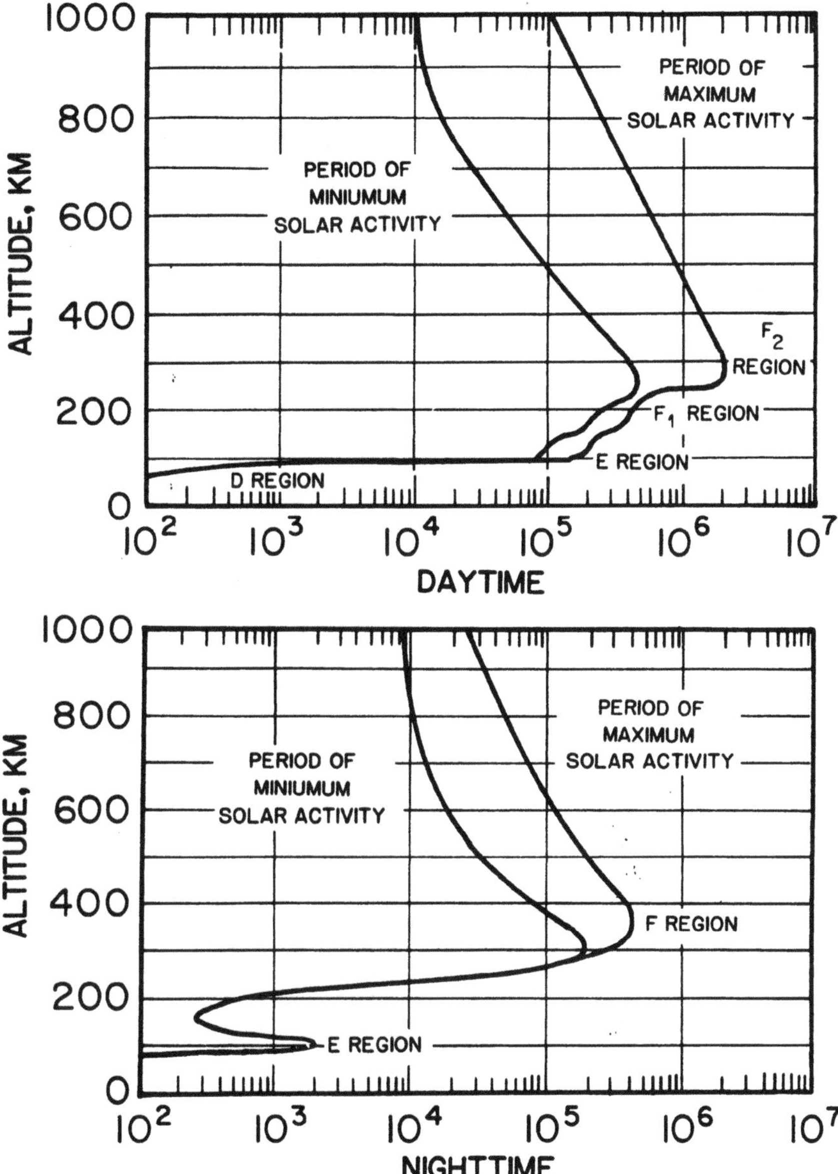

Fig. 3: Diurnal variation of ionospheric electron density over a solar cycle.

SOLAR FLARES

The behavior of a flare-active Sun presented more difficult puzzles, most notably the connection between solar flares and sudden ionospheric disturbances (SID). Various forms of SID are identified as follows:

1) SWF, sudden shortwave radio fadeout or blackout. Radio transmission on 5 to 20 MHz fades with the onset of D region absorption.

2) SCNA, sudden cosmic noise absorption. Cosmic radio noise at 18 MHz is strongly absorbed in passing through the D region.

3) SPA, sudden phase anomaly. Very long radio waves are reflected from the lowered ionosphere at the base of the D region.

4) SEA, sudden enhancement of atmospherics at 27 kHz. Increased signal strength from lightning noise results from enhanced reflectivity of the D region at its base.

5) Magnetic crochets. A sudden augmentation of the geomagnetic field is superimposed on the normal diurnal variation.

Because the visible manifestation of a flare is so clear in the hydrogen red line, it led most solar physicists to see a connection between the resonance line H L-α at 1216 Å, which is responsible for the normal D region, and the increased ionization that is produced by the flare. But a simple analysis of the radiation enhancements that are required, immediately rule out the ultraviolet process. The increase in electron density that occurs during shortwave fadeout is roughly a factor of 2.5, 4.5, 7.0 and 9.4 for flares of importance 1,2,3 and 3+, respectively. The radiation flux from the small area of the flare would need to increase by factors of 6,20,50 and 90, respectively. Such increases are virtually impossible astrophysically.

Analysis of the flux enhancements required to explain the SPA confirm the estimates based on radio fadeout. The SPA measurements are made with very low frequency radio wave (16 kHz) reflections from the base of the ionosphere. When a flare begins, the phase changes rapidly to indicate a drop in the reflecting height. The altitude change may be as large at 16 km in a strong flare. If Lyman alpha were the sole source of ionization, the flux increase required to explain a drop of two scale heights would be 100 times the normal emission.

Attempts to explain the other SID phenomena by Lyman alpha enhancement fail for the same reasons x-rays offer a much better solution. At a wavelength of 2.5 Å, x-rays have the same atmospheric absorption coefficient as Lyman alpha, but the x-ray ionization efficiency is several thousand times as high because x-rays ionize all the constituents of the atmosphere, whereas Lyman alpha can ionize only the trace of nitric oxide. Virtually all the energy delivered in Lyman alpha is dissipated in dissociation of molecular oxygen. Approximately 10^{-36} erg cm^{-2} s^{-1} of 2.5 Å x-rays would suffice to produce a normal D region. Shorter wavelength x-rays would penetrate to the heights associated with SID phenomena.

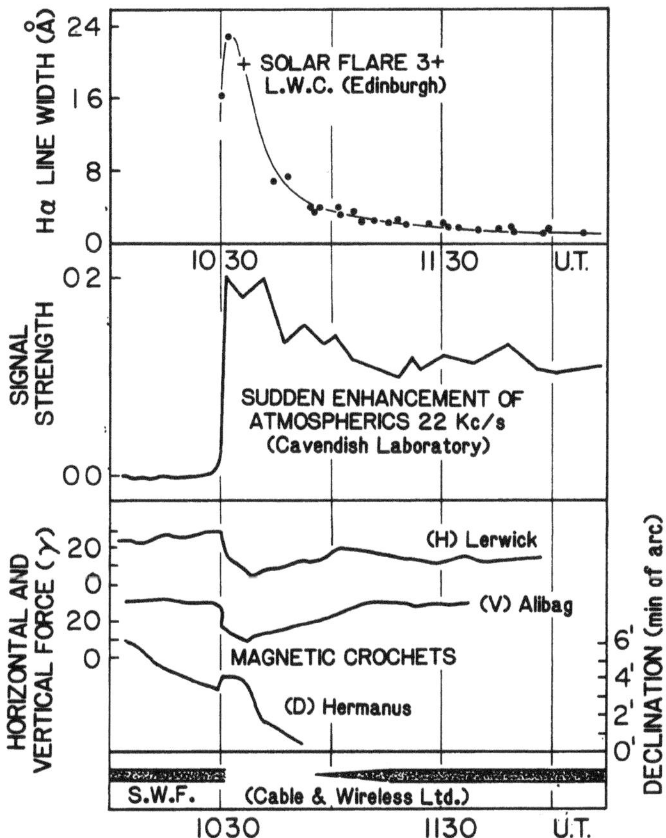

Fig. 4: The development curves of a solar flare and its related ionospheric disturbances. (November 19, 1949).

Rockoons

The observational challenge of detecting flare x-rays was taken up during the IGY. Because the onset of a flare is unpredictable even within a few minutes, liquid fuel rockets that required an hour or more of preparation before firing were unsuitable, and, more importantly, it was impractical to tie up an Aerobee launching tower for days with a rocket in place while waiting for a flare to erupt. In the early 1950's, Lt. Lee Lewis in the U.S. Office of Naval Research conceived of launching a solid propellent rocket suspended from a balloon in the stratosphere. Experiments were begun by James van Allen with a Deacon rocket hung on a Skyhook balloon and floated at sea at a height of 25 km, where it could be fired by radio command. The combination of rocket and balloon was called "Rockoon" and it afforded a cheap mode for cosmic ray studies.

The Rockoon appeared to be a practical solution to the rocketry problem for the study of solar flares. An NRL proposal for a naval expedition as part of the International Geophysical Year (IGY) was approved, and the NRL rocket team obtained the support of the USS Colonial, a Landing Ship Dock with a flight deck from which the Rockoon could be launched. The ship embarked on Project San Diego - Hi in 1956 and sailed to a launch area about 400 miles off the coast of California. The plan was to release a Rockoon each morning on ten Successive days. If evidence of the start of a flare was obtained, the rocket was instantly fired. Out of ten tries, one flare was caught in the act and gave convincing evidence of flare x-rays.

A year later Rockoons went out of style and were replaced by two-stage rockets. In place of the balloon a Nike booster propelled the Deacon to the height of the stratosphere where it was ignited to carry on up into the ionosphere. The rockets were fired from a simple rail launcher that the NRL team set up on San Nicholas Island off Point Mugu. In 1958 and 1959, a series of these rockets were launched at times of flares and gave confirmation of the dependence of SIDs on the intensity and wavelength of flare x-ray emissions. By the end of the IGY, the satellite era was upon us, and the new means of long term observations came into use. Instead of occasional rocket measurements of solar x-rays and ultraviolet, the Sun could be monitored continuously. Over the course of a decade the NRL series of ten SOLRAD satellites gave a full solar cycle map of x-ray activity.

The Electrojet

Let us return to the story of "Rockoons" as they were applied to other studies of the upper atmosphere in conjunction with ground based magnetic observations. For the IGY some 190 magnetic observatories were implemented, scattered widely over the globe. For in-situ measurements it was necessary to employ rockets that could be launched from shipboard stations moveable about both sides of the magnetic equator. As early as 1949, the U.S.S. Norton Sound went to sea with three Aerobee rockets instrumented by Fred Singer and his colleagues with fluxgate magnetometers to detect the electrojet. The rockets were launched from the Pacific Ocean some 1000 miles west of Peru. Singer's group thought they had detected an electrojet between 58 and 65 miles, but the measurements were marginal.

By the beginning of the IGY, the proton precession magnetometer had been developed for rocket flight, and Van Allen's team set out to fly them on Rockoons. In August and November 1957 they attempted 54 Rockoon firings while cruising from near Thule in northern Greenland to off the coast of Antarctica. In August, they launched eighteen Rockoons from the U.S.S. Plymouth Rock; four misfired, three suffered telemetry failure, and three more failed because instrument components froze. Altogether, only seven flights could be described as good or partially successful. A month after the return of the Plymouth Rock, the electrojet research team switched to Loki II rockets for their Rockoons and left Boston for Antarctica aboard the icebreaker U.S.S. Glacier. Between October 14 and 20, 1957, six successful shots were fired along a line running north-south 650 miles across the expected path of the electrojet. The results were puzzling; not one but two current layers were penetrated. Some of the results were attributed to "sporadic-F" in the upper parts of F region, apparently consisting of discrete clouds of ionization.

Since the days of Rockoon observations, sporadic E and Spread F have come to be much better understood; although there are still many puzzling details to be unraveled. Sporadic E reflects high-frequency radio waves that normally escape the ionosphere. Signals can be received as far away from the source as 2000 km on a single hop. In summer, severe interference may affect television signals. Sporadic E and Spread F will be the subject of intensive study in the equatorial regions in the International Space Year (ISY) during 1992.

Ionospheric Electron Loss Processes

While solar radiation provides the input function for ionization and dissociation of the atmospheric constituents, the equilibrium electron density depends equally importantly on the rates of recombination for atoms and molecules at various heights. The slowest process is radiative recombination of electrons and atomic ions,

$$x^+ + e \rightarrow x + h\nu$$

Molecular ions are neutralized by dissociative recombination,

$$xy^+ + e \rightarrow x' + y'$$

Prime symbols indicate excited atomic states. Theoretical values of the radiative recombination coefficients are $4.8 \cdot 10^{-12} cm^3 s^{-1}$ for H^+ and He^+ and $3.7 \cdot 10^{-12} cm^3 s^{-1}$ for O^+. For the dissociative recombination coefficients of N_2^+, O_2^+ and NO^+, the values are somewhat uncertain but approximately $10^{-7} cm^3 s^{-1}$ for all three at 300 K.

To get observational evidence of the recombination coefficients, ionospheric scientists have resorted to eclipse records of the rate of decay and recovery of the ionosphere. During the course of a total eclipse the amount of solar radiation obscured from the earth diminishes by different amounts, depending on wavelength. The ratio of the instantaneous solar flux at any given wavelength to its uneclipsed value is called the eclipse function. For visible light in the course of a total eclipse, the eclipse function is unity at first and fourth contact and zero at second and third contacts, almost exactly equal to the unobscured fraction of the solar disk.

In the ultraviolet and x-ray regions the function behaves irregularly. Isolated regions of enhanced brightness in x-rays and ultraviolet may lie high enough above the disk that they are not eclipsed. In Lyman alpha, as in visible radiation, the eclipse is very nearly total. X-rays are never completely obscured. In measurements made at the eclipse of March 7, 1970, the Lyman alpha flux at totality was reduced to only 0.15 percent. At the same time, solar x-rays in the bands 2 to 8 Å, 8 to 20 Å and 44 to 60 Å, the major sources of E region ionization, gave residual fluxes at totality of 5,7, and 16 percent, respectively. Clearly the heights of origin were well above the chromosphere, in the lower corona.

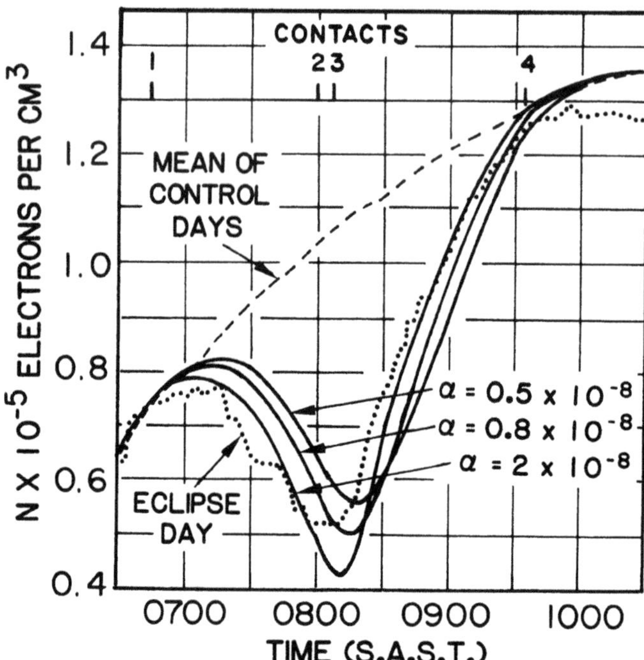

Fig. 5: Theoretical and observed peak electron densities for the E-region during an eclipse.

Thermosphere and Exosphere

Satellite dynamics, beginning with the small Vanguard test-sphere in 1958 and the Echo balloon in 1959, provided valuable information on the high neutral atmosphere. In the exosphere above 1500 km, hydrogen begins to dominate the gas composition. The gas concentration is so low that an atom cannot make a complete orbit of the Earth without colliding with another atom. Echo was intended to be a reflection target for radio communication, but such passive radio reflectors were quickly superseded by active transponders. Echo was so large and easily visible however, that it provided valuable information about the density of the atmosphere from the effect of drag.

From 1959 to 1962 the drag on Echo seemed to indicate an unexpectedly high density at 1500 km if the principal constituent were atomic oxygen. To drive so much oxygen to 1500 km required a thermospheric temperature greater than 2000 K. Neither could an oxy-hydrogen mixture satisfy the requirement; the hydrogen content would have to exceed the generally accepted models by an order of magnitude. Marcel Nicolet solved the puzzle by introducing helium into the model atmosphere to match the observed drag without exceeding a temperature of 1500 K. His model required a helium bulge between altitude regimes which were first predominantly oxygen and at higher levels, hydrogen.

The mass spectrometer aboard the Explorer XVII satellite confirmed Nicolet's explanation in 1963. The major atmospheric constituent between 500 and 1000 km was helium but the concentration at any particular time varied with solar activity. Near solar maximum helium dominated the composition from 1100 to 5000 km because of higher thermospheric temperature. Calculation of the heat flow through the thermosphere above 200 km indicates a nearly height independent temperature. Oxygen, helium and hydrogen separate out by thermal diffusion in agreement with Nicolet's helium bulge prediction. Hydrogen moves upward most rapidly and escapes into interplanetary space. At the level of the mesosphere, solar ultraviolet radiation provides a continuous source of replenishment by dissociation of water vapor. As it diffuses upward the hydrogen cloud is detectable as geocorona 50,000 miles high.

Radioactive decay of radium and thorium in the ground supplies helium to the atmosphere. From estimates of the abundance of isotopes in the basalt and granite of the Earth's crust the expected outflow of helium is about 10 atoms per square-centimeter column per second. It follows that only one million years are needed to fill the terrestrial atmosphere to its normal helium contents and that the rate of escape from the exosphere to maintain the equilibrium balance requires a thermospheric temperature of 1500K.

Satellite drag is a sensitive indicator of the absorption of solar energy by the atmosphere. When solar activity is high, the atmosphere expands in response to the heating of the absorbing levels of the atmosphere. The expanded atmosphere exerts greater drag on satellites and forces them down to lower altitudes. According to Kepler's law of orbital motion, the speed of the satellite must increase as its orbit shrinks. It then loses energy faster to drag, and its demise becomes inevitable.

Airglow

Astronomers had discovered the airglow early on as a faint background radiation in stellar spectra. It is barely visible to the human eye even on moonless nights in the countryside away from city lights. The most prominent airglow emission came in the green line of atomic oxygen at 5577Å. When it was observed that the green light grew stronger from the zenith to the horizon, it became clear that it originated in the atmosphere in the neighborhood of 100 km. By the time of initiation of the IGY it was known that the red lines of oxygen (6300Å, 6460Å) glowed well above 150 km. Rocket studies identified the yellow light of sodium at about 80 km. Specific other colors are attributed to molecular oxygen, atomic and molecular nitrogen, potassium and lithium. The strongest emission comes from hydroxyl, just beyond the limit of visual perception in the near infrared. Altogether, these airglow emissions add up to about one-tenth the intensity of starlight. The general explanation for these various emissions was that solar radiation ionized atoms and dissociated molecules during the day at characteristic levels of the upper atmosphere. At night they recombined and radiated. But the patterns of emission were very

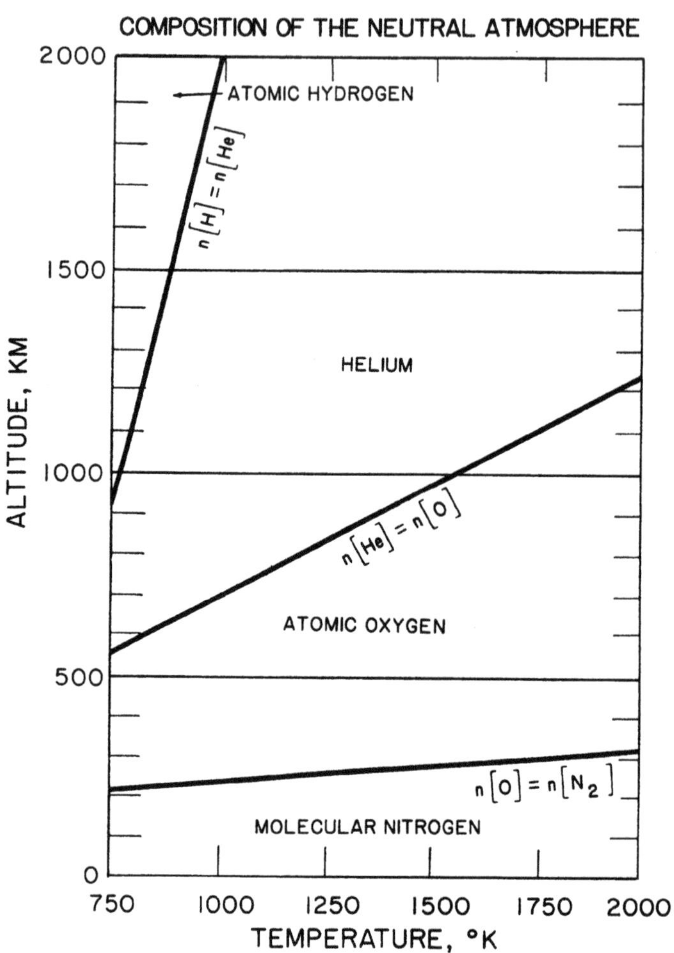

Fig. 6: Predominant atomic species at various altitudes. Temperature depends on time of day and the phase of the solar cycle.

strange, varying from one part of the sky to another in fast moving clouds at speeds of several hundred km per hour. Speculations were that wave motions created the patterns, that they were related to sporadic E and the Electrojet.

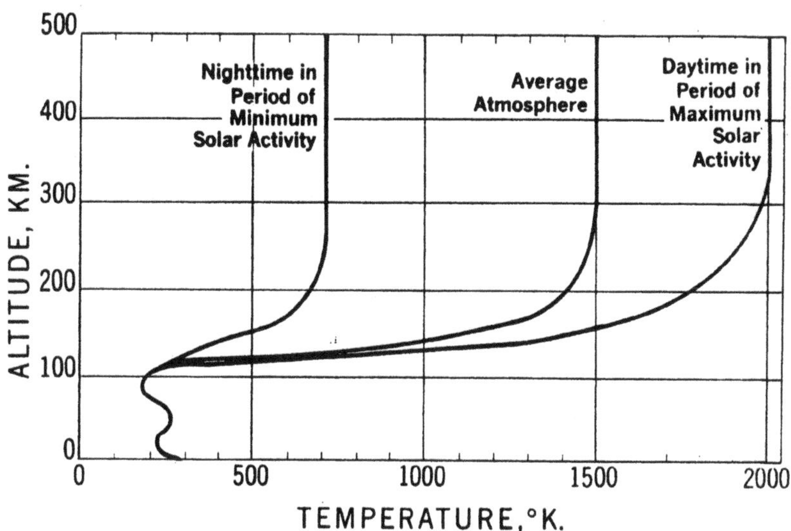

Fig. 7: Variation of thermospheric temperature over a solar cycle.

In 1955, the NRL group attempted a rocket survey of the extreme ultraviolet radiation of the night sky from an Aerobee rocket. Above 70 km the hydrogen Lyman-alpha detector was swamped with radiation that exceeded all the visible airglow. In 1957, a more refined scan showed that the Lyman alpha from outside the Earth's shadow cone was being backscattered from exospheric hydrogen to the dark side of the Earth. At first it was speculated that the scattering hydrogen atoms must be interplanetary, but further measurements and theory identified the hydrogen with a giant geocorona reaching tens of thousands of km beyond the Earth.

The clearest picture of ultraviolet glow was photographed with an ultraviolet camera placed on the moon during the Apollo-16 mission. To the long wavelength side of Lyman alpha, the image in the atomic resonance line of atomic oxygen at 1314Å was so bright that the entire sunlit hemisphere was suffused with its light. Spectacular arcs extended into the nighttime hemisphere, excited by the recombination of oxygen ions with electrons in the F region of the ionosphere. The arcs were aligned with the double humped peak of the ionospheric F region where it brackets the geomagnetic equator.

Efforts to study atmospheric dynamics via artificial airglow seeded with the aid of vapor releases from rockets began in the 1950's. David Bates, after the evidence of the free sodium at 90 km from rocket measurements, proposed that sodium be ejected from a rocket at about 80 km. Vapor atoms excited by sunlight would fluoresce the resonance D lines at twilight to from a visible cloud. The first sodium cloud was produced by ejection of sodium pellets in a mixture of Al + Fe_2O_3 from an Aerobee rocket in 1955. At about the same time development of the French Veronique rocket had proceeded to flight test. It had a performance capability similar to the Aerobee but no substantial funding for scientific payload development was provided. French scientists seized on the vapor release experiment as a way to get good science at a minimal cost. The first successful cloud release was achieved in March 1959 from Hammaguir and the temperature of the thermosphere was deduced. At 102 km, the turbopause was discovered from evidence of a sharp limit between a lower lying turbulent eddy structure of the atmosphere and an overlying laminar regime.

As early as 1931, Sidney Chapman and Vincent Ferraro proposed that sudden fluctuations in the magnetic field might be attributed to the impact of solar plasma clouds on the Earth's magnetic field. Chinese astronomers a thousand years earlier had noted that comet tails were driven away from the Sun. Until relatively recently it was thought that radiation pressure swept back the gas from the comet head into the thin, elongated tail. But comet tails are so tenuous that radiation pressure can have little effect.

In 1951, Ludwig Biermann suggested that the ionization in comet tails was the result of charge exchange between a solar wind that had not yet been observed and cometary gas. The acceleration of comet tails, in sudden jumps, he attributed to momentum transfer from solar wind electrons and protons for which he required a density of about 1000 per cc. The idea of charge exchange and a high number density of particles in a solar wind were soon shown to be wrong, and theorists chose instead to attribute the acceleration to gross plasma and magnetic field interaction between a thinner solar wind and the comet tail.

Biermann later proposed to imitate conditions in cometary plasma at high altitudes by creating artificial clouds of alkaline earth metals such as calcium, strontium, and barium that have low ionization potentials and resonate in the visible spectrum. In May 1963, two Centaure rockets were launched successfully from Hammaguir with barium burners in their nose cones. The clouds were spectacular, but unfortunately displayed no evidence of barium atoms or ions, only useless barium oxides.

When the reaction used in the burners was changed to combining barium with copper oxide, the results were highly successful. The gas produced with this mixture was almost pure barium with a small percentage of strontium and the photoionization time was short - about 30 seconds, indicative of a large production of ions. Direct visible evidence of the conversion of atoms of barium into ions was expressed by the color change from the green atomic resonance to the purple resonance lines of ions. The cloud, initially spherical, quickly transforms to an elongated shape as the electrons and ions drift along the magnetic field lines. The entire conversion from atoms to ions takes place in about 2 minutes. Subsequent releases introduced the use of lithium and europium.

A later technique for producing barium clouds resorted to pyrotechnics. A shaped charge was covered with a thin surface of barium metal. By pointing the shaped charge in the direction of magnetic field lines, a high velocity (12 to 14 km/s) gas jet was directed along the lines of force when the charge exploded. If the explosion occurred above collision dominated altitudes (400 km) the luminous track traced the magnetic field lines to very high altitude. Historically, barium clouds gave the first glimpse of large scale magnetospheric convection as early as 1960.

More recently the barium cloud method has been performed with great sophistication in a U.S. – U.K. – German satellite mission called AMPTE for Active Magnetospheric Particle Tracer Explorer. Two charges of about 2 kg of barium vapor were released outside the Earth's magnetic field in December 1984 and July 1985. The German Ion Release Module (IRM) was instrumented with a complete set of instruments to measure magnetic field, electric and magnetic waves, the three dimensional velocity distribution of ions and electrons, and the charge, mass and energy distribution of the ions. Similar instruments were carried on the United Kingdom Sub-satellite nearby. Optical observations from aircraft and from the ground completed a versatile ensemble of instruments for a comprehensive space laboratory.

It is interesting to note that Iosif Shklovsky, in the Soviet Union, proposed early on the use of vapor releases to form artificial comets. Sodium vapor was released from the second Soviet moon rocket in 1957 and fluoresced strongly enough to provide useful tracking of the rocket.

Solar Wind

Eugene Parker, in 1958, first proposed that a solar wind must result from simple hydrodynamic expansion of the solar corona. It is interesting to recall how little attention had been paid to Bierman's ideas and how few scientists seemed to accept the notion of a solar wind. Although Chapman had speculated that the outer limits of the solar corona stretched to the distant planets he did not describe a flow of wind. About two years after Parker's theory of a wind was published, the Soviet Lunik 2, that impacted on the moon, reported tentative evidence of a solar wind. Subsequently, Lunik 3 and a Soviet Venus probe together with NASA's Explorer 10, gave further confirmation. But it was only after the results reported from Mariner 2 in 1962 that the evidence was widely accepted.

The general picture emerged of a solar wind – magnetosphere interplay in which the wind flows around the dayside to the nightside where the Earth's magnetic field is drawn out into an extended magnetotail. The lines of the magnetotail are stretched in the direction of the solar wind like a wind sock far beyond the distance to the moon. The magnetotail serves to store energy that is extracted from the interaction between the solar wind and the magnetosphere. On occasions, violent magnetospheric substorms release the energy into the auroral zones. At play in the substorm are enormous electric currents. The interaction between the wind and the magnetosphere resembles a cosmic dynamo that drives the currents through space across and along magnetic field lines. Part of the electrical flow is in the form of a current sheet down the center of the tail, dividing it into two lobes of oppositely directed magnetic fields that direct the current into the auroral ovals.

The Earth's magnetic shield is so effective that only 0.1 percent of the mass of the solar wind that hits it manages to penetrate. Penetration is believed to be a process of magnetic field merging and reconnection. The most revealing space probes have been the International Sun-Earth Explorers. In 1977, NASA and ESA launched a pair of satellites into nearly identical orbits. As the two spacecraft chased each other around the magnetsophere, they sensed the position and movement of the bow shock and magnetosheath about 130,000 km above the Earth. Where magnetic merging of the wind and the magnetosphere occurred on the sunward side, the magnetosheath was peeled back toward the dark side of the Earth, hundreds of thousands of km into the magnetotail. As merging of the fields progressed, the field lines were sharply bent and particles caught inside the bends were accelerated as though projected by a sling-shot.

The Global Electric Circuit

To complete the description of the space environment the electrical properties must be connected to the surface of the Earth via a global electric circuit. Early in the eighteenth Century, atmospheric scientists observed a weak permanent electrification of the atmosphere even in fair weather, with the Earth negatively charged and the air positively. The gradient near the earth's surface was about 100 volts per meter. It was believed that the Earth retained a primordial charge from the time it was formed.

At the beginning of the 20th Century most scientists attributed the ionization to the emanation of radioactivity from the ground. In 1912, Victor Hess tried to measure the variation in conductivity with altitude by flying his electroscope on a manned balloon to a height of 5 km. At that height he believed the air would shield his detector from ground level radiation. Instead he found no decrease with height from which he deduced the ionizing radiation of great penetrating power was arriving from outside the atmosphere. He had discovered cosmic rays for which he received the Nobel Prize 24 years later. The cosmic rays provided continuous ionization that supplied enough current to neutralize the Earth's negative charge.

Some 65 years ago, C.T.R. Wilson, inventor of the cloud chamber, proposed that thunderstorms created a generator that maintained the Earth's negative charge. Thousands of storms are in progress around the world at any moment. Beneath a thundercloud negative charge is carried to ground, while above the cloud positive charge flows to the ionosphere. The charge spreads around the Earth rapidly and comes down to ground through the atmosphere in fair weather regions. Global lightning maintains a potential difference of hundreds of thousands of volts between the ground and the ionosphere. The global electric circuit has been compared to a giant leaky capacitor, with the highly conducting ground and ionosphere representing two oppositely charged plates and the lower atmosphere playing the role of an insulating dielectric.

Thunderstorms sustain the charge separation. Most clouds develop and dissipate without substantial production of lightning. Cloud electrification depends on convective turbulence. Strong electrification comes with strong convection, generally beginning with rapid horizontal and vertical development of the fair-weather cumulus cloud into a cumulonimbus cloud. When vigorous updrafts and downdrafts occur, charge undergoes separation in the thundercloud and gives rise to lightning. With about 1000 lightning flashes per second, the integrated current amounts to from 1000 to 2000 amperes and the potential difference 200kv to 400kv.

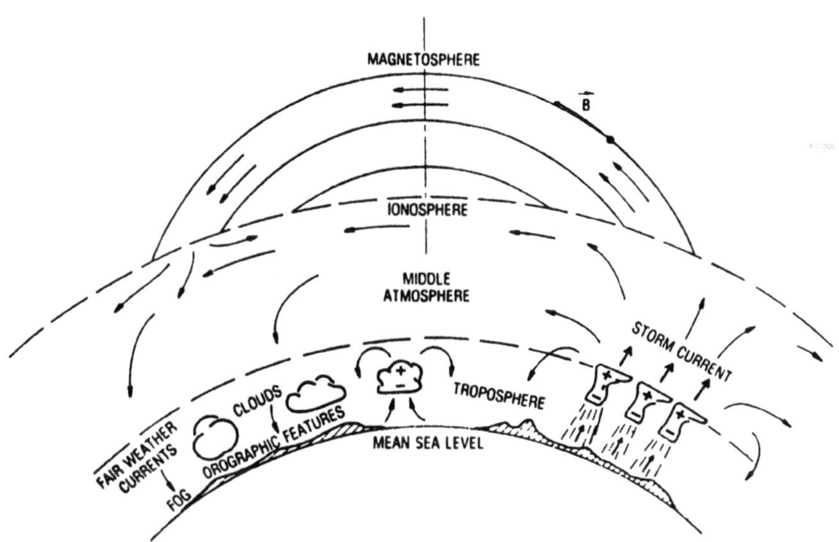

Fig. 8: Thunderstorms create conduction currents and maintain global electric circuit.

Since the earlier studies of atmospheric electricity there have been many attempts to ascertain a solar-terrestrial relationship between ground circuits, electric fields and thunderstorm frequency that would correlate with solar activity and auroras. Ground based, aircraft-and balloon-borne instruments have hinted at atmospheric electrical responses to solar flares, sunspot cycle variations, geomagnetic activity and auroras. It is difficult, however to separate the induced variations in the global atmospheric electric circuits from solar and upper atmosphere variability because they are superimposed on complex electrical variations associated with meteorological and anthropogenic influences in the lower atmosphere.

It is still puzzling in detail how the lower atmosphere global electric circuit connects with the intense current systems flowing in the high upper atmosphere. The ionospheric wind dynamo at about 100km carries horizontal currents of about a hundred thousand amperes on the Earthward side. At high magnetic conjugate latitudes and at about the same altitude, the solar wind magnetospheric dynamo drives currents as high as a million amperes. How all these current systems are linked to solar and auroral variability remains a challenging problem in atmospheric science.

* * * * * * * * * * * * *

This paper discussed only a sampling of diverse phenomena in the Sun/Earth system from the perspective of historical development of the science. Most of the phenomena are understood in broad principle but are very complex in detail, and much more research will be required to unravel all the interactions.

General References

Chapman, S. and Bartels, J., Geomagnetism, Oxford University Press, New York (1940)

Ratcliffe, J. H. ed., Physics of the Upper Atmosphere, Academic Press, New York (1960)

Friedman, H., "Solar Observations Obtained from Vertical Sounding", Reports on Progress in Physics XXV (163), 164-217 (1926)

Proceedings of the International Conference on the Ionosphere, The Institute of Physics and the Physical Society, London (1963)

Dessler, A. J., "Solar Wind and Interplanetary Magnetic Field", Reviews of Geophysics 5 (1), 1-41 (1967)

Akasofu, S.I. and Chapman, S., Solar-Terrestrial Physics, Clarendon Press, Oxford, England (1972)

OVERVIEW FROM A SYSTEMS PERSPECTIVE

C. K. PURVIS
NASA Lewis Research Center
Cleveland OH 44135
USA

ABSTRACT. The interactions between space systems and their orbital environments are discussed in overview. Attention is focussed on those environment factors appropriate to earth orbital vehicles. Ways in which space systems interact with their environment and effects on system performance and lifetime are described. Approaches to hazard assessment and protection are noted.

1. Introduction

Design of any system demands consideration of the environment in which it must operate, to ensure proper system function, reliability and lifetime. Space systems are no exception to this rule. Indeed, because of the difficulty and cost of repair, environmental compatibility is particularly critical in space.

For present purposes, the "space environment" to be considered is that of near-Earth space, i.e., orbits ranging in altitude from Low Earth Orbit (LEO) to geosynchronous (GEO) and beyond, and including all inclinations. The natural near-Earth space environment is complex and dynamic. Its features are determined partially by the characteristics of the Earth itself, partially by the interactions between the Earth and the Sun, and partially by processes occurring in interplanetary and interstellar space. Environmental factors include neutral atoms and molecules, plasmas and charged particles, fields, particulates and electromagnetic radiation. Many of the environmental components vary with position in orbit (e.g., altitude, longitude, latitude), local time, season and level of solar activity. The presence and activities of orbital systems can significantly modify the "lower energy" environment components such as neutral gases, ionospheric plasmas and electric fields, so that the local environment may be rather different from the natural one. The local environment is what the system actually interacts with, so a comprehensive treatment of environmental compatibility requires an integrated approach comprising consideration of environment, system, subsystems and effects.

R. N. DeWitt et al. (eds.), The Behavior of Systems in the Space Environment, 23–44.

2. Environment Factors and Effects on Systems

The interrelationships among the system, its subsystems or payloads and the environment are illustrated in cartoon form in Fig. 1. One begins by considering the natural environment at the orbital location(s) of interest and the characteristics of the system in its various modes of operation to identify how they may interact to cause the local environment to differ from the natural one. One then examines the interactions between the system and its local environment to determine the effects on the system, its subsystems and payloads. The importance of any particular effect is judged by its impact on the system's ability to perform its mission. Thus the requirements placed on various subsystems will depend on the goal of the system as a whole, as will the relative importance of various environmental interactions and their effects.

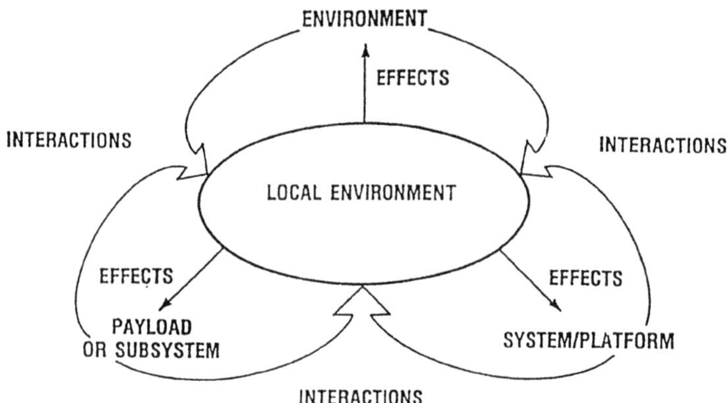

Fig. 1: Environmental interactions and effects concept.

The various environment components, or factors, interact in different ways with space systems and produce a variety of effects. Because the physical processes of interaction depend on the type of environment factor under consideration, it is usual to "sort" environmental interactions and effects by environment factor. Following this approach, a list of environment factors and their effects on space systems in near-Earth space is provided in Table 1.

In what follows, each of these environment factors and its effects are discussed briefly, and strategies for mitigating undesirable effects noted. Finally, the system perspective is discussed briefly. This material is intended to be introductory rather than encyclopedic; later papers in this volume should address the environments, interactions and effects, and application to specific systems in more detail.

Table 1: The terrestrial space environment and effects on space systems.

ENVIRONMENT FACTOR	EFFECTS
SUNLIGHT AND EARTHSHINE	HEATING, THERMAL CYCLING, MATERIAL DAMAGE, SENSOR NOISE, DRAG, TORQUES, PHOTOEMISSION, [POWER]
GRAVITY*	ACCELERATION, TORQUES, [STABILIZATION]
B AND E FIELDS*	TORQUES, DRAG, SURFACE CHARGES, POTENTIALS
NEUTRAL ATMOSPHERE*	DRAG, TORQUES, MATERIAL DEGRADATION, VACUUM, CONTAMINATION, HV BREAKDOWN
PLASMAS*	CHARGING, INDUCED ARCING, PARASITIC CURRENTS, SYSTEM POTENTIALS, ENHANCED CONTAMINATION, ES AND EM WAVES (NOISE), PLASMA WAVES AND TURBULENCE, CHANGE OF EM REFRACTIVE INDEX
FAST CHARGED PARTICLES	RADIATION DAMAGE, SINGLE EVENT UPSETS, ARCING, NOISE
METEOROIDS AND DEBRIS	MECHANICAL DAMAGE, ENHANCED CHEMICAL AND PLASMA INTERACTIONS, LOCAL PLASMA PRODUCTION
SYSTEM GENERATED	SYSTEM DEPENDENT: NEUTRALS, PLASMAS, FIELDS, FORCES AND TORQUES, PARTICULATES, RADIATION

*LOCAL ENVIRONMENT STRONGLY INFLUENCED BY SYSTEM

2.1 SUNLIGHT AND EARTHSHINE

Figure 2 shows an overview of the solar electromagnetic spectrum from gamma ray through radio wavelengths [Refs. 1,2]. Some 99.5% of the Sun's radiant energy is in the 1200 Angstrom to 10 μm wavelength range (ultraviolet (UV), visible and infrared (IR)). Flux levels in this range are relatively constant and do not vary much over the 11-year solar activity cycle. In contrast, flux levels in both the short (gamma-ray, x-ray, extreme UV) and long (radio) wavelength ranges show strong variations (up to about 100X) with solar activity. Indeed, the solar flux at $\lambda = 10.7$ cm is used as a measure of solar activity.

The solar spectrum in Earth orbit in the UV through IR range is well approximated by blackbody radiation from a 5760°K object. The total radiant energy per unit area and time (integrated over wavelengths) measured at 1 AU is called the Solar Constant. The currently adopted value of the Solar Constant is 1371± 5 W m^{-2} [Ref. 1].

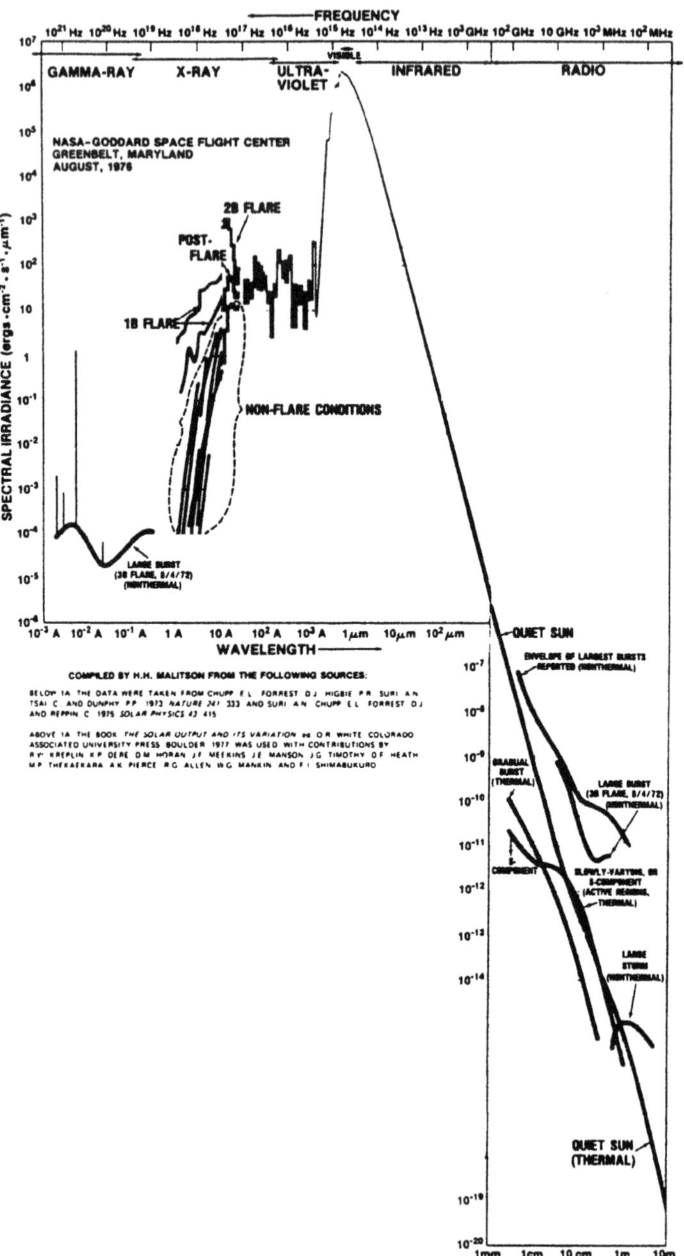

Fig. 2: The solar spectrum [Refs. 1, 2].

The Earth and its atmosphere reflect sunlight and emit thermal radiation. The thermal spectrum as seen from orbit is a composite of the surface spectrum and that from the atmosphere (at wavelengths where the atmosphere is opaque). The surface spectrum is approximated by a 288°K blackbody curve, while the radiation from the atmosphere can be approximated by a 218°K blackbody. Reflected solar radiation (albedo) exhibits strong local variations with surface type and cloud cover. The fraction of total energy reflected (global albedo) is 0.39.

One evident consequence of sunlight and earthshine for space systems is heating. Spacecraft surface materials are frequently selected on the basis of their thermal properties (absorptance and emissivity), because controlling the temperature both inside and on the surfaces of satellites is a basic requirement for proper functioning [Ref. 3]. Earthshine can make an important contribution to the thermal balance of a spacecraft in LEO because of the large solid angle subtended by the Earth. In addition to overall temperature control, it is necessary to consider the effects of thermal expansion of materials. Bonds between materials with different coefficients of thermal expansion will be placed under stress as the temperature changes. The thermal cycling which results from repeated eclipse passage is of particular concern for long term fatigue failure of such bonds (e.g., solar cells and circuits on lightweight plastic substrates). Differential expansion can also result in flexing of large structures. These effects are most significant for LEO spacecraft which pass through eclipse most frequently.

Material degradation due to irradiation by the higher energy portions of the solar spectrum (UV, X, γ) is a significant concern for spacecraft surface materials and vacuum UV stability is a basic requirement for material acceptability. Degradation may affect optical, thermal, electrical and mechanical properties of materials and coatings [Ref. 4].

Specular and diffuse reflection of sunlight results in radiation pressure drag and torques on spacecraft [Ref. 5]. These are small compared to aerodynamic drag and torques at LEO, but become significant at higher altitudes. Photoelectrons released by (mainly UV) photon impact (current density typically about 10^{-9} A/cm^2) play an important role in some plasma/electromagnetic interactions, notably in GEO spacecraft charging. Incoming photons can also be a source of noise in some types of sensors, which must be protected from direct exposure.

Finally, one effect of solar radiation on materials, the production of electron-hole pairs in solids, and particularly in semiconductor diode junctions, is the basis for the most common power source on Earth orbital systems, solar cell arrays!

2.2 GRAVITY

The Earth's gravitational field is a consequence of its mass and mass distribution. The gravitational acceleration is often expressed as the gradient of a scalar potential which is written in terms of a harmonic expansion in spherical coordinates [Ref. 1]. The "monopole" term dominates, with the most significant deviation from sphericity being the flattening at the poles (or equivalently, the equatorial bulge). The coefficients of the various terms in the harmonic expansion for the gravitational potential have been obtained by observing the orbits of low altitude satellites.

In addition to its role in determining spacecraft orbits, the gravitational field plays a role in determining spacecraft orientation via gravity gradient forces and torques [Ref. 6]. These tend to cause a spacecraft to align its "long axis" (actually the axis of minimum moment of inertia) with the local vertical, i.e., in a radial direction. They also produce a tension in the spacecraft structure along the local vertical and produce restoring torques on a misaligned system. These forces and torques are generally quite small, and oscillation periods consequently long. However, gravity gradient is a useful passive method of spacecraft stabilization.

There has been much recent interest in conducting material processing experiments in the "microgravity" environment of space. Indeed, the acceleration felt by a "black box" in a space system will be much less than that due to gravity at Earth's surface. However, it is not, of course, zero. The acceleration felt will depend on the box's location in the spacecraft relative to the system's center of mass, the system's mass distribution, and forces due to system operations and other disturbances such as drag.

2.3 MAGNETIC AND ELECTRIC FIELDS

Figure 3 shows a schematic illustration of the Earth's magnetosphere. Near the Earth, i.e., within a few thousand kilometers of the surface, the magnetic field is essentially dipolar; further out the field becomes greatly distorted. The overall shape of the outer magnetosphere is determined by the interaction between the Earth's magnetic field and the solar wind, and the various plasma currents in the magnetosphere. The Earth's geomagnetic dipole axis is tilted at about 11.5° from its spin axis, and also somewhat displaced from the geographic center so that the "end" of the geomagnetic dipole nearest the geographic north pole (north end of the spin axis) is located in Greenland (geographic coordinates: 78.5°N, 291°E) [Ref. 7]. It may be noted that this is the Earth's south magnetic pole (the north pole of a compass needle points toward it). Because of this tilt and displacement, the Earth's geomagnetic equator, its geographic equator and the ecliptic define three different planes.

Magnetic field effects on systems include torques due to spacecraft magnetic moments and current flows and development of induced potentials due to motion of the spacecraft through the earth's field. Magnetic torques will be exerted on spacecraft having residual net dipole moments, uncompensated current flows or permeable materials onboard, and on spinning spacecraft having charged surfaces. It is customary to "de-Gauss" spacecraft before launch to minimize net moments, to design solar array circuits to minimize net moments due to current flows, and to design attitude control systems to compensate anticipated magnetic torques.

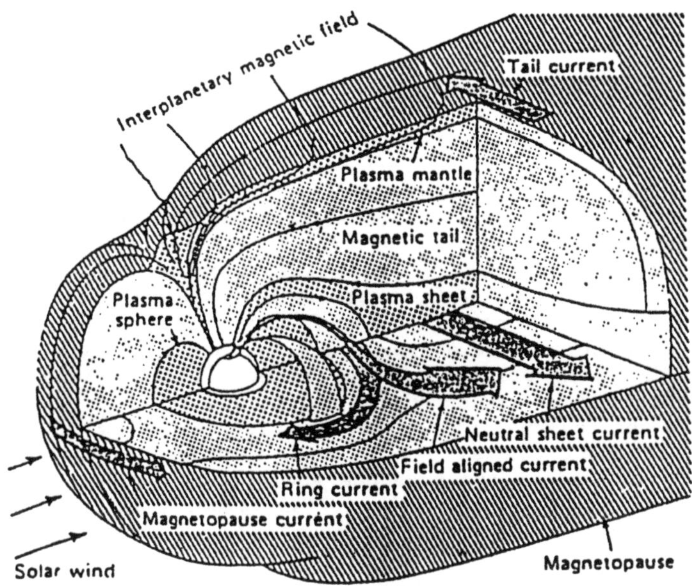

Fig. 3: Schematic of the Earth's magnetosphere.

In the spacecraft's reference frame, its motion through the magnetic field is seen as an electric field and causes a potential to develop across the spacecraft in the direction perpendicular to both **B** and the spacecraft's velocity vector, $\mathbf{v}_{s/c}$, of magnitude $\phi = (\mathbf{v}_{s/c} \times \mathbf{B})\cdot\mathbf{L}$, where L is the spacecraft's dimension. This potential is largest in LEO where both $v_{s/c}$ and B are largest, and zero at GEO. A value of ϕ of about 0.3 V/m is typical at LEO. Very large systems can develop substantial potentials which can drive currents through large spacecraft structures. This is the operational principle of the electrodynamics tether [Ref. 8]; for other systems, e.g., the Space Station, such structural currents are undesirable and to be minimized in design.

There are electric fields present in the magnetosphere, associated with the various magnetospheric current systems. The total voltage drops across the magnetosphere are large, but the fields are small, of order tens of millivolts per meter [Ref. 1], and from the perspective of space system environment interactions, the electric fields tend to be dominated by system-induced potentials. Such potentials include system voltages in the power system and instruments (where exposed to the environment), charging of surfaces by ambient plasma populations, and the motionally induced potentials discussed above. System produced potentials drive plasma interactions in the ionosphere and may result in enhanced chemical reactions (O^+ erosion) and surface sputtering. A charged spacecraft may experience Coulomb drag [Ref. 9] in addition to aerodynamic drag. Additionally, charged surfaces can experience electrostatically enhanced rates of contamination.

2.4 NEUTRAL GASES

Atmospheric density and pressure fall off rapidly with altitude [Refs. 1, 10] so that even at very low orbital altitudes (say 300 km), the vacuum is "hard" by surface standards. Nonetheless, there is sufficient gas present to cause significant effects. A typical number density value at 300 km is 10^9 cm^{-3} (compare to 10^{19} cm^{-3} at sea level). The kinetic temperature of gases at orbital altitudes is high (order 1200°K) compared to surface gas temperatures of order 300°K, and the composition is quite different. Residual atmospheric densities, temperatures and composition show strong variations with solar activity levels and also vary diurnally. Atomic oxygen (O) is an important constituent; it is generally the dominant constituent at altitudes below about 500 km, and dominates to altitudes beyond 1000 km during periods of high solar activity [Ref. 10]. Other important constituents include H, He, N_2, O_2 and Ar.

Because the residual atmosphere is relatively "low energy" - 1100°K corresponds to about 0.1 eV - the local neutral environment is often significantly altered by the presence and operations of a space system. In LEO, a spacecraft's along-track orbital velocity is large compared to the atomic and molecular thermal speeds, so ram and wake regions are formed. Outgassing of materials in the space vacuum is a source of contaminant neutrals, and operation of spacecraft thrusters for attitude control and reboost, as well as water dumps from manned systems, leaks etc., can create high neutral densities.

Neutral gas interactions produce a number of effects on space systems. Atmospheric drag and torques due to the motion of a spacecraft through the residual atmosphere are important considerations for satellite lifetime on orbit and for attitude control [Ref. 6]. Vacuum stability, particularly in the presence of solar UV, is a basic requirement for spacecraft material. Rapid degradation of materials on ram-facing surfaces due to oxidation by the atomic oxygen in the residual atmosphere was brought to light after early Shuttle missions and is an important area of recent research [Refs. 11, 12, 13]. Another phenomenon associated with ram-facing surfaces and of current research interest is "glow" [Ref. 14]. Contamination of surfaces, largely by deposition of spacecraft-produced gases, is another important concern [Ref. 15]. Finally, as higher voltages are employed on spacecraft, designers must allow for the possibility of gas breakdown in regions of high electric fields. Residual atmospheric gas densities are sufficiently low that this is largely a matter of avoiding production of large system-generated neutral clouds in such regions.

2.5. PLASMAS

In addition to the residual neutral atmosphere, ionized gases, or plasmas, abound in near-Earth space. For present purposes, the term plasma is used to refer to those ionized gases in the magnetosphere with energies of $\leq$ 100 KeV, which do not produce significant "radiation" effects. Of primary interest here are the relatively low energy/high density ionospheric plasmas and the higher energy/lower density plasmas associated with geomagnetic substorm activity.

The ionospheric plasma is generally considered to be overall neutral, i.e., containing equal concentrations of electrons and ions, and has characteristic temperatures of the order of 0.1 eV. Plasma density varies with altitude and exhibits peak densities of order 10^6 cm^{-3} at about 300 km (magnetic) equatorial altitude (see Fig. 4). The densities, temperatures and ionic composition vary diurnally, seasonally and with solar activity [Ref. 1]. The motion of the ionospheric plasma particles is greatly influenced by the Earth's magnetic field and the ionospheric plasma co-rotates with the field. Because of the magnetic field's influence, the particle densities also vary with latitude, being larger near the magnetic equator than near the poles at a given altitude.

The other plasma environment of particular interest for spacecraft interactions is that associated with geomagnetic substorm activity. These plasmas are characterized by higher temperatures (5 - 20 KeV) and lower densities (≤ 1 cm^{-3}) than the ionospheric plasmas, as illustrated in Fig. 5 [Ref. 16]. Such plasmas surround spacecraft in near-geosynchronous orbits in the pre-midnight to dawn local time region, and are carried down magnetic field lines to produce higher-energy particle streams in the auroral zones.

A space system in a plasma environment comes into "electrostatic" equilibrium with the plasma by acquiring surface charges and system potentials such that the net current is zero [Refs. 17, 18]. This process occurs point-by-point for insulation, taking local fields, plasma sheath structures and surface charged particle emissions such as secondary

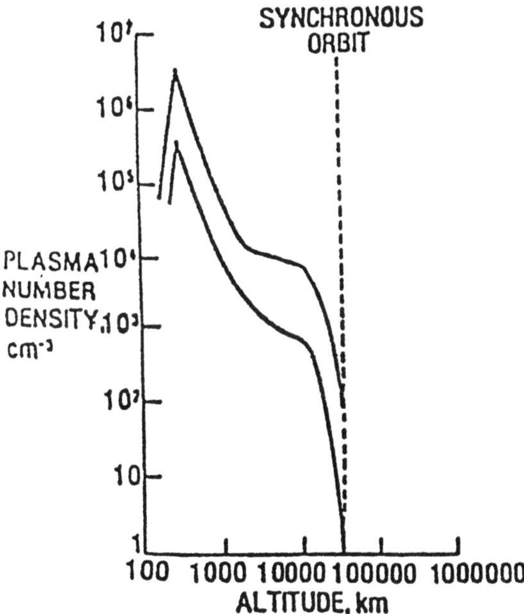

Fig. 4: Densities of low energy plasma in the plasmasphere.

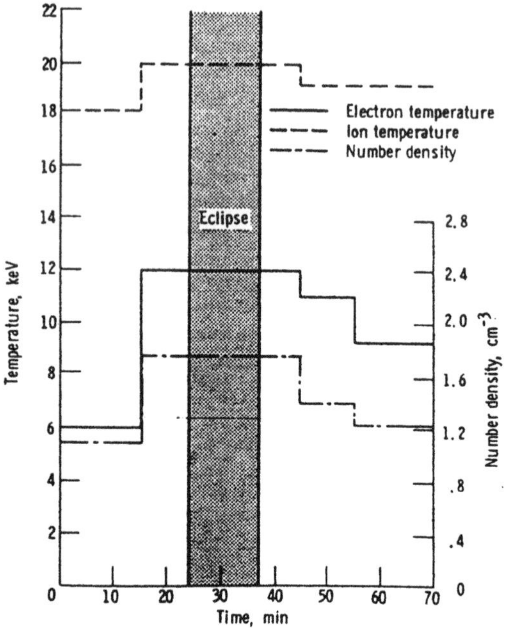

Fig. 5: Model of geosynchronous substorm plasmas (temperature and density).

electrons into account, and globally for conductors. Motionally induced potentials and system-generated voltages must also be considered, as must all sources of current from the natural environment (electrons, ions, photoelectrons) and from the system itself. This equilibration process establishes the system and surface potentials relative to plasma reference. The equilibrium is generally treated as quasistatic but is dynamic in the sense that it changes with any changes in current between the system and the environment.

The charging response of a surface in a plasma may be illustrated by considering the simple case of an insulating surface element and the current densities which must be balanced to obtain net zero current to it, as illustrated in Fig. 6 [Ref. 19]. Current densities to the surface are those due to environmental electrons and ions, secondary and backscattered electrons produced by these primaries, leakage current through the bulk insulator, and, if sunlit, photoelectrons. Other current densities which can be considered are secondary ions due to ion impact (generally a small contribution) and surface leakage currents. In general, all the current densities are functions of the surface potential (and the local electric fields due to adjacent surfaces). For an isolated surface in a plasma, i.e., ignoring photoelectrons and leakage currents and assuming electric fields at the surface due only to its own potential, the "rule of thumb" is that the surface will charge negatively to a potential of the order of the electron temperature. The surface charges negatively because the electrons are much less massive than the ions and consequently their flux is much larger, assuming that the electron and ion temperatures are approximately the same.

For purposes of analyzing plasma-systems interactions, geosynchronous substorm plasmas and ionospheric plasmas represent two distinct regimes. In the hot, tenuous substorm plasmas, the Debye lengths, or distances over which the plasma will rearrange itself to screen a (small) charge or potential, are hundreds of meters, large enough so that in general the space system may be considered small by comparison. Space charge effects are negligible, so Laplace's equation may be used to compute potential distributions around the spacecraft and current densities from the plasma may be computed via orbit-limited Langmuir probe theory. System voltages are generally small relative to the plasma temperatures, so the natural environment drives the interactions electrically. The system may also be considered to be at rest relative to the magnetic field and the plasma. Environmental plasma current densities are very low, the electron current density being typically of order 10^{-10} - 10^{-9} A/cm^2. Because of this, photoelectron currents from sunlit surfaces, with typical current densities of order 10^{-9} A/cm^2, play an important role. Thus, in sunlight, the shadowed insulating surfaces of a spacecraft immersed in a substorm plasma charge negatively as described above, while the sunlit surfaces remain near zero potential until the negative potential on the shaded surfaces becomes large enough to form potential barriers over the sunlit side which suppress the emission of the low energy (kT ≈ 2 eV) photoelectrons, allowing the entire spacecraft to begin charging negatively [Ref. 20]. This process results in the development of kilovolt-level differential potentials

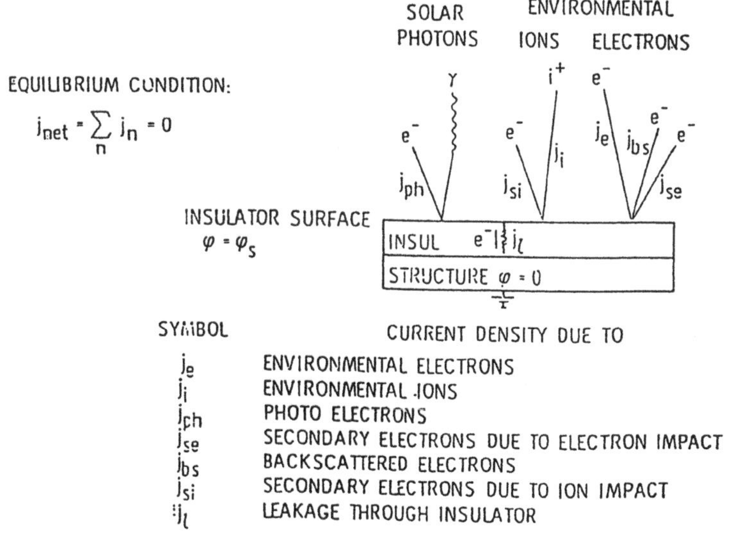

Fig. 6: Charging response: simple case of isolated surface.

between various surfaces, and large local electric fields. The higher energy tail of the electron energy distribution (20 - 100 keV particles) can also penetrate several microns into insulating materials, depositing "buried" charge layers with corresponding high electric fields within these materials. Arc discharges due to differential charging and/or breakdowns due to buried charge layers are believed responsible for anomalous behavior of many geosynchronous satellites [Ref. 21].

By contrast, in the relatively cold dense ionospheric plasmas, Debye lengths are of order mm to cm, so space systems in this regime are large by comparison. Space charge in this regime plays an important role both in sheath formation and in determining current densities to surfaces, necessitating the use of Poisson's equation to compute potentials and space-charge-limited formulations for sheath thickness and current densities. System voltages are generally large compared to plasma temperatures. The system is also in motion relative to both the plasma and the magnetic field. At LEO, the spacecraft along-track orbital speed (≈7.5 - 8.0 km/sec) is large compared to the ion thermal speeds (0.5 - 1.0 km/sec) but small compared to the electron thermal speeds, so that plasma ram and wake phenomena are important. For large systems the motionally induced potential is also important.

In this regime, the initial focus of plasma-system interactions analysis is determination of the electrical floating potentials of the system. Because in this case the system is regarded as high voltage, overall current balance requires that some portions of the system are positive relative to plasma to collect electrons while others are negative to collect sufficient ions to attain net zero current. In general a larger area must be negative, again because of the small mass (large mobility) of electrons relative to ions. The calculation is complicated by the motional effects, the need to account for system-generated plasmas and in some cases the need to account for the effects of secondary emission from insulating surfaces in determining currents to adjacent conductors [Ref. 18]. Concerns include potentials of the system (and in particular the system electrical reference or "ground" potential) relative to plasma, parasitic currents flowing through the plasma, currents driven through structures and arcing of negatively-biased areas, including solar arrays, to the plasma [Ref. 22].

The situation of low altitude, high inclination spacecraft presents some special concerns, because they pass through the auroral zones. At high latitudes, the substorm plasmas, flowing down the magnetic field lines, result in kilovolt-energy streams of electrons and ions at low altitudes in the auroral zones. These impinge on high inclination spacecraft as they traverse these zones and can result in various charging-related phenomena [Ref. 23].

Short term effects of plasma interactions on systems include surface material charging, parasitic currents through plasmas or structures, reference potential shifts relative to plasma, arcing between differentially charged surfaces or from the spacecraft to space with associated current pulses and EMI, and breakdown of thin insulating films covering conductors at voltage. Long term effects include electrostatic enhancement of surface contamination, sputtering induced by ions accelerated into biased surfaces, and enhanced degradation of surface materials due to electrical stress and/or breakdowns.

2.6. FAST CHARGED PARTICLES

Energetic ($\geq$ hundreds of keV) particles in the Earth's environment include the trapped particles (electrons and ions) in the radiation or Van Allen belts, solar flare protons and Galactic cosmic rays [Ref. 24]. Because of their high energies, these particles penetrate into materials on spacecraft and cause a variety of detrimental effects which must be considered in spacecraft design. The trapped particles are an integral part of the Earth's magnetosphere; their motion is dominated by the Earth's magnetic field and comprises spiraling about the magnetic field lines, bouncing or being reflected (mirrored) in the northern and southern hemispheres and drifting in longitude [Ref. 25]. Two zones, the inner and outer are usually identified. Both protons and electrons exhibit peaks in flux in the inner zone at magnetic equatorial altitudes of about 1.5 - 2.0 Earth radii (R_E, measured from the Earth's center). The proton fluxes dominate, with proton energies to hundreds of MeV. This proton population is quite stable. The outer zone exhibits peak fluxes of electrons at around 5 R_E, with energies to about 7 MeV. The electron populations are quite dynamic and much more variable than the protons. Models of the energetic trapped particle populations are maintained by the National Space Sciences Data Center (NSSDC) at NASA's Goddard Space Flight Center [e.g., Refs. 26, 27].

Energetic protons (10 MeV to ~1 GeV) associated with solar flares impinge on Earth's magnetosphere sporadically. Earth orbital spacecraft are somewhat protected from these by the magnetosphere, but systems in high inclination orbits or at very high altitudes can be strongly affected [Ref. 24].

Galactic cosmic rays originate outside the solar system. They comprise protons (85%), alpha particles (14%) and heavy nuclei [Ref. 1], and have energies in the 100 MeV - plus range. The fluxes of these particles with energies below about 10 GeV are modulated via interaction with the heliosphere (solar wind and interplanetary magnetic field) so that their flux is reduced during periods of high solar activity [Refs. 1, 24].

Energetic particles interact with materials to produce displacement of atoms via momentum transfer and to produce ionization in the materials. Displacement disrupts the lattice and produces bulk damage in solids; nuclei are far more efficient in producing displacement than are electrons. Bulk damage is of most concern for semiconductors whose electrical properties can be severely affected [Ref. 24].

Both electrons and nuclei produce ionization in materials. In most cases, it is the cumulative effect of ionization, known as radiation dose (a measure of energy deposited per unit mass of material) which is of chief concern, though the rate at which the energy is deposited (dose rate) is important in some cases [Ref. 24]. Semiconductors, semiconductor devices and dielectrics are subject to this type of damage, which can degrade electrical and mechanical performance.

Degradation of materials due to displacement and ionization processes is generally referred to as radiation damage. Materials, and particularly semiconductor devices, are carefully screened for appropriate radiation resistance for use in space applications.

In addition to the long term effects of energetic particle impingement, there are a group of effects, known as single event phenomena, which present particular hazards to semiconductor devices and are the consequence of a nuclear particle passing directly through a junction and depositing charge there [Ref. 28]. These include: Single Event Upset (SEU) in which the impinging nucleus produces sufficient charge to change the logic state of the device, but the circuit is such that it may be reset by command; Single Event Latchup (SEL) which is the same except that the affected logic switch cannot be reset; and Single Event Burnout (SEB) in which the particle, impinging on the gate region of a power FET results in its failure and shorting of the power supply. Evidently, some level of SEU may be tolerable, but SEL and SEB must be avoided! The single event phenomena present increasing hazards as technology "improves" to produce devices which are smaller and lower energy, because such devices are vulnerable to lower energy impinging particles than their "less sophisticated" counterparts.

As a final note, energetic particles impinging upon a spacecraft produce secondary radiation in passing through the spacecraft's materials. This includes Bremsstrahlung or "braking radiation" produced as electrons are slowed down, and secondary protons and neutrons produced by impinging nucleons. These may be of concern for some applications [Ref. 24].

2.7. METEOROIDS AND DEBRIS

The term "meteoroids" refer to the solid particulates of extraterrestrial origin present in interplanetary space in the Earth's vicinity. There are a number of models for the flux of meteoroids as a function of mass, one of which (due to Grun et al. [Ref. 29]) is shown as Fig. 7. The range of masses is extensive; the models are given in terms of cumulative flux in number m^{-2} sec^{-1} versus mass in gm. The model shown in Fig. 7 is for the interplanetary flux at 1 AU. In order to use such a model to analyze meteoroid impacts on Earth orbital spacecraft it is necessary to modify the interplanetary flux to account for the presence of the Earth and the velocity of the spacecraft in orbit. Meteoroids also exhibit a distribution in velocities which is presumed isotropic in direction, but whose speeds change between Earth and interplanetary space [Ref. 30]. The range of velocities is from 11.1 km/sec (escape velocity) to about 72.2 km/sec at the top of the Earth's atmosphere, with an average of about 16.9 km/sec. In interplanetary space, velocities range down to zero, and the average also shifts down, to about 12-13 km/sec.

Modifications to the interplanetary flux of meteoroids for Earth orbital applications must account for gravitational focussing (which also accounts for the change in the velocity distribution) and geometry factors due to shielding of the flux by the Earth (these depend on altitude and the orientation relative to the Earth of the spacecraft surface of interest). Gravitational focussing generally increases the flux as the Earth is approached according to the formula for the focussing factor χ [Ref. 31].

$$\chi(H) = 1 + v_{esc}^2/v_\infty^2$$

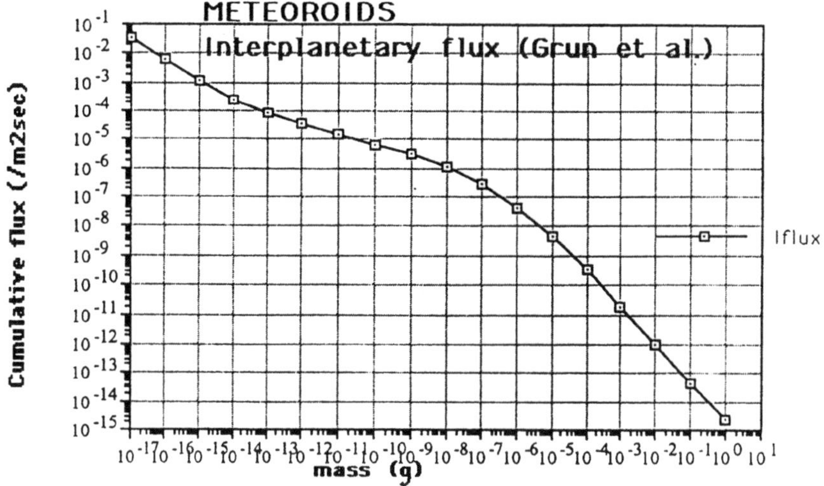

Fig. 7: Model of interplanetary meteoroid flux [Ref. 29].

where H is the altitude, v_∞ is the particle velocity in interplanetary space and v_{esc} is the escape velocity at H which is given by

$$v_{esc} = (2GM/(R+H))^{1/2},$$

with G the gravitational constant, M the Earth's mass and R the Earth's radius. It may be noted that in general for a particle at altitude H,

$$v(H) = (v_\infty^2 + v_{esc}^2(H))^{1/2},$$

which accounts for the velocity distribution modification implied by gravitational focussing. The Earth shielding factor (ESF) for a particular surface will depend on altitude and surface orientation [Ref. 32]. It is always ≤ 1. The adjustments χ and ESF do not account for the orbital motion of the spacecraft through the (isotropic) velocity distribution of meteoroids at altitude H. This requires an additional adjustment (akin to a ram/wake calculation) to the flux which depends on spacecraft velocity in orbit and surface orientation relative to this velocity [Ref. 32]. In addition to the meteoroids there are solid "particles" in Earth orbit of manmade origin. About 5% of these are active spacecraft; the rest include inactive spacecraft, spent stages, fragments and smaller particulates resulting from collisions, explosions and operations of spacecraft [Ref. 33]. These are not truly part of the natural environment but are a part of the orbital environment of increasing concern as a hazard to space systems, and are referred to as orbital debris. Most of these are in orbits below about 2000 km altitude and in all inclinations, with a smaller "ring" near geosynchronous orbit. The larger debris objects (> 10 cm diameter) are tracked by the US Systems Command; in this size range they far exceed natural meteoroids in flux. There is evidence from Skylab, Solar Max and other sources for the presence of debris particles in small mass ranges as well [Ref. 33]. Analysis of impact data from LDEF, currently underway, should help to resolve the debris fluxes in the smaller mass ranges [Ref. 34]. The most recent model for debris flux versus mass has been developed by Kessler et al. [Ref. 35]. Because the debris population is in orbit about the Earth, its velocity distribution is quite different from that of the meteoroids; average velocities are lower, and the velocity of impact is strongly dependent on spacecraft surface orientation (in a given orbit, debris particle speeds are comparable to spacecraft orbital speeds).

The obvious concern for spacecraft from both meteoroid and debris impacts is mechanical damage. Aside from the possibility of a collision with some large debris object (probably catastrophic, but statistically still reasonably unlikely so far!), the concern is for long term material degradation from repeated small impacts, and for protection of critical elements which might be vulnerable to a single impact. Examples of the latter would be puncture of a pressurized vessel or a critical electrical element. There is some evidence that particulate impact of insulation separating power system components at different voltages can produce permanent short circuits, particularly at higher ($\geq$ 100 V) voltages [Ref. 36]. In impacting a material at high velocity, a particulate vaporizes (as does some of the impacted material) and produces a dense localized plasma which can conduct large currents if an electric field is present. Such currents if sustained by external voltage sources (e.g., the power system) can result in more extensive damage and permanent short circuits.

Particulate impacts can also produce synergistic effects which may be hazardous. These include enhanced rates of chemical (e.g., O erosion) or plasma interactions due to damage of thin protective coatings and impact-stimulated release of buried charge in insulators such as surface materials of geosynchronous or polar orbiting spacecraft.

2.8. SYSTEM GENERATED ENVIRONMENT

The contribution of the system to its local environment depends on characteristics of the system such as size, power/voltage levels, type of operations and/or payloads and so forth. The level of disruption of the natural environment also depends on that environment itself in the sense that it is the "lower energy" environment factors which are typically most readily changed by the system's presence and activities. Environmental factors likely to be affected include neutral gases, low energy plasmas and fields.

A system may affect its local neutral gas environment via its motion through the ambient neutral population in LEO, producing ram and wake phenomena. It may contribute directly to the neutral population via outgassing, operation of thrusters, effluent dumps, chemical gas releases from experiments and so forth.

Low energy plasma populations in LEO are similarly affected by spacecraft motion and releases of ionized gases, including ionized fractions of thruster plumes, as well as plasma components produced by charge exchange of ambient ions with spacecraft-released neutrals. These plasma populations are also strongly affected by the electric and magnetic fields generated by spacecraft power systems, surface charges and/or motion.

Surface charging and the resulting electric field configurations play an important role in geosynchronous spacecraft charging because they control the behavior of the low energy secondary and photoelectrons which constitute an important part of the current balance [Ref. 20]. In geosynchronous, as in LEO, any spacecraft-generated source of charged particles will alter the current balance and therefore the charge state of the spacecraft.

As noted earlier, the local acceleration (perceived "gravity") onboard a space system will be affected by thruster operations, internal motions and so forth. Spacecraft sources of radiation include secondary radiations and onboard nuclear devices.

A comprehensive treatment of environmental effects for any given environment factor requires assessment of the spacecraft-generated components, which become part of the local environment. A fully comprehensive environment compatibility assessment at the system level demands consideration of effects for all the environment factors.

3. System Design for Environment Compatibility

A conceptual flow chart for the system design process is shown as Fig. 8. One begins with a mission, a stated goal or set of goals for the system to fulfill. The mission may be an operational application such as a communications platform, scientific data gathering (e.g., measurements of the space environment itself or space-based observations of Earth or other celestial objects), or it may be technology demonstration. It may be manned or unmanned. Some systems such as the Space Station have a multiplicity of intended uses; this renders the design process more complex. The mission, once defined, determines the system's orbit or location in space, and defines certain system characteristics such as general size, power level, lifetime and complexity of required operations.

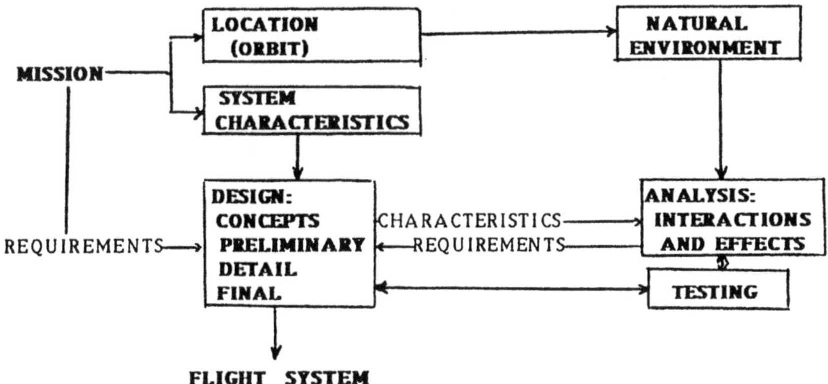

Fig. 8: Conceptual flow chart for system design process.

The orbital location (and for some factors, mission duration) in turn determines the natural environment in which the system must operate. The system characteristics lead to a conceptual design which is refined in stages (preliminary, detailed, final) for each subsystem and for the system as a whole. As part of this design/refinement process, analyses are made to the system's interactions with its environment and the consequences for performance. Such analyses use as input various specifications for the environment factors and the system characteristics, and produce requirements on the system design which are incorporated into design refinements. Testing is conducted at various stages to aid in the analysis and as proof of concepts and design demonstration. At each stage the mission requirements are reviewed to ensure that the system will fulfill the mission. In cases where requirements conflict, trade studies must be done, with satisfaction of the mission objectives as the deciding factor.

Design approaches to control environmental impacts include selection of materials and coatings, selection of electrical components and circuits, protection of critical elements (e.g., from impact or radiation), hardening, design and placement of subsystem elements to minimize interactions or interference and restrictions on operations of some subsystems. In some cases, control of the local environment is indicated, such as providing cryogens or pressurized containers. Use of plasma sources to control spacecraft potentials has been considered for some systems. Other active environment control measures may also be useful.

Requirements stemming from one set of environmental compatibility concerns are frequently different from, and sometimes in conflict with those stemming from another set. In such cases subsystem or system level trade studies are necessary to identify the optimum design choice. As an example, a material with desirable thermal and vacuum stability properties may be unacceptable from an electrical viewpoint (too conductive, too insulating or unstable under atomic oxygen bombardment.) Then another material must be sought, or perhaps a new material or coating developed for the application.

The usual approach to environmental interactions has been to treat each environment factor separately, performing analyses and conducting tests, and placing requirements on materials, electronics, subsystems and so forth as indicated. The requirements levied by the individual analyses must then be integrated via system requirements documentation. As missions and systems become more complex, this approach becomes more tedious and subject to conflict. In recent years there has been increasing effort to develop more comprehensive engineering analysis tools to aid in system-level design trade studies. Examples of this trend include ESABASE [Ref. 37] in Europe and EPSAT [Ref. 38] in the US.

4. Conclusion

The natural environment of near-Earth space is complex and dynamic and contains many components, some of which are strongly influenced by the system's presence and operations. Environmental compatibility is an essential element of system design, with the relative importance of various environmental effects ultimately determined in each case by their impact on the system's ability to perform its mission. Traditionally, environmental effects on systems have been considered as a set of separate entities, with each environment factor treated separately in terms of analysis and testing, and resulting in separate sets of requirements on materials, electronics, subsystems, etc. which are then integrated at the system level via requirements documents. As missions and their associated systems become more complex, such an ex-post-facto approach becomes inefficient at best. Development of comprehensive environmental compatibility analysis tools and inclusion of environmental compatibility assessments as an integral part of the design process from the inception of mission/system definition would appear to be the direction for future progress.

The intent of this paper has been to introduce, in overview form, the various environment factors and some of their effects on Earth orbital systems, and to suggest a "systems" perspective from which to view these environment interactions and impacts. The coverage has been intentionally brief. It is anticipated that the following papers in this volume will provide details of the environment, the interactions and effects, and summarize experience with some actual space systems from the environment effects perspective.

References

[1] R. E. Smith and G.S. West, compilers (1983), Space and Planetary Environment Criteria for Use in Space Vehicle Development, 1982 Revision (Volume 1), NASA TM 82478.

[2] O.R. White, ed., The Solar Output and its Variation, Colo. Assoc. Univ. Press, Boulder (1977).

[3] D.F. Gluck, Space Vehicle Thermal Testing; Principles, Practices and Effectiveness, NASA/SDIO Space Environmental Effects on Materials Workshop, NASA CP 3035, Part 2, p. 395 (1989).

[4] W.S. Slemp, Ultraviolet Radiation Effect, ibid, p. 425 (1989).

[5] W. J. Evans, Aerodynamic and Radiation Disturbance Torques on Satellites Having Complex Geometry, **Torques and Attitude Sensing in Earth Satellites**, S. F. Singer, ed, Academic Press, p. 83 (1964).

[6] R.F. Fischell, Passive Gravity Gradient Stabilization for Earth Satellites, ibid., p. 13 (1964).

[7] S.B. Gabriel and H.B. Garrett, An Overview of Charging Environments, NASA/SDIO Space Environmental Effects on Materials Workshop, NASA CP 3035, Part 2, p. 495 (1989).

[8] M. Martinez-Sanchez and D.E. Hastings, A Systems Study of a 100 KW Electrodynamic Tether, Tethers in Space, Advances in the Astronautical Sciences, Volume 62, P.M. Bainum, I. Bekey, L.

Guerriero and P.A. Penzo, eds, American Astronautical Society, p. 341 (1987).

[9] S.F. Singer, Forces and Torques Due to Coulomb Interaction with the Magnetosphere, **Torques and Attitude Sensing in Earth Satellites**, S.F. Singer, ed, Academic Press, p. 99 (1964).

[10] **US Standard Atmosphere**, NOAA-S/T 76-1562, US Government Printing Office (1976).

[11] B.A. Banks and S.K. Rutledge, LEO Altitude Atomic Oxygen Simulation for Materials Durability Evaluation, proc. Fourth European Symposium on Spacecraft Materials in Space Environment, CERT, Toulouse, France, 6-9 September, p. 371 (1988).

[12] D. Morison, R.C. Tennyson, J.B. French, T. Braithwaite,, M. Mosian and J. Hubert, ibid., p. 435 (1988).

[13] R.C. Tennyson, Atomic Oxygen and its Effect on Materials, this volume.

[14] B.D. Green, The Spacecraft Flow: A Review, proc. Fourth European Symposium on Spacecraft Materials in Space Environment, CERT, Toulouse, France, 6-9 September, p. 477 (1988).

[15] E.N. Borson, Spacecraft Contamination Control and the Selection of Materials, ibid., p 275 (1988).

[16] C.K. Purvis, H.B. Garrett, A.C. Whittlesey and N.J. Stevens, Design Guidelines for Assessing and Controlling Spacecraft Charging Effects, NASA TP 2361 (1984).

[17] E.C. Whipple, Potentials of Surfaces in Space, Rep. Prog. Phys., 44, p. 1197-1250 (1918).

[18] I. Katz, Current Collection by High Voltage Spacecraft, this volume.

[19] C.K. Purvis and D.C. Ferguson, Surface Phenomena in Plasma Environments, NASA/SDIO Space Environmental Effects on Materials Workshop, NASA CP 3035, Part 2, p. 511 (1989).

[20] C.K. Purvis, The Role of Potential Barrier Formation in Spacecraft Charging, **Spacecraft/Plasma Interactions and Their Influence on Field and Particle Measurements**, A. Pedersen, D. Guyenne and J. Hunt, eds, European Space Agency, EAS-SP-198, p. 115-124 (1983).

[21] G.L. Wrenn and A.J. Sims, Spacecraft Charging in Geosynchronous Orbit, this volume.

[22] D. Hastings, Arcing on High Voltage Power Systems in Low Earth Orbit, this volume.

[23] E.J. Daly, Plasma Interactions at Low Altitude, this volume.

[24] **The Radiation Design Handbook**, ESA PSS-01-609 (draft), European Space Agency, Paris (1989).

[25] A.L. Vampola, The Space Particle Environment, NASA/SDIO Space Environmental Effects on Materials Workshop, NASA CP 3035, Part 2, p. 367.

[26] D.M. Sawyer and J.I. Vette, AP-8 Trapped Proton Environment, NSSDC WDC-A-R&S 76-06.

[27] M.J. Teague, K.W. Chan and J.I. Vette, AE-6, A Model of Trapped Electron for Solar Maximum, NSSDC WDC-A-R&S 76-04 (1989).

[28] W.A. Kolasinski, Effects of Space Radiation on Electronic Microcircuits, NASA/SDIO Space Environmental Effects on Materials Workshop, NASA CP 3035, Part 2, p. 383 (1989).

[29] E. Grun, H.A. Zook, H. Fechtig and R.H. Giese, Collisional Balance of the Meteoritic Complex, Icarus, 62, p. 244 (1985).

[30] H.A. Zook, The State of Meteoritic Material on the Moon, Proc. Lunar Planetary Science Conference - LPI Houston, p. 1653 (1975).

[31] E.J. Opik, Collisional Probabilities with the Planets and the Distribution of Interplanetary Matter, Proc. Royal Irish Acad., p. 165 (1951).

[32] T.J. Stevenson, K. Sullivan, P.R. Ratcliff and J.A.M. McDonnell, The Space Particulate Impact Environment on NASA's Long Duration Exposure Facility, unpublished manuscript, USS/Physics Laboratory, Univ. of Kent at Canterbury (1991).

[33] D.J. Kessler, Orbital Debris Environment and Data Requirements, NASA/SDIO Space Environmental Effects on Materials Workshop, NASA CP 3035, Part 1, p. 281 (1989).

[34] J.A.M. McDonnell, The Near Earth Space Particulate Environment, this volume.

[35] D.J. Kessler, R.C. Reynolds and P.D. Anz-Meador, Orbital Debris Environment for Spacecraft Design to Operate in Low Earth Orbit, NASA TM 100471 (1989).

[36] K. Bogus, F. Krueger, M. Rott, E. Schneider and H. Thiemann, Micrometeoroid Interactions with Solar Arrays, proc Fourth European Symposium on Spacecraft Materials in Space Environment, CERT, Toulouse, France, 6-9 September, p. 675 (1988).

[37] M. Frezet, ESABASE Extension to Spacecraft Charging, Final Report to ESTEC, June, 1985 (1985).

[38] G.A. Jongeward, R.A. Kuharski, E.M. Kennedy, K.G. Wilcox, N.J. Stevens, R.M. Putnam and J.C. Roche, Environment-Power System Analysis Tool Development Program, in Current Collection From Space Plasmas, N. Singh, K.H. Wright and N.H. Stone, eds, NASA CP 3089, p. 352 (1990).

INTRODUCTION TO THE SPACE ENVIRONMENT

R. W. NICHOLLS
Centre for Research in Earth and Space Science
York University
4700 Keele St., North York, Ontario, Canada M3J 1P3

ABSTRACT. *Near-Earth Space* is the principal space environment in the context of the Advanced Study Institute. It includes the (≅ 1000km thick) terrestrial atmosphere, in which most contemporary space activity takes place, and the magnetosphere. For planetary missions interplanetary space and the atmospheres of other planets are also important parts of the space environment. The near-earth space and planetary space environments are greatly influenced by the effects of solar particulate and electromagnetic radiation, and the effects of planetary gravitational and magnetic fields.

This paper introduces the concepts of space activities and the space environment. It also reviews the properties, constitution, structure and energy exchange processes of the atmosphere with emphasis on neutral species.

1. Introduction

This NATO Advanced Study Institute is devoted to a discussion of the Behaviour of Systems in the Space Environment. In the introductory presentations we therefore first review the nature and properties of the environment in which space systems operate to provide a scientific context for the discussions of space systems which follow.

Contemporary space activities of many nations may be broadly classified into the following categories:

- Space Applications
  Use of Communications and Earth Observations Satellites.

- Space Technology
  Fabrication and engineering of components of launch systems, space vehicles, payloads, communications and data handling systems.

R. N. DeWitt et al. (eds.), The Behavior of Systems in the Space Environment, 45–66.

- <u>Space Science</u>
  Use of Space Systems to perform research:
  IN SPACE (e.g., microgravity research in space laboratories)
  ON SPACE (e.g., study of local space plasmas)
  FROM SPACE (e.g., astronomical and earth observations from observational satellites)

Apart from the use of communications satellites and some planetary probe activities, most of the above space applications and space science activities are carried out within the 1000km thick terrestrial atmosphere. A few extend into the magnetosphere and the solar-terrestrial region of near earth space, and some occur in interplanetary space. These are the regions of the space environment with which we are mainly concerned in this Advanced Study Institute.

The principal influence to which near-earth space is subjected is the energy flux from the sun. It has particle (solar wind) and electromagnetic radiation (from the X-ray to microwave) components. Solar properties and radiation are reviewed in following contributions.

The magnetic field of the earth exerts a profound influence on the charged particles in the solar wind. It gives rise to the magnetosphere and the radiation belts in near earth space. Further, the terrestrial magnetic field exerts a great influence on the ionospheric layers of the high atmosphere. The properties and phenomena of the magnetosphere and of the ionospheric are also reviewed in following contributions.

The properties of interplanetary space, and the properties of cosmic rays and energetic particles, all of which contribute to the space environment are also reviewed in later lectures.

Part of the local near-earth space environment is particulate, with natural (micrometeors) and anthropogenic (space debris) components. Both of these have to be understood, for they can cause collision damage to space systems. They too are each reviewed below in the hypervelocity and impact physics contributions.

Typical space instrument platforms, and their nominal altitudes are as follows.

INSTRUMENT PLATFORMS	ALTITUDES (km)
Ground (upward looking)	0
Aircraft	5-10
Balloons	25-30
Sounding Rockets	100-200
Earth Observations Satellites	300-700
Astronomical Satellites (various)	500; 600; 30;000; 80,000
Shuttle	300
Space Station/Mir	700

The near-circular orbits of many spacecraft thus lie in the upper atmospheric region of near-earth space. Spacecraft with more eccentric orbits traverse not only the upper atmosphere but also the more remote regions of near-earth space. Consequently many spacecraft experience the physical conditions of the mesosphere, the thermosphere and the exosphere of the terrestrial atmosphere as well as the magnetosphere in some cases. The conditions are determined by neutral and charged particle interactions, particulate collisions, electromagnetic radiation over a very broad spectral range, and electric and magnetic fields. We therefore devote much of this presentation to an overview of the space environment of the terrestrial atmosphere.

The orbits of other spacecraft (e.g., geosynchronous satellites, magnetospheric satellites and planetary probe satellites) are much further from the earth, and they experience the physical conditions of the magnetosphere and of interplanetary space.

It should also be remarked that unlike typical laboratory circumstances, at satellite orbital speeds of say 10 km/sec, the kinetic energy of collision between the satellite surface and an environmental particle, such as an O-atom, is about 8ev which can give rise to significant physico-chemical effects. These are discussed in later presentations.

An excellent contemporary reference to some of these topics is Handbook of Geophysics and the Space Environment [Ref. 23, 3rd edn] produced by the Air Force Geophysics Laboratory. The earlier editions of this fine reference work [Ref. 30, 31] and also the Satellite Environment Handbook [Ref. 22] are also of great scientific and historical interest.

2. Structure of the Terrestrial Atmosphere

We are concerned here with the general properties of the terrestrial atmosphere as it currently exists. It is beyond the scope of this brief presentation to discuss the long term evolution of the atmosphere into its current state [see e.g., Ref. 19, 32].

The atmosphere is a complex multicomponent compressible fluid. At different altitudes it is influenced by, and exhibits the following phenomena:

- the mechanics of compressible fluids and their mass transport (e.g., winds, meteorology and climate)
- atmospheric (photo) chemistry; the production of atomic and molecular neutral and ionized species and their energetic reactions (e.g., aeronony)
- atmospheric ionization and electromagnetics (e.g., ionospheric phenomena and radio propagation)
- atmospheric luminosities (e.g., the aeronomy of the aurorae, airglow, shuttle glow, etc.)

Continuum, gas kinetic and particle ballistic models have been developed to describe the behaviour of various atmospheric layers.

Many ground based, aircraft and balloon based, sounding rocket and shuttle based, and satellite based instrument packages, which depend on diverse technologies have been developed over the past few decades to study atmospheric phenomena by remote sensing and by immersion sensing methods. The interpretation of such work requires an extremely multidisciplinary approach, calling for the skills of meteorologists, climatologists, atmospheric scientists, modellers, physicists, chemists, equipment designers, engineers, data analysts, etc. Further, while a general understanding of the structure, composition and processes of the atmosphere exists, that understanding is still rapidly evolving. New phenomena such as the polar ozone "holes" [Ref. 3, 4, 15] continue to be revealed and are intensively studied.

In addition to the various editions of the Handbook of Geophysics cited above, other excellent texts on atmospheric properties and aeronomic processes exist. For authoritative discussion of some of the topics briefly reviewed below, the material in Ratcliffe [Ref. 27], Whitten and Poppoff [Ref. 33, 34], Banks and Kockarts [Ref. 1], Chamberlain [Ref. 8], Brasseur and Solomon [Ref. 2] and Rees [Ref. 28] is highly recommended.

The upper limit of the atmosphere is not sharply defined. It is the region where the number density of particles is indistinguishable from that of the magnetospheric medium into which it merges. Its location and properties are time dependent and vary with the strength of solar activity. For our purposes it is convenient to adopt about 1000km as the altitude of the fringe region of the atmosphere from where the light mass components H and He escape into space. In that region the collisional mean free paths are on the order of 10^6 m. The gas laws do not strictly apply, and molecular trajectories are ballistic.

The properties of the 1000km thickness of the atmospheric environment are reviewed below, for that is the region in which many space systems are operationally immersed. The lower regions of the atmosphere are electrically neutral. The upper layers have significant populations of electrons and positive ions in the various layers of the ionosphere, the plasma properties of which are significant.

The principal species of the atmosphere and the variation of their concentration with altitude is displayed in Figs. 1a and 1b. It is clear from these figures that the concentration of each of the most abundant components (N_2, O_2, A, CO_2) appear to decrease exponentially with altitude to about 86km, and that their relative concentrations do not change much up to this altitude. The presence of photochemically produced species like O, OH, NO, H is also clear. Above 86km the variation of species abundance with altitude is somewhat more complex.

The principal atmospheric regions (troposphere, stratosphere, mesosphere, thermosphere and exosphere) are characterized by the atmospheric temperature profile. This is illustrated in Fig. 2. The "pauses" where $dT/dz=0$ were suggested by Chapman to delineate the boundaries between regions. The physical conditions in each of the regions, and the characteristic processes which take place in them vary from region to region. The temperature profile of a region is a manifestation of its heating budget which depends upon absorption and emission of radiation and the related photo-dissociation, photo-ionization processes and mass transport processes.

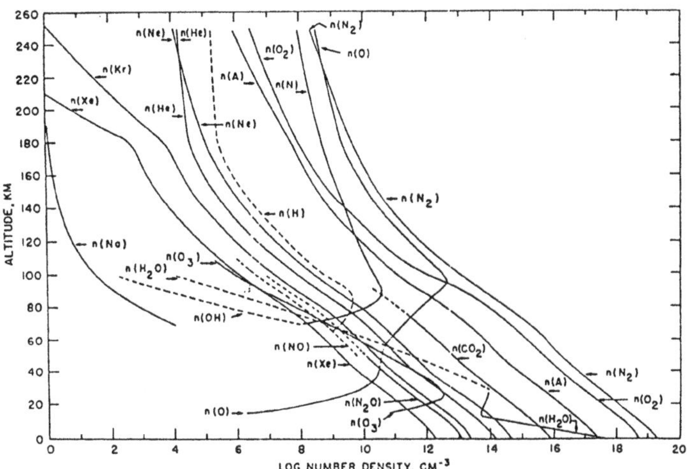

Fig. 1a: Vertical Distribution of Atmospheric Constituents from Surface to 25 km [Ref. 30, 8-2].

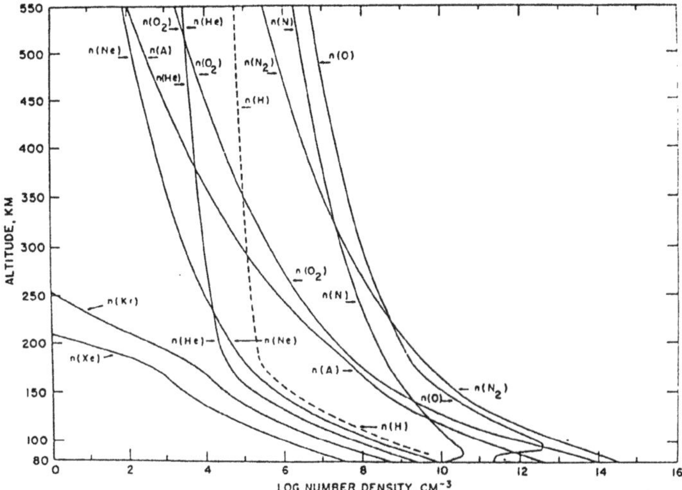

Fig. 1b: Vertical Distribution of Atmospheric Constituents from 80-500km [Ref. 30, 8-3].

Energy input in a layer depends upon the product of number density of absorbing species (N_i), frequency distribution of radiation density $\rho(\nu)$ (or intensity $I(\nu)$) at the altitude concerned, and the cross-section ($\sigma(\nu)$)

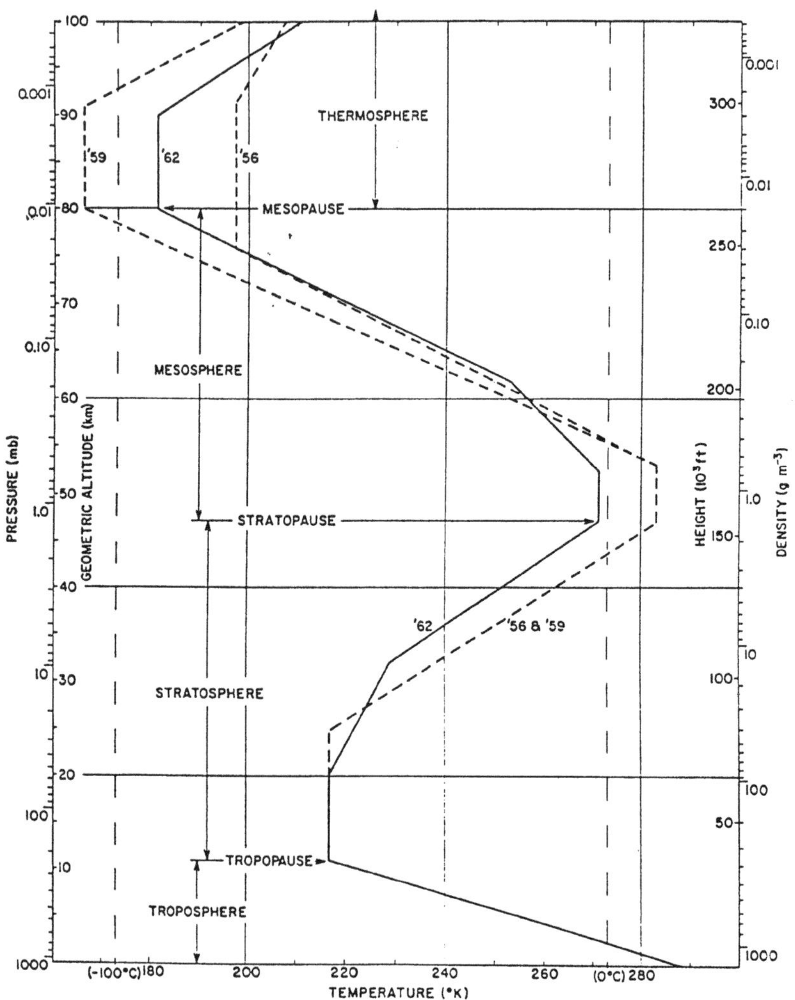

Fig. 2: Temperature profile with Atmospheric Layers 2-1 HdG 1965.

for the absorption process. The profile of a layer at altitude z is then controlled by $N_i(z)\ I(\nu,z)\ \sigma(\nu,z)$. By extension of these concepts, Chapman [Ref. 11, 12] developed a celebrated phenomenological model of atmospheric layer formation in which account is taken of radiation absorption leading to photodissociation or photoionization. Its applications to the formation of the ionospheric layers is briefly reviewed elsewhere [Ref. 31, 2-1].

2.1. ATMOSPHERIC LAYERS

The troposphere, which occupies the important lowest ≃ 10km of the atmosphere, contains about 75% of the atmospheric mass at any point (depending on latitude). It is the realm of climate, weather, meteorology, aircraft flight and, of course, life. It maintains important environmental interactions with all parts of the earth's surface. The planetary boundary layer where viscous phenomena and wind shears are emphasized lies in the lowest kilometer of the troposphere. The universal gas law

$$P = \rho R^*T/M_o = NR^*T/N_A = N\ k\ T \tag{1a}$$

holds well in the troposphere. At altitude z, P is the total pressure, ρ is the mass density, R^* is the universal gas constant, k is Boltzmann's constant, T is the absolute temperature, M_o is the mean molecular weight, N is the total number density of molecules, and N_A is Avogadro's constant.

The partial pressure P_i of species i of number density n_i is:

$$P_i = n_i\ k\ T \tag{1b}$$

and

$$P = \Sigma\ P_i\ . \tag{1c}$$

The vertical equilibrium of an atmospheric slab of vertical thickness dz which supports a pressure change of -dp is given by:

$$-dP = g\ \rho\ dz \tag{2a}$$

where g is the altitude-dependent acceleration of gravity $MG/(R_o+z)^2$. M and R_o are, respectively, the terrestrial mass and radius, and G is the universal constant of gravitation.

From Eqs. (1a) and (2a)

$$P/P_o = \rho/\rho_o = N/N_o = \exp\ [-z/H]\ , \tag{2b}$$

where

$$H = R^*T/gM_o \tag{2c}$$

is the so-called "*scale height*". It is the altitude change over which the pressure, mass density and number density changes by 1/e (≅0.46). The concept of scale height has many applications in the aeronomy of planetary atmospheres. It is in general z-dependent, for both T and M_o vary with altitude.

Equation (2b) is the familiar "barometric" equation. Its excellent representation of atmospheric circumstances to an altitude of about 86km is seen in the linearity to this altitude of the semi-logarithmic plots of abundance of major species in Fig. 1. The semi-logarithmic plots of Figs. 3a and 3b display the variation of P, ρ, N, T, M_o, mean free path,

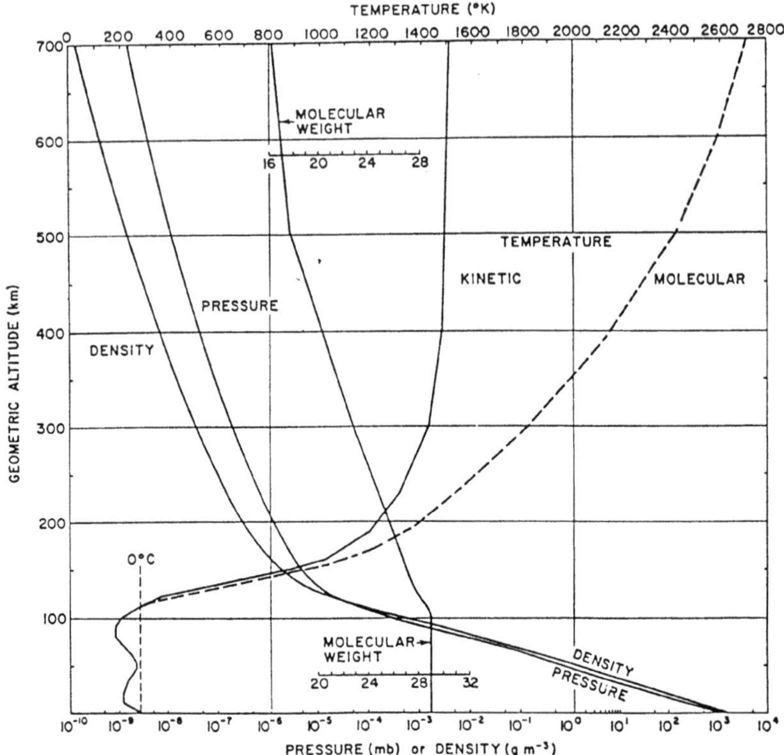

Fig. 3a: Profiles to 700km of temperature, pressure, density and molecular weight from U.S. Standard Atmosphere 1962 [Ref. 31, 2-2].

collison frequency and particle speed, with z up to 700km. Note the linearities of P, N, and ρ in these plots up to 86km.

The reason that the barometric equation so well represents atmospheric conditions up to 86km is seen in Fig. 3a which displays the variation of mean molecular weight M_o of atmospheric constituents with altitude to 700km. M_o is constant (28.964kg/kmol) from the surface to about 86km, which is the altitude of the onset of the thermosphere. Over this range of altitudes the effects of diffusion and of photochemistry have a minimal influence on M_o. The relative proportions of the major constituents is unchanged. The decrease of M_o with altitude in the thermosphere above 86km is due to photodissociation and photoionization of major species in the thermosphere.

Photochemical effects also occur in the stratosphere and the mesosphere with the consequent formation of important minor constituents, as is seen in Fig 1a. This has a very small effect on the mean molecular weight in those regions because the number density of new species (principally O and O_3) formed in the stratosphere are about 7 orders of

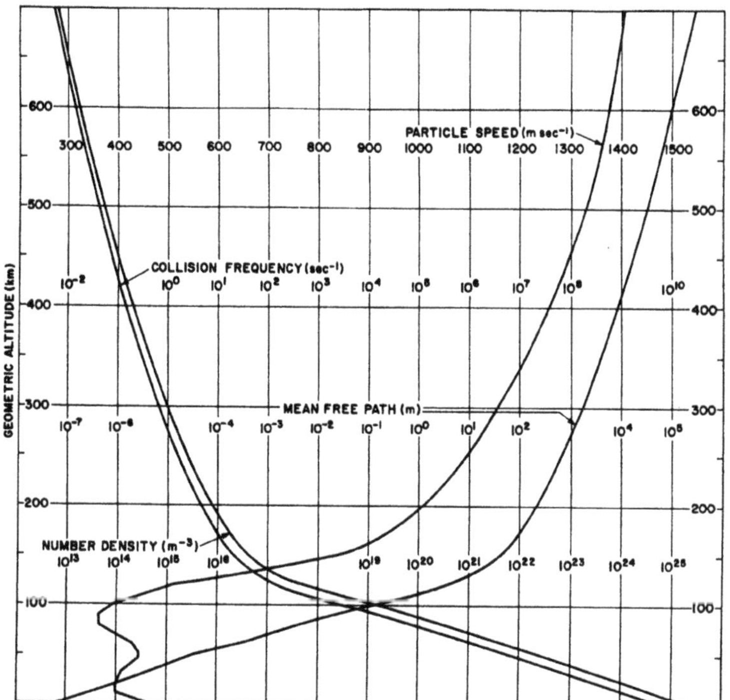

Fig. 3b: Profiles to 700km of number density, collision frequency, mean free path and particle speed from U.S. Standard Atmosphere 1962 [Ref. 31, 2-3].

magnitude below the number densities of the principal atmospheric constituents, (N_2, O_2, A, CO_2) which control M_o.

In the troposphere, the constant decrease in temperature with height of between 6 or 7 °K/km is called the (adiabatic) lapse rate. In Fig. 2a we recognize the tropospause at about 10km. Above this is the stratosphere through which the temperature rises as a result of absorption of solar radiation and the formation of ozone which is discussed at length in section 2.2. The stratosphere which terminates at the stratopause near 50km, is followed by the mesosphere (middle atmosphere) in which many complex photochemical processes occur, and throughout which the temperature falls. The mesosphere terminates at about 86km which is the location of the mesopause. Above the mesopause is the thermosphere, across which the temperature rises dramatically. There are few opportunities in this region for heat produced by photodissociation and photoionization processes to dissipate, for the collision frequency is so small (e.g., ≅ 1 sec^{-1} at 300km), and the collision mean free paths are so large (e.g., ≅ 1km at 300km). Figures 3a and 3b display the variation of collision frequency and mean free paths to 700km. Diffusion is important at these altitudes.

Gravity wave phenomena are observed through the mesosphere and thermosphere [Ref. 21, 29].

The positive temperature gradient in the thermosphere is related to heating arising from absorption of ultraviolet sunlight with consequent photodissociation of oxygen molecules into "hot" O atoms. Atomic oxygen rapidly becomes a permanent component of the thermosphere with increase in altitude. The temperature gradient decreases in the region of 500km at the thermopause. Its location is very dependent upon solar activity.

The fringe region of the atmosphere above the thermosphere is called the exosphere in which molecular trajectories are ballistic. Throughout the thermosphere photodissociation causes the mean molecular weight to decrease with increasing altitude as indicated in Fig. 3a. Further, photoionization processes give rise to the various layers (D, E, F, etc.) of the ionosphere, the onset of which lies at the top of the mesosphere. Most of the ionospheric layers lie in the thermosphere. It is reviewed more fully in a later chapter [Ref. 25].

The atmospheric layers identified above lie in three broad regions: *The Homosphere, The Heterosphere and The Exosphere* characterized by the physical conditions which exist in them.

The HOMOSPHERE occupies the bottom 86km of the atmosphere. It includes the troposphere, the stratosphere and the mesosphere. In this region the atmosphere is well mixed, collision frequencies are high, mean free paths are small (see Fig. 3b), the perfect gas law and the barometric equation hold well, and the relative fractions of the principal components (N_2, O_2, A, CO_2) are constant. The mean molecular weight is also constant (28.9644 kg/kmol, see Fig 3a). The region is homogeneous. Different photochemical processes indeed occur in the three regions, but these relate to the chemistry of the minor components as seen in Fig. 1. The scale height in this region depends solely on T/g, and thus upon the variation of T with altitude.

The HETEROSPHERE lies above the homosphere. It contains the thermosphere. In the heterosphere, from 86km to above 500km, depending on solar activity, photochemical reactions and diffusive separation of components ensure a steady decrease in mean molecular weight with increase in altitude (see Fig. 3a). Mean free paths are large and collision frequencies are small (see Fig. 3b). The region is thus heterogeneous. Mixing effects of winds do not influence the variation of species with altitude. Photodissociation of O_2 causes O to be a very important component of the heterosphere. This can have a significant effect on space systems at those altitudes. Photoionization effects in the heterosphere give rise to the high concentrations of electrons in the F regions of the ionosphere [Ref. 25]. In the heterosphere the scale height depends on T/Mg. Its increase with altitude is controlled both by the increase in temperature and by the decrease in mean molecular weight with increase in altitude.

The EXOSPHERE is the region above 500km in which the light species, mainly H and He whose scale heights are large, are the principal components. The collision frequencies are so small ($\cong$ 1 per minute) that atoms of H and He escape from the atmosphere on ballistic trajectories. The number density of principal components of the exosphere are illustrated in Figs. 4a and 4b as a function of height.

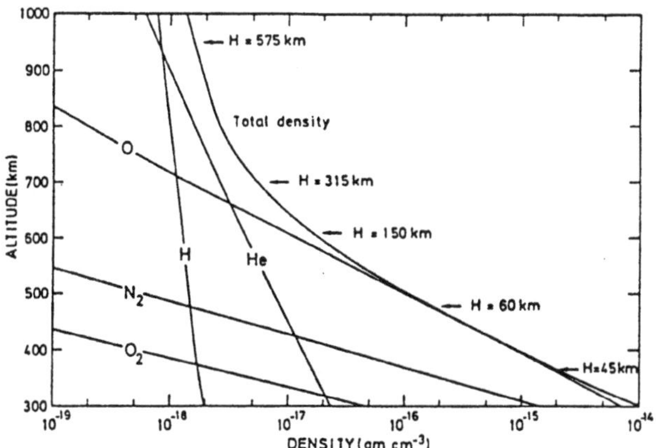

Fig. 4a: Atmospheric density and contributions from components between 300 and 1000km. Scale heights at various altitudes are also indicated [Ref. 1, p10].

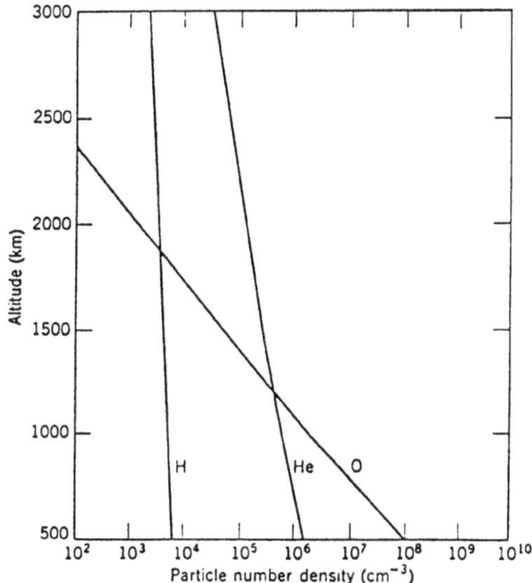

Fig. 4b: O, H, and He number densities in the exosphere [Ref. 34, p111].

A study of the vacuum UV spectrum of the airglow reveals that the earth is surrounded by a cloud of H which emits Lyα at 1216Å. This suggested the use of the term geocorona to describe the outer regions of the exosphere, for Lyα is also a major spectral emission from the solar corona.

2.2. IMPORTANCE OF PHOTODISSOCIATION PROCESSES IN ATMOSPHERIC STRUCTURE

It is clear from Fig. 1 that the allotropes of oxygen, O, O_2 and O_3 are persistent and important upper atmospheric species variously in the troposphere, the stratosphere, the mesosphere and the thermosphere. The very significant effects of atmospheric O on space systems will be discussed later in this Advanced Study Institute. Thus in this section a discussion is given, in an atmospheric context, of the formation of O and O_3 from photodissociation of O_2.

The potential curves of the more important electronic states of O_2 are displayed in Fig. 5. From the standpoint of long path atmospheric absorption phenomena, the electronic absorption transitions (band systems) which involve these states are [Ref. 20]:

$h\nu + O_2\ (X^3\Sigma_g^-) \rightarrow O_2(a^1\Delta_g)$ (Infra-red Atmospheric Absorption System)

$\rightarrow O_2(b^1\Sigma_g^+)$ (Atmospheric Absorption System)

$\rightarrow O_2(A^3\Sigma_u^+)$ (Herzberg I System)

$\rightarrow O_2(B^3\Sigma_u^-)$ (Schumann-Runge System)

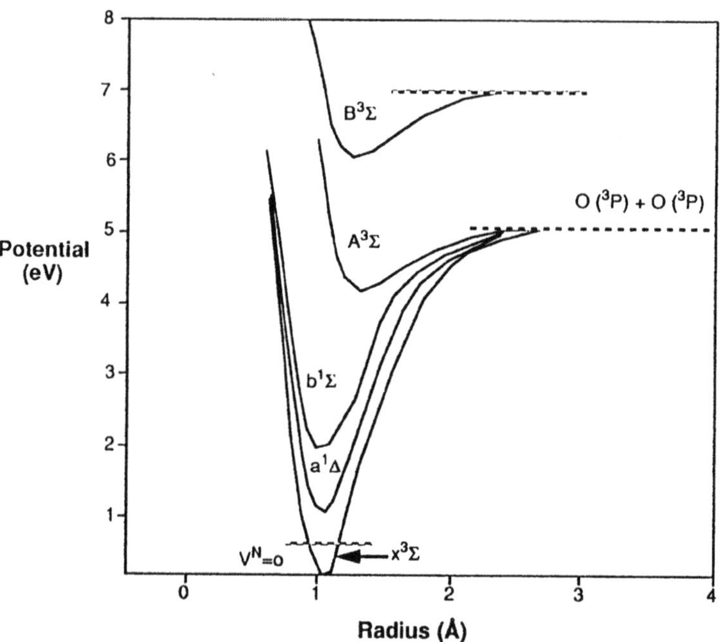

Fig. 5: Potential Curves of O_2 electronic states.

The strongest IR Atmospheric Absorption bands lie in the near IR at 1.2683 μ (0,0) and 1.0674 μ (1,0). The system is "forbidden" by electric dipole selection rules. The resulting metastability of the long lived $a^1\Delta_g$ state has important implications in the energetics and chemistry of the troposphere. The $b^1\Sigma_g^+$ state is less metastable, and bands of the atmospheric absorption system are readily observed in long path lengths of O_2. The strongest are at 6969Å (0,0) and at 6895Å (1,0). These bands are indicated in Fig. 6 which is a realistic line-by-line spectral synthesis of long path atmospheric extinction from satellite altitudes [Ref. 24].

The Schumann-Runge [Ref. 18], and Herzberg I [Ref. 14, 7] systems play very important roles in the oxygen photochemistry of the stratosphere, mesosphere and thermosphere. Because of the significant change Δr_e in equilibrium internuclear separations r_e between the X and the respective A and B states, the absorption progressions from $X^3\Sigma_g^-(v''=0)$ to both the A and the B states involves a set of bands attached to a shorter wavelength photodissociation continuum (see Figs. 5, 7).

Solar radiation at wavelengths of the Schumann-Runge continuum is strongly absorbed in the thermosphere and in the mesosphere. It does not penetrate lower into the atmosphere. The Schumann-Runge continuum transition is therefore the main source of O atoms in those regions. Solar

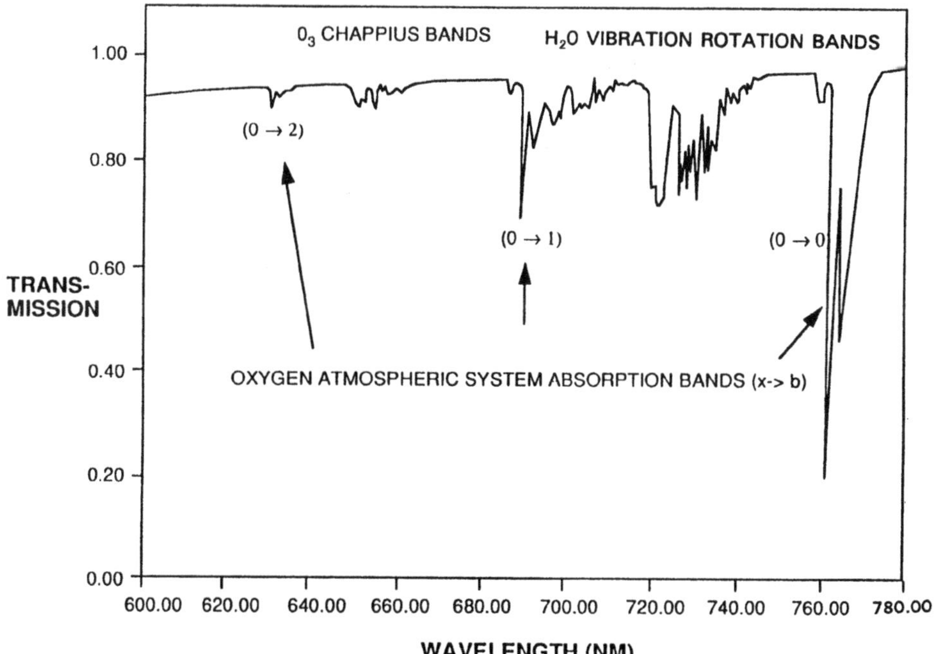

Fig. 6: Spectral synthesis of atmospheric transmission spectrum between 600 and 700nm. Absorption features of the (0,0); (0,1); and (0,2) Bands of the Atmospheric system, O_3 Chapuis bands and H_2O vibration rotation bands are indicated.

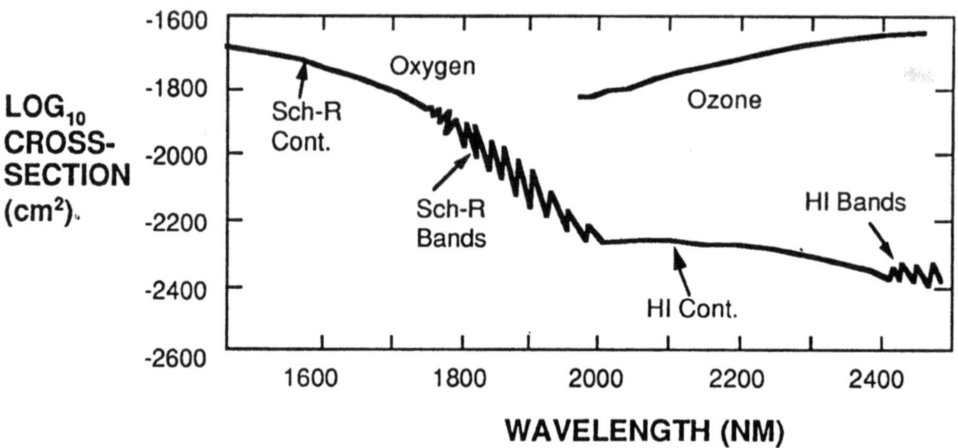

Fig. 7: Spectral synthesis of absorption cross section of O_2 and O_3 between 150 and 200 nm [after Ref 5].

radiation at wavelengths of the Herzberg continuum is less strongly absorbed. It penetrates to the stratosphere where it is the main source of stratospheric O atoms which give rise, after some chemistry, to the ozone layer.

Absorption of solar radiation into the continua occurs when Franck Condon "vertical" transitions from $X^3\Sigma$ (v=0) terminate respectively on the A or B potential curves at a location ($r < r_e$) above the respective dissociation limits. In these cases the bound-free transitions

$$h\nu + O_2(X^3\Sigma_g^-) \begin{cases} O(^3P) + O(^1D) & : \text{S-R [Upper state } O_2(B^3\Sigma_u^-)] \\ O(^3P) + O(^3P) & : \text{H I [Upper state } O_2(A^3\Sigma_u^+)] \end{cases}$$

occur [Ref. 26]. The Herzberg I transition is modestly forbidden as an electronic dipole transition for it breaks the - <-/-> - selection rule [Ref. 20, p. 277]. Its bands and photodissociation continuum are thus weaker than the Schumann-Runge bands and continuum (see Fig. 7). The formation of $O(^1D)$ in the Schumann-Runge continuum has important implications in the aeronomy of the mesosphere and thermosphere which are well discussed by Banks and Kockarts [Ref. 1], Whitten and Poppoff [Ref. 34], Ratcliffe [Ref. 27] and by Rees [Ref. 28].

In the case of the B state curve-crossing, predissociations which occur at vibrational levels near v'=4 and above enhance the effect. Figure 7 displays a semi-logarithmic plot of the absorption cross section for the O_2 molecule (from $X^1\Sigma$ v"0) and the O_3 molecule between 2500Å and 1500Å. The O_2 cross sections were computed using a realistic line-by-line code which we have developed [Ref. 5, 6, 24]. The O_3 data were measured by Griggs [Ref. 17].

In this figure, bands of the v"=0 progression of the Herzberg I system lie between 2500Å and 2429Å. They merge into the Herzberg I photodissociation continuum at 2430Å. This continuum is overlaid between 2000Å and 1751Å with bands of the v"=0 progression of the Schumann-Runge system. These bands merge into the Schumann-Runge photodissociation continuum at 1751Å. This continuum runs to about 1350Å with a maximum at about 1425Å. In Fig. 7 the segment of the continuum between 1751Å and 1500Å is displayed.

Thus the two absorption processes which respectively give rise to photodissociation of O_2 in the atmosphere are:

Herzberg Continuum - *Stratosphere*

$h\nu + O_2(X^3\Sigma_g^+) \rightarrow O(^3P) + O(^3P) \quad (\lambda < 2430Å)$

Schumann-Runge Continuum - *Mesosphere* *Thermosphere*

$\rightarrow O(^3P) + O(^1D) \quad (\lambda < 1751Å)$

Solar irradiation of the atmosphere is the principal cause of the photodissociation processes. It also stimulates photoionization processes which give rise to the ionosphere. These are discussed below [Ref. 25].

The solar radiation which illuminates the terrestrial atmosphere is the superposition of a) the illumination from the solar photosphere which radiates an approximately $T \cong 6000°K$ black body Fraunhofer continuum crossed with chromospheric absorption lines, and b) the atomic and ionic line emission and very high temperature ($\cong 10^6$ °K) continuous spectrum of the solar corona [Ref. 31, Chaps 15 and 16; Ref. 23, Chaps 1 and 2].

The high number density of O in the thermosphere indicated in Figs. 1a and 1b is principally due to the effects of solar radiation in the Schumann-Runge photodissociation continuum. Smaller contributions are made by solar radiation at the wavelengths of those Schumann-Runge bands where predissociation occurs, from Hydrogen line radiation at the Lyman α wavelength (1216Å), and from the Herzberg photodissociation continuum, in order of decreasing importance, as illustrated in Fig. 8.

Similarly the relative importance of radiation at these wavelengths which penetrates to various levels of the mesosphere is indicated in Fig. 9. In this figure it will be noted that solar continuous radiation in the Herzberg continuum has a dominating effect below about 65km as the atmosphere becomes more and more opaque to shorter wavelengths.

At stratospheric altitudes solar radiation in the Herzberg continuum is the only available photodissociative radiation which penetrates to these levels. This is shown in Fig. 10 which also shows O_2 photodissociation rate from Schumann-Runge predissociated band wavelengths in the lower mesosphere (where radiation reaches) and from solar radiation in the Herzberg continuum in the stratosphere.

$$J_{O_2}(\text{Lyman } \alpha) = 3 \times 10^{-9} \quad \text{sec}^{-1}.$$

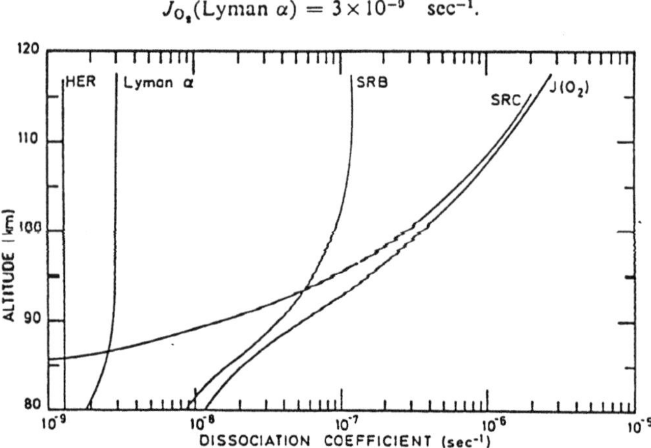

Fig. 8: Photodissociation coefficient of O_2 [$J(O_2)$] in the lower thermosphere with an overhead sun, together with relative contributions of the Herzberg continuum (HER), Lyman α, Schumann-Runge Bands (SRB) and Schumann-Runge Continuum (SRC) [Ref. 1].

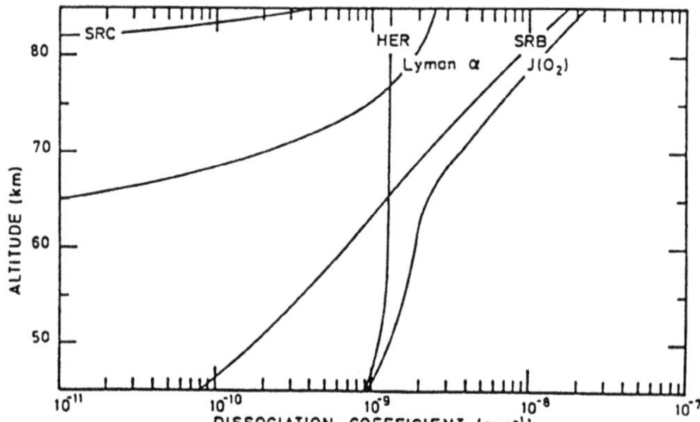

Fig. 9: Photodissociation coefficient of O_2 [$J(O_2)$] in the mesosphere with an overhead sun, together with relative contributions of Herzberg continuum (HER), Lyman α, Schumann-Runge continuum (SRC), and Schumann-Runge Bands (SRB) [Ref. 1].

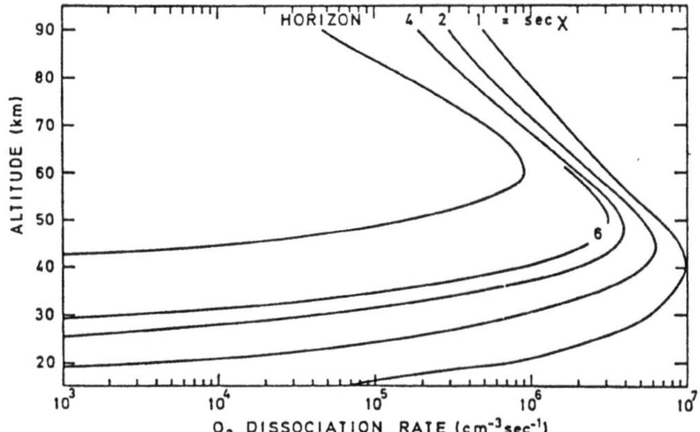

Fig. 10: Photodissociation rate of molecular oxygen for various solar angles in the mesosphere and the stratosphere [Ref. 1].

Stratospheric absorption of solar radiation into the Herzberg continuum gives rise to the formation of O atoms (both in the ground and ^3P state). Chapman [Refs. 9, 10, 13] suggested that the three body recombination reaction:

$$O + O_2 + M \rightarrow O_3 + M$$

is responsible for ozone formation in the stratosphere. The collision partner M is needed to balance angular momentum in the collision. M is probably an O_2 molecule.

The photochemical processes which take place in the stratosphere result in reaction products with enhanced kinetic energy. This is the source of stratospheric heating and the temperature rise to the stratopause.

The ozone which is formed in the stratosphere exhibits three electronic transitions, which in absorption give rise, respectively, to the Hartley bands (2000Å-3000Å), the Huggins bands (3000Å-3600Å) and the Chappuis bands (4500Å-8200Å) [Ref. 17]. The intensity profiles of these absorption transitions are displayed in Figs. 11a, 11b, and 11c).

The Huggins and Chappuis bands, which lie in the visible spectra region, are quite weak. The UV Hartley band absorption however is very strong. It is the source of the UV opacity of the ozone layer over a broad spectral range. It prevents solar radiation with wavelengths below about 3000Å to reach the surface of the earth. It thereby acts as a shield of the earth's surface from damaging UV solar radiation. The fragility of the ozone layer to anthropogenic chemical attack by CFCs and oxides of nitrogen is of great environment concern [Ref. 3, 4, 15].

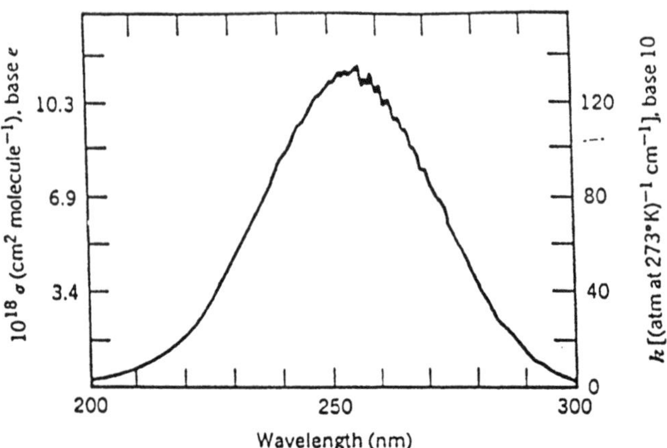

Fig. 11a: Absorption coefficient of the O_3 Hartley bands [Ref. 17].

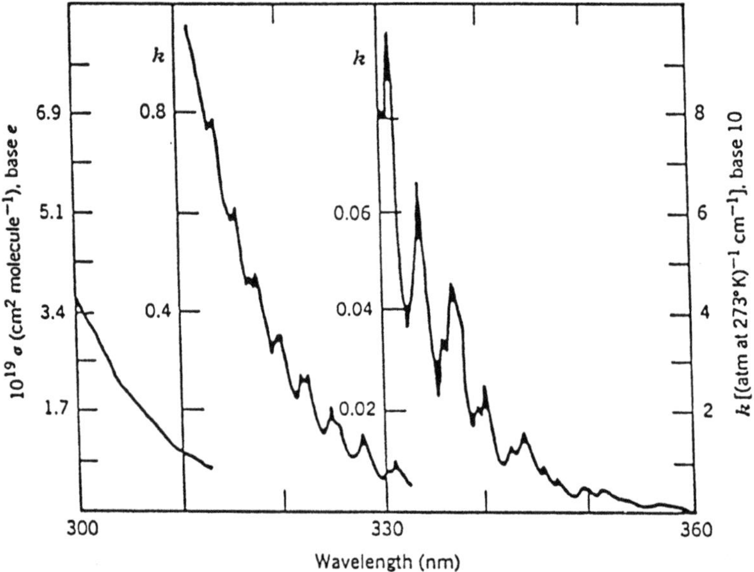

Fig. 11b: Absorption coefficient of O_3 Huggins bands [Ref. 17].

Most optical remote sensing of the state of the ozone layer are based on the spectroscopic properties of all three of these transitions. It is a very important contemporary activity in atmospheric science [Ref. 16].

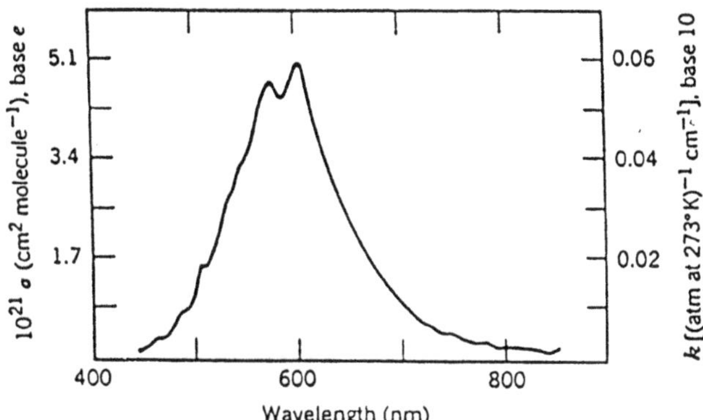

Fig. 11c: Absorption coefficient of the O_3 Chappuis bands [Ref. 17].

3. Summary and Conclusions

A brief summary has been made here of some of the properties of the local near-earth space environment of the neutral terrestrial atmosphere in which many contemporary space activities take place. No account has been given of luminous phenomena such as the aurora and the airglow, nor of the luminosities which occur around spacecraft in the mesosphere, often called the O-glow. The atmospheric constitution, structure and processes are summarized in Fig. 12.

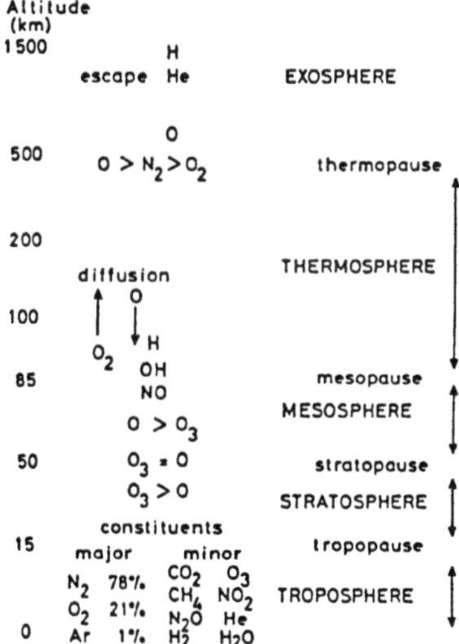

Fig. 12: Summary of atmospheric composition and nomenclature [Ref. 1].

Acknowledgements

This work has been assisted by research grants from the Natural Sciences and Engineering Research Council of Canada, the Atmospheric Environment Service of Canada, the Defense Research Establishment, Valcartier, Quebec, the Institute for Space and Terrestrial Science, and the Canadian Network for Space Research.

References

[1] P.M. Banks and G. Kockarts, Aeronomy (Parts A and B), Academic Press, New York (1973).

[2] Guy Brasseur and Susan Solomon, Aeronomy of the Middle Atmosphere, D. Reidel, Dordrecht Holland (1984).

[3] W.H. Brune, J.G. Anderson, D.W. Toohey, D.W. Fahey, S.R. Kawa, R.L. Jones, D.S. McKenna, and L.R. Poole, 'The potential for ozone depletion in the arctic polar stratosphere', Science 252, 1260-1266 (May 30, 1991).

[4] F.E. Bunn, F.W. Thirkettle, and W.F.G. Evans, 'Rapid motion of the 1989 Arctic Ozone Crater as viewed with TOMS data', Canadian Journal of Physics 69, 1087-1092, "Special Issue on Space Science" (1991).

[5] M.W.P. Cann, R.W. Nicholls, W.F.J. Evans, J.L. Kohl, R. Kurucz, W.H. Parkinson, and E.M. Reeves,. 'High resolution atmospheric transmission calculations down to 28.7 km in the 200-243nm spectral range', Applied Optics 18 (7) 964-977 (1979).

[6] M.W.P. Cann, J.B. Shin, and R.W. Nicholls, 'Oxygen Absorption in the spectral range 180-300nm for temperatures to 3000°K and pressures to 50atm, Canadian Journal of Physics 62 (12), 1738-1751 (1984).

[7] M.W.P. Cann and R.W. Nicholls, 'Spectral Line Parameters for the O_2 Herzberg I Band System, Canadian Journal of Physics 69, 1163-1165, "Special Issue on Space Science" (1991).

[8] J.W. Chamberlain, Theory of Planetary Atmospheres, Academic Press, New York (1978).

[9] S. Chapman, 'A theory of upper atmospheric ozone', Memoirs of the Royal Meteorological Society 3, 103-125 (1930a).

[10] S. Chapman, 'On ozone and atomic oxygen in the upper atmosphere', Philosophical Magazine 10, 369-383 (1930b).

[11] S. Chapman, 'The absorption and dissociative or ionizing effect of monochromatic radiation in an atmosphere on a rotating earth', Proceedings of the Physical Society of London 43, 26-45 (1931a).

[12] S. Chapman, 'The absorption and dissociative or ionizing effect of monochromatic radiation in an atmosphere on a rotating earth part II: Grazing incidence', Proceedings of the Physical Society of London 43, 483-501 (1931b).

[13] S. Chapman, 'Some phenomena of the upper atmosphere', Proceedings of the Royal Society A 132, 353-374 (1931c).

[14] V. Degen, S.E.H. Innanen, G.R. Hebert, and R.W. Nicholls, "Identification Atlas of Molecular Spectra (series 1) 6: The O_2 $A^3\Sigma_u^+$ - $X^3\Sigma_g^-$ Herzberg I System, Centre for Research in Earth and Space Science and Department of Physics, York University, North York, Ontario Canada M3J 1P3, ppvii + 31 (1968).

[15] W.F.G. Evans, F.E. Bunn, and A.E. Walker, 'Global ozone trends from a re-analysis of TOMS data', Canadian Journal of Physics 69, 1103-1109, Special Issue on Space Science (1991).

[16] W.B. Grant, 'Ozone Measuring Instruments for the Stratosphere', (Volume 1, Collected works in Optics), Optical Society of America, Washington, D.C (1989).

[17] M. Griggs, 'Absorption coefficient of ozone in the ultraviolet and visible regions', Journal of Chemical Physics 49, 857-859 (1968).

[18] G.R. Hebert, S.E.H. Innanen, and R.W. Nicholls, Identification Atlas of Molecular Spectra (series 1) 4; The O_2 $B^3\Sigma_u^-$ - $X^3\Sigma_g^-$ Schumann-Runge System Centre for Research in Earth and Space Science and Department of Physics, York University, North York, Ontario Canada M3J 1P3, pp vii + 30 (1967).

[19] A. Henderson-Sellers, 'The origin and evolution of planetary atmospheres', Adam Hilger Ltd. Bristol (1983).

[20] G. Herzberg, Molecular Spectra and Molecular Structure II The Spectra of Diatomic Molecules, Van Nostrand, New York (1950).

[21] C.O. Hines and D.W. Tarasick, 'On the detection and utilization of gravity waves in airglow studies', Planetary and Space Science 35, 851 (1987).

[22] F.S. Johnson (ed.), Satellite Environment Handbook, Stanford University Press (1961).

[23] A.S. Jursa (Scientific Editor), <u>Handbook of Geophysics and the Space Environment</u>, Air Force Geophysics Laboratory, Air Force Systems Command, United States Air Force (Note additional copies can be obtained from National Technical Information Service, 5285 Port Royal Road, Springfield, VA 22161 (1985).

[24] R.W. Nicholls, 'Realistic synthesis of atmospheric and interstellar extinction spectra', Journal of Quantitative Spectroscopy and Radiative Transfer 40 (3), 275-289 (1988).

[25] R.W. Nicholls, 'The upper atmosphere and the ionosphere', this volume, p 103 (1991a).

[26] R.W. Nicholls, 'Franck-Condon Factor Sum Rules for transitions involving bound and unbound states', Journal of Quantitative Spectroscopy and Radiative Transfer 45 (5), 261-265 (1991b).

[27] J.A. Ratcliffe (ed.), Physics of the Upper Atmosphere, Academic Press, New York (1960).

[28] M.H. Rees, Physics and Chemistry of the Upper Atmosphere, Cambridge University Press, Cambridge (1989).

[29] D.W. Tarasick and C.O. Hines, 'The observable effects of gravity waves on airglow emissions', Planetary and Space Science 38, 1105-1119 (1990).

[30] United States Air Force Geophysics Research Laboratory (1957, 1960), <u>Handbook of Geophysics</u> (1957, private distribution, Revised Edition), MacMillan, New York.

[31] S.L. Valley (Scientific Editor), Handbook of Geophysics and Space Environments, Air Force Cambridge Research Laboratories, Office of Aerospace Research, United States Air Force (1965).

[32] J.C.G. Walker, Evolution of the Atmosphere, MacMillan, New York (1977).

[33] R.C. Whitten and I.G. Poppoff, Physics of the Lower Ionosphere, Prentice Hall, New York (1965).

[34] R.C. Whitten and I.G. Poppoff, Fundamentals of Aeronomy, John Wiley, New York (1971).

THE SUN - ITS ROLE IN THE ENVIRONMENT OF THE NEAR EARTH SPACE

VICENTE DOMINGO
Space Science Department of ESA
ESTEC, Noordwijk (NL)

ABSTRACT. The basic characteristics of the Sun are described as the background that determines the permanent features of the near Earth environment. The Sun transmits its influence to the Earth essentially by gravitation, electromagnetic radiation and particle emission (solar wind and energetic particles). Leaving gravity apart, the average characteristics and the variability of the different components of the electromagnetic radiation as a function of wavelength are discussed; and for particle emission, the characteristics of the solar wind, including its changes with solar cycle and the properties of the energetic particles sporadically emitted by the Sun, are analyzed.

A brief summary of the structure of the Sun and of the main processes that take place in the solar interior and in the solar atmosphere is presented, as well as the origin of the different components of the radiation and particles that influence the near Earth environment. Particularly important for noticeable effects in the environment are the transient phenomena caused by the so called solar activity: solar flares, coronal mass ejections, 22-year magnetic cycle, etc. Finally the status of the prediction of environmental changes in the near Earth space is discussed; the understanding of the solar phenomena is crucial for the prediction of solar events that induce changes in the space near the Earth.

1. Introduction

We shall consider as near-Earth space environment that part of space that lies around the Earth, between some 300 km above its surface and 2 million km. In this region operate the low-earth orbit satellites (a few hundred km height), the geostationary orbit satellites (36000 km height); this includes most of the application satellites, and also the less frequent scientific satellites that are located in highly eccentric orbits around the Earth as well as those that are placed around the Sun-Earth Lagrangian points.

An object in the near-Earth environment will be affected directly by the Sun's emissions and indirectly by the modifications induced by the Sun's emissions in the elements filling this environment (i.e., gases, plasmas, electric or magnetic fields). In this paper we will review the

R. N. DeWitt et al. (eds.), The Behavior of Systems in the Space Environment, 67–101.

solar emissions, in particular their variations, and will include also a short description of the solar structure as this is important to predict how the environment may change.

2. Basic Properties of the Sun and Their Variations

The following is a list of the most important characteristics of the Sun in terms of physical quantities:

Distance: The mean distance between the Sun and the Earth, also known as an Astronomical Unit (AU) is 1 AU = 149 597 870±2 km. The distance varies between 1.47×10^8 km at perihelion in January, and 1.52×10^8 km at aphelion in July.

Mass: $m_0 = (1.989 \pm 0.002) \times 10^{30}$ kg

Radius: $r_0 = (6.9626 \pm 0.0007) \times 10^5$ km

Total mean irradiance at mean distance $S = 1367 \pm 2$ W/m^2

The mean irradiance at 1 AU or "solar constant" has been observed to vary of the order of 0.15% over periods of days, due to solar active regions effects, and 0.1% over the duration of one solar cycle (cycle 21).

Rotation: the Sun does one rotation in about 27 days, as seen from the Earth. The period of rotation is dependent on the solar feature that one observes to make the rotation measurement, and on solar latitude.

2.1. SOLAR INTERIOR STRUCTURE

A schematic cross-section through the Sun is shown in Fig. 1. Energy is generated in the solar core by nuclear reactions. The primary reaction is the fusion of hydrogen nuclei to form helium nuclei ($4H^1 \rightarrow He^4$). For every gram of hydrogen entering the reaction, 0.007 g is converted into energy. To produce the solar luminosity, about 4.3×10^9 kg/s must be converted to energy.

The energy generated in the core is transferred outward by radiation. Because of the high densities, the radiation is absorbed and re-emitted many times on its outward journey. Radiation that began in the core as high energy gamma rays is degraded by these successive absorptions and re-emissions until they finally emerge as visible radiation characteristic of the solar surface. However, the energy is not carried all the way to the surface as radiation. As it moves outward from the core, the temperature, density, and pressure drop. As the temperature drops, free electrons can be trapped by atoms into bound states, causing an increase in the opacity. Thus radiation becomes less effective in transporting energy. A large temperature gradient results and convection becomes the primary energy transport mechanism. Helioseismological measurements indicate that this transition occurs at 0.71 solar radii from the center. Photospheric observations clearly show that this convection zone reaches the solar

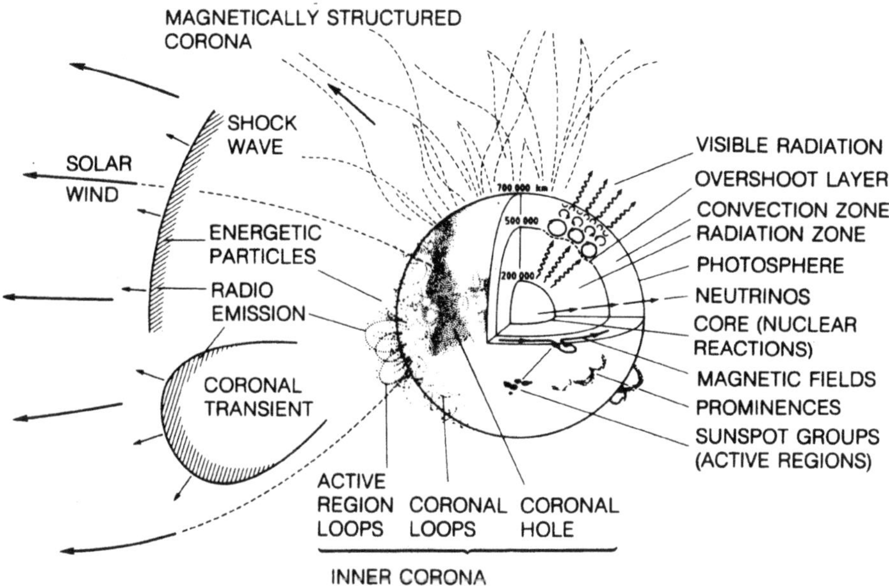

Fig. 1: Schematic solar structure and phenomena.

surface, and that at least 3 or 4 scales of convection are present. Near the solar surface, radiation can escape into space, and it again becomes the primary energy transport mechanism.

The above description of the solar interior results from applying a standard stellar model to the known solar boundary conditions. Standard stellar models are based on the application of the fundamental physics principles, although their application requires a certain amount of assumptions as the physical conditions are significantly different from those in the laboratory and they are not well known, because of the observational difficulties. So far we have not mentioned rotation or magnetic field; but we will come back to them when we discuss variability (paragraph 2.4).

2.2. SOLAR ATMOSPHERE

The solar atmosphere is divided into several "layers" for convenience of description and analysis. The boundaries between the layers are generally defined in terms of temperature (Fig. 2).

The photosphere - Even though the whole Sun is gaseous, we can define clearly a solar surface, the photosphere where the atmosphere begins. Throughout the solar interior the thermally emitted photons are reabsorbed by gas that is opaque to them. The opacity in the upper convection zone is due mainly to the absorption of the photons by negative hydrogen ions. The mean free path or absorptivity is highly dependent on temperature and

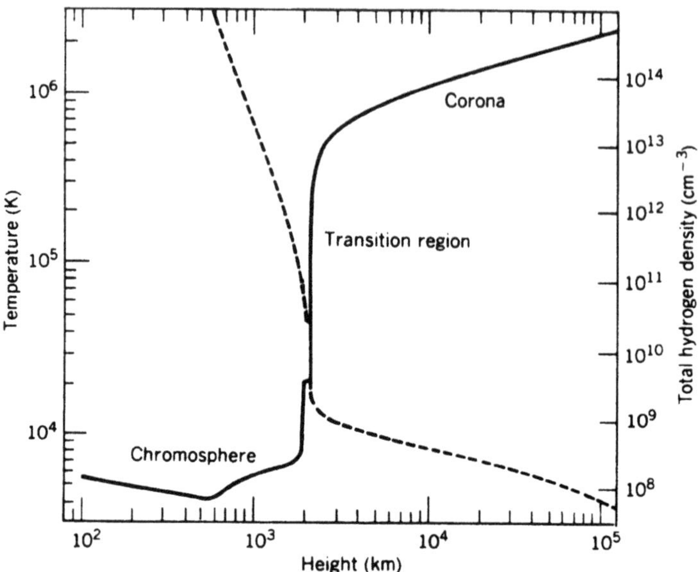

Fig. 2: Plots of solar atmosphere temperature (solid line) and density (dashed) from a model of the quiet network [Ref. 1].

density. The solar surface is the level above which most of the photons are radiated without further interaction. Essentially the mean free path of the photons becomes longer than the mass of material left above. From a practical point of view one defines the level zero of the solar atmosphere as the level at which the electromagnetic radiation of 500 nm has an optical depth equal to 1. This is the level from where the probability of a photon escaping is 1/e. This radiation into free space produces a cooling of the material that in turn induces convection. Since the gas below the photosphere is opaque, it behaves as a black body that radiates a continuum spectrum. If it is assumed that the emitted radiation is in equilibrium and the Stephan-Boltzmann law (energy flux = σT^4) is used to calculate an effective temperature for the whole emitting layer, a value of 5780 K results.

The energy that is radiated by the photosphere constitutes almost all of the energy emitted by the Sun into space and is centered in the visible part of the spectrum. The photosphere is only a few hundred kilometers deep. The visual structure of the photosphere reflects the turbulent structure of the convection zone. The observation of the photosphere shows the presence of ever forming "granules" that are manifestations of the underlying convection. The presence of the magnetic field in this region appears to accommodate itself to the emerging structures from the convection zone.

The chromosphere - The transparent region above the photosphere (about 2000 km) is a region where the density decreases very rapidly with height. The temperature that reaches its minimum at the bottom of the chromosphere (about 4200 K) is seen to increase first slowly and then rapidly in what is called the transition region into the corona at 10^6 K. For practical purposes one defines the chromosphere as the region above the photosphere with temperatures between the temperature minimum and 25000 K. As a consequence much of the radiation emitted by the chromosphere lies in the extreme ultra-violet (EUV). One aspect that is worth mentioning is that this variation of average properties with height, particularly when we are above the photosphere, must be considered as a way to describe particular regimes of plasma and has only a statistical meaning. In fact both the chromosphere and the corona are highly structured, and it is possible to find structures of chromospheric plasma amidst the coronal gases. The chromosphere is the region where the structures of the photosphere are carried to by means of the magnetic fields, which become dominant in this region with respect to the hydrodynamic forces. A particularly important structure is the chromospheric network, that has enhanced magnetic fields, and is visible because of its resulting higher temperature. Characteristic of the chromosphere are the spicules, that are brilliant jets of luminous gas that rise to approximately 1000 km and then fade in 2 to 5 minutes.

Transition region - A very thin layer where the temperature rises from 25000 K to 10^6 K. This region, though not very important from the point of view of overall radiation flux, is of great interest for scientific research of the solar atmosphere because the observation of the Sun in the emission lines of the ions that are generated at the transition region provide a very sharp definition of the boundary between chromospheric and coronal structures (Fig. 3).

The corona - From the chromosphere one arrives through the thin transition region into the outermost part of the solar atmosphere that reaches out into interplanetary space - the corona - with temperatures between one and two million degrees. The corona appears to be structured mainly by magnetic fields, and its dynamics are driven by magnetically-induced motions, sudden-releases of energy and by the explosive expansion into the surrounding vacuum of space. The heating of the corona is not completely understood; several mechanisms of mechanical, electrical and magnetic energy dissipation are proposed with limited observational support. Propagating waves and oscillations originating in the photosphere, the chromospheric spicules, and dissipation of Alfven waves propagating upwards along magnetic field lines are among the mechanisms that are proposed to explain the coronal heating. The temperature reaches up to 2×10^6 K in the lower corona and decreases slowly above that level.

Again we must remember that the corona is not a layered region but a rather structured region where the temperatures, densities and mass flows vary in space and time. Different regions like streamers or radially-elongated bright areas, loops, coronal holes or dark areas show small variations in temperature (on the order of 30%) from area to area, but large variations in density up to one or two orders of magnitude.

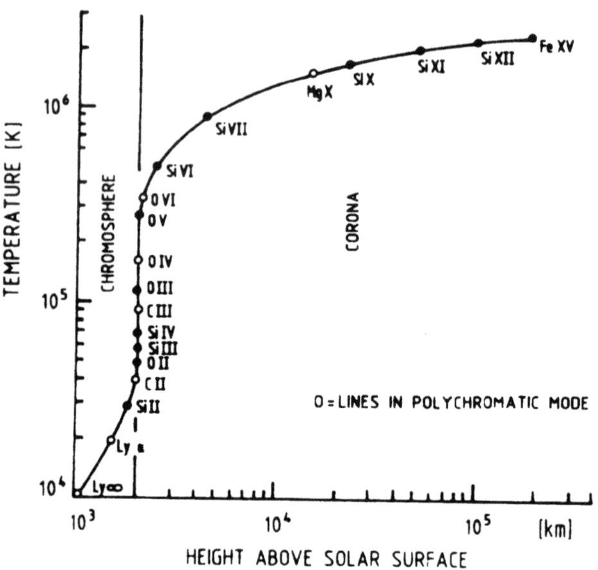

Fig. 3: Temperature as a function of height in a mean solar model atmosphere. Dots indicate the temperature of formation for some atomic species (open circles: observed with the HCO spectrometer) [Refs. 2, 3].

2.3 SOLAR ACTIVITY AND THE SOLAR CYCLE

The fact that the Sun rotates, that it contains a magnetic field, together with the convection zone dynamics are probably the cause of what is called solar activity. That is what suggests the "dynamo theory" that attempts to describe the so-called 11 year cycle of the Sun. A relatively small magnetic axial dipole can be measured in the Sun. The dipole reverses polarity every 11 years, thus making the overall cycle last 22 years.

The development of the activity cycle can be described as follows [Refs. 4-6]. The Sun rotates differentially; that is, the rotation rate is highest at the equator and decreases with increasing solar latitude such that the rotation rate at the poles is about 20% slower than at the equator. As viewed from the Earth, the equatorial rotation rate is about 27 days. At the start of the 11-year solar cycle, the configuration of the solar magnetic field is similar to the Earth's, with positive polarity at the north pole of the Sun and negative polarity at the south (see Fig. 4). The magnetic field lines run primarily from north to south. The differential rotation begins to stretch the field lines as more rapidly rotating equatorial regions run ahead of the regions at higher latitudes. The stretched field lines below the solar surface become twisted and intensified until bundles of twisted field lines become buoyant and pop

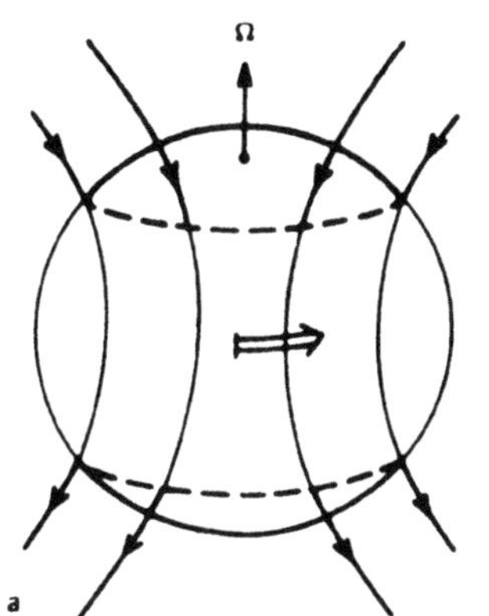

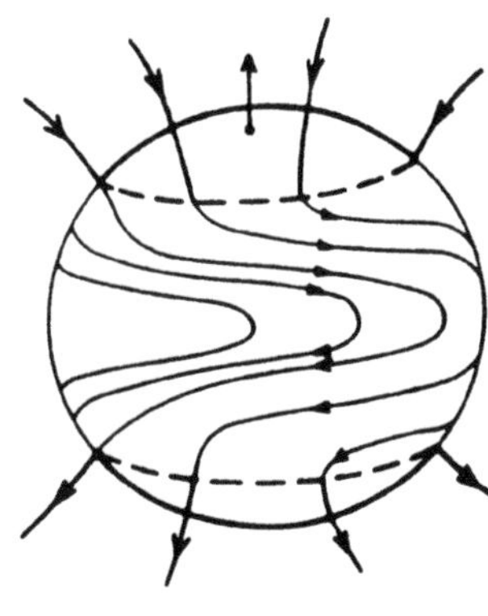

Fig. 4: Illustration of the Babcock model of the solar magnetic cycle. (a) Initial poloidal field before winding up by differential rotation. (b) Predominantly toroidal field generated by differential rotation with equatorial acceleration [Refs. 5, 7].

through the surface forming magnetic bipolar regions (Fig. 5). Polarity is positive where one end of each flux bundle passes through the surface and negative at the other end. Motions of the gas at the solar surface cause the surface fields to break up with the following polarities drifting toward the poles where they mix with the opposite polarities there (Fig. 6). This cancels existing polar fields and eventually replaces them with fields of the opposite polarity. Eventually (after about 11 years), the global field reverses, forming a situation similar to that at the start, but with opposite polarities at the solar poles (negative polarity at the north pole, positive polarity at the south pole). This sets the stage for a similar cycle of winding up of the fields, etc. Thus, the 22-year magnetic cycle has two maximum periods of strong magnetic fields at the surface, periods when there are many sunspots and extensive amounts of solar activity.

The magnetic fields that reach the solar surface produce other structures besides sunspots. Magnetic loops extend above the surface connecting regions of opposite magnetic polarity and become filled with hot plasma with temperatures of 1.5 to 3 x 10^6 K. The plasma heating appears to be caused by quasi steady-state dissipation of magnetic energy stored in the twisted magnetic-field lines. These loops are part of the corona. They emit EUV and X-ray radiation, which can be observed with instruments flown above the EUV/X-ray absorbing layers of the Earth's upper atmosphere (Fig. 7). Since the amount of EUV and X-ray radiation emitted by the Sun depends on the fraction of the solar surface covered by strong magnetic fields, the EUV and X-ray energy received by the Earth depends on solar rotation (since the magnetic fields are not uniformly distributed) and upon the solar cycle.

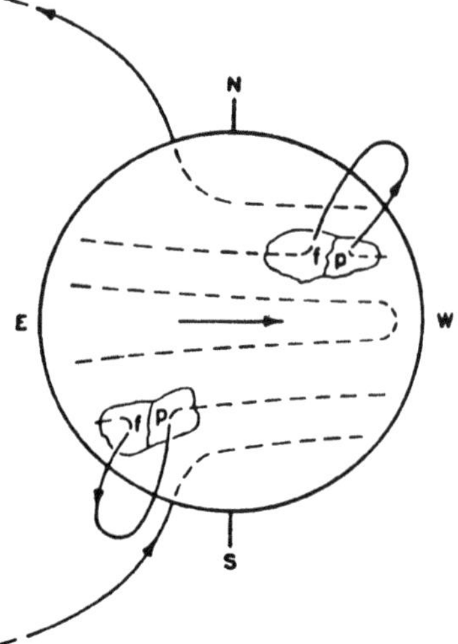

Fig. 5: Bipolar magnetic regions are formed where buoyant flux loops of the submerged toroidal field are brought to the surface. These continue to expand, and the flux loops rise higher into the corona. The letter f and p denote magnetic polarities following and preceding the direction of solar rotation [Refs. 5, 7].

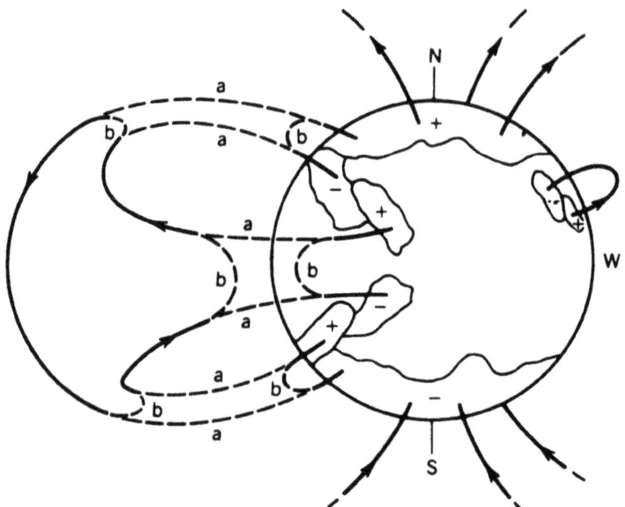

Fig. 6: The expanding lines of force above older bipolar magnetic regions move out to approach the lines of force of the main dipole field. Severing and reconnecting gradually occur, and a portion of the main field is neutralized. Also, a large flux loop of low intensity is liberated into the corona. Continuation of the process results in the formation of a new main dipolar field of reversed polarity [Refs. 5, 7].

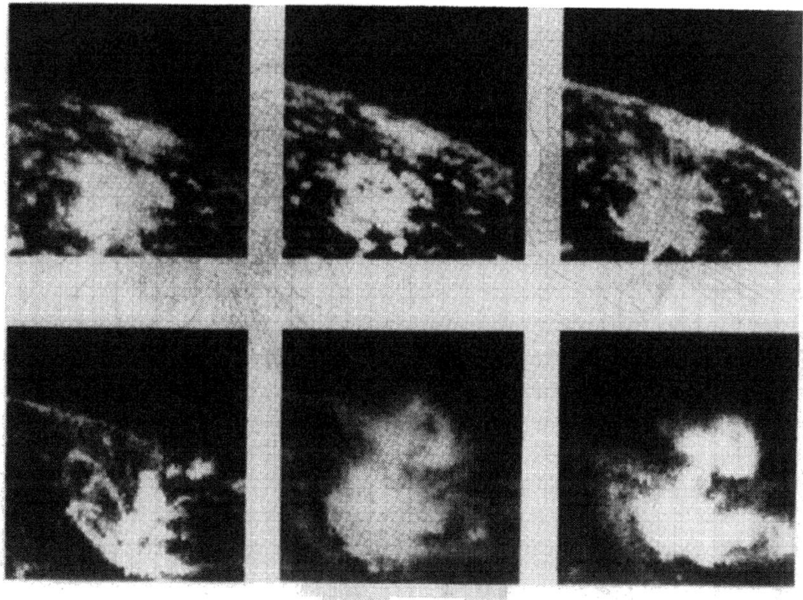

Fig. 7: EUV emissions from loop structures associated with an active region. The spectroheliograms are taken in integrated spectral lines of the ions Ly-α, C III, O VI, Ne VII, Mg X and Si XII, corresponding to plasma temperatures of 10^4 to several 10^6K. The field of view is 5 x 5 $(\text{arcmin})^2$ and the images were built up by a raster scan with a 5 x 5 $(\text{arcsec})^2$ picture element. These observations were obtained with the Harvard College Observatory EUV spectrometer flown on Skylab.

In some areas of the Sun, the magnetic fields are sufficiently weak that the gas pressure of the hot coronal plasma exceeds the pressure of the magnetic field, breaking open the magnetic loops, and allowing the plasma to flow out into interplanetary space to form the solar wind. The most prominent sources of steady-state solar wind are coronal holes, large magnetic unipolar regions with open magnetic configurations (one end of each field line is rooted in the Sun, the other is carried far out into space by the solar wind). The large dark areas in the X-ray photograph of the Sun (Fig. 8) are coronal holes. The bright areas are regions with closed magnetic-field configurations (magnetic loops with both ends of the field lines attached to the solar surface). Regions with the strongest magnetic fields have the brightest X-ray emission. Coronal holes are sources of high speed solar wind (mean speed of about 700 km/s). Small open areas within magnetically complex regions appear to be the source of the low-speed wind (mean speed of about 340 km/s). Magnetically complex regions appear to have predominantly closed configurations and are characterized by moderate to bright X-ray emissions (see Fig. 8).

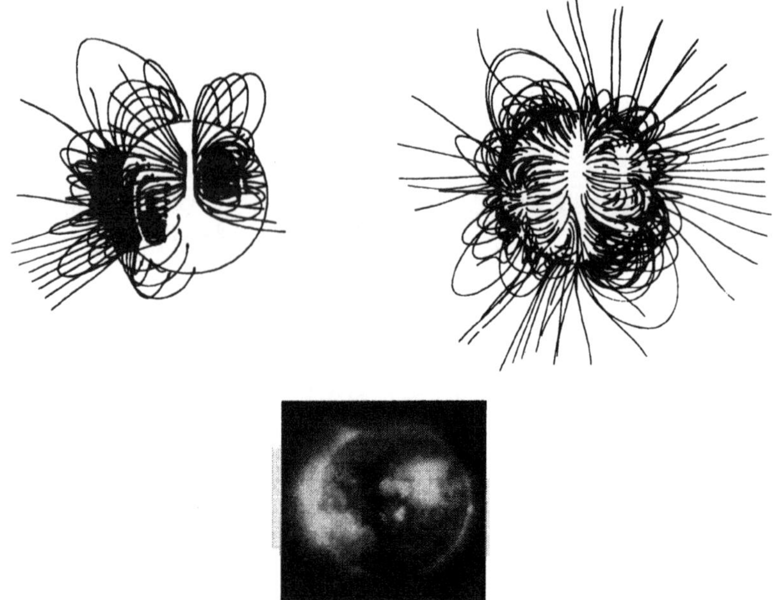

Fig. 8: Comparison of the observed X-ray corona and calculated magnetic fields. X-ray image from the American Science and Engineering telescope on Skylab. By permission of L. Golub [Ref. 7].

Occasionally, energy stored in the twisted magnetic-field lines is released suddenly. As a result of the dissipation of magnetic energy by an as yet ill-defined mechanism, electrons and protons are impulsively accelerated, probably in the coronal regions of the loop. These particles subsequently propagate along the magnetic field lines and interact with the ambient plasma in the legs of the loop and at the footprints. Hard X-rays, γ-rays, microwave radio emission, and neutrons are produced during this "impulsive" phase. At the loop footprints, the particle beam is thermalized; its energy is converted to heat in the denser plasma of the lower corona, transition region, and chromosphere. Although this excited plasma emits radiant energy at a variety of wavelengths, radiative losses do not remove the excess thermal energy fast enough, and explosive "evaporation" of chromospheric material occurs. This material has been heated to temperatures of tens of millions of degrees K, and cools slowly by radiation (of soft X-rays) and conduction during the "gradual" or cooling phase as it fills the large loop structures [Ref. 8] (Fig. 9). The bursts of X-ray and EUV radiation travel outward into interplanetary space at the speed of light and, among other things, heat and ionize gas in the upper terrestrial atmosphere. Some of the high-energy accelerated particles may also escape the Sun and, if conditions are suitable, some of these particles may reach the Earth. Flares that occur near the west limb

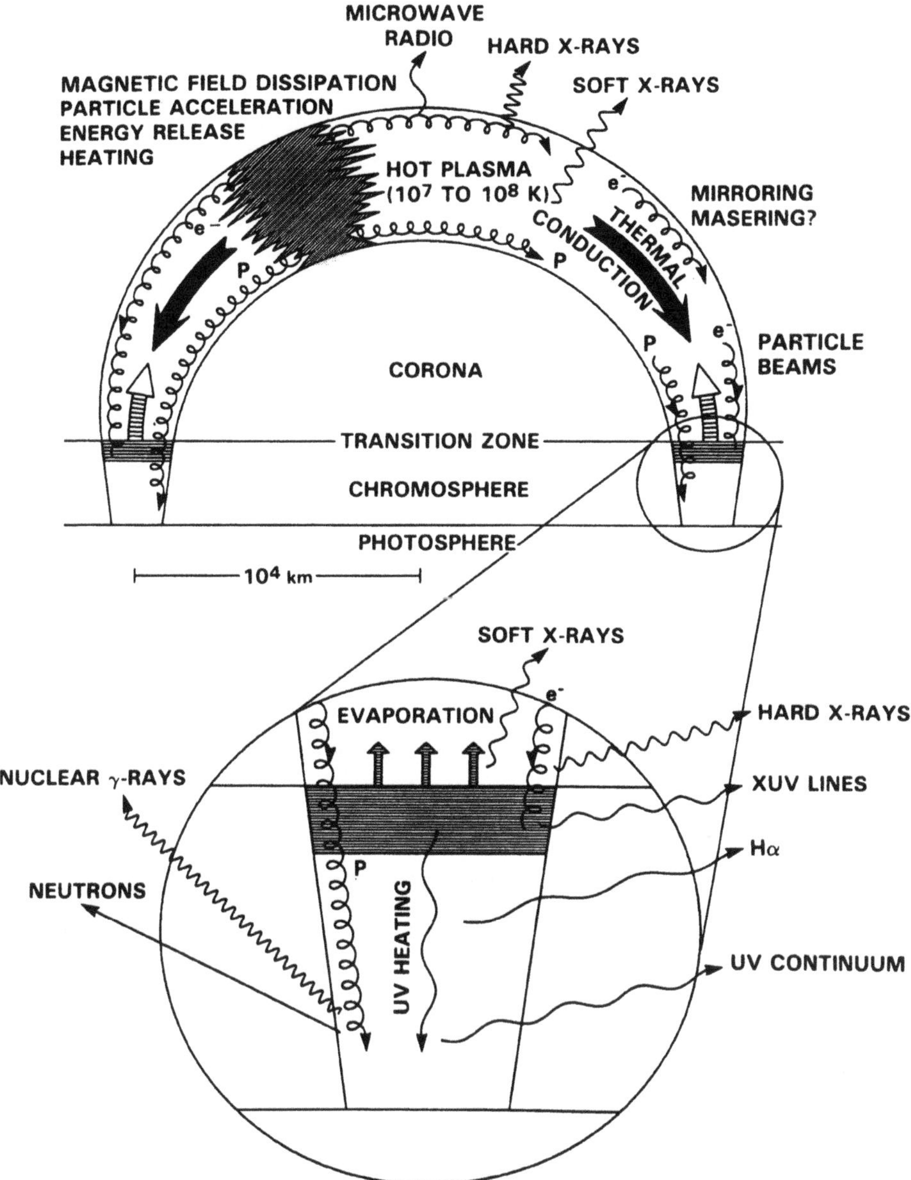

Fig. 9: Schematic diagram of the possible flare scenario described in the text. Although a single loop structure is shown in cross-section, much more complicated field geometries are common [Ref. 8].

of the Sun tend to have more favorable conditions for this due to the configuration of the interplanetary magnetic field, which spirals out from the rotating Sun much like the water from a rotating water sprinkler. The solar magnetic-field lines are carried away from the Sun by the solar wind. Finally, the disruption of the coronal magnetic field may blow off large amounts of coronal gas, which produces what is known as a coronal mass ejection. If this plasma cloud (enhanced solar wind) is directed toward the Earth and impacts on the Earth's magnetosphere, it affects conditions in the magnetosphere and upper atmosphere [Ref. 4].

3. Solar Emissions - Variability and Influence in the Near Earth Space Environment

We will now see how the structure of the Sun and its variations affect the radiation emitted by the Sun. Any aspect of the solar radiation and particulate emissions influence the near-earth environment, but in most cases we will focus our attention on the variations and consider that the mean value constitutes the background.

3.1. ELECTROMAGNETIC RADIATION - THE DIFFERENT PARTS OF THE SPECTRUM - ORIGIN IN THE SUN, VARIABILITY AND RELATIVE IMPORTANCE IN THE NEAR EARTH SPACE ENVIRONMENT

The spectrum of solar emitted electromagnetic radiation has been measured over a very large span of wavelengths (see Fig. 10). The spectral irradiance varies over the spectrum by more than 20 orders of magnitude, but the effects are not only related to the relative intensity but mainly to the sensitivity of the system we are dealing with. We will only deal with those aspects of solar emissions that are known to influence the near-Earth space environment.

3.1.1 Solar Irradiance (Solar Constant)

The total solar irradiance measured at the top of the Earth atmosphere, at its mean distance to the Sun, is known as the solar constant. Its absolute value is known with an error of about 0.15%. It has been observed to vary up to 0.15% with a characteristic time of a few days due to the presence of sunspots on the solar surface (see Fig. 11). It has also been observed to vary about 0.1% during about one solar cycle (it has been measured only during the last cycle) (Fig. 12).

The solar irradiance affects the systems in space, through radiative heating and by radiation pressure, and in both cases the observed variability of about 0.1% is well within the noise of the thermal characteristics of the space systems. This variation can be compared with the annual variation in irradiance of 6% due to the eccentricity of the Earth orbit around the Sun.

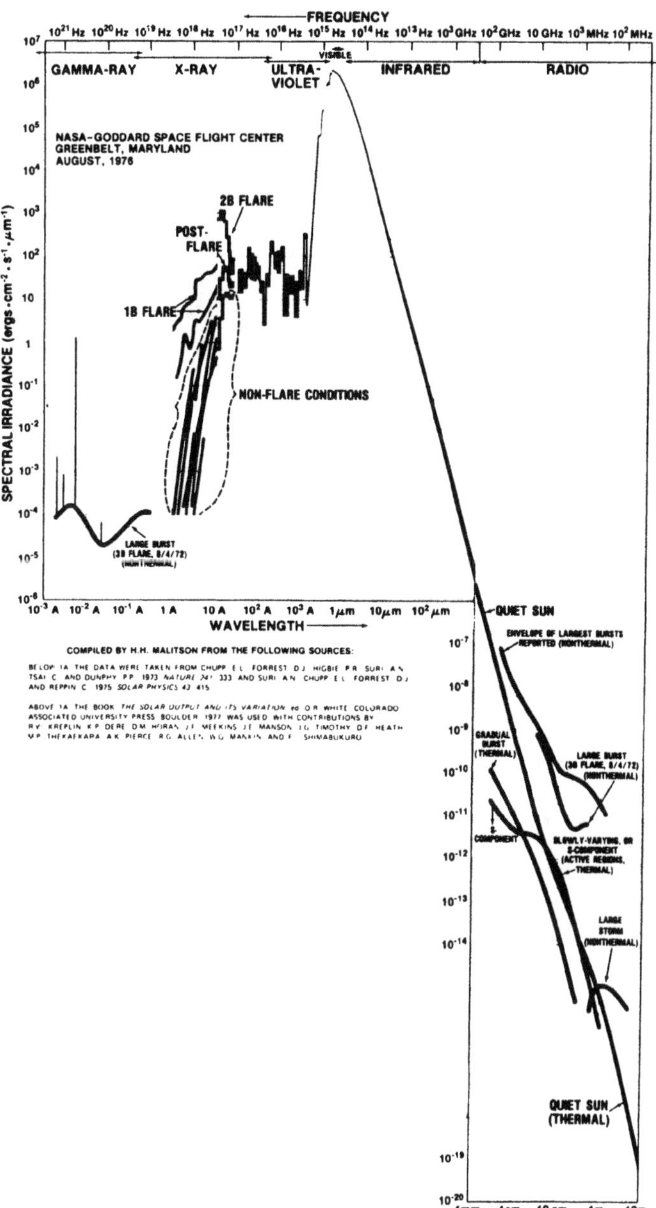

Fig. 10: Spectral distribution of solar irradiance [Ref. 9].

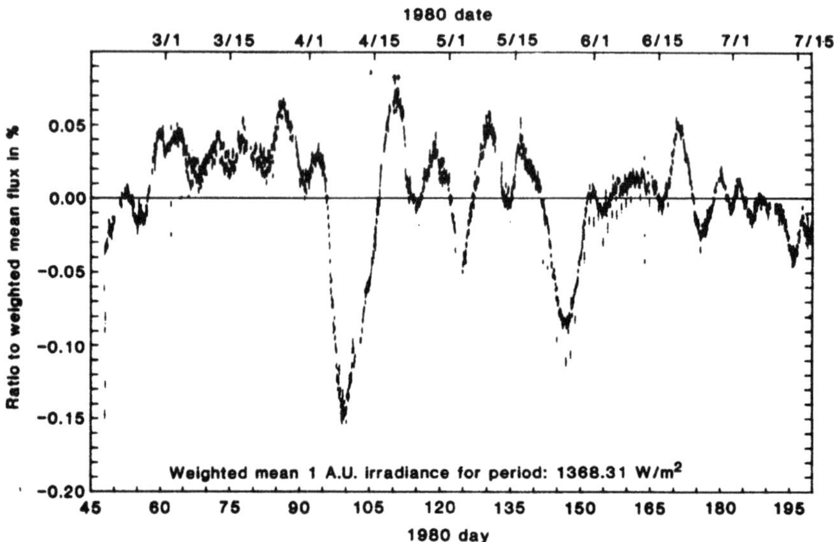

Fig. 11: Total solar irradiance at 1 AU, shown as a percentage variation about the weighted mean for the first 153 days of SMM [Ref. 10].

A particular type of solar variations, solar oscillations, are the oscillations that have been observed in the Sun. These oscillations are generally observed in the form of velocity oscillations by Doppler measurements or as oscillations of the photometric intensity. They are the manifestation of standing acoustic waves, driven by pressure, or gravity waves. These standing waves are resonant in cavities such as the interior of the Sun, limited by the photosphere, or the chromospheric cavity limited by the temperature minimum and the transition zone. The most prominent of these oscillations have been observed as photospheric surface oscillations, with a peak amplitude around a 5-minute period. These kind of variations, which are of great interest for the study of the solar structure (helioseismology), are of no relevance for the near-Earth space environment.

3.1.2 The Radio Spectrum

Figure 13 illustrates the solar radio spectrum, which begins in the microwave region at $\lambda = 1$ mm. The thermal "quiet-Sun" spectrum continues smoothly from the infrared into the radio region.

Superimposed on the quiet solar radio emission is a spectrum of great variability. The S-component (for slowly variable) is correlated to the solar 11-year activity cycle, its spectrum is approximately thermal, and its flux normally is 1 or 2 orders of magnitude below the quiet-Sun flux. On the other hand, there are rapid bursts of radio emission, on time scales

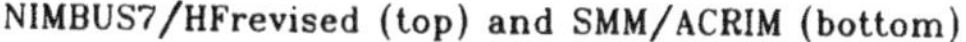

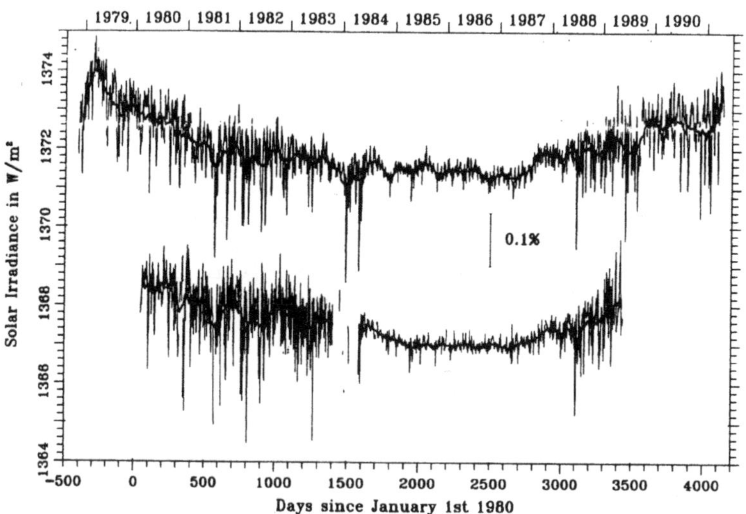

Fig. 12: Total solar irradiance observations by the ACRIM I experiment on the Solar Maximum Mission (SSM) and by the cavity pyrheliometer on NIMBUS 7. Points are the mean values for a day's observations [Refs. 12-14].

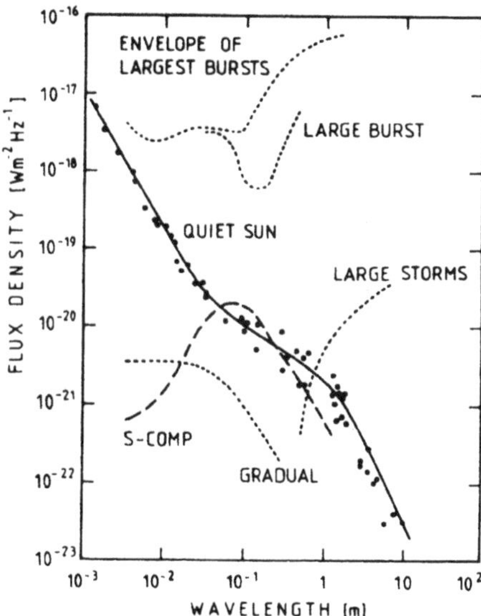

Fig. 13: Solar radio emission. Dots and solid curve: quiet Sun; dashed: slowly varying component; dotted curves: typical rapid events [Ref. 2].

of seconds to days. During such events the flux may exceed the quiet-Sun level by several orders of magnitude with substantial deviations from the thermal spectrum. The frequency of occurrence of radio bursts is again correlated to the 11-year cycle. Many of the bursts are associated with flares or with eruptive prominences. Their study has been used to interpret the propagation of shocks through the corona.

- Near-Earth space environment effects -

The radio bursts have an indirect effect on the space environment because during their peak of activity that may last from about 10 minutes to about 2 hours (Fig. 14) they may disturb the radio communication with spacecraft. They also constitute predictive information about disturbances that will reach the near-Earth environment in the following hours or days.

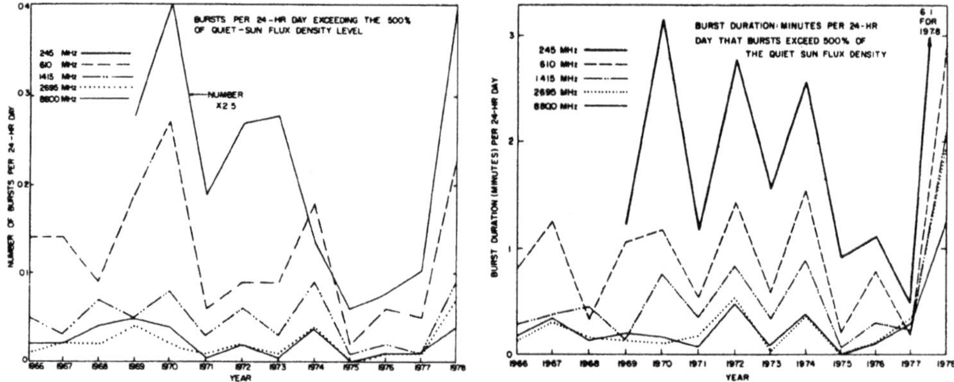

Fig. 14: a) The number of radio bursts per day that exceeded 500% of the quiet-sun flux-density level for the years 1966-1978. b) The number of minutes per day that solar burst radiation, at various frequencies, exceeded 500% of the quiet-sun flux density [Ref. 11].

We should mention the index known as $F_{10.7}$, the solar 10.7-cm radio flux, i.e, the brightness of the Sun as observed at a wavelength of 10.7 cm. It is a measurement of the slowly varying radio continuum, and it is, after the sunspot number, the most widely used indicator of solar activity. It is used as a proxi indicator of the level of EUV radiation emitted by the Sun. The 10.7 cm flux originates predominantly from the upper chromosphere and the lower corona.

3.1.3 Ultraviolet Radiation

The solar ultraviolet spectral irradiance is shown in Fig. 15. The sharp decrease near λ = 210 nm and the continuum which follows towards shorter wavelengths are due to the ionization of Al I. The whole band between

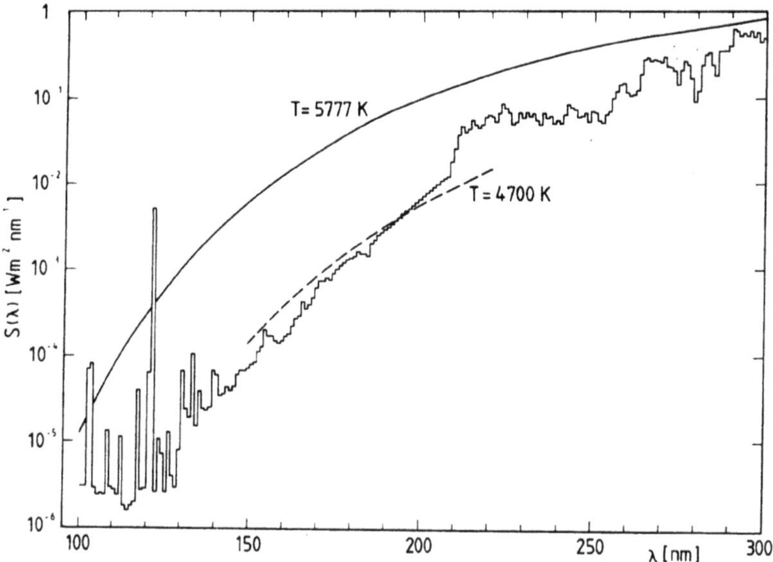

Fig. 15: Solar spectral irradiance in the ultraviolet, averaged over 1-nm bands. The solid and dashed curves are black-body spectra [Fig. 2].

200 nm and 150 nm is approximately represented by a brightness temperature of 4700 K.

Below 150 nm, emission lines dominate the spectrum. Most prominent is the Lyman-α line of hydrogen, a line of about 0.1 nm width centered at 121.57 nm; the average irradiance in this line alone is 6 mW/m^2, and this is as much as in the whole spectrum below 150 nm besides Lyman-α.

Increasingly towards shorter wavelengths, the ultraviolet irradiance is variable. For example, at time scales comparable to the Sun's period of rotation (27 days) changes of up to 25% are observed at 120 nm. These changes are partly true temporal variations and partly a manifestation of the non-uniform distribution of the sources in the solar atmosphere; these pass across the visible hemisphere as the Sun rotates. Still larger in amplitude - up to a factor 2 - is a variation which correlates with the 11-year activity cycle. However, for the total luminosity of the Sun the mentioned variations are of minor importance: the bands from 300 nm to 330 nm, 210 nm to 300 nm, and 150 nm to 210 nm contribute only 1.5%, 1%, and 0.01%, respectively [Ref. 2].

- Near-Earth space environment effects -

Solar radiation at wavelengths shorter than about 320 nm is totally absorbed in the Earth's upper atmosphere. This ultraviolet region is the principal source of energy in the upper atmosphere and controls the neutral and ion composition, temperature, and photochemistry in the stratosphere, mesosphere, and thermosphere. The absorption of solar UV in the Earth's atmosphere for an overhead Sun is illustrated in Fig. 16, which gives the altitude at which the rate of absorption at a particular wavelength is at maximum. At these altitudes, solar radiation is reduced by a factor of $e^{-1} = 0.37$ of its value above the Earth's atmosphere. The primary atmospheric constituents that absorb the radiation in the different wavelength regions are also given in the figure [Ref. 11].

It is worth noticing that above 200 nm the spectral irradiance is relatively constant, the radiation comes mainly from the upper photosphere and the lower chromosphere. Moreover, though this radiation is important for atmospheric chemistry, it does not affect significantly the space environment.

3.1.4 EUV and X-rays

Figure 17 shows a spectral snapshot in the EUV region of the electromagnetic radiation from three different parts of the solar atmosphere. Notice that a prominence consists of gas at chromospheric temperature. We have seen that both the chromosphere and the corona are strongly conditioned by the magnetic field structure that evolve with the solar activity. Therefore it is to be expected that the radiation that is emitted by the chromospheric and coronal structures will vary with the solar cycle, and also that the radiation that reaches the Earth will be influenced by the solar rotation as the structures rotate with the Sun.

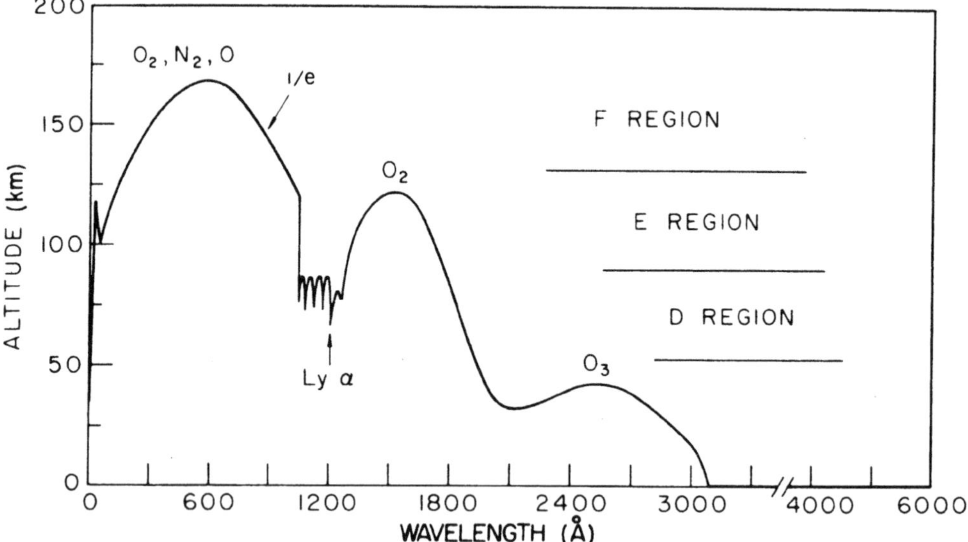

Fig. 16: The altitude at which the rate of absorption of solar UV radiation is at maximum. The principal atmospheric constituents that absorb the radiation in the different wavelength bands are indicated [Ref. 11].

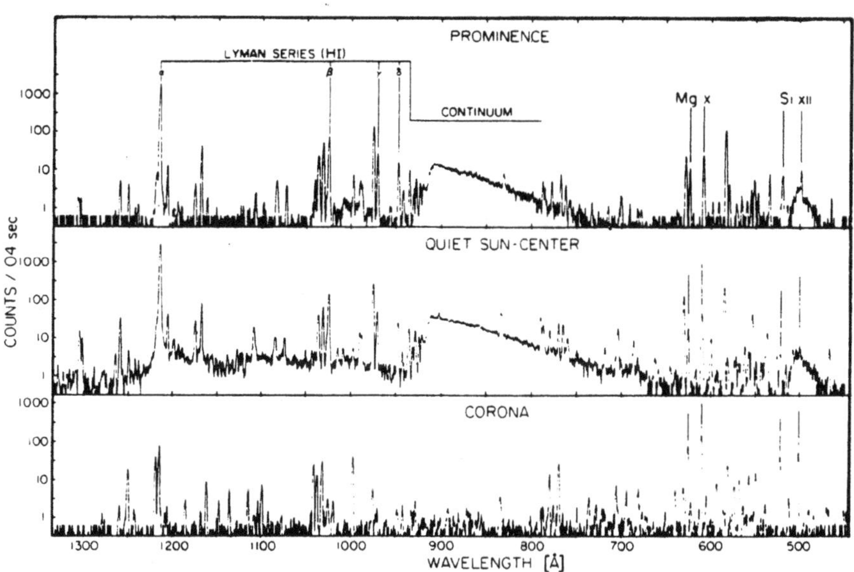

Fig. 17: Solar spectrum in the extreme UV [Refs. 2, 3].

The EUV irradiance (below 180 nm) variation is illustrated in Fig. 18, where the ratio of the irradiance near solar maximum during January 1979 to the irradiance near solar minimum during July 1976 for solar cycle 21, is plotted for 2.5 nm intervals of wavelength. The figure shows that the variation decreases with increasing wavelength toward 180 nm. At wavelengths longer than about 230 nm the ratio approaches unity, and therefore solar activity has a negligible effect on the spectral irradiance at these longer wavelengths. The large ratios that are apparent at wavelengths below 50 nm originate from highly ionized atomic species produced in the solar corona. Individual lines that fall within the 2.5 nm intervals can vary by a significantly higher ratio than that plotted in Fig. 18 since the plotted ratio represents the irradiance averaged over several lines that fall within the wavelength interval. For example, the ratio of the coronal emission line of Fe XVI at 33.54 nm increases by a factor greater than 100 from solar minimum to maximum, although the ratio of the averaged irradiance in the interval 32.5-35 nm is a factor of 12.

Figure 19 illustrates the relative variation of the solar spectral irradiance for several 5-nm intervals in the 140-175 nm range, along with the hydrogen Lyman-α line, at 121.6 nm. The variations for the Fe XV 28.4 nm line emission and for two intervals of the 17-20.5 nm range are shown in Fig. 20. These data were obtained on the AE-E satellite [Ref. 15].

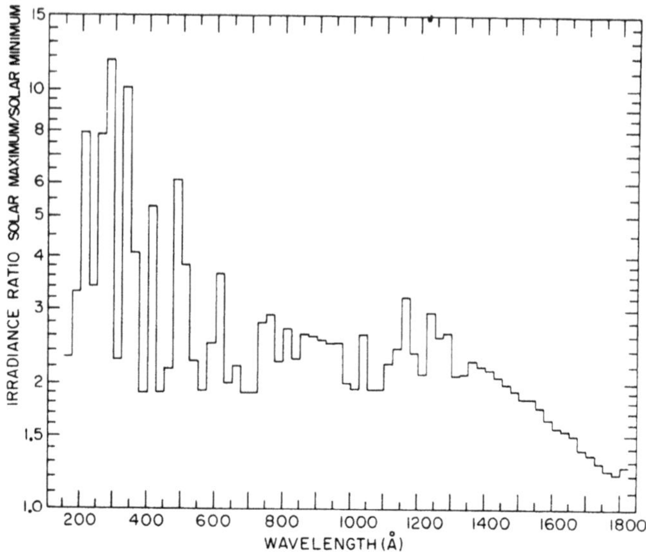

Fig. 18: The ratio of the solar spectral irradiance near solar maximum during January 1979 to the irradiance near solar minimum during July 1976 for solar cycle 21. The ratios are plotted for 25 Å intervals of wavelength [Refs. 11, 15].

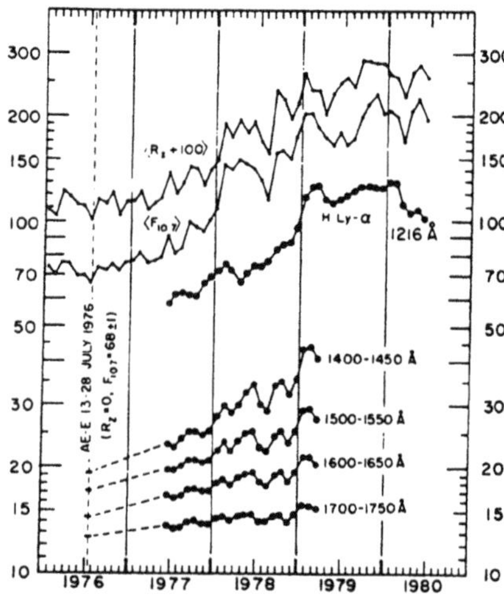

Fig. 19: Irradiance variations at wavelengths from 1216 to 1750 Å from AE-E satellite measurements during solar cycle 21 [Refs. 16, 17].

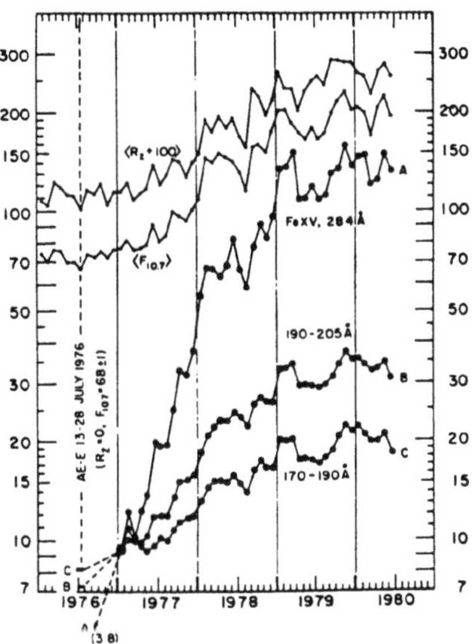

Fig. 20: Irradiance variations of some solar coronal emissions from AE-E satellite measurements during solar cycle 21 [Refs. 16, 17].

Each point represents the average of all measurements for the month. For comparison, the monthly averages of the sunspot number R_Z and the 10.7 cm radio emission are included in the figures. These data illustrate that the variation of the irradiance with solar activity between 1977 and 1980 depends upon the region of production of the radiation in the solar atmosphere. The variation of the H Lyman-α line that is produced in the lower chromosphere is a factor of approximately 2.3, while the variation of the Fe XV line produced in the corona is approximately 8.0.

- Near-Earth space environment effects -

The EUV radiation affects the space environment in two ways. First directly, because many materials used in space systems are sensitive to EUV radiation, and second, indirectly by modifying the space environment near the Earth. Because the EUV flux is not measured regularly, several models have been developed using proxi ground based measurement of 10.7 cm radio flux ($F_{10.7}$) and HeI 1083 nm equivalent width [Ref. 18] to infer the EUV fluxes.

A number of solar space telescopes working in the EUV part of the spectrum have shown effects of severe degradation in a short period of time. Figure 21 shows the degradation observed in the transmittance of a telescope flown on board OSO-8 for the study of the Sun in several spectral lines between 100 and 400 nm. Although there is no proven explanation of the effect, one assumes that it is probably due to the interaction of EUV

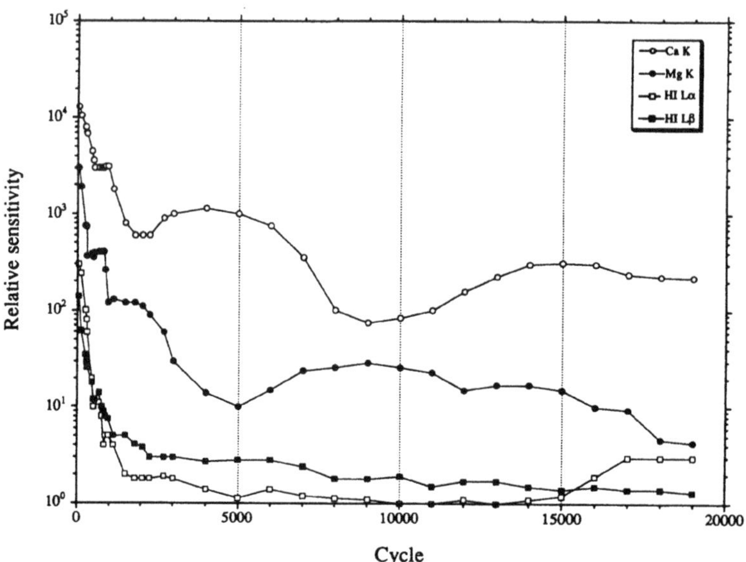

Fig. 21: Relative sensitivity change, as a function of time in orbit, for channels CaII K (393.5 nm), MgII K (279.6 nm) and HI Lyman-α and Lyman-β (121.6 and 102.5 nm) in a telescope aboard OSO.8 [Ref. 19]. Cycle = orbital revolution, about 15 cycles per day.

radiation with organic contaminant materials that are deposited on the optical surfaces. Photopolymerization of hydrocarbon components is the most important effect proposed to explain the degradation.

As for the space environment at low altitude, high thermospheric temperatures are maintained against thermal diffusion primarily by the absorption of EUV radiation by atmospheric constituents (mainly atomic oxygen). Because the solar flux of EUV varies strongly over the 11-year solar cycle, thermospheric temperatures likewise vary strongly over the solar cycle. In an average solar cycle, solar EUV and thermospheric temperatures increase by about a factor of two from solar cycle minimum to solar cycle maximum. Because density is mainly an integrated effect of temperature, the change in density over the solar cycle is amplified with respect to the temperature change. Changes in temperature cause changes in composition and these changes act to mitigate the direct thermal effects. The net effect of temperature and composition is to give order-of-magnitude changes in density over the solar cycle. The atmospheric drag on satellites is proportional to atmospheric density. The effects on satellite lifetimes are profound. A fairly typical satellite initially at 500 km would have a lifetime of about 30 years under typical solar cycle minimum conditions and only about 3 years under solar maximum conditions (Fig. 22). Large increases in the rate of change of the orbital period from solar minimum to solar maximum are possible. The fractional increase

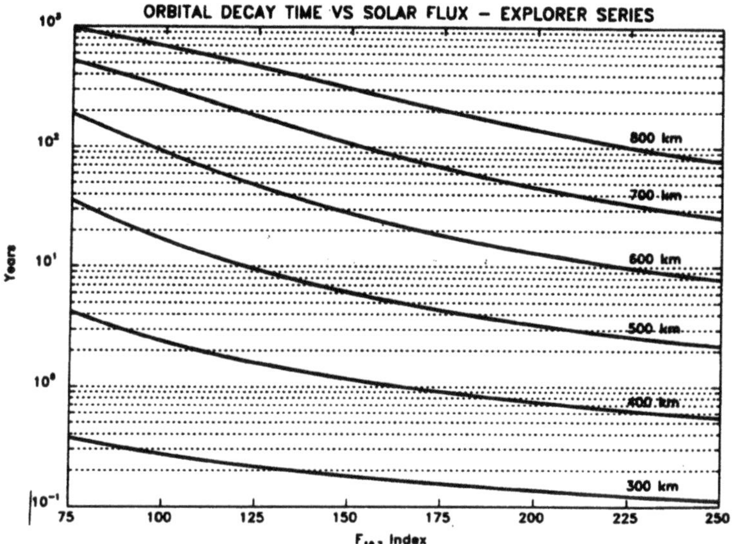

Fig. 22: Satellite lifetimes vs $F_{10.7}$ values for circular orbits for various initial altitudes [Ref. 16].

in the highly reactive species atomic oxygen can exceed two orders of magnitude [Ref. 16]. The variation of EUV produced ions at geostationary orbit is another observed effect (Fig. 23).

3.2 PARTICULATE EMISSION - SOLAR CYCLE RELATION

At a not well determined distance from the solar surface, the solar atmosphere explodes into a supersonic wind. The solar wind is a fully ionized, electrically neutral plasma that carries a magnetic field and streams outward from the solar corona at all times. The wind is highly variable both in time and in space and fills the interstellar region around the Sun.

Since the conductivity of the solar wind is extremely high, there is essentially no diffusion of the plasma transverse to the magnetic field and the field is said to be "frozen in" the flow. The plasma carries the magnetic field with it into space. If the Sun were not rotating the field would extend straight outward but the rotation causes the field to appear wound in a spiral or "garden hose" shape. The characteristics of the solar wind plasma have been measured by many spacecraft both in the neighborhood of the Earth and throughout the solar system. Table 1 lists the average characteristics at 1 AU, and gives an idea of their variability (σ = standard deviation).

The solar wind velocity ranges between 300 and 700 km/s 90% of the time. At the Earth's position, the velocity vector is radial. Solar wind variations observed on time scales of the order of days are found to be

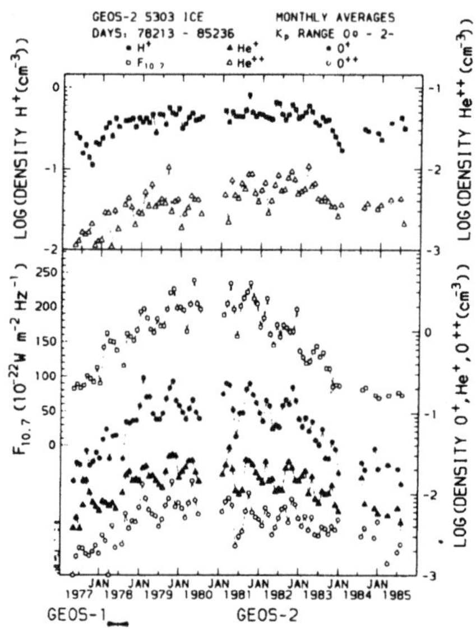

Fig. 23: Monthly averages of ion densities and the $F_{10.7}$ cm index from 1977 to 1985, at geostationary orbit [Ref. 20].

Table 1: Plasma Characteristics of Various Types of Solar Wind Flows.

	Average		Low Speed		High Speed	
Parameter	Mean	σ	Mean	σ	Mean	σ
N (cm^{-3})	8.7	6.6	11.7	4.5	3.9	0.6
V ($km\ s^{-1}$)	468	116	327	15	702	32
NV ($cm^{-2}s^{-1}$)	3.8×10^8	2.4×10^8	3.9×10^8	1.5×10^8	2.7×10^8	0.4×10^8
N_α/N_p	0.047	0.019	0.038	0.018	0.048	0.005

caused by the large shocks associated with coronal transient phenomena (flares, coronal mass ejections) and by high-speed streams associated with co-rotating coronal holes (Fig. 24).

When one averages the solar wind parameters over longer time scales, one finds that the average properties of the solar wind as observed near the Earth show little change along the 11-year solar cycle. But it is the structure of the small scale changes and the latitudinal distribution of the solar wind characteristics that change with solar cycle, as is to be expected from the changes in the solar corona described above. They induce changes in the Earth's magnetosphere and in the galactic cosmic ray modulation.

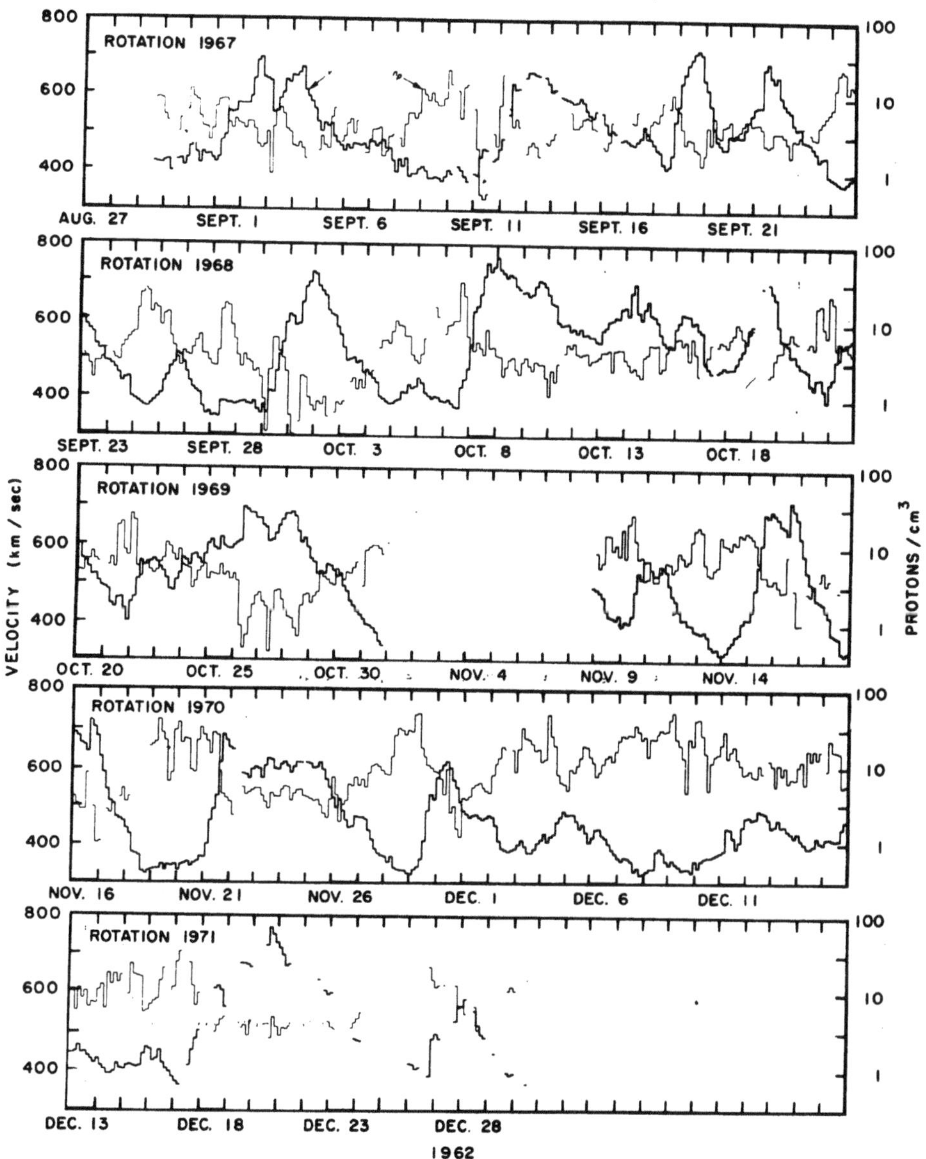

Fig. 24: Three-hour averages of the solar wind proton density (light line) and flow speed (heavy line) from mariner 2 on its flight toward Venus in 1962. Each panel presents 27 days of data. Note the velocity stream structure and the tendency for recurrence of streams on each solar rotation [Ref. 11, 22].

- Near-Earth space environment effects -

The only direct effect that has been considered for the solar wind to produce in space systems outside the magnetosphere is the consequence of the solar-wind ions impinging upon optical systems pointing at the Sun. They can produce sputtering effects on the mirror or filter external layers, or bubbles or ion implantations. But there are important indirect effects:

a) Through the Earth's magnetosphere

The magnetosphere is the result of the interaction of the solar wind with the Earth's magnetic dipole. The description of the Earth's magnetosphere is the subject of another paper in this book [Ref. 21]. We will just indicate that any change in the characteristics of the solar wind has an effect on the state of the magnetosphere: the velocity and density of the solar wind produce changes in the dimensions by hydrodynamic pressure, and the direction of the magnetic field embedded in the solar wind changes its connectivity with the Earth's magnetic field.

Solar wind shocks induce geomagnetic storms that generate large currents at different levels of the magnetosphere from its external boundary down to the lower atmosphere. Geomagnetic activity most often measured by the magnetic index Kp or by the ring current magnetic index Dst is directly linked to solar wind characteristics and variations. Since most of the spacecraft spend all their life inside the magnetosphere, the environment in which they are is affected by the solar wind. The following are a couple of examples showing the effects of solar wind activity on magnetospheric activity and on spacecraft.

Figure 25 shows how the surface of a spacecraft that goes through Earth eclipse at high altitude charges depending on whether at that moment the magnetosphere is quiet or disturbed. At high altitude the charging of the spacecraft surface is produced by hot plasma electrons (>1-2 keV) and becomes effective in the eclipse periods because at these moments photoemission, that acts as a voltage limiting mechanism, is interrupted. During geomagnetic activity the hot plasma density becomes larger on the night side of the magnetosphere. At geosynchronous altitude, major geomagnetic storms produce an increase of very energetic electrons, and energetic electrons penetrate into bulk dielectrics in satellites, inducing thick-dielectric charging (e.g., in cable insulation and printed circuit boards) that can exceed the breakdown potential of the dielectric (Fig. 26).

b) The modulation of galactic cosmic rays

Charged particles populate the inter-stellar wind with a spectrum of energy that reaches up to 10^{20} eV. The solar wind creates a cavity around the Sun, the heliosphere, and modulates the penetration of the charged particles in such a way that they are observed in lesser quantity at the Earth orbit during maximum of solar activity and in larger quantity during minimum of solar activity (Fig. 27).

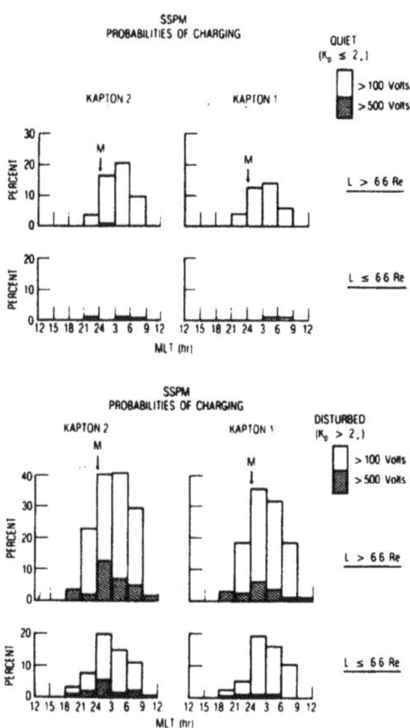

Fig. 25: Histograms of the probabilities of charging versus magnetic activity, for a spacecraft at high altitude in the night magnetosphere [Ref. 23].

High energy cosmic rays may cause errors in memory contents and in microelectronics. Spacecraft operating in low Earth orbit at low inclinations are largely protected from galactic cosmic rays by the Earth's magnetic field. All the other spacecraft are exposed to galactic cosmic rays.

3.2.2 Solar Energetic Particles - Flares

There are two main sources of energetic particles of solar origin in the near-Earth environment: the flares and the interplanetary shocks. The most important are the chromospheric events known as flares that result from the sudden release of very large amounts of energy described in section 2.4 above. Solar flares produce electromagnetic emission, accelerate electrons and ions, and if conditions are favorable, inject these particles into space. Generally the very large particle events with particles accelerated to very high energies are the consequence of solar flares.

Unlike solar electromagnetic radiation, both the onset time and maximum intensity of the solar particle flux depend on the heliolongitude of the flare with respect to the detector location in space. This directionality results because particles will move most easily along the interplanetary magnetic field-direction. The interplanetary magnetic-field topology is determined by the solar wind outflow and the rotation of the

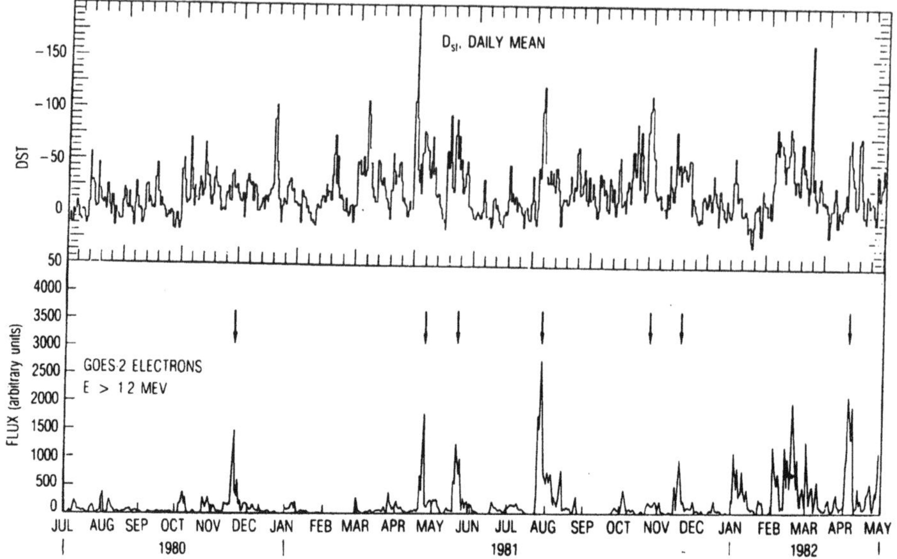

Fig. 26: Comparison of D_{st} with energetic electrons measured by GOES. The lower trace is a plot of the E_e > 1.2 MeV electron flux as measured at geosynchronous orbit by the GOES weather satellites from mid-1980 to mid-1982. The arrows indicate times when the sun-sensor anomalies ascribed to thick dielectric charging were experienced on several geosynchronous satellites. The upper panel is a plot of D_{st} for the same time period [Ref. 23].

Sun, which during "quiet" conditions can be approximated by an Archimedian spiral (Fig. 28).

Energetic solar particles reach the orbit of the Earth within a few minutes if the particles have very high energies, or within hours for the lower-energy particles. The enhanced solar plasma associated with the flare usually propagates to the Earth within one or two days when it causes aurora and geomagnetic disturbances. The propagation time and intensity of the energetic particles from the flare to the Earth depends on the propagation of the particles, first by diffusion in the solar corona from the flare site to the root of the magnetic-field lines and then by transport along interplanetary magnetic-field lines. Figure 29 shows typical time scales for the particle phenomena in a solar flare. Since the electromagnetic radiation propagates unhindered from the Sun to the Earth, it may be used to predict within tens of minutes the probable arrival of energetic particle fluxes (Fig. 30).

The frequency of the solar flares is correlated with the cycle of solar activity in general, but the kind of correlation depends upon the parameter that one observes. It has been shown [Ref. 25] that for flare-generated proton events (proton energy greater than 30 MeV) the annual integrated flux is very small for the years around sunspot minimum (when

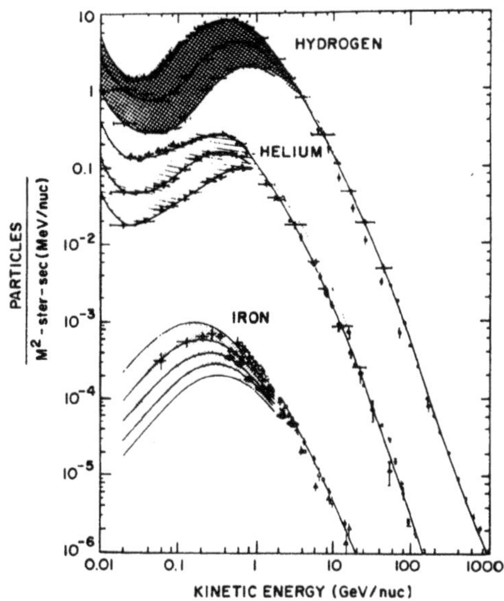

Fig. 27: Primary cosmic ray differential energy spectrum. The upper envelope indicates the solar minimum spectrum while the lower envelope indicates the solar maximum spectrum. The shaded area indicates the range of the solar modulation over a solar cycle. The hydrogen spectrum in this figure has been multiplied by a factor of 5 so the modulated portion of the spectrum avoids merging with the top of the helium spectrum [Ref. 24].

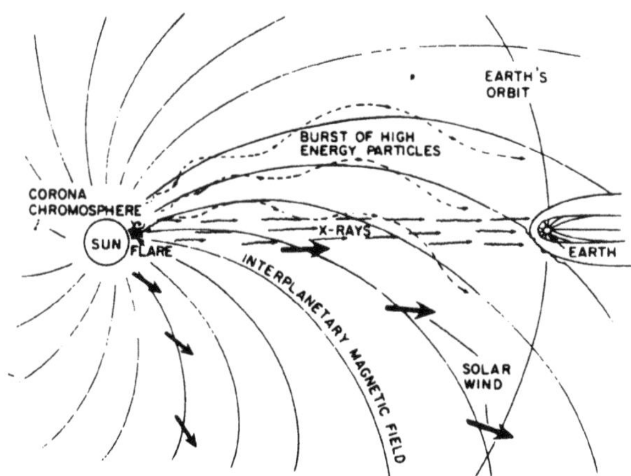

Fig. 28: Characteristics of the idealized structure of the interplanetary medium [Ref. 24].

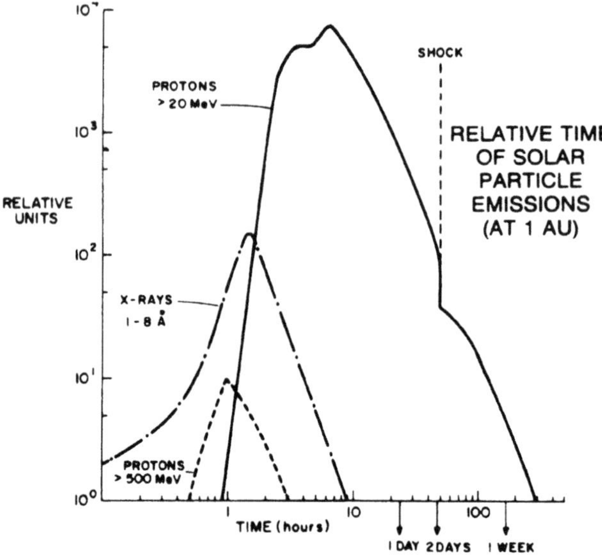

Fig. 29: Time scale of solar emissions arrival at 1 AU (notice the logarithmic scale) [Ref. 24].

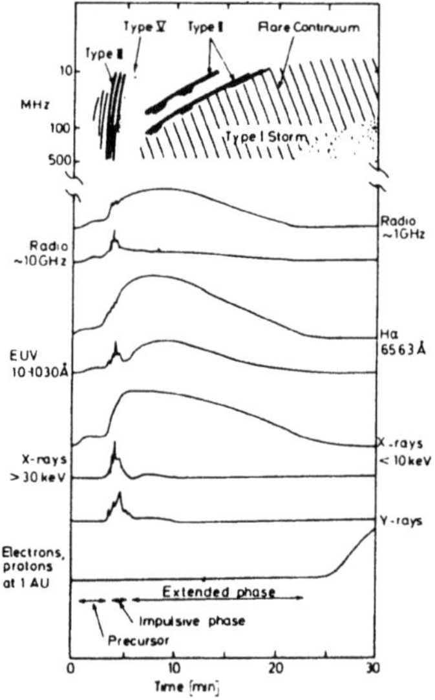

Fig. 30: Electromagnetic and particle radiation for a typical flare [Ref. 26].

the year integrated sunspot number is less than 35), and it is independent of sunspot number for integrated sunspot number greater than 35. One can say that it behaves like an on/off system (Fig. 31).

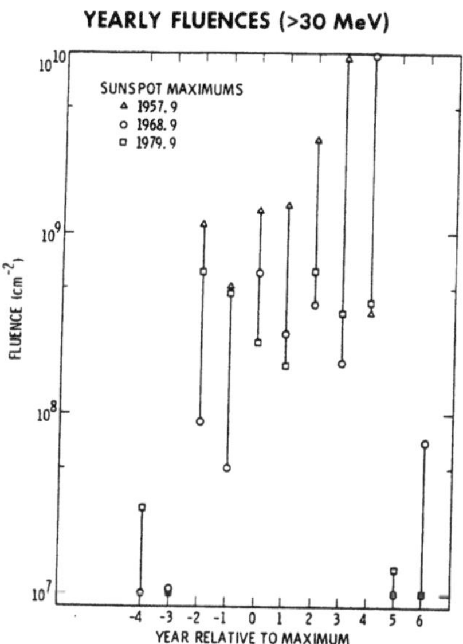

Fig. 31: Integrated yearly fluences of protons of energy greater than 30 MeV versus year relative to sunspot maximum [Ref. 25].

Another source of energetic particles are interplanetary shocks. Almost any shock wave propagating through the corona and the solar wind will generate measurable fluxes of ions and electrons accelerated by the effect of the shock propagation through the magnetized plasma. Some of these shocks are originated by coronal mass ejection, associated with disappearing filaments.

- Near-Earth space environment effects -

The main source of high energy particles that encounters a spacecraft outside the inner Earth-radiation belts are the proton events generated by solar flares. The effects of high energy particles on the spacecraft systems are discussed in several papers in these proceedings. Typical effects are: decrease in the efficiency of the solar cells, single event upsets (SEU) in the microprocessors, degradation of CCD detectors, etc. Large solar proton events are a cause of major concern for out-of-the-Earth magnetosphere human travel.

4. Predictions

We have seen the many effects that changing solar activity have on the near-Earth space environment. Predicting future solar activity is therefore an important subject for spacecraft designers and space mission planners.

There are two main kinds of predictions that are interesting: one is the overall average level of solar activity, the other, prediction of single events, such as proton flares or geomagnetic storms. For the prediction of the solar activity level, the predictors are based on the application of statistical methods to the history of the solar activity. Figure 32 gives an indication of the degree of success in predicting the level of solar activity, represented by the number of sunspots, 12 months in advance. For the prediction of individual solar originated disturbances in the near-Earth environment, the methods are purely phenomenological. Based on continuous observations of the Sun and continuous monitoring of the solar radiation from spacecraft and from radiotelescopes, it is possible to see, for instance, when a flare has occurred, the type, and relative importance of the event, then on the basis of accumulated knowledge of similar cases, provide a warning with an indication of the probability that a disturbance will arrive at the Earth in the following minutes or hours (see Fig. 29). The observation of the evolution of solar active regions on the surface of the Sun allows the prediction of probable disturbances within time scales of days.

5. Summary

We have seen that the solar emissions toward the Earth vary in many ways that are capable of modifying the near-Earth environment. Particularly important are:

- The variations, with the solar cycle, of the extreme ultraviolet radiation affect both space systems directly, via effects on the surface materials, and the space environment at low orbits, because it heats and modifies the upper Earth atmosphere and ionosphere.

- The variations of the solar wind caused by the rotation of the coronal structures from which it originates and by shocks propagating through it induce large changes in the population of particles and plasmas at all levels in the Earth magnetosphere, where most satellites reside.

- The solar flares with their emission of electromagnetic radiation bursts, particularly at X and EUV wavelengths, that disturb the ionosphere, the radio bursts that disturb radio transmission, the energetic particles that may damage the space systems, and the associated solar wind shocks that produce large magnetic storms near the Earth.

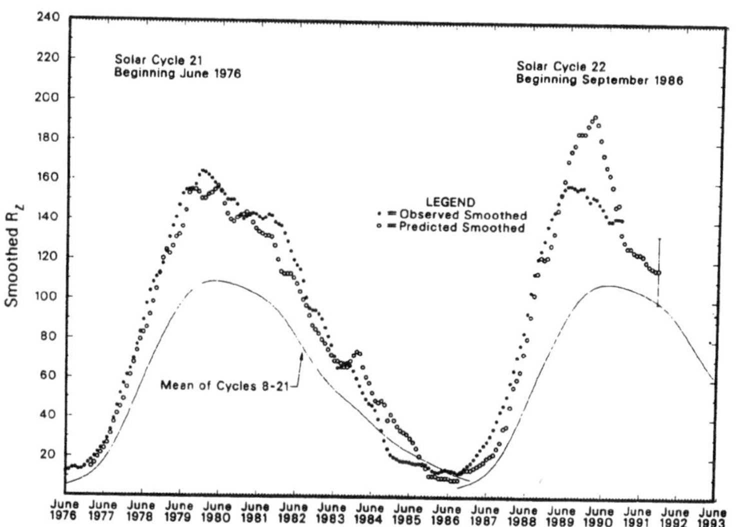

Fig. 32: Observed and one-year-ahead predicted sunspot numbers [Ref. 27].

6. Bibliography and Credits

The reader is referred to books such as The Sun [Ref. 2]; Solar Astrophysics [Ref. 7]; Handbook of Geophysics and Space Environment [Ref. 11]; and the special section on 'Solar Cycle Effects,' published in the Journal of Spacecraft and Rockets in November-December 1989, for a more in-depth study of the subject. Many parts of this paper are a transcript of these publications.

References

[1] A.H. Gabriel, 'A magnetic model of the solar transition region,' Phil. Trans. Royal Soc. London, 281, 339 (1976).

[2] M. Stix, The Sun, An Introduction, Springer-Verlag, Berlin (1989).

[3] E.M. Reeves, M.C.E. Huber, and J.G. Timothy, 'Extreme UV spectroheliometer on the Apollo Telescope Mount', Applied Optics, 16, 837 (1977).

[4] G. Withbroe, 'Solar activity cycle: History and predictions,' J. Spacecraft and Rockets, 26, 394 (1989).

[5] H. Babcock, 'The solar magnetic cycle,' Astrophys. J., 133, 572 (1961).

[6] R.W. Noyes, The Sun, Our Star, p. 109, Harvard University Press, Cambridge (1982).

[7] P. Foukal, Solar Astrophysics, Wiley, New York (1990).

[8] J.B. Gurman, ed., NASA's Solar Maximum Mission: A Look at a New Sun, p. 6, NASA/GSFC (1987).

[9] O.R. White, The Solar Output and Its Variations, Colorado Associated University Press, Boulder (1977).

[10] R.C. Willson, S. Gulkis, M. Janssen, H.A. Hudson, and G.A. Chapman, 'Observation of solar irradiance variability,' Science, 221, 700 (1981).

[11] A.S. Jursa, ed., Handbook of Geophysics and Space Environment, Second Edition, Air Force Geophysics lab., USAF (1985).

[12] Fröhlich, C.

[13] R.C. Willson and H.S. Hudson, 'The Sun's luminosity over a complete solar cycle,' Nature, 351, 42 (1991).

[14] D.V. Hoyt, H.L. Kyle, J.R. Hickey, and R.H. Maschhoff, 'The Nimbus-7 solar total irradiance: A new algorithm for its derivation,' J. Geophys. Res. 97, 51 (1992).

[15] H.E. Hinteregger, 'Representation of solar EUV fluxes for aeronomical applications,' Adv. Space Res., 1, 39 (1981).

[16] R.L. Walterscheid, 'Solar cycle effects on the upper atmosphere: implications for satellite drag,' J. Spacecraft and Rockets, 26, 439 (1989).

[17] H.E. Hinteregger, 'AE-E experiences of irradiance monitoring for 1250-1850 A,' in Proceedings of the Workshop on Solar UV Irradiance Monitors, NOAA/ERL, Boulder (1980).

[18] W.K. Tobiska and C.A. Barth, 'A solar EUV flux model,' J. Geophys. Res., 95, 8243 (1990).

[19] P. Lemaire, 'Sensitivity changes in the CNRS ultraviolet spectrometer aboard OSO-8', Journal, 15, 237 (1991).

[20] M. Stockholm, H. Balsiger, J. Geiss, H. Rosenbauer, and D.T. Young, 'Variations of the magnetospheric ion number densities near geostationary orbit with solar activity,' Annales Geophysicae, 7, 69 (1989).

[21] S. Cowley, 'The Magnetosphere', these proceedings (1993).

[22] M. Neugebauer and C.W. Snyder, 'Observations of the solar wind,' J. Geophys. Res., 71, 4469 (1966).

[23] A.L. Vampola, 'Solar cycle effects on trapped energetic particles,' J. Spacecraft and Rockets, 26, 416 (1989).

[24] D.F. Smart, and M.A. Shea, 'Solar protons events during the past three cycles,' J. Spacecraft and Rockets, 26, 403 (1989).

[25] J. Feynman, T.P. Armstrong, L. Dao-Gibner, and S. Silverman, 'Solar proton events during solar cycles 19, 20 and 21,' Solar Physics, 126, 385 (1990).

[26] D.J. McLean and N.R. Labrum, ed., Solar Radiophysics, p. 55, Cambridge University press, Cambridge (1985).

[27] Solar-Geophysical Data, Solar-Geophysical Data Prompt Reports, p. 29, NOAA, Boulder, Colorado (1991).

THE UPPER ATMOSPHERE AND THE IONOSPHERE

R. W. NICHOLLS
Centre for Research in Earth and Space Science
York University
4700 Keele St., North York, Ontario, Canada M3J 1P3

ABSTRACT. A review is given of the structure, properties and composition of the D, E, F1, F2, Topside and Plasmasphere components of the terrestrial ionosphere. It is a unique part of the near earth space environment, ionized layers which are formed in the mesosphere, the thermosphere and the exosphere as a result of selective photoionizations by X-ray and extreme ultraviolet components of the sunlight. The primary photoionizations are followed in many cases by altitude-specific ion chemistry reactions such as dissociative recombination. Ionospheric properties thus depend strongly on the solar activity cycle and time of day.

1. Introduction

In a previous paper [Ref. 6] a general discussion was given of the structure and constitution of the atmosphere, which is an important part of near earth space. The controlling influence of solar irradiation on atmospheric conditions was described in general terms. The emphasis of that presentation was principally on neutral species, such as O_2, and on important processes such as photodissociation (and its products) induced by solar radiation as indicated below.

$$h\nu + O_2 \rightarrow O + O$$

Many atmospheric properties, particularly below 100km, involve such neutral species. By contrast, in the upper atmosphere (roughly between 60 and 1000km including the mesosphere, the thermosphere and the exosphere) charged species such as:

$$O^+,\ O_2^+,\ O^-,\ O_2^-,\ NO^+,\ N_2^+,\ CO_2^+,\ He^+,\ H^+,\ e^-$$

are prevalent. They arise from the combined effects of primary photoionization processes such as:

$$h\nu + N_2 \rightarrow N_2^+ + e$$

R. N. DeWitt et al. (eds.), *The Behavior of Systems in the Space Environment*, 103–116.

often stimulated by absorption by the atmosphere of solar EUV and X-ray photons, followed in most cases by secondary ion chemistry reactions such as:

$$N_2^+ + O \rightarrow NO^+ + N$$

This example of ion chemistry is given because of the important role played by NO^+ in the ionosphere. Although NO is a species of low natural atmospheric abundance, its photoionization cross section is such that it is preferentially formed in the D, E, and F ionospheric layers of the mesosphere and the thermosphere. It plays important roles in ionospheric processes.

Figure 1 displays the solar emission spectrum recorded between 1300Å and 270Å on the NASA OSO 3 satellite. Much of the radiation originates in the upper chromosphere and corona of the sun. The strong features at 1216Å (H Lyman α), 1026Å (H Lyman β), 977Å (CIII), 304Å (He II) are responsible for much of the photoionization in the ionosphere.

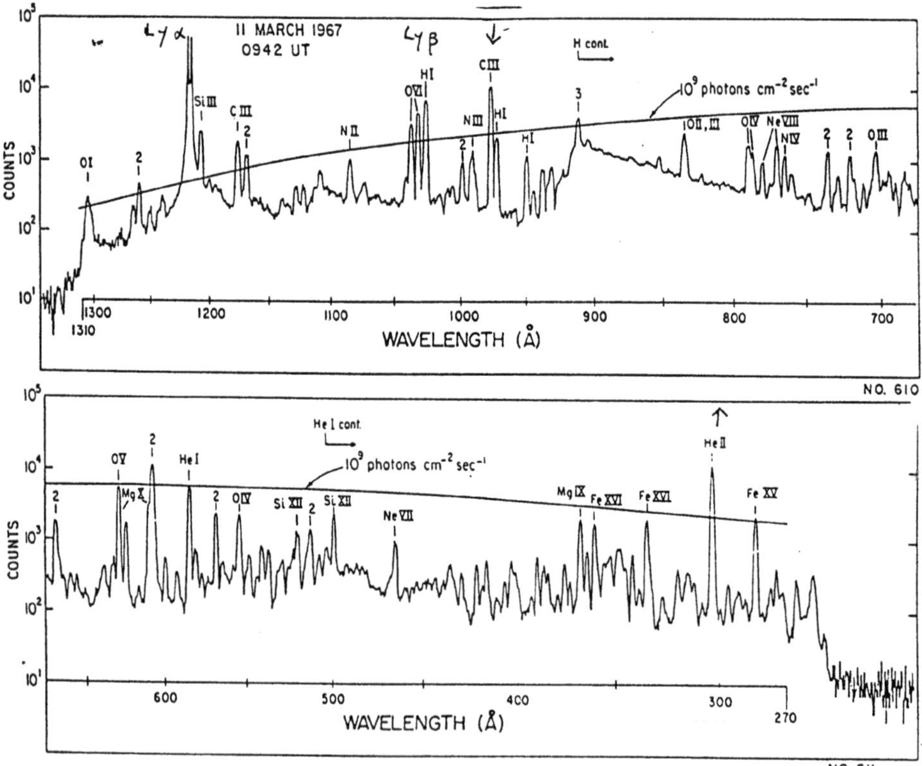

Fig. 1: Solar UV Spectrum between 1300Å and 270Å observed from the OSO 3 satellite [Ref. 5].

The solar UV irradiance (λ < 3200Å) is a very small fraction (1-2%) of the total solar irradiance but its absorption in the atmosphere is a major source of energy. It drives atmospheric photochemistry and temperature profiles throughout the stratosphere, the mesosphere, the thermosphere, and the exosphere.

Figure 2 displays the broad band extinction properties of the atmosphere for solar UV and X-radiation. An altitude profile at which the solar radiation is attenuated by 1/e ($\cong$0.368) is shown as a function of wavelength. The principal absorbers are indicated as well as the levels of the ionospheric layers. Below 50km much of the extinction arises from absorption by neutral species. At ionospheric altitudes, nearly all of the extinction comes from photoionization processes.

Many ion-chemistry reaction cycles, which involve positive and negative ions, electrons and neutrals, take place in the high atmosphere. The eventual equilibrium between ionization and recombination reactions gives rise to the equilibrium concentrations of charged and uncharged species. These are very time dependent, on diurnal and longer scales, driven as they are by solar radiation.

Charged and neutral species thus exist in the upper atmosphere which therefore exhibits some plasma properties in these regions. The resulting charged layers (D,E,F1 F2) of the ionosphere are found in the mesosphere, the thermosphere and the exosphere. The purpose of this chapter is therefore to summarize briefly the structure, constitution and principal properties of the ionosphere which is an important part of the space environment to which various space systems are exposed.

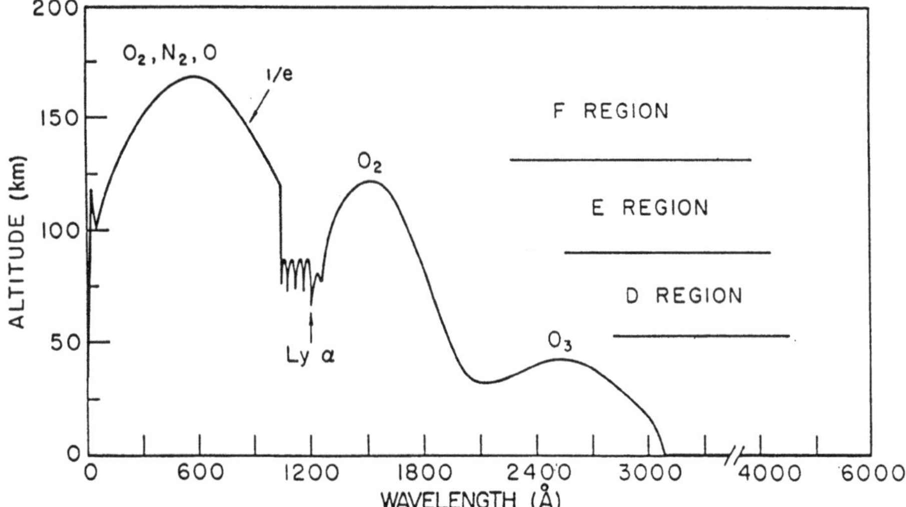

Fig. 2: Atmospheric extinction of the solar UV radiation. A broad band 1/e extinction altitude profile is shown between the X-Ray region and about 3000Å. Principal primary absorbers and altitudes of ionospheric regions are indicated [Ref. 5].

It is not the purpose of this chapter to provide a complete review of the aeronomy of the ionosphere on which there is a very rich literature. A compact historical review of the discovery of the ionosphere, which was stimulated by interest in long distance short wave radio propagation, and the related phenomenology of radiowave reflection properties of ionospheric layers, is given by Whitten and Poppoff [Ref. 12, 13]. Other excellent reviews on the formation and properties of the ionosphere can be found in the various editions of the Handbook of Geophysics [Refs. 5, 10, 11], in Banks and Kockarts [Ref. 1], in Whitten and Poppoff [Ref. 12, 13], and in Ratcliffe [1960, Chapter 9]. An excellent summary ion chemistry of the thermosphere is given in appendix 5 of Rees [Ref. 9].

2. The Structure of the Ionosphere

The ionosphere, the presence of which was first inferred as an atmospheric radio-wave reflecting layer to explain world wide short wave radio communication, consists of the ionized layers D, E, F_1, F_2 between 70 and 1000km as summarized in Figs. 3 and 4.

Figure 3 shows that the ion/electron number densities in the E and F regions are between 10^4 and $>10^6$ /cc. It is seen in the same figure that these number densities are less than the number densities of neutrals at the same altitudes by some orders of magnitude. The ionospheric regions thus lie in a dilute plasma. The principal positive ion species of the ionospheric plasma layers, and the photoionizing radiation which produce them are indicated in Fig. 4.

A simple theory of the refractive index of a collisionless ionospheric plasma (in the absence of a magnetic field) [Ref. 5, Chapter 10] predicts that the critical frequency, the plasma frequency at which total reflection of radio waves will occur is given by:

$$f^2 = \frac{e}{4\ \pi^2 \epsilon_0 m}\ N_e \tag{1a}$$

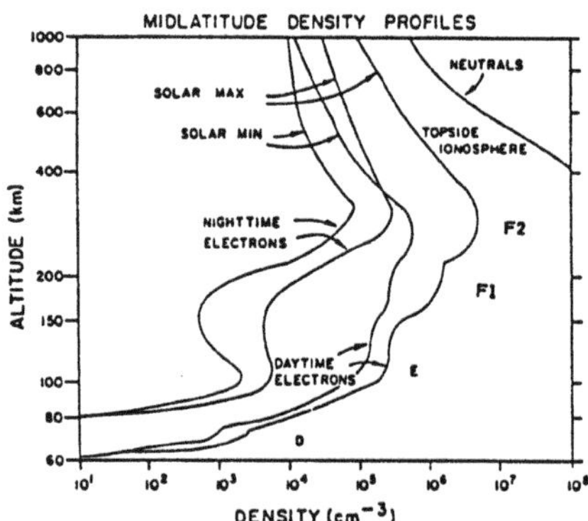

Fig. 3: Ionization altitude profiles and ionospheric layers for solar maximum and solar minimum conditions [Ref. 5].

Layer	Altitude(km)	Major Component	Production Cause
D	70–90 km	NO^+, O_2^+	Lyman Alpha, x-rays
E	95–140 km	O_2^+, NO^+	Lyman Beta, Soft x-rays, UV Continuum
Fl	140–200 km	O^+, NO^+	He II, UV Continuum (100–800Å)
F2	200–400 km	O^+, N^+	He II, UV Continuum (100–800Å)
Topside F	> 400 km	O^+	Transport from Below
Plasmasphere	> 1200 km	H^+	Transport from Below

Fig. 4: Daytime midlatitude ionospheric layers: Altitudes, major ionic components and production cause [Ref. 5].

The symbols are conventional. ϵ_0 is the permitivity of vacuum and N_e is electron density. Thus.

$$f = 0.009\ [N_e]^{1/2}\ \text{MHz}\ ;\ N_e = 1.24 \times 10^4\ f^2\ \text{cm}^{-3} \qquad (1b)$$

Therefore, for ionospheric electron densities between 10^4 and 10^6 cm^{-3} typical plasma frequencies lie between about 0.9 and 9 MHz. The scanning frequencies of ionosonde reflection equipment thus commonly lie between 0.5 and 20 MHz in the short wave radio band.

Figures 3 and 4 identify D, E, F_1 and F_2 layers across the mesosphere and thermosphere. In addition "topside F" and plasmasphere layers are identified in the upper thermosphere and in the exosphere.

The D region is the lowest ionospheric layer (70-90km). It is only present during daylight hours. It is produced by very penetrating radiation and particles from extra-terrestrial space. It is a weak plasma imbedded in a neutral atmosphere. Many attachment, detachment and charge exchange processes take place in it. Its charge density peaks between 90 and 78km depending on geophysical circumstances. It is the most complex ionospheric region, and its variations have a great effect on absorption and reflection of low frequency radio signals.

The E region (95-140km) peaks at about 110km. At sunset the electron concentration rapidly drops to an equilibrium density which is retained during the night. Ionization of molecules is very important in this region. Electron loss in this region takes place by dissociative recombination.

The F region has two components F_1 and F_2. The F_1 (140-200km) region peaks at about 200km and is absent during the night. The F_2 (200-400km) region peaks at about 300km during the day and somewhat higher at night.

In the F_1 region (140km-200km) the principal ions are O^+ and NO^+. Solar radiation of λ < 800Å stimulates the main primary ionization. The main electron depletion mechanism is dissociative recombination. Ion interchange is needed for this.

In the F_2 region (200km-400km), solar radiation of λ < 800Å is responsible for much of the ionization to form O^+ and N^+. That is because the principal neutral constituents O and N have respective ionization potentials of 13.6 ev and 15.58 ev.

The so-called Topside region (400km-1200km) lies above the F_2 region peak. In it the O^+ and H^+ ions are transported from below and their density decreases with altitude. Magnetic field lines of 60° (or less) geomagnetic latitude are closed. They arise from a point in the southern hemisphere and terminate at a point in the northern hemisphere. Above 1200km they constrain hydrogen ions moving out from the topside region and prevent them from drifting off into space. The region above about 1200km and the 60° geomagnetic latitude line which holds these trapped ions is called the plasmasphere. Its upper boundary, the 60° fieldline, is called the plasmapause. Magnetic lines of higher latitude in the approximately dipole field are not closed. The Topside ionosphere and plasmasphere are well illustrated in Jursa [Ref. 1].

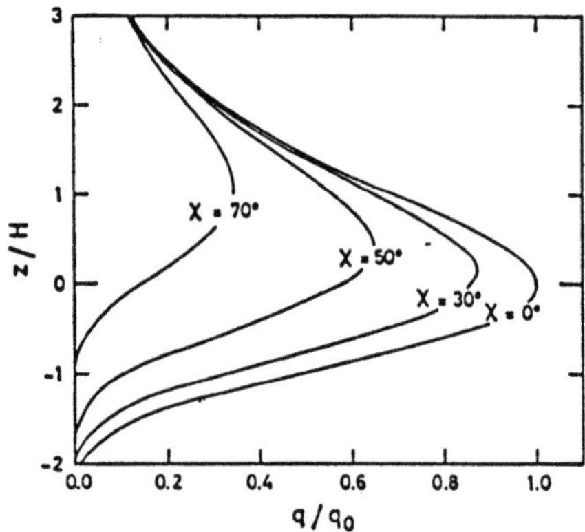

Fig. 5: Normalized ion production profiles for a Chapman layer in an idealized isothermal atmosphere for different solar zenith angles x [See Eq. (5) in which ξ is the solar zenith angle] [Ref. 1].

3. Ionospheric Processes

The physical conditions existing in each of the ionospheric regions depend upon the balance between primary ionization and secondary recombination effects. The primary ionizations arise from solar X-ray and EUV radiation together with the effects of galactic cosmic rays throughout the whole ionosphere. At different altitudes, these effects play different roles. The more highly absorbed less penetrating radiations are the principal ionizing agents in the high atmosphere. Lower in the atmosphere, less highly absorbed and more penetrating radiation is responsible for most of the ionization. The recombination effects depend upon species which are present and their collision frequencies.

3.1. THE CHAPMAN MODEL

Chapman [Refs. 3, 4] developed an idealized phenomenological theory for atmospheric layer formation arising from photodissociation and photoionization processes in an exponential atmospheric slab, characterized by a scale height, and in which solar radiation is absorbed. This absorption results in photodissociation or photoionization. The photodissociation effects are important in the neutral stratosphere and mesosphere. The photoionization effects are important in the ionosphere.

The aspects of his model which describes ionospheric layer formation through such primary ionizations in a simple one component slab atmosphere is briefly reviewed below. It provides good physical insight into the morphology of the ionospheric layers. It has been more recently augmented to take account of a multicomponent atmosphere and a spherical earth [Refs. 1, 2, 8, 12. 13].

The Chapman theory and the associated Chapman function leads to expressions for q(z) the primary ionization rate at altitude z which accounts for layer formation. Its formulation uses three concepts: Extinction/Absorption, Scale Height, and Energy Deposition as follows.

Extinction. At a given wavelength and for a given species of a concentration N(z) at altitude z, and ionization cross section σ cm^2 per particle, the extinction -dI of a flux of radiation I along an oblique atmospheric path dl (= dz secξ where dz is the altitude increment and ξ is the zenith angle) is:

$$-dI = \sigma N(z) I dl = \sigma N(z) I dz \sec\xi \qquad (2a)$$

Integrating this from altitude z to the fringe of the atmosphere (∞) we obtain the Beer extinction law equivalent:

$$I(z) = I_\infty \exp [- \sigma \sec\xi \int_z^\infty N(z) dz] \qquad (2b)$$

Scale Height. In simple cases [see Ref. 6, section 2.1] we can write N(z) in terms of a scale height H by Eq. (3a)

$$N(z) = N_o \exp [-z/H] \qquad (3a)$$

where

$$H = \text{scale height} = RT/M_o g \qquad (3b)$$

and M_o is the mean molecular weight

Energy Deposition. The energy deposition dI per unit cross section along oblique path increment dl (=dz sec ξ) is I σ N dl = I σ N dz sec ξ. If one electron-ion pair is produced per W(ev) of this deposited radiation, then the ionization rate q per unit volume at z is:

$$q(z) = \frac{1}{W} \frac{dI}{dz} \cos\xi \tag{4a}$$

Thus from Eq. (2a) we can write

$$q(z) = \frac{\sigma N(z)}{W} I(z) \tag{4b}$$

which from Eqs. (2b) and (3a) becomes

$$q(z) = \frac{\sigma N_o I_\infty}{W} \exp\left[- \frac{z}{H} \sigma N_o \sec\xi (\exp(-z/H) \right] \tag{4c}$$

This is one version of the Chapman function.

The maximum ionization rate q_m occurs at the altitude z_m, where dq/dz (or d(lnq)/dz) = 0 at z_m. Thus:

$$z_m = H \ln [H \sigma N_o \sec\xi] \tag{4d}$$

This is equivalent to unit slant optical depth. From Eqs. (4c) and (4d), the maximum ionization rate q_m is:

$$q_m = q(z_m) = I_\infty \cos\xi / eWH \tag{4e}$$

Here e is the base of natural logarithms. From Eqs. (4c) and (4e)

$$q(z) = q_m \sigma N_o W e \cos\xi \exp\left[- \frac{z}{H} - \sigma N_o \sec\xi (\exp-\frac{z}{H})\right] \tag{5}$$

which is another form of the Chapman function. It is often expressed in non-dimensional terms. Figure 5 displays one set of Chapman functions which illustrate how individual ionospheric layers are formed.

3.2. IONOSPHERIC LAYERS

3.2.1. *The D Region (70-90km)*. The primary ionization in the D region is caused by very penetrating radiation whose photoionization cross-section is sufficiently small ($\sigma < 10^{-19}$ cm$^2 \cong 10^{-3}$ Å^2) that it has not been completely absorbed at higher altitudes. The sources of this radiation include:

- * Solar X-Rays (λ < 10Å)
- * Solar UV (λ < 1750Å)
- * H Lyman α (λ = 1216Å)
- Galactic cosmic rays (BEV)

The intense H Lyman α emission line from the solar corona penetrates to the D region as it happens to lie at the same wavelength as a deep transparency window ($\sigma \cong 10^{-20}$ cm^2) in the vacuum ultra violet absorption spectrum of O_2. The X-rays ionize N_2 and O_2. Lyman α ionizes NO and other low ionization potential metallic species in the D region.

Many complex processes take place in the D region in which the collision frequency is high (10^4- 10^6 sec^{-1}). The primary ions formed are N_2^+, O_2^+ and NO^+. Ion chemistry resulting from collisions with these ions rapidly produce many other positive and negative ion species [Refs. 1, 9]. The D region is only present during daylight hours when the solar primary ionization source is active. A typical positive ion distribution in the D region is shown in Fig. 6.

3.2.2. *The E Region (95-140km)*. The primary ionization in the E region arises from radiation whose absorption cross section is somewhat larger ($\sigma < 5 \times 10^{-18}$ cm$^2 \cong 5 \times 10^{-2}$ Å^2), but not sufficiently large as to be absorbed by higher altitude regions. Sources of this radiation include:

- Solar soft X-rays in the region 31Å < λ < 100Å
- Solar EUV continuum in the region λ > 800Å
- Solar H Lyman β (λ = 1025Å)
- Solar CIII (λ = 977Å)
- Solar Lyman continuum (λ < 910Å)

Lyman β and the CIII line both ionize O_2. The Lyman continuum ionizes OI. In this region ionization of molecular species is predominant, with the formation of N_2^+, O_2^+, and to a lesser extent O^+ and NO^+. Electron-ion losses occur through dissociative recombination processes such as:

$$NO^+ + e \rightarrow N^* + O^*$$

$$O_2^+ + e \rightarrow O^* + O^*$$

$$CO_2^+ + 3 \rightarrow CO^* + O^*$$

$$N_2^+ + e \rightarrow N^* + N^*$$

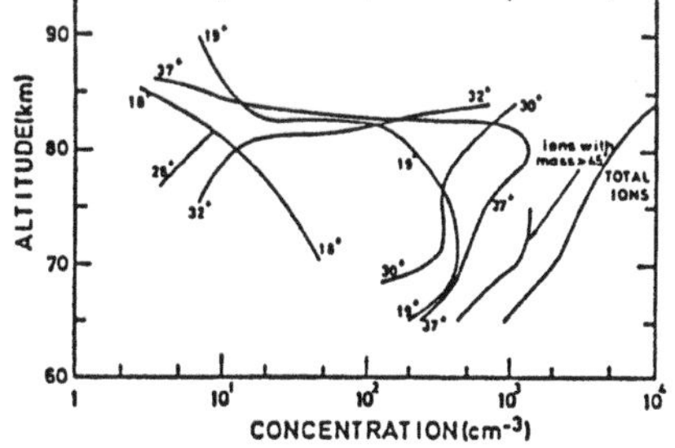

Fig. 6: Altitude distribution of the positive ion composition of the D region molecular weights of individual components are indicated [Ref. 1].

where N* etc may be optically excited species. After the relevant ion chemistry has taken place, O_2^+ and NO^+ are the most abundant species in the E region as shown in Fig. 7.

3.2.3. *The F_1 Region (140-200km)*. The F_1 region peaks near 200km and is absent during the night. Ionization in the whole F region is caused by solar radiation between 100Å and 800Å region where a large photoionization cross section $\sigma > 10^{-17}$ sq. cm $\cong$ 0.1Å^2 for atmospheric species exists. The principal components of this ionizing radiation are lines of HeII and the solar X-ray continuum. The principal positive ions initially formed in this region are N_2^+ and O^+. Complex ion chemistry between these and other constituents leads to NO^+ being the most abundant ionic end product, followed closely by O^+. This is illustrated in Fig. 8.

3.2.4. *The F_2 Region (200-400km)*. At altitudes up to 200km, and including the F_1 region, collision rates are sufficient to ensure that the balance between photochemical and ion chemical processes determine the quasi-equilibrium structure of each layer. At higher altitudes however, the diffusion of ions through the neutral gas becomes a very important factor which has to be taken into account. Above 250km, diffusion rates and recombination (chemical) loss rates of ions become comparable. Diffusion exerts control on ion and electron number densities. By diffusion, charged particles move long distances before they are influenced by loss processes. The collisional mean free path at 200km is about 1 km. Diffusion of charged particles across magnetic field lines is of course inhibited and this has an effect on the ionic dynamics in the high atmosphere. At very high altitudes diffusion is the dominating process. The competition between diffusion and recombination gives rise to the upper F_2 peak of the ionosphere, at about 300km.

As indicated in section 3.2.3, the predominant ionizing radiation between 100Å and 800Å is identical to that for the F_1 region. The principal ions formed are O^+ and N_2^+ and O_2^+, and the predominant ion which contributes to this part of the space environment is $O^+(^4S)$.

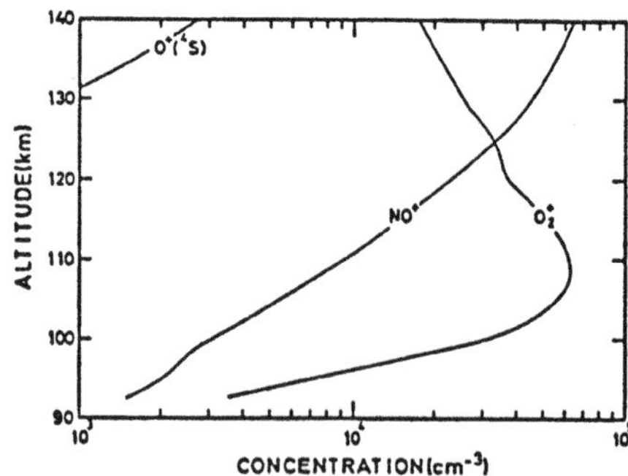

Fig. 7: Altitude distribution of E region ions at a solar zenith angle of 60° [Ref. 1].

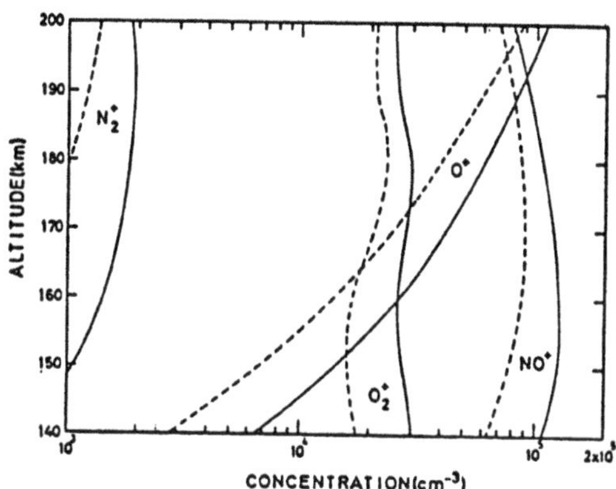

Fig. 8: Altitude distribution of F_1 region ions for a 1000°K thermosphere at low solar activity. Solar zenith angle = 60° for dashed curve and 0° for solid curve [Ref. 1].

Figure 9 displays the results of calculations of the altitude distributions of O^+ ions between 200 and 450km in F_2 region, based on the assumptions of two typical photoionization rates for the $O + h\nu \rightarrow O^+$ reaction. The smaller of the two rates (1.5×10^{-7} sec^{-1}) corresponds to low solar activity and thus a relatively low solar UV flux. The larger of the two (3.0×10^{-7} sec^{-1}) corresponds to high solar activity with which is associated a higher solar UV flux.

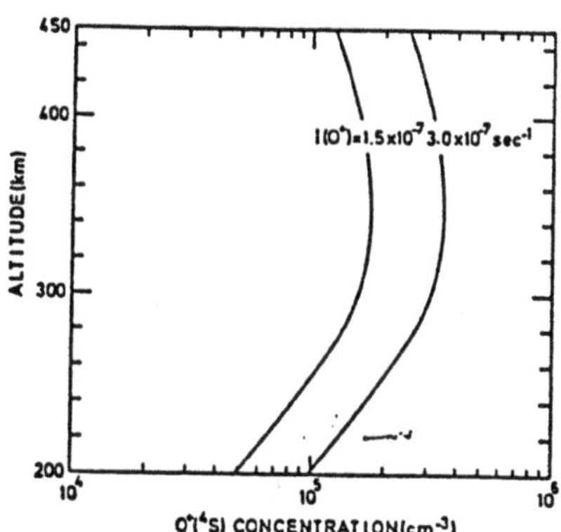

Fig. 9: Altitude distribution of F_2 region O^+ ions for diffusive equilibrium in a 1500°K model thermosphere, with overhead sun calculated on the basis of two photoionization rates (1.5×10^{-7} sec^{-1} typical of low solar activity and 3×10^{-7} sec^{-1} typical of high solar activity [Ref. 1].

Electron loss in this region occurs through dissociative recombinations:

$$e + N_2^+ \rightarrow N + N$$

$$e + O_2^+ \rightarrow O + O$$

It can also occur through a somewhat more complex set of reactions involving NO^+ as an intermediary:

$$O^+ + N_2 \rightarrow NO^+ + N$$

$$NO^+ + e \rightarrow N + O$$

$$O^+ + O_2 \rightarrow O_2^+ + O$$

3.2.5. *Topside Ionosphere and Plasmasphere (z > 400km).* The dominant ion in the topside region (400km < z < 1000km) is O^+ which is transported from the F_2 region. H^+ ions formed in the F_2 region by the almost totally resonant charge exchange reaction

$$O^+(^4S) + H(^2S) \rightarrow O(^3P) + H^+$$

are also transported into the topside region where their abundance increases with altitude until they become the dominant species of the exosphere (z > 1000km) as indicated in Fig. 10. In regions above the topside ionosphere, where the magnetic trapping of ions discussed in section 2 occurs, H^+ is the dominant specie of the plasmasphere. Ion motion at these large altitudes is largely ballistic.

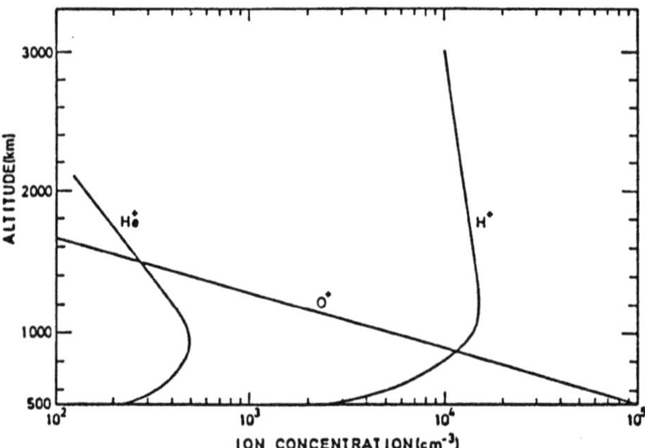

Fig. 10: Altitude distribution for topside ionosphere and plasmasphere ions in diffusive equilibrium with a 750°K thermosphere [Ref. 1].

4. Conclusion

In keeping with the "space environment" theme of the Institute, this brief review of the upper atmosphere and the ionosphere has emphasized the structure, formation, and constitution of the ionized layers of the upper atmosphere. In the mesosphere and the thermosphere, they are embedded in the surrounding neutral atmosphere to form a dilute plasma. In the exosphere they are the dominant components.

Among the important ionospheric phenomena which have not been treated here, are mass motion and dynamics, winds, current sheets, radio wave propagation effects, auroral effects and coupling with the magnetosphere. Information on these topics will be found in other presentations to the Institute and in the monographs listed among the references.

Acknowledgements

This work has been assisted by support from the Natural Sciences and Engineering Research Council of Canada, the Atmospheric Environment Service of Canada, the Defense Research Establishment, Valcartier, the Institute for Space and Terrestrial Science and the Canadian Network for Space Research.

References

[1] P.M. Banks and G. Kockarts, Part B, Aeronomy, Academic Press, New York (1973).

[2] J.W. Chamberlain, Theory of Planetary Atmospheres, Academic Press, New York (1978).

[3] S. Chapman, The absorption and dissociative or ionizing effect of monochromatic radiation in an atmosphere on a rotation earth. Proceedings of the Physical Society of London 43, 26-45 (1931).

[4] S. Chapman, The absorption and dissociative or ionizing effect of monochromatic radiation in an atmosphere on a rotating earth, Part II. Proceedings of the Physical Society of London 43, 483-501, (1931).

[5] A.S. Jursa, (ed.) Handbook of Geophysics and the Space Environment, Air Force Geophysics Laboratory, Air Force Systems Command, United States Air Force (Note additional copies can be obtained from National Technical Information Service, 5285 Port Royal Road, Springfield, Virginia 22161) (1985).

[6] R.W. Nicholls, Introduction to the Space Environment, Proceedings of this ASI, p 45 (1991).

[7] J.A. Ratcliffe, Physics of the Upper Atmosphere, J. Wiley, New York (1960).

[8] J.A. Ratcliffe and K. Weeks, The Ionosphere Physics of the Upper Atmosphere, Chapter 9, pp 377-470 (1960).

[9] M. Rees, Physics and Chemistry of the Upper Atmosphere, Cambridge University Press, New York (1989).

[10] United States Air Force Geophysics Research Laboratory (1957, 1960), Handbook of Geophysics, (1957 private distribution, 1960 Revised edition), MacMillian, New York.

[11] S.L. Valley (Scientific Editor), Handbook of Geophysics and Space Environments, Air Force Cambridge Research Laboratories, Office of Aerospace Research, United States Air Force (1965).

[12] R.C. Whitten and I.G. Poppoff, Physics of the Lower Ionosphere, Prentice Hall, New York (1965).

[13] R.W. Whitten and I.G. Poppoff, Fundamentals of Aeronomy, John Wiley, New York (1971).

THE NEAR EARTH PARTICULATE ENVIRONMENT

J.A.M. MCDONNELL
Unit for Space Sciences
University of Kent at Canterbury
Canterbury, Kent CT2 7NR, U.K.

ABSTRACT. Satellite orbits in proximity to the gravitationally attracting planet Earth are immersed in an active swarm of particulates, now both natural *and* man-made. High velocities relative to spacecraft can lead to extensive impact damage, and in the developing scenario of human exploitation of space, characterization and recognition the particulate environment forms a design driver for space structures with a high area-time product. Concern for the increasing role of man-made *space debris*, at centimetre scale dimensions and above, has led to increased activity in both tracking and modelling. In the size range deemed "unprotectable" regarding space station design, databases have been developed; below this size, space debris and natural particulates vie for dominance and currently form an exciting area for study and debate. The return of NASA's Long Duration Exposure Facility (LDEF) provides a case study for the understanding of their roles and illustrates how radically the spacecraft pointing direction will determine the penetration rate of space structures. The paper reviews knowledge and current understanding of the two populations as inputs to the space environment models.

1. Perspective of the Particulate Environment

Knowledge of the population of interplanetary space and the near Earth space environment precedes the space age. Prior to the launch of satellites, knowledge of the potential component of a natural dust cloud about the Earth led to NASA's strong support for micrometeorite detectors. The Earth's Dust Belt [Ref. 1] was short lived [Ref. 2] and the hazard aspect relaxed to the effect of unbound natural interplanetary meteoroids from the comet or asteroid families. Only a modest gravitational enhancement of this population would be expected in Low Earth Orbits (LEO) due to the attraction of the Earth.

Fields of study and lines of evidence on the distribution of matter maintaining the LEO particulate environment – which must indeed be replenished – are shown in Table 1. Characterization of the information available from such fields of study has been reviewed on numerous occasions. Recent reviews of significance are offered by Grün, et al.

R. N. DeWitt et al. (eds.), The Behavior of Systems in the Space Environment, 117–146.

Table 1: Evidence for the presence of particulates in the near Earth environment. Differing techniques are vital to determine the properties of this complex mix of particulates covering a size regime from submicron to centimetre dimensions.

PHENOMENON	TECHNIQUE	APPLICATION
Zodiacal Light	Terrestrial and in-situ optical sensing	Global properties of Solar System
Meteor trails	Optical and radar tracking	Meteoroid orbits, stream classification
Cosmic spherules	Pacific sediments, deep ice and stratospheric dust retrieval	Microanalysis, and petrology isotopic studies
Fireballs	Terrestrial recovery of meteorites	Mineralogy and orbits of particular asteroids
Impact craters	Terrestrial crater identification	Impact bombardment history
Tektites	Impact crater ejecta finds	Identifiers of impact events
Impacts on spacecraft and sensors	Penetration and Impact plasma	Current environment and dynamics

[Ref. 3] focussing on the interplanetary environment; in LEO space, in-situ satellite techniques and measurements have been reviewed by McDonnell [Ref. 4]. It is in the near-Earth space that we now have opportunity for dramatic developments in understanding by study of the data returned from NASA's Long Duration Exposure Facility (LDEF). We shall review evidence derived from key space experiments on LDEF and examine how they shape our understanding of the roles and especially the interplay of natural particulates and space debris. First, however, we need to understand some basic properties of the characterization of particulate distributions; such properties define the relative importance concerning penetration frequency, the scattering area and the mass distributed within the population.

2. Characterization and Properties of the Flux Distribution

The cumulative flux distribution, illustrated in Fig. 1 and instanced in measurements returned from LDEF (Section 3, et seq.) is plotted on a logarithmic flux-logarithmic mass scale. A straight line plot in this plane would be characterized by the form:

$$\log \Phi(m) = \alpha \log m + \log c \quad (1)$$

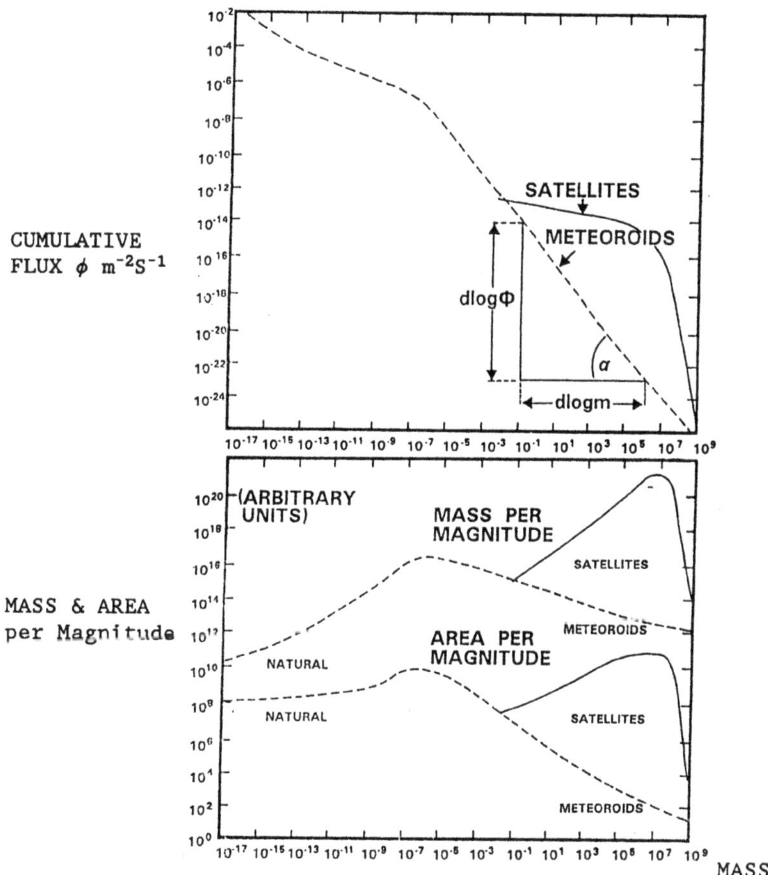

Fig. 1: The cumulative mass distribution in LEO orbit [Ref. 28]. The slope α of the logarithmic plot determines which masses are significant in terms of the frequency of encounter, and the mass or area within a distribution. Figure 1(b) illustrates the mass per decade interval and (b) the area per decade interval derived from Eqs. (4) and (9).

where α = log cumulative slope $= \dfrac{\partial \log(\Phi(m))}{\partial \log(m)}$ (α is negative). It may be extended to cover arbitrary forms of the distribution Φ(m). We note that Φ(m) refers to the particles greater than a limiting value, and is a *threshold* measurement, e.g., Φ(m)= number per $\text{metre}^2\text{sec}^1$ of particles of mass m or *greater*; the "average" particle differs by an amount dependent on the slope of the cumulative distribution α. Only at very "high" values of the (negative) slope α, does the average and the threshold mass converge.

This average mass may be shown to be given by m $\alpha/(\alpha+1)$ and has validity for $\alpha < -1$; whereas the most frequent mass is close to m, the average mass can be much larger. For $\alpha = -1.2$, e.g., the value is 6.m.

The number of particles $\partial\Phi(m)$ in a small interval ∂ log m or ∂ m may be found (see Appendix) using: $\partial \log\Phi(m) = \alpha\, \partial \log m$ and $\partial\Phi(m) = \alpha\, \Phi(m)\, \partial m/m$. The *flux* is defined as the number of particles passing through a flat surface, which is accessible to particles approaching from 2π steradians, but because of geometric target foreshortening it is equivalent to a detector of isotropic angular sensitivity open to π steradians.

The logarithmic flux distribution is useful for plotting the size distribution over a wide range of particulate fluxes and mass but is inappropriate for computation of, e.g., (1) the mass of particles in a given range, (2) the area of particles in a range, or (3) the total area of spacecraft erosion. Such properties are required for evaluation of environmental effects but equally important for understanding relationships in the source and sinks of particulates; these properties require us to first identify the differential mass distribution.

Exponentiating equation (1) we find

$$\Phi(m) = c\, m^{\alpha} \qquad (2)$$

which shows that our flux distributions are typically of power law form. The differential flux $d\Phi(m)$ is given by

$$\phi(m) = \alpha c . m^{\alpha-1}\, dm. \qquad (3)$$

The (differential) mass index $s = \alpha - 1$ is, for example, -2 for a value of the cumulative slope of $\alpha = -1$.

We can now determine the total number of particles over a range m_1 to m_2 in the flux distribution by integration of Eq. (3). This could have been obtained from the difference of the two cumulative fluxes $\Phi(m_1)$ and $\Phi(m_2)$ (Eqs. 1 or 2). But for the *mass* within such an interval m_1 to m_2 we must integrate (see Appendix) which yields the result:

$$= \int_{m1}^{m2} \phi(m) . m . dm$$
$$= \frac{\alpha}{\alpha+1} . \Phi(m) . m^{\alpha} . (m_2^{\alpha+1} - m_1^{\alpha+1}) \qquad (4)$$

This solution is valid except for $\alpha = -1$, unfortunately a value fairly typical of many natural distributions, e.g., of meteoroids impacting the Earth's atmosphere. For this special case $\alpha = -1$, we find the total mass:

$$= \alpha . \ \Phi(m) . m . \log_e \left(\frac{m_2}{m_1}\right) \qquad (5)$$

In the case where m_1 and m_2 represent intervals on the logarithmic mass scale as for example *decade* intervals where $m_2 = 10m_1$, $\log(m_2/m_1) = 1$, then $\log_e m_2/m_1 = \log_e 10 = 2.30$, and the total mass $= \alpha\ c\ \log_e \frac{m1}{m2} = \log_e 10 . \alpha\ c = \log_e 10 . \alpha\ \Phi(m)\ m$ per decade. Thus the mass of particles in any particular logarithmic mass interval is uniform over the range for $\alpha = -1$ since $\Phi(m) = c.m^{-1}$. The value of $\alpha = -1$ is then a "turning point" regarding properties of the distribution. To illustrate the properties for $\alpha < -1$ and $\alpha > -1$, we integrate over a wide range of masses (i.e., for $m_2 >> m_1$) for α representing a 'high' slope and for α representing a 'low' slope.

<u>For $\alpha < -1$</u>, integration of equation (4) yields a total mass

$$= \frac{\alpha}{\alpha+1} . \Phi(m_1) . m_1 \tag{6}$$

We see that the mass is determined only by the <u>lower</u> limit m_1, i.e., is 'concentrated' there. For $m_1 = 0$, we have infinite mass, but in practice (reflecting the need to constrain the mass to the finite sources of mass available) the distribution "falls off" (by a decline in the slope α). This conservation of mass - mass budgeting between sources – we shall see is essential to understanding the sources and sinks of particulates. We must, in fact, account for the total mass budget in interplanetary dust, comets, asteroids and the planets as a closed solar system. Mass is redistributed as it is interchanged between various sources and sinks but its flow can be charted. We must also find in e.g. the space debris scenario, that the mass of fragmentation products should never exceed the mass of source objects – a useful constraint!

For distributions with low slopes, the mass is concentrated near the largest mass (m_2) of any interval, and is given by:

$$Total\ mass = -\frac{\alpha\Phi(m_2)\,m_2}{\alpha+1} \tag{7}$$

We must complement the above relationships by extending it to the <u>area</u> of flux distribution, e.g., representing what the scattered or reflected light from such a distribution would be. We take the geometric cross section ($A(m) = 4\pi r^2$ where r is the particle radius) but could readily extend it to incorporate the optical scattering function for light scattering calculations or penetration parameters for calculating the area of impact erosion:

$$\int_{m1}^{m_2} \phi(m) \, . A(m) \; dm = \int_{m1}^{m_2} \frac{\alpha \phi(m)}{m\alpha} \left[\frac{3\pi^{1/2}}{4\pi\rho} \right]^{2/3} . \log_{e10} \tag{8}$$

The critical value of α concerning *area* is found to be $\alpha = -2/3$ and yields for over the interval m_1 to m_2 an area per magnitude:

$$2.32\alpha \; C \left(\frac{3\pi^{1/2}}{4\rho} \right)^{2/3} = \alpha \; \frac{\Phi(m)}{m^{2/3}} \cdot \left[\frac{3}{4\pi\rho} \right]^{2/3} . \log_{10} \tag{9}$$

For high slopes ($\alpha < -2/3$) the area is concentrated in the smaller particles but only for $\alpha > -2/3$ is the area of large particles dominant. We can find situations (e.g., α between -2/3 and -1) where the mass is dominated by large particulates and yet the area is dominated by the smaller of the particulate populations.

Knowledge of the sources and sinks of matter in the solar system we are currently exploring is derived from our planetary vantage point at 1 A.U. heliocentric distance. In the inner solar system we see a relative high density of small particulates, but whose free lifetime, due primarily to the relativistic Poynting Robertson drag, is only some 10^4 years. The same particulates demonstrate, however, in terms of mineralogy, ages of some 10^9 to 10^{10} years perhaps predating the solar system [Ref. 5]. Figure 2 illustrates schematically these routes for processing of matter in the solar system. The particulates must be *liberated* in very recent times, and we see comets and asteroids as potential sources; comets are evidently prolific sources in generating their dust tails. IRAS data also shows asteroids have dust trails, but of larger dimension. Self grinding within this cloud, which is probably in equilibrium over timescales some 10^7 years, contributes a further (and mixed) source of particulates nearer the Sun: this can be shown [Ref. 6] to be generated where the *area* of distribution is high - namely between 30 to 100 microns particle diameter. The available *mass* in the distribution is shifted towards slightly higher masses in line with Eqs. (4) and (9). We see that the size range 100 μm to 1000 μm, namely the lower end of the sporadic meteoroid population, provided the dominant source and is largely generated from comets. Characteristics of the orbital elements of these different sources intercepted in LEO space will be used to determine properties which we can compare with our observations, LDEF especially.

Figure 1(a) showed us the cumulative flux distribution $\Phi(m)$ for objects encountered in Earth orbit. We shall later examine the relative importance of natural and space-age particulates from modelling and measurement, but use the relationships we have developed to derive from this data the mass per decade magnitude Fig. 1(b) and the cross-sectional

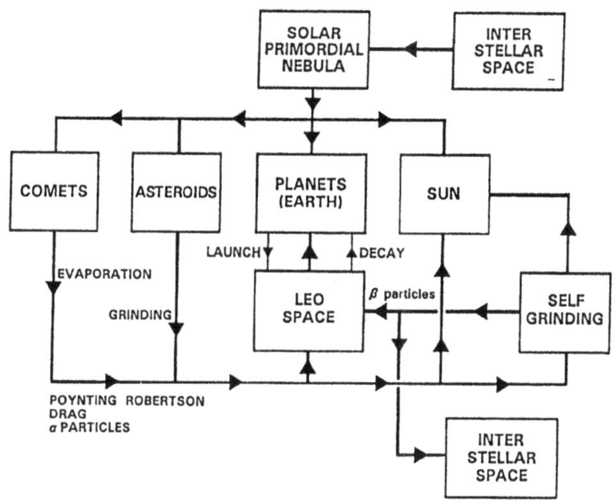

Fig. 2: Sources and sinks of particulates contributing to the LEO particulate environment.

area per decade magnitude Fig. 1(c). Examining these we can readily appreciate that the satellite population has the *potential* to exceed (during a collisional break-up scenario) the natural population of interplanetary meteoroids; we note that the collision frequency with satellites is (fortunately!) low but the capability of exceeding the natural meteoroid hazard at small masses is quite real.

In the same way concerning the natural population of meteoroids, we see that their mass per magnitude enables them to provide a ready source of smaller particles by e.g., self-grinding in interplanetary space. We identify from 1(c) which particles dominate the interplanetary scattering-the Zodiacal Light.

3. Measured Distributions

We look at a few of the range of impact related experiments on LDEF; this set marks a dramatic increase in the definition of impact measurements and damage assessment in the LEO environment. LDEF's vital role in the reformulation of the environmental definition arises from three factors:

a. Large area x time product (110 metres^2 x 5.75 years);

b. Retrieval from orbit to permit laboratory study using sophisticated analytical tools;

c. Stabilization (in geocentric co-ordinates) of the spacecraft attitude.

Key experimental results are now described.

3.1 LDEF MICROABRASION FOIL PACKAGE (MAP)

Hypervelocity Impact perforations from the first surface of the Microabrasion Foil Experiment (MAP), which comprises multilayer micron dimension foils, have been reported by McDonnell, et al. [Refs. 7, 8]. Typical hypervelocity impact features (shown in Fig. 3) are detected optically in foils of different thickness, and plotted as a function of f_{max}, the ballistic limit. Sample fluxes from several thousand perforations are shown in Fig. 4. Representing the cumulative penetration rate for independent areas of aluminum (and brass, but not shown), the data clearly demonstrate the high East (LDEF orbital ram direction) flux. Asymmetry between the North and South fluxes is significant; error bars, representing the statistical uncertainty $\sqrt{N}$ for each surface are small with the exception of the lower values west fluxes (N is shown). The angular offset of LDEF (8° towards the North, Fig. 2) reduces the apparent magnitude of the real bias towards the north for the small particles. The bias is clearly size dependent and reverses (again significantly) for penetration thicknesses of > 20 microns. As yet these features cannot be explained! We may perhaps have to involve anisotropy in Earth bound space debris orbits (despite the expected randomization of orbits for small particles) and also (for larger particulates) anisotropy of the interplanetary environment. We have not yet, however, performed the separation of space debris and natural particulates! We must examine first other data sources emerging from the LDEF analysis.

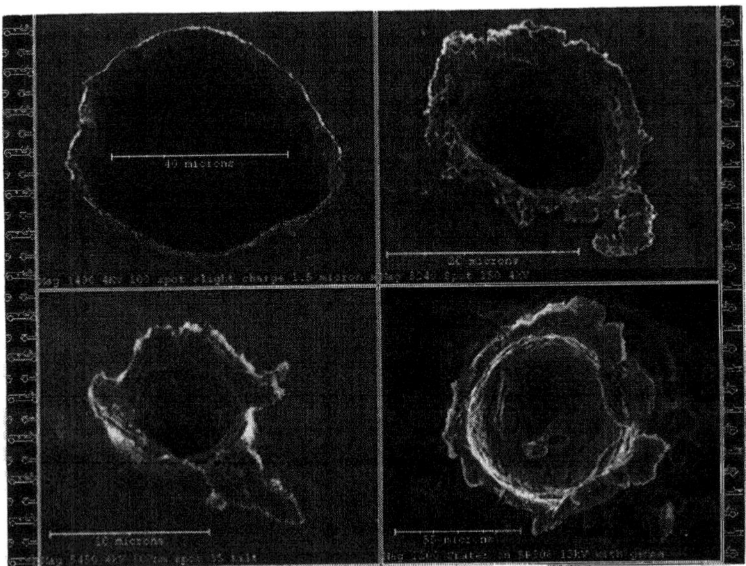

Fig. 3: Perforation of a 5 micron LDEF MAP foil by a hypervelocity space particulate. A crater diameter 60 microns would be caused by a particle of some 40 microns at 17 kms^{-1} velocity.

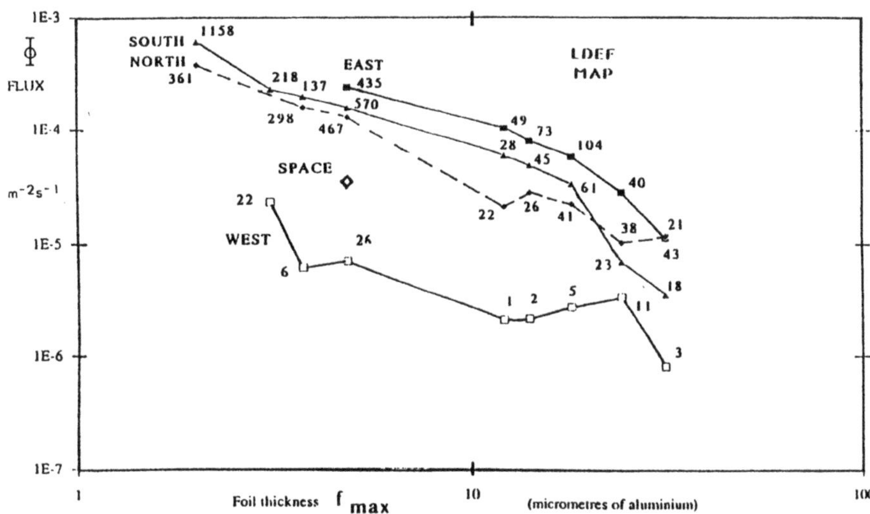

Fig. 4: Penetration rates of differing foils as a function of thickness for five pointing directions on the MAP experiment. Penetration counts are shown and error limits, given by root N, where N is the number of perforations above the ballistic limit, are generally insignificant except for the (lower) West flux; the difference between North and South is significant as indeed is the reversal of this trend. The excess flux observed on the East Ram direction is consistent with an orbital population but only for the smaller particulates (~ 5μm diameter). At larger dimensions the flux anisotropy is consistent with natural (unbound) particulates as the dominant but not necessarily exclusive source.

3.2 LDEF ULTRA HIGH COSMIC RAY COVER (UHCRE) EXPERIMENT

The Cosmic Ray Experiment [Ref. 9] provided via its thermal covers a valuable meteoroid detector. Analysis of impacts is currently being performed at Canterbury [Ref. 10]. The programme is complementary to that being conducted at NASA JSC on another sub-set of the total of 18 m^2 of detector surface. Eleven trays on LDEF were used for carrying the UHCRE experiment, located on nine of LDEF's twelve peripheral faces. Nine segments are currently under study; two chosen for the pilot study reported here were from rows four and ten, opposing rows and near to the EAST and WEST pointing directions.

The blankets were scanned for features using the "Large Optical Scanning System" (LOSS) which employs a method of photometric detection with a resolution of approximately 3 microns at its highest magnification. The process is automated with computer based software which searches for features and logs their position and optical size. In the study, after initial scanning, each feature was returned to for verification as an impact site. Three-colour stereo CCD images were then taken of each impact site and logged in a database, set up for use by the scientific community, where specific craters or types of craters may need to be selected for study.

With the thermal blankets we are not able to yet define accurately the penetration sensitivity due to their unusual structure; prior to their proper calibration they are considered to be all Teflon sheets 200 microns thick, and having an equivalent thickness of 170 μm aluminium.

After taking images of each feature these images are subsequently processed to extract data including information on hole size, the diameters of the rings observed around the perforations. A plot of some source data from the Thermal Blankets, leads to the distribution of hole sizes, shown in Fig. 5. On leading (Row 10) and trailing (Row 4) shows an East to West ratio varying from 4 to 12. This ratio is significant in the identification of orbital particulates, and its low value argues against them at this size. The data is incorporated with other impact experiments on LDEF into the model used in Fig. 6.

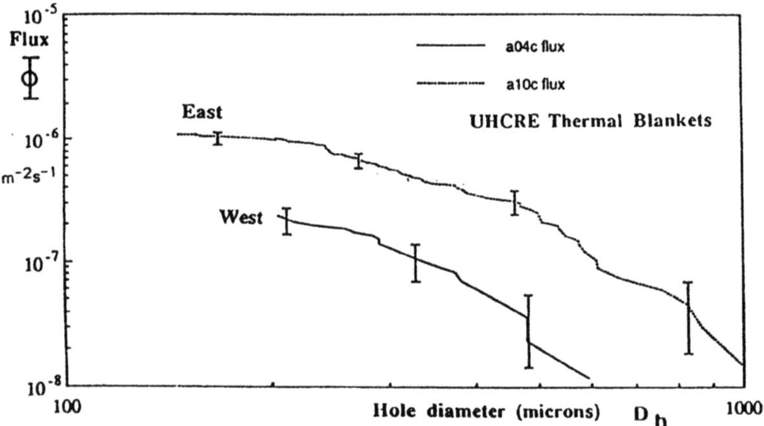

Fig. 5: Penetration hole diameter distributions for the (UHCRE) thermal blanket covers, and leading (Row 10) and Trailing (Row 4) faces of LDEF. The composite of 120 μm Teflon and 80 microns paint may be plotted at an equivalent thickness of 170 μm aluminium.

3.3 SDIE EXPERIMENT

Representing an area comparable to the cosmic ray covers, the SDIE experiment [Ref. 11] offers for higher masses an effectively "infinitely thick" target regarding penetrations. No perforations of 2.5 mm aluminium were detected. Valuable comparisons can be made between the impact effects on such thick targets and the thin MAP and UHCRE data, where the larger particles clearly penetrate. Opportunity for the "calibration" of thick target cratering and foil perforation relationships abounds; such work was reported at the Workshop "Hypervelocity Impacts in Space", Canterbury, 1991.

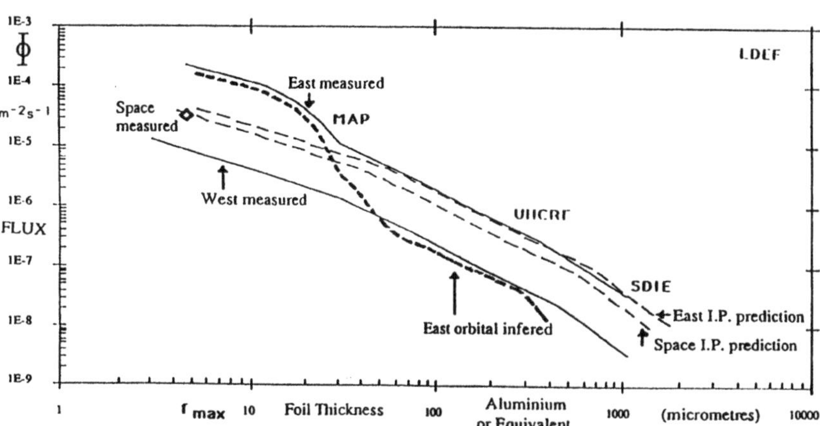

Fig. 6: The application of modelling of particulate dynamics to LDEF West and Space flux data yields a particulate velocity V_{PE} of 17.4 kms^{-1}. It is further transformed to the West Face of LDEF, which shows that the prediction based on interplanetary unbound particulates is in good agreement with measured West fluxes at medium masses. At smaller masses, however, the presence of orbitals is demonstrated on the East face (dashed line).

Data is acquired primarily in terms of crater diameter measured at the original detector surface level. Accompanied by depth to diameter relationships of 0.50 determined by Humes [Ref. 12] for the larger craters (D_c > 500 μm) we can convert diameter distributions (D_c) to penetration depth distributions (T_c) and relate to foil penetration penetration thickness limit (f_{max}). The ratio of f_{max}/T_c has been reported for thin foils as 1.17 [Ref. 13]. Humes [Ref. 12] infers a higher value near to 1.7. Although remembering that this calibration must in future be refined by the LDEF data set, we note that the log-log representations of data render this factor less critical regarding our inspection and intercomparison of the different data sets at this stage.

3.4 INTERPLANETARY DUST EXPERIMENT (IDE)

Measuring the smallest particles on LDEF, of submicron dimensions, the IDE experiment [Refs. 14, 15] offers vital characteristics for tracking down the source of (at least the very small) particulates. We will see (section 5) that the modelling particle dynamics to flux ratios from the MAP data demonstrates the presence of orbital particulates, but only for the particulates causing perforation of the aluminium foils of 20 micron thickness and lower. This trend is supported by the IDE experiment data; experiment features which provide a special perspective to understanding the LDEF data-set are:

(i) Time-tagged impact detection (but only for the first year)

(ii) Six directional flux measurements (4 peripheral, Earth and space pointing).

The time tagged information shows a surprising unique structure; it shows that the orbital population is neither isotropic nor constant! As perceived from LDEF's orbit, periodicity in the variations is seen and shows the interception of orbits on successive periods, dismissing the concept that microparticulates should have diffused into an isotropic cloud! This selective distribution in orbits – where the spatial density momentarily exceeds the average by many orders of magnitude – has been interpreted as evidence of space debris [Ref. 15] and alternatively as compatible with natural particulates captured into Earth orbits either directly (via aerocapture) or indirectly by particulate atmospheric disruption followed by capture (aero-fragmentation capture [Ref. 16]). Chemical evidence of impact residues must provide more definitive answers, and we could very possibly, given the extent and high definition of data from LDEF, distinguish the presence of both contenders to this orbital population. There could indeed be both components present. Although we await also, the chemical identification of the residues of larger impact craters, we must accept that these are generally dominated by natural meteoroids. Our modelling demonstrates this by returning a velocity of 17.4 km/sec, well in excess of orbital values. On the west face, analysis of 15 impact residues of the Frecopa experiment [Ref. 17] also reported 15 out of 15 studies as chondritic in composition, hence certainly not space debris. Additionally though thick target residue analysis [Ref. 18] has yielded a small fraction which are compatible only with man-made space hardware.

3.5 SUMMARY FLUX CHARACTERISTICS

We are thus able to generate a flux distribution from these data sets. We refer them not to particle mass but the equivalent thickness of aluminium perforated; it is the flux *ratios* which are significant! Corresponding to the lowest (West), highest (East) and the space pointing directions, we now have a measure of the anisotropy of the average LDEF preferred flux. We will later wish to account for the 8° effect of LDEF as more data is made available [Ref. 19]. The model fit extends from micron to millimetre dimensions, covering 9 orders of magnitude in particle mass. This is our basis for modelling.

4.0 Modelling of LDEF Data and the LEO Environment

4.1 EFFECTS OF SPACECRAFT EXPOSURE ATTITUDE

The diversity of flux rates for different exposure attitudes on LDEF (flux anisotropy) will be incorporated into engineering models and guidelines and used directly for spacecraft exposures similar to LDEF. The data refers to velocity at mean altitude of 465 km and in an orbit of inclination

28.5°. To apply the data to any other satellite exposure requires an understanding of the particulate orbits. We must develop modelling of particulate orbits, incorporating the relative motion of the satellite (LDEF) and a supposed geocentrically referenced cloud, in order to understand what this LDEF anisotropy means. We may then understand better the sources of particulates. Such modelling of the Interplanetary micrometeoroid environment as encountered by LDEF has been developed [Ref. 9]. Due to LDEF's fast precession rate the average interplanetary flux is assumed random [Ref. 20]. Our model calculates the relative fluxes and impact velocity of particulates arriving at each of LDEF's 14 faces. Such ratios are calculated at a particular (arbitrary) initial mean geocentric particulate velocity V_{PE} which may be related to an interplanetary Earth approach velocity V_∞.

The model results are then applied firstly to the MAP data, which is deployed on five different pointing directions of LDEF; other LDEF impact data is incorporated at larger particulate dimensions. The thin foil perforation flux distributions measured on each of these faces (North, South, East, West and Space), are supposed to be caused by either natural micrometeoroids (unbound orbits) or man-made space debris (orbital). It can be shown that the unbound micrometeoroids are able to encounter all faces of LDEF, but for the orbital particulates, there can be only a very small fraction on the West, Space and Earth faces: the only possible exception to this would be impacts from highly eccentric orbits. We see, therefore, that a study of the West and Space fluxes is the key in deconvolving the naural micrometeoroids and orbital fluxes.

To understand the modelling approach we first imagine that LDEF is stationary, with respect to the Earth; it will encounter an isotropic Interplanetary flux and with the exception of Earth shielding, all LDEF faces will detect the same flux. If, though, the spacecraft is travelling at its orbital velocity through this geocentrically preferenced isotropic cloud, detection efficiency will modify the measured fluxes of each face. These effects are *mass sensitivity* and *detection geometry*. The geometry effect is the 'sweeping-up' of particles onto the forward faces. The apparent flux on the moving detector relative to the stationary detector can be determined for each face and are termed the 'K' factors. The motion of LDEF also increases the mean impact velocities on the forward faces relative to those of the stationary spacecraft, and likewise reduces the velocities on the rear faces, giving rise to the mass sensitivity effect. Here, because particulate foil penetration (or crater depth) depends upon the impact velocity, the forward faces will receive more perforations than the rear faces at a particular size because of the increased number of particulates of smaller size. Therefore, the forward faces not only *encounter* more particles than the rear faces due to the geometry effect, but also *detect* more particles due to the mass sensitivity enhancement. This underscores the importance of LDEF's flux ratios.

A flux transformation from the West face to the Space face (Fig. 6) is then composed of a horizontal (sensitivity) shift to account for the mass sensitivity (the relative increase in foil thickness at constant mass), and a vertical (flux) shift to account for the geometry effect along with Earth shielding (the flux ratio at constant mass). The flux ratio at constant foil thickness can then also be defined if the slope of the foil thickness perforation distribution is known. Both these mappings are dependent upon the initial geocentric meteoroid velocity, and therefore, so also is the relative transformed flux. This transformation is then a prediction of the foil penetration flux that would occur on the Space end face. By performing this at an arbitrary velocity V_{PE}, we can find a solution for the geocentric particle velocity which generates the *observed* Space flux from the (transformed) West flux. This is calculated to be 17.4 km s^{-1}, for the MAP data, which compares favorably to mean micrometeoroid velocities derived from observations of meteor streams as measured as 16.9 kms^{-1} by Kessler [Ref. 21], as 19.2 kms^{-1} by Dohnanyi [Ref. 22], as 16.5 kms^{-1} by Erickson [Ref. 23] and 15.2 km/s by Southworth and Sekanina [Ref. 24]. The Earth approach velocity V_∞ is deduced as 13.6 km second. We shall later examine the implications of this modelling for the LDEF data set in the context of Interplanetary models by Grün, et al. [Ref. 3] and Cour-Palais [Ref. 25] (Fig. 10).

4.2 ANGULAR DISTRIBUTION OF THE FLUX AND DETECTOR GEOMETRY EFFECTS

In order to extract as much information as possible from LDEF, the NASA LDEF ("A Team") analyzed the entire frame of the spacecraft prior to the removal of the experimental trays for distribution and analysis. They discovered, on the aluminium clamps, flanges and tray lips, a total of 768 impacts of diameter 0.5mm and above, with the number of impact sites ranging from only 3 on the trailing West face (03), to 87 on the East (10) face. The main advantage of using data from the LDEF frame is that there are effectively 24 side-facing directions instead of the normal 12, providing more opportunity for resolving any possible effects of a fine structure from a debris component. Plotted in Fig. 7 are sample results for impacts on the clamps, flanges and tray lips [Ref. 26] for large impacts (ca. 500 μm). Any data point from such an analysis represents the highly smoothed effect of a flat plate detector. In an effort to understand the true incident distribution, we must examine the efforts of such smoothing and the possible de-convolution. The "Errors" represented by a spike in this data arise from the statistics of the small number of large impacts.

Each data point measured on an LDEF surface is the result of an impact on a flat-plate detector by a microparticle incident from an unknown direction. A particle striking a particular face could also have done so on any of the other (nearby pointing) faces LDEF, but in each case the probability of it occurring would have been different, as well as the incident angle. Noting that such factors will lead to a deterministic and, known, smoothing function, then it may be possible to calculate the flux of particles coming from each specific range of directions by the process of mathematical deconvolution. We note two main effects (1) Geometrical Sensor Area, which is proportional to the cosine of the angle from the normal (α) and (2) a penetration sensitivity which follows a $(\cos \alpha)^{0.88}$ dependence, according to empirical relationships [Ref. 27]. This has led to a partial deconvolution of the data; the results show that the distribution of large craters is not caused by the "butterfly" distribution expected for orbital impacts [Ref. 28]. The LDEF results for large

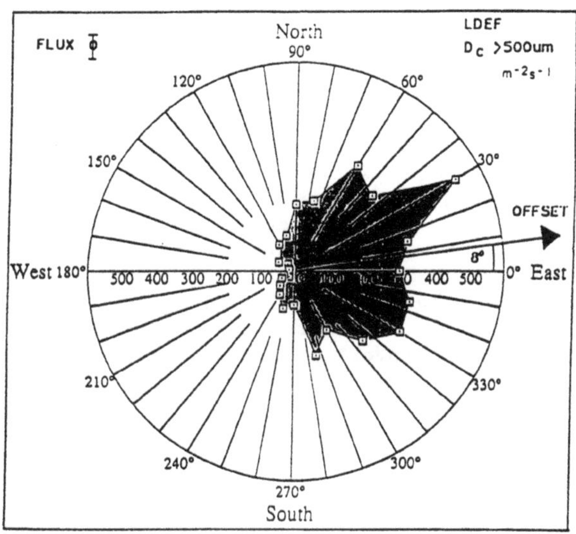

Fig. 7: Radial plot of impact flux for LDEF's peripheral pointing directions.

particles differ markedly from the MAP micro foil data where orbital particulates are significant.

4.3 CURRENT SPACE DEBRIS STUDIES

NASA has promoted space debris studies with considerable success [Ref. 29]. Following this lead, ESA's Space Debris Working Group concluded in the ESA SP-1109 that "For future manned missions (e.g., the Columbus programmes and Hermes) it is essential to establish a European database of all space debris data......"

In response to these findings a concerted effort was started by an advisory group leading to a document complementing NASA's volume. An initiative was effected to obtain various *tools* to support the findings and to monitor the space debris environment; these are also coordinated through the Space Debris Advisory Group. Tools developed include DISCOS - The Database and Information System Characterizing Objects in Space, designed and developed with ESA and implemented under contract at the University of Kent at Canterbury. It is now on-line at ESOC (European Space Operations Centre) Darmstadt, Germany. Also available is ESABASE, a risk assessment modelling tool comprising both micrometeoroid and space debris models on-line.

DISCOS takes data on tracked objects only. These are mainly satellites, payloads and fragmentation debris that are tracked by USSPACECOM. There are four main *space objects* data sources; namely, USSPACECOM Two-Line Elements, NASA's Satellite Situation Reports, RAE Table of Earth Satellites and Teledyne Brown Engineering History of On-Orbit

Satellite Fragmentation. There are also two Solar and Geomagnetic Activity Data files; namely, NASA's Long Term Solar and Geomagnetic Activity predictions and ESOC SOLMAG data. The database is implemented in Oracle and *users* access the main data files which contain hybrid information from the four *space objects* data sets via SQLPLUS.

DISCOS can be used for collision dynamics and probability, time evolution models, re-entry predictions and risk assessment, especially for manned missions, strategies for collision avoidance and launch window calculations. DISCOS is used as a primary data source for space debris and has been used to estimate impact probability calculations on a spacecraft such as LDEF (Long Duration Exposure Facility). Data for larger masses (as shown in Fig. 1) gives the distribution of satellites and fragments in Earth orbit and can readily be generated from DISCOS.

ESA is continuing to invest in space debris monitoring and control as underlined by recent contracts awarded for characterizing impact probability and distributions on a spacecraft, e.g., the new LDEF data at smaller masses; it is also promoting studies via SIRA (UK) in collaboration with the University of Kent for the study of optical tracking techniques for space debris.

5. Sources and Sinks - Current Viewpoints

Modelling the natural particulate components on LDEF, which dominate on the West and Space faces, leads us to the expectation that mid-sized particulates detected by all faces of LDEF are dominantly natural; by deduction of the particle velocity from the comparison of the Space and West pointing detectors we are able to deduce the interplanetary velocity and flux distribution by incorporating effects of Earth shielding and gravitational Earth focussing. This compares well (Fig. 8) with the interplanetary flux distribution deduced by Grün, et al. [Ref. 3]. At larger masses, we see a significant divergence, namely an enhancement in flux above accepted meteoroid fluxes. If confirmed, it will demand explanation; it might well be explained by impact comminution products of the terrestrial satellite population. Previous modelling [Ref. 29] predicted a dominance by debris above 1mm particulate size, and we could be seeing here in the LDEF data first evidence of collisional increase.

Previous evidence from the Solar maximum Satellite data [Ref. 30] indicating a micropopulation of debris some 1 to 2 orders of magnitude above the natural population have been shown [Ref 31] to be exaggerated by invoking inappropriate penetration formulae [Ref. 27]. The effects of this are seen in Fig. 9, where the originally published data is shown adjacent to the new interpretation using the McDonnell-Sullivan penetration formula. Table 2 shows various penetration formulae which have been used to decode crater and penetration measurements in space. Taking note of the need to incorporate the effects of target strength and density and the wide range of dimensions on LDEF, we are forced to conclude that the McDonnell-Sullivan relationship should be used at this stage. It is probable that the LDEF data itself will offer scope for the examination of the parametric dependencies within this relationship.

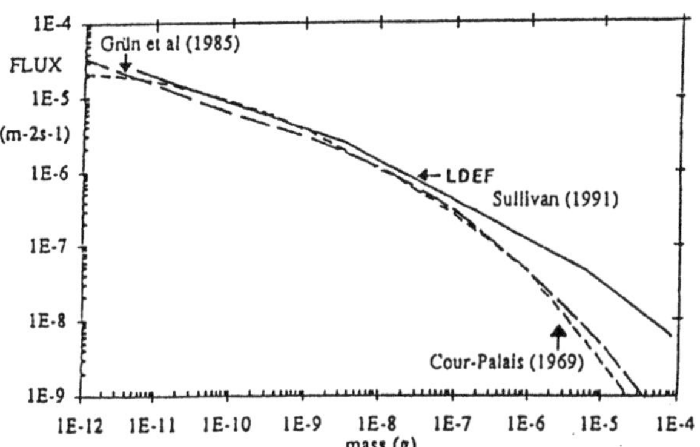

Fig. 8: Fluxes measured from LDEF data corrected for gravitational enhancement and Earth shielding compared to the interplanetary model of Grün, et al. [Ref. 2] and Cour-Palais [Ref. 25]. Agreement supports the interplanetary origin of the West and Space pointing LDEF data.

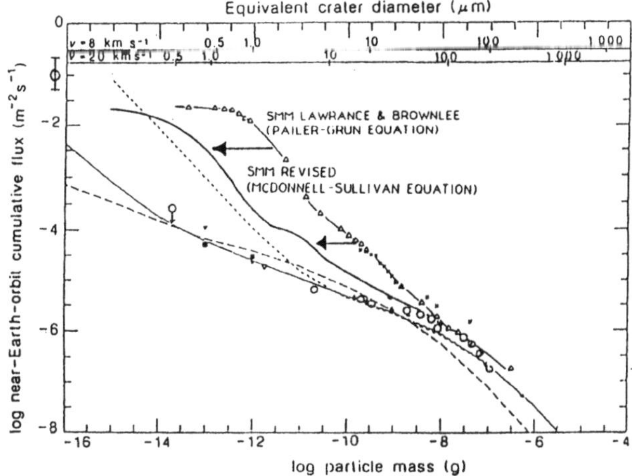

Fig. 9: Data from Laurance and Brownlee [Ref. 30] interpreted using the penetration formula as published [Ref. 27] and more recent interpretations [Ref. 31]. Although the excess "debris" component is considerably reduced a significant excess remains which needs addressing and resolving in terms of local spacecraft contamination or orbital particulates.

To get a viewpoint of the current status of understanding of the sources and sinks, we begin with the interplanetary flux [Ref. 3] (Fig. 8). "Mainstream meteoroids" (above 10^{-6}g mass) characterized by a cumulative index of $\alpha = -1.2$ are produced, and must be continuously replenished [Ref. 32] by cometary sources in heliocentric orbits. Asteroidal sources

Table 2: Penetration formulae used to decode hypervelocity impact data. The need to incorporate target strength and density plus particle density velocity and dimensional scaling underscores the value of the McDonnell-Sullivan formula in the decoding of foil perforation data from LDEF.

$$\frac{f}{d} = 1.023\ d^{0.056} \left(\frac{\rho_P}{\rho_T}\right)^{0.476} \left(\frac{\sigma_{Al}}{\sigma_T}\right)^{0.134} V^{0.664} \qquad \text{[Ref.31]}$$

$$\frac{f}{d} = 0.79\ V^{0.763} \qquad \text{[Ref. 4]}$$

$$\frac{f}{d} = 0.57\ d^{0.056}\ \varepsilon^{-0.056} \left(\frac{\rho_P}{\rho_T}\right)^{0.5} V^{0.875} \qquad \text{[Ref. 36]}$$

$$\frac{f}{d} = 0.772\ d^{0.2}\ \varepsilon^{-0.06}\ \rho_P^{0.73}\ \rho_T^{-0.5}\ (V\cos\alpha)^{0.88} \qquad \text{[Ref. 27]}$$

$$\frac{f}{d} = d^{0.056}\ \rho_P^{0.52}\ V^{0.875} \qquad \text{[Ref. 37]}$$

$$\frac{f}{d} = 0.635\ d^{0.056}\ \rho_P^{0.5}\ V^{0.67} \qquad \text{[Ref. 25]}$$

are not precluded if in short period heliocentric orbits, i.e., perigee within 1 A.U. We see that the Pioneer 8 and 9 data [Ref. 6] shows that both the velocity and directional characteristics of the interplanetary flux changes at mass 10^{-11}g. We see a swing at smaller masses to a solar flux direction, dominated by a higher velocity. We can to a first order separate out the bound and unbound heliocentric orbits. The *unbound* orbits at these very small masses represent those particles expelled from comets or collisionally generated near the sun between the impact of differing meteoroid streams. On liberation they find a gravitational field which is reduced by radiation pressure to the extent that their orbits are hyperbolic. Figure 10 shows the effects of a separation into this "bimodal" population, separated not so much by their nature but by virtue of their fate in terms of orbital elements. Otherwise they may well be of

similar characteristics, namely cometary fragments of nature similar to the IDPs recovered from the Earth's stratosphere [Ref. 33].

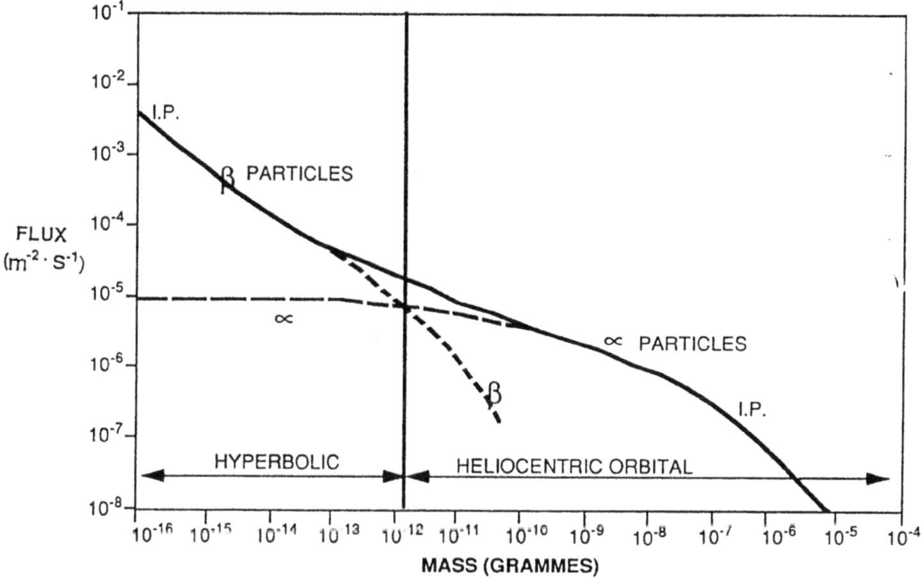

Fig. 10: Interplanetary flux (IP) data [Ref. 3] resolved into an orbital component (α particles) and hyperbolic β particles [Ref. 6].

Noting that the interplanetary "bimodal" flux at 1 A.U. is incident, and without significant gravitational enhancement, on the lunar surface, we can compare this with lunar crater distributions. Review data [Ref. 34] is used in this instance to be representative of the differing data which have been published. We are still left in some considerable difficulty, however, regarding the absolute calibration of lunar rock exposure ages. A normalization is made therefore to the Grün interplanetary fluxes at larger masses where the distribution slope is high and the existence of secondary craters rendered insignificant. Figure 11 shows this operation, which leads us to readily identify the excess lunar microcrater flux at smaller craters, which had been suspected by numerous investigators (e.g., [Ref. 35]).

We are now able, and this is especially interesting in our study of the LEO environment, to collate more reliably summary evidence in Earth orbit (Fig. 12). We see LDEF as compatible for medium masses with the interplanetary flux, but gravitationally enhanced. This situation applies to all faces of LDEF. At smaller masses, however, the peripheral faces, especially the East through to North and South, show an "excess", namely the component we attribute to orbital particulates. The SMM spacecraft (become of "randome" exposures) sampled this orbital excess for only a fraction (a quarter) of its exposure, but due to the higher flux value it would have seen perhaps equal mixes of orbital and interplanetary particulates at this mass. Although we have seen, at the smaller masses

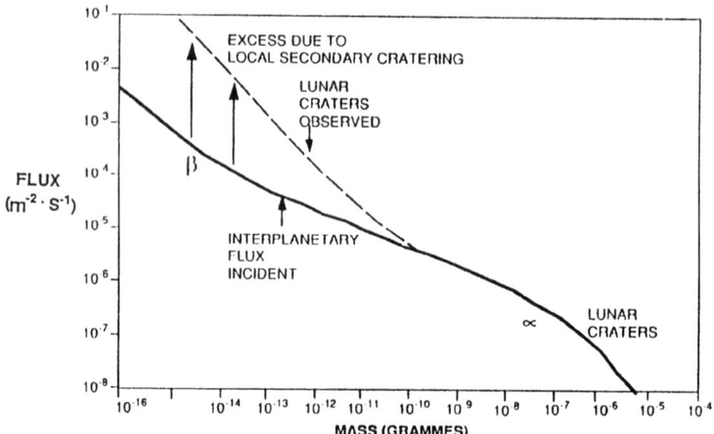

Fig. 11: Interplanetary particulate flux data compared to lunar crater impact data, using arbitrary flux rate scaling for the Lunar data normalized for the larger masses where primary craters must dominate. It shows clearly the existence of secondary craters at small sizes which are generated by the larger (α) particulates.

detected by the craters on the aluminium louvers on SMM, that the flux was exaggerated at a particular mass scale, we do still see a continuing excess above the interplanetary level as we go to the smallest masses. This would well comprise both locally generated secondary craters on SMM (due to the proximity of the solar cell structures) and a real flux of orbital particulates now seen on the LDEF MAP and IDE experiments.

6. Issues in the LEO Particulate Environment

We see now that the LEO environment poses a design challenge for the high reliability space systems and for extended lifetime missions. The impact of particulates has hitherto been able to be ignored in the small satellite area, perhaps less due to ignorance than because of higher failure from other mechanisms. Milestones in particulate environment definition were clearly the NASA meteoroid model for the near Earth Environment and for interplanetary space [Ref. 25]. We see now in the light of more recent review interplanetary measurements [Ref. 3] that this remains a good model, certainly for isotropic or random mission exposure to the meteoroid flux. New knowledge is, however, being generated: in deep space currently this is via the Galileo, Ulysses and Hiten missions; in near-Earth space, LDEF's return has provided unprecedented opportunity for the definition and understanding of environmental modelling. Yet we still see such understanding the sources of particulates as vital to the extrapolation of such data and its application to new missions.

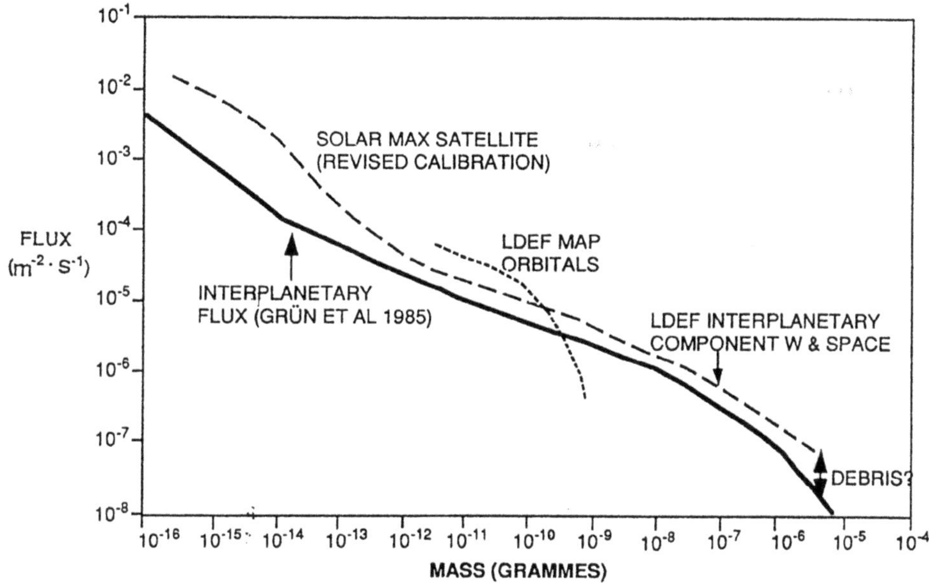

Fig. 12: LEO flux distribution data compared. The detection of LDEF of an orbital population at smaller sizes from the MAP data [Ref. 7] could be explained in terms of space debris or natural orbital component. Chemical evidence will be needed to resolve this matter. The interplanetary flux dominates for all faces between particle diameters of say 5μm and 500μm but a possible space debris component at larger diameters could be indicative of a contribution from space debris comminution. The source of the microparticle orbital population is suspected to be, but as yet unconfirmed, space debris.

LDEF's data set generally comprises *average* exposures, but in certain instances, *time resolved* data. Though we see the meteoroid role as dominant, orbital particulates are clearly demonstrated by statistical grounds invoking dynamic modelling and by direct evidence such as the chemical data only recently emerging. The intermix of interplanetary and orbital particulates is size dependent, and this no doubt holds clues to the origin of the particulates. This mixing ratio is also orientation sensitive, and design criteria for spacecraft can be clearly seen to benefit from appropriate exposure attitude relative to the orbit vector. Characteristics of the impacting particulates and their impact parameters are therefore also altitude dependent and prescriptions for meteoroid efficient shielding will therefore also differ. The time dependent data available (such as the IDE experiment) is generally restricted to the "insignificantly small" sub-micron particulates; in terms of hazards their significance could though be vital to our understanding! Are they orbital debris from rockets and impact comminution or natural fragments from meteoroid fragmentation? Comprising the tail of the satellite fragmentation distribution they would give a significant handle on problems relating to the definition of a suitable in-orbit comminution distribution which cannot be readily measured at larger dimensions. At medium masses, the dominance of the meteoroid population may mask the significant damage they could cause in the millimetre size range and yet be a population which could be increasing with time.

LDEF's data set will be understood and resolved only by a multi-disciplinary attack from the research field. From the application of dynamical modelling of orbits, from crater morphology, from the mass-analysis and the isotopic analysis of crater residues, and from the time history of some of this information, sufficient dimensions are surely available to resolve the major aspects of the source and sinks of particulates. The answers, as yet unresolved, are now on the detector surfaces in the laboratory and yet do vitally need the continued thrust of investigations of international status.

Acknowledgements

The research is supported by the Science and Engineering Research Council (SERV), U.K. and by the Strategic Defense Initiative by its Office of Innovative Science and Technology through contract N60921-86-A226 with the Naval Surface Warfare Center. Substantial collaboration is acknowledged from the LDEF MAP Co-Investigator Team, especially contributions from S. Deshpande, S. Mullen, K. Sullivan and M. Neish. Special thanks are due to D.J. Gardner for discussions and assistance in proof corrections. Manuscript preparation by Alision Rook.

References and Bibliography

[1] C.W. McCracken, et al., 'Direct measurements of interplanetary dust particles in the vicinity of Earth', Nature, 192, 441-442 (1961).

[2] C.S. Nilsson, 'Some doubts about the earth's dust clouds', Science, 153, 1242-1246 (1966).

[3] E. Grün, et al., 'Collisional balance of the meteoritic complex', Icarus 62, 244-272 (1985).

[4] J.A.M. McDonnell, 'Microparticle studies by space instrumentation', In Cosmic Dust, Ch. 6, J. Wiley & Sons (1978).

[5] K. Nishiizumi, et al., 'Exposure history of individual cosmic particles', Earth and Plan. Sci. Letters (1990).

[6] O.E. Berg and E. Grün, 'Evidence of hyperbolic cosmic dust particles', Space Research 13, 1047-1055 (1973).

[7] J.A.M. McDonnell, S.P. Deshpande, S.F. Green, P.J. Newman, M.T. Paley, P.R. Ratcliff, T.J. Stevenson, and K. Sullivan, 'First Results of Particulate Impacts and Foil Perforations on LDEF', Adv. Space Res. (in press) (1990).

[8] J.A.M. McDonnell and K. Sullivan, 'Hypervelocity Impacts on Space Detectors: Decoding the Projectile Parameters', workshop on Hypervelocity Impacts in Space Abstracts Volume, UKC, July 1991 (1991).

[9] D. O'Sullivan, A. Thompson, C. O'Ceallaigh, V. Domingo, and K.P. Wenzel, "The Long Duration Exposure Facility (LDEF) Mission 1 Experiments, A High Resolution Study of Ultra-Heavy cosmic-Ray Nuclei (A0178)", NASA SP-473.

[10] M.T. Paley, 'The large optical scanning system', Proc. Hypervelocity Impacts in Space, University of Kent, Canterbury (in submission) (1991).

[11] Humes (1980).

[12] Humes (1991).

[13] J.A.M. McDonnell, 'Factors affecting the choice of foils for penetration experiments in space', in Space Research X, North Holland, p. 314 (1970).

[14] Singer, et al. (1984).

[15] J.D. Mulholland, et al., Proc. 1st LDEF Science Conference, National Aeronautics and Space Administration Langley Research Center, Hampton, VA, submitted for publication (1991).

[16] J.A.M. McDonnell and P.R. Ratcliff 'Aerocapture and Aerofragmentation Capture' Proc. Symposium Asteroids, Comets & Meteoroids III, University of Northern Arizona, Flagstaff (1991).

[17] J. Borg, 'Micrometeoroid analysis on board FRECOPA Payload', Proc. Hypervelocity Impacts in Space, University of Kent, Canterbury (in submission) (1991).

[18] See, et al., Presentation LDEF 1st Post Retrieval Symposium, Orlando, FL (1991).

[19] D.H. Niblett, 'Survey of flux Data Derived from impacts on LDEF surfaces', Workshop on Hypervelocity Impacts in space Abstracts Volume, UKC, July 1991.

[20] H.A. Zook, Lun. Plan. Sci. Conf. (Abstracts volume), pp. 1577-1578 (1991).

[21] D.J. Kessler, AIAA J., 2337-2338 (1969).

[22] J.S. Dohnanyi, Bellcom TR66-340-1 Bellcom Inc. (1966).

[23] J.E. Erickson, J. Geophys. Res. 73 pp. 3721-3726 (1968).

[24] R.B. Southworth and Z. Sekanina, NASA CR-2316 (1973).

[25] B.G. Cour-Palais, NASA SP-8013 (1969).

[26] M. Zolensky, D. Atkinson, M. Allbrooks, C. Simon, M. Finckenor, J. Warren, E. Christensen, F. Cardenas, 'Meteoroid and Orbital Debris Record of the Long Duration Exposure Facility, Proc. Hypervelocity Impacts in Space, University of Kent, Canterbury (in submission) (1991).

[27] N. Pailer and E. Grün, 'The penetration limit of thin films', Planet. Space Sci. 28, 321-331 (1980).

[28] Space Debris, ESA SP-1109 (Report of the Space Debris Working Group) (1988).

[29] D. Kessler, et al., 'Cosmic dust and space debris', Space Res., Vol. 6, No. 7 (1986).

[30] Laurence and Brownlee (1985).

[31] J.A.M. McDonnell and K. Sullivan, 'Whence the LEO Particulates? LDEF's Data and New Penetration Formulae Reshape the Arguments on the Balance of Sources', Workshop on Hypervelocity Impacts in Space Abstracts Volume, UKC, July 1991 (1991).

[32] F.L. Whipple, On maintaining the meteoritic complex. In Zodiacal Light and the Interplanetary Medium. NASA SP 150, pp. 409-426 (1967).

[33] D.E. Brownlee, et al., 'Interplanetary dust; a new source of extraterrestrial material for laboratory studies, Proc. Lunar Sci. Conf. 8th, 149-160 (1977).

[34] L.B. Le Sergeant d'Hendecourt and P.L. Lamy, On the size distribution and physical properties of interplanetary dust grains. Icarus 43, 350-372 (1980).

[35] H.A. Zook, 'Meteoroid directionality on LDEF: Asteroidal versus cometary sources and how to obtain an effective velocity for beta meteoroids', Workshop on Hypervelocity Impacts in Space, University of Kent, Canterbury, (in submission) (1991).

[36] R.H. Fish and J.L. Summers, "The effect of material properties on threshold penetration," Proc. 7th. Hypervelocity Impact Symposium, 6, pp. 1-26 (1965).

[37] R.J. Naumann, "The Near-Earth Meteoroid Environment", NASA TN D-33717 (1966).

[38] W.C. Carey, J.A.M. McDonnell, and D.G. Dixon, 'An empirical penetration equation for thin metallic films used in capture cell techniques', Proc. Properties and Interactions of Interplanetary Dust. eds, R.H. Giese, P. Lamy, and D. Reidel, 291-295 (1984).

[39] S.P. Deshpande and S.F. Green, 'Can objects in geostationary transfer orbits (GTO) affect the debris population at low earth orbit (LEO)?', Hypervelocity Impacts in Space, University of Kent, Canterbury (in submission) (1991).

[40] S.F. Green and J.A.M. McDonnell, 'A numerical model for characterization of the orbital debris environment', Hypervelocity Impacts in space, University of Kent, Canterbury (in submission) (1991).

[41] H.J. Hoffman, H. Fechtig, E. Grün, and J. Kissel, 'First results of the micrometeoroid experiments 215 on HEOS 2 satellite', Planet. Space Sci. 1975, Vol. 23, pp. 215 to 224, Pergamon Press (1975).

[42] J.A.M. McDonnell, 'Particulates detected in the Near-earth space environment aboard the Long Duration Exposure facility (LDEF): Cosmic or Terrestrial? Proc. IAU Colloq. 126. Kyoto, Japan, (in press) (1990).

[43] D. McKnight, 'An assessment of recent satellite break-ups on the near earth environment', poster paper NATO ASI (1991).

[44] D. McKnight, 'Debris Growth: Isolating Sources and Effects. (poster paper NATO ASI, theses proc.) (1991).

[45] S. Mullen, 'The taxonomy of impact of the LDEF UHCRE thermal blankets', Proc. Hypervelocity Impacts in Space, University of Kent, Canterbury (in submission) (1991).

[46] S. Mullen, 'Penetration data from the LDEF UHCRE thermal blankets: reduction to incident mass distributions', Proc. Hypervelocity Impacts in Space, University of Kent, Canterbury (in submission).

[47] NASA Space Debris Book.

[48] H. Klinkrad, 'Advances in Space Research', Proc. XXVII COSPAR (1991).

APPENDIX
RELATIONSHIPS IN PARTICULATE DISTRIBUTIONS

A (straight line) plot in the log flux-log mass plane in which our cumulative flux distributions are plotted is characterized by the form:

$$\log \Phi(m) = \alpha \log m + \log c$$

where α = log cumulative slope = $\dfrac{\delta \log \Phi(m)}{\delta \log m}$. $(\alpha < 0)$. (A1)

$\Phi(m)$ refers to the particles greater than a limiting value, and is a *threshold* measurement, e.g., $\Phi(m)$ = number per metre^{2}sec^1 of particles of mass m or greater than m. The number of particles $d\Phi$ in a small logarithmic interval $\partial \log m$ may be found using: $\partial \log \Phi(m) = \alpha\, \partial \log m$. By differentiating or exponentiating we find: $\partial \Phi(m) = \alpha\, \Phi(m)\, \partial m/m$ = number of particles in a mass interval. Exponentiating Eq. (A1) we find

$$\Phi(m) = c\, m^{\alpha} \tag{A2}$$

which shows that our flux distributions are typically of power law form. The differential flux is $d\Phi(m)$ is given by

$$d\phi(m) = \alpha c . m^{\alpha-1}\, dm. \tag{A3}$$

The (differential) mass *index* $s = \alpha - 1$.

We can determine the total number of particles over a range m_1 to m_2 in the flux distribution by integration of $\phi(m)$. We could have obtained this form from the difference of the two cumulative fluxes $\Phi(m_1)$ and $\Phi(m_2)$, but $\phi(m)$ is needed for other manipulations.

Mass within a flux distribution.

For the *mass* within such an interval m_1 to m_2 we integrate the function $\phi(m)$ m:

$$\text{Total mass} = \int_{m1}^{m_2} \phi(m)\,.m.\,dm$$

$$= \int_{m1}^{m_2} \alpha\; c\; m^{\alpha-1}\; m\; dm$$

$$= \alpha \frac{c}{\alpha} + 1\; [m^{\alpha+1}]_{m_1}^{m_2}$$

$$= \frac{\alpha}{\alpha+1} c.\,(m_2^{\alpha+1} - m_1^{\alpha+1}) \qquad \text{(A4)}$$

Mass for $\alpha = -1$.

This solution is valid except for $\alpha = -1$, a value fairly typical of many rational distributions, e.g., of meteoroids impacting the Earth's atmosphere. For the case $\alpha = -1$, we find

$$\text{Total mass} = \int_{m1}^{m_2} \alpha c\; m^{\alpha-1} m\; dm$$

$$= \alpha\; c \int_{m1}^{m_2} \frac{d\,m}{m}$$

$$= \alpha\; c\; [\log_e m]_{m_1}^{m_2}$$

$$= \alpha\,.\,c\; \log_e\left(\frac{m_2}{m_1}\right) \qquad \text{(A5)}$$

In the case where m_1 and m_2 represent intervals on the logarithmic mass scale as for example *decade* intervals where $m_2 = 10m_1$, $\log (m_2/m_1) = 1$, then

$\log_e m_2/m_1 = \log_e 10 = 2.30$, and the total mass $= \alpha\; c\; \log_e \dfrac{m_1}{m_2} = \log_e 10.\alpha c$

$= \log_e 10.a\; \Phi(m).m$ per decade. Thus the mass of particles in any particular logarithmic mass interval is uniform over the range for $\alpha = -1$ since $\Phi(m) = c.m^{-1}$.

Mass for $\alpha < -1$ (high slope)

Integration and substitution yields: (A6)

$$\text{Total mass} = \frac{\alpha\,.\,c}{\alpha + 1}\; \frac{1}{m_1 - (\alpha+1)} = \frac{\alpha}{\alpha+1}\,.\,\Phi(m_1)\,.\,m_1$$

We see that the mass is determined only by the *lower* limit m_1, i.e., is concentrated there, for what are termed "high" slopes in the flux distribution.

Mass for $\alpha > -1$ (low slope)

The mass is concentrated near the largest mass and is given by total mass:

$$-\frac{\alpha c}{a+1} \cdot \frac{1}{m_2-(\alpha+1)} = -\frac{\alpha \Phi (m_2)\, m_2}{\alpha+1} \tag{A7}$$

Area within a distribution

For the integrated area of a flux distribution, e.g., representing the scattered or reflected light from such a distribution, we take the geometric cross section (A)= πr^2 where r= particle radius. We can readily extend it if necessary to incorporate the optical scattering function for light scattering calculations. We find the geometric cross section area:

$$= \int_{m1}^{m_2} (number\ X\ Area)\, dm$$

$$= \int_{m1}^{m_2} \alpha\ c\ m^{\alpha-1} . m^{2/3} . \left(\frac{3\pi^{1/2}}{4\rho}\right)^{2/3} .\ dm$$

$$= \int_{m1}^{m_2} \alpha\ c \left(\frac{3\pi^{1/2}}{4\rho}\right)^{2/3} . m^{\alpha-1/3} . \partial m.$$

$$= \int_{m1}^{m_2} \alpha\ c \left[\frac{3\pi^{1/2}}{4\rho}\right]^{2/3} . m^{\alpha-1/3} . dm$$

$$= \left[\frac{\frac{3\pi^{1/2}}{4\rho}}{\alpha+2/3}\right] . c .\ [m^{\alpha+2/3}]_{m_1}^{m_2} \tag{A8}$$

Area for $\alpha = -2/3$

The critical value of α concerning area is found to be $\alpha = -2/3$ and yields for $\alpha = -2/3$ over the interval m_1 to m_1:

$$\text{area per magnitude} = 2.32\alpha\, c\left(\frac{3\pi^{1/2}}{4\rho}\right)^{2/3}$$

$$= \alpha\, c.\left[\frac{3\pi^{1/2}}{4\rho}\right]^{2/3}.\log_e 10 \qquad \text{(A9)}$$

<u>Area for $\alpha < -2/3$ (high slopes)</u>

Integration limits applied to the general expression for area yields:

$$\text{Area} = -\left(\frac{3\pi^{1/2}}{4\rho}\right)^{2/3}\frac{c}{(\alpha + 2/3)}.m_1^{a+2/3}$$

$$= -\frac{\left(\frac{3\pi^{1/2}}{4\rho}\right)^{2/3}}{(\alpha + 2/3)}.\ \Phi(m_1).m_1^{2/3} \qquad \text{(A10)}$$

<u>Area for $\alpha > -2/3$ low slopes</u>

$$\text{Area} = -\left(\frac{3\pi^{1/2}}{4\rho}\right)^{2/3}.\frac{c}{(\alpha + 2/3)}.m_2^{a+2/3}$$

$$= -\frac{\left(\frac{3\pi^{1/2}}{4\rho}\right)^{2/3}}{(\alpha + 2/3)}.\ \Phi(m_2).m_2^{2/3} \qquad \text{(A11)}$$

<u>Average mass incident</u>

The average mass is found by the weighted flux x mass distribution. This function yields that of the integrated mass within the distribution normalized to the total flux. We find for a limiting threshold mass m

$$\text{average mass} = \frac{\alpha}{\alpha + 1}.m. \qquad \text{(A12)}$$

This has validity only for $\alpha < -1$, i.e., for high distribution slopes. This is reasonable because for $\alpha > -1$, the integral of the mass within the distribution is infinite. However, this point serves to illustrate the fact that unless the distribution slope is high, the average mass is considerably greater than most frequent mass.

<u>Average area in distribution</u>

This is, as in the case of average mass, valid only for $\alpha < -2/3$ and yields a value for a limiting size of m or greater:

$$-\frac{\left(\frac{3\pi^{1/2}}{4\rho}\right)}{\alpha + 2/3} . m^{2/3} = A(m) \frac{-3}{(\alpha + 2/3)} \tag{A13}$$

It is thus closer to the threshold than is the average mass due to the higher area to mass function for small particulates.

Arbitrary forms of flux $\Phi(m)$

Where $\Phi(m)$ is not conveniently expressed by an equation, and may represent an experimentally determined function, perhaps even with an associated noise or error, the mass and area functions may be determined by using sampled values of $\Phi(m)$ and the slope $\alpha(m)$. The number of particles in a small interval d log(m) is given by dN = exp(log $\Phi(m)$ - $\alpha(m)$ dlog m) - $\Phi(m)$. Hence the mass in that same interval is:

$$m.dN \tag{A14}$$

and the area in that same interval is:

$$A(m)\ dN = \left(\frac{3\pi^{1/2}}{4\rho}\right)^{2/3} m^{2/3}\ .dN \tag{A15}$$

This technique is used to generate the data shown in Figs. 1(b) and (c). The number of samples required per decade is determined by the accuracy required and the quality of the data.

THE MAGNETOSPHERE AND ITS INTERACTION WITH THE SOLAR WIND AND WITH THE IONOSPHERE

S. W. H. COWLEY
Blackett Laboratory
Imperial College
London SW7 2BZ
United Kingdom

ABSTRACT. A tutorial discussion is presented of the Earth's magnetosphere, and its interaction with the solar wind at its outer boundary, and with the ionosphere at its inner boundary. Topics covered include magnetospheric cavity formation, magnetic reconnection, plasma convection, structure of magnetosphere plasma populations, substorm dynamics, ionospheric convection, magnetospheric and ionospheric current systems, particle precipitation and the aurora.

1. Introduction and Theoretical Background

1.1. OBJECTIVES

Spacecraft and ground-based observations over the past thirty years have revealed the Earth's magnetosphere to be a highly complex, structured, time-dependent plasma system which interacts strongly with the solar wind at its outer boundary (the magnetopause), and with the ionosphere at its inner boundary. Nevertheless, it is the thesis of this article that the observed properties of the magnetosphere and its interactions can be understood at a zeroth order level in terms of the combination of a number of relatively simple elements which obey a small number of simple physical rules. The picture which emerges from this discussion is intended primarily to form a framework within which to appreciate the context of more detailed studies.

1.2. THEORETICAL BACKGROUND

Although the understanding of plasma systems at a detailed level is a notoriously complex business, only two simple theoretical ideas are required to meet the "zeroth order" objectives stated above; namely, the concepts of "frozen-in" transport of the magnetic field and the plasma, and the force exerted by the field on the plasma. These we will first outline.

1.2.1. *Frozen-in Flow.* We will first consider the motion of individual charged particles in an electromagnetic field. The simplest case is particle motion in a uniform steady magnetic induction **B**, in which case

R. N. DeWitt et al. (eds.), The Behavior of Systems in the Space Environment, 147–181.

ions and electrons gyrate transverse to the field in circles of radius $V_\perp/\Omega$, where $V_\perp$ is the particle speed transverse to B and Ω is the gyro frequency ($\Omega = qB/m$, where q is the particle charge and m its mass), while moving with uniform speed $V_\parallel$ along them. The general motion is thus a helix. If we now add a uniform electric field **E**, perpendicular to **B**, then to this helical motion will be added a particle drift, V_D which is perpendicular to both **E** and **B** and given by

$$V_D = ExB/B^2 \tag{1}$$

It is important to observe that this "ExB drift" is quite independent of the mass and charge of the particles, such that the addition of the transverse electric field is entirely equivalent to a transformation of the frame of reference. For example, if we take our original system with a uniform **B** and no **E**, and then transform to a frame moving with velocity $-V_D$ transverse to **B** relative to this system (some arbitrary velocity), then it is clear that in the new frame all the particles will drift transverse to **B** with the same velocity V_D. This motion in the new frame is, however, accompanied by the existence of an electric field in this frame given by the non-relativistic transformation $E = -V_D x B$. Substitution of this **E** into (1) then shows that the transverse drift in this frame is just the "**ExB** drift" in the electric field which automatically appears in this frame by virtue of the frame transformation.

The zeroth-order motion of the particles thus consists of the sum of a gyratory motion about the field (which gives no net transport), together with flow along the magnetic field, and a transverse **ExB** drift given by (1). If we now consider a more general field configuration in which **E** remains perpendicular to **B**, but in which both fields vary in space and time (but only slowly compared with the gyroradius and gyroperiod of the particles respectively), then this motion has a very simple and important property. If we consider a set of particles whose gyrocentres at some instant of time are all located on one magnetic field line, then at any other instant of time (earlier or later) their gyrocentres will also be located on one magnetic field line, as illustrated in Fig. 1. The reader is referred to elementary plasma physics texts for a formal proof that this is indeed the case.

This result is of central importance in giving a simple understanding of the relationship between the motion of the plasma and the behavior of the electromagnetic field. However, there are two completely equivalent ways of thinking about it. First, we may focus on the motion of the plasma and consider the magnetic field lines to be transported along by the flow "frozen" into the plasma. As the flow twists and bends so the magnetic field is also twisted and bent (but not with impunity, due to the force which the field then exerts on the plasma as we will shortly discuss). This way of thinking about the above result is appropriate when the energy of the flow dominates the energy of the field in the system, as is the case, for example, in the solar wind. Alternatively, we may think about the field lines themselves as moving (with the **ExB** drift), and carrying the plasma particles along with them. This way of thinking about the result is appropriate when the field energy exceeds the flow energy, as is the case, for example, in the inner dipolar parts of planetary magnetospheres.

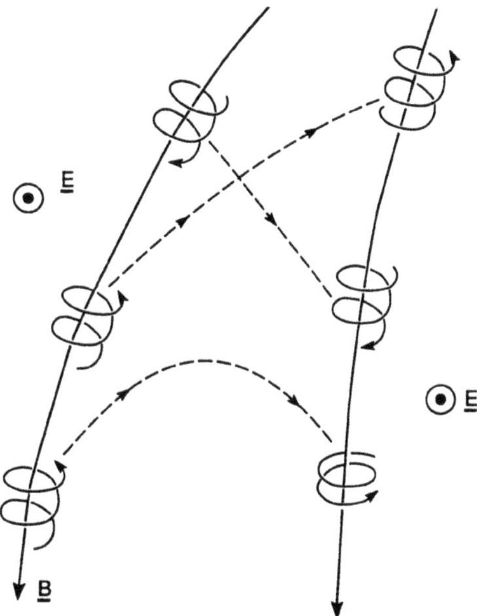

Fig. 1: Sketch showing the motion of charged particles in a non-uniform magnetic field **B** in the presence of a perpendicular electric field **E**. On the left we show some charged particles whose centers of gyration lie at some time on one field line. After a certain time they have moved due to the **ExB** drift and to motion along the field. The "frozen-in" theorem then tells us that their gyrocentres still lie on a magnetic field line, as shown.

An additional point needs to be made about the motion of the particles along the magnetic field, which under certain circumstances is not unrestricted. The gyration of the charged particles results in the formation of a loop of electric current, which is consequently associated with a magnetic moment given by $\mu = -(mV_\perp^2/2B)\mathbf{b}$, where **b** is the unit vector along **B**. In a non-uniform magnetic field there consequently exists a net force on the particle $\mathbf{F} = (\mu.\nabla)\mathbf{B}$ which is directed along the field such as to repel the particle from regions of high field strength. On moving along the field into such a region this force will act to reduce and reverse the field-aligned motion of the particle, causing it to "mirror" and move out of the region of high field strength. The overall motion along the field is governed by the constancy of the total particle speed and the particle magnetic moment (i.e., $V_\perp^2/B$, the "first" adiabatic invariant). Consequently, in a magnetospheric dipolar magnetic field geometry the particles will "bounce" rapidly along the field lines between mirror points in the northern and southern hemispheres, moving back and forth across the field strength minimum at the dipole equator. This oscillatory behavior is the motion which lies behind the formation of trapped particle "radiation belts" in dipolar planetary magnetospheres.

Finally, we should point out that the three-component motion discussed above represents only a zeroth order approximation which is strictly valid only for plasma particles of low energy. In a weakly inhomogeneous and time-dependent field, particles of finite energy will also experience drifts transverse to **B** which are proportional to the energy of the particle and to the non-uniformity and time-dependence of the field, and whose sense also depends on the sign of the charge (these are the "grad B", "curvature", and "polarization" drifts etc.). These drifts (together with the magnetization of the plasma) are responsible for the currents transverse to **B** which are carried by the plasma in such a field. As a result of these drifts, however, the frozen-in transport of the field and the plasma represents only a zeroth-order approximation for a plasma of finite temperature. Any initial collection of particles threaded by a given field line will gradually spread apart across the field, ions and electrons in opposite directions. The rate of this spreading depends upon the degree of inhomogeneity and time-dependence in the field relative to the particle gyroradius and gyroperiod. In the limit that the field scale lengths or times approach the latter values then the above picture of particle motion breaks down, and with it the concept of "frozen-in fields". In the following sections we will therefore need to remember that while the "frozen-in" concept is valid in fields with large spatial scales and long time scales, it is suspect in structures with small scale lengths or short time scales.

1.2.2. *The Force of the Field on the Plasma.* The force exerted by the electromagnetic field on an individual charged particle is given by the sum of the electric force $q\mathbf{E}$ and the magnetic force $q\mathbf{V}\times\mathbf{B}$. However, when these forces are summed over some finite volume of plasma, the electric force sums to zero since the plasma as a whole will be charge neutral, while the total magnetic force per unit volume sums to $\mathbf{j}\times\mathbf{B}$, where $\mathbf{j}$ is the electric current density in the plasma. This force can also be written wholly in terms of the magnetic field $\mathbf{B}$, since $\mathbf{j} = \text{curl } \mathbf{B}/\mu_0$ (neglecting the displacement current in the Ampere-Maxwell equation). Use of a vector identity then enables us to write the magnetic force per unit volume as the sum of two terms as follows

$$\mathbf{j}\times\mathbf{B} = (\mathbf{B} \cdot \nabla)\mathbf{B}/\mu_0 - \nabla(B^2/2\mu_0) \quad . \tag{2}$$

The first term represents the effect of the tension in the magnetic field and produces a force on the plasma like that which would be exerted by bent rubber bands. The plasma motion under the action of this term is such as to reduce the bending of the field. The second term is the magnetic pressure term, and shows that the force per unit area exerted by the field on the plasma is $B^2/2\mu_0$. The concepts of magnetic tension and magnetic pressure will be used extensively in the discussion below.

2. Formation of a Magnetospheric Cavity

In this section we will use the two concepts introduced above to discuss the formation of a magnetospheric cavity in the solar wind surrounding a magnetized planet, such as the Earth. This was the first quantitative problem to be solved in solar-terrestrial physics, by Chapman and Ferraro working at Imperial College in the early 1930s. However, at that time these authors did not know that the solar wind is a continuous outflow, nor did they know about the interplanetary magnetic field, so they considered what would happen if a front of unmagnetized ionized gas should emerge from the Sun and interact with the Earth, the latter previously residing in a vacuum. This is the Chapman-Ferraro problem.

2.1. THE CHAPMAN-FERRARO PROBLEM

The solution of the Chapman-Ferraro problem is simple in principle, and is illustrated in Fig. 2. If we assume that plasmas and magnetic fields are perfectly frozen together as in the above discussion, then by the same token the Earth's magnetic field will not be able to penetrate the

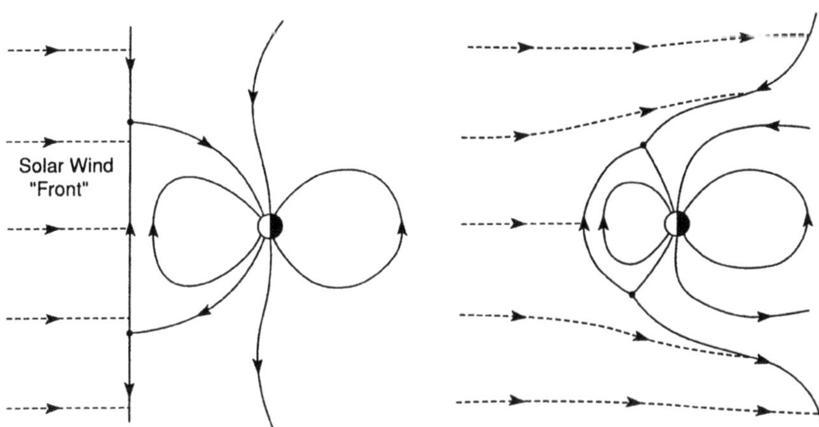

Fig. 2: Illustration of the behavior of the field and plasma in the hypothetical "Chapman-Ferraro" problem, in which a "front" of unmagnetized perfectly conducting plasma expands outwards from the Sun and interacts with the Earth's magnetic field. The solid lines show the magnetic field lines and the dashed lines the plasma streamlines. Initially the plasma sweeps up the magnetic field and compresses it on the dayside, as shown on the left of the figure. However, the increasing magnetic pressure eventually brings the plasma to rest near the dayside equator, while away from the equator the flow is deflected around the Earth, such that a magnetospheric cavity is formed when the front is well downstream.

expanding "front". Consequently, the field will be swept up by the plasma and compressed on the dayside as shown on the left hand side of the figure. An electric current will flow in the surface of the plasma (the "Chapman-Ferraro" current) which will switch off the magnetic field in the interior of the plasma (though mathematically the solution for the field is most easily obtained using image dipole techniques). However, the field compression clearly cannot continue indefinitely since an increasing pressure will be exerted on the plasma "front", which will eventually bring the plasma to rest at a stagnation point on the equator on the dayside of the Earth. Elsewhere the flow will continue, but it will be deflected around the planet by the magnetic pressure force, as shown on the right hand side of the figure.

Clearly, when the plasma "front" is well downstream from the Earth, the magnetic field will be confined to the interior of a cavity surrounding the planet; this being the planetary magnetosphere. The scale size of the cavity is determined by the condition for pressure balance at the stagnation point; the field pressure on the magnetospheric side of the boundary must balance the dynamic pressure of the solar wind on the other. A simple calculation shows that for the Earth the boundary (termed the magnetopause) should lie at a geocentric distance of about 10 R_E, and this is indeed where it is observed. Within this simple picture the cavity will also eventually close on the downstream side at a much greater distance, due to the effect of the small but finite thermal pressure of the solar wind. One final point is that due to the fact that the solar wind flow is supersonic (with a Mach number typically between 5 and 10), a stand-off shock will form ahead of the magnetospheric obstacle, across which the flow will be slowed, compressed and heated. The region of turbulent plasma which lies downstream from the shock is called the magnetosheath.

2.2. GENERALIZATIONS

The simple picture of an unmagnetized solar wind interacting with a vacuum planetary dipole field may readily be extended to include field and plasma from both sources, as illustrated in Fig. 3. Here we now include an interplanetary magnetic field which is convected outwards from the Sun, frozen into the solar wind, and which is compressed in the slowed flows of the dayside magnetosheath and draped over the magnetospheric cavity. We also include an interior source of planetary plasma (e.g., from the planet's ionosphere), which in the absence of other effects will corotate at the planetary rotation period, as enforced by atmospheric drag on the ions in the lower collisional part of the ionosphere. The key point, however, is that if we assume that the plasmas and fields are perfectly "frozen" together, then there will always be a strict division between the plasmas and fields of solar and planetary origin. In general, space will be broken up into cells containing the plasma and field from the different sources, the location of the boundaries between them being determined by pressure balance. These boundaries will also constitute current sheets, since the magnetic field will in general change abruptly across them, both in magnitude and direction.

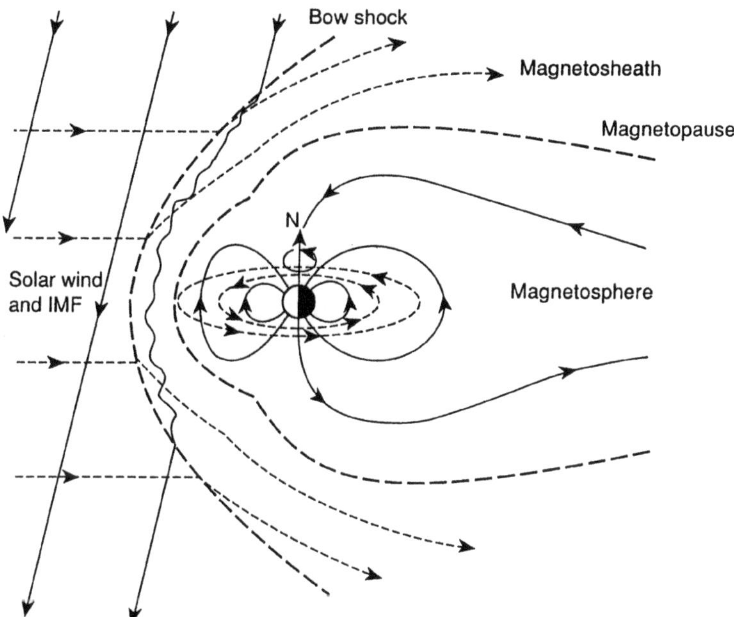

Fig. 3: Sketch of the magnetosphere system expected on the basis of the "frozen-in" flow approximation, in the noon-midnight meridian plane. The arrowed solid lines indicate magnetic field lines, the arrowed dashed lines the plasma streamlines, and the heavy long-dashed lines the principal boundaries (bow shock and magnetopause). The magnetopause represents a perfect boundary between planetary field and plasma (originating from the planetary ionosphere), and the solar wind and interplanetary magnetic field. Inside the magnetospheric cavity the plasma circulates with the planetary rotation period. Outside, a bow shock lies upstream of the magnetopause, across which the plasma is slowed, compressed and heated. The turbulent plasma from the shock is termed the magnetosheath downstream [Ref. 1].

3. Reconnection and the "Open" Magnetosphere

The picture derived above does give an appropriate initial view of the way in which magnetospheres are formed in the solar wind around magnetized planets. However, it is by no means the full story. We derived this picture on the basis of the perfect freezing together of field and plasma, and we pointed out in Section 1 that this would be valid provided that the scale length of the field is large (compared with the typical particle gyroradii). However, application of this approximation to the solar wind-planetary field problem has automatically produced a system which includes a small scale length, namely the magnetopause current sheet which separates the two regimes. While the frozen-in approximation will therefore be appropriate over very nearly the whole volume of the magnetosphere and its solar wind environs, it is liable to break down in one crucial region; namely, the magnetosphere-magnetosheath interface where the primary coupling between the planetary and solar plasmas takes place. This turns out to have consequences which are crucial to the physics of the Earth's magnetosphere, as we will now go on to discuss.

3.1. MAGNETIC RECONNECTION

When there exits a large magnetic shear across the magnetopause current sheet (i.e., when the magnetosheath field differs substantially in direction from the northward-pointing field on the terrestrial side of the boundary), breakdown of frozen-in flow leads to the effects shown in Fig. 4. The magnetic fields on either side of the boundary may diffuse relative to the plasma into the current sheet from either side, thus forming an x-type magnetic field geometry leading to the production of magnetic field lines which connect across the boundary between the two plasma regimes. This process is called magnetic reconnection, since we may picture the initial field lines as having been broken and "re-connected" across the boundary. The magnetic tension effect will then cause the current sheet plasma and the reconnected field lines to contract along the current sheet away from the site of reconnection, as illustrated in the figure, thus allowing further field lines from the initial populations to diffuse into the sheet and become reconnected.

The theory of magnetic reconnection was pioneered by Dungey in the early 1950s, and applied by him to the solar wind-magnetosphere coupling problem in the early 1960s. The initial effect of this process is to form "open" magnetospheric field lines which thread across the magnetopause boundary, and which are connected into the solar wind at one end and to the Earth's polar regions at the other. The consequences of this are three-fold. First, and most important, the existence of these open tubes allows an efficient transfer of solar wind momentum into the magnetosphere as these open field lines are transported downstream by the solar wind flow. Second, the open tubes also form a direct magnetic pathway for the transfer of plasma particles across the boundary between the magnetosphere and the solar wind. Third, as the reconnected field lines contract along the boundary away from the site of reconnection, the plasma on these tubes is accelerated and heated. In the sections below we will discuss each of these three key effects, starting in the next section with the convective flow which is generated inside the magnetosphere.

3.2. CONVECTIVE FLOW IN DUNGEY'S "OPEN" MAGNETOSPHERE

The geometry of the magnetic field and the associated flow of the plasma which occurs in a steady-state version of the "open" magnetosphere is shown in Fig. 5. Here magnetic reconnection at the dayside magnetopause produces open field lines connected to the Earth's polar regions which are carried antisunwards by the solar wind flow, and stretched out on the nightside into a long magnetic tail. As they are carried antisunwards, the near-Earth portions of these open tubes sink towards the center of the

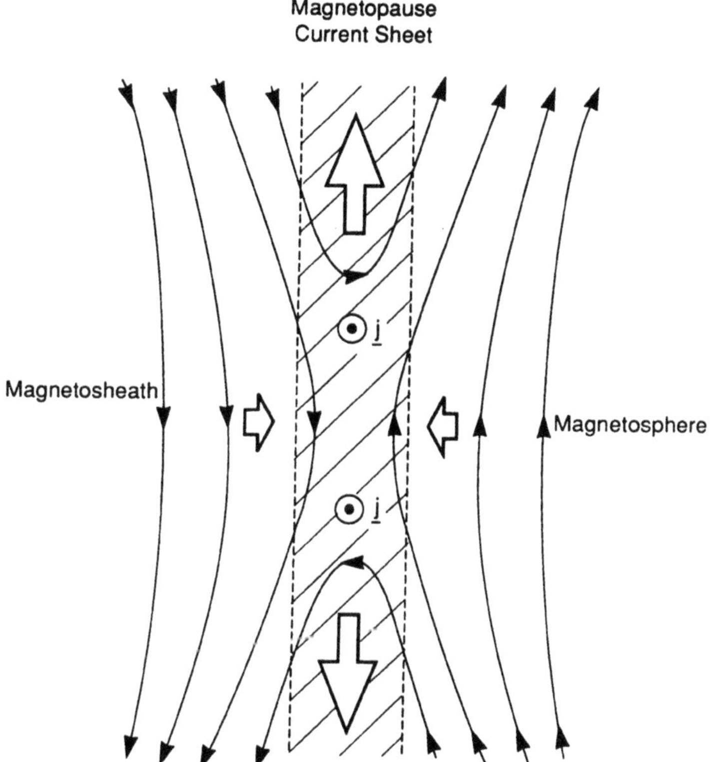

Fig. 4: Sketch showing the effect of magnetic reconnection at the magnetopause current sheet (hatched region), where the current density j points out of the plane of the diagram (circled dots). The arrowed solid lines show the magnetic field lines, while the large arrows show the direction of plasma motion. Field lines diffuse into the current layer from both sides and reconnect, producing "open" field lines which thread through the boundary from the magnetosheath to the magnetosphere. The magnetic tension of the field lines results in rapid field and plasma flow along the current sheet away from the site of reconnection [Ref. 1].

tail, and then reconnect again at a x-type field configuration on the nightside. In the region tailward of the tail-reconnection region the "disconnected" flux tubes flow back out into the solar wind, while Earthward of the reconnection region the newly closed field lines flow back towards the dayside magnetopause, where the cycle repeats. The reconnection process thus results in a large-scale cyclical flow being driven in the magnetosphere, with open tubes moving antisunwards over the Earth's poles, and closed tubes moving sunwards again through the central magnetosphere. Overall the flow cycle time is 6-12 hours, of which a given

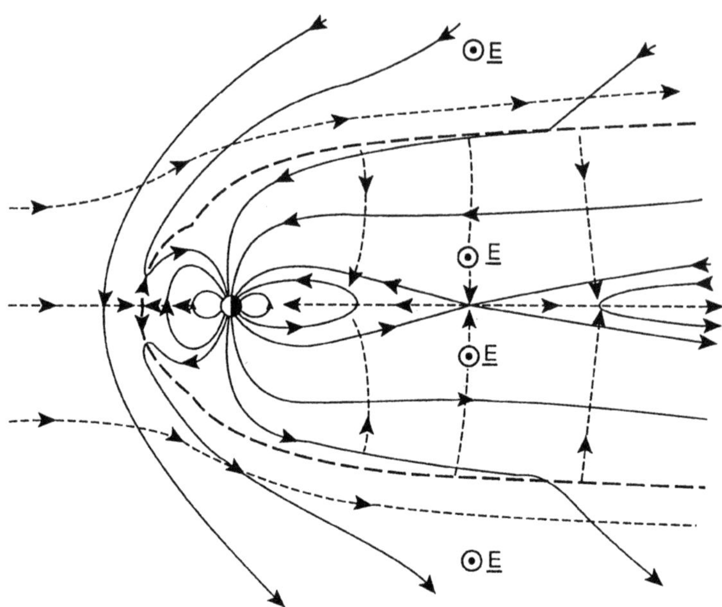

Fig. 5: Sketch of Dungey's "open" magnetosphere in the noon-midnight meridian plane, where arrowed solid lines indicate magnetic field lines, arrowed short-dashed lines the principal plasma flows, and the long-dashed heavy line the magnetopause. The circled dots marked **E** indicate the electric field associated with the flow, pointing out of the plane of the diagram [Ref. 1].

field line is open and being stretched out downtail for 2-4 hours, and closed and flowing back to the dayside for 4-8 hours. The former interval allows us to make a simple estimate of the length of the magnetic tail, as the product of the solar wind speed times the time for which a field line remains open. This estimate gives a value of around 1000 R_E, as originally obtained by Dungey, much greater than the distance to the dayside magnetopause (about 10 R_E). However, a "tail" of distended disconnected flux tubes will persist to much greater distances.

Four additional points are in order. The first is that not all of the flux tubes in the magnetosphere take part in the solar wind-driven convection cycle. Specifically, a "core" of flux tubes extending in the equatorial plane typically to about 4 R_E instead approximately corotate with the Earth. The flux tubes taking part in the convection cycle flow around outside this central "core" from the nightside to the dayside. The central region therefore has different plasma characteristics to the remainder of the magnetosphere (it corresponds to the plasmasphere, as will be discussed later), and has to be treated separately. It may be noted that while this corotation region corresponds only to a small volume of the magnetosphere, it actually corresponds to a majority of the Earth's surface, up to a magnetic latitude of about 60°.

The second point is that the solar wind-driven flow system is not steady, but varies on few-minutes to few-tens-of-minutes time scales. The flow is excited whenever closed magnetic flux is reconnected at the dayside magnetopause and transferred to the tail, and correspondingly when open flux in the tail is reconnected and transferred back to the dayside; these two processes acting only in a loosely coupled fashion. The first of these processes, the dayside reconnection rate, depends principally on the direction of the interplanetary magnetic field (IMF). When the IMF points to the south, opposite to the direction of the Earth's equatorial field (as shown in Fig. 4), equatorial reconnection is rapid and drives a large flow. However, when the IMF points north, equatorial reconnection at the dayside magnetopause is weak or absent, and the flow driven from this source decays to small values. The flow within the system thus depends significantly on the north-south component of the IMF. Flow is also driven by nightside reconnection, and this is also variable, occurring in bursts lasting a few tens of minutes corresponding to intervals of geomagnetic and auroral disturbance (substorms), as will be discussed further below. However, the precise conditions under which rapid tail reconnection is triggered remains the subject of study and debate.

The third point is to remind the reader that the magnetospheric flows are also associated with a magnetospheric system of electric fields, given by $\mathbf{E} = -\mathbf{V}\times\mathbf{B}$. In the noon-midnight meridian cross-section shown in Fig. 5 this electric field points everywhere out of the plane of the diagram as shown, though of course it is not everywhere of equal strength (the steady-state magnetic field lines are electric equipotentials, so the steady-state electric field is stronger near the Earth where the field lines converge). In the outer magnetosphere the electric field is typically a few tenths of a mV m^{-1}, corresponding to a cross-magnetosphere voltage of several tens of kV. The magnetospheric voltage also relates directly to the rate of transfer of magnetic flux from the dayside to the tail and back again. From Faraday's law a cross-magnetosphere e.m.f. of 1 volt is equivalent to a flux transfer rate to 1 Wb s^{-1}.

The fourth point is that solar wind momentum may also be communicated to the magnetosphere by processes other than reconnection, thus also exciting large-scale cyclical flow. In particular, solar wind particles flowing adjacent to the magnetopause may be scattered across the boundary, e.g. by plasma waves, carrying their antisunward-directed momentum with them, thereby setting closed magnetospheric flux tubes into motion towards the tail. Eventually these flux tubes must flow back towards the dayside again through the central magnetosphere, thus completing the flow cycle. It is important to emphasize that such a cycle of closed flux tube flow can co-exist with similar flows excited by reconnection on a continuous basis; they are by no means mutually exclusive. The main question then concerns the relative importance of these flows as quantified by the flux transfer rates involved, or equivalently by their contributions to the cross-magnetosphere voltage. Observations in the magnetosphere and ionosphere have shown that the voltage associated with the "viscous" process is typically about 10 kV, while that associated with reconnection is variable between essentially zero and 200 kV (with typical values between 50 and 100 kV), depending upon the direction of the IMF and the state of magnetospheric disturbance. Thus "viscosity" may play a significant role during quiet periods and when the IMF points north, but in general reconnection plays the dominant role. It is for this reason that we focus mainly on the reconnection cycle in this exposition.

3.3. ENERGY EXCHANGES BETWEEN FIELD AND PLASMA

The structure and properties of magnetospheric plasmas depends upon three main factors. The first is the nature of the plasma sources and sinks. The second is the nature of the motions which transport the plasma from the sources to the sinks. The third is the energy exchanges between the plasma and the field which take place during that transport. In the Earth's magnetosphere the principal plasma sources and sinks are the solar wind and the ionosphere, and the main transport process is the time-dependent cyclic convection which we discussed in the last section. In this section we now address the third major issue concerning plasma-field energy exchanges.

A simple way to approach this problem arises from noting that the energy input from the field to the plasma per unit volume per unit time is given simply by ($\mathbf{j}.\mathbf{E}$). In a steady state conservation of energy (Poynting's theorem) is then expressed as

$$\mathrm{div}\,\mathbf{S} + \mathbf{j}\,.\,\mathbf{E} = 0, \tag{3}$$

where $\mathbf{S} = \mathbf{E}\mathrm{x}\mathbf{B}/\mu_o$ is Poynting's vector which describes the energy flux (energy per area per time) in the electromagnetic field. Thus where $\mathbf{j}.\mathbf{E}$ is positive, the plasma gains energy from the field and such a region is a sink of $\mathbf{S}$ (div $\mathbf{S}$ negative), while where $\mathbf{j}.\mathbf{E}$ is negative the plasma loses energy to the field and such a region is a source of $\mathbf{S}$. Now we pointed out above in our discussion of Fig. 5 that the electric field points everywhere out of the plane of that diagram (from dawn to dusk across the magnetosphere), so to discuss energy exchanges we need only to consider the sense of the electric current. This is shown in a similar cross-sectional view in Fig. 6, where the dashed lines show the Poynting vector flow and the circled symbols the flow of the current as indicated by curl $\mathbf{B}$. Circled dots indicate regions of current flow out of the plane of the diagram, parallel to $\mathbf{E}$, which are consequently regions where energy flows from the field to the plasma. Conversely, circled crosses indicate regions of current flow into the plane of the diagram, antiparallel to $\mathbf{E}$, which are consequently regions where energy flows from the plasma to the field.

It can be seen from Fig. 6 that current flows out of the plane of the diagram both in the dayside magnetopause (Chapman-Ferraro) current sheet and in the center plane of the tail, so that plasma is energized in these regions. These are the current sheets where reconnection is taking place and where the plasma is consequently being heated and accelerated by the field tension away from the x-type regions. Examinations of the sense of the field tension effect on the plasma gives an equivalent alternative way of discussing its acceleration and energization. However, current flows into the plane of the diagram over the tail lobe magnetopause, so that this is a region where the plasma loses energy to the field. The field tension

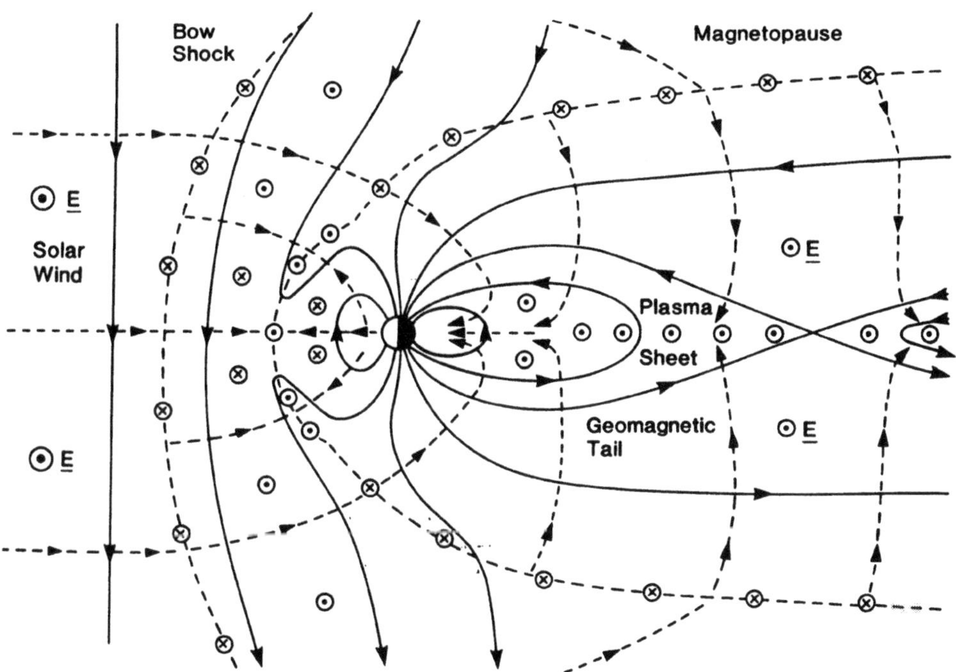

Fig. 6: Sketch of the open magnetosphere in the noon-midnight meridian plane showing the principal field-perpendicular currents and the flow of electromagnetic energy. The solid lines show magnetic field lines, the arrowed short-dashed lines the flow of electromagnetic energy (the Poynting flux **S**), and the unlabelled circled symbols the direction of the field-transverse current flow. The circled dots indicate current flow out of the plane of the diagram, and the circled crosses current flow in [Ref. 2].

effect opposes the tailward flow of the magnetosheath plasma in this region as the open tubes are convected downstream. The energy extracted from the magnetosheath plasma flows into the tail lobes a Poynting flux, and is then fed into the plasma in the center of the tail. This is the dominantly important energy exchange process which powers convection-dominated magnetospheres such as the Earth's. The acceleration and heating which takes place at the dayside magnetopause is of secondary significance.

4. Structure of Magnetospheric Plasma Populations

Having discussed plasma sources, transport and energy exchanges we are now in a position to go on to look at the structure of the magnetospheric plasma populations to which these processes give rise. We will approach this topic by first considering the behavior of individual ions in the flow before generalizing to the plasma as a whole. We also assume that to a first approximation the plasma electrons simply follow the ion behavior such as to maintain charge neutrality.

4.1. SOLAR WIND SOURCE

The shocked solar wind (magnetosheath) plasma adjacent to the magnetopause consists of protons (and a few percent alpha particles) with energies of a few 100 eV, together with electrons of energy ~ 100 eV, flowing round the magnetopause with speeds of a few 100 km s^{-1}. In this section we will consider the plasma populations to which this source gives rise, beginning with a discussion of the motion of individual ions starting in the magnetosheath.

4.1.1. *Two Proton Trajectories.* In Fig. 7a we sketch the trajectories of two representative solar wind protons which impinge on the magnetopause, one on the dayside (proton A in the figure), the other further downtail (proton B). Proton A starts at point 1 on the sketch in the dayside magnetosheath adjacent to the magnetopause, and convects with the flow into the magnetopause current sheet. There it will be accelerated by the field tension effect away from the site of reconnection (assumed to be near-equatorial) to energies of about 1 keV, and will either be reflected off the current sheet back out into the magnetosheath, or will be transmitted across it to subsequently move along the newly opened field lines down towards the Earth (we assume the latter here for sake of argument, as shown by point 2 on the trajectory). At point 3 in the sketch the particle is mirrored in the high field strength region near the Earth and moves back up the field lines again, but it does not move back to the dayside magnetopause due to the continual tailward motion of the open flux tubes during its time of flight to the Earth and back (the **ExB** drift of the field lines is indicated by the short arrows in the sketch). Instead the proton moves back out towards the tail lobe magnetopause. Whether it reaches it or not depends on the field-aligned speed of the particle compared with the tailward speed with which the open flux tubes are being stretched down-tail by the magnetosheath flow. If the field-aligned speed of the proton is greater than the magnetosheath speed then the particle will "catch up" with the "end" of the field line at the tail magnetopause, and will move back out into the magnetosheath. Conversely, if its field-aligned speed is less than that of the magnetosheath (as assumed here for sake of argument), then it will remain within the tail and will instead convect with the field line towards the tail center plane (point 4); the distance down-tail where the particle reaches the central current sheet is given simply by $(V_P/V_{SW}) \times L_T$, where V_P is the field-aligned speed of the ion, V_{SW} the speed of the solar wind (essentially equal to the magnetosheath speed well downstream from the Earth), and L_T the length of

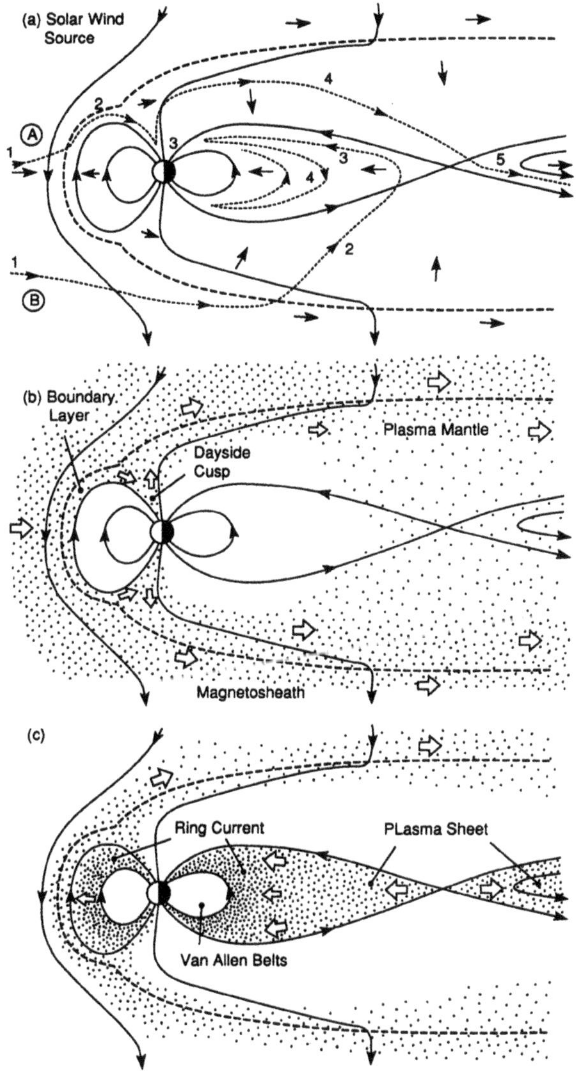

Fig. 7: (a) Trajectories of two typical magnetosheath ions which impinge on the Earth's magnetopause, one at the dayside magnetopause (ion A), the other at the tail magnetopause (ion B). The short arrows indicate the direction of the **ExB** drift. (b) Plasma populations in the outer magnetosphere of magnetosheath origin. The large arrows indicate the direction of plasma flow. (c) Plasma populations resulting from particle acceleration in the tail current sheet.

the tail estimated above using Dungey's argument (~ 1000 R_E). Thus ions with smaller down-tail speeds convect into the central current sheet closer to Earth than those with higher speeds, since the higher the speed, the further the ions travel down the tail in the time that the field lines remain open. What then happens to the particle in the central current sheet depends on the location of the x-type reconnection region in the tail, and whether the particle reaches the current sheet Earthward of this region, or tailward of it. Present evidence indicates that in the central tail and under normal conditions this region lies typically ~ 100 R_E from Earth, so that only ions with relatively low field-aligned speeds reach the current sheet Earthward of this location. We thus assume (again for sake of argument) that our ion reaches the current sheet tailward of the reconnection region. There it will be accelerated by the tension in the field away from the Earth, and will move back out into the solar wind along the "disconnected" field lines at speeds typically between 500 and 1000 km s^{-1} (point 5 on the trajectory).

Proton B starts from a point which has a somewhat greater magnetospheric "impact parameter" than proton A (point 1) and instead impinges on the tail lobe magnetopause. We again assume that the ion is transmitted across the current layer into the magnetosphere, now losing a small amount of energy in the process (probably only a few tens of eV) due to the retarding action of the field line tension. We assume that it will then have a somewhat smaller down-tail speed than proton A, consequently convecting into the central tail current sheet closer to the Earth than the latter (point 2). We also assume for sake of argument that it convects into the region lying Earthward of the tail reconnection region, so that instead of being accelerated tailward by the field tension and being lost to the solar wind as before, it is instead accelerated Earthwards by the field into the region of closed field lines (point 3). Typical post-acceleration energies are a few keV, corresponding to Earthward flows along the closed field lines of several hundred km s^{-1}. Subsequently, however, the particle will mirror near the Earth and return to the current sheet, but at a point well Earthward of its initial position due to the Earthward contraction of the closed field lines during the ion's time of flight. There it will be further accelerated (but to a lesser extent than before), followed by a second mirroring on the opposite side of the current sheet and a second return to the center plane etc. (point 4). The particle thus becomes trapped on the closed flux tubes and convects in towards the Earth bouncing between mirror points. Subsequent accelerations after the first can be approximately described in terms of conservation of the bounce ("second") adiabatic invariant (the integral along the field between the mirror points of the field-parallel velocity). The particles may be lost during transport to the dayside by precipitation into the ionosphere resulting from wave-particle scattering, or by charge-exchange with the Earth's atomic hydrogen geocorona (of which, more later). If neither occurs, then the particle will be lost to the magnetosheath once the field line on which it is moving becomes open as a result of reconnection at the magnetopause.

4.1.2. *Boundary Layers, Dayside Cusp and Mantle.* We now generalize our discussion of the two solar wind ions to consider the related plasma populations. In this section we consider the populations of relatively unprocessed magnetosheath plasma which are observed flowing in the anti-sunward direction in the outer parts of the magnetosphere. We discuss separately below the accelerated populations which arise from the current sheet interaction in the tail center plane.

In Fig. 7b we sketch the three connected populations involved. First we have the boundary layer of modestly accelerated magnetosheath plasma which lies on open field lines adjacent to the dayside magnetopause. This layer has a typical thickness of a few 1000 km, with densities similar to magnetosheath values (of order 10 cm^{-3}), but somewhat higher temperatures and bulk speeds. In addition to the layer formed by reconnection, boundary layers of magnetosheath particles which have entered by other "viscous" processes are also observed, as mentioned above. These boundary layers tend to be somewhat broader, have a somewhat lower density, higher temperature, and a bulk speed which is less than that of the adjacent magnetosheath plasma. These boundary layer populations flow down the field lines to the Earth to form the dayside cusp plasma, and while the majority of the particles mirror and flow out again, a small fraction precipitates into the dayside ionosphere forming the most important component of the dayside auroral zone. The particles on open boundary layer flux tubes then flow up into the outer tail lobe, forming the plasma mantle population adjacent to the tail magnetopause surface. This population is augmented by additional entry across the magnetopause surface into the tail lobe. In the near-Earth tail this "mantle" layer is typically a few R_E thick, compared with a total tail radius of about 20 R_E, but increases in width further away from the Earth due to convection of the lobe field lines towards the tail center plane. Since ions with the smallest down-tail speed arrive at the center plane closest to Earth, a velocity gradient exists across the mantle, with the slowest (field-aligned) particles being found in its inner regions, and with the velocity (and density) increasing towards magnetosheath values as the magnetopause is approached.

4.1.3. *Plasma Sheet and Ring Current.* We now consider what happens when the lobe plasma convects into the tail center plane and is accelerated. Although we are considering the solar wind source explicitly here, we note that similar arguments also apply to any ionospheric plasma which flows into the current sheet on the nightside, as we will discuss further below. Ions from either source will be accelerated in the current sheet away from the reconnection region to speeds of 500 to 1000 km s^{-1}, corresponding to energies of several keV (it is found that the electrons are accelerated in this region from a few tens of eV in the lobe to a few hundred eV in the central region, probably by a wave-particle process). As indicated in Fig. 7c this accelerated plasma is termed the plasma sheet, irrespective of the side of the reconnection region on which it is located. Tailward of the reconnection region the plasma sheet forms a jet several thousand km wide in which the plasma and field lines flow unrestrained back into the solar wind. A similar jet of Earthward-flowing plasma also forms Earthward of the reconnection region, but is terminated in the near-Earth region of closed field lines by the return flux of particles mirrored from the Earth. In this near-Earth region the Earthward jet is then confined to the outer surface layer of the plasma sheet (called the plasma sheet boundary layer), while in the central plasma sheet, where mirrored particles are present, the plasma is observed to have become thermalized.

In the action of their being accelerated in the tail current sheet, the plasma sheet particles automatically produce just the correct electric current which is required to flow between the two lobes of the magnetic tail (note that an acceleration requires a cross-system displacement of the ions in the direction of the electric field, and of the electrons in the opposite direction, such as will produce a current in the appropriate direction). In the magnetotail proper this current closes around the lobe magnetopause to produce an overall current flow which has the shape of the greek letter "theta" when viewed along the tail axis. However, as the plasma sheet plasma convects with the field lines towards and around the Earth in the quasi-dipolar region of the magnetosphere, and onwards to the dayside magnetopause, it continues to carry a current due to the grad B and curvature drifts of the particles, which now closes westward around the Earth. The hot plasma in the quasi-dipolar region is consequently termed the ring current, and the energy of the plasma which it contains can be monitored by measurement of the magnetic perturbation produced on the ground. Of course, in a steady state this plasma cannot enter the "corotating core" in the inner magnetosphere, as indicted in Fig. 7c. Finally, this figure also shows that the ring current plasma which does not precipitate into the atmosphere during the flow from the tail to the dayside eventually is lost across the boundary, forming a layer of energetic particles outside the magnetopause which flows downstream with the magnetosheath. The particles which do move into the atmosphere, together with those from the dayside cusp, form an annular band of precipitation circling each pole which corresponds to the auroral zone.

4.2. IONOSPHERIC SOURCE

Initially we expect the ionosphere to form a low-energy source of light ions for the magnetosphere, consisting mainly of protons flowing out with speeds of a few tens of km s^{-1} (corresponding to an energy of a few eV). This picture may generally be appropriate both at low latitudes in the corotating region, and at high latitudes in the polar cap. However, in the auroral zone in between additional processes occur to complicate the picture. First, ionospheric ions are heated transverse to the magnetic field by ion cyclotron waves in the topside ionosphere, producing "pancake" distributions in velocity space. The mirror effect acting on these ions then accelerates them out of the ionosphere where their distribution folds along the field to produce velocity space "conics" above the acceleration region. This process accelerates not only protons but also heavier ions, principally singly charged oxygen, with smaller amounts of singly-charged helium and sometimes singly charged nitrogen and nitric oxide, etc. The energy of these particles is typically of order 10 to 100 eV, corresponding to protons with speeds around 100 km s^{-1} and oxygen ions with speeds of a few tens of km s^{-1}. The dayside cusp ionosphere has been found to be a particularly intense source of these ions (fluxes around 10^9 cm^{-1} s^{-1}), such that this region has been termed the "cusp ion fountain" (typical fluxes at ionospheric heights are about a factor of ten less). Second, it has been found that upward-flowing beams of field-aligned ionospheric ions (protons and oxygen) are formed above discrete auroral regions, possibly by the action of field-aligned electric fields. The ion energies in this case are somewhat higher, of order a few keV, corresponding to protons with speeds of several hundred km s^{-1} and oxygen with speeds of ~ 100 km s^{-1}. Broadly speaking, therefore, in the region of open field lines the ionosphere represents a source of protons (the "polar wind") and O^+ ions (mainly from the cusp ion fountain) which flow out at speeds of a few tens of km s^{-1}. The auroral zone on closed field lines at lower latitudes then represents a spatially structured source of protons and O^+ over a range of energies from a few tens of eV to a few keV, corresponding to outflow speeds of a few hundred km s^{-1} for protons and several tens of km s^{-1} for oxygen. In the next section we consider the trajectories of typical ions in this outflow.

4.2.1. *Ionospheric Ion Trajectories*. Trajectory A in Fig. 8a shows the path of a low-speed (few tens of km s^{-1}) ion flowing out of the open field line region of the ionosphere (corresponding to a "polar wind" proton or a "cusp ion fountain" O^+). The ion flows out into the open tail lobe, but due to its low field-aligned speed it does not reach far down the tail in the time that the field line on which it is moving remains open. Typically these ions will therefore move into the tail current sheet within a few tens of R_E of Earth. On reaching the current sheet the ions will be accelerated by the field tension to speeds of several tens to several hundreds of km s^{-1} as for solar wind ions, forming a component of the plasma sheet.

Trajectory B similarly shows the motion of a low-speed ion which flows out into the quasi-dipolar closed field line region, on the dayside. This ion will remain of low speed, bouncing slowly between mirror points, as it convects with the field lines towards the dayside magnetopause current sheet. When the field line reconnects the behavior of the ion depends on which portion of the field line the particle is located at that time. If the particle is located on the portion of the field line lying between the Earth and that adjacent to the magnetopause it will remain of low energy and flow outward into the tail lobe, subsequently behaving like trajectory A. If, however, it is located in that portion of the field line lying adjacent to the magnetopause, as shown, then it will be accelerated in the current sheet to speeds of a few hundred km s^{-1} (about 1 keV for protons, 10 keV for O^+), subsequently moving either into the magnetosheath or, as shown, down the cusp and into the plasma mantle (depending on the field-aligned ion speed).

Finally, trajectory C shows the motion of a higher-speed (few hundred km s^{-1}) proton moving out into the closed field line region from the nightside auroral zone. Such particles are directly injected into the plasma sheet and subsequently convect back to the Earth with the flow, being further accelerated as they do so. Eventually these particles are lost either by precipitation back into the atmosphere, by charge-exchange with the geocorona, or by outflow at the magnetopause.

4.2.2. *Ionospheric Plasma Populations*. In Figs. 8b and 8c we show the ionospheric plasma populations to which the above behavior gives rise, divided, roughly, into low-speed populations (tens of km s^{-1}), and

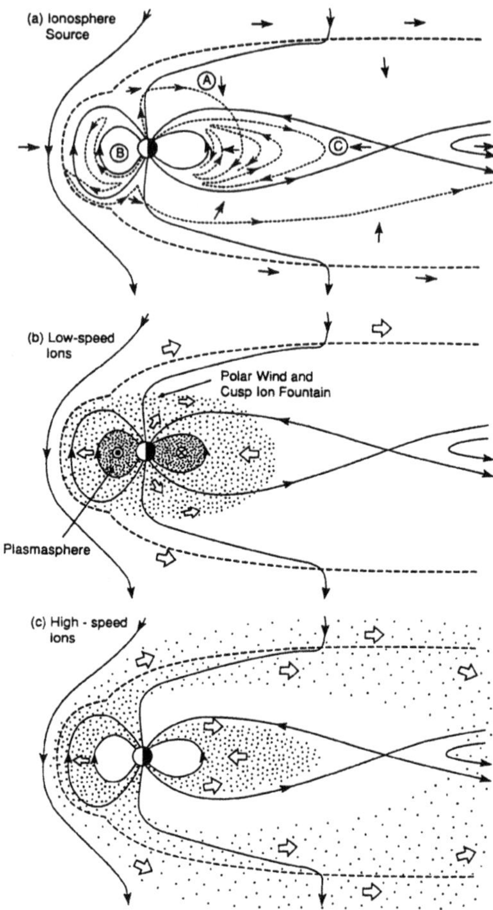

Fig. 8: (a) Trajectories of three typical ionospheric ions which flow out from different regions with different speeds. Ion A is a low-speed ion (few tens of km s^{-1}) which flows out from the open field line region and is then accelerated in the tail. Ion B has a similar initial speed, but flows out of the closed field line region on the dayside and is subsequently accelerated in the dayside magnetopause (and tail current sheet further down the system). Ion C is a high-speed proton (few hundred km s^{-1}) which flows out from the nightside auroral zone. (b) Populations of low-speed (few tens of km s^{-1}) ionospheric ions in the magnetosphere. (c) Populations of high-speed (few hundred km s^{-1}) ionospheric ions in the magnetosphere and environs, formed by direct injection from the auroral zone, and by acceleration in the tail and dayside magnetopause current sheets.

high-speed populations (hundreds of km s^{-1}) respectively, of both protons and O^+. The low-speed plasma will generally be found throughout the near-Earth region, forming field-aligned low-density tailward-flowing streams in the tail lobes extending to a few tens of R_E, and a low-energy component in the plasma sheet and ring current. In the figure we assume for definiteness that the outer limit of these flows lies within the region of closed plasma sheet field lines, Earthward of the tail reconnection region. The high-speed component is formed from ions accelerated near the Earth and directly injected into the plasma sheet and ring current, as well as from initially low-speed plasma which has been accelerated in the tail current sheet and dayside magnetopause. As shown in Fig. 8c these ions will populate the plasma sheet, ring current, boundary layer, dayside cusp and plasma mantle, as well as escaping into the magnetosheath. Overall, it is estimated that the relative contributions of the ionosphere and solar wind to the hot plasma regimes (plasma sheet and ring current) are approximately equal.

Finally, we consider the central "core" of flux tubes which corotates with the Earth. Here the ionosphere again forms a source of low energy (few eV) light ions (mainly protons), and in a steady state this region is filled to equilibrium levels (hundreds to thousands cm^{-3}) with such cold plasma. This region is therefore termed the plasmasphere, as indicated in Fig. 8b. The plasma density falls to much lower values (few cm^{-3}) outside this region, in the ring current and inner plasma sheet, because of the heating and loss of plasma which takes place on these flux tubes during the solar wind-driven convection cycle. Thus as we move across the boundary (the plasmapause) from the plasmasphere into the ring current, the density abruptly decreases while the average energy of the plasma abruptly increases.

4.3. SUMMARY

The overall structure of magnetospheric plasma populations may finally be appreciated by mentally superposing the various components shown in Fig. 7 and 8. The final picture is complex; plasma from solar wind and ionospheric sources form spatially structured convecting populations of protons and oxygen spanning a range of energies from a few eV (in the plasmasphere) to over 100 keV (in the inner ring current). However, from the approach adopted above we are able to see that the overall complexity is built from the superposition of a number of simpler components whose basic properties can be understood in quite simple terms.

5. Magnetospheric Dynamics

Much of the discussion in the above sections was based on the assumption that the convective flow driven by the solar wind in the magnetosphere is of a steady nature, as was necessary for simplicity of argument. In section 3.2 we already pointed out, however, that this is not the case. Flow in the magnetosphere is driven firstly by coupling processes at the magnetopause which transfer flux from the dayside to the tail, and secondly by tail processes which result in its return. These two sets of processes act only in a loosely coupled fashion, and both are time-dependent.

5.1. DAYSIDE RECONNECTION

The rate of dayside reconnection depends principally on the direction of the interplanetary magnetic field (IMF). The most important component in this regard is the north-south (z) component, though the east-west (y) component may also play a weaker role. (Here the x axis points from the Earth to the Sun, the x-z plane contains the Earth's magnetic dipole vector, and y completes the right-handed set.) On average "equatorial" reconnection is strong when the IMF is strong and southward-pointing and weak when it points north, and there is a corresponding modulation of magnetospheric flows as flux is transferred from the dayside to the tail. However, even when the IMF is steady the reconnection at the dayside magnetopause often appears to take place in a series of pulses lasting for ~2 minutes and separated by ~8 minutes. These pulses are known as "flux transfer events", and they must give rise to corresponding pulses in the boundary layers, dayside cusp and mantle, as well as in the corresponding dayside flows. The origin of these reconnection pulses is the subject of active research at the present time.

Two additional effects associated with the orientation of the IMF should be mentioned. The first is that when a y-component of the interplanetary field is present, there is an associated east-west magnetic tension in the open flux tubes which results in an asymmetric addition of the open tubes to the tail lobes. For example, when the y-component is positive, open tubes in the northern hemisphere are pulled towards dawn while open tubes in the southern hemisphere are pulled towards dusk, and vice versa when the y-component is negative. There is a corresponding set of dawn-dusk flow and plasma asymmetries in the dayside cusp, polar cap and mantle plasma, as well as knock-on effects in the region of closed field lines.

The second effect is that although equatorial reconnection may cease when the IMF points north, large magnetic shears will then exist across tail magnetopause, poleward of the cusp, between the draped northward IMF and the previously opened flux tubes of the tail lobes. The ensuing reconnection between these fields does not alter the amount of open flux in the system, but "stirs" the open flux into circulatory motion by transferring flux from one side of the tail lobe to the other.

5.2. NIGHTSIDE RECONNECTION

The processes and conditions which control the rate of nightside reconnection remains one of the great unknowns in magnetosphere physics. Nevertheless, the fact that the amount of open flux in the tail generally varies only between rather narrow limits (between ~4 and ~12×10^8 Wb) indicates that the nightside reconnection rate must follow that at the dayside, at least in an average sense.

Let us then consider what happens when the dayside reconnection rate is augmented following a sustained southward turn of the IMF. New open flux is created at the dayside and transferred to the tail, but the tail reconnection rate clearly cannot respond immediately since the information that the change has occurred is communicated at a finite speed. For example, if the tail reconnection region lies at 100 R_E down-tail, the information about the new open flux will not arrive for at least 30 minutes, this being the time taken by the new open tubes to reach this location at usual solar wind speeds of ~400 km s^{-1}. During this interval, therefore, the amount of open flux in the system will grow, and the tail will expand. The dayside magnetopause is correspondingly "eroded" inwards.

The evidence is, however, that before the pre-existing tail reconnection region can adjust to accommodate to the change in the dayside reconnection rate, the near-Earth tail becomes unstable with the consequences shown in Fig. 9. During the above "growth phase" the plasma sheet is observed to thin near the Earth, and a new x-type reconnection region forms within it, as shown in the upper sketch in the figure. Reconnection in this region is sufficiently rapid that after a few minutes the old plasma sheet becomes disconnected from the Earth, forming a closed-loop "plasmoid" which travels tailwards out to the magnetosphere at speeds of ~500 to 1000 km s^{-1} (lower two panels). Continued rapid near-Earth reconnection over the next ~30 minutes then closes some fraction of the open flux in the lobes, resulting in a reduction in the tail field and an "injection" of plasma and closed flux into the ring current. In the nightside ionosphere the region of bright aurora moves poleward (the "expansion phase") as open flux is closed, with the production of hot plasma. After about 30 minutes the tail reconnection rate generally slackens, and the reconnection region moves out into the distant tail so that initial conditions are restored, this corresponding to the "recovery phase". Overall, the above sequence of events is termed a "magnetospheric substorm", and typically several may occur in the course of one day. A full-blown "magnetic storm" occurs when a sequence of substorms takes place (due to interplanetary conditions) with sufficient rapidity and over a sufficiently long period (several hours) that the magnetosphere is highly disturbed with the ring current built up to very large energies.

6. Magnetosphere-Ionosphere Coupling

In section 4 we considered one aspect of magnetosphere-ionosphere coupling; namely, the fact that the ionosphere forms a spatially structured source of proton and heavy-ion (mainly oxygen) plasma for the magnetosphere, at energies between 1 eV and ~1 keV. In this section we will examine the background to these effects, concentrating on the impact which the magnetosphere has on the ionosphere. This is basically of two kinds. The first is that the solar wind-driven convection is communicated to the ionosphere and drives a twin-vortex pattern of flows at high latitudes (in the auroral zone and polar cap), which is superimposed upon corotation. This flow is important in transporting ionospheric ionization e.g. between the dayside and nightside ionospheres. However, the flow is opposed by ion-neutral collisions in the lower ionosphere, principally at altitudes between ~100 and 200 km, which leads both to the frictional heating of the gas and to the excitation of neutral (thermospheric) winds. The ion-neutral drag is also communicated to the magnetosphere and solar wind

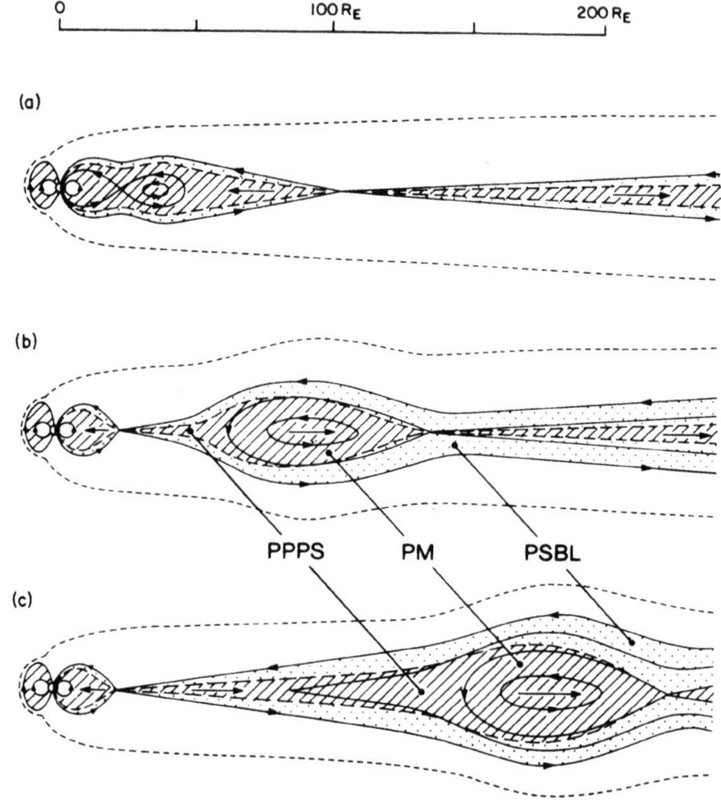

Fig 9: Diagrams showing the instability which occurs in the geomagnetic tail during a magnetospheric substorm. A new x-type field region forms in the near-Earth plasma sheet (upper panel), followed by plasmoid (PM) formation and down-tail propagation (middle and lower panels). Solid lines are magnetic field lines, the outer dashed line the magnetopause, and the hatched region the plasma sheet. Arrows indicate the direction of plasma flow. "PSBL" indicates the plasma sheet boundary layer containing energetic ions and electrons, and "PPPS" the "post-plasmoid" plasma sheet. [From Richardson et al., *J. Geophys. Res.*, **94**, 15189-15520, 1989].

via a large-scale current system. Secondly, as we have already noted, the structured magnetospheric plasma precipitates into the atmosphere, forming another source of energy input, causing the atmosphere to glow (i.e., producing the aurora), creating ionization, and consequently also modifying the ionospheric conductivity. The ionization is mainly produced in the lower ionosphere by precipitating electrons with energies between a few hundred eV and a few keV, at the expense of about 30 eV of energy per ion-electron pair produced. In this section we will discuss these effects in more detail, starting, however, with a brief discussion of the structure of the Earth's atmosphere and quiet-time ionosphere.

6.1. EARTH'S ATMOSPHERE AND IONOSPHERE

In Fig. 10 we show typical altitude profiles of the density of the principal constituents of the neutral atmosphere and of the ionosphere, as well as the variation of the temperatures of these constituents, from the Earth's surface up to 1000 km. At the Earth's surface the atmosphere is composed of ~78% molecular nitrogen (N_2), ~21% molecular oxygen (O_2), together with ~1% of other gases such as the inert noble gases, water vapor, etc. The temperature at the surface is ~300 K. As we move upwards through the atmosphere the temperature lies between ~200 and ~300 K, the variations of which delineate the several regions of the lower, middle, and upper atmosphere (the troposphere between 0 and 15 km (10 km near the poles) which is heated from the ground and where the temperature falls with height, the stratosphere between 15 and 50 km where the temperature rises due to the absorption of solar near-UV radiation, the mesosphere from 50 to 85 km where the temperature again falls, and the thermosphere above 85 km where the temperature again rises). For our purposes, however, we can regard the atmospheric temperature as being roughly constant with height up to ~100 km. It is also found that the atmospheric composition remains essentially unchanged up to an altitude of ~100 km, while the density decreases as exp $(-z/H)$ due to gravity. In this expression $H = kT/mg$ is the scale height, equal approximately to 10 km, such that the atmospheric density falls by a factor of 10 with every ~20 km of height (k is Bolzmann's constant, T the temperature (300 K), m the mean molecular mass (29 amu), and g the acceleration due to gravity).

Above 100 km, however, three related changes take place which are all due to the action and absorption of solar far-UV and X radiation. First the temperature rises from a minimum of ~190 K at the top of the mesosphere at 85 km (the mesopause), to ~1000 K at an altitude of ~300 km, above which the neutral air temperature remains approximately constant. The important further consequence is that such a temperature profile strongly stabilizes the turbulent motions which act to mix the gases, which keep the composition constant at lower altitudes. Above ~100 km mixing becomes unimportant and instead the density of each atmospheric constituent decreases as exp $(-mgz/kT)$ according to its own specific mass m. As a consequence the lighter constituents come to dominate the atmosphere at sufficient altitudes. An additional factor in this transition from mixing to gravitational separation (i.e., from homosphere to heterosphere) comes from the decreasing density and consequent decreasing collision rate of the gas particles with height. As a result, the gas particles will separate

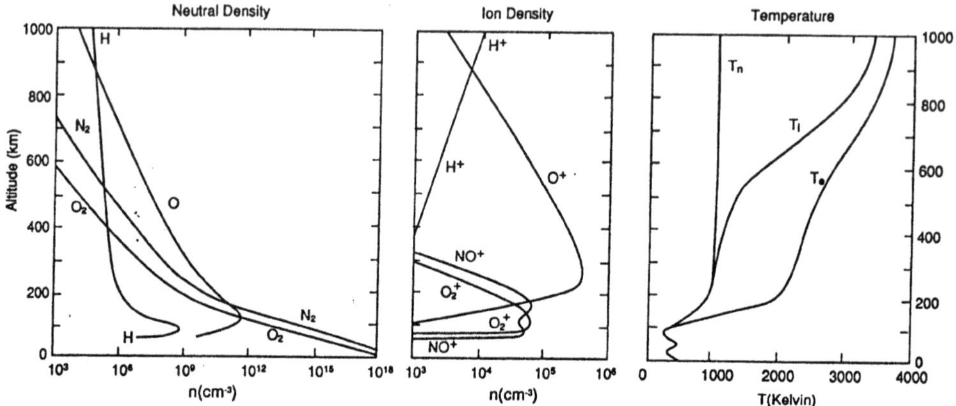

Fig. 10: Typical altitude profiles between 0 and 1000 km of (a) the number density of principal neutral constituents; (b) the number density of principal ionic constituents (note the change of density scale relative to (a)); (c) the temperature of the neutral atmosphere (T_n), ions (T_i), and electrons (T_e).

under gravity with increasing rapidity at increasingly high altitudes, thus requiring increasing levels of turbulence if the constituents are to remain mixed. Instead, the turbulence is suppressed by the temperature gradient in the thermosphere, so that gravitational separation takes place above this region.

The second major effect above ~100 km is that the atmospheric molecular species O_2 and N_2 are photodissociated into atoms. However, atomic nitrogen is rapidly returned to molecular form by reactions with O_2 which in turn produce more atomic oxygen (O). The main effect, therefore, is to convert oxygen from molecular to atomic form (n[O] exceeds n[O_2] above ~120 km), while nitrogen remains predominantly molecular. Since the mass of O is 16 while that of N_2 is 28, gravitational separation then results in the atmosphere becoming dominated by O at altitudes above about 200 km. Eventually the lightest species of all, helium and then hydrogen come to dominate at the highest altitudes, above ~1000 km. Thus it is that while the atmosphere is almost wholly composed of nitrogen and oxygen at the surface, it is instead dominated by atomic hydrogen at the highest altitudes (the hydrogen originates from the photo-dissociation of water vapor and methane lower in the atmosphere). Further, above about 500 km the particle mean free paths become comparable with and exceed the atmospheric scale height, such that above this altitude collisions become unimportant. Instead, the neutral atmospheric particles in this region move on ballistic trajectories in the gravitational field, either escaping from the Earth or eventually falling back, depending upon their velocity (this collision-free region is termed the exosphere). In fact this atomic hydrogen atmosphere (or "geocorona") extends with appreciable densities to many R_E from Earth, throughout the near-Earth magnetosphere.

The third consequence of X-ray absorption in the upper atmosphere is the most important for our purposes, that is that the neutral atmosphere becomes partially ionized, forming the ionosphere. Three principal regions are identified whose ionic species basically reflect the underlying changes in the neutral atmosphere described above. The D region is located at altitudes between 60 and 90 km with ion densities which are too low to show on our diagram ($\sim 10^3$ cm^{-3} by day and ~ 10 cm^{-3} at night). It is characterized by the presence of complex photochemical reactions, NO^+, water cluster ions, and the presence of negative ions rather than electrons. The E region lies between 90 and 150 km where molecular ions predominate, principally O_2^+ and NO^+. Above this in the F-region (150 to a few hundred km) O^+ dominates, with a peak in the density at altitudes between ~250 and ~300 km. Finally at the highest altitudes, above ~1000 km, protons become the dominant ion species (this region is sometimes referred to as the protonosphere). Thus, as pointed out in section 4, the initial expectation is that the ionosphere should form a source of low temperature (~1 eV) hydrogen plasma for the magnetosphere. However, as we also pointed out, acceleration processes in the auroral zone result in the outflow of protons and O^+ with higher energies. It can also be seen by comparing neutral and ion densities in Fig. 10 that the ionosphere is a weakly ionized plasma. In fact the number density of the neutral hydrogen geocorona exceeds that of the magnetospheric plasma out to distances of around 10 R_E, thus encompassing most of the near-Earth magnetosphere.

Finally we should note that the above represents only a simplified initial picture of the ionosphere. In reality the ionosphere also depends on time of day (as does the solar irradiance – the profile shown in the figure is for the dayside, densities decrease by a factor ~10 at night), on the solar cycle (since the solar irradiance at relevant wavelengths is greater at solar maximum than at solar minimum – the profile shown is for solar minimum, densities increase by a factor ~10 at solar maximum), and on latitude, in particular due to the high-latitude convection and precipitation effects which we mentioned briefly above, and which we will now go on to discuss in more detail.

6.2. IONOSPHERIC CONVECTION, PRECIPITATION AND CURRENTS

6.2.1. *Ionospheric Convection.* In Fig. 11a we show a view of the high-latitude ionosphere looking down upon the north pole, with noon (12 MLT) at the top, dusk (18 MLT) to the left, midnight (24 MLT) at the bottom and dawn (06 MLT) to the right. The dashed line in this figure is the boundary of open field lines, with closed field lines lying at lower latitudes (see Fig. 5). The solid lines with arrows then indicate the ExB flows of the ionospheric plasma (corotation subtracted), which to a first approximation consists of an antisunward flow in the region of open field lines and a return sunward flow on closed field lines at lower latitudes, thus forming a twin-vortical convection cycle consistent with the

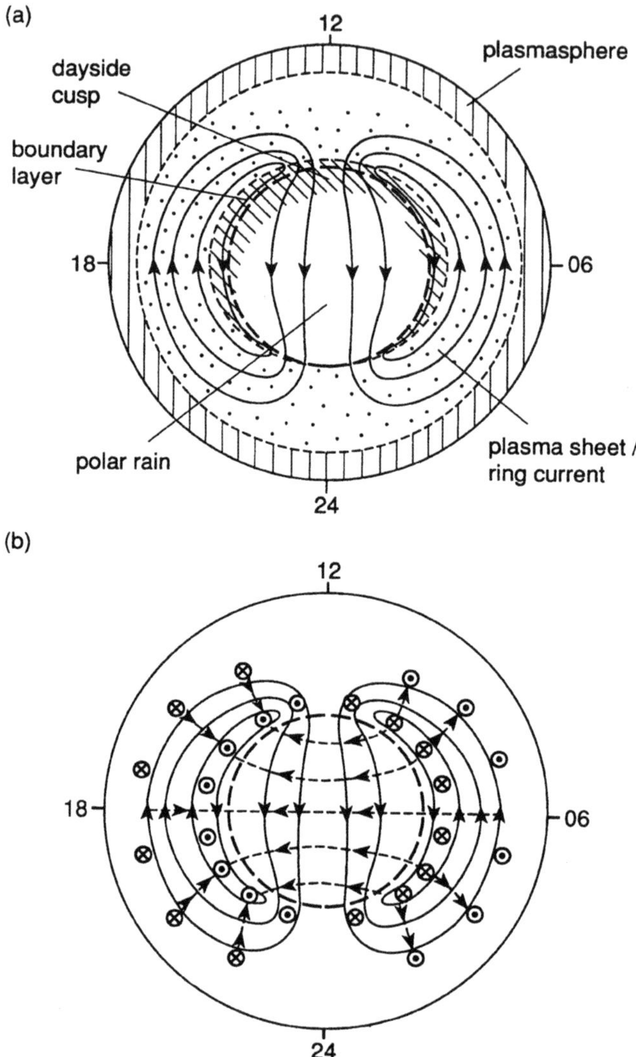

Fig. 11: View looking down on the northern high-latitude ionosphere showing the plasma **ExB** drift paths (arrowed solid lines) and (a) the main magnetospheric plasma precipitation zones, (b) the main ionospheric current systems. The interior heavy dashed circle in each sketch shows the boundary between open field lines at high latitudes and closed field lines at lower latitudes. In (b) the arrowed short-dashed lines show the ionospheric Pedersen current flow along **E**, while the circled dots and crosses show the upward and downward-directed closure field-aligned currents respectively. The ionospheric Hall current flows in the direction opposite to the **ExB** drift paths [Ref. 1].

discussion in section 3.2. However, it will be noted that the boundary between sunward and antisunward flows near dawn and dusk does not quite coincide with the boundary of open and closed flux. This is a consequence of the "viscous" coupling near the boundary, mentioned above, which results in closed field lines adjacent to the equatorial magnetopause being convected antisunwards. The inner light dashed line marks the inner (lower latitude) edge of this boundary layer. The outer light-dashed line marks the boundary between the region dominated by solar wind-driven convection (the high-latitude ionosphere) and that dominated by corotation (the mid-latitude ionosphere) i.e. it corresponds to the plasmapause.

Under usual conditions the boundary of open field lines lies at a magnetic latitude of ~75° during the day and ~70° at night, while the plasmapause lies at a latitude of ~60°. The voltage associated with the flow (i.e., the voltage across the entire region of antisunward flows at high latitudes) is typically ~50 kV. However, the ionospheric flow varies as the corresponding magnetospheric flow varies (with the IMF and with the substorm cycle), and the open/closed field line boundary also moves in latitude as the amount of open flux in the system changes (according to the difference in dayside and nightside reconnection rates). In addition, the flow pattern also exhibits dawn-dusk asymmetries which are associated with the y-component of the IMF, while additional circulatory cells may appear in the open flux region when the IMF is northward due to reconnection between the IFM and the tail lobe (see section 5.1). Like the flow in the magnetosphere, therefore, the ionospheric flow is also a very variable phenomenon.

6.2.2. *Magnetospheric Precipitation.* Fig. 11a also shows the correspondence between the twin-cell flow system and the patterns of particle precipitation. Four principal regions are identified. First, the region at lowest latitudes indicated by the vertical hatching maps along the magnetospheric field lines to the plasmasphere. Consequently there is little energetic particle precipitation into the ionosphere in this region under usual conditions. Second, at higher latitudes the dotted area in the region of sunward flows maps to the ring current and plasma sheet. Magnetospheric particles precipitate into this region with energies ~1 to 100 keV, producing the "diffuse aurora" and augmenting the ionization at E-region heights, particularly on the nightside. Third, the diagonal hatching in the region poleward of the flow reversal maps to the magnetopause boundary layers and dayside cusp where magnetosheath plasma (few 100 eV) precipitates into the ionosphere. This region, together with that mapping to the plasma sheet/ring current, forms the circumpolar "auroral zone". Fourth, the unshaded region at the highest latitudes maps magnetically to the tail lobe and plasma mantle. Here the precipitation from the cusp dies out into a weak structureless drizzle of magnetosheath particles, termed the polar rain. These particles produce little auroral luminosity, so that the central region of open field lines is optically dark. This region is often referred to as the "polar cap". We should point out, however, that this diagram does not indicate the precipitation which gives rise to the structured curtain-like "discrete" auroral forms which often stretch for several thousand km around the poleward portion of the auroral zone. These aurorae are produced by spatially structured few-keV electron precipitation, where it is believed that the electrons are accelerated out of the lower-energy magnetospheric population by field-aligned electric fields operating at altitudes of a few thousand km. Further discussion will be given below, after we have considered the magnetosphere-ionosphere current system.

6.2.3. *Physical Origins of Ionospheric Currents.* The relative drift between ions and electrons which gives rise to electric currents in the ionosphere are due to the collisions which occur between ions and neutral atmospheric particles in the E-region, principally between altitudes of ~100 and ~150 km. These currents are therefore intimately associated with ionosphere-atmosphere frictional drag in a manner which will be elucidated below. The motion of charged particles in the presence of collisions depends upon the collision frequency (the reciprocal of the mean free time between collisions) compared with the gyrofrequency qB/m. When the collision frequency is small compared with the gyrofrequency (i.e. when the particle executes many gyrations between collisions), the motion simply reduces to the **E**x**B** drift transverse to **E** and **B** which we have discussed throughout this article. This behavior applies to electrons above an altitude of ~80 km, so that the electron flow is simply **E**x**B** drift throughout the whole of the important part of the ionosphere (upper D region, E-region and F-region). This condition also applies to the ions in the F-region and above, and since the ion number density is equal to the electron number density throughout, there is no net current in these regions. Below ~150 km, however, the ion motion is effected by collisions. If we imagine an ion starting from rest after suffering a collision, it will first accelerate in the direction of the electric field associated with the flow before turning in the magnetic field and starting to **E**x**B** drift. The mean position of the ion in the subsequent motion is therefore displaced in the direction of **E** relative to its starting position, and this happens every time a collision takes place. As a consequence the ions now have mobility in the direction of the electric field; this corresponds to the take-up of energy by the ions from the electric field and its conversion into atmospheric heating via the collisions. There is a corresponding current flow along **E** termed the Pedersen or "direct" current, given by $\mathbf{j}_P = \sigma_P \mathbf{E}$, where σ_P is the Pedersen conductivity (remember that the energy input from the electromagnetic field to the plasma is $\mathbf{j}.\mathbf{E}$ watts m^{-3}). In addition, however, the collisions also disrupt the **E**x**B** drift of the ions, so that the ion mobility in this direction is reduced. Consequently, since the electrons continue to flow with the full **E**x**B** drift velocity, there will exist a net current in the direction opposite to **E**x**B**. This is termed the Hall current, given by $\mathbf{j}_H = \sigma_H \mathbf{b}\mathrm{x}\mathbf{E}$, where σ_H is the Hall conductivity and **b** is the unit vector along the magnetic field. There is no dissipation of electromagnetic energy associated with the Hall currents.

The Pedersen conductivity peaks in the E-region at ~130 km where the ion collision frequency becomes equal to the gyrofrequency. Under this condition the ions typically only complete part of a gyration between collisions, so that the mobility along **E** is very large; the average drift in this direction peaking at half the **E**x**B** drift speed. At that point the ion drift in the **E**x**B** direction is also reduced by a half. However, the Hall current peaks below this altitude, at about ~110 km, where the ion collision frequency is large compared with the gyrofrequency, so that the ions are collisionally tied to the neutral atmospheric particles and hardly drift at all in any direction (in the neutral air rest frame). The Hall conductivity is reduced below this altitude by the declining ion and electron number densities in the lower E-region.

For many purposes, however, a detailed knowledge of the altitude profile of the current is not required, rather it is important to know only the total current integrated with height through the ionosphere. The height-integrated current components are then $J_P = \Sigma_P E$ and $J_H = \Sigma_H bxE$, where Σ_P and Σ_H are the height-integrated Pedersen and Hall conductivities, respectively. These are variable quantities, depending on the ionization produced by solar illumination and electron precipitation, but are usually of order 10 mho. Given typical high-latitude electric fields of ~50 mV m^{-1} associated with flows of order 1 km s^{-1}, we find typical height-integrated currents of ~0.5 A m^{-1}. Integrating over the total area of the polar cap, for example, then gives a total current flow of order a few MA. A simple calculation of the expected magnetic effect on the ground, using Ampere's law, then indicates that the perturbation field should be a few hundred nT, about 1% of the Earth's main field at the surface, and easily measurable. However, as we will indicate below, the majority of the magnetic effect seen on the ground is due to the Hall currents; the Pedersen currents, while of a similar magnitude in the ionosphere, produce only a small magnetic effect on the ground.

6.2.4. *Ionospheric Current Circuits.* In Fig. 11b we show the overall pattern of ionospheric currents in a view looking down on the northern ionosphere, in the same format as Fig 11a. The heavy dashed line again indicates the boundary between open and closed flux tubes and the solid arrowed lines the **ExB** drift streamlines. Since the Hall currents flow in the direction opposite to the **ExB** drift, the latter solid lines also indicate the pattern of Hall current flow, in the direction opposite to the arrows shown. In principle these currents could therefore close wholly in the ionosophere by flowing round the **ExB** streamlines. However, this will only be the case if the ionospheric conductivity is uniform. As pointed out above, however, the E-region ionization and hence conductivity is enhanced by electron precipitation in the auroral zone compared with the polar cap. Consequently, the integrated auroral zone Hall currents will be larger than those in the polar cap, such that continuity requires current closure in the magnetosphere via currents flowing along the magnetic field lines. The field-aligned current (FAC) must be directed out of the ionosphere in the poleward part of the nightside auroral zone, and into the ionosphere in the poleward part of the dayside auroral zone (not indicated in the diagram). During substorms the nightside east-west auroral zone flows (and the associated north-south electric fields) are particularly intense, and the conductivities are also augmented by intense precipitation from the newly-formed hot plasma sheet population in the tail. The nightside auroral zone Hall current flow is thus particularly intense under these conditions, forming the eastward and westward "electrojet" currents in the pre- and post-midnight sectors respectively. The magnitude of these currents is ~1 MA over a latitudinal strip a few degrees wide, leading to characteristic magnetic effects (called "magnetic bays") on the ground underneath of magnitude several hundred nT.

The arrowed short dashed lines in Fig. 11b then show the electric field lines associated with the flow, and hence also the pattern of Pedersen currents. These currents flow from dawn to dusk across the polar cap, reversing in sense in the auroral zone. Consequently it can be seen that in principle these currents cannot close in the ionosphere, but instead must close in the magnetosphere via a system of field-aligned currents. The Pedersen currents converge on the dusk flow reversal between the auroral zone and polar cap, where the FAC must consequently flow out of the ionosphere as indicted by the circled dots in the figure. Similarly the Pedersen currents diverge away from the flow reversal region on the dawn side, where FAC must consequently flow into the ionosphere as shown by the circled crosses. These FAC, located at and equatorward of the boundary between open and closed field lines, are termed the "Region 1" currents. Another set of FAC must similarly lie at the equatorward boundary of the auroral zone as shown, which flow in at dusk and out at dawn. These currents are called the "Region 2" FAC. Together, the extended sheets of field-aligned current formed by the Region 1 and Region 2 systems produce magnetic perturbations in the region between them which are oppositely directed to the plasma flow (i.e., sunward over the polar cap and antisunward in the auroral zone). The Pedersen currents flowing in the ionosphere then act simply to "switch off" these magnetic perturbations in the region under the ionosphere; in the limit of a uniformly conducting ionosphere the magnetic perturbation field under the conducting layer is identically zero.

The physics behind the closure of these currents in the magnetosphere can be thought of in terms of simple circuit theory, in which a generator (the flowing solar wind or magnetospheric plasma) is connected across a load (caused by frictional drag in the ionosphere) by currents flowing along highly conducting wires (the field-aligned currents). The simple circuit associated with the polar cap currents is shown in Fig. 12a, which is a view in the dawn-dusk meridian plane looking towards the Sun, so that the plasma flow is directed out of the plane of the diagram (circled dots). The arrowed solid lines show magnetic field lines, while the current circuit is shown by the arrowed long-dashed line. The Pedersen current flows from dawn to dusk (left) in the polar cap ionosphere, up the dusk Region 1 field-aligned current to the magnetopause, from dusk to dawn (right) in the magnetopause, and back down to the ionosphere in the dawn Region 1 field-aligned curet. The magnetosheath plasma acts as the generator (**j**.**E** is negative in the magnetopause), the ionosphere as the load (**j**.**E** is positive in the Pedersen current), and electromagnetic energy flows from the former region to the latter in the space between them as a Poynting flux (short-dashed lines marked **S**) which is associated with the electric field of the flow and the magnetic perturbation of the overall current system. As indicated above, the current loop creates a sunward-directed perturbation field over the polar cap, and this, combined with a dawn-to-dusk convection **E**, produces a downward-directed component of **S**. This is shown explicitly in Fig 12b, which shows a "side" view of the system in the noon-midnight meridian plane. The perturbation field causes the field lines to tilt over in the space between the magnetopause

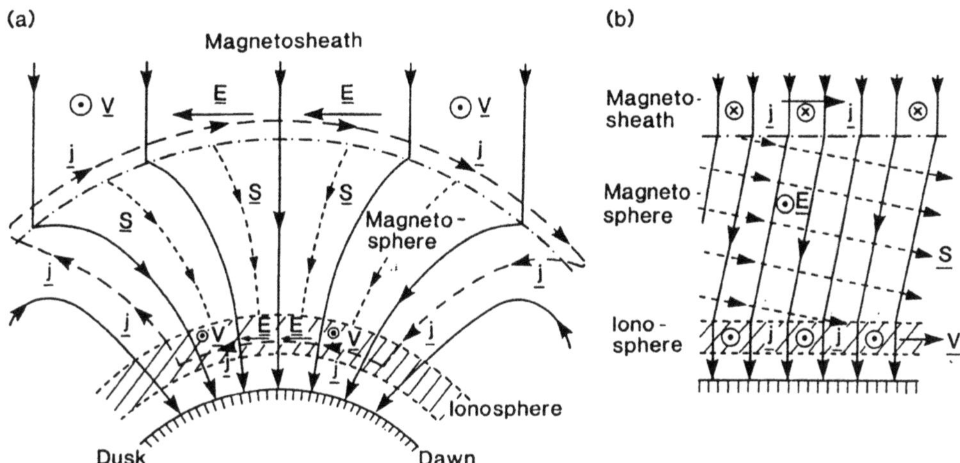

Fig. 12: Sketch of the "Region 1" current system which is associated with the transmission of solar wind energy and momentum into the ionosphere. Sketch (a) shows a view looking towards the Sun with the flow in the magnetosheath and ionosphere pointing out of the plane of the diagram, while sketch (b) shows a side view looking from dusk to dawn, with the plasma moving from left to right. In each diagram the solid lines show the magnetic field lines, the arrowed short-dashed lines the Poynting flux (S), while in (a) the current system is shown by the arrowed long-dashed line, and in (b) by the circled symbols [Ref. 2].

and the ionosphere as shown. The Poynting vector **S** then has a downward component into the ionosphere, equal in magnitude to the ionospheric Joule heating rate per unit area $J_P.E = \Sigma_P E^2$. In force terms it is clear that the field tension acts to slow the magnetosheath and to maintain the ionospheric flow against frictional atmospheric drag. In between, the field lines tilt over as a result of this drag while being "pulled" by the magnetosheath flow at the other end.

Another similar, but more complicated, current system is associated with the closure of the auroral zone Pedersen currents. In this case the currents close via the Region 1 FAC and the magnetopause on the poleward side, and the Region 2 FAC and the ring current plasma on the equatorward side.

6.2.5. *Discrete Aurorae*. The discrete auroral forms mentioned above are associated with spatially structured regions of upward-directed FAC, and are consequently particularly prominent in the dusk flow reversal region where the Region 1 current flows upward out of the ionosphere. Downward currents are readily carried by cold electrons flowing out of the ionosphere, but cold ionospheric ions can only carry a small upward-directed field-aligned current. Consequently, upward field-aligned current must generally be carried by precipitating hot magnetospheric electrons, and a substantial voltage (few keV) must generally exist along the field lines to accomplish this. The downward-accelerated electrons create the discrete aurora in the E-region, while ionospheric ions are also accelerated upwards to create a source for the magnetospheric plasma at few keV energies, as previously mentioned in section 4.2. The field-aligned voltages can be produced by perturbing the equipotentials of the electric field associated with the flow so that they no longer lie parallel to field lines e.g. by closing off above the ionosphere the equipotentials across the dusk flow reversal between the auroral zone and polar cap. In this case part of the magnetosheath voltage falls along the FAC rather than across the ionosphere, and correspondingly, part of the downward-directed Poynting flux is absorbed in the region of field-aligned voltages rather than in the ionospheric Pedersen current layer. The energy is transformed into the accelerate motions of the auroral electrons and the outflowing ionospheric ions.

Further Reading

[1] S.W.H. Cowley, The plasma environment of the Earth, Contemp. Phys., 32, 235-250 (1991b).

[2] S.W. H. Cowley, Acceleration and heating of space plasmas: basic concepts, Ann. Geophys., **9**, 176-187 (1991a).

[3] J.W. Dungey, Interplanetary field and the auroral zones, Phys. Rev. Lett., **40**, 47-48 (1961).

[4] S.I. Akasofu and Y. Kamide (editors), "The Solar Wind and the Earth", Terra Sci. Publ. Co., Tokyo (1987).

[5] G. Beynon, R. Boyd, S.W.H. Cowley, and M.J. Rycroft (editors), "The Magnetosphere, the High-Latitude Ionosphere, and their Interactions", Roy. Soc. Publ., London (1989).

[6] S.W.H. Cowley, Plasma populations in a simple open model magnetosphere, Space Sci. Rev., **26**, 217-275 (1980).

[7] J.K. Hargreaves, "The Upper Atmosphere and Solar-Terrestrial Relations", Van Nostrand Reinhold Co., London (1979).

[8] E.W. Hones, Jr (editor), "Magnetic Reconnection in Space and Laboratory Plasmas", Geophys. Mono. 30, AGU Publ., Washington DC, USA (1984).

[9] M. Lockwood, S.W.H. Cowley, and M.P. Freeman, The excitation of plasma convection in the high-latitude ionosphere, J. Geophys. Res., **95**, 7961-7972 (1990).

[10] G.K. Parks, "Physics of Space Plasmas", Addision-Wesley Publ. Co., Redwood City, CA, USA (1991).

[11] E.R. Priest (editor), "Solar System Magnetic Fields", D. Reidel Publ. Co. (1985).

COSMIC RADIATION

TITUS MATHEWS
Department of Physics
The University of Calgary
Calgary, Alberta
Canada T2N 1N4

ABSTRACT. A brief description of primary cosmic radiation, its charge composition and energy spectra is given followed by an account of secondary radiation in the atmosphere. Geomagnetic effects, cut-off phenomena, latitude and east west effects and asymptotic directions of particles are explained. 11-year solar cycle variation and Forbush decreases are illustrated with Calgary neutron monitor. Solar flare ground level events are discussed using the recent series of events. Finally, a very brief account of theoretical explanations of cosmic ray modulations in the heliosphere are given.

1. Introduction

Since its discovery by Hess in 1912, cosmic radiation has been a subject of intense investigation. As a result we know a great deal about its nature, composition and variations in intensity with time. The name remains an appropriate one, for its origin cannot be attributed to a single source or even several sources with certainty. The sun produces particles of relativistic energies sporadically, but its contribution to cosmic radiation of galactic origin is not significant. A good part of our knowledge has come from ground based, balloon and rocket borne instruments before the sputnik era. The advent of earth satellites and space probes has enabled us to study the radiation outside the earth's atmosphere – the primary cosmic rays, especially those of lower energies.

Several aspects of cosmic radiation continues to be studied vigorously of which the high energy astrophysics perhaps receives the greatest attention now. The discovery of cosmic x-rays and gamma-rays during the mid-sixties added a new dimension to the study of origins of cosmic rays. The discovery of solar wind and interplanetary magnetic field has contributed much to the understanding of solar modulation of galactic cosmic rays reaching the earth. The sporadic release of relativistic particles by the sun has called attention to study the acceleration of particles to cosmic ray energies and their propagation in the inner solar heliosphere. Extended voyages of humans to space calls for renewed investigation of biophysical effects of cosmic radiation. Cosmic rays were used as a source of energetic particles to study high energy nuclear

R. N. DeWitt et al. (eds.), The Behavior of Systems in the Space Environment, 183–204.

interactions and many elementary particles were first discovered as part of cosmic radiation in the atmosphere. However this field of study must now be considered as truly in the realm of accelerators, though particles of energy much higher than a TeV will be available only in cosmic rays for the foreseeable future.

In this lecture, I will attempt to give an account of the general features of cosmic radiation at the earth, its charge composition and energy spectrum outside the atmosphere, its transformation as it passes through the atmosphere, the effect of geomagnetic field on cosmic radiation and variations of its intensity with time. Cosmic ray intensity at the earth is strongly influenced by solar activity. Solar influence extends to a large region surrounding the sun. Figure 1 shows a conceptual model of the heliosphere through which galactic cosmic rays must pass before reaching the earth. This model draws heavily from our understanding of the magnetosphere which, to some extent at least, have been subjected to verification by in-situ measurements.

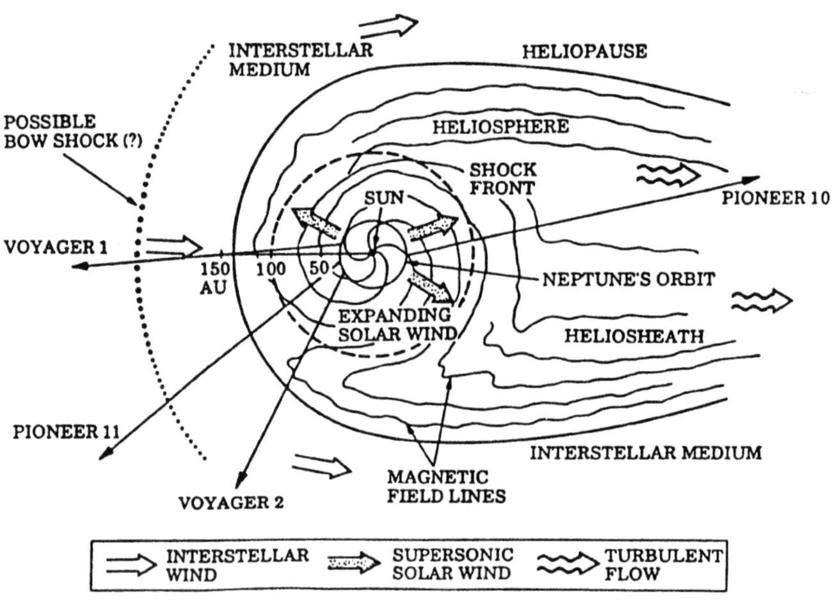

Fig. 1: Conceptual model of the heliosphere.

2. Primary Radiation

The radiation arriving at the top of the atmosphere consists mostly of relativistic protons (85%) and helium nuclei (14%), nuclei of heavier elements accounting for about 1%. Energy of these nuclei range from a few tens of MeV to over 10^{20} eV. Figure 2 gives the energy spectra of protons, alpha particles and heavier nuclei. At energies above 10 GeV, the differential spectra may be expressed as a power law with an exponent of -2.6. At lower energies the spectra are strongly influenced by solar modulation and hence both the flux and shape of the spectra are dependent on the 11-year cycle of solar activity. Examination of the rigidity spectra of protons and helium nuclei given in Fig. 2 shows significant differences. Detailed study of these differences can be used to understand the diffusive transport of galactic cosmic rays in the heliosphere. For a detailed account of primary cosmic rays [see Ref. 1].

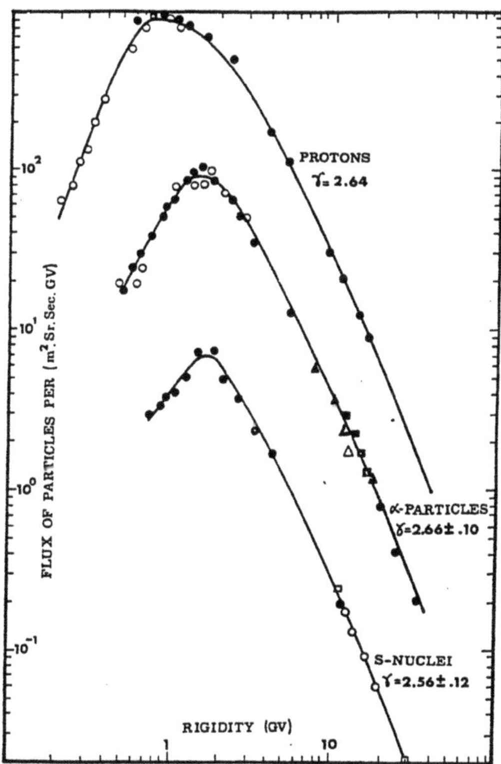

Fig. 2: Energy spectra of primary Protons, Helium nuclei and nuclei of heavier elements at solar activity minimum.

The bulk of the cosmic rays reaching the earth is galactic in origin. At higher energies (>10^{14} eV) the spectra steepens somewhat with spectral exponent decreasing from -2.6 to about -3.2. This steepening is attributed to the leakage of cosmic rays from the galactic plane. At energies >10^{16} a significant fraction of the cosmic rays is thought to be of extra-galactic origin.

Detailed examination of the charge composition of the primary nuclei show that heavier nuclei are relatively more abundant in cosmic rays than in the universe, with the exception of Li, Be, and B which are found to be about five times more abundant. This is an argument used in favor of considering explosive stars, supernovae, etc., which are rich in heavy elements, as the source of cosmic rays. As they wander through the interstellar space, they undergo nuclear interactions and re-accelerations, and both nuclear composition and energy spectra undergo changes. Over abundance of Li, Be, and B results from the interstellar nuclear reactions. By careful study of the isotropic composition of primary nuclei, it has been estimated that the radiation arriving at the earth has travelled through 2-6 g of material. Typical residence time of cosmic ray nuclei in the galaxy is about 200 million years, which is very short compared to the age of the galaxy. If the intensity of galactic cosmic radiation remained about the same as now, it must be generated in the relatively frequent occurrences of supernova events.

Relativistic electrons form only a negligible (1%) fraction of cosmic radiation. Study of this small electron component is very important because of its negative charge and its charge to mass ratio. The negligible presence of electrons in galactic cosmic rays must be attributed to energy loss of electrons in interstellar magnetic fields. Charge measurements of galactic electrons show that only 10% is positive and hence only a small fraction of the electron component originates in the interstellar medium.

High energy photons -- x-rays and gamma rays -- were the subjects of considerable attention during the last two decades. They also cannot be considered as a major component of cosmic rays, for the total energy received at the earth through them is not significant in comparison with cosmic ray nuclei. However, they are important for the information that can be obtained about possible sources of cosmic rays. They are not affected by interplanetary or interstellar fields. Point sources and an approximately isotropic background radiation has been observed in both x-rays and gamma-rays.

3. Cosmic Rays in the Atmosphere

As it passes through the atmosphere, primary radiation undergoes numerous nuclear interactions and what reaches ground level is mostly the secondary radiation and their decay products. Figure 3 illustrates the cascade of nuclear interactions which takes place in the atmosphere. The number of primary nuclei surviving decreases exponentially with atmospheric depth. The mean free path for collision of nucleons is about 80 g/cm^2. The secondary particles undergo both nuclear interaction and decay. These competing processes create a multiplication effect initially until the average energy falls below what is required for multiplication. Thereafter the intensity of radiation decrease gradually. The counting rate of a detector as it goes up in the atmosphere will increase gradually until it reaches a maximum, then decrease until it becomes a steady value corresponding to the primary radiation. The intensity maximum is about 15,000 to 20,000 m above sea level.

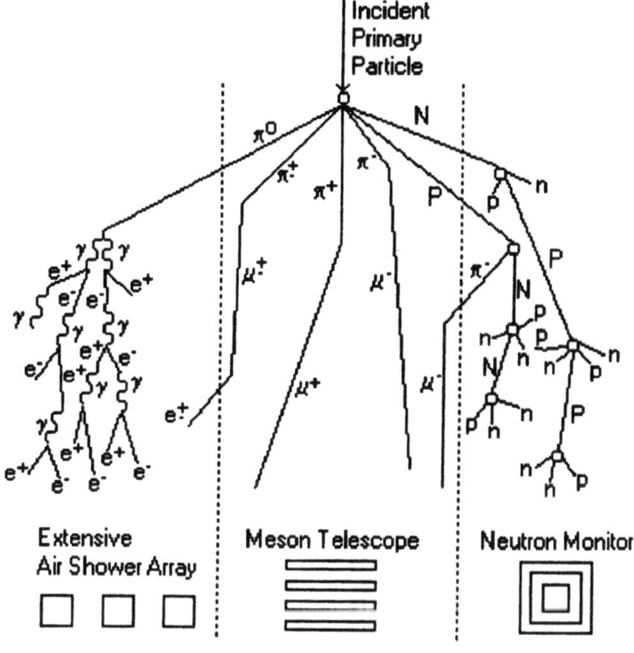

Fig. 3: Cascade of nuclear interactions of cosmic rays in the atmosphere.

The secondary radiation deep in the atmosphere consists of three components (i) nucleons -- both neutrons and protons, (ii) electrons -- both negative and positive, and (iii) muons -- the energetic decay products of pions and kaons produced in nuclear interactions. For continuous monitoring purposes, the nucleonic components are detected with neutron monitors and muon components with ionization chambers or counter telescopes. Muons can be detected deep underground.

4. Geomagnetic Effects

That the primary cosmic radiation is made of charged particles was first inferred from the effect of the geomagnetic field on the intensity of cosmic rays. Two aspects of the geomagnetic effects will be discussed here. The first one is the cut-off phenomena and the other the deflection of particles in the magnetosphere as they traverse the geomagnetic field. The motion of energetic charged particles in a magnetic field is conveniently described in terms of its rigidity ($P = pc/Ze$, p = momentum, c = speed of light in vacuum and Ze = charge), the unit of its measurement being the volt.

The geomagnetic field, in a first approximation, is that of a tilted, off-centric dipole. Motion of charged particles in such a field is complicated. Since the advent of high speed computers, it has become possible to trace the trajectory of particles in such a field and display them on video screens. However, much that we know about geomagnetic effects were first learned from analytical study of the problem by Stormer and by Vallarta and Lemaitre nearly sixty years ago.

(a) Geomagnetic cut-off

The analytical studies have shown that the earth is accessible only to particles of rigidity above a minimum, the cut-off rigidity. It depends on the geomagnetic latitude as well as the direction of arrival of particles at the point of observation. At a given latitude and direction of arrival, the hemisphere above can be divided into a region from which no particle of rigidity below a certain value can come and another region from which all particles above a certain rigidity is permitted. The former is called the Stormer cone and the latter the main cone. The region between the two is the penumbra, in which some particles are allowed and some not.

The cut-off rigidity in the vertical direction corresponding to the Stormer cone and the main cone are shown in Fig. 4. At the equator and at high latitudes the two cones give the same values, but at intermediate latitudes they differ significantly. The fraction of the particles allowed in the penumbral regions varies with latitude in a complicated way.

Figure 5 gives lines of equal cut-offs over the globe. The eccentricity of the geomagnetic field produces a variation of cut-off rigidity along the equator, it being about 13 GV over South America and 17.5 GV over Indonesia. The effects of geomagnetic anomalies as well as the consequence of the tilt of the geomagnetic dipole is illustrated in this diagram.

The intensity of radiation arriving at the earth depends on geomagnetic latitude, it being least at the equator and greatest at the poles. The directional effect is strongest at the equator; the cut-off rigidity ranging from 10 GV for arrival from the western horizon to 60 GV from the eastern horizon, the vertical cut-off rigidity being about 15 GV.

Figure 6 illustrates the east-west effect measured with a muon telescope at the equator for three different zenith angles. The observed effect is greatest at the largest zenith angle, as is to be expected; but it is less than expected from the primary spectra (20% versus 83%). This is a consequence of the generation of secondary radiation in the atmosphere; the specific yield, i.e., the number of secondary particles produced by a primary nucleon depends on energy and atmospheric depth. The product of specific yield and primary spectra is called the response function of an instrument and it is dependent on the atmospheric depth.

Figure 7 illustrates the latitude effect of the nucleonic component at sea level for two levels of solar activity. The intensity of the nucleonic component at the equator is 55-60% of that at high latitude while the latitude effect is only 8% in muon component. This illustrates the

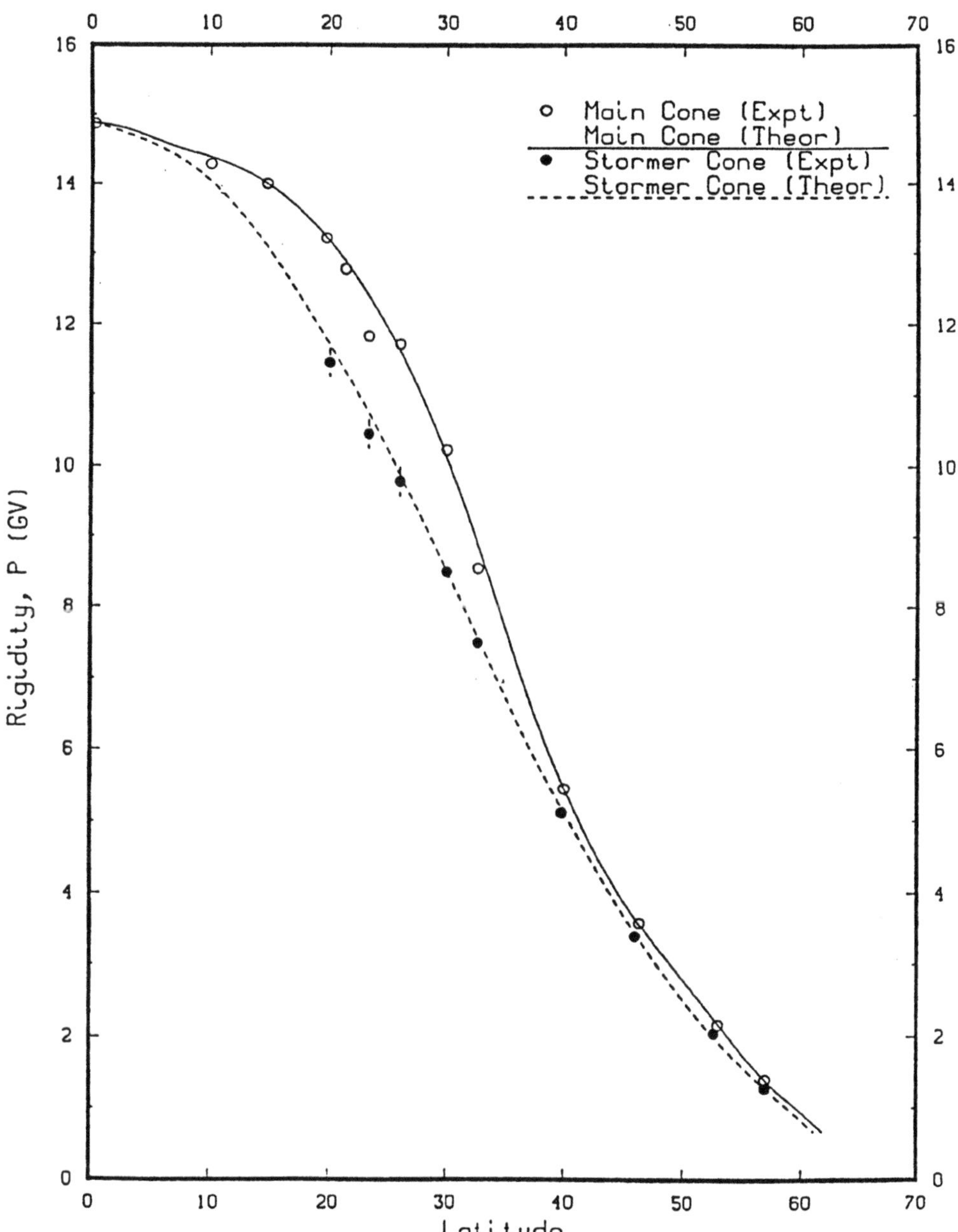

Fig. 4: Vertical cut-off rigidity as per Stormer cone and main cone.

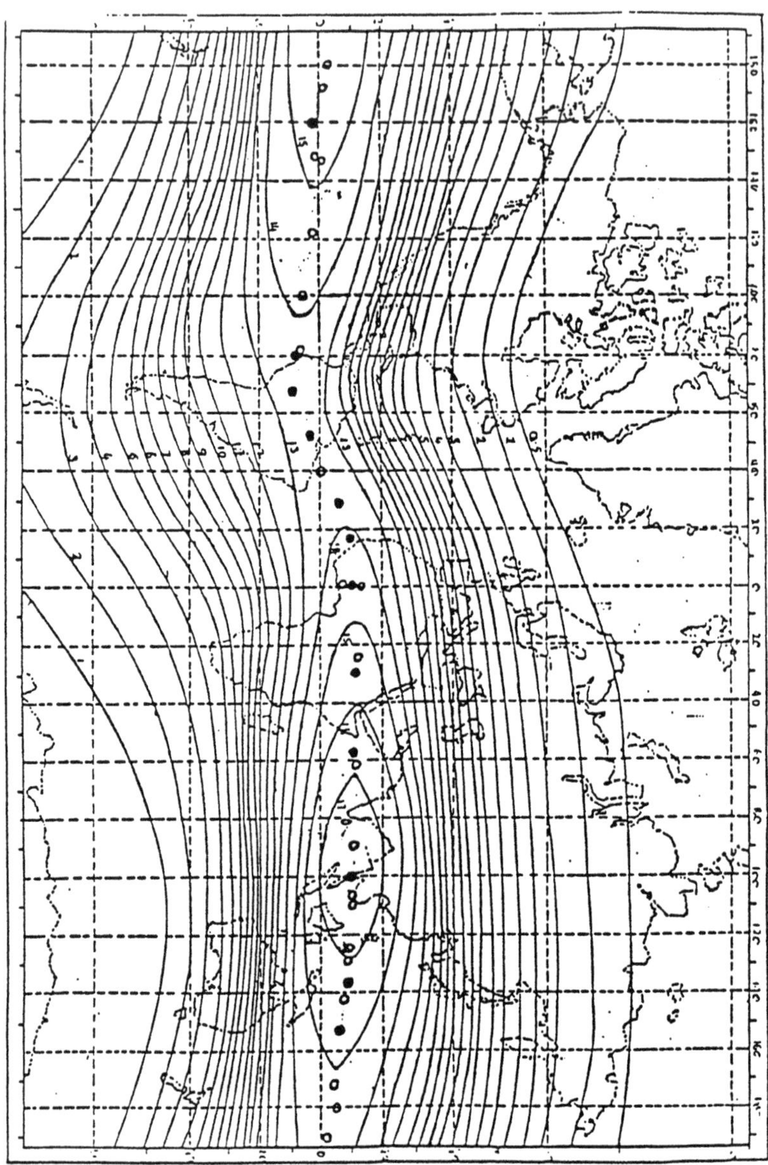

Fig. 5: Lines of equal vertical cut-offs.

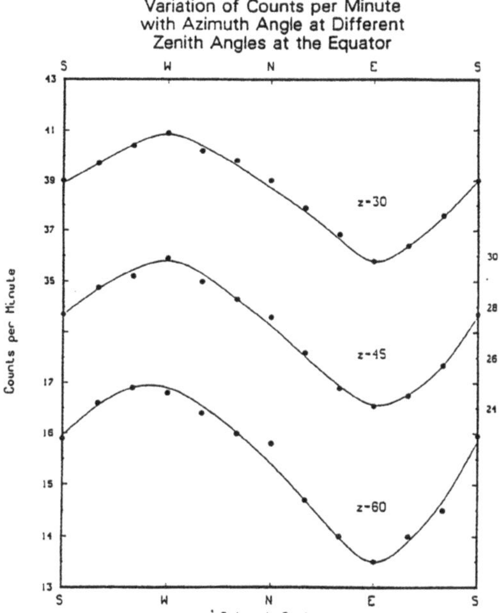

Fig. 6: East-West effect measured at equator for three different zenith angles.

difference in the response functions of the different instruments. It may also be noticed that as one approaches the poles, the intensity does not increase continuously. This is because the primary radiation must be of a certain minimum energy to produce secondary particles at sea level. This minimum is known as the atmospheric cut-off. The atmospheric cut-off of a neutron monitor is about 1 GV where as that of a muon telescope oriented vertically is about 0.45 GeV. The atmospheric cut-off of an instrument depends on its altitude and unlike the geomagnetic cut-off, it is not sharp and well defined.

(b) Asymptotic directions of approach

The term asymptotic directions is used to indicate the direction in which a particle is travelling before it comes under the influence of the geomagnetic field. It depends on the rigidity of the particle, the geomagnetic coordinates of the point of arrival on the earth and the direction of arrival at that point. Particles of very high rigidity compared to the cut-off rigidity are deflected little, whereas those near the cut-off rigidity are deflected a great deal.

One consequence of this is that particles of different rigidities arriving at a particular station originate in different parts of the sky. The part of the sky from which a station records most of the cosmic radiation is called the asymptotic cone of acceptance and is dependent on

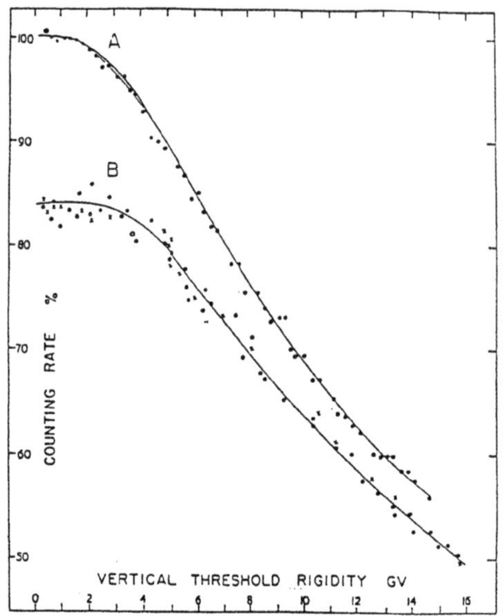

Fig. 7: Latitude effect for nucleonic component at two levels of solar activity.

the response functions of the instrument. The concept of asymptotic cones is essential to understand the anisotropic features of intensity changes observed during Forbush decreases and solar flare ground level events. Figure 8 shows the asymptotic cones of acceptance for typical solar flare generated radiation.

5. Cosmic Ray Intensity Modulations

Continuous recording of the intensity of secondary radiation, both its nucleonic and muonic component has been carried out for several decades at a number of stations and is continuing even today. Much of what we know about the intensity-time variations of relativistic particles at the earth has been learned from the records of earth based monitoring instruments. Several types of intensity variations have been found to occur, of which only three, viz., the solar cycle variations, Forbush decreases and solar-flare ground level events, will be described here. At times, long-lived solar activity centers produce solar flares repeatedly or long lasting high speed streams and produce a 27-day variation corresponding to the rotation period of the sun. A small (0.2-0.5%), persistent daily variation is also seen in cosmic ray intensity. Venkatesan and Badruddin [Ref. 2] has given a detailed account of cosmic ray modulations.

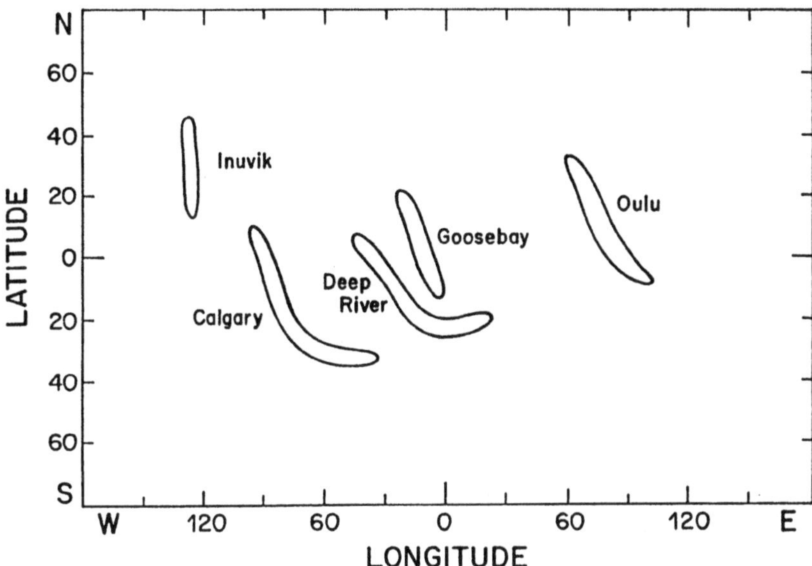

Fig. 8 Asymptotic cones of acceptance of several neutron monitors for solar flare particles.

Figure 9 is a record of the nucleonic intensity records of a neutron monitor in Calgary from 1964. Such records of neutron monitors at other stations go back to 1954 and of ionization chambers (muon component) to 1936. This figure illustrates the 11-year variation in cosmic ray intensity which is out of phase with sunspot activity.

When the intensity records over shorter periods are examined, other types of variations are also found. Figure 10 is the normalized daily counting rates of the Calgary neutron monitor for the year 1990. Apart from gradual changes over many days, the figure also shows shorter term variations, in particular sharp decreases in intensity lasting a few days, coinciding, more or less, with geomagnetic storms and also even shorter lived, sharp increases in intensity following some large solar flares. Geomagnetic storm-time variations are called Forbush decreases (After Scott E. Forbush who first observed these world-wide events). Solar flare events observed with ground based instruments are called "ground level events" (GLEs).

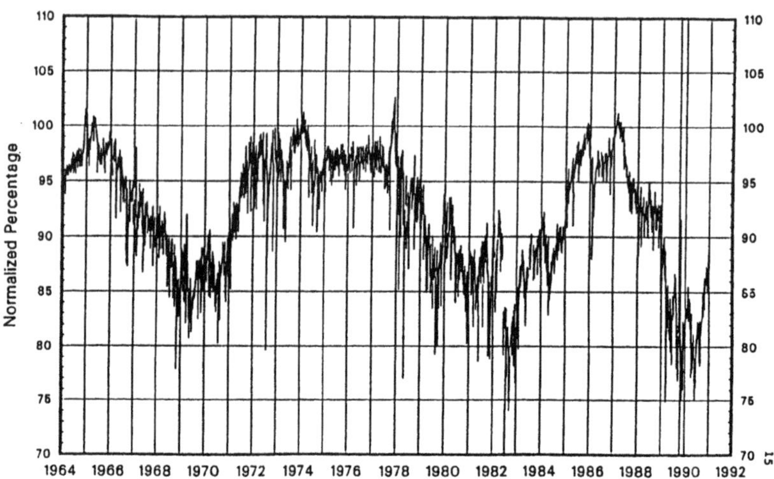

Fig. 9: Normalized daily counting rates of Calgary neutron monitor from 1964 to 1990.

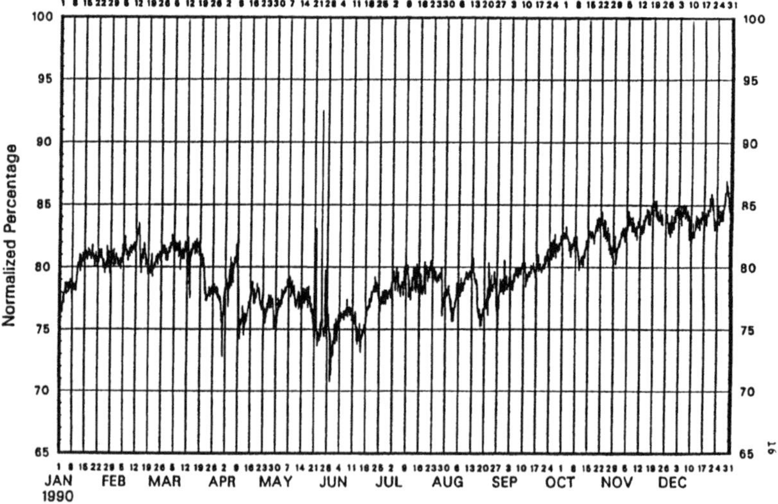

Fig. 10: Normalized daily counting rates of Calgary neutron monitor for 1990.

(a) **Solar cycle variations**

The variation observed in the intensity of a high latitude neutron monitor over a solar activity cycle is 15-25%. It is typically about half of that at the equator, showing that the variation is energy or rigidity dependent. The magnitude and the profile of the variation differs from cycle to cycle; for example, the duration of high intensity (95% of maximum) was quite prolonged (6 years) during the seventies, but during both the sixties and the eighties, it was much shorter (about 3 years). Solar cycle variations have been observed with balloons and satellite-borne instruments and variations in energy spectra of different components measured.

The solar cycle variation clearly is due to the modulation of the intensity of galactic radiation by solar activity. Since the orientation of the large scale solar field reverses at the time of high solar activity, the cycle-to-cycle differences in the 11 year modulation effect may be related to the orientation of the solar dipole field.

(b) **Forbush decreases**

Figure 11 shows a large Forbush decrease. Forbush decreases typically show a sharp reduction of 3-10% over a period of a day or two and gradual, exponential recovery over a period of 7 to 10 days. There is considerable variation from event to event both in terms of the magnitude and of the profile. Figure 12 shows the variation in recovery time of all the Forbush decreases observed during the last two solar cycles. Some of the large

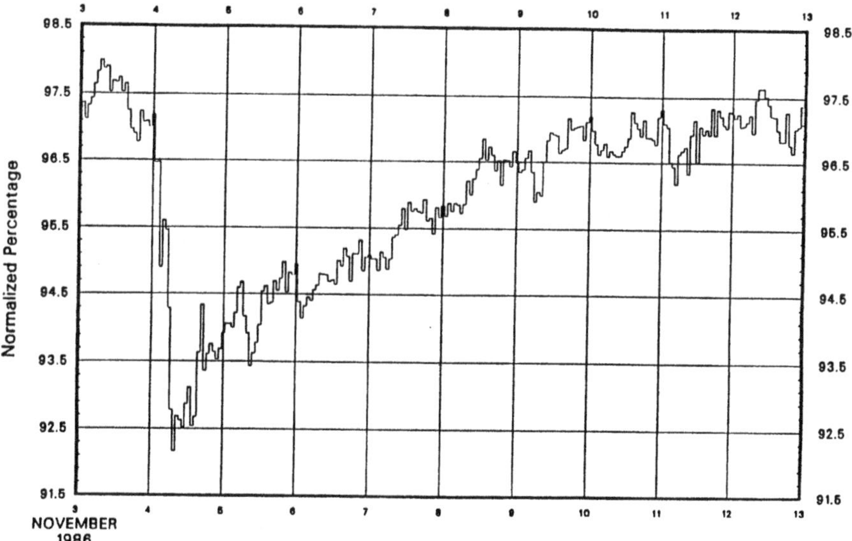

Fig. 11: Normalized hourly counting rates of Calgary neutron monitor during a Forbush decrease.

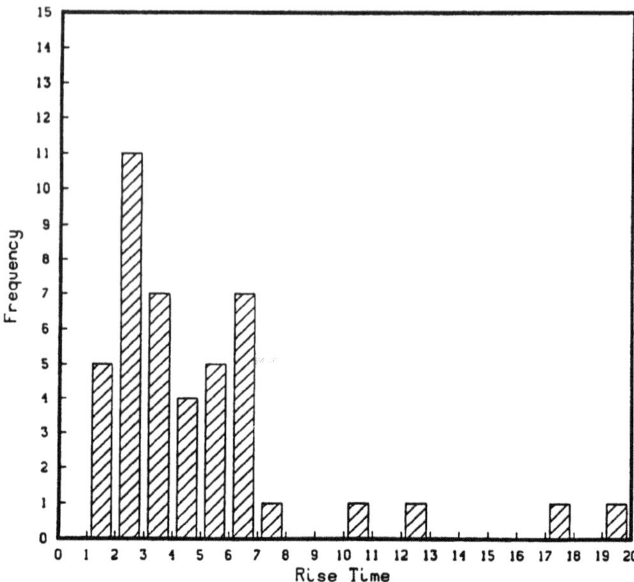

Fig. 12: Frequency distribution of recovery times from Forbush decreases. Risetime given in days.

events have been observed deep underground indicating that even particles of 100 GeV energy can be affected. During such large events cosmic rays of a few GeV may be completely removed from near the earth.

Forbush decreases are clearly associated with solar disturbances which produce travelling interplanetary shock waves; but the mechanisms which cause the decreases in intensity are not fully understood. The travelling interplanetary shock waves can act as a barrier and sweep out galactic cosmic rays producing the sharp decreases in intensity. As the radiation diffuses back into the volume swept by the travelling disturbance, the cosmic ray intensity recovers gradually in an exponential manner. When such decreases are repeated within a few days or weeks, as it happens when the solar activity reaches its peak, they tend to produce a cumulative effect which may partially account for the 11 year variation. Some events are thought to occur as a result of the earth entering magnetic clouds within which the intensity of cosmic rays is significantly less than outside. Such events show a very different time profile, an example of which is shown in Fig. 13.

It has been suggested that the solar cycle variation results from the cumulative effects of Forbush decreases. Forbush decreases begin near the sun and propagate outwards. Forbush decreases have been observed by instruments on deep space probes with appropriate delays to confirm this. Shock waves expand as they move outwards, and eventually they envelope the whole inner heliosphere and also pile up at great distance from the earth.

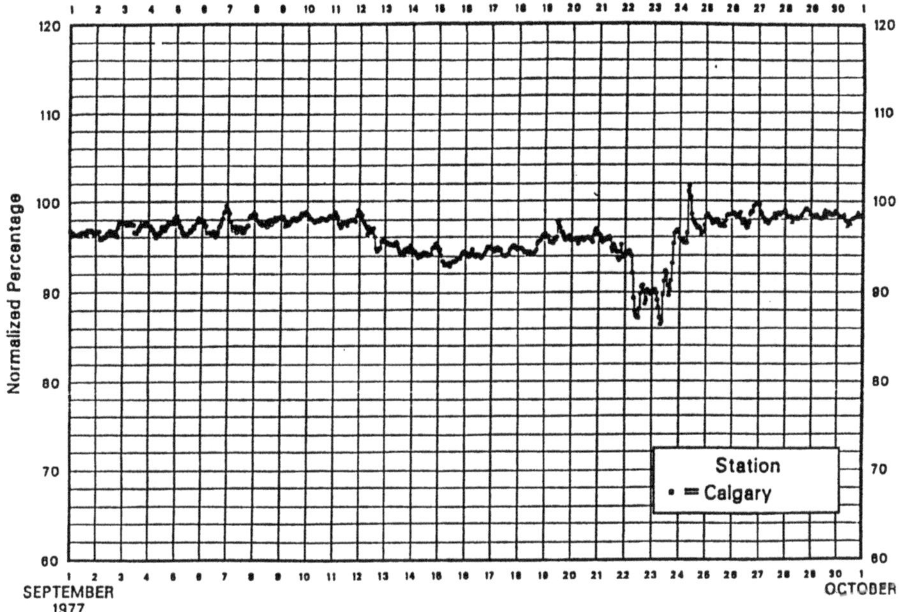

Fig. 13: Example of a Forbush decrease without an exponential rise.

The hysteresis effect, i.e., the tendency of cosmic ray recovery to lag behind the solar activity, has been shown as evidence. In Fig. 14 the number of Forbush decreases per year is plotted along with monthly average counting rates of the neutron monitor in Calgary. It is seen that though there is a correlation, it is not exact.

It has been pointed out that the orientation of the solar dipole field plays a significant role in the solar cycle modulation. At the height of the solar activity cycle the solar field reverses itself. The tilt of the solar field for the years 1976 to 1986 is also shown plotted along cosmic ray data. It is seen that there is a good correlation in this case too. The propagational behavior of galactic cosmic rays is quite different before and after reversal. For example after the 1970 reversal, the cosmic ray intensity rose quite rapidly, but after the 1981 reversal the rate of rise was much less.

6. Solar Flare Ground Level Events (GLE)

The occurrences of solar flares during which energetic particles are released into interplanetary space is very important from the point of view of the radiation levels in space. Such events were once thought to be very rare occurrence, but it turned out to be a consequence of the threshold energy for detection by the instruments used. Until 1956 only five such events were detected over a 20 year period when ionization chambers, with

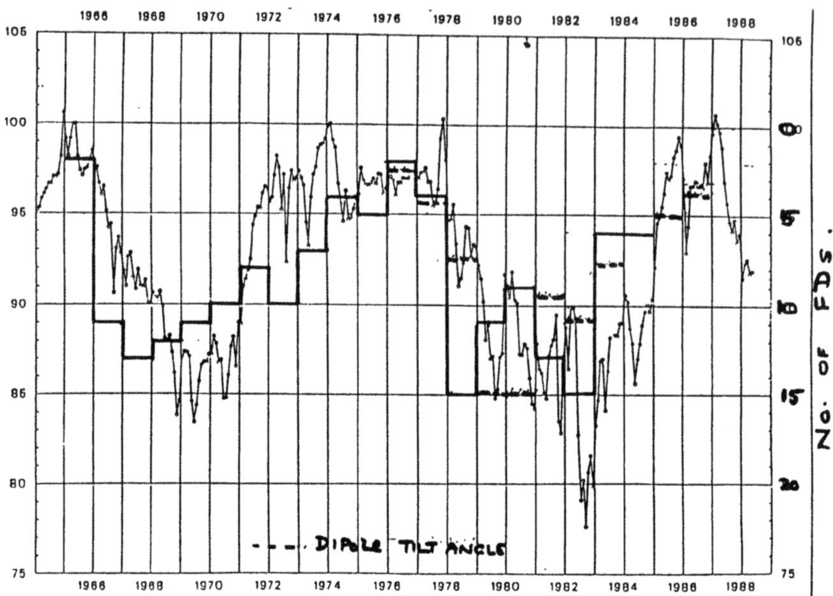

Fig. 14: Number of Forbush decreases per year along with monthly counting rate of Calgary neutron monitor and tilt of solar dipole field.

a threshold energy of 4 GeV, were the main instruments used. Neutron monitors have a threshold of about 0.5 GeV and they have detected some two dozen events over the period from 1956 to 1976. Riometers can detect solar flare particles of energy >10 MeV arriving over the polar cap and have detected more than 80 events in the same two decades. Satellite-borne instruments have no such threshold energy for detection and have shown generation of energetic particles in solar flares to be a fairly common phenomena, though generation of relativistic particles may be relatively infrequent.

A very good account of relativistic solar cosmic rays has been given by [Ref. 3]. I will only discuss some of the principal features of the phenomena, before going onto illustrate them with the most recent events. During the span of a few months in 1989-90 about 10 events were detected. The current cycle of solar activity seems to have been the most active one since continuous monitoring with neutron monitors began.

Solar flare ground level events (GLEs) can be described typically as a rapid (a few minutes to an hour) rise in counting rate of a monitor within a few minutes of an optical solar flare and then a gradual, usually exponential, decay to normal levels over a few hours. Particles released by flares on the invisible side have also been recorded at the earth. When the sources region and time of injection can be identified unambiguously, the features of GLEs observed at the earth can be used to understand the propagation of relativistic particles in the inner heliosphere.

The propagation of these particles are obviously influenced by the interplanetary magnetic field and the irregularities in them. Particles released near the foot of the field line connecting the earth and the sun arrive at the earth first. Cosmic ray monitors with asymptotic cones facing this field line see rapid increase and maximum intensity; those facing the opposite direction observe slow or delayed increase and the lowest intensity. Flare particles released near 60 degrees west of central meridian have the highest probability of detection at the earth. Anisotropy, i.e., the differences in the time profiles of intensity increases observed at different stations, can be used to determine the pitch angle distribution of solar flare particles and its variations with time. This can be related to the presence of irregularities or scattering centers in the interplanetary magnetic field. Widely varying conditions with scattering mean free paths as low as 0.025 AU to nearly scatter free propagations have been observed.

Another measurable parameter is the energy or rigidity spectrum and its variation with time. The rigidity spectrum, if expressed as a power law, typically has an exponent of -5, which is significantly different from that of galactic cosmic rays. Once the spectral exponent is determined one can estimate the integral flux of solar flare particles in space. Typical values for integral flux of particles of rigidity above 1 GV was 10^4 per m^2 s sr, though for the large flare of 23 February 1956, it was estimated to be 10^7 particles per m^2-s-sr. Only on few occasions have the particles with energy greater than 5 GeV been detected.

Figure 15 shows the intensity records of the Calgary neutron monitor for two months of 1989; the most remarkable feature of which is the number of GLEs. The largest of these was on 29 September 1989 which showed a maximum increase of about 405% in Calgary, which is about 1100 meters above sea level. It was an energetic event in that the increases were observed at the equator, and it has been estimated that particles of energy up to 25 GeV were produced. The site of the parent solar flare was estimated to be 15° behind the western limb.

The September 29 event was quite complicated. Figure 16 shows the records from several Canadian neutron monitors. The increase remained anisotropic for several hours, and the pitch angle distribution varied significantly during the time. However at higher rigidities (>6 GV) it appeared to be a normal event as the records of muon monitors show (Fig. 17). The records from a set of inclined telescopes in Ottawa showed that the rigidity spectrum above 6 GV had a power law exponent of -5; but the records of neutron monitors indicated a value of -2.9. This makes it difficult to estimate the integral flux of relativistic particles in space during this event, which could have been between 10^4 particles per m^2-s-sr.

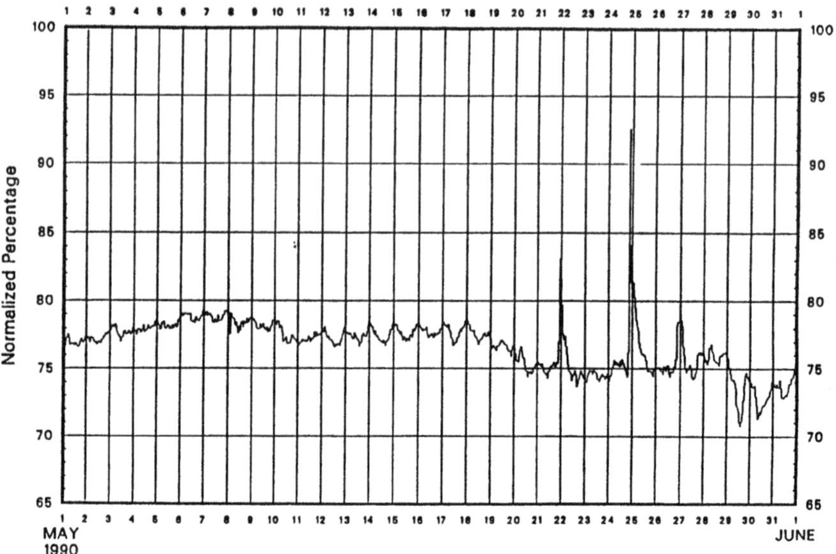

Fig. 15: Intensity records of Calgary neutron monitor for four months in 1989.

Figure 18 illustrates an unusual feature observed during one of the recent events. Only the monitors in Calgary, McMurdo and the South pole observed this sharp, short lived anisotropy. This was so because the pitch angle distribution was sharp and the asymptotic directions of these stations were almost at right angles to the spiral field direction. This was an unusual example of almost scatter-free propagation, controlled by narrow flux tubes or field lines which have been distorted from the usual garden hose directions.

7. Cosmic Rays in the Heliosphere

A conceptual model of the heliosphere, the region in which the effect of solar wind and the frozen-in interplanetary magnetic field influence the propagation of cosmic rays is affected, was shown in Fig. 1. The first attempt to explain the intensity variations of cosmic rays as a consequence of diffusion through the heliosphere was due to [Ref. 4]. According to this the galactic cosmic rays diffuse inward through the irregularities in interplanetary magnetic field conveyed outwards by the solar wind. The strength of irregularities as well as the solar wind speed change with solar activity, resulting in the quasi-periodic modulation of the intensity of cosmic rays reaching the earth.

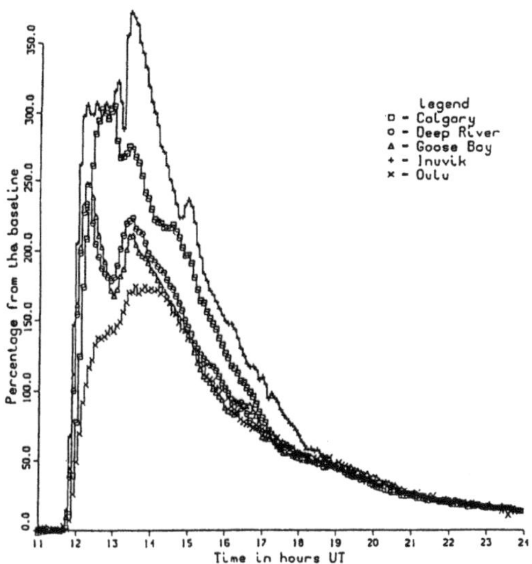

Fig. 16: The solar flare ground level event of 29 September 1990.

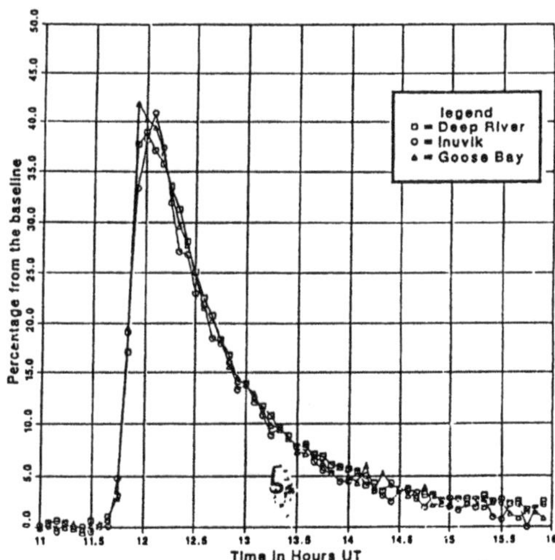

Fig. 17: Muon telescope records during the 29 September 1990.

For the simple case of isotropic diffusion the rate of change of density of particles is written as:

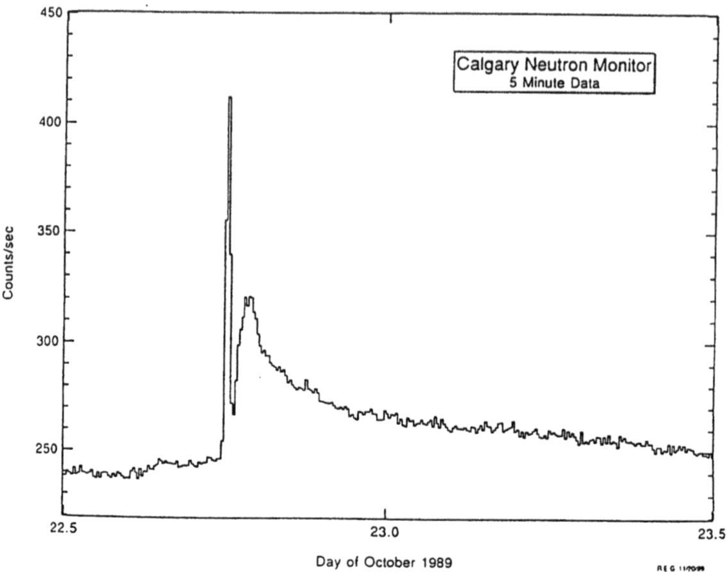

Fig. 18: Solar flare ground level event of 22 October 1990.

$$\frac{\partial n}{\partial t} = \nabla \cdot (K\nabla n) - \nabla(Vn)$$

where diffusion coefficient is given by $K = 1/3\lambda v$, λ being the collision mean free path, v the particle velocity, and V the solar wind velocity. In steady state, the particle density at a distance from the sun is given by:

$$\frac{n(r)}{n(\infty)} = \exp\left(-\frac{3V}{v}\int_x^R \frac{\partial r}{\lambda(r,p)}\right)$$

where R is the boundary of the heliosphere. The diffusion coefficient has been estimated using various assumptions about the radial dependence of mean free path, its dependence on particle rigidity etc., and compared with those estimated from observation of interplanetary field fluctuations. The estimated values have been shown to be reasonable.

One consequence of the convection-diffusion is that, even at solar minimum, the intensity of cosmic rays at the earth is significantly less than that outside the heliosphere. The residual modulation will depend on the radius of the heliospheric cavity. Such a boundary has not been encountered by the deep space probes, even at 60 AU. The radial gradient observed is only 1-2% per AU.

A more refined equation which takes into account anisotropic diffusion, convection and adiabatic deceleration of cosmic rays in interplanetary space is:

$$\frac{1}{r^2}\frac{\partial}{\partial r}\left(r^2UV - r^2K\frac{\partial U}{\partial r}\right) = \frac{1}{3r^2}\frac{\partial}{\partial r}(r^2V)\frac{\partial}{\partial T}(\alpha Tu)$$

where U(r,T) is the differential number density, $\alpha = (T + 2E_0)/(T + E_0)$ with E_0 the rest mass, T the kinetic energy, K(r,T) is the effective diffusion coefficient, and V(r) is the solar wind speed. The Differential intensity is given by $J = vU/4\pi$, where v is particle speed. Numerical solutions of this equation with acceptable values of solar wind speed and effective diffusion coefficient have shown that many features of primary nucleon and electron spectra can be well accounted for. For example, the peak in the energy spectra of nucleons and the positive spectral index below is well accounted for as the energy loss of galactic cosmic rays due to adiabatic deceleration. This equation is valid only for quasi-stationary situation and does not explain the dynamic nature of long and short term modulations. Further they neglect the effect of particle drift of particles in the large scale interplanetary magnetic field.

More recent theoretical work [Ref. 5] take into account the drift of particles in the spiral field as well as the orientation of the large scale field. Numerical solutions of the equations seem to be able to explain solar cycle modulation including the cycle-to-cycle variations. According to these, galactic cosmic rays diffuse primarily along the ecliptic plane encountering the interplanetary disturbances during one half of the 22 year solar cycle. During the other half of the cycle, they reach the earth mainly through the polar regions of the heliosphere. This accounts for the cycle-to-cycle difference in the profile or solar cycle variations. The orientation of the solar dipole field is a very important factor.

The role of Forbush decreases in producing long term variations is still not clear. In fact the physical mechanism responsible for Forbush decreases is not clear. Is it due to scattering by shock waves or due to adiabatic deceleration? The former will be rigidity dependent whereas the latter will be energy or velocity dependent. Relative changes in proton and helium spectra during Forbush decreases may identify this; but such measurements are yet to be made.

Acknowledgements

I wish to acknowledge the assistance of my colleagues at the University of Calgary in preparing this paper.

References

[1] P. Meyer, "Cosmic rays in the galaxy", Annual Review of Astronomy and Astrophysics, 7, 1-38 (1969).

[2] D. Venkatesan and Badruddin, "Cosmic-Ray Intensity Variations in the 3-Dimensional Heliosphere," Space Science Reviews, 124-194 (1990).

[3] S.P. Duggal, "Relativistic Solar Cosmic Rays", Reviews of Geophysics and Space Physics, 17, 1021-1057 (1979).

[4] E.N. Parker, "The passage of energetic charged particle through interplanetary space," Planetary and Space Science, 13, 9.

[5] J. Kota and J.R. Jokipii, Astrophysical Journal, 285, 573 (1983).

THE INTERPLANETARY ENVIRONMENT

JEAN-CLAUDE MANDEVILLE
Department de Technologie Spatiale
CERT-ONERA
BP 4025
31055 Toulouse Cedex
France

ABSTRACT. The knowledge of environment parameters is necessary for the establishment of design requirements for space vehicles and relevant equipments. With the development of planetary and interplanetary missions it is important to assess the space environment not only in the vicinity of the earth but also to describe its variation in the interplanetary space and in the vicinity of the planets. The aim of this paper is to give an overview of the different components of the space environment for an interplanetary mission taking into account its possible variations in the vicinity of the major planets. Emphasis will be made on the description of the particulate environment. Past and future planetary exploration is also summarized.

1. Introduction

The exploration of the solar system is an active and growing international enterprise. Our earth is just one of the many bodies that circle the sun, and with planetary exploration we have become much more aware of our own planetary environment. Many space missions are no longer limited to the vicinity of the earth and are devoted to the study of all the regions of the solar system. If the sun is the main driving mechanism for the interplanetary environment, there is however important variations in the vicinity of the various planets. Of the factors of the environment in which a spacecraft must operate, vacuum, corpuscular radiation, weightlessness, ionized gases, is the presence of solid matter which constitute particulate debris. Their presence attested, for an earth observer, by the occurrence in the night sky of the zodiacal light or the phenomenon of meteors.

Assessment of the natural environment in early stages of a space development program will be valuable in developing a space vehicle with a minimum operational sensitivity to the environment. Environment data are necessary for the establishment of design requirements for space vehicles and associated equipment and for the design of testing facilities on the ground.

R. N. DeWitt et al. (eds.), The Behavior of Systems in the Space Environment, 205–232.

2. Rationale of Planetary Exploration

Astronomy is one of the oldest established sciences. Modern man has made and continues to make more and more sophisticated astronomical measurements from ground based and orbiting observatories. The origin and nature of the solar system bodies have long been the subjects of human research. More or less consciously this research has been made through a systematic approach:

Planetary investigation was firstly based on ground telescopic observation using more and more sophisticated instruments: visual, photographic, radar observation techniques using the resources of the spectrophotometry and polarimetry. This first approach has been very valuable in many cases, and for instance, the design of the first lunar probes was based primarily on results from high quality telescopic observations.

The second step of planetary exploration came with the design of planetary missions of increasing complexity: Spinning spacecraft were only used in the beginning, then 3-axis stabilized probes were used, sometimes coupled with scanning platforms and sometimes sending probes onto the surface of planetary bodies for close-up investigation.

A complete range of interplanetary probes has been used:

- fly-by: Mariner, Pioneer, Helios
- orbiters: Mariner, Pioneer, Voyager, Giotto, Vega, Pioneer Venus, Venera
- unmanned landers: crash landing (Ranger)
  soft landing: unmanned (Surveyor, Luna, Viking),
  manned (Apollo Missions)
- samples return (Luna, Apollo)

Selection on targets has been also systematic, taking into account the various opportunities (technical and astronomical) and launch windows (multiple fly-by); the moon, being the closest and the best known celestial body has received the complete range of spacecrafts and missions.

The most investigated objects have been so far (American and Russian spacecrafts almost exclusively):

The inner or terrestrial planets:

- Mercury: fly-by with Mariner 10 and 11 (1974-1975)
- Venus: fly-by with Mariner 10 and 11, orbiter and lander with Pioneer (1978-1987), Venera (USSR)
- Mars: orbiters, namely with Mariner 4 and Mariner 9
  landers with Viking 1 and 2 (1976-1982)
  fly-by with Phobos (USSR)

The outer planets and their satellites:

- Jupiter: Pioneer 10 and 11, Voyager 1 and 2 (1979)
- Saturn: Voyager 1 and 2 (1980-1981)
- Uranus: Voyager 2 (1986)
- Neptune: Voyager 2

The comets, fly-by: Giacobinni-Zinner (ICE)
Halley: Giotto (1986 ESA), Vega (USSR), Sagikate (Japan)

Present and future missions are in progress or planned for:

- Venus: Magellan in orbit since August 1990, extensive radar mapping
  Jupiter and Galilean satellites: Galileo launched in 1990
- Jupiter: Galileo fly-by
  Cassini fly-by
- Saturn: Cassini/Huyghens (1995), orbiter and entry probe on Titan
- Mars: balloon mission (USSR), rover (Esa-Nasa)
- Moon: orbiter
- Asteroids and Comets: CRAF, Cassini, Rosetta

3. Components of the Environment

The *Sun* is the main driving mechanism for the interplanetary environment. Up to a distance of 35 AU (1 AU = 150 million km) the near earth space environment is only a peculiar case of the interplanetary environment modified locally by the interaction of the magnetic field and of the atmosphere. The environment in the vicinity of the planets is in the same manner modified by the possible occurrence of a planetary magnetic field or an atmosphere.

There are many various components of the space interplanetary environment:

- Vacuum
- Sun and planetary gravitational attraction
- Sun electromagnetic radiation: IR, visible, UV (described elsewhere in this volume)
- Charged particles: solar wind, solar cosmic rays, galactic cosmic rays (described elsewhere in this volume)
- Interplanetary magnetic filed
- Meteoroids and micrometeoroids, interplanetary dust

Planetary environments:

- Gravitational attraction: modification of particulate environment, surface gravity, gravitational focusing, Roche's limit
- Presence of magnetic field: magnetosphere
- Presence of atmosphere: ionosphere, modification of solar radiation
- Ring of particles, ejecta, local population
- Variation of temperature in the atmosphere
- Surface morphology
- Optical properties of atmosphere
- Polarization

- Albedo
- Spectral reflectance
  Chemistry and composition of atmosphere

Mechanisms of degradation of materials exposed to the space environment are many and various. As far as the solid particulates are concerned, the main degradation is caused by mechanical failure of the exposed material under the high dynamical load produced upon impact at very high velocity. Extent of damage depends primarily on the size of impacting body: either complete loss of spacecraft or progressive loss of performance. Of peculiar concern are long duration or manned missions.

4. Overview of the Solar System

4.1. ORIGIN AND FORMATION

In the current view the process begins, with the collapse of the core of a dense molecular cloud forming a protostar and a surrounding accretion disk. The protostar grows by the direct infall of material onto its surface and by accretion from the inner boundary of disk; the gas and dust that remains in the disk provides the raw material from which planets may later form. As the accretion phase ends, 4.5 billion years ago, the dust in the protoplanetary disk settles to the midplane, coagulates and forms planetesimals which may ultimately accumulates into planets. Finally, only planets, satellites, asteroids and comets are left, although the occasional collisions of these larger bodies produced disks of planetary debris and left prominent impact features on most planets.

4.2. DESCRIPTION OF MAIN BODIES

Our solar system consists not only of the Sun and nine planets (and their own satellites) but also of thousands of small bodies which occasionally become more spectacular than the brightest planet: brilliant comets, asteroids wandering in the vicinity of earth orbit, meteor flashing across the sky, meteorites striking the earth. All these objects, whose size spans from several hundred km in diameter for the largest asteroids or planetary satellites to sub-micron particles are mainly leftovers from the formation of the solar system.

Detailed description of the main planetary objects is out of the scope of this paper, for reference the main properties of the planets are summarized on Table 1. Some data of interest for a planetary mission will be given however below:

- The *Moon* has no atmosphere and no magnetic field: no major modification of the interplanetary environment is expected in its vicinity. The surface morphology is characterized by the occurrence of impact craters of every size, nature of the soil surface is regolithic (special surface created by the continual

Table 1: Planetary data [Ref. 12].

Object	Diameter (km)	Mass (g)	Mean density (g/cm³)	Visual geometric albedo	Rotation period (days)	Obliquity	Revolution period (days unless noted otherwise)	Semimajor axis (A.U. for planets; 10³ km for satellites)	Orbit inclination (with respect to ecliptic for planets; planetary equator for satellites)	Orbit eccentricity
Sun	1,391,400	1.987 (33)	1.4	–	25.4	7°.25	–	–	–	–
Mercury	4,864	3.30 (26)	5.5	0.10	58.6	<7°	0.2408 yr	0.387	7.0	0.206
Venus	12,100	4.87 (27)	5.2	0.586	243*R*	~179°	0.6152 yr	0.723	3.39	0.007
Earth	12,756	5.98 (27)	5.52	0.39	1.00	23°.5	1.000 yr	1.000	0.00	0.017
Mars	6,788	6.44 (26)	3.9	0.15	1.02	25°.0	1.881 yr	1.524	1.85	0.093
Jupiter	137,400	1.90 (30)	1.40	0.44	0.41	3°.1	11.86 yr	5.203	1.31	0.048
Saturn	115,100	5.69 (29)	0.71	0.46	0.43	26°.7	29.46 yr	9.54	2.49	0.056
Uranus	50,100	8.76 (28)	1.32	0.56	0.45*R*	97°.9	84.0 yr	19.18	0.77	0.047
Neptune	49,400	1.03 (29)	1.63	0.51	0.6	28°.8	164.8 yr	30.07	1.78	0.008
Pluto	5,800	6.6 (26)	6?	0.13	6.4	?	284.4 yr	39.44	17.17	0.249
Moon	3,476	7.35 (25)	3.34	0.115	27.3	6°.7	27.3	384	18–29°	0.055
Phobos	18 × 22	?	?	0.06	?	?	0.319	9	1°.1	0.021
Deimos	12 × 13	?	?	0.06	?	?	1.26	23	1°.6	0.003
(1) Ceres	800	1.2 (24)?	4.5?	0.1?	0.38	?	1,681	2.767 A.U.	10°.6	0.079
(2) Pallas	490	?	?	0.1?	0.4?	?	1,684	2.767 A.U.	34°.8	0.235
(3) Juno	250	?	?	0.2?	0.30	41°?[b]	1,594	2.670 A.U.	13°.0	0.256
(4) Vesta	490	2.4 (23)?	3.9?	0.25?	0.22	25°?[b]	1,325	2.361 A.U.	7°.1	0.088
JI Io	3,500	8.0 (25)	3.6	0.92	1.769	?	1.769	422	0°.03	<0.01
JII Europa	3,100	5.0 (25)	3.2	0.83	3.551	?	3.551	671	0°.5	<0.01
JIII Ganymede	5,000	1.65 (26)	2.5	0.49	7.155	?	7.155	1,070	0°.2	<0.01
JIV Callisto	4,900	1.02 (26)	1.7	0.26	16.689	?	16.689	1,883	0°.3	<0.01
JV	170?	?	?	?	?	?	0.498	181	0°.4	0.003
JVI	130?	?	?	?	?	?	250	11,470	28°	0.158
JVII	44?	?	?	?	?	?	260	11,740	26°	0.206
JVIII	12?	!	?	?	?	?	737	23,500	33°*R*	0.40
JIX	14?	?	?	?	?	?	758	23,700	25°*R*	0.27
JX	14?	?	?	?	?	?	255	11,850	28°.5	0.135
JXI	16?	?	?	?	?	?	692	22,560	16°.5*R*	0.207
JXII	12?	?	?	?	?	?	631	21,200	33°*R*	0.16
SI Mimas	900?	3.7 (22)	0.1?	0.49	?	?	0.942	186	1°.5	0.020
SII Enceladus	550	7.2 (22)	0.8?	0.54	?	?	1.370	238	0°.02	0.004
SIII Tethys	1,200	6.6 (23)	0.7	0.84	?	?	1.887	295	1°.1	0.0
SIV Dione	820	1.03 (24)	3.6	0.94	?	?	2.737	377	0°.02	0.002
SV Rhea	1,300	1.5 (24)	1.3	0.82	?	?	4.518	527	0°.3	0.001
SVI Titan	4,850	1.37 (26)	2.3	0.21	15.95	?	15.95	1,222	0.3	0.029
SVII Hyperion	350?	?	?	?	?	?	21.28	1,481	0.5	0.104
SVIII Iapetus	1,150	1.5 (24)	1.9	?	?	?	79.33	3,560	14.7	0.028
SIX Phoebe	260?	?	?	?	?	?	550.4	12,950	30*R*	0.163
SX Janus	370?	?	?	?	?	?	?	160?	~0	~0
UI Ariel	1,470?	?	?	?	?	?	2.520	192	0.0	0.003
UII Umbriel	960?	?	?	?	?	?	4.144	267	0.0	0.004
UIII Titania	1,760?	?	?	?	?	?	8.706	438	0.0	0.0024
UIV Oberon	1,600?	?	?	?	?	?	13.463	586	0.0	0.0007
UV Miranda	550?	?	?	?	?	?	1.414	128	0.0	<0.01
NI Triton	3,800	1.4 (26)	4.9	0.36	?	?	5.877	353	20.1*R*	0.0
NII Nereid	540?	?	?	?	?	?	360	5,600	27.5	0.76

gardening of top layer by meteoritic impacts). mass density is 3.34 g/cm^3, surface gravity is 102 cm/s^2, surface temperature 102°K – 384°K, albedo 0.051 – 0.176, sunlight is 1370 W/m^2.

- *Venus* has a very dense atmosphere (surface pressure of 90 bars) composed mainly of CO_2, clouds formed by micron sized droplets of sulfuric acid are prominent. An ionosphere is present, but no magnetosphere because of the absence of magnetic field. The surface morphology is dominated by plate tectonics and craters from impact and volcanic origin. As shown by the S/C Magellan impact features are abundant despite the presence of a thick atmosphere. The cloud layer is opaque to most of the solar spectrum. Due to a strong greenhouse effect (CO_2) the surface temperature is very high (740°K). The mass density is 5.2 g/cm^3, surface gravity 877 cm/s^2.

- *Mercury* has very faint atmosphere (He,H,Ar), a small magnetic field exists with evidence of a small magnetosphere extending to 1.5 planetary radii, but without trapped particles. The surface morphology is similar to the surface of the moon with the presence of a regolith and numerous impact craters. The solar flux is high at 0.38 AU: 9260 W/m^2. The mass density is 5.4 g/cm^3 and surface gravity 370 cm/s^2.

- *Mars* has a thin atmosphere (surface pressure of 5 mbar), composed mainly of CO_2; meteorological phenomena are present with variations of atmospheric pressure, temperature, occurrence of winds responsible for dust storms. Most of these phenomena are driven by season condensation and evaporation of CO_2 on the polar regions. The surface morphology is dominated by a large variety of features: craters and cratered terrains, fossil fluvial features, tectonic evidence, volcanoes and riles, terraced polar regions, regolith. There is no evidence of magnetic field and magnetosphere. Average solar irradiation is 590 W/m^2 and surface temperature varies between 130°K and 300°K. The mass density is 3.93 g/cm^3 and surface gravity 372.5 cm/s^2. Two satellites, probably captured asteroids, circle the planet (Phobos 13x10 km, Deimos 7.5x6 km). The surface of these satellites is heavily cratered.

- *Jupiter* is the largest planet of the solar system (1/1000 of solar mass), its mass density is only 1.32 g/cm^3, thus implying a relatively small solid core (iron-silicate), surrounded by liquid hydrogen. The atmosphere of Jupiter is composed primarily of H_2 (90%) and He; its outer layer is characterized by a complex cloud structure, mainly aqueous NH_3 solution and solid NH_3. The atmosphere is dynamically very active: zonal currents, equatorial features, great red spot. A thick ionosphere (3500 km) is present. Jupiter has a strong and complex magnetic field (4.2 G at 1 jovian radius), its effect is felt as far as Saturn. It is represented commonly as an offset tilted dipole. Consequently the magnetosphere of Jupiter is very large, with a magnetopause extending to 50-100 jovian radii. Charged particles in the vicinity of Jupiter are of major concern for space missions around the planet; more details are given in [Refs. 1, 14].

Jupiter is circled by a large number of natural satellites, the most important being the four galilean satellites: Io, Europa, Ganymede and Callisto. Io has active volcanoes, Callisto is heavily cratered [Ref. 35]. A ring of particles has been discovered by one of the Voyager missions. Details will be given later (Section 5.6).

- *Saturn* is the second largest of outer planets. Saturn is gaseous outside its core (metallic hydrogen) and so, there is no surface. Its atmosphere is composed of H_2 (94%) and He (about 6%). The presence of a magnetic field implies the presence of a magnetosphere as measured by the Pioneer and Voyager s/c. In terms of particle intensities the magnetosphere of Saturn is between Jupiter and Earth in general character [Ref. 1]. Several satellites, some of them with a large number of craters (Iapetus), orbit the planet.

  The best known and most impressive feature of Saturn is its rings. The Voyager imager showed their enormously detailed structure, consisting of hundreds of small rings. There is probably a continuum of body sizes from the resolved satellites to the ring particles. More details will be given in section 5.6.

- *Uranus* has also a distinctive ring system composed of nine faint structures. At this distance from the sun (19 AU) the solar wind progressively merges into the interstellar medium (proton and electron number density 0.014 cm^3) and the solar flux is only 3.8 W/m^2. There is evidence for a magnetic field and a magnetosphere. Fifteen satellites circle the planet.

- Most of the data on *Neptune and Pluto* are still largely inaccurate, details are given on [Ref. 1].

Because they are directly connected with the particulate environment more details will be given in the following paragraphs on asteroids, comets and meteorites.

4.3. ASTEROIDS

They are small planets orbiting the sun mostly between the orbit of Mars and the orbit of Jupiter. Inclination on the ecliptic plane is generally low. The largest is Ceres (discovered in 1801), more than 2000 asteroids have been catalogued. An information source is mainly the Tucson revised index of asteroid data (TRIAD), a computer file from the University of Arizona, based on telescopic observations [Refs. 28, 29] (see Table 2).

Orbits of asteroids are classified according to semi-major axis, eccentricity, and inclination. It is possible to distinguish several

Table 2: Dynamic properties and other data for selected asteroids [Ref. 1].

Number	Name	Year of Discovery	Approxi-mate Diameter (km)	Opposi-tion Magni-tude	a	Orbital period (year)	e	i	Type
1	Ceres	1801	102	7.79	2.767	4.6	0.0802	10.60	C
2	Pallas	1802	583	8.46	2.773	4.61	0.2394	34.82	U
3	Juno	1804	249	9.0	2.671	4.36	0.2574	13.02	S
4	Vesta	1807	555	6.5	2.361	3.63	0.0889	7.14	U
5	Astraea	1845	116	11.3	2.577	4.13	0.1862	5.33	S
6	Hebe	1847	206	9.5	2.424	3.77	0.2019	11.65	S
7	Iris	1847	222	9.2	2.386	3.69	0.2309	5.47	S
8	Flora	1847	160	9.7	2.201	3.27	0.1567	5.88	S
9	Metis	1848	168	10.0	2.387	3.69	0.1233	5.60	S
12	Victoria	1850	135	10.8	2.334	3.57	0.2190	8.38	S
15	Eunomia	1851	261	9.5	2.642	4.30	0.1870	11.76	S
18	Melpomene	1852	164	9.8	2.296	3.48	0.2176	10.15	S
20	Massalia	8152	140	10.2	2.409	3.74	0.1426	0.68	S
192	Nausikaa	1879	99	11.0	2.403	3.72	0.2445	6.87	S
324	Bamberga	1892	256	11.3	2.685	4.39	0.3346	11.30	C
387	Aquitania	1894	113	10.7	2.74	4.53	0.2383	17.97	S
433	Eros	1898	20	7.2	1.458	1.76	0.2230	10.83	S

families, separated by gaps: the Kirkwood gaps, which are orbital periods harmonic of the Jupiter period (Fig. 1). The main families are:

Apollo-Amor, Hungarias, Floras, Nysa.
Main Belt, Eos, Hilda, Hyriyamas.
Earth crossers: Aten (3) a< 1AU
Apollo (23) a>=1AU7
Amor(10) 1.07 <a<2 AU
Mars crossers.
Trojans: Cluster near two of the Lagrangian points of 3-body stabilities in the Sun-Jupiter system (1000).

Some asteroids are on exceptional orbits: Chiron with a semi-major axis of a=13.7 AU, Pallas with an inclination of i = 34.8°. The total mass of asteroids is: $3\ 10^{21}$ KG (1/20 mass of the Moon). Among the largest are: Ceres (D = 987 km), mass: 1.10^{21} kg, Pallas (D = 538 km), Vesta

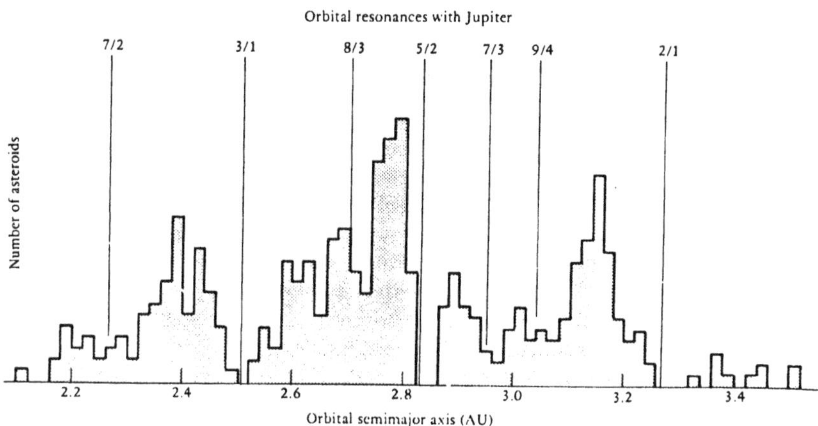

Fig. 1: Distribution of asteroid orbits.

(D = 544 km). The asteroids are divided in several classes, according to Visible and IR spectra and to polarimetry measurements [Refs. 30, 36] (Fig. 2):

C similar to carbonaceous chondrites
S similar to stony iron meteorites
M similar to nickel iron enstatite chondrites
E similar to enstatite achondrites
R and U various types or unclassified

The size distribution is given on Fig. 3. Asteroids are one of the main sources of meteoroids and interplanetary dust. The probability of collision of earth-crossers with earth is one collision of one object with 1 km in diameter every 300000 years.

4.4. COMETS

The comets are the most distinctive and peculiar members of the solar system [Refs. 14, 37]. They are supposed to represent the most primitive objects. They move around the sun in elliptical orbits (often nearly parabolic), their aphelia can be very large (50000 AU), their perihelia is close to 1 AU (see Table 3 and Fig. 4).

- 100 short period comets (<200 years) have been discovered
- 600 are long period comets

They become visible when they are at a distance smaller than 3-4 AU from the earth. They are composed of a small nucleus (1 – 10 km), of a large gas cloud (coma) and a very long dust and gas tail, when close to the sun. Their origin is believed within the Oort's cloud, located at the boundary

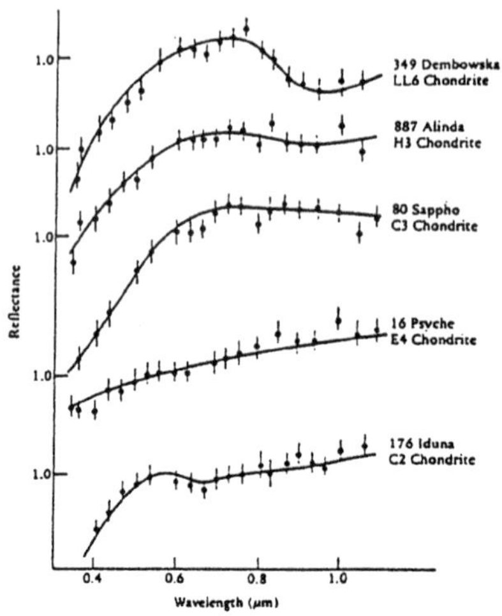

Fig. 2: Typical reflectance spectra of asteroids compared to meteorites.

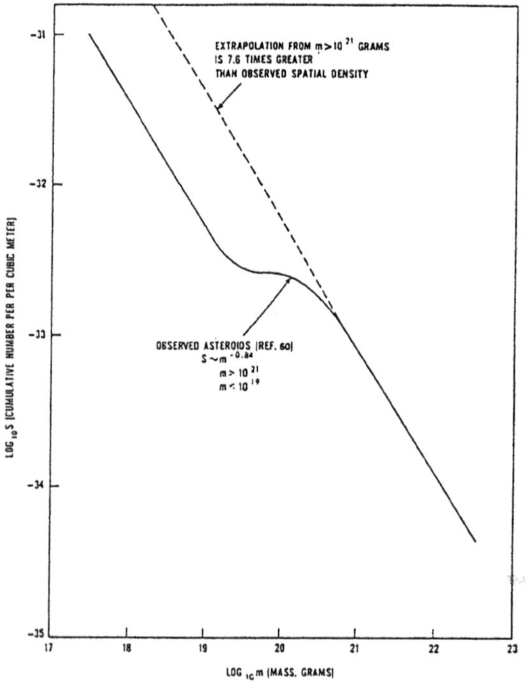

Fig. 3: Mass distribution of asteroids at 2.5 AU [Ref. 3].

Table 3: Cometary data.

Name	Time of Recent Peri- helion	Period (year)	q (AU)	Q (AU)	e	Arg. of Peri- helion	Node	i	No. of Appear- ances
Encke	1974.32	3.30	0.338	4.09	0.847	185.9	334.2	12.0	51
Grigg- Skjellerup	1972.17	5.12	1.001	4.94	0.663	359.3	212.7	21.0	13
Tempel-2	1972.87	5.26	1.364	4.68	0.549	190.9	119.3	12.5	16
Giacobini- Zinner	1972.59	67.52	0.994	5.98	0.715	171.9	195.1	31.7	10
Borelly	1974.36	6.76	1.316	5.84	0.632	352.7	75.1	30.2	10
Halley	1910.30	76.09	0.587	35.33	0.967	111.7	57.8	162.2	27
Tempel- Tuttle	1965.33	32.91	0.982	19.56	0.904	174.6	234.4	162.7	4
						152.8	138.7	113.6	1
Swift- Tuttle	1862.64	119.98	0.963	47.69	0.960				
			0.537		0.99619	354.15	223.96	90.04	1
Bennett (1970 II)	Mar 20.0446 1970		0.142425		1.0000	37.82	257.76	14.30	1
Kohoutek (1973 XII)	Dec 28.4307 1973								

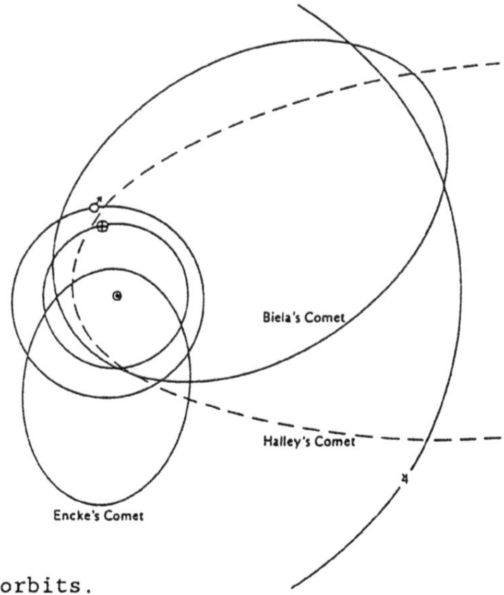

Fig. 4: Cometary orbits.

of the influence of the sun's gravitational attraction (sphere of 50000 AU radius), the total number could be as large as 10^{11} comets [Refs. 12, 13, 14]. They are injected by perturbations caused by passing nearly stars, into the inner solar system and then subject to perturbations from Jupiter. Many comets have been observed two or more times:

P/Encke has the shortest period (3.8 years).
P/Halley has been observed in 86 BC and has been seen 28 times, the last being in 1986 (period of 76 years).

New data on their composition result from observation of comet Halley by Giotto and Vega spacecrafts in 1986 [Ref. 37]:

- The nucleus is a solid conglomerate of water ice with dust grains embedded in it; the sandbank model, widely accepted until now, has to be rejected if Halley is a typical comet. The nucleus is irregularly shaped (15x8x7.5 km) and it is extremely dark (4% albedo).
- Near perihelion the production rate of gas is 18 tons/s and of dust 20 tons/s. Less than 10-20% of the surface of the nucleus look active.
- The composition of the gases is the following: 80% H_2O, 10% CO, 2% CH_4, 1.5% CO_2, 2% NH_3, 2%, 1% H_2CO, <0.1% HCN.

Very small dust particles were detected in-situ (10^{-17} g): probably fluffy particles composed of rock forming silicates coated with an organic component (CHON particles). The abundance of elements is similar to solar abundance, obviously Halley is less differentiated than C1 chondrites. Most of the very small particles are lost through solar radiation pressure and Poynting-Roberston effect and does not contribute to the population of stream particles.

5. Interplanetary Dust

Interplanetary dust particles (IDP) means all solid bodies ranging in diameter from sub-micron (10^{-18} g) to tens of centimeters (10 kg). The total mass of the IDP is only 10^{20} g (10^{-14} of the total of solar system), the flux close to the earth is about 10^{-12} g/m^2/s. They are partly left-over from the formation of the solar system and partly the result from the disintegration of asteroids and comets; a local ring population (around outer planets) could be linked to the formation of the planetary satellites [Ref. 45]. The population is in constant evolution, with however a distribution roughly uniform within the ecliptic plane from 0.5 to 30 AU. Local concentrations are possible within the asteroid belt, the streams of particles (cometary or asteroidal), the Lagrangian points. A modification of the environment is noticed in the vicinity of the large planets (interaction with the magnetic field and atmosphere, gravitational focusing).

5.1. METHODS OF OBSERVATIONS

None of them is able to provide all parameters of interest to describe the status of a dust particle. However various methods complement each other.

5.1.1. Zodiacal Light. The zodiacal light is a cone of light extending along the ecliptic plane [Ref. 19]. It is caused by sunlight scattered by interplanetary dust particles. The major scatterers (80%) are particles in the range of 5 to 100 μm (10^{-10} to 10^{-6} g). Smaller particles are inefficient Mie scatters; larger particles have a small spatial density. The sun F corona forms the inner part of the zodiacal light [Ref. 39]. Its intensity has been measured from earth and from spacecrafts. The satellite Helios, in the inner solar system [Refs. 19, 21] measured a variation with distance from the sun (between 0.3 AU and 1 AU) as:

$$N(r) = r^{-n} \quad \text{with } n = 1.5$$

M. Hanner [Ref. 40] proposed a similar spatial distribution outside 1 AU from Pioneer 10 measurements, with a possible enhancement in the asteroid belt and a cut-off at 3.3 AU. The 3-D distribution is still poorly known [Ref. 39]. The optical depth is $10^{-6} - 10^{-7}$, mass density is about 10^{-19} kg/m^3, and the total mass is estimated at $10^{-16} - 10^{-17}$ kg. Assuming a mass loss of 10 tons/s (by PR effect and by mutual collisions) the lifetime should be limited to typically $10^4 - 10^5$ years, unless there is an adequate source for replenishment.

5.1.2. Collection of Particles. Normally, collection of particles is only possible at low velocities relative to the collecting device. Nevertheless development of capture cells flown on retrievable payloads is encouraging (LDEF, MIR, Salyut, EURECA.

- Collection in the stratosphere by high altitude aircrafts [Ref. 41] of particles decelerated in the atmosphere is interesting for particles in the 10-100 μm range.
- Detection on board spacecrafts.

Several methods have been used:

- Impact ionization detectors (small particles, $10^{-16} - 10^{-11}$ g) could provide data on mass, velocity (rise time of signal), and incident direction of particles. If coupled with a mass spectrometer [Refs. 10, 21] the sensor also yields information on chemical composition.
- Penetration detectors: reliable and simple in design (capacitor, pressurized cell, PVDF).
- Acoustic sensors: reliable, detection area could be large, widely used.

All of these sensors have been used in many spacecraft, in LEO or interplanetary missions (Explorer, Pioneer, Pegasus, Giotto, Vega) [Refs. 5, 6, 7, 8].

5.1.3. Lunar Microcraters. The Moon is continuously hit by interplanetary dust particles. Consequently the surface of samples collected on the moon is covered by many high velocity impact craters with sizes ranging from several millimeters to a fraction of micron. Actually extensive data have been obtained by lunar microcraters counts [Refs. 4, 42]. Limitations are due to possible contribution from ejecta coming from nearby impacts, with a possible overestimate of small size craters, and to an inaccurate determination of the exposure of the rock at the lunar surface (uncertainties in the solar flare track record), see Fig. 5. Physical and chemical properties of impacting particles could be inferred from the geometry of impact craters and analysis of residus [Refs. 42, 46].

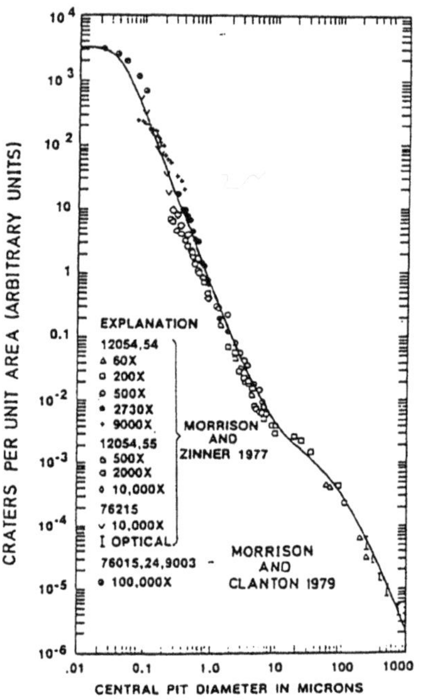

Fig. 5: Lunar microcraters distribution [Ref. 43].

5.1.4. Meteors. When large interplanetary dust particles hit the upper earth atmosphere, they partly melt and evaporate and ionize the surrounding medium. The phenomena can be detected visually (or photographically) and by radar techniques. Particles larger than 10^{-7} g can be detected and if observations are made simultaneous in different locations, the orbit of the body can be computed. In some rare occurrence fragments of meteorites are found on the ground. Until now five chondrites have been recovered, and their orbits determined: aphelia are situated with the asteroid belt, thus implying that asteroids could have a source of chondritic meteorites and possibly of cosmic dust [Refs. 11,18].

5.1.5. Meteorites. Any sizable fragment of a meteor which survive passage to the earth's surface is called a meteorite. Numerous meteorites have been found, recognized by various non-terrestrial characteristics. The earth collects about 10^5 kg per day; evidence of fall of large bodies is given by meteorite craters (Meteor Crater, Arizona, 1.5 km in diameter, one event every 150000 years (?)). Meteorites are classified in: Stones (Chondrites 84%, Achondrites 8%), Stony-irons (1%) and Irons (7%). Chondrites are characterized by the occurrence inside their matrix of small rounded inclusions, whose age of formation is similar to the age of the solar system (4.5 by). Chemical abundances are similar to the composition of the sun, except for volatiles elements. Parent bodies of most of chondrites are obviously asteroids or comets, as shown by orbital determination of some recovered meteorites, Fig. 6. Comparison of reflectance spectra of meteorites and asteroids is used for identification of different types [Ref. 36].

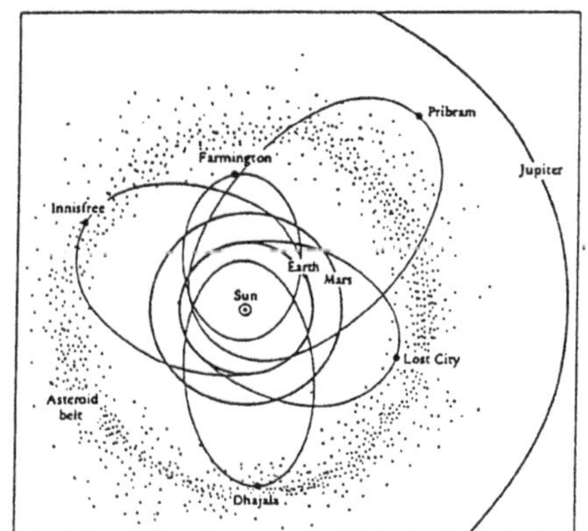

Fig. 6: Meteorite computed orbits [Ref. 13].

5.2. FORCES ACTING ON INTERPLANETARY DUST PARTICLES

In the same manner as other celestial bodies the dust particles generally orbit the sun under the action of several forces:

5.2.1. Gravitational:

- Solar gravitational attraction is the dominant force acting on dust particles:

$$Fg(r) = GMm/r^2$$

$G = 6.67\ 10^{-11}\ kg^{-1}m^{-2}s^2$ is the gravitational constant
$M = 1.99\ 10^{20}$ kg is the solar mass

- Smaller gravitational perturbations (resonance, lagrangian points) are caused by the planets. Significant orbit changes could occur inside a distance of a planet if:

$$D = a_p \, (m_p/2M)^{1/3}$$

a_p = mean distance of the planet form the sun
m_p = mass of the planet

- The sphere of influence has a radius of 0.01 AU for Venus, Earth and Mars and 0.5 AU for outer planets.

- ORBITS: The manner of planetary motion was first established in the seventeenth century by Kepler (Fig. 7):

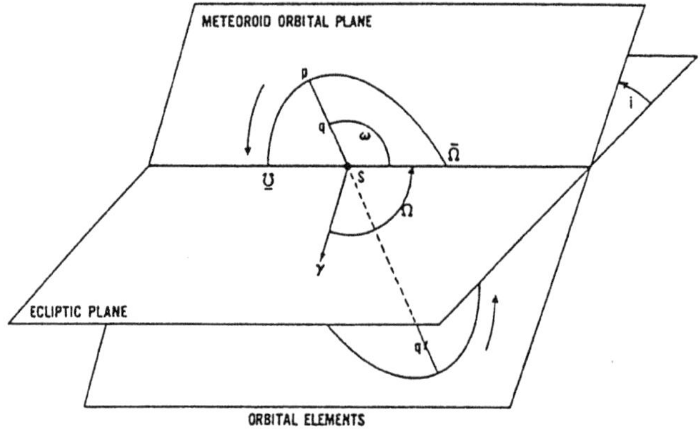

p = POINT OF PERIHELION
q = PERIHELION DISTANCE, ASTRONOMICAL UNITS
q' = APHELION DISTANCE, ASTRONOMICAL UNITS
$a = \frac{1}{2}(q + q')$ = SEMI–MAJOR AXIS, ASTRONOMICAL UNITS
$e = \frac{q' - q}{q' + q}$ = ECCENTRICITY

S = SUN
γ = HELIOCENTRIC POSITION OF THE VERNAL EQUINOX
i = INCLINATION OF METEOROID ORBITAL PLANE, DEG
$\bar{\omega} = \Omega + \omega$ = LONGITUDE OF PERIHELION, DEG
Ω = LONGITUDE OF ASCENDING NODE, DEG
$\bar{\Omega}$ = ASCENDING NODE
$\mho$ = DESCENDING NODE
ω = ARGUMENT OF PERIHELION, DEG

Fig. 7: Definition of orbital elements.

- Each planet (meteoroid) moves in an ellipse with the sun at one focus.

$$r = a(1-e^2)/1+e \cos\theta$$

a = semi-major axis, e = eccentricity

- The lines between the sun and planets sweeps out equal areas in equal amounts of time.

  $dA/dt = \text{constant}, A = \text{area}$

- The ratio of the cube of the semimajor axis to the square of the period is the same for each planet (meteoroid).

  $a^3/p^2 = \text{constant}, \; p = \text{period}$

- VELOCITY EQUATION: This useful equation give the velocity as a function of position:

  $v^2 = GM\,(2/r - 1/a)$ M = mass of body at focus
  G = gravitational constant

- Roche's limit: Roche's limit is defined as the critical distance at which a body with no tensile strength would be torn apart by tidal forces. This notion is important for the discussions of planetary rings. The Roche's limit for two touching particles is:

  $r_R = (2M/m)^{1/3} dr$, dr = dimension on satellite body

5.2.2. Non gravitational:

- Solar radiation pressure [Ref. 23]: Scattering or absorption of solar radiation by particles induces a force directed radially outwards, for spherical particles its value is:

  $F_r = Q_{pr}\ \pi\ s^2\ F_o/c$

  s is the radius of the particle
  F_o the solar irradiance ($W\ cm^{-2} sr^{-1} mm^{-1}$)
  Q_{pr} (=1 for s>1mm) an efficiency factor

- The ratio of Fg/Fr generally used to evaluate the contribution of the solar radiation pressure is given by:

  $\beta = 5.7\ 10^{-7}\ Q_{pr}/\rho s$

The solar radiation pressure could be very efficient for particles smaller than 1 mm at 1 AU.

- Poynting-Roberston effect [Ref. 22]: For a moving dust particle, the incident solar radiation is displaced by an angle v_t/c from the radial direction (v_t is the tangential component of the orbital velocity). The particle motion causes scattering and thermal emission to have a forward component. The result is a braking force and a loss of energy and the particle slowly settles into a smaller and smaller orbit, spiralling towards the sun. For a particle already in circular orbit at heliocentric distance r, the time to spiral into the sun is given by:

$$t_{pr} = 700\ s\ \rho\ r^2/Q_{pr}$$

t is in years (within the range of 10^4 years for a typical dust particle.)
s in μm is the radius of the particle
ρ in g/cm^3 is the density of the particle
r in AU

- Lorentz force [Refs. 24, 25]: The motion if IPD could be also influenced by interplanetary or planetary magnetic fields if they are electrically charged (potential taken by small particles is typically 10-100 volts). However the magnitude of this force is small and its effect a matter of debate.

$$F_1 = q\ v \times B$$

- A further cause of orbital perturbation is the mutual collisions [Ref. 4] between interplanetary particles which result in fragmentation, destruction or change in velocity (magnitude and direction) of the particles.

5.3. PHYSICAL PROPERTIES

Most of the data come from particles collected in the upper atmosphere [Ref. 41], observation of meteors [Refs. 18, 11], and study of impact microcraters [Refs. 6, 17]. Shape is complex but roughly equidimensional. Collected dust has a density in the range 1-3 g/cm^3; observation data indicate that the density is lower for large particles and higher for small particles. Such a density is consistent with a cometary origin.

- Albedo: from zodiacal light measurements [Ref. 19] a value of A = 0.09 is obtained. This value is characteristic of the high carbon content of the collected particles (chondritic material).
- Composition: It is possible to define among collected particles, three categories:
- Chondritic particles (60%), aggregates of smaller grains (0.1 mm), evidence of chondritic abundance for major elements (primitive matter).
- Small spherules (30%) composed of iron-sulfur-nickel could be the result of melting during entry.
- Mafic silicates, olivine and pyroxene rich (10%).

The results are consistent with spectroscopic measurements made by Helios [Ref. 21]. Data obtained upon the Halley fly-by on cometary dust indicate a very dark surface (A=4%) and silicate type particles coated by an organic component. New data are expected from the analysis of remnants inside impact craters formed on materials exposed to space and retrieved (LDEF, MIR) [Refs. 6, 8, 17], see Table 4.

Table 4: Composition of interplanetary dust particles.

CHEMICAL COMPOSITION OF IPD PARTICLES BY WEIGHT
NORMALIZED TO SILICON (14-6)

Element	Fraction
Na	0.04
Mg	0.74
Al	0.06
Si	1.00
S	0.40
K	0.01
Ca	0.07
Cr	0.02
Mn	0.03
Fe	1.25
Ni	0.08

SUMMARY OF DATA:

- Mass: 10^{-16} g to 1g
- Density: 0.2 g/cm^3 to 8 g/cm^3, with the following possible distribution:

$m < 10^{-6}$ g	$\rho = 2$ g/cm^3
$10^{-6} < m < 0.01$ g	$\rho = 1$ g/cm^3
$m > 0.01$ g	$\rho = 0.5$ g/cm^3

- Average velocity: 20 km/s
- Shape: Equidimensional

5.4. FLUX-MASS MODELS

- Flux: Omnidirectional and is given in terms of integral flux: number of particles/m^2/s of mass m or greater against a randomly tumbling plate (2π sr field of view).

The model proposed in the NASA SP-8013 (1969) [Refs. 2, 3] is still largely used:

- Average cumulative sporadic flux mass model is given as:

$10^{-6} < m < 1$ g $\quad \log N_{sp} = -14.41 - 1.22 \log m$

$10^{-12} < m < 10^{-6}$ g $\quad \log N_{sp} = -14.339 - 1.584 \log m - 0.063 (\log m)^2$

N_{sp} is the number of particles of mass m or greater per m^2 per second and m is the mass in grams. The model is valid at 1 AU from the sun and must be corrected for earth focusing.

A slightly different model from Grün, et al. [Ref. 42] has been proposed for the space station environment definition (B.J. Anderson) in NASA SSP 30425 [Ref. 44]:

Interplanetary flux at 1 AU is given as:

$$F_{ip} = (c_1 m^{0.306} + c_2)^{-4.38} + c_3 (m + c_4 m^2 + c_5 m^4)^{-0.36} + c_6 (m + c_7 m^2)^{-0.85}$$

where:

$c_1 = 2.2\ 10^3$ $\quad c_5 = 10^{27}$
$c_2 = 15$ $\quad c_6 = 1.3\ 10^{-16}$
$c_3 = 1.3\ 10^{-9}$ $\quad c_7 = 10^6$
$c_4 = 10^{11}$

- The variation with heliocentric distance is given as $R^{-1.5}$ (R in AU)
- Rapid decrease with distance from ecliptic plane.

Figure 8 shows current model (Grün) of the interplanetary flux at 1 AU.

- MODIFICATION OF FLUX NEAR A PLANET:
  - For planetary shielding:

    shielding factor : $sf = (1 + \cos e)/2$

    with $\sin e = R_p / (R_p + H)$

    R_p = planetary radius
    H height above atmosphere

  - For gravitational focusing:

    $G = 1 + (.76\ R\ V_p^2 r_p) / Ve^2 r$

    where: r_p distance from planet center
    r distance from planet center
    R distance of the planet from the sun in AU
    V_p and V_e escape velocities from planet and earth

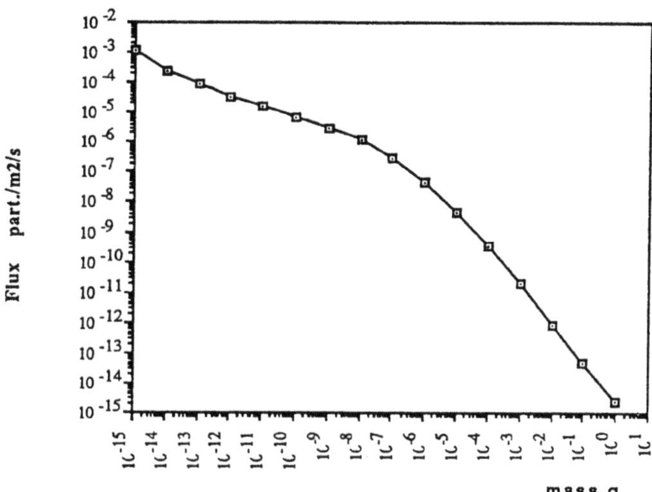

Fig. 8: Interplanetary flux at 1 AU.

5.5. VARIATION WITH HELIOCENTRIC DISTANCE

Before the Pioneer 10 and Pioneer 11 missions, no measurements have been made of the concentration of meteoroids beyond 1.6 AU. Results from observation of cometary meteors implies a variation of the spatial density of cometary meteoroids as R^{-1} to $R^{-1.5}$ (R is the distance from the sun in AU). Results from Pioneers [Ref. 27] show that the spatial density of meteoroids larger 10^{-9} g is essentially constant between 1 and 18 AU (penetration flux $10^{-6}/m^2/s$). There was no increase of particles in the size range within the asteroid belt. The increase of penetrations observed at Jupiter orbit ($10^{-3}/m^2/s$ at $3R_J$) is in agreement with the increased predicted by gravitational focusing for meteoroids in near circular orbits. At Saturn encounter the same increase was observed, probably the result of impact of ring particles, not meteoroids (ring E, at 3.1 R_S, penetration rate $10^{-3}/m^2/s$). The small meteoroids between the asteroid belt and Jupiter orbit are believed to be in eccentric orbits inclined with respect to the ecliptic plane.

5.6. SPORADIC METEOROIDS AND STREAMS

Noticeable increases in the average rate of meteor activity have been observed at regular intervals during the year. These increases are caused by the passage of the earth through a stream of particles traveling in similar heliocentric orbits. These streams are associated in general with comets, but some of them could be associated with asteroids [Refs. 18, 31, 32, 33, 34]. For a mission of short duration in LEO, or an interplanetary mission, it is necessary to take into account the crossing of the different streams by the spacecraft. (See Table 5.)

Table 5. Cometary streams. [Ref. 2]

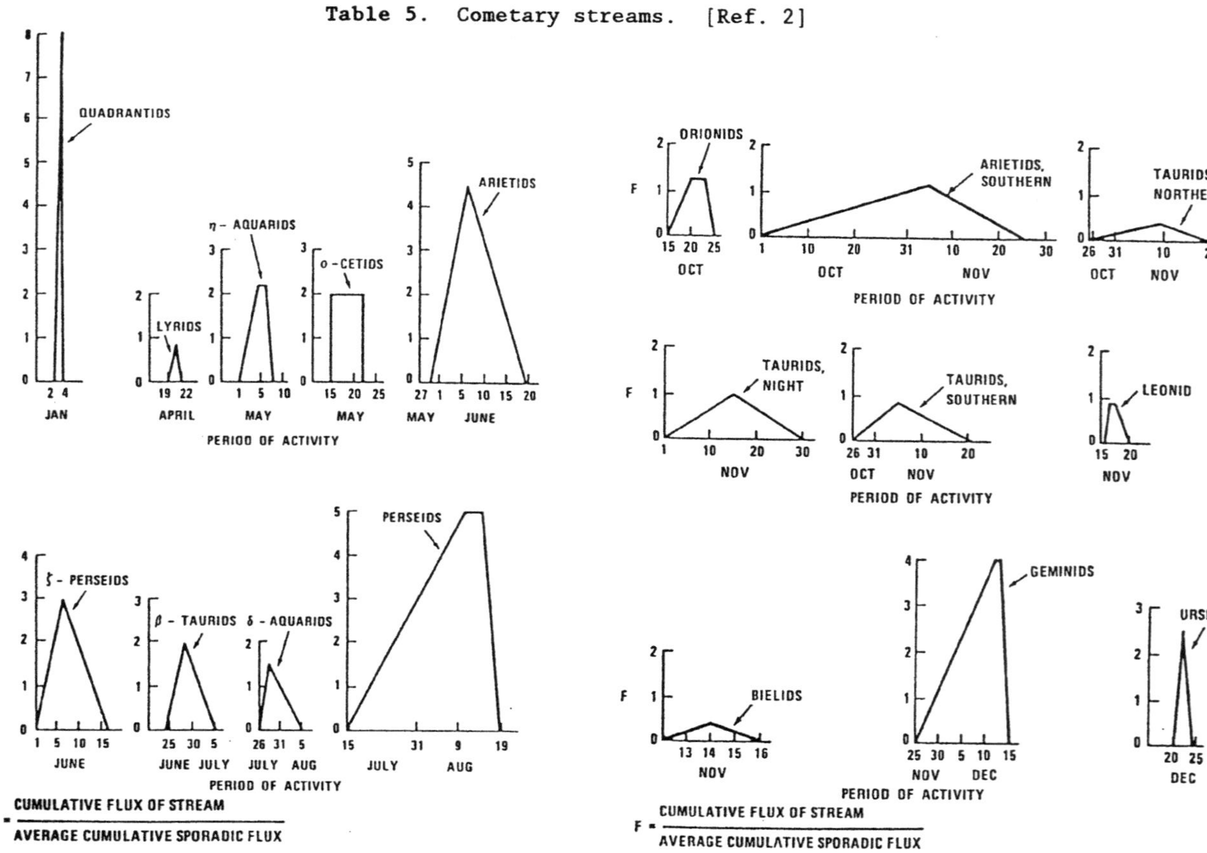

Data on the size distribution of particles inside individual streams are scarce, mainly for the particles smaller than 10^{-6} g (not visible from meteor activity); however, a model has been proposed in the NASA SP-8013 monography [Refs. 2, 11, 18, 33, 38].

for $10^{-6} < m < 1$ g:

$$\log N_{st} = -14.41 - \log m - 4. \log (V_{st}/20) + \log F$$

N_{st} is the number of stream particles of mass m or greater (per m^2/s)
V_{st} is the velocity of each stream in km/s
F stream flux enhancement factor

For a peculiar mission, taking into account the orbital elements of each stream and the orbit of the spacecraft, it is possible to compute the flux enhancement due to the crossing of the meteor streams. As the flux is no longer omnidirectional, flux and impact velocity should be computed for the various faces of the spacecraft. Full details on computation are given in [Ref. 16].

5.7. RINGS OF PARTICLES

Three of the outer planets have prominent ring systems, probably particles left-over from planetary formation. The inner planets are devoid of any rings, perhaps because they are too near the sun for icy materials to have condensed on any such remnants (see Table 6).

Jupiter. A ring of material has been discovered by Voyager 1, its outer edge extends 129000 km from the center of the planet. It is 800 km wide and less than 30 km thick. Optical depth of the ring material is within the range of 10^{-3} in the thermal infrared. Most of the particles must be micron sized particles; they are very dark (carbonaceous material).

Saturn. Its rings system is one of the most prominent feature of Saturn. It is extremely complex as shown by Voyager missions. Rings A and B were discovered by Huyghens in 1659 and separated by the Cassini division. Several other rings have been added, see Fig. 9. They are divided in numerous ringlets, and beyond the present resolution of 100 meters, yet finer structure may exist. Diffuse radial markings, the spokes, are thought to result from interactions between the ring particles, the magnetospheric ions and electrons and the magnetic field of Saturn. Most of the intricate features of the rings result from a variety of gravitational interactions (between particles themselves and between satellites). The satellites could be an important source of ring particles; the Roche's limit criteria is of primordial importance for the formation of the rings. It lies outside the rings for Jupiter and Uranus and at the distance of the Cassini division for Saturn. Ring particles for Saturn could have diameters between 5 mm and 5 m. Water ice is one of the major components of particles.

Table 6: Data on ring particles.

Feature	Distance (km)	Distance $(R_s)^b$	Period (hours)	Comments
Cloud tops[c]	60,330	1.000	10.657	Near 100-m bar level
D ring inner edge	~67,000	1.11	4.91	Extremely small optical depth
C ring inner edge	73,200[d]	1.21	5.61	
B ring inner edge	92,200[d]	1.53	7.93	
B ring outer edge	117,500[d]	1.95	11.41	Inner edge of Cassini division
A ring inner edge	121,000[d]	2.01	11.93	
Encke division	133,500[d]	2.21	13.82	Outer edge of Cassini division
A ring outer edge	136,200[d]	2.26	14.24	About 200-km wide
F ring	~140,600	2.33	14.94	Three narrow components
G ring	~170,000	2.8	19.9	Seen only in forward-scattered light
E ring inner edge	~210,000	3.5	27.3	
E ring maximum	~230,000	3.8	31.3	Near orbit of Enceladus
E ring outer edge	~300,000	5.0	46.6	

Uranus. The nine rings of Uranus are very narrow (between 3 km and 100 km). They are very dark, with an albedo lower than 2.5% (carbonaceous material). Most of the particles should be smaller than 1 meter in radius.

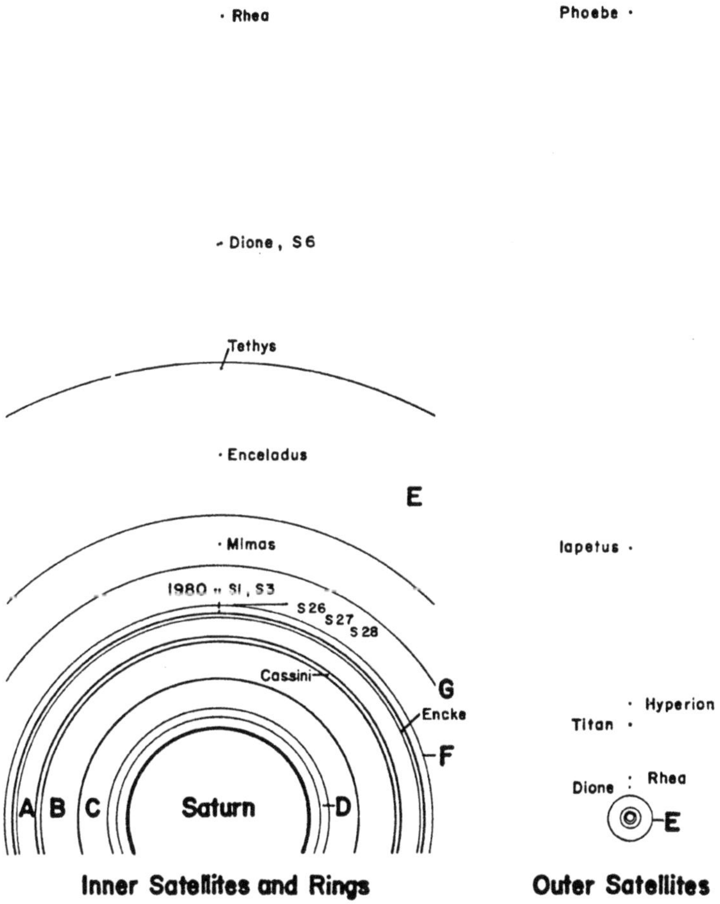

Fig. 9: The rings of Saturn [Ref. 1].

6. Concluding Remarks

Modeling of interplanetary space is a complex and very wide task. As far as the interplanetary dust particles are concerned, in the recent years valuable data have been gathered from the many space experiments devoted to their study. Many uncertainties still remain in the modeling, and it is unlikely that a unique model could be valid for the different regions of the solar system [Ref. 45]. From the engineering point of view, realistic simulation of effects on spacecraft of very high velocity particles is still a challenge, as it was the case during the design of the Giotto encounter with Halley's comet.

References

[1] R.E. Smith and G.S. West, Space and planetary environment criteria guidelines for use in space vehicles development, NASA TM-82501 (1983).

[2] B.G. Cour-Palais, Meteoroid environment model, NASA SP 8013 (1969).

[3] D.J. Kessler, Meteoroid environment model, NASA SP 8038 (1970).

[4] Ch. Leinert and E. Grün, Interplanetary Dust, In Physics and Chemistry in Space, Springer (1988).

[5] J-C. Mandeville and J.A.M. McDonnell, Micrometeoroid multiple foil penetration and particle recovery experiments on LDEF, in IAU Symposium 90 proceedings, D. Reidel, Dordrecht (1980).

[6] J.A.M. McDonnell, et al., First results of particule impacts and foil perforations on LDEF, XXVIII Cospar 1990, to be published (1990).

[7] J.L. Warren, et al., The detection and observation of meteoroid and space debris impact features on the Solar-max satellite, Proc XIXth Lun. Plan. Sci. Conf. 641-657 (1989).

[8] J-C. Mandeville, Aragatz Mission Dust Collection Experiment, in Adv. Space Res, vol. 10, pp. 397-401 (1990)

[9] D.J. Kessler, R.C. Reynolds, and P.D. Anz-Meador, Orbital debris environment, NASA TM 100471 (1989).

[10] W.M. Alexander, et al., Review of direct measurements of interplanetary dust from satellites and probes, Proc. Third Intern. Space Sci. Symp. W. Priester, ed., J. Wiley, p. 891 (1963).

[11] S.F. Singer and J.E. Stanley, Submicron particles in meteor streams, in IAU Symposium 90 proceedings, D. Reidel, Dordrecht (1980).

[12] W.K. Hartmann, Moon and Planets, Bogden and Quisley (1972).

[13] M.Y. McSween, Meteorites and their parent planets, Cambridge Univ. Press (1987).

[14] B.W. Jones, The solar system, Pergamon press 91984).

[15] C.W. Allen, Astrophysical quantities, The Athlone Press, Univ. of London (1955).

[16] G. Melkonian, et al., Computation of micrometeoroid hazards, application to Mars mission, Proc. ESA workshop on space environment analysis, ESA WPP-23 (1991).

[17] J.C. Mandeville, Debris and micrometeoroids: flight data from MIR and LDEF, Proc. ESA workshop on space environment analysis, ESA WPP-23 (1991).

[18] A.H. Delsemme (ed.), Comets, asteroids, meteorites: interrelations, evolution and origin, The Univ. of Toledo Press (1977).

[19] C. Leinert, Zodiacal light - a measure of the interplanetary environment, Space Science Reviews, 18, pp. 281-339 (1975).

[20] L.B. LeSergeant d'Hendecourt and P.L. Lamy, On the size distribution and physical properties of interplanetary dust grains, Icarus, 43, pp. 350-372 (1980).

[21] E. Grün, et al., Orbital and physical characteristics of micrometeoroids in the inner solar system as observed by Helios 1, Planetary and Space Science, 28, pp. 333-349 (1980).

[22] S.P. Wyatt and F.L. Whipple, The Poynting-Robertson effect on meteor orbits, Astrophysical Journal, 11, pp. 134-141 (1950).

[23] J.A. Burns, et al., Radiation forces on small particles in the solar system, Icarus, 40, pp.1-48 (1979).
[24] G. Consolmagno, Lorentz scattering of Interplanetary dust, Icarus, 38, pp. 398-410.
[25] J. Millet, et al., On the electrostatic potential of interplanetary grains: influence of the thermoionic effect, Astron. and Astrophysics, 92, pp. 6-12 (1980).
[26] J.E. Stanley, et al., Interplanetary dust between 1 and 5 AU, Icarus, 37, pp. 457-466 (1979).
[27] D.H. Humes, Results of pioneer 10 and 11 meteoroid experiments: interplanetary and near saturn, JGR, 85, pp. 5841-5882 (1980).
[28] T. Gehrels (ed.), Asteroids, Univ. Arizona Press (1979).
[29] B. Zellner (ed.), Icarus, 40, 3.
[30] M.J. Gaffey and T.B. McCord, Asteroid surface materials: mineralogical characterization from reflectance spectra, Space Sci. Rev., 21, pp. 556-628.
[31] A.F. Cook, A working list of meteor streams, in Evolutionary and physical properties of meteoroids, IAU coll. 13, NASA SP-319 (1971).
[32] B.G. Mardsen, Catalogue of cometary orbits, sixth edition, IAU, SAO Cambridge (1989).
[33] J.D. Drummond, Earth approaching asteroid streams, Icarus, 89, pp. 14-25 (1991).
[34] C.I. Lagerkvist, et al. (eds.), Asteroids, comets and meteors II, Uppsala Univ. Press (1986).
[35] J.C. Mandeville, et al., Reflectance polarimetry of callisto and the evolution of the galilean satellites, Icarus 41, pp. 343-355 (1980).
[36] A. Dollfus, et al., Reflectance spectrophotometry extended to uv for terrestrial lunar and meteoritic samples, Geochimica et Cosmochimica Acta, 44, pp. 1293-1310 (1980).
[37] 20th ESLAB Symposium on the exploration of Halley's comet, ESA SP-250 (1986).
[38] V.S. Sykes, Cometary and asteroids sources of interplanetary dust, Proc. IAU coll. 126, Kyoto, in press (1990).
[39] I. Mann and B. Kneissel, Interplanetary dust close to the sun, Proc. IAU coll. 126, Kyoto, in press (1990).
[40] M.S. Hanner, On the albedo of interplanetary dust, Icarus, 43, pp. 373-380 (1980).
[41] D.E. Brownlee, Microparticles studies by sampling techniques, in Cosmic Dust, J.A.M. McDonnell, ed., pp. 295-336, Wiley, Chichester (1978).
[42] J.B. Hartung, et al., Lunar surface processes: report of the 12054 consortium, Proc. Lunar Planet Sci. Conf. 9th, pp. 2507-2537 (1978).
[43] E. Grün, et al., Collisional balance of the meteoritic complex, Icarus 62, pp. 244-272 (1985).
[44] B.J. Anderson, Meteoroid and Orbital debris, in NASA SSP 30425 (1990).
[45] R.G. Storm, The solar system cratering record, Icarus 70, pp. 517-535 (1987).
[46] J.C. Mandeville, Microcraters on lunar rocks, Proc. Lunar Sci. Conf. 7th, pp. 1031-1038 (1976).

[47] B. Zellner, et al., The asteroid albedo scale.I, Laboratory polarimetry of meteorites, Proc. Lunar Sci. Conf. 8th, pp. 1091-1110 (1977).

ATOMIC OXYGEN AND ITS EFFECT ON MATERIALS

R. C. TENNYSON
University of Toronto Institute for Aerospace Studies
4925 Dufferin Street
Ontario Canada M3H 5T6

ABSTRACT. This lecture addresses the issue of atomic oxygen and its effect on the erosion of spacecraft materials. The nature of the reaction with materials is discussed together with observations from various shuttle experiments, the Long Duration Exposure Facility and ground based atomic oxygen test facilities. In addition, synergistic effects due to UV radiation are examined. Protective coatings and their limitations arising from pin hole defects and micrometeoroid/debris impacts are also considered. Finally, design implications are addressed in terms of material lifetime predictions.

1. Atomic Oxygen Environment

The space environment is characterized by the presence of very low pressure (vacuum), with various atomic species in low concentrations, charged particles, temperature extremes, electromagnetic radiation, micrometeoroids and man-made debris. To some extent, all of the factors influence the design of satellites, depending upon orbital altitude. For this discussion, "space" is defined as beginning at the lowest altitude (nominally around 225 km) that permits a satellite to continuously orbit the Earth. It is evident from Fig. 1 [Ref. 1] that various atomic species exist in the low earth orbit region (LEO), with charged particles populating the higher altitudes (>10^3 km). Note the influence of solar activity on the concentration levels. Figure 2 [Ref. 2] presents in more detail the concentration levels of the various gaseous species in LEO and it is clear that atomic oxygen (AO) is the dominant constituent. Although the AO concentration value may appear to be insignificant compared to the particle concentration at the Earth's surface (~10^{19} a/cm^3), the actual flux of atoms impinging on an orbiting vehicle is quite high (~10^{14} a/cm^2-s) due to the satellite orbital velocity of ~8 km/sec (see Fig. 3, [Ref. 2]). This corresponds to a mean AO energy of ~4.8 ev (Fig. 4, [Ref. 3]).

Space systems operating in a low Earth orbit environment for long periods of time require materials which are stable in that environment. Since the U.S. Space Shuttle began operations, it has been determined that

R. N. DeWitt et al. (eds.), The Behavior of Systems in the Space Environment, 233–257.

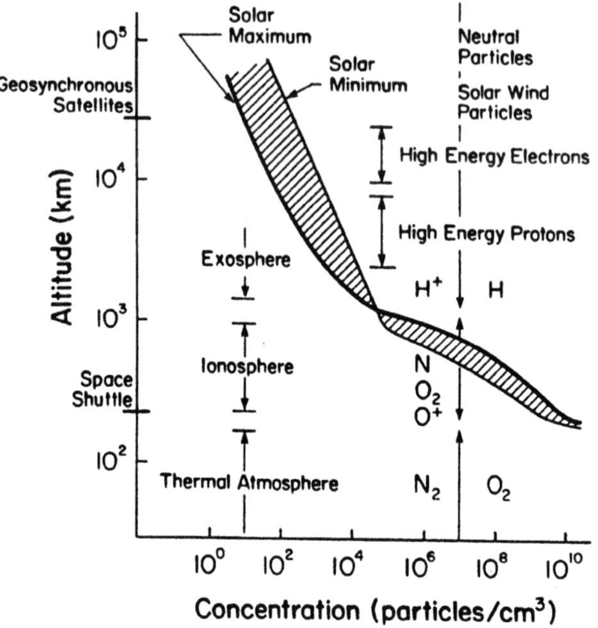

Fig. 1: Particle concentration levels as a function of attitude. [Ref. 1].

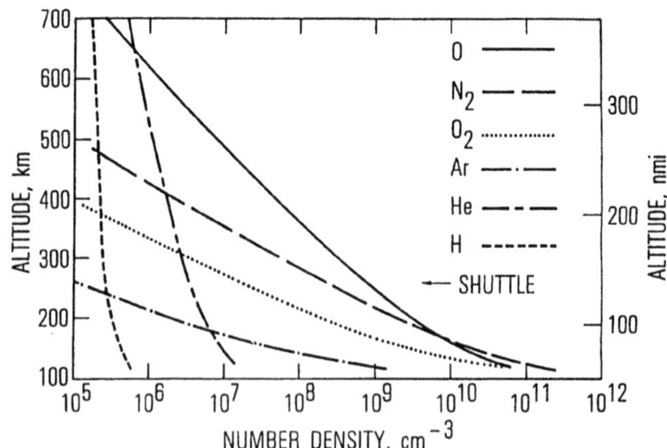

Fig. 2: Particle densities in the low earth orbital environment [Ref. 2].

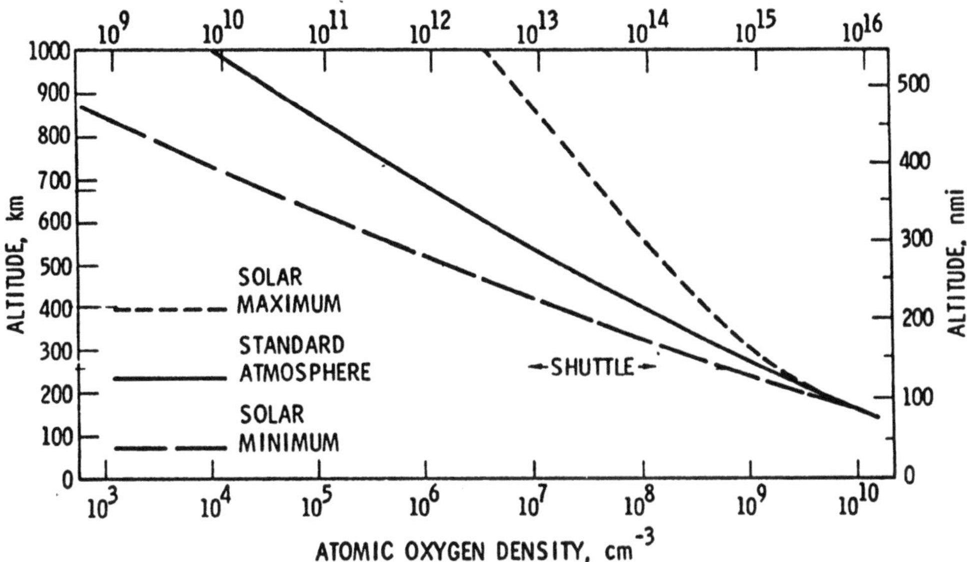

Fig. 3: The atomic oxygen flux on a spacecraft orbiting at 8 km/sec. [Ref. 2].

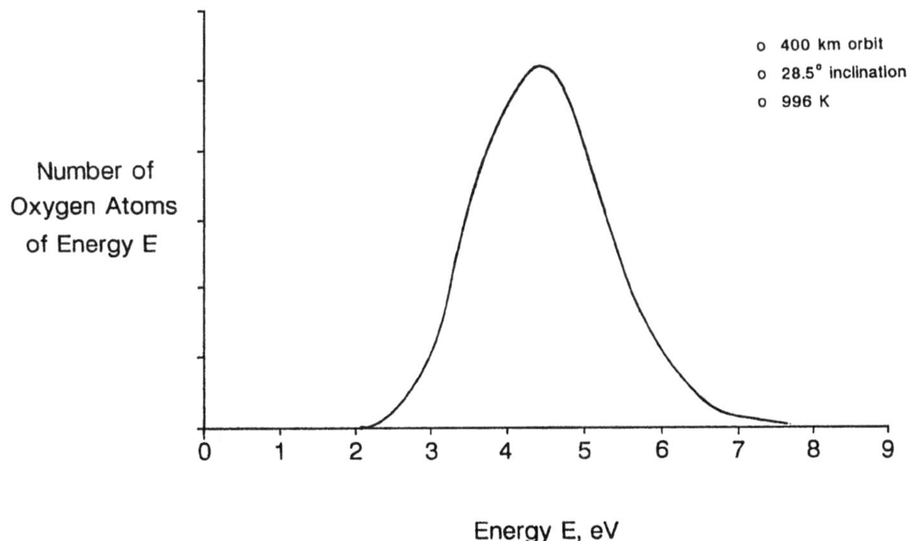

Fig. 4: Typical atomic oxygen energy distribution in low earth orbit. [Ref. 3].

AO has caused significant erosion of many materials [enhanced in some instances by the presence of ultraviolet (UV) radiation], particularly polymers and polymer-based composites. These materials are of special interest because of their applications to space structures, space robots and manipulator arms (e.g., Canadarm on the Shuttle and the Mobile Service System for the Space Station), solar arrays, thermal blankets and second surface mirrors for example. Experience has shown that even a short-term exposure to this environment can have harmful effects on spacecraft surfaces, particularly if they are composed of or covered with organic materials. Organic polymers are especially important to the design of present and future generations of spacecraft power systems. However, because of atomic oxygen, irreversible degradation of material properties can occur (optical, electrical, thermal and mechanical).

2. Atomic Oxygen Reaction with Materials

Atomic oxygen (ground state neutral) is produced by dissociation of O_2 by the vacuum UV radiation (VUV, 1000~2000 Å where 1 Å = 10^{-8} cm) i.e., O_2 + hν → O + O as shown in Fig. 5. Subsequent interaction of AO with a material can lead to erosion or growth of an adherent oxide. As illustrated in Fig. 5, erosion is caused by the formation of volatile oxides (ex: CO) or a non adherent oxide such as occurs with silver (Ag).

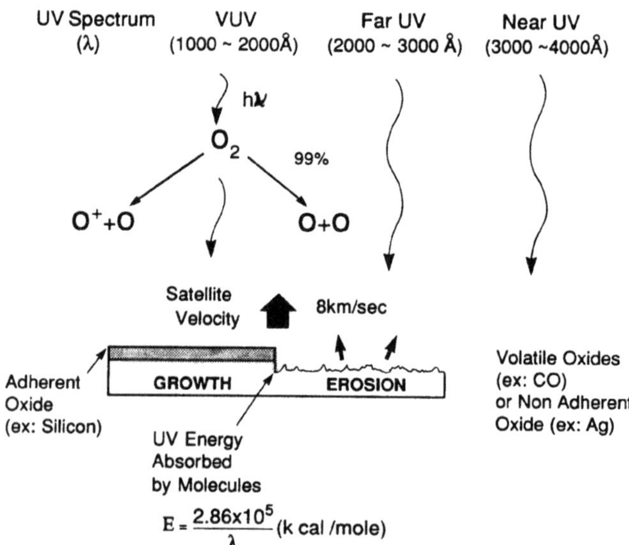

Fig. 5: Reaction of atomic oxygen and UV radiation with a satellite surface.

On the other hand, it is also possible for AO to promote the growth of stable adherent oxides (ex: with silicon coatings). Erosion can also be enhanced by the interaction of UV radiation, in which molecular bonds can be broken through the absorption of UV energy, given by the relation [Ref. 4].

$$E = \frac{2.86 \times 10^5}{\lambda} \quad (K\ cal/mole)*$$

where λ = wavelength (Å).

* Note: the conversion to electron volts (ev) is 1 ev ≅ 23.06 Kcal/mole.

It is clear from data available on chemical bond energies shown in Table 1 [from Ref. 4] that several chemical bonds can in fact be broken by the shorter wavelength VUV radiation.

Various reaction mechanisms have been proposed [Ref. 5] to account for material degradation due to AO. These include:

- abstraction of atoms directly from compounds,
- attachment of O atom to a compound with subsequent elimination of excited state molecule,
- insertion of O atom between two bound atoms with possible portion of molecule departing from the compound as a radical.

Table 1: Chemical Bond Energy* [from Ref. 4].

Bond	Bond Energy (K cal./mole, 25°)
C-C	82.6
C=C	145.8
C≡C	199.6
C-N	72.8
C=N	147
C≡N	212.6
C-O	85.5
C=O aldehydes	176
C=O ketones	179
C-S	65
N-N	39
N=N	100
Si-O silicones	106**

* All values are deduced from aliphatic compounds and are taken from T.L. Cottrell, "The strengths of Chemical Bonds," Butterworths Scientific Publications, London (1958).

** Doubtful value.

To illustrate how erosion can occur, consider the potential energy diagram for two carbon atoms (C) interacting with an incident oxygen atom (O) having a velocity V_z and energy E_z higher than the farfield repulsive level (see Fig. 6). The attractive (V^+) and repulsive (V^-) energy curves are shown intersecting at some distance R_c. The erosion model being considered here requires the O atom to first bond to the carbon atoms. The probability of this occurring is given by the probability that the O atom will cross the curves quantum mechanically at R_c at an energy E^*, thus occupying the lowest energy level shown at $z = 0$. Once bonded to the carbon atoms, it is possible to consider a subsequent impact at this bond site by another O atom having sufficient energy to break a C–O bond, releasing the volatile C≡O gas and leaving a vacancy in the carbon structure. This combination of chemical bonding – physical removal of a volatile is shown schematically in Fig. 7. The strength of the carbon bond (D_o) is about 7.4 ev. If an incident O atom strikes the surface with an energy < 7.4 ev, such that physical sputtering of a carbon atom off the surface does not occur (recall in LEO that the mean energy for AO is ~ 4.8 ev), then there exists a probability that it will chemically bond to two adjacent carbon atoms as shown. The strength of the C-O bond is ~13.1 ev. Noting that the strength of the C≡O gas phase bond is ~ 11.4 ev, then an incident oxygen atom with an energy >1.7 ev can result in the departure of the volatile C≡O with a vacancy site opened in the carbon surface. Calculation of the curve crossing probability function is complex, and to-date the author is unaware of any theoretical predictions of erosion rates using this model.

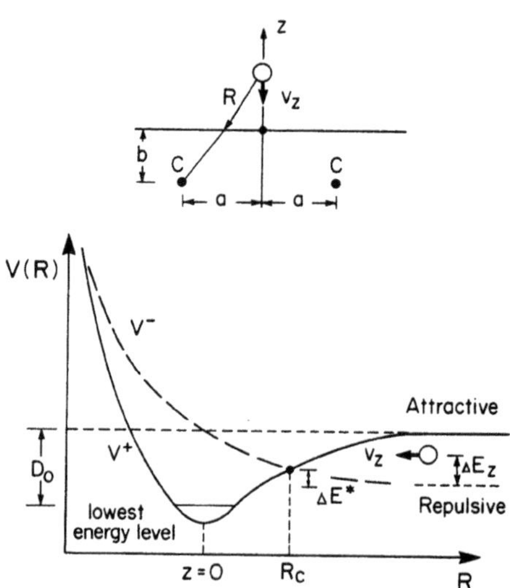

Fig. 6: Potential Energy Diagram for Oxygen Atom and Carbon Bonding.

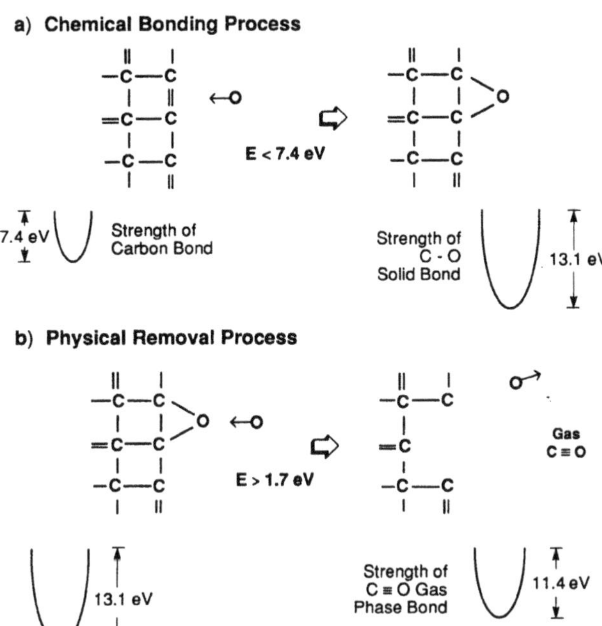

Fig. 7: Atomic Oxygen Erosion Process Graphite Example.

3. Atomic Oxygen Flux on Spacecraft Surfaces

Atomic oxygen flux ($\emptyset$) incident on a spacecraft in the ram direction (i.e. in the direction of the velocity vector) can be calculated as a function of orbital altitude from the relation,

$$\emptyset = NV \text{ (atoms/cm}^2 - \text{s)}$$

where N = number density (atoms/cm^3)

V = velocity (cm/s)

For example, Fig. 2 shows that at an altitude of ~ 400 km,

$N \approx 10^8$ a/cm^3 and for

$V = 8$ km/s then

$$\emptyset \approx 8 \times 10^{13} \text{ a/cm}^2 - \text{s}$$

assuming a standard atmosphere condition (i.e. not taking into account variations in AO density due to solar activity).

However, corrections should be considered due to orbital inclination and atmospheric motion as well as the random Maxwellian thermal motion of the oxygen atoms which correspond to about 0.49 km/s and 0.41 km/s, respectively [Refs. 3, 6]. In addition one must take into account the angular position away from the ram direction to correct for incident flux as well. As a first approximation one can modify the incident flux using the equation

$$\emptyset = NV \cos \alpha$$

where α = angle of incidence relative to the velocity vector.

Because of the random thermal velocity component it has been shown [Refs. 3, 6] that AO can impinge on surfaces at $\alpha > 90°$. Figure 8 [from Ref. 6] illustrates that α can reach about 105°. Variations of flux with altitude and angle of incidence are shown in Fig. 9 [Ref. 6]. Erosion data from ground based simulator tests illustrate this angular dependence, as shown in Figs. 10 and 11 for Kapton® and Teflon® (thin film polymer dielectric materials).

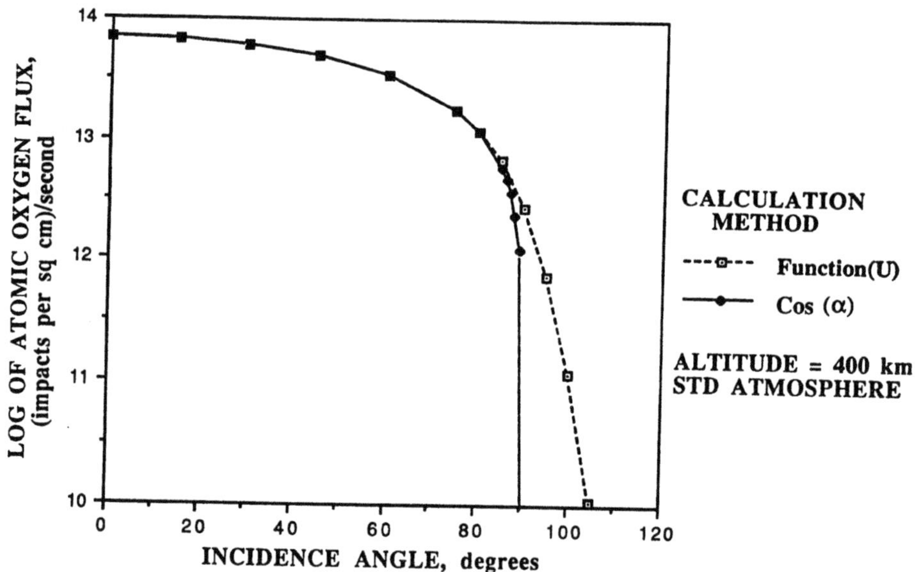

Fig. 8: Effect of Molecular Thermal Velocity on Atomic Oxygen Flux [Ref. 6].

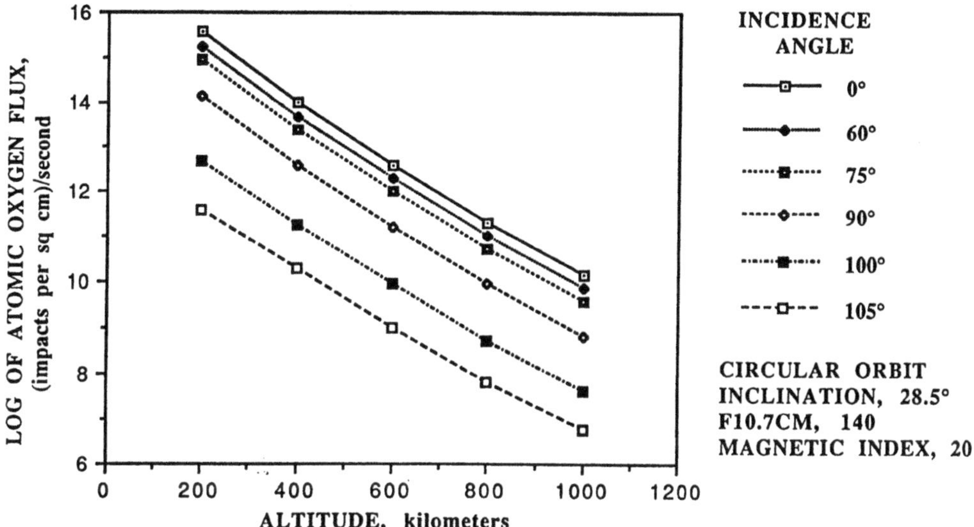

Fig. 9: Atomic Oxygen Flux, Average Conditions [Ref. 6].

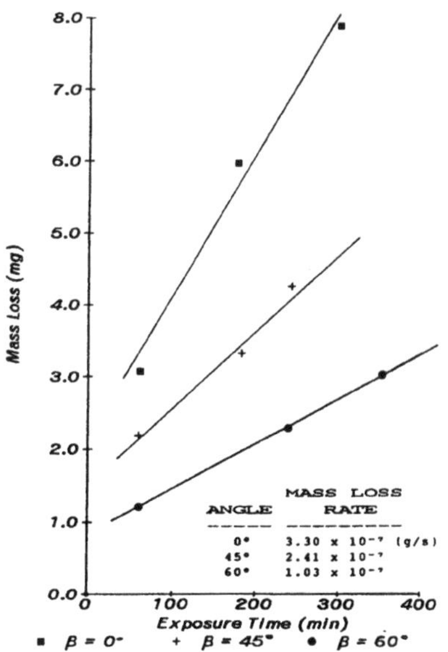

Fig. 10: Effect of AO Beam Angle KAPTON 500HN.

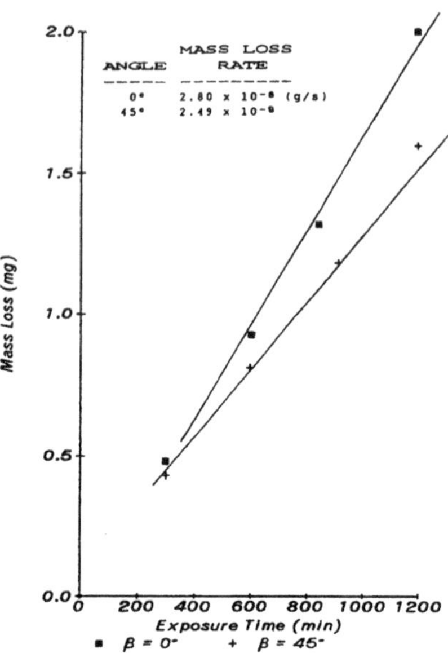

Fig. 11: Effect of AO Beam Angle FEP TEFLON 500C.

4. Material Erosion Data

The characteristic used to quantify the susceptibility of a material to erosion by atomic oxygen is the "erosion yield" (R_e, also referred to as "reaction efficiency"). This parameter is defined as

$$R_e = \frac{\text{Volume of Material Lost}}{\text{Total No. Of Incident O Atoms}} \qquad (\text{cm}^3\ /\text{atom})$$

R_e can be calculated using the relation, $R_e = \dfrac{\Delta m/\rho}{\phi t\ A}$

where Δm = mass loss (g)

ρ = material density (g/cm^3)

ϕ = incident AO flux (atoms/cm^2– s)

t = exposure time (s)

A = exposed surface area (cm^2)

Note that the product $\phi t = F$ yields the total fluence of oxygen atoms.

4.1. SPACE FLIGHT DATA

Reaction of AO with a variety of materials was reported by Leger [Ref. 7] as early as 1982, for STS missions 1-3 and in 1984 for flights 3-8 [Ref. 8]. However, it should be noted that laboratory test data from plasma asher experiments were available as early as 1965 demonstrating the effect of atomic oxygen on polymers [Ref. 9]. The most extensive data base available to-date has been compiled in Ref. 10. For design purposes, classification of this data into erosion yield ranges is given in Table 2. To illustrate the effect of AO on the surface morphology of materials during the erosion process, Figs. 12 and 13 present SEM (scanning electron microscope) photographs of Kapton®, graphite/epoxy and Teflon® samples taken from STS – 41G and the Solar Max satellite.

Perhaps the greatest single source of materials degradation data will emerge from the on-going analysis of the NASA Long Duration Exposure Facility (LDEF) which spent 5.8 years in LEO and was retrieved in January 1990.

Table 2: Classification of Erosion Yield Data [Ref. 10].

Erosion Yield Range (10^{-24} cm^3/atom)

.01 – .09	.1 – .9	1.0 – 1.9	2 – 4	> 4
Al_2O_3	*Polysiloxane/ Kapton (ARO Kapton)*	*Various forms of Carbon*	*Kapton H Polymide*	*Silver*
Al coated Kapton	*Siloxane/ Polymide*	*Epoxies*	*Polycarbonate Resin*	
Diamond	*Polysilane/ Polymide*	*Polystyrene*	*Polyester*	
ITO/Kapton	*401-C10 (flat black)*	*Polybenzimidazole*	*Polysulphone*	
Si Ox/Kapton (aluminized)	*Z-306 (flat black)*		*PMMA*	
Silicones	*Urethane (black conductive)*		*Mylar*	
RTV-615 (clear)	*Apiezon Grease*		*Polyethylene*	
Fluorpolymers	*Tedlar (white)*		*Tedlar (clear)*	
Teflon FEP	*Osmium (bulk)*		*Z-302 (glossy black)*	
Magnesium Fluoride on Glass			*Various forms of Graphite/Epoxy*	
Molybdenum			*Kevlar/Epoxy*	

A schematic of this satellite is shown in Fig. 14, together with the location of the author's composite experiment. The actual fluence of AO impinging on this experiment was estimated at 1.2 x 10^{21} atoms/cm^2, even though it was located 90° from the leading edge (82° from the velocity vector due to an 8° angle of yaw).

Despite this location, substantial erosion of the outer resin (epoxy) layer of a composite laminate can be seen in Fig. 15, comparing the unexposed region (a) with the exposed area (b). The triangle shaped patterns are associated with the AO impacting the surface at 8° above the plane (i.e., grazing incidence). For the fluence level encountered, only the top resin layer was eroded as evidenced in the cross-sectional SEM photographs shown in Fig. 16. Photograph (a) shows the resin layer and interface with the potting compound while photograph (b) clearly shows the absence of this resin layer.

It was also observed that AO reflected off adjacent structures and reacted with the back faces of the composite laminates (i.e. non exposed). Figure 17 illustrates the texture obtained which is characteristic of "normal incidence" erosion, i.e. not directional like the exposed face shown in Fig. 16.

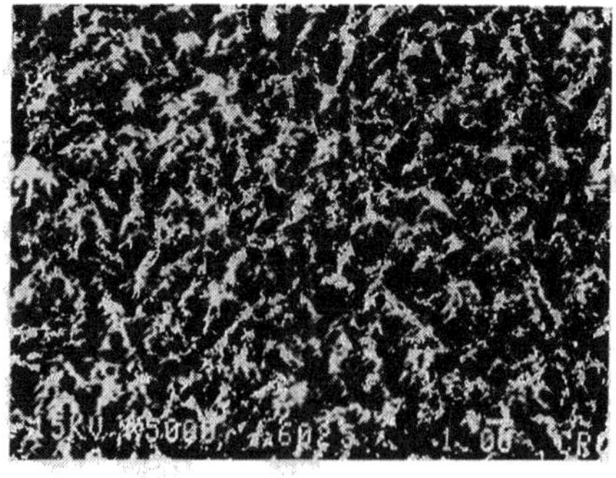

Kapton® (x5000)

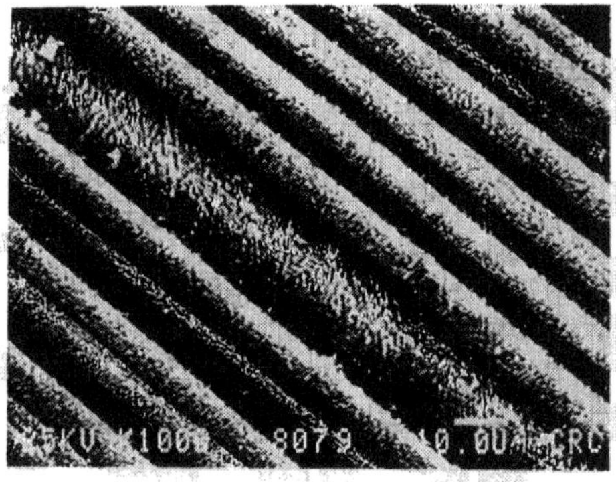

Graphite/Epoxy (x1000)

Fig. 12: SEM Photographs of Exposed Materials from STS-41G-ACOMEX Experiment (Courtesy of D.G. Zimcik, Canadian Space Agency).

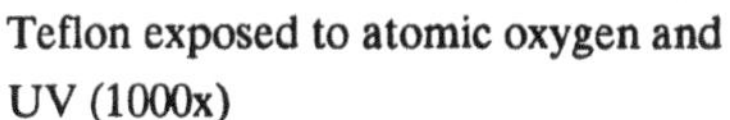

Teflon exposed to atomic oxygen and UV (1000x)

Detail of Teflon exposed to atomic oxygen and UV (5000x)

Fig. 13: SEM Photographs from SOLAR MAX Satellite (from B. Santos-Mason, [Ref. 11]).

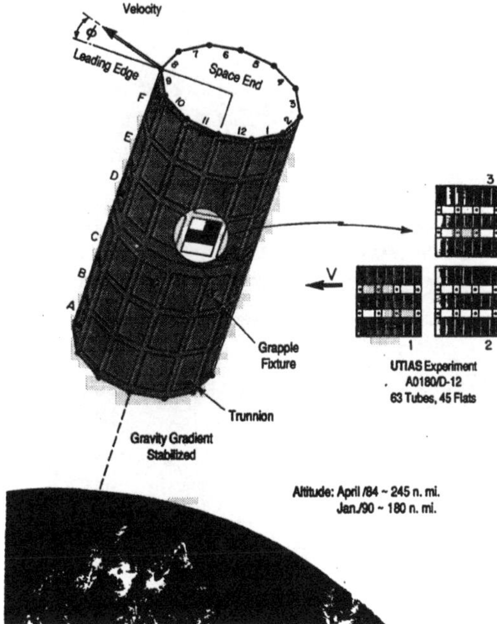

Fig. 14: Schematic of LDEF showing location of author's composite experiment.

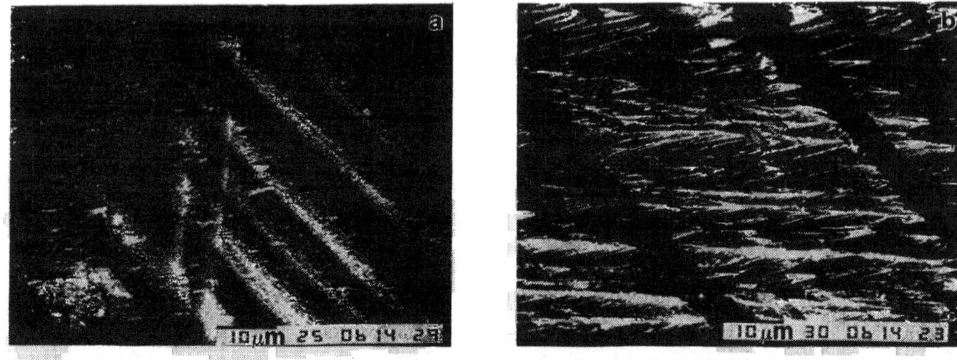

Fig. 15: SEM Photographs of Graphite/Epoxy LDEF Sample
a) 3mm from end fixture (note boundary between unexposed/exposed regions)
b) 6 mm from end fixture-exposed region.

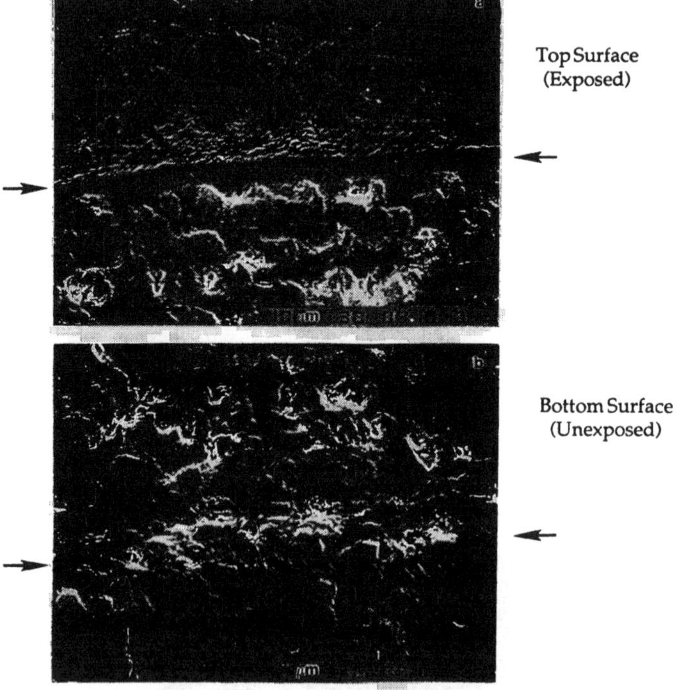

Fig. 16: Cross-Sectional SEM Micrographs of Graphite/Epoxy LDEF Sample (F15) (7 mm from end fixture, out of shadow region). Arrows indicate sample interface with potting compound.

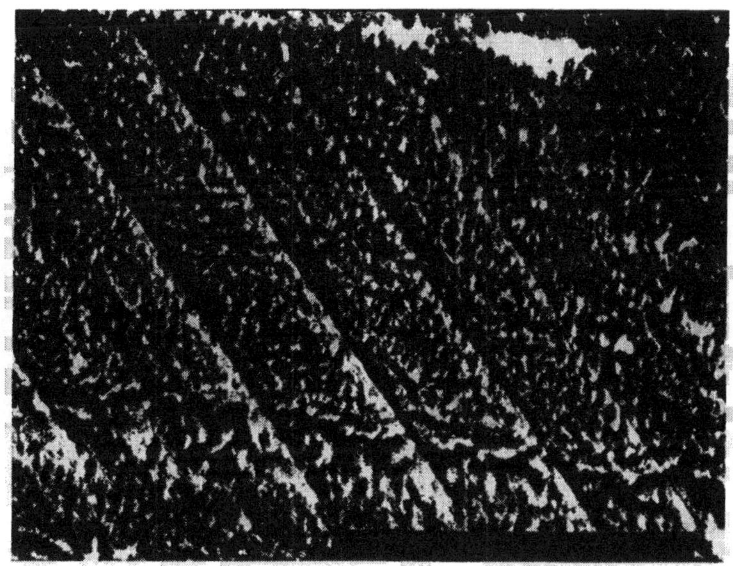

Fig. 17: SEM Photograph of Area Subjected to Reflected AO on Unexposed Face of Graphite/Epoxy Laminate (x2000).

Composite tubes were also flown on LDEF and this provided an opportunity to study the erosion profiles as a function of angle of incidence. Figures 18 and 19 present cross-sectional SEM photographs at different angular positions around the tube (note $\alpha = 8°$ corresponds to the 'ram' direction for LDEF). One can readily see the difference in erosion depth and material loss when one compares $\alpha=30°$ to $\alpha=90°$ (i.e.: grazing incidence). Very little erosion occurred at 90° whereas substantial loss of the composite reinforcing fibers can be seen at $\alpha=30°$.

4.2 GROUND BASED AO/UV SIMULATION TESTS

A recent review of AO ground based test facilities can be found in Ref. 12. One such facility resides at the University of Toronto Institute for Aerospace Studies (UTIAS) and has been used extensively to investigate AO reaction with thin film polymers, polymer matrix composites and various coatings [Ref. 12]. Table 3 [Ref. 14] presents simulator data on erosion yields (R_e) for several materials compared with flight data from Ref. 15. Figure 20 contains SEM erosion photographs for two thin film polymer materials tested in the UTIAS AO facility; Kapton® and Mylar®. It is known that UV radiation can accelerate the AO erosion process. Tests conducted clearly demonstrate enhanced erosion rates for FEP Teflon® as evidenced in Fig. 21.

Erosian Pattern at α =30°

Erosian Pattern at α =65°

Fig. 18: Cross-sectional SEM photographs at indicated angular positions around the tube (α = 8° is LDEF ram direction).

Erosian Pattern at α =75°

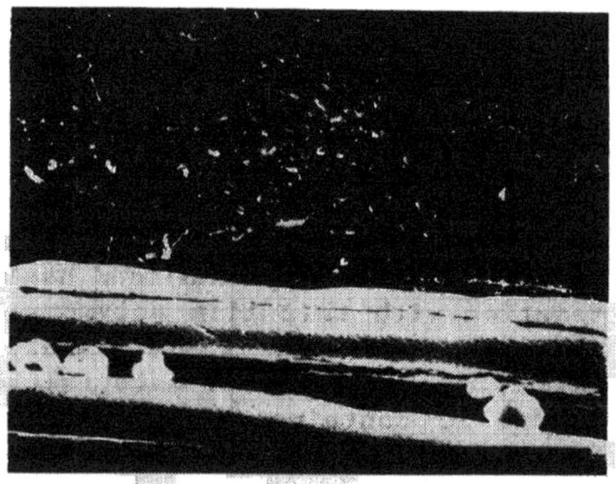

Erosian Pattern at α =90°

Fig. 19: Cross-sectional SEM photographs at indicated angular positions around the tube (α = 8° is LDEF ram direction).

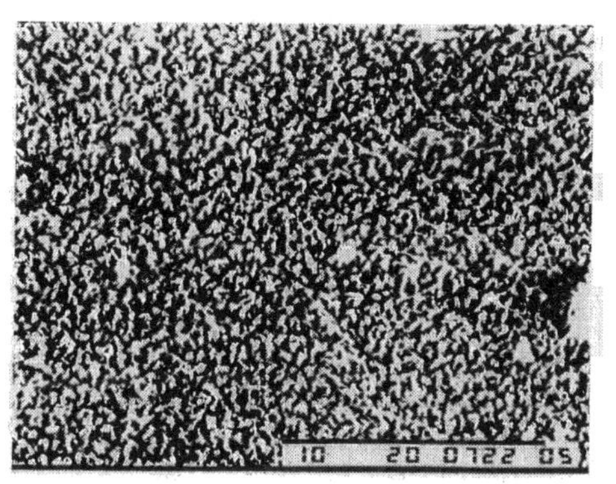

Kapton® (x5000)

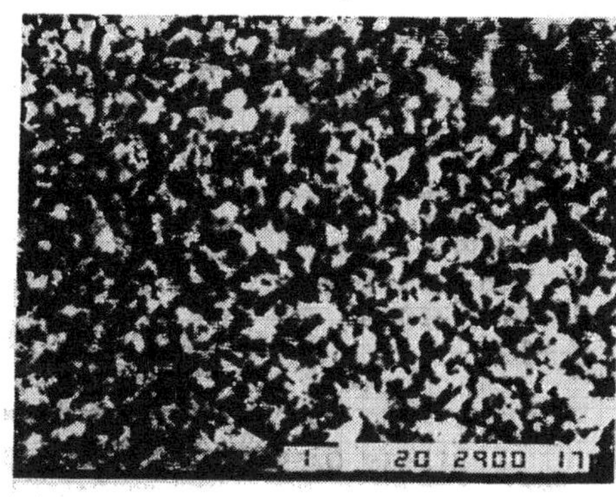

Mylar® (x10000)

Fig. 20: SEM Photographs of Materials tested in UTIAS AO Simulator ($F \sim 10^{20}$ atoms/cm^2).

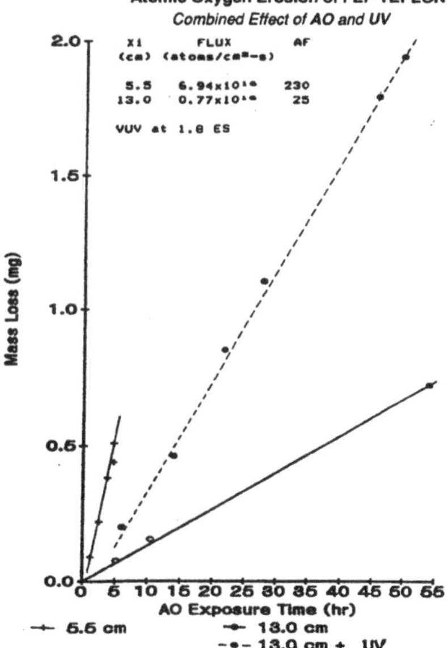

Fig. 21: Atomic Oxygen Erosion of FEP TEFLON

Table 3: Comparison with LEO Flight Data [Ref. 14].

Material	Flight Data [Ref. 15]		AO Simulator
	R_e (cm^3/atom)	$R_e/(R_e)$ Kapton	$R_e/(R_e)$ Kapton
Kapton-H	$3x10^{-24}$	1	1
Polyethylene	3.7	1.23	0.987
Mylar	3.4	1.13	1.360
Tedlar	3.2	1.07	1.260
Pyrolytic Graphite	-	-	0.318
HOPG*	-	-	0.478
Carbon	0.9~1.7	0.30~0.57	-
Teflon FEP	<0.05	<0.017	0.017~0.031
Graphite Epoxy			
1034C	2.1	0.70	-
5208/T300	2.6	0.87	-
AS4/3501-6	-	-	1.037
PEEK			
APC-2/AS4	-	-	0.738

* Highly Oriented Pyrolytic Graphite

4.3. NOMOGRAM FOR CALCULATING MATERIAL THICKNESS LOSS

To assist the reader in estimating material thickness loss for a given AO fluence level, a nomogram has been prepared (Fig. 22). Generally, the design engineer knows the altitude and time in orbit required for a specific satellite application and mission. The intersection of these two lines defines a fluence level, as shown in Fig. 22 for a particular example. One then follows the constant fluence curve until one intersects the specific material erosion curve (see Table 2 for data). Moving horizontally from this point of intersection gives the thickness loss in microns (10^{-6} m). This value corresponds to the worst case, i.e. the "ram" direction, assuming a standard atmosphere.

Any corrections for solar activity can be estimated from Fig. 3. Finally, angle of incidence correction off the ram direction can be applied to the "thickness loss" using the nomogram table.

5. Protective Coatings

One method of protecting materials from AO erosion is to employ barrier coatings having the following properties:

- the barrier must, of course, be resistant to atomic oxygen bombardment;
- it should be flexible, abrasion-resistant, and allow adhesive bonding;

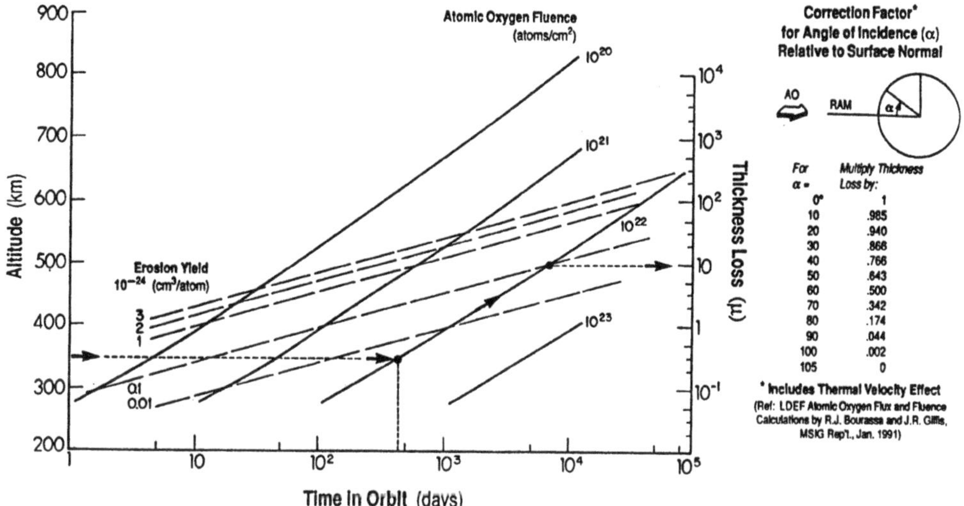

Fig. 22: Nomogram for Calculating Atomic Oxygen (AO) Fluence and Material Thickness Loss as a Function of Orbital Altitude, Time in Orbit and Angle of Incidence (8 km/s Ram, Std. Atm.).

- it should be VUV tolerant, but not alter the substrate's properties;

- finally, surface conductivity should be high in order to prevent the build-up of harmful potential gradients that might result from charging.

As an example, amorphous silicon coatings (a – Si:nH) were tested in the AO beam facility. These coatings were deposited on quartz crystals and in-situ mass loss measurements made as a function of exposure time by observing their frequency shift [Ref. 6]. Results of these tests are shown in Fig. 23. The "control" sample was exposed only to the vacuum environment.

Although there is initial mass loss for all samples, this is associated with outgassing and "cleaning" of the surface. However, it is readily apparent that the coating stabilizes with no subsequent mass loss after this initial phase of outgassing.

There has been considerable research undertaken over the past few years to develop diamond films and coatings. They provide superior properties to other materials for certain applications (ex: hardness, wear

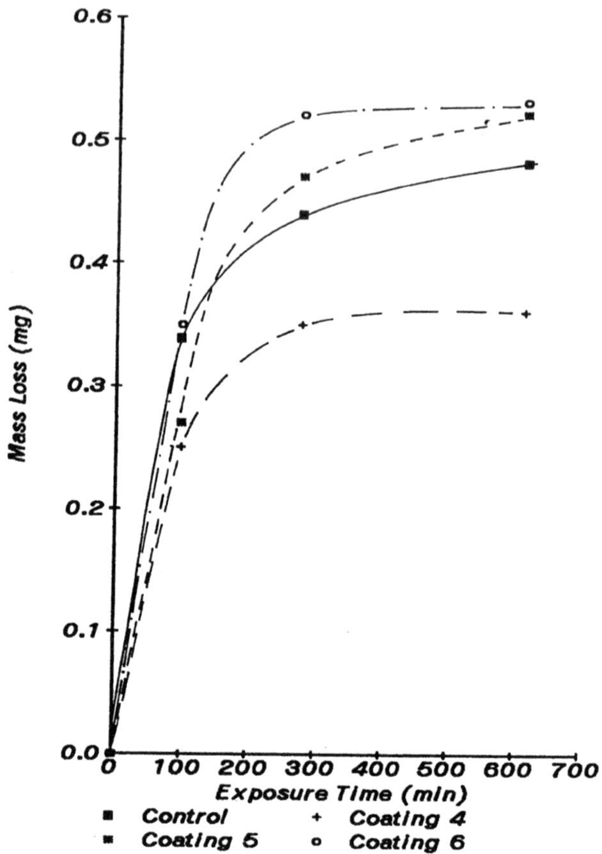

Fig. 23: Mass Loss of Si:n-H Coatings Exposed to Atomic Oxygen.

resistance, etc.), and it was decided to study their resistance to AO. Consequently, a thin diamond coating (Fig. 24a) was applied to a substrate and tested in the AO beam facility at UTIAS. It w , observed that the surface darkened considerably and SEM analysis (Fig. 24b) indicated striations on the crystal faces. Furthermore, some mass loss was also recorded, resulting in an erosion yield equal to ~1.2% of that for Kapton®.

One of the difficulties associated with coatings is the presence of pinhole defects. In addition, micrometeoroids and debris will also impact space structures and because of their high energies, cause local craters and cracking in the coatings. The nature of the interface bond between the coating and substrate will play an important role in determining the extent to which the coating will flake off. This will be exacerbated in the presence of AO which can penetrate through to the substrate and cause

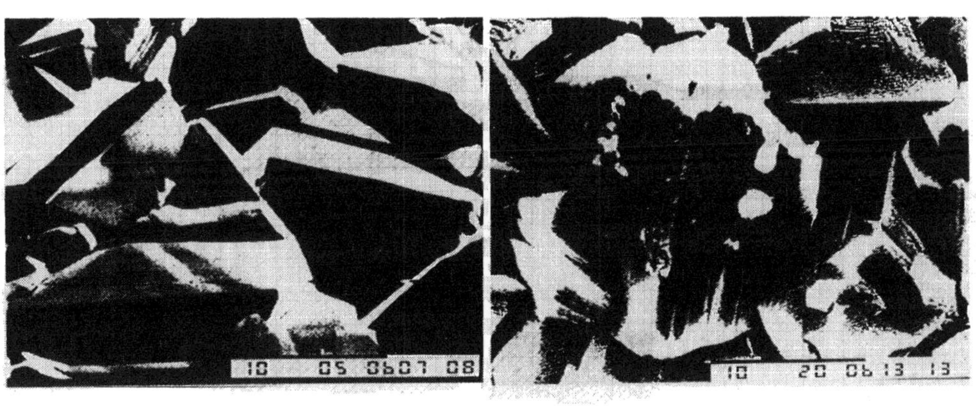

(a) Unexposed (x3500) (b) Exposed (x3500)

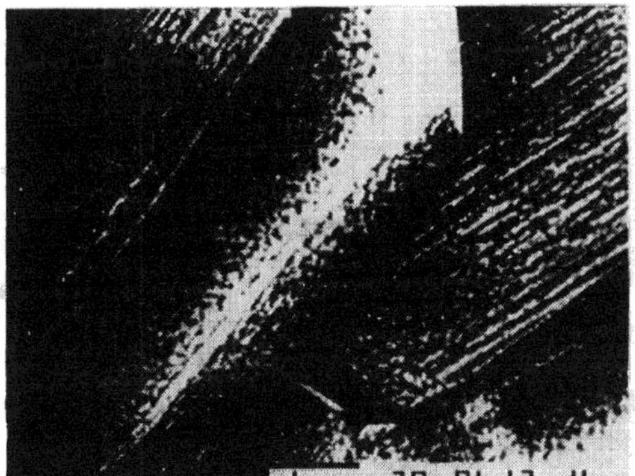

(c) Exposed (x15000)

Fig. 24: SEM Photographs of Diamond Film Before (a) and After (b,c) Exposure to Atomic Oxygen.

erosion. The illustrations shown in Fig. 25 and Fig. 26 [from Ref. 17] depict the effects of AO impinging on a reactive substrate through such a defect. Directional effects are quite clear and the potential for serious undercutting due to swept AO impingement is evident. The actual growth of undercutting was measured as a function of time in a plasma asher experiment reported in Ref. 10. Figure 27 presents these results for several defect sites.

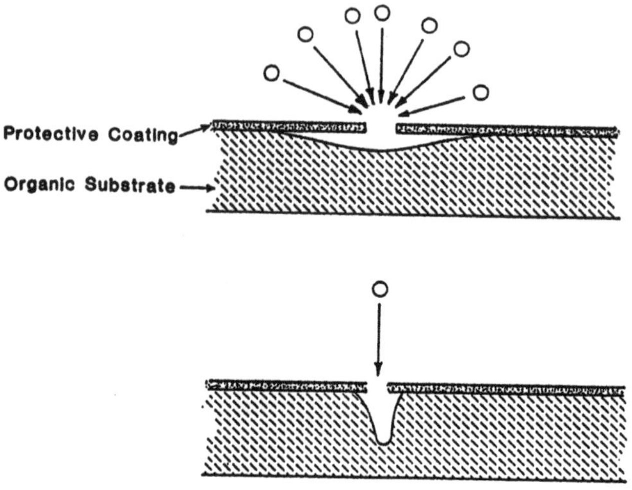

Fig. 25: [Ref. 17].

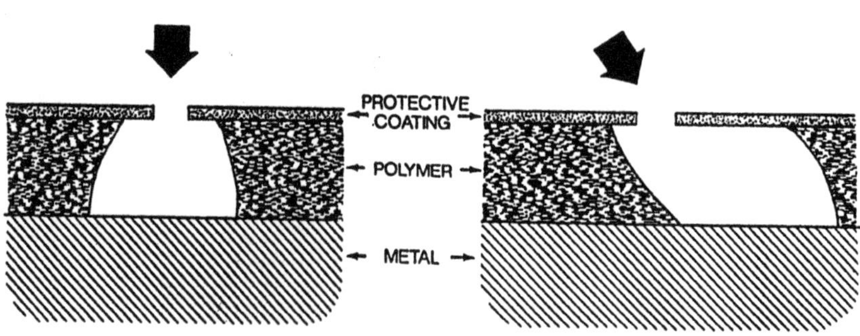

Fig. 26: [Ref. 17].

6. Effect of AO on Optical Properties

Reference 5 contains a summary of optical properties measured on a variety of materials subjected to the AO environment. The most significant changes observed in solar absorbtance and reflectance occurred on exposed organic surfaces.

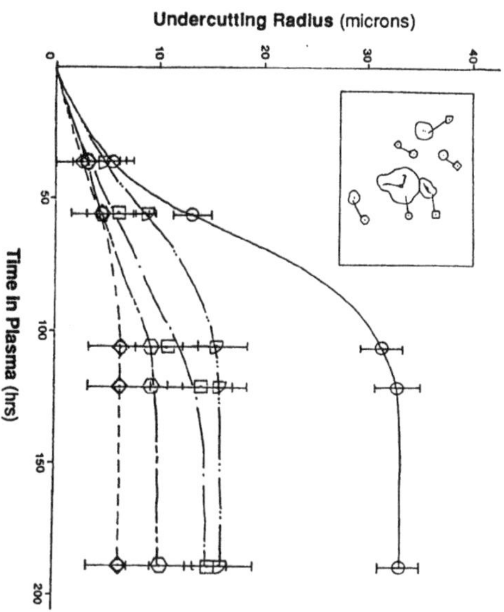

Fig. 27: Atomic Oxygen Undercutting of Defects [Ref. 10].

Acknowledgements

The University of Toronto Institute for Aerospace Studies' (UTIAS) research program on atomic oxygen effects on materials was supported by the Strategic Defense Initiative Organization's Office of Innovative Science and Technology (SDIO/TNI) through contract number N60921-91-C-0078 with the Naval Surface Warfare Center and the Ontario Institute for Space and Terrestrial Science. The author is indebted to the following people who have contributed to the UTIAS program: Dr. W.D. Morison, UTIAS; DR. M.R. Wertheimer, Ecole Polytechnique; Dr. R.H. Prince and Dr. E. Bourdon, York University (CRESS).

References

[1] P.C. Van der Waal, "Effect of Space Environment on Materials", Royal Netherlands Aircraft Factories Fokker, Dept. RV-22 (1968).

[2] D.R. Peplinski, "Satellite Exposure to Atomic Oxygen in Low Earth Orbit", NASA CP 2340 (1984).

[3] B.A. Banks, "Atomic Oxygen", LDEF Materials Data Analysis Workshop, NASA CP 10046, (July 1990).

[4] W.S. Slemp, "Ultraviolet Radiation Effects", Proc. NASA/SDIO Space Environment Effects on Materials Workshop, NASA CP 3035, Vol. 2 (1989).

[5] B.A. Banks, S.K. Rutledge, J.A. Brandy, and J.E. Merrow, "Atomic Oxygen Effects on Materials", Proc. NASA/SDIO Space Environment Effects on Materials Workshop, NASA CP 3035, Vol. 1 (1989).

[6] R.J. Bourassa, "LDEF Atomic Oxygen Exposure", Aerospace Systems Technologies Dept. MS.2-5321.1, Boeing Defense and Space Group, Proc. LDEF MSIG Meeting Seattle, WA, (July 1990).

[7] L. Leger, "Oxygen Atom Reaction with Shuttle Materials at Orbital Altitudes", NASA TM 58246 (May 1982).

[8] L. Leger, J.T. Visentine, and J.F. Kuminecz, "Low Earth Orbit Atomic Oxygen Effects on Surfaces", Proc. AIAA 22nd Aerospace Sciences Meeting, Paper # AIAA-84-0548, Reno, Nev. (Jan 1984).

[9] R. Hansen, J. Pascale, T. De Benedictis, and P. Rentzepis, "Effects of Atomic Oxygen on Polymers", J. Polymer Sci., Part A, Vol. 3 (1965).

[10] B. Banks, "Atomic Oxygen", Proc. LDEF Materials Data Analysis Workshop, NASA CP 10046 (July 1990).

[11] B. Santos-Mason, "Preliminary Results of SMM Exposed Aluminized Kapton and Silvered Teflon:, Proc. SMRM Degradation Study Workshop, NASA Pub. 408-SMRM-79-0001 (May 1985).

[12] S.L. Koontz, "Atomic Oxygen Effects on Spacecraft Materials - The State of the Art of Our Knowledge", Proc. NASA/SDIO Space Environmental Effects on Materials Workshop, NASA CP 3035, Part 1 (1989).

[13] R.C. Tennyson and W.D. Morison, "Atomic Oxygen Effects on Spacecraft Materials Degradation", Proc. Materials Degradation in Low Earth Orbit, Edit. V. Srinivasan and B.A. Banks, Pub. TMS, ISBN No. 0-87339-152-7 (1990).

[14] R.C. Tennyson and W.D. Morison, "The Effect of Atomic Oxygen on the Degradation of Spacecraft Materials", Proc. Canadian Society of Mechanical Engineers, Engineering Forum (June 1990).

[15] L. Leger, J. Visentine, and B. Santos-Mason, "Selected Materials Issues Associated with Space Station", Proc. 18th International SAMPI Technical Conf. (Oct. 1986).

[16] E.B.D. Bourdon, R.H. Prince, W.D. Morison, and R.C. Tennyson, "Real-Time Monitor for Thin Film Etching in Atomic Oxygen Environments", Proc. Int. Conf. on Metallurgical Coatings and Thin Films, ICMCTF-1991, San Diego, CA (April 1991).

[17] B.A. Banks, S.K. Rutledge, B.M. Auer, and F. Di Filippo, "Atomic Oxygen Undercutting of Defects on SiO_2 Protected Polyimide Solar Array Blankets", Proc. Materials Degradation in Low Earth Orbit, Edit. V. Srinivasan and B.A. Banks, Pub. TMS ISBN No. 0-87339-152-7 (1990).

ULTRAVIOLET AND VACUUM-ULTRAVIOLET RADIATION EFFECTS ON SPACECRAFT THERMAL CONTROL MATERIALS

A. E. STIEGMAN and RANTY H. LIANG
Jet Propulsion Laboratory
California Institute of Technology
4800 Oak Grove Dr.
Pasadena, CA 91109

Ultraviolet (UV) and vacuum ultraviolet (VUV) light has been found to contribute significantly to materials degradation in earth orbiting spacecraft. UV/VUV radiation is ubiquitous, occurring at all orbital altitudes and, while the effects of degradation sources such as atomic oxygen may be more pronounced at certain altitudes (*e.g.* low earth orbit), damage due to UV/VUV radiation will be present in all orbiting vehicles. Of the materials used in spacecraft construction thermal control materials (*e.g.* blankets and paints) show the most significant degradation due to UV/VUV exposure. This degradation manifests itself as surface erosion in polymers and as an increase in absorptivity (darkening) in both polymers and thermal paints. It is this change in absorbance that compromises the thermal control properties of the materials. The temperature of the spacecraft is proportional to the ratio of its solar absorptivity (α_{solar}) to its emissivity (ϵ).

$$T_{spacecraft} \propto \alpha_{solar}/\epsilon$$

Typically, emissivity of thermal control coatings does not change. Therefore, if the absorptivity of the thermal control materials increase with time then there will be a concomitant increase in the temperature of the spacecraft. Indeed, data from both flight experiments and laboratory tests accumulated over the past ten years indicates that problems related to VUV exposure can be severe. For example, organic materials, specifically aluminized Kapton and silverized Teflon thermal blanketing materials recovered from the Solar Maximum Mission satellite, showed severe degradation. The thermal blankets, which were in low-earth orbit for four years and two months, showed degradation characterized by extensive pitting and erosion with the most severe decay occurring on surfaces exposed to *both* UV/VUV and atomic oxygen. It is clear that long term space exposure of these materials will greatly compromise their thermal control integrity.

Dramatic materials degradation was also observed on the Long Duration Exposure Facility (LDEF) satellite. Consistent with what was seen on the Solar Maximum satellite the Kapton® and Teflon® thermal blanketing materials showed severe degradation. The most severe degradation was found for samples attached to the leading edge of the satellite which received VUV *and* oxygen atom exposure however, the trailing edge, which experienced

R. N. DeWitt et al. (eds.), The Behavior of Systems in the Space Environment, 259–266.

predominantly VUV impingement, also showed very significant amounts of erosion. Furthermore, thermal control paints applied to the trailing edge showed severe darkening. The VUV induced changes in these material can significantly change their absorptivity thus compromising their thermal control characteristics.

The most comprehensive flight data collected on UV/VUV degradation of thermal control materials was recovered from the ML 12 Thermal Control Coatings Experiment aboard the Spacecraft Charging At High Altitudes (SCATHA) satellite. The ML 12 experiment consisted of 16 calorimetrically mounted thermal control coatings for continuous determination of solar absorptivity and temperature controlled quartz crystal microbalances for continuous monitoring of weight changes that may be associated with decomposition or contamination deposition. The experiment was mounted on the minimum contamination "belly band" of the spinning SCATHA satellite. The experiment, which was launched February 2, 1979 to an orbit of 27,600 x 43,300 km provided continuous data in excess of 10 years. The results of this extensive flight experiment, analyzed and reported by Aerospace Corp., provided an unequivocal demonstration of the degradation of thermal properties by UV/VUV radiation

The changes in solar absorptivity (α_s) as a function of time in orbit for Silver Teflon® and aluminized Kapton® thermal blankets are shown in Figs. 1 and 2 respectively.

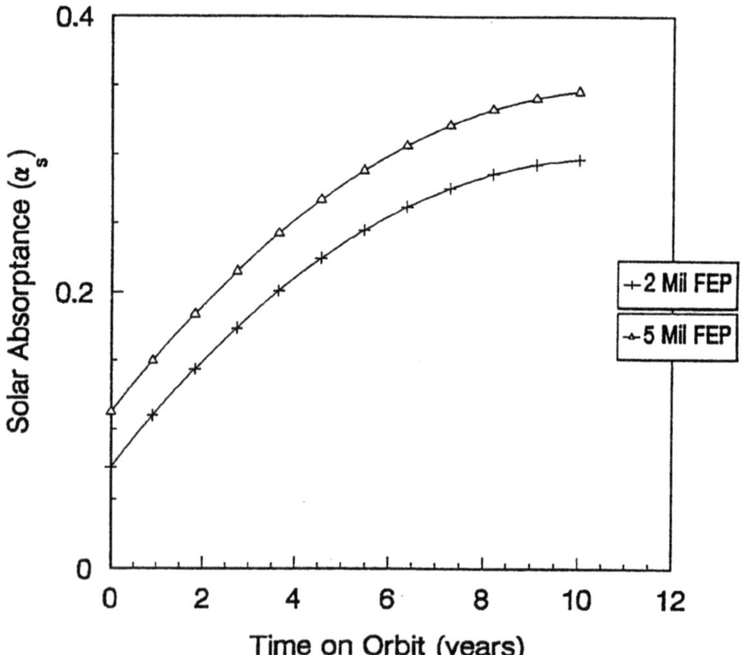

Fig. 1: Changes in solar absorptivity of silver Teflon® with orbit duration.

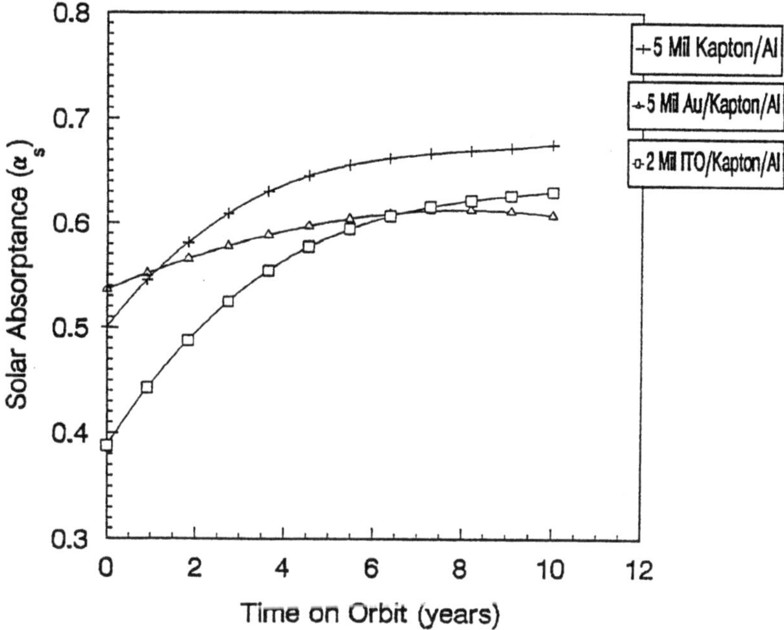

Fig. 2: Changes in solar absorptivity of aluminized and ITO coated Kapton® with orbit duration.

Clearly, the solar absorptivity of these thermal blanketing materials shows a sharp rate of increase early in the mission. For all of the samples the absorptivity appears to plateau and approach a constant value which is, for some of the material, greater than a factor of two higher than it was initially. This increase in absorptivity can result, assuming a constant emissivity, in a considerable increase in spacecraft temperature. It is worth noting that the polished gold and aluminum reflectors flown on SCATHA showed virtually no change in absorptivity for the duration of the mission suggesting that the absorptivity changes observed in the thermal blankets was due to radiation induced changes in the materials themselves and *not* from deposited contaminants.

A fundamental understanding of how polymers degrade under UV/VUV radiation can be gained by analyzing the microscopic photophysics and photochemistry of organic systems in the solar spectral region.

The spectral region in which organic polymers show significant absorption of radiation extends from about 300 nm in the near W to about 100 nm in the VUV. The solar irradiance in this spectral region is shown in Fig. 3. In the region of highest energy, 100-150 nm, the major component (~ 80%) of the irradiance is due to the Lyman-alpha line of atomic hydrogen occurring at 121 nm. At wavelengths higher than 150 nm the flux increases dramatically with by far the greatest flux occurring in the 200 to 300 nm region. While the contribution to the solar flux in the 100-150 nm range

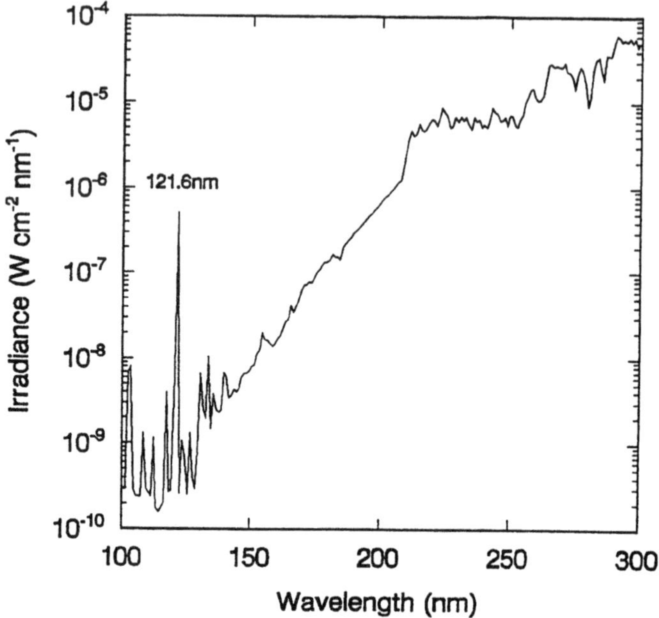

Fig. 3: Solar spectral irradiance.

is lower, damage due to this region may be significant due to the high energy. Conversely, the low energy region (150-300 nm) of the spectrum has a sufficiently high flux that the contribution of this spectral region to materials degradation may be significant. A primary goal of this study is to ascertain which spectral regions, and which photophysical processes associated with those regions, represent the major degradation pathways in organic polymeric materials.

For organic molecules the absorption of light in the region of the spectrum from 100 to 300 nm results in electronic excitations yielding excited state species that are, potentially, highly reactive. It is the reactivity of these energetic states on a molecular level that leads to degradation on a macroscopic scale. The energy of these excited states is dependent on the wavelength of the exciting light while their population is dependent on the photon flux at that wavelength. In general we can, a *priori*, define a number of primary photophysical processes, all potentially leading to degradation. These process can be divided roughly into two categories corresponding to two wavelength regions of the exciting light: the low energy range from 180-300 nm and the high energy region from 100-180 nm. The photoprocesses associated with these regions are shown in Fig 4.

The low energy part of this spectral range of about 220-300 nm (the high flux region of the solar spectrum), corresponds to electronic excitations into the lowest excited states of the molecule. For small aromatic molecules (e.g. benzene or naphthalene) and polymers such as polystyrene or polyvinylnaphthalene this corresponds to $\pi \rightarrow \pi^*$ electronic transitions.

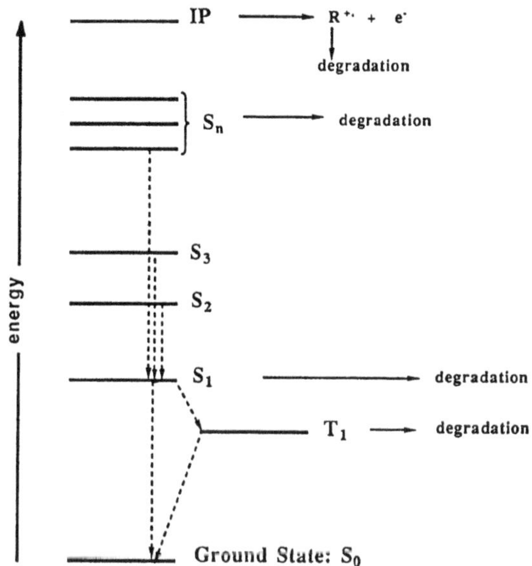

Fig. 4: Energy state diagram showing photophysical and photochemical processes in organic materials.

Using as an example the naphthalene molecule, the lowest energy state is a triplet state, T_1, which has a natural lifetime of approximately 10 microseconds. The next highest state, the singlet S_1 associated with the same transition, lives for 100 nanoseconds decaying either to the triplet state or back down to the ground state. The higher energy states, $S_{2,3}$ or $T_{2,3}$ decay rapidly to these lower two states. Due to the lifetime of the low energy excited states it is generally thought (Kasha's rule) that photochemical processes come only from these states. Photochemical processes which can occur are shown below.

$$A \xrightarrow{h\nu} A* + B \longrightarrow C + D \tag{1}$$

$$A \xrightarrow{h\nu} A* + \longrightarrow E \tag{2}$$

In Eq. 1 the excited chromophore reacts with a second molecule to produce new species. The reaction can be an electron transfer process (oxidation or reduction) or an atom or group transfer process. The resulting product, C or D in Eq. 1, may be volatile small molecules and/or structurally altered polymer chains. These species may be unstable and react further to continue the degradation process. In the space environment the incoming molecule may be energetic species such as charged particles or atomic oxygen.

The excited molecules may also undergo unimolecular processes such as bond cleavage or bond rearrangement (Eq. 2). Examples of unimolecular photochemical processes in organic molecules are CO extrusion from excited benzyl ketone functional groups and ether group cleavage which can produce rearrangement products (Fries rearrangement) or the formation of free radicals. These species are often energetic and can further decay contributing to overall degradation.

It has recently been shown that VUV irradiation accelerates the degradation of Teflon and Kapton caused by atomic oxygen exposure. This cooperative effect is due to the reaction of atomic oxygen with the easily oxidized products of VUV photolysis (species E of Eq. 2).[4]

The spectral range between 180-220 nm corresponds primarily to excitations into higher singlet states (the S_n manifold of Fig. 2). In general these high energy states are extremely short lived, decaying quickly to the S_1 and, subsequently, the T1 state. For example, rates of internal conversion from the higher energy states of naphthalene to the S_1 level are about 10^{12} sec^{-1}. These high energy states may also convert directly to the ground state.

The extremely short lifetime of these upper excited states tends to largely preclude any bimolecular photochemistry (Eq. 1). Unimolecular photochemistry (Eq. 2), however, may be competitive with the fast rates of internal conversion. The high-energy excited states produced in this excitation region may rearrange structurally, cleaving bonds and producing reactive high energy structures or radicals which will subsequently degrade.

In the high energy region of the spectrum, 100-180 nm, a number of new processes occur. Excitation in this wavelength region produces higher excited states of the molecule. This includes the $\pi \rightarrow \pi^*$ continuum, the promotion of "core" electrons into higher energy orbitals, and Rydberg states. For example, in the naphthalene molecule $\pi \rightarrow \sigma^*$ and $\sigma \rightarrow \pi^*$ transitions occur at about 125 nm while the onset of Rydberg type transition starts at about 180nm. As mentioned, the decay of any of these states to the lowest energy excited state is expected to be extremely fast; however, unimolecular processes may be competitive with relaxation.

Photoionization of the molecule also occurs in this region. This involves the ejection of an electron by an energetic photon resulting in a radical cation as shown in Eq. 3.

$$A \xrightarrow{h\upsilon} A^{\cdot +} + e^- \tag{3}$$

In naphthalene photoionization occurs at 152 nm with transitions also at 139 nm and 124 nm. It is clear from this that much of the solar VUV, especially the Lyman-alpha line at 120 nm, will be highly ionizing. It is likely that the highly reactive species produced by this photoprocess will be a major pathway to degradation.

From this analysis it is clear that materials with fast decay rates (radiative or nonradiative) directly to the ground state which bypass reactive excited states (S_1 or T_1) will be inherently more resistive to photodegradation. A common way to accomplish this is the addition chromophores (usually as small molecule dopants) which absorb the radiation preferential to the polymer; common W stabilizers operate in this manner. Alternatively, the polymers selected for thermal control applications in space should be the selected with consideration to their photophysics and photochemistry with an emphasis placed on materials with few reactive excited states.

The particulars of the photochemistry should also be considered when performing accelerated aging studies. The commonly used "multi suns" approach assumes that an increase in intensity will accurately duplicate an extended exposure. However, if the $S_1 \rightarrow T_1$ (Fig. 4) rate is appreciably greater than the $T_1 \rightarrow$ "degradation" then there will be a buildup of T_1. This triplet buildup will decay *both* through the degradation pathway and intersystem crossing to the ground state ($T_1 \rightarrow S_0$) giving a nonlinear response which invalidates the primary assumption of the "multi-suns" approach to accelerated testing. An alternative to this approach is to use one sun of radiant exposure but to accelerate the aging process thermally. By heating the sample the rate constants for the photochemical processes will proportionally incense preventing the build up of the T_1 state.

References

[1] Teichman, L.A.; Stein, B.A. e. "NASA/SDIO Space Environmental Effects on Materials Workshop"; Nasa CP-3035, Part 1, 1989.

[2] Santos-Mason, B. in "Proceedings of the SMRM Degradation Study Workshop" 408-SMRM-79-0001, 1985.

[3] "LDEF-69 Months in Space"; Arlan S. Levine editor, NASA Conference Publication 3134 part 2.

[4] Stiegman, A.E.; Brinza, D.E.; Laue, E.G.; Anderson, M.S.; Liang, R.H. *Journal of Spacecraft and Rockets* 29, 150-151, **1992**.

[5] Stiegman, A.E.; Brinza, D.E.; Anderson, M.S.; Minton, T.K.; Laue, E.G.; Liang, R.H. "An Investigation of the Degradation of Fluorinated Ethylene Propylene (FEP) Copolymer Thermal Blanketing Materials Aboard LDEF and in the Laboratory" JPL Publication 91-10.

[6] Hall, D.F. "Current Flight Results from the P78-2 (SCATHA) Spacecraft Contamination and Coatings Degradation Experiment" in *Proceedings of an International Symposium on Spacecraft Materials in Space Environment*, Toulouse, France 8-12 June 1982, page 143 and references therein.

[7] Heath, D.; Thekaekara, M. "The Solar Output and Its Variation", ed. White, O.R., Colorado Associated University Press, (1977) pp. 193-212.

[8] Turro, N.J. "Molecular Photochemistry" (Benjamin/Cummings Publishing, Menlo Park, CA 1979).

[9] Koch, E.E.; Otto, A.; Radler, K. *Chem. Phys. Lett.* 1972, 16, 131-135.

CONTAMINANT-SENSITIVE SYSTEMS IN THE SOLAR RADIATION ENVIRONMENT

F. R. KRUEGER
Engin.-bureau Dr.Krueger
Messeler Str. 24
D-W-6100 Darmstadt 12
West Germany

ABSTRACT. Many system components outgas organic contaminants being hazardous to other system components. Such sensitive system components are, e.g., mirrors for sun or laser light, charge or photon detectors, mass spectrometer targets or areas. Outgasing components are, e.g., adhesives, isolation materials, paints, plastics, or lubricants. Condensible outgas products are dangerous in the solar radiation field if they are able to polymerize due to solar radiation action. Solar protons produce free radicals in any organic, and UV photons in those organics which absorb UV light considerably. As a consequence such organic contaminants adsorbed at sensitive system surfaces are subject to polymerization; thus being hindered to desorb again. Finally, systems are spoiled due to the joint action of outgasing and solar irradiation. Cross sections and kinetics of this radiation processes are given, and a number of outgas determinations of system components are presented. Due to a chemical identification of all the outgas condensibles, even traces, a proper consideration of endangering is possible. Thus a basis for contamination budget calculations is given.

1. Introduction

A new mode of damage of systems in the solar environment has been recently discovered; namely, radiative enhanced contamination. It is well known that the solar radiation itself can cause a lot of damage: Interaction of solar protons with solid surfaces causes disturbance of the bulk crystal order structure, UV-photons may induce chemical reactions, thus, e.g., bleaching paints and so on.

On the other hand, organic contaminants may be hazardous themselves; e.g., being adsorbed on sensitive surfaces like detectors they may spoil their actions considerably. Generally systems in space are protected against both hazards. Protons can be shielded electrostatically where necessary, photons shielded, and some cleanliness measures are always taken during employment.

However, solar spectrometers and cameras suffered in most cases from malfunctions after some months or even weeks operating in the solar environment (especially the mirrors became blind). The reason was

R. N. DeWitt et al. (eds.), The Behavior of Systems in the Space Environment, 267–271.

apparently due to this new damage mode. One can show, that normal contamination, e.g., of mirrors, with not too refractive species being harmless due to an alternating ad- and de-sorption become hazardous in the solar radiation field. Namely, several desorbed species undergo radical reactions by means of UV-photons and protons, thus often leading to a polymerization hindering the contaminants from desorbing again. As a consequence thick contaminant layers can be built up with jeopardizing properties. Another radiation action is dehydrogenization of the contaminants thus creating pi-bonds which in turn enhance the UV-absorption and the above harmful reactions.

2. Radiation Processing of Organic Layers

That solar radiation can "process" any organic layers is well known in cosmo-chemistry. Namely, condensed gases in the interplanetary environment are polymerized, dehydrogenized, and dehydrated by solar photons [Ref. 1] and protons [Ref. 2]. Both process types are widely studied in the lab, too.

2.1. UV-RADIATION PROCESSING

Photons of a quantum energy larger than approximately 4 eV are able to perform the processes described above. In the distance of 1 A.U. from the sun, the flux density of these solar UV photons is about 20 mW/cm^2 which is about 2×10^{16} $photons/sec/cm^2$.

In order to judge, how fast organic layers are processed, we need a quantum efficiency of the process which is material dependent. In many cases 10% may be a good guess. Exemptions are due to materials containing much oxygen which quench the radicals to, say, 1% efficiency; contraries are due to materials which are subject to chain reactions (like UV-triggered adhesives), in which cases the efficiency may even exceed 100%. Calculating further the 10% example, 2×10^{15} absorbed $photons/sec/cm^2$ are active. For example, with an absorption coefficient of 3000/cm in 10^4 sec, any optically thin layer (below 3 μm in this case) is processed thoroughly.

Most of these processes are dehydrogenation processes and 1/10 are polymerization processes [Ref. 3]. This is sufficient to process such an organic layer in the above time, and 10^{-4}/sec is thus a typical kinetic coefficient for layer processing. One has to compare this value with the desorption kinetic coefficient of the contaminants. If the desorption rate is considerably lower, all adsorbed species are thus trapped by "processing" during sun shine-forming thick adlayers! If it is higher, a branching ratio applies. Solar electrons also can compare in processing with photons in very thin contaminant layers (a few monolayers).

2.2. PROTON PROCESSING

The solar (about 1 keV) proton flux is about 4×10^8 $protons/sec/cm^2$ in 1 A.U. distance from the sun. In LEO orbits, however, it is shielded by the plasma, whereas in GEO orbits it may apply. As the chemical efficiency per unit absorbed energy is comparable (or even somewhat higher for polymerization) to photon absorption, the above proton flux resembles an energy flux like 10^{11} UV-photons/sec/cm^2. However, the typical absorption lengths are only 10 nm. Consequently, a contamination of some ten monolayers completely absorb the proton energy. In this case (UV-transparent, thin contamination layers) proton processing action equals photon processing action.

3. Experimental Investigation of Outgasing

A lot of components and systems, as foreseen to be implemented in SUMER on board SOHO, had been investigated concerning outgasing. As we have seen, outgasing species which can easily be re-adsorbed again at solid surfaces are dangerous, especially if they are active in UV-absorption and ready for polymerization, if free radicals are produced. Those rather refractive species are generally not found in residual gas measurements and thus not determined. CVCM-measurements could only determine the quantity of condensibles, regardless of whether they are harmless (like water, CO2) or not.

3.1. OUTGAS DETERMINATION METHOD

In order to minimize the time consumed for measurements, some kinetic rules had been applied to determine long-term-outgasing. Namely outgasing was measured at several elevated temperatures, and an Arrhenius plot for each single outgasing species allowed the extrapolation at system temperature. The slope of the plot provided the evaporation enthalpy. With a mass spectral (MS) identification all the thermodynamical data of each outgas product are known, especially with the adsorption isotherme the condensibility at any temperature.

The outgas products were separated by gas chromatography (GC) using a silica gel column which separates mainly due to equilibrium vapor pressure. By subsequent flame ionization detection (FID) determination of the outgas concentration in time, the kinetic coefficients could be determined.

With unacceptably outgasing substances or systems, baking and curing was tried and outgasing determined again after a while. So curing procedures could be recommended which improve outgas properties without attacking sensitive components.

3.2. RESULTS

A very interesting class of substances are adhesives. Especially epoxy resins showed a large variety of qualities. Some were acceptable (e.g., 3M, EP310), some medium, some unacceptable for cleanliness purposes depending, however, a lot from curing processes (some even heavily worsened under long-term heating, in contrary to data sheets!). Loctite, polyamides, polyimides, polyurethanes, PTFE, for instance, were studied. Sometimes the chemical fine structure is important (polyamides); sometimes solvents may cause a problem (polyimides) if not properly removed; sometimes baking is important to improve acceptance (Loctite).

More difficult to investigate are, of course, systems, like motors and actuators, because it is essential to drive them during outgas tests. Special outgas chambers with platinum wire feedthroughs had been built to maintain this task. As a result, e.g., a stepper motor which was acceptable for torque, weight, and power demands completely failed in that outgas tests due to several outgasing system parts. Part by part this motor could be improved by altering many of its elements and subsequently testing, thus finitely resulting in an overall acceptable motor for clean space applications. Also other systems, like encoders, plugs, and ball-bearings had been tested with different results.

4. Discussion

Any system in the solar environment with sensitive elements exposed to sun light and radiation should be protected against this type of hazard. With the above rates of UV- and proton- "processing" maximum allowable adsorption rates of contaminants can be calculated for each purpose, depending on defined system lifetime.

Thus a contamination budget can be established. Namely the maximum partial pressure of each possible contaminant at the sensitive area is calculated by the flux at the source (outgas rate) and the vacuum impedances in between and to the outer space. Such calculations are examplified by Krueger [Refs. 4, 5].

Also system environmental contamination should be taken into account, as plume action and spacecraft outgasing, in order to optimize the lifetime of sensitive components as (laser) mirrors, detectors, cameras, and so on.

5. Acknowledgements

Thanks are due to the Max-Planck-Institut of Aeronomy in Katlenburg-Lindau/FRG, Dr. K. Wilhelm, which supported these investigations. The analytical work has been performed in close cooperation with Prof. Dr. E. Schmid, Inst.f.Analytical Chemistry of the Univ. of Vienna/Austria.

6. References

[1] J. Mayo Greenberg, In: 'Comets', ed. by L.L. Wilkening, University of Arizona Press, p. 131 (1982).

[2] G. Strazzulla, 'Modifications of Grains by Particle Bombardment in the Early Solar System', Icarus 61, pp. 48-56 (1985).

[3] G. Foti, In: 'Laboratory Simulation on Organic Cometary Material', ed. by J. Kissel and G. Strazzulla, Catania/Italy, pp. 154-161 (1987).

[4] Franz R. Krueger, 'Cleanliness Theoretical and Semiempirical Study for SUMER on SOHO', Report to Max-Planck-Inst.f.Aeronomie, Katlenburg/FRG. (1989).

[5] Franz R. Krueger, 'Cleanliness Experimental Studies'; 5 study reports to Max-Planck-Inst.f.Aeronomie, Katlenburg/FRG (1990-91).

HYPERVELOCITY IMPACT PHYSICS - PLASMA DISCHARGE PHENOMENA ON SOLAR GENERATORS

F. R. KRUEGER
Engin.-bureau Dr.Krueger
Messeler Str. 24
D-W-6100 Darmstadt 12
West Germany

ABSTRACT. Due to semiempirical values known from impact ionisation experiments and theoretical calculations the time-dependent charge, pressure and density of impact particles (mass .01 to 100 mg, speed 1 to 15 km/s) are calculated. Impacting onto a sandwich layer, the plasma may cause a short-circuit between layers on different electric potential (e.g., with solar arrays). The time-dependent complex impedance of the plasma is calculated. As a result, it can be decided whether in a given electric circuit an arc discharge is ignited after impact and stabilized by the current, or not. It is shown that particle impact plasmas cannot be simulated by laser pulses with any respect, because impact is very active in cratering, but weak in plasma formation; whereas lasers behave vice versa. It is also reported on a series of experiments on particle impact as well as pulsed laser irradiation on solar cells and bus bars with and without voltage applied, which confirmed and extended the theoretical results.

1. Introduction

There is a lot of experience accumulated concerning the damage and plasma formation effects (on the one hand due to particulate impact, on the other hand due to pulsed giant laser irradiation).

The ranges of the particle parameters are that of mass in between some fg and several tens of mg, and that of velocity in between 1 and 60 km/s. In most cases smaller masses refer to larger velocities, and vice versa. The Technical University of Munich/W. Germany (TUM) deals with masses up to some μg in the velocity regime up to 25 km/s; the Ernst-Mach-Institute in Freiburg/W. Germany (EMI) is able to accelerate masses even in the mg regime up to 15 km/s.

Pulsed laser arrangements with pulse duration in the 10 ns regime for comparison must be able to produce flux densities in the range of 0.3 to 100 GW/sq-cm on focal spot areas comparable with those of the particles to be simulated. The region of 0.3 to 10 GW/sq-cm suits for the simulation of plasma formation; that from 3 to 100 GW/sq-cm for cratering.

With focal spots up to 10 μm in the low flux density regime, the LAMMA (R) -1000 instrument is suitable. For simulating larger particles,

R. N. DeWitt et al. (eds.), The Behavior of Systems in the Space Environment, 273–290.

higher-power lasers are needed, as being established at the University of Kaiserslautern/W. Germany. With such laser experiments a properly adjusted light optics for exact focusing and best reproducibility and a good ion and electron optics are badly needed.

From these experiences one can deduce which behaviour is expected with solar cells, with their special geometry and materials, if they would be bombarded with micrometeoroid particles. General semi-empirical formulae, as being derived from the results of those experiments, are used for calculating these expectation values. Furthermore, it was investigated under which conditions laser irradiation may simulate these impact phenomena, as semi-empirical formulae and experience values are also known in that case.

The theoretical calculations of the impact plasma and crater formation may be an additional help for investigating the relevant solar cell damage and short-circuit processes, which is done. Finally, it is reported on particulate impact experiments, and in comparison, pulsed laser irradiation studies, which confirmed the value of the theoretical and semiempirical outline.

2. Impact Damage Phenomena

2.1. PARTICLE IMPACT

When a particle impacts a solid with velocities above some km/s, a central crater is formed; the material mostly being evaporated or otherwise removed. At the crater surrounding edge the material may be damaged and fragments may be ejected. Which fraction of the material totally removed is due to central crater formation and which is due to peripheral ejection, mainly depends on the velocity of the impinging particle and also on the brittleness of the target material bombarded. The smaller the velocity and the more brittle the material is, the more peripheral effects are important relative to the central ones.

2.1.1. *(Primary) Central Effects.* The central effects are often called "primary", however, especially with ionization phenomena, with which the centrally emitted ions are called "primary ions". In the very context of mechanical effects let us speak about "central" phenomena only.

It is a general observation that the central crater volume formed during impact depends linearly on the mass m and quadratically on the velocity v of the impacting particle; thus being proportionally to its kinetic energy. As an example the proportionality factor can be given as follows: Imagine that a certain fraction cE of the kinetic energy E is used for heating up and evaporating the crater material. So we can formally write:

$$c \cdot 0.5 \cdot mv^2 = c\ E = \rho\ V\ c_p\ \Delta T \quad (1)$$

with ρ being the density of the material, c_p its heat capacity, and ΔT formally a temperature rise. For determination of the crater volume we can just rewrite (1) as follows:

$$V = c' \cdot m \cdot v^2/(\rho \cdot c_p) \quad (2)$$

with $c' = 4 \cdot 10^{-4}/K^{-1}$. For practical purposes we may write:

$$V/cm^3 = 0.2\ m/g \cdot (v/\ km/s)^2 \quad (3)$$

The product of density and heat capacity ($\rho\ c_p$) is nearly the same for glass, silicon and silver (2 $J/cm^3/K$), the materials of interest. So we can roughly estimate the crater volume produced by particle impact. The material quality of the impacting particle is of minor importance with sufficiently high velocities.

The geometric form of the crater is more or less a half-sphere with craters of some mm radius. Larger craters are flatter (moon craters!), smaller craters are deeper (microscopic impact channels!). The ratio of the radius r of the crater surrounding circle to the depth d of the crater can be written as

$$r/d = 1. \cdot (r/mm)^{.084} \quad (4)$$

for most practical purposes involving homogeneous material.

How layers of different material in the impact zone will behave is a subject for experimental investigation and is important for solar cell geometries.

2.1.2. *(Secondary) Peripheral Effects.* The effects occurring at the periphery of the crater are called "Secondary" due to historic reasons (see 2.1.1).

With brittle material like glass a lot of damage can be caused outside the central crater. Small fragment pieces are often ablated, and other damage like cracks may happen. Also, with metals and semiconductors fragment pieces are ejected from the rim zone of the crater; these pieces are therefore often called "ejecta". With lower velocities, say 3 km/s, even more material may be such ejected as pieces rather than being excavated in the crater volume. With 10 km/s ejecta play a minor role, except for brittle glasses. With very high velocities, say 50 km/s, ejecta are not observed at all.

A rough estimation is possible giving the range of destruction around the central crater; namely, by semi-empirical values about the "catastrophic disruption" of a non-metallic (ie., more or less brittle) material. A catastrophic disruption in the disintegration of a body occurs when no single crack piece is larger than one half of the original body. There is a general formula, well consistent with empirical data, (deducible from percolation theory) giving the specific energy E/m (energy per unit mass) which has to be applied kinetically to the body of mass in order to catastrophically disrupt it, namely

$$E/m = 1\ J/g \cdot (d/cm)^{-1/2} \quad (5)$$

Percolation theory gives a model-dependent exponent slightly different from 1/2.

The proportionality factor may be as low as 0.1 J/g for very brittle material. Generally a solar cell will be far too large to catastrophically disrupt after impact of a micrometeoroid; however, we may estimate that range around the crater containing the maximal mass m in (6) with the impact energy E in which we expect severe damage. For macroscopic cracks resulting with v = 10 km/s, we find some 4 mm crack range; with v = 3 km/s some 1 mm!

Near layer boundaries from brittle to ductile materials or v.v. ejecta and crack formation may behave very unexpectedly. This is subject to a detailed experimental investigation at particle accelerators with solar cells themselves.

2.2. LASER IRRADIATION

Very much like particle impact cratering, with laser irradiation the matter ablated is a linear function of the light energy absorbed. However, it depends on wavelength and some material qualities which fraction of the laser energy offered is actually absorbed. The efficiency of absorption is best with UV-radiation and material absorbing in this regime. Up to 80% of the offered one may be absorbed in such cases. For wavelengths in the near infrared with only little natural absorption (as being valid with several dielectrics of large electronic band gap) absorption has to be initialized by non-linear multi-photon processes. This is often the case with Nd-YAG lasers. In these cases only 10% of the energy may be absorbed (or even none) if the total flux density is too small to trigger multi-photon absorption.

In the 1060 nm Nd-YAG (12 ns pulse length) the following practical relation concerning the ablated volume V is valid for metals and semiconductors:

$$V/\mathrm{mm}^3 = 0.03 \cdot E/\mathrm{J} \tag{6}$$

with E being again the total incident energy of the laser. With glass this proportionality factor may be somewhat lower; however, in that case more energy may be transferred into cracking or similar damage.

Concerning the crater geometry the situation is very much different than it is with particle impact. Whereas the crater volume and size after impact mainly depend on the total kinetic energy of the impinging particle, the crater size after pulsed laser irradiation depends separately on the focal spot size and the flux density. With low flux densities in the 1 GW/sq-cm (some mJ on 100 μm focal spot) regime the crater covers the spot size; however, it is very flat. With higher flux densities the crater becomes deeper; however, the depth seldom exceeds the diameter. Instead, the diameter increases, too.

At this point, we have to define carefully what is understood with "spotsize" of a laser. Whereas the cross section area of a particle is well defined, that of a laser is not. For details one is referred to [Refs. 1, 2].

With layered material structures, e.g., a metal or semiconductor under a transparent dielectric cover, damage effects due to laser irradiation do happen, which are totally different from those due to particle impact. In that case the laser light may be transmitted through the cover (e.g., glass), being predominantly absorbed in the metal or semiconductor layer. In that case the crater is produced in the second layer, and the first layer explodes due to the related increase of pressure. Evidently, this picture by no means simulate a particle impact whatsoever.

2.3. COMPARISON PARTICLE/LASER

Some aspects of damage due to particle impact on homogeneous materia can be simulated by means of a pulsed laser beam. The laser beam energy has to be typically a factor of 1.5 to 3 larger than the kinetic energy of the simulated particle (due to much energy being consumed for plasma heating, and with dielectrics also due to insufficient absorption), in order to create a comparable crater volume. The focal spot of the laser has to be at least smaller than the impact crater diameter expected, otherwise the crater becomes too wide and flat rather than semi-spherical. Nevertheless, the crater forms will differ anyway. Especially the peripheral effects cannot be simulated by laser irradiation. The formation of ejecta, as observed with low particle velocities, do not have a counterpart in laser irradiation, although by chance also fragmenting and cracking occurs with laser irradiation. These effects, however, severely lack of reproducibility. The contingency of these phenomena do not allow them for simulation.

Solar cells and arrays are composed of several layers of different materials. The damage behaviour of such stacks seems to be very different in particle impact or laser irradiation, respectively. Little is known about it at all; however, some effects as discussed point to a non-comparability of both damage processes. So, it may be worthwhile just to investigate particle impact damage on its own.

Craters, holes, and other damage may be applied by any method whatsoever, especially with a laser. Doing this may be useful, in order to study subsequently other phenomena (like discharges) at these damage sites. For such a purpose laser damage surely is a useful technique. "Simulation" of impact damage, however, may be possible only in a few very special cases.

3. Impact Plasma Formation Phenomena

3.1. PARTICLE IMPACT

Very much like damage phenomena, the formation of electric charges can be subdivided into a central and a peripheral phenomenon. Centrally a plasma is formed consisting of "primary" positive and negative ions and electrons. The ejecta are, at least in part, charged, too; these peripherally ejected charged particles are called "secondary" ones.

3.1.1. *Primary (Central) Effects.* The total charge emitted centrally and measured in at least several mm distance from the point of impact in the vacuum is the residue to the primary impact plasma, probably after a lot of recombination taking place in the very first moments after being generated. This is discussed later.

This plasma, as investigated in a wide particle mass and velocity regime, seems to be more or less neutral; i.e., nearly the same number of positive and negative charge carriers are present. The positive charges are, of course, positive ions of the particle and target material, most of them being singly charged. Only with large particles in the regime considered here (.1 to 1. mm and low velocities above, say, 8 km/s), doubly charged positive ions are expected; however, not yet measured.

The negative charge carriers are in part negative ions and electrons. Whereas with very high impact velocities (above 25 km/s) the electrons are predominant, in the velocity regime considered here both types of charge carriers are important. Metallic particles impacting on metallic surfaces produce nearly only electrons. However, dielectrics also create negative ions, the more polar the material is. The presence of polar surface layers on the target easily produce a lot of negative ions. In these cases the electrons are negligible.

The total "Residual" charge (as defined above) produced and measured in particle impact mainly depends on mass and velocity of the impacting particle, and only little on projectile and target material. A general semi-empirical formula has been established describing the residual charge quantity at least for metallic particles impacting on metallic surfaces; namely,

$$1/\mathrm{Cb} = 1.3 \cdot 10^{-4} \cdot (m/g)^{0.95} \cdot (v/km/s)^{3.5} \qquad (7)$$

This yield of charges may be down a factor of 5 lower, if the target is made of dielectric material.

One easily notes that the charge formation is not linear in the kinetic energy of the impacting particle. There was a lot of speculation which reason would be due to the mass- and velocity-exponents differing from 1 and 2, respectively. The lower mass-exponent points to the participation of surface phenomena in charge production, and the extremely high velocity-exponent (which must converge to 2 for very high v due to energy conservation) to a threshold phenomenon in energy density. More insight into this behaviour was brought by numerical plasma evolution calculations (as discussed later), which showed that generally less than 1 percent of the initially produced plasma survives as the residual one measured. Anyway, with higher velocities the larger energy flux density allows a larger part of the kinetic energy to be used for electronic excitation and thus plasma formation. With low velocities (probably below 1 km/s) no charges are produced at all, as the energy density is not sufficient. This may explain the threshold-like behaviour.

Up to now the initial charge distribution concerning types, temperatures, and densities in the initial state of creation within the impact channel is unknown. Only some theoretical calculations allow some estimation from the residual plasma back to the initial one.

3.1.2. *Secondary (Peripheral) Effects*. It is well known from the study of all groups that in addition to the central impact plasma "ejecta", charges are produced. They are found in oblique angles, are electrically rigid (not much deflected by electric fields), and carry much momentum. When impinging on walls, they create secondary charge emission. It is obvious that the nature of these charges is due to ejected target pieces, as already discussed, with only little specific charge.

With smaller impact velocity the relative contribution of "secondary" charges dramatically increase. Generally, below v=5 km/s they carry most of the charge produced. This behaviour can be deduced easily from the items discussed above; namely, whereas the "primary" ion formation behaves with velocity as $v^{3.5}$, the ("catastrophically") damaged surface regime behave as $v^{1.6}$. If one assumes the number of ejecta being proportional to the area damaged, and the charge per ejectum constant, "secondary" ion formation also behaves as $v^{1.6}$; thus, the ratio of the number of secondary to primary ions behaves nearly like the velocity-square. That is exactly what one observes.

Consequently, the measurement of secondary charges, well discriminated from the primaries, is useful to determine the out-of-crater surface damage. However, the important inner-volume out-of-crater damage cannot be determined by charge measurements.

3.2. LASER IRRADIATION

During irradiation most of the laser energy is used for plasma heating, except at low laser flux densities near threshold. With the Nd-YAG laser (pulse length 20 ns) typical plasma formation threshold flux densities are approximately:

for semiconductors	(like carbon or silicon)	E_0 = 1.0 GW/sq-cm
for metals	(like silver or copper)	E_0 = 1.5 GW/sq-cm
for dielectrics	(like glass or capton)	E_0 = 4.0 GW/sq-cm.

With laser flux densities more than a factor of 3 above threshold, more than one half of the laser energy is consumed for plasma heating, especially speeding up the ions produced. As a result the ions produced possess high kinetic energy and are highly charged. With 10 GW/sq-cm up to quadruply charged ions can be observed with mean kinetic energies up to 1 keV. Singly charged ions have less than 100 eV mean energy: doubly and triply in between. The total number of ions per unit focal area produced behave in a peculiar way as a function of flux density above threshold; namely, (exactly for carbon, estimated to be valid for other material approximately - with the different threshold E_0)

$$N(+)\ /\text{mCb/mm}^2 = .58\ (\ln E_1/E_0) \quad (8a)$$
$$N(++)\ /\text{mCb/mm}^2 = .12\ (\ln E_1/E_0) \quad (8b)$$
$$N(3+)\ /\text{mCb/mm}^2 = .017\ (\ln E_1/E_0) \quad (8c)$$
$$N(4+)\ /\text{mCb/mm}^2 = .003\ (\ln E_1/E_0) \quad (8d)$$

E_1 = Laser flux density. Probably the total positive charge is compensated by an equal negative charge.

One can show that with higher flux densities - a factor of 30 above threshold - the part of total energy consumed for plasma formation reaches its maximum, and then decreases again. This behaviour is very much different than that with particle impact, apparently.

3.3. COMPARISON PARTICLE/LASER

The plasma formation characteristics of particle impact and laser irradiation is very dissimilar, especially when comparing with damage.

With low impact velocities the plasma formation efficiency is very low. A 10 μg particle impacting with 2 km/s contains 2 mJ kinetic energy and produces about 25 nCb total charge. A 2 mJ laser pulse on a .01 sq-mm focal spot, as simulating the above particle, produces about 2 μCb total charge; i.e., nearly one hundred times more. The laser crater would be a .1 mm wide and 5 μm deep flat one: the impact crater would be of similar volume - perhaps a little bit more narrow and deeper. Such a particle impacting at 10 km/s (a factor of 25 more kinetic energy) creates 7 μCb charge (roughly a factor 300 more). The "simulating" laser of 50 mJ may create 20 μCb charge, a much smaller difference than before.

Assuming a 100 μg particle at 2 km/s producing approximately .25 μCb (the focal spot being a factor of 2 larger for simulation) the 20 mJ laser may produce about 15 μCb charge (the crater is not so flat anymore). That particle impacting at 10 km/s creates about 65 μCb charge. The "simulating" laser of 500 mJ may create about 150 μCb charge (an even smaller difference than before.)

3.4. COMPARISON PARTICLE/LASER WITH RESPECT TO THE COMBINED PROCESS

It is obvious already now that there are no laser conditions found, under which the combined process of mechanical damage and plasma formation of particle impact can be simulated.

Either the crater volumes are comparable, in this case the laser will generally produce a more dense and hotter plasma; or the plasma clouds are comparable, in this case the laser will generally not produce a sufficient damage. Moreover, especially in cratering, a series of impact phenomena cannot be simulated by laser irradiation at all. Those which are mainly due to the mechanic shock wave which stems from the enormous momentum transfer do not have a direct counterpart in laser irradiation. The secondary particles from the rim of the excitation zone and some destruction processes outside the impact crater are examples.

An even more serious difference is due to the behaviour of layered material. Whereas with impact the density and shock velocity differences are relevant for the effects to be considered, with laser irradiation light absorptivity and other electronic parameters are relevant. As solar cells are actually layered materials, this note seems to be important.

In plasma formation little is known about the initial plasma conditions. Anyway, even with the residual plasma often measured, one is aware about a lot of differences; namely, a laser-generated plasma is much hotter electronically as well as kinetically, because it is secondarily heated by the laser beam. One easily finds 4-fold charged species in the residual plasma, whereas with impact one even does not find 2-fold charged IIa-element ions. The energy distribution of impact generated ions depends on the impact velocity. With the velocities discussed here some tens of eV are the upper limit; however, with laser irradiation one has to be aware of some 100 eV ions. What this does mean for the initial plasma in the impact zone when comparing with the initially produced laser plasma is discussed in the next chapter.

Before we discuss the consequences for short-circuit studies, some remarks are given concerning the plasma evolution processes. This is in order to have access to the initial plasma, whenever only the residual is measured or otherwise known, as no direct initial plasma measurements are foreseen. It is actually the initial plasma which determines the conditions for discharges following the (impact or laser) dissipation.

4. Theoretical Aspects of Plasma Evolution

4.1. PARTICLE OF IMPACT PLASMA

In order to discuss the theoretical aspects of the evolution of the impact plasma, first of all one has to distinguish at least 3 different regimes of impact velocity, which are:

- the mechanical deformation regime $1 \leq v \leq 5$ km/s
- the "cold" plasma regime $5 \leq v \leq 15$ km/s
- the "hot" plasma regime $15 \leq v \leq 50$ km/s

The impact plasma apparently behaves in a very different manner when comparing these regimes.

In the low velocity regime (v = 1-5 km/s), most, by far, of the charged particles emitted are due to surface emission phenomena. The surfaces of the projectile and target material are deformed (and with the higher velocities also melted), thus being rapidly heated. As a consequence a number of atoms and molecules are desorbed from both surface regions, and are in part ionized due to their ionization potential. Positive as well as negative ions are produced; however, only with pure and clean metals are the negative charges mainly electrons. Theoretically this "plasma" formation can be treated well by the concept of desorption by rapid dissipation processes [Ref. 3]. In these cases only rather few cold (some eV) ions per unit surface area are produced, which do not collide any more. This is clearly demonstrated by examining the mass spectra, which are dominated by intact molecular ions due to the material of the projectile as well as of the target, in addition to atomic alkali (+) and halide (-) ions due to contamination.

Consequently, in this regime the residual plasma can be considered to be the same as the initial one, except for expansion only. Thus the total charge measured in some cm distances can be traced back to its origin in the region of particle-target interaction and in the time interval of some tens (or, with larger particles, hundreds) of nanoseconds after the beginning of the impact.

The phenomena are much complicated within the second regime of medium velocity (v = 5-15 km/s). With such impact speeds most or all of the particle evaporates, and some of the target material does also. Because typical acoustic propagation velocities in solid matter are around 5 km/s, with impact speeds so high the treated material suffers from supersonic shocks, which are indeed non-linear compression waves. The actual compression of the solid matter causes an additional effect in plasma formation; namely, the electronic states are changed and elevated. The "Fermi-sea" of electrons can in part flow out into the vacuum thus leading to free electrons and residual positive ions. The kinetics of these processes are still under heavy discussion. Nevertheless, one observes the fluorescence light and can easily attribute lines to the presence of, say, three-fold charged ions. However, as this highly excited matter in its initial phases is still extremely compressed, many reactions do occur further on, which are especially recombination processes. Anyway, the residual ions one observes are only singly charged.

Consequently, in this second regime the residual plasma can be considered as a mixture of desorbed ions (especially from the surrounding target material) and of residual plasma ions. It is difficult to estimate, how much the initial charge was larger than the residual, and how the spatial distribution actually was. The situation is complicated especially due to the two different sources of charges (desorption and compression plasma) which give rise to different initial distributions of ion types and spatial as well as temporal origin. Measuring the types and energies of the residual ions, one can at least distinguish them: The excess of the number of positive ions over that of negative ones is a measure of the number of residual compression plasma charges. The energy distribution of the latter ones is correlated with the impact velocity v (roughly a mean of v/3 ion velocity), whereas that of the desorption ions is independent of v. One may guess that the initial number of charges in the compression plasma is some ten times larger than the residual number, the positive charges being ions; however, almost all negative charges are electrons (except for VIa- and VIIa-element containing species - e.g., oxygen and chlorine, which also produce negative ions).

For completeness, let us quickly discuss the high-velocity regime, over 15 km/s. Apparently around 15 km/s the collision plasma undergoes a phase transition. Desorption is an effect quantitatively negligible for charge formation here. Whereas the residual plasma with lower velocities is governed by polyatomic (small molecular) ions, at higher velocities the situation dramatically changes. Most of the ions become atomic ones, and "secondary" effects are nearly absent. Theoretically, in the initial plasma even four-fold ionized atoms are produced, although only singly charged ones (even for IIa-elements) are experimentally found in the residual plasma. From these calculations [Ref. 4] one may infer, that the initial plasma contains some hundred times more charge than the residual does and seems to be extremely hot.

4.2. LASER IRRADIATION PLASMA

Plasma formation by a laser beam needs free electrons in the material irradiated, if the wavelength of the light is near the visible regime. In contrast, far-UV photons are able to create free electrons out of any matter, and far-IR radiation does not excite electrons, but creates photons (excites the lattice).

As Nd-YAG lasers (if any) seem to be most suitable for impact simulation, their wave length of 1060 nm is taken as a basis for the following discussion. Only metals and some semiconductors possess enough free electrons in order to absorb single photons directly from the light field. With dielectrics and many semiconductors, laser light absorption has to be initialized by free electron production via multi-photon processes. This determines the threshold energy for plasma formation; the band gap energy determines the number of photons which have to be absorbed at once in a multiphoton-process to create one free electron. This process is thus highly non-linear. Once enough free electrons are produced, they act further absorbing light by inverse Bremsstrahlung (low wavelengths) or light-field induced acceleration (high wavelengths).

Contrary to particle impact, in which the matter is initially compressed hot and then cools by expansion, with laser irradiation, the matter forms an expanding gaseous state, and further energy is transmitted by absorbing more light from the laser beam, shining for nanoseconds.

Consequently, the thermodynamic state of the matter during laser plasma formation differs completely from that during impact plasma formation. Namely, whereas the initial state of the matter after particle impact is a state with high pressure, high atomic temperature, and medium electronic temperature, that after laser irradiation is a state of medium pressure, medium atomic temperature, and high electronic temperature. Thus the dissipative (non-adiabatic) processes in the very beginning of the plasma evolution are very much different, until the plasma is such diluted that adiabatic expansion starts. An illustration and comparison of both effects is given in Fig. 1.

5. Electric Circuit – Complex Impedance Calculations

5.1. PARTICLE IMPACT

Let us now discuss which short-circuit phenomena are expected due to the particle impact phenomena. With the damage we have to exhibit those conditions with which galvanic short-circuits are possible. With the plasma formation we have to consider the electric impact channel conditions in terms of complex impedances, charge carrier properties, and so on, in order to exhibit short-circuit conditions due to permanent arc-discharges, or the ignition of transient discharges possibly resulting in permanent ones further on.

5.1.1. *Mechanic Effects – Galvanic S-c's.* A galvanic short-circuit becomes possible if an impacting particle is able to connect two conductors by destroying the isolating layer in between and bending and/or melting the conducting layers such that interconnection is possible.

If the probe consists of a layered sandwich structure (as it generally will be the case) the impacting particle has at least to destroy the insulating cover (if present), the uppermost conductor and the separating insulator, in order to cause such a damage. The lower conductor needs not necessarily be affected. This depth condition of damage gives lower mass and speed limits, with which such short-circuits are possible.

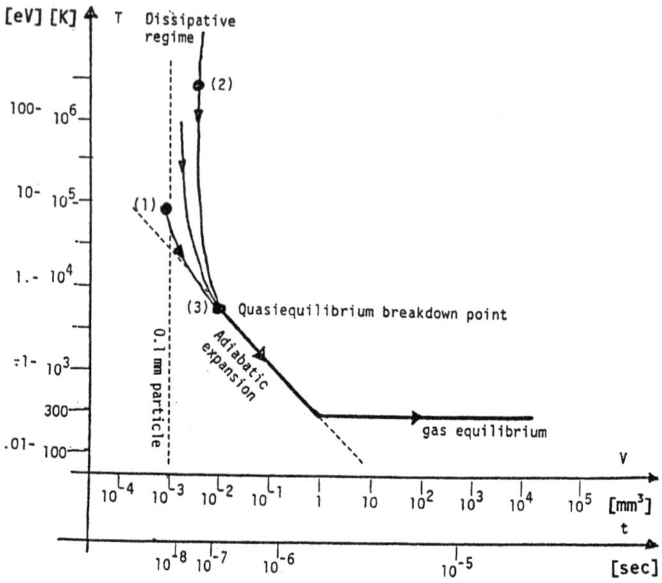

Fig. 1: Typical state diagram and evolution scheme of the thermodynamics of particle (.1 mm) impact – starting at (1)–, and laser irradiation simulating – starting at (2)–.
(1): Particle impact leads to a high-compression state of comparably low "temperature"; the dissipation takes place mainly in the atomic system – the electronic system remains cold.
(2): Laser irradiation leads to a low-compression state of comparably high "temperature"; the dissipation takes place mainly in the electronic systems – the atomic system is heated later.
(3): "Comparable" laser parameters provided, both evolution paths may arrive at the same state, where the quasiequilibrium of the initially created plasma ceases. From this point on adiabatic expansion is the joint process of further evolution.
Note that the initial plasma at least within the first 100 ns may differ completely depending on the excitation; however, similar conditions may be expected nevertheless in some mm distance after some μs!

With high velocities (above 8 km/s) and damage depths larger than necessary it is expected, that the dissipation is high. As a result the region in between the conductors may vaporize thus preventing them to connect. Also a secondary dissipation, as an arc discharge, may vaporize such an interconnection thus healing the short-circuit. So the thickness d of the material down to the surface of the second conducting layer is relevant for the onset of galvanic short-circuits. Numerical values are found in the literature.

5.1.2. *Electric Effects – Arc Discharge S-c's.* Let us now consider the conditions of discharges caused by an impact plasma created within an impact channel of length d and cross section area a. For sake of simplification we assume a cylindrical channel with a voltage applied in between top and bottom. The evolution of Debye-length D, pressure p and temperature T is given in Fig. 2. Thus the plasma acts much alike a parallel plate capacitor of plate distance d and area a, filled with a dielectric (thus forming a capacitance C), which has a certain conductance (thus forming a resistor R).

The voltage U applied causes a current I, both varying in time t due to the complex impedance $Z = R + 1/\omega C$, with ω being the main Fourier component of the a.c. current. The Fourier spectrum, however, is given by the above channel impedance behaviour as well as by the source impedance. The upper Fourier limit is given by $\omega_u = 2\pi/RC$; the lower Fourier limit by $\omega_i = 1/t$, with t the total measuring time interval.

First let us deduce the real part $R = \mathrm{Re}\, Z$ of the plasma impedance Z. The electric current density j is given by

$$j = \sigma E \tag{9}$$

with σ being the specific conductivity and E the electric field strength. The conductivity itself is given by

$$\sigma = e_0\, n\, b \tag{10}$$

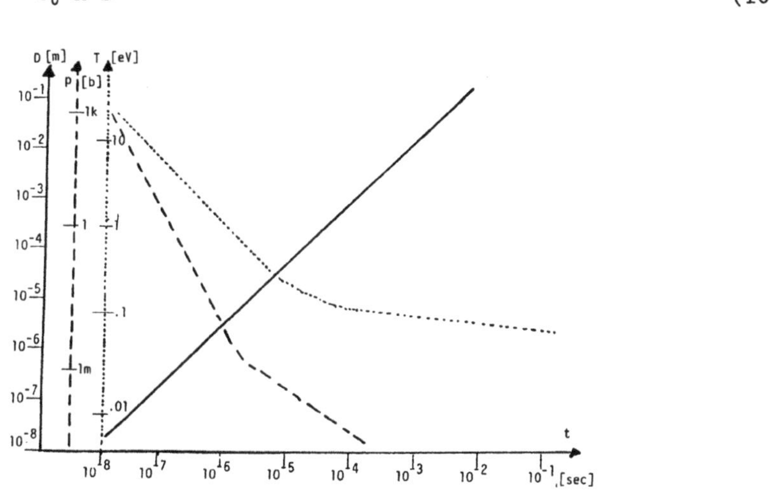

Fig. 2: Evolution scheme of the plasma of a typical impact particle (.1 mm, 5 km/s). Debye-shielding length D, plasma pressure p, and temperature T. After about 0.1 msec the external impedance determines the possibility of an arc discharge. Namely, we are in the Raether-spark or the Townsend discharge ignition regime, depending on the actual impact parameters. The pseudopark is irrelevant.

with e_0 being the elementary charge, n the charge carrier density and b their mobility (velocity per unit field strength). So equation (9) becomes

$$j = e_0 \, n \, b \, E \tag{11}$$

The mobility b is given by

$$b = 1/2 \, (e_0/m) \, (\lambda/c) \tag{12}$$

with m the mass of the charge carrier (here the free electrons), λ the free path length with

$$\lambda = 1/ns \tag{13}$$

and c the mean thermal velocity of the charge carriers. s is the cross section of the atoms hindering the electrons from flying freely. In principle, also the ions have to be considered as charge carriers, however, they contribute only little to the conductivity.

Inserting equations (12) and (13) into equation (10), one gets a highly remarkable result, namely:

$$\sigma = 1/2 \, e_0^2 \;\; 1/(mcs) \tag{14}$$

The conductance comes out to be independent on the plasma density; it depends only on the thermal velocity, i.e., the square root of the plasma temperature, which varies only within one order of magnitude. As a typical value of that conductance one gets $\sigma = 7000 \; \Omega^{-1} \; m^{-1}$. This is around that of graphite, much less than that for pure (undoped) semiconductors. However, this is only valid as long as the field E is able to penetrate whole the plasma length d. With dense plasmas the Debye shielding length D may be smaller in which case the specific conductance decreases roughly by the factor D/d. With our dense primary impact plasmas this effect is very large so that they behave like insulators during the first few microseconds.

The above conductance is not valid for plasmas too dilute, with which the continuous charge carrying process breaks down; namely, when the free path length λ increases beyond d. In that case the specific conductance decreases by the factor d/λ. For a typical impact crater dimension of a few tenth of a mm, the resulting resistance R, which is the real part of the impedance Z, is given in the Fig. 3.

In order to calculate the imaginary part of the impedance we have to determine the capacitance of that very capacitor. The relative dielectric constant of the plasma is very high, however; with Debye-shielded dense plasmas only a minor part of the thickness D zone near both "capacitor plates" acts as dielectrics. One may determine its dielectric constant by the dielectric influence law:

$$a \, \varepsilon \varepsilon_0 E = Q \quad a < 1 \tag{15}$$

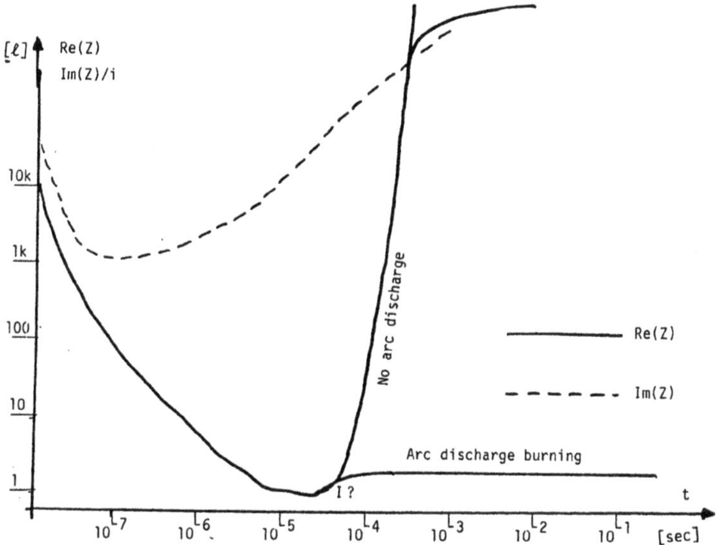

Fig. 3: Evolution of the complex impedance of the plasma of the impact as in Fig. 2, useful for electric circuit dynamic calculations. Real and imaginary part are given separately. Depending on the current which can flow due to the external impedance after about 0.1 ms it is decided whether an arc discharge is ignited or not.

with Q the part of the plasma charge resident in the volume aD. The whole plasma channel acts in that limit as two serial capacitors of this kind. Tens of pf (perhaps 1 nF) are likely for short intervals.

After expansion up to a Debye-length D = d, the channel acts as one capacitor of still high capacitance, now with Q of equation (15) the total residual charge in the channel (1 pF and lower). When the plasma is gone, the vacuum dielectric constant remains with a typical capacitance of some fF. A rough plot of the resulting imaginary impedance is given in the Fig. 3.

Just for comparison the figure also shows the typical behaviour of the Debye-shielding length D, the plasma pressure p and its temperature T. It should be mentioned that the discharge region may behave like the Raether-spark regime or the Townsend regime; namely, the parameter pd of the Paschen-function would allow several tens of Volts for ignition only, as the gas is already ionized! However, it can be assumed, that a discharge is already burning when these regimes are encountered; the pseudospark regime is not encountered.

An arc discharge is stable if the voltage does not drop further then the main ionization voltage of several Volts, with a typical current of an Ampere (Current density j = 500 A/sq-cm). Thus the source impedance determines whether the discharge gets distinct or not.

5.2. LASER IRRADIATION

It is the aim of this chapter to compare, to which extent laser irradiation may serve for simulating short-circuit phenomena themselves, as already discussed in detail with particle impact. So it is not necessary to discuss laser processes in detail anymore, but only how far the analogy is valid.

5.2.1. *Mechanic Effects.* In order to create a damage in a sandwich structure down to the surface of the second conducting layer, generally a far larger laser pulse energy is needed, when comparing with the kinetic energy of the simulated micrometeoroid. As a result the hole burnt in will be wider than that of the impact. Even more important is the fact that the light energy dissipation depends heavily on the sandwich structure, as it was already discussed. Consequently, the mechanical damage will exhibit a much different geometric structure than that caused by the particle impact. A comparison seems thus not possible.

An additional mechanic effect may be caused due to the huge plasma created by the laser pulse. Secondary effects like arc discharges may considerably alter the mechanic structure in a different manner than particle would have done. Taking all these effects into account, one feels completely unable to simulate galvanic short-circuits by pulsed laser irradiation.

5.2.2. *Electric Effects.* The situation is not as bad with the simulation of electric effects by plasma creation in impact channels already created, either by a previous impact or by any other procedure. Although the initial plasma quality differs considerably when comparing those created by impact with those by irradiation, one can find appropriate laser parameters for simulation as well. As the impedances of those plasmas are rather insensitive to the detailed plasma quality, a very similar short-circuit behaviour is expected.

The prompt onset of arc discharges, or the delayed Raether spark ignition may as well be studied by pulsed lasers. For this sake laser fluences producing the same order of magnitude of plasma densities for comparison are sufficient. Such fluences can be deduced from the semi-empirical formulae already given. This procedure has the additional advantage that a correct focusing is possible with multiple use at the same spot, as the laser damage is comparably small. So the reproducibility of the discharge induced short-circuit phenomena can also be studied.

6. Experimental Investigations

6.1. PARTICLE IMPACT EXPERIMENTS

A solar array simulator (SAS) was provided by ESTEC which delivered the high voltages (up to 130 V as built up with Columbus power plant) and with maximal currents up to 9 A. By means of this SAS impact studies could be performed with sandwich structures (bus bars, solar cells, etc.) under voltage conditions comparable to those found in space. The transient voltage and current could thus be measured down to the 100 nsec time regime. In another series of experiments the complex impedance of the impact plasma was simulated by static resistors and capacitors and a fast mercury switch. By comparing the actual impact conditions with simulation experiments, the theoretical predictions of onset of arc-discharges and arc complex impedances were well verified.

Measurements have been performed with the Munich Electromagnetic Dust Accelerator Facility. Due to the small particles (see ch. 1) permanent arc discharges could not be ignited (as expected), however, transient short-cuts were detected. Accelerating larger particles with the Freiburg Gas Gun. Permanent arc discharges could also be produced at 130 V. With lower voltages, say 70 V, permanent arc discharges could not be made, in accordance to theory.

6.2. LASER IRRADIATION EXPERIMENTS

The solar array structure possesses more and less sensitive points for impact. Particles, however, cannot be deliberately directed to a mm-precise point, in contrary to laser shots. So laser experiments were performed with the Kaiserslautern Giant Pulse Laser Facility. As theory predicts a laser pulse producing a short-cut at voltages above 100 V always ignites a permanent arc discharge.

7. Results and Discussion

Although the mechanic and electric effects of hypervelocity impact, especially with sandwich targets under voltage, are rather complicated, a semi-empirical description of the main important properties could be given, and proved by a series of experiments on solar array structures. Although a straight-forward simulation of hypervelocity impact by means of pulsed laser irradiation is not possible, laser experiments are informative. Laser shots resemble the "worst case" of plasma formation, i.e., no particle impact creating a comparable damage will produce more or hotter plasma. So laser shots create the most electric damage per unit energy dissipated possible, and electric withstanding laser pulses of comparable energy always withstand hypervelocity impacts.

For strategic defense initiative purposes one may state that if mechanical destruction is needed, debris or other particle impact is best. However, destruction of electric systems is best maintained by laser pulses: Power plants are much affected, and the laser plasma is able to induce huge destructive electric high-frequency pulses in the electronics, which are outstanding.

8. Acknowledgements

Thanks are due to ESTEC which provided support of the investigation of that impact phenomena under ESA-Contr. No. 7418/87/NL/IW(SC) and for their continuous interest of K. Bogus and A. Robben. The main contractor was PTS, Freiburg/FRG. The particle impact experiments were performed at

Technical University, Munich/FRG, under the guidance of M. Rott; at Ernst-Mach-Institute, Freiburg/FRG, under the guidance of E. Schneider. The laser irradiation experiments were performed at the University of Kaiserlautern, under the guidance of K. Rohr. Thanks are due to all these gentlemen and to my former collaborator D. Maas for their skillful experimental work and many critical discussions.

9. References

[1] F.R. Krueger, D. Maas, M. Rott, W. Reschauer, E. Schneider, K. Weber, and H. Thiemann, 'Investigation of Micrometeoroid Interactions with Solar Arrays', Final Report to ESTEC. Contr.-No. 7418/87/NL/IW(SC), (1989a).

[2] F.R. Krueger, D. Maas, M. Rott, K. Rohr, and K. Bogus, 'Simulation of Micrometeoroid Impacts by Laser Pulses' In: Proc. of the European Space Power Conf., Madrid, ESA SP-294, pp. 837-842, (1989b).

[3] F.R. Krueger, 'Thermodynamics of Ion Formation by Fast Dissipation of Energy at Solid Surfaces', Z. f. Naturforschung 38a, pp,. 385-394, (1983).

[4] F. Arnaudeau and G. Munck, 'Technical Notes on Giotto Residual Ionization', Rungis/France. ESTEC-Contr. 83/5775 (1984).

[5] K. Bogus, A. Robben, F.R. Krueger, D. Maas, M. Rott, W. Reschauer, E. Schneider E., K. Weber, and H. Thiemann, 'Micrometeoroid Impacts on Silver Bus Bar Targets at Elevated Voltages', In: Proc. of the European Space Power Conf., Madrid, ESA SP-294, pp. 820-836 (1989).

[6] R. Dinger, K. Rohr, and H. Weber, 'Ion Distribution in Laser Produced Plasma on Ta-Surfaces at Low Irradiances', J. Phys. D 13, pp. 2301-2307, (1980).

[7] J. Kissel and F.R. Krueger, 'Ion Formation by Impact of Fast Dust Particles and Comparison with Related Techniques', Appl. Phys. A 42, pp. 69-85, (1987).

[8] F.R. Krueger and J. Kissel, 'Experimental Investigations on Ion Emission with Dust Impact on Solid Surfaces', ESA-SP-224, pp. 43-48, (1984).

METHODOLOGY FOR IMPACT DAMAGE ASSESSMENT

T. J. STEVENSON
Unit for Space Sciences
University of Kent at Canterbury
CT2 7NR, United Kingdom

ABSTRACT. There is now an abundance of data describing the phenomenology of impact damage to spacecraft, provided by the recovery of various surfaces in recent years, notably the Long Duration Exposure Facility (LDEF). While these surfaces represent a range of materials and constructions commonly used in spacecraft engineering, the data extracted from their examination cannot be utilized directly where failure mode analysis warrants an assessment of damage and secondary effects. This is particularly true where complex structural geometries, unusual or highly variant stabilization regimes or orbits other than low inclination LEO are involved.

This paper describes the means by which conservative but realistic design guidelines may be derived from existing and projected knowledge of this aspect of the space environment.

1. Introduction

The tools which are enabling mankind to exploit the operational properties of artificial satellites must include a detailed knowledge of the hazards to be encountered during their missions. The means of design, construction, launch and operation of spacecraft are costly, and their appropriation must be constantly justified. The justifications of usefulness and prestige will always feature in this process, but increasingly there is a need to demonstrate reliability and longevity.

Many factors determine the lifetime of a space based system, and it could be said that, so far, only a few mission failures have occurred due directly to a lack of knowledge of the operating environment. This is because most launch, orbital and re-entry environments are not only well characterized, they are analytic (in the mathematical sense) and can be easily applied to a wide range of circumstances.

If one was to categorize the mechanisms which operate, often synergistically, to threaten failure in spacecraft systems, one might divide them into continuous and episodic. Continuous environments, such as Solar UV, are well characterized simply because they are always present to measure. Episodic environments that can be reproduced, such as launch

R. N. DeWitt et al. (eds.), The Behavior of Systems in the Space Environment, 291–298.

structural loads, are also well known, but those that occur unpredictably or rarely, such as the effects of the more extreme forms of Solar activity or the impact of micrometeoroids, are characterized in proportion to how often they have been experienced. A lack of experience manifests itself in poor statistics. One further characteristic, unfamiliar to those used to geological or cosmological timescales, is the potential for non cyclic variability - short term change - introduced through the activities of Mankind.

The subject of this contribution is one such environment; that of hypervelocity impact, its causes and effects. Having emphasized the uncertainties that we no doubt suffer under when attempting to anticipate the impact hazard to space systems, we will describe the state of knowledge and its future.

2. The Existing Data

Throughout the space age, and in fact preceding true orbital flights, there have been attempts to measure the cosmic dust believed to be found between planets. While one might look at a mass flux distribution, appreciate the steep increase in spatial density with decreasing size and then walk outside on a clear night to see a meteor without too long a wait, it remains true that damaging meteoroid impact on spacecraft surfaces are still a rarity anywhere in our part of the Solar System - cometary coma interiors excepted. The relatively low spatial densities resulted in large errors in the early measurements, particularly those originating from sounding rocket flights.

2.1. THE DATA SOURCES

These can be conveniently divided into near Earth measurements and interplanetary measurements on the expectation that the respective environments differ.

Table 1 summarizes some important sources.

2.2. DETECTOR TECHNIQUES AND THE LIMITATIONS OF THE DATA

This is not a comprehensive survey of detector techniques, but is intended to point out some of the pitfalls in using data derived from sources such as these.

2.2.1. *Penetration measurements*. Extracting knowledge of the mass of the impactor from the hole it makes in a thin sheet requires detailed information regarding the penetration mechanics of the materials involved. Empirical formulae relating the size of hypervelocity impact features, both perforations and craters, are given by various authors, such as Pailer and Grün [Ref. 1] and Carey, McDonnell and Dixon [Ref. 2]. Unfortunately, these involve the velocity and density of the impactor which is rarely, if

Table 1: Major data sources by mission.

Mission/Experiment	Reference	Orbit	Primary Measurement
Explorers 16 and 23		LEO	TM Foil perforation
Ariel II		LEO	TM Foil perforation
STS-3 MFE	McDonnell et al (1984)[3]	LEO	LM Foil perforation
Solar Maximum	Laurance and Brownlee (1986)[4]	LEO	LM Crater metrology and chemistry
LDEF	Various - Kinard (1990)[5]	LEO	LM Foils and craters
Apollo windows			LM craters
HEOS			
Helios	Grün[6]	IPE	TM Impact plasma
Ulysses	Grün (1983)[7]	IP	TM Impact plasma
Lunar materials	Morrison and Zinner[8]	-	LM Craters

Key: LEO - Low Earth orbit, low inclination, near circular, less than 100 km. IP - Interplanetary. IPE - Interplanetary but restricted to near the ecliptic. TM -Telemetered data (surfaces not recovered). LM - Laboratory measured.

ever, measured directly and independently. Thus, assumptions must be made if the analysis of a single site is to proceed, but where a number of sites can be attributed to the same source, knowledge of the distribution of velocities might be applied. This is done in the case of LDEF analysis, where a velocity distribution, derived from observations of visual and radar meteors, is used to narrow down the range of masses associated with any given size of perforation. The limitation in this technique is that the observed meteors are generally many orders of magnitude more massive than those that impacted LDEF, and the velocity distribution may differ at these sizes.

The flux attributed to the mass observed is derived from the area of the detector, its field of view and the pointing of that field of view. Complications arise due to uncertainties in the surface area sensitivity and exposure, the problem of impacts on nearby surfaces causing secondary impacts on detectors, the presence of the Earth or other planetary bodies gravitationally enhancing the flux, and the directions of detectable particles being unknown or very restricted.

2.2.2. *Impact plasma measurements.* The mass calibration of these detectors does not depend so strongly on velocity, and indeed detectors do have the capability to measure the impact velocity directly, if it is appreciably charged, and indirectly by the risetime of the plasma signals. Dietzel et al [Ref. 9] laid down the basis for the mass calibration of this detector which has been used several times in interplanetary missions.

Transformation to a directional or isotropic flux is hampered by the same set of geometrical problems outlined above.

3. The Origins of the Detected Particles

3.1. THE INTERPLANETARY FLUX

The interplanetary flux is made up of several component populations. The rarest likely to be material from outside of the Solar System, the interstellar particles. These have never been unambiguously observed, and will only be discriminated by isotopic composition measurements and on the grounds of incident velocity, which will be high. The flux is likely to be isotropic. Another element in the interplanetary mix will be primordial material, originating from the ablation of comets, which has had little chance of thermal or shock modification at planetary surfaces. These particles are small, and well distributed. The third component is planetary ejecta, formerly part of a planet or asteroidal surface, whose size range extends up to asteroids themselves, and are concentrated in the plane of the ecliptic.

3.2. THE NEAR EARTH FLUX

A crude assessment of the data shows a difference between interplanetary and low Earth orbit measurements; this can be attributed in some way to the presence of the Earth. The increase in flux near Earth seems to be due to two factors, gravitational enhancement of the interplanetary flux and space debris from intensive near Earth space operations.

The issue of space debris is now well explored, except with respect to the exact hazard posed. Sources of the material are of course all related to space faring activities, not all of which are honorable.

The era is nearly over where jettisoning of covers and other mission related items without regard to the longevity of their subsequent orbits was routine practice. Such items may, at one time, have made up the majority of nonfunctioning objects in Earth orbit, but since then many satellite breakups have been well documented, and the inheritance of a number of these is a complex of fragments in a wide range of orbits. Most of these objects, whose existence is inferred from fragmentation modelling and experiments, are undetected from the ground.

A celebrated example of a catastrophic breakup was the explosion of an Ariane third stage some months after burnout, caused by the detonation of residual propellant. This event, taking place in the upper part of the LEO region, gave rise to hundreds of trackable objects and probably thousands of smaller particles. Several workers have provided the means by which estimates of the lifetime of this population with this single common source may be made and some results of this are summarized in Table 2 below.

Table 2: (Taken from [Ref. 10]) Estimates using the methods of two authorities of the number of the presently trackable fragments of the V16 Ariane third stage breakup of 1986 remaining in orbit on the dates shown.

Year	from King-Hele (1987)[11]	from Reynolds (1983)[12]
2000	125	127
2010	97	102
2020	78	79
2030	67	64
2040	57	56
2050	53	47
2100	34	25

4. Models and Methods for Hazard Assessment

4.1 AN ISOTROPIC FLUX MODEL

Most investigators will publish a cumulative mass flux model showing their new data against a background of previous measurements and functional fits such as that of Grün et al [Ref. 6]. Figure 1 is an example of this, from

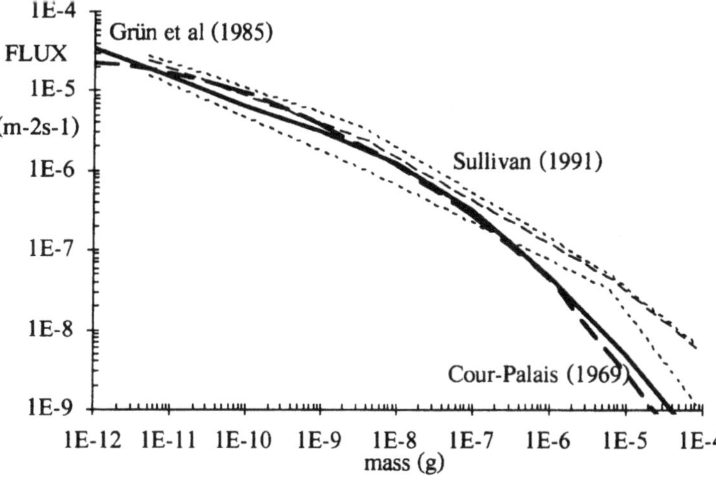

Fig. 1: A cumulative mass flux plot derived from LDEF data and transformed into an inertial frame. Bold curves are: dotted; from Cour-Palais [Ref. 14], solid; Grün et al [Ref. 6]. Feint long dashed line is LDEF data corrected for near Earth effects, and with a particle velocity and density of 17.4 km s^{-1} and 1 respectively. Short dashed lines are error contours derived from numbers of impacts plus and minus their square root. Penetration equation employed is that of McDonnell et al [Ref. 13], and the figure is from Sullivan [Ref. 15].

McDonnell et al [Ref. 13]. Despite the difficulties of comparing data from detectors of differing technologies such as those outlined above, these are in good agreement. This plot represents therefore, the number of particles intercepted by a tumbling flat plate of $1m^2$ in one second whose mass exceeds the value given by the curve – it is a cumulative mass flux plot.

This model or discrete data points from real data sets may be used to arrive at a crude approximation of the damage accruing to a spacecraft. The accuracy of an assessment arrived at by this means is enhanced by similarities between the assessed situation and that of the model; that is, frequent changes in attitude and an orbital regime similar to those of the measurements. Data is frequently transformed to account for the shielding of the Earth and the orbital motion of the detector, and this must be accounted for where necessary. There are few circumstances where simple area-times-flux calculations are appropriate, because all spacecraft move, and the flux is not isotropic.

4.2. CURRENT PRACTICE: NASA SSP 30425, REV A, CHAPTER 8 – METEOROIDS AND ORBITAL DEBRIS

This document reflects the need for improved impact damage assessment in an era of large LEO structures, namely Space Station Freedom. It must contain the current understanding of the environment, while still being of utility to design engineers. This it does admirably, including as it does the Grün [Ref. 6] interplanetary measurements and the space debris model of Kessler et al. [Ref. 16]. This last model has received some attention from those seeking to improve it such as Maclay et al. [Ref. 17], this in the area of the eccentricity of debris particle orbits.

Close study of the document and knowledge of the sources of the data used reveals that the analysis and assessment that results will not aid short term operations, as time resolved information is not present. This can only come from measurements and modelling of the natural and space debris particle distributions. This has begun in earnest with the analysis of data from the Interplanetary Dust Experiment of LDEF [Ref. 18] and the modelling work of Jehn [Ref. 10] and others.

5. Impact damage assessment

5.1. DAMAGE SCALE

The damage occurring at an impact site is described empirically in several places in the literature, and these descriptions are being continuously reviewed in light of the analysis of retrieved spacecraft surfaces [Ref. 2 and NASA SP-8042]. Limitations in ground based accelerator facilities have, so far, restricted the validation of these models largely to regimes where either the impactor density is wrong, the velocity is too low, or some parameter is not well determined.

5.2. DAMAGE SENSITIVITY

Present guidelines give very small probabilities for large body impacts even on the total surface area of Space Station Freedom. There is some concern that a growth in space debris, particularly through a cascade effect initiated by a series of collisions generating a large population, will increase these probabilities in the millimeter to centimeter range where pressurized module protection is necessary. The sizing of bumper shields and the positioning of vulnerable surfaces is being driven by debris growth models inherent in SSP 30425.

There are many other mission threatening failure modes such as those identified by Thorner [Ref. 19] however, and these must be assessed. The nature of the mass distribution, that is, steeply increasing number densities with decreasing mass, leads an assessor to look at the product of damage sensitivity and total exposed area. Solar arrays are prime candidates for this scrutiny, particularly with respect to the synergism between the impactor energy and the electric fields present at the impact site.

References

[1] N. Pailer and E. Grün, Planet, Space Sci. 28, pp 321-331 (1980).

[2] W.C. Carey, J.A.M. McDonnell, and D.G. Dixon, **Properties and** Interactions of Interplanetary Dust (Guise R.H. and Lamy, P., Editors), pp 291-295 (1984).

[3] J.A.M. McDonnell, W.C. Carey, and D.G. Dixon, Nature 309, pp 237-240 (1984).

[4] M.R. Laurance, D.E. Brownlee, Nature 323, pp 136-138 (1986).

[5] W.H. Kinard, D.J. Carter, and J.L. Jones, NASA SP-473, (1984).

[6] E. Grün, H.A. Zook, H. Fechtig, R.H. Giese, Icarus 62, pp 244-272 (1985).

[7] E. Grün, H. Fechtig, R.H. Giese, J. Kissel, D. Linkert, J.A.M. McDonnell, G.E. Morfill, G. Schwehm, and H.A. Zook, ESA SP-1050, pp 227-243 (1983).

[8] D.A. Morrison and E. Zinner, "Distribution and flux of micrometeoroids", Phil. Trans. R. Soc. London A, 285, pp. 379-384, (1977).

[9] H. Dietzel, G. Eichorn, H. Fechtig, E. Grün, H.J. Hoffman, and J.J. Kissel, Phys. (E), Scientific Instruments 6, 209 (1973).

[10] R. Jehn, ESA Journal 15, 1 pp 63 (1991).

[11] D.G. King-Hele, Satellite Orbits in an Atmosphere, Blackie and Son Ltd, Glasgow/London (1987).

[12] R.C. Reynolds, et al., J. Spacecraft 20, 3, pp 279-285 (1983).

[13] J.A.M. McDonnell, et al., Proc. 25th Lunar and Plan. Sci. Conf (1991 in press).

[14] B.G. Cour-Palais, NASA SP-8013 (1969).

[15] K. Sullivan, Proc. Workshop on Hypervelocity Impacts in Space, University of Kent at Canterbury (July 1991, in press).

[16] D.J. Kessler, R.C. Reynolds, P.D. Anz-Meador, NASA TM 100471.

[17] T.D. Maclay, R.A. Madler, R. McNamara, and R.D. Culp, Proc. Workshop on Hypervelocity Impacts in Space, University of Kent at Canterbury (July 1991 in press).

[18] F.S. Singer, J.E. Stanley, P.C. Kassel, J.J. Wortman, J.L. Weinberg, G. Eichorn, J.D. Mulholland, and W.J. Cooke, Proc. 21st Ann. Mtg. Amer. Ast. Soc. Div. Dyn. Ast. (1990).

[19] G. Thorner, ESA ESTEC WP 1511 (1988).

PROTECTION AND SHIELDING

J. BOURRIEAU
CERT-ONERA/DERTS
2 Avenue E. BELIN
31055 TOULOUSE Cedex
FRANCE
Tel. (33)61 55 71 11

ABSTRACT. The effects of radiation on spacecraft materials and electronic devices are directly related to the energy transferred by incident particles to the components (atoms, orbiting electrons and nuclei) of the matter encountered. Such events result in the formation of electron hole pairs and point defects (interstitials, vacancies) which, in turn, give rise to macroscopic damages. Induced secondary radiations (prompt or delayed emission) add their effects to those produced by primary incident particles.

The first part of the paper describes the various sources of particles encountered in space (galactic cosmic rays, trapped protons and electrons, protons and heavier ions from solar events) and their characteristics on various orbits. Secondary radiations (neutrons, bremsstrahlung and recoiling nuclei) are also defined.

The different methods for simulating particle transport and examples of radiation damage modellings are then described. Results for absorbed doses and dose equivalents, including shielding effectiveness considerations, are given and compared to in-flight data.

In the same way the methods for the forecast of single event effects (upset and latch-up) induced on electronic devices by heavy ions (direct process) and by protons (indirect process) are summarized. Some results are given and compared to in-flight measurements.

1. Introduction

Since 1958, when the Explorer and Pioneer satellites discovered the presence of ionizing particle belt around the Earth, a larger number of satellites have been launched by different countries. Many malfunctions in the experiments flown were observed, which could be attributed to environment effects. Due to the growth of mission duration for space systems, and to the increasing use of sophisticated on-board electronics, prediction of the damages on materials and devices in the orbital environment is becoming more and more cunning.

The primary sources of energetic particles are the Earth's radiation belt, the Galactic Cosmic Rays (GCR) transiting ions and the sporadic

R. N. DeWitt et al. (eds.), The Behavior of Systems in the Space Environment, 299–351.

emissions of energetic electrons, protons and heavier ions during solar events (Solar Cosmic Rays – SCR).

The interactions of these primary particles with the spacecraft structures and atoms of the high atmosphere lead to:

- slowing down, absorption and diffusion of primaries,
- prompt or delayed emissions of secondaries, depending on the crossed material and on the nature and energy of impinging particles (Bremsstrahlung, Cerenkov radiations, X and gamma rays, light fragments and recoiling nuclei).

All these radiations are able to induce materials and system damages; these effects can be:

- transient and flux dependent (background noise in detectors, Single Event Upset, material charging...),
- permanent, depending on the fluence received and on the exposure rate if recovery occurs.

The hazard level depends mainly on the orbit and epoch (solar activity effect), on the location inside the spacecraft (shielding effect) and on the consequences at the system level of the failure occurring in a particular device.

For manned missions, the main characteristics of a biological system response to the radiations are its high sensitivity and its capability to restore the damages. Except for heavy exposure (and the central nervous system) effective recovery occurs within some months through tissue production, but modified cells able to duplicate may be created, increasing the risk of late diseases. Radiobiological effects fall into two categories, non stochastic and stochastic damages.

The first, early or late effects (skin erythema, sterility, cataracs..), result from changes occurring in large number of cells. They appear above thresholds; the occurrence and severity of symptoms increase with the exposure.

Stochastic late effects are characterized by the fact that the nature and severity of responses are always the same (cancers, genetic mutations..) and therefore do not depend on the received dose. It is the probability of disease, appearing several years after irradiation, that increases with the exposure level.

Heavy irradiation, inducing crew capability reduction or early death, are improbable in space; such situation can only be encountered during very large solar events in unprotected locations (Extra Vehicular Activities – EVA, outside the magnetosphere).

On the other hand, late effects can have a major impact on manned space flights on geostationary (GEO) or polar Low Earth Orbits (LEO), and during interplanetary travels. During such missions, without specific shielding design and mission planning, the doses can reach the limits specified by the Space Science Board Committee on Space Medicine. In addition the ALARA principle (As Low As Reasonably Achievable) is always applied in order to maintain the risk per mission at the lowest possible level.

2. The Radiation Sources in Space

The Sun is supplying nearly the totality of the energy distributed on the Earth and in its environment. The radiation fluxes are directly or indirectly determined by the solar activity and its fluctuations.

On the one hand, the Sun is a source of a permanent flux of particles, the solar wind. These particles of low energy are easily stopped by the spacecraft and by the magnetospheric shielding. On the other hand, solar flares are sporadic sources of protons and heavy ions with energies up to a few GeV; they are one of the most severe hazards for spacecraft. That is why we must improve our ability to predict solar flare occurrence and proton fluxes.

2.1. THE SOLAR CYCLE

On the photosphere surface, dark spots can often be seen. They are easy to observe and have been studied for centuries. They often cluster by pairs showing opposite magnetic polarities. Sometimes they form large groups with magnetic field showing very complex structures; these are called active centers. They can have a long life duration and can be seen during several rotations. The solar flares burst in these areas. It is a detailed observation of the optical and magnetic structures of the active centers which can allow in some degree the prediction of the occurrence of a solar flare.

The sunspot observation has revealed the Sun rotation at least at the photospheric level. This rotation takes place around an axis nearly perpendicular to the ecliptic plane (83 degrees). The rotation is faster in the equatorial plane (26 days) than at higher latitude regions (29/30 days). The differential rotation acting on the solar dipolar magnetic field can explain the formation and the evolution of the sunspots and their fields. The Sun rotation combined with the solar wind emission force the interplanetary magnetic field lines into an Archimedean spiral shape. This structure plays an important role in the propagation of high energy charged particles from the Sun to the Earth at the time of solar flares.

The number of sunspot present on the solar disk varies according to an 11 year cycle. During cycle maximum (about 7 years), tens of spots may be simultaneously present on the disk, and during minimum (about 4 years) long periods without spots may be observed. The maximum amplitude varies considerably from a cycle to another (Fig. 1) [Ref. 1].

During a whole cycle in a group of two spots, the leading spot always shows the same polarity but will show the reverse polarity in the following cycle. In the same way, the Sun dipolar magnetic field reverses of polarity at each new cycle. So, on the magnetic point of view, the solar cycle should be of 22 years.

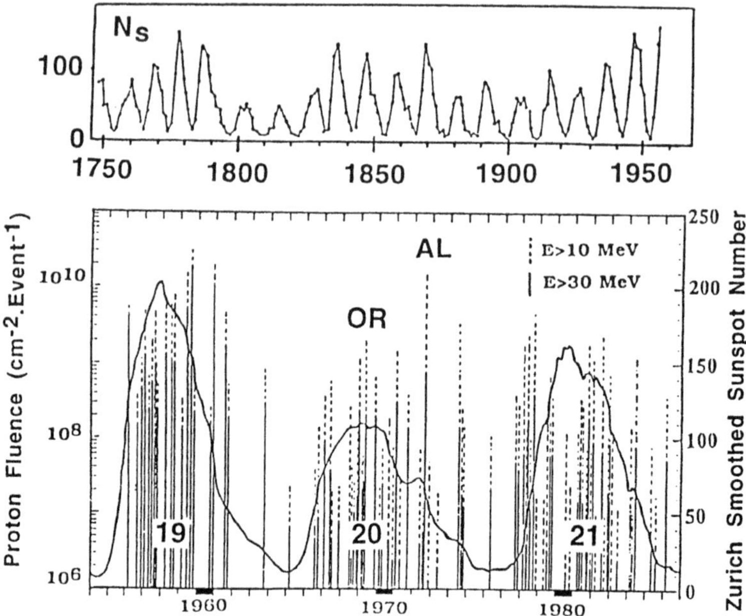

Fig. 1: Solar activity variations during the last centuries (above) Solar proton events and Zurich smoothed sunspot number variations versus time during cycles 19 to 21 (bottom).

2.2. THE SOLAR FLARES

2.2.1. General. A solar event is an explosive phenomenon due to reorganization of instable and strong magnetic fields, occurring above active centers in the low solar atmosphere. A sudden and large optical brightening (optical flare) is seen seldom in white light but generally in emission lines such as H_α.

Flares produce heating, particle acceleration and both photon and mass ejection. Shock waves also propagate through the interplanetary medium. When magnetic fields are open, particles (protons, electrons and heavy ions) may escape into the interplanetary space. They travel along the interplanetary magnetic field lines. After interactions with the medium (including reacceleration by shock waves), they are observed in the Earth's environment. Flares generate natural phenomena such as interplanetary shock waves and geomagnetic storms. Because of the spiral shape of the interplanetary magnetic field (Fig. 2), the solar longitudes around 50 west are more directly connected with the Earth. Solar flares are very frequent in the optical range (up to 10^4 per year according to Fig. 3). Proton events are far less frequent (a few per year according to bottom part of the same figure) [Ref. 2].

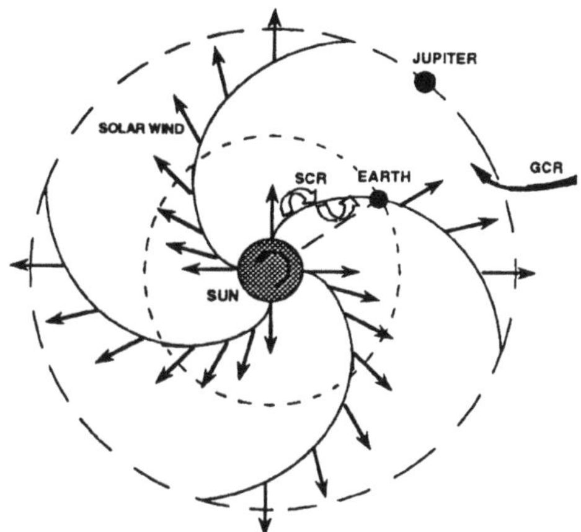

Fig. 2: Archimedean spiral shape of the interplanetary magnetic field.

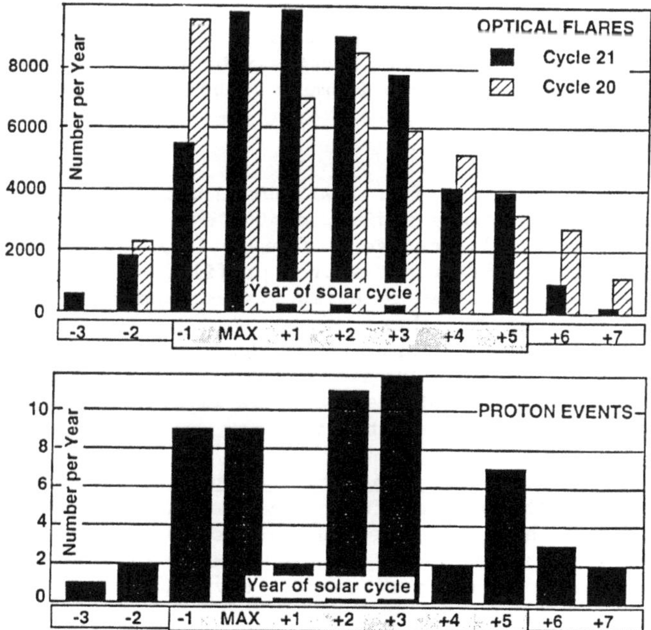

Fig. 3: Number of solar optical flares and proton events per year.

2.2.2. Proton Events

2.2.2.1. General Characteristics. The most important proton energy range to be studied when radiation protection is concerned, is above 10 MeV. Below this limit, most of the protons are easily stopped even with a thin shielding.

Temporal Distribution

The number of events follows the solar activity modulation (Fig. 1); solar events occur mainly during the seven years of the maximum activity period; the risk is negligible during the four years of the cycle minimum.

For cycle 19 (1954-1964) the proton fluences are deduced from indirect ground-based observations (Fig. 1). Thus the measurements are not as reliable for this period as for the next cycles.

Because the most active sunspot groups rapidly recover their ability to give other flares, most proton events occur in groups: for example all very large events were concentrated within only five weeks in the past 30 years.

Delay Between Flare and Proton Event

With the proton events observed during last cycles, it is possible to give precise values of the time sequence of events and to study statistical aspects of the rising phase. The first protons arrive on average 1.5 hours after the flare itself; the maximum of the flux is reached on average 20 hours after the flare, with an observed shortest delay of about 2 hours. During the first six hours, the maximum is only observed in 10% of the events; thus in most cases this leaves enough time for proper actions [Ref. 2].

Proton Event Time Profiles

Flares have a duration of a few hours, but because of particle diffusion in the interplanetary medium, proton events have a duration of a few days. The exponential increase of flux is followed by a much longer decrease. Figure 4 [Ref. 3] give temporal profiles for typical periods: a single proton event is related to a few flares, and the first event is often immediately followed by others.

The proton event time-profile is dependent upon the solar longitude of the flare. When the flare occurs in the eastern part, the delay is much longer. Thus the localization of the flares is important to determine the timing of actions for radiation protection.

2.2.2.2. Analysis of the Flare Prediction Methods. In addition to the shielding provided by the spacecraft and by the Earth's magnetic field, predictions of flares help to lower the risks of radiation in space. Available methods are summarized on Fig. 5. Months or weeks in advance, it will be possible to know the periods of activity, by studying the laws of emergence of solar active regions.

A few days in advance, the observation of the complexity of the magnetic field topology provides an estimate of the flare probability (Fig. 6). The method has been operational since 1965 in the warning centers throughout the world.

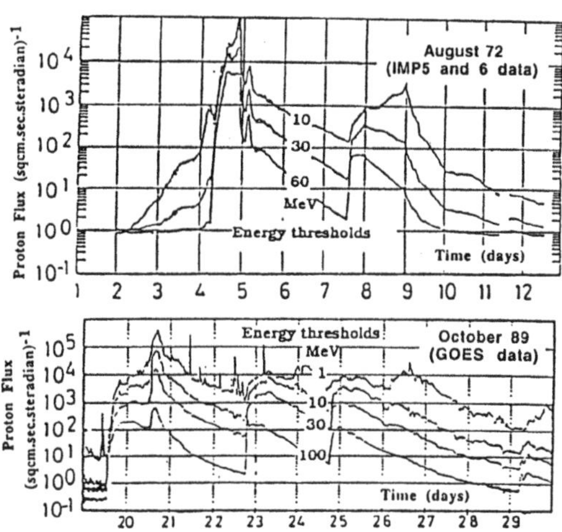

Fig. 4: Very large proton events (August 72 and October 89) integral flux variations versus time.

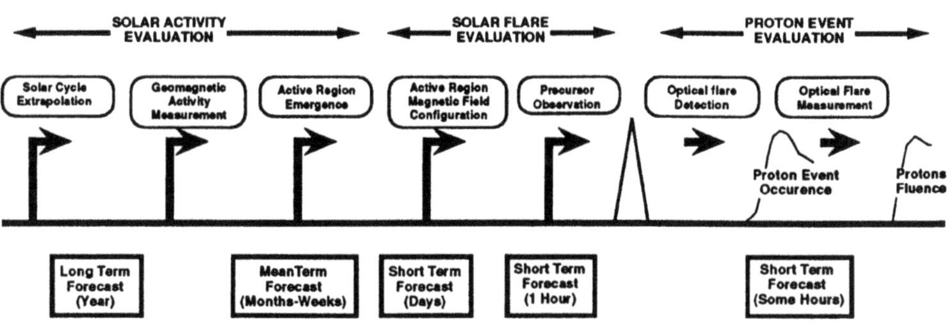

Fig. 5: Proton event forecasting method.

A few hours before a large flare, precursors may be observed, in particular in the X-ray range. Nevertheless, in the absence of appropriate real-time data, the method is not presently operational.

Modelling of the Time Intensity Profile Expected at Earth

When the flare is observed on the Sun, the delay of proton maximum arrival is sufficient, often, to make appropriate decisions. Thus, methods have been developed to predict delay and proton flux from flare observations, in particular in the radio and X-ray ranges.

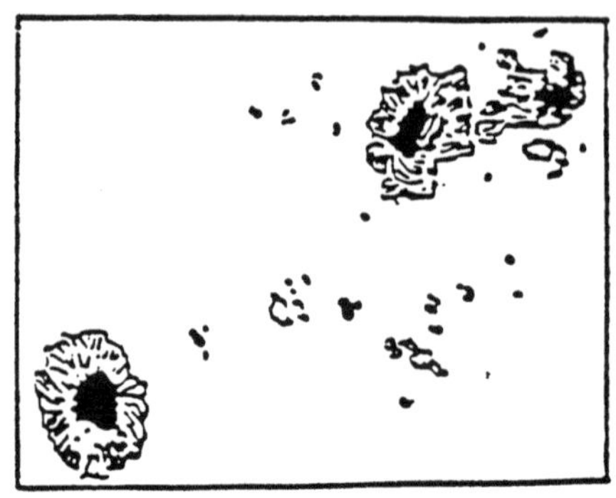

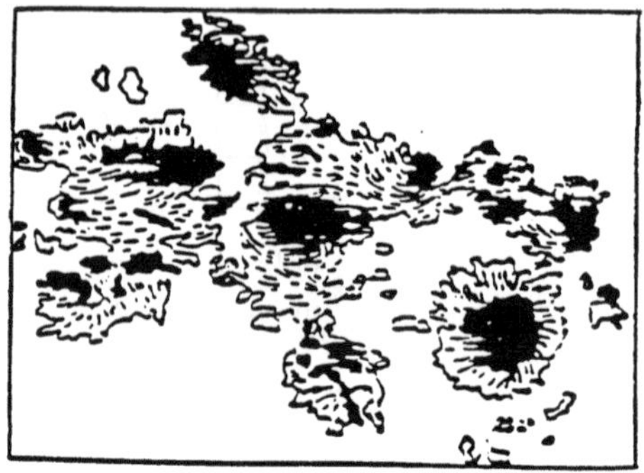

Non Flaring Center **Flaring Center**

Fig. 6: Active center configurations.

A procedure used by US Air Force is available [Ref. 4] to generate a computerized time intensity profile of the solar proton intensity expected at the Earth after the occurrence of a significant flare on the Sun. It is a forecasting tool and it aims at a prediction of the solar proton flux at Earth for the 300 hours following a flare. The scheme of the time intensity profile at Earth is illustrated in Fig. 7.

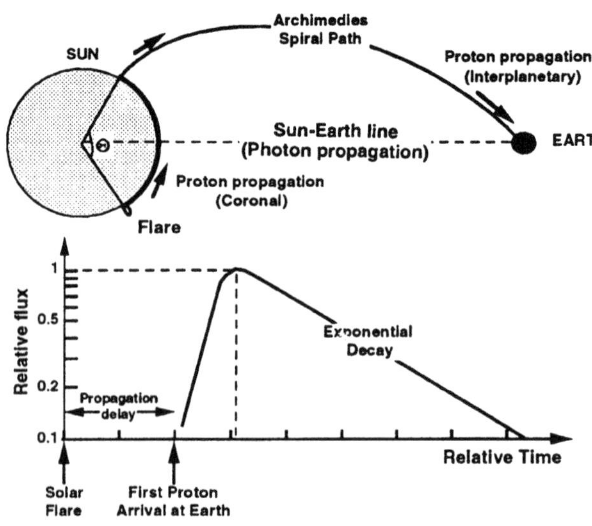

Fig. 7: General characteristics of proton event time profile.

In the model the flare localization is given by optical observations; the shape of the spectral signature of the radio emission is used as indicator that solar protons are being released in the corona; the proton flux emitted is deduced from the radio burst and X-ray intensity measurements. Then the attenuation and the diffusion time in the corona from the flare to the foot of the archimedian line connecting the Sun to the Earth are computed, as well as the transit time along the line for each proton energy. The shape of the rising phase of the proton flux and its maximum are then obtained.

In the following part of the evaluation an exponential decrease is assumed, with a decay constant depending on the solar wind speed, the distance along the archimedian spiral path and the angular distance travelled in the corona. With this model, the delay from the flare observation to the proton flux peak and the fluence (E>10 MeV) may be predicted with standard errors corresponding respectively to a factor 2 and 10.

Probabilistic Model for Long Term Forecast - SOLPRO Code

This code [Ref. 5] is designed for a statistical evaluation of proton fluences to be encountered at one AU outside the magnetospheric shielding. A probabilistic analysis of the solar cycle 20 fluxes allows an evaluation of the probability of exceeding mission fluence levels as a function of mission duration.

Flares are supposed to occur only during the seven years of the active part of the solar cycle, and only this period of time is considered to determine the mission duration. The fluctuations of the solar cycle frequency are not included and uncertainty of about one year is expected at the beginning of each period.

The code includes a distinction between two kinds of flares:

– ordinary (OR) events, corresponding to relatively weak fluence levels (some 10^8 protons per centimeter and per flare) and occurring about ten times per year,

– anomalously large (AL) events, inducing high proton fluence (may be more than 10^{10} protons per cm^2 and per flare) as those of July 1959, December 1960, August 1972 and October 1989.

It must be noticed that only the August 72 AL event had been studied in flight, and that available data apply only to the energy range from 10 to 100 MeV. For the two others in 1959 and 1960 their characteristics have been deduced from ground measurements (neutron monitor mainly). Following these remarks the August 72 flare is used as reference in the code (Fig. 8).

The number of AL flares and the flux due to OR events (Fig. 8) are given versus flight duration during the maximum activity period and confidence level Q. Following recommendations of E.G. Stassinopoulos, Q = 0.75 is used for the computations of AL event number. Some remarks must be made concerning the validity of the model:

- the variations linked to the cycle amplitude modulation are not included (Fig. 1);

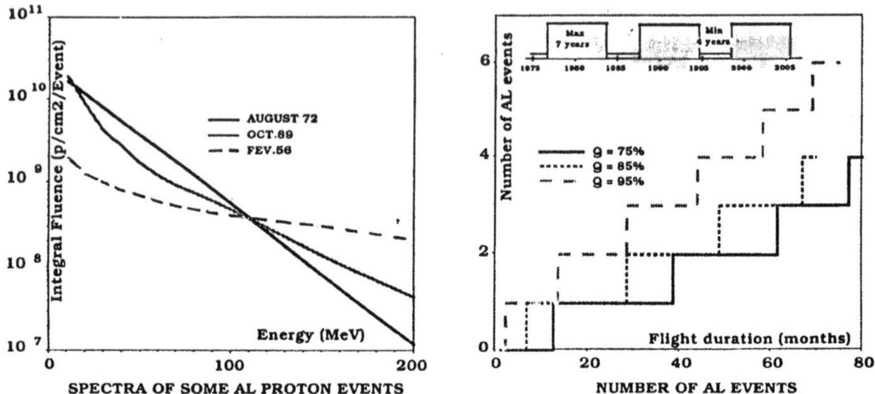

Fig. 8: SOLPRO model of AL proton events.

- the probabilistic analysis of the AL events, based on a reduced number of such flares, is not well established;
- the August 72 flare, used as reference, cannot be considered as a worst case.

More recently, a complementary study covering the solar events observed during the period 1956-1986 had been performed [Ref. 6]. The results of this analysis show that the distributions of the proton fluence (E>10 and 30 MeV) per flare, including the AL event defined by SOLPRO, follow a log normal distribution. Nevertheless the risk associated to the solar proton flares obtained with both models are not fundamentally different (Fig. 9).

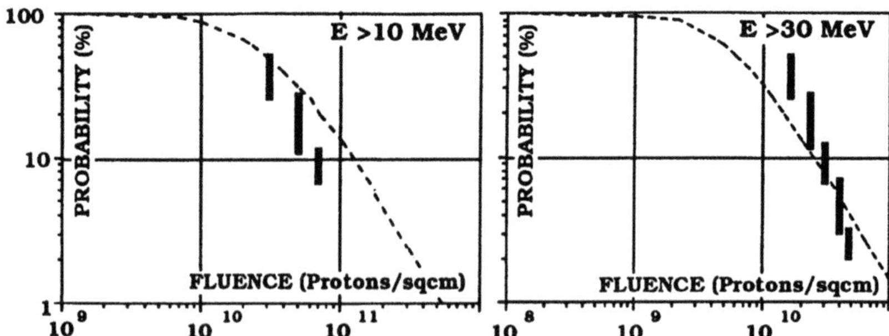

Fig. 9: Comparison of the solar proton fluence estimates Feynman's (dashed line) and SOLPRO (bars) results.

2.2.3. *Heavy Ion Events*. During some flares high atomic number and Energy (HZE) particles (the Solar Cosmic Rays – SCR) are observed, adding to the continuous Galactic Cosmic Rays (GCR see her under) background.

The main difficulty in SCR flux predictions comes from the fact that they are sporadic and highly variable in composition, energy, charge state of the ions and fluxes. SCR are not completely ionized and during their path, they have only few interactions with matter; their charge state at the Earth's orbit level is approximately the one they have during the flare. Therefore, the ionization state varies from event to event and is dependent on the temperature of the photosphere or the solar corona at the moment of the flare.

The composition of solar energetic particles is highly variable from event to event. It is also dependent on the time and the energy during the flare. The heavy ions energies are generally in the 1-50 MeV/nucleon range, but spectra vary from flare to flare and during the event (see Fig. 10 [Ref. 7] showing the energy distribution observed at the peak of the Sept. 24, 1977 event, often considered as a worst case).

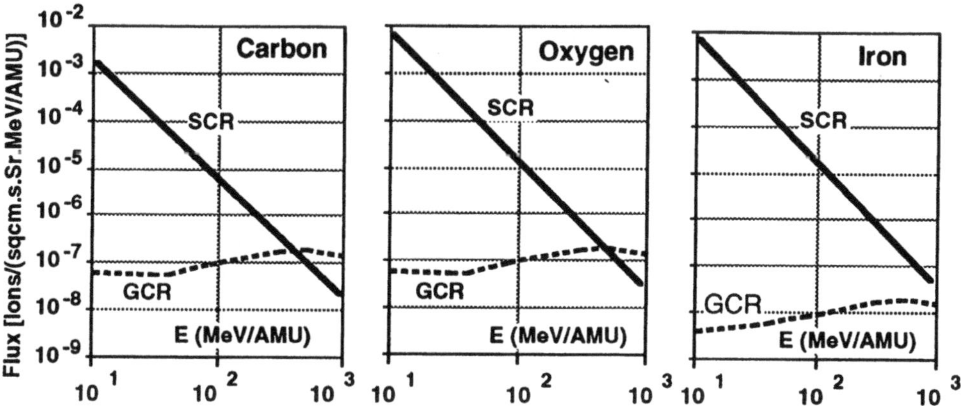

Fig. 10: Comparison between GCR (solar maximum period) and SCR spectra (09/24/77 event at peak).

The SCR component dominates the GCR fluxes only during 2% of the time, and about five events per year may be considered during solar maximum periods. For evaluation of the Single Event Effects (SEE) induced in the electronic systems, models of SCR had been developed [Refs. 8 and 9].

2.3. GALACTIC AND EXTRAGALACTIC COSMIC RAYS

GCR consists primarily of high energy ions with origins outside our solar system. Atoms are emitted by astronomical objects like stars (star flares) but also by novae or supernovae. Approximately, cosmic rays are composed by 87% of protons, 12% of α particles (He nuclei) and only 1% of nuclei with $Z > 2$. P. Meyer [Ref. 10] gives the relative abundance of the elements from H to Ni normalized to that of carbon (Fig. 11). Differences between cosmic ray abondances (R_c) and local composition (R_1) deduced from abondances in the solar system, are explained by the ion spallation during their path in the interstellar medium. The heaviest elements of the Mendeleiev table are also present in cosmic rays but they are in very low quantities.

Due to the interactions of cosmic rays during their paths, the ions are completely stripped and their sources cannot be identified; the particles which reach the vicinity of the Earth's magnetosphere have an isotropic distribution. If we observe the time dependence of the fluxes, a modulation in intensity is detected. This variation correlates with the solar activity (Fig. 12) [Ref. 11]. This phenomenon is due to the solar wind carrying away a turbulent magnetic field up to ten A.U; diffusion of the ions by this magnetic field grows during the solar maximum period inducing a decrease of the fluxes in their low energy parts (Fig. 11).

2.4. MAGNETOSPHERIC SHIELDING

2.4.1 The Magnetosphere. The Earth is surrounded by the magnetosphere (Fig. 13), which is a magnetic cavity created by its own magnetic field, almost dipolar, within the plasma stream issued from the Sun. The geomagnetic field is the sum of two components, the core and the crustal fields.

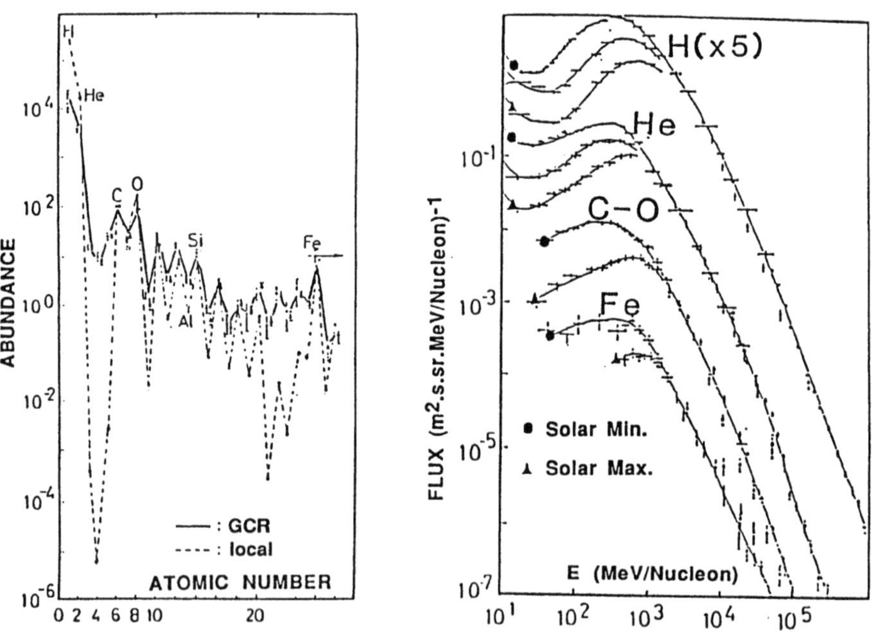

Fig. 11: Galactic Cosmic Rays characteristics composition (left) and spectra (right).

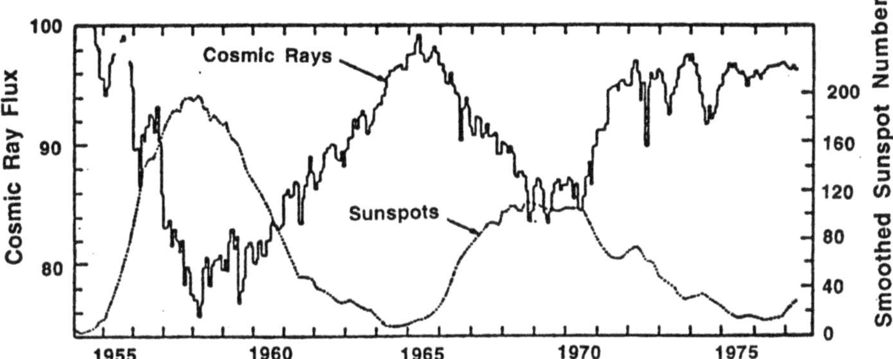

Fig. 12: Solar activity modulation of GCR.

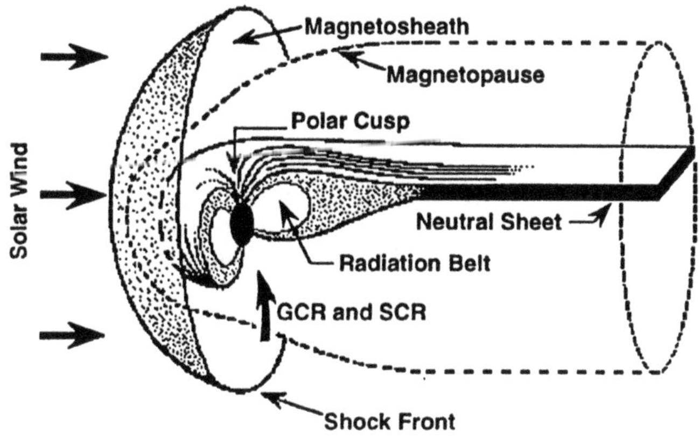

Fig. 13: Structure of the magnetosphere.

The main source, from the molten core of the planet is dipolar; the corresponding dipole is shifted (about 500 km towards West Pacific) and tilted (about 11°) of the rotation axis. Mass and wave motions in the molten core are probably the origin of long term variations, the secular drift, of intensity (about – 27 nt/year), of orientation (south pole drifts about 0.014°/year towards West), and of the shift (about 4 km/year towards West Pacific).

The second term, the crustal field, is induced by magnetized materials from the Earth's surface to depths of the order of 20 km. Its contribution to the ambient field is frequently large near the Earth's surface; the linked irregularities may be neglected for the radiation budget in orbit.

In the magnetosphere the field is the sum of these internal components with fields carried by the solar wind and induced by charge motions in the magnetosphere. These external sources vary with the local time (diurnal variations), the season (tilt effect) and, depending strongly on the solar activity, fluctuate rapidly mainly during magnetic storms.

Up to 4 Earth's radii, internal sources prevail, the field may be considered as dipolar; at this limit external sources modify the dipolar field. The region of space in which magnetic forces due to the field compression balance the solar wind dynamic pressure is the magnetopause (Fig. 13); this frontier depends of course on the solar activity. On the day side, the magnetopause is located at about 10 Earth's radii; in the anti solar direction it extends over more than 500 Earth's radii forming the magnetotail, from which charged particles can reach the Earth's vicinity during geomagnetic instabilities (magnetic storms). In this structure, at high latitudes in the polar cups, transiting particles coming from outside the magnetosphere (GCR and SCR) can move down and hit the high atmosphere.

For studies of charged particle motions in the magnetosphere the field is characterized by two parameters, the geomagnetic coordinates B and L:

- B is the field strength at a fixed point,
- L, a parameter defined by Mc Ilwain [Ref. 12], is approximately equal to the geocentric distance R (in Earth's radii) of the field line point in the equatorial plane.

NB: Assuming a dipolar magnetic field of moment M, with the geomagnetic latitude λ and the radial coordinate, the field line crossing the equatorial plane at geometric distance R_o and the field strength are given by:

$$R = R_o \; . \; Cos^2\lambda$$

$$B = \frac{M}{R^3} \; (1 + 3 \sin^2\lambda)^{1/2}$$

Or, using the Mc Ilwain parameter $L = R_o = R / \cos^2\lambda$:

$$R = L.\cos^2\lambda$$

$$B = \frac{M}{L^3}$$

For the actual geomagnetic field (disturbed by solar wind and magnetospheric currents in large L regions) the geometric interpretation of L become increasingly invalid for equatorial distance greater than 4.

2.4.2. *Magnetospheric Shielding Effect*

<u>Rigidity and Cutoff Rigidity</u>

The earth's magnetic field interacts with the charged particles of cosmic rays and modify their path. The radius of curvature, r, of the particle trajectories is given by:

$$r = R/B = P.c/(Z_{eff}.e.B)$$

with:

- P: the momentum of the particle,
- B: the strength of the magnetic component,
- $Z_{eff}.e$: the charge of the particle,
- c: the light velocity,
- R: the rigidity of the particle.

Particles which are able to reach a certain position in the magnetosphere must have rigidities exceeding a particular value, the cutoff rigidity or geomagnetic cutoff. In the case of galactic cosmic rays, the ions are completely "stripped", $Z_{eff} = Z$, but not in solar flares (we see here why the knowledge of the charge state of ions emitted during a flare is important).

As a first approximation, the earth's magnetic field can be assumed as a perfect dipole shifted of 500 km in the West Pacific direction with regards to the earth's center. Using this assumption, STORMER [Ref. 13] has calculated the geomagnetic cutoff in dependence on the geographic coordinates and the particle incidence:

$$R_c = 59.6 \times r_0^2 \times \frac{\cos^4\lambda}{(1 + \sqrt{1 - \sin\gamma \times \cos^3\lambda})^2}$$

with:

- R_c: the magnetic rigidity cutoff (expressed in GV),
- r_o: the Earth's radius,
- r: the distance from the dipole centre,
- λ: the geomagnetic latitude,
- γ: the direction from which the particle arrives with respect to local East in dipole coordinates.

The geomagnetic cutoff varies drastically along orbit (maximum in the equatorial plane, negligible at high latitudes). Large field perturbations during magnetic storms change the cutoff conditions.

On LEO, satellites are shielded against cosmic rays coming from the opposite side of the Earth. Particles arriving from the West are deflected towards the Earth, while those arriving from East are deflected away, resulting in a large East – West anisotropy of SCR and GCR fluxes on LEO.

In various models [Refs. 8 and 14] including vertical cutoff data [Ref. 15], the Earth shadowing and the state (quiet or stormy) of the magnetosphere determine the shielding effect along the orbit versus particle rigidity. Using these results and the solar protons, SCR or GCR characteristics outside the magnetosphere, the codes determine the energy spectra of the transiting particles received by the spacecraft (Figs. 14 and 15) [Ref. 16].

The effect of the magnetosphere state (quiet or stormy) can be neglected; variations of about 2% are seen on fluxes in LEO. The stormy condition (maximum exposure) is generally used in the evaluations. The magnetospheric protection can be considered as complete for solar protons, GCR and SCR on LEO for inclination lower than 50°, and the altitude effect is negligible in the range for exposed orbits (Figs. 14 and 15). Consequently strategies involving an altitude decrease during proton flares are not useful.

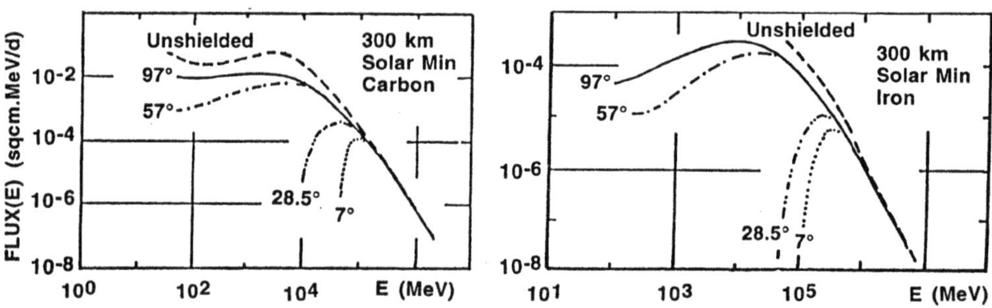

Fig. 14: Examples of magnetospheric shielding effect on GCR encountered on LEO.

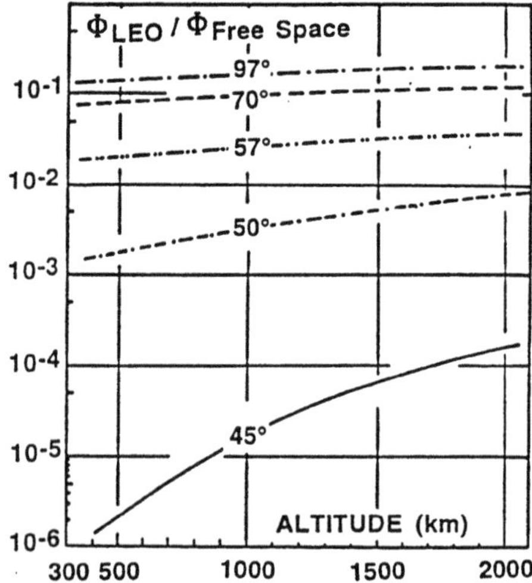

Fig. 15: Solar event transmission ratio (Aug. 72 – E>10 MeV) on LEO.

On geostationary orbit (GEO) the magnetospheric shielding, negligible for heavy ions, reduces considerably the solar proton flux up to some tens of MeV [Ref. 17], depending on the parking longitude.

2.5. TRAPPED PARTICLES IN THE RADIATIONS BELTS

2.5.1. Generalities

2.5.1.1. Origins. Charged particles in the magnetosphere have two origins:

1) The first is the albedo neutron source; cosmic rays striking the atmosphere cause nuclear reactions, among the reaction products are also fast neutrons; some of them leave the atmosphere; as their lifetime is about 13 minutes some neutron decays occur in the magnetosphere producing protons and electrons. This process is the main source of trapped particles in the inner belt (see here after).

2) The second origin providing particles mainly in the outer belt is a drift and acceleration of particle from the solar wind through the magneto-tail during magnetic storms.

2.5.1.2. Trapping. Charged particle trapping results (Fig. 16) from:

- cyclotron gyration around field lines,
- bouncing motion between the magnetic mirrors in regions of higher field strength towards the poles,
- drift around the Earth, due to the radial field gradient, respectively Eastward and Westward for negative and positive charges.

so that charged particles are confined in a magnetic drift shell.

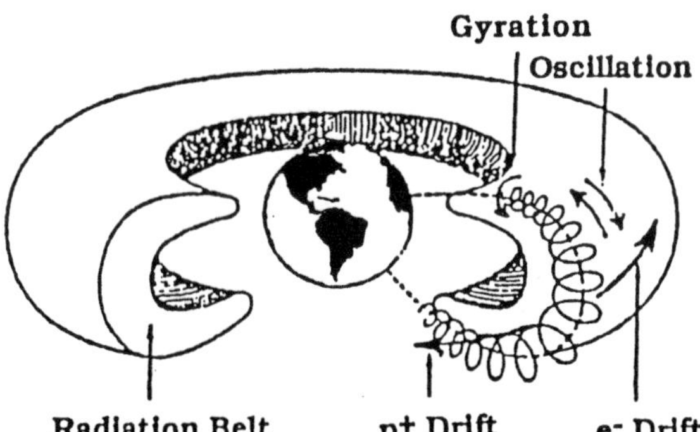

Fig. 16: Trapped particle motions in the Earth's magnetic field.

Magnetic mirroring results from the conservation of the particle magnetic moment, E_n/B, where E_n is the energy associated with the motion perpendicular to the field line. With α the angle between the particle velocity vector and the field line (α: pitch angle) at a point where the field strength is B, this mean in turn that $B/\sin^2 \alpha$ is constant along the field line. Mirroring occurs when $\alpha = 90°$, for large B value (depending on the nature, energy and, injection pitch angle of the particle). If the mirroring point altitude is too low, the particle precipitates in the atmosphere and disappears.

Interactions with atoms of the residual atmosphere or between particles modify their velocity leading to partial loss particles. Due to asymmetry of the field (day-night effect) the drift around the Earth can lead from a trapped state in a region to untrapping conditions in another part of the magnetosphere (pseudo-trapping).

2.5.1.3. Trapped Particle Distributions. For hazard problems electrons and protons are the only particles to be considered as trapped in the magnetosphere; nevertheless small alpha particle fluxes and probably heavier ions of low energy are also observed. Many measurements of electron and proton fluxes in space had been performed during the years 1960 to 1975; they are used by the NASA National Space Science Data Center to establish models of the trapped radiations in the magnetosphere. They are continually being constructed with different degrees of sophistication. The more recent, AP8 for protons and AE8 for electrons, describe (for solar minimum and maximum period) the time averaged particle fluxes versus energy in particular volume of space using (B,L) coordinates. Figure 17 depicts the energy integrated flux contours for protons and electrons in polar section of idealized dipole (the field line equation $R = L. \cos^2 \lambda$ is used). The trapped particles appear distributed in rings, the radiation belts (Van Allen belts) around the Earth.

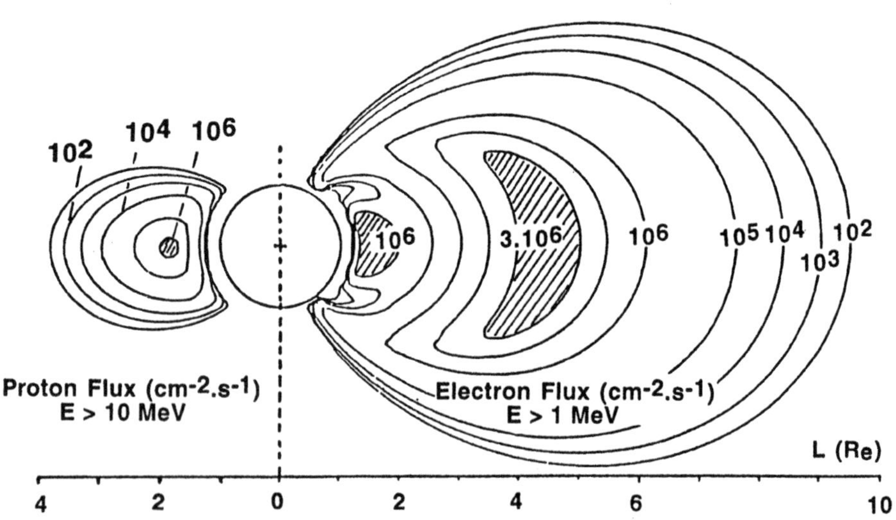

Fig. 17: Electron (right) and proton (left) flux contours (geomagnetic polar section).

Energetic electron belts are distinguished into "inner zone", with its maximum in the equatorial plane at about L = 1.5, and "outer zone" centered at L = 4. The outer belt extends beyond the geostationary orbit, up to about L = 10 and is brought to low altitudes by geomagnetic field lines at high latitude, forming the polar horns around the poles. Outer zone has peak fluxes exceeding those of the inner belt by about one order of magnitude and spectra extend to energies higher (about 7 MeV) than in the inner zone (less than 5 MeV).

The volume occupied by energetic protons depends inversely on their energy and consequently cannot be assigned to inner and outer zone. For energetic protons (E>10 MeV) the flux peak is centered at about L = 1.5 in the equatorial plane, and polar horns do not exist. In LEO, the offset and tilt of the geomagnetic axis, referred to rotation axis, brings the belts to low altitudes upon the south atlantic. In this region (the South Atlantic Anomaly - SAA) high proton and electron fluxes are observed as well as high electron fluxes at high latitude corresponding to the polar horns (Fig. 18) [Ref. 18].

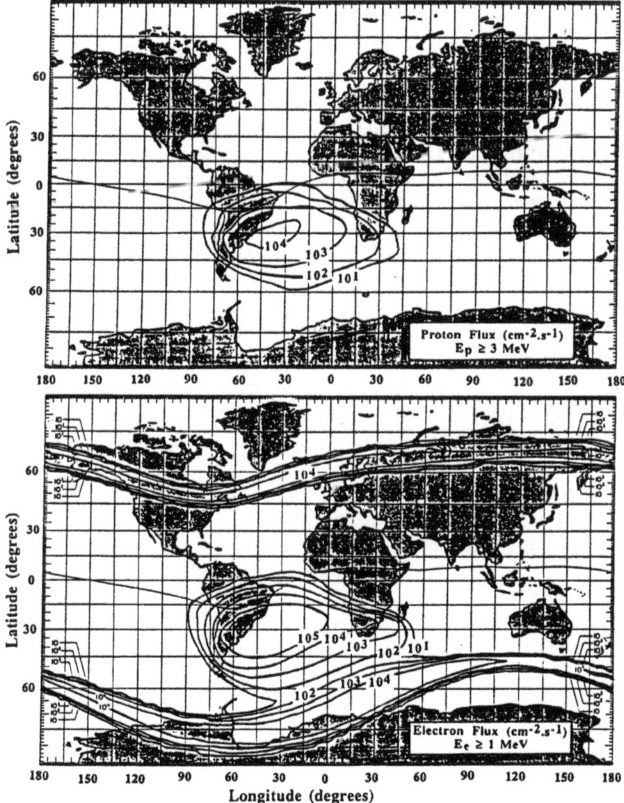

Fig. 18: Proton and electron integral flux contours (altitude 500 km).

2.5.1.4. Time Variations of Trapped Particle Fluxes. The particle fluxes observed are the consequence of various processes leading to supply (particle injections during magnetic storms, neutron albedo) and loss (precipitation in the Earth's atmosphere). They can be

- solar activity effect on GCR fluxes, or on atmospheric and magnetospheric extensions,
- solar flares and magnetospheric storms,
- secular drift of the internal field.

Long Term Variations

The most straightforward magnetospheric modification is the one linked to the long term Earth's magnetic field evolution. Due to a secular field drift, one has to bring up to date the magnetic field values used to compute the particle fluxes at a given time. Applied to the Shuttle's low orbit, this procedure yields false results and overestimates the proton fluxes by a factor 6 during the solar minimum and up to ten during the maximum [Ref. 19].

The origin of this error is well understood now: at first sight, the field decrease and the drift toward the Pacific result in a lowering of the magnetosphere which reaches more and more low altitudes, mainly in the SAA, increasing fluxes on LEO. In fact this description neglects the atmospheric absorption increase on the lowest part of the particle trajectories. Actually the models AE8 and AP8 take into account the losses related to these atmospheric interactions, but for the conditions that were present at the time of the measurements (1964 to 1972). This leads NASA to recommend at least provisionally, a procedure using the 1964-1972 values of the magnetic field.

Other long term variations come from the solar activity effect linked:

- on one hand, to the atmospheric absorption rise during solar maximum (increase of the atmospheric density at high altitudes)
- on the other hand, to the source parameter modulation (increase of cosmic ray fluxes and decrease of magnetospheric tail injection during solar minimum).

The AP8 and AE8 models consider these two solar activity periods. On LEO, this leads to a decrease of electron fluxes and an increase of proton fluxes during the solar minimum.

Medium Term Variations

The measurements carried out on DMSP (840 km, polar) show the short and medium term stability of the internal zone, whereas the external zone electronic population can undergo as much as two orders of magnitude variations in a few days (Fig. 19) during magnetic storms.

Short Term Variations

The magnetospheric deformation (day – night effect) under the solar wind pressure can be neglected for low L values. Beyond three Earth radii, the magnetospheric asymmetry prevents the use of (B,L) coordinates which describe a perfect dipolar magnetic field.

Occasionally at low altitudes and large geomagnetic latitude (50-70 degrees), a sudden increase of energetic protons and electron fluxes, due to trapped particle precipitations, occurs. In practice, this phenomenon, lasting some hundred seconds, dose enhancement is negligible.

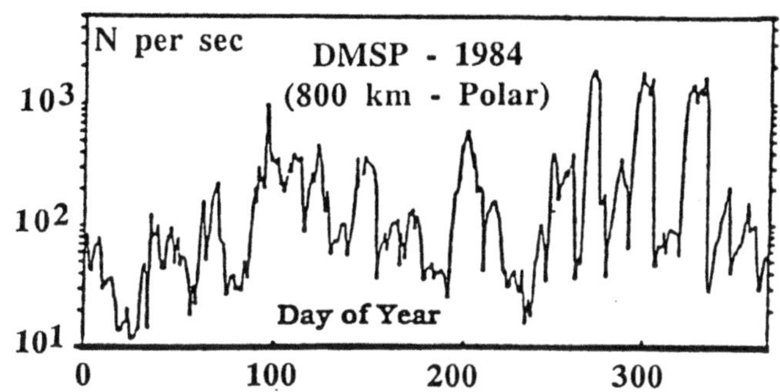

Fig. 19: Outer zone electron counting rate variations VS time.

2.5.1.5. Anisotropy: Angular Distribution of Trapped Particles. At a given place in LEO, protons arriving from the East are gyrating around a magnetic field line being at lower altitudes than for protons coming from the West. The gyration radii length can be a few tens of kilometers long and gives rise to preferential atmospheric attenuation for arriving protons from the East. This anisotropy has often been observed and can reach one or two orders of magnitude.

Trapped particles encountered at low altitudes are near their mirroring points and the angular distribution is directional; more particles arriving nearly perpendicular to magnetic field lines.

2.5.2. Trapped Particle Fluxes

2.5.2.1. Trapped Particle Flux Forecast Method. Codes are available for evaluation of trapped particle fluxes using AP8 and AE8 models [Ref. 20]. With the usual orbit parameters, the code computes the trajectory in geographic coordinates system (altitude, longitude, latitude) versus time. In each step the geomagnetic coordinates (B,L) are determined and used to interpolate the flux values in the AE8-AP8 models. The outputs give integrated omnidirectional fluxes by square centimeter and day, differential fluxes, peak flux per orbit and point by point versus threshold energy. The computations are performed for each solar activity period, during the maximum with AP8 max and AE8 max modes, and in solar minimum with AP8 min and AE8 min.

2.5.2.2. Trapped particle Fluxes - General Results

Solar Activity Modulation

As seen before, the solar activity changes the terms of the balance in the radiation belts and the trapped particle fluxes. At fixed LEO, the solar maximum period corresponds to a growth of electron fluxes and a drop of proton exposure (Fig. 20) [Ref. 21]. This effect is important for protons at low altitudes, due to the atmospheric absorption. On GEO important effect of the solar activity, not included in the models AP8 and AE8, is observed on electron induced doses (Fig. 21) [Ref. 22].

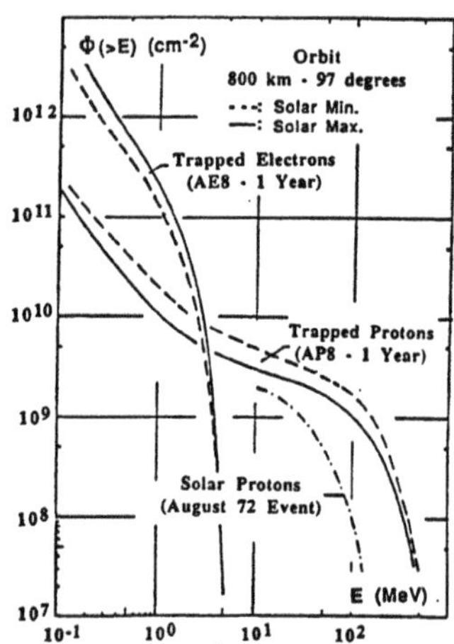

Fig. 20: Trapped particle spectra VS solar activity on LEO.

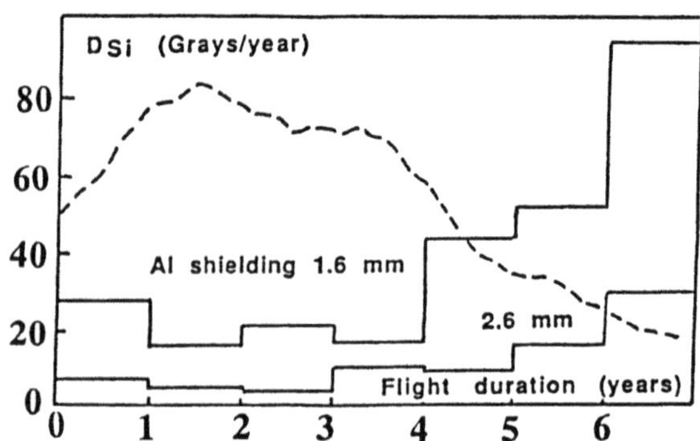

Fig. 21: Annual dose variations VS solar activity on GEO.

Orbit Altitude and Inclination Effects on LEO

Figure 22 depicts the integral proton and electron daily averaged fluxes on various LEO; trapped particle fluxes increase with altitude at any inclination in the studied range. Electron fluxes increase with inclination, due to SAA and polar horn (over 45 degrees) crossings. Harder spectra can be observed (energies up to 7 MeV) in the outer belt for inclined orbits, when polar horn crossings occur. At fixed altitudes, energetic proton fluxes (over 10 MeV) increase with inclination up to 20-40 degrees (depending on altitude); the maximum is for complete SAA crossing, beyond this limit fluxes decrease slightly (less than a factor 2) up to polar orbits. At any altitudes studied, the 7° inclination orbits can be considered as trapped radiation free.

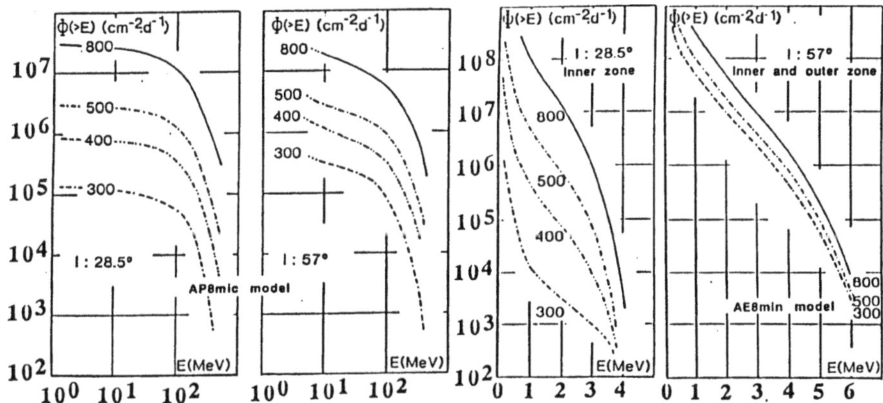

Fig. 22: Trapped particle fluxes on various LEO.

Flux Variations Versus Time Along the Orbit on LEO

Due to the satellite motion and the repartition of trapped particles on LEO (SAA and polar horns), other things being equal, the instantaneous fluxes encountered by the spacecraft change along the orbit (Fig. 23) [Ref. 16]; these results point out that, for the Extra Vehicular Activities (EVA), exposure reduction can be expected by mission planning.

2.5.2.4. Problems and Accuracy. AP8 and AE8 are steady state models and, of course, short term fluctuations (day/night effect, magnetic storms...) cannot be obtained. It must be pointed out that large flux variations observed in the outer zone of the radiation belts (high latitude region on LEO) disappear in the inner region where the geomagnetic field is less submitted to the external conditions.

Recent measurements from STS show that the use of updated field model leads to an over-estimate of trapped proton fluxes on LEO during solar maximum; following NASA recommendations, computations are performed with the geomagnetic field for 1972.

AP8 and AE8 models come from old measurements (before 1972), updated models must be developed from the results of the outstanding in flight experiment performed on board CRRES. The remarks above explain the uncertainties allowed by NASA [Ref. 17] for the long term flux forecast:

- about a factor 2 up/down for trapped protons,
- about a factor 2-3 for inner zone electrons,
- maybe 10 for outer zone electrons.

2.6. SECONDARY RADIATION

Interactions of primary particles, protons, electrons and heavy ions, with the spacecraft structures and, in some degree, with low atmosphere, induce various kinds of secondary radiations (prompt or delayed emissions). The contribution of these secondaries to the total exposure depends mainly on the nature and spectra of primaries as well as on the thickness and nature of the materials encountered.

In practice, neutron and gamma ray albedo from the atmosphere can be neglected and we must only take into account the spacecraft induced radiations. These are mainly:

- delayed emission linked to material activations by protons and heavier ions, may be a long term problem for space station;
- prompt emissions of photons (bremsstrahlung from electron slowing down and gamma rays from exited nuclei), secondary recoiling nuclei, neutrons and protons from nuclear reactions induced by proton and heavier ions.

Bremsstrahlung photons are penetrating radiations (their spectra extend up to the primary electron energy and the structural shielding efficiency is weak). They predominate on primary electron exposure inside aluminium shields thicker than $2 g/cm^2$. Emission increases as Z^2 (Z: atomic number of the crossed matter); high Z materials must be prohibited in the outer side of the spacecraft.

Protons and heavy ions induce nuclear reactions in materials, so they are sources of energetic (up to some hundred MeV) protons and neutrons, light fragments and recoiling nuclei. Generally they can be neglected in total dose effects, but single events, due to these secondaries, may be a problem for electronic devices.

3. Radiation Effects – Doses and Displacements

3.1. GENERALITIES

3.1.1. Particle Interactions with Matter. The space radiations interact with matter to produce direct effects, ionization and displacements, as well as to induce various kinds of secondaries which act themselves in turn. The primary point defects created by the radiation may be permanent or transient, vanishing by complementary defects created by the radiation recombinations (electron – hole, vacancy – interstitial) or forming secondary defects by clustering (divacancies,..) and by association with trapping centers of the material.

Primary and secondary defects may be active, inducing changes in the macroscopic properties of materials or devices (transmission losses in irradiated glasses, electrical drifts in electronic devices). These damages, increasing with the exposure, can produce long term failures in the considered system.

The macroscopic property changes are related with the active defect concentration, itself a function of the density of energy (or absorbed dose) transferred to the material, in some sensitive part (charges in the oxide of MOS for instance). The relation between dose and the defect density may be simple if they are induced by ionization without recovery and clustering. It can be more complicated if the radiation efficiency depends on its nature and energy (atom displacements).

In practice, for the evaluation and modelling of the radiation effects, it is necessary to know the mean dose repartition inside the material as well as often the nature and energy of particles in the sensitive volume. Considering the fluxes encountered and the damages that can be induced in space, the radiations of concern are electrons, protons, heavier ions and energetic photons.

Neutron induced damages are generally negligible, except in some specific situations such as spacecraft with a nuclear power source and radiobiological hazards on-board heavy shielded vehicles exposed to high proton or heavy ion fluxes (interplanetary mission for instance). The interaction between the projectile and the target (atom, electron or nucleus) depends on the nature (mass and charge) and energy of the impinging radiation as well as on the mass and charge of the target.

Neglecting photonuclear reactions, photons interact with atoms through the photoelectric effect (dominating at low energy, up to some tens of keV), Compton scattering (the main effect between some tens keV and some MeV) and pair production (energy threshold: 1.02 MeV). In all these cases energetic free electrons are produced.

The principal interaction, source of the major part of the slowing down of charged particles, is the Coulomb scattering by atomic electrons, causing both excitation and liberation (ionization of the atom target) of the involved electron. Coulomb scattering can transfer sufficient energy to the atoms to displace them from their position in the lattice. Nuclear scattering (elastic or inelastic) transfers a part of the projectile energy to the target nucleus and can eject the atom from its site.

Nuclear reactions, involving the absorption of the projectile by the sample and emission of other particles, produce energetic recoiling nuclei. In turn the energetic secondaries slow down by excitation and ionization and can displace other atoms if they are sufficiently rapid. The maximum energy transferred to a target of mass M at rest by a projectile (mass: m, energy: E) is given by:

$$\Delta E = 4EmM/(M+m)^2$$

Assuming a displacement threshold T_d (minimum energy that must be furnished at the target for its ejection), for each kind of projectile there is an energy threshold E_s. Above it, the considered particle can induce displacement so that the energy threshold is given by

$$E_s = T_d(M+m)^2 / (4mM)$$

For instance in silicon ($T_d = 12.9$ eV) the energy thresholds for protons and electrons are respectively equal to about 100 eV and 160 keV. Particles having energy above this threshold induce a number of displacements that can be computed using the displacement cross sections (Fig. 23) [Ref. 23].

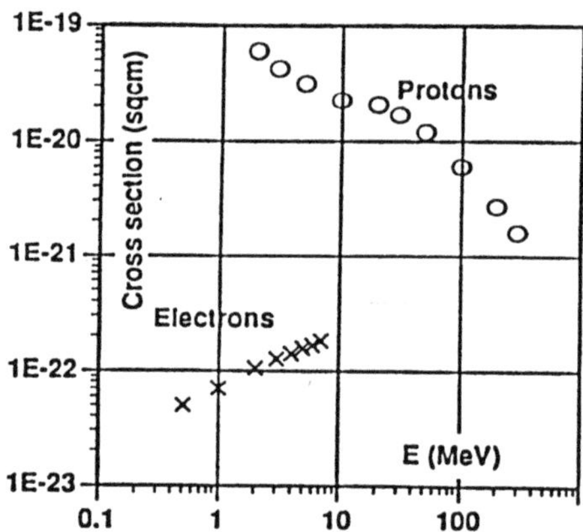

Fig. 23: Displacement cross section vs energy for electrons and protons in Silicon.

3.1.2. Linear Energy Transfer – Stopping Power. A fundamental measure of particle – matter interactions is the Linear Energy Transfer (LET). It is defined as the ratio dE/dL (particle energy loss/trajectory length on which dE occurs). The LET values do not only depend on the particle species and energies but also vary with the material composition. But in a large range of atomic number (Z about 6 to 30), the stopping power induced by ionization (the main process involved in particle slowing down for energies encountered in space) can be considered as independent of the nature of material when the trajectory length is expressed in g/cm^2 (Exception: Hydrogen and its compounds are more effective shieldings). The majority of materials used in space follow this rule and the shielding effect evaluations are performed with aluminium to determine internal radiation exposures.

3.1.3. Absorbed Dose and Dose Equivalent. The absorbed dose is the density of energy deposited by radiations in materials. Its unit is the Gray (1 Gray = 1 Joule per kilogramme of matter); the rad is often used (1 rd = 1 cGy). For a specific radiation, the absorbed dose depends on the nature of the medium crossed and this information must be given; a notation such as D_{Si}: (absorbed dose in silicon) can be used.

If the nature and the energy of particles modify the material response to a fixed absorbed dose, then quality factors or equivalent effectiveness must be defined with regard to the radiation and damage. Dose equivalent for biological effects includes this aspect in a quality factor QF:

$$\text{Dose equivalent} = \text{QF} \times \text{absorbed dose.}$$

The unit of dose equivalent is the Sievert (Sv) corresponding to 1 Joule per kilogram; the rem is also still use (1 rem = 1 cSv). The quality factor is a legal and conservative concept defined by the International Commission on Radiological Protection from a compilation of radiobiological effectiveness observed. Equal to 1 for gamma rays, the quality factor (QF) is only function of the LET in water (Fig. 24) [Ref. 24] for charged particles. For neutrons recommended values are respectively 2.3 and 20 for thermal and fast neutrons.

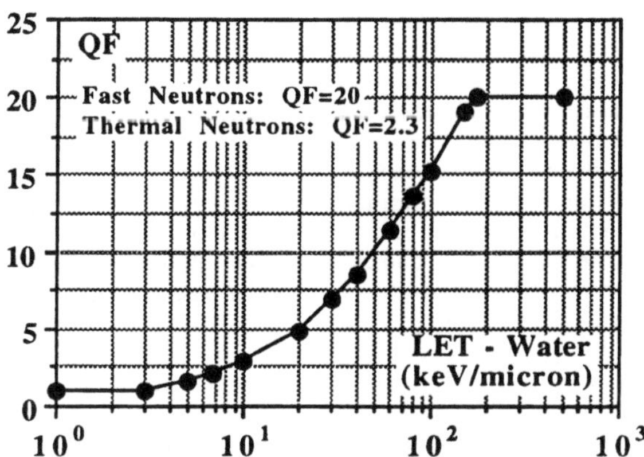

Fig. 24: Quality factor variations with the radiation's LET.

3.2 CODES FOR PARTICLE TRANSPORT COMPUTATION

The calculation of radiation transport in matter can be carried out with a variety of codes ([Refs. 25 and 26] for instance). They often use Monte Carlo method in order to describe individual interactions (Energy loss, scattering...) of each particle during its transport. Most of these codes use simple and idealized shielding configurations as finite or semi infinities slabs, cylinders, solid spheres or spherical shells; usually aluminum is chosen as reference shielding material and equivalence for others defined as seen before. Only a few codes (ex.: NOVICE), including mathematical definition of all shieldings (geometry and nature) exist and are able to follow the particle transport in the 3D spacecraft structure.

Generally they can be used for all radiation types and exposure conditions such as: unidirectional, isotropic or omnidirectional fluxes, continuous spectra or monoenergetic primaries. They furnish at each point along the trajectory in the crossed material, the location, the energy and the direction of the studied particle (primary or secondary). To this part, forming the core of the computation, may be attached various modules in order to determine, for each of the followed particles, and summing their contributions:

- the absorbed dose or dose equivalent profiles,
- the displacement profiles,
- the SEE rates induced by protons or heavier ions,
- the secondary fluxes, doses,... (photons, protons, neutrons, recoiling nuclei),
- the amount of activations,
- the absorbed charges and induced electric field (deep charging of dielectrics).

3.3. EXAMPLES OF APPLICATIONS

3.3.1. Absorbed Dose and Dose Equivalent Evaluations. Dose computations can be performed for the shielding material or for other detecting media (for instance: Tissue for dose equivalent in biological sample or Si for absorbed dose in electronic devices). Often, dose evaluations inside the vehicle follow the four steps described here under.

In the first step, considering an infinite thickness, plane shield irradiated by the chosen environment, the energy, direction and location of each particle furnish the density of the transferred energy, the LET in the material and the quality factor. The absorbed dose and the dose equivalent induced by the considered particle are computed. The sum of the contributions of all particles gives the doses profiles versus depth x.

In a second step analytical expression, assuming straight – ahead approximation and isotropic fluxes furnishes the dose on the detecting media at the center of solid sphere (radius x) or spherical shell (thickness x). The dose profile obtained versus x, for the detecting medium inside the reference shielding material, depends only on the radiation environment characteristics. It can be considered as dose specification for a particular mission (orbit, epoch, duration).

The third step is a sectoring analysis of spacecraft structures in order to obtain the shield thickness distribution (percentage of solid angle for each thickness) for specific locations inside the vehicle.

The last step furnishes the dose at the fixed point by weighing the environmental dose by the thickness distribution.

Considering 3D exact computations results obtained for various shield geometries [Ref. 27], the straight-ahead approximation induces an overestimate of doses for scattering radiations (primary electrons, bremsstrahlung,...); the values obtained with slabs are lower than those computed for spherical shells by approximately a factor 2 and 6 for solid sphere. These results can be considered as limits for the geometrical effects; the sectoring analysis associated with the solid sphere computation appears as an adequate overestimate for the dose estimates inside spacecraft.

Examples of dose equivalent results at the center of solid sphere are shown on Figs. 25 and 26 [Ref. 28] for LEO. The computation (including self shielding by the astronaut's body) of the shielding distribution, S(x), around the considered point inside the spacecraft is then performed. Weighting the environmental dose by S(x), we obtain the dose equivalent in the considered organ. Comparison between doses inside HERMES [Ref. 28] and in flight measurements on board the Shuttle [Ref. 29] may be seen on Fig. 27. The difference, less than a factor 3, in spite of the non equivalence between the shielding distributions, gives an idea of the uncertainties that can be expected on LEO.

Examples of computed absorbed dose profile and comparison with in flight obtained results on polar LEO (DMSP - [Ref. 30]) and GEO [Ref. 22] are shown on Fig. 28. The gap, mainly for weak shielding thicknesses, between measurements and predictions on GEO, probably comes from an overestimate of the trapped electron flux. It can be also pointed out that in GEO the measurements show the important effect of the solar activity which is not foreseen with the AE8 model.

3.3.2. Neutrons and Recoiling Nuclei Productions under Proton Exposure. The radiobiological effectiveness of neutrons is high. ICRU-defined quality factor for these particles reachs approximately 23; consequently secondary neutron production can be a problem for manned missions on orbits exposed to high trapped protons fluxes (altitude upper than 500 km) and solar protons fluxes (inclined orbits or interplanetary travels). In another way, secondary recoiling nucleii and light fragments are sources of Single Event Effect (SEE) under trapped proton exposure in the SAA and in the polar regions during solar events (see for instance UOSAT2 results here under).

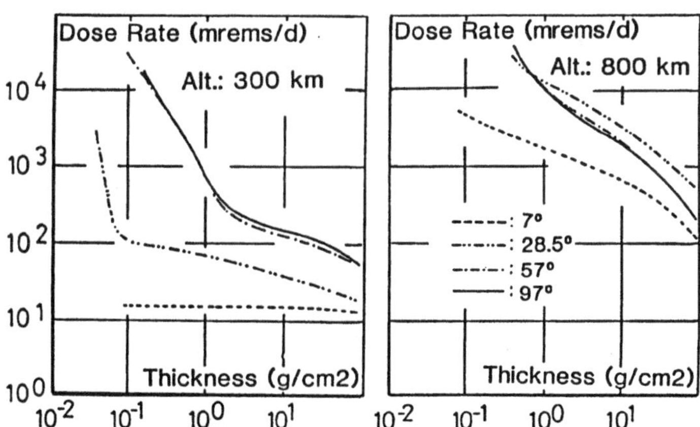

Fig. 25: Dose equivalent rates induced on LEO by GCR and trapped particles (Al solid sphere shielding).

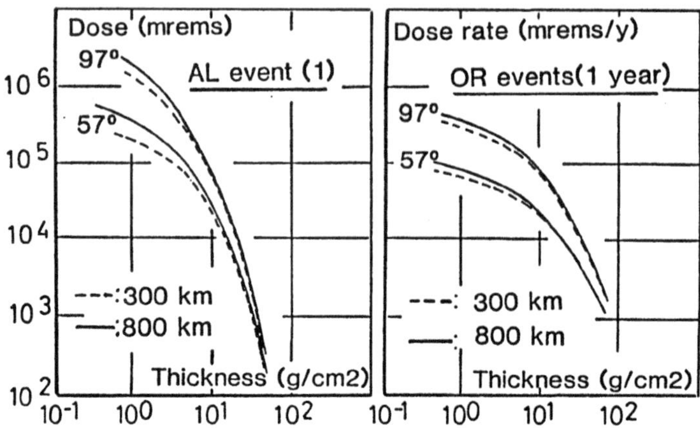

Fig. 26: Dose equivalent rates induced on LEO by OR solar events (1 year) and the August 72 event (Al solid sphere shielding).

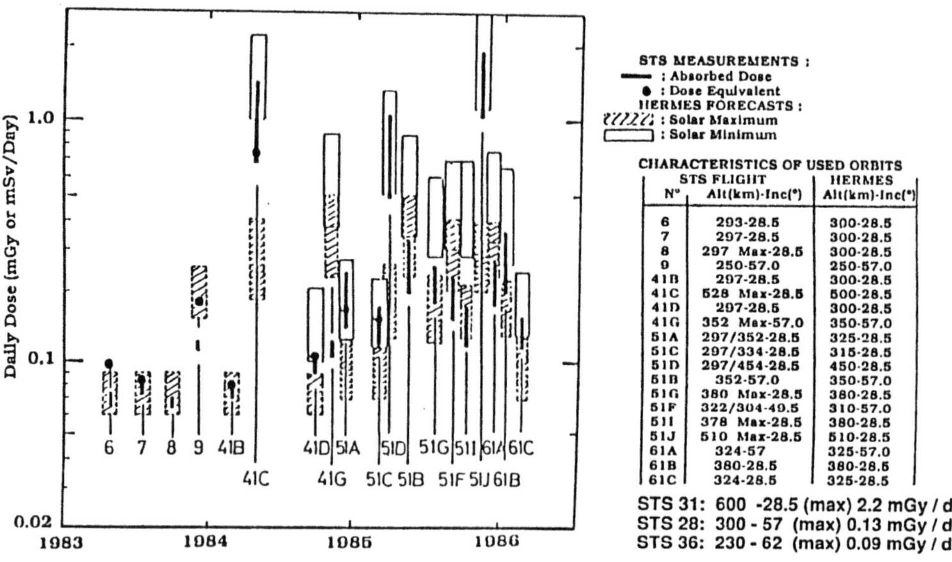

CHARACTERISTICS OF USED ORBITS

STS FLIGHT N°	STS FLIGHT Alt(km)-Inc(°)	HERMES Alt(km)-Inc(°)
6	293-28.5	300-28.5
7	297-28.5	300-28.5
8	297 Max-28.5	300-28.5
9	250-57.0	250-57.0
41B	297-28.5	300-28.5
41C	528 Max-28.5	500-28.5
41D	297-28.5	300-28.5
41G	352 Max-57.0	350-57.0
51A	297/352-28.5	325-28.5
51C	297/334-28.5	315-28.5
51D	297/454-28.5	450-28.5
51B	352-57.0	350-57.0
51G	380 Max-28.5	380-28.5
51F	322/304-49.5	310-57.0
51I	378 Max-28.5	380-28.5
51J	510 Max-28.5	510-28.5
61A	324-57	325-57.0
61B	380-28.5	380-28.5
61C	324-28.5	325-28.5

STS 31: 600 -28.5 (max) 2.2 mGy / d
STS 28: 300 - 57 (max) 0.13 mGy / d
STS 36: 230 - 62 (max) 0.09 mGy / d

Fig. 27: Measured (STS) and computed (HERMES) mean daily absorbed dose or dose equivalent.

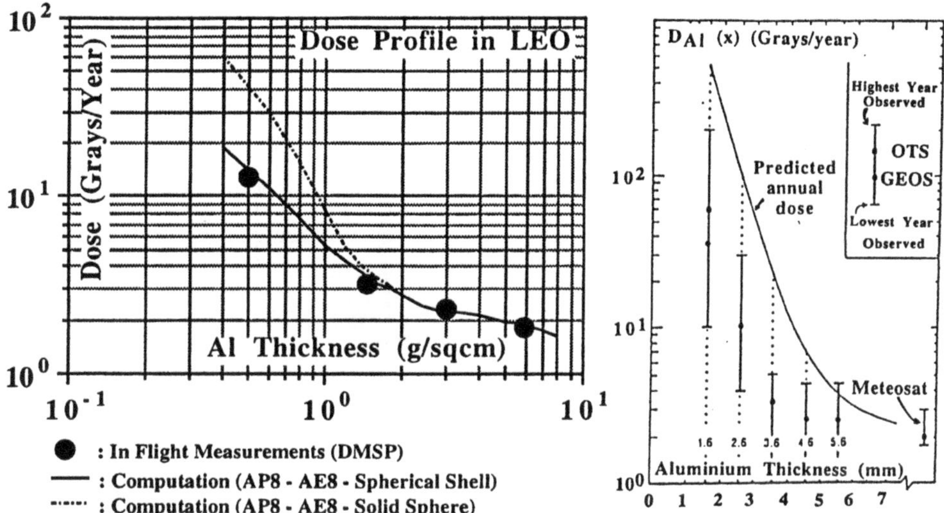

Fig. 28: Absorbed dose profiles computed and measured in flight on LEO (left) and GEO (right).

A specific module has been developed at DERTS for the modelling of the involved process. In this module a Monte-Carlo method is used to follow the interactions between incident protons and nucleons inside the nucleus [Ref. 31]. Its outputs are the angular distributions and energy spectra of recoiling nucleii and secondary nucleons [from cascade and evaporation (Fig. 29)]. Using the Dose/Flux ratio versus neutron energy defined by Haffner [Ref. 32] and the quality factor given in the ICRU Report 40, the neutron induced dose equivalents are then obtained (Fig. 30) [Ref. 33].

In practice on LEO the secondary neutron dose equivalent is not a major problem; it represents less than ten percent of the total dose directly induced by GCR and protons inside representative spacecraft shieldings. The same module furnishing energy and angle distributions of secondaries is also useful for the determination of the proton induced SEE rates (see here under).

3.3.3. Examples of Radiation Damage Modelling

3.3.3.1. Material Swelling Under Irradiation. One aspect of the radiation damage to glasses is the swelling induced in mirrors exposed to radiations in space. The heterogeneity of the dose repartition (absorbed dose decreases from the irradiated surface to the rear side) induces support deformations if the radiation effect is a permanent density change in the irradiated part. In isotropic material the result is a parabolic bending of the surface (convex if density decreases or concave if material shrinking occurs). This effect may be a long term limiting factor for optical systems in space.

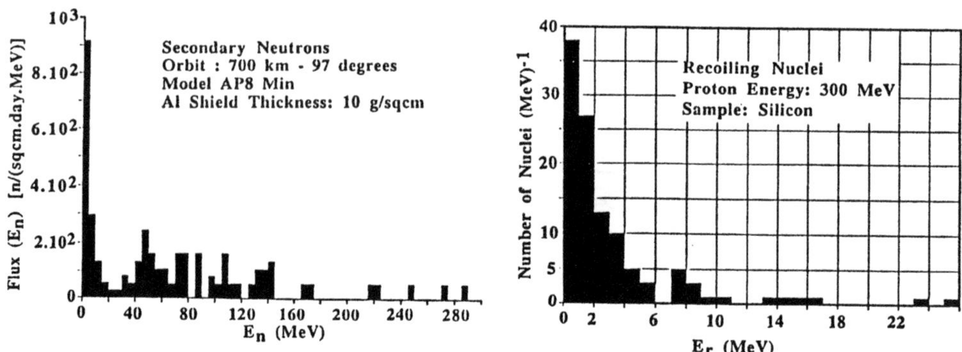

Fig. 29: Examples of neutron and recoiling nucleus spectra induced under proton exposure.

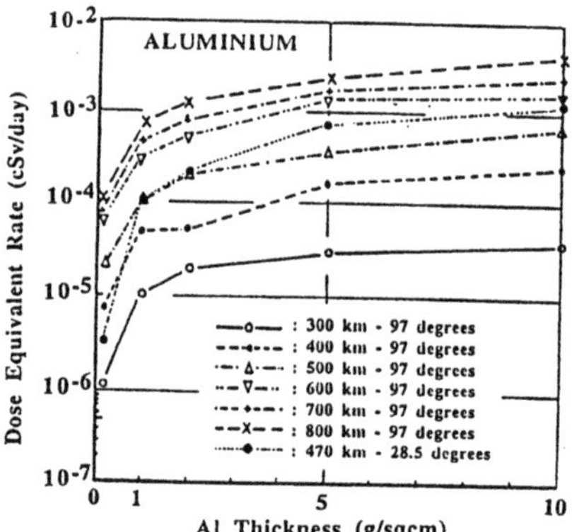

Fig. 30: Neutron induced dose equivalent on LEO.

Taking into account the absorbed dose profile induced in the irradiated material, a swelling model of circular samples had been developed; the mean linear expansion or shrinking coefficient of the irradiated material is given by [Ref. 34]:

$$a_1 = kC.e^3/[x_d R^2 (e-x_d)]$$

with:

C: the measured rise of the chamber,
e and R: the thickness and the radius of the sample,

x_d: damaged depth (can be computed or measured if coloration occurs),

k: a parameter depending on the dose distribution form (k = 0.83 to 1.03).

From the experimental studies carried out on mirror support materials (Cervit, Zerodur, Pyrex, and various silica), it appears that the linear expansion (or shrinking) coefficient is given by:

$$a_1(i) = \alpha_i \cdot D_m^{0.5}$$

with:

D_M: the mean absorbed dose in the irradiated part of the material,

α_i: a constant characteristic of the material sensitivity.

On the vitreous ceramics (Cervit and Zerodur) the studies, performed with various fluence rates, particle types and energies, show that these radiation characteristics are weightless factors (Fig. 31).

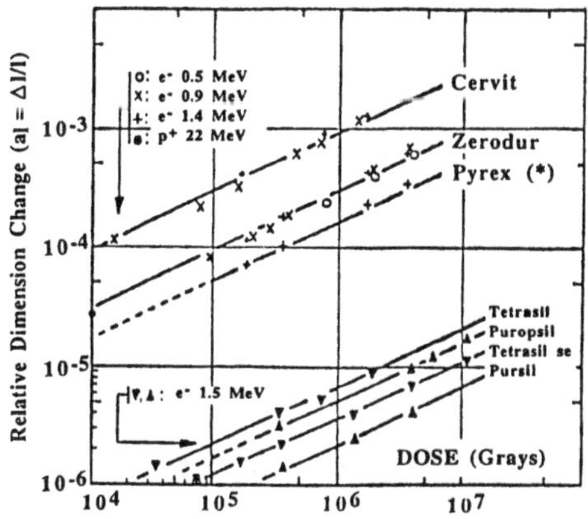

Fig. 31: Linear shrinking (or Expansion*) coefficient variations with the absorbed dose in mirror support materials.

3.3.3.2. <u>Thermal Control Coating (TCC) Damages</u>. The optical property changes of the TCC exposed to radiations in space is an important problem for the thermal control of the satellites. For these directly exposed materials, only UV radiations and low range particles (protons from solar wind and the low energy part of the trapped particles spectra) must be considered.

For the radiation damage modelling [Ref. 35], it is assumed that the degradation is due to absorption bands (absorption coefficient variation induced by radiations: $\Delta\mu(\lambda)$, width: w) appearing in the irradiated material. The number of active defects is proportional to the product $p(\lambda) = W\,\Delta\mu(\lambda)$ and then we must obtain a simple relation between the absorbed dose and $p(\lambda)$.

The studies have been performed on various TCC as paints and Secondary Surface Mirrors (SSM) irradiated by low energy protons and electrons. In the model the absorbed dose profile computations give the mean dose D_m and the irradiated (damaged) thickness x_d of the sample. The absorption coefficient variations and the band width are deduced from the measurements of the spectral reflectance with a model of the photon propagation on diffusing (paints) and non-diffusing (SSM) media.

Results obtained for aluminized Kapton (SSM) and S13G white paint are shown in Fig. 32. For the SSM, bibliographical data and DERTS experiment results are used for the model validation; the density of active defects (producing an absorption band at 600 nm) is proportional to the mean absorbed dose; the nature and energy of particles and fluence rate are weightless parameters.

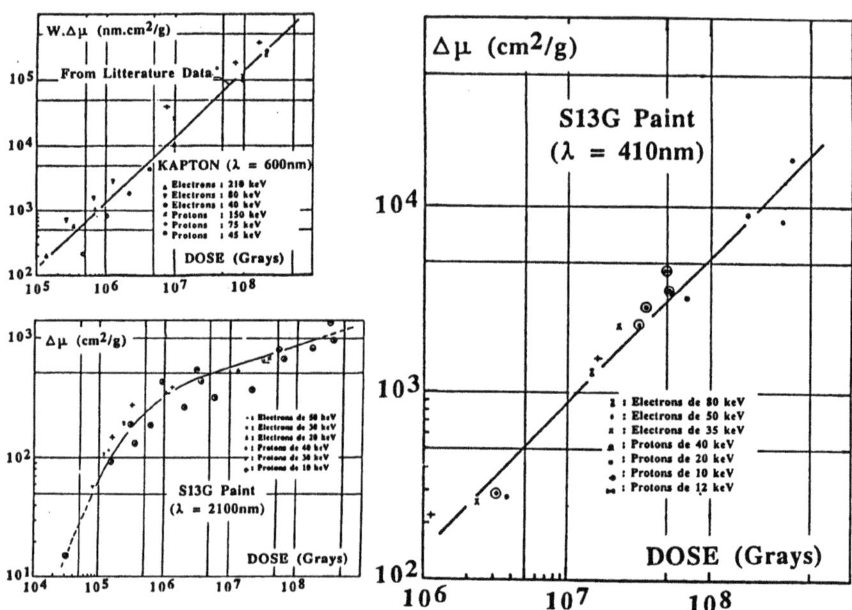

Fig. 32: Optical damages induced in Thermal Control Coatings (Kapton and S13G paint) by low energy particles.

For the S13G paint two absorption bands are observed (410 and 2100 nm). The growth of the active centers does not depend on the particle nature and energy; it depends only on the absorbed dose.

3.3.3.3. Material Activations. Activations are not generally a problem in terms of total dose, except maybe for manned missions on-board vehicles staying for long periods on exposed orbits, as the space station for instance. On the other hand activations occurring in a radiation detector can induce constricting background noise on low level measurements.

For example, computations have been performed [Ref. 36] for the X-ray telescope SIGMA on-board SIGNE 3 (orbit altitude 500 km – inclination 50°). Scintillation detectors (NaI-Tl and CsI-Tl crystals) are used and activation induced background noise appears under trapped proton exposure (during SAA crossings). The activation cross sections are obtained using Rodstan [Ref. 37] and Waddinton [Ref. 38] relations.

Taking into account the shielding effect by the telescope structure, the production rate, and the period of the radioisotopes, the number of disintegrations per second inside involved crystals is computed by integration along the trajectories. Fig. 33 depicts some results obtained for:

- a high disintegration rate product (O^{19} -half life 29 sec.). In this case the disintegration rate follows the trapped proton fluxes variations during SAA crossings. The base line, corresponding to GCR protons without magnetospheric shielding, is overestimated, and in practice one can take advantage of background free periods to perform low level radiation measurements;
- a long life (13 days) isotope (I^{126} in CsI crystal). In this case the time between SAA crossings is negligible compared to the half life, and the disintegration rate increases slowly up to a saturation value (depending on the cross section, the fluxes at the sample level and the half life of the radioisotope), reached after approximately three months.

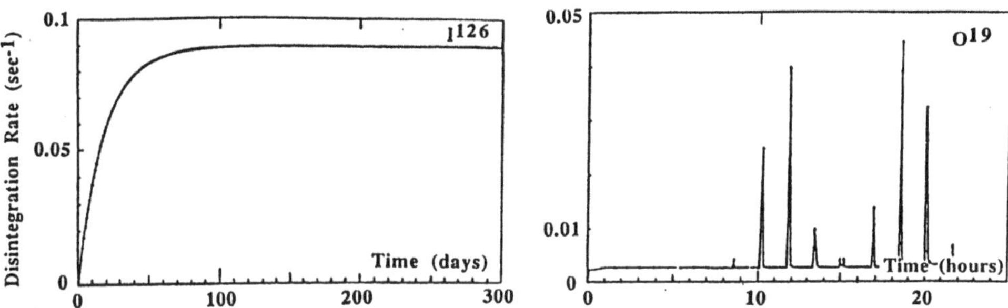

Fig. 33: Activation of Cal (TIC) and CsI (TIC) crystals and LEO (500 km – 50 degrees), followed products I126 and O19.

4. Single Event Effects – SEE

4.1 GENERALITIES

SEE occurs in a specific node of an electronic device when the collected part of the charge deposited in the sensitive volume by the incident ion (direct process) – or by a group of secondaries coming from a proton induced nuclear reaction in its vicinity (indirect process) – exceeds a characteristic value, the critical charge QC of the node. The collected charge depends on the amount of charge generated in the node along the particle's trajectory. Taking into account the proportionality between the number of electron-hole pairs created and the total energy lost, the above conditions may lead to:

- SEE occurs if the energy transferred to the node exceeds a critical energy E_c.

Or, because the amount of energy lost, assuming a constant stopping power in the node, is given by the product of the LET and the trajectory length in the node:

- only ions with sufficient LET (greater than a LET threshold L_S) are able to induce SEE in the considered node.

Due to a large amount of electronics to be used, SEEs may be a difficulty that must be considered. They are a current problem in VLSI devices and will become increasingly important because of the trend to reduce the device size, to enlarge the complexity of the circuitry and to raise the number of memory bits on board.

Direct effects due to primary heavy ions from GCR or SCR, and indirect phenomena occurring under proton exposure must be studied. Both source of SEEs (Latch up and Upset – SEL and SEU) have been identified in flight. On GEO the GCR and SCR contributions appear clearly on board:

- METEOSAT 2 [Ref. 39]: Less than 0.2 SEU per day in quiet period by GCR, about 50 SEUs induced during the 10/19/89 solar event, with a rate reaching about 20 SEUs per day at the peak of the proton flux;
- ETS 5 [Ref. 40]: near 200 SELs and 70 SEUs during the same event, with rates up to, respectively, 250 and 40 SEEs per day; to be compared with about 4 SELs and 1 SEU per day induced by GCR during quiet periods.

On LEO, measurements with UOSAT 2 (700 km – 98 degrees) [Ref. 41] and MOS 1 (910 km – 99 degrees) [Ref. 42], show the contributions of GCR, SCR and trapped protons to the SEE rates. On-board MOS 1 the daily averaged SEU rate increase from 1 during quiet period up to 9 during the October 89 event. On-board UOSAT 2 the daily averaged SEU rate, due to GCR and trapped protons, observed during quiet periods, less than 9, increase up to about 20 SEUs per day during the October 89 event.

For both satellites the South Atlantic Anomaly (SAA) appears clearly on the map of SEU location, identifying trapped protons induced anomalies; the GCR contribution ranges between 20 and 50%, depending on the device, during quiet periods on-board UOSAT 2 (Fig. 34). For this satellite the plot of the latitude of occurrence as function of day show also the magnetospheric shielding effect on GCR and SCR; the SEE rate during the October event is negligible up to 45 degrees and approximately 80% of the GCR induced SEEs occur at greater latitudes (Fig. 34).

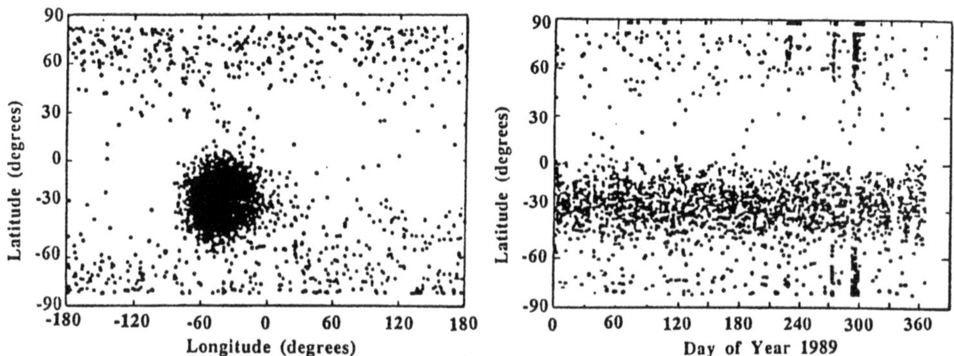

Fig. 34: SEU locations on the TMS4416 RAMs (UOSAT2, 01/89 – 01/90). Left: latitude / longitude plot. Right: Latitude / time of occurrence plot.

For a fixed device the anomaly rates corresponding to these sources depend on:

- the fluxes encountered; they are function of the mission (Orbit, epoch),
- the shieldings around the device, which defines the fluxes impinging the circuit,
- the device sensitivity; obtained from experimental results or models, which gives SEE rates for specific particles (energy, incidence and type) hitting the device.

This defines the various steps of the SEE rate forecast:

- evaluation of the fluxes encountered outside the vehicle,
- determination of the shielding effect by structures,
- measurements and (or) modelling of the device sensitivity,
- estimate of the SEE rate for the fixed device taking into account its location inside the spacecraft and its sensitivity to each ion type.

4.2. PHYSICAL PROCESSES INVOLVED

4.2.1. Heavy Ion Interactions with Matter. The energy of the projectile is transferred by Coulomb interactions to the electrons of the atoms. These secondaries (delta rays) have very low energies and short ranges (for instance: less than 0.1 μm in silicon for secondary produced by a 10^2 MeV iron ion). The LET depends on the energy of the ion and increases with its atomic number (Fig. 35). In silicon the energy to create one "electron – hole" pair is equal to 3.6 eV; high charge density levels can be observed along the trajectory (6.10^3 pairs/μm for 100 keV protons, 10^5 for 100 MeV iron ion) in a narrow channel around the ion trajectory.

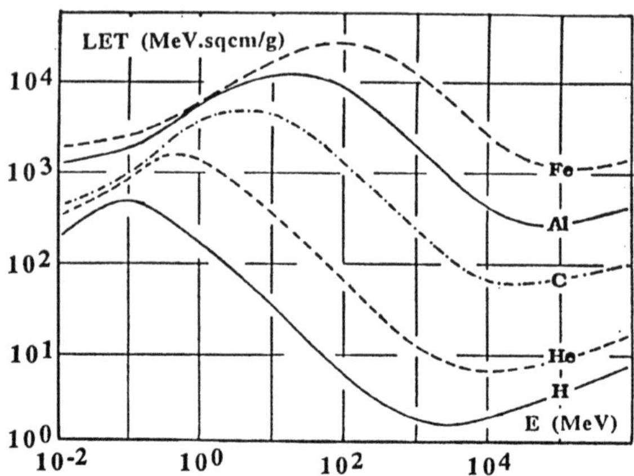

Fig. 35: Variation of the LET on silicon with the energy of ions.

4.2.2. Effects on Electronic Devices. In an electronic device, information storage is performed via charges, voltage state, or current pulse in localized nodes (particular transistor's junctions – Fig. 36). A charge pulse induced in such node can modify its state (upset) if the among of charge exceeds a critical value, the critical charge, Q_{crit}, of the node.

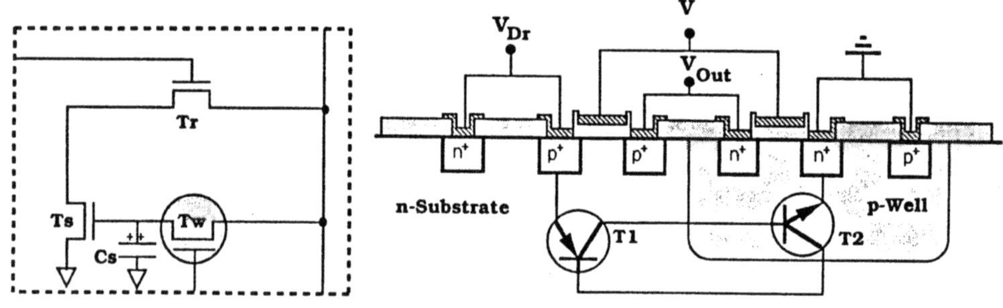

Fig. 36: Sensitive node in a DRAM and pnpn parasitic structure (SCR) in a RAM CMOS bulk.

In some types of devices (bulk CMOS memory cell for instance – Fig. 36) a parasitic pnpn (Silicon Control Rectifier – SCR) structure exists. Under normal conditions, the breakover voltage for this parasitic SCR is higher than the power supply voltage and latch-up does not occur. However, if a sufficient charge pulse is injected into the interior junctions of the SCR, the breakover voltage can be lowered during a short period of time under the power supply voltage, allowing the device to go into latch-up.

The critical charge in the more recent technologies (Fig. 37) can be lower than 0.1 pC, corresponding to an amount of energy deposited in the node lower than about 2.2 MeV in silicon (obtained with Ne and heavier ions at their LET's maximum on a one micrometer trajectory length).

When an ion crosses a p-n junction with a depletion width a at an angle of incidence θ, the charges induced along the path (a/cosθ) in the depletion region are separated by the electric field inducing a prompt charge pulse Q_1 (Fig. 38). In addition the bending of the field

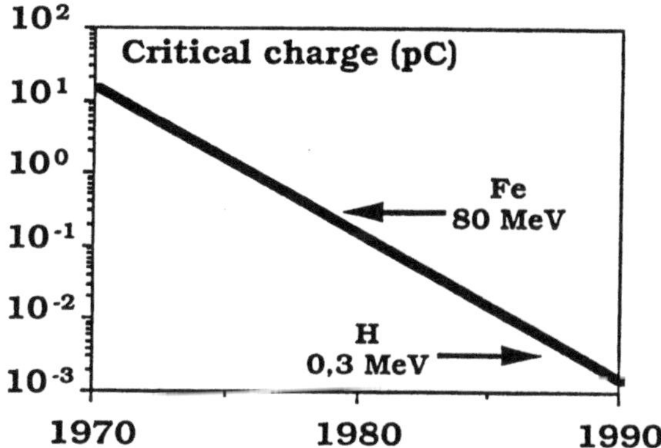

Fig. 37: General shape of the critical charge evolution during the last twenty years. (Charges induced per μm are shown for H ane Fe.)

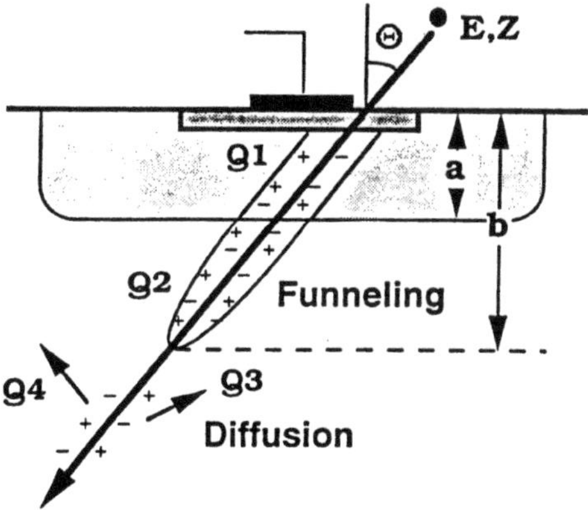

Fig. 38: Mechanism involved in heavy ion induced SEE (direct process).

line into the substrate, due to the high induced charge density, allows prompt ($<10^{-9}$ second) collection of charges Q_2 from an extension (length X_f) of the depletion region along the track (funneling effect). Also minority carriers diffusing from the substrate to the depletion region of the cell can be collected giving additional delayed (10^{-7} second) charge pulse Q_3 at the node (neighboring cells can be affected by a part of this diffusing charges).

Taking into account that 3.6 eV are necessary to create an electron-hole pair, the collected charge is related to the ion's LET by:

$$Q_c = \alpha[k_1 L(E,Z)a/\cos\theta + k_2 L(E,Z)X_f] + k_3 Q3$$

with:

- α: correction from charge to energy, α = 22.5 pC/MeV,
- k_i: reciprocal of the collection efficiency for the considered charge,
- $L(E,Z)$: LET of the ion Z at the energy E, assumed constant in the node.

When the amount of collected charge, Q_c, exceeds the critical charge of the crossed cell, then a single event (upset or latch-up) may be observed if it is a sensitive node. Neglecting the contribution of the diffusion to the collected charge results in

$$Q_{crit} \leq \alpha L(E,Z)[a/\cos\theta + X_f]$$

4.3. DIRECT EFFECTS BY HEAVY IONS

Primary heavy ions (GCR and SCR) can induce sufficient ionization around their track in the sensitive volume to produce SEE by a direct process. The susceptibility of the devices under exposure to these particles is characterized by the cross section variation versus the LET of ions.

4.3.1. LET Threshold and SEE Cross-Section. Considering monocinetic ions (energy, E; atomic number, Z) impinging on a device under normal incidence, SEE can occur if they cross the collection volume of the sensitive cells, and if their LET , L(E,Z), is greater than a minimum value, the LET threshold L_s, given by

$$L_s = Q_{crit}/[\alpha(a + X_f)] = Q_{crit}/(\alpha a')$$

Assuming that n identical cells exist in the device and that they are right-angled parallelepipeds, the sensitive surface of the device, its SEE cross section σ, is given by:

$$\sigma = nb'c'$$

b', c' and a' (the dimensions of the collecting volume perpendicular to the beam and its thickness) are the extension, by funneling and diffusion, of the information storage junctions. These dimensions are functions of L(E,Z); consequently the cross sections increase more or less rapidly with the ion's LET, from L_s up to a saturation level corresponding to the value defined by the relation above (Fig. 39) [Ref. 43]. Various types of collection volumes can exist in a device, so the total cross section is the sum of the cross section corresponding to each node.

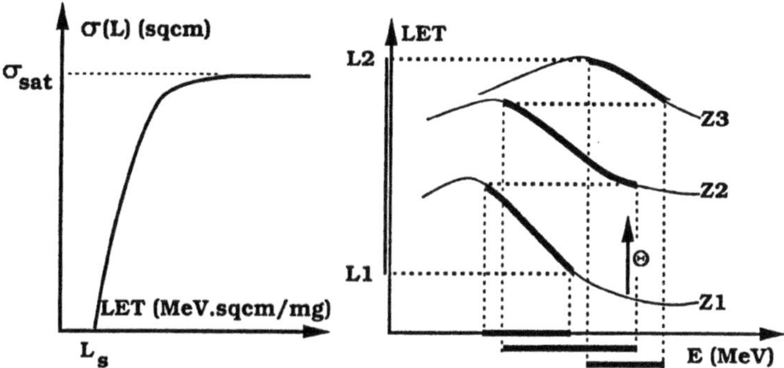

Fig. 39: Shape of the cross section variation with the ion's LET (left) and methods to change the LET during tests with accelerators (right) [Leff = $L(E_i, Z_i)/\cos\theta$].

In order to determine SEE rates induced in space by heavy ions (GCR and SCR) its necessary to know (see her under) the cross section variation $\sigma(L)$ with the LET; experimental measurements must be performed under energetic ion beams.

4.3.2. Measurements of the Device Sensitivity. Cross section measurements are performed with ion beams furnished by high energy accelerators. Considering the domain of the ion's LET encountered in space, up to about 100 MeV cm^2/mg, a large range of LETs must be studied for each device (obtained experimentally by changing (Fig. 39) the ion specie (Z), energy (E) or incidence (θ)).

The cross section is then defined, from the experimental results, by:

$$\sigma[L(E,Z,\theta)] = \frac{N\cos\theta}{\int_0^T \phi[L(E,Z,t)]\, dt}$$

with:

N: the number of SEE measured during a period of time T,
$\Phi[L(E, Z, t)]$: the instantaneous ion's flux.

Some results of SEU (upset) and SEL (latch-up) cross section measurements are shown on Fig. 40.

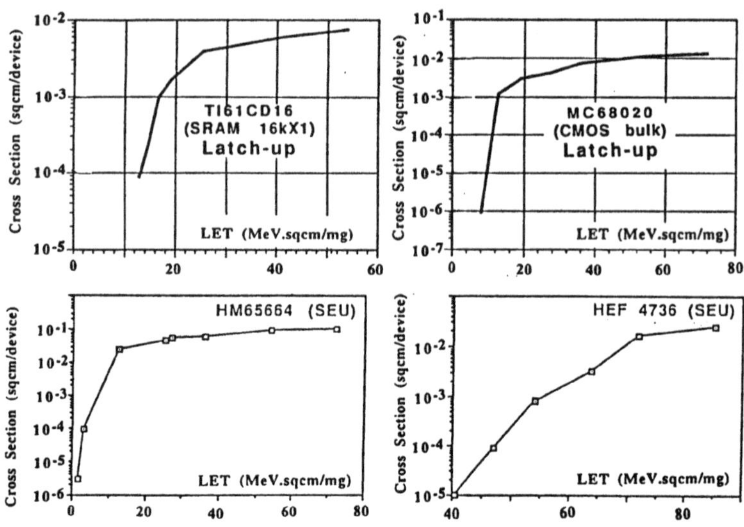

Fig. 40: Examples of SEE cross section measurements results.

Various problems can be observed during these experimental measurements:

- the LET must be constant in the volume (sufficient ion's energy),
- the same LET obtained (Fig. 39) with light E_1, Z_1) and heavy ions (Z_2, E_2) can give different results, due to recombination increase linked with the higher charge density in the channel around the track of the ion of lower atomic number Z_1 [Ref. 44],
- the LET variation with the angle of incidence ($L_{eff} = L / \cos\theta$) is not valid if funneling occurs; in the same way, the cross section correction, $\sigma(\theta) = \sigma.\cos\theta$, is not valid for compact nodes when the lateral surface of the node is not negligible.

Cost and disposal are the main disadvantages of the accelerator facilities; in order to reduce the cost of the device sensitivity measurements, often the fission fragments furnished by radioactive sources (Cf252 for instance) are proposed. In practice they can be used only for preliminary screening tests and to control, before the tests with the accelerator, that the SEE detecting facility works properly.

4.3.3. Forecast of SEE Induced in Space by Heavy Ions. The flux and fluence of each type of ion (GCR and SCR), received by the satellite during a fixed mission, depend on the orbit inclination and altitude as well as on the date of the flight. In practice the effects of these parameters are well known and worse cases may be defined for the various components of the radiation environment.

The solar event probability follow the solar cycle; it is assumed that SCR emissions occur only during the seven years of maximum activity. Primary fluxes crossing the spacecraft structures undergo interactions; they modify energy and angle distributions. Study of the transport must be performed in order to obtain the radiation characteristics (energy and direction) of each ion species inside the vehicle, at the device level.

These outputs are then used to determine, if necessary, the LET spectra of the various sources (GCR and SCR) versus the shell thickness x. Then various ways may be followed to determine SEE rates; in all available codes the inputs to be used are experimental measurements of the device sensitivity, such as the cross section per device or per bit, $\sigma[L(E,Z)]$.

a) In the CREME software [Ref. 8], assuming that sensitive nodes are right angle parallelepipeds of known dimensions (a', b' and c'), and that the exposure is isotropic, one computes:

- A: the average projected area of the node,
- S_{max}: the maximum path length of the ion track in the node,
- F(S): the differential path length distribution.

Then the error rate per bit for the considered silicon device (k) is given by:

$$n_k = 22.5\pi A Q_c \int_{L_{min}}^{L_{max}} F(S)\, F(>L)\, \frac{dL}{L^2}$$

where:

- F(>L): Integral LET spectrum of the considered environment inside the spacecraft,
- L_{max}: maximum of the LET range,
- L_{min}: 22.5 Q_c/S_{max}: minimum of the LET range,
- L_s: LET threshold under which the cross section is equal to zero.

In this method one assumes:

- isotropy of fluxes at the device level. It is not the case mainly for SCR inside the spacecraft,
- constant LET of ions in the node, for a particle near the end of its range that can result in energy loss in the node exceeding the ion energy,
- dimensions (a', b' and c'), including diffusion and funneling, are known. It is often difficult to have these data.

b) Other routines use Monte Carlo simulation to follow the ion's transport from the outside of the shielding up to the device. Then impact, trajectory length and energy loss ΔE in the sensitive node [dimensions a', b', c' and $\sigma(L)$ known] can be determined for each impinging ion. The SEE rate can be then obtained by weighting the spectra of the energy losses by the cross section.

In simplified computations [Ref. 45], taking into account the incidence and the energy (E, Θ) of the ion impinging the sensitive volume, one computes its effective LET, $L_{eff} = L(E,Z)/\cos\Theta$. Neglecting the limitation of the trajectory length in the node by its lateral sides, SEE occurs if L_{eff} is greater than the LET threshold L_s. Finally the computation furnishes the SEE number, $N_{SEE}(>L)$, per square centimeter of sensitive surface versus LET, for the considered ion source and various shielding thicknesses. Figure 41 depicts these results obtained for GCR on SPOT orbit (800 km, 97 degrees) at center of aluminium spherical shells.

Then for any device (k) experimentally characterized [cross section $\sigma_k(L)$ measured], the foreseen SEE rate in flight is given by:

$$N_k = \int_{L_s}^{L_{max}} \frac{\partial [N_{SEE}(>L)]}{\partial L} \sigma_k(L)\, dL$$

with:

- L_s: the LET threshold of the device,
- L_{max}: maximum of the range studied.

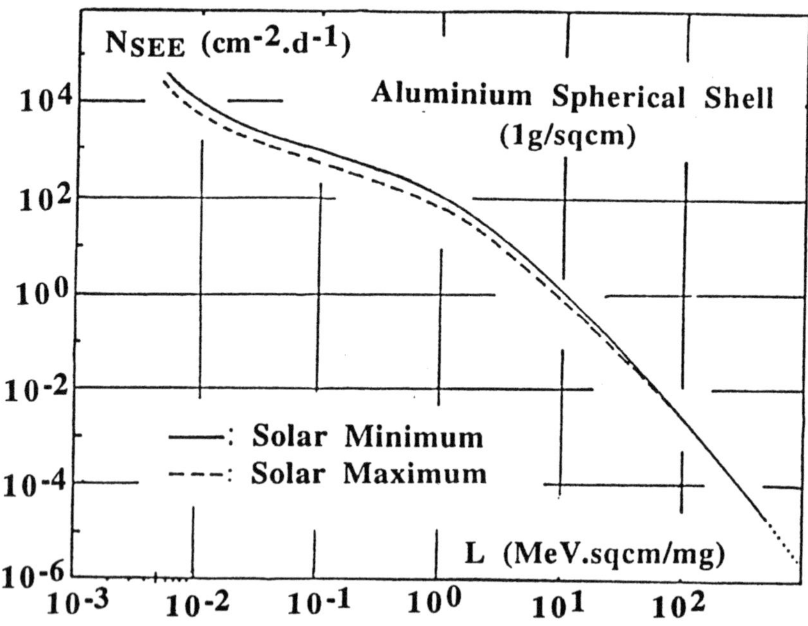

Fig. 41: SEE rate induced by GCR on polar orbit at 800 km versus LET.

The advantage of this method is that only cross section measured results are necessary; but, due to the unrestricted path length, the $N_{SEE}(>L)$ plot must be considered as an upper limit giving an overestimate of the SEE rate, mainly for devices which have compact nodes and high LET thresholds.

4.4. INDIRECT PROCESS, PROTON INDUCED SEE

Under proton exposure a nuclear reaction in the node vicinity induces secondary ionizing radiations (protons, light fragments and recoiling nucleus). The sum of the charges produced along their tracks in the node may exceed the critical Q_c; then indirect SEE occurs.

4.4.1. SEE Cross Section

Experimental Measurements

As for heavy ion induced effects experimentation may be performed in order to characterize the device (k) sensitivity. In this case the measured parameter is the cross section per bit (Fig. 42) [Ref. 46] versus proton energy, $\sigma_k(E_p)$. The domain of energy that must be studied ranges from some MeV up to about 500 MeV. Few accelerator facilities can cover this domain.

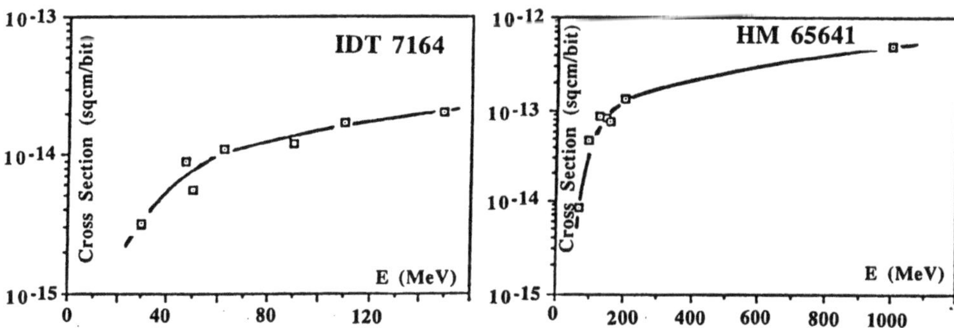

Fig. 42: Examples of measured cross sections for SEU induced by protons (indirect process) versus proton energy.

Bendel's Cross Section Model

A tentative calculation had been performed by Bendel [Ref. 47] in order to obtain a semi-empirical description of the cross-section variations with the proton energy. The single parameter characterizing the device sensitivity is its energy threshold A (the minimum energy of protons to induce SEE).

The advantages of this approach are:

- the cross section is obtained from measurements at low energy (less than 50 MeV); accelerators covering this range may be easily found,

- for specified environment and shielding the SEE rate depends only on A.

Nevertheless this method furnishes doubtful results, and in order to increase the precision of the model, a second parameter has been introduced [Ref. 48] taking into account experimental results obtained at some hundreds of MeV.

Nuclear Reaction Modelling for Cross Section Computation

Taking into account that direct process cross section must be determined to define the sensitivity to GCR, and assuming that the critical charge and the node dimensions may be then obtained, an alternative approach may be developed. It must compute the locally transferred energy by the nuclear reaction products (Fig. 43) [Refs. 31 and 49].

In the DERTS routine the first step is a study of the proton induced reaction in order to define the characteristics of the secondary ionizing fragments. The process is divided into cascade and evaporation stages, both simulated by using a Monte Carlo method.

The nucleus model is that of Hofstadter [Ref. 50] with a Fermi distribution of the charges. Relativistic computation of the interactions is used, and one excludes collisions leading to nucleon energies less than the Fermi level.

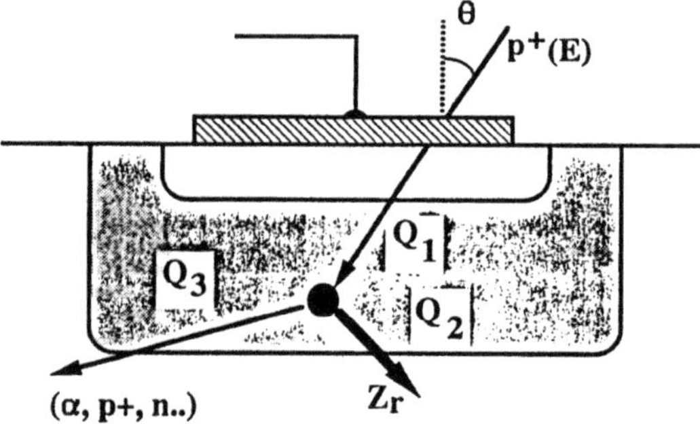

Fig. 43: Mechanism involved in proton induced SEE (indirect process).

The total proton-nucleon and neutron-nucleon cross sections are fits of experimental data obtained by Chen [Ref. 51]; the differential cross sections with respect to the angle distributions are those defined by Metropolis [Ref. 52].

Primary proton and knocked on nucleon trajectories are followed inside the nucleus up to the capture (if the energy drop down a fixed cut-off value) or up to the output of the target. The residual nucleus is left in an excited state; then evaporation occurs. During this process the deexcitation by gamma ray emission is neglected. The secondary nucleon evaporations are assumed isotropic and computed using density levels defined by Weisskopf [Ref. 53] and Dostrovsky [Ref. 54] and pairing energies given by Cameron [Ref. 55]. The evaporation process is stopped when the excitation energy falls sufficiently low.

Taking into account energies and directions of all involved particles (primary proton and secondaries from cascade and evaporation), then the kinetic energy and direction of the recoiling nucleus is obtained as well as its LET. Repeated computations for some thousands of primary protons furnish finally, the energy and angle distributions of secondary nucleons and recoiling nuclei LET spectrum. Considering the mass losses of the target and the secondary nucleon characteristics, good agreement is found with results reported by Metropolis for monocinetic protons up to about 300 MeV.

For higher energies differences appear due to pion and light fragment production neglected in the computation. With regard to the shape of proton spectra in space, the consequences for SEE rate forecast should be weak. These results, given for monoenergetic protons interacting with the target nuclei, must be corrected by the nuclear reaction probability; this is performed by using cross section model developed by TSAO and SILBERBERG [Ref. 56].

Neglecting charges produced by primary and secondary protons involved in the reaction, the LET spectra obtained may be then used for the computation of the SEE cross section versus proton energy. In this last step the node (assumed right angled parallelepiped) dimensions a', b' and c' as well as measurements under heavy ion flux of the SEE cross section, $\sigma(L)$, must be available. If necessary the saturation cross section for high LET may be used for a rough evaluation of the surface dimensions of the node, with:

$$b' = c' = \sigma_{sat.}^{0.5}$$

Averaged trajectory length in the node and solid angle under which the sensitive volume is seen from the nuclear reaction location are computed assuming isotropic exposure; then, correction of the recoiling nuclei LET spectra, taking these values into account, is performed. Finally the proton induced SEE cross section at fixed energy is obtained by integrating the product of the heavy ion induced cross section by the corresponding LET spectrum of recoiling nuclei. Examples of results obtained for the Fairchild 93L422 SRAM and Intel 2164A DRAM and comparisons with other models and experimental measurements are shown on Fig. 44.

4.4.2. Computation of Proton Induced SEE in Space. Assuming that the preliminary studies are achieved:

- the SEE cross section, $s_k(E_p)$, is defined from experimental measurements, Bendel model or nuclear reaction computation,
- the flux forecasts had been performed for trapped and solar protons on the specified orbit,
- the radiation transport in the shielding materials has been studied, giving the spectra of trapped and solar protons at the device's level, $\Phi(E,x)$.

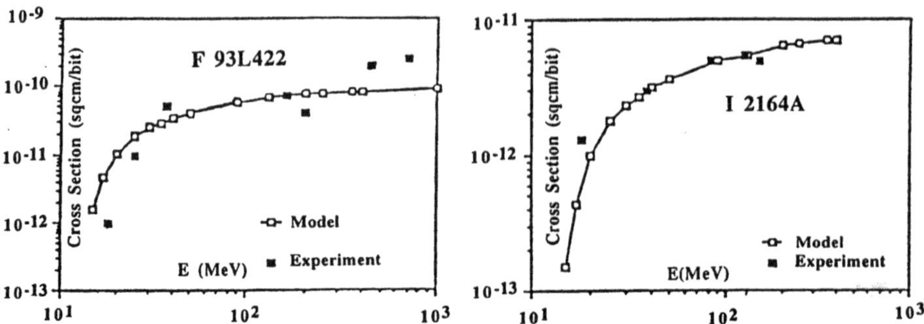

Fig. 44: Comparison of the computed (nuclear reaction model) and measured cross sections for SEE induced by protons.

Then a rough evaluation (angle distribution of particles neglected) of proton induced SEE rate for the device (k) is given by:

$$N_k = \int_0^{E_{max}} \Phi(E,\ x)\,\sigma_k(E)\,dE$$

Some results obtained with the Bendel model are shown on Fig. 45 for trapped and solar protons (August 72 event); for a characterized device (A and number of bits defined), the SEE rate can be directly obtained from the curves.

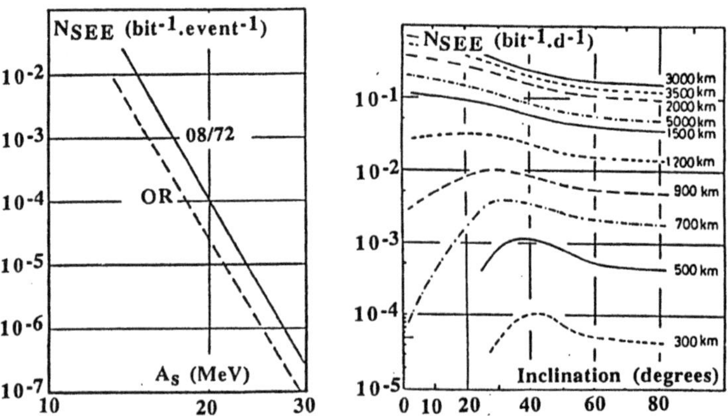

Fig. 45: SEE rate (Bendel model) by proton events (450 km – 64.5°, shielding 37 mm Al), and trapped protons (A = 15 MeV, B = 24).

If nuclear reaction modelling is used, more precise computation, including the angular distribution of the protons impinging the device, may be performed. SEE rates foreseen by using the nuclear reaction model, and measured in flight on board UOSAT 2 [Ref. 57] on HM6554 and HM6516 memories are given on Table 1. All the results are found in good agreement with experiment, showing that nuclear reaction modelling is an effective method, to be used for proton induced SEE cross section computations and for the SEE rate forecast in flight.

Table 1: Computed (nuclear reaction model) and measured (UOSAT2) SEU rates.

DEVICE	HM6516	HM6564
FLIGHT	2×10^{-7}	2.6×10^{-7}
MODEL		
Solar Min.	7×10^{-7}	2.9×10^{-7}
Solar Max.	2.2×10^{-7}	10^{-7}

References

[1] J.H. Adams, et al., "The effects of solar flares on single event upset rates", IEEE Trans. Nucl. Sc., Vol. NS31 n°6 p. 1212 (Dec 1984).

[2] P. Lantos, "Etude de la prédiction des éruptions solaires", Rapport final du contrat HPP/CNES/88/5164/00, Pub. Observatoire de Paris (Oct 1988).

[3] NOAA SESC, "Preliminary report and forecast of solar geophysical data".

[4] D.F. Smart and M.A. Shea, "PPS76 – A computerized "Event mode" solar proton forecasting technique", Boulder Solar – Terrestrial Prediction Workshop Proceedings (Donnelly Ed.) (1979).

[5] E.G. Stassinopoulos, "SOLPRO: a computer code to calculate probabilistic energetic proton fluences", Pub. NASA-NSSDC 75-11 (1975).

[6] J. Feynman, et al., "New interplanetary proton fluence model", J. Spacecraft, Vol. 27, n°4, p. 403, (Jul 1990).

[7] D.L. Chenette, et al., "The solar flare heavy ion environment for SEU: a summary of observations over the last solar cycle 1973-1983", IEEE Trans. Nucl. Sc. Vol. NS 31 n° p. 1217 (1984).

[8] J.H. Adams, et al., "Cosmic ray effects on microelectronics", Part 1: NRL memo report 4506 (1981); Part 2: NRL memo report 5099 (1983); Part 4: NRL memo report 5901 (1986).

[9] J. Bourrieau, "Effets des radiations – particules piégées, éruptions solaires et rayonnement cosmique sur les composants électroniques", Cours Int. de Technologie Spatiale" Qualité, composants et expertise" Cnes Toulouse, Mars (1988).

[10] P. Meyer, et al., "Cosmic Rays – Astronomy with energetic particles", Physics Today p. 23 (1974).

[11] NASA TM 82.478 p. 130.

[12] C.E. McIlwain, Coordinates for mapping the distribution of magnetically trapped particles", J. Geophys. Res. 66, 3681 (1961).

[13] C. Störmer, "The polar aurora", Charendon press (1955).

[14] J. Bourrieau, "COSTRANS: Code de calcul des taux de SEE induits par les ions lourds sur orbite", Pub. ONERA/CERT/DERTS à paraitre.

[15] D.F. Smart and M.A. Shea, "Geomagnetic transmission function for a 400 km altitude satellite", 18th Int. Cosmic Ray Conf. Bangalore – Vol. 3, p. 419 (1983).

[16] L. Lemaignen, et al., "Protection Against Radiations In Space", Final Presentation – Esa Contract 6988/86 (Nov 1987).

[17] E.G. Stassinopoulos, "The space radiation environment for electronics", Proc. IEEE, Vol. 76 n°11 (Nov 1988).

[18] E.G. Stassinopoulos, Publ. NASA SP 3024 (1970).

[19] P.D. McCormac, "Radiation dose and shielding for the space station", Paper IAF/IAA – 86 – 380, 37th Cong. of the IAF Innsbruck (Oct 1986).

[20] J.P. Philippon, "Evaluation théorique et calcul des flux de particules dans la zone stable de la magnétosphère", Thèse d'université n° 258 (March 1970).

[21] J. Berry, et al., "Description of the earth's radiation environment for future European manned space missions" – TN 1001, Contrat ESA n°6988/86/NL/PP (SC).

[22] A. Holmes Siedle, et al., "Calibration and flight testing of a low field pMOS dosimeter" IEEE Trans. Nucl. Sci., Vol. NS32 n°, p. 4425 (1985).

[23] C. Carel, "Phénomènes généraux liés aux contextes radiatifs, RADECS 89 "Radiations: Effets sur les composants et systèmes", 11th Sept 1989 - Montpellier (France), Short course published by the Université des sciences et Techniques du Languedoc Montpellier 2 France.

[24] NCRP, "Guidance of radiation received in space activities", NCRP report 98 (1989).

[25] E.J. Daly, "The evaluation of space radiation environment for ESA projects", ESA Journal 1988, Vol. 12, p. 229.

[26] J. Bourrieau, "Interest for detailed particle - matter interaction simulation in space environment effect evaluation", ESA - Space Environment Analysis Workshop (Oct 1990).

[27] T.M. Jordan, "Electron dose attenuation kernels for slab and spherical geometries", AFWL TR - 81 - 43 (Nov 1981).

[28] L. Lemaignen, et al., "Study of biological effects and radiation protection to future european manned space flights", Final report - ESA contract 6988/86/NL/PP(SC).

[29] E.V. Benton, "Summary of radiation dosimetry results on US and Soviet manned spacecraft", Cospar (paper VII-7) Toulouse (1986).

[30] M.S. Gussenhoven, et al., "New low-altitude dose measurements", IEEE Trans. Nucl. Sc. Vol. NS34 n°6, p. 676 (1987).

[31] T. Bion, "Modélisation des effets singuliers induits dans les composants électroniques par les protons rapides de l'environnement spatial", Thèse de doctorat n° 36, Ecole Nationale Supérieure de l'Aéronautique et de l'Espace - Toulouse (1989).

[32] J.W. Haffner, "Radiation and shielding in space", Ac. press NY (1967).

[33] J. Bourrieau, "Evaluation des flux de neutrons secondaires sur orbites basses (Vols habités)", Publ/ONERA/CERT/DERTS, CR/VH/04 (1989).

[34] J. Bourrieau, et al., "Effect of space charged particle environment on optical components and materials", "Space Materials in Space Environment", ESA SP - 145 (1979).

[35] J. Bourrieau, et al., "Effect of radiations on polymers and thermal control coatings", "Space Materials in Space Environment", ESA SP - 145 (1979).

[36] J. Bourrieau, "Activation des scintillateurs centraux de l'expérience d'astronomie X embarquée sur SIGNE 3", Publ. ONERA/CERT/DERTS, CR/SIGN/04 (1977).

[37] G. Rodstam, "Systematic of spallation yields", Z. Nat. 21a, p. 1027 (1966).

[38] C.J. Waddington, "The fragmentation of cosmic ray nuclei in interstellar hydrogen", University of Minnesota.

[39] R.A. Bull, et al., "Meteosat P2 Latch-up experiment - Monthly report n° 16, October 1989", Pub. Satellites Inst. Limited n°SIL/CR/9190/16 (Jul 3, 1990).

[40] T. Goka, et al., "The on-orbit measurements of Single Event Phenomena by ETS.V Spacecraft", Space Environment Analysis Workshop, Estec (Oct 1990).

[41] R. Harboe Sorensen, et al., "The behavious of measured SEU at low altitude during periods of high solar activity", Space Environment Analysis Workshop, ESTEC (Oct 1990).

[42] Y. Shimano, et al., "The measurement and prediction of proton upset", IEEE Trans. Nucl. Sc. NS36, 6, 2344 (Dec 1989).

[43] J. Bourrieau, "L'environnement spatial (flux, dose, blindage, effets des ions lourds)" RADECS 91 "Radiations: Effects on Components and Systems", 9th - 12th September 1991 - Montpellier (France), short course to be published.

[44] R. Robinson, et al., "Anomalies due to single event upset", 28th Aerospace Sciences Meeting - Reno (Jan 1990).

[45] J. Bourrieau, "L'environnement spatial et ses contraintes pour les systémes électroniques" Journées d'Information Electroniques (JIE90) CEA - Saclay - (Fevrier 1990).

[46] J. Buisson, et al., "Etude des effets dynamiques des ions légers sur les composants électroniques-Collaboration CNES-CNEA", Pub. CEA/DTA/LETI/DEIN/SIR 91/02 (Janvier 1990).

[47] W.L. Bendel, et al., "Proton upsets in orbit", IEEE Trans. Nucl. Sc. Vol. NS30 n°6, p. 4481 (1983).

[48] W.J. Stapor, et al., "Two parameter Bendel model calculations for predicting proton induced upset", IEEE Trans. Nuc. Sc. Vol. 37, n°6, p. 1966 (Dec 1990).

[49] T. Bion, et al., "A model for proton induced SEU", IEEE Trans. Nuc. Sc. Vol. 36, n°, p. 2281 (Dec 1989).

[50] R. Hofstater, Rev. Mod. Phys. 28, 214 (1956).

[51] K. Chen, et al., "A Monte-Carlo simulation of intranuclear cascades" Phys. Rev. n° 166, 949 (Février 1968).

[52] M. Metropolis, et al., "Monte-Carlo calculations of intranuclear cascades", "I: Low energy studies", Phys Rev. 110, 185, "II: High energy studies and pion processes", Phys. Rev. 110, 204 (1958).

[53] W.F. Weisshopf, Phys. Rev., 52, 295 (19730.

[54] I. Dostrovsky, et al., "Monte-Carlo calculations of nuclear evaporation processes, applications to low energy reactions", Phys. Rev. 116, 683 (1959).

[55] A.G.W. Cameron, Can. Jl. Physics, 36, 1040 (1958).

[56] R. Silberberg, et al., "Improved cross sections calculations for astrophysical applications", Ast. Jl. Sup. Ser. N° 58, 873 (1985).

[57] C.I. Underwood, et al., "Observation and analysis of single-event upset phenomena on board the UOSAT-2 satellite", ESA Space Environment Analysis Workshop (Oct 1990).

PHYSICS OF ENERGETIC PARTICLE INTERACTIONS

C. DYER
Defence Research Agency
Space Technology Department
Farnborough

ABSTRACT. This paper is designed to serve as a bridge between preceding papers describing the space radiation environment and ensuing papers dealing with specific effects. Particles of interest include both primary particles present in the space environment (such as protons, heavy ions and electrons) and secondaries (such as neutrons, mesons, X-rays and gamma rays), which are generated by interactions in spacecraft materials. In addition neutrons, X-rays and gamma-rays are frequently used in ground testing to simulate the effects of primary radiation, and are emitted by nuclear power sources.

Relevant reactions are reviewed for all the above and include interactions with atomic electrons leading to ionization and excitation, and nuclear reactions leading to scattering, displacement and the production of secondary particles and induced radioactivity. The terminology used to quantify the interactions is defined and includes stopping power, linear energy transfer, range and reaction cross-section. The dependencies of these on incident particle and target material parameters are described and relationship is made with the parameters used to describe effects such as dose and equivalent displacement damage fluence. Thick target situations require radiation transport calculations to deal with the build-up of secondaries. Such methods are described and illustrated by calculations performed for the Gamma Ray Observatory. In such a heavyweight spacecraft effects are magnified some twentyfold by the build-up of secondaries. Hence predictions for such a system are a good test of the codes. Comparison with early data is very encouraging.

1. Particles of Interest

1.1. PRIMARIES

Energetic particles in the natural environment arise from cosmic rays, solar particle events and the radiation belts. Penetrating cosmic rays and their observation by ground-based monitors have been reviewed by Mathews [Ref. 1] in these proceedings. While particle energies extend to 10^{20} eV,

R. N. DeWitt et al. (eds.), The Behavior of Systems in the Space Environment, 353–381.

from the point of view of space systems it is particles in the energy range 1 to 20 GeV per nucleon which have most influence. The majority are protons (85%) but the influence of alpha particles (14%) and heavier nuclei (1%) is disproportionate to their number as they are more efficient at producing secondaries and are more heavily ionizing. Sporadic solar particle events populate interplanetary space for several days at a time and can penetrate to low earth orbit at high latitudes. Particle energies are typically up to several hundred MeV per nucleon but occasionally reach several GeV. Protons again dominate but significant fluxes of alphas and heavier nuclei contribute to effects and their proportion varies from event to event. An additional "anomalous" component of cosmic rays comprising helium, nitrogen, oxygen and neon ions of relatively low energy (1 to 30 MeV per nucleon) occurs from time to time. These appear to be singly ionized so that they can penetrate to low latitudes and are thought to be picked up from the interstellar medium and accelerated in the solar wind. Energetic particles trapped in the geomagnetic field comprise electrons with energies up to several MeV and protons with energies up to 600 MeV. While energetic electrons are present out to ten earth radii, the energetic protons are confined within three earth radii.

1.2. SECONDARIES

Energetic primaries can interact with nuclei of spacecraft materials and the atmosphere to give a range of secondary products. For primary protons and ions these secondaries include neutrons, protons, mesons and fragments of the incident or target nucleus. Particles knocked directly from the nucleus in the intranuclear cascade can have energies approaching that of the incident particle, while the residual excited nucleus can evaporate neutrons and light fragments and de-excite by gamma-ray emission with characteristic energies up to 20 MeV. Following this evaporation the residual nucleus is usually neutron-deficient, and hence radioactive, with decay products comprising beta-particles and gamma rays of a few MeV in energy emitted according to their half-lives, which range from microseconds to millions of years. Mesons are very short-lived and decay into gamma rays and electrons or positrons. Energetic electrons and positrons produce X-rays of up to the incident particle energy by the process of bremsstrahlung.

1.3. MAN-MADE

Reactor and radioisotope sources are used to power space systems where solar power is insufficient, and these produce neutrons and gamma rays of several MeV. These radiations are also emitted by exoatmospheric nuclear bursts, which in addition emit fission fragments that subsequently decay to enhance the electron levels in the radiation belts. Neutrons, gamma rays and fission fragments, while not primary space radiation, are frequently used as economical methods for ground testing. Clearly the physics of the relevant interactions must be adequately understood for this read-across to be valid.

1.4. PARTICLE AND INTERACTION CATEGORIES

From the point of view of interactions and effects it is convenient to divide the above particles into the following categories.

1.4.1. *Charged Particles* comprise protons, ions and electrons. They all interact by the Coulomb force with atomic electrons to give ionization and with nuclei to give scattering and displacement. Electron scattering on the nuclei gives bremsstrahlung radiation. Protons and ions also interact with nuclei via the short-range strong force to give nuclear reactions.

1.4.2. *Neutrons* are uncharged and interact via the strong force with the nuclei to give both scattering and transmutation reactions.

1.4.3. *Photons* are quanta of electromagnetic radiation and are usually called X-rays when produced by electromagnetic processes (e.g., bremsstrahlung) and gamma rays when produced in nuclear processes. They are indirectly ionizing via electrons and positrons generated in discrete interactions with atomic electrons and nuclei.

2. Ionizing Radiation

2.1. DIRECTLY IONIZING RADIATION

Energetic charged particles can impart sufficient energy by Coulomb interactions to overcome the binding energy of atomic electrons and create electron-ion pairs. This process is always accompanied by excitation of electrons to higher, but bound, orbitals. In the solid state, electrons are raised in energy to the conduction band to give electron-hole pairs. Energy-loss by ionization is a continuous process so that there is a range at which all energy has been lost. However for electrons there is considerable scattering in direction so that their range is less clearly defined.

2.2. INDIRECTLY IONIZING RADIATION

Uncharged particles, such as neutrons and photons, can produce directly ionizing charged secondaries. In this case range is not a valid concept as the particles undergo discrete interactions leading to exponential attenuation.

2.3. RADIATION UNITS

2.3.1. *Unidirectional Particle Flux* is defined as the number of particles crossing unit area perpendicular to their velocity per unit time. For n particles per m^3 of velocity v $m\ s^{-1}$,

$$\text{Flux}(\phi) = nv\ m^{-2}s^{-1}$$

2.3.2. *Omnidirectional Particle Flux* is defined as the number of particles entering a sphere of unit cross-sectional area per unit time.

2.3.3. *Particle Fluence* is the integral of flux with respect to time.

2.3.4. *Particle Energy* is commonly expressed in units of electron-volts (eV).

$$1 \text{ eV} = 1.602\text{x}10^{-19}\text{J}$$
$$1 \text{ MeV} = 1.602\text{x}10^{-13}\text{J}$$

2.3.5. *Specific Energy Loss* (S) is the removal of energy from the particle per unit path length of material.

$$S = dE/dx \text{ in MeV/cm or MeV/m}$$

2.3.6. *Stopping Power* is defined to be S/ρ, where ρ is the density of the medium. Units are MeV/(g cm^{-2}) etc. It is common practice to express shield thicknesses in units of density x thickness (i.e., g cm^{-2} etc). This normalises to the same number of atoms for a given material and approximately normalises to the same number of electrons between different materials (i.e., those with comparable Z/A, see below).

2.3.7. *Linear Energy Transfer* is defined as the energy deposited in the material per unit (density x pathlength). For heavy particles LET is nearly identical to the stopping power, but it can differ when some of the energy lost is not deposited locally.

2.3.8. *Absorbed Dose* is defined as the energy deposited as ionization and excitation in unit mass of material. Units are rads and grays.

$$1 \text{ rad} = 100 \text{ ergs/g}$$
$$1 \text{ gray} = 1 \text{ Joule/kg}$$
$$1 \text{ gray} = 100 \text{ rads}$$

$$\text{Dose (grays)} = \text{fluence } (m^{-2}) \text{ x LET (MeV } m^2/\text{kg) x } 1.602 \text{ x}10^{-13} \text{ (J per MeV)}.$$

It might be hoped that the absorbed dose would uniquely define effects. Unfortunately there are further dependencies on, for example, dose rate and particle type.

2.3.9. *Dose Equivalent* is an attempt to account for this in biological systems by multiplying absorbed dose by a Quality Factor which measures the dependence of radiobiological damage on particle type. A dependence on local density of ionization leads to quality factors of up to 20 for heavy ions and fast neutrons. For absorbed dose in rads, dose equivalent is in rems, while in the mks system the unit is the sievert.

2.3.10. *Energy Loss Per Electron-Hole/Electron-Ion Pair* (w eV) is the average energy required to produce an electron-hole pair (solids) or electron-ion pair (gases). This quantity is greater than the band gap or ionization potential (approximately double for insulators) due to the sharing of energy loss with excitation. Examples are given in Table 1.

Table 1.

Material	w (eV)
Air	33.7
He	42
Si	3.6
SiO_2	17
Ge	2.8
GaAs	4.8

2.4. STOPPING POWER DEPENDENCE FOR IONS

Consider a particle of charge Z_i and velocity v in a medium of atomic number Z_m and atom number density N_m atoms/m^3. This particle will be in the vicinity of each electron for a time proportional to 1/v. The momentum transferred is proportional to force x time, ie to Z_i/v. The kinetic energy transferred is proportional to the square of the momentum transfer and as there are $N_m Z_m$ electrons per unit path length,

$$S \propto (Z_i^2/v^2) \times N_m Z_m$$

As $N_m = N_A \rho/A_m$, where N_A is Avogadro's Number,

$$S/\rho \propto (Z_i^2/v^2) \times (Z_m/A_m)$$

For hydrogen Z_m/A_m is 1, while for other elements the ratio is approximately 0.5 and decreases gradually with increasing Z. Hence hydrogen is the most efficient shield and while other elements have comparable S/ρ, lighter elements are in fact more efficient.

A fuller treatment leads to the Bethe-Bloch equation, but the above enables a simple understanding of the curves presented in Fig. 1. It can be seen that slower ions are more heavily ionizing but their stopping power is eventually limited at low energies by their capturing electrons to become neutralised. At the high energy end their velocities approach the speed of light and they become "minimum ionizing". Also noteworthy is the strong dependence on atomic number which means that high-Z ions can deposit sufficient energy to give single event upsets [Ref. 2]. Stopping powers and LETs are well summarised in the literature [e.g., Refs. 3,4].

2.5. STOPPING POWER AND BREMSSTRAHLUNG FOR ELECTRONS

Similar considerations apply to the stopping power for electrons. However their lighter mass means that they are strongly scattered and also at energies of concern here they are relativistic and hence minimum ionizing. Another consequence of their low mass is the emission of bremsstrahlung

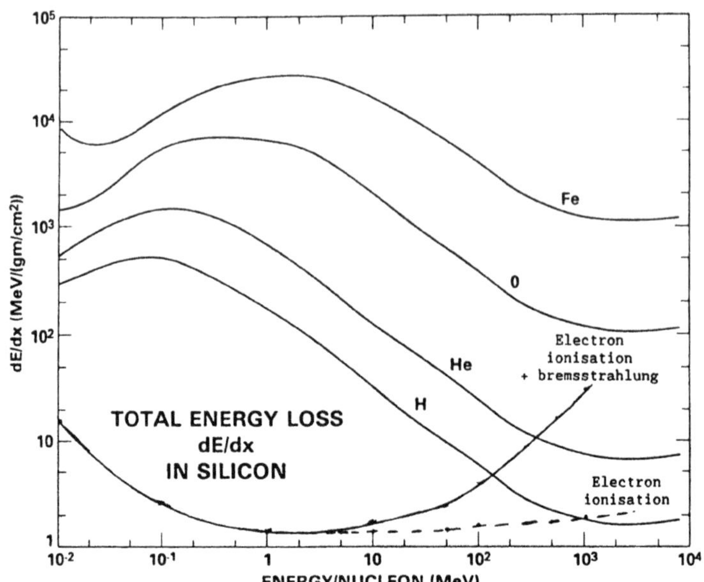

Fig. 1: Ionisation stopping power dependence on energy for ions in silicon compared with ionisation and bremsstrahlung for electrons.

(braking radiation) due to acceleration by the electric field of the atomic nuclei. Rate of radiation is proportional to the square of the acceleration (i.e., to Z_m^2) and increases with particle energy. The emitted radiation is in the form of penetrating X-rays with energies up to that of the electron and these are difficult to shield against. Hence low-Z materials are preferred as electron shields. Stopping-power curves for ionization and bremsstrahlung by electrons are included in Fig. 1 and tabulations may be found in the literature [e.g., Ref. 5].

2.6. PARTICLE RANGES

A consequence of the dependence of stopping power on particle velocity is that the rate of energy deposition and ionization increases as the particle slows down so that a Bragg peak is observed at the end of the track. Fig. 2 (taken from [Ref. 6]) gives a typical curve of ionization density vs depth for an ion. There is a fairly sharp cut-off at the end of track where the ion is finally stopped and neutralised and this gives a well defined range, although there is a small amount of scattering leading to straggling. For electrons scattering is very significant and the shape of the attenuation curve is very different with range less clearly defined (see Fig. 3 taken from [Ref. 7]). Range vs energy curves for protons are presented for low energies in Fig. 4 (taken from [Ref. 7]) and at high energies in Fig. 5 (taken from [Ref. 6]). Note the change in vertical scale

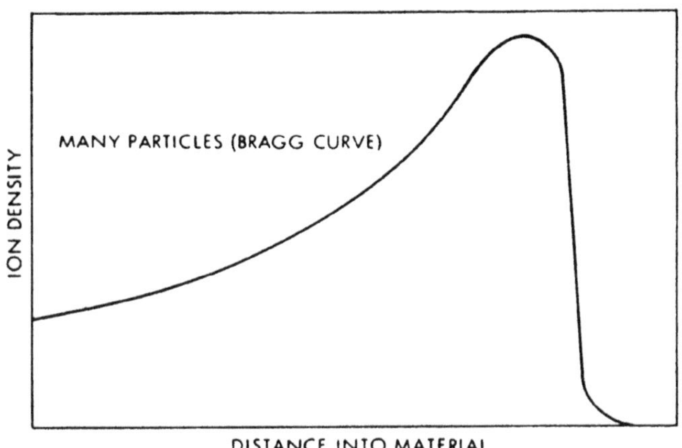

Fig. 2: Representative Bragg curve of ionization density imparted by an ion vs depth in material.

from $\mu g/cm^2$ in Fig. 4 to g/cm^2 in Fig. 5. The higher shielding efficiency of low-Z materials can be clearly seen. Fig. 6 (taken from [Ref. 8]) compares the ranges of electrons and protons. The electrons are minimum ionizing and hence more penetrating up to those energies at which bremsstrahlung dominates. It must be remembered that even if the electrons are stopped the bremsstrahlung radiation must be shielded against.

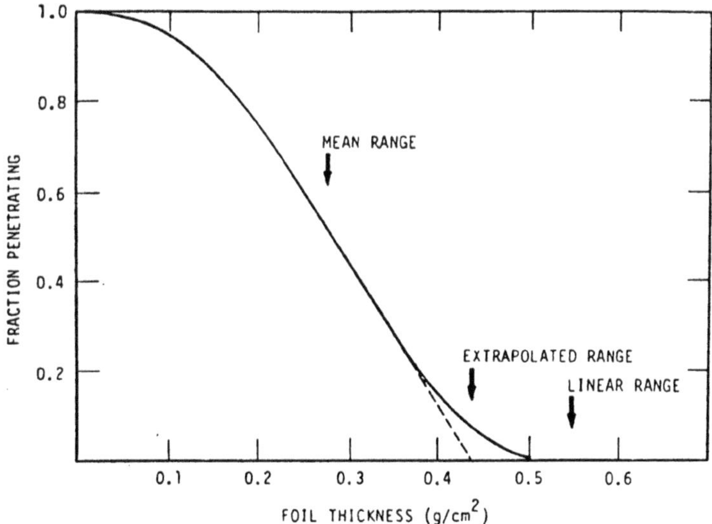

Fig. 3: The penetration of 1 MeV electrons through aluminum shows a less clearly defined range due to scattering.

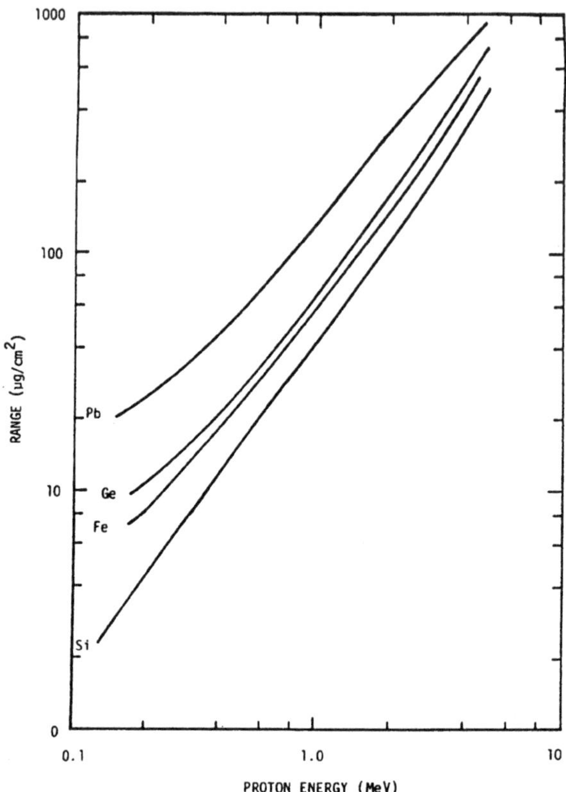

Fig. 4: Curves of range vs energy for low energy protons in Si, Fe, Ge, and Pb.

2.7. EFFECTS ON SPACE SYSTEMS

Accumulated dose leads to degradation of space sytems (e.g., the threshold voltage shift of CMOS transistors), while large linear energy transfers by individual particles can lead to single event upsets and latch-up. Both these phenomena are extensively discussed by Stephen [Ref. 2]. A dose vs depth curve is illustrated in Fig. 7 for a 850 km, sun-synchronous orbit. It can be seen that the electron contribution is more easily shielded than the proton contribution due to their different energies (1 MeV cf 100 MeV). In addition bremsstrahlung dominates electron ionization after 7 mm. Single event phenomena are characterised by integral LET spectra (i.e., particle fluxes greater than a certain LET) as illustrated for a Space Shuttle orbit in Fig. 8. Devices have a certain LET threshold for upset and the integral flux above this threshold may be used to assess upset rates. Comparisons for a wide range of orbits are presented by Bourrieau [Ref. 9].

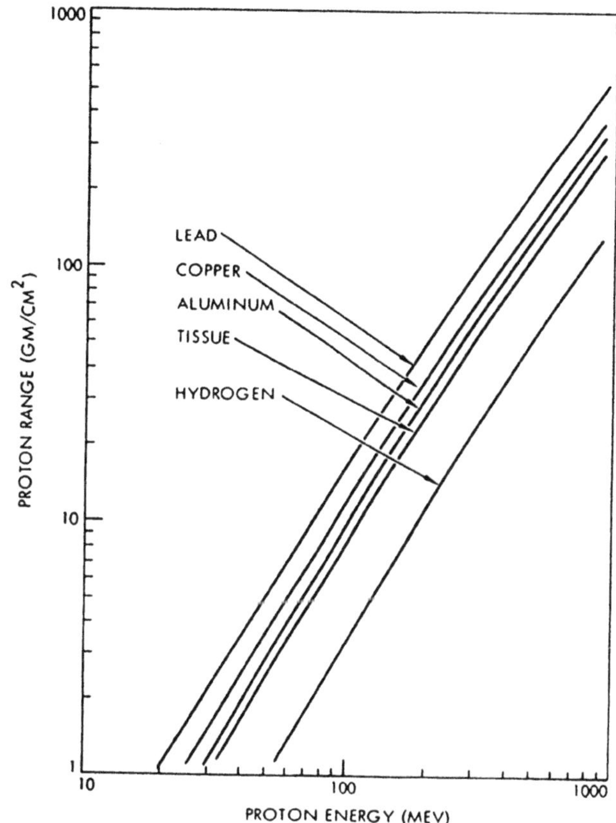

Fig. 5: Curves of range vs energy for energetic protons in hydrogen, tissue, Al, Cu, and Pb.

2.8. PHOTON INTERACTIONS

X-ray and gamma-ray photons give indirect ionization via three discrete interaction mechanisms.

2.8.1. *Photoelectric Ionization* dominates at lower photon energies and involves the absorption of the photon on a tightly bound orbital electron (usually K or L shell). The kinetic energy imparted to the electron is the difference between photon energy and binding energy. The latter is released either as a fluorescent X-ray or as Auger electrons.

2.8.2. *Compton Scattering* dominates at higher energies and can be considered as the elastic scattering of a photon on a free electron (the binding energy is neglected). Conservation of energy and momentum can be

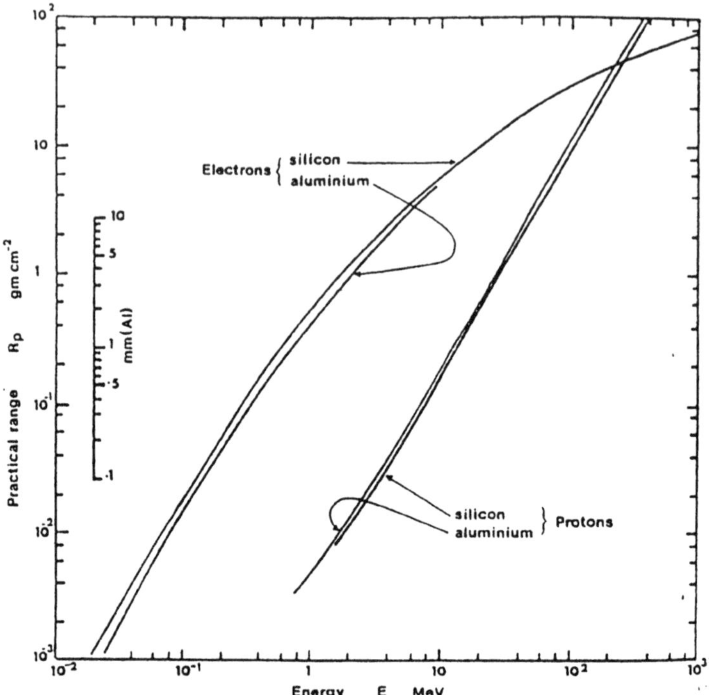

Fig. 6: Curves of range vs energy for electrons in silicon and aluminum compared with protrons.

used to derive the photon energy as a function of scattering angle. Energy transfer to the electron is a maximum for backscatter (180°) but is never total, the backscattered photon energy tending to 256 keV (half the electron rest-mass energy) at high values of incident photon energy (see Fig. 9 taken from [Ref. 7]).

2.8.3. *Pair Production* can occur for photons of greater than 1.022 MeV (twice the electron rest-mass energy) and is the materialisation of the photon into an electron-positron pair by interaction with the intense electric field of the atomic nucleus. The heavy nucleus has negligible recoil energy and the kinetic energy imparted to the electron and positron is the difference between the photon energy and their total rest mass (1.022 MeV). This rest mass reappears as two 0.511 MeV photons when the positron is brought to rest and annihilates an ambient electron.

2.9. PHOTON ATTENUATION

The discrete interaction probabilities of photons with electrons and nuclei are described in terms of reaction cross-sections and lead to exponential attenuation. Similar treatments and terminology apply to the nuclear reactions of particles.

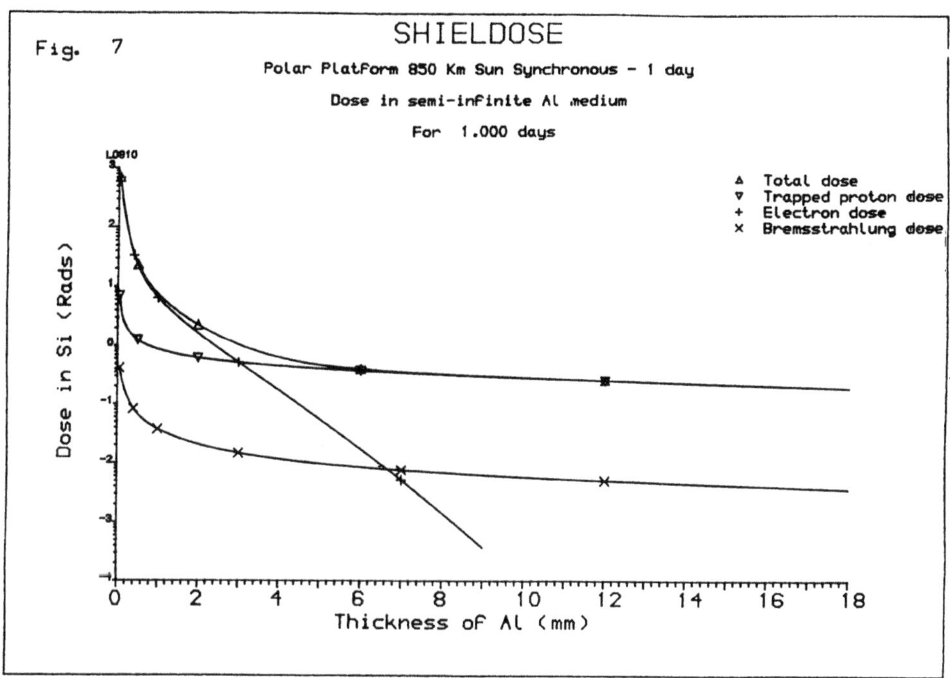

Fig. 7: Daily dose in silicon vs depth in a semi-infinite aluminum shield for radiation-belt electrons and protons encountered in 850 km, near-polar orbit.

2.9.1. *Microscopic Cross-Section* (σ) is the effective area presented to the incident beam by the target or nucleus for the particular reaction. The unit commonly employed is the barn (an approximate measure of the cross-sectional area of a nucleus).

$$1 \text{ barn} = 10^{-24} \text{ cm}^2 = 10^{-28} \text{ m}^2$$

For N_m target nuclei per unit volume the effective target area is $N_m\sigma$, and hence the reaction rate per unit volume is $N_m\sigma\phi$.

A thickness of material dx leads to removal of a flux

$$d\phi = -N_m\sigma\phi \, dx$$

so that on integration $\phi(x) = \phi(0) \exp(-N_m\sigma x)$

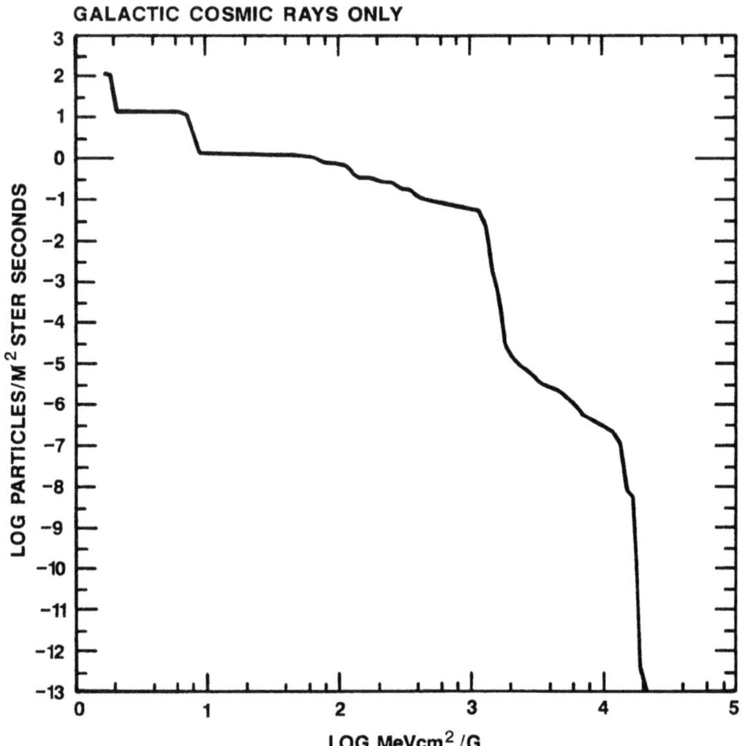

Fig. 8: Integral linear energy transfer spectrum for a low inclination, low altitude space shuttle orbit.

The product $N_m\sigma$ is the probability of interaction per unit path (m^{-1}) and is commonly called the linear attenuation coefficient (μ) for photons. In neutron physics this quantity is termed the macroscopic cross section (Σ).

The reciprocal of μ or Σ has the dimensions of length and represents the distance required to attenuate the flux by a factor of e. It is also equivalent to the average distance between interactions or "mean free path."

By dividing the linear attenuation coefficient by the density of material the more fundamental mass attenuation coefficient (m^2/kg) is obtained.

$$\mu/\rho = N_A\sigma/A_m$$

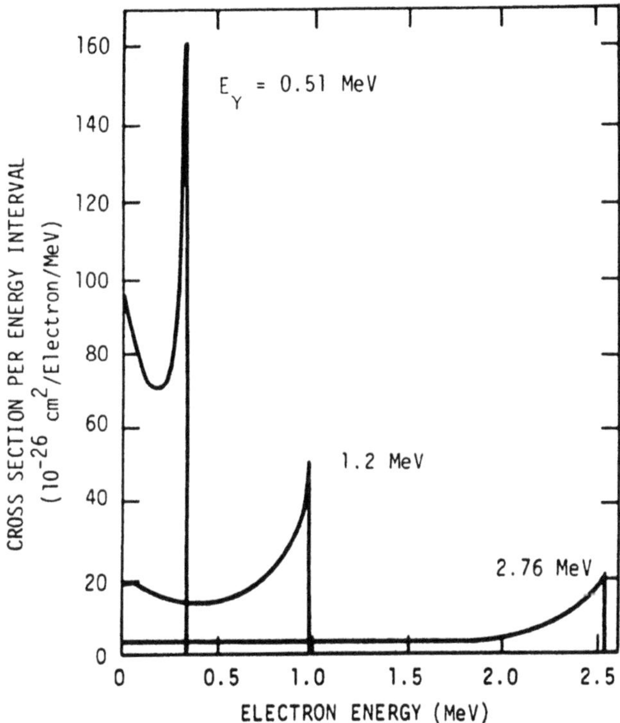

Fig. 9: Energy spectrum of Compton electrons produced by primary photons of energy 0.51, 1.2, 2.76 MeV.

2.9.2. *The Dependence of Attenuation on Material* is expressed in the following relationships.

i) Photoelectric $\sigma \propto Z_m^4$; $\mu/\rho \propto Z_m^4/A_m$

ii) Compton $\sigma \propto Z_m$; $\mu/\rho \propto Z_m/A_m$

iii) Pair Production $\sigma \propto Z_m^2$; $\mu/\rho \propto Z_m^2/A_m$

Thus both photoelectric and pair production processes increase in probability very strongly with the atomic number of the material, whereas for Compton Scattering the dependence is weak for elements heavier than hydrogen and operates in the converse sense. Hydrogen is approximately twice as efficient in giving Compton Scattering compared with other elements. In general high-Z materials are favoured for photon attenuation. In space systems it is common to deploy low-Z materials sufficient to stop the bulk of primary electrons with maximum efficiency and minimum bremsstrahlung and then to follow this with high-Z material to attenuate the bremsstrahlung photons.

2.9.3. *The Dependence of Attenuation on Photon Energy* is expressed in Fig. 10 taken from [Ref. 6]. The photoelectric effect probability falls off rapidly with increasing energy, the Compton process probability is relatively flat, while the pair production probability increases with energy above the threshold. The regions of domination of the various processes are mapped in Z,E space in Fig. 11 taken from [Ref. 10]. It can be seen that for silicon the Compton process dominates over most of the energy range of interest.

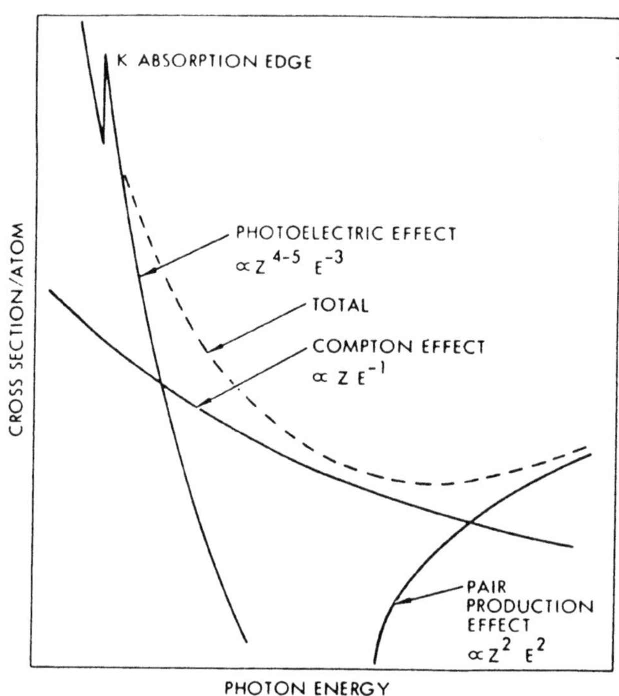

Fig. 10: Photon interaction cross-sections as a function of photon energy and atomic number.

2.9.4. *Dose Enhancement* occurs at the boundaries between materials of different Z when there is a significant discontinuity in the mass attenuation coefficients. This is less severe for MeV photons when Compton scattering dominates but can be very marked for low energy photons when the photoelectric effect dominates. For example 150 keV photons show an enhancement factor of about 20 in regions of silicon adjacent to a layer of gold. For this reason care must be taken in performing total dose testing using low energy X-ray machines and to avoid scattering to low energies of more energetic sources such as Cobalt-60.

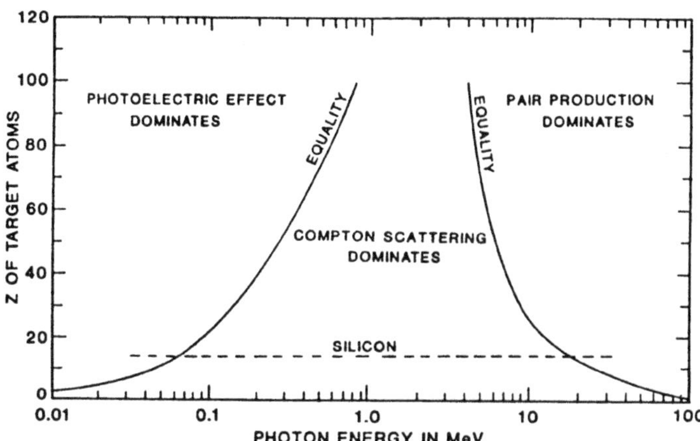

Fig. 11: Illustration of the relative importance of the three types of photon interaction as a function of Z and photon energy. The solid lines correspond to equal interaction cross-sections for the neighbouring effects. The dashed line corresponds to silicon.

3. Particle Interactions With Nuclei

3.1. INTERACTION CATEGORIES

3.1.1. *Neutrons* are uncharged and interact via the strong force which is short range.

3.1.2. *Elastic Scattering* conserves kinetic energy and leads to lattice displacement damage. There is no threshold energy for the reaction but there is a threshold to displace an atom from its lattice site.

3.1.3. *Inelastic Scattering* produces nuclear excitation at the expense of kinetic energy and results in gamma rays from the subsequent deexcitation. There is a threshold energy related to the first excited state of the nucleus.

3.1.4. *Capture* leads to gamma-ray de-excitation and a radioactive neutron-rich product. There is no threshold and cross-sections increase with decreasing neutron energy.

3.1.5. *Transmutation* is the general name given to a wide range of reactions which lead to particle emissions and radioactive products. These generally have thresholds related to the binding energies of the emitted particles.

3.1.6. *Heavy Charged Particles* interact via the strong force to give similar reactions to neutrons. However they must have sufficient energy to overcome the Coulomb barrier and this raises the energy threshold for reactions and makes radiative capture much less likely. In addition the interaction at a distance via the Coulomb force leads to scattering and lattice displacements.

3.1.7. *Electrons* interact via the Coulomb force only to give scattering and bremsstrahlung. Their low mass compared with nuclei leads to less efficient lattice displacement damage.

3.1.8. *Interaction Probabilities* are expressed by microscopic and macroscopic cross-sections (see section 2.9.1). The total cross-section is the sum of a number of partial cross-sections for the various reactions. Example cross-sections for neutron reactions on silicon are given in Figs. 12 and 13 taken from [Ref. 7].

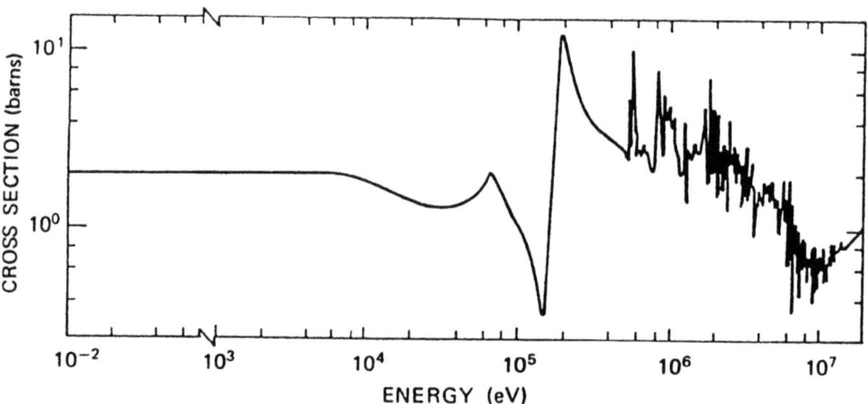

Fig. 12: Cross-section for elastic scattering of neutrons on silicon as a function of energy.

3.2. DISPLACEMENT DAMAGE

3.2.1. *The Mechanism for Displacement Damage* is the recoil energy imparted to lattice atoms by both scattering and transmutation reactions. If this is sufficient to overcome the lattice binding energy the atom is displaced to an interstitial position leaving a vacancy (Fig. 14 taken from [Ref. 11]). If the recoil energy is great enough a cascade of defects is produced. The alteration of the ideal lattice structure produces energy levels within the band gap and changes electrical parameters such as minority carrier lifetime, carrier concentration and mobility. Reduction of minority carrier lifetime is important in reducing the gain of bipolar transistors, the efficiency of solar cells and charge-collection efficiencies in radiation detectors.

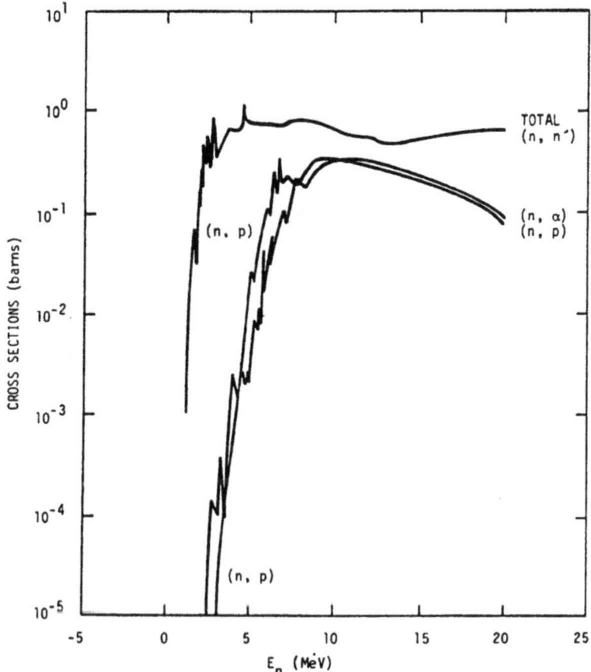

Fig. 13: Cross-sections for neutron inelastic scattering (n,n′) and transmutation reactions, (n,p) and (n,α) on silicon.

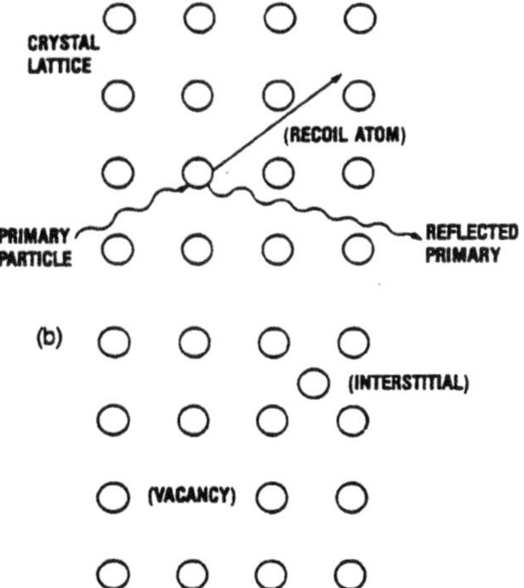

Fig. 14: Schematic of atomic displacement damage in crystalline solid showing an atomic displacement event and the resulting defects (simple vacancy plus interstitial).

3.2.2. *The Kinematics of Elastic Scattering* provide an important insight into the relative displacement damage given by different mass particles. From simple "billiard ball" kinematics the kinetic energy of a mass 1 particle of initial energy E_0 scattering elastically through an angle from a nucleus of mass A is

$$E = E_0(A^2 + 2A\cos\theta+1)/(A+1)^2.$$

The minimum energy following scatter through 180^0 is then

$$E_0(A-1)^2/(A+1)^2,$$

and the corresponding maximum recoil energy of the target nucleus is

$$E_0 - E = 4E_0A/(A+1)^2.$$

For neutrons and protons scattering on hydrogen total energy transfer is possible while for high-A targets the recoil energy is inversely proportional to A. For incident electrons the recoil energy is reduced by the relative mass of the electron (a factor of 1860). The threshold recoil energy for lattice displacement is about 13 eV in silicon so that incident particle thresholds for damage are 100 eV for protons and 190 keV for electrons.

3.2.3. *The Dependence of Displacement Damage on Particle Type and Energy* is illustrated by measurements of carrier lifetime degradation presented in Fig. 15 taken from [Ref. 12]. It should be noted that Cobalt-60 damage arises from the secondary electrons generated by Compton scattering. Considerable success has been achieved in correlating radiation damage with calculations of the nonionizing energy loss as presented in Fig. 16 taken from [Ref. 12]. At low energies protons interact by Coulomb scattering and show a 1/E dependence as for ionization. The lower probability of displacement compared with ionization can be noted from comparison of Fig. 16 with Fig. 1 and is due to the lower probability of nuclear encounters compared with orbital electron encounters. At higher energies nuclear reactions dominate and proton effects are comparable with neutron effects which give nuclear reactions only. Electrons are less efficient and show a threshold due to their lower mass. Another important consideration is the fraction of recoil energy going into ionization as compared with nonionizing displacements. This is the Lindhart fraction as presented in Fig. 17 taken from [Ref. 7]. At low energies the recoil atoms keep their orbital electrons and scatter elastically while at higher energies they become stripped ions and hence more effective at ionization.

3.2.4. *Equivalent Displacement Damage Fluence* is used to express the displacement damage in a spectrum of radiation by the equivalent fluence of monoenergetic radiation which would give the same effect. For space systems, and in particular solar arrays, it is usual to use the equivalent 1 MeV electron fluence. Fig. 18 presents the degradation of solar cell parameters with 1 MeV electron fluence as measured at RAE by Goodbody et al. A lot of electronics is tested for nuclear weapons and reactor effects

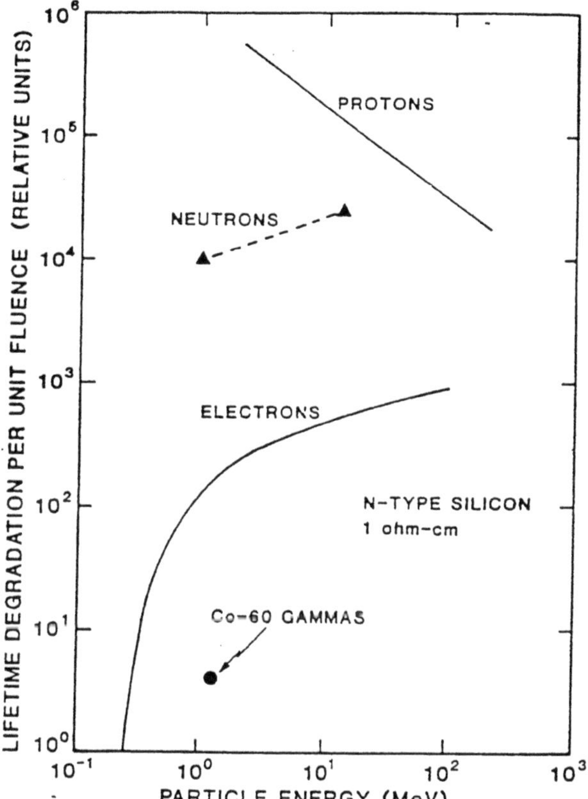

Fig. 15: Dependence of displacement damage on particle type and energy as measured by the degradation of carrier lifetime in silicon.

and here the results are expressed in terms of equivalent 1 MeV neutron fluence.

The equivalent fluence is defined as

$$\phi_{equiv} = \int_0^\infty \phi(E)\, D(E) dE / D(1MeV)$$

where $\phi(E)$ is the particle spectrum and D(E) the displacement damage as a function of energy.

3.3. SPALLATION REACTIONS

Spallation is the removal of nucleons from the nucleus by energetic particle interactions. The term fragmentation is used when a large nuclear

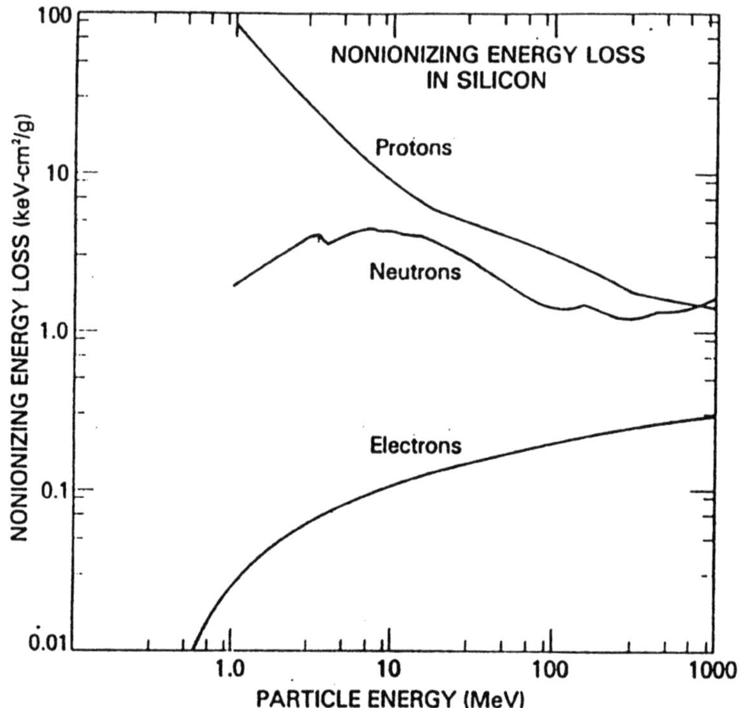

Fig. 16: Calculated nonionizing energy loss in silicon for protons, neutrons and electrons.

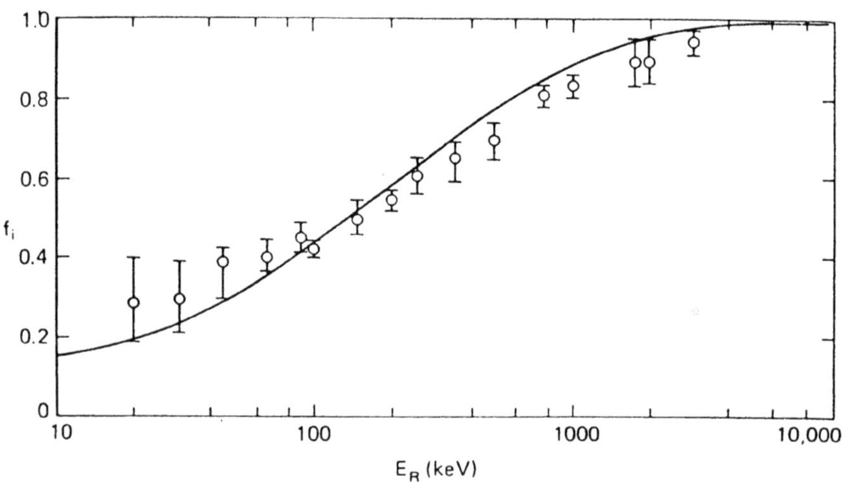

Fig. 17: Calculated Lindhart fraction (f_i) of recoil energy (E_R) going into ionization compared with data points.

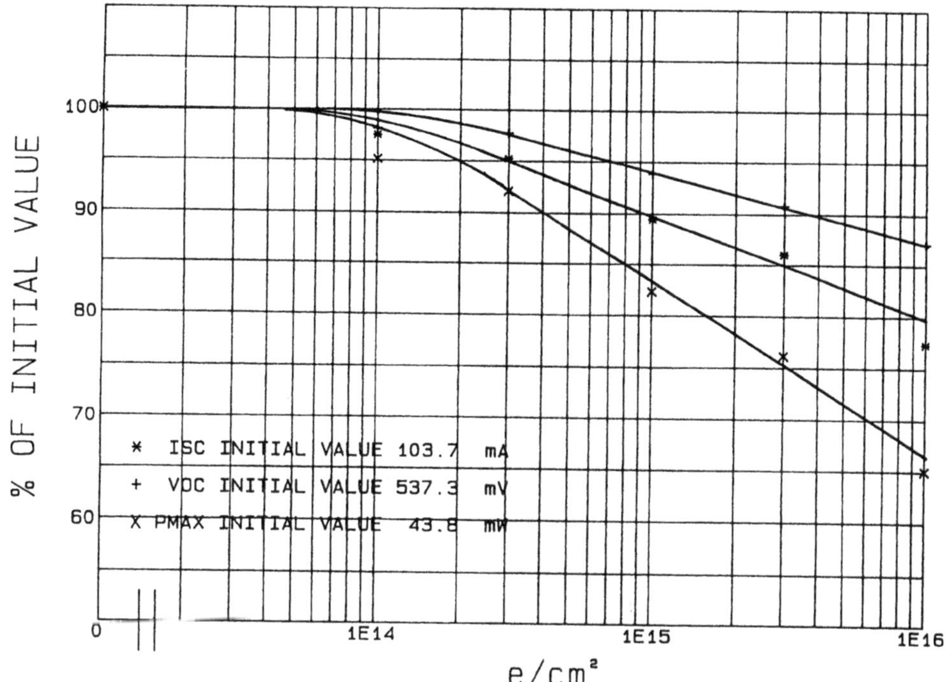

Fig. 18: Measured degradation of short circuit (Isc), open circuit voltage (Voc) and maximum power as a function of 1 MeV electron fluence for a silicon solar cell.

fragment is split off while the term fission is reserved for those cases where the nucleus splits into two comparably-sized nuclides. Such reactions are considered to occur in two stages. Firstly there is the prompt (10^{-23}s) knocking out of nucleons by interaction with the incident particle in an intranuclear cascade. The residual nucleus then de-excites on a much longer timescale (10^{-13}s) by the evaporation of neutrons, protons and light fragments and usually results in a radioactive product.

3.3.1. *Spallation Cross-Sections for Protons* are of considerable interest in a number of applications. Total cross-sections have the form shown in Fig. 19 taken from [Ref. 13] and scale with the geometric cross-section of the target nucleus (ie as $A^{2/3}$). Numerous significant spallation products are produced (hundreds for high-A targets) and only a few of their partial cross-sections have been determined experimentally. Semi-empirical fits to the limited data are used to predict the wide range of unknown cross sections. A five-parameter fit proposed by Rudstam has been considerably extended by Silberberg and Tsao [Ref. 14] to give useful predictions over a wide range of nuclides. An alternative technique employs Monte Carlo simulations of the intranuclear cascade and evaporation processes (e.g., Bertini et al [Ref. 15]).

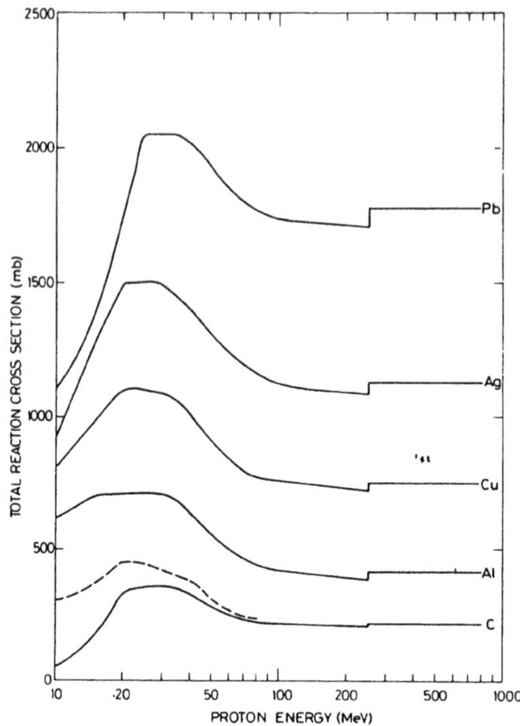

Fig. 19: Total reaction cross-sections as a function of proton energy for a number of elements. The step at 250 MeV is non-physical and was used for computational convenience while the dashed line for carbon includes inelastic excitation of the first excited state of the carbon nucleus at 4.43 MeV.

3.3.2. *Secondary Particle Production* is important in determining the radiation in thick shield situations. Of particular significance are secondary neutrons for which the multiplicity increases with energy of the incident particle and atomic number of the target. There exist a number of parametric fits to the simulation results for yields and spectra [Ref. 16], and these are used as input data for radiation transport calculations.

3.3.3. *Single Event Upsets* can result from the localised ionizing energy depositions of the recoil nucleus and the evaporated particles. An example reaction is illustrated in Fig. 20 taken from [Ref. 17]. This mechanism leads to significant upset rates in regions of high proton fluxes such as the inner radiation belt, as well as during solar flares in geomagnetically exposed orbits.

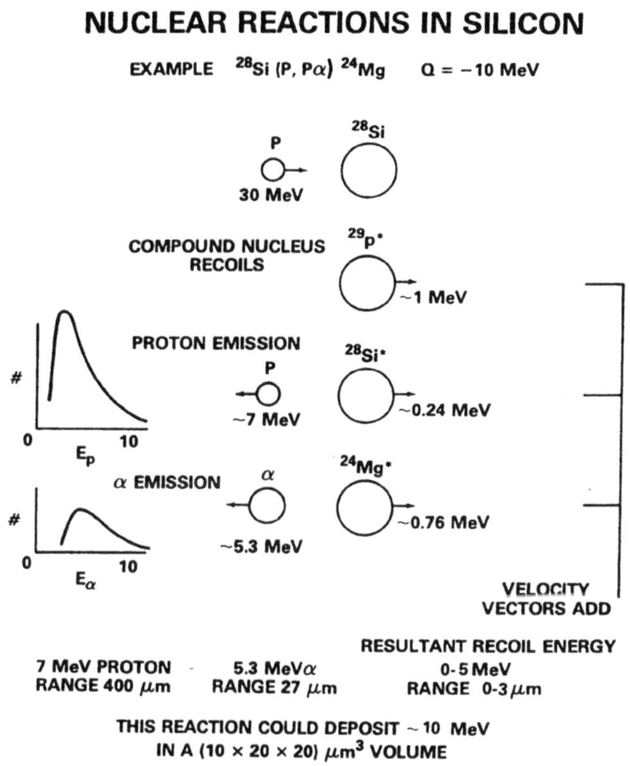

Fig. 20: Example of a proton reaction in silicon leading to a localised energy deposition.

3.3.4. *Induced Radioactivity* results from the wide range of radioactive spallation products and leads to sensitivity-limiting background counts in gamma-ray detectors deployed in space for astrophysical observations, planetary spectroscopy or the detection of hostile objects (see [Ref. 18] for a number of papers on this topic). Calculations of such backgrounds involve the summation of the contributions of many nuclides (approximately 100 for sodium iodide scintillators [Ref. 19]).

4. RADIATION TRANSPORT CODES

The above discussions point to the complexity of radiation interactions with matter and the importance of radiation scattering and the production of secondary particles, for example X-rays from electron bremsstrahlung and neutrons from proton reactions. In addition there is cross-coupling between particle types, for example proton reactions produce mesons which in turn lead to electron-gamma showers. Adequate treatment of radiation effects and shielding requires the development of powerful computer codes embracing all the relevant physics. A schematic of the requirements for complete treatment of space radiation is given in Fig. 21.

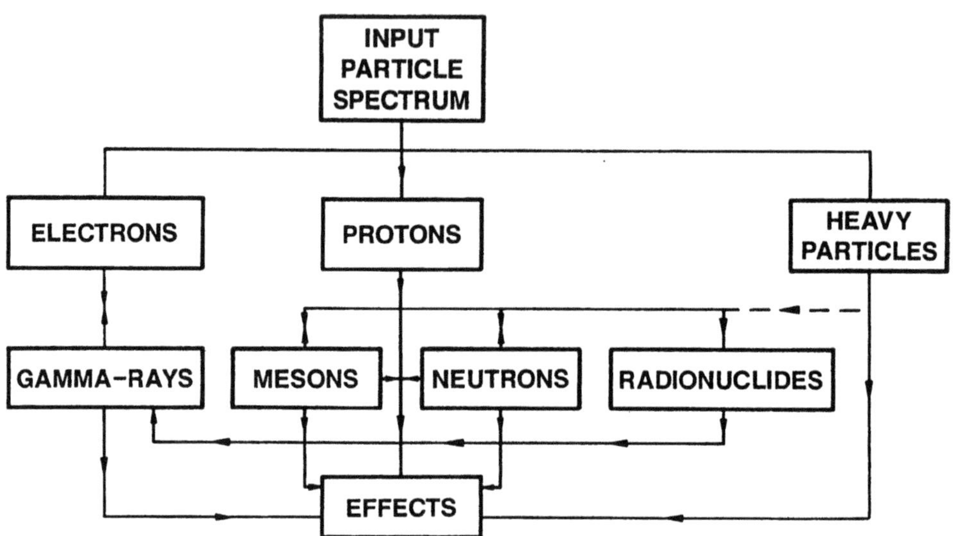

Fig. 21: Schematic of the radiation transport codes required to treat space radiation effects.

4.1. MONTE CARLO CODES

These employ sampling of probability distributions by means of random number generation to simulate the physical processes. They are required for the full treatment of complex geometries, anisotropies, simultaneous events etc, but have the disadvantage of long computer run times to give adequate statistical accuracy in the results. Examples commonly used in the space business are ETRAN [Ref. 20] for electron-gamma transport and the Integrated Radiation Transport Suite (IRTS) as operated at DRA Farnborough. The latter is a suite of three codes (LHI, MORSE and EGS) modified to share a common geometry description and output format. These are capable of covering most of the physics expressed in Fig. 21 with the exception of ions heavier than atomic number 10.

4.1.1. *The Light Heavy Ion* (LHI) code [Ref. 21] is an extension of the Oak Ridge High Energy Transport Code [Ref. 22] and uses an extended version of

Bertini's intranuclear cascade model [Ref. 15] to cover the interactions of nucleons, mesons, and light compound nuclei (A<10) with atomic nuclei of all elements. Protons are transported down to 15 MeV by which energy their residual range is negligible. Neutrons may be produced by the intranuclear cascade or evaporation processes. Those produced at less than 15 MeV are followed by the MORSE code. Fig. 22 shows a schematic of the input and output to the code including interface with space environment codes which generate particle spectra and anisotropies.

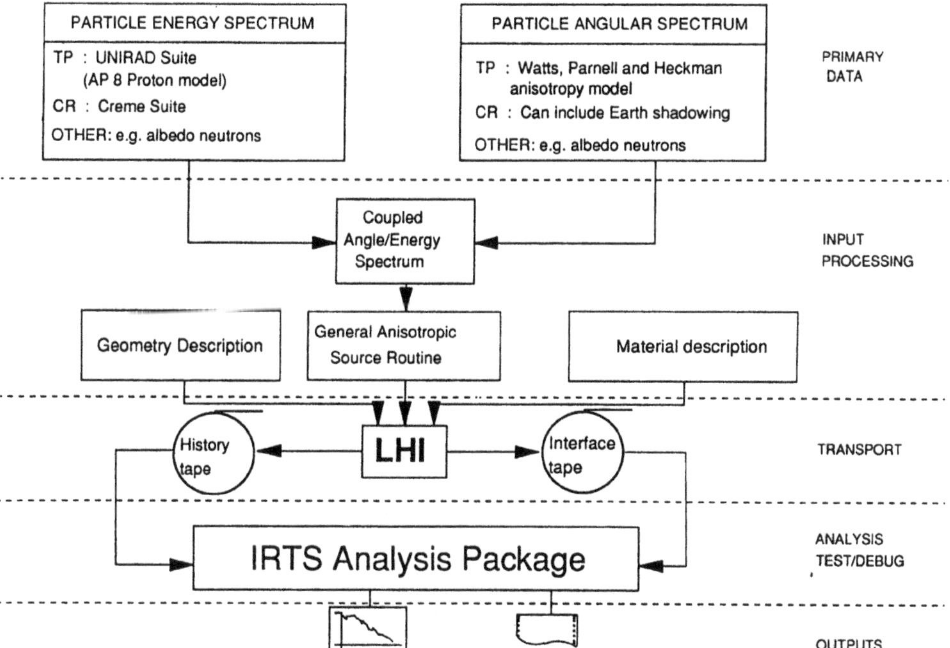

Fig. 22: Schematic interface between the Light Heavy Ion Code (LHI), space environment codes and the Integrated Radiation Transport Suite.

4.1.2. *MORSE* is the Multigroup Oak Ridge Stochastic Experiment [Ref. 23] which is used within IRTS to transport low energy neutrons only, the gamma rays being written to an interface tape for subsequent transport in EGS. The primary limitation on the utilisation of MORSE is the availability of cross-sections for the materials of interest.

4.1.3. *EGS* is the Electron Gamma Shower code [Ref. 24] which models the interactions of electrons, positrons and photons in the energy regime from several thousand GeV down to 10 keV for particles (1 keV for photons).

Interaction cross-sections for any material may be generated by the material preprocessor program PEGS. Interface may be made with gamma rays generated by MORSE or LHI (from photonuclear de-excitation or meson decay).

4.2. DETERMINISTIC CODES

These are used for faster engineering treatment when certain approximations are valid. An example is SHIELDOSE [Ref. 25] which uses ETRAN results in semi-infinite slabs to generate dose-depth curves for general electron spectra and for a limited range of geometries (slab, spherical or cylindrical). For protons and neutrons the Baryon Transport Model (BRYNTRN) uses parameterisation of secondary yields and the straight-ahead approximation to generate doses and fluences as a function of shielding depth [Ref. 16].

4.3. EXAMPLE STUDY FOR THE GAMMA RAY OBSERVATORY

A particularly challenging example for the IRTS is the NASA Gamma Ray Observatory launched in April 1991. At seventeen tons this is the heaviest scientific spacecraft ever orbited and materials include a large portion of high-Z materials such as sodium iodide and tungsten used in gamma ray detectors. Many particle paths therefore involve several particle interaction lengths through elements having a high multiplicity for secondary production. Simulations have been performed to determine the radioactivity induced the Oriented Scintillation Spectrometer Experiment [Refs. 26, 27] which comprises four identical telescopes, each with a 13" x 4" central NaI crystal actively shielded by CsI and NaI and collimated by tungsten. The Monte Carlo simulations show that particle cascades enhance the background over that due to primaries by a factor 5 for inner-belt protons and by a factor of 19 for the more energetic cosmic rays. Such simulations allow event tagging and these have been used to show that an important component of background arising from the neutron-capture product I-128 is predominately caused by neutrons produced within the same detector and is not amenable to shielding. The influence of particle anisotropies in the South Atlantic Anomaly has been included and shows that the relative orientation of spacecraft and detector can make a factor 2 to 4 difference in the induced radioactivity. The source term for spallation products has been combined with the response functions of seventy radionuclides in NaI, as computed using photon transport codes, to give a preflight prediction of the background spectrum resulting from cosmic rays. This gives very good agreement with flight data (Fig. 23) and provides encouragement in the use of these complex codes.

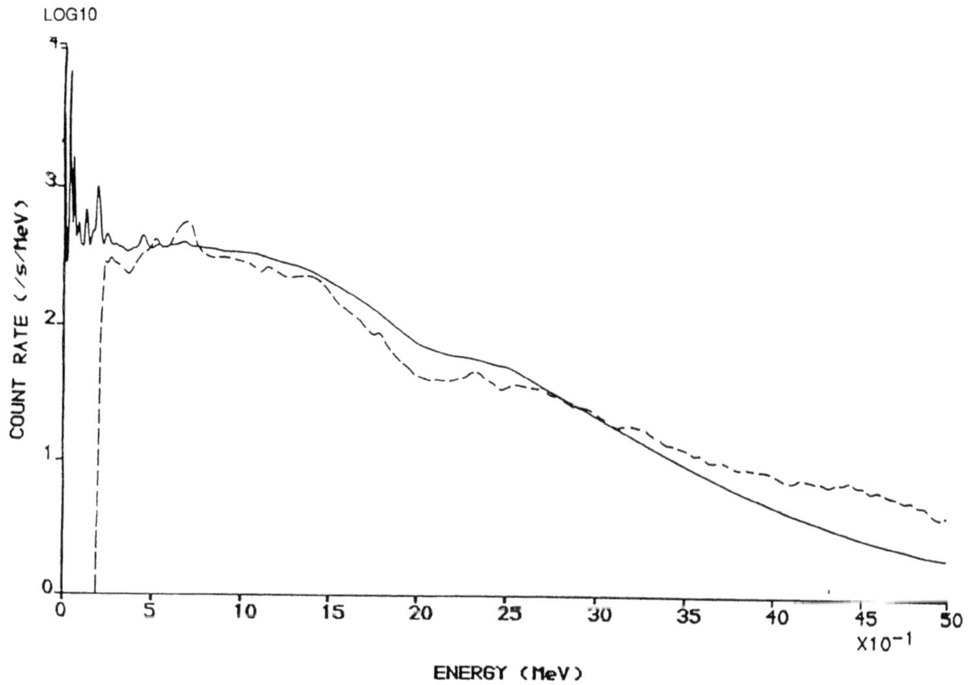

Fig. 23: The background spectrum observed in the GRO/OSSE instrument after 12 days in orbit is compared with a prelaunch prediction using the IRTS codes.

REFERENCES

[1] Mathews, Titus "Cosmic radiation and energetic solar flare particles", These proceedings (1991).

[2] Stephen, J., "Hazards to electronics", These proceedings (1991).

[3] Barkas, Walter H., and Berger, Martin J., "Tables of energy losses and ranges of heavy charged particles", NASA SP-3013 (1964).

[4] Ziegler, J. F., "Handbook of stopping cross-sections for energetic ions in all elements," Pergamon Press (1980).

[5] Berger, Martin J., and Seltzer, Stephen M., "Tables of energy losses and ranges of electrons and positrons", NASA SP-3012 (1964).

[6] Haffner, James W., Radiation and Shielding in Space. Nuclear Science and Technology, Vol 4. Academic Press New York and London (1967).

[7] Van Lint, V. A. J., Flanagan, T.M., Leadon, R.E., Naber, J.A., and Rogers, V.C., Mechanisms of Radiation Effects in Electronic Materials, Vol 1., John Wiley and Sons, New York (1980).

[8] Product Assurance & Safety Department, European Space Research and Technology Centre, The Radiation Design Handbook, ESA PSS-01-609 (1989).

[9] Bourrieau, J., "Protection and shielding," These proceedings (1991).

[10] Srour, Joseph R., "Basic mechanisms of radiation effects on electronic materials, devices and integrated circuits," Tutorial Short Course, IEEE 1983 Nuclear and Space Radiation Effects Conference (1983).

[11] Mclean, Barry F., and Oldham, Timothy R., "Basic mechanisms of radiation effects in electronic materials and devices," Tutorial Short Course, IEEE 1987 Nuclear and Space Radiation Effects Conference (1987).

[12] Dale, Cheryl J., Marshall, P.W., Burke, E.A., Summers, G.P., and Wolicki, E.A., "High energy electron induced displacement damage in silicon," IEEE Trans. Nucl. Sci., NS-35, 6, 1208-1214 (1988).

[13] Measday, D.F., and Richard-Serre, C. "Loss of protons by nuclear reactions in various materials," CERN 69-17 (1969).

[14] Silberberg, R., and Tsao, C.H., "Partial cross-sections in high-energy nuclear reactions, and astrophysical applications," Astrophysical Journal Supp., 220, 315-368, (1973).

[15] Bertini, H.W., Guthrie, M.P., and Hermann, O.W., ORNL-RSIC CCC-156, (1971).

[16] Wilson, John W., et al, "BRYNTRN: a baryon transport model," NASA TP-2887 (1989).

[17] Petersen, Edward L., "Single event upsets in space: basic concepts," Tutorial Short Course, IEEE 1983 Nuclear and Space Radiation Effects Conference (1983).

[18] Rester, A.C., and Trombka, J.I. (eds) Proceedings of the Conference on High Energy Radiation Background in Space, AIP Conference Proceedings 186, (1989).

[19] Dyer, Clive S., Trombka, J.I., Seltzer, S.M., and Evans, L.G., "Calculation of radioactivity induced in gamma-ray detectors during spaceflight," Nuclear Instruments and Methods, 173, 585-601, (1980).

[20] Seltzer, S.M., "An overview of ETRAN Monte Carlo methods of coupled electron/photon transport calculations," in Proceedings of the Course on Monte Carlo Transport of Electrons and Photons Below 50 MeV, International School of Radiation Damage and Protection, Ettore Majorana Centre for Scientific Culture, Erice, Italy, Plenum Press, (1987).

[21] Armstrong, Tony W., and Colborn, Bonnie L., "A thick-target radiation transport computer code for low-mass heavy-ion beams," Nuclear Instruments and Methods, 169, 161-172 (1980).

[22] Armstrong, Tony W., and Chandler, K.C., Nucl. Sci. Eng., 44, 104, (1971).

[23] Emmett, M.B., ORNL-4972, (1975).

[24] Nelson, W.R., Hirayama, H., and Rogers, SLAC-265, (1985).

[25] Seltzer, S.M., IEEE Trans. Nucl. Sci., NS-26, 4896, (1979).

[26] Johnson, W.N., et al, "The oriented scintillation spectrometer experiment (OSSE): instrument description," in Proceedings of the Gamma Ray Observatory Science Workshop, NASA/GSFC, W.N. Johnson (ed), 2-22 to 2-38, (1989).

[27] Dyer, C.S., Truscott, P.R., Sims, A.J., Comber, C., and Hammond, N.D.A., "Predictions of radiation backgrounds for GRO/OSSE," in Proceedings of the Gamma Ray Observatory Workshop, NASA/GSFC, W.N. Johnson (ed), 4-521 to 4-528, (1989).

RADIATION DAMAGE TO SURFACE AND STRUCTURE MATERIALS

ALAIN PAILLOUS
CERT/DERTS
2, Avenue Edouard Belin
31055 Toulouse Cedex, France

ABSTRACT. Spacecraft materials are subjected to the various components of the space environment (ultraviolet radiation, electron and proton fluxes, vacuum and residual atmospheric species, thermal cycling...). They can be degraded severely in space: changes in thermo-optical properties, decrease of mechanical performance (cracking, delamination, variation of thermal expansion...), etc. The total mission times are expanding, currently exceeding ten years, thus requiring prediction of the long-term performance of materials. The understanding of basic radiation damage processes is essential in order to enable one to select the simulation techniques to be used for that purpose. Damage mechanisms for polymeric materials, paints, and optical surfaces is summarized. Some important concerns for establishing effective and valid accelerated irradiation test procedures are addressed: does rate, equivalence of effects, recoveries... Special attention is paid to some synergistic effects possible in space conditions: a few examples of synergy between UV and ionizing particles, contamination and radiation, radiation and thermal cycling are presented. Some results of accelerated testing on materials employing combined exposure conditions are also given for organic and inorganic base paints, polymeric films, and composite materials.

1. Introduction

Spacecraft materials are submitted to the various components of the space environment (vacuum and residual atmospheric species, thermal cycling, ultraviolet radiation, electron and proton fluxes). Due to the remote operation of spacecraft for ever growing periods of time (ten years for current missions, up to thirty years for the future Space Stations), it is mandatory to predict the long-term performance of materials to be used on board during a specific mission. The purpose of this paper is to define the ultraviolet and high-energy particle radiation environment, to synopsize its impact on the stability of thermal control coatings and structural materials, and to give some attention to some problems of the space simulations techniques to be employed.

R. N. DeWitt et al. (eds.), The Behavior of Systems in the Space Environment, 383–405.

2. Review of Experimental Work

A detailed description of the space radiation environment is the subject of other lectures. The reader is referred for more information on the subject of UV radiation and high energy particle fluxes encountered in Space, to other papers of this series of ASI lectures. We can only summarize a few features.

The total electromagnetic intensity of the sun emission is 1356 W m^{-2} at 1 astronomical unit and at air mass zero (AMO). 99.9% of the energy is emitted between 0.1 and 10 μm, with a maximum at approximately 0.45 μm. Considering the material damage aspects, the UV part of the sun emission is the only which is to be considered; therefore only radiations with wavelengths less than or equal to 0.4 μm and corresponding to only 118 W m^{-2} are of concern. The UV radiation is often divided into: (a) the near-UV region from 0.2 to 0.4 μm (this is corresponding to 6.2 to 3.1 eV for the energy of photons), and the vacuum-UV region below 0.2 μm. Although the intensity of the vacuum-UV is negligible (0.11 W m^{-2}) compared to the total UV intensity, one must consider that it has potential effects due to the complete absorption in all materials and the very high energy of photons. "UV equivalent sun-hours" (UV-esh) are generally used in order to define the solar exposure[1].

The high energy particle environment is highly complex; the composition, the energy and fluxes vary both locally and with the time (namely during the solar cycle). Several components must be distinguished:

- cosmic galactic radiation; 90% protons, He^{2+} and heavy ions; energies ranking from 10^{-2} to 10^{10} GeV; very low fluxes,
- solar wind; 96% protons, He^{2+} oxygen ions, electrons; approximately 1 keV for protons, 1 eV for electrons; 2×10^{8} protons cm^{-2} s^{-1},
- solar flare ions; 95% protons, ions; 1 to 100 MeV; highly variable fluxes,
- earth-trapped protons and electrons; proton energies <30 MeV in inner belt, <1 MeV in outer belt; electron energies <5 MeV; fluxes depending on altitude, magnetic latitude, solar activity (specially at GEO), longitude (South-Atlantic anomaly),
- auroral fluxes; protons and electrons; proton energies ranking from 80 to 800 keV; electron energies ranking from 2 to 20 keV; fluxes depending on solar activity.

The exact particle fluxes encountered during one mission depend on the trajectory or orbit, and on time (sun activity). From the point of view of possible damage to materials, in the Geosynchronous Earth Orbit (GEO), one must consider more specially the trapped electrons and protons, and also some particles variable with the sun activity (low energy electrons during substorms, high energy protons). In Low-Earth orbits (LEO), the major radiations are those from electrons and protons trapped in the belt; but in high inclination orbits, low energy electrons can be

---

[1] 1 esh is equal to 118 (W m^{-2}) x 3600 (s), i.e., 4.25 10^{5} J m^{-2}.

received with variable fluxes near the poles during magnetic substorms. Beyond the earth's magnetosphere, the solar wind effects become predominant, but a possible effect of the solar particle events should be weighed also. More details are given in the lecture of J. Bourrieau "Doses and Shielding".

3. The Degradation by Ultraviolet and high Energy Radiations

3.1. MECHANISTIC ASPECTS

Electromagnetic and particle radiations constitute predominant factors in the ageing process of materials in space. Several categories of radiation have to be considered: (i) the solar ultraviolet radiation (i.e., electromagnetic radiation with wavelength $\lambda < 400$ nm), (ii) the energetic electrons, protons and ions which can be encountered in the various zones of space, and (iii) the bremstrahlung radiation induced by interaction of the high energy electrons with materials on board spacecraft.

From the point of view of their chemical action, it is often simpler to consider only two radiation types corresponding to two different energy ranges of photons or particles: (a) the UV radiation, and (b) the so-called "ionizing" radiations. The first type induces photo-chemical reactions in organic materials; the second is associated with high energy radiation chemistry, because the energies taken into consideration ($E > 15$ eV) are superior to those necessitated for the formation of ions from molecules - see Fig. 1.

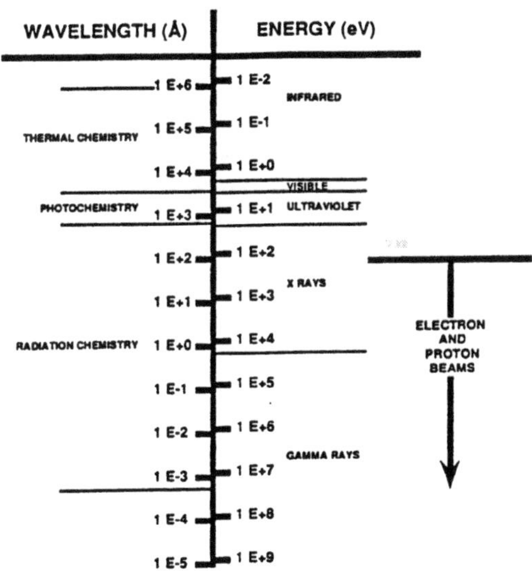

Fig. 1: Scale of radiation energy.

3.1.1. *Primary reactive species created by radiation.*

3.1.1.1. *The case of Ultraviolet Radiation.* Upon UV irradiation, energy is transferred to the molecules or atoms. This transfer is insufficient to allow ionization (that is to say to remove one electron from the molecule), but is enough to allow a transition of electrons from the so-called "ground-state" energy level to an upper "excited" level. The excited states which can be obtained have a life time, an energy and a well-defined structure; diagrammatically:

$$AB \rightarrow AB^*$$

During this excitation process, an electron passes from a bonding orbital (n, σ, π) to an antibonding orbital (σ^*, π^*)[2]. As a general rule, UV radiation can only produce: (a) transitions which take place without change in multiplicity, (b) transitions obeying certain rules of symetry. Excited states of multiplicity different from that of the ground-state can be found in the matter irradiated by UV; in such a case an excited state is obtained in the first place with the same multiplicity as the ground state, then an isoenergetic transition allows an inter-system crossing with a change in multiplicity. Such a process is described in Fig. 2 which shows that the absorption of a photon excites the molecule from its ground state S_0 to the single state S_1 which, in turn, is radiationless converted to the triplet state T_1.

3.1.1.2 *The Case of Ionizing Radiations.* During the interaction of a charged particle with matter, the molecules which are close to the trajectory can be excited or ionized [Refs. 3, 7].

In the first case, the charged particles which move rapidly in the material can only produce optically allowed transitions, that is to say, transitions without a change in multiplicity. On the other hand, particles moving slowly, and electrons in particular, can give triplet states directly from a singlet ground-state; in addition the neutralization of the positive ions (see below) by electrons in the presence of other ions leads to excited singlets and triplets.

In the case of ionisation $AB \rightarrow AB^+ + e-$, the electron liberated as such by ionization can have in itself sufficient energy to leave the zone of primary ionization, and therefore can produce ulterior ionizations and excitations throughout the course of its own trajectory. Electron δ is such a case (see Fig. 3).

---

[2] Let us recapitulate that an orbital defines the spatial distribution of an electron for a given state of energy. For example in a diatomic molecule, the presence of electrons between nuclei makes the chemical bond possible: the bonding orbitals are characterized by the presence of electrons between the nuclei under consideration, whereas the anti-bonding orbitals are characterized by a weak electronic cloud density between the nuclei (structure with low stability).

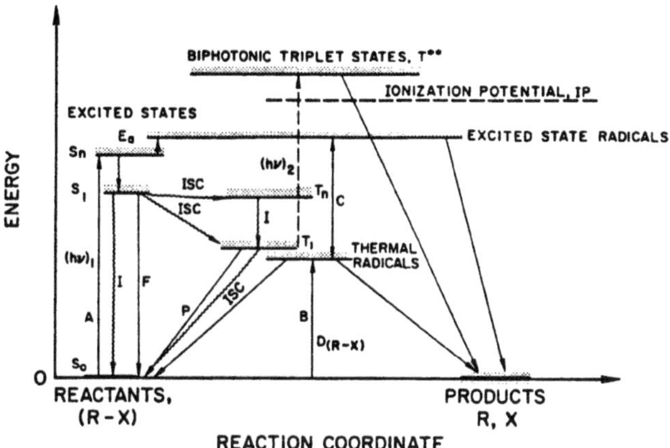

Fig. 2: Generalized energy level diagram (From [Ref. 44]).

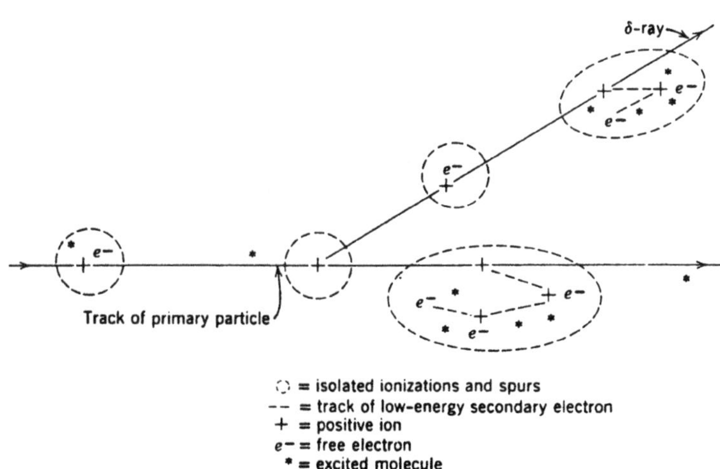

Fig. 3: Distribution of ions and excited molecules in the track of a fast electron (From [Ref. 3]).

High energy electromagnetic radiations (for instance the γ-radiation generated in the bremstrahlung process) produce a similar effect because the absorbed energy is transferred to the secondary electrons and positions which dissipate it during their trajectories. The nature of the created species (ionic or excited) responsible for ulterior chemical evolutions is essentially the same, no matter which type of radiation is used (X, γ, electrons and protons). It is not at all the same spurs. These spurs are very close together along a proton (or heavy ion) track forming a continuous column of ions and excited species; as a consequence of that, energy is quickly transferred to the material and the penetration range is short. In the case of electron irradiation, the spurs are widely spread; the range is far larger and less definite because of the large scattering of δ-rays. Typical penetration ranges for 1 MeV particles are one millimeter for electrons and only 20 μm for protons.

The ratio of the energy lost by the particles in ionizing and exciting the matter is approximately equally divided between these two processes in organic polymeric materials [Ref. 4]. It must be emphasized in addition, that heavy particles (including the protons), besides, can provoke atomic displacements in ordered solids (metals, ionic crystals, glasses) resulting in a vacancy at the initial position and an interstitial atom in a position not occupied in the original lattice.

3.1.2. *Ulterior evolution of ions and excited species.* The excited molecules and ions created during the primary action are unstable and reactive: they can react among themselves or with the surrounding molecules to produce radical products[3] [Ref. 7] which are the most important intermediaries in reactions affecting the materials and sometimes give rise to new chemical groups and molecular products (see an example in Fig. 2). These radicals are often involved in chain reactions. In polymeric materials each reactive intermediary or radical created, can diffuse around the point of creation in a restricted volume (of several atomic distances). Such a diffusion is allowed by the free-motion of molecular segments, and the affected volume depends on the irradiation temperature, the cristallinity of the polymer and the existence of a glass transition. A schematic diagram showing some possible reaction mechanisms in matter irradiated with UV and high-energy charged particles is depicted in Fig. 4.

As to the physico-chemical alterations sustained by the irradiated material, one can say that they depend directly on the quantity of the intermediate reactive species that are created [Refs. 4, 5, 6]. This quantity is in itself a direct function of the quantity of energy lost by radiation in the matter, that is to say, the absorbed energy dose (often expressed in rads or in Grays[4]). In reality the distribution of the absorbed energy dose is never uniform.

---

[3] Free radicals are atoms or molecules having one available electron which permits an ulterior chemical bond formation. For example they are produced when a molecule is split at the level of a covalent bond, and in such a way that each fragment of this molecule carries away one of the electrons which participated in this bond: $R - S \leftrightarrow R\cdot + S\cdot$

[4] 1 Gray (Gy) = 1 J kg^{-1}; 1 rad = 100 ergs g^{-1}.

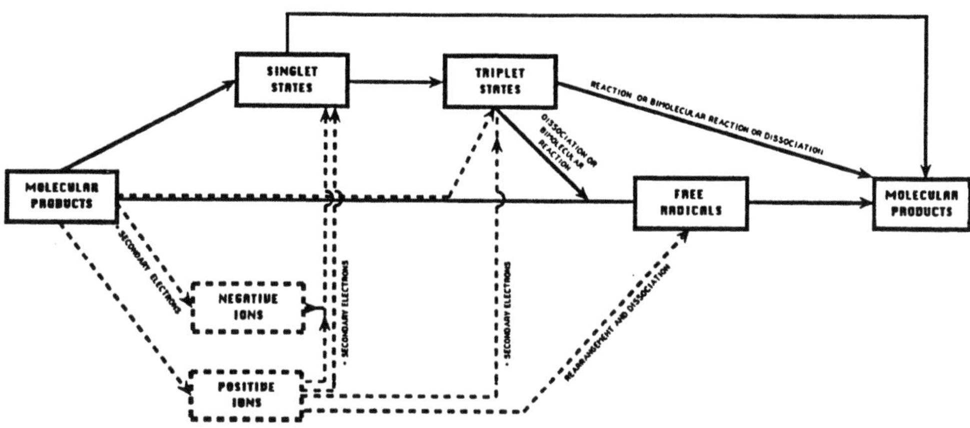

Fig. 4: Reactions of ions, excited molecules and free-radicals. The reactions common to Photochemistry and High-energy Radiation Chemistry are shown in heavy lines.

3.1.3. *Chemical changes*. The main molecular changes induced in polymers as the final step of irradiation processes can be divided into two categories: crosslinking and scissions [Refs. 4, 5, 7]. Crosslinking is a process whereby two separate long chain molecules can be linked together by recombination of two radicals formed on these molecules; an increase in the average molecular weight is thus observed. Scission is a process in which the polymer suffers a breaking of the macromolecular chain; this involves a reduction of the average molecular weight. Typical reaction schemes are shown in Fig. 5. On most of the polymers currently used in space technology, both cross-linking and degradation occur simultaneously; however, in certain cases depending on the polymer structure, either the crosslinking or the degradation can be the dominant feature of the irradiation mechanism [Ref. 5].

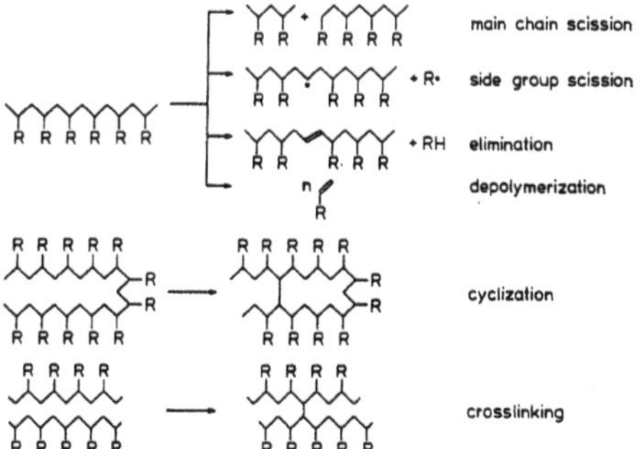

Fig. 5: Scission and crosslinking of macromolecular chains (From [Ref. 8]).

3.2 EFFECTS OF IRRADIATION

Various effects of irradiation, which are of importance concerning spacecraft materials, are depicted in Fig. 6 (adapted from J. Dauphin [Ref. 9]).

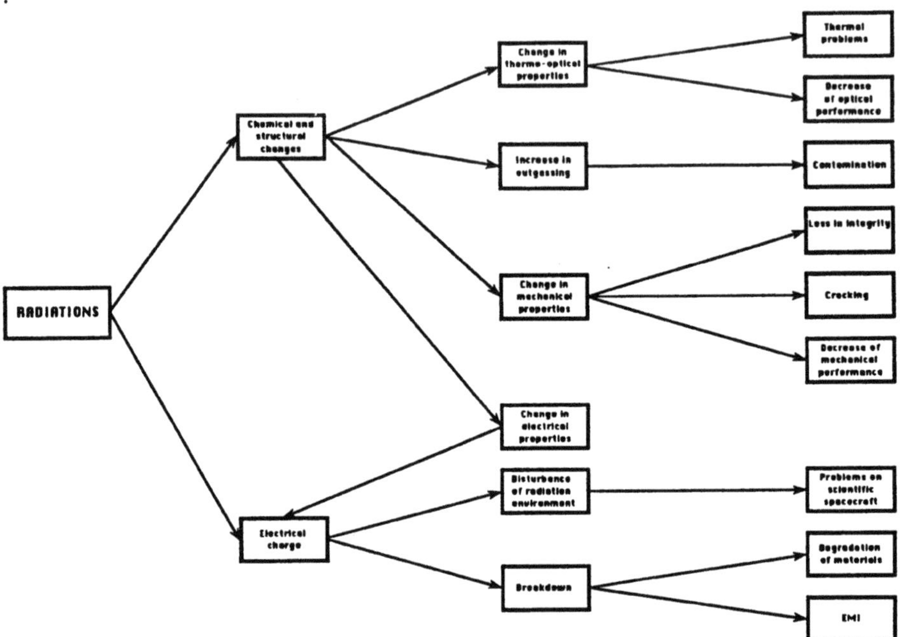

Fig. 6: Effects of irradiations on spacecraft materials (Adapted from [Ref. 9]).

The degradation of mechanical properties of polymers results from a competition between scission and crosslinking. Chain scission usually leads to a degradation of properties such as tensile strength and elasticity; crosslinking can read to increase in ultimate tensile strength and embrittlement; net changes in properties depend upon the relative importance of these processes [Refs. 10, 11]. When chain scissions occur, low molecular weight products are formed in the material; under vacuum these volatile products can be released, becoming potential contaminants for spacecraft vulnerable coatings and optics onto which they can be subsequently condensed [Ref. 12].

The formation of blisters on metallic surfaces bombarded with protons has been reported; it is highly temperature-dependent. This phenomenon can be attributed to the agglomeration of hydrogen gas at the lattice imperfections, grain boundaries and vacuum deposited film interfaces: such agglomerations which are obtained for fluences exceeding $5x10^{16}$ protons cm^{-2}, occur near the predicted penetration depth of protons [Ref. 23].

Optical absorption increases are generally observed as the result of irradiations performed under vacuum. Color centers are formed in ionic crystals, glasses and metal oxide pigments. These centers are due to lattice defects [Ref. 14] (see Fig. 7); the absorption of optical photons by these defects can be attributed to a possibility of excitation of a trapped charge (electron or hole) from its ground state to a higher excited state. This can only occur for a limited frequency range for any specific center; therefore, the optical absorption spectrum of a crystal containing these defects will show an absorption band which is characteristic of that particular center. Most polymeric resins and films also exhibit pertinent color changes due to the formation of chromophoric groups (like conjugated insaturation) or the trapping of free radicals [Refs. 3, 13, 14].

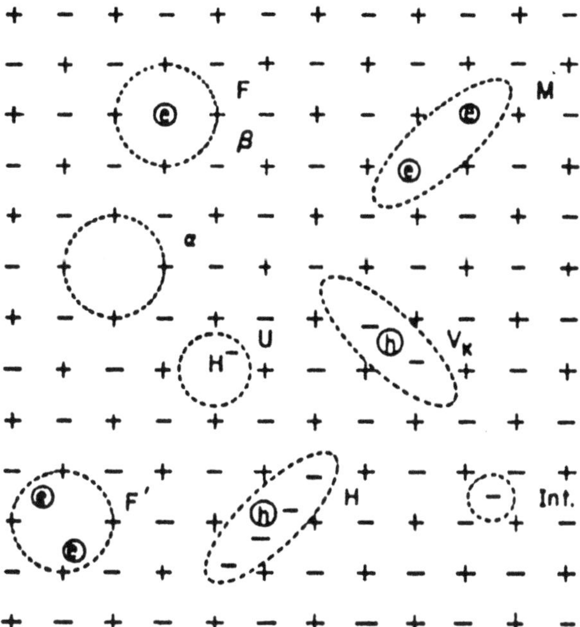

Fig. 7: Schematic representation of some of the prominent colour centres in ionic crystals (From [Ref. 14]).

Lastly, some important surface or bulk electrical conductivity changes are often experienced. They can have some implication on spacecraft charging which is a complex process in which electrical charges, imparted to satellite surfaces by energetic protons and electrons, play also a role. Spacecraft charging is discussed later during this Symposium. One can note that materials can be degraded by bulk electrical breakdown or surface discharge treeing when bombarded with charged particles [Refs. 24, 35].

4. The Simulation of the Space Radiation environment.

4.1 SOME PROBLEMS

4.1.1. *Vacuum irradiations and in-situ measurements.*

There is much evidence that the presence of oxygen molecules strongly modifies the effects of radiation in organic materials [Refs. 3, 15]. The free radicals (and also some triplet states) created are quickly peroxidized: **R· → R – O – O·**. Sometimes radical recombinations which could have lead to crosslink are prevented by the rapid peroxidation reactions which take place [Refs. 3, 5]. The peroxides are chemically active, often involved in chain mechanisms; therefore, there are important changes in both the molecular weight and the molecular branching, as compared with those obtained upon irradiation carried out under vacuum [Ref. 5]. Moreover, a variety of hydroxyl and carbonyl compounds are obtained together with peroxides and other oxidation products upon irradiation in air. Also, many free radicals remain trapped in the bulk of a material irradiated under vacuum. When re-exposed to air post-irradiation, they undergo similar oxidative reactions with products differing in abundance from the products of irradiation in air, but similar in kind, i.e. ketones and alcohols.

An instantaneous bleaching of the radiation-induced coloration is also observed quite often when vacuum-irradiated white paints are exposed to air. This bleaching is due to the reabsorption of oxygen at the pigment interface (during vacuum irradiation, oxygen is removed by photodesorption, leaving oxygen vacancies in which electrons can be trapped, providing optical absorption) [Refs. 13, 16-18]. An example of the air recovery of a paint is given in Fig. 8.

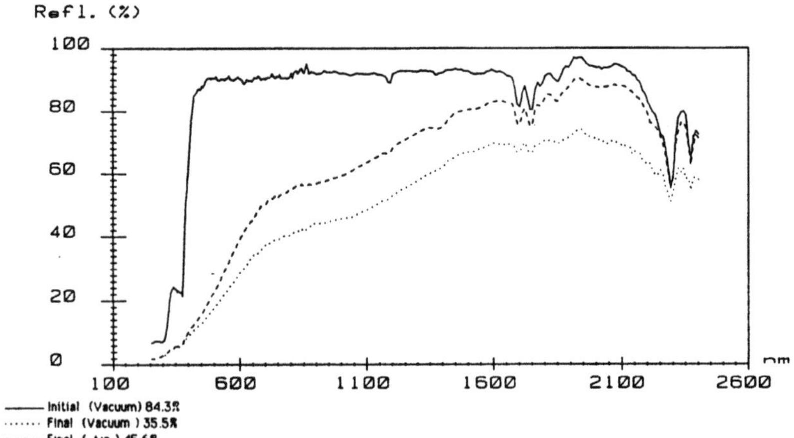

Fig. 8: Air recovery of the SG11 FD white paint irradiated in vacuum with electrons protons and UV. The irradiation conditions were equivalent to 7 years in GEO.

There are some evidences to suggest that return of irradiated specimens to air affect the mechanical properties of polymeric films or composites which have been irradiated in vacuum [Refs. 12, 19-22]. This makes *in vacuo* irradiations as well as *in situ* measurements unquestionably compulsory when space simulation irradiations are made.

4.1.2 *Specificity and synergy of UV and particle irradiations.*

It is apparent from Fig. 4 that the species which play a role during radiolysis and photolysis are qualitatively and quantitatively different. Photolysis only gives rise to a few of the general chemical mechanisms which can be contemplated for the high-energy radiations. Furthermore the UV radiation in principle is more selective in its attack because of the transfer of photon energy to specific chemical groups corresponding to optical absorption bands: a monochromatic source of light can provoke a unique and well defined excited state [Ref. 3]. Also, the defect spurs (specially those densely packed along the proton tracks), which are created by particle radiations, are not produced under UV radiation leading to defects widely spread and uniformly distributed in the planes perpendicular to the light beam.

One therefore expects different effects from radiolysis and photolysis, even for comparable energy doses imparted to matter. Moreover, simultaneous ultraviolet and ionizing radiations (as expected in space environment) can modify the reaction processes which could be followed during an irradiation using only one type of irradiation, or two types of irradiations sequentially carried out. In simultaneous conditions, the additivity of the effects is not observed. Such phenomena may be specially important at low temperature where there is a greater stabilisation of trapped species [Ref. 10]. Several examples of discrepancies between the results of combined irradiations have been reported for polymers [Refs. 10, 12, 24, 25]. For instance [Ref. 25], the polysulfone P1700 shows significantly more effects in sequential irradiations than in a simultaneous one (Fig. 9); in particular, the effect in electrons followed by UV is more than twice that of simultaneous irradiations. The difference likely has to do with radicals from the previous irradiation present at the start of the second; the reason for the large difference between UV/electrons versus electrons/UV is not understood. Lexan℠ polycarbonate shows little effects from UV irradiation; however, it suffers an apparent synergy effect when it undergoes simultaneously 85 keV electron and UV radiation. For a Mylar℠ film protected by an UV inhibitor, the inhibitor is apparently degraded during electron irradiation; this is most readily seen (Fig. 10) when an electron test is followed by an UV irradiation, but it is also seen to a lesser extent in simultaneous irradiations.

Similar experiments have been reported for white paints [Refs. 13, 26]. Where charged particles can physically displace atoms from lattices, UV cannot. The two radiations produce effects which in most cases are dissimilar because of the atomic displacements. The charged particle damage appears to be far more complex and spectrally distributed than that due to UV radiation alone. The spectra from combined irradiation do not

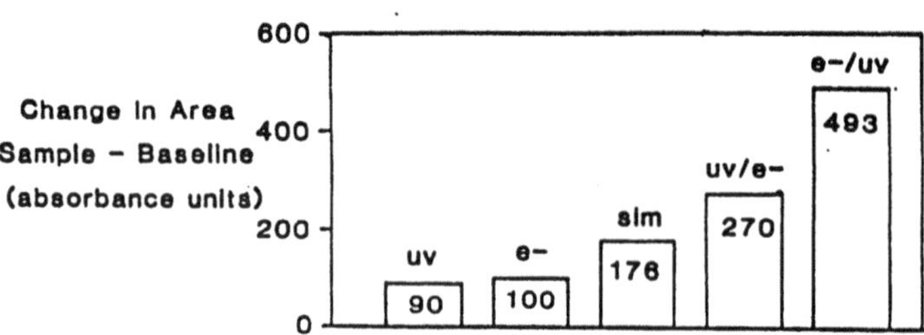

Fig. 9: Effects of simultaneous and sequential combined irradiations for P 1700 polysulfone irradiations (From [Ref. 25]).

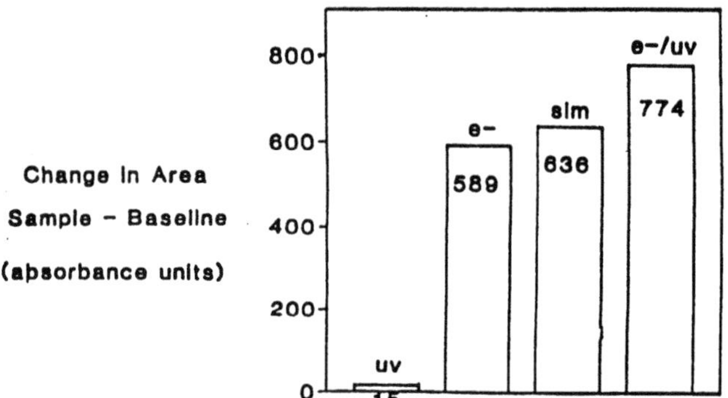

Fig. 10: Effects of simultaneous and sequential combined irradiations for Mylar irradiations (From [Ref. 25]).

resemble either [Ref. 13]: in some cases the total effect is less in a combined environment than in the individual environments only (Fig. 11). One explanation is that certain types of defects created by charged particles are radiatively unstable; in a combined environment they are bleached by the UV radiation.

This behavior has been ascertained several times in combined irradiations performed at DERTS: for various white paints and polymers[5], a partial damage recovery is observed when an ultraviolet irradiation step follows either a particle irradiation step or a simultaneous particle/UV step; the solar reflectance increases in comparizon with the value reached just after the particle irradiation (see Figs. 12 and 13). This increase is noticed for at least 20 days under vacuum. Samples irradiated by particles and only exposed to vacuum do not exhibit reflectance changes as confirmed by a 15-day exposure to vacuum of all samples after the last particle irradiation step [Refs. 27, 28].

---

[5] Polyimide KAPTON; White paints S13G/LO, PSG120, Z93, PSZ 184 (see also [Ref. 29])

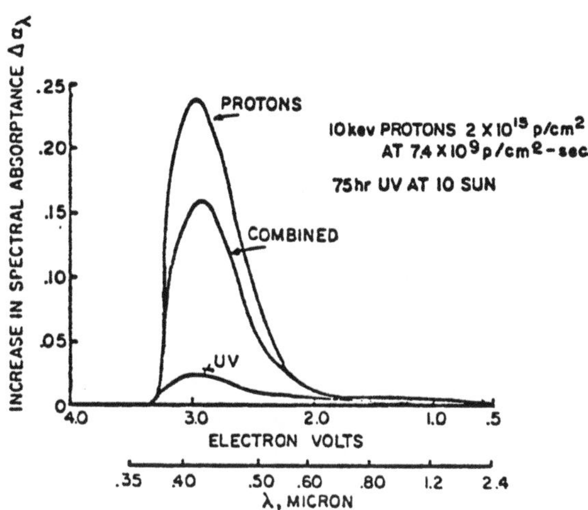

Fig. 11: Increase in spectral absorptance of $ZnO-K_2SiO_3$ coatings exposed to combined and individual environments at 298°K (From [Ref. 13]).

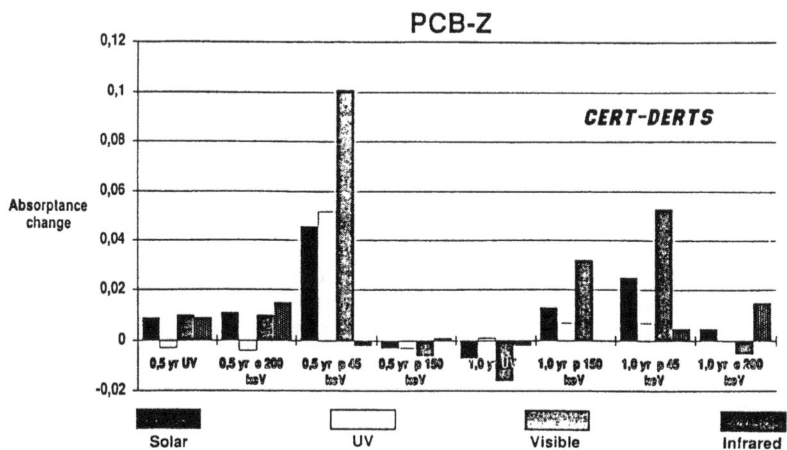

Fig. 12: Effects of sequential irradiations on the solar absorptance of a PCB-Z white paint; the irradiation conditions where those for the north and south faces of a GEO satellite.

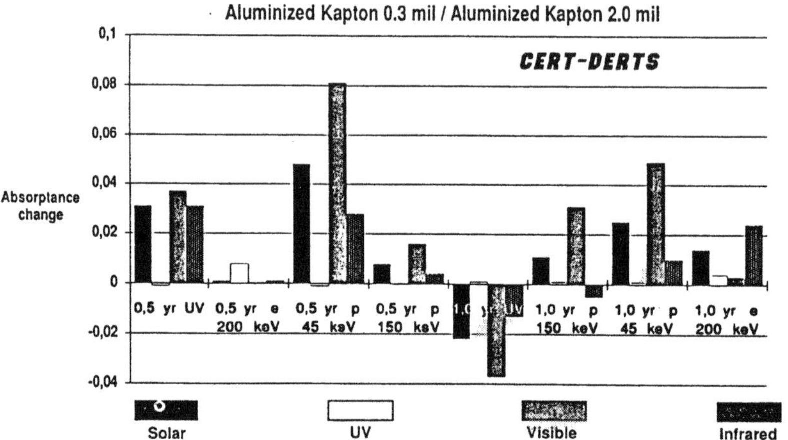

Fig. 13: Effects of sequential irradiations on the solar absorptance of an aluminized Kapton SSM; the irradiation conditions where those for the north and south faces of a GEO satellite.

All these observations are strongly prompting to carry out simultaneous irradiations rather than sequential combinations. When UV irradiations take a longer time than particle irradiations, it seems advisable to end the irradiations with a UV illumination step in order to avoid an overestimate of damage.

4.1.3. *Dose rates.*

A general study of the validity of the reciprocity law (which implies that the effects are equivalent for a same dose whatever the does rate is) is still missing, specially when combined irradiations are performed. The prevailing idea is that the law is valid as long as irradiations with electrons alone (or irradiations with protons alone) are carried out within a ratio of two orders of magnitude in the dose rates. This is substantiated by electron irradiations performed at DERTS on a polyethyleneterephtalate film between 2×10^{10} and 10^{12} cm^{-2} s^{-1} at 600 keV for the same dose [Ref. 19], and experiments at Boeing between 4×10^{8} and 1.7×10^{12} cm^{-2} s^{-1}, on aluminized Kapton, several white paints and surface coatings [Refs. 29, 30]. This may be supported also by measurements on ZnO paints for protons between 10^{10} and 10^{12} cm^{-2} s^{-1} [Ref. 31]. A few anomalies which have been noticed can be explained by thermal effects when the cooling of the samples is not achieved adequately (see for instance [Ref. 32]).

Such parasitic thermal effects are much more difficult to surmount in the case of UV irradiations using high intensities of light. If an appropriate filtering of the visible and infrared parts of the Xenon sources used for solar simulation is achieved, the reciprocity law seems validated for irradiations performed between one and four "UV-suns[6]". Some studies carried out at DERTS on a few material types have shown that, in such experimental conditions, UV irradiations between 1.0 and 5.0 suns lead to similar thermal control coating degradations (33,34]. An example is given in Fig. 14 for a white paint (High Temperature APA 2474) irradiated up to 1000 equivalent sun hours. Comparable results have been obtained with a Chemglaze II white paint, a silvered FEP film with an ITO layer and a Betacloth fabric. For all these materials, the effect of the solar UV irradiation intensity seems to have only a minor effect on the overall solar absorptance changes which are computed from the spectral data obtained after UV irradiation. Of course, low dose rates (i.e., similar to those experienced on spacecraft) are expected to give the best results in space simulation experiments.

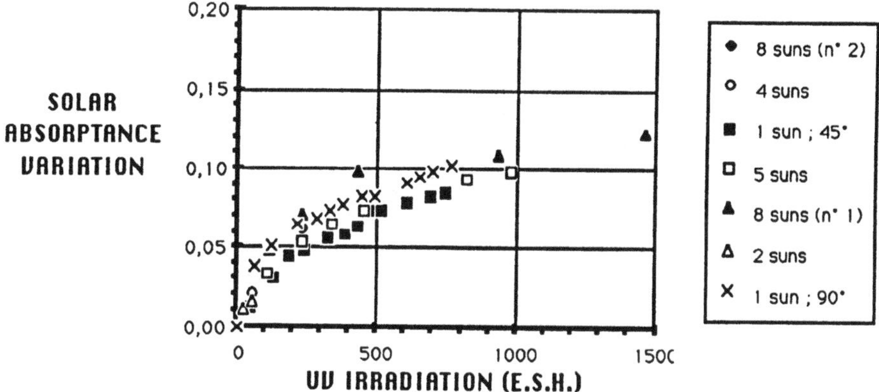

Fig. 14: Solar absorptance variation of an APA 2474 paint for various beam intensities of UV irradiations.

[6] One UV-sun corresponds to 11.8 mW cm^{-2} of the radiation emitted by the sun at wavelength shorter than 400 nm.

4.1.4. *Temperature*

For polymeric materials molecular chain motions are hindered for temperatures below the glass temperature transition T_g. Above T_g radical and/or ion recombinations are possible; they can modify the chemical reactions and therefore the degradation ultimately obtained. The color centers can be affected by temperature changes, also, (recombinations hole/electron, interstitial atom/vacancy,...) but these phenomena are also time-dependent. Temperature variations also are linked to electrical conductivity changes which can be of concern for material damage evaluation [Ref. 24]. The deterioration of a FEP film in an actual low temperature (-30°C) space environment will approximately occur at a rate 3-10 times that observed at elevated temperature (+50°C) [Ref. 10]. Some inconsistencies between laboratory and flight data (from LDEF), related to the degradation of the solar reflectance of several materials, could well be explained by the differences between the conditions of deep temperature cycling experienced in space and the fixed temperature chosen for standard laboratory irradiation tests [Ref. 36]. Of course thermal cycling and irradiations can act synergistically on bonded or composite materials which can suffer differential strains at the interfaces; the micro-cracking of fiber-reinforced organic composites is a well-known example of this problem [Ref. 37].

4.2 PRINCIPLES OF IRRADIATION TESTS

The experiments generally aim at reproducing the combined effects of vacuum, temperature, protons, electrons and UV radiation. They are expected to give information on the behaviour of materials as a function of the elapsed time in orbit.

The energies and fluxes of particle irradiations are selected so as to reproduce the energy dose distribution which is obtained within materials as a consequence of the exposure to space radiations. The principles of the evaluation of the absorbed energy doses within a material, from the flux particle models currently available, are described in one of these ASI lectures[7]. The doses have to be computed for each exact mission, taking into account the orbit, the time (duration, and also date when the particulate fluxes of interest are expected to vary with sun activity), possible shielding effects by spacecraft parts, function required for the material on board, etc... If the considered materials are used for structures, their bulk properties are important and the simulation conditions will be optimized for the penetrating particles (namely the electrons). For thermal control coatings, one will address the degradation of surfaces, and the test conditions will have to favour the proton and ultraviolet radiations effects. Fig. 15 shows the dose distribution within a PSB white paint resulting from the polyenergetic proton and electron fluxes of GEO space environment. It is compared to that obtained for each of the three particle irradiations selected for the simulation at

---

[7] J Bourrieau, 'Doses and Shielding'.

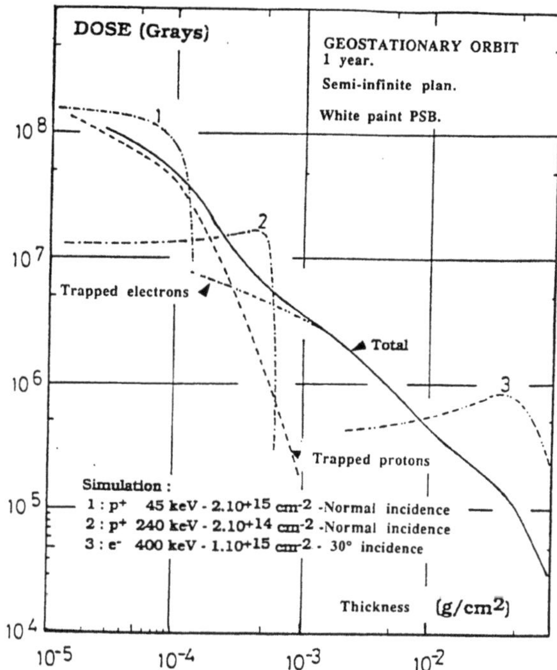

Fig. 15: Dose repartitions within a PSB white paint resulting from the polyenergetic proton and electron fluxes of GEO environment and from three particle irradiations selected for space environment simulation.

laboratory in order to match the space doses accumulated for one year mission. The following values have been utilized [Ref. 46]:

protons 45 keV : 2.0×10^{15} particles cm^{-2} (normal incidence)
protons 240 keV : 2.0×10^{14} particles cm^{-2} (normal incidence)
electrons 400 keV : 1.1×10^{15} particles cm^{-2} (30° incidence).

4.3. FACILITIES

At the DERTS UV radiation, electrons and protons are simultaneously directed onto the samples which are maintained under a high vacuum (less than 10^{-6} Torr) for all the irradiation time as well as for the measurement periods. The UV-irradiation is uninterrupted (except during the measurement periods) while the high energy particle irradiations are achieved by steps (with the UV-beam on). *In situ* measurement techniques are associated to the irradiation tests which are performed under vacuum. Dedicated chambers are used: chiefly, SEMIRAMIS for the measure of optical reflectance in the solar range [Ref. 27], and MIRASITU for the measure of elastic modulus and damping of organic matrix composites [Ref. 38]. These chambers are connected to two Van de Graaff accelerators; one delivers protons with energies ranging from 40 keV to 2.5 MeV, the second gives electrons of 600 keV to 2.5 MeV. The scattering of electrons through an aluminum foil and an electrostatic rastering of the proton beam allow uniform irradiation of a 12 x 12 cm sample array. The samples receive ultraviolet radiation (in the spectral domain 200-380 nm) by the means of short arc Xenon sources, equipped with interferential filters eliminating the visible and infrared parts of their spectrum. UV-irradiations are carried out at 1-4 suns.

Typical laboratory conditions call for a 20-40 day exposure duration in order to simulate one year in space. As the operational life of spacecraft tends to become longer (more than 10 years for GEO operational satellites, 30 years for the Space Stations Freedom and Columbus), the duration of tests is considerably extended with correlative requirements for the test facility reliability.

5. Some Results

5.1. EFFECTS OF CONTAMINANT LAYERS

SEMIRAMIS has been used for a variety of experiments related to the ageing of polymeric materials and thermal control coatings in space simulation conditions. A few results are given in [Ref. 27] from which Fig. 16 is taken out as an example.

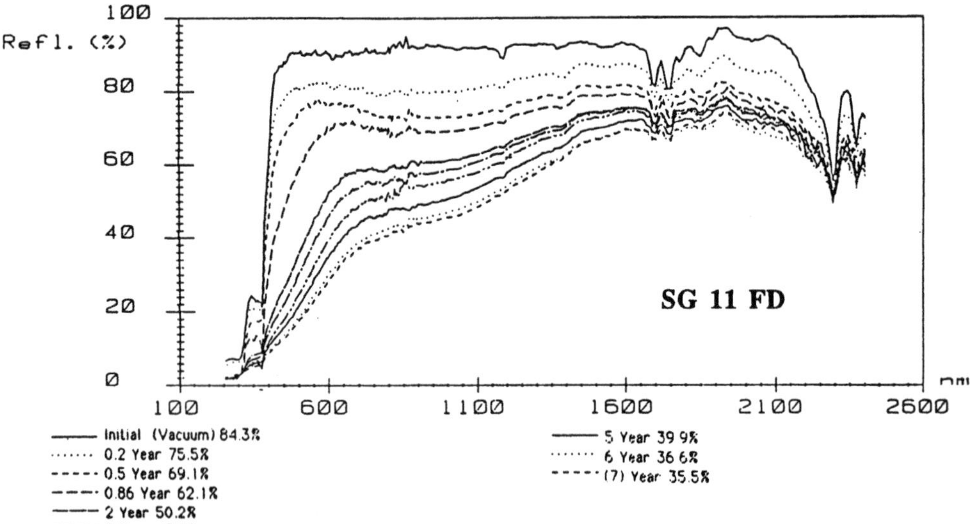

Fig. 16: Variation with time of the optical reflectance decrease of the SG11 FD white paint irradiated in vacuum with electrons protons and UV. Irradiation conditions of a 7-year GEO environment.

Several studies highlight the strong effect of contamination layers in enhancing the degradation of optical properties of surfaces upon exposure to the space environment. Upon UV and particle irradiation low molecular weight contaminants are fixed by polymerization on surfaces; the polymerized films adhere strongly to the substrate and are heavily discolored. Conductive and non-conductive OSRs have been submitted to various thicknesses (50 to 500 Å) of a contamination layer deposited *in situ* at the start of a 5-year GEO environment simulation test. After simultaneous irradiation with UV and particles, the measured variation of optical absorptance was three to four times greater for the OSR having received 500 Å of contaminant (VCM product from RTV 566) than for a clean OSR [Ref. 39], (see Fig. 17). The generality of the synergistic effect of radiations and contamination on optical properties and also the secondary electron emission, has been further studied using other contaminants (from RTV and epoxy adhesives, from pump oils), and using other substrates (paints, aluminized FEP Teflon®, PSG 120 paint, silica fabrics) [Refs. 40, 41]. The observations which have been made strongly support the fact that contamination is the most likely cause of the excessive solar reflectance degradation often experienced in GEO by OSRs [Ref. 42].

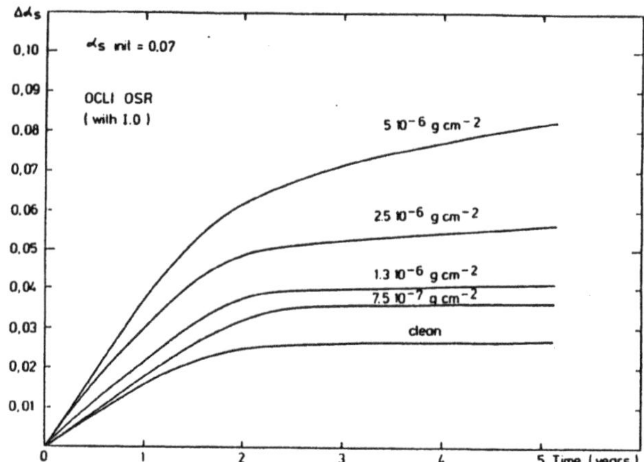

Fig. 17: Variation with time of a PPE OSR precontaminated with variable amounts of a silicone contaminant, then irradiated in vacuum with electrons, protons and UV. Irradiation conditions of a 5-year GEO environment.

5.2. IRRADIATION OF ORGANIC-MATRIX COMPOSITES

The synergistic effect of thermal cycling and irradiations has been demonstrated on carbon fiber reinforced materials using thermoset matrices. In these, residual strains can develop during the cure cycle. They are released by microcracking when thermal cycling occurs. The microcracks are produced in the matrix, and they extend parallel to the fiber direction. This microcracking is more intense for carbon/epoxy materials which are irradiated with electrons [Refs. 37, 38] (see Fig. 18).

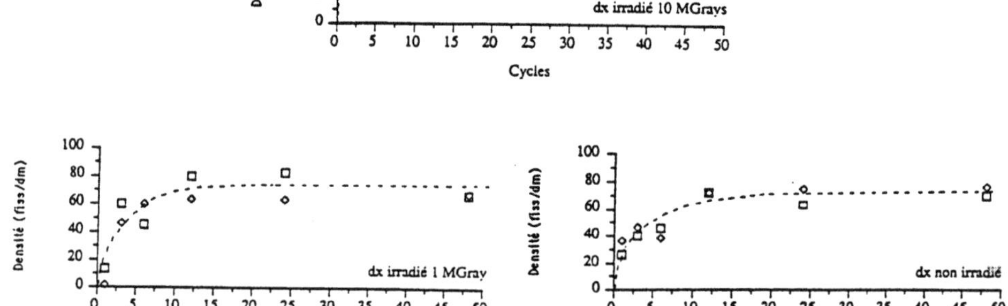

Fig. 18: Effect of the electron irradiation dose on the microcracking density of T300/914 composite $(0_2,90_2)_s$ (from [Ref. 38]).

The combined effects of radiation and thermal cycling on the dimensional stability of polymer matrix composites is twofold [Ref. 44]: (a) matrix plasticization at elevated temperatures can cause permanent residual strains in composites that are exposed to such temperatures, and (b) matrix embrittlement at low temperature enhances matrix microcracking due to thermal cycling, and causes additional residual strains and changes in the CTE.

6. Conclusion

After extended periods of time in space, ultraviolet and high-energy particle radiations significantly affect the performance of materials. However, the degradations which are obtained in space (some of them have been clearly demonstrated by the recent retrieval of samples and pieces of hardware exposed on Solar Max, LDEF, or MIR) are evidently the result of synergy effects between the natural and man-made components of the space environment. The mechanisms of radiation damage seem understood in so far as one type of radiation is present. This is seldom the case in actual space conditions: UV, electrons, protons, atomic oxygen, thermal cycling, debris impacts act simultaneously and synergistically. The origin of actual damage is very complex. As already stated by Schwinghamer in 1980, far more work will be required in order to "become able to theoretically predict the true response of a material in a combined environment exposure. The present understanding of the behaviour of engineering materials exposed to complex environmental stresses largely precludes the quantitative prediction of the time-dependent property changes; hence, there is still a great need for simultaneous environment testing".

References

[1] Turro, N.J., 'Molecular Photochemistry', W.A. Benjamin, New York, 1967.
[2] Kan, R.O., 'Organic Photochemistry', McGraw Hill Book Company, New York, 1966.
[3] Spinks, J.W.T., and Woods, R. J., 'An Introduction to Radiation Chemistry', John Wiley & Sons, New York, 1964.
[4] Charlesby, A., 'Atomic Radiation and Polymers', Pergamon Press, Oxford, 1960.
[5] Chapiro, A., 'Radiation Chemistry of Polymeric Systems', Interscience Publ., New York, 1962.
[6] Schmitz, J.V., 'Testing of Polymers', Chapter 5, Interscience Publ., New York, 1966.
[7] Wilson, J.E., 'Radiation Chemistry of Monomers, Polymers and Plastics', Marcel Dekker, New York, 1974.
[8] Schnabel, W., 'Polymer Degradation', Hanser International, München, Wien, 1981.
[9] Dauphin, J., 'Selection and Control of Materials for Space Applications', ESA Journal, 9, pp. 53-59, 1984.
[10] Judeikis, H.S., 'Space Radiation Effects on Teflon Films', Report SAMSO-TR-79-070, 1979.
[11] Gillen, K.T., and Clough, R.L., 'Occurrence and implications of radiation dose-rate effects for material ageing studies', Radiation Phys. Chem., 18, pp. 679-687, 1981.
[12] Schwinghamer, R.J., 'Space environmental effects on materials', NASA-TM-78306, 1980.
[13] Zerlaut, G.A., Harada, Y., and Tomkins, E.H., 'Recent advances in thermal control materials research', A72-31806, 1972.
[14] Compton, W.D., 'Radiation effects in ionic crystals', The interaction of radiation with solids pp. 386-420, North-Holland Publishing Company, Amsterdam, 1963.
[15] Bolt, R.O., and J.G., 'Radiation Effects on Organic Materials', Academic Press, New York, 1963.
[16] McCargo, M., Greenberg, S.A., and Douglas, N.J., 'A study of environmental effects upon particulate radiation-induced adsorption bands in spacecraft thermal control coatings pigments', AIAA 69-642, AIAA 4th Thermophysics Conference, San Francisco, CA, June 16-18,1969, 1969.

[17] Zerlaut, G.A., Harada, Y., and Tomkins, E.H., 'UV irradiation of thermal coatings in vacuum', NASA SP-15, p. 391, 1965.
[18] Breuch, R.A., and Greenberg, S.A., 'Space radiation damage to the thermal radiative properties of materials', Proceed. 3rd Int Heat Transfer Conf, Chicago, Ill., p. 246, 1966.
[19] Paillous, A.J., and Fayet, Ph., 'Some problems concerning the simulation of plastic films degradations by ionizing radiations', AST/IES/AIAA 5th Space Simulation Conference, Gaithersburg, MD, 1970.
[20] Brown, B.L., Thomasson, J.F., and Kurland, R.M., 'Space radiation effects on Spacecraft materials', Proceed. 24th SAMPE Symp. and Exhib., San Francisco, p. 1021, 1979.
[21] Ferl, J.C., and Long, E.R., Jr., 'Low energy electron effects on tensile modulus and infrared transmission of a polypyromellitimide film', NASA-TM-81977, 1981.
[22] Kurland, R.M., and Thomasson, J.F., 'Radiation testing of materials, in-situ versus ex-situ effects', NASA CR-3475, 1981.
[23] Gillette, R.B., 'Proton Radiation effects on solar concentrator reflective surfaces', AIAA Paper 67-341, AIAA Thermophysics Conf., New Orleans, Louisiana, 1967.
[24] Koitabashi, M., and Al., 'Degradation of spacecraft thermal shields materials by irradiation of energetic particles', Proceed. Int. Symp. on Space Technology May 27, June 1984, Tokyo, Japan, pp. 577-582, 1984.
[25] Kieffer, R.L., and Orwoll, R.A., 'Space environmental effects on polymeric materials', N-89-15255, NASA-CR-184648, 1989.
[26] Fogdall, L.B., Cannaday, S.S., and Brown, R.R., 'Electron energy dependence for in-vacuum degradation and recovery in thermal control surfaces', AIAA Paper 69-643, AIAA 4th Thermophysics Conference, San Francisco, CA, 1969.
[27] Marco, J., Paillous, A., and Levadou, F., 'Combined radiation effects on optical reflectance of thermal control coatings', Proceed. Fourth European Symp. on Spacecraft Materials in Space Environment, 6-9 September 1988, CERT, Toulouse, France, 1988.
[28] Marco, J., Millan, Ph., Paillous, A., Sablé, C., and Siffre, J., Note CERT-DERTS, Final Report on ESTEC Contract 5781/83/NL/AB, Study Nr2 'Sequential irradiations and Synergy', 1985.
[29] Fogdall, L.B., and Cannaday, S.S., 'In situ electron, proton and ultraviolet radiation effects on thermal control coatings', Boeing, Final report on Contract NAS 5-9650, 1970.
[30] Brown, R.R., Fogdall, L.B., Cannaday, S.S., 'Electron-ultraviolet radiation effects on thermal control coatings', AiAA Paper 68-779, 1968.
[31] Bourrieau, J., Paillous, A., Romero, M., Note CERT-DERTS, Final Report on ESTEC Contract 2515/75/HP, 'Dégradation de revêtements de contrôle thermique souls l'effet des rayonnements ultraviolets et particulaires', 1976.
[32] Ferl, J.E., and Long, E.R., 'Infrared spectroscopic analysis of the effects of simulated space radiation on a polyimide', IEEE Tans on Nucl Sc., NS-28, p. 4119, 1981.

[33] Marco, J., Millan, Ph., Paillous, A., Sablé, C., Siffre, J., Note CERT-DERTS, Final Report on ESTEC Contract 5781/83/NL/AB, Study Nr 1 'Increase in intensity of UV irradiations', 1985.

[34] Marco, J., Millan, Ph., Paillous, A., Sablé, C., Siffre, J., Note CERT-DERTS, Final Report on ESTEC Contract 5781/83/NL/AB, Supplement to Study Nr 1 'Increase in intensity of UV irradiations', 1985.

[35] Fogdall, L.B., and Cannaday, S.S., 'Radiation effects on thermal control coatings, polymers and optical materials', Proceed. Intern. Seminar Simulation and Space, CERT, Toulouse, France, 1973.

[36] Guillaumon, J.-C., and Paillous, A., 'Experiment AO 138-6: Thermal Control Coatings', Proceed. First Post-retrieval LDEF Meeting, 3-7 June 1991, Kissimee, Florida, 1991.

[37] Tenney, D.R., 'Structural materials for space applications', NASA Conf. Public. 3035, NASA/SDIO Space Environmental effects on Materials Workshop, June 28 – July 1 1988, NASA-LaRC, Hampton, VA., 1980.

[38] Plottard, P., and Durin, C., 'Space environment effects on Carbon/Epoxy materials', ESA SP-289, Proceed. Intern. Conf. Spacecraft and Mechanical Testing, Noordwijk, The Netherlands, p. 535, 1988.

[39] Marco, J., and Paillous, A., 'Long term tests of contaminated OSRs', ESA SP-232, Proceed. 3rd European Symposium on Spacecraft Materials in Space Environment, ESTEC, Noordwijk, The Netherlands, pp. 245-253, 1985.

[40] Lévy, L., and Paillous, A., 'Effect of contamination on the charging behaviour of silica fabrics', ESA SP-178, Proceed. 2nd European Symposium on Spacecraft Materials in Space Environment, CERT, Toulouse, France, pp. 89-101, 1985.

[41] Marco, J., Millan, Ph., Paillous, A., Sablé, C., and Siffre, J., Note CERT-DERTS, Final Report on ESTEC Contract 5781/83/NL/AB, Study Nr 3 'Etude des procédures d'application de contaminants', 1985.

[42] Marvin, D.C., Hwang, W.C., Arnold, G.S., and Hall, D.F., 'Contamination-induced degradation of solar array performance', N88-9052, 1988.

[43] Tenney, D.R., 'Space radiation effects on dimensional stability of composites', Proceed. Fourth European Symp. on Spacecraft Materials in Space Environment, 6-9 September 1988, CERT, Toulouse, France, pp. 133-144, 1988.

[44] Reinisch, R.F., Gloria, H.R., and Androes, G.M., 'Photoelimination reactions of macromolecules', Photochemistry of Macromolecules, pp. 185-217, Plenum Press, New York, 1970.

[45] Townsend, L.W., and Wilson, J.W., 'Interaction of space radition with matter', IAF/IAA-90-543, 41st Congress of the International Astronautical Federation, Dresden, Germany, 1990.

[46] Marco, J., 'Irradiations ultraviolettes puis particulaires sur des peintures', Note CERT/DERTS CR/REV/34, Rapport final sur Comande CNES 50th du 9/7/90, 1991.

HAZARDS TO ELECTRONICS IN SPACE

JAMES H. STEPHEN
Consultant
35 Appleford Drive
Abingdon
Oxon OX14 2BZ
UK

ABSTRACT. Semiconductor structures, especially VLSI devices such as microprocessors and solid state memories, have a vital role in the exploitation of space covering research, communications and manned space flight. The sensitivity of semiconductors to nuclear radiation was appreciated before the first satellite was launched, but the pronounced effect of the space radiation was not anticipated. As a consequence the behavior of semiconductors in space has been intensively studied since the early sixties for both defence and civil requirements. These studies have had to include the rapid developments in VLSI devices which nowadays are the backbone of modern electronic systems.

The purpose of this lecture is to describe the effects of space radiation on microelectronic devices building on the information in the accompanying lectures. The overall influence of ionizing radiation due to charged particles, electronics and photons interacting with the spacecraft materials causes a steady degradation of the electronic performance of the devices leading to ultimate failure. In addition, the passage of a charged particle (e.g., of cosmic origin) through the semiconductor material releases a pocket of charge which is capable of disturbing the normal operation of the circuit element producing a single event upset (SEU). These two aspects of space radiation will be discussed together with a brief reference to the effects of neutrons. Nuclear weapon effects will not be discussed.

The hazards of space radiation on silicon based microelectronics will be discussed by initially considering the effects of p-n junctions, bipolar and MOS transistors and finally on typical VLSI devices which are currently in common use in space.

The suitability of a microelectronics device for a particular application depends on the space environment, which can be calculated from mathematical models, and the sensitivity of the device which is a function of its technology, fabrication methods, radiation hardening, if any, and packaging. Although there is information on radiation effects in a number of databases, it is frequently necessary to perform terrestrial radiation testing to assess if a component is suitable for the specific space application. The above topics will be addressed in the lecture, together with a description of the methods for reducing the hazards of space

R. N. DeWitt et al. (eds.), The Behavior of Systems in the Space Environment, 407–435.

radiation. Lastly, a description will be given of the methods currently being used on satellites to measure the space radiation for microelectronic applications.

1. Introduction

This lecture outlines the problems encountered in employing electronic devices in a space environment. The accompanying lectures have described in detail the space radiation environment, and it is the purpose here to describe the interaction between the radiation and the semiconductors commonly used in spacecraft, mostly based on silicon, though gallium arsenide is also being used. The hazards due to the radiation are described, followed by methods of minimizing the hazards and the techniques used for the ground simulation of the radiation conditions. The need for ground simulation arises from the fact that silicon integrated circuits can vary considerably in their behavior to radiation due to their increasing complexity, design variations and fabrication methods.

Looking back over the last half century of space exploitation and microelectronics development, one is struck by the prodigious advances made in these two fields of science since the end of the War. Microelectronics progressed relatively slowly after the invention of the transistor with one of the main applications being the transistor computer. With the development of the silicon planar technology, the field opened out from computers to include telecommunications, military applications including space, transport and domestic applications with the digital watch to the calculator and the video recorder. Logic families were the first to benefit by the development of integrated circuits and later computer chips were realized incorporating system oriented designs such as the microprocessors and memories. Over an interval of 30 years the microelectronics has advanced from one transistor per package to over 64 million transistors in the current VLSI random access memory chips.

The research and development into microelectronics has been powered by a number of driving forces some of which are mentioned above. An enormous industry has followed the invention of the transistor in 1947 and in 1991 it shows no signs of slackening; it is a key technology for the Information Age in the next century.

Space technology has benefitted fully from the advances in microelectronics from the earliest times (Telstar) to the very complex satellites now in orbit (Galileo). The majority of the devices flown in space are based on the MOS technology which was initially thought to be radiation-hard as it was a majority carrier device. This optimism vanished when it was shown experimentally that the surface conduction could be affected by stored charge in the gate and passivating oxides. A very considerable amount of effort has been expended in the understanding of this phenomenon and how it can be avoided. This research has continued for over 30 years and is aimed at a rapidly moving target due to the microelectronics development. [See reference books and IEEE Trans on Nuclear Science (1960-1992).]

Radiation hard circuits have been produced by a number of specialist companies for specific applications. These designs tend to lag behind the current commercial and military designs because of the modifications required to the chip layout and fabrication methods. To overcome this problem it is common practice for many satellite applications to select commercial or military circuits and experimentally test their radiation behavior. A number of data banks exist which list test data on particular devices from single transistors to VLSI devices.

The effects of nuclear radiation on microelectronics have been extensively studied using terrestrial sources of radiation, but very little has been done using the space radiation itself, for obvious reasons. However in recent years some experiments are being flown to study the effects on microelectronic components (see data bases).

This lecture has been planned to cover the effects of space radiation on semiconductors, principally silicon, the hazards to microelectronic components, means for reducing these effects, and methods of radiation testing.

2. Effects of Space Radiation on Semiconductor Materials

2.1. BRIEF REVIEW

The space radiation experienced by spacecraft is of two distinct types if the effects of nuclear explosion are not included. Firstly, there is the internal radiation environment in the spacecraft due to shielding of the high energy electrons, protons and cosmic rays (total dose effect). Secondly, there are the high energy cosmic particles which can deposit a small charge in a microcircuit causing it to temporarily or permanently malfunction (single event upset (SEU)). Previous lectures have dealt with the theory and simulation of space radiation conditions from low earth orbit to a geostationary orbit and interplanetary flight. In order to assess the effect of the radiation on the electronics in a particular spacecraft, an estimate must be made of the expected radiation conditions in order to be assured that the correct types of microelectronic components are chosen [Refs. 1,2,3,4,5].

Energy is deposited in semiconductor materials, including insulators such as silicon dioxide, in two ways: by the production of electron-hole pairs and by atomic displacements. These two mechanisms produce different types of degradation in the microelectronic devices.

The effects of high dose rate gamma-ray radiation and high neutron fluxes which could result from a nuclear explosion are not discussed. The result of operating in a low level fast neutron flux are mentioned to cover the case of a satellite with an RTG (radioisotape thermal generator, e.g., Galileo).

2.2. TOTAL IONIZING DOSE EFFECTS

Total dose effects are due to the highly penetrating ionizing radiation produced by the interaction of the high energy electrons and protons with the materials of the satellite body. The intensity of the radiation decreases with increasing thickness of the satellite skin affording some shielding (Fig. 1) [Refs. 1,5].

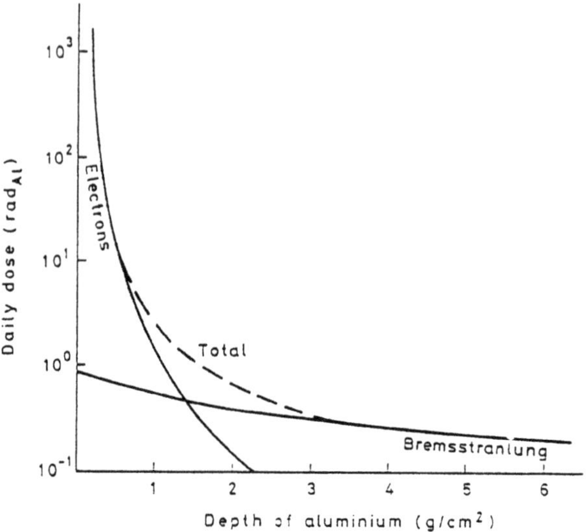

Fig. 1: The daily dose in geostationary orbit.

The radiation produces further ionization on the semiconductor (silicon) and its associated insulator (SiO_2) resulting in the release of high energy electrons which in turn loose energy by the release of electron-hole pairs and ultimately heat when the e-h pairs recombine producing a phonon. The deposited energy is expressed in rads (Si or SiO_2) i.e., radiation absorbed dose. One rad is equivalent to 100 ergs/gm in cgs units or one gray (Gy) (equal to 100 rads) equal to 1 joule per kilogram.

The carrier generation constant for silicon (3.6 eV is required for the formation of one e-h pair) is

$$g = 4.2 \times 10^{13} \text{ electron-hole pairs per cm}^3\text{-rad (Si)}$$

$$= 4.2 \times 10^{13} \times 1.602 \times 10^{-19} \text{ coulombs per cm}^3\text{-rad (Si)}$$

$$= 6.73 \times 10^{-6} \text{ coulombs per cm}^3\text{-rad (Si)}$$

The current produced in a typical transistor in an integrated circuit for a total dose of 100 Krads per year is so small that it has negligible effect on the operation of the circuit. There is a completely different situation when a large ionizing dose is deposited in tens of nanoseconds as very large photo currents can be produced.

The silicon has no mechanism for storing the charge produced by the ionization for more than a few milliseconds, but this is not the case for the silicon dioxide insulating layers used for passivating the silicon surface and in the gate oxide in MOSFETS. Silicon dioxide can be treated as a "semiconductor" with a wide bandgap of about 8 eV. The ionization produces electron-hole pairs in the SiO_2. The electrons due to their high mobility in SiO_2 are rapidly removed leaving a fraction of the positive charges trapped in the SiO_2. If an electric field is present across the SiO_2 directed towards the Si-SiO_2 interface, the positive charges will accumulate near the interface. The fraction trapped for "soft" oxides is between 0.2 and 0.5, and these oxides can be expected to slowly anneal. Hard oxides have a trapped fraction of less than 0.1 which is a desirable situation for space microelectronic components (Fig. 2) [Ref. 13].

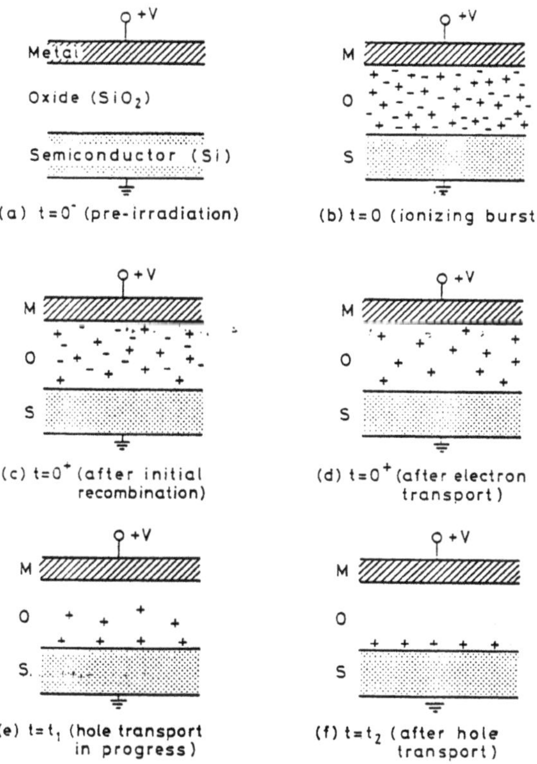

Fig. 2: Illustration of recombination, transport, and trapping of carriers in SiO_2 films [Ref. 13].

Another important effect of the ionizing radiation is to produce interface traps at the Si/SiO_2 interface. They are amphoteric being donor type (positive) in p channel MOSFETS and adding to the trapped oxide positive charges. In n-channel MOSFETS the traps are acceptor type (negative) thus compensating the trapped positive oxide charges (Fig. 3).

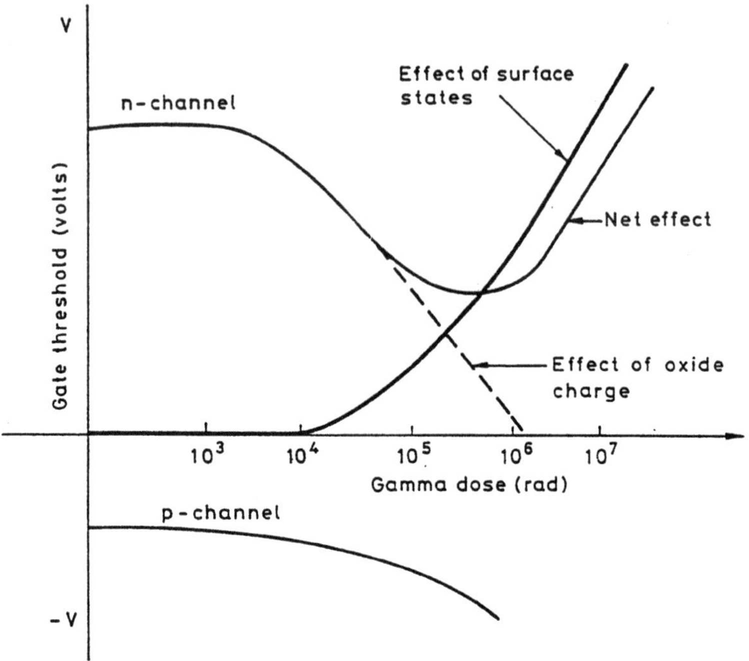

Fig. 3: Typical gate threshold shift for n- and p-channel MOSFETs.

The process for the production of the interface states is not fully understood, but it is known that the wet oxides (grown in steam) have more interface states than in carefully grown dry oxides which are usually used for gate oxides.

The change in the charge compensation in the n-channel MOSFET due to annealing has the effect of moving the operating gate voltage more positive than its unirradiated value. The phenomenon is known as "rebound" and it can seriously affect the operation of the MOSFET especially when used in a complementary pair (Fig. 3) [Ref. 13, Book:MA and Dressendorfer Chap 4].

The dose rates experienced in space are normally very low, being between a few hundred and several thousand rads (Si) per year depending on the spacecraft's orbit. The total ionizing dose received during the operational life of a satellite, which can be up to twenty years, can exceed 100,000 rads (Si). As annealing of the stored positive charge in the oxide is likely to occur over this length of time, leaving the interface states uncompensated, it is essential to conduct low dose rate irradiations to explore the overall effect. Figure 4 illustrates the dependence of the dose to failure on the dose rate. Some experimental approaches are discussed in Section 5.

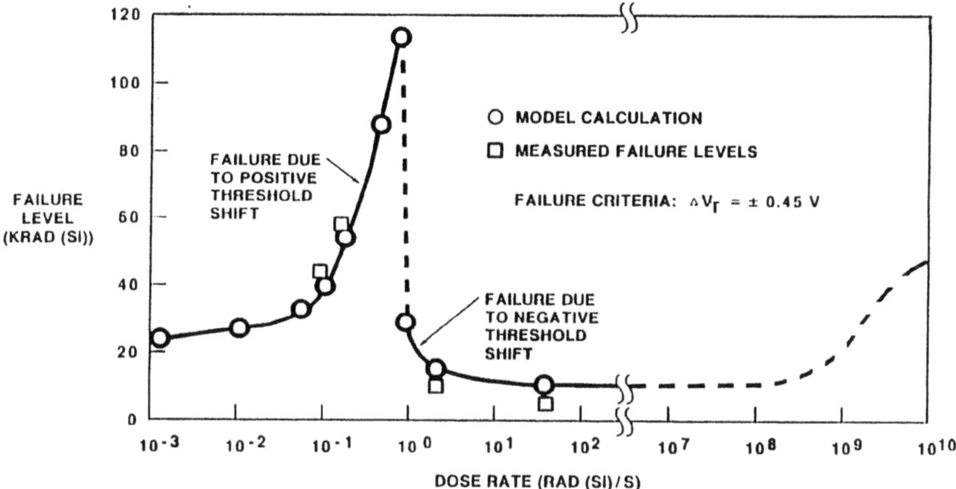

Fig. 4: Total dose required to cause failure as a function of the dose-rate at which the radiation is delivered. [Ref: Ma and Dressendorfer (1989)].

2.3. SINGLE PARTICLE EFFECTS

When charged particles, electrons, protons or heavy ions pass through silicon, they loose energy by electronic loss and nuclear interactions. The electronic loss appears as a cloud of electron-hole pairs along the path of the particle (Fig. 5). The term linear energy transfer (LET) refers to this phenomenon, and its magnitude is dependent on factors such as the particle type and its energy. The electronic charge is deposited in a very short time (picoseconds), and the sudden injection of charge (10^{-14} to 10^{-12} coulombs) into a microcircuit can temporarily disrupt its operation producing a single event upset (SEU); an effect which is particularly serious in dense solid state memories (DRAMS and SRAMS). A related phenomenon known as "latch-up" occurs in complementary MOS circuits in which configuration of the circuits on the chip produces thyristor type structures which can be switched on by a transient pulse from a particle. The latch-up is serious as it paralyses the chip and can cause very large currents to pass and burn-out conductors on the chip. More recently a very similar phenomenon has been observed in power MOSFETS. The passage of a particle through the transistor structure produces a local avalanche which causes the parallel parasite bipolar structure to conduct and carry a very heavy current leading to local burn-out and destruction of the transistor. The SEU was predicted in 1962 by Wallmark and Marcus [Ref 6] and observed in static rams in 1976 [Ref 16].

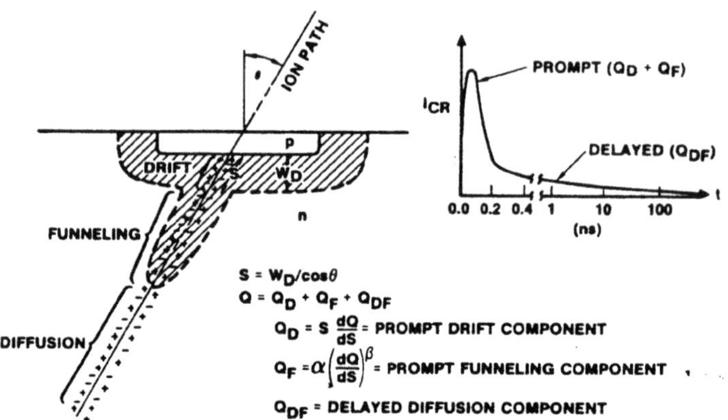

Fig. 5: Cosmic ray induced current pulse showing prompt components due to drift from the depletion region and charge funneling and the delayed component due to diffusion.

The sensitivity of a circuit to SEU or latch-up is measured in terms of a cross-section obtained by irradiating the device under test with a measured particle fluence and observing the number of events under operating conditions. Precautions are taken to limit the latch-up currents. The technique is discussed later.

2.4. DISPLACEMENT DAMAGE

Apart from the ionization effects just described, the radiation of silicon and SiO_2 produces lattice damage effects in which silicon atoms are displaced from their usual lattice positions to take up interstitial sites and leaving a vacancy behind. These damage sites are electronically active with energy levels in the silicon bandgap. These sites have a very pronounced influence on the carrier recombination and generation (i.e., r-g centres) and the carrier mobility. They seriously affect the leakage properties of p-n junctions, the current gain in bipolar transistors and the mobility of carriers in the channel regions of MOSFETS.

Displacement damage is produced by high energy electrons ($E > 180$ keV), protons and all the heavier particles. The damage can occur along the path of the particle by elastic scattering which causes ionization and displacements provided sufficient energy is transferred to the target atom. Inelastic scattering, in which the target atom captures the particle and then emits another particle (e.g., an alpha particle), causing further ionization and displacement due to the recoil of the transmuted target atom.

It is reported that nuclear reactors are being studied for supplying the energy for interplanetary spaceflight. Such a reactor system will involve the emission of fast neutrons and gamma rays from which the crew will have to be shielded. The control and measurement of the reactor system will involve microelectronics in relatively unshielded positions in the vicinity of the reactor core. It is appropriate therefore to mention here that the effect of fast neutron bombardment is to produce damage very similar to that produced by protons. The changes in the electronic properties of the silicon are therefore essentially the same. Figure 6 illustrates the relative damage effects of ^{60}Co gamma-rays, electrons, protons and neutrons.

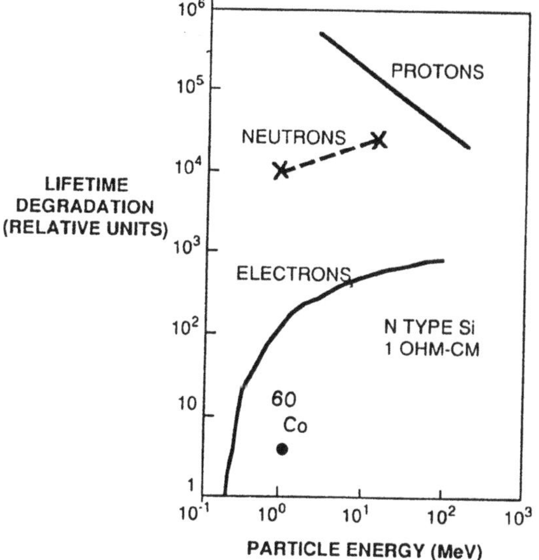

Fig. 6: Comparison of the effects of various particles on carrier lifetime in silicon.

3. Hazards to Semiconductor Devices in Space

3.1. INTRODUCTION

The hazards to microelectronic devices in the space environment are discussed in this section under the three basic headings of ionizing radiation, single particle effects and displacement damage. The discussion is confined to silicon, SiO_2 and Si_3N_4.

3.2. EFFECTS OF IONIZING RADIATION

The effects of low dose rate ionizing radiation on the following devices will be discussed;

p-n junction diode (signal diode and radiation detector)
Bipolar transistors

Junction field effect transistors
MOSFETS p and n channel including CMOS pairs, CCDs, power MOSFETS and RADFETS
Integrated circuits based on the above basic devices
Bipolar, digital, analogue (ADCs etc.)
MOS VLSI, digital, microprocessors, memories/DRAM, SRAM, EPROM and EEPROMS.

3.2.1. *P-N Junction Diode*. The junction diode is a fundamental component of the great majority of semiconductor devices. The energy deposited by the ionizing radiation produces a photocurrent, but as the dose rate is very low, the current is insignificant when compared with the normal leakage current. This is not the case for very high dose rates (>10^6 rad/sec) when the photocurrents can be large and affect the performance of the circuit. It is possible to use the photocurrent generated in large area photodiodes to measure the flux of the radiation.

In the planar technology the perimeters of p-n diode structures are protected by a layer of passivating silicon dioxide. The positive charges produced in the oxide can induce an n-type layer in the silicon and extend the area of the diode by modifying the carrier concentration at the surface. The net result is an increase in leakage current and a softening of the breakdown voltage. The effect can be reduced by the incorporation of guard bands around the diode perimeter.

3.2.2. *Bipolar Transistors*. The effect of the ionizing radiation is similar to that discussed above for the diode. The positive charges on the oxide can produce a conducting path from the base to the emitter which directs base current away from the main base current path resulting in a loss of gain of the transistor. This effect is particularly noticeable in "super beta" transistors which operate with very small base currents. TTL, ECL and power transistors have much larger base currents and are less affected as the diverted current is a smaller fraction of the base current.

It is reported that those transistors which incorporate a recessed field oxide (to planar the structure on the wafer) have been observed to fail at the low dose of 5 Krads(Si) while other bipolar structures can operate up to 1 Mrad(Si) [Ref. 21].

3.2.3. *Junction Field Effect Transistor*. The main effect of ionizing radiation on junction field effect transistors is to increase the gate to channel leakage current and contribute to the noise currents. This is important as JFETS are frequently used in the first stage of low noise amplifiers, both discrete and integrated designs.

3.2.4 *MOSFETS*. The main long term hazard faced by MOSFETS is the accumulation of trapped charge in the oxide layers. Positive trapped charge at the gate oxide silicon interface will be in opposition to the intended negative bias and so will shift the threshold voltage in a negative direction. Hence n-channel MOSFETS will be turned off and p-channel MOSFETS turned on as the charge accumulates. If the radiation takes place with a positive bias on the gate, the electrons are swept out of the oxide leaving a larger fraction of trapped holes.

The mechanism of charge trapping is illustrated in Fig. 2. At $t = 0$ a burst of ionization generates electron-hole pairs throughout the oxide layer. Immediately after the burst some recombination occurs together with the movement of the electrons out of the oxide due to their high mobility compared to the holes (10^6 times greater). Some holes become trapped at the interface. Additional interface states can be produced by the radiation directly, though this process is not fully understood. Interface state effects become important for accumulated doses of over 10^5 rads (SiO_2). In an n-channel MOSFET the interface states are negative and compensate the positive oxide charge causing the threshold voltage to increase (Fig. 3), while in a p-channel MOSFET the oxide charge and the interface states add making it necessary to increase the gate bias to turn the transistor on. As the gate is biased negative the holes will be attracted to the gate, and hence fewer holes will be trapped at the interface. The shift in threshold voltage ΔV_{ot} is given by:

$$\Delta V_{ot}\ (E_{ox}E) = \frac{q}{C_i} K_g(E) f_y(E_{ox},E) f_T(E_{ox}) d_{ox}^2\ D_{ox}(E)$$

where

q = electronic charge
C_i = Farads/cm (EE_o)
K_g = Energy dependent charge generation coefficient
f_y = Field dependant fractional charge yield
E = Energy of the radiation
E_{ox} = Field in the oxide
f_T = Fraction of holes trapped
d_{ox} = oxide thickness
D_{ox} = ionizing dose (rads (SiO_2))

This equation is the often quoted "thickness squared" dependence of V_{OT} on dose. The factor f_T is very dependent on the processing steps during fabrication.

The performance of the complementary MOS (CMOS) pair (Fig. 7) is seriously affected by the movement of the threshold voltages as it alters the operating conditions, and the output voltage from the pair can drop from the supply voltage V_{DD} to half V_{DD} with the result that the pair will not be able to drive succeeding stages.

The positive trapped charge in the field oxide, especially in the recessed oxide structure (birds beak) where the gate metal rises off the gate oxide onto the field oxide, can cause leakage currents from source to drain which increase steadily with dose and finally push the device out of specification (Fig. 8). The radiation damage can be annealed out by simply allowing the device to stand at room temperature. Biasing and raising the temperature can increase the rate of recovery. At the low dose rates in space, the rate of oxide charge generation and the annealing rate tend to cancel one another, with the degree of compensation very dependent on the characteristics of the oxide. The steady build-up of interface states will produce the positive rebound with a resultant overcompensation.

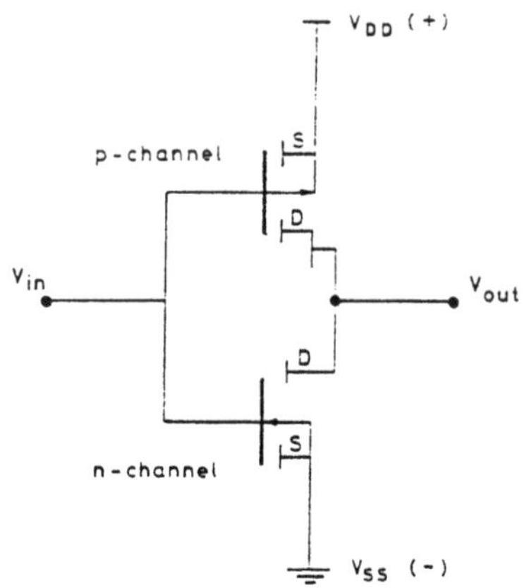

(a) Schematic a CMOS inverter circuit

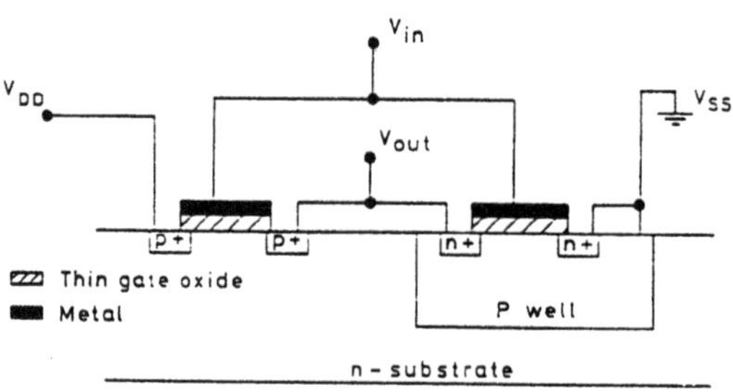

(b) A typical CMOS inverter cross-section

Fig. 7: A CMOS inverter cell: (a) schematic; (b) cross-section.

It has been shown experimentally that the positive charges in the oxides can be annealed by UV irradiating the device to release electrons from the silicon into the oxide which then are captured by the positive charges. The interface states are not affected by the process. The above remarks apply equally well to discrete MOSFETS such as the power MOSFET and integrated circuits.

There is one important MOSFET known as the RADFET which makes use of the charge storage in the gate oxide to act as an integrating dosimeter. The above equation is a guide to the design on the RADFET. The RADFET comprises a p-channel transistor in which the oxide charge and interface

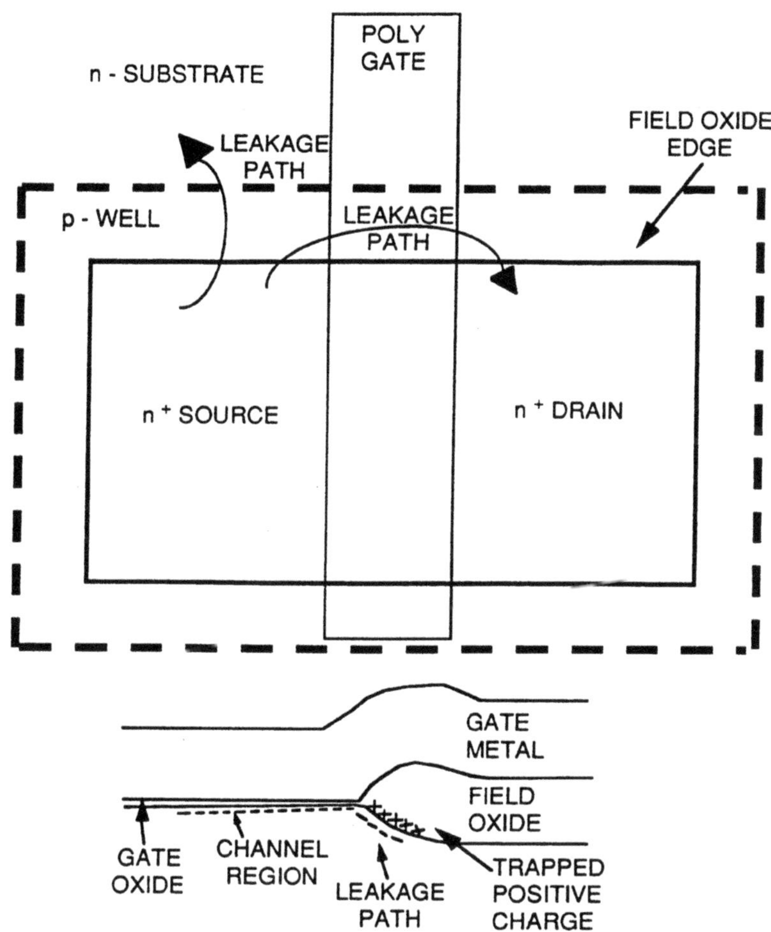

Fig. 8: Possible radiation-induced leakage paths under field oxide in an n-channel MOS transistor structure. [Ref: Ma and Dressendorfer (1989)].

states aid one another and increase the stored charge. In order to increase the sensitivity, a very thick gate oxide (>500 nanometers) must be used. Typical sensitivities are a delta V_T of 0.5 volts for a dose of 1000 rads (Si) with a large positive bias across the oxide [Refs. 18,19].

3.2.5. *Integrated Circuits.* Bipolar integrated circuits (e.g., TTL and ECL) exhibit performance degradation between 10^5 and 10^6 rads (Si) with some circuits working well above 10^6 rads (Si). The input circuits of operational amplifiers using super-beta transistors may show effects at lower doses. Recently it has been reported that bipolar circuits using the recessed oxide structure have failed at doses as low as 5 krad (Si). The recessed structure is shown in Fig. 9. The stored charge in the recessed oxide produces a channel between emitter and collector. It is unfortunate that the recessed oxide technique, which allows higher density and speed to be obtained, introduces an increased susceptibility to ionizing radiation. Caution must be exercised in the choice of bipolar integrated circuits.

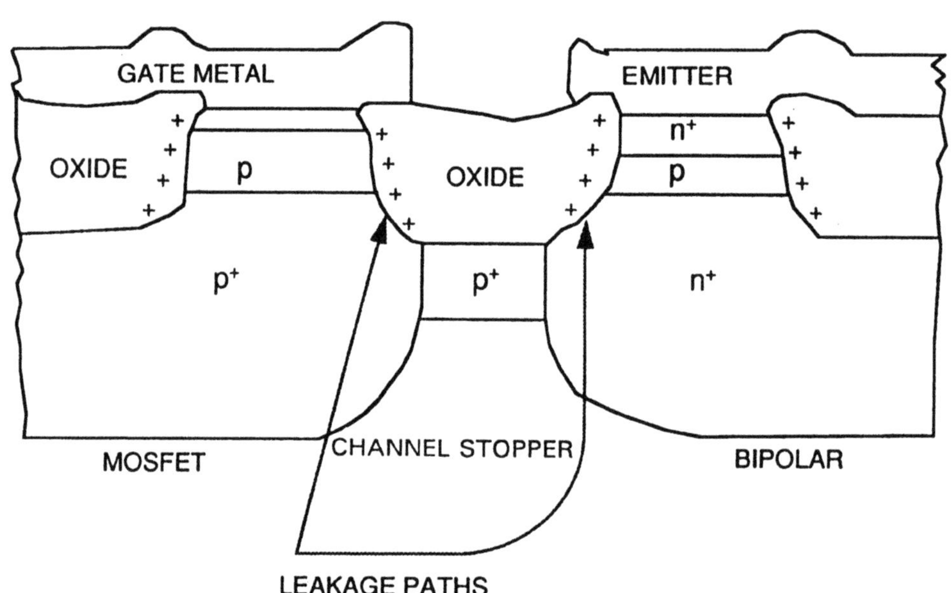

Fig. 9: Recessed oxide structure [Ref. 21].

MOS integrated circuits are the mainstay of digital circuitry in space, with CMOS being the preferred technology for microprocessors (e.g., 80C86, transputer etc.), memories (SRAMS, EPROMs, ROM etc.) and associated circuitry (e.g, logic arrays). The transistors in the MOS integrated circuits will suffer the changes in threshold voltages which can produce functional failure. The gate oxide thickness in typical VLSI transistors is about 30 nm (300 angstroms) which reduces the stored oxide charge due to the delta $V_T = K\ t_{ox}^2$ relationship.

However this is only part of the problem as the charge stored in the field oxide, especially in the vicinity of the transistors (Figs. 8 and 9), produces parasitic transistor action which increases the leakage currents and leads to parametric failure at relatively low doses (10 to 20 krad(Si)). In many cases the device (e.g., a memory chip) will continue to function, but it will be drawing an excessive current from the satellite system. Special design features are required to avoid the parasitic effect, and these are not usually incorporated in commercial designs, as they cost silicon real estate. It is possible to hardened to 100 krad (Si), which is adequate for most of the space applications.

3.3 EFFECT OF ENERGETIC PARTICLES ON DEVICES

As mentioned earlier, the passage of a particle through a semiconductor produces a highly ionized track of electron-hole pairs which result in the very sudden injection of charge into elements of an integrated circuit, such as a memory. This transient produces a single event upset (SEU) or a latch-up. The following devices will be considered: (i) p-n diode (as a detector of particles), (ii) Integrated Circuits (CMOS, CMOS/SOS), and (iii) Power MOSFETS.

3.3.1 *p-n Diode.* Large area p-n diodes make very suitable detectors of the high energy particles as experienced in space. The diodes have been used very successfully in detector systems used in the Concorde airliner, the UOSAT3 satellite (University of Surrey) and now in the US Shuttle.

The ionization produced by the particle is collected as an electronic charge which is directly related to the particle LET. A charge sensitive amplifier is used to integrate the charge from the diode and produce an output voltage directly proportional to the charge. In practice ten diodes are used each with an area of 1 cm^2 and about 200 microns thick.

As an example the LET, in $MeV/gm\ cm^{-2}$, can be calculated in terms of charge as follows:

$$1\ MeV/gm\ cm^{-2} = 10^6 \times 3.6^{-1} \times 2.33 \times 10^{-4} \times 1.602 \times 10^{-19}$$

$$= 1.04 \times 10^{-17}\ coulombs/micron$$

where 3.6 ev is the energy for e-h pair generation and 2.33 is the density of silicon (gm/cm^3). Hence the charge produced in a p-n diode with a path length of 200 microns is 2.08×10^{-15} coulombs for a LET of 1 $MeV/gm\ cm^{-2}$.

The range of LET values to be expected in space are given in Table 1 for protons and iron ions (see earlier lectures.) It should be noted that the LET decreases with increasing energy if it is beyond the Bragg peak. If the particle is stopped inside the diode, all the energy released as ionization, is collected.

Table 1. Energy Loss in Silicon

PROTONS			IRON (^{56}Fe)		
ENERGY MeV/ Nucleon	LET MeV/ gm cm^{-2}	CHARGE Coulombs/ Micron	ENERGY MeV/ Nucleon	LET MeV/ gm cm^{-2}	CHARGE Coulombs/ Micron
0.2	0.44	4.53×10^{-15}	0.179	18	1.85×10^{-13}
1.0	0.18	1.86×10^{-15}	0.89	27	2.78×10^{-13}
5.0	0.06	1.34×10^{-16}	1.79	28	2.88×10^{-13}
10.0	0.034	8.12×10^{-17}	8.93	18.5	1.9×10^{-13}
100.0	0.006	6.18×10^{-17}	17.9	13.0	1.34×10^{-13}

3.3.2 *Integrated Circuits.* Transient upsets produced by cosmic particles in space take two forms, the single event upset (SEU) and latch-up. Both upsets can lead to the corruption of data in a solid state memory chip (SRAM) or the latch-up in CMOS circuits which can lead to permanent damage due to melting on the chip.

It was postulated nearly thirty years ago that ions could deposit enough charge in an integrated circuit to upset its operation [Ref. 6], and in the mid-seventies it was recognized that errors in a satellite could be attributed to a single event upset [Ref. 16]. The phenomenon was brought forcibly to the attention of the semiconductor memory manufacturers when it was shown that alpha particles emitted from naturally occurring uranium and thorium in the materials of the chips and its package [Ref. 25] could cause "soft" errors.

The dynamic RAM cell stores its information as a charge in a potential well below a capacitor plate with a positive charge on it putting the well into deep depletion. If the alpha particle or any cosmic particle deposits enough energy, the electrons produced can be captured by the well, changing its logic sense. DRAMS tend to be more sensitive to particle bombardment than other memory devices and for this and other factors they are not commonly used in satellite systems.

The static RAM cell stores its information on flip-flops (i.e, latches), usually formed by connecting two inverter pairs in a cross-coupled configuration. The information is stored as 1's or 0's, depending on which inverter is conducting, and it is stored for as long as the power is applied to the cell. A cosmic particle can deposit sufficient energy to cause the flip-flop to change state and change a zero to a one or visa versa.

It is customary to measure the SEU sensitivity as a cross-section. This can be calculated from the number of upsets per bit (or device) divided by the number of particles/cm^2 over the same time interval. As the energy loss of the particles increases, due to many factors, the energy crosses a threshold value L_c for an upset. Above this threshold the upset rate increases rapidly until all the particles produce upsets, i.e. saturation.

The SEU sensitivity varies very considerably from device to device and even within one device type from a particular source. The memory chips fabricated in CMOS on SOS have very small cross-sections; whereas some commercial memories of the same capacity have 10^2 to 10^3 times larger cross sections. It is difficult to estimate the sensitivity of particular devices, and the best way of proceeding is to conduct physical testing.

An examination of the CMOS inverter pair with its adjacent n and p channel transistors form a parasitic thyristor or "silicon controlled rectifier" which under normal conditions remains in the "off" state (Fig. 10). The thyristor can be triggered to the conducting or "on" state by a voltage transient and also by the ionization from the passage of a particle through the inverter pair structure. It is known as a latch-up and can only be terminated by removal of all power to the chip. Very large currents (many tens of milliamps) can flow causing excessive load on the power system, loss of stored information due to the reduced supply voltage and the strong possibility of melting of the tracks and bonding wires in the chip, effectively destroying it. The occurrence of a latch-up is very serious, and if there is any doubt about the susceptibility of a CMOS chip to latch, it should be tested. The sensitivity to latch-up is expressed as a cross-section for the device.

CMOS RAM
LATCHUP MECHANISM

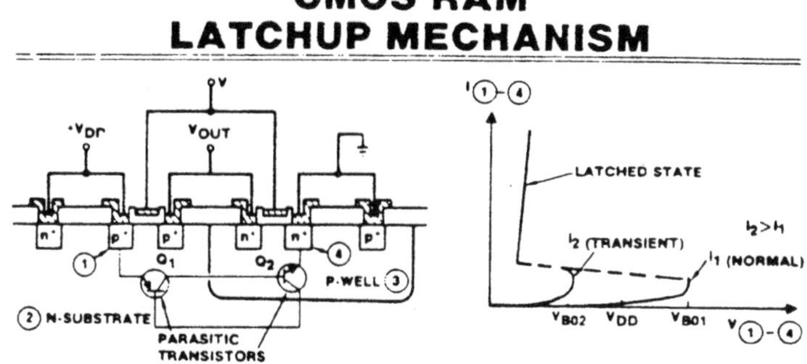

- PARASITIC pnpn PATH RESULTS FROM CMOS FABRICATION
- WHEN $\alpha(Q_1) + \alpha(Q_2) \geq 1$, STABLE LOW IMPEDANCE STATE OCCURS
- α CAN BE INCREASED BY CURRENT INJECTED FROM SINGLE PARTICLE IONIZATION, YIELDING SMALLER V_{BO} (90MeV Fe => 0.28pC/μm => ~ 3ma/μm)
- WHEN V_{BO} REDUCED TO LESS THAN V_{DD}, LATCHUP OCCURS

Fig. 10: Illustration of CMOS latchup mechanism and analogy to SCR.

Another important transistor incorporating a parasitic bipolar transistor structure is the power MOSFET. The power MOSFET is a VLSI component comprising a very large number, about 50,000 on average, individual MOSFETS connected in parallel and each capable of carrying about 1 mA. Associated with each n-channel MOSFET is a parasitic n-p-n bipolar transistor, whereas it's a p-n-p for a p-channel MOSFET (see Fig. 11). The passage of an ionizing particle through one of the transistors when in the "off" state can produce local avalanching which in turn can force the n-p-n structure into conduction and diverting all the current from the external supplies into a small region of the power MOSFET. Local heating occurs leading to melting and destruction of the whole transistor as the source and drain electrodes become joined together. It is known as a single event burn-out (SEB).

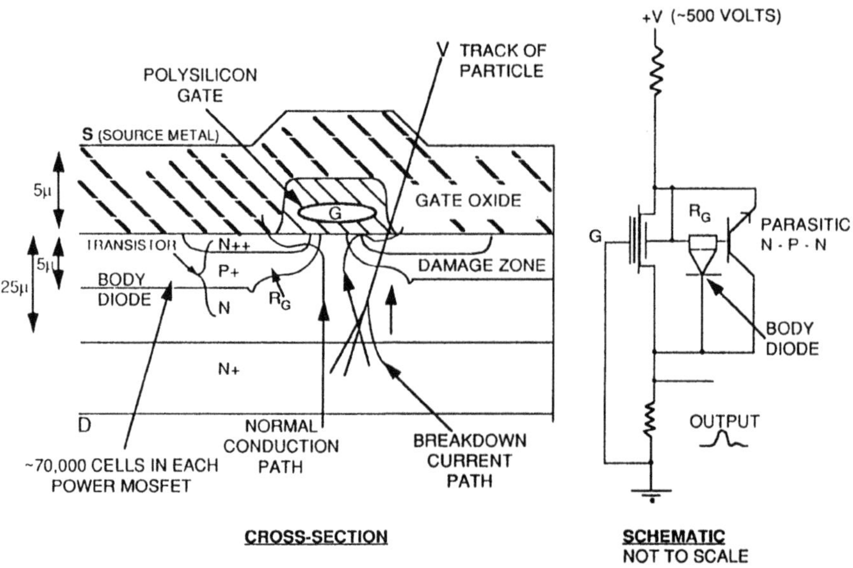

Fig. 11: Power MOSFET.

The sensitivity of the power MOSFET depends on its method of fabrication, the operating conditions and the external circuitry, particularly the amount of energy stored in the external circuitry available for "zapping" the transistor. Power MOSFETS are being manufactured which are claimed to be free from this form of destructive breakdown (single event burn-out). If it is intended to use readily available military or commercial devices, it is prudent to test samples from the batch intended for the satellite.

3.4. SUMMARY

In summary, cosmic particles have been shown to be capable of depositing sufficient charge in silicon devices to:

(i) Corrupt data stored in dynamic or static RAMS, microprocessors and logic arrays,

(ii) Initiate latch-ups in CMOS devices, and

(iii) Initiate a destructive latch-up in power MOSFETS.

3.5. CONTRIBUTIONS FROM DISPLACEMENT DAMAGE

3.5.1. *Introduction*. As this paper is concerned with the normal space environment, displacement damage effects are only likely to become significant when the total dose exceeds about 100 krad (Si). The predominant effects are discussed below under the device headings.

3.5.2. *p-n Diode*. The most important effects are an increase in the number of recombination-generation (r-g) centres producing an increase in the r-g current component, and a reduction in the switch-off time. Switching diodes have been shown to work up to massive doses of ionizing radiation, e.g. 100 Mrad (Si), provided the increased leakage current can be tolerated. Solar cells in their exposed position makes them very vulnerable to radiation damage with a consequent loss of efficiency.

3.5.3. *JFETS*. In common with the junction diode the JFET exhibits an increase in its gate leakage current, a decrease in transconductance and an increase in the 1/f noise. The JFET is a very suitable component for use in very hazardous environments.

3.5.4. *Discrete Transistors*. The appearance of r-g centres in the base region of the bipolar transistor results in a degradation of the current gain and a reduction in the base width. Wide base, low frequency bipolar transistors are of major concern.

Power MOSFETS are affected by changes in the resistivity of the drain region which increases the "on" resistance. The r-g centres will also increase the source-drain leakage current. The effect of displacement damage on the single event burn-up has not been investigated, but it is likely that the decrease in carrier lifetime may reduce the occurrence of SEB.

3.5.5 *Integrated Circuits*. Bipolar integrated circuits, such as op-amps, will have their performance degraded in a similar manner to the individual bipolar transistor. In the op-amp the open loop gain will be degraded, while the input bias current increases, the magnitude depending on the type of amplifier. The use of lateral p-n-p transistors and the super beta n-p-n transistors and its low operating currents are factors which increase the sensitivity to displacement damage.

MOS integrated circuits use majority carrier devices which are not significantly affected by displacement damage, but they may well exhibit an increase in the leakage currents. Charge storage on a node, as in a SRAM, may decay more rapidly and hence require an increase in the refresh rate. The charge storage in the gate oxide as described earlier (3.2.4) is a much more serious effect.

4. Techniques for Reducing the Hazards of Space Radiation

4.1. INTRODUCTION

The usual methods employed to reduce exposure to nuclear radiation are to reduce the time of exposure, keep as far as possible from the source and, lastly, to employ suitable shielding. It is not possible on a satellite to employ such methods as the sources of radiation are very energetic, and in some cases are omnidirectional, and the time of exposure is the duration of the mission, which can be up to twenty years. It is possible to shield effectively from the low energy electrons and protons in the earth's magnetosphere by using a few millimetres of aluminum. The low Z number materials such as carbon, aluminum, glass (i.e., in circuit boards) and hydrocarbons (i.e., in epoxy resins) are preferred as they minimize the Bremsstrahlung emission. Shielding in the spacecraft can be partly achieved by placing the microelectronic components adjacent to large masses, such as batteries and transformers. Localized shielding can be achieved by positioning small shields of tungsten or tantalum on the critical components.

It is impossible to shield against the energetic cosmic particles as they are capable of penetrating many meters of material. In fact shielding can make the situation worse because a reduction in energy <u>increases</u> the LET until the energy is less than the Bragg peak when the LET decreases.

It is difficult to shield against neutrons because of the mass of material required. The most satisfactory solution is to place the source of neutrons as far as possible from the sensitive circuitry.

4.2. TECHNIQUES FOR REDUCING THE HAZARD

4.2.1. *Introduction*. If the suggestions above for minimizing the hazards are not adequate, it will be necessary to rely on radiation hardened devices or commercial devices that have been tested to be sufficiently hard to the radiation. Radiation hardened components are supplied by a number of companies (e.g., Harris, United Technologies, GEC-Marconi, Matra Harris, IR). The range of devices offered is restricted and understandably not of the very latest technology. Information on the radiation performance of microcircuits can be found in a number of databases which have been compiled by JPL (NASA), Hahn-Meitner Inst. (ESA) [Refs 1,2,3].

4.2.2. *Device Design*. As a guide the following techniques can be employed in radiation hardening against total dose and SEU effects.

4.2.3. *Total Dose.* The microcircuits should be made with thin gate oxide (say 25 nm, Delta $V_T = K\ t_{ox}^2$), and the recessed oxide structure should be avoided. The growth conditions during the oxide formation and cooling must be carefully controlled to ensure a stable V_T. The field oxide must be processed to have a high threshold voltage to avoid surface conduction effects.

4.2.4. *SEU Sensitivity.* The main objective in the design is to reduce the amount of charge injected into the microcircuit component. This can be done by reducing the sensitive volume by the use of epitaxial silicon or, better still silicon-on-insulator (SOI) such as silicon on sapphire or SIMOX (ion implanted SiO_2 layer). In addition, the sensitivity of SRAM flip-flop memory elements can be reduced by incorporating resistors in the cross-coupling leads, at the expense of a loss in speed.

All the above methods entail additional costs with the result that some radiation hard microcircuits can cost between 10 and 100 times the cost of the equivalent commercial equivalent device. There is understandably a strong desire amongst designers with limited budgets to "make do" with the commercial item provided it meets their total dose requirements.

5. Terrestrial Testing Techniques

5.1. INTRODUCTION

It is clear from the earlier section that there is a need to simulate as closely as possible the expected conditions in space for a particular orbit and then to subject microelectronic components to these conditions. It is convenient to test for total dose effects and SEUs separately though combined exposures have been successfully attempted. It is desirable that the correct dose-rate be used at the correct temperature. The devices should preferably be powered up, a condition which is essential in the testing of SEU, latch-up and single event burnout in power MOSFETS.

5.2. TOTAL DOSE TESTING

A number of gamma-ray sources (e.g., ^{60}Co, ^{137}Cs and ^{96}Sr) can be used for simulating the total dose effect. The space industry has chosen ^{60}Co as it is widely available in a large range of strengths. The half life of ^{60}Co is 5.27 years, and it decays by emitting a weak electron followed by the emission of two gamma-rays of energies 1.17 MeV and 1.33 MeV. They are emitted with equal intensity and can be regarded as a single gamma-ray of 1.25 MeV. The specific gamma-ray constant of ^{60}Co is

$$\text{given by } 1.32\ R(m^{-2}\ h^{-1}\ Ci^{-1})$$

where R is the exposure in roentgens, (rads (Si) = 0.869 roentgen). For example 1 Curie (Ci) of ^{60}Co has a radiation output of 1.32 R/hour at 1 meter (assuming a point source). All testing is carried out in suitable shielded cells.

5.3. TOTAL DOSE DOSIMETRY

There are various techniques available for the detection and measurement of the dose-rate and total dose from ionizing radiation. The commonly used methods are the ion chamber (IONEX) for dose rate and for total dose, thermoluminescent dosimeters and sensitized perspex. A gamma alarm is used for safety reasons to indicate the presence of the radiation and the dose rate in the radiation cell. A mercury thermometer is kept in the cell to monitor the temperature.

5.4. TEST EQUIPMENT

The devices under test (DUT) can be tested outside the cell and then irradiated under bias if required and measured at intervals out of the cell. A more satisfactory arrangement is to measure in situ at the end of cable which then allows for a computer controlled system. The Hewlett Packard 4145A or B Parameter Analyzer has been used for measuring irradiated devices backed up by electronic digital volt/ammeters and digitally controlled power supplies.

It is usual to make up special jigs for a number of devices of each component type for biasing and measuring during irradiation and any annealing.

5.5. SINGLE EVENT UPSET TESTING

5.5.1. *Introduction.* High energy particle accelerators such as cyclotrons, tandem Van de Graff and linacs are used to produce a flux of high energy particles which are directed into the silicon chip under test. The irradiations are performed with the DUT under vacuum with the device package lid removed. The DUT is connected by a multiway cable to the computer system alongside the vacuum chamber [Ref. 7].

5.5.2. *Ion Sources.* Ion accelerators are few in number and are committed to other physics experiments limiting their availability and making them costly to use. One interesting simulator is the man-made element ^{252}Cf which undergoes spontaneous fission emitting high energy particles with large values of LET. ^{252}Cf is used for determining limiting cross-sections and sensitivity to latch-up. The characteristics of ^{252}Cf are [Ref. 10]:

Effective half life	2.639 years
Alpha decay	96.91% (6.08 MeV and 6.12 MeV)
Fission decay	3.09%

It can be shown that one microcurie of ^{252}Cf will produce

3.6×10^4 alphas/second
2.2×10^3 fission particles/second
4×10^3 fast neutrons/second.

Table 2 lists some typical ion species and their interaction with silicon.

Table 2: Typical Ion Species used for SEU Testing

Ion Species	Energy MeV	Initial LET MeV/mg/cm^2	Approx Range in DUT Microns
He (alphas)	5	0.6	25
Be	25	1.3	55
O	150	2.3	179
Ar	160	15	41
Fe	260	48	52
Kr	150	40	22
^{252}Cf fragments	90 – 140	43	15

A comprehensive listing of LET and range for all the elements in silicon and other elements are published by [Ref. 24]. If testing with alpha particles is required for examining sensitive devices such as DRAMS and CCD's, Americium-241 is a very suitable source. There is no isotope source that emits protons spontaneously, and the only source of high energy protons is an ion accelerator.

5.5.3. *Source Calibration Method.* The particle flux from an ion accelerator or ^{252}Cf fission source is measured at the position of the DUT using a surface barrier detector suitable for heavy ions (e.g., Ortec type BF-030-100-60). The output from the detector is amplified by a charge integrating circuit and fed to a multichannel analyzer. The particle count per unit of time can be recorded as well as the energy deposited in the detector.

5.5.4. *Experimental Method.* The measurement of SEUs requires the DUT to be operating under the normal specified conditions. The DUT must be driven by an external computer system so that data can be transferred to and from the DUT and checked and the number of events recorded. The SEU cross-section is given by

$$\text{Cross section } (cm^2) = \frac{\text{No. events}}{\text{No. particles}/cm^2}$$

The majority of devices require special interface circuit boards to couple them into the computer and through the vacuum seal into the evacuated test chamber. Problems arise due to operating the DUT at the end of two or three meters of multiway cable, but in most cases they can be overcome without degrading the performance of the DUT. Precautions must be taken during SEU testing to guard against latch-up. In some computer systems when a latch-up occurs the power is removed from the DUT to remove the latch-up. The event is recorded and normal testing continued. It is usually essential to write special software to test devices such as microprocessors, memories and logic arrays [Refs. 7,9,10].

The measurement of latch-up and single event burn-out in power MOSFETS using ^{252}Cf requires a simple circuit in which the DUT forms the switch in a monostable circuit (Fig. 12). When the latch-up occurs a limited current from a capacitor flows through the DUT and ceases when the voltage across the DUT has decayed to a value which can no longer sustain the SCR action. The capacitor recharges to its initial value (e.g., V_{DD}). Care is taken to ensure that the minimum of energy is stored in the capacitor, and the initial current surge is limited by a resistor. An electronic counter is used to register the number of events. The system can be left running for long periods when using a ^{252}Cf source [Refs. 7,8,9,23].

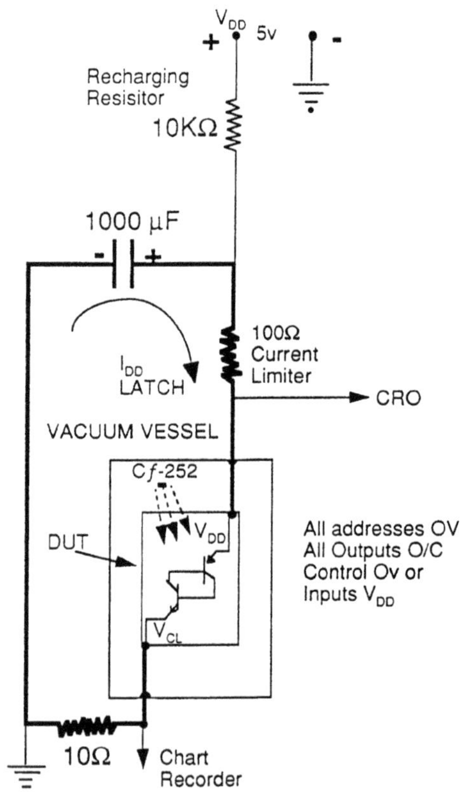

Fig. 12: Simplified latch-up test circuit [Refs. 9, 23].

6. Radiation Testing in Space

6.1. INTRODUCTION

The widespread use of microcircuits in satellites with many varied orbits makes it very desirable to measure the radiation environment in and around the vehicle so that ground testing will more readily represent the space conditions. This approach should help confirm the various predicted conditions. It would be of assistance if small radiation test circuits could be incorporated in modern satellites and the data transmitted to the ground as part of the housekeeping data.

6.2. METHOD OF MEASURING RADIATION EFFECTS IN SPACE

It is necessary to measure total dose and SEU events. The detectors should be small and easy to operate in the satellite.

6.2.1. *Total Dose.* The Geiger-Muller tube and the RADFET have been used for measuring gamma radiation. In the case of the GM tube, which has the disadvantage of requiring a high voltage, the dose is obtained by counting the number of pulses from the tube. The GM tube is very sensitive and it can have a limited life depending on the count rate.

The RADFET is not as sensitive but quite satisfactory for total doses exceeding 100's rads to over 20,000 rads. The sensitivity can be changed by altering the gate oxide thickness. In common with MOSFETS, the RADFET is prone to loose stored charge due to annealing. Temperature measurements should be made adjacent to the above detectors so that temperature corrections can be made [Ref. 19].

6.2.2. *Single Event Upset.* A number of approaches are being tried for detecting the charge deposited by a cosmic particle in a silicon device. These are:

(i) Detecting flips in a large array of static RAMS [Ref. 9].

(ii) Special flip-flops designed to be sensitive (JPL).

(iii) PIN diode array coupled to a charge amplifier and a simple pulse height analyzer. (See 3.3.1) [Ref. 14].

6.2.3. *Current Experiments.* Current equipments to measure various radiation effects are:

(i) CRRES (US)

(ii) CRUS (US)

(iii) CREAM (UK RAE) on Concorde, UOSAT3 and soon in the Shuttle [Ref. 22].

7. Summary

Microelectronic systems are essential for the conduct of space research and exploitation. It is therefore necessary for the microelectronic components to perform reliably during the life of satellite. The two important sources of radiation are the ionizing radiation produced by electrons and protons interacting with the satellite structures and SEUs produced by high energy particles.

In response to the space radiation hazard there is a need to predict the radiation environment in the satellite's orbit and then to simulate the effect on the ground. In the case of ionizing radiation the dose rate may be very low so that testing becomes lengthy and methods for producing a satisfactory answer in a shorter time are required. Methods are being researched to this end.

Many space system designers, especially those doing research with a limited budget, are being compelled to use commercial microcircuits with little or no radiation data available on them. Simulated testing of these parts becomes very important.

The radiation hardness of all the microcircuits in a satellite system should be verified before a detailed design is undertaken to avoid a later expensive redesign. The importance of the microcircuits in the space programme makes it necessary to obtain experimental information on radiation behavior from space to support the predictions and simulations used at present.

The purpose of these lecture notes has been to highlight the hazards experienced in space by microcircuits and to direct the reader to the wealth of published material on the subject.

8. Acknowledgements

The assistance of colleagues at the Harwell Laboratories and Dr. G. Wrenn and Dr. C. Dyer of RAE Farnborough is gratefully acknowledged.

9. References

Books

[1] V.A.J. van Lint, et al., Mechanism of Radiation Effects in Electronic Materials, Vol 1., John Wiley and Sons, New York (1980).

[2] G.C. Messenger and M.A. Ash, The Effects of Radiation on Electronic Systems, Van Nostraud Reinhold Company, New York (1986).

[3] T.P. Ma and P.V. Dressendorfer, Ionizing Radiation Effects in MOS Devices and Circuits, John Wiley and Sons, New York (1989).

[4] S.M. Sze, Physics of Semiconductor Devices, 2nd Edition, John Wiley and Sons, New York (1981).

[5] N.J. Rudie, Principles and Techniques of Radiation Hardening, Editions I and II, Western Periodicals, North Hollywood, California (1980, 1989).

[6] ESA, The Radiation Design Handbook. Published by ESA Publications Division, ESTEC, Netherlands, also A. Holmes-Siedle and L. Adams (1992), "Handbook of Radiation Effects", Oxford University Press (Spring 1992) (1987).

Journal Articles

The great majority of articles and papers on radiation effects in semiconductors are published in the December issues (No. 6) of the IEEE Transactions on Nuclear Science, Volumes NS-1 (1960) to NS-39 (1992). In addition the IEEE Nuclear and Space Radiation Effects Conference Short Course Notes 1982 to 1992 are a further source of reference material.

Review Articles

[1] E.G. Stassinsopoulos and J.P. Raymond, "The space environment for electronics", Proc IEEE, Vol 76, 11, p 1423 (1988).

[2] J.R. Srour and J.M. McGarrity, "Radiation effects on microelectronics in space", Proc IEEE, Vol 76, 11, p 1443 (1988).

[3] S.E. Kerns and B.D. Shafer, Editors, "The design of radiation-hardened ICs for space: A compendium of approaches", Proc IEEE, Vol 76, 11, p 1470 (1988).

[4] R.L. Pease, A.H. Johnston, and Azarewicz, "Radiation testing of semiconductor devices for space electronics", Proc IEEE, Vol 76, 11, p 1510 (1988).

[5] R.D. Rasmussen, "Spacecraft electronics design for radiation tolerance", Proc IEEE, Vol 76, 11, p 1527 (1988).

[6] J. T. Wallmark and S.M. Marcus, "Minimum size and maximum packing density of nonredundant semiconductor devices", Proc IRE, p 286-298 (1962).

[7] A.E. Waskiewicz, J. W. Groninger, V.H. Strahan, and D.M. Long, "Burn-out of power MOS transistors with heavy ions of Californium-252", IEEE Trans Nuclear Science, Vol NS-33, 6, p 1710.

[8] J. H. Stephen, T. K. Sanderson, D. Mapper, J. Farren, R. Harboe-Sorensen, and L. Adams, "Cosmic ray simulation experiments for the study of single event effects and latch-up in CMOS memories", IEEE Trans Nuclear Science, NS-30, 6, p 4464 (1983).

[9] T. K. Sanderson, D. Mapper, J.H. Stephen, J. Farren, L. Adams, and R. Harboe-Sorensen, "SEU measurements using Californium-252 fission

particles on CMOS static RAMS, subjected to a continuous period of low dose rate Co-60 irradiation", IEEE Trans Nuclear Science, NS-34, 6, p 1287 (1987).

[10] D. Mapper, T.K. Sanderson, J.H. Stephen, J. Farren, L. Adams and R. Harboe-Sorensen, "An experimental study of the effect of absorbers on the LET of the fission particles emitted by CF-252", IEEE Trans Nuclear Science, NS-32, 6, p 4276 (1985).

[11] P.S. Winokur, F.W. Sexton, G.L. Hash, and D.C. Turpin, "Total-dose failure mechanisms of integrated circuits in laboratory and space environments", IEEE Trans Nuclear Science, NS-34, 6, p. 1448 (1987).

[12] J. Farren, J.H. Stephen, D. Mapper, T.K. Sanderson, and M. Hardman, "Low level radiation testing of microelectronic components. Part 1, Review of current technology and proposed experimental programme", AERE Harwell Report R-11025 (1984).

[13] M. Hardman, J. Farren, D. Mapper, and J. H. Stephen, "Low level radiation testing of microelectronic components. Part 2, Preliminary studies of gamma dose rate effects on CMOS converters using the LORAD facility", AERE Harwell Report R-11679 (1985).

[14] D. Mapper, J.H. Stephen, J. Farren, B.P. Stimpson, D. J. Bolus, and A.M. Ellaway, "CREAM – a Cosmic Radiation Effects and Activation Monitor for Space Experiments". Part 1, Design and construction of experimental packages for flight on a NASA shuttle mission". AERE Harwell Report R-12916 (1987).

[15] A.H. Johnston and S.B. Roeske, "Total dose effects at low dose rates", IEEE Trans Nuclear Science, NS-33, 6, p 1487 (1986).

[16] D. Binder, E.C. Smith, and A.B. Holman, "Satellite anomalies from galactic cosmic rays", IEEE Trans Nuclear Science, NS-22, p 2675. (1975).

[17] A.H. Johnston, "Super recovery of total dose damage in MOS devices,", IEEE Trans Nuclear Science, NS-31, 6, p 1427 (1984).

[18] A.G. Holmes-Siedle, "The space charge dosimeter: general principles of a new method of radiation detection", Nuclear Instruments and Methods, Vol 121, p 169 (1974).

[19] A.G. Holmes-Siedle and L. Adams, "RADFET: A review of the use of a metal oxide silicon device as an integrating dosimeter", Radiation Physics and Chemistry, Vol 28, p 235 (1986).

[20] T.R. Oldham, A.J. Lelis, H.E. Boesch, J.M. Benedetto, F. B. McLean and McGarrity, "Post-irradiation effects on field oxide isolation structures", IEEE Trans Nuclear Science, NS-34, 6, p 1184 (1987).

[21] R.L. Pease, R.M. Turfler, D. Platteter, D. Emily, and R. Blice, "Total dose effects in recessed oxide digital bipolar microcircuits", IEEE Trans Nuclear Science, NS-30, 6, p 4216 (1983).

[22] C.S. Dyer, A.J. Sims, J. Farren, and J. H. Stephen, "Measurements of solar flare enhancements to the single event upset environment int he upper atmosphere", IEEE Trans Nuclear Science, NS-37, 6, p 1929 (1990).

[23] J.H. Stephen, et al., "Investigation of heavy particle induced latch-up using a CF-252 source in CMOS, SRAMS and PROMS", IEEE Trans Nuclear Science, NS-31, 6, p 1207 (1984).

[24] Littmark and J. F. Zielger, "Handbook of range distributions for energetic ions in all elements", New York, Pergamon Press (1980).

[25] T. C. May and M. H. Woods, "Alpha-particle-induced soft errors in dynamic memories", IEEE Trans Electron Devices, Vol Ed 2b, p 2 (1979).

Data Bases

[1] JPL.

[2] ESA Radiation Data Base, ESA, ESTEC, Netherlands (1989).

[3] Hahn-Meitner Institute, Data compilation of irradiation tested electronic components (2nd edition), HM1-B353, Hahn-Meitner Institute, Berlin, Germany (1987).

PLASMA INTERACTIONS AT LOW ALTITUDES

E. J. DALY • D. J. RODGERS
European Space Agency
ESTEC, Noordwijk, The Netherlands

Dedicated to the memory of Bernt Maehlum.

ABSTRACT. The low altitude environment is normally characterized by the cold dense plasma of what is mainly an ionospheric domain. Although it is apparently a more benign environment than that experienced at high altitudes, spacecraft can experience a number of important plasma-related effects. These interactions are discussed and their impacts are identified for future space systems such as the International Space Station and its co-orbiting platforms, shuttle vehicles, and polar-orbiting platforms. Significant surface charging is not normally possible in such a low-temperature plasma. At high latitudes spacecraft cross the auroral oval. During magnetospheric disturbances, hot electrons precipitate down field lines and can cause high levels of surface charging. This has been observed on polar orbiting (DMSP) satellites. The velocity of a spacecraft is faster than the local ion sound speed and this results in the creation of wakes and ion-free void regions. Charging of surfaces or other bodies is expected to reach higher levels within such wakes. In addition, the large plasma disturbance represented by the wake leads to plasma fluctuations and enhanced electrostatic noise levels. Spacecraft carry a local contaminant environment with them or produce one through venting, thruster firing, etc. These neutral contaminants can be ionized through charge-exchange with the ambient plasma and then form a local pick-up ion environment. Enhanced wave activity and surface contamination can result.

1. Introduction

The next few years will see the development and flight of several large space systems which will operate in low altitude orbits. (By 'low altitude' we mean altitudes below ~1000 km, also referred to as LEO: Low Earth Orbit.) The largest planned system is the International Space Station, but in addition there are plans for the European Columbus free-flying man-tended laboratory and for various shuttle-deployed platforms such as ESA's Eureca and Japan's SFU. These missions are intended primarily for 28° inclined orbits. There are also plans around the world for a considerable increase in the exploitation of the polar orbit with large platforms such as NASA's EOS, ESA's Polar Platforms, and

R. N. DeWitt et al. (eds.), The Behavior of Systems in the Space Environment, 437–465.

continuation of the ESA ERS and CNES SPOT programs. To these we must add the large reusable transportation systems – the US Space shuttle (and extended-duration EDO), Europe's Hermes 'spaceplane', Japan's HOPE, the CIS's Buhran and finally the continuing use and development of the CIS's MIR spacestation. This list is not exhaustive and is no doubt subject to 'revision' but it does indicate important trends: to larger systems, to greater sophistication, and to greater reusability. These inevitably increase the system susceptibility to problems induced by the space environment in general.

In this paper we deal with plasma interactions with such systems and in particular with non-active aspects of the interaction. 'Active' generally means that one deliberately perturbs the plasma; for example, by emitting a charged-particle beam, applying large potentials (e.g., high voltage solar arrays), deploying tethers, etc. These interactions are dealt with by other authors in these proceedings. We concentrate on three basic areas: (i) the mesothermal motional interaction leading to wake creation; (ii) spacecraft charging in polar orbits; and (iii) contaminant-plasma interactions. The first two areas will be given the greatest weight. In discussion of these topics, an attempt will be made to review flight data, ground-based experiments and computational activities. The space available does not allow a complete review, and some aspects are necessarily abbreviated. Therefore, the reader is referred to previous reviews of the subject by Stone [Ref. 46], Murphy and Lonngren [Ref. 24], Wright [Ref. 51], Raitt [Ref. 35] and Martin, et al. [Ref. 21], and to proceedings of several relevant meetings [Refs. 10, 30, 34, 38].

2. Introduction to Plasma Interactions At Low Altitudes

The low-altitude plasma is normally cold and dense. This has two important consequences. Firstly, natural electrostatic charging of a body to more than a few volts is not normally possible. In a non-drifting plasma, the more-mobile electrons tend to cause surfaces to charge negatively to a degree sufficient to repel electrons and accelerate ions until their currents balance and an equilibrium potential is reached. This is a common characteristic of floating Langmuir probes in laboratory plasmas. The floating potential is linked closely to the characteristic energy of the electron distribution (temperature if the distribution is Maxwellian). Typically a surface might charge to about three times the electron temperature (expressed in volts). So in a cold plasma, high-level charging is not possible. In space, the motion of the spacecraft through the ionosphere, leads to enhanced ion currents and electrons and solar photons induce the emission of secondary electrons from a surface. Consequently it is even more difficult for a low-altitude spacecraft to reach high potentials. We will see later that this statement is untrue when an energetic component is added to the electron distribution, such as encountered in auroral regions. The second consequence of a cold dense plasma is that potentials on bodies are more effectively screened from the surroundings. The perturbation to the potential in space around a body is limited to a region known as the *sheath*. The scale length for the size of a sheath is the Debye length. If a small potential perturbation ϕ_0 is established in the plasma, the potential around it falls off as:

$$\phi = \phi_0 \exp(-|x| / \lambda_D)$$

where λ_D is the Debye length [Ref. 5], given by:

$$\lambda_D = \left(\frac{\varepsilon_0 K T_e}{e^2 n} \right)^{1/2} \qquad (1)$$

Here, K is Boltzmann's constant, T_e is the electron temperature in Kelvin, n is the ambient plasma density, e is the electron charge and ε_0 is the permittivity of free space. Figure 1 shows the altitude-dependence of the plasma density, electron temperature and Debye length at low altitude. It is apparent that the millimeter-scale of Debye length is orders of magnitude smaller than typical spacecraft dimension. This is therefore a *thin sheath* regime.

It is important to note that Debye screening is a phenomenon associated with small potential perturbations, when Poisson's equation can be linearized. High level charging or systems with high applied potentials have to be treated with non-linear theory. The diode theories of Langmuir and Blodgett and Child and Langmuir become appropriate [*Katz*, these proceedings]. The Child-Langmuir sheath thickness then defines the extent of the perturbation to the local plasma:

$$D_{CL} = 1.26 \lambda_D \, (e\phi / KT_e)^{3/4}.$$

This can be considerably larger than the Debye length and comparable to the spacecraft size. Note that the Child-Langmuir sheath distance depends on the body potential whereas the Debye length does not. The importance of the sheath stems from the effect it has on current collection: it modifies the effective size of a body as seen by the local particle population. The short Debye length at low altitude is in contrast to the large Debye lengths at high altitudes where most of the time sheath and space-charge effects may be neglected; that is, by contrast, a *thick-sheath* regime [*Wrenn*, these proceedings].

3. Motion of a Body through Plasma

At low altitude, a spacecraft's velocity (body velocity) v_b is *mesothermal*, i.e. it is faster than the ion thermal speed but slower than the electron thermal speed:

$$v_{\theta i} < v_b < v_{\theta e}.$$

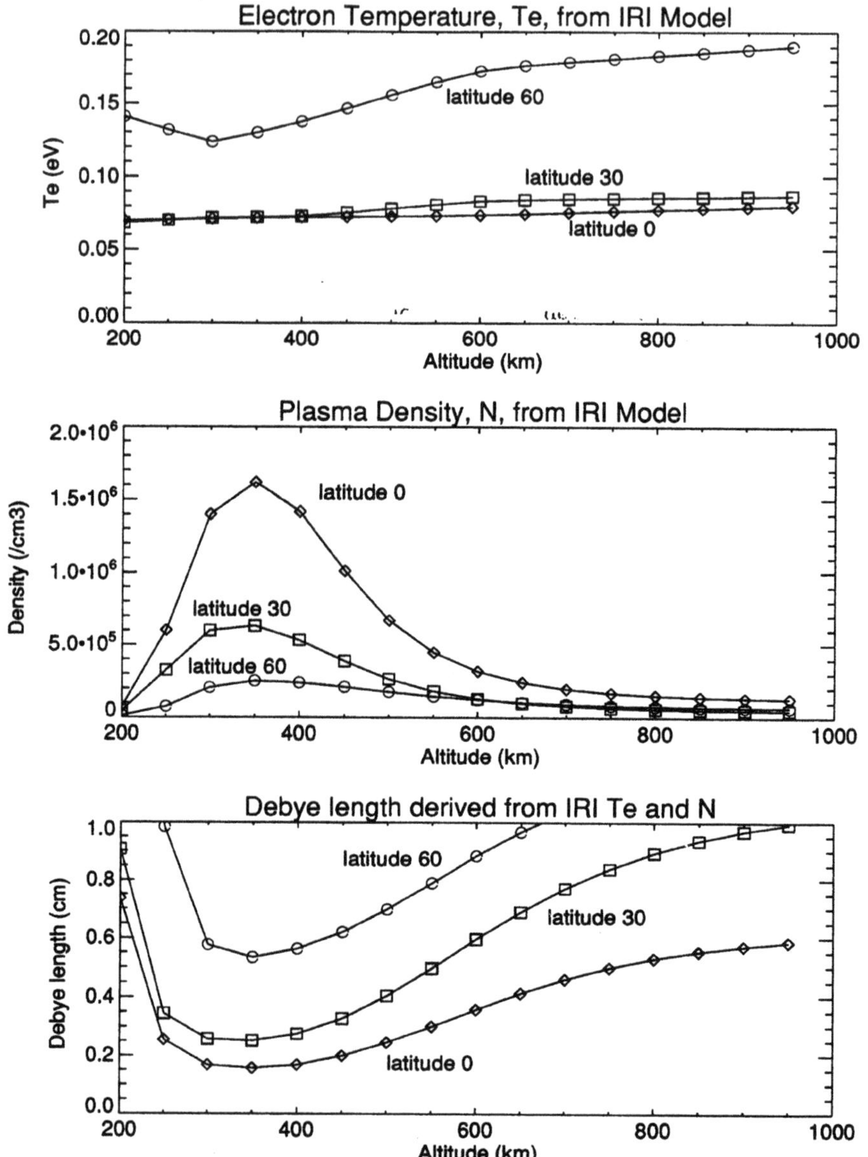

Fig. 1: The variation of plasma density, electron temperature and Debye length with altitude for latitudes 0°, 30° and 60°. Derived from the IRI ionospheric model.

The spacecraft orbital velocity is given by:

$$v_b^2 = \mu \, (2/r - 1/a)$$

$$= \mu/r \text{ for a circular orbit,}$$

where μ is the Earth's gravitational constant ~$4\text{x}10^{14}$ m^3/s^2, a is the orbit's semi-major axis and r is the satellite's radial position. This gives between 7.5 and 8 km/s at low altitude. The above expression shows that the velocity can be up to $\sqrt{2}$ higher at the perigee of a highly eccentric orbit compared to a circular orbit with r=a. The velocity is also larger for orbits around large planets which have a higher gravitational constant.

A particle's thermal velocity is given by:

$$v_\theta = (2KT/m)^{1/2},$$

where T is its temperature and m is the particle mass. (When T is expressed in electron volts (eV), this reduces to $5.9\text{x}10^5\sqrt{T}$ m/s for electrons and $1.4\text{x}10^4\sqrt{(T/A)}$ m/s for ions; A is the ion mass in amu.) Temperatures of electrons and ions in low Earth orbit are typically below 0.1 eV and the dominant ion species is oxygen (A=16). So the ion thermal speed is around 1 km/s while the electron thermal speed is around 150 km/s, confirming that a spacecraft moves mesothermally.

A further important velocity is the ion acoustic, or sound speed,

$$v_s = (KT_e \, / \, m_i)^{1/2}$$

which is dependent on the electron temperature T_e and the ion mass m_i. The reason for this mixing of electron and ion properties (T_e and m_i) is that motions are often a result of electron attempts to move by virtue of their relatively high thermal speeds while they are restrained to behave collectively by the ions whose motions are hampered by their inertia. The electrons are the "driving force" in establishing electric fields to which ions then respond. This equation is strictly only applicable in the cold ion, T_i=0, limit. This behavior also leads to an important plasma wave mode, the ion acoustic wave [Ref. 5]. At low altitudes, since ion and electron temperatures are similar, the ion sound speed is similar to the ion thermal speed. Motions are therefore also *supersonic* with respect to the ions. The body *Mach number*, defined as

$$M = v_b \, / \, v_s,$$

has a value of between 6 and 8.

4. Plasma Wakes

The speed inter-comparisons made above imply that an ion void will form 'behind' a body (on the anti-velocity vector side) since the ion motion relative to the body has only a small component perpendicular to the spacecraft velocity vector. This is referred to as the *near-wake* region. (We will use *longitudinal* to refer to directions parallel to the velocity vector and *radial* to refer to directions perpendicular to the velocity vector.) Figure 2 shows the near-wake void schematically, along with other interesting features of the interaction.

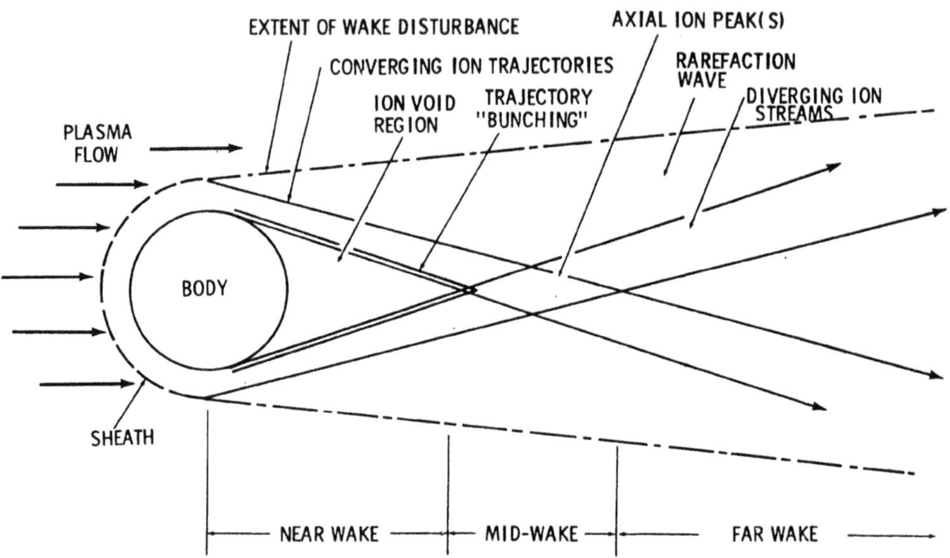

Fig. 2: Schematic of motionally-induced interaction of a body in low Earth orbit [Ref. 46].

When the plasma flowing relative to the spacecraft impinges on the edges of the body, a rarefaction wave is established which propagates radially away from the body at the ion sound speed v_s. As the spacecraft moves, a cone is produced by the boundary of this rarefaction region, which is known as the *Mach* cone by analogy with aerodynamics. The angle that this boundary makes with the velocity vector is the *Mach angle*:

$$\theta_M = \tan^{-1}\left(\frac{1}{M}\right) = \tan^{-1}\left(\frac{(KT_e/m_i)^{1/2}}{V_b}\right) .$$

For a body at or near the local plasma potential, the filling of the wake void is due to a combination of ion thermal motions and motion related to the ion sound speed (that is, driven by electron electrostatic effects). The void is a region of negative potential because of the mobility of the electrons; a barrier is established to exclude them and encourage ions. Wright [Ref. 51] provides a discussion of this wake filling process. An approximate closure length L is:

$$L \approx \frac{V_b}{V_s} R$$

where R is the object radius perpendicular to the flow direction. Filling of the void is an equivalent problem to expansion of a plasma into a vacuum, and so-called *self-similar* solutions are found [Ref. 51]. Here, quasi-neutrality is assumed, electrons obey the Boltzmann law (they are in equilibrium with the electric field):

$$n_e = n_0 \exp\,(e\phi\,/KT_e)\,,$$

and resulting ion (plasma) motions are related to the sound speed. Estimation of the void closure length in practical situations is complicated by potentials established on the spacecraft which exert a force on the ions. Normally these potentials are negative so that ions are further attracted into the void. This is especially true in cases of severe charging in polar orbit, discussed further below, where hundred or thousand-volt potentials can be established. The resultant electric field leads to a focusing of ions. This focusing is strongest in an axially symmetric geometry. Ions are deflected towards the axis at the body's sheath edge and continue to experience the field as they move close to the edge of the void. Near the axis, they encounter ions deflected from other parts of the body. The convergence of ions leads to a region of enhanced ion density and in the case of axial symmetry to an axial peak in density. Some ions cross the axis while others are deflected by the positive space-charge in a near-axial direction. This region is the *mid-wake*. Ions repel each other in this region and further 'downstream', in the *far-wake* region, the ion streams diverge. These features are seen in both computational and ground-based experimental simulations [Ref. 23]. Figure 3 shows the results of a two-dimensional 'particle-in-cell' numerical simulation of the wake of a charged disk [Ref. 36]. The features discussed above are clearly visible. Figure 4 shows the results of an experimental simulation with similar parameters [Ref. 22].

Detailed in-space measurements have yet to map fully such morphology, but many features of wakes were confirmed by measurements made around the Shuttle by the Plasma Diagnostics Package, PDP, as a bay/arm-mounted monitor on STS-3 [Ref. 25] and as a free-flying sub-satellite on Spacelab-2 [Refs. 49, 26]. This experiment performed the first in-situ measurements of plasma wakes behind a large object (around $1000\lambda_D$ wide) in LEO. It had the advantage of being capable of exploring the mid- and far-wake regions out to around 240 m from the shuttle. Density depletions and the Orbiter's Mach cone were clearly observed. Some evidence for the ion stream focusing was also seen. Stone, et al. [Ref. 47] report that the observed filling

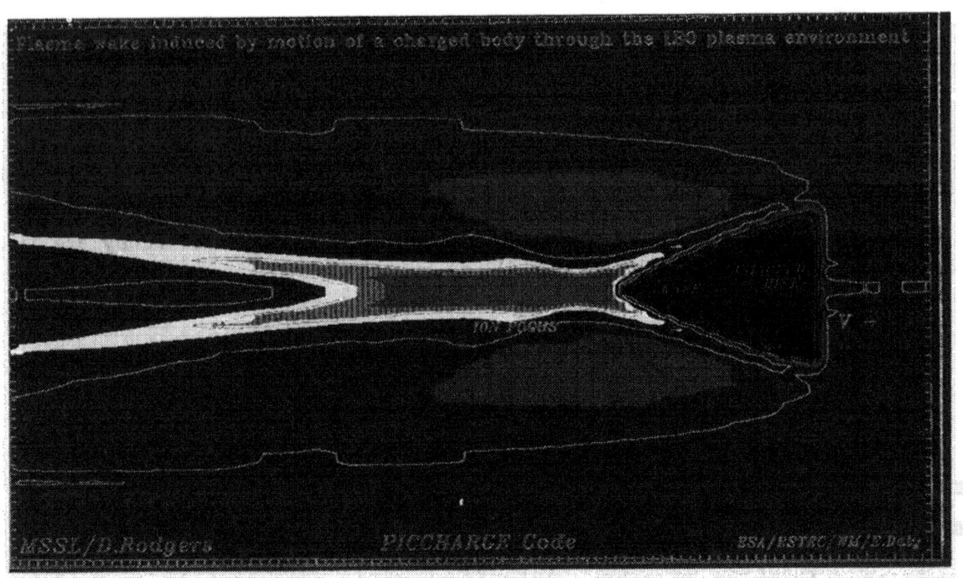

Fig. 3: Two-dimensional numerical simulation of a plasma wake, made using the PICCHARGE code [Ref. 36]. Plasma is incident on a charged disk from the right. This shows the Mach cone, the void region and ion focusing to an axial peak.

of the Shuttle wake was consistent with the self-similar expansion mentioned above. The PDP also measured enhanced electron temperatures in the rarefied void region. Previous small satellites, such as Ariel 1 [Ref. 14], Explorer 31 [Ref. 39] and AE-C [Ref. 40] have observed the near-wake void and other wake features in a variety of parameter regimes. Features such as enhanced electron density and axial ion peaks were reported. These observations have been summarized by Stone [Ref. 46] and by Martin, et al. [Ref. 21].

Murphy and Katz [Ref. 27] used the three-dimensional POLAR computer code to compare shuttle measurements with numerical predictions of the wake morphology. POLAR also uses a self-similar model for expansion into the wake void. The agreement was generally good, but the work indicated the importance of including the contaminant-generated plasma which is currently missing from the POLAR model.

The major ion current to a body moving mesothermally is the *ram ion current*. Incident ions can also be reflected and scattered, with energies which can exceed the Critical Ionization Velocity (CIV) of the ambient neutral gas [Ref. 45].

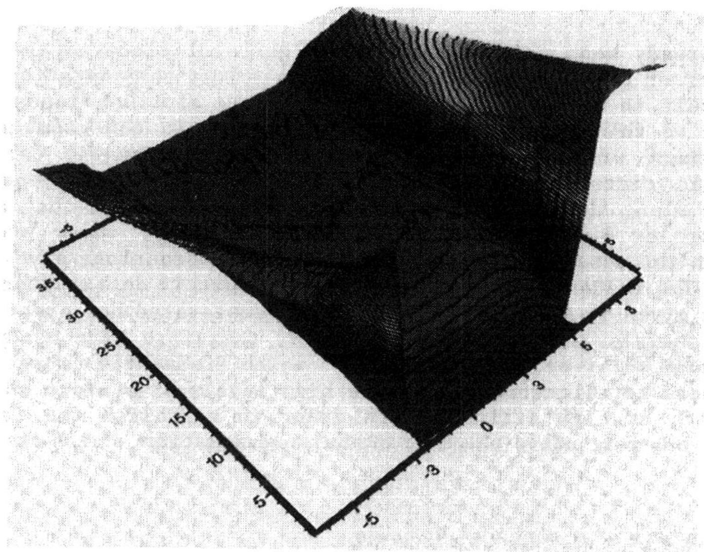

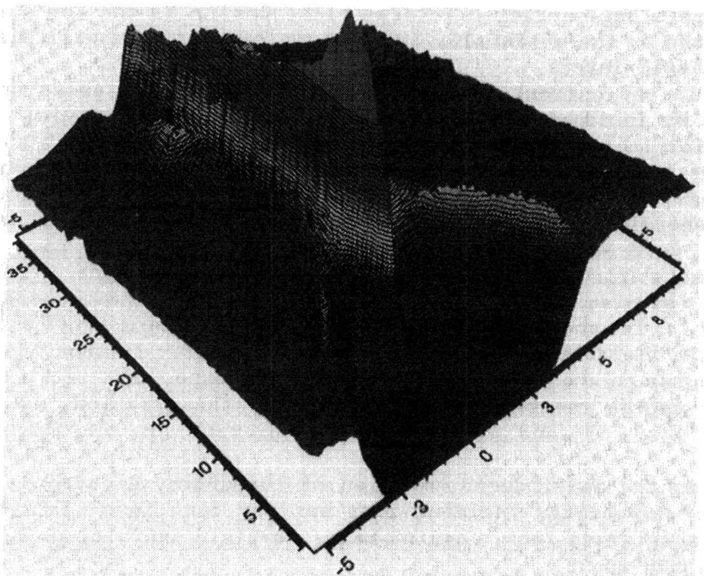

Fig. 4: Experimental observation of wake phenomena made in the NDRE facility. Plasma is flowing from the lower right onto a disk. Upper part shows the ion density void with no potential on the disk. Lower shows closing of void and axial peak due to $-100\mathrm{x}T_e$ potential on disk.

5. The Possibility of High-Level Charging in Polar Orbits

It has already been pointed out that high-level electrostatic charging, which is often seen on high altitude orbits (close to geostationary orbit), is difficult to induce in the cold dense low altitude ionosphere. The exception to this is in the auroral regions where the geomagnetic field lines connect with the plasma sheet of the geomagnetic tail. During geomagnetic disturbances, electrons in this region are energized and propagate along the field lines towards the poles where they are further accelerated by B-field-aligned potentials [Cowley, these proceedings]. They reach the upper atmosphere, where their interactions give rise to the aurorae. The reader is referred to the introductory description of auroral processes given by Akasofu [Ref. 1]. The accelerated hot electrons, which can have energies in the range 10-20 keV, therefore have ready access to low altitude polar-orbiting spacecraft which frequently cross the auroral oval. These hot electrons can cause high levels of surface charging. In common with at high altitudes, in order to establish the level of the charging, one must solve the current balance equation, which at equilibrium is:

$$J_e + J_i + J_{se} + J_{si} + J_\gamma + J_R = 0 \tag{2}$$

where J_e is the ambient electron current and includes both cold ionospheric and precipitating energetic auroral electrons; J_i is the ion current which is dominated by the motionally-induced ram current. We will discuss these in more detail later.

J_{se} is the current of backscatter and secondary emission electrons generated by incident electrons. Many materials have a yield of greater than one for incident electrons of energies from below 1 keV to a few keV. This leads to 'threshold energies' for spacecraft charging [Wrenn, these proceedings] because a sufficient fraction of the incident electrons need to have energies above a certain value before there is a net charging current. Cold ambient electrons produce few secondary electrons. The major effects arise with precipitating energetic electrons. True secondary electrons (as distinct from backscattered electrons) have energies of only a few eV. In the LEO regime this leads to Larmor gyration in the geomagnetic field with radii of a few centimeters. If the field is nearly perpendicular to the surface, electrons can escape. But if it is inclined, electrons can be returned to the surface by the gyration, as illustrated in Fig. 5 [Ref. 18] and their ability to prevent surface charging will be impaired.

J_{si} is the ion-induced emission of low-energy electrons. This will be low for low-energy incident ions but may come into play when charged surfaces accelerate ions. Again, magnetic field inhibition of the emitted electrons may occur.

J_γ is the current of low-energy photoelectrons which result from the incident solar UV flux. This current density is of the order of 20 $\mu A/m^2$. Some photoemission may result from Earth albedo illumination.

J_R is the conducted current to other parts of the spacecraft, when there are potential differences.

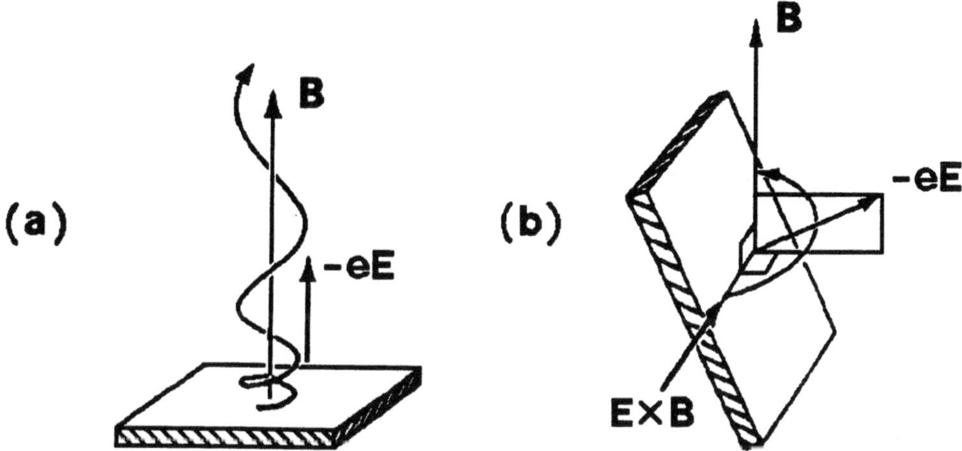

Fig. 5: The inhibition of low-energy secondary emission caused by a magnetic field [Ref. 18]. Electrons can only escape if the angle between *B* and the surface is sufficiently large.

In section 2 we said that the spacecraft motionally-induced ram current was large. So at the heart of the polar charging question is whether the electron current, J_e, is sufficient to exceed the ion current, J_i, and if it is, whether the charging time-scale is short compared with the transit-time of a satellite through the oval (typically a minute or so). This latter question depends on the satellite capacitance and on the charging current. We summarize some of the major differences between high-altitude and low-altitude charging processes in Table 1.

Table 1: Comparison of plasma characteristics at high and low altitude which are important for spacecraft charging.

High Altitudes	*Low Altitudes*
- densities are low ($\sim 10^6$ m^{-3})	- densities are high ($\sim 10^{11}$ m^{-3})
- sheaths are large — no space-charge effects	- sheaths are small — space-charge effects
- motional effects are negligible	- motional effects important
- environment quasi-isotropic	- environment strongly anisotropic
- storm-time temperatures high ($T_e \sim$ 10–20keV)	- energetic auroral distributions ($E_e >$ 10keV)
- events last several minutes	- passage through auroral arcs in seconds
- charging currents $\sim 2\mu$A/m^2	- charging currents $\sim 100\mu$A/m^2

The ion ram current ($J_i = e\ n_i\ v_b$) is about 1 mA/m^2 ($n_i = 10^{12}$ m^{-3}) under quiet conditions. During aurorae, the ambient density can be depleted and n_i can drop to below 10^9 m^{-3} yielding currents as low as 1 μA/m^2. In a severe aurora, the electron charging current is about 100 μa/m^2. The ambient (i.e., nonprecipitating) thermal electron current can be comparable to or higher than this but because these electrons are cold, any significant potentials developing will exclude them from the problem. These and other parameters are summarized in the following table:

Table 2: Some illustrative plasma characteristics in the auroral zones.

Ambient ion temperature	0.1 eV
Amb. electron temperature	0.1 eV
Ambient plasma density	10^{11} m^{-3} (quiet)
	10^9 m^{-3} (disturbed)
Ram ion current, J_i	1 mA/m^2 (quiet)
	1 μA/m^2 (disturbed)
Precipitating el. current, J_e	100 μA/m^2 (disturbed)
Debye length, λ_D	1 cm
Larmor radius of cold el.	2 cm (cold); 4 m (hot)

In considering whether a satellite will charge, one must consider changes to the sheath around it as potentials begin to build, and the effect that this has on the collection of current. When the concern is natural negative charging, we must ask if ion collection is enhanced by sheath modification of the effective collection area to an extent sufficient to offset the charging effects of precipitating electrons. Parks and Katz [Ref. 31] (also Katz and Parks [Ref. 15]) developed a simple model to demonstrate the possibility of high level charging in auroral regions. Their model considered the ram ion current and the precipitating auroral electron current as unidirectional currents onto a sphere. Apart from this, their model was not geometrical. It did not, for example, consider the wake generated by the body or differential charging caused by electrons being incident on specific parts of the surface. Despite these admitted shortcomings, the model was able to demonstrate the physical mechanisms involved in the phenomenon. The model was based on Langmuir-Blodgett theory of a sheath surrounding a large ($R_0 \gg \lambda_D$) spherical 'probe' at high ($|e\phi| \gg KT_e$, KT_i) potential. Parks and Katz indicate that when the sphere is attracting and collecting ions, the effective collection radius R_E is given by an increasing function of potential:

$$\frac{R_E}{R_0} = f\left(\frac{e\phi}{K\Theta_i}\left(\frac{\lambda_D}{R_0}\right)^{4/3}\right)$$

giving

$$R_E \approx R_0 + \frac{2\sqrt{2}}{3}\lambda_D\left(\frac{e\phi}{K\Theta_I}\right)^{3/4} \qquad (3)$$

for $R_E/R_O \leq 1.05$, where R_O is the sphere radius and Θ_i is an equivalent isotropic temperature assigned to the mono-energetic streaming ram ions, such that the currents entering the sheath in the two cases are equal. This requirement yields:

$$k\Theta_i = \pi/4\ E_i$$

where E_i is the equivalent streaming energy of the ram ions. The effective radius Eq. (3) is clearly related to the Child-Langmuir sheath thickness for high-voltage sheaths. [The reader will recall that the sheath thickness is only represented by λ_D in low-potential (linear) circumstances.] The above variation in ion collecting area is clearly more significant for small spacecraft. Collecting area is a function of body potential, as mentioned previously.

Precipitating electrons are represented by Parks and Katz [Ref. 31] by an equivalent Maxwellian distribution and the electron collection is determined simply by the Boltzmann relation:

$$J_e = J_{e0}\ \exp(e\phi/KT_e),$$

where J_{e0} is the electron thermal current at the sheath edge. This current is collected over an area of πR_O^2, whereas the ion current is collected over πR_E^2. In reality, the electron distribution is strongly non-Maxwellian, as will be shown in the next section.

Even allowing for secondary emission, the model shows that a large spacecraft can reach high potentials in the precipitating electron environment before ion collection can become sufficient to balance the electron current. The results of Katz and Parks [Ref. 15] are shown in Fig. 6, where the equilibrium potential of the sphere is shown as a function of the ratio of net precipitating electron to net ram ion currents for two sphere radii. The net currents include secondary emission terms. Table 2 shows that ratios of 10 are easily achievable in the absence of secondary emission. Since secondary emission for electrons typically falls off above a few KeV for many materials, its influence is not great. Even with precipitating-to-ram ratios as low as 2-3, potentials of the order of a kilovolt are predicted for a 5 m sphere, as shown by Fig. 6. The importance of the ratio of hot electron flux to total ion flux is further emphasized in the review of Laframboise and Parker [Ref. 19] (albeit with slightly different definitions).

The high auroral electron currents (J_E) lead to correspondingly fast charging rates:

$$dV/dt = JA/C \sim J_E R_0/4\varepsilon_0 \sim 10^6\text{-}10^7\ V/sec$$

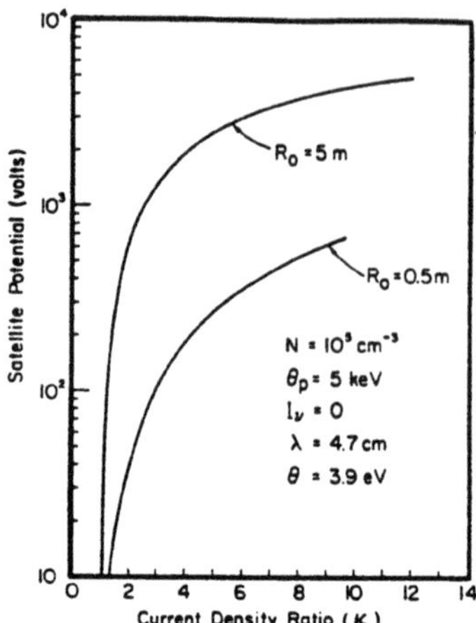

Fig. 6: The results of the analysis of Katz and Parks [Ref. 15] showing the potential reached by a spherical body as a function of the precipitating electron to ram ion current ratio. This shows that kilovolt levels are possible, especially for larger structures.

for absolute charging, and

$$dV/dt \sim J_E\, d/\varepsilon \sim 10^7 d\ V/\mathrm{sec}$$

for differential charging of a dielectric coating of thickness d meters. For a 50 μm Kapton layer, this gives ~500 V/sec. Crossing severe auroral arcs can take several seconds, corresponding to an orbital arc of 10-100 km [Ref. 12], allowing high level absolute and differential charging to be established.

Note that the Parks and Katz model does not consider differential charging between different parts of the satellite, nor explicitly anisotropic effects which are very important. If one considers the wake structures presented in section 3, it will be realized that surfaces or bodies in the void region will be subject to low initial ion currents. Precipitating electrons will cause high-level charging which will be followed by ions being pulled into the void and onto charging surfaces. Apart from being highly non-Maxwellian in energy distribution, the electron incidence is highly anisotropic and will lead to large asymmetries. Further complication results from the fact that negative potential regions established in the wake represent a potential barrier, inhibiting low-energy secondary emissions from surfaces. Polar charging is clearly a highly complex phenomenon, ideally requiring explicit treatment of anisotropy and geometry.

Katz, et al. [Ref. 16] and Rubin and Besse [Ref. 37] considered analytically the charging of bodies (such as an astronaut performing EVA) in the wake of the Shuttle Orbiter during an auroral event, considering ions pulled onto the body. They find that kilovolt charging levels are attainable and also show that high charging currents are not required. Work with the three-dimensional POLAR computer code [Ref. 6] also showed EVA astronaut charging [Ref. 13], and significant differences between astronaut and nearby shuttle potentials, a situation which is clearly dangerous.

6. The Observation of High-Level Charging in Polar Orbits

The first direct observations of high-level charging in polar orbits were made from the DMSP satellite [Ref. 12]. DMSP had an environment-monitoring package which included electron and ion plasma instruments; measuring fluxes of both species in the energy range 30 eV to 30 keV. The electron sensor was able to monitor the occurrence of high-energy electrons during auroral events. The sensors measured particle fluxes from the zenith which, in the auroral oval, are nearly in the field-aligned (B) direction (i.e., precipitating). DMSP also had a Langmuir probe and a retarding potential ion analyzer (RPA) which monitored the ambient (i.e., cold) plasma characteristics.

Figure 7 shows a spectrogram from the electron and ion sensors, reproduced from Yeh and Gussenhoven [Ref. 52]. In these plots, the horizontal axis represents time (UT), with tick-marks every minute, and the vertical axis represents particle energy. The color-coding depicts particle flux, as shown by the color scale. The upper panel shows the time-evolution of electron fluxes. The spectrograms show the entry of the satellite into the auroral oval shortly before 14000 UT. The electron spectrogram between about 14000 and 14060 shows two events having typical *inverted-V* spectrogram shapes. That is, the intense fluxes at high energy which are 'beam-like' stand apart from the low-energy part of the spectrum; the spectrogram shows them as a bright band at high energy. They evolve in time such that the energy of the 'beam' increases to a maximum value and then decreases again, producing an inverted-V signature [Ref. 8]. Coincident with these particular inverted-V events, the ion sensor saw fluxes of accelerated ions as can be discerned on the lower panel by pink-blue features. The first one is a brief burst of ions of energies up to 500 eV and the second was longer-lived with energies up to 100 eV. These are interpreted as ions which were initially cold but are accelerated in the electric field of the spacecraft. Their energies indicate the spacecraft potential with respect to the local plasma. Therefore, the spacecraft is inferred to have charged to -500 V and -100 V respectively during these two events. Many examples of charging of DMSP have been observed and the published record is -700 V [Ref. 52] although -1400 V has recently been reported [Ref. Rubin]. These observations are not able to say anything about differential charging on the satellite, but relate to the absolute satellite charging level, or at least to the level at the sensor.

Yeh and Gussenhoven [Ref. 52] have performed a statistical survey of environments seen by the DMSP satellite and their relationships with charging levels. Various models of the electron energy spectrum are given,

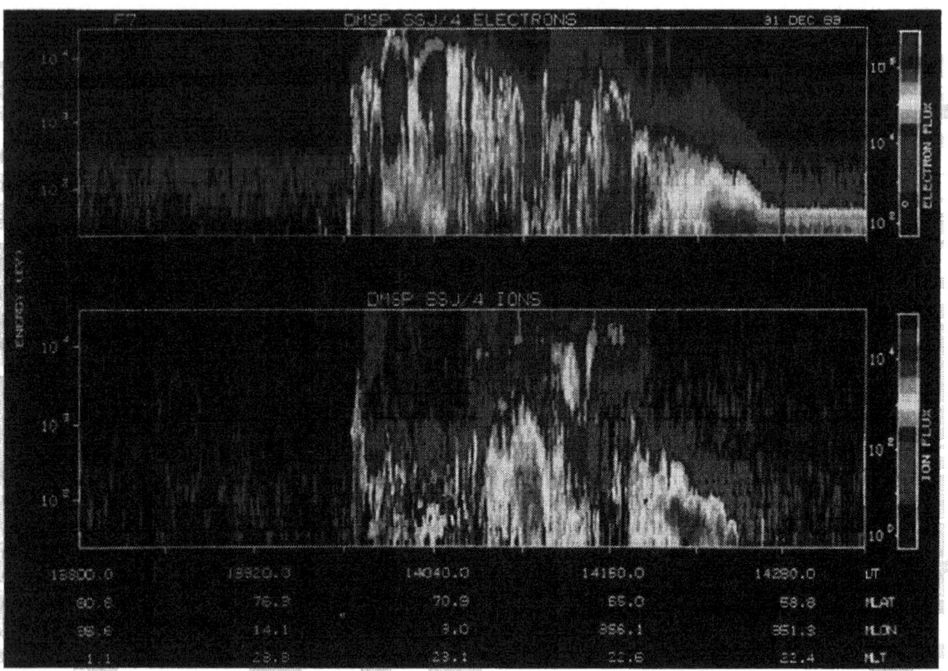

Fig. 7: Spectrograms from DMSP electron and ion detectors, showing inverted-V auroral electron structures (upper panel) accompanied by accelerated ion features (lower panel) at 14000-14060 OUT which indicate spacecraft charging [Ref. 52].

corresponding to different observed spectral shapes and reflecting the origin of the electrons. These are shown in Fig. 8.

- Type 1 has a 'warm' Maxwellian distribution up to a threshold energy and a 'hot' Maxwellian above. The distributions are unaccelerated, so that it is assumed that high-potential structures aligned with the geomagnetic field $\boldsymbol{B}$ are absent. Since the majority of the electron current in this type of spectrum is carried in the low-energy portion, this shape is not usually associated with high-level charging events. Severe examples of this environment have a warm component of T_e~5 kev and corresponding densities of 6×10^7 m^{-3} and 1.2×10^6 m^{-3}. The transition to hot Maxwellian occurs at around 12 keV.

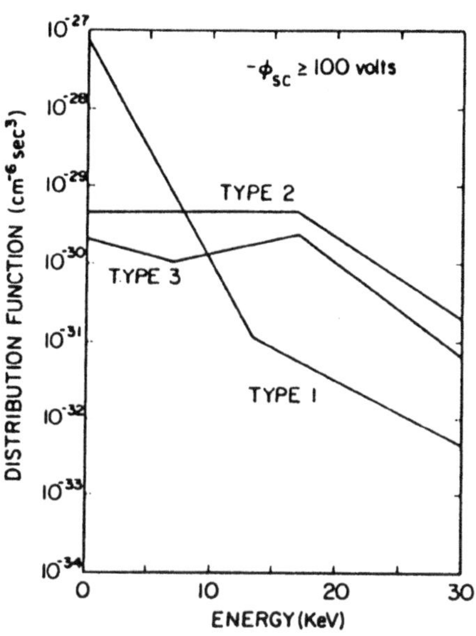

Fig. 8: Shapes of the model electron energy spectra for auroral charging events used in the analysis of Yeh and Gussenhoven [Ref. 52]. Type 2 is most commonly associated with charging events.

- Type 2 has a constant flux at low-energy and an accelerated Maxwellian above the acceleration energy. So this case represents a situation where there is a significant acceleration of electrons along *B* and relative depletion at low energies. This is termed a *modified inverted-V*.

- Type 3 is representative of a *typical inverted-V*. Above a high-energy peak threshold energy, it is an accelerated Maxwellian. Below the threshold, two exponentials model a dip in fluxes which occurs around about 5 keV. In high-level charging cases, the peak threshold, which is also the acceleration energy, is in the region of 16 ± 5 keV.

Yeh and Gussenhoven provide tables of fit parameters for each model for low- and high-level charging. The highest proportion of charging events occur with type 2. Over 40% of events with charging magnitudes ≥145 V have type 2 spectra. In these cases, the transition between low and high-energy parts is at an electron energy of 17.5 ± 4.5 keV and the Maxwellian (accelerated by this transition energy) has a temperature of 4 ± 2 keV. The average total electron current for type 2 events is about 100 $\mu A/m^2$, with some 35% being due to high-energy electrons. This contrasts with only 5% for type 1.

In reviewing previous work, Yeh and Gussenhoven indicate that the directional distribution of type 2 spectra are likely to be *U-shaped*. That is, the flux peaks in directions along and perpendicular to *B*, with a dip in fluxes from intermediate directions.

It should also be added that DMSP was in darkness during charging events, so that the photoemission current J_γ in Eq. (2) was zero, and that the orientation of the satellite was such that secondary emitted electrons could escape along *B*. Furthermore, data from the ambient plasma instruments on DMSP show that the local plasma density becomes considerably depleted during charging events. From the discussion in section 2 it will be clear that this leads to reduced ion current but also to a considerably increased Debye length. This can probably exceed a meter. Simultaneous low energy ambient electron temperatures are not available, but close to a charging event Burke et al. [Ref. 2] report ambient electron temperatures of about 0.4 eV. Yeh and Gussenhoven demonstrate that there is indeed the expected correlation between the charging level and the ratio of hot electron to ram ion fluxes mentioned in the previous section.

Computationally, the regime tends somewhat towards that at high altitudes, and the POLAR code, used for low altitude spacecraft-plasma interaction simulation, has difficulty coping with the large sheaths generated. Cooke, et al., [Ref. 7] suggest that the use of an orbit-limited current collection model then becomes desirable. Nevertheless, their work shows that the POLAR code successfully converges with in-flight DMSP observations and, furthermore, shows that considerably higher potentials can be expected on larger polar-orbiting spacecraft.

As far as mitigation of charging is concerned, one effective passive way is to use high secondary emission materials, such as ITO, on thermal blankets, provided that there is an escape path through the magnetic field. Careful attention to grounding of surfaces is also important and should prevent differential charging taking place and ensure that emissions from one part of the satellite benefit the rest of the satellite.

7. Plasma Waves and Noise

The interaction of a supersonic body with a plasma inevitably leads to the generation of various types of plasma waves. We will briefly review some basic plasma wave modes but refer the reader to introductory plasma physics texts such as Chen [Ref. 5] for details. The simplest and most fundamental type of oscillation in a plasma is electron oscillation at the *plasma frequency*, ω_p. In this case, electrons displaced from equilibrium respond to the generated electric fields by oscillating at a frequency.

$$\omega_p = (n\, e^2/\varepsilon_0 m_e)^{1/2} .$$

When electrons have a finite temperature, this oscillation propagates in the form of an electron plasma wave:

$$\omega^2 = \omega_p^2 + 3/2\; k^2 v_e ,$$

where v_e is the electron thermal speed and k is the wave number. The presence of a magnetic field leads to a hybrid oscillation which is a combination of the plasma oscillation and the cyclotron gyration. The resulting mode is called the *upper hybrid* oscillation:

$$\omega_h^2 = \omega_p^2 + \omega_c^2$$

where ω_c is the electron cyclotron frequency ($=eB/m_e$). Again propagation requires finite electron temperature.

Ion waves occur at lower frequency, due to the large ion mass, and are a common result of low altitude spacecraft-plasma interactions. The most important electrostatic wave is the *ion acoustic* wave. Here, the ion fluid experiences both pressure-gradient and electrostatic restoring forces. Light electrons respond easily to the electrostatic field. The result is a wave described by [Ref. 5]:

$$\omega^2 = k^2\ (KT_e + 3KT_i)/m_i\ (=k^2 v_s^2 \ for\ T_i=0)\ .$$

With cold ions, the group velocity of this wave is the ion sound speed.

Again, we consider the effect of a magnetic field. If the wave propagates almost exactly perpendicular to B, electrons can no longer respond to the electric field in the same way — they must obey the full equation of motion. The result [Ref. 5] is the *lower hybrid* resonance frequency:

$$\omega_l^2 = \Omega_c\ \omega_c$$

where Ω_c and ω_c are the ion and electron cyclotron frequencies respectively. If the wave propagates partially in the direction of B, electrons can move along the field-line in response to electric fields. The *electrostatic ion cyclotron* wave then results:

$$\omega^2 = \Omega_c^2 + k^2\ v_s^2\ .$$

These are just some of the simplest plasma oscillations and wave modes. Electromagnetic and hydromagnetic waves are also possible, a particularly important one being the Alfven wave which propagates along B, with a perturbation $\delta B \perp B$. Propagation is at the Alfven speed:

$$v_A = B/(nm_i\mu_0)^{1/2}\ .$$

Spacecraft moving supersonically through the ionosphere, severely perturbing the ambient plasma and creating wakes, change the ambient plasma distribution from approximately isotropic Maxwellian to one with regions of local non-neutrality, counter-streaming populations, non-Maxwellian distributions and sharp density discontinuities. All of these features may lead to wave activity. Plasma instabilities that can lead to wave formation include the *streaming instability* where one plasma species passes through another with a different velocity and the *kinetic instability* where the plasma is not perfectly Maxwellian. Another source of wave activity, *ion pickup* due to the charge-exchange ionization of neutral gas, is associated with the immediate vicinity of some spacecraft, particularly manned missions. The importance of such interactions is that they transfer energy from one form to another, e.g., from the bulk kinetic energy of a flowing plasma population into thermal energy. During this process energy is also converted into electromagnetic interference which may be problematic.

Because the magnetic field is strong in low earth orbit, both ion and electron wave activity is expected to be dependent on the strength and orientation of the magnetic field. Coherent wave activity is expected around the lower hybrid frequency if propagating perpendicularly to the magnetic field and at the ion gyro-frequency if propagating at smaller angles. Parallel to the magnetic field, oscillations at the electron plasma frequency are expected. In practice, however, broadband turbulence seems to be more common.

A number of in situ studies of spacecraft-plasma interactions have been carried out by the Ariel I, Explorer 31, AE-C and Gemini-Agena 10 satellites as well as the Space Shuttle. Most recent data on waves in spacecraft wakes come from the PDP Plasma Diagnostics Package [e.g., Ref. 28] in the wake of the Shuttle. On STS-3 in 1982 [Refs. 42, 43], the PDP was activated as it lay in the payload bay. The VLF electric field wave measurements showed broadband wave activity up to about 300 KHz in the local plasma, with a peak between 300 and 500 Hz [Ref. 43]. It is difficult to specify how much of the wave activity seen by satellites is caused by the wake structure and how much by ion pick-up. When the Shuttle turned so that the payload bay was in the wake, turbulence was strongly reduced. Large increases in turbulence were seen during thruster firings. A possible source of some of this turbulence lies in streams of ions reported by Stone et al. [Ref. 48]. These were seen at large angles (up to 50°) to the ram direction. During the same mission, the PDP was picked up by the Shuttle arm. This lifted the PDP 10 m away from the Shuttle and, according to Murphy, et al. [Ref. 25], the highest wave activity was at the boundary of the void region. This apparent dependence of waves on wake structure was not seen by Siskind, et al. [Ref. 45] who reported that wave activity was strongest in the ram direction. Samir, et al. [Ref. 41] have concluded from this that these waves were not a wake effect. Below the lower hybrid frequency (ω_h), these waves appear to be mainly electrostatic and unpolarized [Ref. 28]. Above ω_h, there was evidence of polarization which increased at higher frequencies. In 1985 the shuttle Spacelab-2 flight carried the PDP again [Ref. 49] when it was allowed to fly free. It explored the shuttle wake up to 250 m downstream of the orbiter. Wave strength was highest at about twice the assumed ion plasma frequency ($\omega_{pi}=\sqrt{(ne^2/\varepsilon_0 m_i)}$). Wave activity was seen within the Mach cone of the shuttle, extending at least 200 m downstream. However, strong turbulence was confined to within 20 m of the spacecraft. Figure 9 shows how the noise level ($\Delta n/n$) is low in the wake and enhanced at the wake

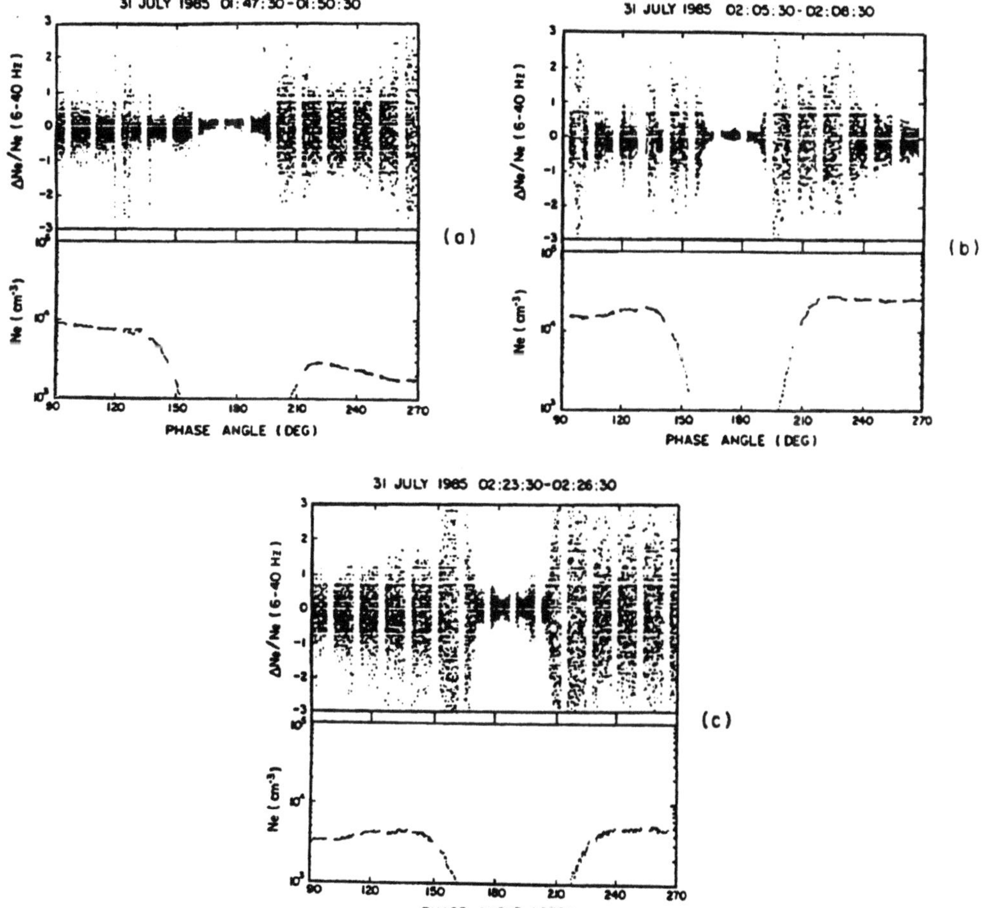

Fig. 9: Low-frequency electrostatic noise level around the wake of the Shuttle, showing enhancement at the wake edges. The wake is indicated by the electron density drop-out in the Orbiter phase angle range 150°-210° [Ref. 49].

edge. Cairns and Gurnett [Ref. 3] show that the magnitude of some waves below ω_h are controlled by the relative orientations of the magnetic field and the velocity vector. They identify the low-frequency waves as Doppler-shifted lower-hybrid waves driven by the pick-up of locally-ionized contaminant water ions. They find no relation between these waves and the wake or thruster firings. Fluctuations are, however, greatest when the velocity vector and B are perpendicular.

Figure 10, from Murphy [Ref. 29], based on the work of Picket, et al. [Ref. 33], shows the relation between the local contaminant environment and the strength of the broadband electrostatic noise on Shuttle. It is clear that emissions from thrusters and from the flash evaporator system give rise to considerable increases in the noise level.

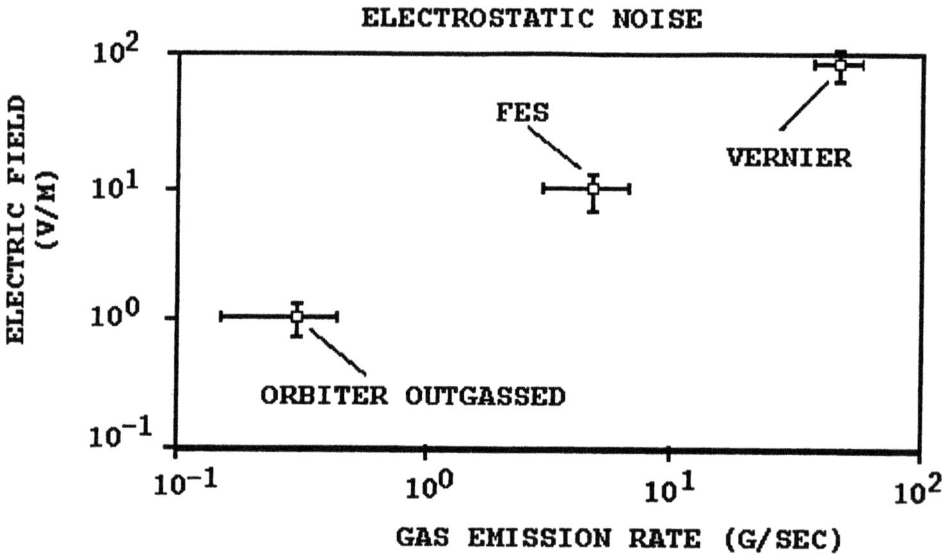

Fig. 10: The relationship between broadband noise level and the emission rate of gasses from the Shuttle [Ref. 29].

Singh [Ref. 44] reviewed the potential sources of electromagnetic waves around the space station and showed that important sources due to the station motion are likely to be Alfven waves, electromagnetic ion-cyclotron waves, ion-acoustic waves and ion Bernstein waves. Wake-induced waves and contaminant-ionization wave effects mentioned above are also expected to be present around the space station.

We can conclude this section by saying that many more in-situ measurements are necessary before a complete picture of the wave environment of a large spacecraft and its co-orbiting contaminant environment can be presented.

8. Contaminant-Plasma-Interactions

Spacecraft carry a local contaminant environment with them in the form of a steady outgassing from surfaces, or produce one through occasional venting, thruster firing, etc. This is especially true on manned missions. Shawhan, et al. [Ref. 42] report a pressure of 10^{-5} Torr in the Shuttle bay at the start of the STS-3 mission. The ambient level of 10^{-7} was reached after nearly 24 hours of subsequent outgassing. These emissions considerably alter the plasma environment [Refs. 42, 33, 32].

The previous section has already indicated that one very important effect of contamination is the production of plasma turbulence. In addition, the contamination of surfaces is a concern for future sensitive payloads, and this can be enhanced as a result of electrostatic re-attraction of contaminant ions. If ions are re-attracted by high voltage surfaces, surface damage can also result [Ref. 9]. Enhanced contamination also increases the likelihood of discharge on elevated voltage systems.

A number of possible processes exist whereby emitted contaminants can be ionized. These include: photoionization from solar UV irradiation, critical ionization velocity phenomena (CIV) where the relative motion of a neutral gas through a plasma causes ionization, and charge-exchange between the ambient plasma and the contaminant gases to form the local pick-up ion environment. The latter mechanism is thought to be dominant.

Figure 11 [Ref. 29] shows schematically the environment around a body in LEO. Ions which are generated in the body's frame of reference (i.e., moving at 8 km/s across B and through the ambient plasma) experience the magnetic field and immediately execute Larmor gyration. One might expect some charge separation, because of oppositely gyrating electrons and ions [Ref. 35]. This would create an equilibrium with zero Lorentz force $E + v \times B$, with the generated polarization electric field allowing an $E \times B$ drift of ions in the original neutral molecule's direction of travel. In fact, this field seems absent and ions are left behind on field lines, gyrating about them and moving along the field line by virtue of their $v \cdot B$ velocity component. In the vehicle frame, this corresponds to a drift in the $-v$ direction consistent with an $E \times B$ drift in the motionally-induced electric field $E + v \times B$. The resulting pick-up ion wake is therefore perpendicular to B, as shown in Fig. 11.

Contaminant ions are predominantly H_2O^+ (which has an ionization potential of about 13.2 eV) created in the charge exchange reaction with ambient O^+:

$$H_2O + O^+ \rightarrow H_2O^+ + O$$

These ions have been observed on Shuttle flights with the PDP mentioned previously and are described in detail by Grebowsky and Schaefer [Ref. 11]. The H_2O^+:O^+ ratio is typically a few percent although, in the absence of venting, the ratio falls as the orbiter outgasses over a week

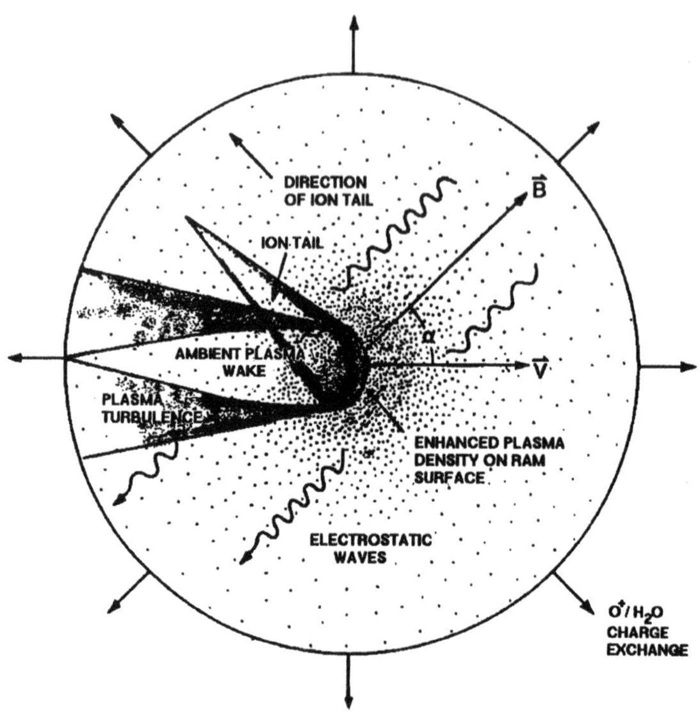

Fig. 11: Schematic of the interactions between the Shuttle and the ambient environment, including charge-exchange plasma contamination, a pick-up ion tail and the generation of turbulence [Ref. 29].

or so. The effects of thruster firings are much more complex and difficult to interpret because of sequential firing of different thrusters, the large emission rate and emission velocity.

The Space Station Freedom will be considerably larger than the Shuttle and of uncertain cleanliness. The planned attitude control will result in frequent thruster firing and venting of Station waste is also expected. This will produce interactions similar to those seen on Shuttle and also reminiscent of interactions of a comet with the interplanetary environment or of the Jovian satellite Io with Jupiter's environment [Ref. 32]. The study of similar spacecraft phenomena is therefore of wider interest. Nevertheless, from an engineering perspective, contamination is a nuisance and spacecraft-plasma-contaminant interactions need to be properly understood and evaluated for future missions.

9. Ground-Based Experimental Simulations

An important contribution to our understanding of spacecraft-plasma interactions can come from carefully controlled experiments in plasma chambers on the ground. Some examples of such experiments are those described by Stone [Ref. 46] Martin et al. [Ref. 23], Chan et al. [Ref. 4] and Lebreton, et al. [Ref. 20]. Martin, et al. [Ref. 21] provide a review of a number of facilities and experiments.

In performing experiments, it is clearly desirable to produce a plasma with the characteristics which are as close as possible to those in space. A streaming plasma with the appropriate stream speed (Mach number) is needed. In addition, electron temperatures and plasma densities must be chosen to produce a Debye length (Eq. 1) small compared to the chamber size. If interactions with a medium-large body are being simulated, the relative size of the body with respect to the Debye length (R/λ_D) needs to be properly represented. When applying potentials to a body, the potential normalized to the electron temperature (eV/KT_e) is often adopted. An important complication to ground-based experiments is the problem of cold charge-exchange ions created in the chamber. These do not possess the stream velocity and can partially fill any wake void structure or contribute an undesired ion current to surfaces in charging experiments. A further problem is that the streaming plasma from a source usually has a divergence of the stream which is obviously absent in space and which affects the morphology of any wakes. It is often difficult simultaneously to simulate all the scaled parameters correctly. If source conditions are altered to change one parameter, other parameters are usually also modified.

Martin, et al. [Ref. 23] report on a coordinated series of controlled experiments in two facilities, with low background pressures, and representative plasma parameters. One chamber, at NDRE, also had the capability of canceling or augmenting the Earth's magnetic field. These experiments resulted in a reasonably good agreement between measured and computed wake characteristics.

10. Conclusions

We have shown that various types of plasma interaction are possible in low Earth orbit. Potentially the most troublesome interaction is that resulting in high-level charging in the auroral zones. There is, however, a shortage of in-flight data on high-level charging events which it would be highly desirable to correct on future polar platforms. Other problems arise from contaminant interactions, wave and turbulence phenomena (which may produce EMI at problematic levels), and effects involving enhanced voltage systems. There is therefore an engineering justification for continued study of these and related interactions – wake studies, for example. Apart from the applied research needs, the plasma interactions are of interest from a general plasma physical viewpoint, with numerous novel interactions, as well as similarities with other interaction in nature which we mentioned previously (comets, natural satellites, etc.).

The subject should receive greater attention with the flight and operation of the Space Station which is a very large body with enhanced voltages, producing a significant contaminant environment. Efforts have been made to define the requirements for monitoring the complex station environment [Ref. 50]. In the interim, there are plans for further Shuttle payloads, including the tethered satellite system (TSS), and possible US and European Spacelab flights, and for free-flyer experiments such as on the Japanese SFU [Ref. 17].

Progress on understanding the interactions will result from coordinated efforts with numerical simulation, ground-based experiments and in-space measurements and experiments. The most difficult to execute are the space-based activities but these are undoubtedly the most crucial.

References

[1] S.I. Akasofu, 'The Dynamic Aurora', Scientific American, p. 54 (May 1989).

[2] W.J. Burke, D.A. Hardy, F.J. Rich, A.G. Rubin, M.F. Tautz, N.A. Safekos, and H.C. Yeh, 'Direct Measurements of Severe Spacecraft Charging in Auroral Ionosphere', in Spacecraft Environmental Interaction Technology 1983, C.K. Purvis and C.P. Pike (eds.), NASA-CP-2359, AFGL-TR-85-0018, p. 109 (1985).

[3] I.H. Cairns and D.A. Gurnett, 'Control of Plasma Waves Associated with the Space Shuttle by the Angle Between the Orbiter's Velocity Vector and the Magnetic Field', J. Geophys. Res. 96, 7591 (1991).

[4] C. Chan, J. Browning, S. Meassick, M.A. Morgan, D.L. Cooke, C.L. Enloe, and M.F. Tautz, 'Current Collection in a Spacecraft Wake: Laboratory and Computer Simulations', in Proceedings of the Spacecraft Charging Technology Conference 1989, R.C. Olsen (ed.), p. 601, Naval Postgraduate School, Monterey, CA (1991).

[5] F.F. Chen, Introduction to Plasma Physics, Plenum Press, New York (1974).

[6] D.L. Cooke, I. Katz, M.J. Mandell, J.R. Lilley, and A.J. Rubin, 'Three-Dimensional Calculation of Shuttle Charging in Polar Orbit', in Spacecraft Environmental Interaction Technology 1983, C.K. Purvis and C.P. Pike, NASA-CP-2359, AFGL-TR-85-0001, p. 205 (1985).

[7] D.L. Cooke, M.S. Gussenhoven, D.A. Hardy, M. Tautz, I. Katz, G. Jongeward, and J. Lilley, 'POLAR Code Simulation of DMSP Satellite Auroral Charging', Proceedings of the Spacecraft Charging Technology Conference 1989, R.C. Olsen (ed.), p. 194, Naval Postgraduate School 1991.

[8] J.F. Fennell, D.J. Gorney, and P.F. Mizera, 'Auroral Particle Distribution Functions and Their Relationship to Inverted Vs and Auroral Arcs', in Physics of Auroral Arc Formation, Geophys. Monogr. Ser. Vol. 25, S.I. Akasofu and J.R. Kan (eds.) p. 91, AGU (1991).

[9] D.C. Ferguson, D.B. Snyder, R. Carruth, 'Findings of the Joint Workshop on Evaluation of Impacts of Space Station Freedom Grounding Configurations,' NASA Tech. Mem. 103717, also in Proceedings of the ESA Workshop on Space Environment Analysis, ESA-WPP-23, ESTEC/WM (1990).

[10] H. Garrett, J. Feynman, and S. Gabriel, (eds.), Space Technology Plasma Issues in 2001, JPL Publication 86-49 (1986).

[11] J.M. Grebowsky, and A. Schaefer, 'Observations of Thermal Ion Influxes about the Space Shuttle', in Current Collection from Space Plasmas, NASA CP-3089, N. Singh, K.H. Wright, and N.H. Stone (eds.) (1990).

[12] M.S. Gussenhoven, D.A. Hardy, F. Rich, W.J. Burke, and H.C. Yeh, 'High-Level Spacecraft Charging in the Low-Altitude Polar Auroral Environment', J. Geophys. Res. 90, A11, 11009 (1985).

[13] W.N. Hall, P. Leung, I. Katz, G.A. Jongeward, J.R. Lilley, J.E. Nanevicz, J.S. Thayer, and N.J. Stevens, 'Polar Auroral Charging of the Space Shuttle and EVA Astronaut', in The Aerospace Environment at High Altitudes and Its Implications for Spacecraft Charging and Communications, AGARD Conference Proceedings No. 406, p. 34-1 (1987).

[14] C.L. Henderson and U. Samir, 'Observations of the Disturbed Region Around an Ionospheric Spacecraft', Planet. Space Sci. 15, 1499 (1967).

[15] I. Katz, and D.E. Parks, 'Shuttle Orbiter Charging', J. Spacecraft and Rockets 20, 22 (1983).

[16] I. Katz, D.E. Parks, D.L. Cooke, M.J. Mandell, and A.J. Rubin, 'Polar Orbit Electrostatic Charging of Objects in Shuttle Wave', in Spacecraft Environmental Interaction Technology 1983, C.K. Purvis and C.P. Pike (eds.), NASA-CP-2359, AFGL-TR-85-0018, p. 229 (1985).

[17] H. Kuninaka, K. Takahashi, and K. Kuriki, 'Space Experiment of High Voltage Solar Array', in Proceedings of the ESA Workshop on Space Environment Analysis, ESA-WPP-23, ESTEC/WM.

[18] J.G. Laframboise, 'Is There a Good Way to Model Spacecraft Charging in the Presence of Space-Charge Coupling, Flow and Magnetic Fields?', in Proceedings of the Air Force Geophysics Laboratory Workshop on Natural Charging of Large Space Structures in Near Earth Polar Orbit, R.C. Sagalyn et al. (eds.), AFGL-TR-83-0046, p. 57 (1983).

[19] J.G. Laframboise and L.W. Parker, 'Spacecraft Charging in the Auroral Plasma: Progress Towards Understanding the Physical Effects Involved', in The Aerospace Environment at High Altitudes and Its Implications for Spacecraft Charging and Communications, AGARD Conference Proceedings No. 406, p. 13-1 (1987).

[20] J.P. Lebreton, Y. Arnal, and R. Debrie, 'Ionospheric Plasma Simulation in a Large Space Simulator', in Proceedings of the Symposium on Environmental Testing for Space Programmes, ESA-SP-304, 163 (1990).

[21] A.R. Martin, D.J. Rodgers, R.L. Kessel, A.D. Johnstone, A.J. Coates, B.N. Maehlum, K. Svenes, and M. Friedrich, Spacecraft-Plasma Interactions and Electromagnetic Effects in LEO and Polar Orbits, Final Report Vol. 1, A Review of Previous Work, ESA Contract 7989/88 ESA-CR(P)-3025a (1990).

[22] J. Troim, private communication, based on work described in Ref. 23.

[23] A.R. Martin, R.A. Bond, P.M. Latham, D.J. Rodgers, R.L. Kessel, A.D. Johnstone, B.N. Maehlum, J. Troim, M. Friedrich, G.L. Wrenn, and A.J. Sims, Spacecraft-Plasma Interactions and Electromagnetic Effects in LEO and Polar Orbits, Final Report Vol. 2, Experimental and Computational Studies, ESA Contract 7989/88 ESA CR(P)-3284 (1991).

[24] G.B. Murphy and K.E. Lonngren, 'A Review of the Findings of the Plasma Diagnostic Package and Associated Laboratory Experiments: Implications of Large Body/Plasma Interactions for Future Space Technology,' in Space Technology Plasma Issues in 2001, JPL Publication 86-49, p. 237 (1986).

[25] G.B. Murphy, J.S. Pickett, N. D'Angelo, and W.S. Kurth, 'Measurements of Plasma Parameters in the Vicinity of the Space Shuttle', Plan. Space Sci. 34, 10, pp. 993-1004 (1986).

[26] G.B. Murphy, D.L. Reasoner, A. Tribble, N. D'Angelo, J.S. Pickett, and W.S. Kurth, 'The Plasma Wake of the Shuttle Orbiter', J. Geophys. Res. 94, A6, 6866 (1989).

[27] G.B. Murphy and I. Katz, 'The POLAR Code Wake Model: Comparison with in situ Observations', J. Geophys. Res. 94, 9065 (1989).

[28] G.B. Murphy, S.D. Shawhan, L.A. Frank, N. D'Angelo, N., J.M. Grebowsky, D.L. Reasoner, and N. Stone, 'Interaction of the Space Shuttle Orbiter with the Ionospheric Plasma', in Proceedings of the ESLAB Symposium on Spacecraft-Plasma Interactions ESA-SP-198 (1983).

[29] G.B. Murphy, 'Contaminant Ions and Waves in the Space Station Environment', in A Study of Space Station Contamination Effects, NASA CP-3002, M.R. Torr, J.F. Spann, and T.W. Moorhead (eds.), p. 19 (1988).

[30] R.C. Olsen (ed.), Proceedings of the Spacecraft Charging Technology Conference 1989, Naval Postgraduate School, Monterey, CA (1991).

[31] D.E. Parks and I. Katz, 'Charging of a Large Object in Low Polar Earth Orbit', in Spacecraft Charging Technology 1980, N.J. Stevens and C.P. Pike (eds.), NASA CP-2182, AFGL-TR-81-0270 (1981).

[32] W.R. Paterson and L.A. Frank, 'Hot Ion Plasmas from the Cloud of Neutral Gasses Surrounding the Space Shuttle', J. Geophys. Res. 94, A4, 3721 (1989).

[33] J.S. Pickett, G.B. Murphy, W.S. Kurth, C.K. Goertz, and S.D. Shawhan, 'Effects of Chemical Releases by the STS 3 Orbiter on the Ionosphere', J. Geophys. Res. 90, A4, 3487 (1985).

[34] C.K. Purvis, and C.P. Pike (eds.), Spacecraft Environmental Interaction Technology 1983, NASA-CP-2359, AFGL-TR-85-0018 (1985).

[35] W.J. Raitt, 'Vehicle Environment Interactions: An Overview', Adv. Space Res. 8, 1, pp. (1)177-(1)186 (1988).

[36] D.J. Rodgers, 'Comparing Computer Simulations of Plasma Wakes in LEO and Laboratory Environments', ESTEC WP-1609 (1991).

[37] A.G. Rubin and A. Besse, 'Charging of a Manned Maneuvering Unit in the Shuttle Wake,' J. Spacecraft and Rockets 23, 1, 122 (1986).

[38] R.C. Sagalyn, D.E. Donatelli, and I. Michael (eds.), Proceedings of the Air Force Geophysics Laboratory Workshop on Natural Charging of Large Space Structures in the Near Earth Polar Orbit, AFGL-TR-0046 (1983).

[39] U. Samir and G.L. Wrenn, 'The Dependence of Charge and Potential Distribution Around a Spacecraft on Ionic Composition', Planet. Space Sci. 17, 693 (1969).
[40] U. Samir, R. Gorden, L.H. Brace, and R. Theis, 'The Near-Wake Structure of the Atmospheric Explorer C (AE-C) Satellite, A Parametric Investigation', J. Geophys. Res. 84, 513 (1979).
[41] U. Samir, N.H. Stone, K.H. Wright, 'On Plasma Disturbances Caused by the Motion of the Space Shuttle and Small Satellites: A Comparison of In-Situ Observations', J. Geophys. Res. 91, pp. 277-285 (1986).
[42] S.D. Shawhan, G.B. Murphy, and D.L. Fortna, 'Plasma Diagnostics Package Initial Assessment of the Shuttle Orbiter Plasma Environment', J. Spacecr. & Rockets 21, 4, 387 (1984a).
[43] S.D. Shawhan, G. B. Murphy, and D.L. Fortna, 'Measurements of Electromagnetic Interference on OV102 Columbia using the Plasma Diagnostics Package', J. Spacecr. & Rockets 21, 4, 392 (1984b).
[44] N. Singh, 'Space Station Induced Electromagnetic Effects', in A Study of Space Station Contamination Effects, NASA CP-3002, M. R. Torr, J.F. Spann, and T.W. Moorhead (eds.), p. 31 (1988).
[45] D.E. Siskind, W.J. Raitt, P.M. Banks, and P.R. Williamson, 'Interactions Between the Orbiting Space Shuttle and the Ionosphere', Planet. Space Sci. 32, 7, pp. 881-896 (1984).
[46] N.H. Stone, The Aerodynamics of Bodies in a Rarified Ionized Gas with Applications to Spacecraft Environmental Dynamics, NASA TP-1933 (1981).
[47] N.H. Stone, K.H. Wright, U. Samir, and K.S. Hwang, 'On the Expansion of Ionospheric Plasma into the Near-wake of the Space Shuttle Orbiter', Geophys. Res. Lett. 15, pp. 1169-1172 (1988).
[48] N.H. Stone, U. Samir, K.H. Wright, D.L. Reasoner, and S.D. Shawhan, 'Multiple Ion Streams in the Near Vicinity of the Space Shuttle', Geophys. Res. Lett. 10, 1215 (1983).
[49] A.C. Tribble, J.S. Pickett, N. D'Angelo, and G.B. Murphy, 'Plasma Density, Temperature and Turbulence in the Wake of the Shuttle Orbiter', Planet. Space Sci. 37, 8, pp. 1001-1010 (1989).
[50] E.C. Whipple and J.N. Barfield, 'Neutral Environment with Plasma Interactions Monitoring System on Space Station (NEWPIMS)', in Proceedings of the Spacecraft Charging Technology Conference 1989, R.C. Olsen, ed., p. 556, Naval Postgraduate School 1991.
[51] K.H. Wright, A Study of Single and Binary Ion Plasma Expansion into Laboratory-Generated Wakes, NASA CR-4125 (1988).
[52] H.C. Yeh and M.S. Gussenhoven, 'The Statistical Electron Environment for Defense Meteorological Satellite Program Eclipse Charging,' J. Geophys. Res. 92, A7, 7705, (1987).
[RUBIN] A. Rubin, unpublished communication, ESTEC (1990, also referred to in Ref. 7.

POTENTIAL THREATS TO THE PERFORMANCE OF VACUUM-INSULATED HIGH-VOLTAGE DEVICES IN A SPACE ENVIRONMENT

R. V. LATHAM
Department of Electronic Engineering and Applied Physics
Aston University
Birmingham V4 7ET, U.K.

ABSTRACT. Most instabilities in a high voltage vacuum-insulated gap are believed to stem from parasitic "cold-cathode" electron emission processes. This paper reviews the fundamental surface studies that have established the vital role played by micro-particulate contamination in promoting a "non-metallic" field-induced mechanism that injects electrons into a vacuum gap; it also discusses the technological procedures that can be adopted to limit the problem. The review concludes with a consideration of the specific factors that pose potential threats to the reliable operation of high voltage devices in space.

1. Introduction

There are many devices and systems that rely for their operation on the ability to support high electric fields, or voltages, between metal electrodes in a vacuum environment. In practice, this requirement frequently poses a serious technological problem, since such an high-voltage (HV) gap is inherently liable to spontaneously "breakdown", and suffer an irreversible catastrophic degradation in its performance. To understand the physical origin of this phenomenon, it has first to be appreciated that transient instabilities (microdischarges), and breakdown events, are both manifestations of a conducting path being established between the electrodes of a vacuum-insulated HV gap. Thus, any physical process that has the potential to generate a microplasma within such a gap must be regarded as a threat to its operational stability.

As illustrated in Fig. 1, there are in principle two configurations involving an HV vacuum gap; the first is where the space between the two electrodes is entirely vacuum, and the second is a "mixed" situation where the electrode gap is bridged by a solid insulator. Although these two regimes are closely related, they are in fact controlled by different sets of physical processes and, as a result, have been traditionally treated as largely independent entities [Refs. 1, 2]. This review will follow this custom and confine itself to a discussion of the "pure" vacuum gap.

R. N. DeWitt et al. (eds.), The Behavior of Systems in the Space Environment, 467–489.

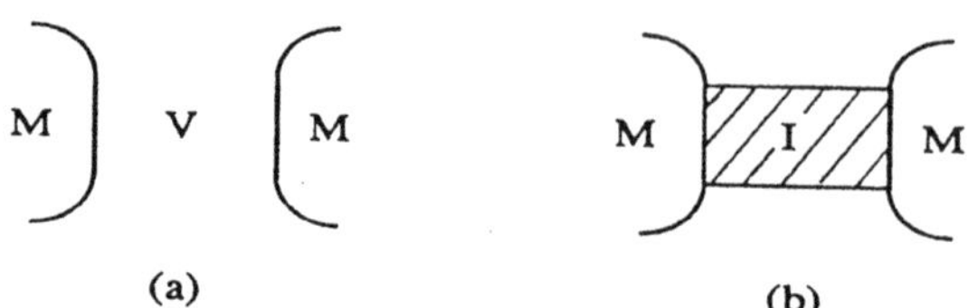

Fig. 1: High-voltage electrode gap configurations: (a) a "pure" vacuum gap and (b) a vacuum gap bridged by a solid insulator insertion.

From an historical perspective, it was traditionally assumed that the insulation of high voltages by a vacuum gap is ultimately limited by two types of electrode surface process which give rise respectively to the well known 'electron emission' and 'microparticle induced' breakdown mechanisms [Ref. 1]. Of these, the former has now assumed the greater importance, and will be the central subject of the present paper. In contrast, the role of microparticles in a HV gap has received less attention of late, and the effects of their presence remain uncertain: in particular, it remains in doubt as to whether they are responsible for the spontaneous creation of emission sites when they attach themselves to electrode surfaces [Ref. 1]. For an assessment of the potential importance of the "ballistic" breakdown mechanism associated with them, the reader is referred to the review of Chatterton [Ref. 3].

With a growing need for higher and more reliable performances from HV devices and systems, notably for space applications, the past decade has seen a sustained effort to understand the physical basis of the microscopically localized electron emission process that occurs randomly on extended electrode surfaces in the typical field range of 10-20 MV/m. In particular, there has been a considerable investment in the development of a range of sophisticated analytical techniques dedicated to the study of this highly specific and localized surface phenomenon [Ref. 4]. The outcome of this programme has been a major advance in understanding, whereby it is recognized that electrons are emitted by a complex field-induced hot-electron emission (FIHEE) mechanism that is promoted by the presence of particulate contamination on electrode surfaces. Not surprisingly, this realization has led both to a major revision in contemporary HV technological procedures, and a changed perspective regarding future progress. The present paper will review the main conclusions drawn from recent fundamental studies, and evaluate their potential impact on contemporary space-orientated HV technology.

2. Parasitic "Cold Cathode" Emission

Metallic conductors, notably copper, are characterized by having a dense gas ($\sim 10^{28}$ m^{-3}) of "free" electrons within the bulk of the material that are responsible for transporting electrical charge. Under the conditions prevailing in most light and heavy current applications, such as electronic circuitry and electrical machinery, these electrons are totally confined within the material by the presence of a surface potential barrier, known as the work function of the metal. There are however other applications in which metal surfaces have to withstand large surface electric fields during the course of their operation. Examples of these would include the HV electrode assemblies of particle accelerators, vacuum switches, capacitors, transmission valves, travelling wave tubes, and electron diodes. In all of these cases, it is essential that none of the "free" electrons leak through the electrode surface to become truly free, since in this state they are able to initiate a range of destructive interactions that usually result in electrical breakdown.

For an ideal single-crystal planar electrode, such a leakage would appear to be very unlikely, since the necessary conditions are not satisfied for supporting the standard electron emission mechanisms that are known to occur from ideal metal surfaces. Thus, there are normally no UV photons available to promote photoelectric emission, the temperature is normally far too low for there to be any significant thermionic emission, and the surface field is normally well below the threshold value of ~3 x 10^9 V/m for metallic field emission to occur. However, as will be explained in the following sections, the real world is far from ideal, and it is possible for "local" conditions to occur on electrodes that promote the cold-emission of electrons under anomously low-field conditions, and the consequent risk of a breakdown (or flashover) of the high-voltage gap. From a technological point of view, such events are frequently catastrophic, since they result in an irreversible degradation in the hold-off performance of the HV gap and the consequent malfunctioning of the associated device or system. Thus, Fig. 2 illustrates the wide diversity of major technologies where progress has been severely limited by this process. For this reason, a long-standing high priority has been given to the development of reliable technological procedures for suppressing the occurrence of parasitic electron emission processes.

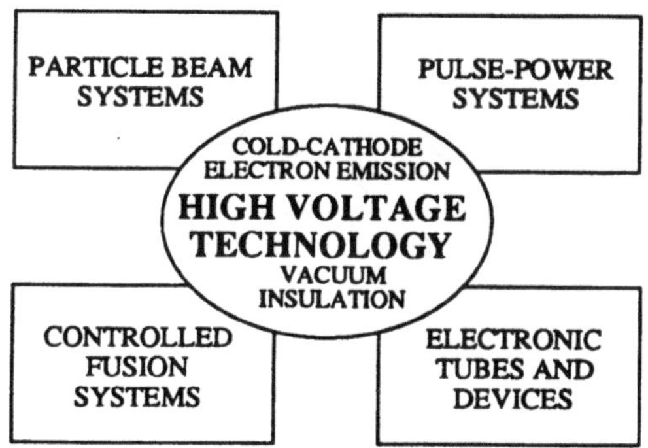

Fig. 2: Major applications of high voltage technology.

2.1. I-V CHARACTERISTIC OF AN HV GAP

If the current-voltage characteristic of a typical operational HV gap is investigated, it will be found to have the general form illustrated schematically in Fig. 3a. For a virgin electrode, there is a threshold field in the range 5-15 MV/m associated with the initial "switch-on" of the prebreakdown current, which subsequently exhibits a reversible and noisy response with increasing field until the gap spontaneously breaks down at fields in excess of 20 MV/m. If however the field is reversed before breakdown occurs, the hysteresis effect shown dotted in Fig. 3a is observed; i.e., where the current decays smoothly to zero, but over an extended field range. The effect of pressure can be somewhat arbitrary, but in general, an increase in pressure results in a lower switch-on field, a higher current and associated noise levels, and a lower breakdown field. In fact, at pressures in excess of $\sim 10^{-5}$ mbar, there have also been reports of an "ignition" effect [Ref. 1], in which the current can switch between two states with an associated hysteresis between them.

On subsequently cycling the field, the gap becomes progressively "conditioned" such that switch-on events no longer occur, and the I-V characteristic becomes approximately reversible. At this stage, it is conventional to characterize the gap in terms of its Fowler-Nordheim (F-N) plot as illustrated in Fig. 3b. Such plots are generally linear at low fields, particularly under low pressure conditions, and have a slope that is determined by the nature of the emission process. The reciprocal of the slope is termed the β-factor, which are typically found to be in the range 100-400.

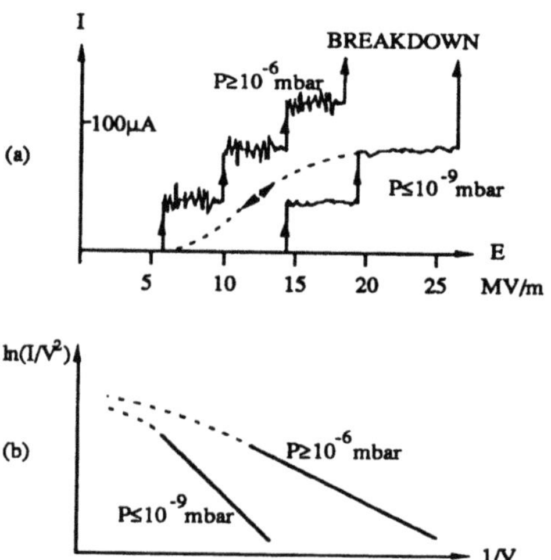

Fig. 3: Illustrating the effect of pressure on (a) the d-c current-voltage characteristic of a typical HV gap, and b) the corresponding F-N representation of the characteristic.

2.2. SPATIAL IMAGING OF "ELECTRON PIN-HOLES"

Although of practical significance, the above type of measurement provides no information about the physical origin of prebreakdown currents, in particular, their spatial distribution over the cathode surface. However, this question can readily be answered by employing a "transparent anode" technique. This instrumental facility, illustrated in Fig. 4, has proved to be a simple but powerful means of studying both the spatial distribution of emitted electrons from planar cathodes, and also, in conjunction with video recording techniques, the temporal evolution of a population of such sites [Refs. 5, 6]. The anode consists of an optically flat glass disc whose surface has been made conducting by the use of a standard tin-oxide coating procedure.

As an illustration of the application of this technique, Figs. 5a and 5b present pairs of simulated time-lapse images that are typically obtained from electrodes under low and high pressure conditions respectively, and at gap currents of several microamps and applied fields in excess of 10 MV/m. From these, it will be seen that in all cases the emission comes from randomly located point emissions centres, i.e., electron "pin-holes", and that the density of these pin-holes increases with increasing pressure. It will also be seen that the spatial distribution of these sites is very much more stable at lower pressures. At higher pressures, sites tend to switch on and off randomly with time, which is one of the processes that accounts for the higher noise level in the gap current under these conditions.

Referring to the pair of images of Fig. 5c, the technique may also be used to study the breakdown event. These show how the initiating arc is spatially centred on one of the prebreakdown emission sites; i.e., supporting the long-held belief that a breakdown event is initiated by the prebreakdown electron emission process. From this pair of time-lapsed images, it will be seen how a breakdown event can frequently result in the creation of a great many additional new sites which, in practical terms, can often correspond to a catastrophic degrading of the insulating performance of an HV gap. As with the increased frequency of the switching

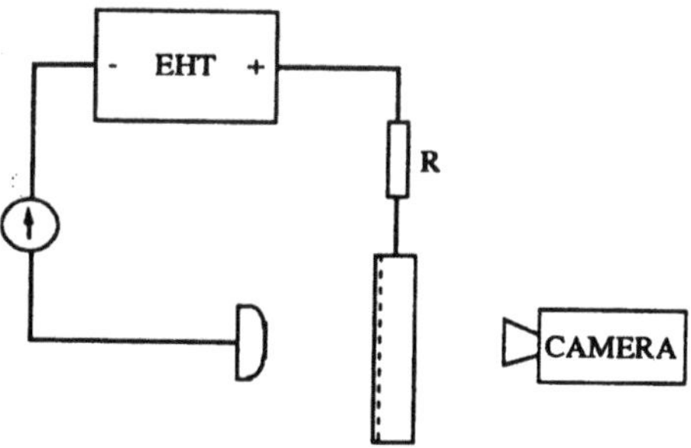

Fig. 4: An illustration of the transparent anode technique for displaying the spatial distribution of "electron pin-holes" on broad-area electrodes.

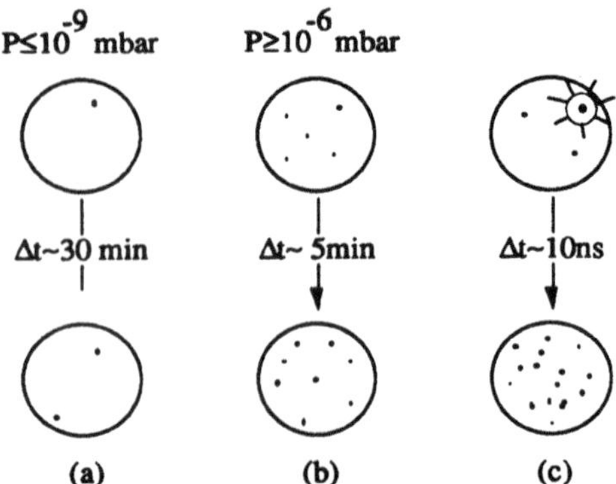

Fig. 5: Schematic representations of pairs of time-lapse images obtained from the transparent-anode images technique. These show in a) and b) how the spatial stability of populations of emission sites are influenced by pressure, and in c) how breakdown events are typically centred on existing emission sites.

phenomenon, the probability of any emission site precipitating a microdischarge or breakdown even also increases with pressure. It should however be noted that both Cox [Ref. 7] and Hurley [Ref. 8] identified at least two types of site: namely, those that become unstable with increasing emission current and initiate breakdown events (type A), and those that remain stable with increasing field (type B).

The versatility of the transparent anode technique can be further illustrated by reference to an important study in which it was used to investigate the initiation of arcs in high pressure SF_6 at fields in excess of 10 MV/m [Ref. 5]. Thus, by first recording the vacuum emission sites of a test cathode, and then recording the first arc when the chamber had been subsequently filled with SF_6 gas at a pressure of 10 bar, it was possible to establish a spatial correlation between the location of the arc and one of the vacuum emission sites. Although the statistical sample of such data was relatively limited, the findings were sufficiently consistent to warrant further investigation to confirm whether the performance of vacuum and HP gaps are controlled by the same basic physical process.

2.3. MICROTOPOGRAPHY OF "ELECTRON PIN-HOLES"

For the in situ study of individual sites, it is necessary to employ some form of scanning probe technique to firstly locate individual emission sites, and then to position the chosen site for subsequent analysis. The pioneering system of Cox [Ref. 9] employed an anode probe hole technique. However, a more convenient approach is to use an arrangement consisting of an anode probe in the form of an electrolytically etched tungsten microtip, in conjunction with the goniometer specimen stage of a scanning electron microscope (SEM). Thus, by using the tip as a scanning field probe, it is possible to isolate an emission site on the axis of the SEM and hence image its topography. At the same time, an auxiliary X-ray or Auger facility can be used to provide information about its material composition.

It was this type of analytical system, illustrated schematically in Fig. 6, that was first used by Athwal and Latham [Ref. 10] to routinely record images which showed that electron pin-holes are invariably associated with the presence of electrically insulated particulate contamination on the electrode surface; i.e., rather than metallic whiskers, as had traditionally been thought. It was also found that only about 1 in 10^6 – 10^8 particles on an electrode surface actually give rise to emission sites in the field range 10-20 MV/m. Subsequently, this type of analytical system was greatly improved by Niedermann, et al. [Ref. 11] to allow the possibility of a comprehensive Auger study of the material composition of a large sample of emitting structures under controlled UHV conditions. This investigation revealed the technologically important information that "active" particles can have a wide range of element composition, with carbon being the most predominant and typically accounting for over 20% of all emitters.

2.4. EMISSION CHARACTERISTICS OF "ELECTRON PIN-HOLES"

In order to gain a physical insight into the nature of the mechanism responsible for prebreakdown emission sites, the automated electron spectrometer facility illustrated in Fig. 7a has been progressively established by Latham and co-workers. As described elsewhere [Refs. 1, 4, 12, 13], this instrument employs a plane-parallel electrode gap that is interfaced with the spectrometer by means of a small anode probe hole and a beam-forming electrostatic lens system. The facility is capable of generating a "site map" of a test cathode, i.e., similar to that

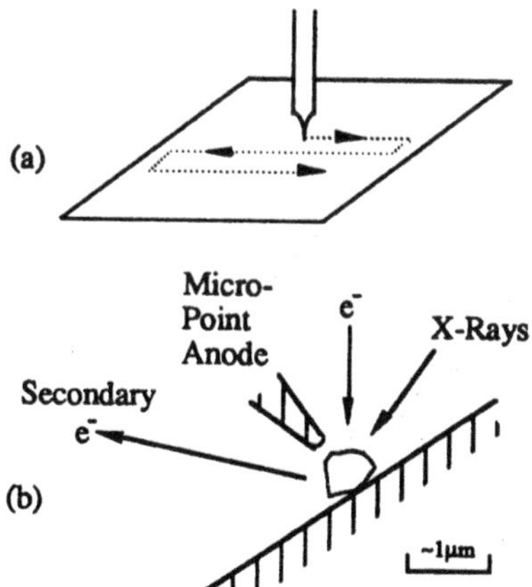

Fig. 6: A schematic illustration of the micropoint anode technique for a) mapping the positions of emission sites, and b) analyzing their topographical and material nature.

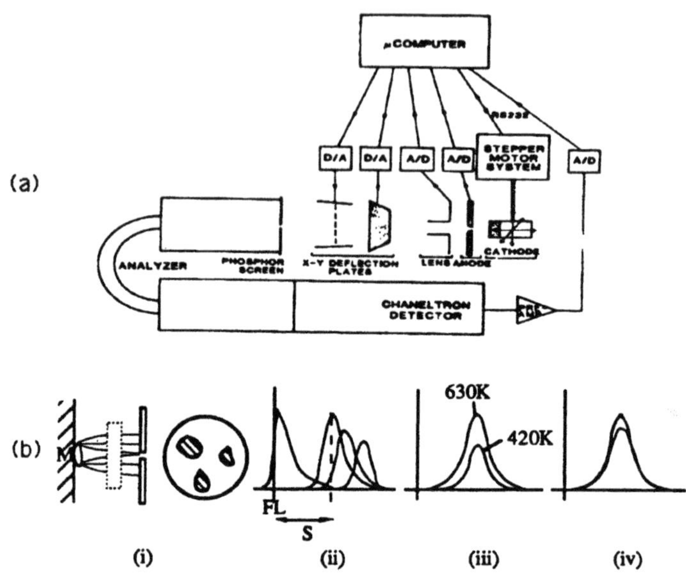

Fig. 7: (a) The electron spectrometer facility, and (b) the major types of information obtained from spectroscopic studies of individual emission processes.

shown in Fig. 5, and then recording the electron energy spectrum of any chosen site, referenced with respect to the Fermi level (FL) of the bulk cathode. A further capability of this system is to record an energy-selective emission image of the site, together with spatially-resolved measurements of the associated energy spectrum of the emitted electrons [Ref. 14]. More recently, this facility has been significantly upgraded to include a computer-controlled data acquisition and processing capability, and a vacuum-interlocked pre-chamber specimen treatment facility.

The key findings to emerge from this study are illustrated in frames (i) to (iv) of Fig. 7b and can be summarized as follows:

(i) The typical projection emission image of a particulate emission site consists of a group of diffuse spots, where the individual sub-sites switch on and off randomly with time with a mean frequency that increases with pressure, e.g., ~1 Hz at a 10^{-8} mbar, and a ~ 10 Hz at ~10^{-6} mbar.

(ii) Each sub-site has its own single-peak spectrum with a broad halfwidth (FWHM) and a characteristic low-current shift S from the FL, where both of these parameters increase nonlinearly with increasing emission current [Ref. 12]; for reference, the type of spectrum obtained from a metallic emitter is shown with its high-energy slope situated on the FL.

(iii) and (iv) The emission spectra are strongly stimulated under constant-gain conditions by both temperature and external uv illumination;

significantly, a temperature increase results in a shift of the spectrum towards the FL, whilst UV irradiation gives no observable shift.

All of these findings point to a "non-metallic" emission mechanism; i.e., quite distinct from that assumed in the traditional Fowler-Nordheim metallic whisker model for explaining prebreakdown electron emission [Ref. 1].

A further important finding is that carbon sites, notably graphite flakes [Ref. 15], give rise to a characteristic type of single or multiple segmental emission image. This emission regime has been studied using an energy-selective imaging capability [Ref. 14] that is built into the integrated analytical facility of Fig. 7a. With this system, it is possible to obtain the type of data shown in Fig. 8, where frame (i) was recorded under "open window" conditions using all the electrons forming the emission spectrum, frame (ii) using only those electrons forming the high-energy slope of the spectrum, and frame (iii) using the electrons forming the low-energy tail of the spectrum. This sequence of images clearly demonstrates that the emission mechanism gives rise to significant spatial variations across individual segments in the energies of the emitted electrons, and has been interpreted in terms of a model involving the coherent scattering of electrons [Ref. 14].

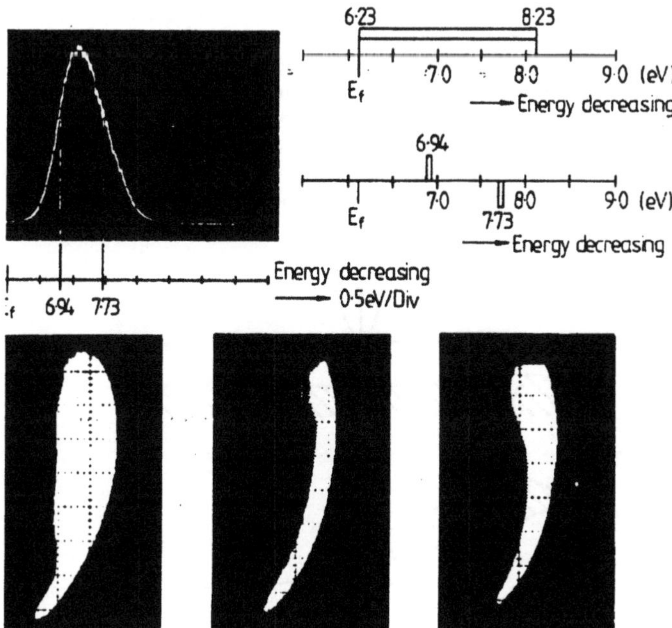

Fig. 8: A segmental emission image recorded by an energy-selective imaging technique to reveal the spatial distribution of electron energies.

2.5. THE ELECTRON EMISSION MECHANISM

On the basis of the findings described in the previous section, two possible models of the emission mechanism have been proposed, with both involving an insulating medium [Refs. 12-14]. The first, the metal-insulator-vacuum (MIV) model, is depicted in Fig. 9a, and represents either an insulating layer or particle I, in intimate contact with a metal electrode M; i.e., corresponding to the experimental situation shown schematically in Fig. 6, or the dielectric coating data to be discussed in a following section. This MIV regime is assumed to emit electrons through a dielectric switching process, in which a conducting channel is electro-formed within the insulating medium. According to this model, electrons are "heated" through several eV and emitted quasi thermionically over the surface potential barrier. The second emission model is depicted in Fig. 9b, and involves a field enhancing "antenna" effect [Ref. 16]. Here it is assumed that a particle or flake (e.g., graphite), which is attached to an electrode surface via a blocking electrical contact, will probe the electric potential distribution at its highest point above the electrode, and hence establish a greatly enhanced electric field across the contact junction. Thus, the necessary conditions are created for the initiation of the dielectric switching process discussed above, with electrons being injected into the vacuum through the M-I-M contact junction [Ref. 14]. By assuming that the emitted electrons are coherently scattered at the flake edge, this model has also been able to provide a qualitative explanation of the segmental emission images shown in Fig. 8.

A quantitative analysis of the MIV model [Ref. 12], based on the band diagram representations of Fig. 10, has shown that it provides a satisfactory explanation for the initial switch-on of sites, the linear and non-linear regions of the typical F-N plot of a gap (or site) characteristic, and the detailed field dependent electron spectral data

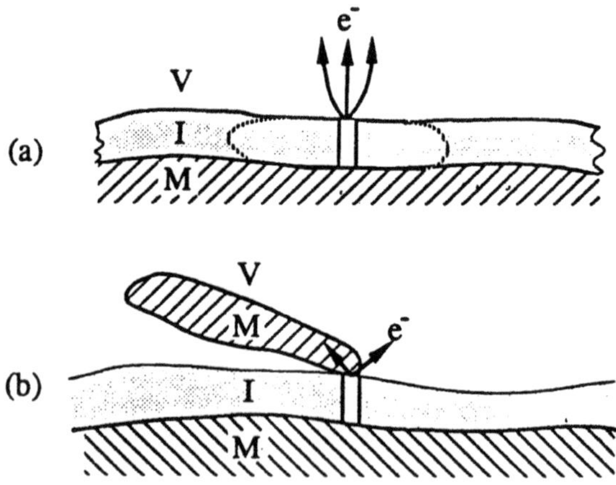

Fig. 9: A schematic representation of emission regimes proposed for (a) the MIV model and (b) the MIMV model.

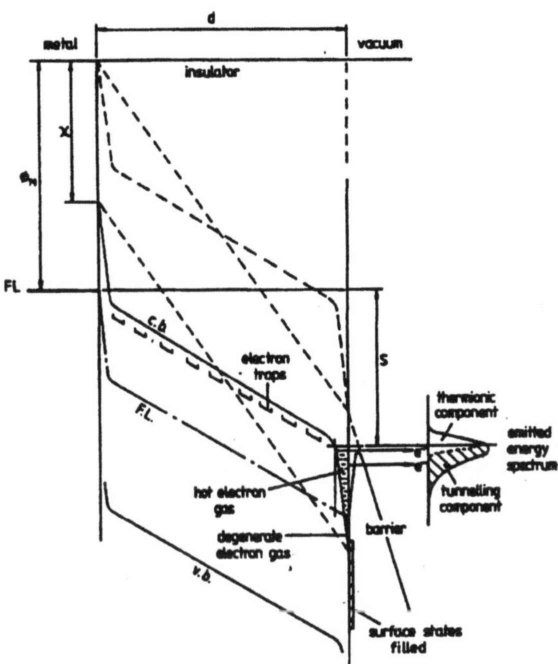

Fig. 10: A detailed band diagram representation of the MIV emission model.

(i.e., shape, half-width and shift). As an example of the predictions of this model, an analysis of emission site data [Ref. 12] reveals a sharp transition between the contact- to bulk-limited conduction regions of the I-V characteristic: indeed, it is this transition that is believed to be responsible for the high-field non-linearities of F-N plots depicted in Fig. 3.

In an alternative theoretical treatment of this type of model, based on the presence of low-density absorbates on an electron surface, Halbritter [Ref. 17] considers the role of hot-electrons in creating internal, self-quenched current avalanches through impact ionization processes. The most significant prediction of this "dynamic" model, as yet not verified experimentally, is that the emitted current will be pulsed with a frequency in the GHz range.

3. Influence of Electrode Processing Procedures

As stated at the outset of this review, the ultimate objective of the fundamental research programmes described above is to provide informed technological advice as to how the insulating capability of vacuum-insulated HV electrodes can be improved. Accordingly, there have been a number of parallel "technical" investigations that have investigated how the basic emission mechanism responds to standard electrode processing procedures.

3.1. GAS CONDITIONING

Apart from the adverse effects of running a gap at an elevated pressure, namely an increase in the incidence of microdischarges and the potential risk of breakdown events, there can also be a beneficial effect, as manifested by the well known and widely used Gas Conditioning process. As illustrated in Fig. 11, this technique involves running a device at a progressively higher field in a helium atmosphere at currents of a few microamps and a pressure of $\sim 10^{-5}$ mbar, and allowing the emission currents to quench to their asymptotic limit at each stage, until the final operating field of the device is achieved. The gas normally used for this process is helium, and the phenomenon has been traditionally interpreted in terms of the blunting of metallic emitters by the sputtering action of high-energy gas ions [Ref. 1].

However, with the emergence of new experimental evidence which questioned the validity of the F-N "metallic" whisker model, it was necessary to reconsider the sputtering interpretation of the gas conditioning phenomenon. In addition, there was no satisfactory explanation of why helium is more effective than other gases. To clarify some of these issues, a detailed investigation has recently been undertaken to determine how both the gas species and temperature influence the conditioning process [Ref. 6].

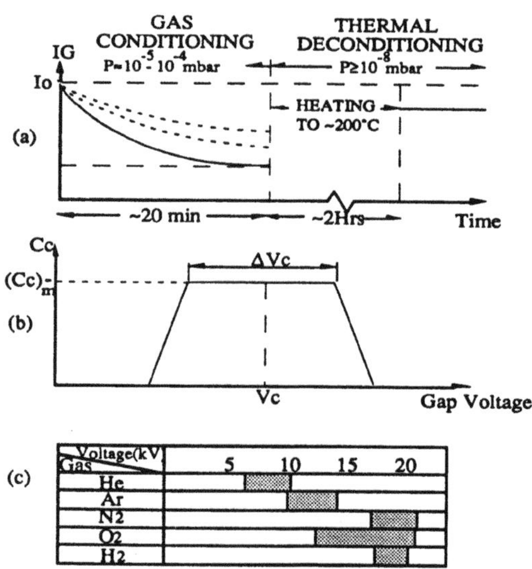

Fig. 11: The general properties of the Gas Conditioning phenomenon, highlighting in a) the "Temperature" effect, and in b) and c) the "Voltage" effect.

The first important finding to emerge from this study is that all of the gases studied do in fact exhibit a positive conditioning effect. However, as illustrated in Figs. 11b and 11c, there is a built in "Voltage Effect", whereby each gas species has a characteristic voltage range over which it is effective as a conditioning gas. From a technological standpoint, this finding indicates that the optimum conditioning procedure for a device might well involve a change of gas as the voltage is progressively raised. The second important finding concerns the permanency of the conditioning, since recent studies have also shown [Ref. 6] that the process can be significantly reversed by low-temperature thermal cycling in the range 200-300C: indeed, as indicated in Fig. 11a, up to 80% of the original current can be retrieved by such a procedure. Furthermore, the dotted lines on this figure illustrate how the effectiveness of the process falls off with increasing electrode temperature. Thus, it is important that gas conditioning is the final treatment process of electrodes, and that the procedure is not used where a device is required to tolerate temperature fluctuations in excess of 100C. A concluding observation on these findings, is that the well known "Current" conditioning technique [Ref. 1] could also be a manifestation of the Voltage effect, whereby the various residual gases progressively have a conditioning effect as the gap voltage is gradually raised.

As to the physical origin of gas conditioning, the dielectric channel model of Figs. 9 and 10 has been shown to provide an "electronic" explanation for the "reversible" nature of the process [Ref. 6]. Thus, referring to Fig. 12, it is assumed that high-energy gas ions are implanted into an emitting channel to create electron traps in the sensitive interface region. It follows therefore that different gas ions will penetrate different distances for a given gap voltage (i.e., the "Voltage" effect), whilst the traps thus formed may be emptied and annealed through

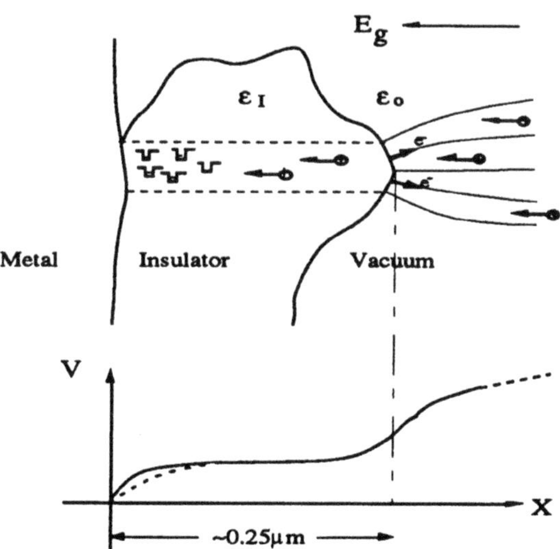

Fig. 12: A schematic representation of the "electronic" explanation for the Gas Conditioning effects.

temperature-enhanced diffusion processes (i.e., the "Temperature" effect). However, it must also be noted that recent studies have shown that there are special electrode conditions where the traditional sputtering mechanism operates [Ref. 18].

3.2. TEMPERATURE CYCLING

As indicated in Fig. 13, the emissivity of a cathode shows a marked increase with its bulk temperature [Ref. 19]. This effect arises from two physical processes: firstly, there is an increase in current from each individual site, and secondly, there is an increase in the total number of emission sites. Recently, the anode probe technique illustrated in Fig. 6 has been used by Niedermann, et al.. [Ref. 11] to investigate the effect of electron bombardment heating on populations of emission sites on niobium electrodes. Thus referring to Fig. 13b, it will be seen that the progressive heating to temperatures up to 750 C results in the stimulation of new emission sites, but that as the temperature climbs to values in excess of 1200 C, all emission sites are generally suppressed.

However, if this same cathode is subsequently recycled through a lower temperature excursion of say only 850 C, it is found that a new generation of sites are formed [Ref. 11]. Furthermore, these new sites are all likely to have a similar material composition (sulphur in the case of Nb electrodes), having been formed by the diffusion of impurity material along the grain boundaries to the electrode surface. To remove these "2nd

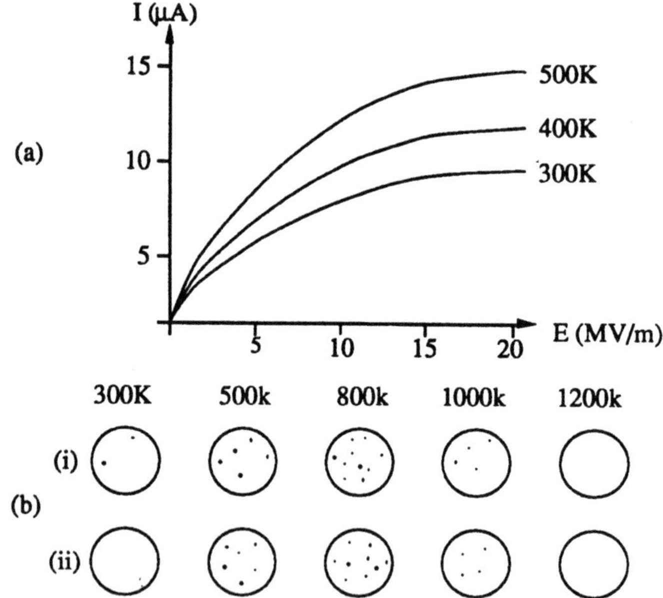

Fig. 13: The influence of electrode heating on (a) the prebreakdown emission current and (b) the emission site density.

Generation" sites, it is again necessary to raise the electrode temperature to values in excess of 1200 C.

In a more recent study [Ref. 19] illustrated in Fig. 14a, a laser probe has been used to investigate the effect on the emission process of heating only the surface region of an electrode: in particular, to determine whether the above effects can be simulated with this technique. Early findings presented in the plot of Fig. 14b indicate that there is a threshold dosing above which the process gives rise to a marked "conditioning" effect. Typically, there is a halving of the β-factor of the electrode, and a reduction in the total number of emission sites at a given field level.

3.3 ELECTRODE COATING

Another apparently attractive approach to eliminating electron pin-holes is to "bury" or "cover" them with a thin over-layer of suitable material [Ref. 1]. Early attempts using evaporated metal films proved unsuccessful due to adhesion problems between the film and electrode under high field

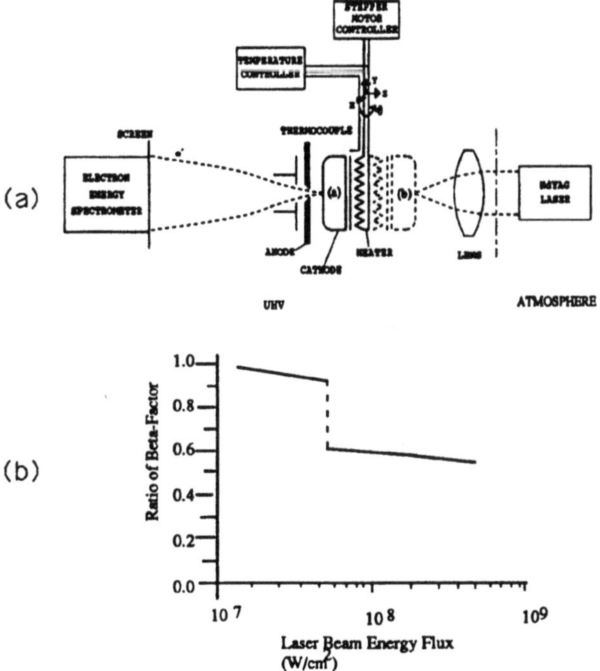

Fig. 14a: (a) The scanning laser probe facility, and (b) a schematic illustration of how the rate of the "after" to "before" β-values of an electrode vary with the laser beam energy flux.

stress conditions; in contrast, attempts at using insulating or semiconducting films have been more successful in this respect, and have given rise to some interesting findings.

Surface Oxidation. The most numerous reports on the use of this process describe how various anodizing techniques can be used to suppress prebreakdown emission and result in an improved voltage hold-off performance of electrodes. The electrode materials most widely used in this context are aluminium [Ref. 20] and niobium [Ref. 21], since their oxides have high electrical resistivities, and hence give rise to a strong insulating barrier at the electrode surface. In contrast, there have also been studies of how individual emission processes are influenced by naturally grown semiconducting oxides. In the case of copper [Ref. 22], it has been shown that, whilst a thick (50nm) air-grown oxide layer has only a relatively minimal influence on the emission site population, it does have a significant influence on the Fowler-Nordheim plot of the I-V characteristic through an accentuation of the high-field non-linearity effect shown in Fig. 3. Also, the presence of such an oxide overlayer has a dramatic influence on the field-dependence of the electron spectral response; i.e., in particular, it gives rise to a marked enhancement of the "shift effect" (illustrated in Fig. 7b) as a result of additional energy-lossy scattering processes in the oxide layer. From the technological perspective, these finding indicate that this type of oxidation cannot be considered as a satisfactory process for "burying" unwanted emission sites.

Dielectric Films. Whilst the coating of electrodes with certain materials, such as Glyptol, has been shown to be effective in suppressing emission, the technique has not been generally adopted as a routine procedure because of its long-term unreliability. Thus, whilst such a film may be initially effective in suppressing emission, there is a serious risk of it subsequently being irreversibly ruptured by the occurrence of an explosive switch-on event. Another potential disadvantage of this procedure is to be found in two recent studies which were deliberately set-up to investigate the emission properties of planar [Ref. 23] and micropoint [Ref. 24] metal-insulator cathode structures. In both cases, the dielectric film led to an order-of-magnitude enhancement in the emission over that obtained with an uncoated metallic cathode at similar field levels. With planar electrodes, the increase in emission current was shown to be a consequence of the switching on of many more emission sites.

Experiments have also been conducted in which electrodes have been coated with a dielectric film that is loaded with a suspension of micron-sized particles [Ref. 25]. These studies showed that such a procedure dramatically enhances the emissivity of electrodes, whereby the switch-on fields of sites can be lowered by nearly an order of magnitude. One important practical implication that follows from this finding is that great care should be taken to ensure that the insulating coatings on high-voltage transformer windings and other components should be free of suspended material.

4. Technological Implications

The single most important conclusion to emerge from the findings outlined in this paper is that, if optimal electrode performances are to be achieved, there is an over-riding need to eliminate particulate contamination from electrode assemblies. It is therefore crucially important to revise the standards relating to the purity of electrode materials, and the electrode manufacture and assembly procedures. In certain "active" applications, such as with particle beam and electron beam diode systems, special precautions may be necessary to protect high-field regions from "internally generated" particulate contamination. By committing the necessary investment, and systematically following this more enlightened approach, it is possible to achieve dramatically improved performances. In this context, special reference will be made in the following section to the experience of the RF superconducting cavity community.

4.1. PREPARATION, ASSEMBLY AND COMMISSIONING OF ELECTRODES

Although there will be an optimal sequence of procedures for each particular application, it is possible to identify a number of general considerations that would apply in all cases. Thus, attention must firstly be paid to the quality of the base electrode material. Secondly, it is necessary to develop an electrode surface preparation procedure that not only removes existing contamination, but also produces an electrode surface with an acceptable microtopography. Thirdly, an electrode must be rinsed, dried and assembled under "clean room" conditions. Lastly, electrodes have to be 'conditioned' before they can be relied upon to operate according to their specifications.

Electrode Material. The choice of this parameter is clearly fundamental, and will invariably be dictated by the particular application being considered. However, there will usually be some flexibility in the choice of the quality of the material. Thus, it is of key importance to establish the elemental nature of the dominant bulk impurities, and whether for example they segregate out along grain boundaries in particulate form. If this is the case, such as happens with carbon in may refractory metals, it has to be recognized that some of these same contaminants will inevitably "appear" in the surface of an electrode and present a potential hazard to its reliable operation. The practical implication of this observation is that it may be necessary to refine the bulk metal manufacturing process, such as was done in the case of the niobium used for RF superconducting cavities [Ref. 26].

Electrode Polishing. Historically, it was conventional to prepare electrodes by some form of mechanical polishing process. However, it is now widely accepted that such a procedure will degrade an electrode surface through the impregnation of contaminant particles from the polishing medium. Accordingly, electrodes are nowadays routinely prepared by using either a "chemical" or "electro-chemical" polishing process: the choice between these two approaches being determined by the constraints of the particular application. In the case of Nb cavities for example, it has been found preferable to use a chemical polishing technique [Ref. 27]. This finding highlights the fact that the performance of an electrode is not necessarily related to the "smoothness" of its surface microtopography. Thus, although electro-polishing produces a smoother Nb surface, a chemical polishing procedure generally results in a more reliable operational surface. It is also essential to recognize that both polishing processes can give rise to "self-contamination", and that consequently it is necessary to build in filtration procedures for removing suspended particulate material from both the gas and liquid media [Ref. 28].

Electrode Assembly. Following the polishing process, and prior to assembly, it is necessary to subject an electrode to a sequence of ultrasonic cleaning and rinsing procedures. These would generally start with a de-greasing agent such as methanol, acetone or trichloro-triflouro-ethelene, and conclude with a de-ionized water rinse. It is also important to ensure that great care is taken in choosing the quality of all agents, and particularly to ensure that all organic material is removed from the final DI-water rinse. In addition, it is crucial for the whole of this operation, together with the subsequent assembly procedure, to be conducted under conditions of clean lamina flow. For more detailed information, the reader is referred to the literature on this very specialized technology [Ref. 27, 28].

In Situ Processing. Having taken the above precautions, there will frequently remain some remnant prebreakdown current from such 'as-assembled' electrodes. To deal with this potential threat to their operational stability, there are two widely used in situ procedures that have been shown to be very effective in passifying the residual emission sites responsible for this current. These are High Temperature Annealing and Gas Conditioning, and are based on the studies described above in Sections 3.1 and 3.2 where their detailed practical implementation, e.g., mode of heating and the choice of gas species, will depend upon the particular device application (see for example the following section). There is also a third in situ procedure, known as "spark knocking", that is used in certain situations to condition electrodes [Ref. 1]. However, this technique tends to be drastic and unpredictable, and results in considerable topographical damage to the electrode surface; accordingly there are limits to the range of its applications.

4.2. ENHANCEMENT OF ELECTRODE PERFORMANCE

A very good example of the technological benefits that can be derived from following the above procedures is the up-grading in the performance of superconducting RF cavities that has been achieved over the past ten-year period. These devices are widely used in particle accelerating and storage systems, and have the general shape shown in Fig. 15a. In the past, they have conventionally been made from copper; however, with the growing demands from nuclear physicists for ever higher particle energies, there has recently been a major effort to enhance their performance by operating them under superconducting conditions with higher accelerating field gradients. To achieve this objective, it has been necessary to invest an enormous technological effort to suppress the onset of field emission, which would otherwise catastrophically degrade their performance.

To illustrate the level of achievement attained with this type of cavity, the reader is first referred to Fig. 15a which shows the trajectories that would be followed by electrons field emitted from the high-field region near the neck of the cavity. It follows from this that the bigger the area of this region, the greater will be the potential incidence of emission sites: it is therefore not surprising that smaller high-frequency cavities have generally achieved higher operating fields than the larger low-frequency cavities. As seen from Fig. 15b, performance levels have typically more than doubled with low-frequency cavities over the past decade, but have approached an order-of-magnitude improvement with the high frequency cavities [Ref. 29]. To illustrate how the various sequences of treatment procedures contribute to the final performance, Fig. 15c summarizes the collated data obtained from an extended series of tests on 1.5 GHz cavities at Cornell University [Ref. 29]. From this data, it will be seen that, in broad terms, a cavity that has been subjected to a final heat-treatment will exhibit an 80% improvement in performance over

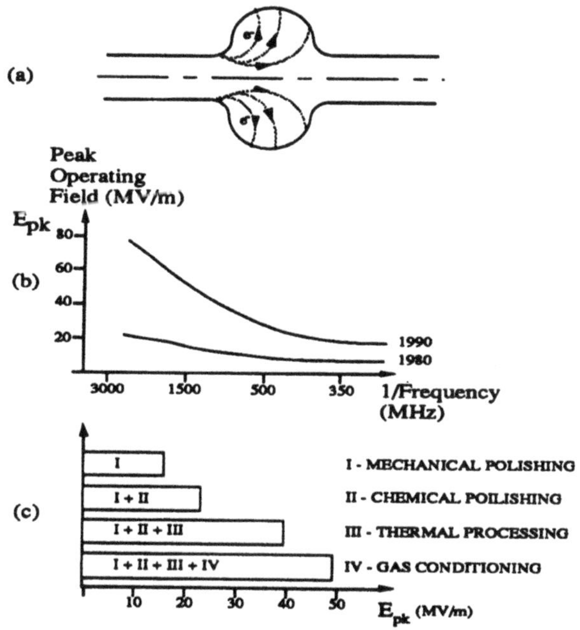

Fig. 15: Illustrating how the performance of superconducting RF cavities has been improved over the past ten-year period as a result of a newly developed sequence of electrode treatment procedures. a) The basic cavity geometry, showing the trajectories of parasitically emitted electrons, b) the improvement in performance, and c) the relative contributions of the individual procedures.

one that has only been chemically processed, and that subsequent He processing results in a further 20% improvement. In fact, peak operating fields in the region of 60 MV/m are now routinely obtained. However, even after this sequence of treatments, there always remain a few persistent emission sites, and it is essential for the future development of RF cavity technology that a method is found for suppressing this residual emission.

5. Future Goals

5.1. GENERAL CONSIDERATIONS

Although the above performance data represent a major technological breakthrough, it has also to be pointed out that, not only are the procedures expensive, but that the high temperatures involved limit their range of applications. In addition, there is a general need to evaluate the effectiveness of this technology in other device applications and, in particular, to confirm whether or not there is a corresponding improvement in the actual breakdown performance of passive electrode applications. More generally, however, there remains an on-going need for developing an alternative technological approach to electrode processing; in fact, the ideal would be for one that is less expensive, and is conducted under near ambient temperatures. For this to be possible, it will be necessary to conduct further fundamental studies aimed at obtaining a better physical understanding of the underlying non-metallic electron emission mechanism, and hence how best to suppress it.

Another important long-term objective must be the development of analytical techniques capable of studying the microscopically localized electron emission processes that occur on broad-area electrodes under fast-rise, pulsed-field conditions. At present, it is not known for example how the d-c MIV and MIM mechanisms discussed in this paper respond to fast-changing fields, or indeed if there are other mechanisms that operate under this type of regime. There is also an urgent need to establish whether a "pure" vacuum gap, and one that is bridged by a solid insulator, share a common breakdown/flashover triggering process; i.e., is triple junction electron emission the same "animal" as that discussed throughout this paper?

5.2 SPACE-SPECIFIC CONSIDERATIONS

In the case of future generations of HV devices that will have to operate reliably over a long periods of time in the harsh space environment, there are an additional set of developmental challenges that have to be added to the "base-line" requirements discussed above. In particular, devices must be able to cope with the effects of solar and cosmic radiation, atomic oxygen, micrometeorite damage, varying electric and magnetic fields, and low-density plasma conditions. As a first step towards reducing the problem to manageable proportions, it would seem inevitable that such HV devices will have to be "encapsulated", rather than "nude", and electromagnetically shielded. However, even with these precautions, electrodes will still have to remain stable under the effects of repeated thermal cycling and residual ionizing radiation. Accordingly, future fundamental research programmes must address these issues, and include studies that simulate the space environment.

Acknowledgement

The author wishes to acknowledge the extended support he has received from the Strategic Defense Initiative Organization's Office of Innovative Science and Technology (through Naval Surface Warfare Center Contract No. N60921-91-C-0078) and Auburn University (Alabama, USA), that has enabled him to maintain an active involvement in this field of research. He also wishes to thank his colleagues Helen Cutler and Andrew Abbot for their help in preparing the diagrams and text of this paper.

References

[1] R.V. Latham, HV Vacuum Insulation: The Physical Basis, London/New York, Academic Press (1981).

[2] H.C. Miller, "Surface Flash-over of Insulators", Proc. XIII DEIV, Paris pp. 233-240 (1988).

[3] P.A. Chatterton, "Recent Developments in Vacuum Breakdown and Discharge Physics Relevant to Accelerator Tube Design", Nucl. Insts. & Meths in Phys. Res., Vol 220, pp. 73-81 (1984)

[4] R.V. Latham, "High-Voltage Vacuum Insulation. New Horizons", IEEE Transactions on Electrical Insulation, Vol. 23 No. 5, pp. 881-894 (1988).

[5] R.V. Latham, K.H. Bayliss, and B.M. Cox, "Spatially Correlated Breakdown Events initiated by Field Electron Emission in Vacuum and High Pressure SF_6", J. Phys. D.: Appl. Phys., Vol. 19, pp. 214-231 (1986)

[6] S. Bajic, A.M. Abbott, and R.V. Latham, "The influence of gap voltage, temperature and gas species on the "gas conditioning" of high voltage electrodes", IEEE Trans. Elec. Insul., Vol. 24 No. 6, pp. 891-896 (1989).

[7] B.M. Cox, "The Nature of Field Emission Sites", J. Phys. D: Appl. Phys., Vol 8, pp. 2065-73 (1975).

[8] R.E. Hurley, "Electrical Phenomena Occurring at the Surface of Electrically Stressed Metal Cathodes. III Current-Voltage Characteristics of Electroluminescent (k-spot) Regions on Broad Area Cathodes", J. Phys. D: Appl. Phys., Vol. 13, pp. 1121-28 (1980).

[9] B.M. Cox and W.T. Williams, "Field-Emission Sites on Unpolished Stainless Steel", J. Phys. D., 10, pp. L5-9 (1977).

[10] C.S. Athwal and R.V. Latham, "A Micropoint Probe Technique for Identifying Field-Emitting Sites on Broad-Area HV Electrodes", Physica, Vol. 104C, pp. 46-9 (1981).

[11] Ph. Niedermann, N. Sankarraman, R.J. Noer, and O. Fischer, "Field Emission from Broad-Area Niobium Cathodes: Effects of High-Temperature Treatment", J. Appl. Phys., Vol. 59, pp. 892-901 (1986).

[12] K.H. Bayliss and R.V. Latham, "An Analysis of Field-Induced Hot-Electron Emission from Metal-Insulator Microstructures on Broad-Area HV Electrodes", Proc. Roy. Soc., Vol. A403, pp. 285-311 (1986).

[13] C.S. Athwal and R.V. Latham, "Switching and Other Non-Linear Phenomena Associated with Pre-breakdown Electron Emission Currents", J. Phys. D.: Appl. Phys., Vol. 17, pp. 1029-1034 (1984).

[14] N.S. Xu and R.V. Latham, "A Spatially Resolved Energy Analysis of Field" Induced Hot-Electron Emission (FIHEE) from MIM Microstructures", J. de Physique, Vol. 47, pp. C7(95-99) (1986).

[15] N.S. Xu and R.V. Latham, "Coherently Scattered Hot-Electrons Emitted from MIM Graphite Microstructures Deposited on Broad-Area Vacuum-Insulated High-Voltage Electrodes", J. Phys. D.: Appl. Phys., 19, pp. 477-82 (1986).

[16] C.S. Athwal, K.H. Bayliss, R.V. Latham, "Field-Induced Electron Emission from Artificially Produced Carbon Sites on Broad-Area Copper and Niobium Electrodes", IEEE Trans. Plasma Sci., Vol. 13, pp. 226-229.

[17] J. Halbritter, "Dynamical Enhanced Electron Emission and Discharges at Contaminated Surfaces", J. Appl. Phys., A39, pp. 49-57 (1986).

[18] A. Zeitoun-Fakiris and B. Juttner, "On the dose of bombarding residual gas ions for influencing prebreakdown field emission in vacuum", J. Phys. D: Appl. Phys., 24, pp. 750-756 (1991).

[19] A.D. Archer and R.V. Latham, XIV'th Int. Symposium on "Discharges and Electrical Insulation in Vacuum", Santa Fe (USA) (1990).

[20] P. Graneau and D.B. Montgomery, "Insulator Flashover Mechanism in Vacuum Insulated Cryocables", J. Vac. Sci. Tech., 13, pp. 1081-7 (1976).

[21] E.L. Garwin and R.E. Kirby, "Surface Studies of Nb, its Compounds, and Coatings", Proc. of the Second Workshop on RF-Superconductivity, pp. 455-504 (1984).

[22] R.V. Latham, K.H. Bayliss, and S. Bajic, "The Influence of Surface Oxidation on Prebreakdown Electron Emission Processes", IEEE Trans. Elec. Insul., Vol 24 (6), pp. 901-903 (1989).

[23] S. Bajic, N.A. Cade, A.D. Archer, and R.V. Latham, "Stimulated cold-cathode emission from metal electrodes coated with Langmuir-Blodgett multilayers", Inst. Phys. Conf. Ser. No 99: Section 3, Bath (1989).

[24] R.V. Latham and M.S. Mousa, "Hot-Electron Emission from Composite Metal-Insulator Micropoint Cathodes", J. Phys. D: Appl. Phys., Vol 19, pp. 699-713 (1986).

[25] S. Bajic and R.V. Latham, "Enhanced Cold-Cathode Emission Using Composite Resin-Carbon Coatings", J. Phys., D: Appl. Phys., 21, pp. 200-4 (1988).

[26] H. Padamsee, "The Technology of Nb Production and Purification", in 2nd Workshop on RF Superconductivity, pp. 339-376 (1984).

[27] D. Bloess, "Chemistry and Surface Treatment", in 2nd Workshop on RF Superconductivity, pp. 409-426 (1984).

[28] P. Kneisel, "Clean Work and its Consequences – Contamination Control Considerations", in 2nd Workshop on RF Superconductivity, pp. 509-532 (1984).

[29] H. Padamsee, J. Kirchgessner, D. Moffat, R. Noer, D. Rubin, J. Sears, and Q.S. Shu, "New Results on RF and DC Field Emission", in 4th Workshop on RF Superconductivity, KEK, Tsukuba, Japan (1989).

SURFACE CHARGING OF SPACECRAFT IN GEOSYNCHRONOUS ORBIT

GORDON L. WRENN and ANDREW J. SIMS
Defence Research Agency, Space Sector
RAE Farnborough,
Hants GU14 6TD, UK

A new satellite is placed into geostationary orbit and successfully commissioned (geosynchronous = orbital period of 24 hours, while geostationary additionally requires the orbit to be circular with 0° inclination). All goes well for a month or so but then the mission controllers are confronted by a number of abnormal situations which are generally classified as 'Operational Anomalies'. For example, on-board systems suddenly change state (maybe turn ON or OFF), housekeeping telemetry channels exhibit out-of-limit values for no obvious reason, attitude control systems and instruments with stepping sequences misbehave.

QUESTION: Do these problems result from surface charging?

It is helpful to classify the various anomalies that have been encountered, using three levels of severity, based upon mission impact and recovery management; typical examples are noted:

Level 1: Annoying (wrong status indication, logic upset)
No performance outage; rectify by ground command intervention.
e.g. Power Bus under-voltage flag set [MARECS]
Battery charge switched ON [METEOSAT]

Level 2: Critical (phantom command, spurious mode switching)
Performance outage; contingency procedure effects recovery.
e.g. Camera 'Spin-Scan' start [GOES]
Radiometer N-S scan stop or jump [METEOSAT]
Emergency Sun Re-acquisition triggered [MARECS]
Antenna Platform Spin-Up initiated [ANIK]

Level 3: Serious (part failure, system shutdown, unknown)
Permanent degradation or mission fatality; blame 'Act of God'.
e.g. Power Surge [DSCS]
Component Failure [GOES]

It is difficult to be sure of the numbers involved, since only a fraction have been reported, but our guess is that over the last twenty

R. N. DeWitt et al. (eds.), The Behavior of Systems in the Space Environment, 491–511.

years you can think in terms of order 10^4 for Level 1, $10^2 - 10^3$ for Level 2, and 10^1 for Level 3.

Since it is usually easier to ask questions than to answer them, let us proceed via a score of subsidiaries.

Q 1: What are the possible reasons for the anomalies?

The following list includes the most common suspects:-

- Telemetry Glitch
- Mission Control Problem
- Part or System Fatigue (Surface Aging)
- Thermal Strain
- Radio Frequency Interference
- Electromagnetic Pulse
- Single Event Upset
- **ELECTROSTATIC DISCHARGE**
- Impact Damage (Micrometeoroid, Space Debris)
- Something Else

Spacecraft engineers have been all too ready to assign the blame for their problems to effects of the environment, rather than any deficiency of their design. In reality, the items at the top of the list are much more likely, and certainly should be eliminated before the latter candidates are considered seriously.

Q 2: Why should we target Electrostatic Discharge?

Although there is no direct proof that a single operational anomaly has been caused by ESD, there is very strong circumstantial evidence.

This is based upon three observations:-

(i) In 1972 Sherman DeForest published a paper [Ref. 1] in which he showed that a geosynchronous spacecraft [ATS-5] could become charged to well over 10,000 volts.

(ii) Initial studies of the anomalies showed that they were not uniformly distributed in local time but appeared preferentially in the midnight to dawn sector. Figure 1 serves to illustrate this by plotting the times of MARECS-A anomalies [Ref. 2] which occurred between 27 January and 9 June 1982 [Ref. 3]; note the absence of events during eclipse.

(iii) Closer inspection disclosed that the frequency of occurrence of such anomalies is linked to the prevalent level of geomagnetic activity.

The implications of these will become clear but, before jumping to any conclusions, you should be warned that MARECS-A was a classic case; anomaly patterns for most other satellites are more complex.

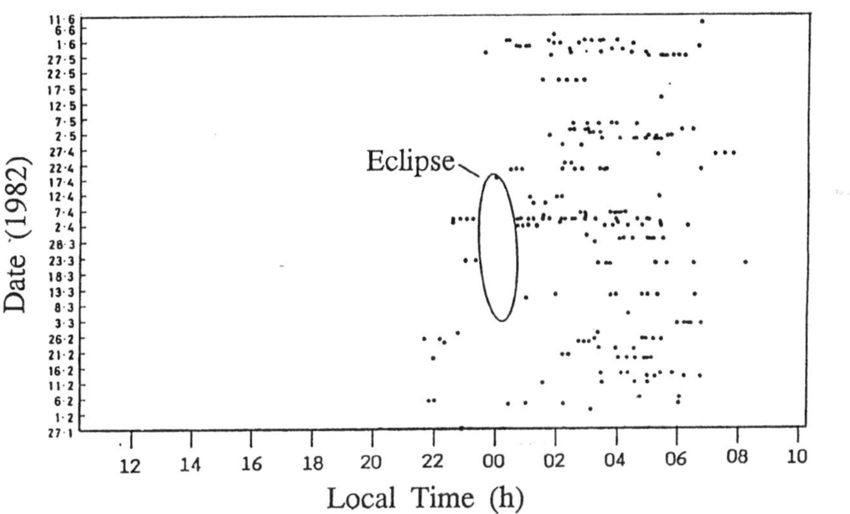

Fig. 1: Date and local time of MARECS-A anomalies, 27 January to 9 June 1982, 217 power bus 'under voltage' status switchings (A108).

Q 3: How is an electrostatic discharge generated?

The following chart describes an accepted scenario:

Spacecraft in Sunlight (i.e. NOT in Eclipse)
+
Some Isolated Surface Element in Shadow
+
Breakdown Path to Structure
+
Subsystem Susceptible to ESD
+
Spacecraft enters 'Hot Plasmasheet' or Spacecraft in Plasmasheet + Substorm
↓
Enhanced Flux of keV Electrons
↓
Threshold Voltage for Breakdown exceeded
↓
ESD → **OPERATIONAL ANOMALY**

Although the process is relatively straightforward there are some qualifiers that need expansion, which leads you to a more detailed analysis of surface charging. First let us dispose of a few more basic questions.

Q 4: keV electrons appear to be the culprits, why is this?

Given a thermal plasma with temperature T, the current density J is proportional to the product of the concentration N and the mean thermal speed $\bar{c}$ where $\bar{c} = \sqrt{8\,k\,T / \pi\,m}$ varies inversely with the square root of particle mass m, and thus the current incident upon a surface tends to be electron dominated (k = Boltzmann constant).

Since negative charge build-up induces a potential on the surface which serves to retard incident electrons, energies of order keV are required to create potentials high enough to trigger break-down. Electrons with energies of ~100 keV or more will penetrate the surface, these could produce ESD via 'deep dielectric charging'; particles with higher energy can trigger Single Event Upsets.

Q 5: What is the significant of Sunlight and Shadow?

Given the intensity of sunlight at 1 AU with normal incidence, the current density due to photo-emission from most surfaces well exceeds that due to incident plasma and inhibits charging. It is now important to introduce a distinction between 'ABSOLUTE Charging' and 'DIFFERENTIAL Charging'. In GEO, near equinox, a satellite can spend up to 75 minutes in Earth eclipse; photo-emission is absent and the whole spacecraft can assume a potential of many kilovolts (-19 kV has been observed). This absolute charging is dramatic in respect of on-board charged particle detection but it rarely seems to cause operational anomalies; in sunlight, absolute charging is limited to a few hundred volts. However, surface elements, which are electrically isolated from each other and the satellite structure, charge differentially – particularly when they are in shadow. It is this effect which more frequently gives rise to damaging discharges.

Q 6: How does the discharge trigger the anomaly?

We can postulate that discharge implies one of three modes – 'Flash-over', 'Punch-through' or 'Blow-off'. In each case, large current transients can be conducted, inductively coupled, or radiated into sensitive electronic systems anywhere on the spacecraft. Our feeling is that flash-over is the common mode as surface breakdown paths develop; blow-off might be expected to prevail for absolute charging but the fact that anomalies are seldom observed in eclipse suggests that it is less important.

Q 7: Can we determine the expected potential of a surface?

For all surfaces, the potential V will move towards a value which achieves a zero net current.

$$C_A\, dV/dt = \underline{J}(V) + \sigma V \qquad = 0 \text{ for } \Sigma\, I = 0 \qquad (1)$$

The capacitance per unit area C_A and conductivity σ are not easily determined but the current density $\underline{J}$ is very difficult, being non-linear and non-local; the inherent negative feedback operates through sheath formation with space charge transfer, and the creation of potential barriers. Figure 2 registers all the current components that must be considered.

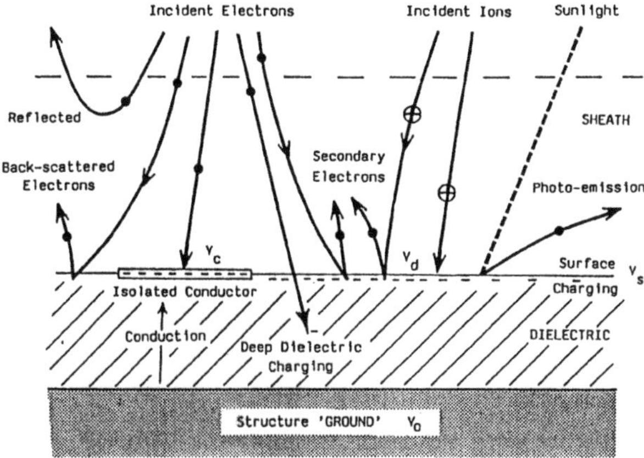

Fig. 2: Currents which control charging at a spacecraft surface - isolated conductor or dielectric layer.

$$J = - J_e + J_b + J_{se} + J_i + J_{si} + J_p \pm J_c \quad (2)$$

While some of these components may be negligible at V = 0, they can still have an important role as $V \rightarrow V_s$ approaching equilibrium

The rate of charging, determined by C_A, is critical because of the dynamic nature of the charging environment, continually changing solar aspect angles (specially for elements of a spinning spacecraft) and the finite eclipse times – surface potentials seldom reach equilibrium

Q 8: Why is Local Time significant?

A satellite at 6.6 Re normally traverses both the plasmasphere and the plasmasheet within 24 hours and Fig. 3 shows their typical configuration in local time. The plasmasphere is a cold plasma which will not support charging while the plasmasheet is characterized by the keV electrons referred to above. Unfortunately the picture is not so simple because the configuration and boundary positions are constantly changing in response to (i) variations in the Interplanetary Magnetic Field (IMF) and solar wind and (ii) substorm activity, during which electrons are injected from the magnetotail region near local midnight. Many of these electrons are pseudo-trapped by the magnetic field and are then subject to gradient and curvature drifts which take them towards dawn.

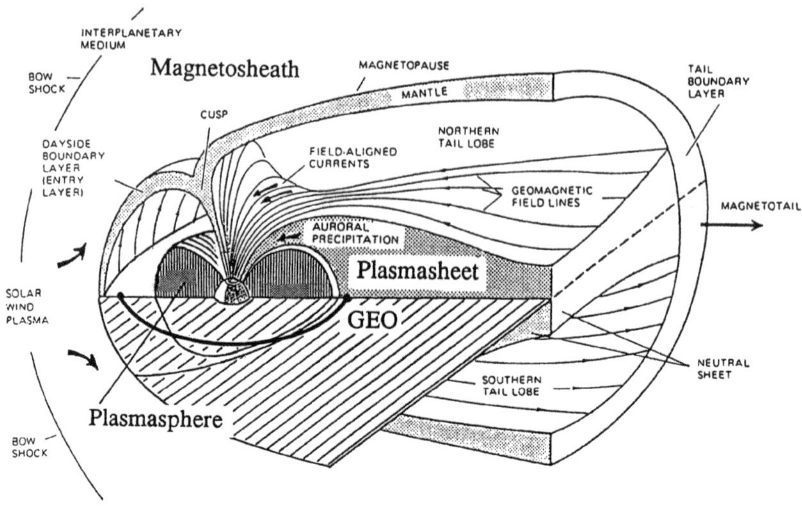

Fig. 3: The magnetosphere showing the geosynchronous orbit with respect to the plasmasphere and plasmasheet.

At 6.6 R_e, their drift speed is between ~ 3.5 x E(keV) degrees per hour at large pitch angle and ~ 1 x E(keV) degrees per hour at small pitch angle [Ref. 4]. (Pitch angle = direction of particle's spiral motion with respect to magnetic field). This means that electrons with energy in the range of 5 to 20 keV can catch a geosynchronous (15 degrees per hour) satellite several hours after an injection, and it accounts for the special interest in the midnight to dawn sector.

Q 9: **Where is the plasmasheet boundary?**

The reason for the boundary is best understood in terms of a simple representation of the equatorial electric field, which results from a radial 'corotational' component plus a uniform 'cross-tail' component from dawn to dusk shown as Fig. 4 [Ref. 4].

Electrostatic potential,

$$U(r,\phi) = C_1/r - C_2\, r \sin\phi \tag{3}$$

where C_1 = 91.5 kV R_e and C_2 ~ 0.5 to 2 kV/R_e depending upon geomagnetic activity.

This predicts a pear shaped plasmasphere pointed towards dusk, in which the cold plasma, produced by photo-ionization of the atmosphere, corotates, whilst outside the U = 0 contour, the plasmasheet particles are constrained by the magnetic field [Ref. 4] but tend to follow the indicated drift paths, after arriving from the tail.

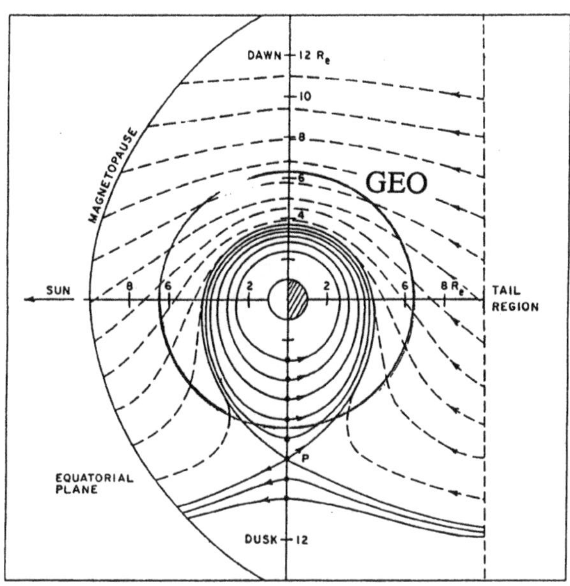

Fig. 4: The equatorial electric field showing equi-potentials which are drift paths for 'zero' energy particles. U = 0, exhibiting a stagnation point at P, separates regions of corotation and convection.

In reality, we can soon appreciate that such a simple model is totally inadequate by looking at Fig. 5, a spectrogram of electrons and protons seen in GEO with ATS-5 at 105°W (LT = UT - 7 h) [Ref. 2]. Here color is added to the conventional grey-scale to indicate pitch angle distribution, approaching blue for field-aligned and orange for 90° particles. The data, for 22-23-24 January 1971 (starting at 18 UT on the 22nd), are wondrous to behold - but certainly not fitting to any simple or static geometry; however, the plasmasheet position is clearly defined with entries at 21 h and 22 h local time. The common practice of modelling particle populations using Maxwellian distributions is clearly far from ideal in this regime.

On neither day are there many electrons with energy exceeding 4.5 kV until 12 UT (03 LT) on the 24th when a high flux of injected electrons catch the satellite, higher energies arriving first. This neatly demonstrates how charging probability tends to peak in two ways (i) spatial and (ii) temporal. The first reflects the usual structure of the plasmasheet seen at GEO - a sharp boundary in the late evening followed by a maximum density and a gradual decay towards the morning hours. The second reflects the magnetospheric disturbances induced by substorm processes, repeated particle injection and energization with movement of the plasmasheet boundary. In fact, hazardous charging only occurs when high fluxes of 5-10 keV electrons are encountered [Ref. 5] in what might

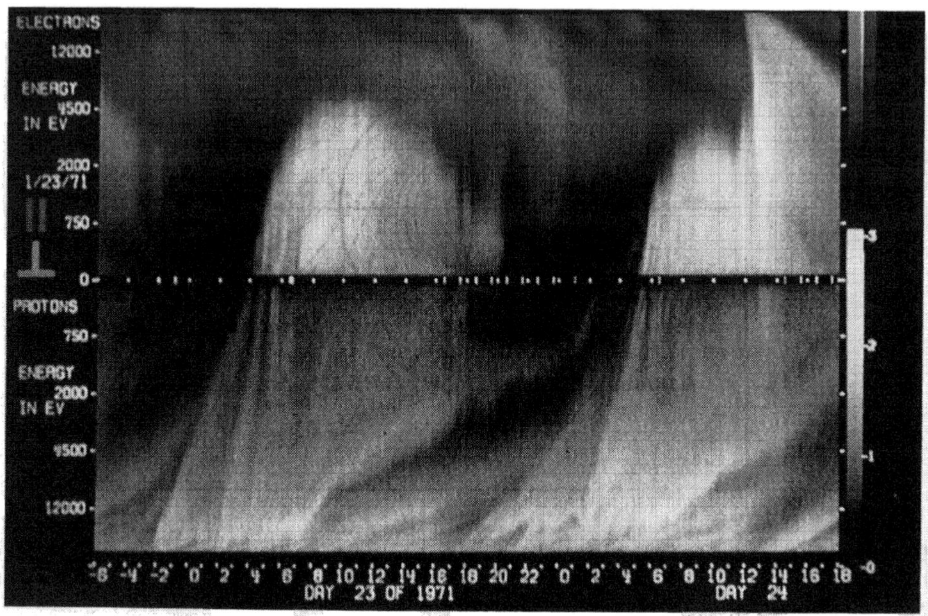

Fig. 5: An ATS-5 spectrogram covering 48 hours in January 1971.

be termed the 'Hot Plasmasheet'; this is often reached an hour or two after an initial plasmasheet entry.

The picture changes from day to day, sometimes dramatically, as the magnetosphere responds to variations in the solar wind and the IMF, these are perceived as 'geomagnetic activity' and monitored by means of ground-based magnetometers. A number of indices are routinely computed in an attempt to quantify this activity [Ref. 6], the most common being the 3-hourly planetary K_p which employs a pseudo-logarithmic scale of 28 values (0o,0+,1-,1o,1+,2-,... .... 8o,8+,9-,9o) from very quiet to very disturbed.

For the duration of the spectrogram K_p shows fairly quiet values:-

Day 22	3o	3o	3o	2–	0+	0+	2–	3o
23	3–	2+	2o	1o	1–	1+	2–	1+
24	2–	2+	1o	2+	2+	1–	3+	4–

The local time of the plasmapause/plasmasheet boundary has been determined as:

$LT(h) = 25.5 - 1.5\ K_p$ ATS-5, Winter 1970-71 [Ref. 7] (4)

$LT(h) = 25.5 - 2.07\ K_p$ METEOSAT-2, 1981-5 [Ref. 8] (5)

The idea that the cold plasma is neatly confined to a pear shaped region is also far from the truth. Note that the ATS spectrograms do not actually go below 50 eV, the detectors are not sensitive to cold plasma. Figure 6 plots the concentration of cold ions seen by GEOS-2 during December 1978 and lists Ap (a daily mean of a linear equivalent of K_p, ranging between 0 and 400) [Ref. 9]. This shows that plasmaspheric ions are generally encountered in the evening but are found all around the orbit at very quiet times. Following a disturbance, cold plasma outside the contracted U = 0 contour is free to convect towards the frontside; the flux tubes can then take several days to refill, the refilling rate depending upon the level of geomagnetic activity [Ref. 10].

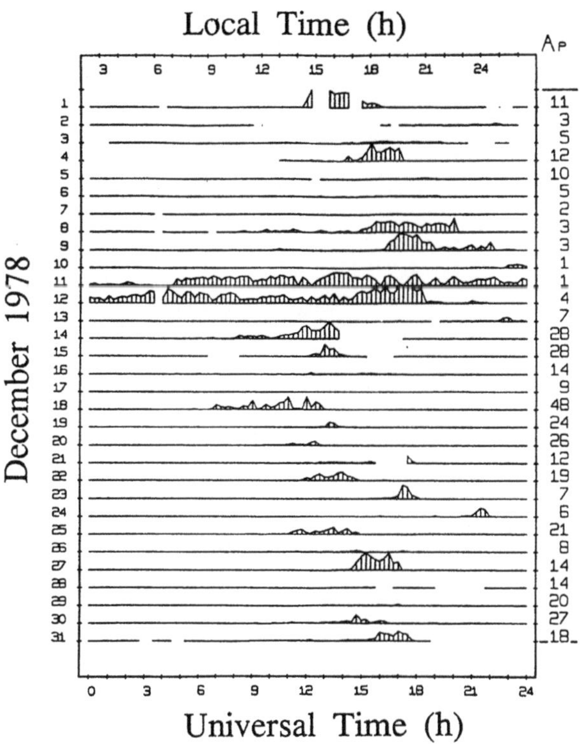

Fig. 6: GEOS-2 measurements of cold plasma at 6.6 R_e for December 1978. Each panel has a linear vertical scale of 0 to 100 cm^{-3}.

Q 10: What does charging look like?

Ref. [2] gives examples of the spectrograms that initially baffled DeForest and McIlwain with evidence of absolute charging in eclipse. Both flux and energy of the electrons were then reduced but the ATS-5 proton detector gave the clearest signature when all protons were accelerated to an energy > 4500 eV [their Fig. 1], saturating some energies and leaving the lower channels empty. Similar data have been used to demonstrate absolute charging up to a few hundred volts in sunlight [Ref. 2]. Evidence for differential charging is more subtle, but it is also clearly provided by the ATS data [Refs. 2,11]. An electron detector alone can sometimes provide conclusive evidence, if the differential charging polarity is right. This was the case with Meteosat-2 eclipse events in which a differential voltage of 650 V, typically developed in about 1 hour, accelerated secondary electrons into the detector, see Fig. 7 [Ref. 8].

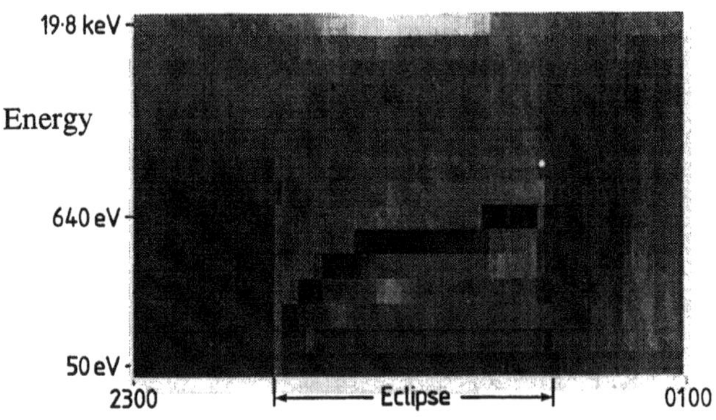

Fig. 7: METEOSAT-2 electron spectrogram covering two hours during the night of 3/4 April 1982: eclipse charging event.

Q 11: What do the current components depend upon?

It is now informative to examine the detail of Fig. 2, i.e., the terms appearing in Eq. (2). J_c is the conductivity term which is proportional to the differential voltage; it can also reflect field-enhancement pre-breakdown. The resistivity of dielectric materials often decreases during long exposure to the space environment.

The photo-emission current J_p depends upon the solar spectrum which is plotted in terms of eV energy on Fig. 8. At 1AU the Solar Constant ~1.4 kW/m^2, but only photons with energy greater than the material work function can generate emission (as the yield curves for Aluminum and Gold show) so that the effective 'constant' is orders of magnitude smaller; J_p is typically 10 to 40 μA/m^2. For a sphere, the cross-sectional area is only 0.25 of the surface area; the establishment of potential barriers can prevent the escape of photo-electrons.

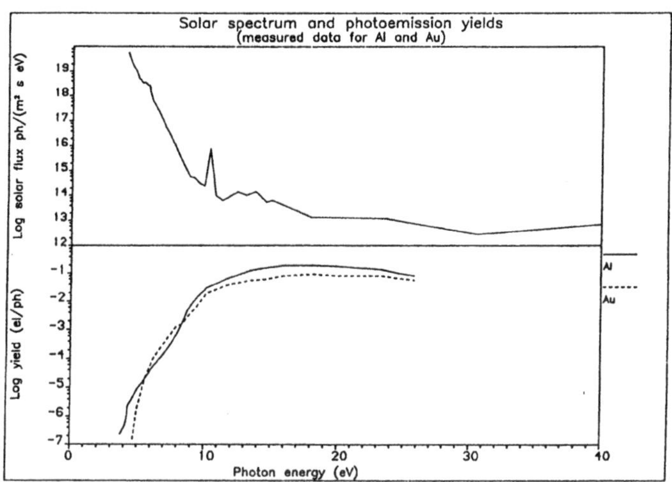

Fig. 8: Solar spectrum and photoemission yields of Aluminum and Gold.

J_e and J_i are integrations of those incident fluxes which have sufficient energy to overcome the voltage bias of the collector. Plasmasheet electrons and plasmasphere protons have already been identified as important but highly variable current components. However, it turns out that the secondary emission terms, J_b, J_{se} and J_{si}, can be critical [Ref. 12].

Q 12: What is the difference between backscatter & secondary emission?

In theory, one can discriminate between incident electrons which are re-emitted from a material (back-scatter) and those which escape following the impact and scattering process. Figure 9 presents a simplistic picture of the interaction, and the contrasting energy distributions of the products; the true secondaries resemble a thermal source corresponding to a few eV. In practice, it is impossible to label the emitted electrons and a partition is arbitrarily set at 50 eV.

For secondary emission, the yield is determined by

$$Yield = \int_0^{X_m} N(E,x) \cdot P(x) \cdot dx \qquad (6)$$

where

N(E,x) = Number of electrons liberated at depth x

$\propto$ Stopping Power, dE/dx

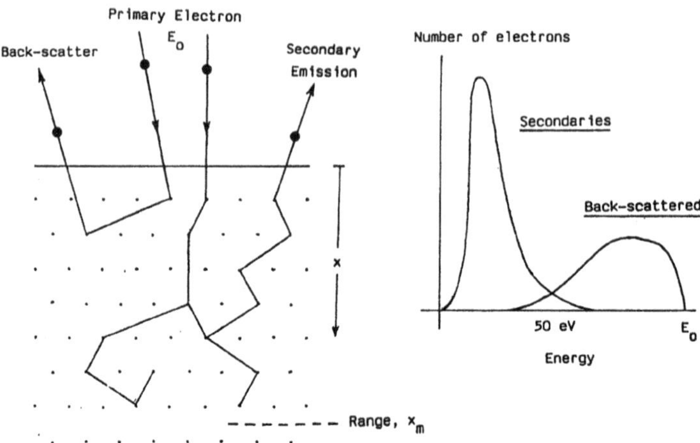

Fig. 9: Electron impact: secondary emission and back-scatter.

Range, $Xm \propto E^n / n$ with $n > 1$

$P(x)$, Probability of escape from depth $x \propto \exp[-\alpha x]$

and for

Electron impact: $dE/dx \propto E^{1-n}$

Ion impact: $dE/dx \propto E^{0.5} / (1+E/Em)$

For normal incidence, this leads to an equation of the form,

$$Yield = \delta(E) = \delta m \cdot (Em/E)^{n-1} \cdot \frac{1 - \exp[-R\,(E/Em)^n]}{1 - \exp[-R]} \tag{7}$$

where

δm = peak yield at E = Em

R = dimensionless Range, $(1 - 1/n) \cdot (\exp[R] - 1)$

Then δm, Em and n can conveniently be fitted to experimental data using simple iteration process.

For back-scattering electrons, the yield is determined by

$$\text{Yield,} \quad \eta(\theta) = \eta(0)\exp[-(1-\cos\theta)\cdot\ln(\eta(0))] \tag{8}$$

where

$\eta(0) = 1 - (2/e)^{0.037Z}$ [Katz et al.]

$\eta(0) = 0.0115\ E^{-0.223}$ polymers [Burke]

$\eta(0) = 0.1\exp[-E/5000]$, $E < 10000$ eV [Shimizu]

$\eta(0) = 0$, $e \leq 50$ eV

Z = atomic number for target, θ = angle of incidence

These expressions establish how the yield depends upon incident energy and angle, and material properties. Figure 10 models the total yield for a representative insulator (kapton) in terms of an Albedo factor,

$$A(E) = 1 - \delta(E) - \eta(E) \tag{9}$$

such that an energy integration of A times the observed electron flux will give a resultant charging current density. If it is assumed that plasmasheet protons have a concentration and temperature similar to the electrons, and that their mean secondary electron yield is ~1, their contribution can be included with the reduction of A by 0.05.

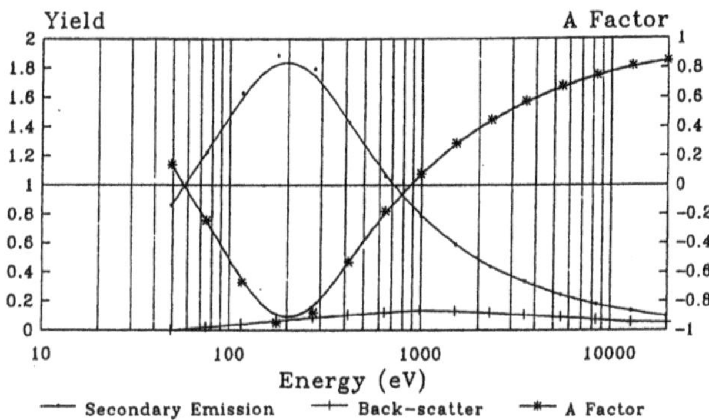

Fig. 10: Electron albedo factor A; assuming yield values of Kapton.

Q 13: Which components determine the current balance?

Let us consider a very simple geometry, represented by a spherical conducting satellite of 1 m radius, with a small isolated kapton patch of thickness 25 μm, see Fig. 11. Imposing a representative plasmasheet environment composed of three Maxwellians,

0.1 cm^{-3} @ 1 eV + 0.9 cm^{-3} @ 600 eV + 1.7 cm^{-3} @ 26 keV electrons
0.1 cm^{-3} @ 1 eV + 1.0 cm^{-3} @ 350 eV + 1.7 cm^{-3} @ 25 keV protons,

the following table of current densities ($\mu A/m^2$) was determined with the EQUIPOT charging code [Ref. 13] to show the relative magnitudes of the components and the net values. The first five correspond to zero bias while the last is taken with V = -7 kV, equilibrium being at V_s = -7.2 KV. The figures give some insight into the variations between materials, and the effects of increased bias.

	V	J_e	J_i	J_{se}	J_b	J_{si}	J_p	J_c	J
Satellite in Sunlight									
ALUM	0	8	0.2	1.5	2	0.4	10*	0	+6
GOLD	0	8	0.2	4	5	0.6	8*	0	+9
Satellite in Eclipse									
ALUM	0	8	0.2	1.5	2	0.4	0	0	-4
Patch in Shadow									
KAPT	0	8	0.2	2	1.4	0.7	0	0	-4
KAPT	-7 KV	5.6	0.9	0.8	0.9	2.1	0	0.3	-0.2
* corrected for sphere									

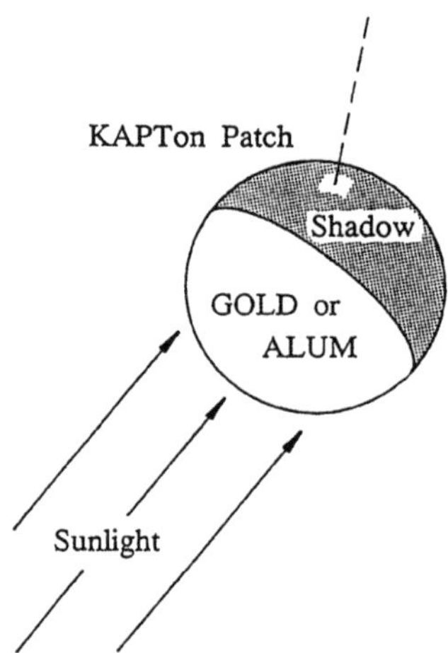

Fig. 11: Simple spacecraft and patch model for differential charging.

Q 14: How fast do the surfaces charge?

Substituting the net current densities into Eq. 1 gives:

	J ($\mu A/m^2$)	C_A ($\mu F/m^2$)	dV/dt (V/s)
Satellite	-4	0.00001	40000
Patch	-4	1	4
	-0.2	1	0.2
Satellite:	$C_A \sim 9 \times 10^{-12}$ /R	$F\ m^{-2}$	Radius R
Patch:	$C_A \sim 9 \times 10^{-12}\ \varepsilon$ / L	$F\ m^{-2}$	Thickness L

During charging, the rate falls off exponentially, but these figures serve to demonstrate that absolute charging is fast whilst differential charging can take several hours. In practice, dV/dt clearly depends upon the detailed form of the J components in addition to parameters such as R, ε and L.

Q 15: How often do serious charging conditions occur?

Meteosat-2 carried a detector for electrons between 50 eV and 20 keV; Fig. 12 shows a grey scale spectrogram covering 24 hours, with plots of integrated current density J_{TOTAL} and J_{NETT}, the latter involving the A factor described above. Given the uncertainty in the modelling and the material properties, J_{NETT} appears to provide a good quantitative index of charging conditions. Binning all data between April 1982 and March 1987, half a solar cycle, it is possible to determine the percentage times for which J_{NETT} exceeds any given threshold. Figure 13 plots such results for 0.1, 0.2, 0.5, 1 and 2 $\mu A/m^2$ as a function of local time, confirming the expected profile and establishing a scale for judging a level of seriousness for any specific system.

Q 16: Are there seasonal or solar cycle trends in charging?

Both the duration of plasmasheet transition at GEO and the frequency of substorm injections are well correlated with the level of geomagnetic activity, as monitored by K_p. Statistically, it is clear that higher K_p tends to occur near equinox and during years close to solar maximum. Charging conditions follow a similar pattern, as does the frequency of ESD related anomalies. Figure 14 illustrates this perfectly by summing the number of spurious switchings per month on Marecs-A for the years 1982 to 1989.

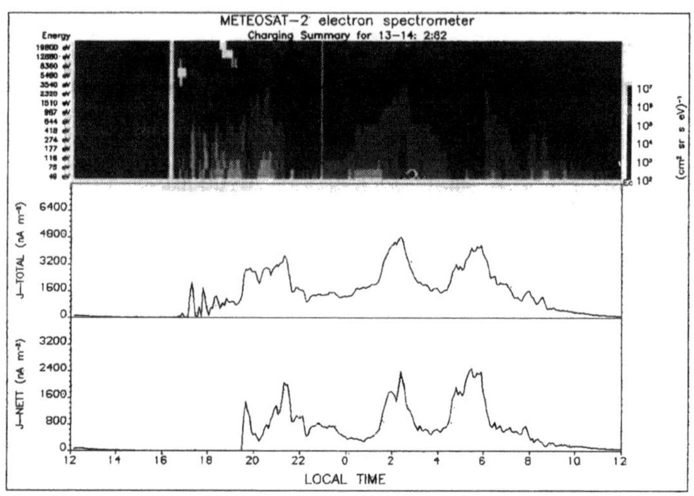

Fig. 12: METEOSAT-2 spectrogram for 13/14 February 1982 with plots of integrated electron and nett charging current densities.

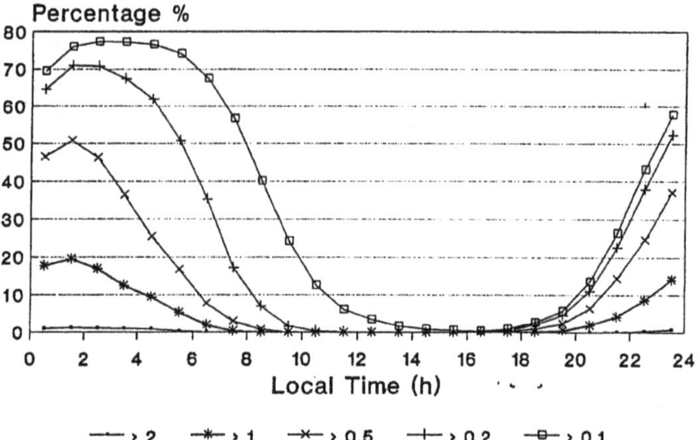

Fig. 13: Percentage of time between April 1982 and March 1987 that METEOSAT-2 observed charging currents greater than 0.1, 0.2, 0.5, 1 and 2 $\mu A/m^2$.

Q 17: Can charging times be inferred from geomagnetic indices?

The study of satellite anomalies is difficult because operational spacecraft never carry particle spectrometers. Sometimes a suitable instrument on a nearby satellite can provide direct evidence of charging fluxes. Any separation in longitude then introduces complication because

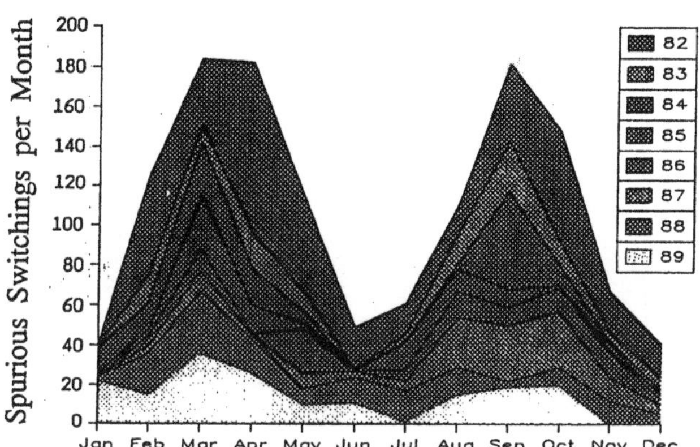

Fig. 14: Monthly rates of spurious switchings on MARECS-A, 1982 to 1989, [K. Derbyshire, ESTEC, private communication, 1991].

events tied to purely spatial structure will be ordered by Local Time, whilst injection arrival delay depends upon the relative drift displacement.

Although charging phenomena can be statistically correlated with K_p, such a 3-hourly planetary index is of little value in the study of individual events [Ref. 6]. Other indices such as AE or a single station K values may be utilized, if available, but really the search for a useful correlation is a difficult and frustrating exercise [Ref. 14].

Q 18: How well are the material properties known?

Satellite surfaces tend to be clean, subject to temperature extremes, and seriously aged by the combined effects of ultra-violet and high energy particle radiation. Properties such as the electron emission yields and surface conductivity are highly sensitive in all these respects. There have been few reliable laboratory measurements for space materials, particularly dielectrics, and these have usually been at room temperature. Particle beam studies have been performed mono-energetically at near normal incidence; sample cleanliness has often been an unresolved problem.

Simulation codes have been used to show that the surface potentials which develop are very sensitive to the key material properties [Ref. 13], and the lack of good data represents a severe limitation at present. Figure 15 illustrates this problem with a spread of curves giving the equilibrium potential of our shadowed patch as a function of the secondary electron emission from proton impact, both the yield at 1 keV (0.2 – 05.) and the energy for peak yield, E_m (100 – 300 keV).

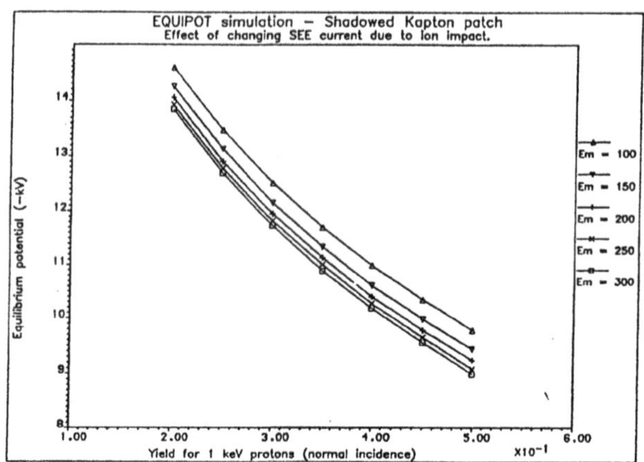

Fig. 15: Results of EQUIPOT simulation for the shadowed Kapton patch, showing the effect of changing SEE current due to ion impact.

Q 19: Can a spacecraft be designed to be immune to charging problems?

The short answer is 'yes', but it can be expensive. NASA has issued excellent guidelines [Ref. 15]. The following protection measures can be considered but, in practice, a degree of compromise will be unavoidable, and the trick is to be able to quantify the risk.

* CONDUCTIVE COATINGS - GOOD GROUNDING
    * Vacuum-deposited Metals or Indium Tin Oxide
    * Thermal Paints
    * Indium Oxide/Polyimide
        + Thermo-optical properties retained
        + High secondary electron emission yields

* 'LEAKAGE' DIELECTRICS
    * Doped Kapton, Carbon-filled Teflon
    * Ion implanted Kapton and Cover Glasses
    * Conductive Adhesives

* DESENSITIZED CIRCUITS
    * Filters and delay components to attenuate transients

* HARD COMPONENTS
    * Radiation hard devices also relatively immune to ESD

* ACTIVE SYSTEMS
    * Cold plasma sources, Electron guns

* CIRCUMVENTION
    * Susceptible systems disabled on substorm alerts

Q 20: **How can the level of immunity be established?**

It is generally impossible to put a fully integrated spacecraft into a vacuum chamber and fire electrons at it, although it has been attempted; ESD injection tests are commonly conducted as a proof of the grounding strategy and sub-system tolerance. Charging tests on critical items like solar cell arrays and thermal blankets are strongly recommended. Computer simulation of overall charging characteristics has been very successfully performed using NASCAP [Ref. 12], a fully three dimensional and dynamic code, which can handle a model of the whole spacecraft and identify barrier sheath development as well as possible discharge sites.

ANSWER:

Given all the reservations implied above, we can now confidently offer an **'irrefutable maybe'**, BUT not all anomalies are due to ESD and not all ESD results from surface charging. Figure 16 analyses 1464 anomalies in GEO in terms of local time and geomagnetic activity; those during very active periods fit the picture of surface charging in the plasmasheet but the others do not [Ref. 16]. Event chronograms for many individual spacecraft also suggest that a different mechanism must be operating, as Manola Romero explains [Ref. 17].

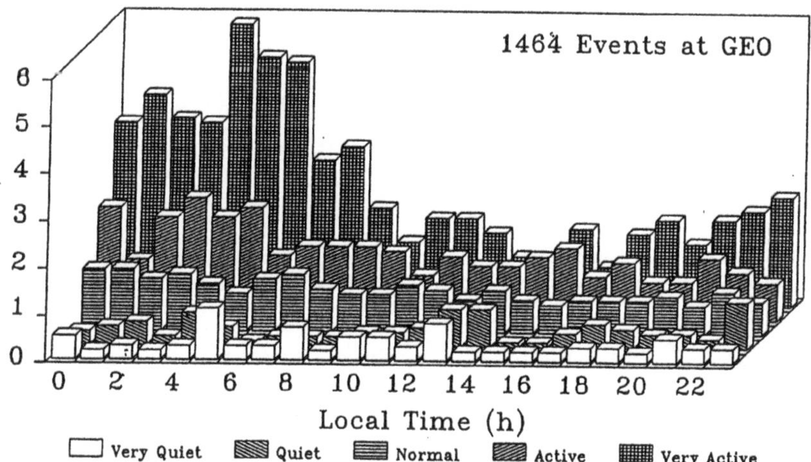

Fig. 16: Diurnal distribution of spacecraft anomalies at GEO for five levels of geomagnetic activity; scale normalized to give a mean of 1 for each local time bin. (NGDC Data Base, Boulder)

Further reading can reasonably be provided by review articles [Refs. 16, 18, 19], specific interests can then be pursued using their reference lists. All references to the High Altitude Gas Grenade Intelligence Satellite have been removed in the interests of NATO security.

References

[1] S.E. DeForest, 'Spacecraft charging at synchronous orbit', J. Geophys. Res., 77, 651-659 (1972).

[2] J.J. Capart, and J.J. Dumesnil, 'The electrostatic discharge phenomena on MARECS-A', ESA Bulletin 34, 22-27 (1983).

[3] J.E. Haines, 'Report on the Marecs-A in-flight anomalies, BAE Report TP 7962 (1982).

[4] J.G. Roederer, 'Dynamics of Geomagnetically Trapped Radiation, Springer-Verlag (1970).

[5] R.C. Olsen, 'A threshold effect for spacecraft charging', J. Geophys. Res., 88, 493-499 (1983).

[6] G. Rostoker, 'Geomagnetic Indices', Rev. Geophys. Space Phys., 10, 935-950 (1972).

[7] B.H. Mauk, and C.E. McIlwain, 'Correlation of Kp with the substorm-injected plasma boundary', J. Geophys. Res., 79, 3193-3196 (1974).

[8] G.L. Wrenn, and A.D. Johnstone, 'Evidence for differential charging on Meteosat-2', J. Electrostatics, 20, 59-84 (1987).

[9] J.J. Sokja and G.L. Wrenn, 'Refilling of geosynchronous flux tubes as observed at the equator by GEOS 2', J. Geophys. Res., 90, 6379-6385 (1985).

[10] X.T. Song, R. Gendrin and G. Caudal, 'Refilling process in the plasmasphere and its relation to magnetic activity', J. Atmos. Terrest. Phys., 50, 185-195 (1988).

[11] R.C. Olsen, C.E. McIlwain and E.C. Whipple, 'Observations of differential charging effects on ATS 6', J. Geophys. Res., 86, 6809-6819 (1981).

[12] I. Katz, M. Mandell, G. Jongeward and M.S. Gussenhoven, 'The importance of accurate secondary yields in modelling spacecraft charging', J. Geophys. Res., 91, 13739-13744 (1986).

[13] A.J. Sims and G.L. Wrenn, 'Sensitivity analysis with a simple charging code', Proceedings of Spacecraft Charging Technology Conference (1989), Monterey, California, 147-158 (1990).

[14] H.L. Lam and J. Hruska, 'Magnetic signatures for satellite anomalies', J. Spacecraft, 28, 93-99 (1991).

[15] C.K. Purvis, H.B. Garrett and A.C. Whittlesey, 'Design guidelines for assessing and controlling spacecraft charging effects', NASA Technical Paper 2361, 43p (1984).

[16] G.L. Wrenn, 'Spacecraft charging effects', Proceedings of the Solar Terrestrial Predictions Workshop (1989), Leura, Australia, 196-205 (1990).

[17] M. Romero and L. Levy, 'Internal charging and secondary effects', ibidem, xxx-xxx (1991).

[18] H.B. Garrett, 'The charging of spacecraft surfaces', Rev. Geophys. Space Phys., 19, 577-616 (1981).

[19] E.C. Whipple, 'Potentials of surfaces in space', Rep. Prog. Phys., 44, 1197-1250 (1981).

SPACE ENVIRONMENT AND EMC/ESD PHENOMENA

J.P. ESTIENNE
Head of the Space Environment &
Electromagnetics Group
MATRA Espace
Toulouse, France

1. Introduction

The orbital medium, where space vehicles are operating (Telecom, observation, scientific satellites, future space station, Hermes) is a very hard environment that must be considered at the beginning of the system design to guarantee a successful mission.

Various space environment constituents lead to system alteration:

- Radiations: trapped particles, solar events, heavy ions, secondary radiation interior to the system
- Meteoroids, debris
- Atomic oxygen
- Thermal conditions
- U.V.

In addition to these constraints, electrical and electromagnetic environments are present, resulting from:

- the local vacuum environment (mainly plasma)
- the system (conducted and radiated pulses, power conditions: PWM, ...)
- The earth (interferences, intentional or non-intentional electromagnetic pulses, ..)

All these sources induce specific effects leading either to degradation of vital functions of the system (permanent effects) or to interferences that occasionally disturb the standard operations.

R. N. DeWitt et al. (eds.), The Behavior of Systems in the Space Environment, 513–564.

The main difficulty lies in the definition of a balanced solution owing to the environmental mission-dependent constraints and the technical and economical ones that control the design of the space system. The integration of these effects and the set-up of a specific procedure acting at system level and levels are very costly and difficult to apply. An economical solution in view of cost and test reduction lies in numerical simulations. This one is generally coupled with tests that act as calibrations. The drawback of this procedure is that a large knowledge in the physics of the phenomena is required as well as a wide experience in various fields such as CAD, numerical techniques, and algorithm design.

In this paper, we will focus our attention on electrical and electromagnetic effects. Those associated to particle degradation of space vehicles can be found in [Refs. 1, 2]. This presentation is oriented around numerical techniques that allow us to represent on computers the electrical and electromagnetic behavior of a space vehicle (EMC, ESD, ...) under system aspect evaluation.

Chapter 2 describes briefly the main causes of defaults which appear on-orbit. Chapter 3 presents the various contemporary numerical techniques which are widely used (Finite differences, Finite volumes, method of moments, ...). Each technique is presented in increasing complexity in order to initiate the reader to these numerical techniques. Application examples follow each numerical technique, demonstrating the potential of such methods on realistic problems.

2. Electromagnetic Phenomena and Effects

On-orbit space environment leads to charging as well as transient discharging phenomena which are mainly observed on geostationary orbit. Trapped particle spectra are reproduced in Fig. 1.

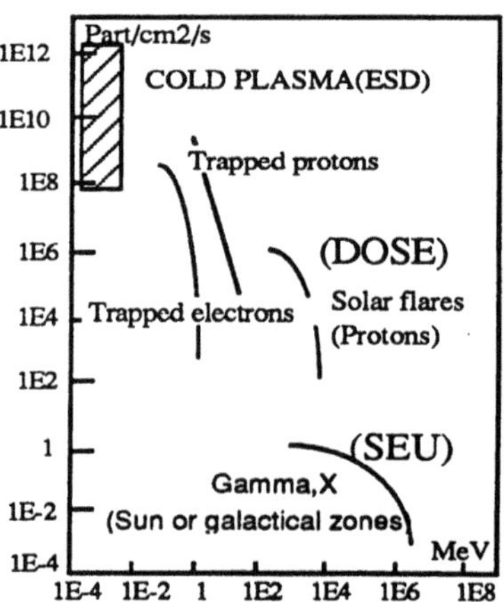

Fig. 1: On-orbit typical spectra.

2.1. ELECTROSTATIC CHARGE

2.1.1. *Surface Charging*. On geostationary orbit, the plasma surrounding the satellite leads to the so-called surface charging effect. This effect was detected in the seventies when more complex telecommunication satellite were launched, with increasingly integrated digital electronic components which are relatively sensitive (MOS for example) to transients. The correlation between the observed events and the local plasma conditions (position of the satellite in the magnetosphere, geomagnetic activity, sun lightning, electrical properties of external satellite coatings,..) have indicated that these anomalies were caused by the plasma environment.

Due to various accumulating phenomena (electrons, ions, protons) on the insulating parts of the satellite (dielectrics, floating metallisation), high voltage gradients are induced. This phenomenon is mainly controlled by the photoelectric and secondary emission and charge space limitation effect which act as balancing effects defining the electrical state of the satellite. This effect is the source of breakdown. We will see hereafter that it is the discharging effect which is the source of events.

In a general way, plasma effects can be treated without regard to the characteristics of the plasma. Due mainly to numerical and pragmatic reasons, low orbit and geostationary orbit are treated differently. This is due to the associated Debye length which is large on geostationary orbit (some tenth of meters) and very low (a few cm to some mm) on low orbits.

The design of spacecraft and protective actions lead then to the use of a system analysis tool able to reproduce various on-orbit situations. This tool must be able to predict, at least in relative units, what is the impact of a modification compared to another. In the space community, the NASCAP tool (See Fig. 2) is the most well known. It solves the POISSON equation in a 3D space, taking into account the surrounding plasma characteristics and various hypotheses specific to geostationary orbit.

POISSON equation:

$$\Delta \Phi = - \frac{\rho(\Phi, E)}{\varepsilon_o} \text{ , where } \Delta = \frac{\partial^2}{\partial x^2} + \frac{\partial^2}{\partial y^2} + \frac{\partial^2}{\partial z^2}$$

Debye approximation:

$$\frac{\rho(\Phi, E)}{\varepsilon_o} = \frac{\Phi}{\lambda^2}, \quad \lambda^2 = \frac{\varepsilon_o \cdot \theta}{n \cdot e}$$

θ = Temperature in eV

It is well known through numerous publications. This is the reason why the simulation technique and the principle of the solving of the Poisson equation, solved with the help of finite elements, is not presented in this article.

In polar zones and on low orbits, the plasma that surrounds the space object is more dense and/or hot. As a consequence, the preceding approximation is no longer valid. Vortex phenomena are generated in the

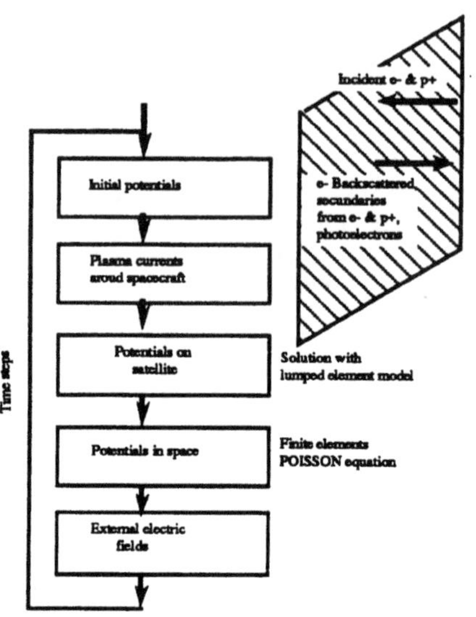

Fig. 2: NASCAP voltage computation.

plasma, due to ram and wake effects. It is then necessary to use in a self consistent manner the screening charge and the associated surface potential around the object: solving the VLASOV equation or Particle In Cell method (PIC). One important point is the presence of high voltage which causes strong losses (high voltage solar generators).

2.1.2. *Bulk Charging.* This effect is associated to the high energetic part of the electron spectrum (trapped particles), typically some hundreds of keV to some MeV. This one constitutes presently one plausible hypothesis that explains events that are not caused by surface charging.

The principle is equivalent to an ionic implantation in a semiconductor. Charges (electrons) crossing the shield (Honeycomb, internal protection, equipment boxes, wire shielding,...) are trapped inside insulating materials (epoxy, alumina of I.C, dielectric of wires, MLI,...). The implantation thickness depends on the energy of the impinging particle, the shielding and the insulating materials. When this charge is trapped inside the dielectric, it undergoes the electrical laws specific to an insulating material (trapping, detrapping recombination, generation,...). The internal potential of the dielectric material increases if the charge accumulating rate exceeds various loss mechanisms (external circuit, constitutive laws of the medium, ...). It is interesting to note that this is an integrating effect with a very high time constant (1000 to 10000 secs, or more) dependent on the insulating "quality" of the material.

Research is under way in this field to clarify the importance of this effect [Ref. 3], understand the basic mechanisms [Ref. 4], develop a system analysis tool [Ref. 4] able to reproduce on-orbit conditions, and define critical situations and various strategies of hardening.

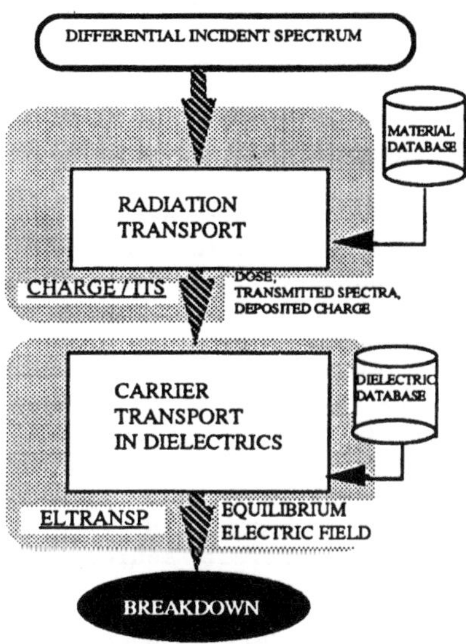

Fig. 3: Bulk charging computation.

This problem is solved in the one dimensional approximation that covers the major part of situations. From pseudo-static formulation of Maxwell equations, one deduces the following relationships:

$$\nabla \times \vec{H} + \vec{J} = \sigma(D, \dot{D}, E)\,(\vec{E} + \varepsilon)\,\frac{\partial \vec{E}}{\partial T}$$

$$\int_S \nabla \times \vec{H} \cdot dS = I_{Tot}.$$

where

D is the Cumulated dose; $\dot{D}$, the Dose-rate;

E, the internal electrical field; σ, medium conductivity; J, electron source current; and

I_{Tot}, the total current inside the medium.

This rather seemingly simple equation is quite difficult to solve due to non linearities (especially for the conduction term) and because of the system's large time constants, of complex conduction models and of the external electrical circuit (Terminal load of a wire for example). Source term evaluation requires the use of an electron transport code (ITS for example: [Ref. 5]) to compute $\vec{J}$. Once the source terms are evaluated, the time and space discretisation are performed and the initial conditions are defined (terminal electrical circuit, interfaces metal-kielectric, ...). All these constraints lead to numerical problems which must be solved carefully [Ref. 6].

2.2. ELECTROSTATIC DISCHARGE

2.2.1. *Surface Discharge.* The surface discharge generates a maximum overshoot current in the range of a few hundreds of amps abd occurs when the voltage exceeds the breakdown potential. At equipment level, surface discharge acts by way of electromagnetic fields of which the amplitude is not well known, and the rise time is probably a few nanoseconds. Replacement currents and associated fields penetrate inside the satellite and couple to target points (antenna, lines). They induce electromagnetic noise which may cause interferences. Without shielding actions, resulting currents and/or voltages are large enough to produce permanent damages on electronics or transients.

The modeling [Ref. 7] itself is simple and consists in the representation of a charged particle source (electrons and eventually ions), controlled by space charge limitation. In vacuum these particles look like a non-linear transient current density $\vec{J}$. Therefore, variations of electric and magnetic fields are produced and governed by the relationship $\vec{J}=F(\vec{E},\vec{H})$ from Maxwell equations. The generated fields tend to establish a more stable equilibrium electrical state. At the beginning of the discharge, ejected charges are accelerated by the electrostatic process and modify the initial field lines. Their movement is self consistent with the resulting total field: Static + dynamic (Charge space limitation effect).

2.2.2. *Bulk Discharge.* Discharges inside dielectrics also cause electromagnetic interferences on space systems. Preliminary results from the CREES experiment (CRESS, work ESA [Ref. 8]) show that dielectric discharges occur on dielectrics exposed to space radiation.

This charge displacement and the dielectric breakdown induce a high transient voltage that is directly transferred by the line to connected electrical interfaces. Bursts from hundreds of volts are mentioned in the literature with very short time duration (few tenths of a nanosecond).

The involved charge is small compared to surface discharge, but this effect is quite important since the resulting transients are transferred directly to the electronics. They are not altered by complex coupling paths as for the surface discharge, which therefore reduce drastically the magnitude of the effect.

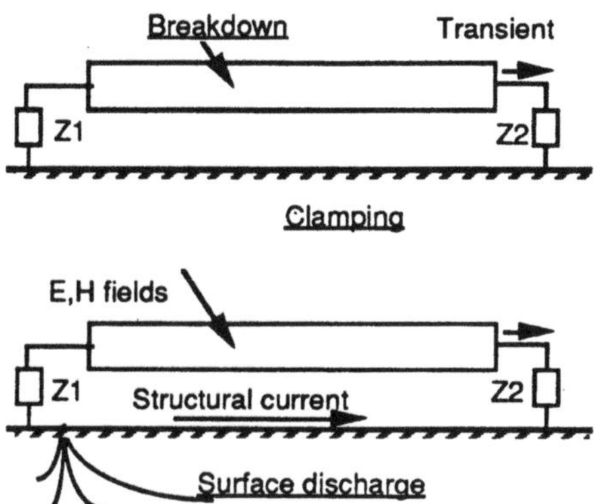

Fig. 4: Coupling sources for ESD.

2.3. E.M.C., SCATTERING, CONDUCTED AND RADIATED EFFECTS

A spacecraft is a very sensitive object which interacts with the space medium (electrostatic discharge for example). Therefore, communications from the vehicle to the external world through antennas generate self electromagnetic noise via:

- the power conditioning from the solar generator
- digital signal lines which are radiating themselves on other lines (cross-talk)
- power switches, power interferences
- radar, antenna
- propagation effects (load matching) and grounding.

In a general way, all these effects are classified as E.M.C. (ElectroMagnetic Compatibility) effects and are defined as sources of E.M.I. (ElectroMagnetic Interference). The cause of an E.M.I. problem may be either within the system one is dealing with, in which case the problem is labelled an intrasystem problem. Other E.M.I. may come from outside, in which case the problem is given the intersystem designation. A very common cause of both intrasystem and intersystem problems is a signal intended for one circuit which also reaches a circuit or circuits for which it was not intended.

Electrical transmission paths constitute a primary means by which the E.M.C. characteristics of equipment are classified. Transmission paths are both conducted (via material: metal, lines, ground, lumped components capacitors, transformers,...) or radiated (via external medium: air or vacuum by means of near field or induction effects) as depicted on the following figure.

Several factors exist simultaneously to create an E.M.I. problem. They are:

- An electromagnetic or electric energy source
- A device sensitive to the type of energy being generated
- A medium for coupling the energy from the source to the sensitive device at a time when the effect is a noticeable degradation of system performance

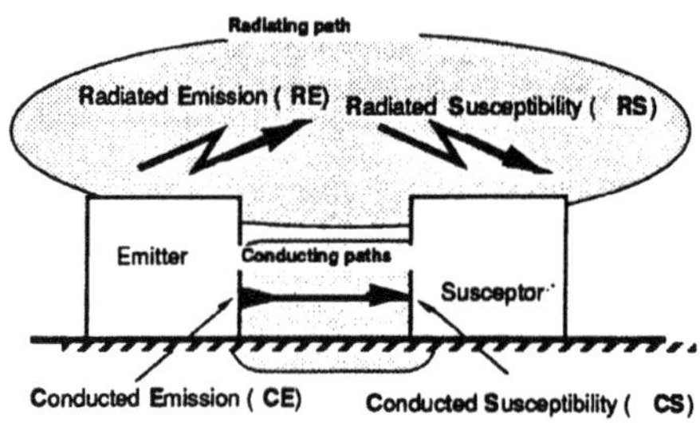

Fig. 5: Transmission paths for E.M.C.

The purpose of the E.M.C. engineer is to insure system or sub-system compatibility. Control is achieved through the use of proven design techniques having a sound theoretical and practical basis. All these interfering effects require testing, shielding and simulation procedures under the following aspects:

- Radiating mode (irradiation of the object with an antenna), and
- Conducting mode (current and/or voltage injection).

The testing procedures are used to reproduce as well as possible an electrostatic discharge or transients which are representative of EMC, lightning stimuli. Modeling aids the engineer and extends the domain of testing results, especially for design, hardening and overtesting which is not acceptable on an object such as a flying satellite model.

2.4. ELECTROMAGNETIC PERTURBATIONS AND COUPLING

The most probable target element is mainly a line that works as detector by means of an antenna effect. A system analysis must take into account these various kinds of perturbations which excite the receiving element. Resulting transients are then evaluated at system or equipment level for:

- electromagnetic coupling,
- dielectric breakdown....

which are dependent on the various electrical configurations within spacecraft (loading of lines, type of line: simple, bifilar, coaxial, bus,...), grounding design.

Associated models are:

- line solver when the loading can be represented as a simple load with R, L, C elements: RADIATE model [Ref. 9]
- an electrical network solver which includes line solver and coupling analysis (MATEMC - [Ref. 10])
- a wire electromagnetic model for coupling in free space (antenna for example) or simple wire above ground place (method of moments) [Refs. 11,12,13]

Each of these methods presents advantages and drawbacks. Physical hypothesis, terminal loads and complexity of the receiving element make selection of the line model.

3. Numerical Methods - Algorithms and Applications

In the frame of system analysis and testing, the use of numerical tools which solve conducted and radiated phenomenon is highly recommended, because:

- Some effects cannot be reproduced on the earth in realistic conditions due to testing and financial constraints, planning or quality assurance, but can be represented on computers
- Simulation can be used as a tool to understand complex physical processes
- Out of scale resolution limits can be supplied by numerical simulations
- Shielding, protective design and verification can be validated numerically
- Immunity of the system is verified with respect to various aggressions (ESD, lightning described under specifications) either in frequency or time domain.

Developing of such tools is a hard task and requires wide investigations in algorithm and numerical techniques. Numerical constraints, mainly computing power and memory resources, are very important in this field of computation. In the majority of cases, the designer searches numerical techniques well adapted to its own computer and easy to implement. This leads to the selection of explicit methods (see hereafter). Implicit methods are mainly obtained when volume or surface integral representation is used (method of moments, finite elements).

The following chapters present numerical techniques which are or will be extensively used in future space programs. Each numerical technique is gradually presented with respect to the level of difficulty to initiate readers to a domain which is very specialized.

The finite difference technique is first presented because it is the oldest one which is widely used and very simple to implement. Some examples applied to EMC/ESD are presented. Finite volumes, a more complex technique is as easy to implement as finite differences, but more powerful in geometrical representation. Basics of the method of moments are then presented under frequency and time domain application. Some typical examples are illustrated with results of codes which are actually running at the MATRA space center. Examples concerning scattering of an object to a impinging wave as well as current injection results are presented.

3.1. MATHEMATICAL MODELS

Simulation requires first to establish the mathematical laws of the phenomenon. These can be described either under differential form or under integro-differential form.

3.1.1. *Differential Formulations*.

Direct formulas describing partial derivative (time, space) relationships are used. The following examples are straightforward and describe the physics under differential mathematical representation:

Example 1: Line equation

Line equations are very well known, they describe the relationship between the line voltage and current as following:

$$\frac{d\{V\}}{dx} = -[R]\,.\,\{I\} - [L]\,\frac{d\{I\}}{dt} + \{V_s\}$$

$$\frac{d\{I\}}{dx} = -[G]\,.\,\{V\} - [C]\,\frac{d\{V\}}{dt} + \{I_s\}$$

{V} and {I} are voltage and current vectors of the line

[R], [L], [G], [C] are the matrices defining the electrical parameters of the line. $\{V_s\}$, $\{I_s\}$ are vectors which represent the stimuli applied to the line (impinging wave, breakdown,....)

Example 2: POISSON equation

$$\frac{\partial^2\Phi}{\partial x^2} + \frac{\partial^2\Phi}{\partial y^2} + \frac{\partial^2\Phi}{\partial z^2} + \frac{\rho}{\varepsilon} = 0$$

Φ potential, ρ charge density

Example 3: Maxwell equations

For electromagnetic applications, the starting point is Maxwell's equations under differential form; the $\vec{E}$ field and $\vec{H}$ field are related to the current density $\vec{J}$:

$$\overrightarrow{\nabla \times \vec{E}} = -\frac{\partial \vec{B}}{\partial t}$$

$$\overrightarrow{\nabla \times \vec{H}} = \vec{J} + \varepsilon . \frac{\partial \vec{E}}{\partial t}$$

Along these equations, the initial state is defined with the following two equations:

$$Div\ \vec{E} = \frac{\rho}{\varepsilon}$$
$$Div\ \vec{B} = 0$$

and the equation of continuity:

$$\nabla . \vec{J} + \frac{\partial \rho}{\partial t} - 0$$

is only a consequence of the above transient equations, but very helpful. Each of these examples is characterized by space and time partial derivative.

3.1.2. *Integral Formulations*. Differential formulae are not well adapted to treat realistic 3D shapes. As described in the next chapters, integral forms are more suitable to represent 3D complex objects as satellites which include very thin open surfaces (solar generator, extensions, ..) and closed surfaces (spacecraft body) as well as wires (lines, antenna, yokes).

Examples involved with Poisson's and Maxwell's equations presented above are transformed to obtain integral-differential representation. Note that with one differential form and a chosen numerical technique (method of moments, finite elements), several integral formulations may be derived. The selection of the representation is controlled by the following criteria: geometrical modeling, boundary treatment, radiating condition, numerical solution of the problem, ...

Example 1: Poisson equation [Ref. 14]

$$\Delta \Phi + \frac{\rho}{\epsilon} = 0$$

$$\left.\begin{array}{ll} \Phi = \Phi_{s1} & \text{on } S1 \\ \dfrac{\partial \Phi}{\partial n} + a.\Phi = \sigma_{s2} & \text{on } S2 \end{array}\right\} \textit{Constraints on S}$$

$$\int_V \psi \left(\frac{\partial^2 \phi}{\partial x^2} + \frac{\partial^2 \phi}{\partial y^2} + \frac{\partial^2 \phi}{\partial z^2} + \frac{\rho}{\varepsilon} \right) dv = 0$$

ϕ: Twice differentiable voltage function

ψ: user defined; Weighing function, once differentiable

After integration, one obtains:

$$\int_V \left(\frac{\partial \psi}{\partial x}\frac{\partial \phi}{\partial x} + \frac{\partial \psi}{\partial y}\frac{\partial \phi}{\partial y} + \frac{\partial \psi}{\partial z}\frac{\partial \phi}{\partial z} - \psi.\rho \right) dv +$$

$$\int_s \psi(a.\phi - \sigma) ds = 0$$

Φ: Once differentiable voltage function

ψ: Weighing function once differentiable

When $\rho = 0$ on the domain noted V, and if we choose ψ such as:

$$\frac{\partial^2 \psi}{\partial x^2} + \frac{\partial^2 \psi}{\partial y^2} + \frac{\partial^2 \psi}{\partial z^2} = 0,$$

then we deduce:

$$\oint_s \left(\psi \cdot \frac{\partial \phi}{\partial n} - \phi \cdot \frac{\partial \psi}{\partial n} \right) ds = 0$$

The previous equation shows the strength of this method which inserts automatically the boundary conditions (Neumann for charge, Dirichlet for potential). This is not the case in finite differences where the engineer must express the Neumann condition through the normal field quantity. Moreover, the treatment is only to be performed on the surface of the object under this formulation which avoids the use of volumic space meshing (only surface of the object is to mesh) and reduces the number of unknowns (potentials or charge) on surfaces in place of all space surrounding potentials.

Example 2: Maxwell equations, E.F.I.E. formulation [Ref. 15]

Using the following relationship between the electric field, the scalar and vector potential:

$$\vec{E}(r,t) = -\nabla\Phi(r,t) - \frac{\partial \vec{A}(r,t)}{\partial t}$$

where retarded scalar potentials Φ and vector potentials $\vec{A}$ are expressed as:

$$\vec{A}(r,t) = \frac{\mu_o}{4\pi}\int_{V(r)} \frac{\vec{J}(r',t-R/c)}{R}\,dv'$$

$$\Phi(r,t) = \frac{1}{4\pi\varepsilon_o}\int_{V(r)} \frac{\rho(r',t-R/c)}{R}\,dv'$$

one deduces:

$$\vec{E}^T(r,t) = \vec{E}^I(r,t) + \vec{E}^S(r,t)$$

where the index T is related to the total resulting field, I defines incident quantities and S defines scattered quantities; that is, quantities only associated with the response of the object to the incident wave. The scattered fields may be defined as follows:

$$\vec{E}^S(r,t) = -\frac{1}{4\pi}\int_{S(r)}\left[\frac{\mu_o}{R}\frac{\partial\vec{J}(r',t')}{\partial t'} + \left(\frac{1}{R} + \frac{1}{c}\frac{\partial}{\partial t'}\right)\left(-\frac{\sigma(r',t')}{\varepsilon}\frac{\vec{R}}{R^2}\right)\right]ds'$$

with

$\vec{J}(r',t') = \vec{n}' \times \vec{H}^s(r',t')$ *the electrical current density*

and

$t' = t - R/c$

$\vec{R} = \vec{r} - \vec{r}'$

Inserting the equation of continuity, another relationship is derived:

$$\vec{E}^S(r,t) = -\frac{1}{4\pi}\int_{S(r)}\left[\frac{\mu_o}{R}\frac{\partial\vec{J}(r',t')}{\partial t'} - \frac{\vec{R}}{\varepsilon\,.\,R^2}\left(\frac{\sigma(r',t')}{R} - \frac{div\,\vec{J}}{c}\right)\right]ds'$$

which can be used to solve at the same time the current and charge density.

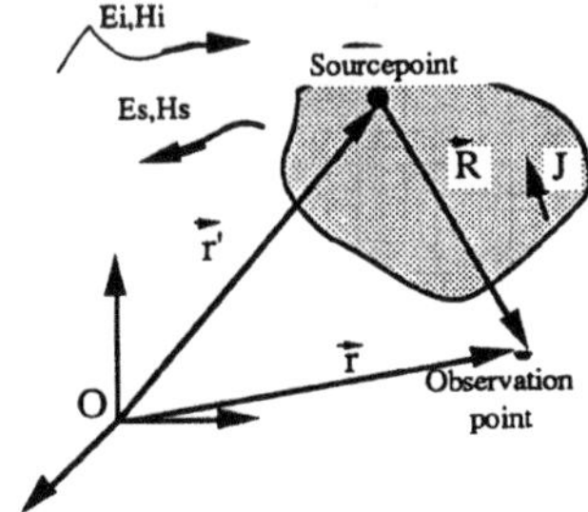

Fig. 6: Scattering object and conventions.

A second order equation can be derived using the Lorentz gauge and the definition of the electric field via scalar and vector potentials. This leads to:

$$\frac{\partial \vec{E}^s(r,t)}{\partial t} = \left\{\frac{1}{4\pi\varepsilon_o}\vec{\nabla}_r\vec{\nabla}_r\cdot - \frac{\mu_o}{4\pi}\frac{\partial^2}{\partial t^2}\right\}\cdot\iint_s \frac{\vec{J}(r',t')}{R}ds'$$

This later equation is fairly complex since the interpolated currents must also be consistent with the terms associated to the second order derivative terms. The only advantage is that the space and time derivatives may be applied after the surface integral is performed.

Example 3: Maxwell's equations, M.F.I.E. formulation [Ref. 6]

Using the magnetic field, and applying the boundary conditions, the following relationships are deduced:

$$\vec{H}^T(r,t) = 2\vec{H}^I(r,t) + \vec{H}^S(r,t)$$

$$\vec{H}^S(r,t) = \frac{1}{2\pi}\int_{S(r)}\left[\left\{\frac{1}{R} + \frac{1}{c}\frac{\partial}{\partial t'}\right\}\left(\vec{J}(r',t') \times \frac{\vec{R}}{R^2}\right)\right]ds'$$

Substituting the equation of continuity, applying the divergence operation and time integration, the charge can be formulated as follows:

$$\sigma(r,t) = 2\vec{n}\cdot\vec{E}^I +$$

$$\frac{\vec{n}}{2\Pi}\cdot\int_{V(r)}\left\{\left(\frac{\sigma(r',t')}{R} + \frac{1}{c}\frac{\partial\sigma(r',t')}{\partial t'}\right)\cdot\frac{\vec{R}}{R} - \right.$$

$$\left.\frac{1}{c\,R^2}\frac{\overrightarrow{\partial J(r',t')}}{\partial t'}\right\}\cdot ds'$$

All these various integral-differential formulations are boundary integral forms, where only surface quantities are computed. From these values, the electromagnetic behavior of the object in space (diffracted fields) can be determined. The procedure of evaluation is the following:

- A series of points r is selected on the surface of the object. A linear system of equations, involving the currents and charges results.

- A point r is then chosen elsewhere for which the radiation due to surface charges and currents is then calculated to obtain the resulting diffracted or total fields.

With this kind of formulation, one is able to solve scattering problems (impinging wave on an object, excitation of the object with a current or voltage source), either in the time or frequency domain. The reader will find the equivalent form in the frequency domain of the above equations by applying the transformation $\partial/\partial t \rightarrow j\omega$. Note that in this case, the charge can be expressed by current via:

$$\rho(r,\omega) = -\frac{\nabla \cdot \vec{J}(r,\omega)}{j\omega}$$

Also interesting to note is that magnetic charges and currents can also be included in the integral formulations. This is well suited to solve accurately aperture problems, effect of thin dielectric sheets on metal, [Ref. 17]. Although when volume treatment is required (particles with charge space limitations, thick dielectrics) these formulations are not well suited because field evaluation is required between each time step.

<u>**Example 4:** Maxwell's equations, volumic, formulations [Ref. 18]</u>

One simple formulation can be found by applying a volume integral to the differential form of Maxwell's equations. This leads to a finite volume representation of Maxwell's equations which rely on volume quantities evaluated for each cell V_i, to flux quantities which are defined on the k surfaces (S_i^k) of the involved volume cell noted i.

$$\iiint_V \nabla \times \vec{E} \, . \, dv = \oint\!\!\oint_s \vec{n} \times \vec{E} \, . \, ds = \int_V -\frac{\partial \vec{B}}{\partial t} \, . \, dv \approx - V \, . \, \frac{\partial \vec{B}}{\partial t}$$

$$\iiint_V \nabla \times \vec{H} \, . \, dv = \oint\!\!\oint_s \vec{n} \times \vec{H} \, . \, ds = \iiint_V \left(\vec{J} + \varepsilon \, . \, \frac{\partial \vec{E}}{\partial t}\right) . \, dv \approx V \, . \left(\vec{J} + \varepsilon \, . \, \frac{\partial \vec{E}}{\partial t}\right)$$

The above formulae come from a space discretisation; it gives information between surface values (known flux) and volume values (fields to evaluate). We need another step which defines surface variables (unknown flux) from volume values (known). This step is obtained via an interpolating technique or a more refined process (Theory of Characteristics). An equivalent formula is derived from edge values to surface values.

Divergential formulation is also well adapted to volumic treatment. It is well suited for the analysis of complex 3D shapes and is defined as:

$$\frac{\partial u}{\partial t} + \frac{\partial F}{dx} + \frac{\partial F}{dy} + \frac{\partial F}{dz} = W \rightarrow$$

$$\iiint_V \frac{\partial u}{\partial t} \cdot dv + \oint\!\!\oint_x F \,.\, ds = \iiint_V W \,.\, dv$$

with:

$$W = \begin{Bmatrix} E_x \\ E_y \\ E_z \\ H_x \\ H_y \\ H_z \end{Bmatrix} \qquad F(u) = \begin{bmatrix} 0 & -\frac{H_z}{\varepsilon_o} & \frac{H_y}{\varepsilon_o} \\ \frac{H_z}{\varepsilon_o} & 0 & -\frac{H_x}{\varepsilon_o} \\ -\frac{H_y}{\varepsilon_o} & \frac{H_x}{\varepsilon_o} & 0 \\ 0 & \frac{E_z}{\mu_o} & -\frac{E_y}{\mu_o} \\ -\frac{E_z}{\mu_o} & 0 & \frac{E_x}{\mu_o} \\ \frac{E_y}{\mu_o} & -\frac{E_x}{\mu_o} & 0 \end{bmatrix} \qquad W = \begin{Bmatrix} \frac{J_x}{\varepsilon_o} \\ \frac{J_y}{\varepsilon_o} \\ \frac{J_z}{\varepsilon_o} \\ 0 \\ 0 \\ 0 \end{Bmatrix}$$

Inserting the above vector and matrix in the divergence formula, the reader will find Maxwell's equations.

Other volumic formulations [Ref. 19], especially formulations well suited to finite element solution, can be found in the literature. This kind of solution is not so easy to implement as finite volumes, because of the implicit nature of the solution and radiation condition to verified for open problems.

One finite element representation is:

$$\nabla \times \vec{H} = \vec{J} + \sigma(D, \dot{D}, E) \,.\, \vec{E} + \varepsilon \,.\, \frac{\partial \vec{E}}{\partial t}$$

$$\int_V \left(\varepsilon \frac{\partial \vec{E}}{\partial t}\Psi - \vec{H}.\nabla \times \Psi\right) dv + \int_V \vec{J} \,.\, \Psi \, dv = 0$$

$$\int_V \left(\mu \frac{\partial \vec{H}}{\partial t}\Psi + \vec{E}.\overrightarrow{\nabla \times \Psi}\right) dv = 0$$

for $\vec{E} \times \vec{n} = 0$ on all boundaries

Ψ is a second order vector function.

3.1.3. *Implicit and Explicit Solutions.* The causality of certain problems forces the solution to be solved as an initial valued problem. In such cases, each field or current quantity (at time t_v) is computed from the previously calculated values ($t_{v-1}, t_{v-2}, \ldots$) and the source terms. For terms involving partial derivative operators, the technique used (next paragraph) will lead to either an implicit or explicit solution:

- Explicit solution: This type of solution will not require matrix inversion or linear system solving. Each localized quantity is computed from previous known quantities in time and space. As Maxwell's equations are hyperbolic [Ref. 20], the stability criterion limits the time discretisation as follow:

$$\delta t \leq \frac{\Delta x}{c}$$

This is valid for Maxwell's equation or line's equations. Finite difference Yee algorithm is one of the famous examples which leads to an explicit solution for Maxwell's equations [Ref. 21].

- Implicit solution: In this case, the differential or integro-differential problem has been transformed in such a way to obtain a linear system. For Maxwell's equations, the resulting matrix is full, very sparse, symmetric, ... (depending on the solving technique). In all cases, the dimension is very large and related to space discretisation. Stability criteria does not exist for implicit solution.

One straightforward implicit solution is the Poisson's equation solved with central finite differences Problems in frequency domain are always implicit.

3.1.4. *Matrix Problems.* Electrostatic or electromagnetic problems generate various types of linear and non-linear systems of equations which, when implicitly solved, may be expressed as:

$$[A] \cdot \{x\} = \{y\}$$

$\{y\}$: Right term (Time dependent or independent)

$\{x\}$: Unknowns (potential, fields quantities, currents, charges, ..)

To express a system of equations as such, the problem must be approached by a space discretisation (either for open problems). This procedure leads to unknown quantities defined on volume centers, surfaces or edges. The dimension of the matrix is defined by the number of cells (surfaces or edges) multiplied by their degree of freedom:

- Maxwell: 6 field components per volume, 2 components of currents on surfaces, ...).

- Poisson: Potentials localized at volume centroid or nodes or surface and/or charge density for surface solving.

Depending on the numerical technique (finite elements, finite difference, ...) and chosen model (interpolating functions, weighing functions, ...), matrix [A] may be very sparse or not, and/or may present interesting properties (symmetric, Toeplitz, real, ...). Each problem leads to a specific case to be solved with the adequate linear system.

Because of the enormous number of unknowns, (typically 10^4 to 10^5 in 3D), a sparse matrix solutions is preferred because only non-zero terms and pointers need to be stored. Standard libraries offer a wide selection of efficient FORTRAN routines that are well adapted to the linear systems. Note that for sparse matrices, it is quite interesting to solve the linear system without inversion. This increases the computing time (solution is to computed at each time step) but saves memory requirements (matrix inversion requires the full inverse matrix). This kind of solution takes full advantage of the sparse matrix and solves large problems on moderate computers.

3.2. FINITE DIFFERENCES (F.D.) METHOD

The basics of finite differences comes from the Taylor expansion of a function with its partial derivative. Centered derivatives are commonly used because they are automatically of 2nd order accuracy. This leads to a more robust algorithm [Ref. 20]. Expansion of the f(x) function around x_0 point gives:

$$f(x_0 + \Delta/2) = f(x_0) + \frac{\Delta}{2} f'(x_0) + \frac{\Delta^2}{8} f''(x_0) + O(\Delta^2)$$

$$f(x_0 - \Delta/2) = f(x_0) - \frac{\Delta}{2} f'(x_0) + \frac{\Delta^2}{8} f''(x_0) + O(\Delta^2)$$

Subtracting each terms of the above equation leads to:

$$f(x_0 + \Delta/2) - f(x_0) - \Delta/2 = \Delta f'(x_0) + O(\Delta^2)$$

From which we deduce:

$$\left.\frac{\partial f}{dx}\right|_{x=x_0} = \frac{f(x_0 + \Delta/2) - f(x_0 - \Delta/2)}{\Delta}$$

Note that the two function values at x_0 are not required, and that the two nearest points are used. Second order accuracy is achieved only with the immediate right and left neighbors. This is the reason why non-physical parasite oscillations can sometimes take place. This relationship constitutes the basis of 2nd order finite difference algorithms denoted "leap-frog algorithm" because 2nd order is achieved through two coupled variables. The above formula can be easily expanded to multiple variable functions and for non uniform meshes.

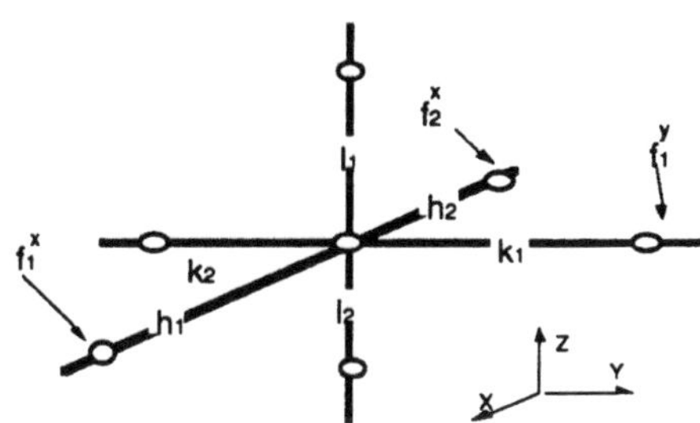

Fig. 7: Centered finite differences mesh.

For a multivariable function f(x,y,z) there gives:

$$\frac{\partial f}{dx}\bigg|_{x=x_o} = \frac{h_1^2.(f_1^x-f_0) - h_2^2.(f_2^x-f_0)}{h_1.h_2(h_1 + h_2)}$$

$$\frac{\partial f}{dx}\bigg|_{y=y_o} = \frac{k_1^2.(f_1^y-f_0) - k_2^2.(f_2^y-f_0)}{k_1.k_2(k_1 + k_2)}$$

$$\frac{\partial f}{dx}\bigg|_{z=z_o} = \frac{l_1^2.(f_1^z-f_0) - l_2^2.(f_2^z-f_0)}{l_1.l_2(l_1 + l_2)}$$

where:

$f_1^x = f(x_o+h_1,y,z)$ $\quad f_2^x = f(x_o-h_2,y,z)$

$f_1^y = f(x,y_o+k_1,z)$ $\quad f_2^y = f(x,y_o-k_2,z)$

$f_1^z = f(x,y,z_o+l_1)$ $\quad f_2^z = f(x,y,z_o-l_2)$

$f_o = f(x_o,y_o,z_o)$

Similarly, second order derivatives are computed and lead to, for a uniform mesh:

$$\left.\frac{\partial^2 f}{\partial x^2}\right|_{x_o,y_o,y_o} = \frac{2}{\Delta_x^2}\left(f(x_o+\Delta,y_o,z_o) + f(x_o-\Delta,y_o,z_o) - 2f(x_o,y_o,z_o)\right)$$

3.2.1. *Application to Poisson's Equation.* Solving Poisson's equation with finite differences requires first a space discretisation. A 2D F.D. meshing is represented hereafter in 2D Carthusian coordinates. Using the expression of second order space derivative at each node of the meshed area (except boundaries), numerical solution of Poisson is defined as follows:

$$\Delta V + \frac{\rho}{\varepsilon_o} = 0$$

$$\rightarrow (V_{i+1j} + V_{i-1j} + V_{ij+1} + V_{ij-1} - 4V_{ij}) = -\frac{\Delta^2}{2}\frac{\rho_i}{\varepsilon_o}$$

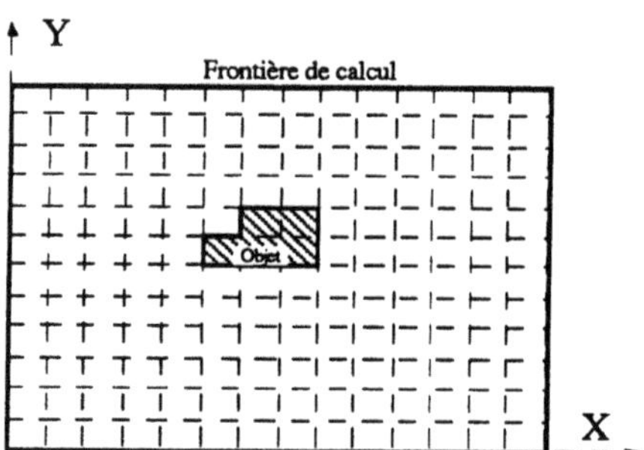

Fig. 8: Example of geometry POISSON problem with F.D. solving.

Boundary conditions are required to obtain a single solution. A Dirichlet condition, (i.e., Voltage value on boundaries) is easy to implement because boundary conditions are the required values for nodes connected to the boundaries. This condition is implemented by imposing the known potentials on the boundaries.

The problem is now expressed as follows:

$$[A] \cdot (V) = (S)$$

where:

(V) represents the voltage vector (All unknown potentials),

(S) Source term which includes known charge density and Dirichlet condition.

[A] Sparse matrix Each row is filled by at least 5 terms. The position of these terms inside the row depends on the chosen node numbering.

An internal numbering may be performed in order to optimize the presence of terms around the diagonal.

$$A = \begin{bmatrix} -4 & 1 & 1 & 0 & 0 & - & 0 & 0 & 0 & 0 \\ 1 & -4 & 1 & 0 & 0 & - & 1 & 0 & 0 & 0 \\ 1 & 1 & -4 & 1 & 0 & - & 0 & 1 & 0 & 0 \\ 0 & 0 & 1 & -4 & 1 & - & 0 & 0 & 0 & 0 \\ 0 & 0 & 0 & 0 & -4 & - & 0 & 1 & 0 & 0 \\ - & - & - & - & - & - & - & - & - & - \\ 0 & 0 & 0 & 0 & 1 & - & -4 & 1 & 1 & 0 \\ 0 & 0 & 0 & 0 & 0 & - & 1 & -4 & 1 & 1 \\ 0 & 0 & 0 & 0 & 0 & - & 0 & 0 & -4 & 1 \\ 0 & 0 & 0 & 0 & 0 & - & 0 & 1 & 1 & -4 \end{bmatrix}$$

Matrix inversion or linear-system solving gives the solution. This method presents some drawbacks:

1: The shape of the object (and/or the boundary) is only represented as a collection of cubes or parallelepipeds. This is due to Finite Difference assumptions.

2: The Neumann boundary condition is not easy to introduce, and leads to problems for irregular shapes.

3.2.2 *Applications to Lines RADIATE Software [Ref. 9].* RADIATE is a system analysis tool developed by MATRA to solve in various configurations transient line problems either in conducted mode or radiated mode. The library includes single line over a ground plane, 2 coupled wires over ground plane, coaxial wire, Bus line, ... with generalized loading (R,L,C). Conducted mode is simulated with a current/voltage source inserted with terminal loads. Radiated mode results from an impinging electromagnetic wave defined as follows:

- User defined at line location

- Aperture diffracted fields at wire location. They are computed according to BETHE theory

Direct time and space line solving are performed with the help of the finite difference technique.

We will consider the case of a single wire over the ground plane, and show that the finite difference algorithm is very easy to obtain. The line equation is written as:

$$\frac{dV}{dx} = -R \, . \, I - L \frac{dI}{dt} + V_s$$

$$\frac{dI}{dx} = -G \, . \, V - C \frac{dV}{dt} + I_s$$

As we have two kinds of variables, time and space are half discretized to produce centered 2nd order accuracy derivatives and obtain an explicit solution.

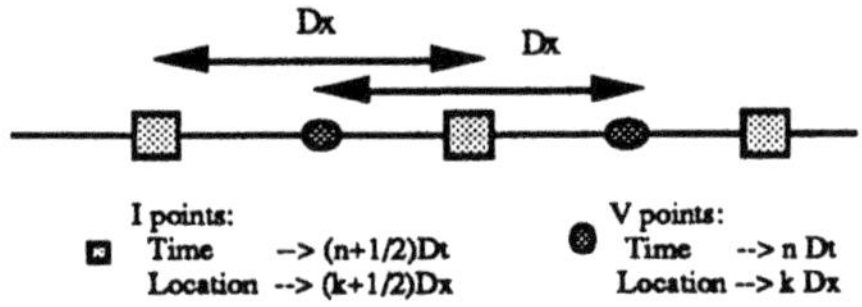

Fig. 9: RADIATE I and V nodes.

A Finite difference algorithm is obtained with the following definitions:

$$\left.\frac{dV}{dx}\right|_{t=n\Delta t}^{x=(k+1/2)\Delta x} = \frac{V_{k+1}^{n} - V_{k}^{n}}{\Delta x}$$

$$= -\frac{R}{2}\cdot\left(I_{k+1/2}^{n+1/2} + I_{k+1/2}^{n-1/2}\right) - \frac{L}{\Delta t}\left(I_{k+1/2}^{n+1/2} - I_{k+1/2}^{n-1/2}\right) + V_{s\,k+1/2}^{\;n}$$

$$\left.\frac{dI}{dx}\right|_{t=(n+1/2)\Delta t}^{x=k\Delta x} = \frac{I_{k+1/2}^{n+1/2} - I_{k-1/2}^{n+1/2}}{\Delta x}$$

$$= -\frac{G}{2}\cdot\left(V_{k}^{n+1} + V_{k}^{n}\right) - \frac{C}{\Delta t}\left(V_{k}^{n+1} - V_{k}^{n}\right) + I_{s\,k}^{\;n+1/2}$$

After insertion of the above formula into the line equation and some mathematical manipulation, an explicit solution is obtained:

at time $t = n.\Delta T$ for line potentials V_k^n at location $x = k.\Delta x$

at time $t = (n+1/2)\Delta t$ for currents $I_{k+1/2}^{n+1/2}$ located at $x=(k+1/2).\Delta$ as a function of the past line values and source terms:

$$I_{k+1/2}^{n+1/2} = \frac{\left(\frac{R}{2} - \frac{L}{\Delta t}\right) I_{k+1/2}^{n-1/2} + \frac{V_{k+1}^{n} - V_{k}^{n}}{\Delta x} + V_{s\ k+1/2}^{n}}{\left(-\frac{R}{2} - \frac{L}{\Delta t}\right)}$$

$$V_{k}^{n+1} = \frac{\left(\frac{G}{2} - \frac{C}{\Delta t}\right) V_{k}^{n} + \frac{I_{k+1/2}^{n+1/2} - I_{k-1/2}^{n+1/2}}{\Delta x} + I_{s\ k}^{\ n+1/2}}{\left(-\frac{G}{2} - \frac{C}{\Delta t}\right)}$$

The stability criterion $\frac{\Delta x}{\Delta t} < c$ is the only constraint of this algorithm which is very robust and powerful.

The above equations are only valid for all points which are not connected to terminal loads.

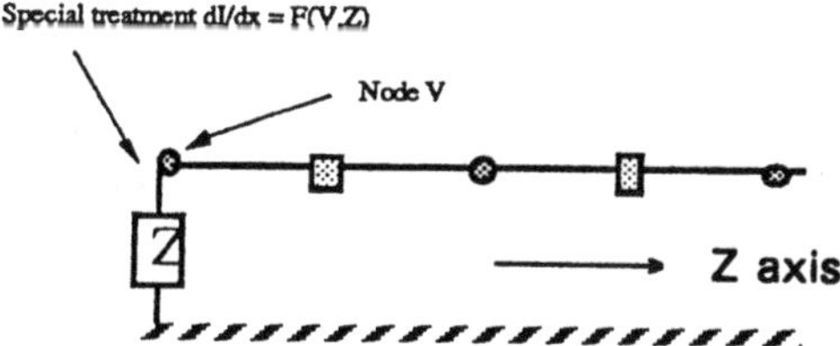

Fig. 10: Boundary Condition in RADIATE software.

The end of the wire requires the selection of an I or V point. Then a specific treatment is applied in order to insert terminal loading condition. (R, L, C, Egen, Igen, Diode).

In RADIATE software, the terminal node is considered as the V point. The space derivative dI/dx for the end of the line is then approximated with a 2nd order Lagrange polynomial. The two nearest current values are required to define weighting factors in the Lagrange polynomial. After some mathematics, one derives a corrected finite difference equation, including the kind of terminal loading. This technique is valid for all elementary elements (R, L, C, voltage/current source) as well as non-linear elements.

These equations are easily generalized to multi-lines. The following examples illustrate the robustness of the algorithm and the interest of such models.

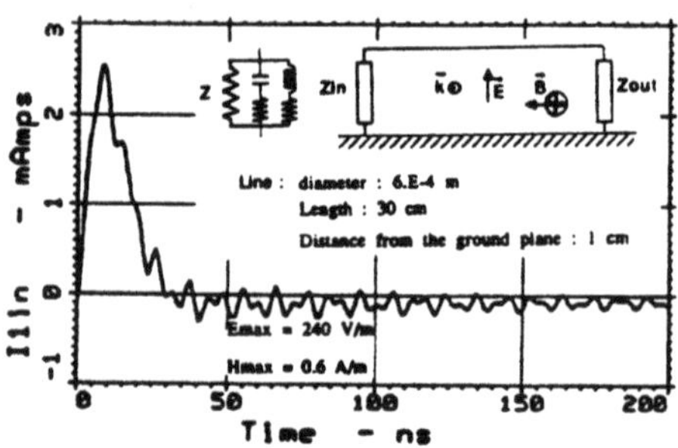

Fig. 11: E, H field coupling on a single wire.

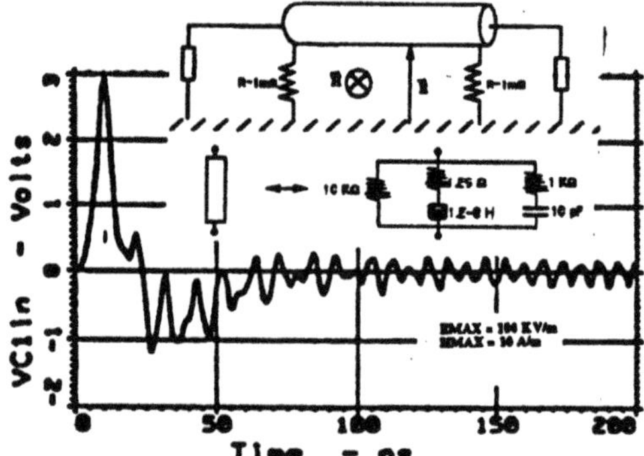

Fig 12: E,H field coupling on a coaxial cable.

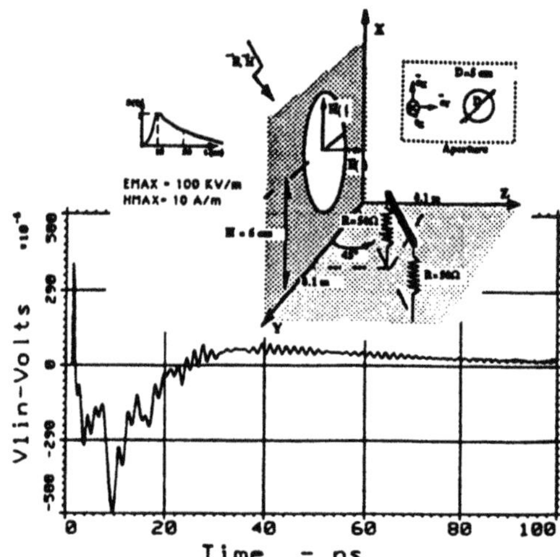

Fig. 13: Coupling of field to wire via an aperture.

3.2.3. *Application to Maxwell's Equations [Ref. 21]*. The same technique as presented for lines is applied to the Maxwell equations. Because of multi-dimensional partial derivatives and time derivative, half-time and space discretisation on electric and magnetic fields is applied to insure 2nd order accuracy. For a fully tri-dimensional problem, space is discretized with elementary cubes or parallelepiped in all directions (Carthusian meshing). The object is represented in this space by these elementary shapes. Axi-symmetrical problems are quite similar and are represented in the R,Z) plane with a collection of rectangles.

Insertion of derivative formulae under finite difference form leads to an explicit solution. Actual electric and magnetic field components are expressed as a function of known quantities in the past and the source term. The numerical scheme is as the previous one, half time and space interval interleaved.

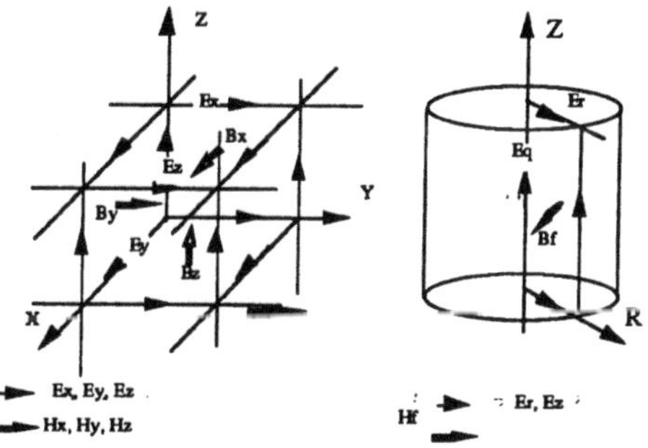

Fig. 14: Nodes E and H for F.D.

Stability criterion limits the time discretisation according to space meshing. For a non-uniform meshing, the, minimum time discretisation must not exceed:

$$\Delta t << \frac{1}{c\sqrt{\frac{1}{\Delta x_{min}} + \frac{1}{\Delta y_{min}} + \frac{1}{\Delta z_{min}}}}$$

For a parallelepipedic meshing, discretized Maxwell's equations are as follows:

$$H_x^{n+1/2}(i,j+1/2,\ k+1/2) = H_x^{n-1/2}(i,j+1/2,\ k+1/2) -$$

$$\frac{\Delta t}{\mu_o}\left\{\begin{array}{l} \frac{h_3^2 + h_4^2}{2h_3h_4(h_3+h_4)}(E_z^n(i,\ j+1,\ k+1/2) - E_z^n(i,\ j,\ k+1/2)) - \\ \frac{h_5^2 + h_6^2}{2h_5h_6(h_{15}+h_6)}(E_y^n(i,\ j+1/2,\ k+1) - E_y^n(i,\ j+1/2,\ k)) \end{array}\right\}$$

$$H_y^{n+1/2}(i+1/2,j,k+1/2) = H_y^{n-1/2}(i+1/2,j,k+1/2) - \frac{\Delta t}{\mu_o}\left\{\begin{array}{l} \frac{h_5^2+h_6^2}{2h_5h_6(h_5+h_6)}(E_x^n(i+1/2,j,k+1) - E_x^n(i+1/2,j,k)) - \\ \frac{h_1^2+h_2^2}{2h_1h_2(h_1+h_2)}(E_z^n(i+1,j,k+1/2) - E_z^n(i,j,k+1/2)) \end{array}\right\}$$

$$H_z^{n+1/2}(i+1/2,j+1/2,k) = H_z^{n-1/2}(i+1/2,j+1/2,k) - \frac{\Delta t}{\mu_o}\left\{\begin{array}{l} \frac{h_1^2+h_2^2}{2h_1h_2(h_1+h_2)}(E_y^n(i+1,j+1/2,k) - E_y^n(i,j+1/2,k)) - \\ \frac{h_3^2+h_4^2}{2h_3h_4(h_3+h_4)}(E_x^n(i+1/2,j+1,k) - E_x^n(i+1/2,j,k)) \end{array}\right\}$$

$$E_x^{n+1}(i+1/2,j,k) = E_x^n(i+1/2,j,k) - J_x^n(i+1/2,j,k) + \frac{\Delta t}{\varepsilon_o}\left\{\begin{array}{l} \frac{h_3^2+h_4^2}{2h_3h_4(h_3+h_4)}(H_z^{n+1/2}(i+1/2,j+1/2,k) - H_z^{n+1/2}(i+1/2,j-1/2,k)) - \\ \frac{h_5^2+h_6^2}{2h_5h_6(h_5+h_6)}(H_y^{n+1/2}(i+1/2,j,k+1/2) - H_y^{n+1/2}(i+1/2,j,k-1/2)) \end{array}\right\}$$

$$E_y^{n+1}(i,j+1/2,k) = E_y^n(i,j+1/2,k) - J_y^n(i,j+1/2,k) + \frac{\Delta t}{\varepsilon_o}\left\{\begin{array}{l} \frac{h_5^2+h_6^2}{2h_5h_6(h_5+h_6)}(H_x^{n+1/2}(i,j+1/2,k+1/2) - H_x^{n+1/2}(i,j+1/2,k-1/2)) - \\ \frac{h_1^2+h_2^2}{2h_1h_2(h_1+h_2)}(H_z^{n+1/2}(i+1/2,j+1/2,k) - H_z^{n+1/2}(i-1/2,j+1/2,k)) \end{array}\right\}$$

$$E_z^{n+1}(i,j,k+1/2) = E_z^n(i,j,k+1/2) - J_z^n(i,j,k+1/2) + \frac{\Delta t}{\varepsilon_o}\left\{\begin{array}{l} \frac{h_1^2+h_2^2}{2h_1h_2(h_1+h_2)}(H_y^{n+1/2}(i+1/2,j,k+1/2) - H_y^{n+1/2}(i-1/2,j,k+1/2)) - \\ \frac{h_3^2+h_4^2}{2h_3h_4(h_3+h_4)}(H_x^{n+1/2}(i,j+1/2,k+1/2) - H_x^{n+1/2}(i,j-1/2,k+1/2)) \end{array}\right\}$$

The h_i coefficients define for each direction the interval between E and H nodes.

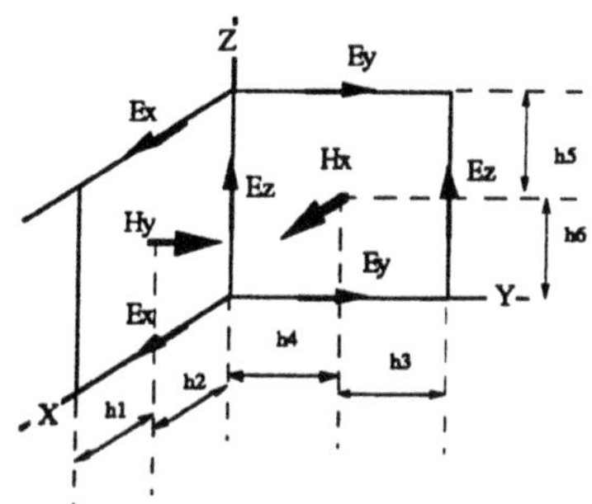

Fig. 15: Variable meshing for Maxwell F.D.

As for space, variable time stepping can be performed. This is a volumic representation of Maxwell's equations. It is well suited to phenomena involving volume treatment as the electrostatic discharge problem or S.G.E.M.P. (nuclear induced system generated electro magnetic pulse) problem. Scattered phenomena are also treated with this method [Ref. 22]. Note that in this case, wires models are often necessary to represent. A specific formulation must then be developed to adapt volume treatment of Maxwell's equations to very thin objects as wires [Ref. 22].

3.3. FINITE VOLUME METHOD (F.V.)

One of the major difficulties concerning the use of numerical methods for industrial purposes lies in the representation of 3D objects such as satellites. The engineer has to mesh complex space, and the application must handle the various elementary shapes of the mesher. Furthermore, complex objects and various boundary conditions must be taken into account.

The finite volume method is very popular in numerical aerodynamics and seems very powerful and adapted to solve volumic electromagnetic problems. Realistic objects as well as accurate boundary conditions can be easily implemented (This is not true for finite differences.), because of the notion of numerical flux defined in the following section.

3.3.1. *Basis of the Method*. The first step of this method is to make a volumic partition of the computational space surrounding the object: cubes, tetrahedrals, pyramids, ... Maxwell's equations are then solved taking into account their finite volume representation on each defined cell.

3.3.2. *Finite Volumes for Maxwell's Equations.*

Method 1: Direct integration:

Direct volumic integration of Maxwell's equations leads to the following relationships:

$$\oint\oint_s \vec{n} \times \vec{E} \cdot ds = - V \cdot \frac{\partial \vec{B}}{\partial t}$$

$$\oint\oint_s \vec{n} \times \vec{H} \cdot ds = V \left(\vec{J} + \varepsilon \cdot \frac{\partial \vec{E}}{\partial t} \right)$$

The time derivative is treated with centered finite differences.

The explicit iterating solution for centroïd field values (noted C) is then obtained as a function of source term (Current density) and quantities denoted flux involving surface field values (noted S).

$$\vec{H}_{C,k}^{n+1/2} = \vec{H}_{C,k}^{n-1/2} - \frac{\delta t}{\mu_o \cdot V_k} \sum_{i=1}^{N_k} \Delta S_i \cdot \vec{n}_i \times \vec{E}_{S,i}^{n}$$

$$\vec{E}_{C,k}^{n+1} = \vec{E}_{C,k}^{n} + \frac{\delta t}{\varepsilon_o} \left(\frac{1}{V_k} \sum_{i=1}^{N_k} \left(\Delta S_i \cdot \vec{n}_i \times \vec{H}_{S,i}^{n+1/2} \right) - \vec{J}_{C,k}^{n+1/2} \right)$$

As for finite differences, this numerical scheme uses an interleaving half time and space interval process to solve first for electric field E and then magnetic fields H. Remarks:

- Cell quantity values require the knowledge of flux values (i.e., surface field values) related to the cell
- An object can be represented with a collection of any elementary 3D shape (cubes, prisms, tetrahedrals, ..). Boundary conditions are automatically implemented on cell surfaces connected to the object. A mixed condition using E and H local field can be easily used.
- Non-structured meshes are allowed.

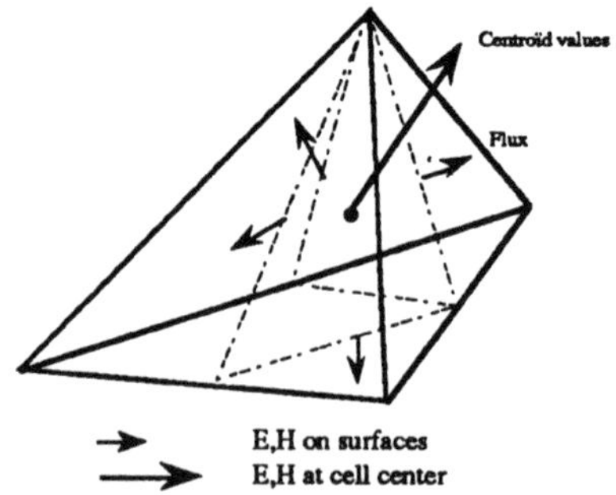

Fig. 16: Flux and cells in F.V.

At this step of computation, we need to evaluate in a self consistent manner surface field quantities. Another computational step must be performed to derive the required values:

$$E^n_{i+1/2} = F(E^n_i, E^n_{i+1}, H^{n-1/2}_i, H^{n-1/2}_{i+1})$$

A simple scheme consists in the evaluation of surface field values through an interpolating process from nearest centroïd values.

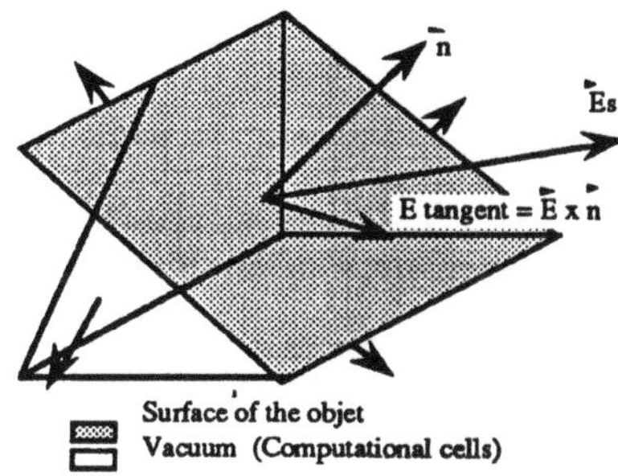

Fig. 17: Boundary conditions for F.V.

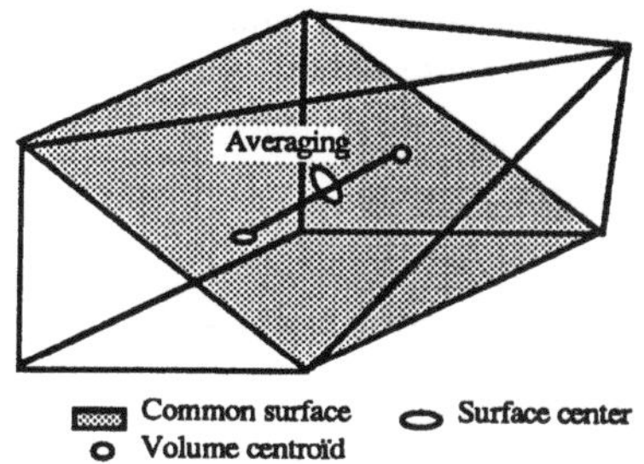

Fig. 18: Computation of flux for F.V.

Method 2 Conservative flux scheme [Ref. 18] Application to electrostatic discharge

This is another method that avoids the drawbacks associated with the preceding method. The method consists of evaluating the surface flux quantities from centroïd values, with a higher order scheme than interpolation. The starting point is the divergential form of Maxwell equations defined as:

$$\frac{\partial U}{\partial t} + \frac{\partial E}{\partial x} + \frac{\partial F}{\partial y} + \frac{\partial G}{\partial z} = W$$

$$E = \begin{Bmatrix} 0 \\ -\frac{D_z}{\varepsilon_o} \\ \frac{D_y}{\varepsilon_o} \\ 0 \\ \frac{B_z}{\mu_o} \\ -\frac{B_y}{\mu_o} \end{Bmatrix} \quad F = \begin{Bmatrix} \frac{D_z}{\varepsilon_o} \\ 0 \\ -\frac{D_x}{\varepsilon_o} \\ -\frac{B_z}{\mu_o} \\ 0 \\ \frac{B_x}{\mu_o} \end{Bmatrix} \quad G = \begin{Bmatrix} -\frac{D_y}{\varepsilon_o} \\ \frac{D_x}{\varepsilon_o} \\ 0 \\ \frac{B_y}{\mu_o} \\ -\frac{B_x}{\mu_o} \\ 0 \end{Bmatrix} \quad U = \begin{Bmatrix} B_x \\ B_y \\ B_z \\ D_x \\ D_y \\ D_z \end{Bmatrix} \quad W = \begin{Bmatrix} 0 \\ 0 \\ 0 \\ -J_x \\ -J_y \\ -J_z \end{Bmatrix}$$

Numerical field evaluation is performed in each cell in two steps:

1: Field computation at cell centroids results from known flux values. Time discretisation is defined as finite differences.

2: Flux computation is derived through the method of characteristics in the 3 directions of space. This is obtained from analysis of eigen values and eigen vectors of the jacobian:

$$\left| \frac{\partial E}{\partial U} - \lambda I \right| = 0$$

Through the characteristic λ, U and E variations are obtained from the jumping conditions. For a specific direction λ, one obtains:

↑↓ *Modes* 0: $\lambda = 0$ $\quad \begin{cases} \xi \times (H_s^+ - H_s^-) = 0 \\ \xi \times (E_s^+ - E_s^-) = 0 \end{cases}$

← *Mode* $+c$: $\lambda = +c.\xi$ $\quad \left\{ B_C^R - B_s^+ = \frac{\xi}{c} \times (E_C^R - E_s^+) \right.$

→ *Mode* $-c$: $\lambda = -c.\xi$ $\quad \left\{ B_s^- - B_C^L = \frac{\xi}{c} \times (E_s^- - E_C^L) \right.$

$$\xi \times \vec{H}_s^- = \xi \times \vec{H}_s^+ = \xi \times \{(\vec{H}_C^R + \varepsilon_o\, c\, \xi \times \vec{E}_C^R) + (\vec{H}_C^L - \varepsilon_o\, c\, \xi \times \vec{E}_C^L)\}$$

$$\xi \times \vec{E}_s^- = \xi \times \vec{E}_s^+ = \xi \times \{(\vec{E}_C^R + \mu_o\, c\, \xi \times \vec{H}_C^R) + (\vec{E}_C^L - \mu_o\, c\, \xi \times \vec{H}_C^L)\}$$

The same computation is performed for the other directions.

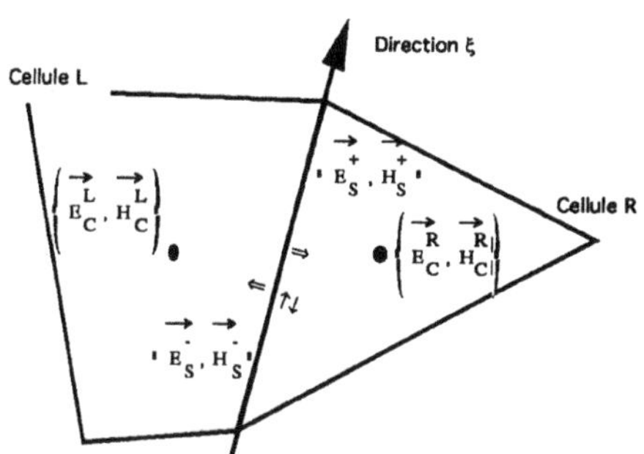

Fig. 19: Flux splitting and field computation.

Such a numerical scheme is on development at TOULOUSE MATRA-space center in order to derive the electromagnetic response of a complex object such as a satellite, to an electrostatic discharge. This technique will allow us to simulate thin objects and to represent as easily as F.D. an open problem. The planned architecture of the software is represented hereafter:

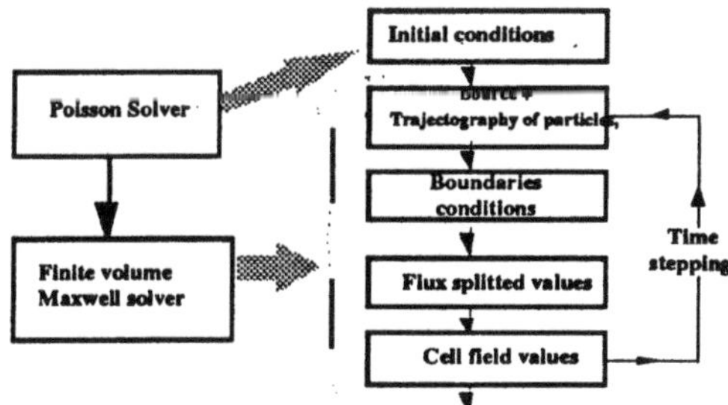

Fig. 20: Electrostatic discharge - simulation.

3.4. METHOD OF MOMENTS (M.O.M.)

3.4.1. *Basis of the Method*. Method of moments (M.O.M.) [Ref. 23] is based on the solution of an integro-differential system through variable expansion. This method is able to solve any formal equation of the form:

$$L \cdot f = g$$

where:

L is any differential, integral or integro-differential operator.

The solution is obtained numerically through a discretisation of the integral and transformation of the integro-differential system into a linear system.

The first operation is the definition of an inner product noted < – | – > and a family of functions: "the basis functions" $| f_i>$. With these assumptions, the unknown function f becomes:

$$f = \sum_{i=1}^{N} a_i \cdot |f_i> \quad , \quad |f_i> = \begin{Bmatrix} f_1 \\ \cdots \\ f_1 \end{Bmatrix}$$

a_i: unknowns to evaluate

Insertion of the development of this function inside the integro-differential system and using linear properties, one obtains:

$$\sum_{i=1}^{N} a_i \, (L \cdot |f_i>) = g$$

To obtain the coefficients, it is necessary to define a matrix. As a consequence, we need to project the resulting above equation on another family of functions (Weighing or testing functions) to build the linear system.

$$< h_i \; | = (h_1 \; \ldots \; h_N) \; .$$

We obtain finally:

$$\sum_{i=1} a_i <h_j \; | \; L \cdot |f_i> = g_j$$

$$g_j = <h_j \; | \; g$$

This can be written as:

$$[A] \cdot (x) = (y)$$

where:

(x) represents the unknown vector of coefficients, a_i,

(y) is a vector which is defined as the projection of source terms (function g) on the weighing functions

[A] is the matrix resulting from the L operator projected onto the weighing and basis functions. By definition, this kind of solution leads to an implicit solution.

The inner product is generally chosen for electromagnetic and electrostatic applications as a surface integral:

$$< f_2 \; | \; f_2 > \rightarrow \iint_S f_1 \cdot f_2 \, ds$$

The selection of weighing functions is derived by the nature of the problem (scalar, vectors, physical quantities) to reproduce accurately the signal. Furthermore, the combination basis functions and testing functions must be easy to implement to:

- allows an easy computation of matrix elements. In particular, singularity treatment (on diagonal of the matrix) should be solved under an analytic form. Infinite integral means that the choice of basis or weighing functions is not well adapted.
- to obtain a matrix with interesting properties (symmetric, Toeplitz, sparse, ...)
- to produce a well conditioned matrix, easy and fast to solve.

More commonly the testing functions used are:

testing function = basis function
→ "Galerkin" method

- Testing/basis functions = Dirac
→ "Point Matching" method
- Basis function = f(r), testing function = Dirac
→ "Collocation" method

Development of functions with local support is generally preferred (term $\Pi(r_j)$ inside the basis function) because they lead to better conditioned matrices.

Method	function term	
	base j	test i
Galerkin (global)	$a_j \cdot f_j(r)$	$f_i(r)$
Least square	$a_j \cdot f_j(r)$	$Q(r) \cdot \frac{\partial \varepsilon(r)}{\partial a_i}$
Point-matching	$a_j \cdot \delta_j(r-r_j)$	$\delta(r-r_i)$
Collocation (global)	$a_j \cdot f_j(r)$	$\delta(r-r_i)$
Collocation (local)	$\Pi(r_j)\cdot\sum_k a_{jk}\cdot f_k(r)$	$\delta(r-r_i)$
Galerkin (local)	$\Pi(r_j)\cdot\sum_k a_{jk}\cdot f_k(r)$	$\Pi(r_j)\cdot\sum_k f_k(r)$

a_j : Constants to evaluate
$\pi(r)$: Gate function
$f_{i/j}(r)$: Basis function
$\delta(r-r_{j:i})$: Testing function (Dirac)
$Q(r)$: Function of r definite positive
$e(r)$: Error equation or residual

3.4.2. *Integral Formulation in Frequency Domain.* For most of the problems, the choice of the E.F.I.E. formulation is more powerful because of its generality. It is more difficult to solve than M.F.I.E. equation [Ref. 24]. M.F.I.E. is only valid for closed objects. It raises numerical problems for open objects, because in this case, currents have to be evaluated on both sides of the object (very thin) producing ill-conditioned matrices for implicit solution.

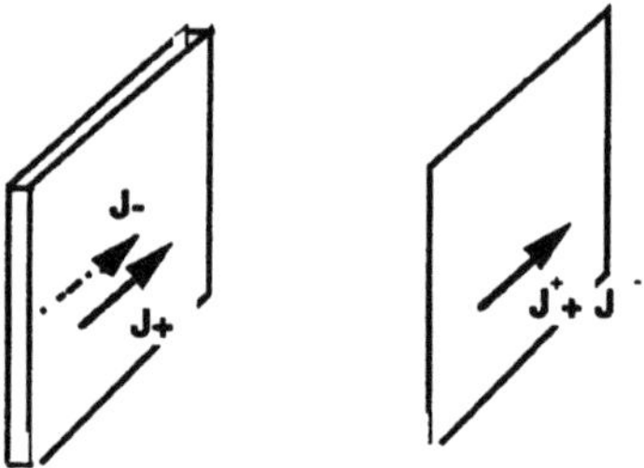

Fig. 21: MFIE and EFIE: Treatment of open objects.

E.F.I.E. avoids this problem, by means of quantities like current and charge which are represented in this equation by the sum of both contributions on each side of the open or closed object.

The following equation is to be solved in frequency domain:

$$E^T(r) = E^I(r) + E^S(r)$$

$$\vec{E}^S(r) = -\frac{j\,\mu\,\omega}{4\pi}\left\{\iint_S \vec{J}(r')\cdot\frac{e^{-j\omega\sqrt{\mu\cdot\varepsilon}\cdot R}}{R}\,ds' + \iint_S \frac{c^2}{\omega^2}\nabla_r\cdot\nabla_r\left(\vec{J}(r')\cdot\frac{e^{-j\omega\sqrt{\mu\cdot\varepsilon}\cdot R}}{R}\right)ds'\right\}$$

$E^T(r)$ is the total electric field in the entire space. The total tangent electric or normal magnetic field on the object will constitute the boundary condition that we will apply. It is well known on the object (Pure metallic condition, impedance condition, ...):

- Ohmic law is a simple example: $\vec{E}^T(r) = \frac{\vec{J}(r)}{\sigma}$

- Perfect conductor: $E^T = 0$

– Surface impedance condition: $\vec{E}^T(r) = \left(\frac{1}{\sigma} + L\omega\right) \cdot \vec{J}(r)$

– Capacitive condition: $\vec{E}^T(r) = - \frac{\nabla_r \vec{J}(r)}{j\ \omega\ C}$

Other conditions can also be implemented:

- Polarization current representing a thin dielectric layer above a conductive surface.
- A source voltage generator (Electric field gap on a wire is a common mean to represent this source)
- Current generator, ...

The basis functions are vector functions that are very powerful to represent easily tri-dimensional objects with complex shapes. With this assumption, unknowns become scalar. The more commonly used functions are the well known RAO and GLISSON functions [Ref. 24]. They are defined for a triangular meshing of the external surface of the object as follows:

On Figure 22, triangles T_n^+ and T_n^- have a common edge noted l_n

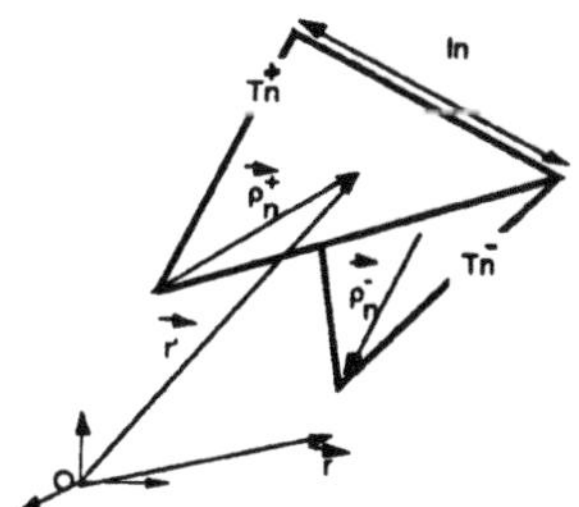

Fig. 22: Patch elements from RAO and GLISSON.

Basis functions f are defined as:

$$J(r') = \sum_{i=1}^{3} J_i^n \overrightarrow{f_n(r')}, \quad \overrightarrow{f_i(r)} = \begin{cases} \frac{l_i}{2A_i^+}\vec{\rho}_i^+ & \vec{r} \in T_n^+ \\ \frac{l_i}{2A_i^-}\vec{\rho}_i^- & \vec{R} \in T_n^- \\ 0 & \vec{r} \notin (T_n^+ \cup T_n^-) \end{cases}$$

The selection of this kind of functions is very well adapted to 3D objects because:

- the basis function is normalized on the ln edge
- it forces continuity of the current component normal to the ln edge on each triangles
- J_n coefficients represent the component of this current in the normal direction of the edge l_n. This property is very well

adapted to insert efficiently an edge boundary condition (open surface with $J_n = 0$)

The same technique applies for basis functions on segments (Piecewise linear functions). A specific function is defined to make connections between surfaces and segments [Ref. 25]. The detail of the computation of the matrix elements and source terms is not reported in this paper because it is well documented [Ref. 26].

Applying as previously defined the method of moments (Galerkin), we obtain a linear system that describes the electromagnetic response of the object at frequency ω:

$$[Z] \cdot \{J_n\} = \{V\}$$

Inversion or direct solution of the linear system gives the $\{J_n\}$ vectors. Surface currents at a patch center can be determined from the definition of surface currents versus basis functions. The computation of the scattered field at any point in space is then achieved from E.F.I.E. equation, since all unknown currents $\{J_n\}$ are now known.

Such a software is operational at MATRA-space center. It is presently validated for simple and complex shapes subjected to voltage/current injection (MAROTS Satellite hereafter represented). Insertion of more complex impedance electrical boundary condition has been implemented and is under testing. The following examples make illustration of the power of this method that is able to represent EMC/ESD problems with complex shapes in various electrical configurations.

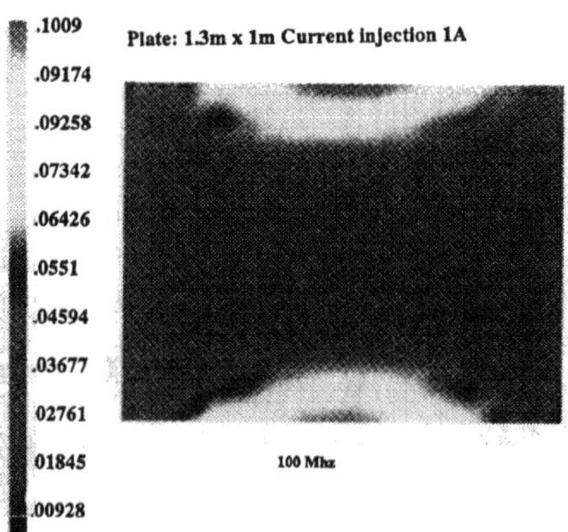

Fig. 23: Current injection inside a plate.

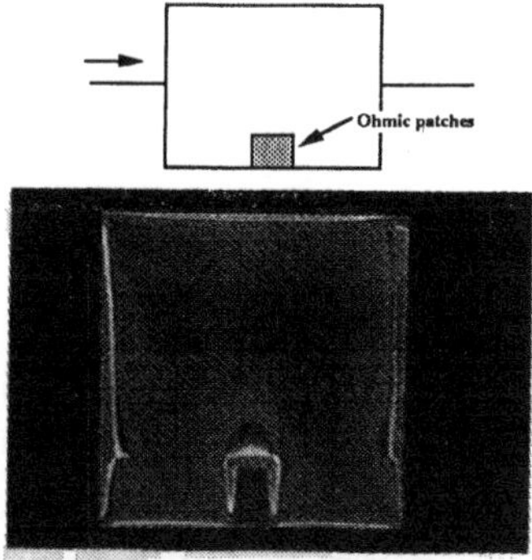

Fig. 24: Ohmic patches on a plate under voltage injection.

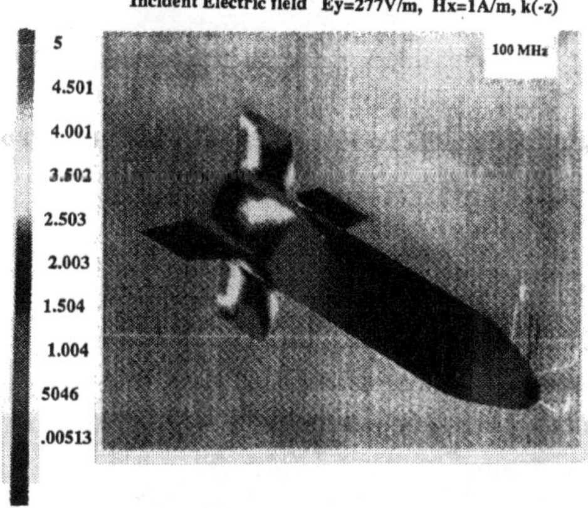

Fig. 25: Missile example: Scattering of an impinging wave.

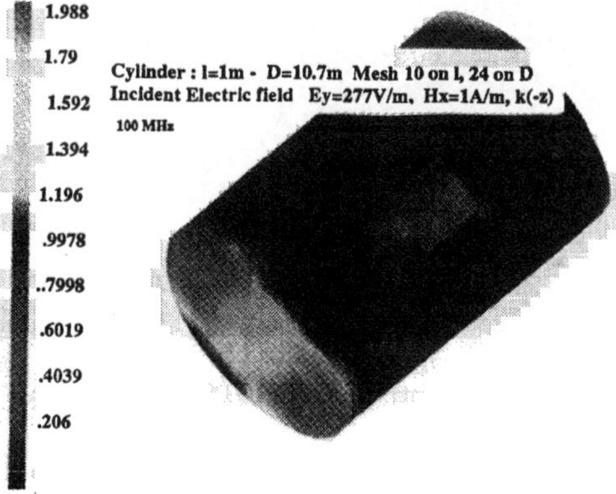

Fig. 26: Scattering of a cylinder with aperture: currents.

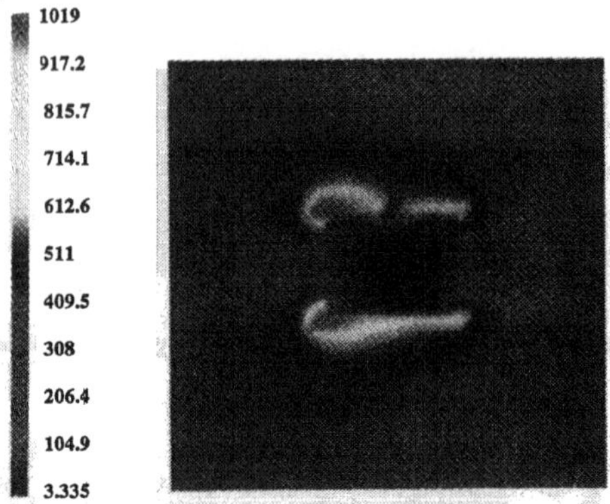

Fig. 27: Scattering of a cylinder with aperture: Total fields.

Fig. 28: Current injection on the MAROTS satellite.

3.4.3. *Integral Formulation in Time Domain.* Time domain solution is more complex than the frequency solution because now, space-time coupled variables are to be taken into account with propagating effects. The basis of the method can be found in [Ref. 27].

The example of a simple wire structure is analyzed. A complex tridimensional shape is treated in the same way owing to function bases that are well adapted to tridimensional treatment as in frequency domain. In another way, a surfacic object can be represented at a first order approximation as a collection of wires connected as a grid.

Electric integral space time formulation: E.F.I.E.

Owing to the hypothesis of the thin-wire model; that is,

- Current uniformity around the axis.
- Thin radius

the electric space-time integral equation is described as a function of resulting current and linear charge density:

$$E_s(r,t) = -\int_{C(r)} \left[\frac{l'}{R} \cdot \frac{\partial I(r',t')}{\partial t'} + \frac{c \cdot \vec{R}}{R^2} \cdot \frac{\partial I(r',t')}{\partial l'} - \frac{c^2 \cdot \vec{R}}{R^3} q(r',t')\right] \cdot dr'$$

l Normalized vector at point r, tangent to the wire

l′ Normalized vector at point r′, tangent to the wire

$\vec{E}_I(r,t)$ Incident field at point r and time t

$\vec{R} = \vec{r}' - \vec{r}$

R Distance from observation point to source point

I(r′,t′) Current at source point at delayed time t′

q(r′,t′) Charge at source point at delayed time t′

dl′ Integration step along the wire

r′ Source point

r Observation point

$t' = t - R/c$ Delayed time

c = light velocity

This equation is valid in all space, except source zones, i.e., the domain internal to the wire with radius a(r).

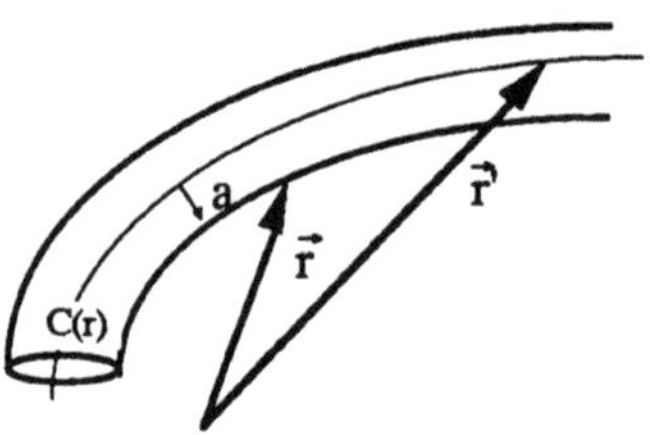

Fig. 29: Wire geometry.

Basis functions

As in the frequency domain, basis functions are chosen. They are defined as:

$$I(r',t') = \sum_{i=1}^{N_s} \sum_{j=1}^{\infty} I_{ij}(r' - r_i, t' - t_j) \cdot U(r' - r_i) \cdot V(t', - t_j)$$

U and V functions are gate functions defining the application interval of this relationship.

$$U(r' - r_i) = U(s_i'') = \begin{cases} 1 & |r' - r_i| \leq \Delta_i/2 \\ 0 & |r' - r_i| > \Delta_i/2 \end{cases}$$

$$V(t' - t_j) = \begin{cases} 1 & |t' - t_j| \leq \delta_j/2 \\ 0 & |t' - rt_j| > \delta_j/2 \end{cases}$$

As the function I_{ij} is only defined on a segment of lengths Δ, the space variable can be transformed as a curvilinear coordinate on wire path. δ_j is the time increment at time step j and Δ_i is the length of the i segment. The function $I_{ij}(r' - r_i, t' - t_j)$ is obtained by means of a chosen interpolating function: In this case, we have chosen 2nd order LAGRANGE polynomials. Second order means that we have at least continuity of 2nd order derivatives.

Inserting this definition, the current and linear charge densities become:

$$I(s'',t'') = \sum_{l=1}^{l=1} \sum_{m=n}^{m=n+2} B_{i,j}^{l,m} \cdot I_{i+1,j+m}$$

$$q(s'',t'') = \sum_{l=1}^{l=1} \sum_{m=n}^{m=n+2} B_{i,j}^{l,m} \cdot q_{i+1,j+m}$$

$$B_{i,j}^{l,m} = \prod_{p=1}^{p=1} l \left(\frac{s_i'' + s_i - s_{i+p}}{s_{i+1} - s_{i+p}} \right) \cdot \prod_{q=n}^{q=n+2} \left(\frac{t_j'' + t_j - t_{j+q}}{t_{j+m} - t_{j+q}} \right)$$

S''_i: Curvilinear coordinate of the i segment center.
and

$$n = \begin{cases} = -2 \text{ if } \Delta R = \dfrac{r' - r}{c} < 0.5 \\ = -1 \text{ if } \Delta R = \dfrac{r' - r}{c} > 0.5 \end{cases}$$

in order to preserve interpolation for small segments.

In all cases, extrapolation must be avoided because the solution obeys causality. This is the key of numerical convergence and stability.

Solving

The summation acting on the E.F.I.E. is discretised in N_S terms (N_S segments). Taking as weighing functions Dirac, one obtains the following equation:

$$-\vec{I} \cdot \vec{E}_I(r,t) = -\frac{\mu_o}{4\pi} \sum_{i=1}^{N_s} \vec{I} \int_{\Delta C_i}$$

$$\left[\frac{\vec{I}'}{R} \frac{\partial I(r',t')}{\partial t'} + c \frac{\vec{R}}{R^2} \cdot \frac{\partial I(r',t')}{\partial r'} - c^2 \frac{\vec{R}}{R^3} \cdot q(r',t') \right] dr$$

$\overrightarrow{l_{i/u}}$: Normalized vector at point i/u

Another equation is to be used, owing to the properties of the continuity equation. The first equation reduces to the following relationship using properties of U and V gate functions inside I(r',t').

$$-\vec{l_u} \cdot \overrightarrow{E_I(r_u, t_v)} =$$

$$-\frac{\mu_o}{4\pi}\sum_{i=1}^{N_s} l_u \int_{-\Delta_i/2}^{\Delta_i/2} \left\{ \frac{l_i}{R_{iu}} \cdot \frac{\partial I_{ij}(s_i'', t_j'')}{\partial t_j''} + c\, \frac{\vec{R}_{iu}}{R_{iu}^2} \cdot \frac{\partial I_{ij}(s_i'', t_j'')}{\partial s_i''} - c^2\, \frac{\vec{R}_{iu}}{R_{iu}^3} \cdot q_{ij}(s_i'', t_j'') \right\}_{t_j = t_v - R_{iu}/c - t_j''} ds_i''$$

with the following definitions:

$$R_{iu} = R_i(s_i'') = \left|\vec{r_u} - \vec{r_i} - s'' \cdot \vec{s_i}\right|$$

$$|s_i''| \leq \Delta_i/2 \ , \ |t_j''| \leq \delta_j/2$$

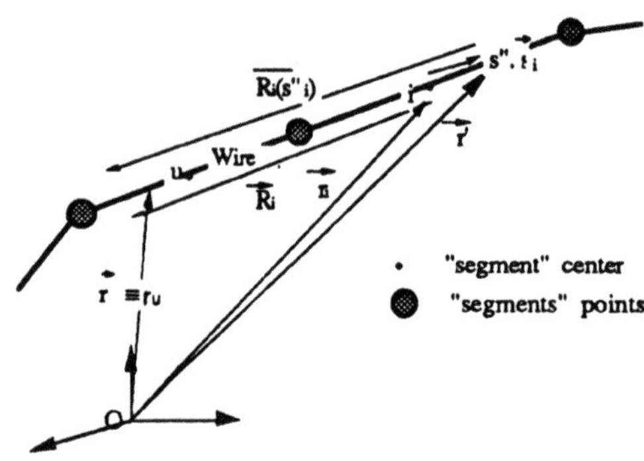

Fig. 30: Points i (source), u (Observation) and local variable s".

Linear System

The two above equations, seeming rather complex, define the linear system. By insertion of the explicit form of current (Lagrange polynomials) and selecting actual involved terms noted v and others that are associated to known past variables (v-1, v-2, ...), then we obtain an iterating linear system from which currents and linear charges are obtained.

$$\sum_{k=1}^{N_s} \frac{\mu_o}{4\pi}\vec{s}_u \sum_{a=0}^{3}\sum_{b=0}^{3} \left\{ \sum_{p=0}^{1} \delta(r_{i,u} \cdot p) \begin{pmatrix} \overline{\overline{F^{1\,a,b}_{\ i,u,1,p}}} \\ \overline{\overline{F^{2\,a,b}_{\ i,u,1,p}}} \end{pmatrix} I^{a,b}_{i,u} \cdot \overline{I}_{k,v} \right.$$

$$+ \sum_{p=0}^{1} \delta(r_{i,u} \cdot p) \begin{pmatrix} \overline{\overline{G^{1\,a,b}_{i,u,1,p}}} \\ \overline{\overline{G^{2\,a,b}_{i,u,1,p}}} \end{pmatrix} I^{a,b}_{i,u} \cdot \overline{q_{k,v}} \Bigg\}$$

$$= \begin{pmatrix} \overline{E^{I}_{u,v}} \\ \overline{E^{I}_{u,v}} \end{pmatrix} - \frac{\mu_o}{4\pi} \vec{s}_u \sum_{k=1}^{N_s} \Bigg\{ \sum_{m=n}^{n+2} r_{i,u} \sum_{a=0}^{3} \sum_{b=0}^{3} \begin{pmatrix} \overline{\overline{F^{1\,a,b}_{i,u,1,p}}} \\ \overline{\overline{F^{2\,a,b}_{i,u,1,p}}} \end{pmatrix} I^{a,b}_{i,u} \cdot \overline{I_{k,v-r_{i,u}+m}}$$

$$- \sum_{m=n}^{n+2} r_{i,u} \sum_{a=1}^{3} \sum_{b=1}^{3} \begin{pmatrix} \overline{\overline{G^{1\,a,b}_{i,u,1,p}}} \\ \overline{\overline{G^{2\,a,b}_{i,u,1,p}}} \end{pmatrix} I^{a,b}_{i,u} \cdot \overline{q_{k,v-r_{i,u}+m}} \Bigg\}$$

With: $\begin{cases} u=1,N_s & for\ \ i=1,N_s \\ k=i+1 & for\ \ 1=-1,0,1 \end{cases}$

Term's F1, F2, G1, G2 and $I^{a,b}$ are constants, only dependent on the geometry of the problem. Variables $I_{k,v}$ and $q_{k,v}$ are unknown currents and charge density on segment k at time v. they are formally obtained from the following linear system as:

$$[A] \cdot \begin{Bmatrix} I^v \\ q^v \end{Bmatrix} = \begin{Bmatrix} E^1 + f(I^{v-1}, \ldots I^1) + f'(q^{v-1}, \ldots q^1) \\ E^1 + g(I^{v-1}, \ldots I^1) + g'(q^{v-1}, \ldots q^1) \end{Bmatrix}$$

This solution is the basis of more complex situations where wire connections, various boundary conditions and symmetries can be implemented (ohmic condition, self or capacitive condition).

This formalism is easily expanded to complex wire geometry with multiple junctions with help of Kirchoff laws [Ref. 29]. The formulation is not changed because junctions are seen as corrections to the original matrix without junctions.

Field Computation

The total field in space is the most important variable, because it is accessible to measurements and constitutes the source for energy coupling. It is obtained in outer source zones as the sum of the eventual incident field and scattered field resulting from previous current and charge density computation.

As in the frequency domain, the scattered electric field is obtained directly from the definition of E.F.I.E. operator and insertion of basis functions. The magnetic field is obtained from M.F.I.E. operator and insertion of basis functions. In this case, M.F.I.E. operator does not fail, either with open surfaces or wires since the observation point does not lie inside or on the source zone.

After some mathematical manipulation, the resulting fields are obtained as a quadrature of charge density and current, weighted with geometrical coefficients. For efficiency and time cost purpose, space zones are splitted into 3 zones:

- Far field zone: $1/R^3 << 1/R^2 \rightarrow \vec{r} \approx \vec{R}$

  Fields are charge density independent, 1/R dependence

- Intermediate zone: $1/R^3 << 1/R^2$

  Fields are charge density independent

- Near zone: No hypothesis

  Current and charge density dependent as depicted

on the following figure.

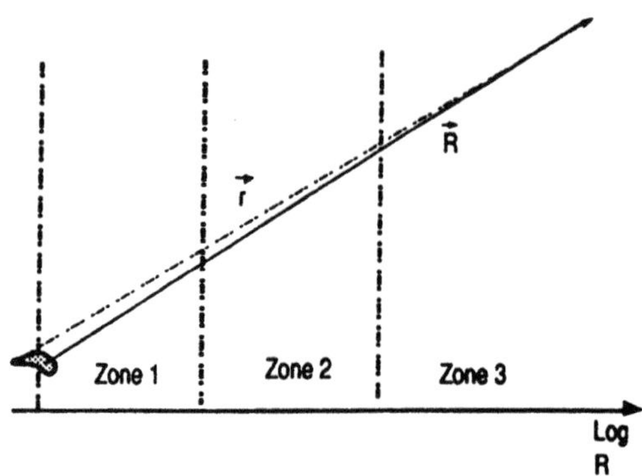

Fig. 31: Field computational zoning.

As an example, far fields at u point express as:

$$\vec{E}_s(R_{ou}, t) = -\frac{\mu_o}{4\pi \cdot R_{ou}} \sum_{i=1}^{N_s} \sum_{l=-1}^{1} \sum_{m=n}^{n+2} \frac{\{F1 \cdot \vec{I}_l\} \cdot I_{i+1, v-r_{ou}+m}}{D_i^{l,m}}$$

Where F1, F2 and D are only geometry dependent.
This result can be used for RCS evaluation for example. The following examples illustrate some of the results obtained by means of this method [Ref. 29] developed at the space-MATRA center.

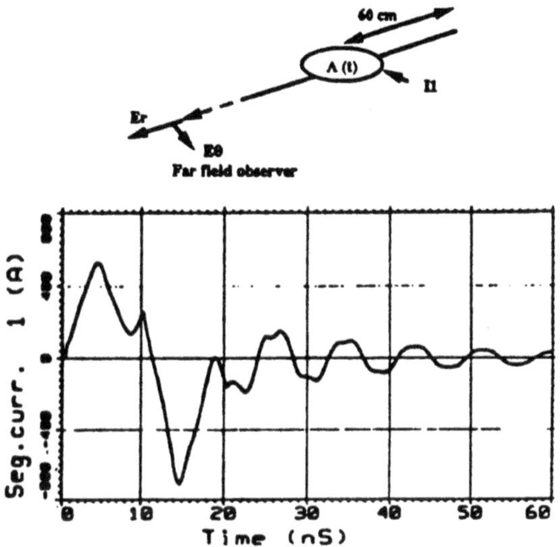

Fig. 32: Example of an antenna (Radiating mode).

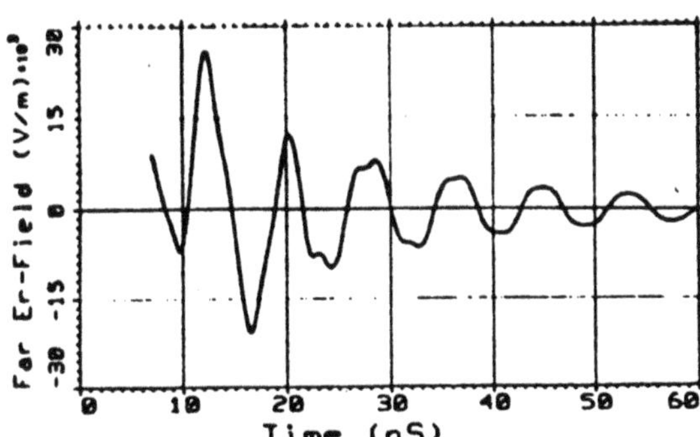

Fig. 33: Far field of the antenna.

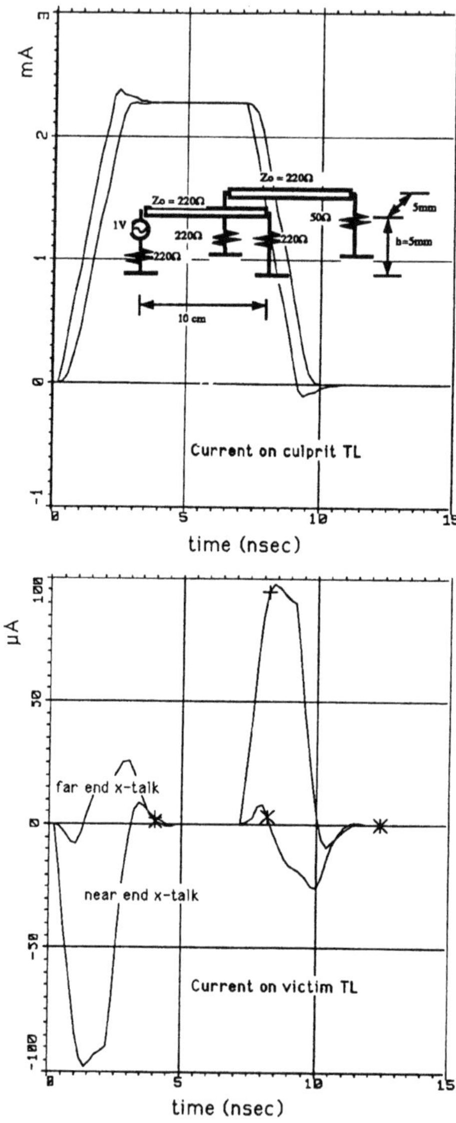

Fig. 34: Example of cross-coupling.

4. Conclusion - Advantages/Drawbacks of These Methods

Needs in the space domain are very important under EMC/ESD aspects. The problems are increasingly critical due to the increasing complexity and sensibility of the systems and their specialty (Long life mission, high cost, no possibility of on-orbit reparation). As a consequence, increasingly engineers must be educated to EMC/ESD science and related simulation technique to anticipate and solve problems.

This problem is more critical for big systems like HERMES, or the space station where ground testing cannot be performed at sub-system or system level. Simulation will be the only means to demonstrate the EMC/ESD immunity of the system.

Two main classes of problems are to solve:

- Problems that require volumic treatment: Electrostatic discharge, static plasma effects. All problems which require a self consistent solution of Maxwell-Vlassov or Poisson-Vlassov equations.

- Fully surface problems where the target object is considered as a scattering object. This is the case for an impinging wave on a structure, current or voltage injection (EMC/ESD testing), field penetration into an aperture, ...

For volumic problems, the treatment of the Maxwell equations by means of finite volumes seems to be the more adapted in view of geometrical representation, mathematical moderate complexity and ability to represent boundary conditions (radiating condition with flux splitting selection, electrical condition). This method is very efficient:

- to reproduce accurately the geometry of the problem

- to apply various boundary conditions (Ohmic materials, surface impedance, dielectrics, multilayers, ...) in a natural way.

In all other cases where a volumic solution is not required, the boundary integral equation solved by M.O.M. is strongly advised because this method allows us:

- to reproduce with high accuracy the geometry

- to limit the number of unknowns (Variables are defined only on the object but in surrounding space for volumic formulation)

- to take automatically the radiating condition without numerical artifact (Absorbing boundary sheets, big external cells to delay the reflected wave, extrapolation 1/r, ...)

The mathematical formalism is more difficult to develop compared to the volumic treatment. A surfacic mesher is required, but this technique allows us to treat very complex situations on moderate computers (VAX, DEC station). The following table summarizes the advantages and the drawbacks of the various methods that are today widely used in EMC/ESD simulation. The method of moments is clearly the more powerful because it is able to reproduce with the minimum of unknowns, complex geometrical situations with a transparent radiating condition.

	M.O.M.	F.D.	F.V.
Operator	Integral	Differential	Integral
Domain	Frequency and time	Time	Time
Pb boundary		infinity pb	Infinity-flux selection
Objects	3D Surfacic mesher	kind: cubic	kind: finite elements
Type Pb	Implicit	Explicit	Explicit
Nb unknowns	1.5 N1	6 N^3	6 N^3
Pb num.	matrix inversion, Complex matrix, Singular integrals	num. instabilities Stability criteria	Stability criteria

N1 = Number of edges on the object

N = Number of elementary volumes representing the vacuum around the object, >>N1

Software based on the method of moments in the frequency as well as in the time domain are today running at MATRA-space center. They are able to represent complex situations as on-orbit satellites. They allow us soon to treat system aspects related to EMC, ESD, lightning, ... especially on the projects that are difficult to test on ground because of their size.

Acknowledgment

The author gratefully acknowledges Miss Baradat for her contribution to running examples and pictures in MOD for the frequency domain, L. Gravelle who helped me in the development of the time domain thin-wire and surface MOM computer program, J. Seille and C. Garres who are the key persons for MATEMC and RADIATE software.

References

[1] J.P. Estienne, J. Lalande, J.F. Pascal, and J. Seille, Conference J.I.E., OSRAY, "Conception de systèmes résistants à l'environnement spatial" (1990).

[2] J.P. Estienne and J.G. Ferrante, RADECS 89: "Méthodes d'analyses et d'optimisations contre les rayonnements pour les véhicules spatiaux", Annales de physique Colloque N°2, Supplément au N°6, vol 14 (Dec 1989).

[3] "Spacecraft internal electrostatic discharge," internal document MATRA (1989).

[4] Internal electrostatic charging," internal document MATRA: Response to A.O ESA 1-2381/90/NL/JG (1990).

[5] "I.T.S.: Integrated Tiger Series of coupled e$^-$/ph monte-carlo code system" Halbleib and Mehlorn, Sandia lab.

[6] J. Seille and J.P. Estienne, "Internal electrostatic charging analysis and simulation: logiciel MATBULK" internal document MATRA.

[7a] J. Seille, "PIC method for ESD:" internal document MATRA (1987).

[7b] "Trajectography of space charge during an ESD c" Contract CNES (1988).

[8] J.P. Estienne, A. Baradat (MATRA), Mattuci (PROEL-italie), "A combined theoretical and experimental anaysis of worst case on-orbit e$^-$ spectrum for bulk-charging" presented at ESA EMC workshop EMC (Dec 1990).

[9a] J.P. Estienne and J. Seille, "Software RADIATE: Coupling of an EM wave on various loaded lines with aperture," Reference V 1.1 89.

[9b] J. P. Estienne, "Modelisation of the response of lines with any kind of loading for an impinging EM wave or an ionising pulse" MATRA doc 139/JPE/030.86.

[9c] J.P. Estienne, J.G. Ferrante, J. Seille, "Computer simulation of diffraction of an E,H field through an aperture and coupling on a transmission line" conference paper, IEEE TURIN (1989).

[10a] G. Garres and J. Seille, "MATEMC user reference manual," MATRA document TS/101/RT/06.87C.

[10b] F. Holderer, J.P. Estienne and J. Seille, "Study of new harness models for MATEMC software - Time and frequency domains" Doc MATRA J. (1989).

[10c] J. G. Ferrante, J. Seille, C. Garres and J.P. Granger, "Approximate methods for EMC circuit analysis in the time domain", conference paper IEEE 1989 TURIN.

[11] "Numerical electromagnetics code (NEC) Method of moments" LLL Technical notes Doc 116, Vol 1, NOCSC, San Diego, CA J (1981).

[12a] H.H. Chao and B.J. Strait, "Computer programs for radiation and scattering by arbitrary configurations of bent wires," Syracuse University (Sept 1970).

[12b] F.J. Dedrick, "WAMP: Wire antenna modeling program," California University (Dec 1973).

[13a] R.M. Bevensee, "A thin wire computer code for antenna or scatters with pulse expansion functions for current," LLL F (1976).

[13b] J.H. Richmond, "Thin wire antenna over a perfectly conducting ground plane," Ohio State University (Oct 1974).

[13c] J.H. Richmond, "Computer program for thin wire structure in a homogeneous conducting medium," Ohio State University (J 1974).

[14] G. Dhatt and G. Touzot, "The finite element method displayed," J. Wiley and Sons, NY.

[15a] R. Perala, T. Rudolph and F. Eriksen, "Electromagnetic interaction of lightning with aircraft," IEEE, Vol EMC24 n°2 (May 1982).

[15b] C.L. Bennett, "Integral equation solution," Sperry Research Center (Oct 1975).

[16] C.L. Bennett, "Space-time integral equation approach to the large body scattering problem," Sperry Research Center (M 1973).

[17] R. Mittra, "Computer techniques for electromagnetics," Springer Verlag, NY.

[18a] J.P. Estienne and G. Picart, "Trajectography of space charge emitted during an ESD," Contract CNES 840/CNES/87/4858/00 – Final Report.

[18b] T.W. Roberts, "The behaviour of flux difference splitting scheme near slowly moving shock wavec," Journal of computational physics, Vol 90/N°1 90.

[19] "Mixed finite elements in R^3 – NEDELEC Numer. Math. 35, 515-341 (1980).

[20] F. Hildebrand, "Finite difference equations and simulations, Prentice Hall, New Jersey.

[21] "Numerical solution of initial boundary value problems involving Maxwell equations in isotropic media," YEE IEEE T.A. and P. Ap 14, pp 302-307 (May 1966).

[22a] "The aribitrary body of revolution code ABORC - A. WOODSn N. Delmer INTEL-RT-8141-028.

[22b] T.A. Tumolillo and J.P. Wondra, "MEEC-3D: A computer code for self consistant solution of the Maxwell-Lorentz equations in 3D," Jaycor IEEE NS, Vol NS14, N°6 (Dec 1977).

[22c] R. Holland, "THREDE code," IEEE T.N.S. NS24 pp. 2416-2421 (Dec 1977).

[22d] R. Holland, L. Simpson and R. John, IEEE T. NS, NS26, pp 4964-4969 (Dec 1979).

[22e] R. Holland, L. Simpson and K. Kunz, IEEE EMC, EMC22, A (1980).

[22f] R. Holland and L. Simpson, "Implementation and optimisation of the thin-strut formalism in THREDE", MRC IEEE, TNS, NS27 N°6 (Dec 1980).

[23] "Field computation by moment method," R.F. Harrington and R.E. Krieger publishing company, Malabar, Florida (1982).

[24] R. Wilton, M. Rao and W. Glisson, "Electromagnetic scattering by surface of arbitrary shape," Syracuse University, NY.

[25] L. Medgyesi-Mitschang, "Radiation from wire antenna attached to bodies of revolution: the junction problem," IEEE A&P Vol AP29, N°3 M (1981).

[26a] S.M. Rao, "Application to the method of moments to electromagnetic scattering by surfaces of arbitrary shape," Rochester Institute of Technology.

[26b] S.V. Yesantharao, "EMPACK - A software toolbook of potential integrals for computational electromagnetics," University of Houston.

[26c] "Computation of skin currents for an impinging wave on satellites," Stage report ENSEEIHT A. Baradat Ref S374/RT/18.89.

[27] M.V. Blaricum and E.K. Miller, "TWTD: A computer program for the time domain analysis of thin wire stuctures," LLL (1972).

[28] R. Mittra, "Numerical and asymptotic techniques in electromagnetics," Springer Verlag.

[29a] J.P. Estienne, "Time wire-frame MATRA simulator," internal document MATRA.

[29b] J.P. Estienne, A. Baradat and J. Seille, "Evaluation of surface currents in time domain for illumination and injection cases," Part 1: Theory Ref S413/NT/31.90.

INTERNAL CHARGING AND SECONDARY EFFECTS

MANOLA ROMERO, LÉON LEVY
CERT-ONERA/DERTS
2, Avenue Edouard Belin
31055 Toulouse Cédex
FRANCE

ABSTRACT. The charged particle environment of spacecraft includes as well very low energy plasma (ionospheric cold plasma), storm particles of a few keV and high energy particles trapped in the magnetospheric belts. Initially, only the so called storm particles were held responsible for surface charging and anomalies. Since 20% of the recorded discharges on SCATHA occurred without surface charging, trapped high energy particles are also a matter of concern and interest.

These particles have of course very different ranges within materials and many electrostatic situations can be produced according to the particles and materials.

Two questions are important to understand what happens when voltages are built and then when discharges occurs:

1) What are the properties of materials and configurations that are meaningful in order to observe high internal fields in the system? For example, what are the various forms of conductivity (bulk, surface, radioactivated...) and how can they be measured in the laboratory.

2) What happens to the charge when the voltage builds up (where are they...) and when the discharge occurs (single and double layers, where do the charges come from during discharges...)?

1. Introduction

Some satellites, such as DSP and MARECS A (see Fig. 1) present a time distribution of events within the 0–6h range which is characteristic of surface charging occurring during substorms. Others, such as Symphonie B and Telecom 1-A [Ref. 1] present a quite different feature, inferring that a different charge-discharge mechanism is involved.

Skynet 2-0B is an example of quasi isotropic time distribution in which the efficient part of the satellite environment is not obviously correlated with local time, as is most of the low energy particles at geostationary orbit. In SCATHA, 20% of the discharge recorded by the TPM (Transient Pulse Monitor) experiment occurred without surface charging, and were equally distributed through the orbit, and not correlated with substorms.

R. N. DeWitt et al. (eds.), The Behavior of Systems in the Space Environment, 565–580.

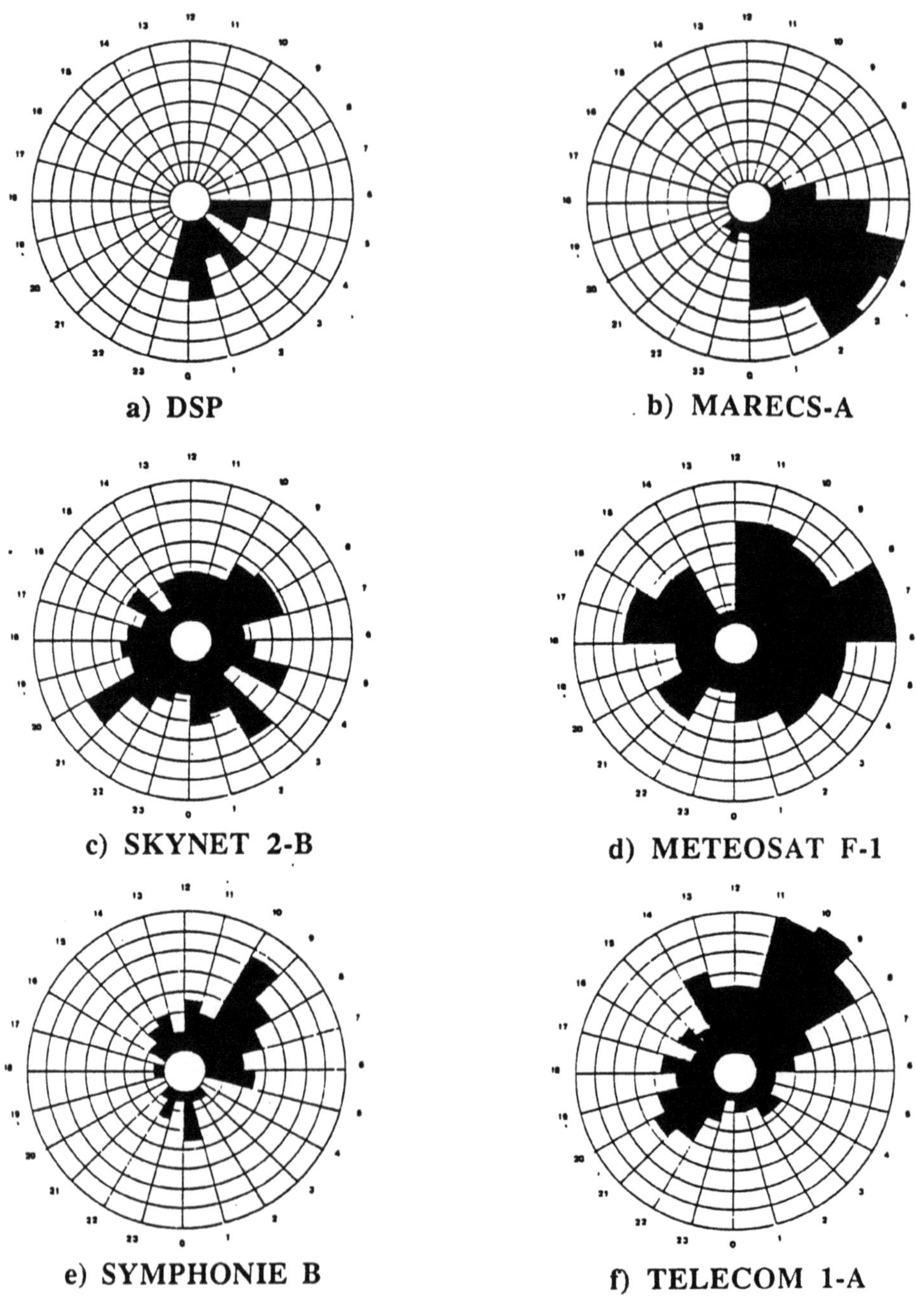

Fig. 1: Chronograms of events/local time for various satellites.

In order to understand these various behaviors, concern and interest are to be given to problems such as:

- what happens when a part of a satellite is exposed to a given environment?
- what happens to the charges when a discharge occurs?

2. Charge Build Up and Discharges

2.1. CHARGE BUILD UP

When a dielectric material is exposed to an electron beam, its surface voltage increases until it reaches either a maximum value or a discharge threshold.

The surface voltage is a consequence of the balance of various currents such as incident beam (depending on the voltage), secondary emission (depending on the material, the energy of the beam and the surface voltage), and leakages depending on the conductivity (bulk and surface) of the material – itself a function of the irradiation on going and the charge repartition – and the geometrical configuration.

The Fig. 2a presents the equivalent RC circuit of the sample. Due to the high time constant of this circuit ($\tau = \rho.\varepsilon$ is about several hours long), the charges embedded are durably trapped within the material, and the voltage due to these charges increases,

$$v(t) = Q(t)/C = \frac{1}{C}\int J(t)dt \qquad (Fig.2b)$$

Hence, the charge repartition has effects both on the resulting current J through the conductivity and the capacitance C of the sample (for instance, a layer of charge embedded in a dielectric can be considered as a capacitance formed by a series of two capacitances respectively between the layer and the front side and the layer and the rear side).

Discharges occur when V(t) reaches a critical value, usually around -10000 Volts. The discharge is a physical process in which a fast transfer of charges produces current transients.

We will describe several of these physical processes; for the sake of simplification we will consider only electron transfer. The geometrical configurations used as a reference are described in the Fig. 3. Samples 1 and 2 are connected to the satellite structure symbolized by its capacitance C_0 and its voltage V_0 with respect to the ambient space. Sample 3 schematizes a small patch of thermal coating in which the front metallization has not been grounded to the structure.

2.2. DIELECTRIC BREAKDOWN

When a voltage is applied to a dielectric and no other opportunity of charge transfer than through the material is possible, at last, when a threshold value Vs is reached, violent charge transfer occurs along a few tracks crossing the material which is highly ionized and even pyrolized along these tracks.

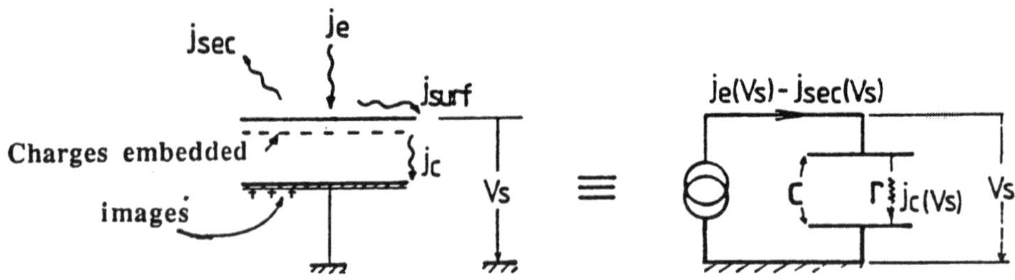

$$C dV_s/dt = j_e(V_s) - j_{sec}(V_s) - j_c(V_s)$$

Fig. 2a: R-C equivalent circuit of sample.

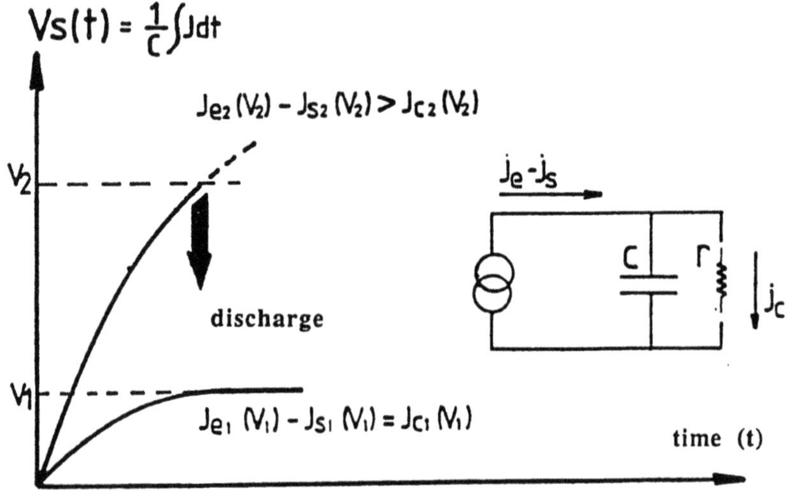

Fig. 2b: Charge build up.

The surface potential is neutralized. The physical processes by which the neutralization occurs are a matter of research and debate. Some models postulate the emission of the embedded electrons. It is more likely that the surface neutralization is the result of the reattraction on the surface of the ions that are produced in a plasma formed in some area of the material [Ref. 2]. Surface and sub-surface channels have been photographed at the S.E.M. [Ref. 3], showing evidence of the transformation of the solid material into plasma. This phenomena is labelled by a in the Fig. 3.

The threshold values of the voltages are usually very high in the lab, and almost none of the cases of charge, measured in the SCATHA

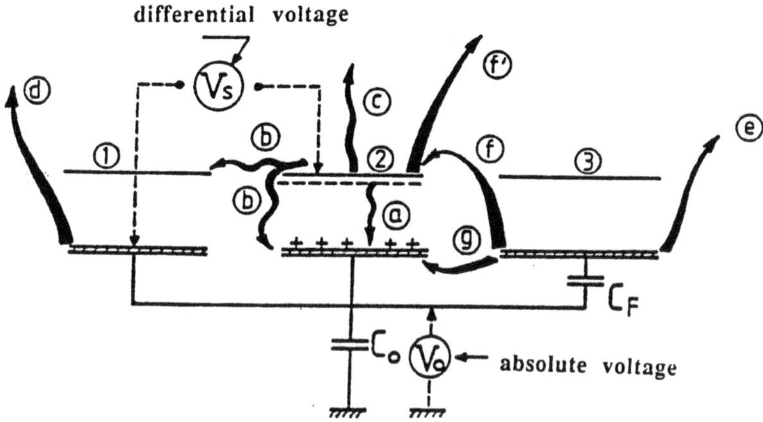

Fig. 3: Three classes of sample mounting, with schema of the electron paths for various kinds of discharge.

experiments were high enough to compare with them (but one, the 10th of June 1980, in which no discharge occurred).

2.3. INVERTED GRADIENT

In some cases, floating metallizations or floating cables can get a negative charge with respect to the nearby dielectrics. Then, edge effects (or field emission) can trigger a discharge which usually – due to the smallness of the floating metallization – does not contain many charges but – due to draining of these charges by the highly conductive metal – does have high maximum currents I and very high dI/dt [Refs. 1, 4]. These discharges are very dangerous, having a high capacity to link with the satellite circuitry.

2.4. FLASH OVER

On an experiment set-up like the following (see Fig. 4), only charges leaving the material (blow-off) can be seen by the measuring system; a probe current around the wire which grounds the holder.

Charges leaving the sample and that would be collected by the holder or that slide along the surface as "flash-over" currents cannot be detected by the probe. Moreover, when the total charge emitted by blow-off is compared to the whole charge lost by the sample during an event (the surface potential probe provides the information), it is observed that more charge is lost during a discharge event than is emitted as blow-off. From this, the flash-over current existence is deduced.

This process supposes that at some point a charge transfer is triggered, for instance through a dielectric breakdown or, more probably, due to edge effects.

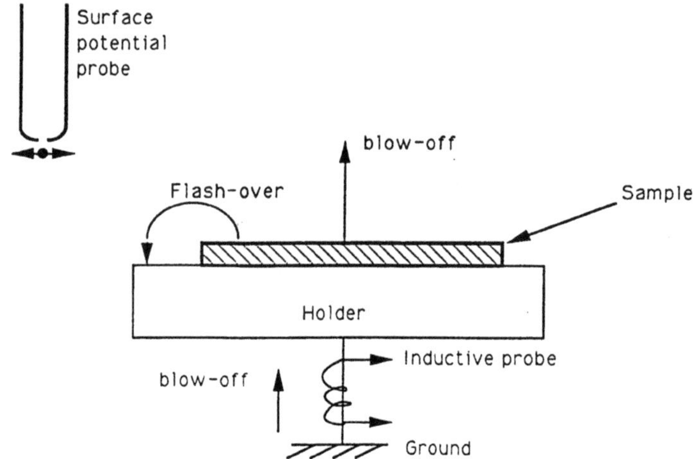

Fig. 4: Sample mounting for flash over test.

At that point the electric field, initially oriented perpendicular to the surface, becomes parallel due to the charge drain. Hence, a high electric field appears and plasma outgassing from the material, or the metallization, or both, is created. This plasma interferes with the surface whilst charges are created by the emission stimulated by the ions of the plasma, a propagation of the plasma at the surface occurs, so called "bush fire arc discharge", and an emitted current is observed.

This process, schematized by b on Fig. 3 and by Fig. 5 was described by Inouye [Ref. 5].

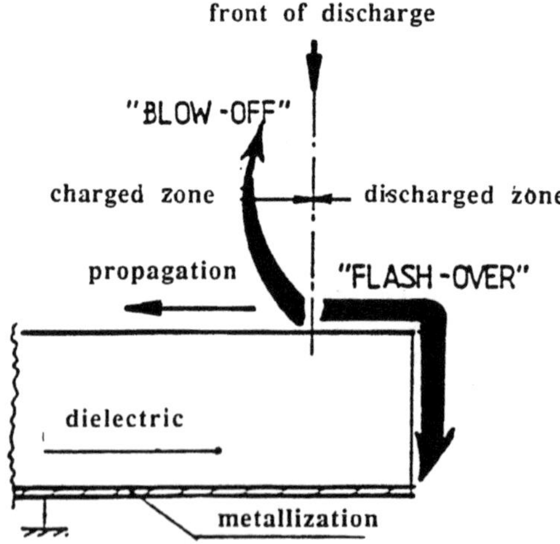

Fig. 5: Propagative discharge according to Inouye.

2.5. BLOW OFF

In that case, when the discharge is initiated, part of the negative charges is expelled toward the space. Different configurations can occur such as (see Fig. 3) dielectric surface to space (c), satellite structure to space (d), floating metallization to space (e), floating metallization to dielectric surface then to space (f,f'), dielectric surface to metallization then to space (b,d)...

The common consequence of all these cases is that the global charge of the satellite is changed and hence its absolute voltage, too, and a space charge is created all around. Then, the probability that electromagnetic interferences occur is increased, due to the fact that a larger surface of the satellite is exposed to the EM emission of the surrounding plasma.

2.6. CONSIDERATIONS UPON THE PRE-DISCHARGE AND THE POST DISCHARGE STATES

It has been stressed that discharges occur in orbit at a voltage low when compared to the dielectric breakdown threshold of the sample. Several ideas have been explored to answer this question when no floating metallizations or inverted gradient are involved.

Meulenberg [Ref. 6] has assumed that the charges could form a double layer (Fig. 6) with electrons embedded, so that an electric field large enough to trigger a discharge could appear between the charge layers. Instead, what was observed is the formation of a double layer after the discharge has occurred.

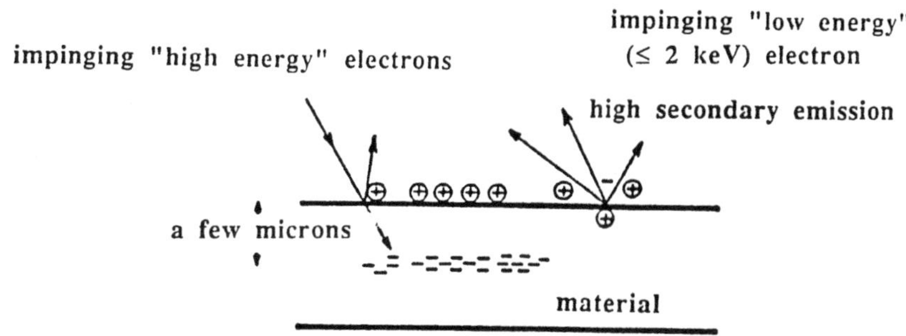

Fig. 6: Double layer according to Meulenberg (ions at the surface and electrons embedded).

The experiments that have been performed to investigate this possibility are [Ref. 2]. The evaluation as a function of the time of an electric surface potential Vs(t) has been studied after it has been submitted to three sequences of charging by electrons and discharging:

a. Charging/cutting off the beam/recording Vs(t).

b. Charging/Discharging by closing briefly a switch/reopening the switch/recording Vs(t) (beam off)

c. Charging up to the discharge obtention/recording Vs(t) (beam off).

Figure 7 gives a schematic and simplified description of the various experiments performed as well as the observed Vs(t) behavior.

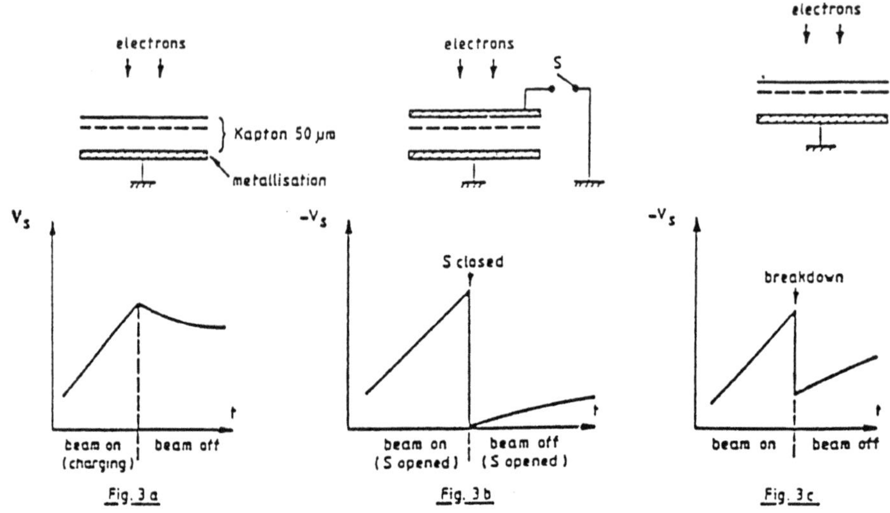

Fig. 7: Evolution (as a function of time) of a dielectric surface potential Vs, submitted to different sequences of charging and discharging.

The first experimental sequence (a) leads to a surface voltage Vs(t) decreasing with time (in absolute magnitude). The second sequence (b) allows the observation of the "return voltage" build up after the surface has been grounded. This "return voltage" build up is an increase of the measured voltage without any new injected charge. It has been observed and modeled by J.M. Siguier et al. [Ref. 2].

The third sequence (c) shows the return voltage build up after occurrence of a blow-off discharge on the surface of the dielectric. The resemblance between sequences (b) and (c) indicates that blow-off produces a positive charge layer on the sample surface and that, afterwards, embedded electrons migrate towards the surface under the effect of the electric field radiated by the positive charge on the free surface.

It must be emphasized that the currents collected during the discharge consist of both electrons and ions blow-off (emitted) by the charged dielectric surface. Each species is observed depending on the direction of the electric field at the sample surface and underlines the importance of the "floating" testing configuration which allows for holder voltage variation.

The magnitude of the surface potential is about −7000 V before the discharge, and falls to −1700 V after the blow-off. It first increases by 1000 V and 200 V according to the incident energy (20 and 30 keV resp.), then starts decreasing monotonically after about 100 mm.

This can be regarded as a "return voltage" build up according to reference 2 combined to a natural decay. Such a behavior can be explained by the two electric fields existing inside the dielectric after the blow-off which drift the embedded charges towards the surface or the backside, depending on its positions. Compensating the "return voltage" build up by the natural decay due to the conduction, we get results very similar to the "return voltage" build up after a short-circuit of the sample.

Using the model of reference 2, we find that for the 20 keV electrons, the amount of embedded charges is q = 4.E-7 C/cm^2 and the depth of the charge centroïd with respect to the free surface is l = 1.4 mm. For the 30 keV electrons we find that q = 4.4E-7 C/cm^2 and l = 4 mm. The agreement between the evaluated embedded charges before discharge and the measured "return voltage" build up is fairly good.

As outlined by P. Coakley et al. [Ref. 7 and 8] it is important to take into account the real orbital spectral distribution of electrons for lab experiment. The table hereby enclosed (extracted from Ref. 7) present the breakdown threshold as observed for various spectra of incident beam.

MATERIAL	MONO-ENERGY (keV)	MULTI-ENERGY	
Kapton perfore 130 mm	-6,5	Discharges suppressed	
Kapton 50 mm	-13	Discharges suppressed	
OSR S 200	-6à-12 according energy	-0.9 -3	-2.8 (a) -2.5 (b)
Mylar 50 mm	-13	-1.9 -2.4	-3.5 (c) -2.2 (d)

SPECTRA:
(a) [25 keV/1 nA.cm^{-2}] + [3 keV/5 nA.cm^{-2}]
(b) [25 keV/0.3 nA.cm^{-2}] + [5 keV/1 nA.cm^{-2}]
(c) [35 keV/1 nA.cm^{-2}] + [4 keV/7 nA.cm^{-2}]
(d) [60 keV/0.3 nA.cm^{-2}] + [5 keV/3 nA.cm^{-2}]

3. Internal Electric Field

The previous considerations stresses the importance of the charge repartition and the subsequent electric field and conductivity.

3.1. PARTICLES SPECTRA AND RANGES

The space spectra indicate current densities which are decreasing functions of the energy. With most of the particles (90% of them) of energies less than 50 keV, deposition will be near the surface (within the first 15 microns).

Some typical spectra, as measured by SCATHA, are reported in the Fig. 8. They represent various kinds of situations: quiet, substorm, post substorm. A simulation spectrum enhanced in the 100-150 keV file that can be obtained on the facility described in paragraph 4 is mentioned so as to point out that all kinds of situations can be taken into account during ground experimentation.

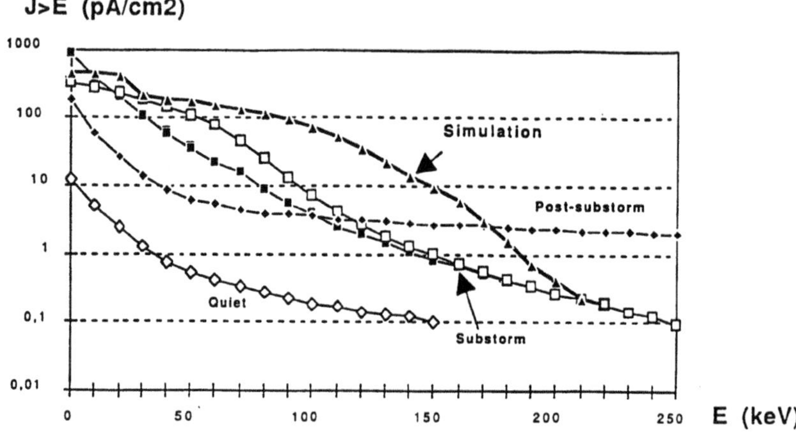

Fig. 8: Electron spectra as observed by SCATHA (+ one spectrum used for ground simulation).

The Fig. 9 presents the range observed in Kapton for monoenergetic electrons impinging normally the surface with energy between 1 to 100 keV. For other materials, except Hydrogen and heavy elements, a rather good approximation of the range is obtained assuming that the penetration is equal for all materials when the depth is expressed in g/cm^2.

3.2. STORED CHARGE, ELECTRIC FIELD, CONDUCTIVITY, LEAKAGE CURRENT

This paragraph does not aim at giving precise results for such material but the guidelines for an eventual computation.

a) *Stored charge, electric field*

The rate of charge deposited in the material can be calculated from the continuity equation, in one dimension:

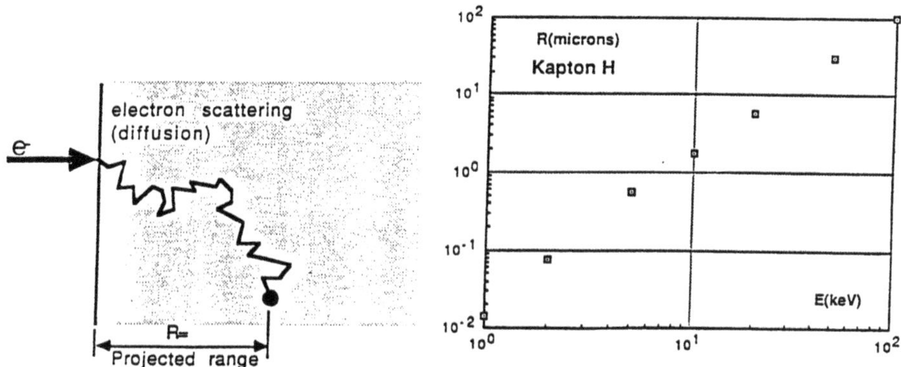

Fig. 9: Energy range relation of electrons in Kapton.

$$- \frac{\partial \rho}{\partial t} = \frac{\partial J}{dx}$$

where:

x = depth in the dielectric,
r = charge density,
J = current density.

The current density appears as the sum of the conduction current Jc (proportional to the electric field) and of the incident current Ji(x)

$$J(x,t) = Jc(x,t) + Ji(x,t)$$

where:

Ji(x,t) = incident current transmission (that could be extracted from the figures given in 3-1), and

$$Jc(x,t) = (\sigma_D + KD^\circ(x))\ E(x,t)$$

where:

σ_D = dark conductivity,
K = radiation induced conductivity coefficient,
$D^\circ(x)$ = dose rate profile (first step).

So, the conductivity is a strongly varying function of depth in the dielectric.

The charge density $\rho(x,t)$ can be determined as:

$$\rho(x,t) = \rho(x,0) + \int_0^t \frac{\partial \rho}{\partial t'} dt'$$

The electric field results from the application of Poisson's law:

$$\frac{\partial E}{\partial x} = \frac{\rho(x)}{\varepsilon} \text{ and } E(x) = E(0) + \frac{1}{\varepsilon} \int_0^x \rho(x).dx$$

where:

V - potential
ε - sample permitivity
E(0) - electric field at the surface (x=o).

b) *Computation of the induced conductivity*
In the irradiated zone, the expression of the radiation induced conductivity will be taken as:

$$\sigma = \sigma_D + KD^{\circ} \text{ or } \sigma = \sigma_o + KD^{\circ\Delta}$$

where:

σ_D - dark resistivity
K - radiation induced conductivity coefficient
D° - dose rate (computed from 3-1). It is a strongly dependent function of the depth (x).
Δ - a coefficient < or ≈ 1 that is suggested in the literature.

In the non-irradiated zone, however, there is also an induced conductivity due to the local field. The conductivity for instance in most polymers and dielectrics is a strong function of the electric field through the Poole-Frenkel effect which allows detrapping by lowering the potential barrier of the traps.

L. Levy [Ref. 9] has developed a method based on the surface potential decay. Results obtained using this method have confirmed that the conductivity depends strongly on electric field. We will assume this field dependence to be the same as that used in the NASCAP code:

$$\sigma(E) = \frac{\sigma_0}{\varepsilon} \left(2 + Ch \left\{ \frac{\beta_F \sqrt{E}}{2\ kt} \right\} \right)$$

where

$$\beta_F = \left(\frac{q^3}{\pi \varepsilon}\right)^{1/2} = \textit{Poole-Frenkel coefficient}$$

σ_o = bulk conductivity of a sample not exposed to any radiation and not subjected to any internal electric field (input data n°3 of the materials properties of NASCAP).
q = charge of the electron
ε = dielectric permitivity

c) *Computation of the leakage current*

The leakage current, computed in the non irradiated area will be computed as:

$$J_L = \sigma(E).E_{(x>R)}$$

with R = maximum penetration depth of the electrons, and such that for x>R, $\rho(\chi) = 0$.

4. Description of an Experimental Array Used for Deep Charging Study

The SIRENE facility of CERT/DERTS is described here as a mere example of what could be an experimental array dedicated to deep charging effect. The main and specific characteristic of SIRENE is to yield an electron beam with a spectral distribution corresponding to a substorm environment with enhanced high energy particle fluxes. In order to do so, two electron guns in the range of 0–30 keV for the lowest energies and of 0–250 keV for the highest energies are used. The spectrum really impinging the sample is refined using a multifoil scattering technique. Any reference spectrum can be obtained with a rather good approximation; the spectrum most used being the spectrum observed on SCATHA September 22, 1982; selected since it was the richest with electron energies up to 100 keV. Figure 8 presents an example of an enhanced spectrum that can be reached.

In order to study sun-aspect related anomalies, transient states produced by shadow-sun passages can be studied simulating photoemission by a proton beam. The table hereby enclosed summarizes the main characteristics of this facility. Tests have been performed on layers of pinched Teflon, metallized and with adhesive at the rear face. Whilst irradiated with a monoenergetic beam of 30 keV, such samples always undergo discharges.

When irradiated with a beam of energy distributed between 0 and 200 keV – with the same intensity of 450 pA/cm^2 – different behaviors are observed depending on the quality of the rear contact and the thickness of the sample (more precisely of the fluctuation of thickness around the nominal value). With a good contact and a rather thin sample, no discharge is observed; underlining the beneficial effect of penetrating particles on bulk conductivity.

OBJECTIVES:
*Study of potential hazard charging outside and inside the satellite *Study of voltage building on spacecraft dielectric materials and equipment samples for geostationary orbit, evaluation of materials properties and protections; *characterization of discharges; *study of solar orientation effect through simulation of photoemission by proton irradiation; *measurement of natural and radiation induced conductivity of materials
FACILITIES:
*Pressure ≈ 10^{-6} hPa; *Spectral distribution of electrons (0–200 keV, 0–1 nA/cm^2) representative of one day of activity tbd as required; *monoenergetic protons within the 10–150 keV range (0–500 nA/cm^2); *sample capacity of 15 x 15 cm^2; *instrumentation with photography, currents and voltages probes (sampling at 1 ns, bandwidth, 300 MHz).

Other tests performed on samples shielded – representative for instance of cables – have shown the detrimental effect meaning the presence of discharges only with an extended spectrum.

These kind of facilities can be used to measure the conductivity of dielectric materials, either natural or radiation induced. To do so, the sample is charged using the adequate electron beam, and the natural decay of the surface voltage is observed with the voltage probe which has the property that it does not alter the charge repartition and does not trigger discharges.

5. Conclusions

Historically, ESD on satellites were considered only as a surface effect and strong correlations with the local instantaneous environment – and more precisely with substorm plasmas – were observed. Nevertheless, for real orbital cases, these correlations were not that easy to establish, due to the superposition of other factors such as sun exposure and condition of coupling of the EM emission of discharge with the internal circuitry of the spacecraft.

Since then, two trends of evolution have been observed. One is due to the fact that the present missions of orbital systems tend to use electronic components integrated on a larger and larger scale, of high cadences, of large bandwidth, with the consequence that these components are more sensible to perturbation and especially to currents – even low – with a high frequency. The other is the progression of the knowledge concerning surface discharges, and the elaboration of conception rules for satellites aiming at a fair reduction of the events observed. By the implementation of the appropriate guidelines, spacecraft are already more immune than they were in the past. The effort has now to be put on the persistent few "non typical" remaining anomalies.

Hence, many of the events presently observed are due to processes involving embedded charges and strongly depending on the properties of the materials, and of the evolution of a rather badly known significant environment – meaning the real environment observed by the real satellite, and not mean values – of electrons from 0 to approximately 300 keV cumulated during several hours, if not days. The processes also imply small internal parts of the satellite, such as floating threads within cables, the presence of which is not always well known. As a result, further advances in the field will be obtained only by careful modeling of the environment and its fluctuations, precise experiments on discharge mechanisms and material properties measurement on adequate facilities, and complete documentation of spacecraft at all steps of their development and orbital life.

References

[1] L. Lévy, D. Sarrail, J.P. Philippon, J.P. Catani, J.M. Fourquet, *"Sur la possibilité de Charge Différentielle de plusieurs Kilovolts dan le secteur Jour de l'Orbite Géosynchrone"*, The Aerospace Environment at High Altitude and its Implications for Spacecraft Charging and Communications (1988).

[2] J.M. Siguier, J. Marco, L. Lévy, D. Sarrail and R. Coelho, *On the "Return Voltage" Buildup Following an Electrostatic Discharge on Space Insulating Materials"*, 13th International Symposium on Discharges and Electrical Insulation in Vacuum (1988).

[3] K.G. Balmain and G.R. Dubois, *"Surface Discharges on Teflon, Mylar, and Kapton'*, IEEE Transactions on Nuclear Sciences NS 26 (1979).

[4] R. Reulet, J.M. Siguier, D. Sarrail and L. Lévy, *"Etude de Charge d'Eléments Internes et Externes de Satellites"* CERT Report CR/CLAQ/22 (March 1990).

[5] Inouye, *"Brushfire ARC discharge model"*, TRW Defence and Space System Group, 1980 NASA CP 2182, p. 133-162 (1981).

[6] A. Meulenberg, Jr., *"Evidence for a new discharge mechanism for dielectrics in a plasma"*, Spacecraft Charging by Magnetospheric Plasma, Ed. A. Rosen, Progress in Aeronautics and Astronautics, Vol 47, p. 237, Martin Sommerfield Series Ed.

[7] P. Coakley, B. Kitterer and M. Treadaway, *"Charging and discharging characteristics of dielectric materials exposed to low and mid-energy electrons"*, IEEE on NS. NS 29(n°6) (1982).

[8] P. Coakley, M. Treadway, N. Wild and B. Kitterer, *"Discharge Characteristics of Dielectric Materials Examined in Mono-, Dual-, and Spectral Energy Electron Charging Environments"*, Spacecraft Environmental Interactions Technology 1983 – NASA Conference Publication 2359 – AFGL-TR-85-0018:511-524 (1983).

[9] L. Lévy, D. Sarrail and J.M. Siguier, *"Conductivity and secondary emission properties of dielectrics as required by NASCAP"*, Proceedings of the 3rd European Symposium on Spacecraft Materials in Space Environment, Noordwijk, ESA SP 232, Nov. 1985.

TETHER PHENOMENA OBSERVED IN THE OEDIPUS-A EXPERIMENT

H. G. JAMES
Communications Research Centre
Department of Communications
Ottawa, Ontario K2H 8S2 - Canada

ABSTRACT. The OEDIPUS-A experiment was undertaken to address three areas in basic ionospheric science: weak field-aligned dc electric fields in the auroral ionosphere; large electrostatic probes; and bistatic wave propagation. An instrumented double payload was launched on a Black Brant sounding rocket from Andøya, Norway on 30 January 1989. An electrically conducting tether between the payload halves was extended to a length of 960 m on a trajectory with an apogee of 512 km. The flight produced novel data in all three basic-science areas. It also illustrated some characteristics of the operation of tethers in space. From among the results, a number of phenomena have been selected which are of general interest regarding the behavior of large electrodynamic structures in the ionospheric plasma.

1. Introduction

The OEDIPUS-A payload was launched from Andøya, Norway during an active auroral display on 30 January 1989. After being placed on a suborbital trajectory by a three-stage Black Brant X rocket, the payload separated into two instrumented subpayloads. An important feature of this experiment was a long insulated conducting tether connecting the two subpayloads. This system was used to make passive observations of the ambient auroral plasma and to stimulate the plasma. The major scientific objectives of this project were:

- to make passive observations of the auroral ionosphere, in particular, the natural magnetic-field aligned dc electric field $\mathbf{E}_{||}$ utilizing a large double probe;
- to measure the response of the large probe in the ionospheric plasma; and
- to seek new insights into plane- and sheath-wave propagation in plasmas.

R. N. DeWitt et al. (eds.), The Behavior of Systems in the Space Environment, 581–603.

Initially it was planned to address all three objectives equally on the OEDIPUS-A flight. After the payload design was compared with available rocket capabilities, it was decided that the project would stress the second and third objectives. Then, when a more powerful rocket became available that could lift the payload to ionospheric altitudes where the relatively weak $\mathbf{E}_{||}$ was certain to be observed, a second flight would take place with the accent on the first objective.

The results from the OEDIPUS-A experiment are providing new insights into the electrodynamic properties of large space structures. These results complement the findings of other completed or planned investigations into spatial tethers. Both the MAIMIK (Ref. [1]) and CHARGE (Ref. [2]) experiments centred around plasma-probe interactions stimulated by on-board energetic electron guns, whereas the OEDIPUS acted as large classic double probe and applied active stimulation by means of electromagnetic fields. The planned USA-Italy Tethered Satellite System (Refs. [3] and [4]) will operate on NASA's Shuttle at low and mid latitudes. It is principally concerned with electrodynamical processes, such as that induced by the $\mathbf{v} \times \mathbf{B}$ potential, obtained on large structures orbiting through a magnetosplasma.

The objective of this paper is to discuss new results from OEDIPUS, both to convey some scientific notions and to comment on the ionospheric plasma as an environment for experiments. The theme that is common to most of the presentation is the role of the sheath surrounding solid bodies in plasmas. All spacecraft orbiting through the ionospheric or magnetospheric plasma have sheaths which, through electrostatic or electrodynamic processes, can affect the operation of the spacecraft. This discussion is intended to provide an introductory understanding of parts of the experiment. References are included to literature that can be pursued for details.

Section 2 deals with the characteristics of the OEDIPUS-A experiment mission and flight conditions. The payload design is overviewed because it reflects the different kinds of measurements and environmental conditions expected. Sections 3 and 4 are devoted to two different results from the experiment. Section 3 relates to the original theme of OEDIPUS: the tethered payload as a double probe for diagnostics of the auroral ionospheric plasma. This part discusses how the requirement for measurement of $\mathbf{E}_{||}$ in the high-latitude ionosphere led to a requirement for active tests of the double probe. Section 4 describes how the data from the transmitter-receiver pair HEX and REX were used in research on sheath waves on the tether.

2. General Description of Experiment

2.1. PAYLOAD SCIENTIFIC INSTRUMENTS

The OEDIPUS-A instrumentation was mounted in two cylindrical housings having the usual diameter for Black-Brant sounding rocket payloads, 43 cm. Each subpayload was about 2 m long, and the total payload mass was 266 kgm. Both subpayloads carried equipment for passive measurement of auroral processes and other equipment for active experiments. In the first category were energetic particle detectors, triaxial flux-gate magnetometers and plasma (Langmuir) probes. One of each was on the forward subpayload and one on the aft. Passive measurements were also carried out by a dc-VLF electric field probe connected to a 5-m dipole on the aft portion. A package called the Tether Control and Monitor (TCM) was located in the aft subpayload, and operated the ensemble both for passive current and voltage measurements on the tether and for active determination of the double-probe response characteristic. Finally a VLF-HF receiver, REX, was connected to the aft 5-m dipole, for monitoring background radio fields. Current-voltage response data recorded by the TCM are the subject of Section 3. Section 4 is concerned with radio frequency data recorded by the REX when it worked synchronously with a low power transmitter HEX on the forward payload.

All of the TCM, HEX-REX and plasma-probe modes were synchronized within a major frame that repeated every 6 s throughout the flight. The major frame was composed of 12 minor frames, each of 500 ms length, during which the TCM, HEX and REX executed one or other of their modes. Temperature-stabilized oscillators provided the master signal for clocks controlling these instruments. On the flight upleg before payload separation, the two oscillators were hard-wire connected for exact phase synchronism. After separation and throughout the remainder of the flight, the stability of the now independent clocks assured event synchronization among the TCM, HEX and REX. This permitted efficient use of the flight time through rapid sweep modes in the three instruments. The other instruments were not mutually synchronized, but rather ran freely throughout the flight.

2.2 PAYLOAD SUBSYSTEMS

Several important subsystems were required for the unique objectives of OEDIPUS A. Four of the most important are described briefly here: an attitude control system for magnetic-field alignment, a tether-spool system, a propulsion separation system and a relative attitude measurement of the separated payloads. Other standard rocket support systems such as the electrical power and telemetry systems are not described.

2.2.1. *Attitude Control System (ACS).* Experiment requirements called for the alignment of the unseparated payload axis to within 1° of the earth's magnetic field **B** prior to fore-aft payload separation at an altitude of about 285 km. NASA designed and supplied the ACS which used a 3-axis magnetometer as an attitude sensor. Argon thrusters provided pitch and roll correction.

2.2.2. *Tether Spool System.* The principal functions of the tether-spool subsystem were to deploy about 1000 m of conducting wire in a controlled manner and to provide a continuous measurement of the deployed wire length to ±1%. The heart of the system was a rotating spool holding 1300 m of 24-guage Teflon-coated wire. Slip rings on one side of the spool communicated signals to and from the rotating spool. A magnetic hysteresis brake on the opposite side applied a constant torque of 7.5 oz-in as the wire was paid out. Rotation of the spool, and hence total length of wire deployed, was monitored by an optical sensor count of notches on the spool flange. Rf isolation of the spool from the payload ground was necessary for experiments with the HEX and REX. This was achieved by mounting the spool assembly on nylon standoffs and by isolating the brake from the spool with nylon couplers. A spool-to-ground resistance greater than 10^{11} Ω was obtained.

2.2.3. *Separation Subsystem.* An Argon-gas separation subsystem was required to provide sufficient relative momentum to separate the two payload halves to the final distance of 958 m. The gas system utilized two matched nozzles mounted 180° apart in the forward payload. Initial separation of the subpayloads was accomplished using four small springs installed around the fore-aft separation plane. Immediately thereafter, the Argon was vented through the nozzles, developing an initial separation thrust of about 30 lb.

2.2.4. *Relative Position Measurement System.* The angle (**T**,**B**) between **B** and the tether wire direction **T** was an important variable in the physical interpretation of the tether physics. It was required to determine the off-axis position of the forward payload referred to the aft payload and the relative roll (spin) angle between the two subpayloads. An image-intensified TV camera was located on the aft payload with its optical axis parallel to the aft payload spin axis. The TV was sensitive enough to detect third-magnitude stars. Light-emitting diodes (LEDs) were mounted on the shell and on the boom tips of the forward payload to serve as targets for the imager. The boom failed to deploy correctly, with the result that its LEDs did not turn on. The body-mounted LEDs provided detectable luminosity at a separation of 1.5 km. Although the attitude analysis is not complete, it is hoped that the LED data will be able to provide additional data for the absolute determination of the aft payload attitude.

2.3. FLIGHT AND PAYLOAD PERFORMANCE

2.3.1. *Trajectory Objectives.* During experiment planning, the variation of (**T**,**B**) with time for launches from several possible launch sites was calculated by prime contractor Bristol Aerospace Limited. The trajectory simulation, including such parameters as azimuth and elevation, was based on known range boundaries and impact dispersion data of the Black Brant X rocket. The Andøya Rocket Range near Andenes, Norway was selected over four North-American sites because (**T**,**B**) remained significantly smaller across the flight compared with flights from the other sites. This is shown in Fig. 1.

The simulation showed that as the flight progresses, the gravity-gradient force tends to align **T** with local vertical. Near flight's end, the **T** direction has been rotated by as much as 6° in elevation. The **B** vector moves in the same direction as **T** only in the Andøya launch case. The computed angle (**T**,**B**) does not exceed 2° during the flight from Andøya, which is several degrees better alignment than in the other trajectories computed for Fig. 1.

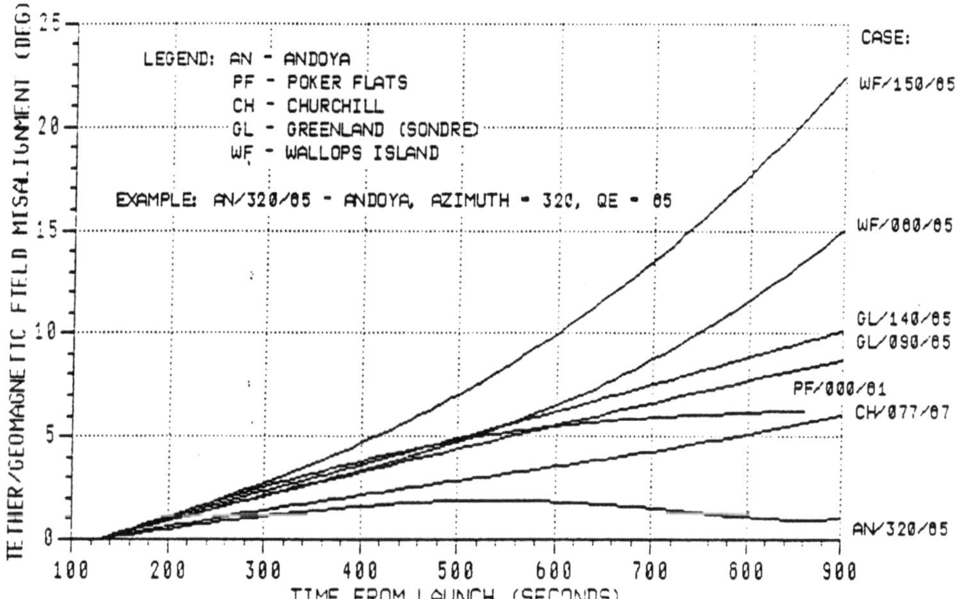

Fig. 1: Values of the angle (**T**,**B**) between the tether and magnetic field directions, computed along payload trajectories for launches from five candidate sites.

2.3.2 *Flight achieved.* The OEDIPUS-A payload was launched using a three-stage Black Brant X rocket at 2114:40 UT on 30 January 1989 from Andøya. An apogee of 512 km was achieved and 560 s of prime data were recorded. At apogee, the payload was at an invariant latitude of 70.9° and a magnetic local time of 00 h 11 min.

Figure 2 identifies some of the major payload events along the trajectory. After deployment of probes and antennas was completed near 164 km, the attitude control system began to align the payload spin axis with the local **B**. The alignment was complete and the spin rate adjusted to 0.7 rps at 285 km. At this time, the tether separation sequence bagan. Full extension to 958 m was achieved a little past apogee. Telemetry data for the tether length confirmed the smooth deceleration of subpayload relative motion by the magnetic brake.

Because of winds, safety rules required the Andøya launch authority to use a lower elevation than the optimal 85° assumed in Fig. 1. Consequently, an apogee of 512 km was obtained instead of the targeted

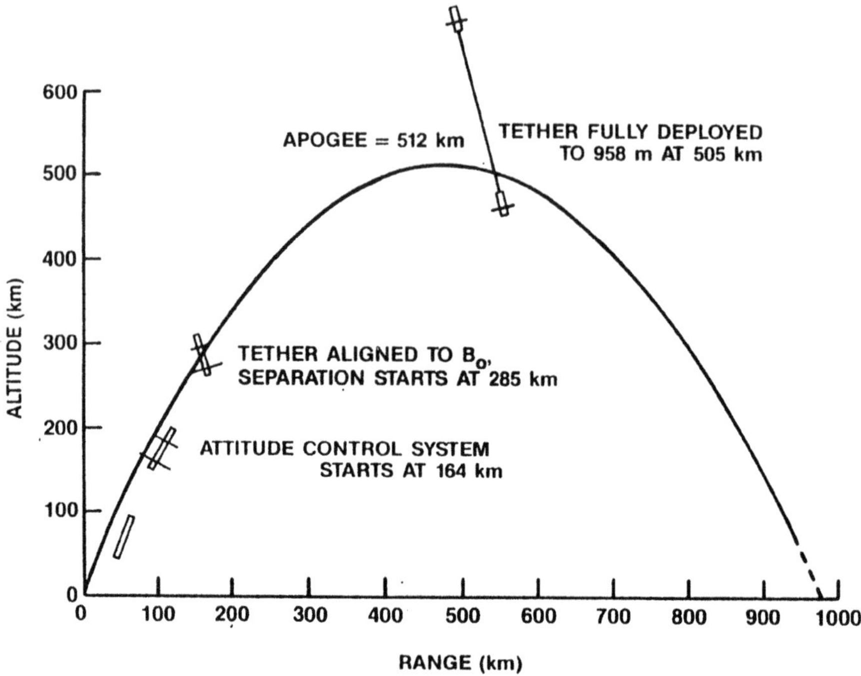

Fig. 2: Sequence of major payload events along the payload trajectory, with their time after launch in seconds: Attitude Control System start, 121; Separation start, 175; Apogee, 408; Tether deployment complete, 450; Re-entry at 80 km, 727.

620 km. An unexpected tip-off skewed the initial payload separation. This was the principal reason that the (T,B) angle went as high as 5° during the trajectory.

2.3.3. *Scientific conditions during flight.* All the scientific instruments in the payload operated as designed. The OEDIPUS-A launch azimuth was 46° west of geographic north toward a widespread aurora of moderate intensity located several hundred kilometers north and west of Andøya over the Norwegian Sea. Ground magnetometers at Andenes showed modest activity during the flight; this was consistent with low ionospheric absorption observed by the Andenes riometer. The Andenes ionosonde indicated a strong E-layer with foE ≈ 3 MHz and weak, diffuse returns from the F region in the 4-5 MHz range.

The ambient ionospheric electron density profile along the trajectory, as measured by both the plasma probes and by rf techniques with the HEX and REX, exhibited fluctuations typical of those expected in the auroral zone. At an altitude of about 290 km on the downleg, around 640 s after launch, a deep density depression was encountered which was apparently related to signatures of geophysical origin in potential-difference measurement and the energetic particle data. After passing that depression, the payload experienced quickly rising densities and passed through an E-region peak density at about 710 s, corresponding to foE = 5.8 MHz at a peak altitude of 135 km. Further information about the geophysical circumstances of the flight are reported by Ref. [5].

3. Double-Probe Studies Using OEDIPUS

3.1 MAGNETOSPHERIC MEASUREMENTS WITH A DOUBLE PROBE

The OEDIPUS experiment was first conceived to measure $\mathbf{E}_{||}$ in the magnetosphere at auroral latitudes. This field arises from the interaction of the solar wind with the earth's magnetic field. $\mathbf{E}_{||}$ fields accelerate charged particles into the auroral regions and are intimately connected with the closure of electrical current between the near and distant parts of the magnetosphere. Understanding of the overall dynamics of the magnetosphere necessarily includes an understanding of $\mathbf{E}_{||}$.

Large electric fields of the order of hundreds of millivolts per meter have been observed in the magnetosphere above a few thousand kilometers altitude. How these fields map down to the densest parts of the ionosphere at altitudes between, say, 100 and 500 km is an open question. In the lower ionosphere, plasma is not tightly "frozen" to field lines, and currents can flow at right angles to **B** to complete electric circuits involving $\mathbf{E}_{||}$'s and parallel currents at higher altitudes. Clear evidence about natural **E** fields in the ionosphere is an important part of the complete picture of the magnetosphere.

Weak $\mathbf{E}_{||}$ values have been observed at ionospheric heights. The reported values have varied significantly in magnitude, from tens of millivolts per meter down to less than 100 μV/m. The measurements of small $\mathbf{E}_{||}$ values in the presence of much larger perpendicular **E** values has made it difficult to interpret measurements of apparent $\mathbf{E}_{||}$.

The OEDIPUS-A experiment payload was designed to improve the observation of small E values. The payload would form a large double probe in which the potential difference would be monitored between its two subpayload parts. The sensitivity of the double probe for $\mathbf{E}_{||}$ detection is determined by the physical separation of the subpayloads. If the noise of the potential difference measurement is of the order of the thermal plasma potential fluctuations corresponding to a typical temperature of 0.1 eV, and if we wish to have a measurement of an **E** value of 1 mV/m to ±10%, then the separation has to be about 1 km.

The Tether Control and Monitor (TCM) instrument on OEDIPUS measured the potential difference between the subpayloads. It was designed with an internal impedance that was high enough (10^{11} Ω) to prevent the double probe from perturbing the measurement.

In order to understand the performance of the payload in the high-impedance voltmeter mode, it was decided to build into the TCM measurement cycle other modes that would provide a full picture of the ensemble as a double probe. Four TCM modes were chosen. These are schematized in Fig. 3. Panel A shows the voltmeter mode. In C and D, the plasma is deliberately shorted and the current between the payloads detected. Panels C and D are very similar, and provide data that are complementary to those from panel A.

The A, C and D panel data are interpreted in the framework of a longstanding model of an electric probe in a plasma, the Langmuir probe. Much theoretical and practical work on probes in plasmas has followed the first research by Ref. [6]. Modern mathematical techniques for predicting the distribution of plasma and fields around dielectric and metallic spacecraft bodies are an outgrowth of the concepts used to explain the Langmuir probe.

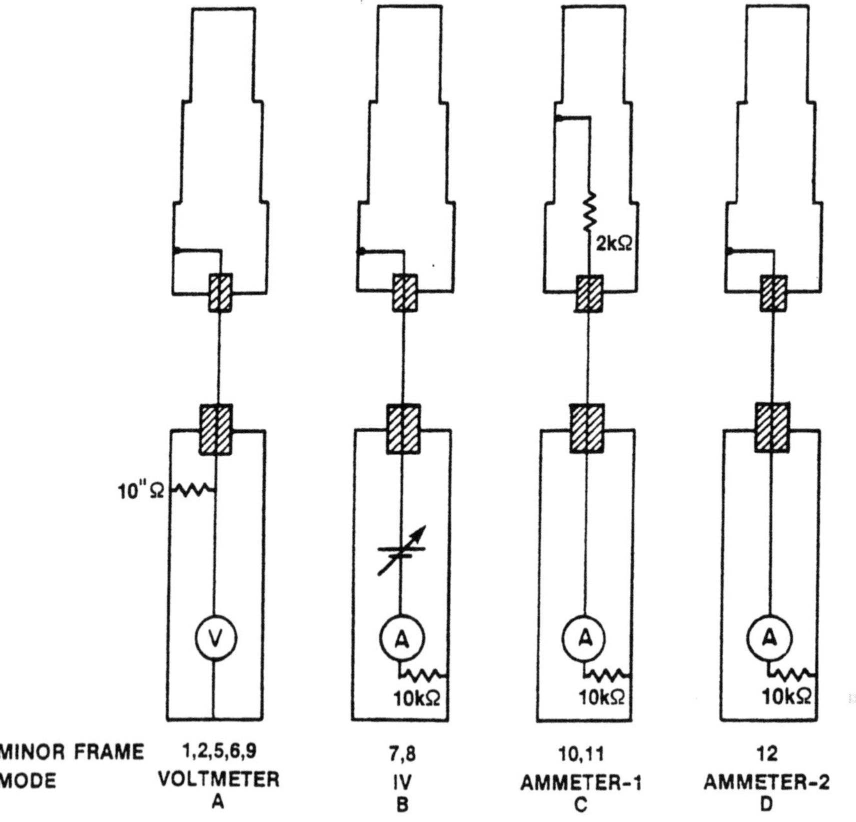

Fig. 3: Schematics of the electrical connections of the Tether Control and Monitor.

Spacecraft-plasma interaction is an important concern in the design of spacecraft. Since the TCM operates the entire payload as a double probe, the TCM data permit a new test of the electrostatic probe theory. The rest of this chapter is devoted to the interpretation of OEDIPUS TCM data recorded in the mode that directly measures the double probe characteristic. This is the current-voltage or "IV" response data recorded using the connection in panel B of Fig. 3.

3.2 IV CHARACTERISTIC OF A LANGMUIR PROBE

Reference [7] has reviewed the quantitative theory of the Langmuir probe. Ref. [8] have worked out the probe current response as a function of applied potential for a cylindrical metallic probe moving through a plasma. An example of the IV curve for OEDIPUS flight parameters is given in Fig. 4.

It is beyond the scope of this article to provide the mathematical development of the probe theory. The probe of Fig. 4 is assumed to be in a flowing equilibrium plasma without a magnetic field. In this diagram, increasing electron current to the probe is plotted along the positive current (y) axis. Langmuir probes can have different shapes. However, their response curves are qualitatively similar to that of Fig. 4.

A Probe IV characteristic has roughly three different regions. On the left side of the diagram at high negative voltages, ions are attracted by the probe and electrons are repelled. Near the probe surface, there are more ions than electrons. The ion excess builds up to balance the negative charge on the probe, establishing an ion sheath outside of which there is no electric field. The flatness of the curve indicates that the dimensions of the sheath change little as the voltage is increased. In this region, called the saturation ion current region, the current magnitude is determined by the combined effect of random thermal motion and ramming of the probe through the plasma.

On the right side of the curve is the saturation electron current region, which is characterized by an electron sheath. The slopes of the curves and the dimensions of the sheaths in the two saturation regions are markedly different because of the difference in the thermal velocities of electrons and ions.

There is a point at about 0.5 volts where there is no field anywhere between the probe and the plasma. This is the space potential. Both charged-particle types move to the probe on account of their thermal speeds. Electrons are more mobile, and the current is carried predominantly by electrons.

As the probe potential is made increasingly negative with respect to the space potential, the curve moves through its transition region. Ions are increasingly attracted and electrons repelled. The electron current here varies exponentially with the ratio of the potential energy to the average thermal energy. The characteristic curve crosses the potential axis at what is called the floating potential, equal in this case to -0.7 V. This is the potential that the probe takes by itself when placed in the plasma. The floating potential has to be negative to balance the electron and ion fluxes. In typical collisionless laboratory- and space-plasma

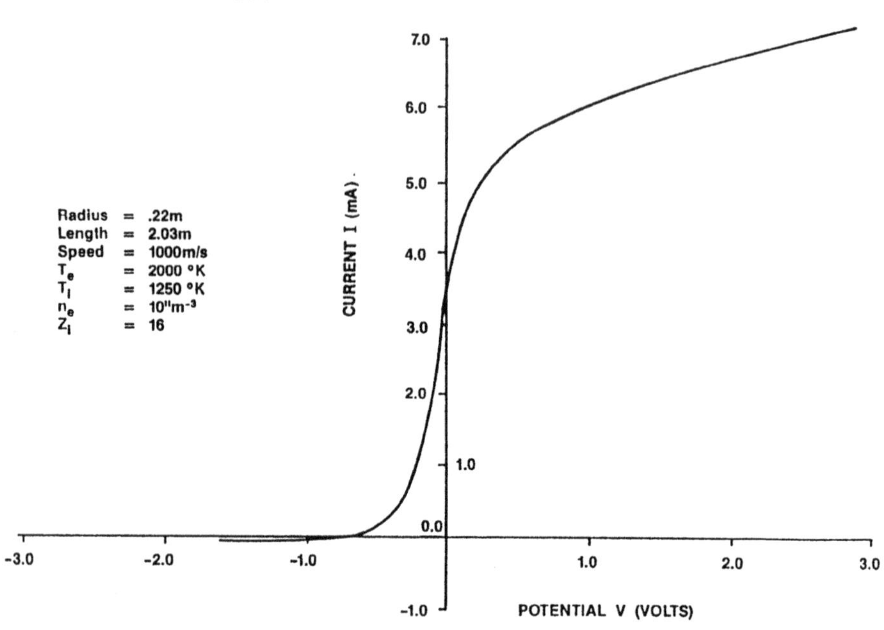

Fig. 4: Current-voltage curve for the aft subpayload considered a cylindrical electrostatic probe in a flowing magnetoplasma, as predicted by the theory of Ref. [8]. Parameters relevant to the OEDIPUS-A experiment have been used.

diagnostic applications, the saturation current regions are used to determine the number densities of the respective species, and the electron temperature is scaled from the transition region.

Sheath thickness is closely linked to the notion of Debye shielding in plasmas (Ref. [9]). Electrons distribute themselves around a charge excess and limit its range to the Debye length λ_D given by

$$\lambda_D = \left[\frac{\epsilon_0 kT}{n_e q_e^2}\right]^{1/2} \tag{1}$$

Where ϵ_0 is the vacuum permittivity
kT is the average electron thermal energy
n_e is the electron density
q_e is the electron charge.

The thickness of the ion sheath around a floating probe is of the order of λ_D. In the OEDIPUS experiment, the plasma frequency was around 2.5 MHz and the electron temperature 2000 °K, which made λ_D about 1 cm.

3.3 DOUBLE PROBE

OEDIPUS is unusual in that the entire subpayload bodies serve as the probe surfaces. This in contrast to most electrostatic probe applications in space where use is made of special probes whose current-collecting surfaces are small compared to the area of the spacecraft upon which they are mounted. With two subpayload bodies of approximately equal area, the configurations in Fig. 3 must be analyzed as double probes. In the double probe, a condition of current balance to the two probes is sought. That is, solutions to both probe's IV characteristics are required such that the net current into one probe equals the net current out of the other.

This balance is demonstrated in Fig. 5. The subpayloads are connected as shown on the left side. The current balance condition is illustrated by supposing for the moment that the two subpayload IV characteristics, top and bottom, are identical and by referring the discussion to the upper characteristic in Fig. 5. One payload operates at potential V_1 and the other at V_2 such that the difference $V_1 - V_2$ equals the applied voltage V, and such that the current drawn by one is exactly minus the current drawn by the other. The symmetrical IV response on the right side is obtained and, to first order, the extremities are just the ion-saturation portions of the individual IV curves.

In Figs. 6 and 7 we come to quantitative results of modelling the OEDIPUS-A double probe reported by Ref. [10]. The IV response curves, observations in solid line, have the opposite shape to the usual form shown in Fig. 4 simply because of the polarity convention chosen by the instrument designers. Fig. 6 shows the double-probe IV characteristic recorded by the TCM at 11 seconds after the payload separation started. The tether length was 20 m. Fig. 7 corresponds to 353 s after separation when the tether was at its final length of 958 m and the payload was on the downleg part of its trajectory.

Reference [10] found optimal theoretical fits (broken-line) to the observations. This yields measurements of the electron density, the electron temperature, the payload velocity and the external potential applied to the tether by natural effects. Good fits to the observations are obtained except at the beginning of the voltage sweep. The authors attribute the observed transient signal to relaxation effects following the step-function application of the starting voltage -8 V.

The external potential obtained in the fitting process is a combination of the effect of the ambient geophysical field and of the potential $T \cdot \mathbf{v} \times \mathbf{B}$ induced in the tether of length T moving at velocity v across the magnetic field lines **B**. The $\mathbf{v} \times \mathbf{B}$ potential is the dominant component of the total external potential. It is largely responsible for the difference between Figs. 6 and 7. It is less than 1 V in Fig. 6, but in Fig. 7 is sufficiently large to shift the double-probe sweep well away from the zero-current point. The Fig. 7 curve is thereby confined largely to the ion-saturation region of operation.

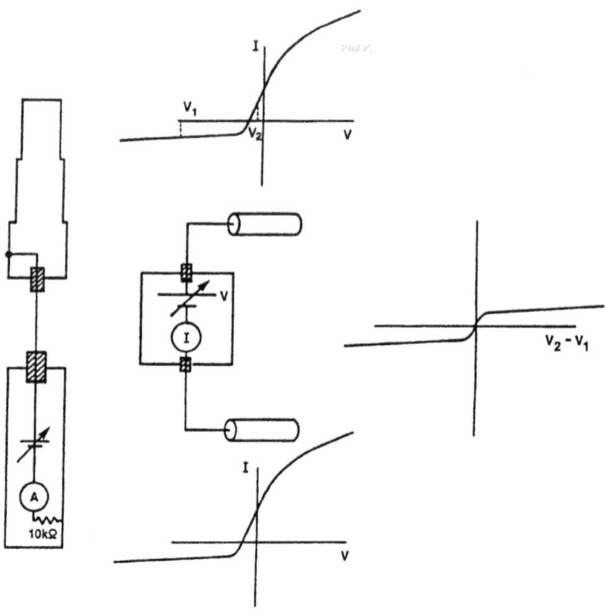

Fig. 5: The synthesis of the IV characteristic of the OEDIPUS-A double probe. This is obtained by solving a current balance equation for the two subpayloads viewed as back-to-back probes. Each subpayload is a cylindrical probe with the IV curves at the top and bottom. The combined payload equivalent circuit is in the middle and its combined IV characteristic is at the right.

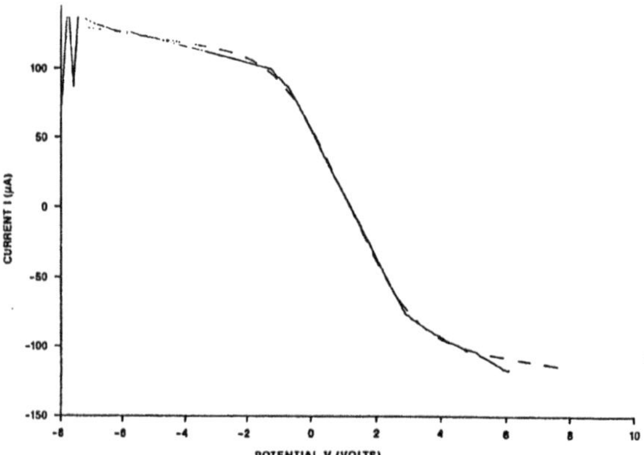

Fig. 6: Observed (solid line) and theoretical (broken line) IV curves for the OEDIPUS-A double probe at 11 s after separation. Altitude = 306 km. Tether length = 20 m.

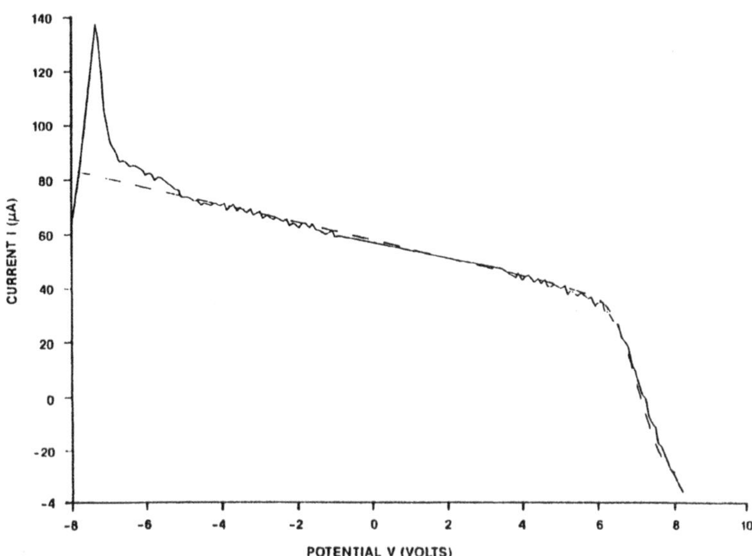

Fig. 7: Observed (solid line) and theoretical (broken line) IV curves for the OEDIPUS-A double probe at 353 s after separation. Compared with Fig. 6, the asymmetry of the curve is caused by the external **vxB** potential induced by the motion of the conducting tether across magnetic field lines. Altitude = 455 km. Tether length = 958 m.

The external potentials obtained from the fitting agree well with the values obtained by the TCM when operating in its voltmeter mode. The temperatures obtained by the Plasma Probe and the TCM are in rough agreement. A comparison of the electron densities measured by the TCM, the plasma probes and an rf technique using the HEX and REX is given in Fig. 8.

The plasma probe and TCM curves are based generally on the same electrostatic probe theory and yield comparable values. These are 1.5 to 2 times smaller than the values from the rf method. This illustrates the widely recognized problem of obtaining absolute density values from Langmuir probes. Possible difficulties in interpreting electrostatic probe behavior in the ionospheric magnetoplasma been mentioned by Ref. [11]. Discrepancies between probe and rf techniques in orbiting spacecraft are not as high, typically 20% (Ref. [12]). It may be that, in sounding rocket payloads, additional processes prevent the probe from performing as assumed in the theory.

Aside from the problem of absolute density measurements with the tethered double probe, the OEDIPUS experiment on the large double probe has indicated that useful measurements of plasma parameters are accessible via the application of the conventional probe theory for flowing collisionless plasma. This may help to put the modelling of spacecraft environments on a better footing in the future.

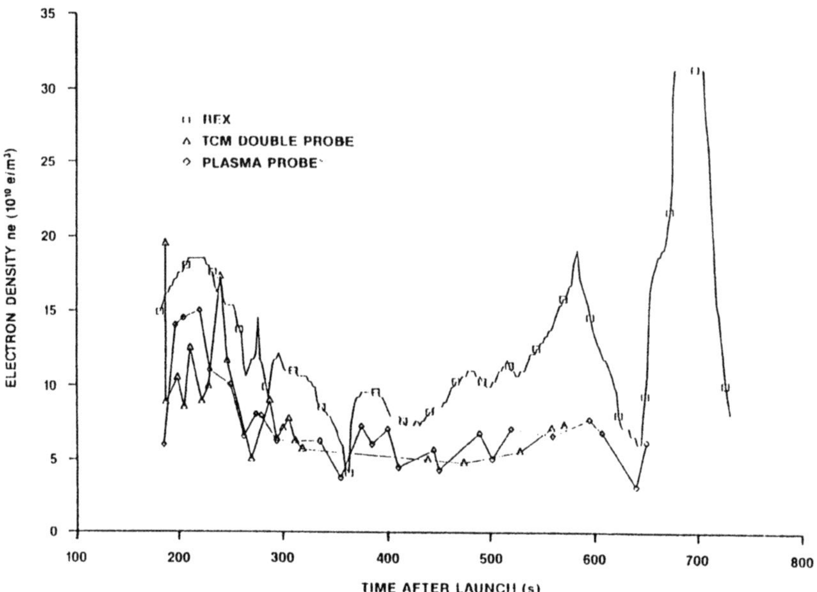

Fig. 8: Electron-density histories during the OEDIPUS-A flight as measured by the Tether Control and Monitor (△), the plasma probe on the aft subpayload (◇) and the HEX-REX (□).

4. Sheath Waves

As already stated in section 3, a solid body in a plasma has an ion sheath around it, on account of the relatively high mobility of electrons. Besides being an important part of our understanding of the double-probe IV response, the sheath plays another role that is examined in the OEDIPUS experiment: the sheath around the tether wire supplies the vacuum gap and the ambient plasma supplies the outer conductor of an effective coaxial rf cable. The electromagnetic waves that are guided by the coaxial cable are called sheath waves (Ref. [13]), and have a TM mode character. Sheath waves have been under investigation for a few decades in the laboratory, but the OEDIPUS project appears to be the first time sheath waves have been deliberately excited and observed in a space setting.

4.1. SHEATH WAVE THEORY

The formal description of the ion sheath in a magnetosplasma and the theory for propagation in the coaxial sheath waveguide have not yet been developed. Hence no theory is available for rigorous comparison with the OEDIPUS results. The theories that most closely approach the OEDIPUS conditions are for a) coaxial geometry in an unmagnetized plasma and b) planar magnetoplasma geometry with propagation, magnetic field direction and the planar surface all parallel.

A particular isotropic plasma theory for sheath waves has been elaborated from the thin-wire moment method of Ref. [14]. Ref. [15] have developed a computer program that is applicable to the analysis of the OEDIPUS data. The payload is modelled as a single wire joining two end plates as shown in Fig. 9. Immediately surrounding the tether wire is a sheath vacuum gap of thickness 1.25 cm and of permittivity ϵ_0. Surrounding the whole structure is isotropic plasma with permittivity

$$\epsilon = \epsilon_0(1-X/U)^{1/2} \tag{2}$$

where $X = (f_p/f)^2$, f_p is the plasma frequency, f is the wave frequency, $U = 1 - j\nu/\omega$, ν is the angular electron-neutral collision frequency and $\omega = 2\pi f$.

The solution of the wave equation yields phase velocities that are somewhat slower than the speed of light c. Propagation has an upper frequency limit $f_s = f_p/\sqrt{2}$. The expression for ϵ has been modified after noting that in the limiting case of **T** $\parallel$ **B**, the rf electric field is perpendicular to **B**. In an anisotropic plasma, the perpendicular component of the permittivity tensor dominates, with its zero at the upper hybrid resonance frequency $f_{uhr} = (f^2_p + f^2_c)^{1/2}$ rather than at f_p, where f_c is the gyrofrequency. Therefore ϵ in equation 2 has been arbitrarily adjusted by replacing X with $(f^2_p + f^2_c)/f^2$ for comparison with the OEDIPUS observations.

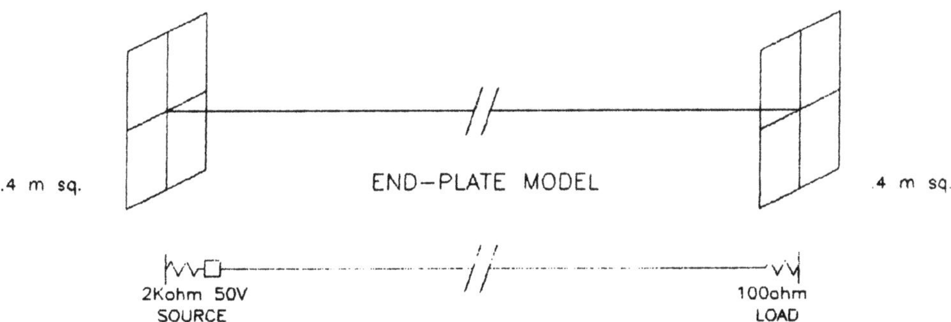

Fig. 9: Wire configuration for moment-method calculation of sheath wave propagation along the OEDIPUS-A tether surrounded by unmagnetized plasma.

Reference [16] analyzed the planar geometry for a cold anisotropic plasma. Their dispersion curves typically show phase and group speeds somewhat smaller than those for plane waves in vacuum at frequencies well below the cutoff f_s. As the frequency is increased to the neighborhood of f_s near $f_{uhr}/\sqrt{2}$, the phase and group refractive indices rise to values greater by about an order of magnitude or more than the vacuum values. When compared with earlier related work cited by Ref. [16], their results confirm the importance of a vacuum sheath in changing the frequency bandwidth of allowed propagation. Also, with f_p and f_c in a ratio similar to that of the OEDIPUS-A experiment, the imaginary part of the phase index indicates increased attenuation around f_c.

4.2 EXPERIMENT DESCRIPTION

Sheath waves were excited on OEDIPUS A by the HEX transmitter and detected by the REX receiver. The HEX is a 2.5-W signal generator located in the forward subpayload. It synthesizes waveforms from 20 Hz to 5 MHz, and also modulates these with rectangular pulse envelopes. The HEX has three different output levels: 1, 50 and 100 V_{rms}. As the major frame proceeds, it connects to a 4.4-m dipole, to the tether or is disconnected.

The REX receiver in the aft subpayload has a bandwidth of 50 kHz and can be centered with 50 kHz spacing anywhere in the range 25 kHz to 5 MHz. During the major frame, the REX input cycled through connections to a 5.5-m receiving dipole, to a tether voltage measurement and to a tether current measurement. Figure 10 shows the different connection schemes of the HEX and REX.

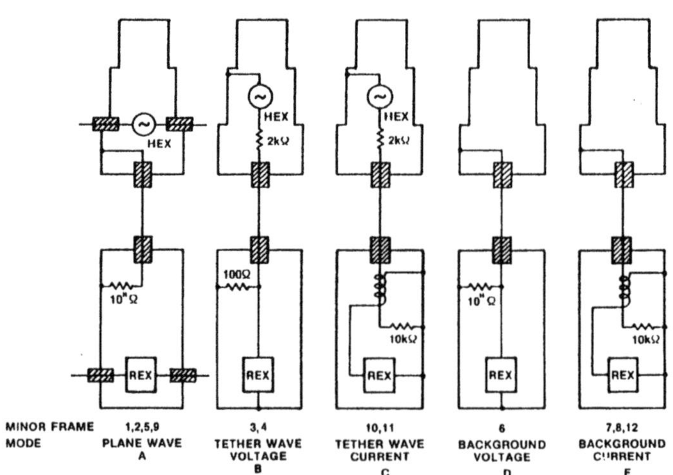

Fig. 10: Schematics of the electrical connections of the transmitter HEX and associated receiver REX.

Connections B and C were designed expressly for tether-wave excitation. Tether signal voltage across the 100-Ω termination is recorded in B whereas it is tether current in C. Connection A, showing HEX and REX connected to the aforementioned dipoles, was intended for plane-wave experimentation. Although some plane wave phenomena have been identified in the connection-A data (Ref. [17]), it turns out that connection A also excites sheath waves.

The present discussion centres around the SH1 frequency mode of HEX and REX in combination with connections B and C. During a 500-ms minor frame, the central frequency f_i of the REX bandwidth was swept in steps of 50 kHz from 25 to 4875 kHz, dwelling on each for 5 ms. The HEX was synchronized to transmit two 300-μs pulses in each band, one at frequency f_i − 12.5 kHz at 0.3 ms and one at f_i + 12.5 kHz at 2.8 ms. The flight data confirmed that the synchronization after separation was obtained to within a few tens of microseconds.

4.3 ANALYSIS OF SHEATH WAVES

Figure 11 is a gray-scale representation of received signal strength from the voltage-measuring mode as a function of both frequency and elapsed time during the flight. A strong passband is observed from zero frequency up to a sharp cutoff between 1.7 and 2.3 MHz. Above the cutoff is a stopband extending upward to a frequency between 3 and 4 MHz, where a fairly strong signal reappears. The signal between 0 and 2.3 MHz is identified as sheath-wave propagation. As already discussed, the sheath wave theoretical cutoff f_s is thought to be somewhere between $f_p/\sqrt{2}$ and $f_{uhr}/\sqrt{2}$. The plasma frequency f_p was measured to fluctuate between 2 and 3 MHz during the flight (Ref. [17]) and f_{uhr} lay about 300 kHz above this. The observed f_s in Fig. 11 is about several hundred kilohertz below $f_{uhr}/\sqrt{2}$.

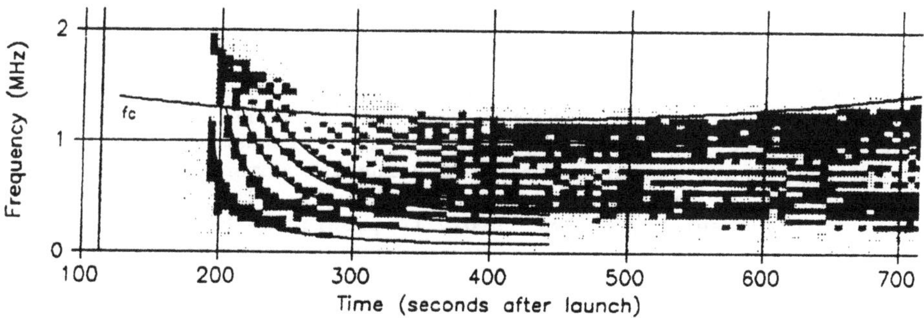

Fig. 11: Flight survey of the strength of signal received by the REX in the tether voltage connection (panel B in Fig. 10), using the SH1 frequency mode. The fringe pattern is interpreted as tether-length resonances of sheath waves.

The curved fringe pattern in Fig. 11 provides strong support for the sheath-wave interpretation of that signal. The pattern is thought to manifest a sheath wave resonance condition between the tether length L and the dominant wave number k. Suppose a sheath wave signal travelling toward the REX end of the tether has phase ϕ. Because of the mismatch at that end of the tether, most wave energy is reflected back toward the HEX end, the phase being changed by $\Delta\phi_1$, by the complex reflection coefficient at the REX end. This wave travels back to the HEX end undergoing a phase change of kL. It experiences another reflection with another phase change $\Delta\phi_2$. It arrives back at the REX end with a total phase change of $2kL + \Delta\phi_1 + \Delta\phi_2 = 2kL + \Delta\phi$. The first fringe from the bottom is interpreted as obeying $2kL + \Delta\phi = 2\pi$. In general fringe maxima occur at

$$2kL + \Delta\phi = 2n\pi \qquad (3)$$

where n is an integer denoting fringe order.

The wave number k is a function of the frequency f and the plasma parameters. Ignoring for the moment the plasma parametrical dependence, the total reflection phase shift $\Delta\phi(f)$ has been found by trial and error as the value which provides a single k solution at constant f for all n fringes. A frequency independent value $\Delta\phi = -1.4\pi$ is found to give a unified solution across most of the band.

This is shown in Fig. 12. The box points were obtained from the wave voltage measurements, panel B in Figure 10, and the triangles from the current measurements, panel C. This constitutes a first-order measurement of the sheath-wave dispersion relation. The waves are seen to propagate somewhat slower than c, in that $k/k_o = (f\lambda_o)/(f\lambda) = c/v_\phi = 1.5$, where v_ϕ is the phase velocity. Furthermore k is mildly dispersive through most of the frequency range observed. This is seen by evaluating the group speed. The slope of the trace $\partial(k/k_o)/\partial f \approx 0.6/2.5 \times 10^6\ \text{Hz}^{-1}$. Expanding

$$\frac{\partial(k/k_0)}{\partial f} = \frac{1}{k_0}\left(\frac{\partial k}{\partial f} - \frac{2\pi k}{ck_0^2}\right) = \frac{1}{f}\left(\mu_g - \frac{c}{v_\phi}\right) \qquad (4)$$

one finds a group index $\mu_g \approx 1.7$ at f = 1 MHz.

At full tether extension L = 958 m, HEX pulses would travel the distance L in $L\,\mu_g/c = 6\ \mu s$. The SH1 mode data are consistent with this, showing pulses that arrive at REX within 10 μs of transmission.

The dispersion relation evaluated using the moment method of Ref. [15] is overlaid as the continuous line in Fig. 12. The good fit to the data confirms the ability of the theory to predict both phase and group speeds of sheath waves.

4.4 THE EFFICIENCY OF SHEATH WAVE PROPAGATION

Sheath waves are remarkable for the efficiency with which they couple the ends of the OEDIPUS payload. While it is difficult to interpret signal levels at the REX end absolutely, the propagation efficiency is demonstrated in two ways. First, maximum signals recorded in the middle

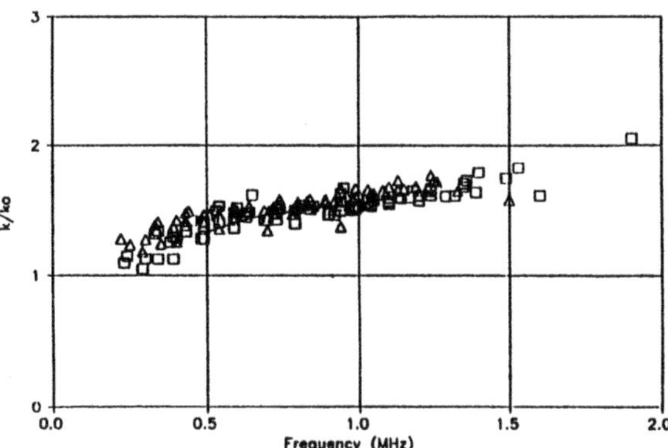

Fig. 12: Sheath-wave-numbers normalized to vacuum wave numbers scaled from various fringes in Fig. 11. The voltage measurements corresponding to panel B in Fig. 10 are plotted with triangles (▵) and the current measurements, panel C, are plotted with squares (▫).

of the sheath-wave band change by less than 10 dB between the start and the finish of the tether deployment. That is, there is little signal attenuation over the 1 km tether. Second, values of the fringe order n (eq. 2) as high as 10 are observed. In other REX data not shown, the decay time for rf pulses is found to be about 50 μs. Since the signal takes about 6 μs to traverse the fully deployed tether, this implies that waves reflect several times before damping out in the waveguide. The multiple fringes are another indication that sheath waves propagate with little loss.

Electromagnetic sheath waves can be efficiently guided along conductors in plasma. The calculations with the model in Fig. 9 show that the signal coupling from one end of the tether to the other is tens of decibels higher than in free space. This is consistent with the observed spectra, of which Fig. 13 is an example. This graph of REX signal power was recorded using the current mode in Fig. 10C at 43 s after the start of payload separation when the tether was 220 m long. The sheath-wave passband signal power in 0-2 MHz is about 50 dB above the stop band starting around 3 MHz. With sheath waves, subpayload coupling is increased over vacuum conditions by 30 dB at 0.5 MHz and 20 dB at 1.5 MHz.

Sheath waves are therefore a significant design issue in the electromagnetic compatibility of avionics at frequencies up to HF on large metal space structures. Fig. 14 shows how the 0-f_s passband may vary on a worldwide scale. Height profiles of $f_s = f_{uhr}/\sqrt{2}$ encountered at four equally spaced locations along a circular low-earth-orbit have been produced with the International Reference Ionosphere (Ref. [18]) and a magnetic field model. The variation of local f_s with Local (solar) Mean Time means that the sheath-wave passband and adjoining stopband will be swept back and forth across the 0 - 8 MHz band during each orbit.

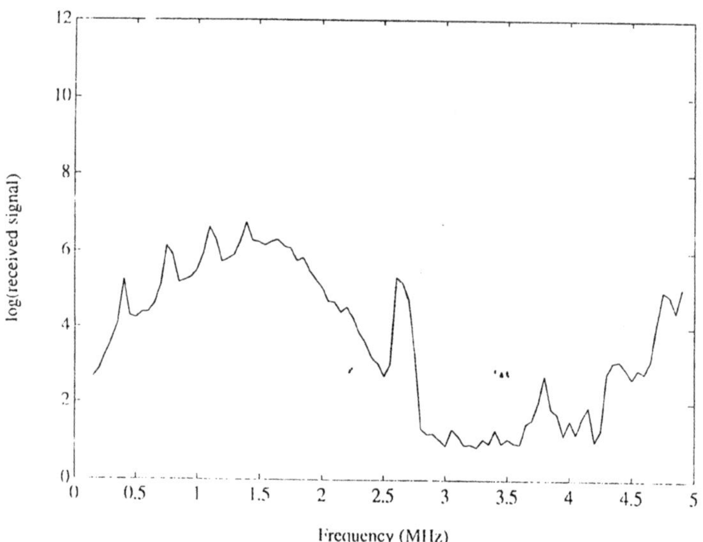

Fig. 13: Frequency power spectrum of sheath-wave current detected by REX using the connection scheme in Fig. 10C. This was recorded at a height of 362 km when the tether length was 220 m, 43 s after payload separation.

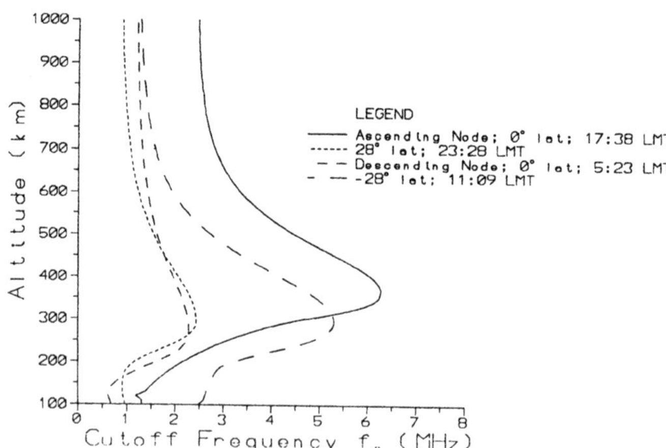

Fig. 14: Predicted altitude profiles of the sheath-wave cutoff frequency $f_s = f_{uhr}/\sqrt{2}$ for four points along a low-earth circular orbit. Date is 15 Jan 1995. Sunspot Number = 40. Orbital inclination = 28°. Orbital height = 296 km.

It is important to note other passbands in Fig. 13 besides the 0-f_s band. A sharp feature at 2.6 MHz and a band rising at about 4.5 MHz are not predicted by the referenced theory. Other REX data show that these bands at higher frequency have comparable levels to the lower band. Their existence may require inclusion of magnetic field and hot-plasma effects in the theory.

5. Concluding Discussion

The OEDIPUS results are good illustrations of the interaction of the ambient ionospheric plasma with spacecraft. The OEDIPUS subpayload bodies can be modelled as cylindrical probes because the response of the double-probe payload agrees to first order with the predictions of existing probe theory. The external potential difference between the two ends induced by the **vxB** effect causes important effects. Any large artificial structure in space will have large induced dc emfs that will affect the electric fields and sheaths around it.

There are other natural energy sources that can perturb the electrostatic fields around a spacecraft in addition to those considered above. These include auroral precipitating particles at energies up to 10 keV and trapped radiation belt particles at somewhat lower latitudes with energies beyond 100 keV. Solar radiation releases photoelectrons from spacecraft metal surfaces, further complicating the description of the dc electric field. In earth orbits below about 300 km, where abundant heavy ions and neutrals are rammed by the spacecraft, atomic and molecular processes are excited near the spacecraft skin. All of these effects are included in elaborate computer-based mathematical codes that are used for modeling spacecraft near environments.

The space plasma sets up a sheath around satellites that guides electromagnetic waves efficiently. The implications for electromagnetic interference and compatibility are clear. Whether positive use can be made of sheath waves, for instance in communications capitalizing on their low dispersion, remains to be seen.

The theoretical analyses of the OEDIPUS TCM and HEX-REX data herein has a common limitation: the theory does not include the magnetic field. The requirement, common to the double probe and coaxial sheath studies, for cylindrical geometry leads to considerable mathematical complexity in the general case when the cylindrical probe/tether and magnetic axes do not coincide. Clarification of the basic physics in this more general case is one of the next big steps for both probe and sheath-wave theory.

6. Acknowledgements

The author is much indebted for useful discussion and other help to R. Godard, K.G. Balmain, C.C. Bantin and G.W. Hulbert whose work on OEDIPUS-A data forms the basis of this paper. OEDIPUS is a collaborative project of the Canadian Space Agency and the national Aeronautics and Space Administration. Bristol Aerospace Limited were the prime contractor for payload design and construction.

References

[1] B.N. Maehlum, W.F. Denig, A.A. Egeland, M. Friedrich, T. Hansen, G. Hölmgren, K. Måseide, N.C. Maynard, B.T. Narheim, K. Svenes, J. Trøim and J.D. Winningham, 'MAIMIK - a high current electron beam experiment from the Andøya rocket range', Proc. of the 8th ESA symposium on European rocket and balloon programmes and related research, ESA SP-270 (1987).

[2] S. Sasaki, K.I. Oyama, N. Kawashima, T. Obayashi, K. Hirao, W.J. Raitt, N.B. Myers, P.R. Williamson, P.M. Banks and W.F. Sharp, 'Tethered rocket experiment (Charge 2): Initial results on electrodynamics', Radio Sci., 23, 975-988 (1988).

[3] C. Bonifazi, 'Tethered Satellite System (TSS) core science equipment', paper presented at International Conference on Tethers in Space, Sept. 17-19, 1986, Arlington, VA, Advances in Astronautical Sciences, 62, American Astronautical Society, San Diego, CA, pp. 173-191 (1987).

[4] T.D. Megna, 'Tethered Satellite System capabilities', paper presented at International Conference on Tethers in Space, Sept. 17-19, 1986, Arlington, VA, Advances in Astronautical Sciences 62, American Astronautical Society, San Diego, CA, pp. 193-205 (1987).

[5] H.G. James, and B.A. Whalen, 'OEDIPUS-A: Space research with a new tether', EOS, Trans. Am. Geophys. Union 84, 499-506 (1991).

[6] H.M. Mott-Smith, and I. Langmuir, 'The theory of collectors in gaseous discharges', Phys. Rev. B28, 727-763 (1926).

[7] F.F. Chen, 'Electric probes', in R.H. Huddlestone and S.L. Leonard (eds.), Plasma Diagnostic Techniques, Academic Press, New York, NY, pp. 113-200 (1965).

[8] R. Godard and J.G. Laframboise, 'Total current to cylindrical conductors in collisionless plasma flow', Planet. Space Sci. 31, 275-283 (1983).

[9] J.L. Shohet, 'The Plasma State', Academic Press, New York, 149-152, 333 pp (1971).

[10] R. Godard, H.G. James, J.G. Laframboise, B. Macintosh, A.G. McNamara, S. Watanabe and B.A. Whalen, 'The OEDIPUS experiment: analysis of the current/voltage data', J. Geophys. Res. 96(A10), 17879-17890 (1991).

[11] J. Margot-Chaker and A.G. McNamara, 'Comparative study of Langmuir probe characteristics in different ionospheric conditions', Planet. Space Sci. 32, 1427-1437 (1984).

[12] J.L. Donley, L.H. Brace, J.A. Findlay, J.H. Hoffman and G.L. Wrenn (1969), 'Comparison of results of Explorer XXXI direct measurement probes', Proc. IEEE 57, 1078-1084.

[13] J. Marec, and G. Mourier, 'Sur la propagation des ondes de surface et la nature des résonances électrostatiques de gaine', C.R. Acad. Sc. Paris B271, 367-370 (1970).

[14] J.H. Richmond, 'Radiation and scattering by thin wire structures in a homogeneous conducting medium', IEEE Trans Antennas Propagat. AP-22, 365 (1974).

[15] M.A. Tilston and K.G. Balmain, 'A multiradius, reciprocal implementation of the thin-wire moment method', IEEE Trans. Antennas Propagat. 38, 1636-1644 (1990).
[16] J.-J. Laurin, G.A. Morin and K.G. Balmain, 'Sheath wave propagation in a magnetoplasma', Radio Sci. 24, 289-300 (1989).
[17] H.G. James, 'Guided Z-mode propagation observed in the OEDIPUS-A tethered rocket experiment', J. Geophys. Res. 96(A10), 17865-17878 (1991).
[18] D. Bilitza, 'International Reference Ionosphere; recent developments', Radio Sci. 21, 343-346 (1986).

RESULTS OF SPACE EXPERIMENTS: CRRES

E.G. MULLEN and M.S. GUSSENHOVEN
PL/GPSP Hanscom AFB, MA, USA 01731

ABSTRACT. Experiments on the Combined Release and Radiation Effects Satellite (CRRES) are gathering the most comprehensive data ever taken to measure the near-Earth space environment and its effects on satellite operations. The data are showing deficiencies in our understanding of space radiation belt models, microelectronic upset frequencies and total dose degradation of system components. We describe the high energy particle detectors and the microelectronic test package used to make the measurements, the data handling techniques used to prepare the data bases, the data comparisons with existing models, and how we are developing new and better models. To conclude the report, we examine the dynamics of cause and effect of spacecraft anomalies using CRRES data from the major solar particle event period of March 1991.

1. Introduction

Almost immediately following the dawn of the space age in the late 1950s, the near-Earth space environment was discovered to have energetic proton and electron populations trapped by the Earth's magnetic field in "Van Allen" belts named after their discoverer [Ref. 1, 2, 3, 4; and references therein]. The realization that satellites and other space systems would have to operate in this harsh radiation environment brought new meaning to the area of scientific pursuit called "Radiation Effects". In the last decade, as computer technology has advanced by leaps and bounds, the advanced microcircuits that now support many of our everyday needs here on Earth are being rapidly transitioned into space systems. With this transition, the radiation effects arena is broadening due to the fact that radiation particles can severely impact the functionality and lifetime of electronic systems made with microelectronic components.

Theorists and modelers have spent countless man-years developing theories and codes to predict satellite anomalies caused by energetic space radiation particles. The anomalies have a variety of causes, many of which are detailed in other parts of this conference proceedings. Five of the most likely causes of environmentally-induced anomalies are: a) spacecraft frame charging, b) material charging and discharging, c) single event upsets in devices, d) total dose degradation, and e) high dose rate

R. N. DeWitt et al. (eds.), The Behavior of Systems in the Space Environment, 605–653.

degradation. The truth test comes when space data are available to test the models. On a fully instrumented satellite such as the Combined Release and Radiation Effects Satellite (CRRES), cause and effect relationships are readily discernible. Before examining these relationships, we show, in Figs. 1-4, examples of CRRES measurements of each of the five candidates for producing anomalies.

DeForest [Ref. 5] first identified spacecraft frame charging using a spectrogram display of electron and proton measurements at geosynchronous altitude. Figure 1 is a spectrogram of data from the Low Energy Plasma Analyzer (LEPA) on CRRES taken on 10 November, 1990 during a charging event. In each panel the color-coded differential flux is displayed in grey-scale as a function of energy (ordinate) and time (abscissa). The LEPA energy range is 100 eV to 30 keV for electrons and 10 eV to 30 keV for ions. Electrons are shown in the top panel, ions in the bottom. The spectrogram covers one full CRRES orbit, from perigee through apogee to perigee. At high altitudes the hot electron population can produce a net negative current to the spacecraft, charging it to negative potentials with respect to the ambient plasma. To balance the negative current, cold (< 10 eV) positive ions are accelerated and collected by the vehicle. They appear in particle detectors as an ion peak at the energy equal to the potential difference between the plasma and the vehicle. In Fig. 1 the peak is readily identified between .1 and 1 keV from 13:45 - 17:45 UT. The electron population that drives the vehicle potential was identified by

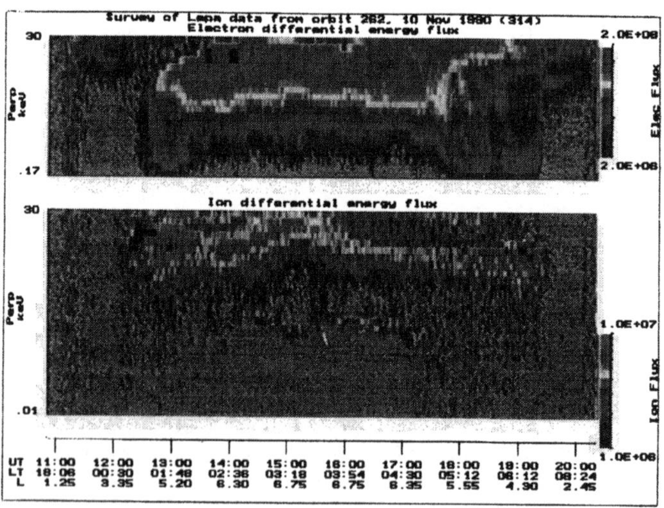

Fig. 1: Electron (top) and ion (bottom) spectrograms giving differential energy flux perpendicular to the magnetic field in [particles (cm^2 s ster)$^{-1}$] in grey-scale as a function of energy and universal time. Local time and L-value are added ephemerides. The data are taken on 10 November 1990 by the Low Energy Plasma Analyzer on CRRES.

Ref. 6 to have energies in the range of 20-50 keV. These electrons appear in the highest LEPA channels concurrently with the charging peak. The charging level decreases as perigee is approached at the end of the orbit because the ion density increases with decreasing altitude, requiring less acceleration to bring in a balancing current.

Material charging and discharging is measured on CRRES with an Internal Discharge Monitor which records pulses in test circuits to several different material and cable samples. The number of pulses per orbit is plotted against orbit number in Fig. 2. The first 700 orbits occurred from 15 July 1990 to 9 May 1991. The pulses are discharges of accumulated charge deposited in the samples over a period of time. The electrons responsible for the charge accumulation have energies greater than those which create frame charging, probably higher than a few hundred keV. The exact range of these causative electrons is still under investigation [Ref. 7, 8, 9; and references therein].

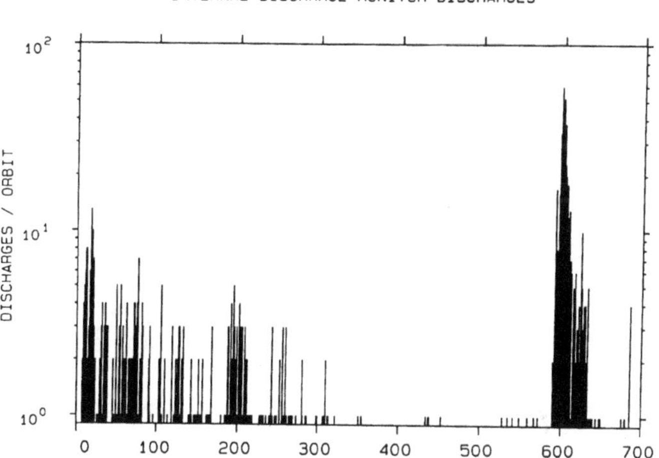

Fig. 2: The total number of discharges per orbit, measured in 16 test circuits in the Internal Discharge Monitor on CRRES, plotted against orbit number from 25 July, 1990 to 7 May, 1991.

"Single Event Upsets" (where a single, energetic particle changes the state of a flip-flop or memory cell location) were non-existent when systems were made entirely from discrete devices. Now state-of-the-art devices show a wide range of upset sensitivity to the protons and cosmic rays ever present in space [Ref. 10, 11, 12, 13, 14]. Although much research has been done over the last few years to better understand and model SEUs, device sensitivity can still only be adequately determined by test. However, we do not know if existing ground test procedures accurately reflect what is happening in space. In ground testing the devices are delidded and tested with protons and ions much lower in energy than what penetrates into and causes SEUs in space systems. The particles which cause SEUs in space are usually so high in energy that shielding the devices by metal enclosures does not alleviate the problem. Single event upset (SEU) rates as measured in the Microelectronics Test Package (MEP) on CRRES are shown as a function of L-shell in Fig. 3 *[Note: L-shell will be used throughout this paper. It identifies magnetic field lines on which particles are confined. On the magnetic equatorial plane and for a dipole field, L is equivalent to distance from the center of the Earth, measured in Earth radii, R_E].* The figure gives the combined upset rates of 35 devices, representing 12 device types which upset the most during the early part of the CRRES mission. The peak in upset rate, near an L-value of 1.5 R_E (R_E is one Earth radius), is also the peak of the inner radiation belt.

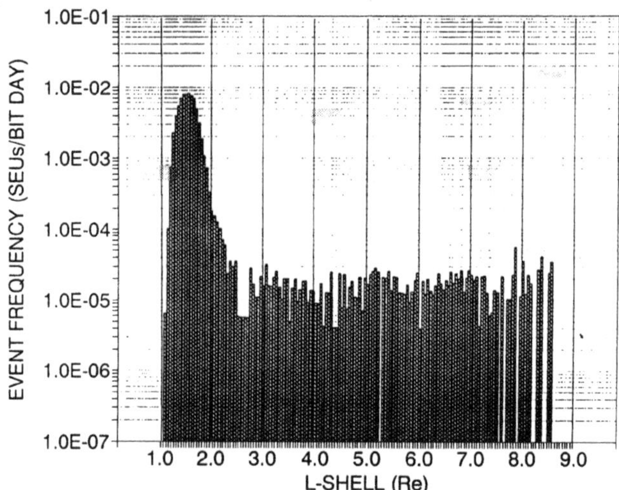

Fig. 3: Single event upset frequency, from 35 proton-sensitive devices in the Microelectronics Test Package on CRRES, plotted against L-value in Earth radii. The data are accumulated from 25 July, 1990 to 22 March, 1991. The peak at an L-value of 1.5 coincides with the heart of the inner radiation belt.

Total dose and dose-rate effects (usually leading to a gradual degradation of the electrical parameters of a device), on the other hand, have always been a problem, even with discrete components [Ref. 15]. Certain new technologies are extremely sensitive to total dose. Millions of dollars are spent every year by members of the space community to produce "radiation hardened" devices that will maintain their functionality for ten years or longer in space. As with SEUs, total dose hardness of a component can only be determined by test, and the standard ground-based ^{60}Co tests may not accurately reflect radiation damage experienced in space from energetic electrons and protons. Test procedures need to be verified with space data and updated to accurately reflect in-space degradation. Figure 4 shows the total dose behind 4 different shielding thicknesses as measured by the Space Radiation Dosimeter on CRRES, as a function of orbit

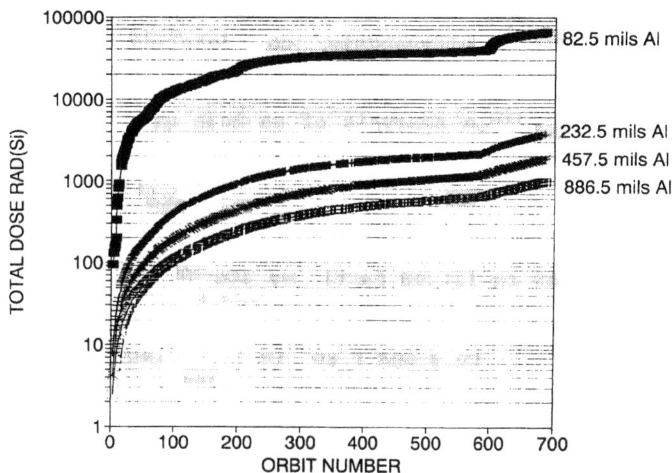

Fig. 4: The total dose measured behind 4 different shielding thicknesses by the Space Radiation Dosimeter on CRRES, plotted as a function of orbit number from 25 July, 1990 to 7 May, 1991.

number. The rate is easily ascertained by the slope of the curves. The measured dose can be directly related to device degradation in the MEP and, thus, to shielding effectiveness.

With the launch of CRRES on 25 July, 1990 at 19:21 UT, a unique opportunity to investigate the high energy space particle environment and its effects on advanced microelectronic components came to fruition. Previous measurements of near-Earth space were used to produce the existing NASA radiation belt models, but even taken together, the sampling of the radiation regions was inadequate. In addition, the early instruments did not have the sophistication in computer control or technology that today's experiments have, and thus could not provide the energy resolution, pitch angle coverage, and magnetic field data needed to map the particles back to the equator or incorporate dynamic and quasi-static radiation processes into the models.

CRRES, in an equatorial orbit, inclined at 18.1° to the geographic equator, with perigee of 350 km and apogee of 33,500 km, cuts through the hearts of the inner and outer radiation belts, spinning at 2 rpm. It provides the platform for 20 experimental packages to characterize the particle, field and wave environment of near-Earth space. With sixteen particle instruments the satellite measures the complete equatorial pitch angle distribution of electrons, protons and ions over the entire energy range of interest. The orbit period is slightly under 10 hours which is short enough to measure the dynamic evolution of radiation belt events.

In addition to the environmental sensors, CRRES has 3 engineering experiments to determine the effects of the radiation particles on advanced microelectronic components, on sample spacecraft dielectric materials and on advanced solar cells. The microelectronic testing represents the first time both the space environment and device upset and degradation parameters have been measured simultaneously to determine cause-and-effect relationships. Previous to CRRES there has been only limited space testing of devices. Much of the in-space performance of chips has come from anomaly analysis or failure analysis of devices used to run operational satellites. Although this data base has been useful, it is not adequate to project lifetimes or upset probabilities of electronic devices or systems in space. Also previous in-space testing cannot adequately characterize chips in severe radiation environments since it has been limited to regions of space (either below or above the major belt particle populations) where few radiation particles are present.

The CRRES space radiation experiments and their affiliated agencies, together with an artist's concept (to scale) of the CRRES satellite and its orbit path through the Earth's outer and inner radiation regimes is shown in Fig. 5. A listing of the instruments and agencies is also included as an Annex. Well over 300 scientists and engineers from around the world are participating in CRRES. Over 50 scientists and engineers participated in the design and testing of just the Microelectronics Test Package.

This paper is organized as follows:

Section 2 gives an overview of the satellite, a listing of the full range of particle, field and wave experiments on CRRES, a short description of the high energy particle detectors to be used in producing the next generation of static radiation belt models, and a complete description of the Microelectronics Test Package.

Section 3 gives a description of how the data are prepared for analysis and any assumptions made during "data selection".

Section 4 gives a first glimpse at what the new radiation belt models will look like and how the CRRES data received to date compares with the existing NASA Radiation Belt Models.

Section 5 gives early on-orbit results from the Microelectronics Test Package (MEP) in terms of SEU sensitivities, total dose degradation, shielding effectiveness, and correlations with particle measurements.

Section 6 is a discussion section in which we use measurements taken during the large solar-magnetic storm event that occurred in March, 1991, to address the impact of CRRES on the Space Radiation Effects Community and preliminary results on spacecraft anomaly investigations.

2. Spacecraft and Experiments

2.1. OVERVIEW

The CRRES spacecraft was built by Ball Aerospace Systems Division (BASD) in Boulder, Colorado under joint sponsorship of the US Air Force and NASA. As shown in Fig. 5, the satellite is in the shape of an octagonal polyhedron. The large solar arrays on top provide the power necessary to run the spacecraft. The spacecraft was designed for a one year lifetime guarantee, but with a 3 to 5 year lifetime goal. The spacecraft has enough station-keeping propellant on board to maintain it on-orbit for 10 years. Data is recorded on satellite tape recorders (2) twenty-four hours a day,

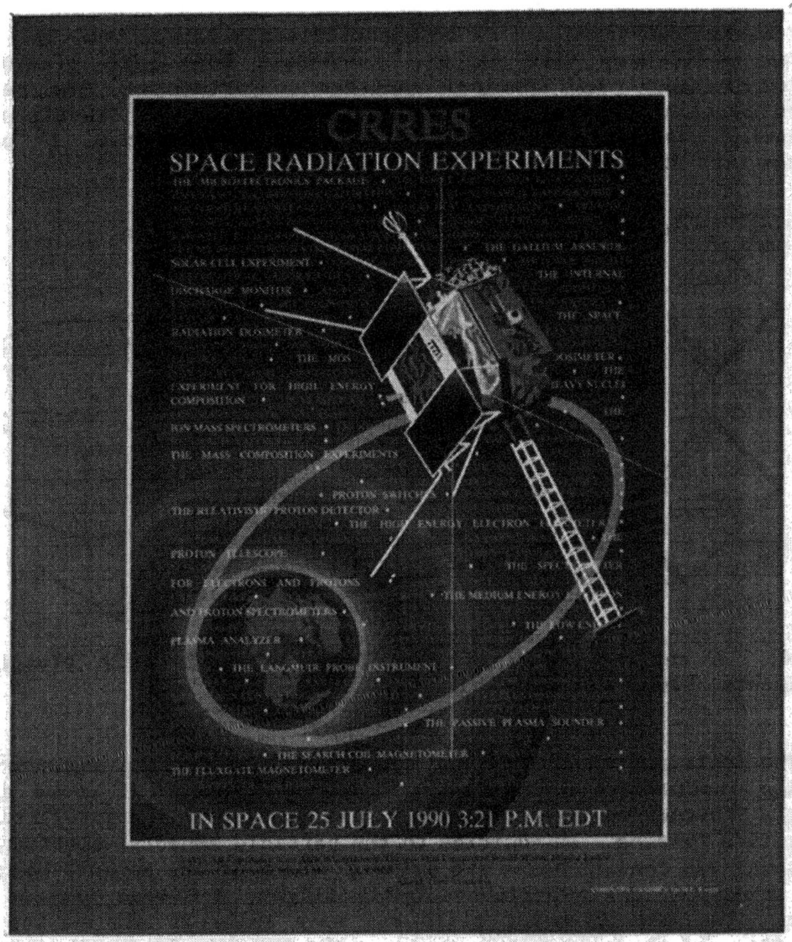

Fig. 5: A schematic diagram of the CRRES satellite and its orbit with a listing of the CRRES experiments and the agencies responsible for them.

7 days a week at 16 kbps. The data is transmitted to the ground daily via the USAF Consolidated Satellite Tracking Center (CSTC) and passed to the data reduction and distribution center at Hanscom AFB, MA.

The main body of the satellite is 37" high and 107" across the top. It has the center cut out for attaching to the launch vehicle. The vehicle weighed nearly 2 tons at launch but will be about 1.5 tons on orbit after releasing about .5 tons of chemicals. On orbit, the spacecraft spin axis is maintained such that the normal to the top surface, containing solar panels, is always within 15° of the solar direction. The spin rate is kept near 2 RPM. All experiments and booms are mounted externally or with external (to the spacecraft) viewing angles on the top, bottom or in 2 side compartments (numbers 4 and 8) on opposite sides of the vehicle. Figure 6 gives a cut-away view of the spacecraft body with placement of experiments, booms and chemical canister compartments. The radiation effects portion of the CRRES mission consists of 23 experiments: 3 engineering test packages, 4 field and wave instruments, 14 particle spectrometers and 2 dosimeters.

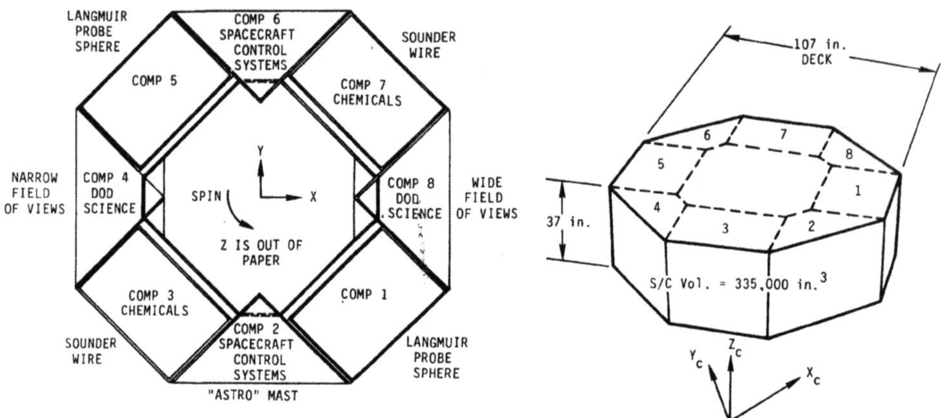

Fig. 6: A cut-away view of the CRRES satellite with placement of experiments, booms and chemicals.

The instrument locations and lines of sight for the engineering and particle instruments are given in Table 1. Figure 7 shows selected particle instruments from Compartment 4 as mounted in the satellite during integration testing. In addition to the particle sensors, electronic data processing and control boxes are evident. Many of the sensors have covers which protected the experiments until flight, and were removed before launch. The two 100-meter field and wave antennas are on top of the satellite, and the magnetometers are on the end of a 20 foot astromat boom deployed from compartment 2. More information on the satellite and experiment configuration is contained in the CRRES System Description Handbook [Ref. 16]. A preliminary description of the CRRES/Spacerad instrument has also been compiled [Ref. 17].

The particle detectors measure the complete particle spectrum of electrons and protons from a few eV to hundreds of MeV, and heavy ions from about 100 eV/Q up to cosmic ray energies of greater than 500 MeV/AMU. Table 2 lists the instruments together with their measurement ranges, energy resolution and angular resolution. Figure 8 shows the energy overlap and cross-calibration ranges for both electrons (8a) and ions (8b). The lower energy particle detectors can produce 3-dimensional distribution functions of both electrons and protons. The instruments are controlled with on-board processors so they can scan in modes to get the best energy, pitch angle and mass discrimination as the satellite orbit moves from one

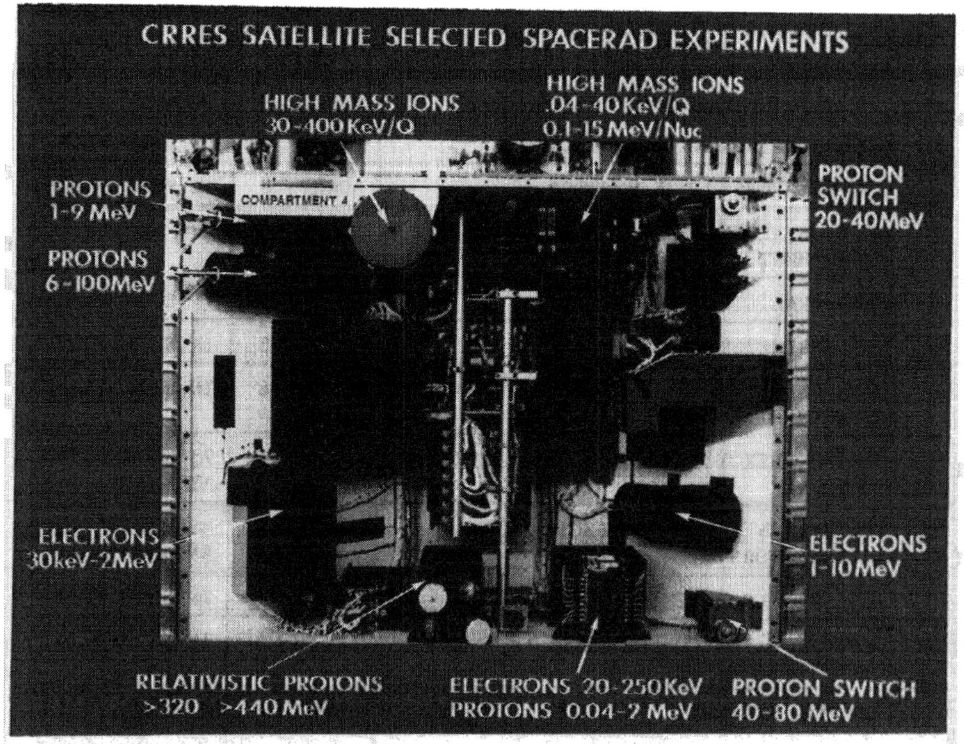

Fig. 7: CRRES experiments in Compartment 4.

very different particle regime to another. Three of the particle detectors that give primary support to the MEP analysis and the radiation belt model upgrades will be discussed in greater detail below. These instruments are the High Energy Electron Fluxmeter (HEEF), the Proton Telescope (PROTEL) and the Space Radiation Dosimeter.

The field and wave experiments on CRRES support the particle measurements in the following ways: a) The measured magnetic field is used to determine the particle pitch angle, that is, the angle between the magnetic field and the direction of the measured particles. With a good magnetic field model the particle pitch angle distribution at one point can be mapped to a distribution at any other point on the field line. This permits a major compression of particle measurements in three dimensions to a two-dimensional empirical model on the magnetic equator. b) The measured power in electromagnetic waves is used to assess pitch angle scattering of the particles into the loss cone where they are no longer trapped, but lost to the upper atmosphere. c) Measurement of the plasma frequency gives an independent measurement of the plasma density which can be used to cross-calibrate lower energy particle detectors. d) The wave spectrum is used to identify wave modes and sources of free energy

Table 1: Instrument locations and line of sight.

INSTRUMENT	LOCATION	LINE OF SIGHT
1. MICROELECTRONICS TEST PACKAGE	COMP. 8	IN SPIN PLANE
2. INTERNAL DISCHARGE MONITOR	COMP. 8	IN SPIN PLANE
3. GALLIUM ARSENIDE SOLAR PANEL EXPERIMENT	TOP	TOWARD SUN
4. SPACE RADIATION DOSIMETER	COMP. 8	IN SPIN PLANE
5. MOS DOSIMETER	COMP. 8	IN SPIN PLANE
6. HIGH ENERGY HEAVY NUCLEI COMPOSITION EXPERIMENT	COMP. 8	IN SPIN PLANE
7. MEDIUM ENERGY ION MASS SPECTROMETER	BOTTOM	75° FROM SPIN AXIS AFT
8. LOW ENERGY ION MASS SPECTROMETER (2 DETECTORS)	BOTTOM	45° & 75° FROM SPIN AXIS AFT
9. MASS ION COMPOSITION SPECTROMETER	COMP. 4	IN SPIN PLANE
10. LOW MASS ION COMPOSITION SPECTROMETER	COMP. 4	IN SPIN PLANE
11. PROTON SWITCHES	COMP. 4	IN SPIN PLANE
12. RELATIVISTIC PROTON DETECTOR	COMP. 4	IN SPIN PLANE
13. HIGH ENERGY ELECTRON FLUXMETER	COMP. 4	IN SPIN PLANE
14. PROTON TELESCOPE	COMP. 4	IN SPIN PLANE
15. SPECTROMETER FOR ELECTRONS AND PROTONS (2 DETECTORS)	BOTTOM	30° & 70° FROM SPIN AXIS AFT
16. MEDIUM ENERGY ELECTRON SPECTROMETER	COMP. 4	IN SPIN PLANE
17. ELECTRON PROTON ANGLE	COMP. 4	25° BELOW SPIN PLANE
18. LOW ENERGY PLASMA ANALYZER	COMP. 8	IN SPIN PLANE
19. HEAVY ION TELESCOPE	COMP. 4	IN SPIN PLANE

responsible for diffusion processes, which to first order control the belt locations and densities. e) Models of the plasma and wave regimes are made and used with X models to produce quasi-static and dynamic radiation belt models. f) Spacecraft charging events will be identified to determine their effect on particle measurements and instrument operations.

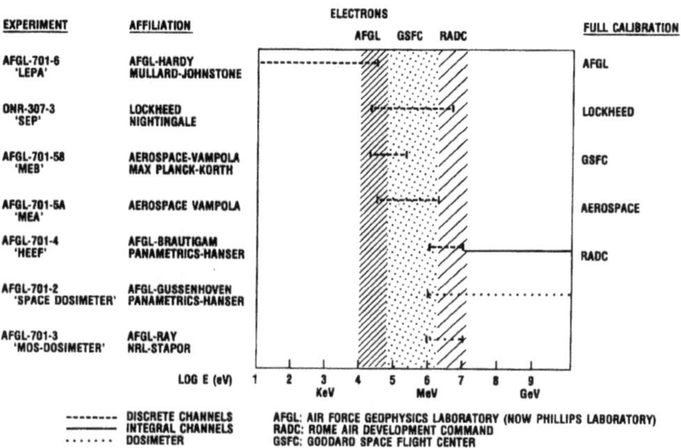

Fig. 8a: Energy ranges of the electron detectors on CRRES, with their affiliations and calibration facilities. Cross-hatching shows instrument groupings for cross-calibrations at the same facility.

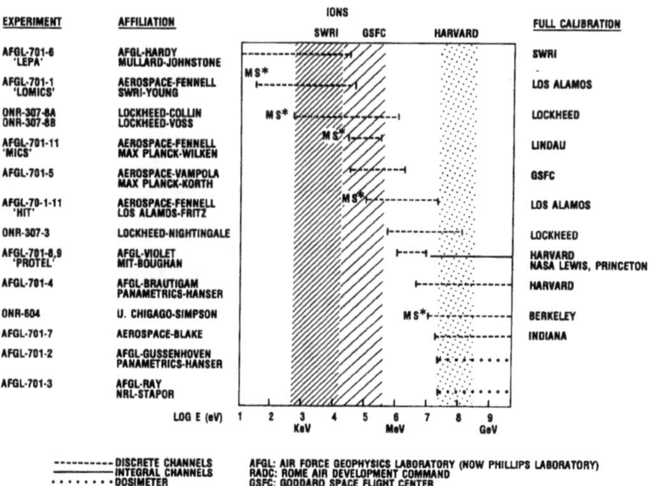

Fig. 8b: Same as 8a, but for ions. MS* designates instruments that differentiate ion mass.

Table 2: Particle detector characteristics.

NUMBER	AGENCY	MEASUREMENT		ENERGY RESOLUTION	ANGULAR RESOLUTION
701-6	AFGL EMMANUEL MULLARD	ELECTRONS PROTONS	10 eV - 30 keV	3%	±2.8° x ±4.0° STANDARD MODE ±2.8° x ±0.5° LOSS CONE MODE
701-5B	AEROSPACE MAX PLANCK	ELECTRONS PROTONS	20 - 250 keV 40 keV - 2 MeV	35% 35%	±2.0° x ±5.0°
307-3	LOCKHEED	ELECTRONS PROTONS	20 keV - 5 MeV 500 keV - 10 MeV	24 CHANNELS* 48 CHANNELS*	6° CONICAL 6° CONICAL
307-8A	LOCKHEED	IONS	0.1 - 32 keV/Q	10%	5° CONICAL
307-8B	LOCKHEED	IONS	20 - 8000 keV-AMU/Q	10%	4° CONICAL
701-11B	AEROSPACE LASL	IONS	40 ev/Q-40 keV/Q	7.5%	±6° x 15°
701-11A	AEROSPACE MAX PLANCK	IONS	30 - 300 keV/Q	10%	1° CONICAL
701-11C	AEROSPACE	IONS	100 keV/AMU - 15 MeV/AMU	<10%	6° CONICAL
701-5A	AEROSPACE	ELECTRONS	30 keV - 2 MeV	18 CHANNEL*	(±3° - ±11°) x ±11°
701-4	AFGL PANAMETRICS	ELECTRONS	1 - 10 MeV	10 CHANNELS*	±7.5° CONICAL
701-8,9	AFGL EMMANUEL MIT	PROTONS	1 - 100 MeV	24 CHANNELS*	±10° x ±10° LOW ENERGY ±12° x ±17° HIGH ENERGY
701-7A	AEROSPACE	PROTONS	20 - 80 MeV	20 - 40 MeV 40 - 80 MeV	180° CONICAL
701-7B	AEROSPACE	PROTONS	>320, >440 MeV	2 CHANNELS	
604	U. CHICAGO	IONS	20 ->500 MeV/AMU	<1%	1° CONICAL

* CHANNEL DEPENDENT ENERGY RESOLUTION

We turn now to discuss in greater detail the three high energy detectors mentioned above and the MEP, since these instruments provide the primary data for analysis in this paper.

2.2. HIGH ENERGY ELECTRON FLUXMETER (HEEF)

HEEF is a solid state spectrometer designed and developed to measure differential electron energy spectra in 10 energy channels from 1 to 10 MeV. The instrument is of a telescopic design with two solid state detectors (each 700 micrometers thick) stacked in front of a BGO scintillating crystal. The front solid state detector has an active area of 100 mm^2 and the back an active area of 50 mm^2. Figure 9 shows a schematic of the sensor head. In order to be counted, an electron must

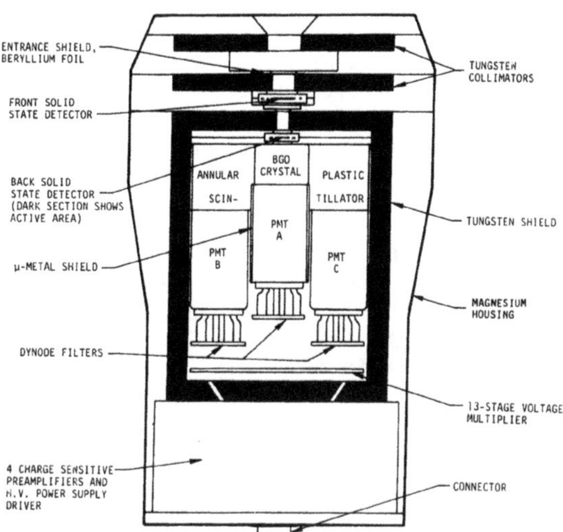

Fig. 9: Schematic diagram of the High Energy Electron Fluxmeter (HEEF).

produce a triple coincidence of pulses in the two solid state detectors and the BGO crystal, and have an anticoincidence with particles detected in the annular plastic scintillator surrounding the BGO crystal. The system works as follows: Each electron between 1 and 10 MeV that enters the detector through the aperture produces pulses in the two solid state detectors and photons in the BGO crystal. The photons produced in the crystal are seen by a photomultiplier tube (PMT-A), optically coupled to the crystal. The photomultiplier tube produces charge pulses proportional to the number of photons seen. The charge pulses seen in coincidence are all pulse-shaped in a shaping circuit. If the 3 pulses all fall within the proper broad pulse height range, the BGO crystal pulses are further analyzed in a pulse height analyzer and placed into the proper energy electron counter bin for transfer to shift registers for satellite readout. The registers are serially read every 0.5 seconds and stored by the satellite data storage and telemetry system. This produces a data base of 2 measurements per second per energy channel

The largest problem with measuring MeV electrons, especially in the CRRES orbit, is contamination due to high energy protons and electron bremsstrahlung. Great care was taken in the detector design to try and eliminate all counts other than the 1 to 10 MeV electrons which come directly in the aperture. Protons that come directly down the aperture are not counted because they have excessive energy loss in the solid state detectors (for protons with energies less than 100 MeV) or in the BGO crystal (for protons with energies greater than 30 MeV). For particles which can penetrate the heavy shielding (greater than 140 MeV protons and greater than 20 MeV electrons), the annular plastic scintillator produces anticoincidence counts which reject the particles being counted. A .006 inch beryllium foil stops electrons less than 140 keV that directly enter the aperture. The tungsten collimators and shield reduce bremsstrahlung, and the magnesium housing reduces bremsstrahlung production from less than 10 MeV electrons. The detector rejection of unwanted particles seems to have worked extremely well in that the data taken on orbit have been remarkably free of background contamination.

2.3. PROTON TELESCOPE (PROTEL)

PROTEL consists of two solid state detector assemblies (sensors) which measure protons from 1 to 100 MeV. The low energy sensor measures protons from 1 to 9 MeV in 8 contiguous energy channels. The high energy sensor measures protons from 6 to 100 MeV in 16 contiguous channels. The two sensors operate much like HEEF in that they consist of detector stacks in a shielded assembly that have to satisfy certain coincidence/anticoincidence conditions to produce particle counts. A schematic of the high energy sensor head is given in Fig. 10. Each system works as follows: a proton with energy between 1 and 9 MeV (6 and 100 MeV) that enters the low energy (high energy) sensor through the aperture produces pulses in the stacked solid state detectors. Two types of detectors are used, surface-barrier and silicon-lithium-drifted. The low energy detector uses five surface barrier detectors: the first 4 in coincidence to determine the input particle energy and the 5th behind an aluminum absorber in anticoincidence to reject particles above 9 MeV. The high energy sensor uses a total of 6 detectors, the first is a surface barrier detector and the rest are silicon-lithium-drifted. The first 5 detectors are used in coincidence to determine the input proton spectra and the 6th in anticoincidence behind a brass absorber to reject particles above 100 MeV. The energy lost in the silicon detectors by incoming protons determines the incoming particle energy. Pulses from individual detectors are pulse height analyzed and fed to an on-board microprocessor which performs the logic analysis to determine when double or triple coincidence pulses (simultaneous energy deposition in two or three detectors) are of the right pulse height to be counted. The processor puts the counts into the proper energy bin and feeds the data to the satellite storage and telemetry system at a rate which provides a 1 sample per second data base.

As with HEEF, contamination from high energy protons presents the biggest problem in producing an accurate data set. Great care was taken in the design to eliminate as much contamination as possible. Each sensor has a collimator made of aluminum in a saw-tooth pattern (see Fig. 10) to reduce forward scattering of particles into the acceptance cone. The sweeping magnet sweeps out electrons that would contribute to higher background levels in the first detectors. The passive shielding configuration of aluminum (Al) and a tungsten alloy (W) is shown in Fig. 10. The high energy sensor also has active anticoincidence rejection via guard rings around the silicon-lithium detectors. The ring is part of the same silicon wafer as the detector but is electrically isolated from it. Pulses detected in the detectors coincident with pulses detected in the guard rings are discarded. For most portions of the CRRES orbit the shielding and anticoincidence rejection do an excellent job, but at the inner edge of the inner proton belt where high fluxes of greater than 100 MeV particles exist, contamination levels can become significant. As will

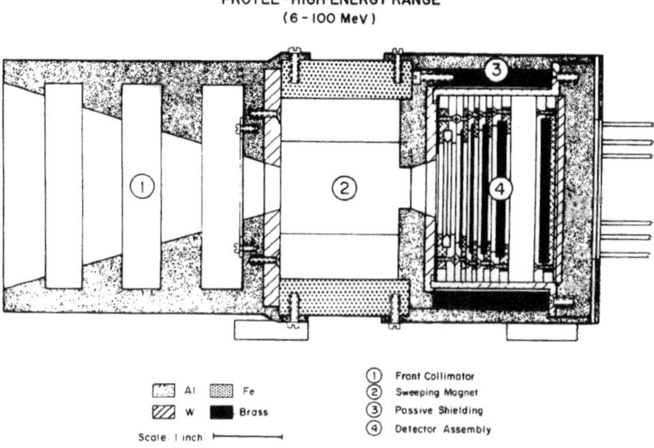

Fig. 10: Schematic diagram of the high energy head of the Proton Telescope (PROTEL).

be discussed below, extensive modelling of the detector is used to reduce/eliminate the data contamination in the inner belt region.

2.4. SPACE RADIATION DOSIMETER

The dosimeter is the work horse of the particle detectors and provides a well-calibrated level of radiation dose for intercomparison with other instruments. It is essentially the same design as the dosimeter flown on the DMSP F7 spacecraft with great success [Ref. 18, 19]. The Space Radiation Dosimeter measures the radiation dose, that is, energy deposited, from both electrons and protons behind four aluminum hemispheres of different thickness. We refer to the four measurements in terms of progressively thicker shielding with 1 being the measurement behind the thinnest shielding. The dosimeter also provides information on the number flux of electrons and protons at energies above the thresholds defined by the shields by counting the number of energy deposits. Figure 11 is a schematic drawing of one of the four sensors of the instrument. The solid state device selected as the active measuring element is a p-i-n, diffused-junction silicon semiconductor with a guard ring, permitting a threshold of 50 KeV to be set for the energy deposition in the device. This allows the detection of both the high energy particles and most of the bremsstrahlung produced in the shield without compromising on-orbit lifetime. The devices used in this detector are 400 microns thick. The Dome 1 detector area is .008 cm^2. The detector areas under Domes 2 and 3 are 0.015 cm^2 and under Dome 4, 1.00 cm^2. Thus, the latter three detectors are planar in nature. The thickness of the domes were chosen to provide electron energy thresholds for the four sensors of 1, 2.5, 5, and 10 MeV, and proton energy thresholds of 20, 35, 51, and 75 MeV. Particles that penetrate the shield and bremsstrahlung produced in the shield will deposit

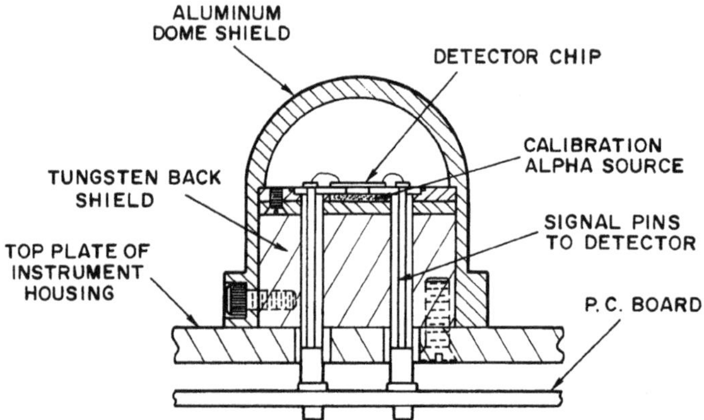

Fig. 11: Schematic diagram of one of the Space Radiation Dosimeter detectors on CRRES.

energy in the device producing a charge pulse. The charge pulse is shaped and amplified. The pulse height is proportional to the energy deposition in the detector. The characteristics of the detector and the threshold are such that energy depositions between 50 keV and 1 MeV give what is termed the low linear energy transfer (LOLET) dose. Depositions between 1 MeV and 10 MeV give what is termed the high linear energy transfer (HILET) dose. The LOLET dose comes primarily from electrons, high energy protons (100-200 MeV), and bremsstrahlung. The HILET dose is primarily from protons below 100-200 MeV. A full listing of detector parameters is given in Table 3.

The dose is directly proportional to the total energy deposited in the detector. The omnidirectional number flux can be estimated by modeling the energy deposition from an assumed distribution and matching the predicted and actual response. The dosimeter is periodically calibrated in-flight with a weak (approximately 0.3 nCi) Am-241 source that emits 5.5 MeV alpha particles. Between the source and the detector is a thin foil that changes the shape of the spectrum such that it peaks near 3 MeV with a half width of 1 to 2 MeV. By having the source in the dosimeter it is possible to measure and correct for any changes in operating efficiency, and thus get accurate long-term dose measurements.

To determine areas of potential SEU upsets in near-Earth space, the dosimeter also has the capability of separately counting very high linear energy transfer (VHLET) deposits in its solid state detectors. The characteristics of the detector and the threshold are such that energy depositions above 40 MeV in detectors 1, 2, and 3, and 75 MeV in detector 4, are counted as very high linear energy transfer dose or "nuclear star" events. The "nuclear star" counts come from high energy proton interactions inside and/or near the sensitive device element or by direct energy deposition from heavier cosmic rays. Unlike the DMSP F7 dosimeter, the detector sizes and thresholds are such that there are no possible path lengths long enough for direct energy deposition from protons to trigger VHLET counts. Counts for each of the 5 measured parameters, LOLET dose,

Table 3: Space Radiation Dosimeter characteristics.

DOME NO.	DOME THICKNESS (mils Al)	DETECTOR AREA (cm^2)	ENERGY DEPOSITION COUNT RANGES LOLET (MeV)	HILET (MeV	VHLET (MeV)
1	82.5	0.008	.050 - 1	1 - 10	>40
2	232.5	0.051	.050 - 1	1 - 10	>40
3	457.5	0.051	.050 - 1	1 - 10	>40
4	886.5	1.000	.050 - 1	1 - 10	>75

DOME NO.	ALUMINUM SHIELDING (gm/cm^2)	DETECTOR THICKNESS (microns)	PARTICLES MEASURED LOLET ELEC. (MeV)	PROT. (MeV)	HILET PROT. (MeV)	VHLET PROT. (MeV)
1	0.57	403	>1.0	>130	20-130	>46
2	1.59	434	>2.5	>135	35-135	>56
3	3.14	399	>5.0	>140	51-140	>69
4	6.08	406	>10.	>155	75-155	>111

LOLET flux, HILET dose, HILET flux and VHLET, are sent to the satellite telemetry system once every 4.096 seconds for each of the 4 sensors.

2.5. MICROELECTRONICS TEST PACKAGE (MEP)

Because the MEP is the critical mission payload for a successful CRRES mission, great care has been taken to develop an optimal experiment. A working team of experts in radiation-hardness-electronics testing was established to a) oversee the design and fabrication of the package; b) select the appropriate devices for both single event upset (SEU) and total dose (parametric degradation and annealing) testing; and c) develop appropriate space test procedures. Each device selected had to be justified by one or more of the selection criteria established by the team for total dose/annealing or SEU or both.

2.5.1 *Device Selection.* The selection criteria for total dose/annealing were:

1. Components are of specific interest to present DOD space systems.
2. Components are of general interest to presently developing DOD systems and may be selected for use in future space systems.
3. Components are common to many existing and planned space systems and usually limit the lifetime due to low radiation tolerance levels.
4. Components are all CMOS, but made with different radiation hardening processes.
5. Components are of the LSI (Large Scale Integrated) bipolar variety.

6. Components are from similar functional groups, but made with varying technologies.

7. Components are manufactured with advanced VLSI processing techniques.

8. Components are radiation-hardened, metal gate CMOS components.

9. Components are used by NASA in instrumented balloons.

10. Components qualify for use in multiply redundant systems.

The selection criteria for single event upsets were:

1. Devices are known to be sensitive to SEU and will thus provide good statistics for long term studies.

2. Devices are or will be modeled for SEU sensitivity and mechanisms, and will be used to test model predictions.

3. Devices are or will be tested on the ground at accelerator facilities so that the validity or usefulness of ground test measurements can be determined.

4. Devices are from specific DOD or NASA systems for which an SEU sensitivity data base would be extremely useful.

5. Devices are of feature size, geometry, or technology such that the effect of these on SEU rate can be distinguished in space.

6. Devices are constructed with new technologies and are not yet in production.

7. Devices represent a variety of function types including RAMs, microprocessors, PROMs, etc.

8. Devices are expected to show a change in SEU sensitivity with total dose exposure.

9. Devices are expected to show a change in SEU sensitivity with changes in operating conditions (voltage, cycle time, source current, etc.).

10. Devices are CMOS that have been SEU-hardened by a variety of methods, in particular those that involve cross-coupled resistors.

11. Devices are CMOS and are known to be sensitive to SEU latchup.

The devices selected for test include random-access memories (RAMs), microprocessors, programmable read-only memories (PROM's), analog-to-digital converters, octal latches, inverters, gate arrays, operational amplifiers, comparators, hexfets, and special test circuits. The special test circuits are devices being developed by JPL and NRL to test different geometries, feature sizes, and technologies on a single chip. These devices should give us much needed insight to device behavior for radiation modeling. The total test package device complement includes over 60 device types and approximately 400 total test devices. Thus, the MEP tests total dose and SEU response in a wide range of device types and technologies important to the entire space community.

2.5.2. *The Flight Instrument and Operations*. The MEP is a radiation hard, microprocessor-based (SA3000) test system for electronic devices. It measures the actual radiation susceptibility of selected advanced technology devices in orbit. It also measures the internal MEP radiation environment using semiconductor dosimeters (56) and silicon and Gallium Arsenide pulse height analyzers (1 each). There are also 46 temperature monitors within the experiment to differentiate dose from thermal effects. The box is shown, with its front cover removed, in Fig. 12. It weighs 106 lbs, has a front face of 45 x 15 inches, is 15 inches deep at its deepest point and draws up to 140 watts of power. It is one of the most complex electronic experiments ever launched. It contains more than 14000 total parts on 40 printed circuit boards. There are a total of 280 printed circuit board layers with more than a quarter of a million interconnects, with some boards having as many as 14 layers. The physical design provides for maximum radiation exposure of devices on the front board since the MEP cover has a thickness of only 10 mils aluminum. The predicted dose for devices on the front board, using the NASA radiation models and an AFWL materials model of the MEP, was 1 megarad during the first year of operation. Because of the high exposure level of the front boards, deep dielectric charging of the board material from penetrating electrons was considered a threat to the proper operation of the MEP. To eliminate this problem, a 'leaky' (in conductivity) paint was developed [Ref. 20] and applied to the front board surfaces to bleed off charge accumulation and prevent discharges inside the package.

Fig. 12: The CRRES Microelectronics Test Package (MEP) with front cover removed.

The box is redundant in all major control units which include the main processor, power supplies, interface control units and device-under-test (DUT) bus systems. The MEP system block diagram is shown in Fig. 13. The system is cross-strapped so a single controller can operate all test devices. The DUTs are in blocks of 1 to 5 chips (depending on power draw), can be individually turned on or off, and are current limited to turn off if preset power levels are exceeded.

The SEU testing is performed once every 2 seconds with 1 of 4 patterns (all zero's, all one's, checkboard one/zero, or checkerboard zero/one). RAMs are tested with all patterns, PROMs with only checkerboard one/zero. If no error is detected, no data are sent. If an error is detected, the

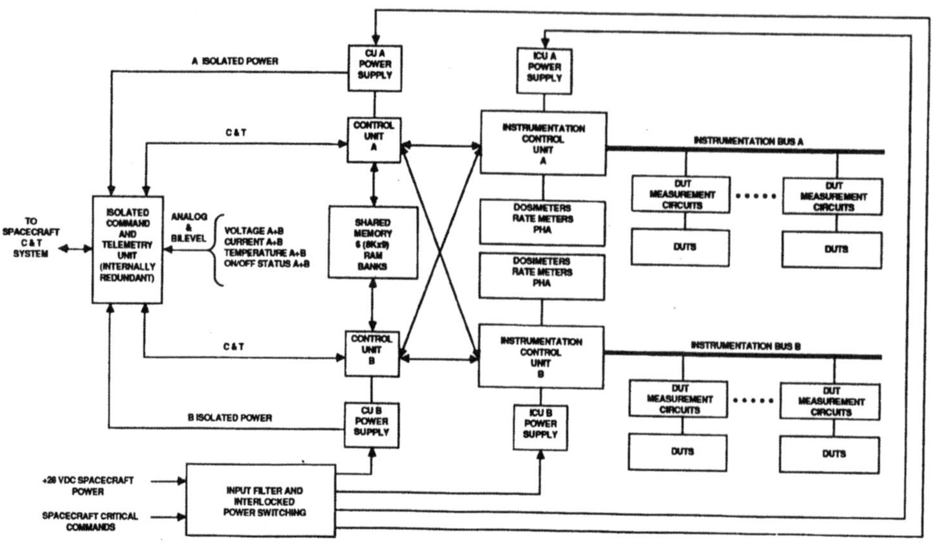

Fig. 13: The circuit diagram for the Microelectronics Test Package on CRRES.

processor immediately rewrites the device and retests. If the error persists it is called a hard error, checked against a permanent fault list, written to the permanent fault list if not already there, and reported in the telemetry as to the block, device, bit address and pattern (to identify whether the bit flipped from 1 to 0 or 0 to 1) of the error. If recovery is seen, the error is called a soft error and is also reported. Soft errors are not put in the permanent fault file. If the number of permanent faults accumulated in a single device exceeds a preset limit, the device will be automatically mapped-around (not tested) for future SEU testing unless overridden by ground command. If more than 5 soft SEUs per 2 seconds are seen during a single device test, the device is put into a compression or histogram mode so that only the number of hits during the accumulation period are reported and not the addresses or bit flip directions. This mode was adopted because of the high number of upsets projected for solar proton events and for the heart of the inner proton radiation belt. (To date the SEU tester has never gone into a histogram mode due to particle upsets.) Special, SEU-soft, 64 K dynamic random access memories (DRAMs), called "ratemeters," were buried deep in the MEP to simulate heavily shielded parts on operational space systems. Only the total number of upsets per device in the two second measuring interval are sent to the spacecraft, and not the addresses of the SEUs.

Total dose testing is performed continuously throughout the orbit, but only once an orbit (about every 10 hours) for each device. The test data are placed into the telemetry stream. Each block of devices has its own independent test circuit which is device dependent. Although the devices remain under bias at all times, they are turned on serially with their tester circuit only at the actual time of testing to reduce power consumption. The data parameters measured are device dependent currents and voltages. These include voltage threshold, output leakage current, open loop gain, input offset voltage, input offset current, input bias current, among others. The MEP also contains some individual device radiation shielding packages called RADPAKs. Dosimeters and non-radiation hardened RAMs are placed in the packages to assess the shielding effectiveness. The packages contain both high Z and low Z material to reduce both direct radiation and bremsstrahlung. The package configurations are the result of extensive radiation transport modeling and are optimized for the types of environments typically experienced by space systems in near-Earth space. The device shielding concept should provide better dose shielding and less weight than box shielding.

The dosimeters scattered throughout the MEP are radiation soft P-channel Metal Oxide Semiconductor (PMOS) transistors, first developed for dosimetry by [Ref. 21]. The basic principle of operation is as follows: When the semiconductor material in the device is exposed to the strong local fields from radiation particles, more electrons than "holes" are lost due to the higher mobility of the electrons. The charge becomes stably trapped in the transistor and changes the operating threshold voltage. Measuring the change in this voltage at a constant current can be directly related to total integrated dose accumulated in the device. The range of these dosimeters is typically 0 - 25 kilorads, but the range can be raised by grounding the gate bias during exposure and only applying voltage to the gate during the short test interval. This was done for 16 of the dosimeters in the MEP extending their range to approximately 50 kilorads. It should be noted however that the relationship between voltage shift and integrated dose is not near-linear as was originally thought. Ground testing of these devices under different operating conditions has shown that assuming a linear relationship will underestimate the dose received, especially at higher total dose levels. Since these devices can only be calibrated by accumulating dose in them, which destroys their usefulness, and no two devices have been found to be exactly the same during testing, there is an error level, not yet fully determined, that must be accounted for to properly interpret the data. These devices are also found to be extremely temperature sensitive as will be shown below in the data discussion section. To get around some of these problems on CRRES, the MEP dosimeters are being cross-calibrated with other particle/dose instruments.

Because the MEP measures so many different types of data, and only records data when things are happening (like an SEU or a total dose parameter), the data are packetized and stored in a buffer within the MEP until they are prompted for transmission to the satellite telemetry system. The data rate for the MEP transmission is 200 bits per second which can be far slower or far faster than the rate at which the buffer is filled. Each data packet is identified by a unique one-byte header which gives both the length and the content of the packet. The data are synchronized with the satellite telemetry system to guarantee the first byte in any major frame is a header. This prevents corruption of the data flow from loss of synch, noise, and so forth, for more than 1 major frame, i.e., a period of 4.096 seconds. Time packets are sent every second, and null packets are sent at the end of a major frame if insufficient space remains to send the next data packet or if no data packets are available in the buffer to be sent. There are over 100 types of data packets. These are described in the CRRES MEP Serial Telemetry Manual [Ref. 22]. Within the data packets are the pertinent SEU, total dose or housekeeping parameters. The data packets include a checksum at the end to guarantee the packet was properly transmitted.

The MEP was extensively tested as a unit before flight and the data from all the testing have been summarized in a report [Ref. 23]. This report is necessary to properly understand the flight data since some "flaky chips" were flown, and some test circuits did not perform exactly as envisioned. Because of the overall complexity of the package, the minor nature of the flaws in comparison to overall mission goals, and the risks associated with disassembling and reassembling the MEP to make changes, we decided to fly the MEP without making all the corrections and test part replacements necessary for a 100% operating package. Duplicity of test parts (4 to 10 for each part type) and test circuits (separate for each bus) meant that some of all part types and well over 90% of all parts were being properly tested before flight.

3. Data Reduction and Data Base Generation

The wide variety of CRRES experiments and program products make it necessary to develop data bases that are user-friendly, transportable (via communication links, storage media, etc.), and versatile. They must accommodate changes in on-orbit operations, calibrations, instrument sensitivity levels, etc. Because the data base is large (500 gigabits per year) and the initial processing time is long, we required a modular data management system with sufficient flexibility at each step that total data base regeneration would not be required if errors were found months or years after launch. Initial agency tape files are created with separate ephemeris, attitude and magnetic field files. From this, the Time History Data Base (THDB) is constructed. It is a data base ordered by time and instrument, in which the raw data are kept at the highest time resolution of the instrument along with calibration and comment files which tell how to convert raw telemetry counts into engineering units. The files are produced on an orbit basis, namely, for one revolution of the satellite around the Earth from perigee to perigee. They are written for easy attachment to the ephemeris files. The files are held on mass storage devices for a 1 year period after which they are rotated to off-line storage as new data become available. An on-line data directory is available to indicate the status of the THDB at any time. Details on the file structures of the THDB are contained in the CRRES THDB Data Reduction Task document [Ref. 24].

Validation of data in the THDB is a major chore. Shortly after CRRES launch, all software used to create the THDB from agency tapes was verified. In addition, ongoing quality checks that remove spurious noise, contamination from unwanted particles, temperature effects, etc., must be performed before the data can be used for analysis. Also calibrations have to be verified with on-orbit data, and cross-calibrations of instruments which overlap in energy have to be performed.

For high energy electrons, three instruments are being used to support engineering type studies: the High Energy Electron Fluxmeter (HEEF), the Medium Energy Electron Analyzer (MEA) and the Electro-Proton-Angle Spectrometer (MEB). The electron energy ranges of these three instruments are 1 to 10 MeV, 30 keV to 2 MeV, and 20 keV to 250 keV, respectively. Much of the cross-calibration of these 3 instruments has already been performed. Figure 14 shows a distribution function for electrons as a function of energy, made using data from the 3 instruments. For MEB, only Detector 1 is used. Each instrument is designated by a separate symbol. The spectrum is made from data averaged over 1/20th R_E centered on L = 6.025 R_E, and 5° in pitch angle, centered on 85°. The data were taken at high altitude and at considerable distance from the magnetic equatorial plane $(B/B_0 = 1.32)$. There were no high energy protons in this region. The distribution function is smoothly varying over the complete energy range and the overlap agreement among the 3 instruments is excellent.

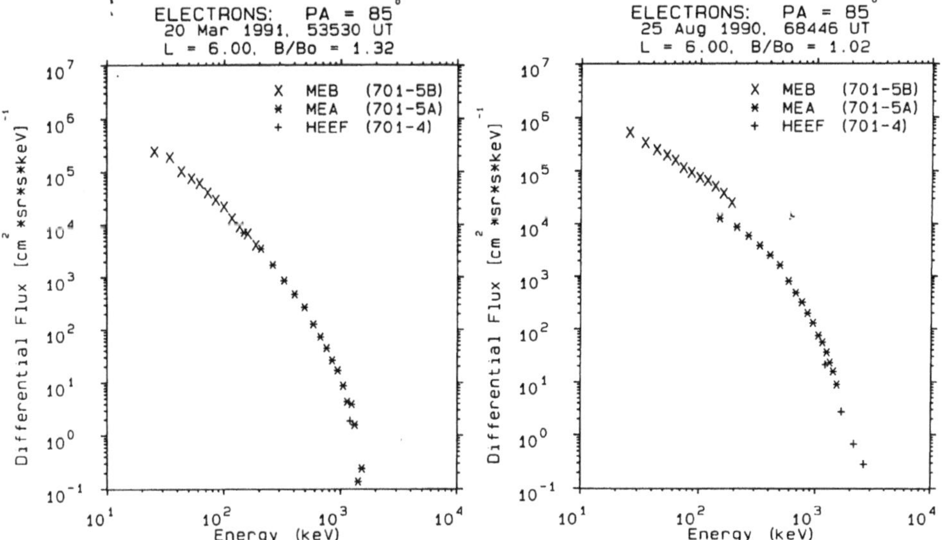

Fig. 14: Electron differential flux as a function of energy, measured by HEEF (crosses), MEA (stars), and MEB (x's) on 20 Mar 1991 at an L-value of 6 R_E.

Fig. 15: Same as Fig. 14, but on 24 August 1990. In both figures the data are averaged over 1/20th of an R_E, centered on L = 6.025, and over 5° in pitch angle, centered on 85°.

This electron spectrum is one of the most steeply falling spectra encountered so far. For energies > 1 MeV the spectrum falls as E^{-N}, where N, the power law index, is -12. In the same format, Fig. 15 gives an electron distribution function taken during an earlier orbit. The spectrum was averaged in the same way as before, taken at the same L-value, but was much closer to the magnetic equator $(B/B_0 = 1.02)$. A weak solar proton event occurred at this time. In this case, the power law index for energies > 1 MeV is -6.5, a much harder spectrum than in Figure 14. HEEF channels up to 3 MeV are well above background in this case. The overlap between HEEF and MEA is not good, as shown by the break in the distribution function. From examination of many other spectra we conclude that electrons with energy greater than 2 MeV can contaminate both of the medium energy electron sensors, MEA and MEB. Corrections for the contamination are still being prepared and must be made if data from these instruments are to be used in the midst of healthy high energy particle populations. The HEEF data are remarkably contamination-fee except at the inner edge of the inner radiation belt where the highest energy protons are trapped. Here some subtraction will be necessary, the exact level still to be determined.

For protons, the primary instruments being used for both engineering studies and radiation belt upgrades are PROTEL and MEB which have energy ranges 20 keV - 2 MeV, and 1 to 100 MeV, respectively. The data from these instruments look to be relatively free from contamination over most of the orbit, except in the inner edge of the proton belt. Figure 16 shows a distribution function of PROTEL and MEB data, averaged in the same manner as the electron data. Three of four MEB detectors were used in the average. The spectrum was taken at high altitude during a moderately large solar proton event. Not only are the data taken at high altitude (L = 6.55

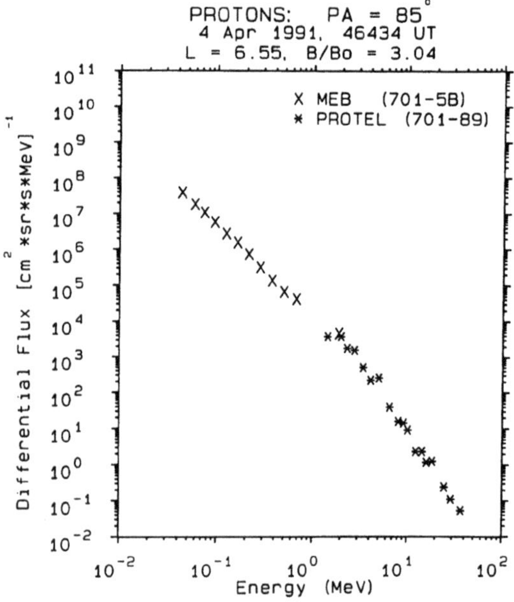

Fig. 16: Proton differential flux as a function of energy, measured by PROTEL (stars) and MEB (x's), on 4 April 1991, during a solar proton event. The data are averaged in the same manner as the electron data.

- 6.60), but are taken nearly as far off the magnetic equator as CRRES can go (B/B_0 = 3.04). Again the function is smoothly varying over the whole energy range, and well-matched near 2 MeV. The power law index for the spectrum above 2 MeV is -4.0. Although there are a significant number of high energy protons in this event, they do not contaminate the spectrum because of the rapid spectral fall-off. At the lower edge of the inner radiation belt, where significant numbers of protons with energy > 100 MeV are trapped, contamination poses a formidable problem. The proton spectrum at this altitude falls off slowly with energy, and the higher energy particles can penetrate detector shielding and produce false counts at lower energies. The false counts can be more than the true counts if a worst-case, flat spectra is assumed. Extensive Monte Carlo contamination modeling of the PROTEL detector has been done [Ref. 25]. The model takes into account all the materials of the detector and the arrival angle of the particle within the detector. Radiation transport codes propagate the particles into the detector with appropriate energy loss, and Monte Carlo statistics are used to determine the probability of a count being measured given the sensitivities of the individual detectors and the coincidences and anticoincidences needed. The modeling indicates contamination levels can be extremely high in the highest PROTEL energy channels, for a spectrum falling at E^{-2} or more slowly. However, two points can be made: 1) Most of the contamination predicted by this model comes from particles above 100 MeV. 2) The accuracy of the model is not directly determinable. Preflight testing of PROTEL to determine high energy proton rejection and on-orbit comparison indicate the contamination is less than the model indicates. The final contamination levels are still being determined and subtraction routines using proton measurements from other instruments, as well as singles counts from the back detector of each PROTEL head, are still in development for the inner belt region. Above the inner belt region contamination is minimal due to the paucity of higher energy protons, even during solar proton events which only extremely rarely contain significant particle populations above 100 MeV.

The Space Radiation Dosimeter data can be handled in a straightforward manner except when the instrument is accidentally turned off or put into a calibration mode for a long period of time. The instrument basically measures LET, and as such, does not differentiate particle type. Although we normally refer to LOLET data as electron data because we use it to measure dose in the outer zones, at the inner edge of the inner belt the LOLET response is dominated by high energy protons. Thus, to properly interpret the data, it is necessary to know what particle regime the measurements come from. Similarly, protons are the primary contributors to the HILET data, but in the presence of high fluxes of > 4 MeV electrons, contributions from this population are seen in the first two detectors, e.g., those with the thinnest shielding. For VHLET at high L-values, in addition to proton stars, direct deposits due to heavy nuclei cosmic rays are also possible. In fact, most of the VHLET counts at altitudes above an L of 2 are most probably due to galactic cosmic rays (except in solar proton events) since this area of space does not contain significant levels of high energy protons. The only real editing done to the dosimeter data is the removal of obvious noise spikes. For total mission dose, taken from the exponent-mantissa counters, and not the four second measurements, interpolations must be made through those time periods during which the instrument was inadvertently turned off by false commands or noise. Figure 17 shows a plot of total dose versus orbit for the first 692 orbits of CRRES behind the thinnest dome. The HILET and LOLET contributions are included. Changes in the slope of the curves reflects changes in dose rate. The large increase in dose rate near orbit 600 is related to proton and electron flux increases commencing around 23 March 1991 which will be discussed below.

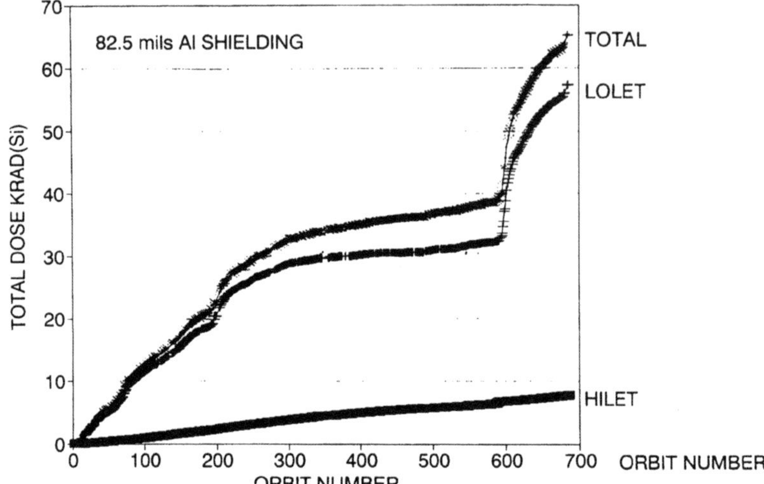

Fig. 17: LOLET, HILET and total dose as a function of orbit number, measured by the Space Radiation Dosimeter on CRRES from 25 July 1990 to 7 May 1991.

Of all the data to prepare for analysis, the MEP data are the most difficult. It was necessary to develop sophisticated software routines to unpack the data and place them into individual device files. The software must recognize each data packet type, determine what parameters are contained in the packet, and place the time of the data measurement with the data. The time must be determined by counting time pulses in the data stream since there are no clock pulses in the instrument data, and satellite time of storage and transmission does not relate to the time the data are taken. The problem is further complicated by time jumps in the spacecraft clock and missing data frames. Before the data can be analyzed, all anomalous behavior must be removed. Among other effects, this includes correcting for temperature changes, removing known bad data from pre-launch test results, removing data from on-board testing idiosyncracies, removing spurious behavior as parts reach end-of-life levels due to total dose accumulation, accounting for tester deficiencies not corrected pre-flight, and editing out data during on-board testing to simulate upsets. Much of the data reduction can only be done by the analysts who were involved in the original conceptual design, and it requires continual referral to circuit diagrams to ensure the proper interpretation of real effects.

Because of manpower and funding limitations, some of the data packet types have still not been edited or even examined for problems nine months after launch. However, much of the data have been edited and some of the results, given below, are new and exciting. To give an idea of one type of correction required, we show in Fig. 18 the temperature and dose from one of the thermistors and one of the MOS dosimeters in the MEP. The temperature fluctuations produce dose variations which are not real. Normalization techniques are used to remove the temperature effects. The MEP data are not contained in the THDB and so portions of the two data bases must be merged in order to perform correlations to determine cause and effect relationships between particles and upsets.

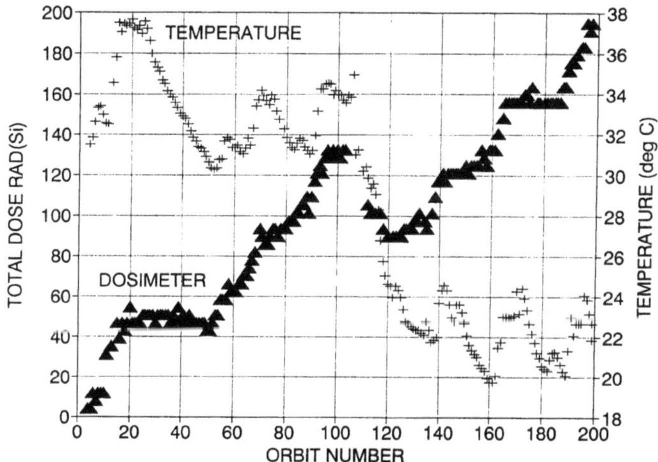

Fig. 18: Temperature and dose measurements from a thermistor and a MOS dosimeter in the MEP, plotted as a function of orbit number from 25 July 1990 to 16 February 1991.

4. CRRES Radiation Particle Analysis

One of the prime products to be produced from CRRES is a new radiation environment model. In final form it should consist of several models both static and quasi-static to accommodate both long term and short term missions. Existing radiation models are only average models and can be expected to give poor fluence estimates for certain orbit requirements or for short term missions. To get a data set statistically large enough to look for spatial versus temporal effects, a minimum of three to five years of data are needed. To address differences between solar minimum and solar maximum, 11 years of data would be necessary. Although CRRES has sufficient hydrazine to last that long in orbit, the severity of the orbit coupled with electrostatic discharge problems on the satellite indicate that a lifetime equal to a full solar cycle is improbable.

At this time 5 radiation models are commonly used: 2 NASA electron models, AE8MAX and AE8MIN [undocumented; Refs. 26, 27]; 2 NASA proton models, AP8MAX and AP8MIN [Ref. 28], and 1 cosmic ray model developed at NRL [Refs. 29, 30, 31]. Comparison of earlier DMSP dosimeter data with AP8MIN and AE8MIN showed that the NASA proton models are generally very good for determining long term dose levels, while the electron models were high by factors of 4 to 10+ [Ref. 18]. Similar model comparisons with CRRES data were done for early orbits [Ref. 32]. The results were basically the same as those for the DMSP orbit. Figure 19 shows a comparison of the AP8MAX omnidirectional flux with CRRES PROTEL flux in the 3 - 4 MeV range, for an orbit on 6 August 1990. The agreement is very good in terms of fluence (approximately equal to the area under the curve). However, the high altitude shapes of the two profiles differ. We will show later that the occurrence of the second belt, although rare, is evidence for proton belt dynamics, hitherto ignored. When averaged over long time periods, the agreement for total flux for the CRRES orbit is excellent.

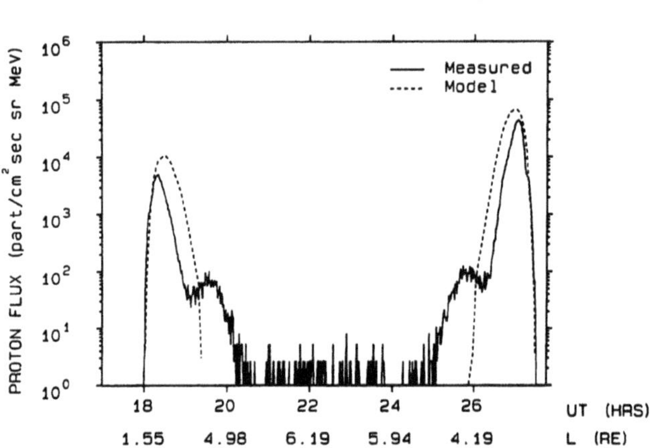

Fig. 19: One-minute averaged proton differential flux from 3 - 4 MeV, as a function of time, for one full CRRES orbit on 6 August 1990. The dashed curve shows the predicted flux from the NASA AP8MAX radiation model for the same orbit.

Figure 20 gives a similar comparison of 2 - 2.5 MeV CRRES electron flux, measured by HEEF, with the AE8MAX flux. Here huge differences are seen between the two plots. Those who defend the model point out that high fluxes of electrons occur episodically, lasting for days, and that these drive the average way up. To examine this hypothesis, we make the same comparison during the large electron event that occurred on 29 March 91. Figure 21 shows the comparison for an orbit on this day. These data are taken near the peak of the largest MeV electron fluxes observed on CRRES to date. For this period the data near apogee closely approximate the model. However, even the measured values at apogee are not sufficiently high to bring an average made over months of data at levels shown in Fig. 20 to the NASA model levels. Other difficulties with the electron model exist. The measured low altitude inner zone electron fluxes differ from

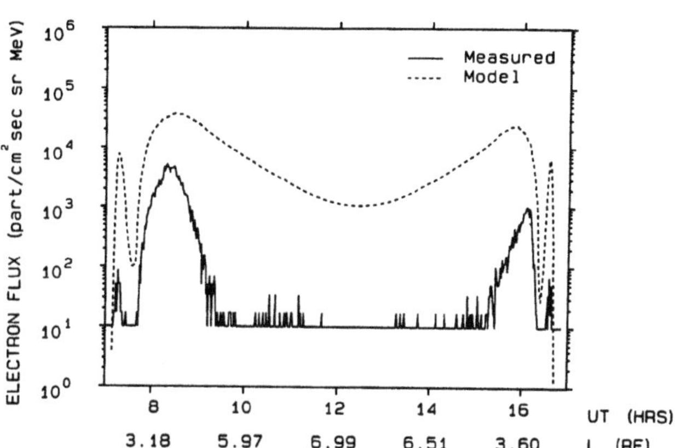

Fig. 20: Same as Fig. 19, only for 2-2.5 MeV electrons and for the NASA AE8MAX model. The data were taken on 28 July 1990.

2.0 - 2.5 MeV ELECTRONS 29 Mar 1991

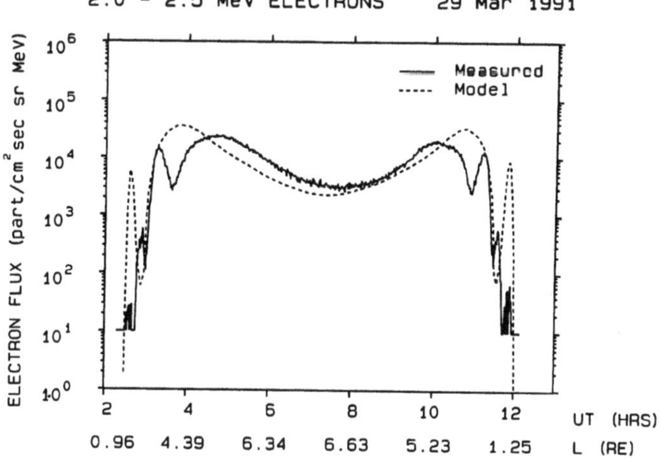

Fig. 21: Same as Fig. 19, only for data taken on 29 March 1991.

the model fluxes by orders of magnitude. This is most likely because early electron inner zone measurements that were used in the model were contaminated by protons. The CRRES electron experiments have far better proton rejection than any other instrument flow to date. Note also that the NASA electron profiles remain high and smoothly varying near apogee, while the CRRES measurements often show a dropout to background levels as in Fig. 20. The difference between measured and model fluence over a full orbit can be more than an order of magnitude for energies near 1 MeV. Because of the consistent discrepancies found when comparing CRRES and DMSP data with the electron models, we do not recommend their use. If they are used, they should be considered a worst case average environment.

Preliminary models are presently being prepared from five data sets, PROTEL, HEEF, MEA, MEB and the Space Radiation Dosimeter. The NASA models are given in terms of omnidirectional integral flux in an L - B/B_0 grid. B is the value of the magnetic field at the satellite, and B_0 the value of the magnetic field when the orbit position is mapped to the magnetic equator. This is also the way we prepare our dose maps. However, we are initially preparing our flux models in terms of differential flux, averaged in 5° pitch angle bins and in the discrete energy widths of the instrument channels. The data are averaged in spatial bins of 1/20 R_E in L. The binned data for each orbit are mapped back to the magnetic equator conserving the first adiabatic invariant and using a combined IGRF (International Geomagnetic Reference Field) and Olson-Pfitzer, 1974 [Ref. 33] quiet magnetic field model. Data from selected orbits are combined. For the preliminary models shown here the selected orbits were those for which the measured magnetic field on CRRES stayed within 5% of the model field throughout, ensuring that the mapping to the equator is valid. We call this the 'Quiet Model'.

Figure 22 shows the preliminary proton model for energies between 100 keV and 66 MeV. Differential flux is mapped in color in an L-shell versus pitch angle grid. The maps show an inward dispersion of peak flux with energy, and normal pitch angle distributions, i.e., peaking at 90° and falling off smoothly into the loss cone. By constructing the model at the magnetic equator at all pitch angles it is possible to map the particle fluxes anywhere along the flux tubes down to the mirror points. Thus, one can determine the fluence, dose, etc., anywhere in the belt region.

Figure 23 shows the corresponding electron model for 340 keV to 4.5 MeV electrons. Again the plot is in pitch angle versus L-shell with the intensity level in color. The data are for the same periods as the quiet proton model. The outer belt fluxes peak over a wide L-shell range, but the inward dispersion of the peak with energy is still evident. The electron pitch angle distributions in the outer belts are nearly isotropic. The flux levels for energies > 1 MeV (HEEF) in the inner belt are much less than given by the NASA models, and the data shown here may even be a little high due to some proton contamination. Below 1 MeV a background correction has been made for proton contamination in the MEA and the resulting model is questionable. It is not yet clear if we will ever have confidence in our high energy inner belt electron levels.

Aside from the integrity of the electron measurements in the inner zones, another type of problem thwarts our static modeling efforts in the outer zones. We illustrate this using LOLET dose maps, the contributions to which, in the outer zones, are only from electrons. Maps were made for two periods, one using successive orbits taken between 27 August and 4 September 1990, and the other using successive orbits taken between 23 September to 3 October 1990. The dosimeter makes an omnidirectional measurement and, thus, does not give pitch angle information. The data for these maps are binned by L value as before, but in B/B_0 bins instead of pitch angle. The IGRF plus Olson-Pfitzer quiet model values are used. The size of the B/B_0 bins vary with B. They are chosen to correspond to each 2° in magnetic latitude if the magnetic field were a dipole. The maps are

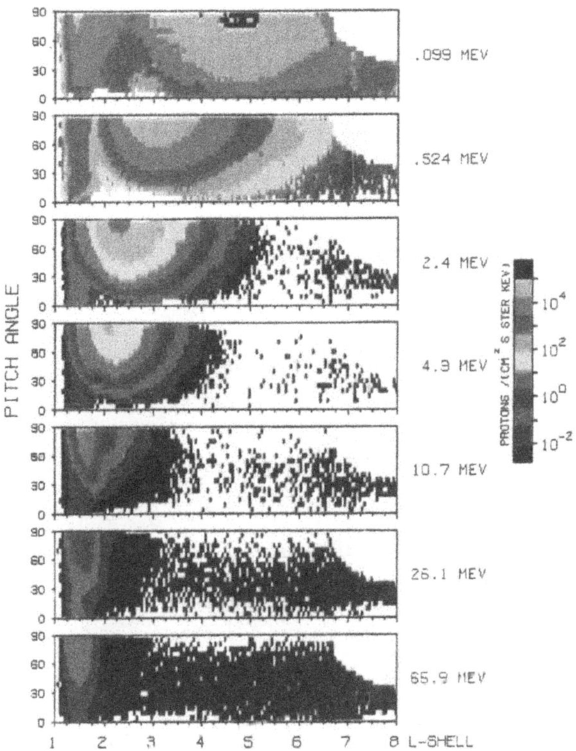

Fig. 22: A preliminary proton radiation belt model, made using CRRES PROTEL and MEB data taken during magnetically quiet periods. The differential flux is shown in color code for seven energy values against a grid in L-value and pitch angle.

given in Figs. 24 and 25. Here the data are color-coded in rads/second and displayed on an L-value vs B/B_0 grid. For the first period and for Channel 1 (> 1 MeV outer zone electrons) the highest dose rate lies between an L of 3 and 4. For the second period the highest dose rate is between an L of 4 and 5; that is, the peak has moved to higher altitudes. For Channel 2 (> 2.5 MeV outer zone electrons) the response in the second period is broad and flat from an L of 3 to 5, which is near double the extent of the previous period, but considerably lower in intensity. These two maps show how dynamic the outer zone electrons are, and because of this, a strictly static model will either overestimate or underestimate the population at a particular altitude unless the conditions at the time of flight are exactly the same as the conditions for the data used in the model.

Another type of model which we generate from the dosimeter data is the average number of VHLET counts as a function of altitude. These averages can be related to single event upsets in proton-sensitive microelectronic devices (see Section 5). Maps of DMSP VHLET dosimeter data taken at 840 km were previously given by [Ref. 34]. Here in Fig. 26 we plot average VHLET counts per second versus L-shell, measured in detector 2, for the

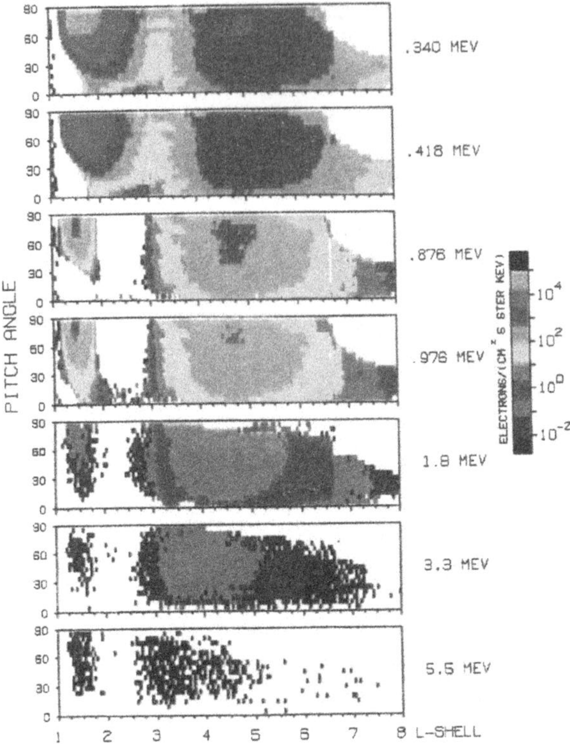

Fig. 23: Same as Fig. 22, but for seven electron energies, the data taken from HEEF and MEA.

first 500 orbits of CRRES. The count rate peaks in the inner belt region near an L of 1.5 R_E. At altitudes above an L of about 2 R_E the rate is fairly constant indicating an equal likelihood anywhere within this altitude range of seeing a VHLET count whether due to a heavy ion, cosmic ray or a star event.

Many CRRES particle event studies are underway at this time with results to be published in the near future. The breadth of the studies ranges from pure data analysis (e.g., particle cutoff determination) to pure theory (e.g., high energy particle interaction with the magnetopause as a scattering center). The event studies are beyond the scope of this paper and will not be addressed here with one exception, which is presented in the discussion section.

5. CRRES Microelectronics Test Package Analysis

The MEP analysis can be divided into two parts, SEU and total dose. Although not mutually exclusive, they will be treated separately until a

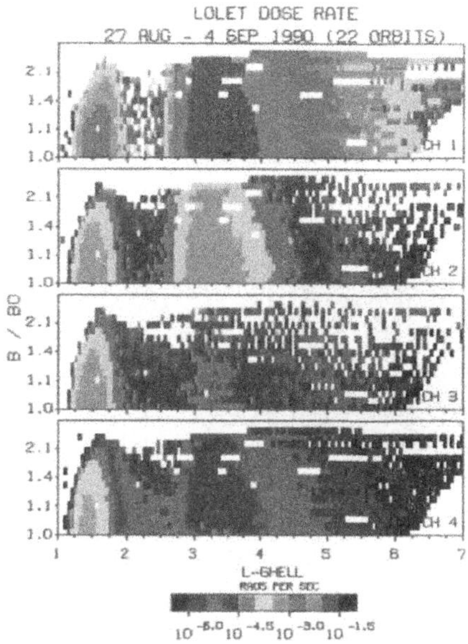

Fig. 24: LOLET dose maps made using Space Radiation Dosimeter data from 27 August to 4 September 1990. The maps are constructed for each of the four shielding thickness.

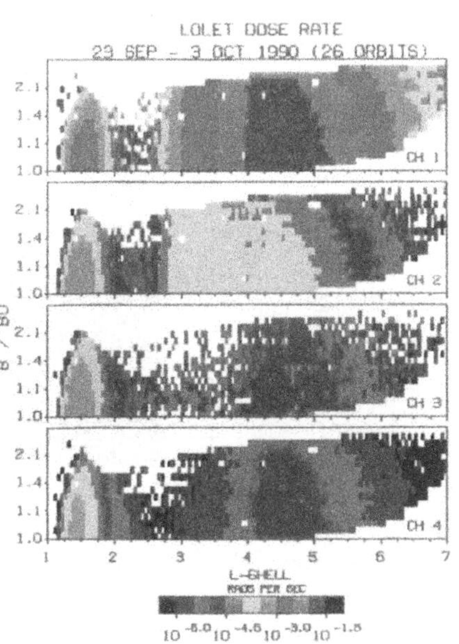

Fig. 25: Same as Fig. 24, but using data from 23 September to 3 October 1990. In each figure the average dose/s in rads is shown in color code against a grid in L-value and B/B_0.

larger data base becomes available. We first look at the SEU data. One of the initial justifications for the CRRES Program was the determination of those near-Earth regions where SEUs are most likely to occur. To examine this, we placed 64,000-bit dynamic rams (DRAMs) in the MEP to get high upset statistics. What we found on orbit, was that these chips upset much less than originally predicted, while other, more sensitive chips, upset at far higher rates. Therefore, to get the best statistical upset map, we combined data from the 13 most-sensitive-to-upset chip types (40 chips, approximately 500,000 bits). This included 5 of the 64 K DRAMs. We showed earlier, in Fig. 3, the upset frequency of these devices (SEUs per bit day) as a function of L-shell. The data were taken from the first 500 orbits during which no major solar proton events occurred; they were sorted into bins whose size was 1/20th R_E, and averaged. The absolute numbers have little meaning since many chip types and technologies were mixed. When sufficient data become available to produce statistically significant numbers for each chip type, then the numbers will give quantitative results which can be better used for comparison to model predictions. However, the plot still shows some very interesting results. First, the upset rate increases to a maximum inside an L value of 1.45 R_E which is in the heart of the inner radiation belt for 55-200 MeV protons.

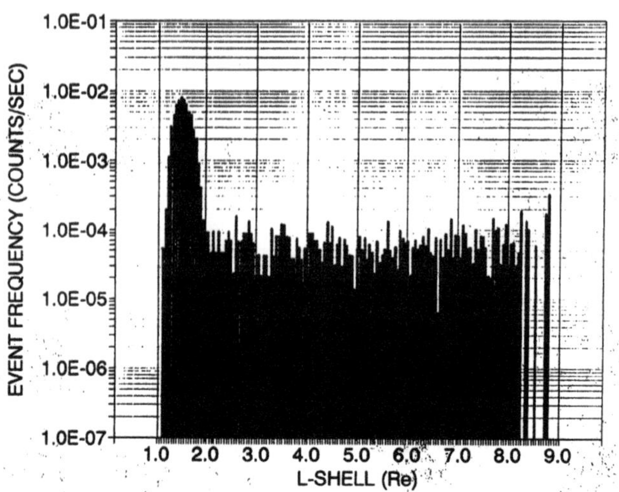

Fig. 26: Average VHLET count rate from Detector 2 of the Space Radiation Dosimeter as a function of L-value. Averages were made using the first 500 orbits (25 July 1990 to 16 February 1991) and binning by 1/20th R_E.

Second, no obvious drop in SEU rate exists from the highest L values down to L = 2 where the high energy proton belt begins. The near-constant SEU rate from geosynchronous altitudes to the outer edge of the proton belt indicates that there is no sharp cutoff of the cosmic ray population that causes the upsets.

To show the value of the Space Radiation Dosimeter VHLET fluxes (star counts) in predicting where SEUs will take place, both the VHLET counts/second for the sum of Domes 2 and 3 (a total of .1 cm^2 of area) and the SEUs described above, converted to SEUs/second, are plotted in Fig. 27. Again, events/second are plotted versus L-value and averages are made in bins of 1/20th R_E. The agreement is remarkably good showing that the dosimeter provides a very good mapping of upset occurrence of proton sensitive chips. It also indicates that nuclear star events in the chips are probably the most important cause of upsets for proton sensitive chips. Outside the belt, at high altitudes, the agreement between the two is so good that it is tempting to identify the VHLET threshold setting of 40 MeV with the energy deposition required for upsets. One must be careful, however, since we do not know the total active area of the 500,000 bits under test and how it compares to the .1 cm^2 of the active dosimeter area. The agreement in absolute level may be only coincidental, and we expect that, in general, a scaling factor will have to be applied to take account of differences in particle energy for producing upsets in specific chips. Determination of exact particle energies needed for upsets is beyond the scope of this paper but will be addressed in future work.

To compare technologies, SEU rates, in errors per bit per day, were calculated for all the RAMs for the first 6 months on orbit. For each device block (3 to 5 chips of the same type, manufacturer, etc.), an upset

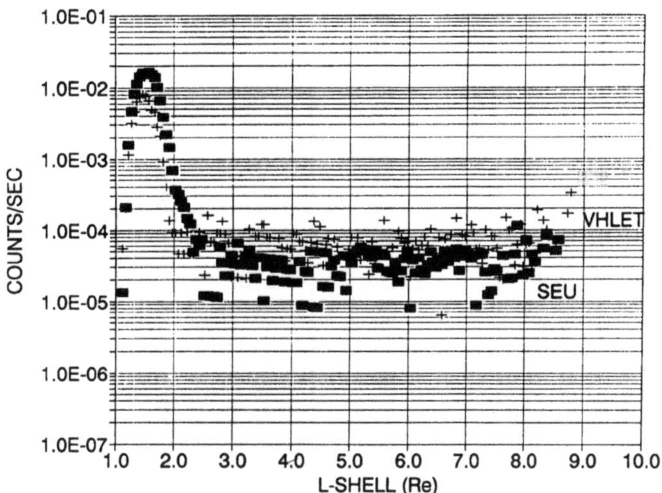

Fig. 27: A comparison of average SEU rate and average VHLET count rate as a function of L value.

rate was determined. The minimum and maximum rates appear in Table 4 for each technology type independent of the number of bits, feature size, manufacturer, etc. For the DRAM, as only 1 device type was used, only 1 value is given. In general, the TTL chips upset the most and CMOS devices the least. The GaAs chip that gave the maximum upset rate was taken from an early pilot line of GaAs RAMs which is no longer producing similar parts. The minimum number is more indicative of the response of this technology. For CMOS RAMs, two upsets (one a double bit hit) have been experienced to date, although one commercial chip block experienced total dose failure in less than 3 months on orbit. These upsets occurred after the first 6 months so they are not included in Table 4. It must be remembered that these numbers are for the CRRES orbit only and cannot be used for other orbits. To obtain prediction rates for these chips in other orbits, the actual orbit dwell times at each altitude must be calculated and weighted by the chip upset rate measured at each altitude as shown in Fig. 3. Only the surface has been scratched in SEU analysis of the CRRES data. Much more analysis is needed before definitive results suitable for space system designers can be obtained.

Compared to SEU analysis, much less has been done to obtain early CRRES total dose results. There are several reasons for this. One is that many of the chips were radiation-hardened to total dose and as yet have not accumulated enough total dose for their behavior to be properly analyzed. The second is that the manpower level available for this effort has been minimal. A third is that the Space Radiation Dosimeter has to be cross calibrated with the dosimeters in the MEP and temperature effects removed from the MOS dosimeters before dose degradation can be properly assessed in the package. While good progress is being made in this area, it is not

Table 4: On-orbit SEU rates.

		ERRORS/(BIT-DAY)	
DEVICE TYPE	TECHNOLOGY	MINIMUM	MAXIMUM
STATIC RAM	GaAs	3.2E-5	1.4E-2
STATIC RAM	NMOS	8.1E-5	1.4E-4
STATIC RAM	TTL	1.0E-3	5.0E-3
STATIC RAM	NMOS/CMOS	4.6E-6	5.7E-6
STATIC RAM	CMOS	0	0
DYNAMIC RAM	SMOS	4.3E-5	
*BASED ON APPROXIMATELY 6 MONTHS OF DATA			

yet completed. The fourth is that dose on parts deep inside the package was less than originally predicted and no significant degradation occurred on buried parts.

Figure 28 illustrates the temperature problem in dose measurement and shows the relative effect of shielding. Here total dose (uncorrected for temperature) in rads from MOS dosimeters at four shielding levels (board 1, board 2, board 3, and RADPAK) is plotted versus orbit number. The dosimeter on the first level has a shielding whose aluminum equivalence is 80 mils thickness. The dosimeter on the second level is behind the 10 mil front cover and one fully loaded circuit board. The third level dosimeter is behind the 10 mil cover and two fully loaded multilayer circuit boards. The dosimeter on the first level saturated in orbit 225. At that time, it had approximately 20 times more dose than the dosimeter on board 2 and 2 orders of magnitude more dose than the dosimeter on board 3. The dosimeter in the RADPAK had much less (> 6 times less) than the dosimeter on board 3. The temperature sensitivity can best be seen in the RADPAK device where the dose is low. The periodic fluctuations are due to temperature variations in the MEP when the spacecraft was reoriented every 2 to 3 weeks to keep the solar panels pointed toward the sun. The large drop near orbit 100 was due to shutting down half the MEP after a power supply problem. After the shutdown, the average temperature dropped approximately 50% from near 40°C to less than 20°C. This also can be seen by a slight decrease in dose of the level 3 dosimeter.

Figure 4, shown previously, gives the total dose behind the 4 domes of the Space Radiation Dosimeter in the same format as Fig. 28 for comparison. The dose behind the thinnest shielding compares very well with the first level MEP dosimeter; dose behind the second dome compares very well with the second level MEP dosimeter; and dose behind the fourth dome compares very well with the third level dosimeter. From calculations, we

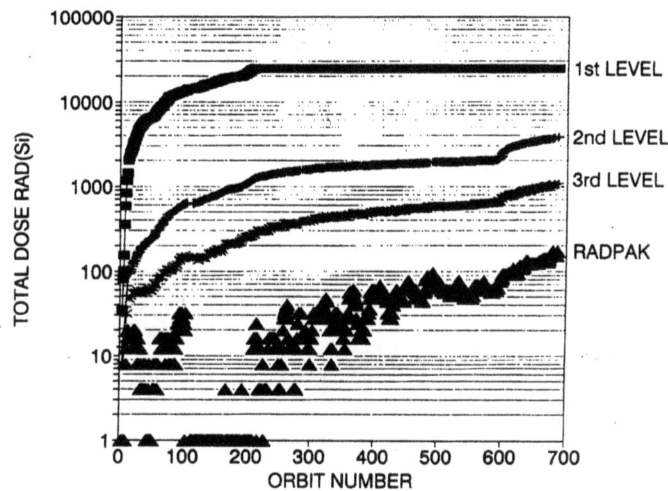

Fig. 28: Total on-orbit dose, in rads, measured by MOS dosimeters at three different board levels and behind RADPAK, as a function of orbit number from 25 July 1990 to 7 May 1991.

know that the MEP cover and MOS dosimeter cover were about equivalent to the Dome 1 thickness (82.5 mils) of the Space Radiation Dosimeter. It was a pleasant surprise that the MEP level 2 dosimeters saw the equivalent of approximately 232 mils aluminum shielding and level 3 dosimeters saw the equivalent of approximately 886.5 mils aluminum shielding. This means we can use the better calibrated and temperature-stable Space Radiation Dosimeter data to determine total dose degradation in the MEP devices. Since the Space Radiation Dosimeter will not saturate, the MEP dose could be properly measured over the entire mission.

The dose data show that for deeply buried parts, passive board shielding can be as effective as about an inch of aluminum, but spot shielding with a combination of high and low Z material is much better. Since the high Z material does not have to be very thick (and therefore does not add much extra weight), it might be built into the box walls and eliminate the need for special shielding internally. Another method would be to put a sheet above and below all the boards containing radiation soft parts. This would add some weight, but would significantly decrease the cost and reduce the delivery schedule necessitated by using radiation hardened parts. Further total dose results will have to await future reports.

6. Discussion

We will end, as we started, with a discussion of spacecraft anomalies, showing the environmental and spacecraft response to the largest solar proton event and the largest high energy electron event that occurred since the launch of CRRES. Both events occurred in the period from 22 March to 1 May 1991. The proton event was associated with a flare which began at 22:42 UT on 22 March. The flare was classified as Importance-3 in brilliance, a level reached in less than 1% of all flares. The flare lasted about 32 minutes which is quite short for an Importance-3 flare. The solar protons measured on the GOES 7 spacecraft in geosynchronous orbit are shown in Fig. 29. Unedited 5 minute average values of the proton flux are given for four energy ranges: 15-44 MeV, 39-82 MeV, 84-200 MeV and 110-500 MeV. All four channels peak at 03:50 UT on 24 March. At 03:42 UT, just before the GOES' protons peak, a Sudden Storm Commencement (SSC) of 184 nT was recorded on the Boulder magnetometer. This is an extremely large SSC indicating the severity of the solar wind shock hitting the magnetosphere at this time. Although confirming data from IMP 8 are not yet available, it is believed the large, rapid increase in protons seen on CRRES were near simultaneous with the shock arrival.

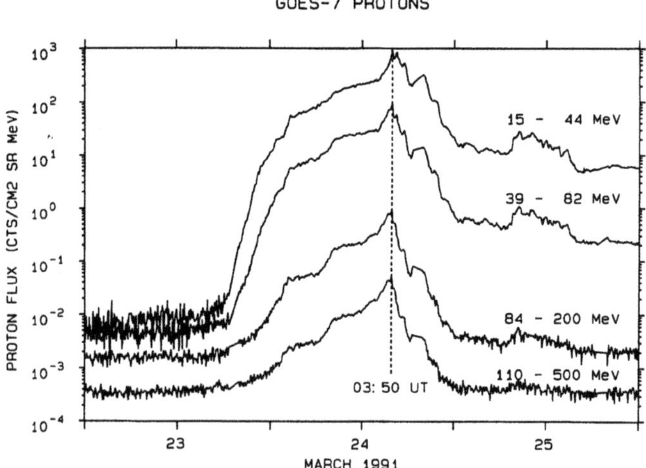

Fig. 29: Five-minute-averaged GOES-7 proton flux in 4 energy channels for the 23-25 March 1991 solar proton event. The dotted line shows the peak fluxes for all 4 channels at 03:50 UT on 24 March, following the Sudden Storm Commencement at 03:42 UT.

As fortune would have it, CRRES was near an L of 2.55 R_E when the shock hit. At first, we thought we had missed the peak of the largest proton event of the CRRES mission, since, at the time, we were well inside predicted cutoff levels for particle entry. The magnetometer on CRRES saw an increase of 77 nanotesla over a 1 minute period. At the same time that the measured magnetic field on CRRES peaked, a peak was measured in the high energy PROTEL channels. All the PROTEL channels from 1 to 80 MeV showed particle increases, but the channels above approximately 20 MeV showed the latest increases in intensity. Above 20 MeV there was no time dispersion with energy. Figure 30 shows the PROTEL energy channels between 19.4 MeV and 80.1 MeV. The plot is for a single CRRES orbit running from perigee to perigee on 23-24 March 91. The one-minute average differential flux in particles/cm^2-sec-ster-MeV are plotted versus UT (L-shell) for the 9 highest PROTEL energy channels marked on the side of the graph. The differential flux scale applies to the 80.1 MeV protons. Each lower energy channel flux level is offset (increased) by 1 order of magnitude for better viewing. Between 18:30 and 19:30 the satellite passes through the steady-state, single-peaked inner proton belt after which the proton flux levels are at background. From 20:04 UT (L = 4.2 R_E) on 23 March, through apogee, to 03:12 UT (L = 3.6 R_E) on 25 March, the satellite encounters solar protons from the solar proton event well in progress. Within 30 seconds of the SSC, at 03:42 UT on 24 March, the protons in all the channels between 20 and 80 MeV showed a large increase in the one minute flux average. In some channels the increase was more than an order of magnitude. At this time the satellite is at an L-value of 2.55 R_E. A blowup of the shock period confirms the near simultaneity of particle increases over these energies. For lower energies, down to 2.4 MeV, the flux peaks were more and more gradually delayed with decreasing energy. The peak for the 2.4 MeV particles occurred about 1 minute and 20 seconds later than the peak at 19.4 MeV.

We follow the event into the next orbit as shown in Fig. 31. The format is the same as Fig. 30. From 04:15 UT to 05:00 UT, the satellite again traverses the inner belt on its way out toward apogee. However, in this pass, the proton fluxes do not fall to background. Instead, the effects of the solar proton event and SSC can be seen extending all the way down into the inner belt region where a second belt (double-peaked inner belt) at higher altitudes is in its formative stage. Near the end of the orbit, as we once again approach perigee, the double belt structure is clearly evident at all energies with the reestablishment of background levels at altitudes just above the double peak. At the 54 MeV level, the peak of the second belt is at an L-shell of 2.6 R_E. Data confirming the existence of the second belt are given by both the Space Radiation Dosimeter and the >200 MeV proton background channel on HEEF. Figure 32, again in the same format as Fig. 30, shows PROTEL data 42 days later, on 5 May 1991. The second proton belt is still present, although the altitude of the flux peak has decreased somewhat (to an L-shell of 2.34 R_E for the 54 MeV particles), probably due to inward diffusion. Thus, once formed, the belt remains in a stable state for a long period. This is the same behavior found following the February 1986 solar proton event [Ref. 35]. Exactly how long the protons remain stably trapped in the second belt will need to be determined in later studies after the belt has died out.

Single event upset (SEU) rates in SEUs/(Bit-Day) as measured in the MEP for the first 500 plus orbits prior to the March 1991 solar proton event were shown as a function of L-shell in Fig. 3. Figure 33 shows a similar plot of the same devices for 92 orbits following the solar proton event.

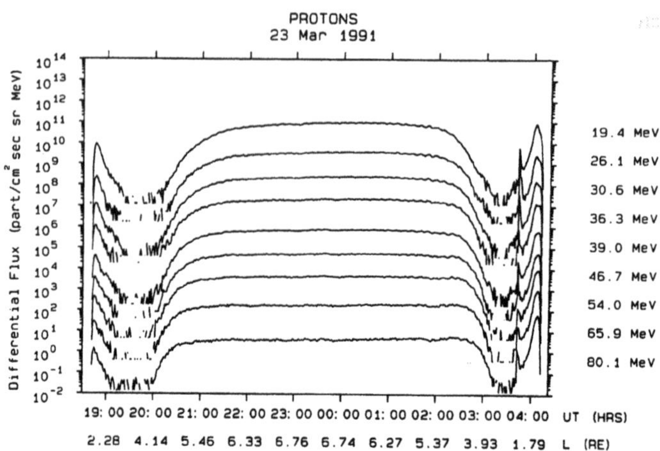

Fig. 30: One-minute average differential flux measurements for 9 energy channels between 19.4 and 80.1 MeV versus UT for CRRES Orbit 587 on 23-24 March 1991. The sharp peak in fluxes at 03:42 UT (L = 2.55 R_E) coincides with the time of the Sudden Storm Commencement.

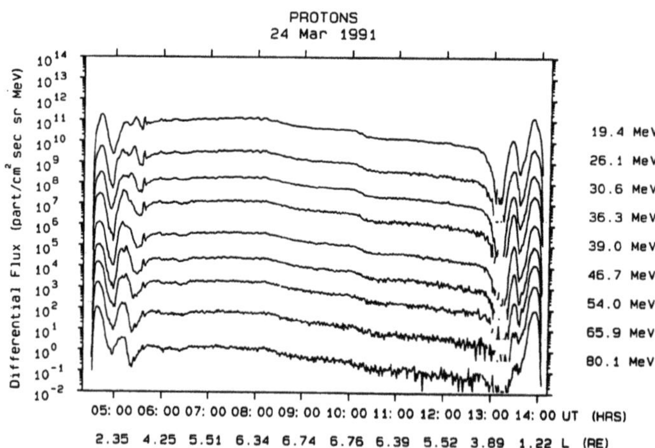

Fig. 31: Same as Fig. 30, but for CRRES Orbit 588 on 24 March 1991 showing the formation of the double-peaked inner proton radiation belt during the early portion of the pass and its stable condition at the end of the pass. The outer peak of the 54 MeV protons is at an L = 2.6 R_E.

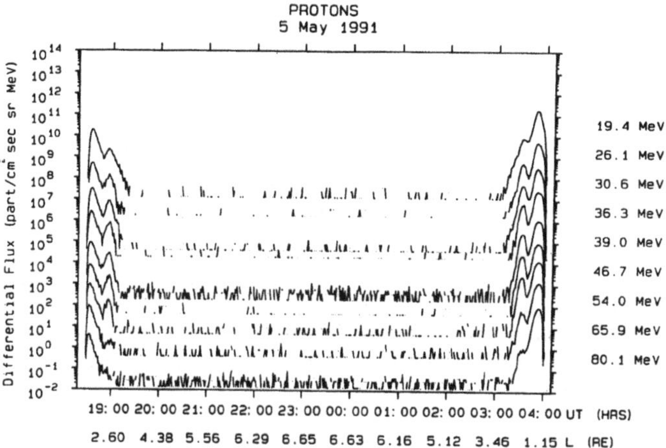

Fig. 32: Same as Fig. 30, but for CRRES Orbit 692 on 5 May 1991. The outer peak of the 54 MeV protons has moved inward to an L = 2.34 R_E.

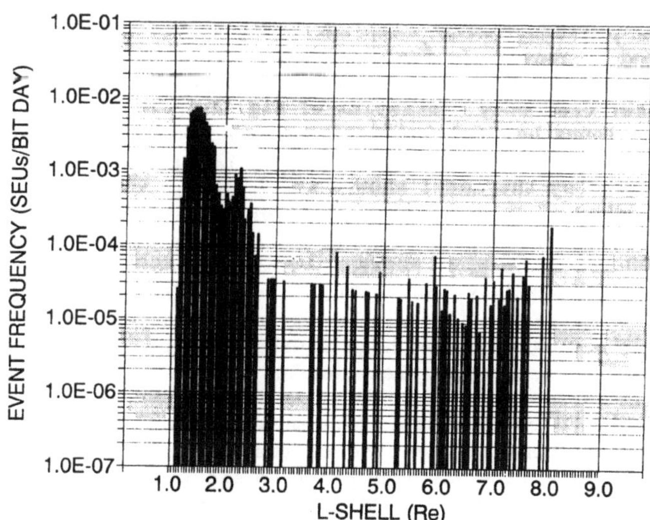

Fig. 33: Single event upset frequency for 35 proton sensitive devices for the first 60 orbits (1 month) following the solar proton event of March 1991. The double belt proton structure is clearly evident in the double-peaked SEU frequency. The dropouts at higher L-values are due to poor statistics.

Here the statistics are not as good due to the paucity of data (fewer orbits) especially at higher L-shells, but in the belt regions the statistics are sufficient to pick out a well separated double-peaked structure. Comparing Figs. 3 and 33, the effects of the double belt structure are very evident in upset rates between L-shells of 1.8 and 2.5. Since this newly formed belt could continue for months, the increased upset rates can have a major impact on spacecraft operating in or through this region of space. These new CRRES results need to be included in an inner belt proton model. The inclusion will not be easy because these events occur infrequently and will have to be treated in terms of the probability that a large shock occur and produce a second belt. The time decay of the second proton belt as a function of particle energy and flux level will also need to be included. The time decay is still to be determined, but it is assumed it will be a function of the original formation level.

In addition to updating the proton belt models, the cosmic ray cutoff models used to determine regions where normal cosmic ray backgrounds and solar protons affect satellite systems also need major rework. The model most often used is called the "Adams Model" which is based on a compilation of data and early studies mostly prior to 1980 [Refs. 29, 30]. The model was made from GeV and higher energy particles and extrapolated down to MeV energies. Earlier studies [Ref. 36] and the CRRES data show that for particles in the energy range that produce SEUs in electronic components in space, the rigidity cutoffs predicted by this model are incorrect. The cutoff altitudes (L-values) are too high both for solar protons that enter into the magnetosphere prior to shocks and for the shock-associated particles that penetrate much, much further into the magnetosphere as shown in Figs. 30 and 31 above. In fact, even galactic cosmic ray particles that create SEUs penetrate much further than the models predict as can be seen in the SEU upset frequency statistics of Fig. 3. The upset frequencies are flat above the inner belt(s) indicating an equal probability of upsets and therefore, an equal accessibility of particles that produce SEUs. The cutoffs predicted by the model show a region above the inner belt where no significant number of upsets from cosmic rays or solar protons would occur.

While the solar protons retreated toward background levels after 24 March, the 100 keV and higher energy electrons reached peak flux levels on the 28th and continued near peak values through the 29th of March before they started to decrease. These levels of electrons were the highest recorded to date on CRRES. The high fluxes were responsible for greatly intensified arcing in the material samples of the IDM as can be seen in Fig. 2. The greatest number of arcing instances occurred between 06:00 UT on the 28th and 12:00 on the 29th. The material samples on CRRES were not the only thing upsetting at the time of the high electron fluxes. Since launch, the CRRES satellite has been prone to electrostatic discharge (ESD) problems. During the period from 28 through 30 March numerous upsets were recorded on CRRES. Among them were phantom instrument state changes, temporary non-functionality of a tape recorder, non-functionality of a digital telemetry unit, and bad status of a power converter unit. Since all the problems occurred after the proton event and during the electron event, it is easy to attribute them to ESD rather than SEU or dose rate since electron dose for buried system components is minimal and SEUs are proton or cosmic ray induced. It turns out that this period is especially useful in studying specific anomalies on the spacecraft, since the flux levels for both the protons and electrons that produced the anomalies were measured, and the anomalies were separated by days so that the causative factors can be clearly separated. In fact, several operational satellites experienced anomalies over the period, with some occurring only during the proton event and some only during the electron event. In both cases, environmental data were used to help identify the specific class of anomalies.

The dose effects for this March 1991 event period were shown earlier in Fig. 17 for thin shielding (82.5 mils of aluminum). For this shielding level, there was as much dose accumulated in the less than 2 week period than had accumulated during the previous 5 months. For thicker shielding (457.5 mils of aluminum) typical of most spacecraft parts, the total dose rate increased by about a factor of 6 starting near orbit 600 as can be seen in Fig. 34. Figure 34 shows LOLET, HILET, and total dose as measured by the Space Radiator Dosimeter versus orbit number over the first 692 orbits.

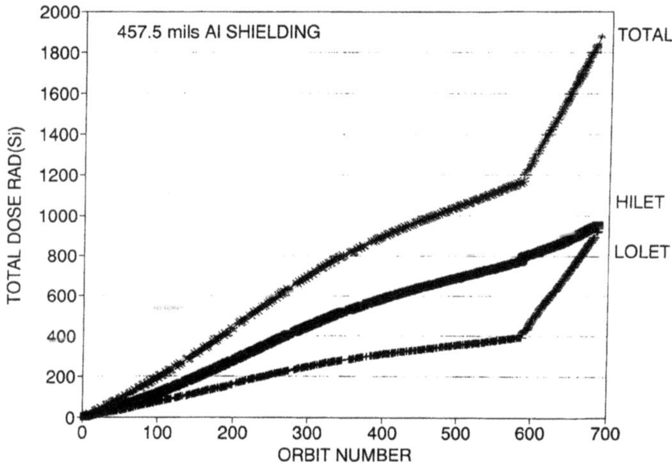

Fig. 34: LOLET, HILET and total dose, measured by the Space Radiation Dosimeter, as a function of orbit number, from 25 July 1990 to 5 May 1991. The increase in dose rate near orbit 600 results from particle increases following the solar flare of 22 March 1991.

The dose rate continues to remain at the increased level over 1 month later. How long this will continue is not known, but it shows that once particles (both electrons and protons) get into the magnetosphere and become trapped, they remain for long periods, and can significantly change the radiation exposure levels of space system components. This further points out the need for dynamic models for both long and short space missions.

In conclusion, CRRES/SPACERAD was launched to better understand the near-Earth radiation environment and how it affects space systems. In the first few months of operation, it has produced a wealth of information that will help identify and correct satellite deficiencies as well as better define the environment in which they must operate. It will also help modelers to better understand why and how devices upset and degrade in space and lead to better radiation test procedures for future space missions. The large solar-initiated event period of 23 March to 1 May 1991 has given us "a worst case" against which to test our models and scientific understanding. If CRRES continues to operate successfully over the next 3 to 5 years as hoped, many, if not most, of the secrets of space anomaly cause, effect and circumvention will be unlocked.

Acknowledgements

The authors wish to thank Ernie Holeman, Don Brautigam, Michele Sperry, Kevin Ray, Kevin Kerns, Carolyn Jordan, Doug Reynolds, Dan Madden, and Mike Violet for developing software and preparing the data presentations shown in the paper. A special thanks goes to Claire Daigle for desktop editing and preparing the final camera ready manuscript and for maintaining her composure while tolerating the authors throughout the lengthy editorial process.

References

[1] M. Schulz and L.J. Lanzerotti, Particle Diffusion in the Radiation Belts, Springer-Verlag, New York (1974).

[2] M. Schulz, 'Energetic-particle populations and cosmic-ray entry,' J. Geomag. Geoelectr., 32, 507 (1980).

[3] W.N. Spejeldvik and P.L. Rothwell, 'The Radiation Belts,' in Handbook of Geophysics and the Space Environment, ed., Adolph S. Jursa, AFGL, USAF, ADA167000, 5-1 (1985).

[4] D.F. Smart and M.A. Shea, 'Galactic cosmic radiation and solar energetic particles,' in Handbook of Geophysics and the Space Environment, ed., Adolph S. Jursa, AFGL, USAF, ADA167000, 6-1 (1985).

[5] S.E. DeForest, 'Spacecraft charging at synchronous orbit,' J. Geophys. Res., 77, 651 (1972).

[6] E.G. Mullen, M.S. Gussenhoven, D.A. Hardy, T.A. Aggson, B.G. Ledley, and E. Whipple, SCATHA survey of high-level spacecraft charging in sunlight, J. Geophys. Res., 91, 1474 (1986).

[7] E.C. Whipple, 'Potentials of surfaces in space,' Rep. Prog. Phys., 44, 1197 (1981).

[8] A.R. Frederickson, 'Electric discharge pulses in irradiated solid dielectrics in space,' IEEE Trans. Elec. Insul., EI-18, 337 (1983).

[9] W.F. Denig and A.R. Frederickson, Deep-Dielectric Charging - A Review, AFGL-TR-85-0123, Environmental Research Paper, No. 919, Air Force Geophysics Laboratory, Hanscom AFB, MA, 01731 (24 May 1985).

[10] D. Binder, E.C. Smith, and A.B. Holman, 'Satellite anomalies from galactic cosmic rays, IEEE Trans. Nucl. Sci., NS-22, 2675 (1975).

[11] T.C. May and M.H. Woods, 'Alpha-particle-induced soft errors in dynamic memories,' IEEE Trans. Elec. Dev., ED-26, 2 (1979).

[12] J.C. Pickel and J.T. Blandford, Jr., 'Cosmic ray induced errors in MOS memory cells, IEEE Trans. Nucl. Sci., NS-25, 1166 (1978).

[13] R.C. Wyatt, P.J. McNulty, P. Toumbas, P.L. Rothwell, and R.C. Filz, 'Soft errors induced by energetic protons, IEEE Tran. Nucl. Sci.., NS-26, 4905 (1979).

[14] G.J. Brucker and E.G. Stassinopoulos, 'Prediction of error rates in dose-imprinted memories on board CRRES by two different methods,' IEEE Trans. Nucl. Sci., NS-38, 913, (1991).

[15] G.J. Brucker, E.G. Stassinopoulos, O. Van Gunten, L.S. August, and T.M. Jordan, 'The damage equivalence of electrons, protons, and gamma rays in MOS devices,' IEEE Trans. Nucl. Sci., NS-29, 1966 (1982).

[16] CRRES System Description Handbook, Rev. E, Ball Aerospace Systems Division, Boulder CO (1990).

[17] M.S. Gussenhoven, E.G. Mullen, and R.C. Sagalyn, editors, CRRES/SPACERAD Experiment Descriptions, AFGL-TR-85-0017, Environmental Research Papers, No. 906, Air Force Geophysics Laboratory, Hanscom AFB, MA (24 January 1985).

[18] M.S. Gussenhoven, E.G. Mullen, R.C. Filz, D.H. Brautigam, and F.A. Hanser, 'New low-altitude dose measurements,' IEEE Trans. Nucl. Sci., NS-34, 676 (1987).

[19] E.G. Mullen, M.S. Gussenhoven, and D.A. Hardy, 'The space radiation environment at 840 km,' in Terrestrial Space Radiation and Its Biological Effects, P.D. McCormack, E.E. Swenberg and H. Bucker, eds., NATO ASI Series, Series A: Life Sciences, Vol. 154, pp 41-60, Plenum Press, New York (1988).

[20] A.R. Frederickson, J.E. Nanevicz, J.S. Thayer, C. Lon Enloe, E.G. Mullen, and D.B. Parkinson, 'Leaky insulating paint for preventing discharge anomalies on circuit boards,' IEEE Trans. Nucl. Sci., NS-36, 2405 (1989).

[21] L. Adams and A. Holmes-Siedle, 'The development of an MOS dosimeter for use in space," IEEE Trans. Nucl. Sci., NS-25, 1607 (1978).

[22] CRRES Microelectronics Package (MEP) Serial Telemetry Manual, SSD-D-CM026, Naval Research Laboratory, Space Systems Development Department, Washington, DC (1 December 1989).

[23] K.P. Ray and E.G. Mullen, Pre-launch Data for the GL701-1A Microelectronics Package, PL-TR-91-2007, Hanscom AFB, MA (1991).

[24] D. E. Delorey, The Combined Release and Radiation Effects Satellite Time History Data Base, Phillips Laboratory Technical report, submitted for publication (1991).

[25] C.A. Hein, Comparison of the PROTEL Contamination Code Predictions with the Harvard Accelerator Calibration Data, GL Technical Memorandum No. 179, Hanscom AFB, MA (2 February 1990).

[26] C.E. Jordan, NASA Radiation Belt Models AP-8 and AE-8, GL-TR-89-0267, Hanscom AFB, MA (30 Sep 1989).

[27] J. Lemaire, M. Roth, J. Wisemerg, and J.I. Vette, Development of Improved Models of the Earth's Radiation Environment, Technical note 1: Model Evaluation, TREND issued at IASB, Printed at MATRA, ESTEC/Contract #8011/88/NL/MAC (28 June 1989).

[28] D.M. Sawyer and J.I. Vette, Models of the Trapped Radiation Environment,' National Space Science Data Center, NSSDC/WDC-A-R&S 76-06 (Dec 1976).

[29] J.H. Adams, Jr., R. Silberberg, and C.H. Tsao, Cosmic Ray Effects on Microelectronics, Part I: The Near-Earth Particle Environment, NRL Memorandum Report 4506, Washington, DC (1981).

[30] J.H. Adams, Jr., J.R. Letaw, and D.F. Smart, Cosmic Ray Effects on Microelectronics Part II: The Geomagnetic Cutoff Effects, NRL Memorandum Report 5099, Washington, DC (1983).

[31] C.H. Tsao, R. Silberberg, J.H. Adams, Jr., and J.R. Letaw, Cosmic Ray Effects on Microelectronics Part III: Propagation of Cosmic Rays in the Atmosphere, NRL Memorandum Report 5402, Washington, DC (1984).

[32] M.S. Gussenhoven, E.G. Mullen, D.H. Brautigam, E. Holeman, C. Jordan, F. Hanser, and B. Dichter, 'Preliminary comparison of dose measurements on CRRES to NASA model predictions,' submitted to IEEE, Trans. Nucl. Sci., 38, 1655 (1991).

[33] W.P. Olson and K.A. Pfitzer, 'A quantitative model of the magnetospheric magnetic field,' J. Geophys. Rev., 79, 3739 (1974).

[34] E.G. Mullen, M.S. Gussenhoven, J.A. Lynch, and D.H. Brautigam, 'DMSP dosimetry data: A space measurement and mapping of upset causing phenomena, IEEE Trans. Nucl. Sci., NS-34, 1251 (1987).

[35] M.S. Gussenhoven, E.G. Mullen, and E. Holeman, 'Radiation belt dynamics during solar minimum', IEEE Trans. Nucl. Sci., NS-36, 2008 (1989).

[36] M.S. Gussenhoven, D.H. Brautigam and E.G. Mullen, 'Characterizing solar flare high energy particles in near-Earth orbits,' IEEE Trans. Nucl. Sci., 35, 1412 (1988).

Annex

EXPERIMENT AGENCY LIST

The experiments, acronyms, numerical designators and affiliated agencies are:

1.	Microelectronics Test Package (MEP) AFGL-701-1A	Air Force Phillips Laboratory Naval Research Laboratory Defense Nuclear Agency Defense Advanced Research Projects Agency Assurance Technology Corporation Telenetics Corporation Fail-Safe Technology Corporation NASA Headquarters NASA Goddard Space Flight Center Microelectronics Working Group Microelectronics Industrial Community
2.	Internal Discharge Monitor (IDM) AFGL-701-1B	Air Force Phillips Laboratory Jet Propulsion Laboratory JAYCOR Inc. Beers Associates
3.	Gallium Arsenide Solar Panel Experiment (GASP) AFAPL-801	Air Force Wright Aeronautical Hughes Aircraft Company
4.	Space Radiation Dosimeter AFGL-701-2	Air Force Phillips Laboratory Panametrics Inc.
5.	MOS Dosimeter AFGL-701-3	Naval Research Laboratory
6.	Experiment for High Energy Heavy Nuclei Composition ONR-604	University of Chicago Louisiana State University
7.	Medium Energy Ion Mass Spectrometer ONR-307-8-3	Lockheed Palo Alto Research Lab Office of Naval Research

	Instrument	Institutions
8.	Low Energy Ion Mass Spectrometer ONR-307-8-1&2	Lockheed Palo Alto Research Lab Office of Naval Research
9.	Mass Ion Composition Spectrometer (MICS) AFGL-701-5B	Max Planck Institut Fur Aeronomie The Aerospace Corporation University of Bergen Rutherford-Appleton Laboratory
10.	Low Mass Ion Composition Spectrometer (LOMICS) AFGL-701-11B	Los Alamos National Laboratory The Aerospace Corporation
11.	Proton Switches AFGL-701-7B	The Aerospace Corporation
12.	Relativistic Proton Detector AFGL-701-7A	University of California San Diego The Aerospace Corporation
13.	High Energy Electron Fluxmeter (HEEF) AFGL-701-4	Air Force Phillips Laboratory Panametrics Inc.
14.	Proton Telescope (PROTEL) AFGL-701-8&9	Air Force Phillips Laboratory Massachusetts Institute of Technology
15.	Spectrometer for Electrons and Protons (SEP) ONR-307-3	Lockheed Palo Alto Research Lab Office of Naval Research
16.	Medium Energy Electron Spectrometer (MEES) AFGL-701-5A	The Aerospace Corporation
17.	Electron Proton Angle Spectrometer (EPAS) AFGL-701-5B	Max Planck Institut fur Aeronomie The Aerospace Corporation University of Bergen
18.	Low Energy Plasma Analyzer (LEPA) AFGL-701-6	Air Force Phillips Laboratory Mullard Space Science Laboratory University of Sussex Amptek Inc.
19.	Heavy Ion Telescope (HIT) AFGL-701-11A	Los Alamos National Laboratory The Aerospace Corporation
20.	Search Coil Magnetometer AFGL-701-13-2	University of Iowa

	Instrument	Institutions
21.	Fluxgate Magnetometer AFGL-701-13-1	Air Force Phillips Laboratory Schonstedt Instrument Company
22.	Passive Plasma Sounder AFGL-701-15	University of Iowa Fairchild Corporation Analytyx Electronic Systems Inc.
23.	Langmuir Probe Instrument AFGL-701-14	Air Force Phillips Laboratory University of California Berkeley Regis College Analytyx Electronic Systems Inc. University of California San Diego Cornell University Weitzmann Corporation

AN OVERVIEW OF THE LONG DURATION EXPOSURE FACILITY: CASE STUDIES FOR THE EFFECTS OF THE SPACE ENVIRONMENT ON SPACECRAFT SYSTEMS[1]

DAVID E. BRINZA
Jet Propulsion Laboratory
California Institute of Technology
Pasadena, California

ABSTRACT. This lecture provides an introduction to the NASA Long Duration Exposure Facility (LDEF). The basic intention for LDEF was to develop a retrievable platform which would provide lower-cost and more frequent access to space for the scientific and engineering community. A broad range of scientific and engineering experiments were integrated into LDEF for the first one-year mission which was deployed from Challenger (STS-41D) in April 1984. Schedule slippage and the Challenger disaster led to the unplanned extended exposure which culminated in the retrieval by Columbia (STS-32) in January 1990. This lecture provides a broad overview of the experiments flown aboard LDEF-1 and preliminary significant results from a spacecraft systems perspective.

Introduction

This lecture will provide a brief overview of LDEF and could only begin to relate a small percentage of the significant scientific and engineering findings to date. The first section of the lecture describes the LDEF Mission 1 concept and configuration. A brief description of the environmental characterization for the LDEF mission is described in Section 2. Selected findings from LDEF experiments are provided in Section 3. A perspective on implications for LDEF on spacecraft systems and some future research directions for space environmental effects is provided in Section 4.

At the outset, the author would like to thank the LDEF Project Office for providing many helpful discussions in relating the vision developed through interactions with the unique and enthusiastic community of LDEF space experimenters. In particular, Dr. William Kinard and Lenwood Clark have been most cooperative in relating the philosophical approach for the LDEF project. Much of the information presented here was gathered by the author during the LDEF de-integration process in early 1990 and in subsequent meetings of the Materials Special Investigation Group and the

---

[1] Presented at the NATO Advanced Studies Institute "Behavior of Systems in the Space Environment" held in Pitlochry, Scotland on July 7-19, 1991.

R. N. DeWitt et al. (eds.), The Behavior of Systems in the Space Environment, 655–667.

First LDEF Post-Retrieval Symposium held in Orlando, Florida in June of 1991.

1. LDEF Mission 1 Description

The Long Duration Exposure Facility (LDEF) represents the first example of an integrated experiment platform designed for deployment, long term space exposure and retrieval for the Space Shuttle Program. The origins of LDEF lie in the design of a very simple low-cost payload which could demonstrate the ability of the Space Shuttle to launch, deploy and retrieve large payloads in low earth orbit (LEO). The predecessor of LDEF was a simple cylindrical payload known as Micrometeoroid Experiment Platform (MEP) designed strictly to characterize the LEO micrometeoroid and debris environment. The proposed experiment was approved as an early Space Shuttle payload. It was noted that with relatively minor modification the structure would be capable of integrating many self-contained experiments for long term space exposure. The platform was designed to permit simple integration of experimental "trays" into a large structure to be launched, recovered, refurbished and reflown on an 18-month cycle. A NASA announcement of opportunity was issued in 1979 for participation on LDEF and over 200 proposals were received. Fifty-seven experiments were selected for LDEF Mission 1 [Ref. 1] covering an extremely broad range of scientific and engineering interests including biology, micrometeoroid and debris, radiation and cosmic rays, materials science and effects on systems due to long term space exposure.

The philosophy for the experiment design for the LDEF project was quite unlike that found in typical space experiments. The planned retrieval and possibilities for reflight encouraged development of low-cost passive and active experimental designs. The LDEF project did not impose requirements upon the experimenters beyond straight forward interface controls and meeting Shuttle safety rules. As a result, the experimenter was able to establish a level of risk consistent with available funding. The ability to perform post-retrieval analysis and refurbishment of hardware to correct any anomalies or examine unexpected phenomena prior to reflight reinforced a low-cost philosophy. This concept was felt to be the wave of the future for space experiments, with NASA, university, industrial, military, other government agencies and foreign participants. LDEF would provide the scientific and engineering community with a facility for long-term microgravity and space environmental exposure planned for frequent access to space.

The LDEF structure is a twelve-sided welded and bolted aluminum cylindrical framework approximately 14 feet in diameter and thirty feet in length. The unpopulated structure weighed approximately 8,000 pounds. The LDEF structure accommodated 70 experiment trays with dimensions of 34" x 50" on the circumference and 14 trays of 34" x 34" on the earth and space ends. The experimenters could select tray depths of 3, 6 or 12 inches with a weight range of 180 to 200 pounds. The populated LDEF structure included experiments listed in Table 1 and weighed nearly 21,400 pounds.

Table 1. LDEF Mission 1 Experiments.

Micrometeroid and Debris	Materials
• Space Debris Impact	• Crystal Growth
• Microabrasion Package	• Atomic Oxygen Outgassing
• Meteoroid Impact Craters	• Atomic Oxygen Interactions
• Dust Debris Collection	• High-Toughness Graphite Epoxy
• Chemistry of Micrometeoroids	• Composite Materials for Space Structures
• Measurements of Micrometeoroids	• Epoxy Matrix Composites
• Interplanetary Dust Experiment	• Composite Materials
• Meteoroid Damage to Spacecraft	• Metallic Materials under Ultravacuum
Power, Propulsion, Thermal	• Graphite-Polyimide and Graphite-Epoxy
• Solar Array Materials	• Spacecraft Materials
• Advanced Photovoltaics	• Polymer Matrix Composites
• Coatings and Solar Cells	• Balloon Materials Degradation
• High Voltage Drainage	• Thermal Control Surfaces
• Solid Rocket Materials	• Textured and Coated Surfaces
• Variable Conductance Heat Pipe	• Radar Phased-Array Antenna
• Low-Temperature Heat Pipe	**Optics and Electronics**
• Transverse Flat-Plate Heat Pipe	• Infrared Multilayer Filters
• Thermal Measurements	• Pyroelectric Infrared Detectors
Space Science	• Metal Film and Multilayers
• Interstellar Gas	• Vacuum-Deposited Optical Coatings
• Ultra-Heavy Cosmic Ray Nuclei	• Ruled and Holographic Gratings
• Heavy Ions	• Optical Fibers and Components
• Trapped-Proton Energy Spectrum	• Holographic Data Storage Crystals
• Heavy Cosmic Ray Nuclei	• ERB Experiment Components
• Linear Energy Transfer Spectrum	• Solar Radiation on Glasses
Life Sciences	• Quartz Crystal Oscillators
• Biostack	• Active Optical System Components
• Seeds in Space	• Fiber Optic Data Transmission
• Student Seeds Experiment	• Fiber Optics Systems
	• Space Environmental Effects

LDEF was placed into orbit at an altitude of 482 km on April 7, 1984 by the Space Shuttle Challenger (STS-41D). The LDEF satellite was gravity-gradient stabilized and was intended to be retrieved after 10 months on orbit. As a result of manifest changes for the Space Shuttle and the Challenger disaster, LDEF was not recovered until January 12, 1990 by Columbia (STS-32). At the time of recovery, the orbit of LDEF had decayed to an altitude of 340 km and had completed over 32,000 orbits over the 69 months in space. Obvious damage had occurred to materials on LDEF as seen in the retrieval photograph (Fig. 1).

Fig. 1: Retrieval of the Long Duration Exposure Facility after nearly six years in Low Earth Orbit.

The extended exposed of LDEF impacted the project in several manners. The leading edge surfaces were exposed to a very high fluence of atomic oxygen, with nearly half of the mission fluence obtained during the last six months on orbit. The external surfaces of LDEF were subjected to substantially greater levels of solar W, particle impacts and ionizing radiation than those anticipated in the original mission plan. In effect, the LDEF satellite had become a collective experiment of greater value than the original set of individual experiments. A small portion of the experiments were adversely affected due to changes in the investigative personnel, loss of pre-flight data and controls, and funding discontinuity. Yet LDEF was unquestionably a treasure chest of space environmental effects data which could provide invaluable information for design of future long-life space assets, particularly Space Station Freedom.

In order to provide a coherent picture of the effects of the unplanned extended duration of the mission, the LDEF Project Office formed four Special Investigation Groups (SIGs). The SIGs drew membership from NASA centers, the Department of Defense, and foreign agencies to brief a broad range of expertise in Materials, Micrometeoroid and Debris, Induced Radioactivity and Systems disciplines. The role of the SIG's was to assess global effects of the extended mission to LDEF on the whole and to provide assistance as requested (and within budgetary constraints) by the LDEF experimenter community in analysis of the individual experiments. The LDEF Project Office also contracted the Boeing Defense and Space Group to perform system and materials analyses in support of the SIG's and investigators.

2. LDEF Environment Definition

The most readily identified effects on LDEF were attributed to atomic oxygen exposure, micrometeoroid and debris impacts, solar ultraviolet radiation and contamination. Some less obvious were effects attributed to thermal cycling and extremes. Effects due to ionizing radiation were indeed subtle upon casual inspection. The LDEF Project Office placed a high priority in quantifying the LDEF environments to permit analysis and interpretation of effects. The LDEF environment definition was accomplished through a team effort, involving LDEF experimenters, the LDEF Project Office and SIGs, and Boeing Defense and Space Group. The following summarizes findings fnr the T.DEF environments.

Visual evidence on the external surfaces of LDEF, particularly the earth and space ends, due to shadowing effects of atomic oxygen, led to speculation that LDEF was slightly misoriented about the yaw axis. More detailed analysis of these shadowing patterns, performed primarily by Dr. Bruce Banks of NASA Lewis Research Center [Ref. 2], and refined by analysis of a pinhole camera [Ref. 3] by Dr. John Gregory of the University of Alabama-Huntsville, indicated that the LDEF attitude was about 8° off of the intended velocity vector in the yaw axis and less than 1° in the pitch axis. This orientation information was important for assessing atomic oxygen fluences and micrometeoroid and debris distributions for the LDEF surfaces. The origin of this slight misorientation remains unexplained as the moments of LDEF were carefully adjusted prior to launch and are believed to lie outside of the preflight predictions for three axis gravity gradient stabilization limits.

The micrometeoroid and debris (M&D) impact distribution was characterized via post-flight inspection of all LDEF surfaces by the M&D

SIG "A-team". In a monumental effort during de integration, over 34,000 impacts were located and catalogued in a computerized database and extensively documented [Ref. 4] by the "A team". Many of the larger or more interesting craters and penetrations were imaged via a digital stereoscopic imaging system and stored on optical disk for subsequent analysis. The impact survey revealed a factor of 10 - 20 greater impact frequency on the leading edge surfaces as compared to trailing edge surfaces. The findings indicate that the Pre-LDEF models underpredict impactors in the small size range (< 5 micron). The mean impact velocity is approximately 17 km/s. Presence of debris impacts on trailing edge and earth end indicate debris exists in highly elliptical orbits, most likely due to orbit transfer vehicles. The largest impactor encountered by LDEF was probably on the order on 1 mm diameter and most large particles were from natural sources. Silvered/teflon thermal blankets seemed to perform effectively as Whipple shield "bumpers". Impact frequency data [Ref. 5] was recorded by the Interplanetary Dust Experiment during the first year of the LDEF mission. This data revealed the presence of persistent debris clouds associated the Shuttle during deployment and a launch event several months after LDEF deployment.

The estimate of atomic oxygen fluences for the surfaces of LDEF shown in Fig. 2 was compiled by Boeing Defense and Space Group [Ref. 6]. The model for estimating fluences included solar activity and geomagnetic parameters for the MSIS-86 model, orbital elements for LDEF from NORAD, effects due to thermal velocity and atmospheric co-rotation and the misorientation of the LDEF satellite. The leading edge fluence of 8.7 x 10^{21} atoms/cm^2 would be expected to erode about 250 microns of Kapton, a result reportedly confirmed on the Interaction of Atomic Oxygen with Solid Surfaces at Orbital Altitudes experiment [Ref. 7]. The effects of LDEF misorientation and atomic oxygen thermal velocities lead to a nearly factor of 100 difference in atomic oxygen fluence for the row 6 and row 12 surfaces which were intended to be nominally parallel to the velocity vector of LDEF. Boeing Defense and Space Group is developing a model to characterize microenvironments for LDEF experiments which will include scattered atomic oxygen as well as direct atomic oxygen and solar fluences.

Contamination effects were easily observed in many locations aboard the LDEF satellite. Heavy, highly localized molecular deposits were noted near cables and connectors as well as near vent locations and various materials. A thin, ubiquitous film was also detected over the LDEF structure. The entire interior surface of LDEF was painted with a diffuse black paint, Chemglaze Z-306, which was not baked. Tests performed at the Air Force Arnold Engineering and Development Center (AEDC) indicate that Z-306 outgassing products would produce a contaminant film on surfaces exposure to solar radiation. An adequate source for the silicone component of the LDEF contaminant film could not be identified aboard the LDEF satellite. The particle contaminant distribution over LDEF was also characterized for the LDEF mission life. LDEF was deployed with a cleanliness level consistent with MIL-STD 1246B Level 1000C to Level 2000C. As a result of orbital exposure, retrieval, ferry flight and deintegration activities, the particle distribution was significantly increased. Results of the study of contaminants aboard LDEF were reported at the First LDEF Post-Retrieval Symposium [Ref. 8].

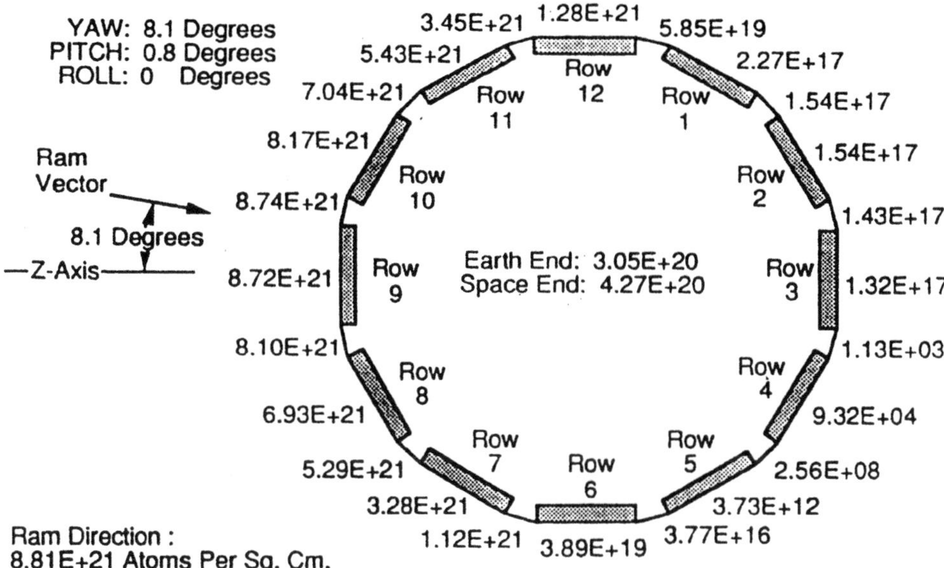

Fig. 2: Atomic oxygen fluences for LDEF surfaces, anomalously high fluences on Rows 2 and 3 are due to unintentional vehicle roll after retrieval.

A thermal model for LDEF was constructed and compared with recorded on-orbit temperatures at various locations. The results were documented to provide a detailed solar fluence [Ref. 9] and time temperature history [Ref. 10] for LDEF surfaces. The space end of LDEF received approximately 14,500 equivalent solar hours (ESH) of direct solar radiation during the mission. The surfaces about the perimeter of LDEF received between 6,000 and 11,000 ESH while the earth end received about 1200 ESH direct solar illumination. The thermal measurements indicated that the LDEF structure experienced temperature extremes ranging from 0°C to 60°C. Experiment surfaces obviously encountered much broader temperature variations, depending upon thermal optical properties, mass and conduction paths of the experiment hardware. A refined thermal model, constructed by NASA Langley Research Center, is accessible to LDEF experimenters to perform individual thermal analyses for experiment components.

The LEO environment is generally considered to be a relatively benign environment with respect to charged particle radiation. LDEF, by virtue of its extended exposure to the space environment, provided an opportunity to examine induced radioactivity about a large three-axis stabilized space structure. Shortly after transfer to the Spacecraft Assembly and Encapsulation Facility (SAEF-2) at Kennedy Space Center, an array of gamma-ray spectrometers from the Naval Research Laboratory were positioned about the LDEF structure. The trunnions and keel pin were also removed and sectioned for gamma-ray measurements. Radionuclei from spallation reactions with aluminum, such as ^{22}Na were detected as expected. A strong anisotropy of proton activation was observed with the west surface (trailing edge) experiencing a substantially greater fluence that the east surface. The more surprising result was the unexpectedly high levels of ^{7}Be detected on the leading edge of LDEF. The ^{7}Be radionucleus is formed by spallation of oxygen and nitrogen at stratospheric altitudes by cosmic rays. The levels of ^{7}Be found on LDEF are far greater than expected by diffusive transport mechanisms, indicating unexpected dynamics within the upper atmosphere which more efficiently transport ^{7}Be to high altitudes. Induced radioactivity and ionizing radiation exposure levels are provided in the report [Ref. 11] from the LDEF Induced Radiation SIG.

3. Selected Findings from LDEF

The primary structure of LDEF was constructed from 6061-T6 aluminum and suffered very little, if any, degradation as a result of extended exposure to the launch and space environments [Ref. 12]. During deintegration, the torques of fasteners for the tray clamps were recorded. A very small percentage of fastener torques were found to lie significantly outside the pre-flight values. Several experimenters reported shearing of smaller fasteners during tray level deintegration. This phenomenon was attributed to galling of the fasteners during installation, particularly in the case of unlubricated fasteners. No cases of metallic cold welding were found aboard LDEF [Ref. 13], including on an experiment designed to present favorable conditions for cold welding [Ref. 14].

Organic matrix composite materials were found to be significantly affected by atomic oxygen erosion on the leading edge. Mechanical properties (modulus and strength) were degraded and dimensional stability affected due to erosional loss of material [Ref. 15]. Composites on the trailing edge or those which were protected by metal films were essentially unaffected. Some composites experienced dimensional changes due to thermal cycling and outgassing over a period ranging from weeks to months [Ref. 16].

Thermal control materials and coatings performance was the subject of several LDEF experiments. A number of white paints were evaluated in varying environments. Chemglaze A-276 was found to be significantly darkened on the trailing edge of LDEF whereas A-276 on the leading edge was found to be white (see Fig. 3). This ram/wake asymmetry is attributed [Ref. 17] to atomic oxygen removal of the urethane binder which is readily darkened by solar ultraviolet radiation. The silicone-based S13G/LO was found to degrade significantly in both ram and wake environments. The ceramic white paints [Ref. 18] such as Z-93 and YB-71 were found to be essentially unaffected aboard LDEF. The chromic acid anodize was found to perform well on LDEF [Ref. 19], the Space Station Freedom sulfuric acid anodize coating for aluminum was not flown on LDEF. The silvered teflon (Ag/FEP) blankets flown on LDEF were obviously affected on the leading edge appearing "fogged" whereas those blankets on the trailing edge appeared unaffected. The fogged appearance of the leading edge Ag/FEP was due to atomic oxygen erosion with approximately 20 microns removed on the leading edge [Ref. 20]. Trailing edge Ag/FEP blankets developed an embrittled layer attributed to solar vacuum ultraviolet radiation [Ref. 21]. Micrometeroid and debris impacts on the Ag/FEP blankets created large delamination zones about the impact site [Ref. 4]. Unexpected degradation occurred for a bonded Ag/FEP film which was attributed [Ref. 22] to the application process which permitted the adhesive to migrate through cracks in the silver layer which subsequently darkened upon solar ultraviolet exposure. A vapor deposited aluminum on Kapton (VDA/Kapton) blanket completely disintegrated on the leading edge, creating vast amounts of aluminum debris [Ref. 23]. The VDA/Kapton blanket on the trailing edge was found to be essentially unaffected.

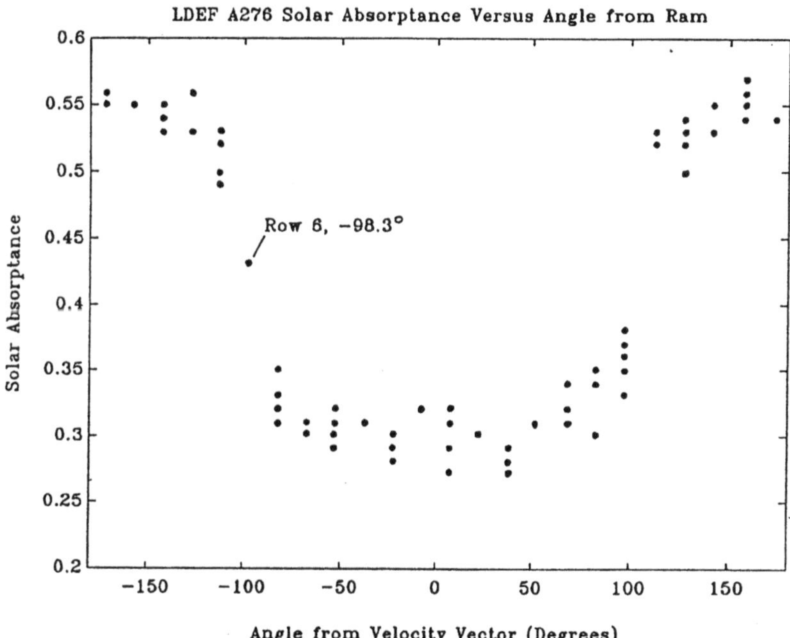

Fig. 3: Plot of A276 solar absorptance versus angle from velocity vector. The partial "scrubbing" of the Row 6 specimen (98.3° from ram) is due to the thermal velocity effects of atomic oxygen.

Optical materials aboard LDEF experienced levels of degradation ranging from essentially unchanged to total loss of performance [Ref. 24]. Many optical test specimens were severely affected by molecular contamination [Ref. 25] with significant loss of spectral transmission or reflectance in the ultraviolet and infrared regions. Several optical materials were damaged by micrometeoroid or debris impacts with a few specimens fractured but many being pitted by small impactors. Several optical coatings were severely degraded by atomic oxygen, thermal cycling or solar ultraviolet radiation.

Solar cells aboard LDEF generally performed quite well. A number of cells were cracked by micrometeoroid or debris impacts. Some cells showed slight performance degradation due to contamination of the covers. In general, the solar cells aboard LDEF showed power production degradation of less than 10% [Ref. 26].

The Experiment Initiate System (EIS), Experiment Power and Data Systems (EPDS), experimenter electronics, batteries, and associated harnessing were planned to be extensively tested by the LDEF Systems SIG [Ref. 27]. The EIS was found to perform nominally in post-flight examination with the exception of one unused indicator relay which suffered from an intermittent condition attributed to particulate contamination within the device. The EPDS generally functioned as expected on orbit with the exception of one of Magnetic Tape Modules in which a relay failure caused a portion of early flight data for the Thermal Control Surfaces Experiment to be overwritten. More detailed analyses of the EPDS was not possible since the pre-flight performance data was purged by the manufacturer. Generally experimenter unique electronics worked well with the exception of one experiment where the failure was attributed to a design problem. LDEF batteries, particularly the lithium sulfur dioxide batteries, performed very well. Some leakage occurred in a lithium carbon monofluoride battery due to deformation of the o-ring seal, however, the batteries performance was not compromised. Harnesses were generally found in good condition. Some connectors used by experimenters were not space rated and outgassed excessively. In spite of relatively widespread use of non-space rated components (i.e. MIL-STD 883B or commercial parts), no anomalies in the electronic and power systems aboard LDEF were attributed to extended exposure to the LEO environment.

4. Implications for LEO Spacecraft Systems, Present and Future

LDEF has provided a great deal of valuable data for spacecraft designers. The integrity of the LDEF structure implies high reliability in welded and bolted metal designs. Advanced high performance composite spacecraft structure need to account for environmental effects due to atomic oxygen and thermal cycling. The materials selection process for thermal control materials and coatings has been aided by the evaluation of these materials on LDEF. Ceramic white coatings performed very well, but flexible white coating did not fare well in LEO. Silver teflon blankets generally performed well but may degrade in very long life missions. Unprotected VDA/Kapton is not recommended for surface expected to encounter significant atomic oxygen fluence. Contamination effects on optical components were demonstrated to significantly degrade performance of certain optics. Micrometeoroid and debris also represents a considerable risk to high-performance large optics. Solar cell performance is expected to suffer little degradation in LEO although micrometeoroid and debris impacts can affect performance. The LEO environment poses little risk to electronic and power systems and the use of fully space qualified (Class S) parts is not seen as necessary provided good design practice is followed. The use of electromechanical devices adds risk to system design and solid state alternatives should be used where practical.

Based on the quantity and quality of data which LDEF is returning, the need for long term space exposure facilities is apparent. Space Station Freedom will provide a platform for conducting space environmental effects research in the twenty-first century. The shorter term thrusts for space environmental effects research should focus on improving our understanding of the space environment by modeling and further space experiments. The space debris environment, in particular, should be characterized more thoroughly and monitored for assessment of debris growth. The need for testing and validation of new materials technologies prior to incorporation into long-life operational space assets will continue to exist. Development of improved ground simulation capabilities will be guided by correlative studies of materials degradation with space exposure experiments. A low-cost long term space exposure platform could also be useful in evaluating long life space mechanisms, high capacity data storage devices, communications technologies, advanced power systems and other advanced subsystem elements prior to use in civilian, military and commercial spacecraft systems.

Acknowledgement

Participation in the LDEF deintegration process and support to the LDEF Materials Special Investigation Group by the author was supported by the Key Technologies Directorate of the Strategic Defense Initiative Organization (SDIO) through the Air Force Wright Laboratory and the Jet Propulsion Laboratory, California Institute of Technology, under a contract with the National Aeronautics and Space Administration.

References

[1] Clark, L.G., Kinard, W.H., Carter, D.J., and Jones, J.L., "The Long Duration Exposure Facility Mission 1 Experiments," NASA SP-473, 1984.

[2] Banks, B., reported at the LDEF Materials Workshop, Hampton, VA, 1991.

[3] Gregory, J.C., "LDEF Attitude Measurement Using a Pinhole Camera With a Silver/Oxygen Atom Detector," First LDEF Post Retrieval Symposium Proceedings, NASA CP-3134, 1992.

[4] See, T.H, Allbrooks, M.A., Atkinson, D.R., Simon, C.G. and Zolensky, M., "Meteoroid and Debris Impact Features Documented on the Long Duration Exposure Facility, A Preliminary Report," Publication #84, JSC #24608, 1990.

[5] Mulholland, J.D., Singer, S.F., Oliver, J.P., Weinberg, J.L., Cooke, W.J., Montague, N.L., Wortman, J.J., Kassel, P.C., and Kinard, W.H., "IDE Spatio-Termporal Impact Fluxes and High Time-Resolution Studies of Multi-Impact Events and Long-Lived Debris Clouds," First LDEF Post-Retrieval Symposium Proceedings, NASA CP-3134, 1992.

[6] Bourassa, R.J. and Gillis, J.R., "Atomic Oxygen Exposure of LDEF Experiment Trays," NASA Contractor Report 189627, 1992.

[7] Gregory, J.C., Christi, L., Raikar, G.N., Weimer, J.J., Wiser, R., "Interactions of Atomic Oxygen with Material Surfaces in Low Earth Orbit: Preliminary Results from Experiment A0114," First LDEF Post-Retrieval Symposium Proceedings, NASA CP-3134, 1992.

[8] Crutcher, E.R., et al, 4 papers in the First LDEF Post Retrieval Conference Proceedings, NASA CP-3134, 1992.

[9] Bourassa, R.J and Gillis, J.R., "Solar Exposure of LDEF Experiment Trays," NASA Contractor Report 189554, 1992.

[10] Berrios, W.M., "Long Duration Exposure Facility Post-Flight Thermal Analysis Orbital/Thermal Environment Data Package," Orbital Environment Data Group, 1991.

[11] Parnell, T.A. "Summary of Ionizing Radiation Analysis on LDEF," available from the LDEF Project Office.

[12] Spear, W.S. and Dursch, H.W., "LDEF Mechanical Systems," First LDEF Post-Retrieval Symposium Proceedings, NASA CP-3134, 1992.

[13] Dursch, H.W. and Spear, W.S, "On-Orbit Coldwelding: Fact or Friction?," First LDEF Post-Retrieval Symposium Proceedings, NASA CP-3134, 1992.

[14] Assie, J.P., "Microwelding (or Cold-Welding) of Various Maetallic Materials under the Ultra-Vacuum LDEF Experiment A0138-10," First LDEF Post-Retrieval Symposium Proceedings, NASA CP-3134. 1992.

[15] George, P.E., and Hill, S.G., "Results from Analysis of Boeing Composite Specimens Flown on LDEF Experiment M0003," First LDEF Post-Retrieval Symposium Proceedings, NASA CP-3134, 1992.

[16] Tennyson, R.C., Mabson, G.E., Morison, W.D., and Kleiman, J., "Preliminary Results from the LDEF/UTIAS Composite Materials Experiment," First LDEF Post-Retrieval Symposium Proceedings, NASA CP-3134, 1992.

[17] Golden, J.L., "Results of Examination of the A276 White and Z306 Black Thermal Control Paint Disks Flown on LDEF," First LDEF Post-Retrieval Symposium Proceedings, NASA CP-3134, 1992.

[18] Wilkes, D.R., Brown, M.J., Hummer, L.L. and Zwiener, J.M., "Initial Materials Evaluation of the Thermal Control Surface Experiment," First LDEF Post-Retrieval Symposium Proceedings, NASA CP-3134. 1992.

[19] Plagemann, W.L., "Space Environmental Effect on the Integrity of Chromic Acid Anodized Coatings," First LDEF Post-Retrieval Symposium Proceedings, NASA CP-3134, 1992.

[20] Hemminger, C.S., Stuckey, W.K., and Uht, J.C., "Space Environmental Effects on Silvered Teflon Thermal Control Surfaces," First LDEF Post-Retrieval Symposium Proceedings, NASA CP-3134, 1992.

[21] Brinza, D.E., Stiegman, A.E., Stazak, P.R., Laue, E.G., and Liang, R.H., "Vacuum Ultraviolet (VUV) Radiation-Induced Degradation of Fluorinated Ethylene Propylene (FEP) Teflon aboard the Long Duration Exposure Facility," First LDEF Post Retrieval Symposium Proceedings, NASA CP-3134, 1992.

[22] Zwiener, J.M., Herren, K.A., Wilkes, D.R., Hummer, L., Miller, E.R., "Unusual Materials Effects Observed on the Thermal Control Surface Experiment (S0069)," First LDEF Post-Retrieval Symposium Proceedings, NASA CP-3134, 1992.

[23] Blakkolb, B.K., Yaung, J.Y., Henderson, K.A., Taylor, W.W., and Ryan, L.E., "Long Duration Exposure Facility (LDEF) Preliminary Findings: LEO Space Effects on the Space Plasma Voltage Drain Experiment," First LDEF Post-Retrieval Symposium Proceedings, NASA CP-3134, 1992.

[24] Hawkins, G.J., Seeley, J.S., Hunneman, R., "Exposure to Space Radiation of High-Performance Infrared Multilayer Filters and Materials Technology Experiment," First LDEF Post-Retrieval Symposium Proceedings, NASA CP-3134, 1992.

[25] Harvey, G.A., "Effects of Long-Duration Exposure on Optical System Components," First LDEF Post-Retrieval Symposium Proceedings, NASA CP-3134, 1992.

[26] Brinker, D.,J., Hickey, J.R., and Scheiman, D.A., "Advanced Photovoltaic Experiment, S0014: Preliminary Flight Results and Post-Flight Findings," First LDEF Post-Retrieval Symposium Proceedings, NASA CP-3134, 1992.

[27] Miller, E.A., Brooks, L.K., Johnson, C.J., Levorsen, J.L., Mulkey, O.R., Porter, D.C., and Smith D.W., "LDEF Electronic Systems: Successes, Failures and Lessons," First LDEF Post Retrieval Symposium Proceedings, NASA CP-3134, 1992.

SPACE POWER EXPERIMENTS ABOARD ROCKETS

E.E. KUNHARDT
Weber Research Institute
Polytechnic University
Farmingdale, New York 11735

1. Introduction

Space Power Experiments Aboard Rockets (SPEAR) is a research program sponsored by the Strategic Defense Initiative Organization, Innovative Science and Technology division (SDIO/IST), and managed by the Defense Nuclear Agency (DNA), whose objectives are: 1) to develop pulse power systems and components that exploit the space-vacuum for insulation (thereby reducing the weight of the payload), and demonstrate their operation in space, and 2) establish the basis for treating problems in the areas of high voltage insulation, "grounding", and operation of high voltage/current systems in the Low Earth Orbit (LEO) space environment.

SPEAR consists of a series of rocket-borne experiments (RBE) and associated supporting effort involving theoretical, computational, and experimental components. Two RBE have been carried out, and a third is presently in preparation. The first and third RBE are being conducted under the direction of Utah State University, whereas in the second Space Data Corporation has had both integration and launch responsibility. The supporting effort is distributed among the following institutions: Auburn University, Maxwell Laboratory, Naval research Laboratory, Polytechnic University, Texas Tech University, University of Maryland, the University of Texas at Arlington, and Westinghouse.

The first RBE, SPEAR I, has addressed basic issues related to the program objectives [Refs. 1-12]. The SPEAR I payload configuration, including a block diagram of the power circuit, is shown in Fig. 1 [Ref. 1]. This configuration was flown on Dec, 1987, in a suborbital flight from the NASA facility at Wallops Island, Virginia. It reached an apogee of 369 km. A sequence of positive bias voltages were applied to the spheres with respect to the rocket body and the resulting current flowing in the circuit measured. Data were gathered for three orientations of the plane, S, of the spheres relative to the direction of the magnetic field of the earth. The I-V characteristics are shown in Fig. 2 [Ref. 1] (the direction of B is given relative to the plane S). From these data and numerical simulations [Ref. 9], the behavior of bi-polar objects in LEO,

R. N. DeWitt et al. (eds.), The Behavior of Systems in the Space Environment, 669–711.

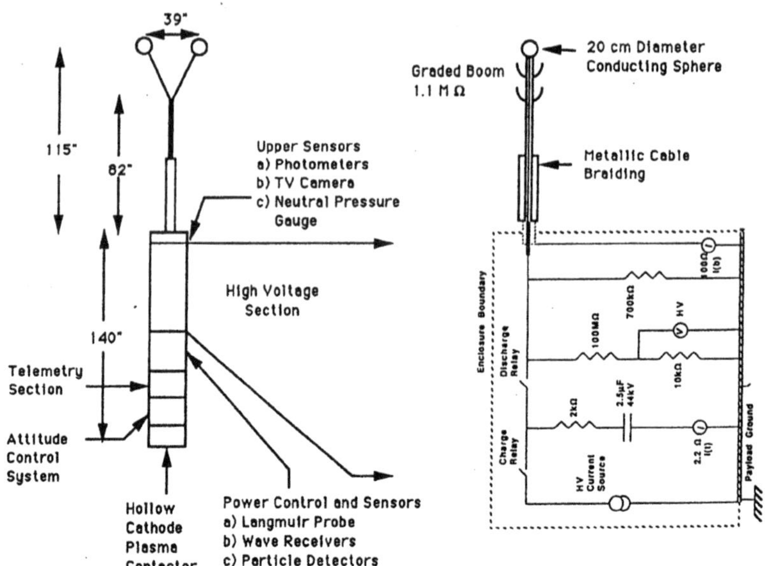

Fig. 1: SPEAR I payload.

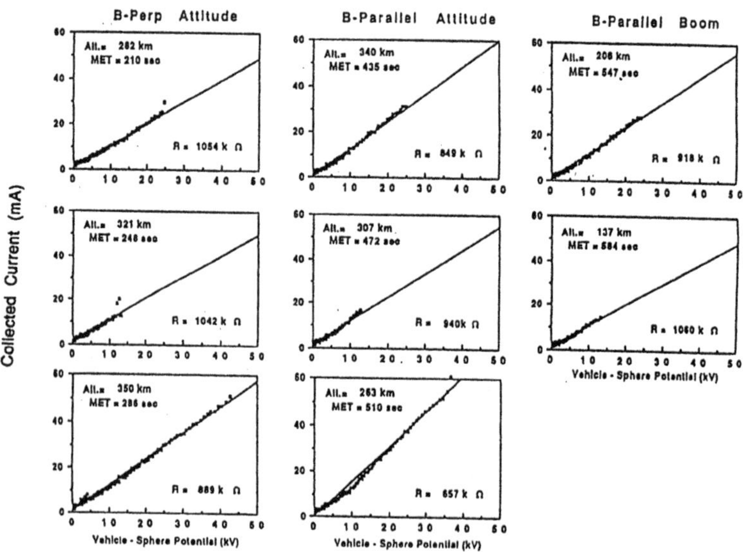

Fig. 2: SPEAR I collected current vs applied sphere voltage.

and in particular the dynamics of interacting sheaths have been elucidated ("Current Collection by High-Voltage Spacecraft", Ira Katz). For voltages up to 46 kV and altitudes above 100 km, no breakdown has been observed under these experimental conditions.

The current-voltage and breakdown characteristics for two electrode systems in test chambers of various sizes and under various background conditions [Refs. 3-8] (including low density neutral and ionized gases) have also been measured. This was done in connection with pre-flight tests of the payload [Ref. 6] and to asses the validity of ground-based ionospheric simulation experiments. Theoretical and computational models [Refs. 9-12] have been developed for determining the current flow to a stressed electrode exposed to LEO, and the threshold voltages for breakdown of the sheath surrounding it, and for breakdown of the medium between two biased electrodes. These threshold voltages have been measured in the chamber experiments using a low density ionized gas as background, and a spherical electrode biased with respect to the chamber wall. The sheath breakdown criterion applies to the regime of charged particle density where the separation, d, between the electrodes (sphere-chamber) is larger than the dimensions of the sheath (denoted by r_s), so that wall effects (due to the interaction of the charge particles with the chamber walls) are minimized. This situation exemplifies that found in SPEAR I. The two-electrode breakdown threshold applies to the density regime where $r_s \gg d$. This type of breakdown can occur between components of partially confined payloads, similar to SPEAR II, when the breakdown threshold voltage is below the operating voltage.

In the second RBE, SPEAR II, the payload consisted of two pulse power systems, selected as models for the power supplies needed for SDIO directed energy and electromagnetic launcher systems. A salient feature of the payload is that critical components in these systems made use of space-vacuum insulation. A block diagram of the payload is shown in Fig. 3. It consists of a high voltage system that delivers a 100 kV, 10 amp (1MW) pulse of up to 50 microsecond duration to the anode-cathode structure of a klystrode RF power tube, and a high current system that drives an electromagnetic launcher operating at 140 kA and 6 kV. A number of space-vacuum insulated components form part of these systems, including a pulse transformer, storage inductor (both designed by Westinghouse), and pulse forming components (designed by Maxwell Laboratory) . The payload was tested in a chamber under conditions simulating the space environment, successfully demonstrating the merit of the design concepts (testing under flight conditions was to take place on 25 July 1990, on a suborbital flight from the White Sands Missile Range; but thirty-five seconds into the flight, the rocket veered off course and had to be destroyed).

The third RBE is in the planning stage and is to investigate issues related to the : a) discharging of charged objects via different "grounding" schemes, b) operation of high voltage solar arrays, and c) the current-voltage and breakdown characteristics of an uni-polar object (single stressed electrode relative to the space-plasma) in LEO (this complements the results obtained in SPEAR I for bi-polar objects. This measurement was not perform in SPEAR I since the cover plate exposing the plasma contactor to the ionosphere could not be opened, and consequently the rocket body could not be brought to plasma potential).

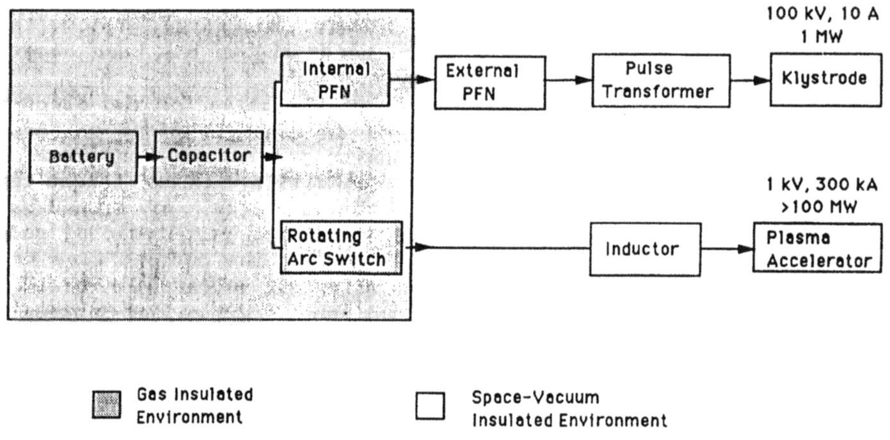

Fig. 3: Schematic Diagram of the SPEAR II payload.

Since each RBE involves a number of issues pertaining to the objectives of the program, instead of using them as framework , this presentation is organized around the body of information that has evolved regarding the insulating properties of the "space-vacuum". This approach allows for a systematic development of the issues and thus exhibits the overall contributions of the program. Moreover, it permits to contrast the state of affairs with regards to insulation for space and ground-based systems, and measure the contributions that the program has made in a relatively short time.

2. Pulse Power Technology In Space

2.1. BACKGROUND

A block diagram of a generic pulse power system, with examples of system components, is shown in Fig. 4. The system contains four major elements: 1) power generator, 2) energy storage components, 3) switches, and 4) the load. The operating time scales and power levels throughout the system are functions of load requirements. Power levels corresponding to SPEAR II are shown in Fig. 3 (similar layouts have been used for Figs. 3 and 4). These factors place requirements on the medium used for insulation and the techniques for integrating the components into an operating system. Since the voltage at which an insulating medium breaks down is a function of the duration of application (increasing with decreasing duration) [Ref. 13], Fig. 4 suggests that the insulation requirements vary along the power train. The general approach for determining these requirements is to use criteria derived from DC application as reference, since it represents the minimum threshold voltage for the failure of the insulation. This paper discusses these criteria in the context of the space environment. In SPEAR II for example, the power generator (battery and capacitors in Fig. 3) and

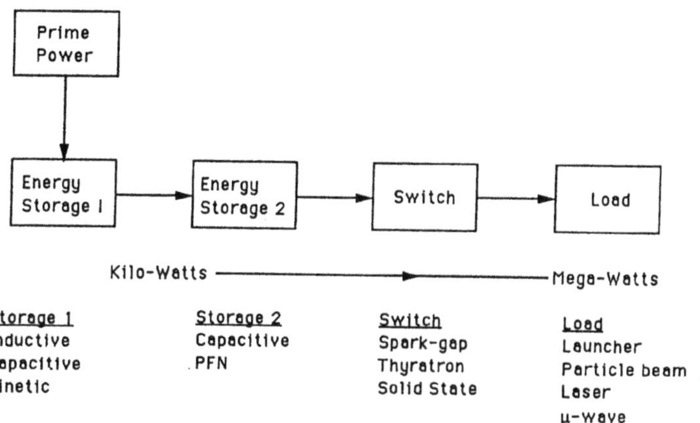

Fig. 4: Schematic diagram of generic pulse power system.

the switches were inside the gas insulated section so that classical insulation principles applied (although the need to minimize weight is always present). The energy storage components used both types of insulation, gas (internal PFN) and space-vacuum (external PFN, Inductor, and pulse transformer), whereas the loads were open to space.

Given that these systems are intended for space applications, there is a compelling need for reducing their size and weight as compared to those intended for earth-based applications. Space-based systems present the designer with the possibility for making optimal use of vacuum insulation to meet the criterion of reduced weight. However a closer look at the parameters defining the "space-vacuum" (see lecture by R.V. Latham p. 467) shows that it has unique dielectric properties, significantly different from those used in earth-based systems. This drive demands an understanding of the electrical properties of materials and methods for controlling electric and magnetic distributions in this environment. As far as its insulating properties are concerned, this medium behaves more like a weakly ionized gas (albeit with extremely low density) than a "vacuum". Moreover, because of the low density and the large volumes which need to be considered, this behavior is significantly influenced by the earth's magnetic field. Thus, as will become evident from the discussion in this paper, to characterize its behavior as an insulator, it is necessary to ascertain both its breakdown characteristics and the loading it represents to a power system due to leakage currents. This can be achieved by obtaining the current-voltage and breakdown characteristics of generic electrode configurations and background conditions, which can then be extrapolated to other situations.

The experimental and theoretical base that serves as a guideline in the use of this medium as insulation has been lacking. Using earth-based systems as a guide, it is recognized that because of the complexity of the problems [arising for example from the interaction of applied fields and the changing local field and material parameters (due to space charge effects)] a special approach is necessary for specifying, with reasonable accuracy, the insulation requirements in a practical system. This approach has two aspects: 1) the theoretical basis for calculating dielectric strengths, leakage currents, and field distributions, given the operating conditions, and 2) an experimental data base for dielectric strengths and leakage currents using model (generic) field distributions (in space and time) similar to those existing in the particular applications. This latter aspect implies more than just the validation of the theory, in that for the specific application, the parameter space may be either unknown or lie outside the range of validity of the theory, and a combination of both aspects may then be necessary.

The SPEAR program has made significant contributions towards both aspects of the space insulation problem. An extensive body of knowledge has evolved which is summarized in sections III-VI without reference to specific contributors. Before doing so, since the theoretical foundations for space insulation start with concepts originating in the parameter space of earth-base systems, this space and the corresponding theoretical foundations are reviewed in Sec. 2.2B. The parameter space associated with the space-vacuum and the models that have been chosen for electrode configuration are discussed in Sec. 2.2C.

2.2. PARAMETER SPACE AND DIELECTRIC STRENGTH OF INSULATING MATERIALS PERTINENT TO EARTH-BASED SYSTEMS

This topic of course is the subject of a number of books and journal articles [Refs. 13-17]. Our only purpose is to present those aspects that will serve to place into context the problem of space-vacuum insulation. There are four insulating media used in earth-based applications: gases, solids, liquids, and vacuum. The theory for gas insulation is the most advanced and the one that is most pertinent. This is because the dielectric strength of the space-vacuum arise from its gas-like characteristics. An important difference between the two medium is evident. For voltages below the breakdown threshold, the pre-breakdown (leakage) currents are essentially zero in a gas (unless the electrodes are highly non-uniform giving rise to corona discharges, see Sec. 2.2B), whereas they may be significant in the space-vacuum. The leakage currents in this case are associated with either current collection from the ambient plasma (see Sec. 2.2C) or with toroidal discharges which are the analog of corona discharges (see Sec. 4). Consequently, in a gas, the problem of insulation reduces to obtaining the breakdown characteristics of the gas in a particular electrode configuration (field distribution).

There are three generic field distributions that form the basis for the theoretical approach to the problem of insulation in earth-based systems. These distributions are created in the following electrode configurations (or geometries): 1) plane-plane, 2) sphere-sphere (the rod-plane geometry is considered as a special case of this geometry), and 3) concentric cylinders. The theory for determining the dielectric strength of a homogeneous insulating gas placed between two plane electrodes is considerably more advanced than for the other configurations. This theory (the Townsend theory) is presented first, followed by those that apply to non-uniform geometries.

A. Uniform Field Breakdown (plane-plane geometry)

The theory of breakdown in uniform fields (Townsend Breakdown) has provided one of the primary tools in the design of earth-based power components and systems, namely the "Paschen law". This "law" expresses the insulating strength of the gas as a function of the parameter space of the medium and inter-electrode separation. The development of the theory is sketched below.

Consider smooth parallel plate electrodes separated by a distance d, with the interelectrode space homogeneously filled with a gas at a density N. The density is sufficiently high that the electron mean free path, λ, is of the order or less than d ($\lambda \leq d$). In the absence of a magnetic field, the avalanche enhanced current, I, resulting from an applied electric field, E=V/d (where V is the voltage between the electrodes) and an external source of current at the cathode, I_0, is determined from the continuity equations for the electron and ion currents, I_e and I_i, respectively. That is,

$$\frac{d}{dx} I_e = \alpha I_e \tag{1a}$$

$$\frac{d}{dx} I_i = \alpha I_i \tag{1b}$$

and the conditions,

$$I = I_e(x) + I_i(x) = I_e(d) \tag{1c}$$

$$I_e(0) = I_o + \gamma I_i(0) \tag{1d}$$

where α is the primary ionization coefficient, and γ is the secondary electron production coefficient due to ion impact on the cathode. The solution to these equations (neglecting space-charge effects so that α can be taken to be constant) is

$$I = \frac{I_o e^{\alpha d}}{1-\gamma(e^{\alpha d}-1)} \tag{2}$$

The condition for which the secondary (γ) process can sustain the current I with I_0=0 is defined as breakdown. From Eq. 2, this is obtained by setting the denominator to zero,

$$\gamma(e^{\alpha d} - 1) = 1 \tag{3}$$

[a more general expression than this equation is obtained by substituting ω/α for γ, where ω is the generalized secondary electron production coefficient ($= \alpha\gamma + \beta + \delta + \ldots$, with β and δ being the electron production coefficients due to ion impact ionization of the gas and photo-ionization, respectively)].

Since α/N and ω/N are functions of E/N, Eq. (3) can in principle be solved for the field (or voltage) that leads to breakdown, E_s (or V_s), in terms of the product Nd. This results in a relationship of the form

$$V_s = V_s(Nd) \tag{4}$$

which is known as the Paschen Law. In certain regimes, the functional relation between a/N and E/N can be approximated by

$$\alpha/N = A\exp(-BN/E) \tag{5}$$

where A and B are constants for a given gas. Assuming that the most important secondary process is the γ-process ($\omega/\alpha \sim \gamma$) and that γ is a slowly varying function of E/N over a wide range of values (this is an experimentally observed fact), then using Eq. (5), Eq. (4) for V_s becomes

$$V_s = B\,Nd/(C_1 + \ln(ANd)) \tag{6}$$

where

$$C_1 = \ln(1 / \ln(1 + 1/\gamma)).$$

From this equation it can be seen that at large values of Nd ("the right hand side of the Paschen curve"), the breakdown voltage, V_s, increases with increasing Nd. Similarly, at low Nd values (" the left hand side of the Paschen curve"), V_s, increases with decreasing Nd. In the transition from low to high values of Nd, the breakdown voltage goes through a minimum (the Paschen minimum), $V_{s\ min}$, at an Nd value denoted by Nd_{min}. With voltages lower than $V_{s\ min}$, it is impossible to cause the breakdown of a uniform field gap. A sketch of the breakdown characteristics in this geometry is shown in Fig. 5. Also shown is the range over which Eq. (6) is valid. The range of Nd below the Paschen regime corresponds to semi-vacuum [Ref. 16] and vacuum breakdown, which involve processes other than the α and ω processes.

Theoretically, Eq. (6) is valid as long as the gas density is uniform and sufficiently high that the various coefficients appearing in the equation can be defined. This occurs in a range of Nd about Nd_{min} such that the mean free path for collision is less than, or of the order of, the inter-electrode separation. Moreover, the avalanche growth ($\exp(\alpha d)$) must be sufficiently low that the evolving space charge does not influence this growth. This is in general true if the space charge field is much less than the applied field at breakdown. At atmospheric pressures, the growth factor

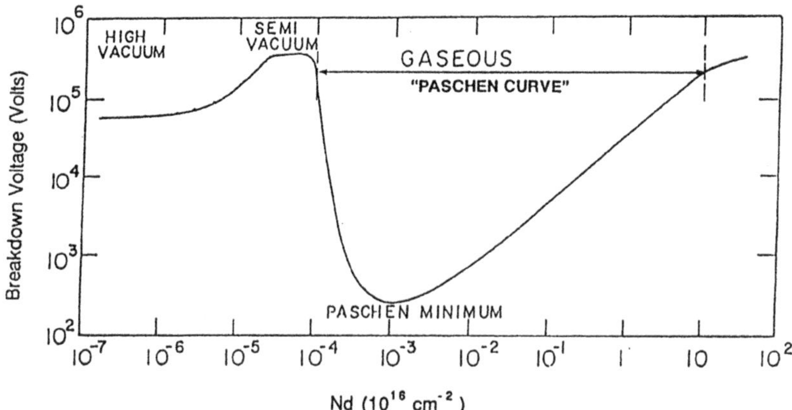

Fig. 5: Sketch of the breakdown voltage vs pd for B = 0, showing Paschen regime.

must be less than 10^7, corresponding to an electron density of 10^{12} cm^{-3}. A number of experimental investigations (of the steady state type) have been conducted to confirm Paschen's law. The results agree well with Eq. (6) for values of Nd up to approximately 10^{20} cm^{-2} and down to approximately 10^{15} cm^{-2} (see Fig. 5). Outside this range, additional factors need to be considered that are not taken into account in Eq. (4). At high Nd, irregularities in the cathode surface cause local field intensification leading to larger than average avalanche gains, and consequently to lower breakdown voltages than predicted from Eq. (4). Similarly, at low Nd, surface irregularities combined with gas inhomogeneities (produced via various mechanisms) result in deviations from Eq. (4).

As mentioned earlier, Eq. (4) is also not applicable when the avalanche gain is sufficiently high that the resulting space charge field becomes of the order of the applied field. In this case, a semi-quantitative breakdown condition can be developed based on the concept of a streamer. The resulting expression may however be viewed as a generalization of Eq. (4), and can be written as

$$\exp\left[\int_0^{x_c} \alpha \, dx\right] = n_c \tag{7}$$

where x_c is the distance along the field line in which an avalanche reaches the critical number, n_c (approximately 10^8), necessary for the initiation of a streamer. It is assumed that if the condition for streamer formation is met, breakdown follows.

In the presence of a transverse external magnetic field, B (as would be the case in space) the breakdown characteristics (Eq. (4)) are affected

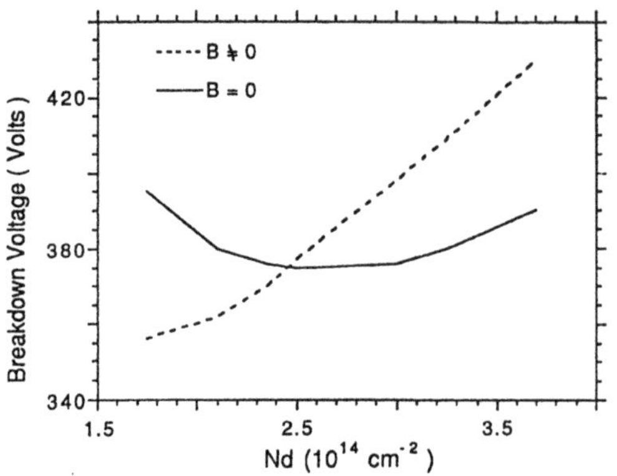

Fig. 6: Sketch of the effect of B on breakdown voltage.

as schematically depicted in Fig. 6 [Ref. 20]. For Nd >> Nd_{min}, a non-zero value for B always increases V_s. Whereas for Nd << Nd_{min}, a small B causes a decrease in V_s, while a large B increases V_s. This effect of the magnetic field on the breakdown characteristics may be incorporated into Eq. (4) via the concept of effective density or field. That is, defining the effective density, N', via the relation [Ref. 18]

$$N' = N\sqrt{1 + (\omega_c t_c)^2} \tag{8}$$

where ω_c is the cyclotron frequency (eB/m), and $t_c = \gamma_c/Nv$, with γ_c being the collision free path at 1 mm Hg pressure and v the average electron velocity. The primary ionization coefficient in the presence of a magnetic field can be obtained from its B=0 value using the relation

$$\alpha_B\,(B/N, E/N) = \alpha(0, E/N') \tag{9}$$

Similar relations hold for the other coefficients appearing in Eq. (4). Thus the breakdown characteristic can still be determined from an expression similar to Eq. (4). However, these concepts are only useful for values of B such that γ_c is less than the cyclotron orbit, 2 r_c, where r_c is the cyclotron radius. There is no theory available outside of this regime.

B. Non-uniform Field Breakdown

The theory for breakdown in the other two generic configurations is not as developed as for parallel-plate electrodes; moreover, the approach is based on either the Townsend or streamer theories for uniform fields, so that the same parameters and concepts developed in connection with these theories are still useful. Thus in practice, where all three types of generic field configurations are encountered, the Paschen law and streamer formation criterion developed (theoretically or experimentally) for these configurations are used for determining insulation requirements (depending on the conditions, as discussed below). Thus Eq. (7) which has been interpreted as an expression for the breakdown condition in both theories, is also applicable to the other generic configurations. In this case, α ($=\alpha_{eff}$) depends on space and the integration in Eq. (7) is taken along the path of highest field strength (highest growth).

For highly non-uniform fields (the degree of non uniformity is measured in terms of the peak to average field ratio), such as point-plane electrodes, manifestations of transient or steady state luminous corona discharges ("partial" breakdown) are observed without the occurrence of "complete" breakdown (where "complete" breakdown refers to the establishment of a low impedance channel that bridges the interelectrode space). The voltage at which a corona discharge develops (the corona inception voltage) is also obtained from Eq. (7). In either case, the mean free path for ionization, λ_i, must be less than the distance, d_c, from the stressed electrode over which $|E(r)| > E_c$, where E_c is the critical field below which electrons cannot gain sufficient energy to cause ionization of the neutral gas. Thus, in non-uniform fields, the voltage that satisfies Eq. (7) may result in either "partial" or "complete" breakdown. Which process actually occurs depends on factors such as the degree of field non-uniformity, gas density, polarity of the highly stressed electrode, and the nature of the gas (for detailed treatment of this subject see [Ref. 21]). As discussed in the next section, corona like discharges are of significant importance to spaced-based applications, since in most situations of interest the field is non-uniform. These discharges are responsible for considerable power losses from high voltage systems and often lead to deterioration of insulation by action of the discharge ions bombarding the insulating surfaces, and other discharge byproducts reacting with it. These discharges also give rise to interference in communication systems.

In the regime $Nd/Nd_{min} >> 1$, an electron avalanche may develop into a streamer within the distance d_c. At breakdown, the streamer propagates across the interelectrode space resulting in the collapse of the voltage between the electrodes. The discharge takes the form of a filamentary channel. At the other extreme, $Nd/Nd_{min} << 1$, electron avalanches can only occur if the effective free path for ionization in the field direction, α_{eff}^{-1}, is at least of the order of d_c. In negative divergent electric fields, this occurs when the applied field is sufficiently high that field-emitted currents can locally heat the electrode and cause gas evolution (by either desorption or vaporization). Subsequently, $\alpha_{eff}d_c >> 1$, leading to ionization growth and breakdown. As for $Nd/Nd_{min} >> 1$, the discharge is filamentary. This mechanism has been studied extensively in connection with vacuum breakdown [Ref. 19]. In this regime, ionization processes occurring in the background (ambient) gas are non-existent and the focus is on processes that result in gas evolution with a significant change in the local gas density, and consequently, the ionization rate. In this sense, "vacuum breakdown" in fact refers to the conditions leading to the transition from a vacuum to a gas background in which ionization growth and breakdown can evolve. The breakdown fields obtained for these conditions are well above those found in the Paschen regime (see Fig. 5).

In positive divergent electric fields, breakdown can occur for $Nd/Nd_{min} \ll 1$ in geometries where electrons can be trapped in orbits whose lengths are at least of the order of λ_i. For this condition, ionization growth can occur before the primary electrons in the high field area are collected, and a partial corona- like discharge can be sustained near the electrode. These discharges are observed for example in the concentric-cylinder configuration when the radius of the inner cylinder is sufficiently small to create the trapped orbits.

Outside the Paschen regime (where the effective pressure or field concepts are applicable), there is no methodology for taking into account the influence a uniform magnetic field, B, on the breakdown condition for non-uniform electric fields. For $Nd/Nd_{min} \gg 1$, where streamers form, it seem plausible to use Eq. (7) with suitably defined coefficients and integrating along a path defined by the application of Newton's equations of motion to the head of the evolving avalanche. For $Nd/Nd_{min} \ll 1$, there have been very few investigations of the breakdown characteristics. Breakdown develops as in the Paschen regime via multiple avalanche growth in association with secondary electrode processes. However, the approach for obtaining the classical Townsend coefficients, α and ω, and applying Eq. (7) has to be modified. For this condition, the dynamics of the electron distribution function in energy at a point r (which is used in obtaining α) and the yield of secondary electrons from the cathode are highly non-local in space due to the near ballistic transport of both electrons and ions. Moreover, the presence of trapped orbits due to either geometric and/or B effects, can further complicate the formulation of a breakdown criterion. This is the situation which exits in the problem of breakdown in the space environment as discussed in the next section.

C. Space-Vacuum as an Insulating Medium: Generic Field Distributions and Parameter Space

Insulation requirements for the components contained within the pressurized region of the SPEAR II payload, for example, can be determined from the theory developed in the previous section. From Eq. (8), note that for a gas density of 10^{19} cm^{-3} (~1 Atm), and B ~ 0.45 Gauss (typical value for flight conditions), $\omega_c t_c \sim 10^{-4}$ so that, N' ~ N, and classical breakdown theory applies. Moreover, the magnetic field would have to increase by more than four orders of magnitude before it has an effect on the effective background density, N'. The determination of the insulation requirements for components exposed to the space-vacuum is more difficult, and general criteria are at present lacking [Ref. 22]. The difference in behavior (as insulators) between this medium and that discussed in Sec. 2.2B arises from the fact that in the former case the background gas is ionized (see lecture by R.V. Latham for further discussions). As a consequence, currents can flow prior to the establishment of a self-sustaining discharge (breakdown). Identifying these currents with pre-breakdown flows, the goal then is to determine the loading that these currents represent to a pulse power system, and the voltage threshold for breakdown (and by comparison, the effect of the background charged particle density on the neutral gas breakdown characteristics).

Three situations are encountered resulting in: 1) breakdown of the sheath surrounding a component, 2) breakdown between components forming the payload (for example between PFN components or PFN and support structure in the SPEAR II payload), and 3) breakdown within a component (for example, the pulse transformer in SPEAR II). This arises from the need to distinguish between breakdown of the space-vacuum, as occurs in the first two situations (analogous that discussed in Sec. 2.2B), and breakdown dominated by surface charging, outgassing, or classical vacuum processes, as occurs in the last situation (and which will not be treated in this paper). In this case, when the background density is high due to outgassing or the electron trajectory between the stressed surfaces is not significantly affected by the magnetic field, then the effective pressure approach (Sec. 2.2B) and conventional vacuum insulation techniques, respectively, are applicable. The effect of surface charging on the breakdown threshold depends on the specific situation, and no general theory is available to determine the breakdown characteristics. The approach that has evolved, and used in SPEAR I and II, is not to directly expose the critical surfaces to the space-plasma.

The difficulty in the analysis of the insulating properties of the space-vacuum, relative to that presented in Sec. 2.2B, stems from the presence of the magnetic field, the low density, and the need to include, in some situations, the space-charge field [recall that in the derivation of the Townsend-Paschen criterion (Eq. (3)), space charge effects have been neglected]. The approach taken, however, is very similar, and the results reduce to those presented in Sec. 2.2B for equivalent conditions. This approach is to determine the steady state I-V characteristic and investigate its behavior with V. The voltage which leads to "infinite" current is defined as the breakdown threshold [the condition leading to Eq. (3)]. As remarked earlier, this threshold represents a lower bound for breakdown under pulsed conditions.

To proceed with the analysis, two regimes have been identified according to the dimensions of the (electrode) sheaths relative to the inter-electrode separation. Each of these regimes consists of three sub-regimes depending on the contribution of the secondary processes to the current in the circuit. These regimes and sub-regimes are quantified below in the context of a generic electrode configuration.

In the spirit of the Townsend theory, generic field distributions, or equivalently, electrode configurations need to be identified that can serve as models for developing breakdown criteria which can then be extrapolated to more complicated situations (as has been indicated in Sec. 2.2B for Townsend breakdown). These generic geometries are essentially the same as those presented in Sec. 2.2B. However, because of the influence of the magnetic field, the finite size of the electrodes, and the low background density encountered in space, electron free paths can be much greater than the size of the electrodes so that the field distributions are invariably non-uniform. Consequently, unlike the analysis of Sec. 2.2B (which is founded on parallel plate electrodes), the corresponding theory must be developed starting from a nonuniform field configuration. Two suitable configurations have spherical-cylindrical and cylindrical-cylindrical electrodes (each pair with radii r_e and r_c, respectively, with $r_e < r_c$) [see Fig. 7]. The reason for these configurations is that it can be used as model for point-plane, and other non-uniform configurations of practical interest. Moreover, it exhibits behavior that are characteristic of more complicated situations while possessing symmetry properties that facilitate its analysis. The presence of the magnetic field of the earth make the analysis at least two dimensional. The goal, however, is to obtain breakdown characteristics that as for Eq. (7) can be applied along a suitable path. In order to make the model more suitable to space applications, the outer cylinder is taken to be only partially enclosing, that is, containing a number of "windows". In the context of the SPEAR II payload, for example, any one of the biased components can be taken as the inner electrode, whereas the remaining components and the supporting frame plays the role of the partially enclosing cylinder.

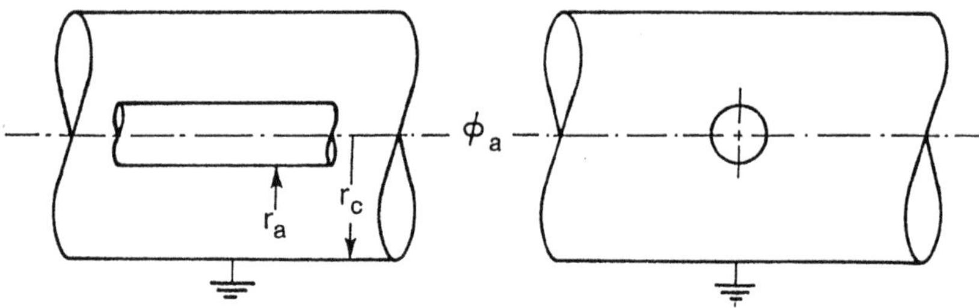

Fig. 7: Generic electrode configurations.

Focusing on the sphere-cylinder configuration (the discussion applies equally to the cylinder-cylinder configuration), it is then possible to quantify the regimes that comprise the parameter space of interest. Let V be the potential (positive or negative) applied to the sphere relative to the outer cylinder, and N, n_e, and n_i be the background neutral and charged particle (electrons and ions) densities, respectively. Consider first the case where r_c is infinite, $B = 0$, and $N/n_e \ll 1$. From the Langmuir-Blodgett results for the I-V characteristics under these conditions [Ref. 23] (see Sec. 4), the sheath surrounding the isolated electrode is spherically symmetric, and has a radial extent, r_{L-B}, that depends on V and the background charged particle density (see Sec. 4). This is a suitable reference model in terms of which the two primary regimes are defined as a function of r_c; that is, by the conditions $r_{L-B} \ll r_c$ and $r_{L-B} > r_c$, respectively (these regimes do not depend on the value of B, and although the character of the sheath for $B \neq 0$ differs from that of the reference model, r_{L-B} can be used as a reference scale in the direction of B). Each of these regimes consists of three sub-regimes which are defined in terms of the ratio of the actual current to the ambient current in the absence of (feedback) secondary processes (see Table I). These processes enhance the current collected by the biased electrode and determine the breakdown thresholds. They are: a) secondary emission from the electrodes due to ion impact (represented by a current $I_{se} = \gamma I_i$), and b) sheath expansion due to the space-charge field, E_{ss}, generated by the secondary electron-ion pairs produced via impact ionization of the background neutrals (this is

Table I. Distinguishing properties of various regions of parameter space.

	Sphere Polarity	Ambient Current Collection	Enhanced Current Collection	Breakdown
$r_{L-B} \ll r_c$	+	$E_{ss} < E$	$E_{ss} = O(E)$	$dI/dV = \infty$
	-	$\gamma\ I_i \ll I_s$	$\gamma\ I_i \sim I_s$	
$r_{L-B} \gg r_c$	±	$\gamma\ I_i \ll I_s$	$\gamma\ I_i \sim I_s$	$\int \alpha_{eff}\, dr = k$

subsequently referred to as the E_{ss} process). The magnetic field has an indirect influence on these secondary processes. Its influence in each of the sub-regimes is on the primary process, impact ionization, and consequently on the magnitude of the avalanche growth. From these considerations, the sub-regimes can be quantified in terms of: r_{L-B}, r_c, I_{se}, and E_{ss}. This is done subsequently, in order of increasing current, for each primary regime.

For $r_{L-B} \ll r_c$, as the voltage is increased from zero, the ambient current collection sub-regime is defined by the condition: for positive bias, $E_{ss} \ll E$, where E is the applied field in the presence of the plasma (with $N/n_e \ll 1$), and for negative bias, $I_{se} \ll I$, where I is the collected current for a negative bias equal to -V. In this sub-regime, the spherical electrode acts as a biased probe collecting a current I determined by the properties of the ambient charged particle density (see Sec. 3). As the voltage increases, the collected current can be enhanced over that provided by the ambient plasma when: a) the neutral density is sufficient that avalanches increase the number of electron-ion pairs, b) the secondary current at either the sheath boundary for positive bias (the E_{ss} process), or at the sphere for negative bias (due to the secondary emission coefficient, γ, becoming greater than one) becomes significant. The change in either magnitude, or dependence on V of the collected current defines the transition to the enhanced current collection sub-regime. This transition is gradual and depends on sphere radius, properties of the neutral gas, and sphere material (on account of γ). Further increases in voltage can lead to a third sub-regime identified by the breakdown of the sheath. This is manifested by an abrupt and large increase in current as the voltage reaches the threshold (See Sec. IV).

The sub-regimes pertaining to the $r_{L-B} > r_c$ regime are defined as for $r_{L-B} \ll r_c$ (in terms of the magnitude of the actual current relative to the ambient current in the absence of electron production). The secondary process that ultimately determines the I-V characteristics for both polarities is the γ process. However, the magnitude of the current flow also depends on the transparency of the cylinder with respect to the ambient plasma (this is a measure of how well the enclosure confines the field. It depends on the potential of the enclosure with respect to the plasma and on the number and size of its "openings"). For low transparencies (such that the field is contained within r_c), the distance over which the electrons multiply is limited to (r_c-r_e) and remains constant with voltage (unlike the $r_{L-B} \ll r_c$ regime, where the boundary expands with voltage). For B ~ 0, the avalanche gain over this distance (for the conditions of interest) is not sufficient to enhance, in concert with the γ process, the ambient current. Consequently breakdown cannot occur via this mechanism (in this case, the breakdown criterion corresponds to that of vacuum). Thus the boundary at r_c inhibits breakdown with respect to a completely transparent boundary ($r_c = \infty$). To achieve breakdown between the electrodes (sphere-cylinder), the gain has to be increased by either increasing r_c or by increasing B. Increasing r_c results in sheath breakdown as discussed in the previous section (neglecting the dynamics at the enclosure openings). On the other hand, for a given r_c, there is a critical B, B_{cr}, which results in sufficient gain to satisfy the breakdown condition between the electrodes. (As discussed in Sec. 5, B_{cr} is the minimum value of magnetic field for which breakdown can occur in a given enclosed system.) Alternatively, for a given B and transparency, there is an upper bound on the size of enclosure which inhibits breakdown (see Sec. 5).

The discharge characteristics (the I-V characteristics for voltages above the breakdown threshold), in particular the space-charge distribution and the magnitude of the discharge current, are also significantly different than for the $r_{L-B} \ll r_c$ case since the sustaining mechanism at the cylinder surface can be a stable source of current over a wide range of conditions. A number of discharge modes have been observed depending on the circuit parameters (see Sec. 5). These discharges do not differ from that which occurs when $n_e = n_i = 0$. The plasma that is formed subsequent to breakdown determines the magnitude of the current and field distribution (for given circuit parameters). This plasma is the same for either condition. These discharge modes are reminiscent of the corona and glow discharge modes obtained in the Townsend regime (see Sec. 5).

The parameter space analyzed above (summarized on Table I) can be identified with the following practical considerations : 1)low magnitude leakage currents due to the conductivity of the plasma representing a minor loading on stressed pulse power components that are exposed to the space environment, 2) "sheath" breakdown which essentially results in the "grounding" of the electrode to the ambient plasma (the impedance of the "connection" is determined by the properties of the resulting discharge which at present are unknown. However, from the discussion of Sec. 4, it is likely to be a corona-like discharge), and 3) breakdown between material electrodes similar to that encountered in the Paschen regime resulting in a glow-like discharge. Note that in the last sub-regime, the medium is no longer useful as an insulator.

In general, the transition from one sub-regime to another occurs at lower voltages as N is increased. A similar behavior is observed with charged particle density. However, the behavior of the breakdown threshold with n_e as the secondary mechanism changes from the E_{ss} to the γ process and as B increases is not completely clear at present (see Sec. 6).

In the following sections, the behavior of the generic two-electrode system in the regime of parameters of interest to the SPEAR program are explored. In the presentation, the inner electrode is taken to be positive, since it exhibits the lowest threshold for breakdown. The procedure used is equally applicable to the negative polarity case (which was the situation in SPEAR II), and results obtained for this polarity are also presented.

3. Ambient Current Collection By Charged Objects in LEO ($r_{L-B} \ll r_c$, no feedback)

This topic has been the subject of a number of lectures in this ASI and consequently the reader is referred to them. In particular the lectures by I. Katz, and D.C Ferguson. Consequently, this section is devoted to a brief discussion of the nature of the trajectories (orbits) of the electrons in the vicinity of a charged spherical object in the ionosphere (i. e. in the sheath surrounding the charged object). This topic forms the basis for a treatment of the ambient I-V characteristics of the spherical object (the generic electrode), and for the determination of the ionization coefficient. It is shown below that the electron trajectories in the field of relevance are chaotic. This fact makes quantitative determination of these characteristics very difficult, although reasonable estimates can be obtained via computer simulation.

The equation of motion for a charged particle (charge $q < 0$, mass m) in a uniform magnetic field, B, in the vicinity of the sphere (radius a) biased to a potential $V > 0$ is

$$m \frac{d}{dt} v = q\, v \times B - q\nabla\left[V \frac{a}{r}\right] \tag{10}$$

where $v = dr/dt$ is the electron velocity. The total energy of the charged particle, and the component of angular momentum parallel to B are constants of the motion. Introducing the cyclotron frequency, ω_c, the characteristic length λ, defined by $\lambda^3 = -qVr_s/(m\omega_c^2)$, and the scaled variables $\tau = \omega_c t$, $\hat{r} = r/\lambda$, the dimensionless equations of motion become [Ref. 24]

$$\hat{v}_x = \frac{d\hat{x}}{d\tau} \qquad \hat{v}_y = \frac{d\hat{y}}{d\tau} \qquad v_z = \frac{d\hat{z}}{d\tau} \tag{11a}$$

$$\frac{d\hat{v}_x}{d\tau} = \frac{\hat{x}}{\hat{r}^3} - \hat{v}_y \qquad \frac{d\hat{v}_y}{d\tau} = -\frac{\hat{y}}{\hat{r}^3} + \hat{v}_x \qquad \frac{d\hat{v}_z}{d\tau} = -\frac{\hat{z}}{\hat{r}^3} \tag{11b}$$

Notice that these differential equations are independent of parameters. As a result the character of the orbit is entirely determined by the dimensionless initial position and velocity of the electron in the six-dimensional phase space of the charged particle. The trajectories are confined to the region

$$V_{eff} \leq \hat{E} \tag{12a}$$

where $\hat{E}$ is the normalized energy and the effective potential, V_{eff}, is given by

$$V_{eff} = \frac{1}{2\hat{\rho}^2}\left[\hat{L}_z + \frac{1}{2}\hat{\rho}^2\right] - \frac{1}{[\hat{\rho}^2 + \hat{z}^2]^{1/2}} \tag{12b}$$

with $\hat{L}_z$ being the angular momentum in the z direction and $\hat{\rho}$ the normalized two-dimensional radial coordinate in planes perpendicular to z. Equipotential contours corresponding to Eq. (12b), for $\hat{L}_z = 1$, are shown in Fig. 8. Electrons coming in from $|z| = \infty$ with a given $\hat{L}_z$, "bounce around" inside this potential before they return to infinity. If they loose energy due to collisions, they fall deeper into the well, and the corresponding well gets closer to the anode (this picture is used in Sec.V for the determination of the ionization coefficient).

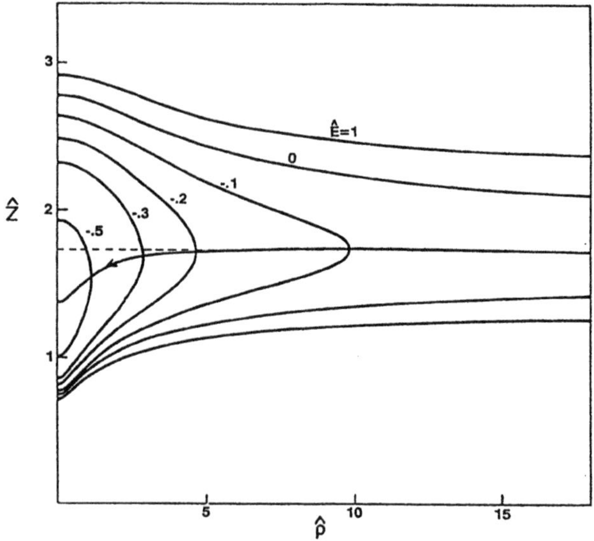

Fig. 8: Contours of constant energy for positive sphere located at the origin.

Equations (11) are the governing equations of motion used to obtain numerical solutions in this study. Two types of initial conditions have been investigated, corresponding to positive (unbounded) and negative (bounded) energy states. Based on the results for the Hydrogen atom [Ref. 25], the motion for some initial conditions is expected to be irregular. To determine the region of initial conditions which lead to these "chaotic" orbits, numerical solutions to Eq. (11) have been obtained and used in constructing Poincare maps for $\rho, \dot{\rho}$ in the z=0 plane. Chaotic orbits lead to maps that fill the entire region allowed by conservation of energy and angular momentum, whereas regular orbits lead to maps in the form of closed lines or points. Calculations have been performed with Eq. (11) and the results converted to dimensional values for the specific case of a 10 cm radius sphere biased at 10 kV. The magnetic field is taken to be 0.4 Gauss. These values correspond to the SPEAR I configuration and lead to normalization constants of $\lambda = 1.521$ m, and $\omega_c = 7.035 \times 10^6$ rad/sec.

The initial velocities have been chosen to correspond to thermal electrons at various altitudes (above 150 km; see Table II for the correspondence between altitude and ionospheric parameters used throughout this paper) and to electrons that have either suffered a collision or have been generated by electron impact ionization of neutrals. The initial positions have been varied to determine the region of space that result in chaotic and regular orbits, and in particular the subset of these trajectories that intercept the sphere (that is, trajectories that are neither trapped nor unbounded). Fig. 9a shows a set of representative Poincare maps with the starting position as a parameter. The solid curve forms the boundary of the permissible region for the energy and angular momentum associated with the initial condition. The nature of the orbit is

Table II. Correspondence between altitude and ionospheric parameters used in this work (obtained from MSIS atmospheric model).

Altitude (km)	Neutral Density (10^{17} m^{-3})	Electron Density (10^{11} m^{-3})	J_o (10^{-4} A/m^2)
120	4.5	0.0065	0.27
150	0.44	0.0071	0.038
160	0.26	0.012	0.068
180	0.1	0.039	0.25
250	0.01	0.75	4.64
300	0.003	1.1	6.9

evident from these maps. Fig. 9b shows the regions of initial positions that result in chaotic, regular, and captured orbits. The conditions that result in chaotic orbits are confined to the region with $z > 0$ and near the electron capture radius. The line of electron capture corresponds to the set of initial conditions for which the lower value of r calculated from Eq. (12) with z=0 is equal to the sphere radius. Trajectories started inside this radius intersected the sphere at some time. This radius is nearly constant with z and is only a few centimeters different from the value obtained assuming that the electron starts from rest at infinity. Results such as those presented are presently being used to determine the current collected by the sphere as a function of bias voltage (the I-V characteristics).

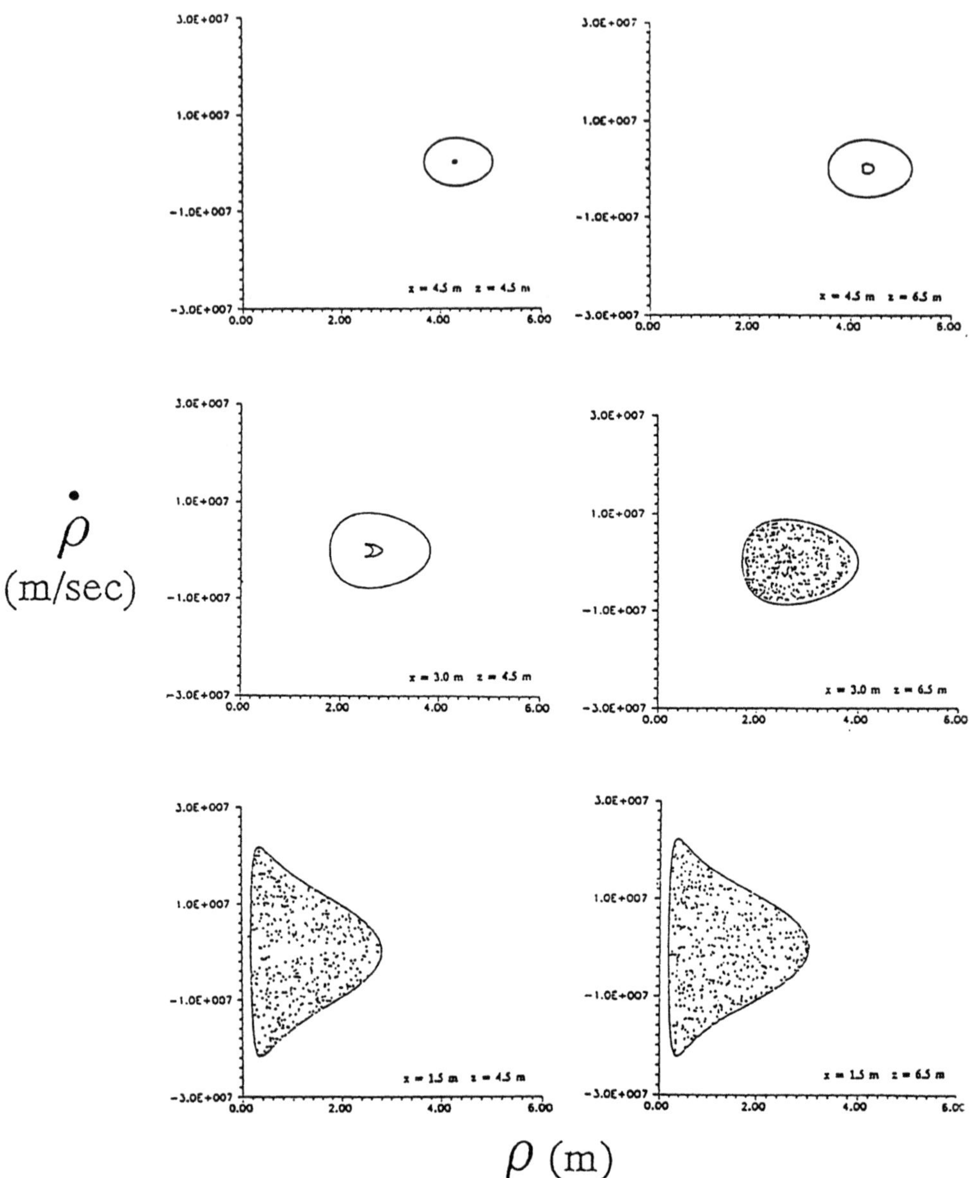

Fig. 9a: Poincare sections at z=0.

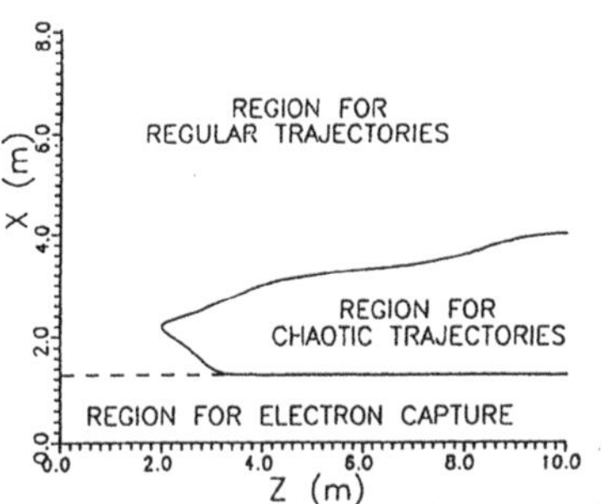

Fig. 9b: Boundaries of chaotic orbits for B=0.4G, V=10kV, and r_e= 10 cm.

4. Enhanced Current Collection and Sheath Breakdown ($r_{L-B} \ll r_c$, with feedback)

For positive bias voltages, when E_{ss} (the space-charge field due secondary electron-ion pairs generated via electron impact ionization of the background neutrals) becomes of the order of E (the applied field in the presence of the background plasma), it is necessary to account for it in the calculation of the I-V characteristics. In this case, the current collected by the spherical electrode at a given bias voltage is enhanced over that obtained in the absence of ionization. The evaluation of the enhancement factor is greatly complicated by the fact that the electron trajectories in the combined field are, as shown in Sec. 3, chaotic. As result, the investigation of enhanced current collection has been restricted to conditions for which the magnetic field effects can be taken into account via an effective pressure (essentially, the influence of the magnetic field on the trajectories is minimal). In this section, the I-V characteristics, assuming that B is negligible, are presented. An approach for including the effect of B is also indicated. The relevance of this regime to situations encountered in space are discussed following the presentation of the results.

It is shown below that as the bias on the sphere increases, a runaway condition develops characterized by the fact that a small change in the bias results in an infinite increase in the collected current. That is,

$$dI/dV = \infty \tag{13}$$

This condition defines the breakdown of the sheath surrounding the sphere. It is equivalent to the condition used in defining Townsend breakdown, Eq. (3) (The Paschen law).

Since the mean free paths are much larger than the scales over which the electric fields changes with position, the energy distributions for electrons and ions are highly non-equilibrium. Consequently, it is not possible to use the continuity equations in the form given by Eq. (1). In this case, it is necessary to account for the energy distribution in the determination of the current enhancement due to ionization and secondary processes. This is carried out below.

Consider first the positively biased electrode, and neglect the presence of the enclosure at r_c. These conditions serve to model the interaction between a biased component and the space-plasma, in the absence of all other components and support structure. Subsequent to this discussion, the influence of the enclosure is considered. As remarked earlier, the analysis for a negatively biased electrode is the same, and results for this case are also presented.

In steady state, for an applied positive potential, V, the sheath surrounding the sphere has dimensions $r_s = r_{L-B}$. The corresponding collected current, $I = I_{L-B}$ ($= 4\pi r_{L-B}^2 J_0$, where J_0 is the thermal current density of the electrons in the space-plasma assuming a Maxwellian distribution. This current is altitude dependent). "Primary" electrons are assumed to enter the sheath boundary at r_s. As the primary electrons travel towards the anode, they generate secondary electron-ion pairs through impact ionization of the neutral component. The density of the secondary electrons can be neglected relative to the primary density (calculations show that the ionization probability for the parameters of interest is in the range of 10^{-2} and therefore the assumption is justified). As the applied voltage is increased, r_s increases with a concomitant increase in the number of primaries that enter the sheath. The mechanism that regulates the change in r_s as a function of the change in the applied potential is the behavior of the space-charge distribution caused by the accumulation of the ions in the sheath (their lifetime in the sheath is greater than that of the electrons). This distribution influences and is influenced by (in a self-consistent manner) the resulting potential distribution, the ionization rate (which depends on particle velocity), and the particle fluxes. This interplay becomes evident from the discussion that follows.

To obtain the potential distribution, the density distributions for primary and secondary particles must be determined, taking into account their energy distribution as a function of position, r. Assuming that the primaries lose little energy, the number of ions produced in a shell at r' is

$$dn_i\,(r') = n_e\,(r')\,\alpha_{eff}\,(r')\,dr' \tag{14}$$

where $n_e(r')$ is the density of electrons at r', and α_{eff} is an average (effective) ionization coefficient which defines the growth in density per unit distance in the r direction [equivalent to that defined in Sec. 2.2B for non-uniform fields]. In steady state, these ions contribute a current at r given by

$$dI_i(r,r') = 4\pi r^2\, e\, dn_i(r')\sqrt{\phi(r) - \phi(r') + \phi_{oi}}\sqrt{2e/m_i}$$

where $\sqrt{\phi(r) - \phi(r') + \phi_{oi}}\sqrt{2e/m_i}$ is the velocity of the ions at r, given that they were generated at r' with a velocity $\sqrt{2e\phi_{oi}/m_i}$, and $\phi(r)$ is the potential at r. From this equation, the ion density at r is

$$n_i(r) = \frac{1}{4\pi e}\int_{r_s}^{r} \frac{dI_i(r,r')}{r^2\sqrt{\phi(r) - \phi(r') + \phi_{oi}}\sqrt{2e/m_i}} \tag{15}$$

The differential current at r' is given by the expression

$$dI_i(r,r') = \delta I_e(r')\, G(r',r) \tag{16a}$$

with,

$$\delta I_e(r') = \int_{r_s}^{r'} G(r',r'')\, \delta I_e(r'')dr'' \tag{16b}$$

The evaluation of this equation, in particular the determination of the kernel G(r',r''), depends on the electron distribution function. For high E/N (low background density), the distribution at r' is assumed to consist of "beams" with energy equal to the potential difference between the point (r'') at which the beam has been created and r' (this assumption implies that the electrons loose little energy as they proceed towards the sphere). With this assumption, G becomes

$$G(r',r'') = a_{eff}(r'-r'') = NQ_i(r'-r'') \tag{17a}$$

where Q_i is the energy dependent ionization crossection. At $r' = r_s$,

$$\delta I_e(0) = I_0 = 4\pi r_s^2\, J_0 \tag{17b}$$

The density of secondary electrons can be obtained in a similar fashion.

Inserting the particle densities into the Poisson Eq., the potential distribution is found to obey the following equation

$$\frac{d}{d}\left[r^2\frac{d}{dr}\,\phi(r)\right] = \frac{J_0\, r_s^2/r^2}{\epsilon_0\sqrt{2e/m_e}}\left\{\sqrt{\frac{m_e}{m_i}}\int_{r_s}^{r}\frac{dI_i(r,r')/I_0}{\sqrt{\phi(r) - \phi(r') + \phi_{oi}}} - \frac{1}{\sqrt{\phi(r) + \phi_{op}}} - \int_{r}^{r_s}\frac{dI_i(r,r')/I_0}{\sqrt{\phi(r') - \phi(r) + \phi_{os}}}\right\} \tag{18}$$

where ϕ_{os}, ϕ_{oi}, and ϕ_{op} are the initial energies of the particles. The subscripts os, oi, and op refer to secondary electrons, ions, and primary electrons, respectively. Eqs. (15)-(18) are a set of integro-differential equation for $\phi(r)$ and the electron and ion currents. The boundary conditions are $\phi(r_s)= \phi_{op}$, $\phi(r_a)=V$, and Eq. (17b). The solution yields I=I(V), where the total current I is obtained by integrating Eq. (17a) over r' from r_e to r_s. This solution deviates from the L-B solution on account of the first and third terms (due to ionization) in Eq. (18). When their contribution is significant, the result is enhanced current collection. A point is reached at which a differential increase in V leads to a runaway of the sheath current. That is to an explosive expansion of the sheath. This situation indicates the breakdown of the sheath. The state of the medium surrounding the sphere for higher voltages cannot be determined from the theory presented. This observation also applies to the collected currents. At present, there is no way to estimate this current.

The results obtained from Eq. (18) for the conditions existing in the ionosphere at 150 and 180 km (using the MSIS-87 atmospheric model, see Table II) and for an anode radius of 10 cm are shown in Fig. 10. The solid curve (curve 1) includes feedback effects, whereas the dash curve (curve 2) does not. The regime of enhanced current collection corresponds to voltages above that for which the two curves separate. The breakdown voltage at 150 km (Fig. 10a) is that voltage for which curve 1 has infinite slope ($dI/dV=\infty$). This occurs at approximately 7.2 kV. No breakdown is observed at 180 km for voltages up to 100 kV, which is well above those used in SPEAR I (this has been found to be the case for all altitudes above 160 km). Typical I-V results for a negatively biased electrode are shown in Fig. 10c for altitudes of 120 and 250 kM. The inner electrode has been taken to be cylindrical with a radius of 0.22 m to model the body of the rocket (in SPEAR I) or the high voltage section (in SPEAR II). For these conditions, the current enhancement due to secondary effects is sufficient to result in breakdown. Moreover, the collected currents are significantly lower than for positive polarity. The values for the breakdown voltage, for positive polarity, as a function of altitude are shown in Fig. 11.

The sensitivity of the collected current, for positive polarity, to changes in the ambient current density, J_0, are illustrated in Fig. 12. The breakdown voltage is found to exhibit a complex behavior with current density in that it can increase, decrease, or remain constant with increasing current density. This behavior arises from the combination of a decrease in the sheath dimensions with increasing J_0 (consequently decreasing the distance over which avalanche growth can take place) and the changes in the ionization coefficient (on account of the potential re-distribution) and the total avalanche growth (on account of the increase in the current density). The behavior obtained for the breakdown voltage depends on the compensation between these competing mechanisms.

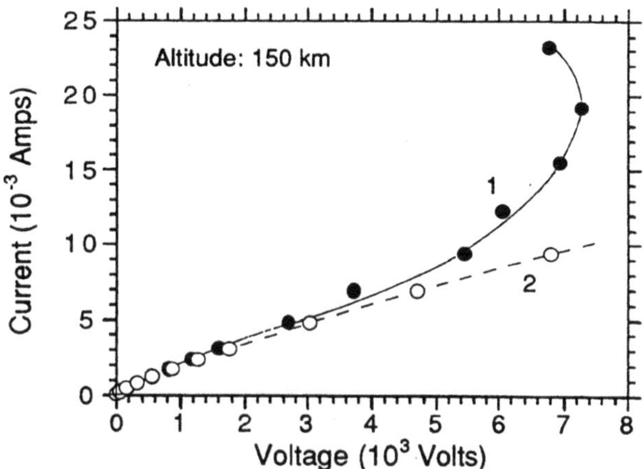

Fig. 10a: Current-voltage characteristics at an altitude of 150 kM showing effect f feedback on the current collected, and the threshold for breakdown.

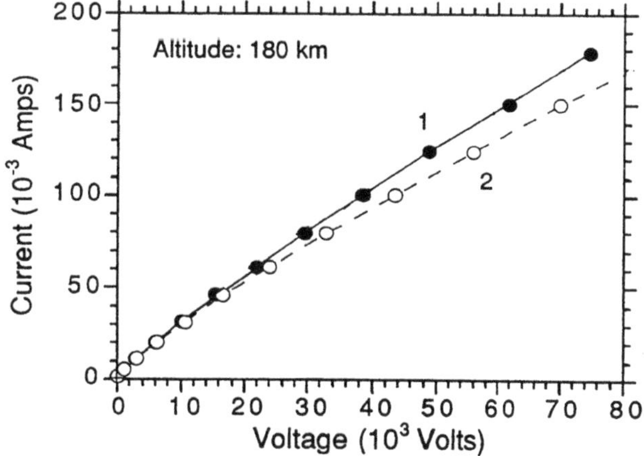

Fig. 10b: Current-voltage characteristics at an altitude of 180 kM.

To include the effects of the magnetic field in lowest approximation, the expressions obtained in Eq. (17a) for the effective ionization coefficient, α_{eff}, and that used for the electron density in Eq. (18) need to be modified. An approach for modifying α_{eff} is discussed in Sec. 5. An approach for modifying the density has not been implemented, although estimates of the density distributions for $B \neq 0$ have been obtained based on the potential well concept discuss in the previous section. More accurate

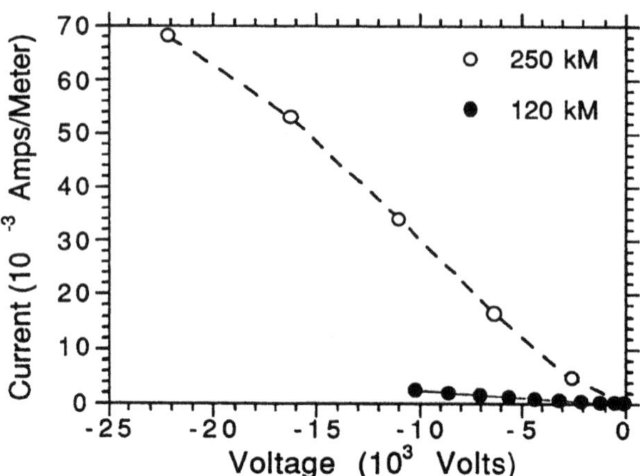

Fig. 10c: Current-voltage characteristics for a negatively biased cylinder (infinitely long).

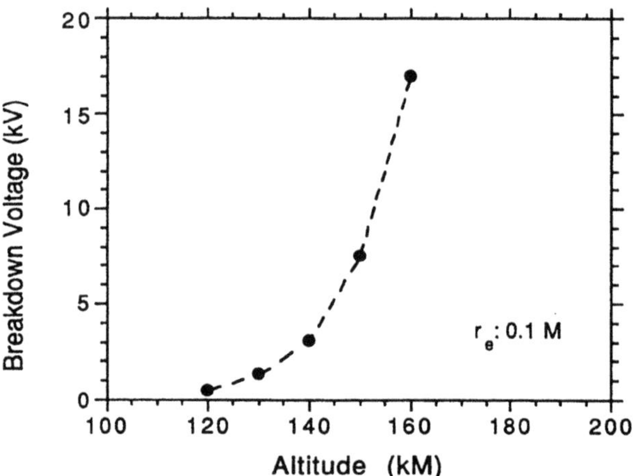

Fig. 11: Breakdown voltage vs altitude.

models are being developed based on this concept. The results show that in the ambient collection regime, the current for B=0 are higher than that for B≠0; however, in the enhanced regime since α_{eff} is higher for B≠0 than for B=0, it is not evident at this time the relative effect B. This is presently being pursued.

For small changes in α_{eff}, the currents obtained from Eq. (18) represent an upper bound on the drain to a pulse power system with a component (of comparable size to that of the sphere) completely exposed to the space environment (that is, $r_{L-B} \ll r_c$) . For the 10 cm sphere example,

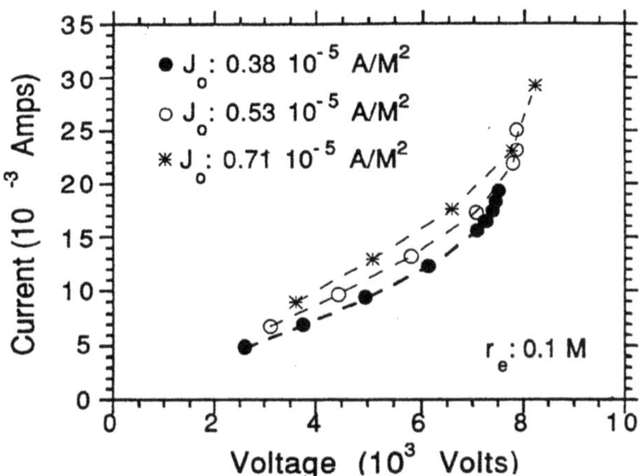

Fig. 12: Sensitivity of breakdown voltage to background current density.

the power drain is of the order of kilowatts at 180 km. Complete exposure of a pulse power component is not expected to be the case since other components tend to provide some shielding relative to the ambient plasma (and consequently modifying the results, as discussed in the next section). This effect has in fact been observed in SPEAR I, where the sheath surrounding the rocket body provided the shielding to the sphere [Ref. 7] (see lecture by I. Katz p. 713). It also represents the prevalent situation in SPEAR II. In this case, the insulation requirement for a particular component is determined in reference to nearby components, accounting for possible effects of the charged particles that leak through their shield (see next section)

An analysis of enhanced current collection and breakdown for negative polarity can be developed from the equations presented in this section for positive polarity. This is presently being carried out.

5. Enhanced Current Collection and Breakdown in Enclosed and Partially-Enclosed Structures ($r_{L-B} > r_c$)

Consider the influence on the results obtained in the previous section for a positive biased sphere of a metallic boundary located at $r_c > r_{L-B}$, but with r_s(at breakdown)$>r_c$. It has been found in the previous section that as the potential increases, r_s increases due to the E_{ss} process. At voltages such that $r_s > r_c$, the dominant secondary process contributing to the current enhancement changes over to the γ process (assuming that the field is to a large extent confined within the enclosure and neglecting the plasma dynamics at the openings), and the distance over which avalanche gain occurs becomes independent of voltage (and equal to $r_c - r_e$). For B=0 (low avalanche gain), this results in the saturation of the current with increasing voltage as compared to the results obtained in the previous section. Thus, the onset of the current enhancement regime is the same as in the $r_{L-B} \ll r_c$ regime except that current saturation occurs when $r_s = r_c$. Moreover, as result of this saturation, (sheath) breakdown does not occur; that is, the boundary inhibits breakdown under these conditions. To achieve breakdown between the electrodes, it is necessary that $B > B_{cr}$ so that the gain is sufficient to satisfy the breakdown condition (see Sec. 6).

The task then is to show the influence of the metallic boundary at r_c on the results of the previous section for B=0, or from a different viewpoint, to assess the effect of the low density plasma on the Townsend breakdown characteristics. A procedure for accounting for the magnetic field is also presented which is implemented in the analysis of the breakdown characteristics.

In the presence of the boundary, the magnitude of the pre-breakdown ambient current depends on the leakage of the sphere field to the ambient plasma. This leakage depends on the degree of enclosure (since the cylinder is assumed to have "windows") and on the bias of the cylinder relative to the ambient plasma. For positive sphere bias and zero cylinder bias relative to the plasma, the current collected is taken to be some fraction (determined by the degree of field confinement) of the thermal current density times the area of the cylinder (the length of the cylinder is assumed to be equal to its diameter). This current decreases with negative bias to the cylinder on account of the shielding by the cylinder sheath. If the cylinder contains a number of windows such that the field distribution deviates significantly from the enclosed case, the collected current for zero cylinder bias is enhanced by the effective area of the leakage field outside the cylinder. Moreover, as the cylinder becomes negative the current collected by the sphere again decreases, although the dynamics of the shielding depend on the details of the windows (on account of the double sheaths that may form at the openings. This dynamics is presently being neglected).

The procedure for determining the collected current, neglecting magnetic field effects, follows from that presented in Sec. 3. Instead of letting the current at the r_s boundary be I_0 [see Eq. (17b)], the secondary emission condition is used, namely

$$I_o = I_s = \int_{r_e}^{r_s} \int_{r_s}^{r''} \alpha_{eff}\,(r' - r'')\,\delta I_e(r'')\,dr''\,\gamma(r')\,dr' \tag{19}$$

where γ is the energy dependent secondary emission coefficient and δI_e is given by Eq. (17a). Since the energy of the ions impacting the cathode is proportional to their place of origin, γ is a function of the position at which the differential beam of ions (δI_i) has been created. For voltages such that $r_s > r_c$, the outer limit of integration in Eq. (19) is set equal to r_c. Equations (18) and (19) can then be integrated to determine the dependence of the current with voltage. The results for a cylinder with $r_c = 4$ m are shown in Fig. 13. As remarked earlier, the current is found to increase with voltage on account of sheath expansion and saturates when $r_s = r_c$. A summary of the results obtained under various conditions is presented in Table III.

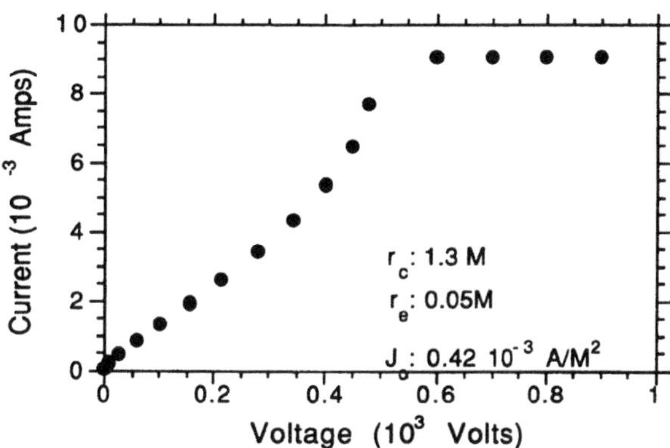

Fig. 13: Effect of a metallic boundary on the current-voltage characteristics.

Table III. Summary of breakdown characteristics of sheath and effect of metallic boundary.

Neutral Density (10^{17} M^{-3})	Electron Density (10^{11} M^{-3})	Breakdown Voltage (Volts)	Sheath Radius (M)	J_0 10^{-4} (A/M^2)
10	5	177	0.5	21.2
5	5	443	0.8	"
2.5	5	1493	1.3	"
10	1	213	0.9	4.23
5	1	483	1.3	"
2.5	1	1327	1.9	"
		r_c = 1.3 M		
5	1	no breakdown up to 5 kV		

To achieve breakdown, it is necessary to increase the gain by making B≠0. To account for B, the following assumptions are made. In the spirit of the Townsend theory, the space-charge effects are assumed to be small up to the breakdown threshold, so that the field distribution is that which exist in vacuum. This assumption greatly simplifies the analysis, since it is not necessary to solve the Poisson Eq. [Eq. (16)]. The procedure follows that used in obtaining the Townsend criterion [Eq. (3)] with two modifications arising from the presence of B and the fact that the secondary current at the cathode is due to ions that have ballistic trajectories [Ref. 26]. The model that is used here for developing the modifications is that of a steady flow of electrons towards the sphere (for positive bias) characterized by a cascade down the effective confining potential [Eq. (12)], and generating secondary ions that travel along ballistic trajectories from their point of origin to the cathode. The energy at the cathode of the group created at r equals the potential at r. The modifications can then be achieved by taking r to be the coordinate along the flow, substituting a_{eff} for α (where a_{eff} includes the effect of the magnetic field as discussed below), and determining the secondary electron current at the cathode from Eq. (19). From these observations, the breakdown condition can be written in the form

$$\int_{r_l}^{r_u} \gamma(r)\ \alpha_{eff}(r)\ \exp\left[\int_{r}^{r_u} \alpha_{eff}(r')dr'\right] dr = 1 \tag{20}$$

where the integration limits r_l and r_u are the boundaries of the region over which a_{eff} is greater than zero [see Fig. 14]. By defining an average secondary emission coefficient Γ, an approximate breakdown condition can be obtained which has the form of Eq. (7) [Refs. 20, 26],

$$\int_{r_l}^{r_u} \alpha_{eff}(r)dr \geq K(\Gamma) \tag{21}$$

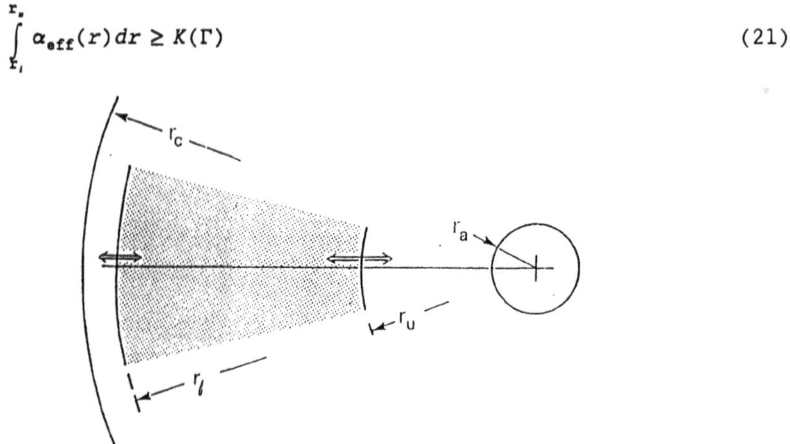

Fig. 14: Boundaries, r_l and r_u, over which avalanching occurs.

where K is an average feedback function . In certain cases, the secondary emission can be greater than 10 so that carrier multiplication as low as 1.1 may be sufficient to satisfy Eq. (21) and cause breakdown. In contrast to the Townsend breakdown, the current growth results more from the γ process than from avalanching in the bulk. From Eq. (18), the breakdown problem has been reduced to the determination of the variables α_{eff}, K, r_l, and r_u. This is done subsequently. To find the effective ionization coefficient along the direction of the electric field (the r direction), knowledge of the electron energy distribution function as a function of r is necessary. Since this distribution function is not known at this time, two approaches have been perused [Refs. 2, 6]. One based on the most probable trajectory and the other on the chaotic nature of the orbits. Both give essentially the same result, namely

$$\alpha_{eff} = P_c \frac{\overline{Q}_i}{\overline{Q}} \frac{1}{\overline{X}(r)} \tag{22}$$

where P_c is the probability that an electron emitted from the cathode enters the avalanche process, Q and Q_i are the average total and ionization collision crossection, and $\overline{x}$ is the mean free path in the r direction (see Ref. 3 for further details).

The lower boundary of the avalanche gain region, r_1, is the radius where $\phi(r) = \eta\, \phi i$, where ϕi is the ionization potential and η is the ratio of the total to the ionization crossection (a modification being pursued is to replace this condition by the location of the bottom of the potential well [Eq. (12)] for an L_z corresponding to this radius). The upper boundary, r_u, is the radius where the secondary electrons produced by ionization are immediately collected by the anode (i. e. they are not in trapped orbits). It is found from the expression

$$r_u = \left\{1 + (8e\ [V - \phi(r_e)]/m\ \omega_c^2\ r_e^2)^{1/2}\right\}^{1/2} r_e \tag{23}$$

where V is the value of the potential at the sphere (the applied potential).

The average feedback function, K, has the general form

$$K = -\ln\ [(G\ P_c\Gamma + 1)/G\ P_c\Gamma] \tag{24}$$

where G is a geometric factor that accounts for the fact that the contribution of the secondaries to the breakdown process is a function of their place of origin. Note that the feedback function is a logarithmic function of Γ, so that the threshold voltage is weakly dependent on cathode material. This is also true of Paschen breakdown.

Thus, from Eqs. (18)-(21), the breakdown characteristics become

$$NP_c \int_{r_e}^{r_u} \frac{\overline{Q}_i}{\overline{Q}} \frac{1}{\overline{X}(r)}\, dr = -\ln\left[\frac{GP_c\Gamma + 1}{GP_c\Gamma}\right] \tag{25}$$

This equation has been solved numerically for an O, O_2 , and N_2-O_2 backgrounds. These backgrounds correspond, respectively, to the dominant species at the altitudes of interest to the SPEAR flights, partially enclosed structures on account of wall recombination, and test chamber experiments. The electron impact ionization crossections have been taken from [Ref. 27].

In contrast to the Paschen curve where the breakdown voltage is presented as a function of Nd, the breakdown characteristics for the space-vacuum are presented as plots of breakdown voltage vs B/N with N as a parameter. The results for a sphere radius of 5.08 cm biased positive relative to a cylinder of 121 cm are shown in Figs.15a and 15b for N_2 and O_2 backgrounds, respectively. The characteristics are found to consist of two branches, intersecting at the critical value of B, B_{cr}. As remarked earlier, breakdown (via the γ mechanism) does not occur for $B < B_{cr}$. This observation is equivalent to the Paschen minimum. The cathode radius has little effect on the lower branch, and a strong effect on both B_{cr} and the upper branch. The anode radius on the other hand has a strong influence on both branches of the characteristics and the opposite effect on B_{cr} than the cathode radius. The dependence of B_{cr} on anode and cathode radius is shown in Figs. 16a and 16b, respectively. These effects can be explained from the combination of secondary processes at the cathode and the dependence, through the electric field, of r_l and r_u on the cathode and anode radii, r_c and r_e, respectively. For fixed r_e and r_c, as V is increased from zero, both r_l and r_u increase from r_e, with r_l increasing faster (thus creating the ionization region shown by shaded area in Fig. 14). Moreover, for $r_c-r_e > r_e$, since the profile of the electric potential in the ionization region (r_l-r_u) is essentially that of an sphere in infinite space, r_l and r_u are influenced mainly by r_e. This explains the strong influence of r_e on the lower branch. For large V (upper branch), r_l approaches r_c, and the potential profile is affected by the cathode; consequently both boundaries influence this branch.

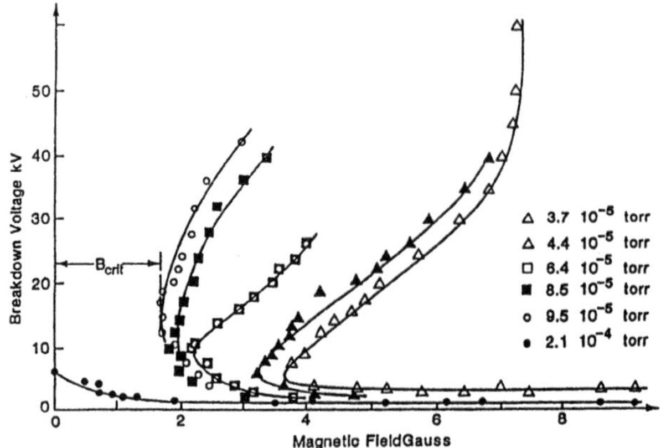

Fig. 15a: Measured breakdown voltage vs B for sphere-cylinder configuration (r_e = 5.08 cm, r_c = 121 cm) in a N_2 background.

An approximate formula for B_{cr} [alternative to an implicit evaluation via Eq. (25)] with respect to chamber parameters can be obtained by noting

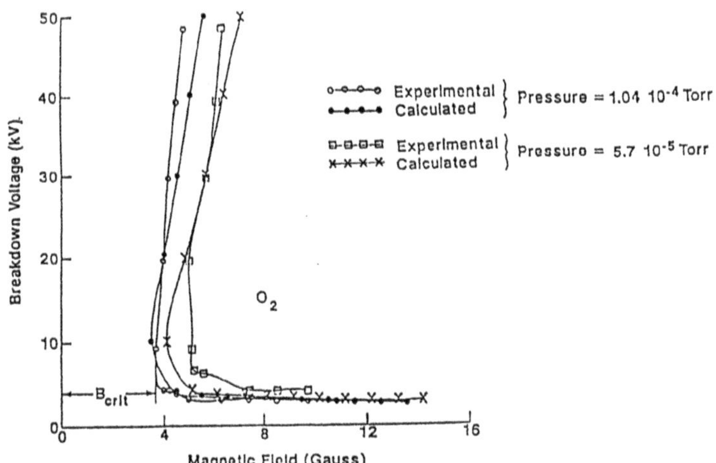

Fig. 15b: Measured and calculated breakdown voltage vs B O_2, for same geometry as Fig. 15a.

that B_{cr} is the value of B at the crossing of the upper and lower branches. When the characteristic confinement radius [Eq. (23)] becomes equal to r_c, the breakdown voltage, V_u, follows the upper branch. Thus,

$$V_u = \frac{3\pi}{32}\frac{\overline{Q}_I}{\overline{Q}}\frac{e/m}{I_n(1+1/\Gamma)}\frac{B^2 r_c^2}{r_e^2} \tag{26}$$

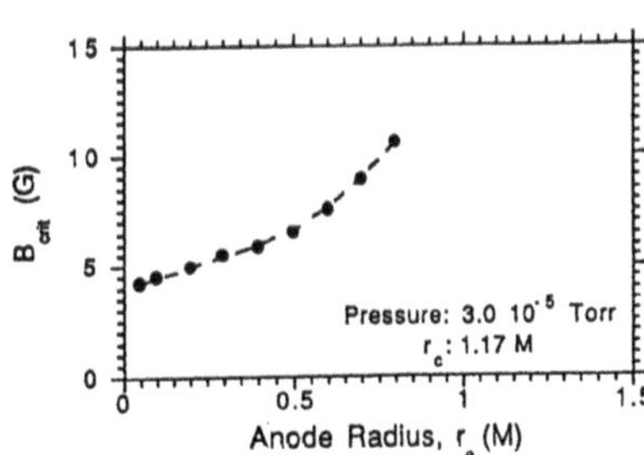

Fig. 16a: Dependence of critical magnetic field, B_{crit}, on anode radius, for constant cathode radius.

On the other hand, when the confinement radius is much smaller than r_c, the breakdown voltage, V_1, follows the lower branch. That is,

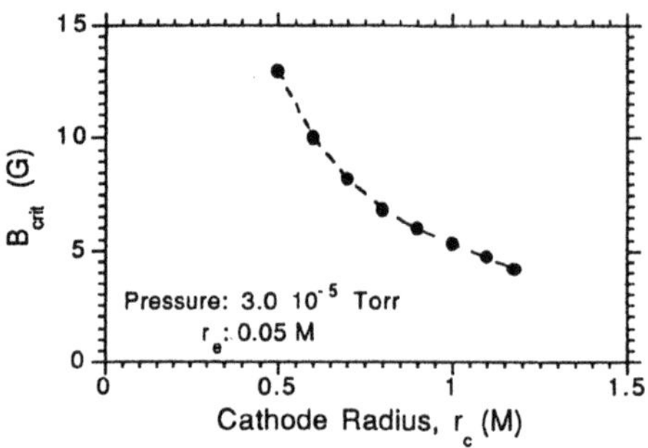

Fig. 16b: Dependence of critical magnetic field, B_{crit}, on cathode radius, for constant anode radius.

$$V_l = \frac{80}{3\pi^2} \frac{\bar{Q}}{Q_i} n \left(\frac{r_c}{r_e}\right) \ln (1 + 1/\Gamma) \; \Phi_i \tag{27}$$

At the crossing of the two branches

$$V_u = V_l$$

and form Eqs. (26) and (27)

$$B_{cr} = \sqrt{\frac{640}{9\pi^2}} \sqrt{\frac{m\Phi_i}{e}} \; \frac{\bar{Q}}{Q_i} \ln (1 + 1/\Gamma) \sqrt{\frac{r_e}{r_c^3} \ln \frac{r_c}{r_e}} \tag{28}$$

where G is assumed to be a weak function of B.

To assess the validity of the theory, the DC breakdown characteristics in the sphere-cylinder configuration have been measured as a function of pressure, magnetic field, anode radius, and plasma density for an O_2 background. The results for a neutral background are also shown in Fig. 15b for the same conditions as the calculations. Similar results have been obtained by Antonaides et.al. [Ref. 6]. The experimental results are in general agreement with the theory. The lower branch of the characteristics has been found to consist of a series of toroidal discharges whose light emission is confined to a region around the meridian plane of the electrode system and whose radial extend increases with sustaining discharge voltage. One of these modes has been observed by Greaves et. al. [Ref. 4]. This series of discharges terminates with the establishment of a glow-like discharge that fills the cylindrical chamber. In the upper branch, only the glow-like discharge is observed.

The I-V characteristics associated with the lower branch of the breakdown characteristics is shown in Fig. 17. The sequence of toroidal discharges preceding the formation of the glow-like discharge are identified with nearly flat portions of the curves. These discharges are the analog of corona discharges. The structure of these curves becomes

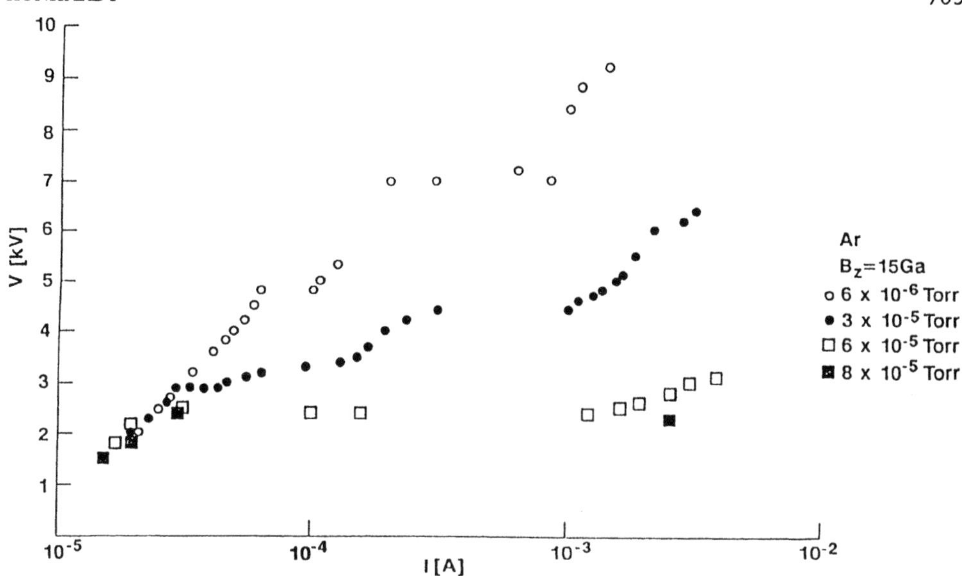

Fig. 17: Voltage-current characteristics associated with the lower branch of the breakdown characteristics. Transitions in the voltage with current indicate onset of a new discharge mode.

less defined at higher pressures, consisting of the first and third mode only, the latter also being somewhat unstable. The transition between the modes are also identified with the sharp increases in anode current, by more than an order of magnitude in some cases. The highest mode develops into the glow-like discharge, which is characterized by much lower sustaining voltage and higher currents. The development of anode current (curve 1) as a function of time for an applied voltage (curve 4) above the sustaining voltage for a particular mode is shown in Figs. 18a, 18b, and 18c. As the applied voltage increases above a particular sustaining mode-voltage, a transition to the next higher mode is observed (see Fig. 18a). Near the threshold for the transition, period doubling oscillations are observed (Fig. 18b), as well as oscillations between modes (Fig. 18c).

To determine the influence of the background plasma ($n_e \neq n_i \neq 0$) on the breakdown characteristics [Eq. (25)], it is necessary to solve Eq. (25) together with the Poisson Eq. [Eq. (18)]. This remains to be done. This influence, however, has been experimentally determined for spheres of both polarities. The results are shown on Fig. 19. As can be observed, the breakdown voltage decreases (significantly) with electron density, a stronger behavior than that of the sheath in the absence of the magnetic field. As remarked in the discussion of those results (Fig. 10), the difference may be due to a larger increase in the gain with magnetic field, which more than compensates for the decrease in sheath dimensions with electron density. This effect essentially lowers the effective value of B_{cr}. This phenomenon remains to be clarified.

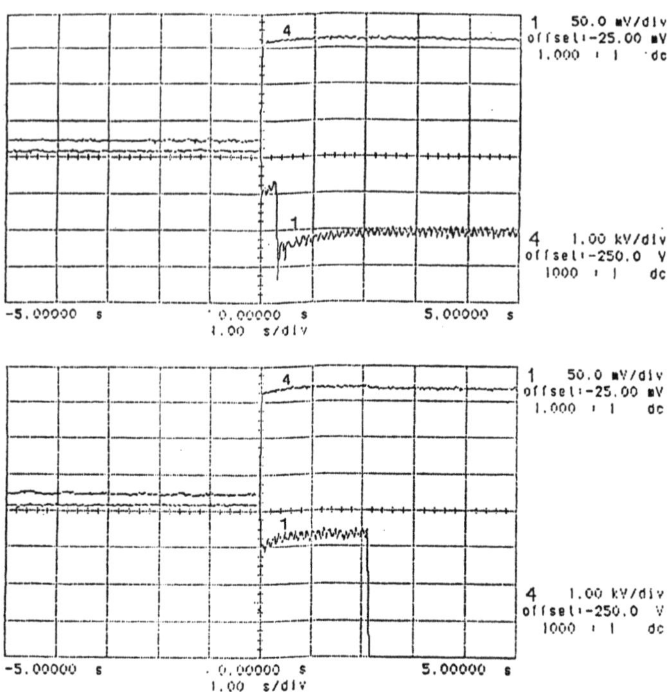

Fig. 18a: Current (1) and voltage (4) waveforms at the anode showing transition between the first and second and second and third toroidal discharge (1mV = 0.625 μA).

6. Applications

The theory, as developed thus far, may be used to determine the integrity of the insulation between components forming a payload exposed to the space environment and between a component and the ambient plasma, and to suggest a design for inductors exposed to the space-plasma. Note that a component relative to its environment can be modelled by either a sphere or a cylinder. For example, the PFN assembly in the SPEAR II payload can be consider (for the purpose of determining its breakdown characteristics relative to the other components and the support structure) as a spherical electrode. Moreover, for all expected flight conditions in SPEAR II, r_{L-B} >> scale length of enclosure, so that the condition $B_{ambient} < B_{cr}$ is a suitable criterion for assessing the electrical integrity of the payload. A third application arises in connection with the payload tests performed in large vacuum chambers. A problem arose when carrying out these tests at the NASA Plum Brook facility with regards to breakdown between the payload and the chamber walls. Mock up experiments were also carried out

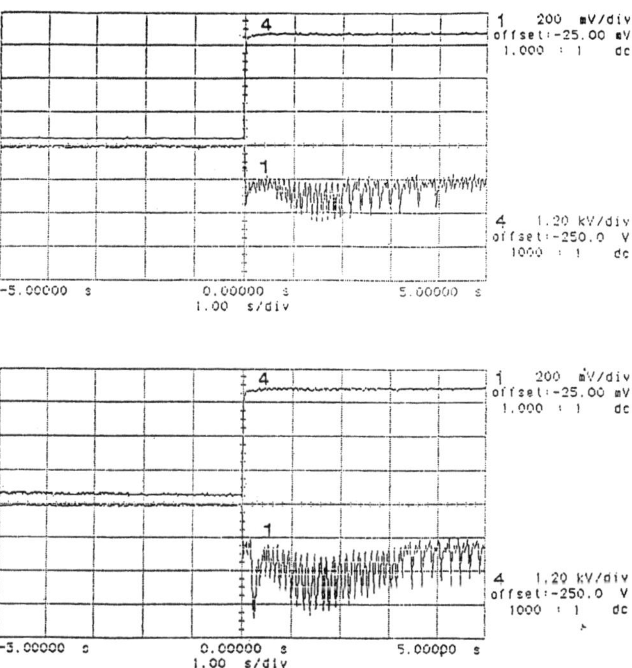

Fig. 18b: Anode current (1) and voltage (4) for an applied voltage near, but below, the threshold for transition between the second and third discharge modes, showing the period doubling instabilities in the current waveform.

in smaller chambers and this breakdown was not observed. These observations are analyzed in this section in the context of the theory presented in the previous sections.

In these cases, if the electric field lines from the object being considered is mainly confined to its enclosure ($r_{L-B} > r_c$, and n_e=n_i=0) and its size is small relative to the dimensions of the enclosure, the sphere-cylinder geometry represents a good approximation and Eq. (22) is directly applicable. The parameter that needs to be determined is B_{cr}, since it determines whether or not breakdown can occur in the chamber. Values for B_{cr} have been obtained for the SPEAR II payload, and the three chambers that have been used in the SPEAR program, namely, Plum Brook, Polytechnic, and the University of Maryland (see Table IV). Each component in a payload, of course, has an associated enclosure (i.e. determined by its immediate surroundings), and consequently, a B_{cr} (there may be two values for B_{cr}, depending on the orientation of the payload relative to the earth's magnetic field and the location of the component inside the enclosure). Instead of proceeding by component, it is only necessary to

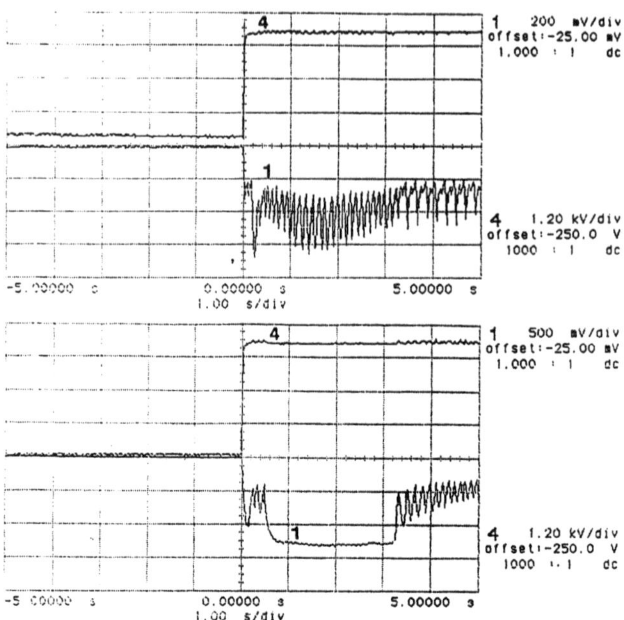

Fig. 18c: Anode current (1) and voltage (4) for an applied voltage near, but above, the threshold for transition between the second and third discharge modes, showing oscillations in current between modes.

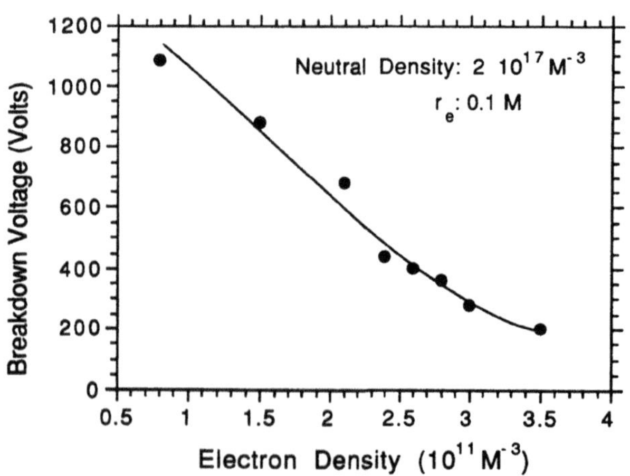

Fig. 19a: Breakdown voltage for positive sphere-cylinder configuration as a function of background electron density.

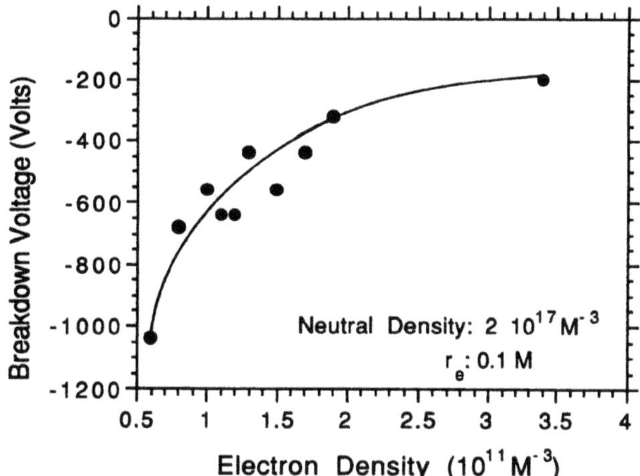

Fig. 19b: Breakdown voltage for negative sphere-cylinder configuration as a function of background electron density.

Table IV. Critical magnetic field iin experimental chambers.

Parameter	LRC	POLY	UM
r_c (M)	6.5	1.21	0.9
r_e (M)	0.2	0.05	0.05
B_{cr} (T)	1.5 10^{-5}	9.06 10^{-5}	1.34 10^{-4}
B_{cr}* (T)	4.3 10^{-5**}	2.5 10^{-4++}	3.8 10^{-4}
r_e (M)		B_{cr}/B_{cr}(LRC)	
0.2	1	8.95	12.75
0.1	1	9.63	14.1
0.05	1	10.1	14.9

LRC Lewis Research Center (Plum Brook)
UM University of Maryland
POLY Polytechnic University

** estimated value
++ measured value

determine the smallest B_{cr}. If this value falls below the ambient magnetic field, then breakdown can occur in the payload. The values of B_{cr} for the cases considered are derived below.

To determine B_{cr}, if the mean electron energy is near the ionization threshold, then in Eq. (28) [Ref. 28],

$$\frac{\bar{Q}}{Q_i} \simeq 2 \frac{Q}{Q_i} \qquad (29)$$

where Q and Q_i are the total respective crossections at 100 eV, since at higher energies this ratio is constant. The pressure dependence of B_{cr} is mostly due to Γ, which can be expressed in the form

$$\Gamma = 8 \sqrt{\frac{2m\phi_i}{e\nu_c} \frac{V r_e}{B^2}} \qquad (30a)$$

where,

$$\nu_c = k p \qquad (30b)$$

where ν_c is the effective collision frequency, p is the pressure, and k is a proportionality constant. This constant can be obtained by approximating the total collision crossection with the expression

$$Q = C \varepsilon^{-0.5} \qquad (31)$$

where C is a fitting parameter. For Nitrogen, k= 1.5 10^{10} sec^{-1} $Torr^{-1}$. For the same gas, pressure and sphere voltage, B_{cr} scales as

$$B_{cr} \sim \sqrt{\frac{r_e}{r_c^3} \ln \frac{r_c}{r_e}} \qquad (32)$$

This scaling expression permits the estimation of the relative behavior of the various components in the SPEAR II payload, and the test chambers.

Consider a component near the center of the payload (such as the high voltage probe, or the PFN in SPEAR II). For an enclosure with dimensions of 0.56 m (the radius of the Aries rocket body) and component size of approximately 0.2 m, the B_{cr}, for a positive bias, is found to be 3.1 Gauss. Consequently, the ambient B (0.45 G) << B_{cr}, and breakdown is not expected to occur (recall that this criterion operation, and that for pulse operation the threshold becomes larger. However, it is not possible, at this time, to assess the change in B_{cr} with pulse length). Since this value for B_{cr} is the lowest for all components, biased either positive or negative, the space-vacuum insulation within the SPEAR II payload is well within the bounds of integrity.

The results for the test chambers is presented in Table IV. The first two lines give the cathode and anode radius used in the tests. The third line gives B_{cr} for a background pressure of 10^{-5} Torr in Nitrogen, and the corresponding lower branch breakdown voltage. The next three lines give B_{cr} relative to the Plum Brook value for three different anode radii. These results show that B_{cr} varies for about an order of magnitude from chamber to chamber for the same conditions. Note that in Plum Brook, the critical field is below the background (Earth's) magnetic field so that breakdown between payload and chamber wall occurs at zero applied B. In fact, breakdown always occurs in the lower branch.

At the other extreme from the conditions discussed above is the case where the background charged particle density is sufficiently large that $r_{L-B} \ll r_c$. This is the situation in the test chambers, for example, when the plasma sources are turned on. In this case, the results found in Sec. IV only provide an upper limit on the breakdown voltage. As noted in Sec. V, the charged particle density, in the presence of a magnetic field, reduces the breakdown voltage from the B=0 case and the trend with density is actually the opposite. However, the values obtained with the B=0 theory remain useful in that if the breakdown criterion is satisfied for this case, it will also be so for $B \neq 0$. For a background charged particle density of 10^5 cm^{-3}, and a neutral density of 10^{11} cm^{-3}, the breakdown threshold from Sec. 4 is found to be approximately 200V, which conditions are not suitable for testing of the payload. The most effective way to inhibit this breakdown is to reduce the avalanche gain by placing metallic boundaries in the r-z plane at the minimum of the potential well [Eq. (13)] that results in the largest gain. This has the effect of increasing the effective B_{cr}. The change in B_{cr}, to B_{cr}*, obtained in the POLY chamber with a rectangular metallic sheet (dimensions: 0.3m, 1.2m) in the r-z plane, about z=0, and 0.08m from the chamber, is shown in Table IV. The change in the B_{cr} of the chambers used for SPEAR tests under equivalent conditions have been estimated and the results given in Table IV. Similar procedures can be used to inhibit breakdown between components of a payload.

Consider, for example, the behavior of a cylindrical inductor (open at both ends) exposed to the space-plasma. On account of the increase in the local magnetic field over ambient during operation, the condition $B \ll B_{cr}$ may be violated. Breakdown can then occur if dB/dt (or equivalently, dI/dt) is sufficiently large to generate voltages above threshold. Note that as the risetime increases, the voltage generated increases, and consequently, likely to approach the breakdown threshold. However, recall from Sec. 2 that this threshold increases with decreasing duration of voltage appliation. Thus, to determine the transient threshold voltage for breakdown requires a dynamic theory, which is not at present available. The technique for inhibiting breakdown by shifting B_{cr}, through the use of properly placed metallic boundaries, can be applied to this situation. An effective way to accomplish this in the context of an inductor is to give it a conical shape. With this geometry, trapped electrons are essentially eliminated ($a_{eff} \sim 0$), and the breakdown criterion corresponds to that of vacuum. Moreover, the conical shape reduces the high stresses created near the ends of the inductor (that is at the point of connection to the circuit) in pulsed operation, by creating a grading in the inter-winding capacitance. This inductor shape was used in the SPEAR II payload.

Acknowledgements

This work has been supported by SDIO/IST through DNA. The work at the Polytechnic University has been done in collaboration with J. Bentson and S. Popovic. I would like to thank H. Cohen for his guidance in the preparation of the manuscript.

References

[1] J. Raitt, SPEAR I Final Report, Utah State University (1988).

[2] D.B. Allred et. al., IEEE Trans. Nuc. Sci. 35,1386 (1988).

[3] E.E. Kunhardt et. al., Proc. XIIIth Int. Symposium on Discharges and Electrical Insulation in Vacuum, Paris, edited by J.M. Buzzi and A. Settier (les Editiones de Physique, Les Ulis, France, 1988) p. 247.

[4] R.G. Greaves, D.A. Boyd, J.A. Antoniades, and R.F. Ellis, Phys Rev. Lett., 64, 886 (1990).

[5] J. Antoniades, R. Greaves, D. Boyd, and R. Ellis, Proc. 28th Aerospace Science Meeting, Reno, Nevada (1990).

[6] J. Antoniades, M. Alport, D. Boyd, and R. Ellis, IEEE Trans. Elect. Ins. 25, 563 (1990).

[7] M.J. Alport, J. Antoniades, D.A. Boyd, R.G. Greaves, and R.F. Ellis, J. Geo. Res. 95, 6145 (1990).

[8] S. Popovic et. al., Proc. XX Int. Conf. Phen. Ionized Gases, Pisa, Italy, edited by V. Palleschi and M Vaselli (Institute of Atomic and Molecular Physics-CNR, Italy,1991)p.1025.

[9] I. Katz et.al., J. Geo. Res. 94, 1450 (1989).

[10] D.L. Cook, and I. Katz, AIAA J. Spacecraft and Rockets 25, 132 (1988).

[11] E.E. Kunhardt et.al., Gaseous Dielectrics VI, edited by L. Christophorou, and M.Pace (Pergamon Press,N.Y., 1991).

[12] A. Syljuasen et.al., Proc. XX Int. Conf. Phen. Ionized Gases, Pisa, Italy, edited by V. Palleschi and M Vaselli (Institute of Atomic and Molecular Physics-CNR, Italy, 1991) p.41.

[13] J.M. Meek and J.D. Craggs, editors, Electrical Breakdown of Gases (John Wiley & Sons, NY 1978) pp. 286-289.

[14] F. Llewelyn-Jones, Ionization and Breakdown In Gases (Methuen, London, 1966).

[15] J. Cobine, Gaseous Conductors (Dover, New York, 1958).

[16] E.E. Kunhardt and L. Luessen, editors, Electrical Breakdown and Discharges in Gases, NATO ASI Series 89a (Plenum Press, New York, 1983).

[17] E. Kuffel and W.S. Zaengel, High Voltage Engineering (Pergamon Press, Oxford,1984).

[18] R. Hackman, J. Appl. Phys. 46,627 (1975).

[19] A. Maitland, J. Appl. Phys. 32, 2399 (1961); G.A. Mesyats, Proc. Xth Int. Conf. Phen. Ionized Gases, Oxford, England (1971); R. Latham, High Voltage Vacuum Technology (Academic Press, New York, 1981.

[20] H.A. Belvin and S.C. Haydon, Aust. J. Phys. 11,18 (1958).

[21] L.B. Loeb, Electrical Coronas (US Press, Berkeley, 1965).

[22] W.G. Dunbar, High Voltage Design Guide: Spacecraft, Air Force Weapons Laboratory, AFWAL-TR-82-2057, Vol. V (1983).

[23] L. Langmuir and K. Blodgett, Phys. Rev. 24,49 (1924).

[24] R. Gajewski, Physica (Utrecht) 47, 575 (1970).

[25] J.B. Delos et.al., Phys. Rev. A30,1208 (1984).

[26] P.A. Redhead, Can. J. Phys. 36,255 (1958).

[27] K. Takayanagi, and H. Suzuki, Crossections for Atomic Processes Vol.I (Research Information Center, Institute of Plasma Physics, Nagoya, Japan, 1978).

[28] W.E. Kappula, and T.S. Stein, Advances in Atomic and Molecular Physics 26,1 (1989).

CURRENT COLLECTION BY HIGH-VOLTAGE SPACECRAFT

IRA KATZ
Maxwell Laboratories, Inc., S-Cubed Division
P.O. Box 1620
La Jolla, California 92038
USA

ABSTRACT. During the past decade, there has been much progress in understanding how spacecraft collect charged particles from the tenuous plasma surrounding the earth. This paper describes the basic theory of steady-state sheaths in collisionless plasmas and its application to spacecraft in the magnetosphere. The first part describes the physical mechanisms, including angular momentum and space charge, which control sheath currents. Next, how spacecraft surface potentials are modified by the plasma currents and the roles of potential barriers and low-energy secondary electrons in determining how currents are distributed among surfaces are briefly discussed. The last topic is how current equilibrium is achieved on geometrically complex spacecraft.

1. Introduction

During the past decade, there has been much progress in understanding how spacecraft collect charged particles from the tenuous plasma surrounding the earth. Basic equations that describe the charging of satellites in geosynchronous orbit during substorms in the magnetosphere [Ref. 1] are well known. The early controversies surrounding electron current collection by sounding rockets [Ref. 2] have been addressed with theory and experiment resulting in basic agreement. While the equations describing the magnetic-limited sheath surrounding high-voltage spacecraft in low-Earth orbit are well known, there still is a debate on which set of algorithms is both sufficient to describe the plasma interactions, and yet practical enough to provide answers in a reasonable time on available computers [Ref. 3]. The upcoming electrodynamic Tethered Satellite System (TSS-1) will have sufficient instrumentation to test the adequacy of the present theories and computer codes. This paper describes the basic theory of steady-state sheaths in collisionless plasmas and its application to spacecraft in the magnetosphere. The major part describes the physical mechanisms, including angular momentum and space charge, that control sheath currents. Following is a brief discussion of how spacecraft surface potentials are modified by the plasma currents and the roles of potential barriers and low-energy secondary electrons in determining how currents are

R. N. DeWitt et al. (eds.), The Behavior of Systems in the Space Environment, 713–729.

distributed among surfaces. The last topic is how current equilibrium is achieved on geometrically complex spacecraft.

The approach employed in this paper is to consider a spacecraft as a large, asymmetrical, high-voltage probe immersed in a magnetoplasma. There has been extensive research into the current characteristics of probes in plasmas [Refs. 4,5,6,7,8,9,10,11,12]. Most of the published work has been for symmetric probes. We discuss the extension of the basic theories to account for geometric asymmetry, the earth's magnetic field, and surface charging.

2. The Orbital Plasma Environment

The ionosphere in LEO (low-Earth orbit) is a cool dense plasma. For sounding rockets, typical ionospheric plasma parameters are

$$n_e = n_i = 10^{11}\ m^{-3}$$

$$\theta_i = \theta_i = 0.1\ eV$$

$$B = 0.4\ Gauss\ . \tag{1}$$

The time and distance scales associated with this plasma are

$$\omega_{pe} = 2 \times 10^{-7}\ sec^{-1}$$

$$\omega_{ce} = 7 \times 10^{6}\ sec^{-1}$$

$$\lambda_D = 0.007\ m$$

$$\lambda_{ce} = 0.02\ m$$

$$\lambda_{ci} = 3\ m. \tag{2}$$

Typical active spacecraft experiments have dimensions of meters, potentials of hundreds of volts or more, and durations as long as seconds. The wide range of time and distance scales in current collection from the ionosphere allows approximations to be made that lead to good estimates of the collected current.

In GEO (geosynchronous-Earth orbit), the plasma density is considerably less but is frequently characterized by higher particle kinetic energies. While in general, the energy distributions found in the magnetosphere are non-Maxwellian, we associate a temperature with them for the purpose of establishing distance and potential scales.

$$n_i = n_i \approx 10^{6}\ m^{-3}$$

$$\theta_e = \theta_i \approx 10^{4}\ eV$$

$$B \approx = 10^{-3}\ Gauss \tag{3}$$

The time and distance scales associated with this plasma are

$$\omega_{pe} = 6 \times 10^4 \ sec^{-1}$$

$$\omega_{ce} = 2 \times 10^4 \ sec^{-1}$$

$$\lambda_D = 7 \times 10^3 \ m$$

$$\lambda_{ce} = 2 \times 10^3 \ m$$

$$\lambda_{ci} = 10^5 \ m \ . \tag{4}$$

The ambient neutral environment in both LEO and GEO is so small that collision lengths are kilometers or greater. For most of the following, collisions with ambient neutrals are ignored. Neutral gases generated by spacecraft, either by thruster firings or outgassing, can play a major role in current collection but is not addressed in this paper.

3. The Physics of Current Collection

The equilibrium state of the plasma surrounding a spacecraft can be described by Poisson's equation and the collisionless Vlasov equation,

$$\nabla^2\phi = -\frac{\rho}{\varepsilon_0}$$ (5)*t300R

$$\rho(\mathbf{x}) = e\left(\iiint f_i(\mathbf{x},\mathbf{v})d\mathbf{v} - \iiint f_e(\mathbf{x},\mathbf{v})d\mathbf{v}\right), \tag{6}$$

where ϕ is the potential and f_i, f_e are the ion and electron distribution functions, respectively. The potential is measured with respect to the unperturbed plasma at great distances. The region of non-zero space charge, ρ, or disturbed plasma surrounding a high potential spacecraft is called the sheath region. Currents to spacecraft surfaces are found from surface integrals over the local distribution functions.

$$j(\mathbf{x}) = e\left(\iiint f_i(\mathbf{x},\mathbf{v})\mathbf{v}\cdot\mathbf{n}\ d\mathbf{v} - \iiint f_e(\mathbf{x},\mathbf{v})\mathbf{v}\cdot\mathbf{n}\ d\mathbf{v}\right) \tag{7}$$

The local particle distribution functions can be related to those in the undisturbed plasma by the Liouville theorem.

$$\frac{df(\mathbf{x},\mathbf{v})}{dt} = \frac{\partial f(\mathbf{x},\mathbf{v})}{\partial t} + \mathbf{v}\cdot\nabla_x f(\mathbf{x},\mathbf{v}) + \frac{\partial \mathbf{v}}{\partial t}\cdot\nabla_v f(\mathbf{x},\mathbf{v}) = 0 \tag{8}$$

If a spacecraft were at the ionospheric potential, the incident particle currents are just the one-sided ambient thermal currents. For typical LEO parameters, Eq. (1), ambient currents are

$$j_{the} \approx 10^{-3} \ amp \ m^{-3}$$

$$j_{thi} \approx 5 \times 10^{-6} \ amp \ m^{-2}$$

$$j_{ram} \approx 10^{-4} \ amp \ m^{-2} \ , \tag{9}$$

where the ram ion current is calculated assuming a 7500 m/sec orbital velocity. In LEO, the plasma environment is mesothermal; that is, the spacecraft velocity is greater than the ambient ion thermal velocity and less than the electron thermal velocity.

$$v_{the} \approx 10^5 \ m \ sec^{-1} > v_{ram} \approx 7500 \ m \ sec^{-1} > v_{thi} \approx 770 \ m \ sec^{-1} \tag{10}$$

For GEO, the current densities are much lower:

$$j_{the} \approx 3 \times 10^{-6} \ amp \ m^{-2}$$

$$j_{thi} \approx 6 \times 10^{-8} \ amp \ m^{-2} \ . \tag{11}$$

In GEO, satellite velocities are much lower than typical particle velocities, so there is little ram or wake effects on particle currents.

Below, we examine physical mechanisms that can limit collection of a plasma particle with impact parameter, b, by a charged sphere of radius, a. The four mechanisms, shown in Fig. 1, are conservation of energy, conservation of angular momentum, conservation of canonical angular momentum in a magnetic field, and space-charge shielding limiting the range of the potential.

3.1. REPELLED SPECIES CURRENTS

For both the ionosphere and the magnetosphere (LEO and GEO) particle currents for repulsive potentials can be estimated directly from Eqs. (7) and (8) if the ambient distribution function, f $(\mathbf{x},\mathbf{v})$, is assumed to be isotropic.

$$\begin{aligned} j(\mathbf{x}) &= e\left(\iiint f_f(\mathbf{x},\mathbf{v})\, \mathbf{v}_f \cdot \mathbf{n}\, d\mathbf{v}\right) \\ &= e\left(\iiint f_0\left(|\,\mathbf{x}\,|,\ \sqrt{\mathbf{v}_f \cdot \mathbf{v}_f + \frac{2eq\Phi}{m}}\right)\mathbf{v}_f \cdot \mathbf{n}\, d\mathbf{v}\right) \end{aligned} \tag{12}$$

For Maxwellian distributions, this reduces to a particularly simple and frequently used expression for repelled current density.

$$j(\phi) = j_{th} e^{-\left|\frac{e\phi}{\theta}\right|} \tag{13}$$

where j_{th} is the one-sided thermal current in the undisturbed plasma. While strictly applicable only for limited cases, Eq. (13) is almost always how repelled currents are calculated. The only significant exception is for ram ion currents, where the current density is assumed constant unless

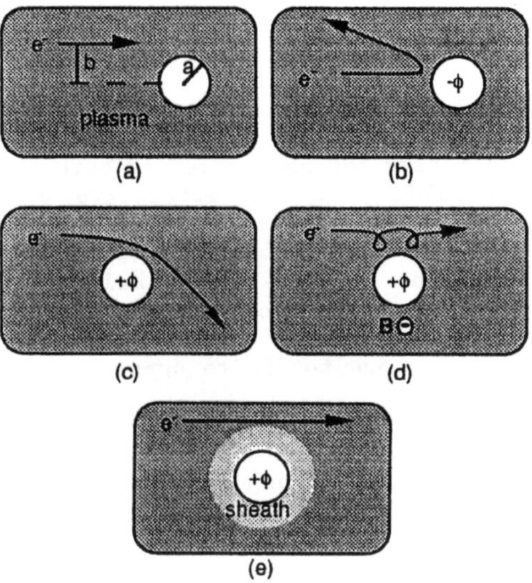

Fig. 1: Physical mechanisms that can prevent plasma particles from being collected by a spacecraft. A particle may have (b) too little energy, (c) too much angular momentum, (d) too much canonical angular momentum, or (e) space charge can shield the potential.

the surface potential can repel all the ions, in which case it is assumed to be zero. This then gives the conditions

$$j_{ram}(\phi) = en_i v_{ram}, \qquad e\phi \leq \frac{1}{2}\, mv_{ram}^2 \tag{14}$$

$$j_{ram}(\phi) = 0, \qquad e\phi > \frac{1}{2}\, mv_{ram}^2$$

In LEO, the energy associated with the ram oxygen ion velocity due to orbital motion is approximately 5 eV.

3.2. ATTRACTED SPECIES CURRENTS

All attracted species particles can energetically reach the spacecraft surface. Examining impact parameters is a good way to estimate the attracted species current that reaches a spacecraft. The total attracted current is for an isotropic plasma,

$$I_{attracted} \approx 4\pi b^2\, j_{ambient}\ , \tag{15}$$

where b is the maximum impact parameter of a particle to hit the surface, or

$$I_{attracted} \approx 2\pi b^2 \, j_{ambient} \tag{16}$$

if current only flows along magnetic field lines. For typical spacecraft dimensions, potentials, and plasma parameters, angular momentum determines the maximum impact parameters for collection of both ions and electrons in GEO. Space-charge shielding of the attractive potential limits ion collection in LEO, while for electrons in LEO either space-charge or magnetic fields can limit the collected current, depending on the particular flight experiment. The discussion below first examines angular momentum and magnetic field restrictions on particle impact parameters for collection, neglecting the charge density in Eq. (5). Then, the restrictions on the current due to finite space charge are considered.

3.2.1. *Angular Momentum.* If the range of the potential were infinite, the maximum impact parameter of particles collected by an attractive potential would be limited by conservation of angular momentum. The initial angular momentum, l_0, for a plasma particle with thermal velocity v_{th} and impact parameter, b, is given by

$$l_0 \approx m \, v_{th} \, b \ . \tag{17}$$

If it were collected by a sphere of radius a and potential ϕ, the maximum angular momentum the particle could have would be its final velocity, v_f, times the sphere radius,

$$l_{max} \approx m \, v_f \, a \ . \tag{18}$$

The maximum impact parameter of a particle that can be collected and still satisfy conservation of angular momentum is found by combining Eqs. (17) and (18),

$$b \le a \, (v_f/v_{th})$$

$$b \le a \, (1 + \phi/\theta_e)^{1/2} \ , \tag{19}$$

using energy conservation. Particles with larger impact parameters will miss the sphere. The collected current, which depends on b^2, increases linearly with potential.

$$\begin{aligned} j(\phi) &= j_{th} \cdot \left(\frac{b}{a}\right)^2 = j_{th}(1 + \phi/\theta_e) \\ I_{orbit} &= 4\pi a^2 j_{the}(1 + \phi/\theta_e) \end{aligned} \tag{20}$$

This type of collection is seen in hot, dilute plasmas, such as the magnetosphere but is observed only for very low potentials in the ionosphere. In LEO, the energy associated with the ram ion velocity due to orbital motion is approximately 5 eV and is used in place of the electron temperature. In that case the above becomes

$$j_{ram}(\phi) = j_{th} \cdot \left(\frac{b}{a}\right)^2 = j_{ram}\left(1+\frac{\phi}{\frac{1}{2}mv_{ram}^2}\right) \tag{21}$$

For an insulating ram-facing surface in LEO, we can find the equilibrium floating potential by balancing ram ions with thermal electrons using Eqs. (13) and (21). Since the ram energy is very large compared with the ambient electron temperature, we ignore the potential term in Eq. (21),

$$j_{the}\exp(e\phi/\theta_e) = j_{ram}(\phi) \approx env_{ram}$$

$$\phi \approx \theta_e \ln\left(\frac{j_{ram}}{j_{the}}\right) \approx -0.2V \tag{22}$$

If the spacecraft were not moving, the equilibrium floating potential would be slightly more negative.

$$j_{the}\exp(e\phi/\theta_e) = j_{thi}(1 + \phi/\theta_i)$$

$$\phi \approx \theta_e \ln\left(\frac{j_{thi}}{j_{the}}\right) \approx -0.5V \tag{23}$$

Spacecraft charging in GEO is a direct consequence of the spacecraft achieving current balance with very high plasma temperatures.

$$\phi \approx \theta_e \ln\left(\frac{j_{thi}}{j_{the}}\right) \approx -40{,}000V \tag{24}$$

At the higher incident kinetic energies, secondary electrons play a major role in reducing potentials from the upper bound in Eq. (24) by more than a factor of two.

3.2.2. *Magnetic-field-limited Currents.* For electrons in the ionosphere, the earth's magnetic field introduces a canonical angular momentum, l_θ, that severely restricts the range of impact parameters that can be collected [Ref. 12],

$$l_\theta = mr^2\left(\frac{d\theta}{dt} + \frac{\omega_{ce}}{2}\right) \tag{25}$$

Conservation of both energy and the above canonical angular momentum limits the maximum impact parameter, b_{PM}, for collection.

$$b_{PM} \le a\sqrt{1 + \left(\frac{8e\phi}{m\omega_{ce}^2 a^2}\right)^{1/2}} \tag{26}$$

For impact parameters larger than b_{PM}, the **v x B** acceleration, which is transverse to the radial direction, makes electrons miss the central sphere. The maximum current that can be collected by a sphere of radius a and potential ϕ is

$$I_{PM} = 2\phi a^2 j_{the}\left(1 + \sqrt{\frac{8e\phi}{m\omega_{ce}^2 a^2}}\right) \tag{27}$$

The 2π in Eq. (27) is because ionosphere electrons flow only along the ambient magnetic field direction. A one-meter sphere at 10,000 volts in a typical LEO environment (Eq. 1) collects less than 0.2 amperes of electrons from the ionosphere due to magnetic limiting. The magnetic field is typically the most limiting condition in electron collection by a symmetrical spacecraft in the ionosphere. Early electron beam experiments aboard rockets reported currents much larger than allowed by the Parker-Murphy theory and lead to speculation that plasma turbulence scattered electrons across magnetic field lines [Ref. 13]. However, later data at altitudes above 250 km show clear evidence of magnetic limiting [Ref. 14]. Recently, Neubert has shown that the high currents collected at lower altitudes were due to ionization of the neutral atmosphere by the electron beams used to modify the spacecraft potential [Ref. 15].

3.2.3. *Space-charge-limited Currents.* The limits described above are predicated on the range of the attracting potential being larger than the limiting angular momentum impact parameter. This condition is clearly violated in the ionosphere for spacecraft at high potentials. A space-charge sheath forms around a high potential spacecraft. This sheath shields the bulk of the plasma from the potential. Since electric fields are very small in the surrounding plasma, it follows from Gauss's law that the sheath space charge balances the spacecraft surface charge,

$$\iiint_{\text{sheath}} \rho d\mathbf{x} + \oint_{\text{spacecraft}} \sigma dS = 0. \tag{28}$$

The sheath dimensions are calculated by solving Poisson's equation (Eq. 5). If the attracted species is very cold compared with applied potential and the geometry is planar, the Child-Langmuir approximation can be used for the sheath, space-charge density,

$$\rho(\mathbf{x}) = \frac{j}{v} = \frac{j}{\sqrt{\frac{2e\phi(\mathbf{x})}{m}}}. \tag{29}$$

The space-charge-limited shielding distance for electrons in one-dimensional planar geometry is

$$l_{SC}^2 = \frac{2.3 \times 10^{-6}\phi^{3/2}}{j_{the}}. \tag{30}$$

The corresponding equation for ions is

$$I_{sc}^2 = \frac{2.3 \times 10^{-6}\phi^{3/2}}{j_{th}\sqrt{\frac{m_i}{m_e}}}. \tag{31}$$

Spherical convergence increases the current density for higher potentials and reduces the shielding distance substantially from Eqs. (30) and (31). The space charge of the attracted electrons or ions in a typical LEO plasma, Eq. (1), shields a ±10,000 volt potential on a one-meter sphere within a ten-meter sheath (Fig. 2). For ions, this is a much shorter distance than either angular momentum limit. For electrons, the magnetic limit is a factor of two less than the space charge limit, thus the current would be closer to the 0.2 amperes from Eq. (28) than the one ampere from the space-charge-limited sheath radius. While the planar sheath dimension obtained from Eq. (31) is a nonsensical 53 meters, Eq. (31) allows one to quickly estimate, when space charge is important. Better estimates can be obtained from tables of results for spherical collection [Ref. 4]. The spherical space-charge-limited sheath radius, R_s, can be found from

$$R_s \approx 0.0235\frac{a^{3/7}\phi^{3/7}}{j_{th}^{2/7}} \tag{32}$$

when R_s is large compared with a, the sphere radius.

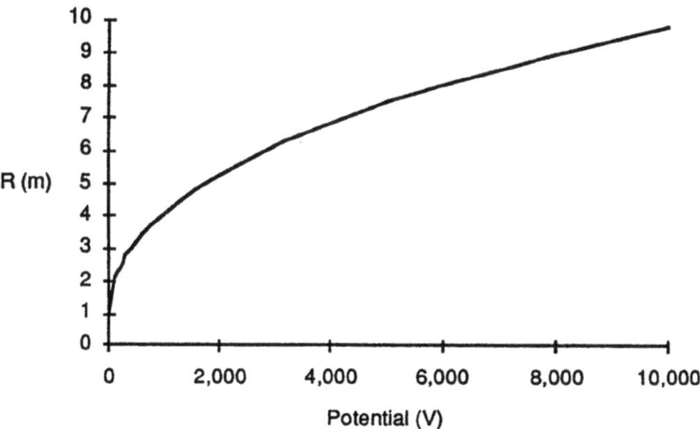

Fig. 2: Space-charge-limited sheath radius around a one-meter sphere in LEO neglecting magnetic field and orbital velocity effects. By neglecting convergence, Eq. (30) would have calculated a 55-meter sheath for the 10,000 volt potential.

Computer models can accurately calculate space-charge-limited sheaths in three dimensions for arbitrarily shaped spacecraft using either an expression like Eq. (29) corrected for convergence [Ref. 16] or self-consistently tracking particles inside the sheath [Refs. 17, 18, 19, 20, 21].

Equation (28) provides insight into the number of electrons in a magnetically limited sheath. Although scattering can leave electrons trapped in the sheath, the number of trapped electrons cannot be substantial unless they generate enough ions to balance their space charge. For sheath dimensions large compared with the effective radius of the spacecraft, the surface charge depends only weakly on the sheath radius. Thus, while the magnetic field may modify particle trajectories and the sheath shape, in the absence of ionization the total number of electrons in the sheath is roughly the same with or without the magnetic field.

3.2.4 *Presheath Effects.* Outside of the space-charge sheath, the ionosphere is perturbed by weak electric fields that focus thermal current to the sheath edge and allow the plasma to satisfy the Bohm criterion at the sheath edge. How this is accomplished in a magnetoplasma is not know. For nonmagnetized plasmas, Parrot et al. [Ref. 22] calculated self-consistent potentials and densities for the quasi-neutral presheath. Their analysis gives a sheath edge potential of 0.5 θ_e and an incident current of 1.45 j_{th}. These results, modified to account for spacecraft motion, are used in most computer calculations.

4. Charge Transport on Spacecraft Surfaces

The previous discussion assumes that the spacecraft surface is entirely conducting and held at a uniform potential. However, for thermal control, most spacecraft surfaces are covered with insulating material. When charged particles land on the spacecraft, surface potentials change due to charge accumulation. The potentials on different surfaces change differently depending on geometry and material properties. Insulating surfaces charge to achieve local current balance, with equilibrium potentials of the order of the electron kinetic energy (Eqs. (22) and (24)). The actual currents locally are often dominated by low energy, ~2 eV, secondary electrons generated by incident electrons, ions, or solar photons. For incident electrons with kinetic energies between 50 electron volts and a few thousand electron volts, secondary electron currents are greater than incident electron currents for most spacecraft materials. Photocurrents are typically 10^{-5} $amp\ m^{-2}$ or more, invariably greater than spacecraft charging currents in GEO. One of the more important phenomena is how potential barriers influence low-energy electron currents. In geosynchronous orbit, potential saddle points can inhibit secondary electron emission and cause daylight charging [Ref. 23]. In the ionosphere, potential barriers to low energy and secondary electrons cause ion focusing for negative potentials and "snapover" for positive potentials [Refs. 24, 25, 26].

For electron collection in LEO, the large secondary electron yields enable insulating surfaces to collect current as if they were conductors.

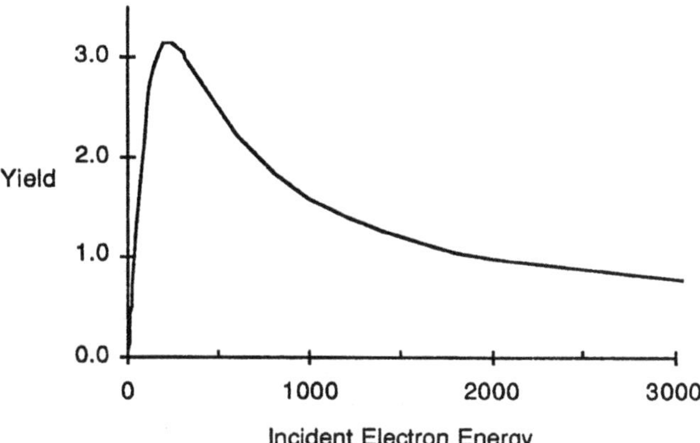

Fig. 3: Secondary electron yield on kapton as a function of incident electron energy.

The incident electrons generate secondary electrons that are attracted to nearby exposed positive conductors. This secondary electron transport increases the effective electron collecting area over the exposed positive conductor area, and the electron currents are correspondingly increased.

For ion collection in LEO, ion-generated secondary-electron currents are of the same sign as the ion currents. They drive insulating surfaces toward plasma potential and steepen the surface electric fields near exposed negatively-biased conductors; for example, arcing on negatively biased solar arrays is caused by the enhanced fields between exposed array interconnects and nearby solar-cell coverslips, which are near plasma potential. Low-potential insulating surfaces surrounding negatively biased conductors reduce the range of the negative potential, and the ion currents are correspondingly decreased.

5. Spacecraft Geometry Effects On Current Collection

Sometimes, sheath dimensions around nonspherical spacecraft are estimated by replacing the spacecraft with a sphere of equivalent surface area. This approach is very problematical at best and should only be used to obtain preliminary estimates for grid dimensions for numerical calculations. Computer models, such as NASCAP/LEO, POLAR, and DynaPAC, accurately calculate in three dimensions the space-charge-limited sheath around arbitrary spacecraft. Estimates, including some geometrical effects are quickly found using EPSAT or the Space Station Environments WorkBench [Refs. 27, 28, 29].

5.1 MULTIPLE SHEATHS

Potentials on spacecraft continue to change until each surface is in current balance, i.e., there is no net current [Ref. 30]

$$0 = j_{net} = j_e + j_{sec} + j_{backscat} + j_i + j_{photo} + j_{conduction} \tag{33}$$

The previous sections examined charge transport inside a single sheath and neglected conduction currents, j_{cond}, between different spacecraft surfaces. For most insulating surfaces, each point on the spacecraft surface individually balances incoming electrons with incoming ions and outgoing electrons. For some insulating surfaces, transport of secondary electrons along the surface may be important.

Spacecraft may have several sheaths of both polarities. Current balance can be achieved in three ways, as shown in Fig. 4 [Ref. 31, private communication]. The first is that conduction and surface transport balance currents between different location on the spacecraft that have their own sheath currents. The upcoming Tethered Satellite System (TSS-1) is an example of a spacecraft where two conducting sections are biased with respect to one another. For TSS-1, the individual sheaths are a few meters in radius and are separated by 20 kilometers. Since the ambient ion current density is much smaller than the electron current density, ion-collecting sheaths must be much larger than the corresponding electron-collecting sheaths if ion currents and electron currents are to balance.

When the two sections are near each other and the potentials are sufficiently high, the space-charge sheaths can overlap as shown in Fig. 4(b). One way of reducing the electron current is for the ion-collecting sheath to be large enough that it envelops part of the electron sheath. Computer calculations of the SPEAR I sheaths showed that current balance was achieved by the large, ion-collecting sheath enclosing most of the electron-collecting sheath [Ref. 32]. This phenomenon will occur only if exposed conductors of opposite polarity are closer than the sum of their space-charge-limited sheath radii. This clearly will be the case for SPEAR III where the ion and electron-collecting sheaths will be separated by only a couple of meters, less than the anticipated ion-collecting sheath around the negatively charge rocket body.

The third way current balance can be achieved is one sheath of one polarity enveloping the opposite polarity sheath. This is frequency the case in GEO when the negative polarity sheath can suppress photoemission. It also occurs in LEO when coverslip potentials shield the sheaths around the edges of positive voltage solar cells.

5.2. GEOMETRY AND MAGNETIC FIELDS

The discussion of magnetic-field-limited currents assumed a spherical spacecraft and spherically symmetric potentials. Analysis of data and computer simulations of the CHARGE-2 and SPEAR I flight experiments show that asymmetry tends to increase the collected currents over that predicted by spherical probe theory.

For estimating the space-charge sheath dimensions, researchers frequently treat the spacecraft as a sphere with the equivalent surface

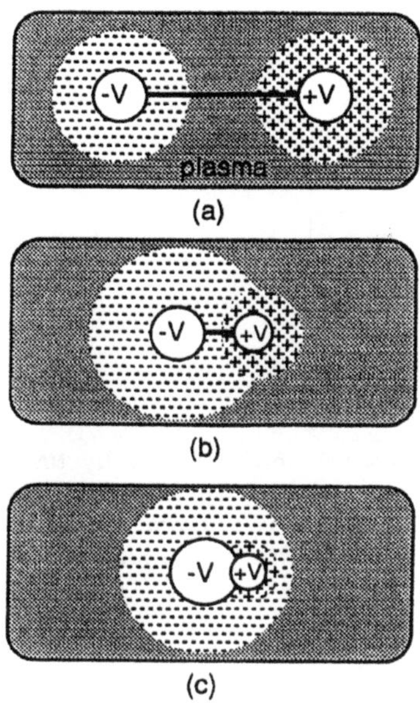

Fig. 4: To achieve current balance, sheath can be either (a) independent, (B) interfering, or (c) completely enveloped [Ref. 31, private communication).

area [Ref. 33]. Some justification for this approximation can be found in Eq. (28), assuming that the sphere has the same average surface-charge density as the actual spacecraft. For magnetic limiting, this approximation is inappropriate, since the limiting mechanism is related to the particle motion normal to the magnetic field. A better estimate would be to use the largest spacecraft dimension normal to the magnetic field, as the effective radius in Eq. (26). Better yet is to perform calculations using computer models that include geometry, space charge, and magnetic field [Ref. 14].

The spherical sheath potentials maximize the magnetic limiting. Magnetic limiting makes particles **v x B** drift around the attracting potential, returning to the same equipotential surface. An interfering sheath, as shown in Fig. 4(b), moves equipotentials into smaller impact parameters; thus, while a particle may initially be unable to be collected, as it circulates around the potential, it may drift into a region where it is closer to the probe, and be collected. This effect was evident on SPEAR I [Ref. 32], as shown in Fig. 5.

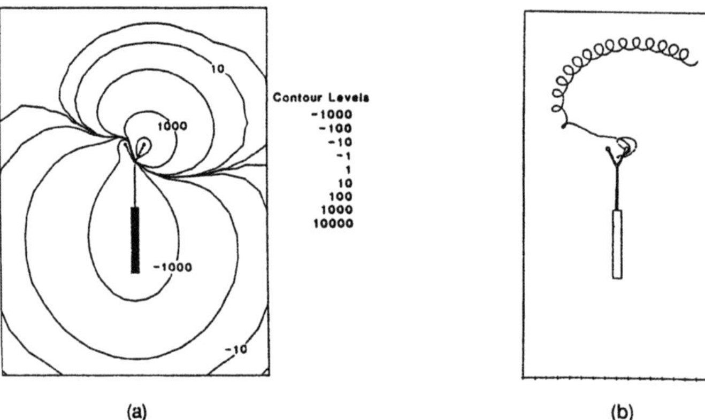

Fig. 5: (a) Potential contours calculated by NASCAP/LEO for SPEAR I with one sphere biased to 46 kV and the spacecraft ground at -6 kV. (b) Path of an electron in the potentials shown in Fig. 5(a). Note that the electrons are collected due to the presence of the ion-collecting sheath.

6. Conclusion

Over the past two decades, a body of knowledge consisting of theories, flight data, and computer models has been created to describe plasma interactions with spacecraft. Many of the phenomena that control plasma currents to spacecraft surfaces, some of which are discussed above, are understood well enough to be compared with flight and ground experiments. Other phenomena, including neutral gas effects in sheaths, plasma double layers, and magnetic-limited electron collection in low-Earth orbit, are presently under active investigation. The next decade promises to be rich with new data and models that will further understanding of spacecraft-plasma interactions.

References

[1] E.C. Whipple, Jr., 'Potential of surfaces in space', Reports on Progress in Physics 44, 1197-1250 (1981).

[2] J.R. Winckler, 'The application of artificial electron beams to magnetospheric research', Review of Geophysics 18, 659 (1980).

[3] J.G. Laframboise, 'Is there a good way to model spacecraft charging in the presence of space-charge coupling, flow, and magnetic fields?', in Proceedings of the Air Force Geophysics Laboratory Workshop on Natural Charging of Large Space Structures in Near Earth Polar Orbit, AFGL-TR-83-0046, 293 (1983).

[4] I. Langmuir and K.B. Blodgett, 'Currents limited by space charge between concentric spheres', Physics Review 24, 49 (1924).

[5] H.M. Mott-Smith, and I. Langmuir, 'The theory of collectors in gaseous discharges', Physics Review 28, 727 (1926).

[6] D.B. Beard and F.S. Johnson, 'Ionospheric limitations on attainable satellite potential', Journal of Geophysical Research 66, 4112 (1961).

[7] F.F. Chen, 'Electric probes', in R.H. Huddleston and S.L. Leonards (eds.), Plasma Diagnostic Techniques, Academic, Orlando, Fla., 113-200 (1965).

[8] J.G. Laframboise and J. Rubinstein, 'Theory of a cylindrical probe in a collisionless plasma', Physics of Fluids 19, 1990 (1976).

[9] J. Rubinstein, and J.G. Laframboise, 'Upper bound current to a cylindrical plasma probe in a collisionless magnetoplasma', Physics of Fluids 21, 1655 (1978).

[10] J. Rubinstein, and J.G. Laframboise, 'Theory of a spherical probe in a collisionless magnetoplasma', Physics of Fluids 25, 1174 (1982).

[11] J. Rubinstein, and J.G. Laframboise, 'Theory of axially symmetric probes in a collisionless magnetoplasma: Aligned spheroids, finite cylinders, and disks', Physics of Fluids 26, 3624 (1983).

[12] L.W. Parker, B.L. Murphy, 'Potential buildup on electron-emitting ionospheric satellites', Journal of Geophysical Research 72, 1631 (1967).

[13] L.M. Linson, 'Current-voltage characteristics of an electron emitting satellite in the ionosphere', Journal of Geophysical Research 74, 2368-2375 (1969).

[14] M.J. Mandell, J.R. Lilley, Jr., I. Katz, T. Neubert, N.B. Myers, 'Computer modeling of current collection by the CHARGE-2 mother payload', Geophysical Research Letters 17(2), 135-138 (1990).

[15] T. Neubert, P. Banks, B. Gilchrist, A. Fraser-Smith, P.R. Williamson, W.J. Raitt, N. Myers, and S. Sasaki, 'The interactions of an artificial electron beam with the Earth's upper atmosphere: effects on spacecraft charging and the near-plasma environment; Journal of Geophysical Research 95, 12209-12217 (1990).

[16] M.J. Mandell, I. Katz, and D.L. Cooke, 'Potentials on large spacecraft in LEO', IEEE Transactions on Nuclear Science NS-29, 1584 (1982).

[17] D.L. Cooke, I. Katz, M.J. Mandell, and J.R. Lilley, Jr., 'Three-dimensional calculation of shuttle charging in polar orbit', in Spacecraft Environment Interactions Technology-1983, AFLG-TR-85-0018, NASA CP-2359, 205 (1985).

[18] J.R. Lilley, Jr., I. Katz, and D.L. Cooke, '**v x B** and density gradient electric fields measured from spacecraft', Journal of Spacecraft and Rockets 23, 656 (1986).

[19] I. Katz, D.L. Cooke, D.E. Parks, M.J. Mandell, and A.G. Rubin, 'A three-dimensional wake model for low Earth orbit', Journal of Spacecraft and Rockets 21, 125 (1984).

[20] M.J. Mandell, T.T. Luu, and J.R. Lilley, 'Analysis of Dynamical Plasma Interactions with High-Voltage Spacecraft - Interim Report,', Maxwell Laboratories, Inc./S-Cubed Division Report SSS-DPR-90-11973 (1990).

[21] M.J. Mandell, T.T. Luu, J.R. Lilley, and D.L. Cooke, 'DynaPAC - A 3-D finite element plasma analysis code', presented at Physics Computing '91 Conference, June 10-14 1991, San Jose, California, Paper BF-26 (1991).

[22] M.J.M. Parrot, L.R.O. Storey, L.W. Parker, and J.G. Laframboise, 'Theory of cylindrical and spherical langmuir probes in the limit of vanishing Debye number', Physics of Fluids 25, 2388 (1982).

[23] M.J. Mandell, I. Katz, G.W. Schnuelle, P.G. Steen, and J.C. Roche, 'The decrease in effective photocurrents due to saddle points in electrostatic potentials near differentially charged spacecraft', IEEE Transactions on Nuclear Science NS-25(6), 1313 (1978).

[24] C.K. Purvis, N.J. Stevens, F.D. Berkopec, 'Interaction of large high power systems with operational orbit charged-particle environments', NASA TM X-73867 (1977).

[25] R.C. Chaky, J.H. Nonnast, and J. Enoch, 'Numerical simulation of sheath structure and current-voltage characteristics of a conductor-dielectric in a plasma', Journal of Applied Physics 52-12, 7092-7098 (1981).

[26] I. Katz and M.J. Mandell, 'Differential charging of high-voltage spacecraft: The equilibrium potential of insulated surfaces', Journal of Geophysical Research 87, 4533 (1982).

[27] G.A. Jongeward, R.A. Kuharski, E.M. Kennedy, K.G. Wilcox, N.J. Stevens, R.M. Putnam, and J.C. Roche, 'The Environment-Power System Analysis Tool development program', in Proceedings of the 24th Intersociety Energy Conversion Engineering Conference, August 6-11, 1989, Washington DC, 371-374 (1989).

[28] R.A. Kuharski, G.A. Jongeward, K.G. Wilcox, T.V. Rankin, J.C. Roche, 'High voltage interactions of a sounding rocket with the ambient and system-generated environments', IEEE Transactions on Nuclear Science 37(6), 2128-2133 (1990).

[29] G.A. Jongeward and V.A. Davis, 'Using EPSAT to analyze vehicle interactions with the space environment', to be published in proceedings for the 1991 Vehicle-Environment Interactions Conference, Johns Hopkins University, March 11-13 1991, Laurel, Maryland (1991).

[30] S.E. Deforest, 'Spacecraft Charging at Synchronous Orbit', Journal of Geophysical Research 77, 651-659 (1972).

[31] W.J. Raitt, Utah State University, Logan, Utah, private communication.

[32] I. Katz, G.A. Jongeward, V.A. Davis, M.J. Mandell, R.A. Kuharski, J.R. Lilley, Jr., W.J. Raitt, D.L. Cooke, R.B., Torbert, G. Larson, and D. Rau, 'Structure of the bipolar plasma sheath generated by SPEAR I', Journal of Geophysical Research 94, 1450-1458 (1989).

[33] N.B. Myers, W.J. Raitt, B.E. Gilchrist, P.M. Banks, T. Neubert, P.R. Williamson, and S. Sasaki, 'A comparison of current-voltage relationships of collectors in the ionosphere with and without electron beam emission', Geophysical Research Letters 16, 365-368 (1989).

THE HUMAN SYSTEM IN SPACE

JANIS H. STOKLOSA, Ph.D.
Manager, Biomedical Research Programs, Life Sciences Division, National Aeronautics and Space Administration, Washington, DC 20546

KATHERINE J. DICKSON
Research Associate, Science Communication Studies, George Washington University, Washington, DC 20006

1. Introduction

Successful human exploration of space depends upon integration and operation of propulsion, spacecraft engineering, communications, life support, and many other systems. Traditionally, aerospace research and development defines optimal, maximal, and critical performance limits that specify a set of operating parameters for each hardware and software system to ensure successful integration into the overall space system. Humans can be treated as a system which is a component of manned space systems. Human performance parameters can be measured and defined, and a critical set of operating parameters for the human system can be determined in order to ensure integration into the overall space system and mission success.

This human system analysis and application has been successfully implemented during National Transportation Safety Board (NTSB) aviation accident investigations and recommendations, and in the National Aeronautics and Space Administration's (NASA) aircraft operations [Refs. 1, 2]. Lack of attention to the human as a key system element has resulted in serious and even catastrophic accidents and incidents in operational non-space systems. Crew communication and coordination problems have been listed as a factor in 85% of all U.S. aviation accidents [Ref. 3], including the United Airlines DC-8 accident in Portland, Oregon on December 28, 1978 in which lack of adequate cockpit communication was cited by the NTSB investigation in the probable cause of the accident. Overreliance on automated systems has also been the cause of several serious aviation accidents, including the Eastern Airlines L-1011 aircraft accident in the Florida Everglades on December 29, 1972, where overreliance on the autopilot contributed to the accident resulting in the deaths of 94 passengers and 5 crewmembers [Ref. 4]. The incident at the Three Mile Island nuclear power plant in Pennsylvania in 1979 was caused by the failure to design systems incorporating human engineering considerations, by inadequate attention to the human-automation interface, and by deficiencies in operator training [Ref. 5]. Problems with group dynamics have also been responsible for the failure of past exploration missions, including crew mutinies during polar expeditions and oceanic voyages.

R. N. DeWitt et al. (eds.), The Behavior of Systems in the Space Environment, 731–778.

Spaceflight operations and research programs have documented unique physiological impacts of the space environment on humans. These include cardiovascular deconditioning, muscle atrophy, loss of bone mineral content, and neurovestibular, hematological, immunological, and endocrinological changes. However, the impact of the space environment on human behavioral and performance relationships is relatively unknown. It is possible to develop an operating envelope that defines critical human performance parameters for specific space missions or sets of operating conditions based on our knowledge of basic behavioral capabilities and human-operations, human-machine, human-environment, and human-human interactions. The integration of these profiles defines the operating envelope for a mission scenario.

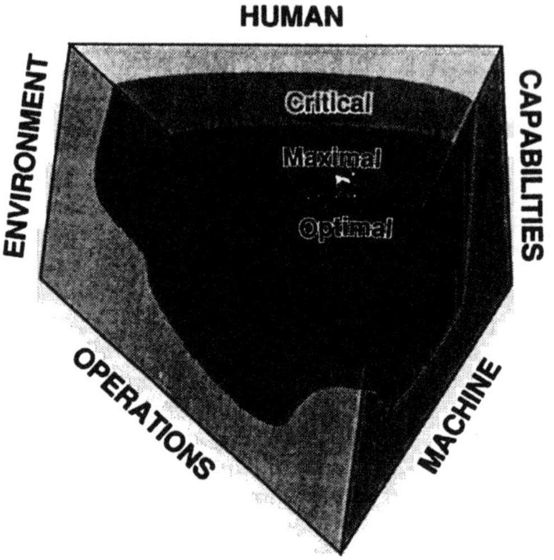

Fig. 1: The space human performance operating envelope defines the optimal, maximal, and critical performance limits.

The human-operations profile focuses on operational and procedural issues related to mission performance. It includes selection and training, mission task analysis, workload, fatigue, sleep, and work scheduling. The human-machine profile focuses on human-equipment interaction dynamics. It involves human-computer interface, automation, work station design, personal equipment, and telecommunications. The human-environment profile focuses on the intrinsic and extrinsic physical factors of the space environment that could affect performance. It is concerned with habitability, lighting, acoustics, vibration, and personal requirements. The human-human profile focuses on crewmember interaction dynamics. This includes crew coordination and communication, small group interpersonal relations, crew composition, and support systems. The behavioral capabilities profile focuses on the intrinsic human capabilities and limitations that are brought to a mission. This involves sensory, perceptual, and cognitive abilities, physical and performance capabilities, and psychophysiological relationships. Information from these profiles will provide a basis for treating the human system in a manner compatible with treatment of other space system components during design and operation of manned space systems, and will be used to establish system requirements, i.e., design requirements, procedures and protocols, and develop technologies that are within the space human performance operating envelope of the mission.

This chapter examines the role of the human as a system in the challenging and complex environment of space. It discusses the importance of space human factors and critical issues posed by space exploration, and provides a brief review of the current space human factors knowledge base. It also describes the NASA Space Human Factors Program, including research activities and the potential benefits to humans on Earth resulting from this research.

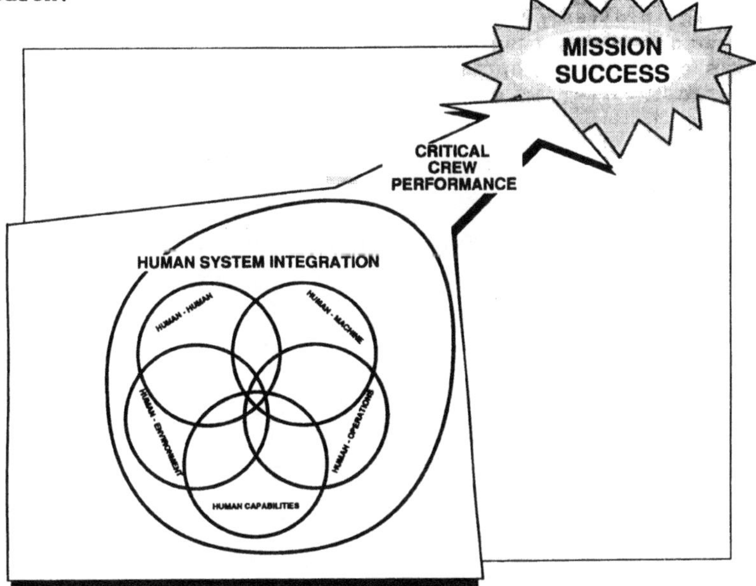

Fig. 2: The space human performance operating envelope defines critical human parameters to ensure mission success.

2. Space Human Factors Issues

More than thirty years have passed since the first human escaped the gravitational force of Earth and hurtled into space. Since then, considerable effort has been expended in assessing the dangers that spaceflight conditions pose to humans. The physiological changes that occur during spaceflight have been and continue to be extensively studied. Behavioral and psychological changes and the psychological needs of space crewmembers have also been studied, but they have not been given the same level of investigation, primarily because of the relatively short duration of most of the flights conducted to date.

Future manned space missions will, however, pose unprecedented challenges for the men and women who undertake them. These missions represent a continuum for the expansion of human civilization and scientific knowledge into the local solar system: space stations where crews will live and work in Earth orbit; a lunar outpost where, for the first time in history, crews will routinely work in a harsh extraterrestrial environment, conducting scientific experiments and eventually supervising complex operations such as in situ resource utilization; a manned mission to Mars, which will involve outbound and return flights of a combined duration of one to three years, including time spent working on the planetary surface; and, eventually, a Mars settlement.

The substantial investment in these missions requires significant advanced planning, founded upon a properly structured scientific knowledge base and a comprehensive picture of the effects of extended stays in space. This planning involves ground-based research and development in all areas of space human factors, from basic to operational, to make certain that key science issues requiring optimal human productivity over extended periods are addressed in order to ensure the success of these missions.

Fig. 3: Crews will work routinely in extraterrestrial environments such as the lunar outpost.

2.1. HUMAN-OPERATIONS

Humans in space play a vital and wide-ranging role in mission operations, from monitoring spacecraft controls to conducting and serving as test subjects in scientific experiments. Because of the importance of crew performance, several factors need to be considered when planning workload, task allocation, and scheduling. The workload for each crewmember must be balanced to allow sufficient time for recreation and rest. Scheduling workload depends on the amount and timeliness of work needed to be done, the amount of workspace available, and the skills required to do this work.

The importance of each of these variables needs to be weighed in planning a program to enhance crew performance. Task allocation between crewmembers, ground control, and machines must be investigated, including issues of crew cross-training. Task analyses of the different phases of a planned mission and human extravehicular activity must be conducted. In addition, methodologies for workload measurement and fatigue assessment must be established to determine adequate task allocation, work-rest schedules, and workshifts.

Circadian rhythms, sleep patterns, and quality of sleep may be altered in spaceflight, affecting the performance and moods (e.g., increased irritability and hostility) of crewmembers. On some Shuttle missions, sleep and work-rest schedules are arranged in two 12-hour shifts to allow continuous coverage, which may result in a shift or alteration in the natural circadian rhythm of some crewmembers. If a phase shift is required in space, operational or workload factors must be adjusted in order to avoid the fatigue and concomitant increase in judgmental and mechanical errors that may arise from sleep disturbances.

Criteria for crew selection and composition are critical for optimal crew mix for specific missions. In addition to a primary requirement of appropriate crew job skills, procedures and protocols for training effective teams in enhanced communication, crew coordination, and interpersonal dynamics also need to be developed.

2.2. HUMAN-MACHINE

Astronauts function in a variety of roles in relation to the spacecraft and its equipment, from being simply a passenger to a highly trained pilot or technician. Automation and human-machine interactions in the space environment need to be better studied and understood. Future spacecraft environments will be composed of a complex system of humans, computers with intelligent systems, robotics, and other sophisticated equipment. The balance between the use of human operators and the use of machines is critical. Automation and use of robotics may eliminate human error in the space exploration system, but it may also reduce creativity and problem-solving techniques that are unique to humans. Other human-machine interactions, such as computer-aided decision-making, need to be investigated to determine the optimal design of the human-machine system. The parameters for levels of automation and robotics and human-computer interface designs need to be fully understood before a manned three-year mission to Mars is undertaken.

Human anthropometric and biomechanic parameters should be considered when designing workstations and related equipment. The ratio of available workspace to crew size needs to be evaluated, especially for long-duration missions. Workstations must be designed to accommodate different body sizes and the human neutral body posture that occurs in weightlessness. Another concern is the design of space suits, gloves, and other equipment for extravehicular activity, to provide adequate protection from the harsh space environment while simultaneously allowing for optimal physical dexterity in performing intricate, critical operations. Individual differences must be considered when designing space suits, especially as crews become more heterogeneous and missions become more specific. Equipment may need to be designed or altered on a mission-specific basis to enable the completion of unique tasks. Compatibility between workstations, equipment, and the crew must be optimized to ensure efficient performance of tasks and crew safety.

2.3. HUMAN-ENVIRONMENT

Earlier spacecraft designs were based primarily on engineering criteria; little thought was given to the personal needs and requirements of the astronauts or cosmonauts. As flight missions increase in length, however, crewmembers will not only be travelling but will actually be living in space. Living quarters and other areas must be engineered for maximum efficiency and comfort, while taking into account the particular difficulties caused by weightlessness and other spaceflight factors. Factors such as privacy, noise levels, temperature, decor, lighting availability, and operation of hygiene facilities need to be understood and optimized in order to ensure productivity in environments that are almost wholly artificial.

Noise is a constant annoyance in spacecraft. Even during the cruise phase of a flight, it can be loud enough to disturb sleep and to interfere with waking activities and conversational speech. Certain kinds of noisy environments tend to adversely affect performance. The effects of noise on perception and performance, adaptation effects, and the fatiguing and annoying aspects of noise need to be determined.

Other habitability factors and personal requirements, such as design of living space, lighting, temperature, and privacy, are also important for maintaining crew health and productivity. Living space design parameters including the use of light, color, windows, decor, and temperature must be studied. Understanding the underlying requirements for privacy and finding

Fig. 4: The crew of Space Shuttle mission STS-51D live in close quarters during the spaceflight.

ways of meeting them, e.g., personal possessions or architectural arrangement, are important. Research is also necessary to establish habitat design requirements and mission operation scenarios.

2.4. HUMAN-HUMAN

Space stations, a lunar base, and a Mars mission will involve small groups of individuals living and working together for extended periods in isolated, confined, and high-risk environments. Extended exploration missions will involve multinational cooperation. Instead of the early homogeneous astronaut corps, crews will be composed of individuals of various cultural and professional backgrounds. As mission length, crew size, and crew heterogeneity increase, the psychological and social factors of a larger number of people working under close confines, for long periods of time, become more important in mission planning. Significant crew psychosocial issues include crew coordination and communication, command and control structure, leadership, and group dynamics and processes. These issues must be investigated in ground-based research to ensure crew stability and productivity during spaceflight.

The space environment is characterized by risk, isolation, confinement, and sensory deprivation. Psychological adaptation will result from prolonged exposure to these stresses. How individual crewmembers react to these stressors and the coping and intervention strategies that are required to counteract them need to be understood. Isolation and confinement, as demonstrated in space analog environments such as submarines and Antarctic camps, have led to a number of physiological and psychological manifestations. Understanding the effects of isolation and confinement is extremely important in alleviating or preventing possible anti-social and counterproductive behavior.

In addition to the stress imposed by long-duration spaceflight, there is an ever-present risk of a crisis onboard the spacecraft (such as the death of a crewmember or failure of the craft to respond to commands) or externally (the threat of an asteroid impact). During planetary exploration, there would be no chance of a timely rescue from an outside source; the crew would be completely isolated from outside help and therefore dependent upon itself. How individual crewmembers and the crew as a whole handle a crisis would be of utmost importance to the success of the mission, the physical and emotional health of the humans aboard, and even the survival of the ship and its crew. Crew support systems, including coping methods, monitoring techniques, and intervention strategies, must be identified to maintain individual and group psychological health, motivation, morale, and cohesion over long periods of time in very difficult circumstances.

Communication is critical to maintaining group cohesiveness. In stressful situations like spaceflight, word choice becomes extremely important to avoid ambiguity and misunderstanding. Stress and interpersonal dynamics may cause crewmembers to become edgy, and tone, speech rate, facial expressions, and other nonverbal cues that modify word usage may be difficult to discern in space. Physical factors such as noise levels and bulky spacesuits may interfere with communication. Reduced atmospheric pressure in a space cabin could alter sound waves such that vocal tone and quality are changed.

2.5. BEHAVIORAL CAPABILITIES

Humans have basic behavioral repertoires and intrinsic physical and psychological capabilities which define the limits of their performance capabilities within any operating system. These capabilities and the effects of the spaceflight environment on them need to be understood to ensure optimal performance and mission success.

The spaceflight environment contains a number of factors, including weightlessness, radiation, acceleration forces imparted during launch and re-entry, spatial and temporal disorientation, and the artificial cabin environment, that induce physiological changes in humans that may lead to decreased performance [Ref. 6]. Among the major physiological changes that occur in space are: (1) cardiovascular deconditioning, including the migration of body fluids to the upper body, resulting in an increased blood volume pumped through the heart; (2) a subsequent decrease in blood volume, including losses in blood volume, red cell mass, and hemoglobin content; (3) loss of bone mineral content, especially calcium, leading to diminished bone strength; (4) muscle atrophy, the loss of muscle tissue mass, strength, and endurance; (5) neurovestibular changes and the occurrence of space motion sickness early inflight; and (6) immunological changes, including decreased lymphocyte activation and endocrine changes. Postflight problems resulting from readaptation to Earth's gravitational force include orthostatic intolerance and additional neurovestibular changes. The effects of these changes on performance inflight are unknown. For example, the alterations in sensory perception and motor control potentially could interfere with task performance. Other physiological changes may also affect the ability of crewmembers to function and to complete their tasks effectively during missions lasting several years. Research is needed to determine if these physiological changes or the measures designed to counteract them have behavioral correlates that might become increasingly important with longer stays in space.

Understanding the functioning of perception, cognitive processes, vision, and motor coordination in the space environment is essential. In order to develop effective systems where humans interact with computer systems and perform a variety of tasks essential for smooth operation of the spacecraft and completion of scientific experiments, a more thorough understanding of the user's perceptual and cognitive abilities and their utilization in the space environment is needed. Methods and tools for defining cognitive tasks also need to be developed. Based upon the crew's capabilities and limitations, the requirements for preflight and inflight training need to be defined.

3. Current Knowledge Base

During the early years of manned spaceflight, life sciences concerns were concentrated on determining the physical needs of the astronauts, particularly in such areas as radiation exposure and physiological effects of microgravity. In contrast, very little attention was focused on behavioral adaptation to reduced gravity or the crews' psychological or social adjustment to spaceflight. As mission durations increased, more emphasis was placed on human factors concerns but research and objective data remain limited.

There is an extensive and comprehensive scientific knowledge base available that is relevant to space human factors. However, much of this base consists of a core of ground-based and aviation research and technology and data from isolated and confined environments, including undersea habitats, submarines, and polar stations. Actual space-based data in human factors are largely sparse, limited, and anecdotal. The specific effects of the space environment on behavioral processes and human performance have not been investigated in any systematic way. Behavioral scientists have had little or no direct access to flight and ground crews, which has resulted in a lack of systematic assessment of individual performance both in training and during missions. The paucity of such data makes it difficult to identify and evaluate factors that may influence human performance.

This review of the current state of available knowledge about human factors in the space environment will cover spaceflight studies and observations of the United States, the former Soviet Union, and other countries, and brief results obtained from analog environments and simulations (e.g., undersea habitats, polar stations, and aviation). It should also be noted that the majority of spaceflight data are from short duration missions.

3.1. HUMAN-OPERATIONS

A number of human factors studies have been conducted to investigate the relationship of sleep and spaceflight with performance. In general, physical performance disturbances have been attributed to abrupt, scheduled changes in sleep-wakefulness cycles from those occurring on Earth. Simulations have revealed that an evaluation of operations and procedures can identify problems that could not be remedied inflight. Work capacity inflight has been studied by the Russians, although there is little conclusive evidence from the reported research.

3.1.1 *Sleep, Circadian Rhythms, and Work-Rest Cycles*. Studies involving observation and measurement of sleep parameters have been conducted during flight. Poor sleep quality and fatigue were reported from the Gemini IV and Apollo 7, 8, 13, 14, and 15 missions [Refs. 7, 8, 9]. During lunar surface stays during the Apollo Program, sleep was generally inadequate. Electroencephalograms were recorded during Gemini 7, although operational problems resulted in limited data being collected [Ref. 10]. Electroencephalograms, electrooculograms, and head motions were recorded during three flights of Skylab. Results from the Skylab studies indicated that no significant changes in sleep latency (time required to fall asleep) were observed except for an increase at the beginning of the 84-day mission. Sleep duration was not significantly changed, but some changes in sleep stage characteristics were observed, though they did not result in performance capability degradation [Ref. 10]. A Spacelab 1 study that recorded eye movements during sleep showed that both the number of eye movements and percentage of rapid eye movement (REM) sleep were elevated early inflight, returning to normal later inflight [Ref. 11]. Thirty percent of U.S. Shuttle astronauts have requested sleep medication inflight, although none had a history of usage on Earth [Ref. 12]. One study revealed that readaptation to Earth's gravitational force is more disruptive to sleep than is adaptation to microgravity [Ref. 10].

During some space missions, crewmembers were required to shift their work-rest cycles by 12 hours, which can cause disturbances in circadian rhythms and lead to performance decrements, excess fatigue, and other negative effects. When sleep schedules were shifted from a 24-hour day set to local time at the launch site during the Soyuz program, cosmonauts experienced some degradation of performance and disturbed sleep [Refs. 13, 14]. During a Russian-Bulgarian mission to the Mir space station, electrophysiological recordings were taken of sleeping cosmonauts. Cosmonauts also answered questionnaires immediately upon waking regarding their sleep quality. Results indicated that changes occurred in sleep architecture or structure, including a decrease in REM sleep. An increase in the amount of time required to fall asleep was also observed [Ref. 15].

Fig. 5: Crewmember sleeps in weightlessness onboard STS-8.

An analysis of work-rest cycles of Salyut 6 and Salyut 7 crewmembers on a daily, weekly, and total-flight-duration basis was conducted. Overall, a decrease in the amount of time spent working and an increase in the amount of time spent sleeping were observed. A change in sleep-wake schedules led to deviations in the functional state of all crewmembers. It was concluded that specific work-rest cycles should be rigorously adhered to for optimum crew workload [Ref. 16].

STS-35 astronauts were exposured to bright light (10,000 lux) prior to launch to determine if this treatment would effectively shift the circadian rhythms of crewmembers required to undergo a 12-hour phase shift for a mission requiring two shifts working around-the-clock. Subjective reports following the flight indicated that during their inverted schedule, the crewmembers were able to sleep soundly during the day and remain alert during the night. Physiological measurements taken from astronauts following an aborted launch attempt that had been preceded by bright light treatment showed that the use of bright light was able to shift the circadian rhythm pacemaker regulating temperature and hormone levels [Ref. 17].

3.1.2 *Work Capacity*. The former Soviet Union has conducted many flight experiments on work capacity and motor coordination. Visual motor coordination and tracking ability investigations were conducted during Voskhod, Soyuz, and Salyut flights using an apparatus called RPS-2M which measures tracking. Studies on Soyuz flights indicated an increase in reaction time for visual motor-reaction in tracking tasks. Cosmonauts flying their second spaceflight had lower error rates than first-time flyers. A Salyut 7 experiment with a visiting crew indicated that tracking errors were elevated early inflight, then decreased as the flight progressed, restoring work capacity to its preflight level [Ref. 18].

Mental activity and work capacity were investigated on the Salyut 6-Soyuz 35 mission and on Salyut 7 using a device, the Balaton, which measures rate and accuracy of information processing. Results indicated an adequate level of performance, although individual differences were large, especially when the task complexity was high. There was a decrease in work capacity early inflight in some crewmembers, but later during the flight all parameters had returned to their baseline levels [Refs. 19, 20]. In the PROGNOZ test conducted onboard Mir, an instrument named Pleven-87 was used to study mental performance and sensory and motor reactions to sound and light stimuli [Ref. 21]. Results indicated that effective psychological performance was maintained throughout the flight [Ref. 22].

Mental workload capabilities, including evaluation of higher mental processes, reflexes and reaction time, and tracking ability were studied during the mission of a Bulgarian cosmonaut using the Pleven-87 and Zora instruments. Results revealed a high mental workload capacity during the flight and no changes in reflex locomotor activity. Improvements were observed in the precision and speed of complex sensorimotor activities in response to visual and audio signals and tracking ability. Time for mental processing, however, was slightly increased: the average duration of a mental operation was extended by 12 to 20 percent [Ref. 23].

Studies of crew movement and equipment stowage and deployment were conducted during the Spacelab Life Sciences-1 (SLS-1) mission, which flew in June 1991. Results indicated that handling of small items was more difficult in microgravity than on Earth. Problems with stowage of equipment included stowage of multiple items in one stowage compartment, excess packaging, discrepancies regarding stowed items, and the distribution of related equipment components across multiple stowage areas. Studies of crew movement through the tunnel connecting Spacelab with the Orbiter Middeck revealed that no difficulties occurred [Ref. 24].

3.1.3 *Analog/Simulation Studies*. In the Skylab Medical Experiments Altitude Test (SMEAT), three astronauts lived and worked in a mission-length simulation. Besides evaluating biomedical equipment in a low pressure environment, the simulation provided an opportunity to examine such human factors issues as Skylab procedures, layouts, data-handling, and communication systems. As a result of this evaluation, many systems and procedures were redesigned. SMEAT was a valuable tool for Skylab designers since it uncovered serious operational anomalies that could not have been remedied inflight [Ref. 25].

In 1977, NASA conducted a 10-day simulation study covering proposed operational activities for Spacelab both inflight and on the ground. The study, called ASSESS, Airborne Science/Spacelab Experiments System Simulation, evaluated a number of objectives related to payload operations procedures. The results revealed a number of inflight integration and check-out problems and underscored the importance of personnel having a clear understanding of their role in the mission [Ref. 26].

3.2. HUMAN-MACHINE

Although much research on space human-machine issues has been conducted on the ground, very little comprehensive research has been done inflight. A comprehensive review of generic requirements for space facilities and related equipment that directly interface with crewmembers can be found in NASA-STD-3000, Man-System Integration Standards [Ref. 27].

Achieving the optimal balance of humans and machines to perform the multitude of required mission tasks is a continuing challenge. Although space missions rely heavily on automation and computers to carry out most functions, unique human intervention has ensured the success of many flights. On both the Apollo 11 and Apollo 16 missions, emergency redirection of the spacecraft was required to avoid hazardous landings on the lunar surface. During a Skylab mission, the Commander and Scientist Pilot performed an extravehicular activity (EVA) to fix a disabled solar panel, thus restoring vehicle operation. Several instances on the space shuttle have demonstrated the crewmembers' ability to perform repair and deployment maneuvers not possible by machines [Ref. 28]. These few examples demonstrate the criticality of humans as operators onboard space missions and the unique capabilities they bring to the space environment.

3.2.1 *Control Devices and Display Technology*. During Shuttle flight STS-29, astronauts completed a questionnaire evaluation of two cursor control devices, a trackball and an optical mouse with a reflective pad, which had been developed in prior ground-based models and KC-135 flight evaluations. The crew reported that the trackball was too large and had too much "play," although the one piece design of the trackball was seen as an advantage over the multiple hardware pieces of the mouse [Refs. 29, 30].

During STS-41, objective performance data were collected by comparing two cursor control devices, a built-in trackball on a Macintosh laptop computer and an Altra Felix™ device. Subjective inflight comments revealed that the ball component of the trackball floated in the housing, causing errors, and that the Felix™ was inadequate for fine movements. Objective data analysis revealed that the Felix™ provides for fast pointing, but both the Felix™ and the trackball resulted in a high error rate [Ref 31].

During Shuttle flights STS-29 and STS-30, astronauts completed an onboard evaluation of the transflective laptop computer display under a variety of lighting conditions (e.g., direct and sun-shafting). The display was reported unacceptable by the astronauts, with off-angle viewing difficulties and image shifting reported as specific problems. During an evaluation onboard STS-41 of an active matrix computer display (not backlit), the crew reported that the resolution of the display was good, but external lighting was essential. This made the display unsuitable for future use. A similar evaluation of an electroluminescent display was completed by STS-41 astronauts. The display was judged far superior to the transflexive displays previously flown, and no problems were reported.

Fig. 6: Crewmember on STS-29 evaluates the laptop computer display on the GRiD laptop computer.

3.2.2 *Analog/Simulation Studies*. In recent years, automation has become an increasingly important issue in aviation. A number of aircraft accidents have occurred that have been caused, at least in part, by factors related to automation. Often, the problem was not a failure of automated equipment so much as human error in its operation and monitoring. Adequate design of the human-machine interface or balance to reduce manual workload of pilots has also been an issue. Studies of automation in MD-80 aircraft indicated that pilots do experience some reduction in total workload, but probably less than that claimed by automation proponents [Ref. 4]. While manual workload is reduced, mental workload is not. With the introduction of automation, some errors are eliminated while others are introduced. One study showed that general aviation pilots made more serious navigational errors in a simulator with an autopilot than without [Ref. 32].

Display devices are another significant human-machine interface issue. A lack of properly designed warning displays was cited as one cause of the Three Mile Island accident [Ref. 5]. As aviation cockpits become more technologically advanced, properly placed displays and use of automatic warnings have grown in importance. For example, the ground proximity warning system has frequently been known to sound false alarms [Ref. 4], which may lead to some pilots ignoring what might be real warning signals.

3.3. HUMAN-ENVIRONMENT

Human factors data were collected during each of the three Skylab missions to provide a knowledge base for designers and program planners for future space missions. The results of these investigations are reported in NASA Technical Memorandum X-59163 [Ref. 33] and in various Skylab Experience Bulletins [Ref. 34]. Information was obtained on several aspects of living in space, including the environment, architecture, mobility and restraint of crewmembers, food and drink, garments, personal hygiene, housekeeping, communication, and off-duty activities. Data were also gathered to understand the crew's capability to perform work in microgravity during long-duration space missions. Four specific areas were investigated, resulting in valuable procedural and design guidelines. These areas were manual dexterity, locomotion, mass handling and transfer, and inflight maintenance. The approach for the Skylab human factors evaluations included subjective techniques, such as questionnaires and interviews, along with objective measures, such as time and motion analysis and sound level measurement. These evaluations resulted in the removal of certain restraint and mobility aids from inclusion in the Space Shuttle.

Fig. 7: Waste management is performed by crewmember during spaceflight.

Currently, requirements for layout and design of the space station maintenance workstations, galley table, and shower are based on guidelines derived from the Skylab investigations.

Noise has been a continuing significant problem on many U.S. spaceflights. An Apollo 7 study revealed systems that both individually and collectively exceeded the NASA-specified acoustic level. Survey results from Shuttle Acoustic Noise studies conducted during the STS-1 and STS-2 missions revealed that some systems exceeded NASA acoustic guidelines but did not interfere with sleep or communication [Refs. 35, 36]. An investigation of human factors environmental issues, including noise and vibration, was conducted during the SLS-1 mission. The noise study utilized noninvasive questionnaires and an automated sound level meter to sample sound pressure levels in six different locations. Preliminary results indicated that noise levels exceeded NASA standards, especially in the Spacelab, and interfered with sleep, concentration, communication, and relaxation. Vibration was noticed by the astronauts, but it did not affect their performance [Ref. 24].

With respect to habitability issues and artificial life support of the spacecraft environment, inflight anecdotal and observational data have revealed that living space is confined, privacy is lacking, and facilities for personal hygiene are limited. In addition, noise levels are high and unpleasant odors abound. Food is restricted in quality and diversity, appetite decreases, and crewmembers become bored with the limited food choices [Ref. 37].

3.3.1 *Analog/Simulation Studies*. In preparation for Skylab, there were two primary efforts to examine human-environment interactions in an environment analogous to spaceflight. Grumman Aerospace Corporation and NASA co-sponsored a submarine mission, headed by Jacques Piccard, to gain some knowledge of what habitability might be like during Skylab missions. A six-man crew traveled undersea from Florida to Nova Scotia in 31 days. Several crew observations about the living conditions were incorporated into Skylab equipment design and procedures. In the mission-length SMEAT simulation, three astronauts lived and worked in an area similar to the living quarters and workstation of Skylab. During this simulation, operational, mechanical, and crew interface difficulties were found in some of the equipment, and they were redesigned prior to the mission. For example, the collection bags in the urine system were found to be often not large enough to hold a 24-hour collection of urine. A complete account of the work conducted in the design of Skylab's habitat is given by Compton and Benson [Ref. 25].

Analog studies have also shown the importance of such habitability issues as illumination and diet. Submarine studies showed that lighting sources must be carefully placed, since the majority of information used by crewmembers is received visually and the light available is almost completely artificial. Taking the design of the vessel into consideration to eliminate glare is also important. Illumination and color have been shown to alleviate depression of crewmembers during submarine missions [Ref. 38]. Studies onboard the undersea habitat Tektite revealed that mealtime was the major social event for the crew and that food preparation was of particular enjoyment [Ref. 38].

3.4. HUMAN-HUMAN

Interpersonal dynamics and other group-related issues have been the subject of inflight observational and anecdotal data. Information about crew interactions and the effects of spaceflight factors such as isolation and confinement also derive from studies conducted in analog environments.

3.4.1 *Interpersonal Dynamics.* Instances of tension have been reported between individual crewmembers and between the crew and ground control personnel [Refs. 39, 40]. These problems appear to occur more frequently with heterogeneous crews, especially with visiting international crews who may present language difficulties, and with extended missions. Problems have resulted from misunderstood communications. One Russian cosmonaut observed that in space each word assumed added importance, and statements were often interpreted wrongly and blown out of proportion [Ref. 41]. Another Russian cosmonaut commented that although conflicts did arise, they were quickly resolved and did not decrease the feelings of friendship that existed between crewmembers [Refs. 39, 42]. Information is also available from space crewmembers on such dynamics issues as crew size and leadership style: cosmonauts have stated that they prefer a shared leadership style [Ref. 40].

Crew selection and training techniques are used to help assure crew compatibility. An example is the former Soviet Union's rigorous training program, involving not just equipment training and preparation for the expected novel spaceflight environmental stresses; they also conduct emotional training exercises designed to foster self-reliance in dangerous situations [Refs. 41, 43]. Training for each mission usually occurs as a group, allowing crewmembers to become accustomed to one another and evaluations of group cohesion and compatibility to be made.

3.4.2 *Psychological Adaptation.* Much of our knowledge on the effects of the stresses of spaceflight on psychological adaptation comes from observational or anecdotal data. Factors such as isolation, confinement, living in a small group, constant danger, physiological changes, and habitability concerns can place a tremendous amount of stress on the individual crewmember. A long list of symptoms, including fatigue, irritability, depression, anxiety, mood fluctuations, boredom, tension, social withdrawal, and motivational changes has been compiled from the observational and anecdotal data [Ref. 44].

Psychological work capacity research on Salyut 6 using both subjective measures (fatigue, mood, and complaints) and objective measures (scope, time, and quality of work performed) identified four psychological stages that occurred throughout the 6-month mission: an acute adaptation stage, which lasted for the first 7 days and represents a period of initial adjustment to the new surroundings and conditions; a period of complete compensation, which lasted up to about the first three or four months of the flight and is a time of stabilization of mental and work capacity; a period of incomplete compensation from four to five months of flight; and the final "breakaway" period during the final month of flight [Ref. 45].

Although there are limited data in this area, it appears that psychological problems can result from the stress of spaceflight on both individual and group levels. There are, however, ways to alleviate or minimize potential difficulties through the use of psychological support measures. The former Soviet Union has actively pursued the use of these measures since the beginning of their space program to optimize crew mental health and well-being and therefore help ensure a successful and productive extended duration flight.

One such psychological support measure is the unmanned Progress ships, which carry new supplies to the space station. These supplies can include fresh fruits and vegetables, special foods, leisure activities, letters and videotapes from friends and family, and items requested by the crew. Communication between crewmembers and persons on Earth is also important. The Russians have used communication with friends, family, celebrities, public figures, and psychologists as a means of psychological support to help maintain the morale of the crew. The communications also serve in a monitoring role, aiding the ground control staff in monitoring the psychological state of the cosmonauts. One means for ground personnel to detect the level of stress of a cosmonaut during these communications is through voice analysis. This method involves analyzing speech pattern, tone, and speed to identify tension; this may cue support personnel to a potentially stressful situation [Ref. 46].

Another means of relieving stress is to make the crew's environment more habitable. Efforts utilized by the Russians to date include the use of pastel colors in the spacecraft interior, recordings of Earthlike sounds such as rain or bird songs, inclusion of windows, and use of lighting [Refs. 41, 47].

3.4.3 *Analog/Simulation Studies.* Spacecraft analog environments are studied to understand elements of interpersonal dynamics of groups living in isolated environments. Analog environments share several features with spacecraft and the space environment, including confinement, isolation, personal discomfort, and potential danger. The occurrence of dysfunctional psychological and psychosocial behavior has been reported during medium- and long-duration submarine missions [Refs. 48, 49, 50 51], in undersea habitats [Refs, 52, 53, 54, 55], and at polar stations [Refs. 56, 57, 58]. Data from these environments have revealed instances of impaired cognition, irritability, hostility, depression, boredom, fatigue, anxiety, miscommunication, and deterioration of individual behavioral adjustment.

Despite these observations, psychological disturbances have not been common and most groups were able to adapt and perform effectively. Several studies have helped identify some of the variables that enhanced or impaired group performance [Refs. 59, 60, 61, 62, 63, 64, 65]. Positive variables were gregariousness, role-sharing, privacy, and good leadership. Negative variables were lack of privacy, lack of social stimulation, lack of objective rewards, inability to communicate with people other than the crew, stagnant environment, family concerns, group adjustment problems, and an absence of sources of emotional satisfaction. The data and procedures derived from these studies are applicable to long-duration space missions in terms of crew selection and training and improved psychological support.

Underwater studies in human factors have been conducted in deep-sea hyperbaric chambers, e.g., on the Ben Franklin [Refs. 66, 67], Sealab II [Refs. 68, 69, 55], and Tektite II [Refs. 70, 71, 72]. Submersible simulator missions ranged from 15 to 60 days and housed both military and civilian personnel. These studies were conducted mostly in the 1960s and 1970s. Comprehensive overviews of these earlier studies on submarines may be found in Christensen and Talbot [Ref. 73] and in NASA CR-2496 [Ref. 74]. Barabasz [Ref. 75] provides a review of human factors data accumulated from polar missions.

U.S. studies have been conducted in land-based simulators built specifically to imitate the conditions to be endured during actual spaceflights. These studies, which were performed in the 1960s and 1970s and lasted between one to three months, tended to emphasize the roles of careful selection and pre-mission training in alleviating some of the anomalies observed within isolated and confined environments. Results were contradictory, however, and no definitive conclusions or recommendations could be drawn from the data [Ref. 73].

Aviation is another spaceflight analog environment that provides information about group dynamics. Approximately 85% of all aviation accidents are caused in part by failures of crew coordination and communications [Ref. 3]. Extensive research has been done in the cockpit on these issues. Training in crew coordination, also known as cockpit resource management training, has been shown to be successful in improving performance [Ref. 76]. Studies have also shown that crews that have flown together previously have fewer accidents than crews that have not flown together, due to better communications between crewmembers [Ref. 77]. According to the U.S. National Transportation Safety Board, the use of cockpit resource management training has played a significant role in preventing aviation accidents [Ref. 78].

3.5. BEHAVIORAL CAPABILITIES

The United States and the former Soviet Union have conducted spaceflight experiments on the effects of the space environment on physical performance criteria, especially sensory-motor disturbances, visual function, illusions, and proprioception. Early studies focused on perceptual and sensory phenomena, partly because over 50% of U.S. and Russian crewmembers experience space motion sickness [Ref. 79]. More recently, emphasis has shifted to perceptual-motor coordination and the effects of long-term exposure to the space environment on the behavioral capabilities of the crew. Most of the published accounts of manned space missions suggest that, with few exceptions, decrements in performance have been transient, occurring early in the mission. The perceptual, cognitive, and psychomotor performance of the crews apparently has been unimpaired in terms of mission completion [Refs. 80, 81, 82, 12]. Many of the reported problems can be related to learning new perceptual and motor skills in weightlessness, as well as to the spatial disorientation and space sickness that occur early in a mission. There have also been a number of contradictory reports on visual function.

3.5.1 *Perceptual-Motor.* The spaceflight environment differs physically from the Earth's environment in a number of ways, one of which is a readjustment of motor and perceptual skills in the weightless environment. In the U.S. Gemini program and during early Russian flights of the Vostok and Voskhod series, crewmembers reported experiencing spatial illusions and vertigo, and feeling as though they were "hanging upside down." These illusions were experienced especially early in flight and during periods of increased motor activity [Refs. 83, 84, 85, 86, 87]. Cosmonauts reported that some suppression of these illusions could be achieved by fixing one's gaze on a stationary object, by not moving, or by relaxation exercises [Ref. 85]. Soyuz 9 cosmonauts observed deterioration of the accuracy of their standard movements during their first few days of orbital flight. Motor responses frequently exceeded nominal levels for a given movement. During nearly all flights there was an initial slowness in accomplishing tasks and a diminished motor coordination during the adaptation to weightlessness. A sensation of slow movement of the hand while writing in weightlessness [Ref. 88] and apparent movement of objects within the visual field [Ref. 86] were also reported. These false sensations are transient, however, and usually of only a few minutes' duration [Ref. 88].

Time and motion studies conducted during the Skylab missions were aimed at determining how well crewmembers could perform specified tasks in microgravity over the course of long-duration spaceflight. It was concluded that the first inflight performance of a task generally took longer than the last preflight performance; performance adaptation was very rapid, however, with behavioral performance continuing to improve from beginning to end of all Skylab missions. Data taken at the end of the Skylab 84-day mission showed no evidence of a performance deterioration that could be attributed to the effects of extended duration spaceflight exposure [Ref. 89].

Proprioceptive illusions were experienced during the performing of deep knee and arm bends, especially immediately upon landing and with the eyes closed, following Shuttle flights STS 41-G and Spacelab 1. The investigators concluded that "the system that compares motor commands with resulting sensory inputs may no longer be able to distinguish correctly between movement of self and movement of the world" [Ref. 90].

A mass perception study conducted during Spacelab 1 found impairments in mass discrimination ability inflight due to the loss of information regarding weight and possibly also to maladaptation to weightlessness. Impairments in mass discrimination were also observed postflight due to maladaptation to Earth's gravity [Ref. 91].

3.5.2 *Neurovestibular.* During Spacelab 1, several neurovestibular experiments were conducted. Electromyographs recording from the gastrocnemius and soleus muscles during a sudden vertical fall in space revealed that the otolith-spinal reflex was inhibited immediately upon entering weightlessness and declined further during the flight, yet it was unchanged from preflight levels when measured postflight [Ref. 92]. Immediately upon returning to Earth, subjects experienced difficulty maintaining their balance after being subjected to falls. This phenomenon returned to normal within a few days [Ref. 93]. A study of positional awareness following flight revealed increased variability of limb position estimation and uncertainty of orientation, compared with preflight values [Ref. 93]. These investigators also studied oscillopsia (apparent displacement of visual targets during passive or voluntary head movement) and found that it was associated with disturbances in the gaze control system [Ref. 93].

Another Spacelab 1 experiment recorded the Hoffmann or H reflex from the soleus muscle in a test of the monosynaptic spinal reflex in conjunction with vertical linear acceleration to assess modification of utriculo-saccular function induced through prolonged exposure to microgravity. Subjects were exposed preflight, inflight, and postflight to sudden, unexpected falls. On the seventh day of flight, the H reflex amplitude showed different characteristics than the preflight and early flight values. Immediately postflight, a rebound effect was observed in the H reflex, but not in self-motion perception. Drops late in the flight were described by the astronauts as sudden, fast, hard, and translational. Late inflight and immediately postflight, the astronauts felt as though the floor was coming up to meet them, rather than as though they were falling towards the floor [Ref. 94].

Studies on STS-8 and STS-11 indicated that self-motion reports and eye reflexes during roll motion showed primary linear translation and reduced ocular counterrolling, compared with preflight levels [Ref. 93]. Data from studies of perceived self motion are consistent with the hypothesis that "signals from receptors that respond to linear acceleration are reinterpreted during adaptation to weightlessness" [Ref. 12].

3.5.3 *Vision*. The majority of sensory input in pilot operations during spaceflight is through the visual system. A form of functional myopia, similar to that experienced by submariners, has been reported. While this may be an appropriate adaptation to confinement in cramped quarters, it can impair distance judgment and spatial vision for distant objects. Data from visual function studies conducted in space and observational data have been somewhat contradictory, with some studies reporting no changes, others claiming increased acuity, and others describing decreased visual function.

The first quantitative vision studies by the U.S. were conducted during the Gemini 5 and Gemini 7 missions. They revealed no significant degradation or improvement in visual function [Ref. 95]. Visual studies conducted on the Apollo missions demonstrated a statistically significant decrease postflight in intraocular tension in all astronauts, compared with preflight values [Ref. 96].

The occurrence of light flashes experienced by astronauts was first reported during the Apollo 11 mission and was studied on several subsequent missions [Refs. 97, 98]. This phenomenon occurred following adaptation to darkness, with eyes either open or closed. It was thought that these flashes were the result of cosmic radiation particles passing through retinal cells. Retinal photography was used following Apollo 15 and Apollo 16 to look for possible retinal lesions or damage to retinal cells. These studies found no evidence of lesions or damage, but they did indicate some increase in retinal vasculature constriction [Ref. 96]. Studies were also conducted on Skylab 4 to measure frequency of visual light flashes in near-Earth orbit and to characterize the mechanism responsible for this phenomenon [Ref. 98].

Visual performance studies were conducted on Shuttle flights STS-5 through STS-8, STS-41B, 41C, 41D, 41G, and 51C. These investigations revealed no significant changes in phoria, eye muscle balance, eye dominance, critical flicker frequency, or stereopsis [Ref. 99]. No statistically significant changes in visual acuity were found [Refs. 100, 99]. Contrast sensitivity tests, however, produced slightly different results. One study found no significant changes in three measures of contrast sensitivity. Another study showed statistically significant individual differences in contrast sensitivity, with crewmembers showing different magnitudes of change at different spatial frequencies [Refs. 100, 99].

A German experiment conducted during Spacelab D-1 examined intraocular pressure in relation to the headward fluid shift that occurs during adaptation to microgravity. This study found that intraocular pressure increased by 30% one hour after launch, and was then followed by a decrease [Ref. 101].

Studies of visual function have also been performed during Russian spaceflights. The optical resolving power of cosmonauts' eyes and the number of eye movements per unit time were measured during Voskhod missions. It was concluded that the functional capabilities of the eye are subject to changes in terms of oculomotor function and visual work capacity in the weightless environment [Ref. 13]. The changes, however, were considered to be slight and insignificant in terms of the visual function required inflight [Ref. 88]. Decreases in visual contrast sensitivity during the first few days in orbit have also been reported.

Studies of spontaneous and evoked eye movement responses during Salyut 6 and Salyut 7 flights showed a suppression, in space, of the slow compensatory eye movements that normally occur with nodding of the head under conditions of gaze fixation. Nystagmic movements in opposite directions also occurred in space. These changes suggest the development of vestibular asymmetry inflight and the disruption of gaze fixation mechanisms in response to spaceflight [Ref. 102].

The Russian experience with visual function in space includes anecdotal reports of increased visual acuity, e.g., being able to see buildings and ships on the Earth's surface from space. Researchers have concluded that these visions constitute optical illusions or mental extensions of actual visual acuity, such as seeing the acute angle of a ship's wake and believing that the ship itself is visible [Refs. 13, 86, 39]. Experiments conducted on spacecraft have not revealed any scientific evidence for increased visual acuity [Ref. 13].

4. NASA Space Human Factors Program

NASA's space life sciences research was initiated in 1960 with the goal of enabling human survival in space. The Life Sciences Division is currently addressing the challenge of space exploration by supporting research to ensure human health, safety, and productivity in space missions. The Space Human Factors Program, a vital component of the Life Sciences Division, is fulfilling a lead role in this challenge by quantifying human adaptive and performance capabilities in the space environment. The mission of the Space Human Factors Program is to develop the knowledge base required to understand the basic mechanisms underlying behavioral adaptation to spaceflight and the capabilities and limitations of the crewmember in the unique environments that will be encountered during future missions. This requires an in-depth understanding of psychological and behavioral adaptation to space and the ways in which adaptive behaviors influence or affect performance. The Program also develops and validates system design requirements, procedures, and protocols that ensure the psychological well-being, safety, and enhanced productivity of space crewmembers.

SPACE HUMAN FACTORS PROGRAM GOALS

Understand the psychological, behavioral, and performance responses to space

Develop requirements, procedures, and protocols to enable safe, productive, and enhanced crew performance

Promote research and technology to improve the quality of life for humans on Earth

SPACE HUMAN FACTORS PROGRAM OBJECTIVES

Determine the acute and long-term psychological, behavioral, and performance responses to space

Determine the critical factors that affect behavior and performance in space and understand the underlying mechanisms

Develop and evaluate design requirements for space environments, equipment, operations, and crew support

Develop effective monitoring techniques, procedures, and protocols for maintaining and enhancing crew performance

Develop and validate ground-based models and/or analogs

4.1. APPROACH

4.1.1 *Methodology.* Over the next several decades, NASA will undertake intensive planning for human space exploration missions. The success of the NASA Space Human Factors Program will depend on the active participation of discipline scientists from the intramural and academic communities, the aerospace industry, and the international space community. The Program, as being planned, is of considerable scope and breadth and will contribute significantly to the development of the existing scientific and technological knowledge base and the development of the advanced requirements and technologies necessary to maintain and enhance human safety, well-being, and productivity in space.

The Space Human Factors Program reflects a multifaceted approach to understanding humans in the space environment, utilizing clinical, personality, social, behavioral, experimental, environmental, systems engineering, and psychophysiological approaches. Approximately one-half of the research funds allocated to the Space Human Factors Program is granted to extramural researchers at universities located within the United States. All research is reviewed for scientific merit and other criteria by experts in the relevant areas. The research involves the collection of qualitative and quantitative data and the development and testing of behavioral and performance models through ground-based studies in laboratories; analogs including the Antarctic and undersea habitat; aviation; mock-ups; and simulations; and during flights on the KC-135 aircraft for parabolic flights, the Space Shuttle, and Space Station Freedom when available.

4.1.2 *NASA Unique Facilities*. Unique resources at NASA's Johnson Space Center (JSC) in Houston, Texas include the Anthropometry and Biomechanics Laboratory (ABL), the Human-Computer Interaction Laboratory (HCIL), the Graphics Analysis Facility (GRAF), and the Remote Operator Interaction Laboratory (ROIL). Other facilities include a Shuttle simulation and training facility, the Weightless Environment Training Facility (WETF), which is a neutral buoyancy tank, and a KC-135 aircraft, that can provide brief periods of weightlessness.

The Anthropometry and Biomechanics Lab (ABL) determines human capabilities in altered gravity environments through the development of a database of static and dynamic anthropometric and biomechanic data. Human strength and motion data are used to develop techniques for performing work in space. The ABL is equipped with state-of-the-art video motion measurement equipment that allows simultaneous collection and analysis of physiological data such as electromyographs (EMG). The ABL also has force dynamometers, which can be used in the KC-135 aircraft and in the WETF, and has several force plates for measuring induced loads.

The Human-Computer Interaction Laboratory (HCIL) performs research on the interactions between humans and computers and on the effectiveness of advanced display and control technologies. It is equipped with a variety of computer systems and prototyping software with which cognitive and perceptual processes and information exchange between humans and computers can be studied. The HCIL is staffed with resources for developing cognitive models and is also equipped with video equipment to conduct usability evaluations and to record and edit video protocols.

The Graphics Analysis Facility (GRAF) provides high fidelity, three-dimensional, dynamic computer graphics modeling of the physical environment, including the human body through computer programs named PLAID and JACK. It predicts human-system interfaces by illustrating human machine positioning, strength, and motion. The images produced by the GRAF are used to analyze payload configurations, crew positions, and fields of view such as payload deployment or movement through Space Station Freedom.

The Remote Operator Interaction Laboratory (ROIL) studies the human interface to remotely operated equipment. It is equipped with two six-degree-of-freedom manipulators, a camera/television switcher console that allows the test conductor to control the images the operator sees, a Datacube computer for image enhancement, and a system for integrating controls and displays from a variety of outside sources.

Fig. 8: The KC-135 aircraft enables researchers to test equipment and theories in a weightless environment.

Specialized resources at NASA's Ames Research Center (ARC) in Moffett Field, California include full-function cockpit simulators and space station mock-ups in the Human Performance Research Laboratory and highly sophisticated aeronautical flight and air traffic control simulators in the Man-Vehicle Systems Research Facility. Other facilities at ARC include the Virtual Interactive Environment Workstation Laboratory (VIEWS), the Spatial Auditory Displays Laboratory, and the joint Army-NASA Aircrew/Aircraft Integration Program (A^3I/MIDAS) Laboratory.

A fully functional Boeing 727-200 simulator is housed in the Man-Vehicle Systems Research Facility. This full-motion simulator provides a representative commercial air transport cockpit environment and provides researchers with input regarding the pilot's perception of the operational environment. It has a dedicated experimenter's control laboratory capable of controlling and monitoring the simulator's operation and a computer-generated visual scene system that provides 4 channel, 120 degree field-of-view, out-the-window visual cues.

The Virtual Environment Facility (VIEWS) is a multi-sensory personal simulator and telepresence device which enables greatly improved situational awareness in complex spatial environments. The VIEWS Laboratory consists of a wide-angle stereoscopic display unit, glove-like devices for multiple degree-of-freedom tactile input, connected speech-recognition technology, gesture-tracking devices, and computer-graphic and video-image generation equipment. The facility enables improved scientific visualization interfaces for exploration of planetary surface data.

Fig. 9: The full-motion Boeing 727 flight simulator at Ames Research Center.

The Spatial Auditory Displays Laboratory consists of a 3-D sound hardware system that places different radio communication streams to separate virtual auditory positions around an individual. This laboratory conducts research into the underlying perceptual principles of auditory displays and the development of an advanced signal processor, emphasizing the role of auditory cues as a critical channel of information in complex spatial environments during periods of high visual workload or when visual cues are limited, degraded, or absent. The laboratory is also used to implement and test auditory display concepts that allow a pilot or crewmember to immediately, accurately, and inexpensively monitor three dimensional information, such as traffic location, through the use of three-dimensional sound.

The A^3I/MIDAS Laboratory uses MIDAS, a prototype human factors computer-aided engineering system, to evaluate aspects of crewstation design and operator task performance. MIDAS contains tools to describe the operating environment, equipment, and mission of manned systems, which are used to create static and dynamic models of human performance and behavior. It provides design engineers and analysts with interactive symbolic, analytic, and graphic components, which permit the early integration and visualization of human engineering principles by using embedded models and principles of human performance and behavior rather than by using human subjects.

4.1.3 *Spaceflight Opportunities*. Several spaceflight activities are currently underway or planned for national and international research in human factors. Assessments of human factors aspects of Shuttle mission operations and crew performance began with the first dedicated space life sciences mission, SLS-1, in 1991 and will continue to be conducted on future Shuttle missions. A mental workload and performance experiment was conducted in January 1992 onboard the International Microgravity Laboratory 1 (IML-1) mission. Human behavior and performance has been designated for the first time as a primary theme of a Shuttle mission, IML-2, which is scheduled for flight in 1994. Human workload, performance, sleep, and circadian rhythms will be comprehensively studied during that mission. Rhesus monkey behavior and performance studies are proposed to be investigated on SLS-3 and SLS-4, scheduled for flights in 1996 and 1998, respectively.

Fig. 10: Cutaway view of the Space Shuttle Spacelab.

As part of the U.S. Congressional mandate focusing on the 1990s as the "Decade of the Brain," a spacelab mission called Neurolab is being planned for 1998. This mission, dedicated to research in the brain and behavioral sciences, would provide access to the unique research environment of spaceflight for basic and applied studies in both human and animal brain and behavior research, and would enhance United States technological leadership and vitality in these fields. Neurolab would allow NASA to further the development of brain and behavior research by providing access to spaceflight to neuroscientists and behavioral scientists in the U.S. academic, government, and industrial communities.

4.1.4 *Interagency Cooperation*. The NASA Space Human Factors Program is proposing several interagency initiatives within the next decade. These initiatives could include research conducted jointly with the National Science Foundation (NSF) in the polar regions, and a collaborative project with the National Oceanic and Atmospheric Administration (NOAA) in undersea habitats. Interagency cooperation has also been initiated with the U.S. Air Force to investigate such issues as crew workload and performance in operational situations.

Under a Memorandum of Understanding between NASA and NSF, an initiative has been developed to conduct joint human factors research in the Arctic and Antarctic. Bases in Antarctica are an analog for extended space exploration in that they are characterized by a high degree of isolation and self-sufficiency, are located in an extremely rugged setting, and are maintained over long periods of time. This initiative will stimulate research leading to an improved understanding of psychological, behavioral, and physiological processes observed under extreme conditions. Research in polar analogs could focus on cognition and performance, small group interactions, stress and adaptation, and human physiology; and could also provide for the development of new or improved data collection and analytical methods.

Fig. 11: The undersea habitat (left) and bases in Antarctica (right) provide analogs for space research.

Cooperative projects are under discussion with NOAA to utilize the undersea habitat, "Aquarius," as a model for a contained space environment to study selected human factors, environmental health, and physiology issues. In human factors, use of this facility could allow researchers to investigate the effects of confinement on crew dynamics, automation, and habitability. The undersea habitat could also serve as a model for studying team dynamics associated with EVA-type activities in space.

4.1.5 *NASA Specialized Center of Research and Training*. In the 1990's, NASA Life Sciences is intending to solicit proposals for an extramural NASA Specialized Center of Research and Training (NSCORT) in human factors. Funding for this Center is expected to be approximately one million dollars per year for five years. The NSCORT will focus on ground-based research to advance basic knowledge, provide a scientific knowledge base, and generate effective strategies for solving specific problems within space human factors. This NSCORT program will be established as an integral part of the Life Sciences Division's Space Human Factors research plan, and will be expected to further the United States' scholarship, skills, and performance in space human factors and related scientific and technological areas. It will also enhance the pool of research scientists and engineers being trained to meet the challenges of the nation's commitment to prepare for future human space exploration missions.

4.1.6 *International Cooperation.* The NASA Life Sciences Program has developed strong international cooperations. Joint Working Groups in life sciences have been established between NASA and the Canadian Space Agency (CSA); the French Space Agency, Centre Nationale d'Etudes Spatiales (CNES); the European Space Agency (ESA); the German Space Agency, Deutsche Agentur für Raumfahrtangelegenheiten (DLR/DARA); the National Space Development Agency (NASDA) of Japan; and the former Soviet Union. The Space Human Factors Program is discussing a number of international collaborative projects. With DLR/DARA, CNES, and ESA, NASA is developing both ground and flight investigations in circadian rhythms, performance, and other human factors issues. The Rhesus Research Facility, being developed jointly by NASA and CNES, will investigate the influence of microgravity and other spaceflight variables on primate behavior. Cooperation with the Russians could include joint activities on Russian Cosmos biosatellites and on the Mir space station.

4.2. CURRENT FUNDED RESEARCH

In response to NASA's programmatic needs, research is currently being conducted in all space human factors areas within the space operating envelope to determine the critical limits of performance and develop design requirements and procedures for the human in the space system. Investigations are being conducted in the following areas: crew selection and training, mission work analysis, sleep and circadian rhythms, workload, small group dynamics, crew communication and coordination, crew composition, automation and computer interaction, EVA glove development, work station design, habitability, noise, stowage and deployment, performance capabilities, perception and cognition, virtual environments, and anthropometrics and biomechanics. A compilation of publications resulting from research funded by the Space Human Factors Program over the past 10 years has recently been published [Ref. 103].

4.2.1 *Human-Operations.* Dr. Schiflett of the Armstrong Laboratory, Brooks Air Force Base, is preparing an experiment planned for the IML-2 Spacelab mission. The experiment, which seeks to understand behavioral and performance issues of crews working and living in space, will utilize six tests from the Unified Tri-Service Cognitive Performance Assessment Battery, accepted by the Department of Defense and NATO. These tests were developed through standardized human performance testing based upon current theoretical models of human performance, and were selected based on space mission task analysis and the hypothesized effects of microgravity. Experimental protocols will examine decision-making, information processing, and perceptual-motor performance in microgravity. The primary objective is identification of the effects of microgravity, separate from those of fatigue and shifts in work/rest cycles, upon specific information processing skills affecting performance.

Armstrong Laboratory at Brooks Air Force Base and JSC staff have also begun a series of studies aimed at developing workload metrics for evaluating alternative system designs, quantifying the effects of partial-g and other environmental stressors on human performance, and developing computer models and simulations of human performance in space. This work, in conjunction with future studies, will serve as the foundation for developing an integrated set of methodologies for the evaluation of human performance and behavior in space.

Drs. Donchin and Kramer of the University of Illinois are examining a number of workload measurement techniques for use in spaceflight. One goal of the study is to identify which psychophysiological indices are predictors of workload performance. Results can then be used to determine the proper allocation of responsibility among operators and automated components of the system, assess the robustness of skill acquisition as a function of workload during training, and predict performance of crewmembers during high-pressure episodes of long-duration missions.

Evaluations of several operational factors are being conducted during several STS missions by Ms. Adam at JSC. Stowage and deployment of equipment, crewmember restraint techniques, and management of loose cables in microgravity are being studied. In addition, evaluations of the effects of the spaceflight environment on overall crew performance, activity scheduling, and translation through the Shuttle-Spacelab transfer tunnel are also being performed.

Sleep duration, sleep quality, and adequately designed work-rest cycles are especially significant in the stressful and unique environment of space where normal work-rest cycles are often shifted for mission purposes. Dr. Czeisler of Harvard Medical School is investigating the use of bright light in resetting the human biological clock. This research involves exposing subjects to 10,000 lux illumination during 8-hour night work shifts and comparing physiological parameters, such as body temperature and plasma cortisol levels, and cognitive performance of test subjects with those of individuals exposed to ordinary room light during the same work shift. Results have shown that bright light treatment adequately shifts these parameters and adapts these individuals to their new work-rest cycle [Ref. 104]. These procedures are being used to shift the work-rest schedules of crewmembers on selected shuttle missions [Ref. 17].

Dr. Monk of the University of Pittsburgh will conduct an experiment on IML-2 on sleep, circadian rhythms, and performance in space. The study will evaluate sleep using polysomnography and sleep diaries. Circadian rhythms will be measured using body temperature, urine volume, and urinary electrolyte and hormone concentration data. Mood and performance will be assessed using subjective measures of global vigor, and through a computer-based performance test. Separate analyses will be performed of the two crew shifts, day-shift and night-shift, to study "shift work adaptation" in addition to "time into mission" effects. This experiment will show how circadian rhythms are altered in space; how this alteration is correlated with sleep dysfunction and impairment in mood, vigilance, and performance; and how a phase shift in the sleep-wake cycle in association with spaceflight impacts the above parameters [Ref. 105].

Selection criteria are required to ensure crew coordination and optimum crew performance in space. "Select-out" criteria, a process by which unqualified individuals are removed from the applicant pool, are already established. NASA is now working on the development and validation of "select-in" criteria: identifying positive crewmember qualities, and selecting for individuals possessing those traits. Researchers Dr. Rose of the MacArthur Foundation and Dr. Helmreich of the University of Texas are investigating characteristics associated with personality and individual and group performance measures among astronauts. Data are being collected on psychological factors, including personality characteristics, psychomotor coordination, verbal and spatial aptitude, and ratings on nine job-related categories such as job knowledge, job performance, leadership, and teamwork. Models are being developed and tested for performance of semi-autonomous work teams. Preliminary results using NASA astronauts indicate that peer and supervisor ratings provide a reasonable prediction of performance [Ref. 106].

4.2.2 *Human-Machine.* Dr. Rudisill and the HCIL staff at JSC are currently conducting research to improve the design of human-computer interfaces during the use of computer-based medical procedures, with the goal of enhancing the human decision-making process. Cognitive models are employed to redesign the interfaces to medical procedures. Preliminary results have shown that human recall of a procedure based on a cognitive modeled interface results in superior performance, compared with traditional medical procedural training [Ref. 107]. Future research will examine human-computer interface issues, including explanation facility, time consuming data input, trust and acceptance variables, terminology diversity, and decision making errors.

Performance of crews during EVA is critical to mission success. Yet, despite this key role, the gloves traditionally used by astronauts during EVA are uncomfortable, lack dexterity, and result in crewmembers performing extended EVAs with a limited ability to perform fine motor functions. Research performed by Ms. Schentrup and Mr. Kosmo at JSC employs advanced laser scanning techniques to develop EVA gloves with enhanced dexterity. This improved measurement technique produces a numerical map of the hand, which is then used to form the innermost layer of the glove. This approach permits more accurate crewmember hand measurements, resulting in a more comfortable, better fitting glove. This technique also leads to decreased fabrication time, reduced overall glove production costs, and improved glove configuration repeatability. This will enable crewmembers to perform tasks in space more efficiently and comfortably [Ref. 108].

Dr. Wiener of the University of Miami is investigating the complex issues associated with the interaction of human crews and automated systems. The goal of this research is to understand, in a real-world setting, how varying levels of automation impact both team performance and group processes. The objectives of this research are to determine human design requirements for automated systems and to develop protocols for training users of automated systems. Analysis of a full-mission simulation is being conducted, in which performance of air transport crews in a highly automated aircraft (MD-88) was contrasted with similar performance in an aircraft equipped with older technology (DC-9). This analysis includes overall team performance, incidence of errors, workload ratings, dynamics of group processes, and the factors that contribute to errors and high workload. This line of research will be relevant to the design of missions and the training of crews for extended spaceflight missions where onboard automation will be necessary to supply the real-time informational support presently provided from the ground [Ref. 109].

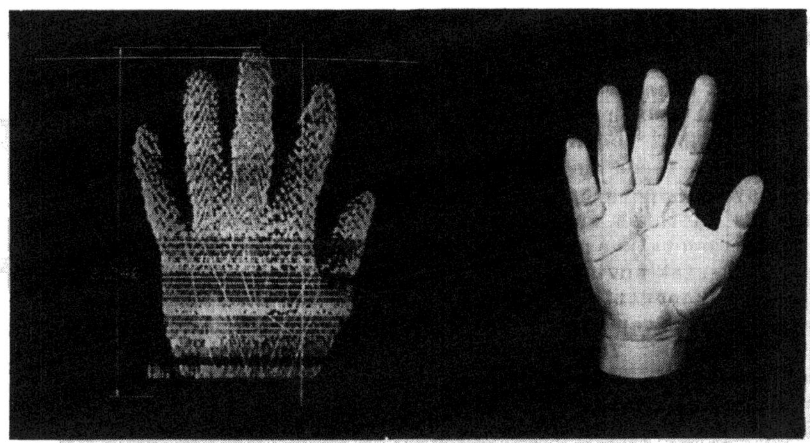

Fig. 12: Laser scanning techniques create a computerized map of the hand to aid in development of the EVA glove.

4.2.3 *Human-Environment*. Human factors evaluations of spacecraft environmental conditions are being conducted by Ms. Adam of JSC during several STS missions. Similar to human factors studies completed on Skylab, the techniques used are non-intrusive with existing operations. Specific objectives include evaluation of the effects of noise and vibration on SLS-1 crewmembers, communication, and mission objectives, and evaluation of maintenance and lighting.

Dr. Stuster of ANACAPA Sciences Inc. is developing design guidelines and operational recommendations concerning significant behavioral, psychological, and sociological issues associated with long-duration space missions from data from analogous conditions that are characterized by extremely long durations, e.g., remote Antarctic research stations and expeditions of research and discovery. Identifying performance measures for remotely monitoring the psychological adjustment of crew personnel to isolation and confinement is also part of this research. This work will provide design guidelines and recommendations for equipment, habitability requirements, and operational procedures to facilitate human performance and productivity in isolated and confined habitats, such as interplanetary spacecraft and lunar outposts [Ref. 110].

4.2.4 *Human-Human*. Dr. Holland of JSC is conducting research to improve our understanding of intercultural crew factors and to identify compatible crew characteristics for long-duration international missions. Research is underway to identify factors that may be unique to the confined, semiautonomous work/habitation environment of Space Station Freedom. In particular, this project proposes to develop a knowledge base of cross-cultural factors which shape the performance of work involved in multinational space missions and to identify the inflight and ground work tasks that can be optimized through cross-cultural training. Crew training approaches that optimize beneficial cultural differences and minimize problematic ones are also under investigation, and procedures and protocols to improve performance are being developed [Ref. 27].

Dr. Kanki of ARC is conducting crew factors research to understand the design and management of the social, organizational, and task factors that can affect crew productivity and well-being. Crew factors research emphasizes crew coordination and communication, examining real task-performing teams in operational environments that are analogous in critical ways to space operations. This includes such harsh or stressful environments as those encountered by aviation crews, undersea saturation divers, and mountaineering teams, as well as larger teams that work in complex, high-risk environments such as air and spacecraft maintenance and launch control operations. Research tools for analyzing audio and video tapes as well as interview and questionnaire techniques have been developed for this research. This research will contribute to the development of training protocols that can aid crews in reducing errors and in coordinating activities for achieving higher group productivity. It also contributes to the development of selection criteria related to group functioning [Ref. 111].

Research being performed by Dr. Helmreich is aimed at isolating the measurable aspects of personality that influence both individual and group performance. Studies focus on personality, demographic variables, and differences in task-related skills which characterize individuals in a group. Studies of social inputs include such variables as group size and group structure, including distribution of authority and tasks. Research issues also include the selection of individual crewmembers and the composition of crews to balance not only technical skills and specialty expertise, but also to enhance performance-related compatibility among crewmembers and leaders. The research has resulted in the development of rating instruments to evaluate crew coordination and individual and team performance of crews. In addition, decomposition and coding of behavior across time was developed to capture central elements in interpersonal communication and group processes that can be related to performance and mission success. The final outcome of these research activities will be the isolation of a group of personality constructs that relate to performance [Ref. 112].

Organizational factors and team performance are being investigated by Dr. Hackman of Harvard University. Data from airline crews are being analyzed to understand team performance in the aerospace environment. This research will provide data regarding the effects of organizational and social influence on the performance of flight crews; the data will be used to develop guidelines for design and operation of effective crews [Refs. 113, 114].

Team problem solving will be required during space missions to solve problems for which no packaged solutions are available. Models of crew performance are needed as a foundation for training and design of performance aids. Research performed by Dr. Orasanu of ARC has led to the development of methods to characterize crew problem solving and communication strategies [Ref. 115]. The focus of Dr. Orasanu's current work is on interactional processes that mediate overall crew performance. These include various aspects of communication, problem solving, decision making, and resource management. Variables that may influence these processes include crew familiarity, crew size, leader personality, gender, fatigue, and other stressors. The results of this work will advance the development of principles for selecting and designing crews, guidelines for preparing crews to cope with unforeseen problems in high stress environments, techniques for resisting effects of stressors, and principles for designing human-centered systems.

4.2.5 *Behavioral Capabilities*. Human performance in space includes the completion of complex tasks, and knowledge of crewmember capabilities and limitations during such operations is critical. Mr. Klute and the staff of the ABL at JSC are documenting crewmember operational reach capability in Shuttle flight suits under varying gravitational loads. Results from these investigations will be used to evaluate the effect of increased gravitational loads on reach capability, the impact of the current suit on crew mobility, and the use of ground-based simulators to conduct realistic inflight emergency training procedures. The ABL collects anthropometric data on all astronaut candidates. A database of over 500 subjects from this astronaut candidate population has been developed. The database is continuing to be developed to include range of motion and strength capability data. The ABL is conducting studies to quantify astronaut impact on the spacecraft microgravity environment from push off loads and exercising, as well as investigations to aid in determining the effectiveness of exercise countermeasures. The ABL is also conducting research to quantify crewmember operational reach and force transmission capabilities under various environmental conditions, including comparing the force transmission capability of high pressure space suit prototypes with that of the current Shuttle EVA suit [Ref. 116]. Results from this study will be used to design the next generation suit for exploration missions.

Mr. Maida and the GRAF laboratory staff at JSC are developing three-dimensional human models, which include data for reach, vision, body mass, and center of mass, using both computed and measured data and the computer models PLAID and JACK. Models will be used to conduct engineering evaluations of human-machine interactions, spacecraft configurations, viewing analysis, task performance scenarios, and anthropometric design criteria. The goal of this research is to develop a validated full-motion computer model of a human and the space environment. These data will benefit design engineers and operation planners in assessing engineering and human factors concerns for spacecraft design and operations [Ref. 117].

Because the vestibular system does not function normally in space, astronauts are forced to rely more heavily on vision for controlling their movements. Therefore a clear understanding of how humans use visual-motion cues to estimate self-motion and depth is critical for anticipating the deleterious effects of microgravity on performance. This fundamental knowledge is also critical for designing visual displays and training paradigms which may minimize or even counteract such effects. Drs. Stone and Perrone of ARC have developed a model of human self-motion estimation that is based on the known properties of motion sensitive neurons in the primate brain [Ref. 118]. They are now using this model to generate simulations of human perception under a number of sparse visual conditions

Fig. 13a: A computer model generated by PLAID at JSC, showing a Shuttle crewmember engaged in extravehicular activity.

Fig. 13b: Corresponding photograph taken in space shows PLAID's ability to realistically model on-orbit tasks.

similar to those encountered during EVA. They are also conducting human psychophysiological experiments to quantify actual human performance in heading and depth estimation tasks under the same sparse visual conditions. Because their model is biologically based and will be psychophysiologically validated, this research will lead to a better understanding of how the human brain processes self-motion information, to the development of a predictive tool for human performance in a host of aerospace environments, and also to the design of more effective displays and visual interfaces for aerospace environments.

Drs. Welch and Adelstein at ARC are using "virtual environments" to examine questions about human perception and performance that would prove either impossible or very difficult to test by means of "real-life" physical environments. This research has four specific objectives: to assess the usefulness of the virtual environment as a tool for the study of perception and performance by examining selected experimental topics in such environments and comparing the result to those obtained in analogous physical environments; to identify the necessary and sufficient stimulus conditions for creating virtual environments that are as functionally similar to physical environments as possible, by directly comparing the responses of subjects to virtual environments and to their physical counterparts; to determine how operations can best use adaptation to overcome the sensory and sensory-motor problems inherent in most virtual environment systems; and to extend knowledge about human perception and performance.

Drs. Rumbaugh and Washburn of Georgia State University are studying rhesus monkeys to assess the effects of microgravity and spaceflight on rhesus behavioral and psychological functioning. The behavioral and performance capabilities of rhesus monkeys parallel those of humans and therefore provide important behavioral model data. An interactive, computer-based testing paradigm has been developed that provides highly motivating activity for the animals, while permitting assessment of their psychomotor functioning, learning, attention, memory, perception, and problem solving. The research will determine the degree to which these performance measures correlate with and predict corresponding physiological indices of adaptation to microgravity. It will also provide environmental enrichment for the monkeys and continuously assess the psychological well-being of rhesus monkeys maintained for spaceflight or ground-based research [Ref. 119]. The ground-based data will form the foundation for inflight assessment of basic task performance in rhesus monkeys planned for the SLS-3 flight.

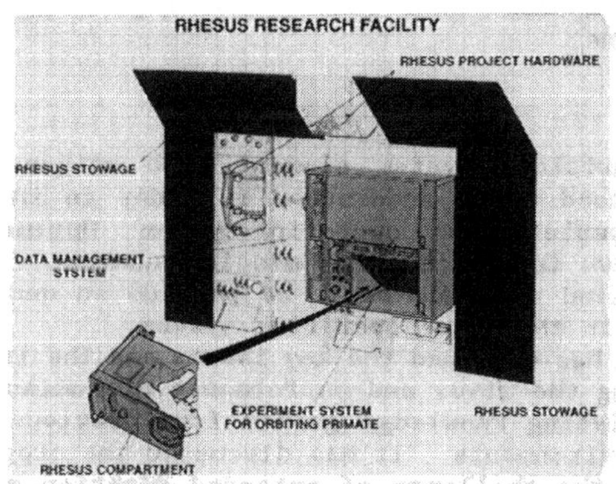

Fig. 14: Diagram showing the facility to be used on the Space Shuttle to study the rhesus monkey.

4.3. EARTH BENEFITS

Research and development activities sponsored by the NASA Space Human Factors Program have been developed to ensure the safety and productivity of humans in space. At the same time, they have resulted in spinoff applications that improve the quality of life for humans on Earth. Certain technologies, tools and techniques, systems, and programs that have been developed for use in space have secondary applications that require little or no change or that can be modified for a new use at a minimal cost. For example, PLAID software has been made available to the Computer Software Management and Information Center (COSMIC). This center, supported by NASA in conjunction with the University of Georgia, makes software developed by the government available to the private sector. PLAID can be used for a variety of modeling and engineering tasks such as equipment design, architectural layout, and traffic patterns. "Virtual environment" computer graphics software is now being used commercially and allows the human operator to explore and interact with artificial environments. This technology and its related systems can be used to create dynamic models of communication networks, traffic control systems, and even recreational games.

Research being conducted on circadian rhythms and the resetting of the human biological clock has direct application to individuals experiencing problems related to work schedules, to situations requiring continuous operations, and to the alleviation of jet-lag difficulties. Investigations into the learning and performance of rhesus monkeys have resulted in technologies that have directly benefitted studies of acquired brain damage and recovery of function in young children. Future space human factors research is expected to provide a variety of advances in the areas of psychological and medical health care, human resources, public safety, housing, training, transportation, communications, recreation, manufacturing and industrial worker productivity, environmental and resource management, and human work and safety in remote, unusual, or stressful conditions.

5. Summary

The human is a critical system element with unique capabilities and limitations that need to be understood in order to integrate the human element into the complete space operating system. Optimal functioning and performance of crews is critical for mission success. The human factors discipline as applied to space seeks to provide an understanding of the role of the human in the space operating system.

This chapter has outlined the key issues and the importance of human factors in ensuring the lives and performance of humans in space and has documented the existing knowledge gained from previous spaceflights and selected analog environments. It has discussed the program that NASA has developed to meet the challenge of extended duration missions and space exploration. The NASA Space Human Factors Program has a two-part goal: to understand the effects of the spaceflight environment on the psychology, behavior, and performance of crewmembers, and to develop the requirements, procedures, and protocols needed to enable safe, productive, and enhanced crew performance in space.

Future missions in space will be milestones in the history of humankind, accelerating science and technology and greatly expanding our knowledge of our home planet, the solar system, and the universe. Research and development sponsored by the Space Human Factors Program will not only ensure the health, safety, and productivity of humans in space, but will also lead to a better life for humans on Earth.

Acknowledgement

The authors wish to thank Dr. Timothy McKay of Lockheed Engineering and Sciences Company, Houston, TX for his valuable input in preparing this document, and his participation in the NASA session of the NATO Advanced Study Institute Program.

Fig. 15: Human factors research on the human in space will help to ensure a better life for humans on Earth.

6. Bibliography

[1] Stoklosa, J.H. Accident Investigation of Human Performance Factors. In: Proceedings of the Second Symposium on Aviation Psychology, Columbus, OH, April 25-28, 1983 (Jensen, R.S., ed.). Columbus, OH: Ohio State University, pp. 429-436, 1983a.

[2] Stoklosa, J.H. The Air Florida and Pan American Accidents - A Further Look. In: Second Aerospace Behavioral Engineering Technology Conference Proceedings P-132. Warendale, PA: Society of Automative Engineers, pp. 11-17, 1983b.

[3] Nagel, D.C. Human Error in Aviation Operations. In: Human Factors in Aviation (Weiner, E.L. and Nagel, D.C., eds.). San Diego, CA: Academic Press, Inc., pp. 263-304, 1988.

[4] Weiner, E.L. Cockpit Automation. In: Human Factors in Aviation (Weiner, E.L. and Nagel, D.C., eds.). San Diego, CA: Academic Press, pp. 443-461, 1988.

[5] Ford, D.F. Three Mile Island. Thirty Minutes to Meltdown. New York, NY: Viking Press, 1982.

[6] Nicogossian, A.E., Huntoon, C.L., and Pool, S.L. (eds.). Space Physiology and Medicine, Second Edition. Philadelphia, PA: Lea & Febiger, 401. pp., 1989.

[7] Berry, C.A. Summary of Medical Experience in the Apollo 7 through 11 Manned Spaceflights. Aerospace Medicine 41: 500-519, 1970.

[8] Goldwater, D. Informal Summary of Shuttle Medical Findings Presented during the Annual Meeting of the Aerospace Medical Association. San Antonio, TX, May 1981.

[9] Strughold, H. and Hale, H.B. Biological and Physiological Rhythms. In: Foundations of Space Biology and Medicine. Volume II, Book 2 (Calvin, M. and Gazenko, O.G., eds.). Washington, DC: NASA Headquarters, pp. 535-548, 1975.

[10] Frost, J.D., Jr., Shumate, W.H., Salamy, J.G., and Booher, C.R. Experiment M133: Sleep Monitoring on Skylab. In: Biomedical Results from Skylab (Johnston, R.S. and Dietlein, L.F., eds.). Washington, DC: NASA Headquarters, pp. 113-126, 1977.

[11] Quadens, O. and Green, H. Eye Movements During Sleep in Weightlessness. Science 225: 221-222, 1984.

[12] Parker, D.E., Reschke, M.F., and Aldrich, N.G. Performance. In: Space Physiology and Medicine, Second Edition (Nicogossian, A.E., Huntoon, C.L., and Pool, S.L., eds.). Philadelphia, PA: Lea & Febiger, pp.167-178, 1989.

[13] Leonov, A.A. and Lebedev, V.I. Psikhologicheskiye Osobenosti Deyatel'nosti Kosmonatov. Moscow: Nauka Press, 1971. (In Russian; English translation: Psychological Characteristics of the Activity of Cosmonauts, 1973. (NASA TTF-727))

[14] Leonov, A.A. Prospects for the Conquest of Space and Psychology. In: Psikhologicheskikh Poletov (Petrov, B.N., Lomov, B.F. and Samsonov, N.D., eds.). Moscow: Nauka Press, p. 28, 1979.

[15] Myasnikov, V.I., Polyakov, V.V., Zhukova, O.P., Ponmareva, I.I., Momand, A.A., Aleksandrova, A.A., and Stoilova, I. The Study of Cosmonauts' Sleep in Flight on Space Station Mir. In: Kosmicheskaya Biologiya i Aviakosmicheskaya Meditsina. Tezisy Dokladov IX Vsesoyuznoy Konferentsii. Kaluga, June 19-21, 1990, pp. 251-252. (In Russian; English Abstract: USSR Space Life Sciences Digest 29: 108-109, 1991.)

[16] Litsov, A.N. and Shevchenko, V.F. Psychophysiological Distinctions of Organization and Regulation of Daily Cyclograms of Crew Activities During Long-Term Spaceflight. Kosmicheskaya Biologiya i Aviakosmicheskaya Meditsina 19(2): 12-16, 1985. (In Russian; English translation: Space Biology and Aerospace Medicine 19(2): 12-18, 1985.)

[17] Czeisler, C.A., Chiasera, A.J., and Duffy, J.F. Research on Sleep, Ciracadian Rhythms and Aging: Applications to Manned Spaceflight. Experimental Gerontology 26: 217-232, 1991.

[18] Khachatur'yants, L.S., Ivanov, Ye.A., and Yepishkin, A.K. The Effect of Space Flight on the Characteristics of Pursuit Tracking. In: Kosmicheskaya Biologiya i Aviakosmicheskaya Meditsina. Tezisy Dokladov IX Vsesoyuznoy Konferentsii. Kaluga, June 25-27, 1986 (Gazenko, O.G., ed.). Moscow: Nauka Press, 1986. (In Russian; English Abstract: USSR Space Life Sciences Digest 15: 58, 1988)

[19] Clark, P. The Soviet Manned Space Program. New York, NY: Orion Books, 192 pp., 1988.

[20] Nechayev, A.P., Ponomareva, I.P., Khideg, Ya., Bognar, L., and Remesh, P. On the Additional Capacities of the Methodology for Studying Human Psychological Work Capacity (Based on Salyut-7 Results). In: Kosmicheskaya Biologiya i Aviakosmicheskaya Meditsina. Tezisy Dokladov IX Vsesoyuznoy Konferentsii. Kaluga, June 25-27, 1986 (Gazenko, O.G., ed.). Moscow: Nauka Press, 1986. (In Russian; English Abstract: USSR Space Life Sciences Digest 15: 56-57, 1988.)

[21] Radkovski, G.Iv. and Getzov, P.St. Study of Cosmonauts' Working Capacity by Means of Psycho-Physiological Methods and Instrumentation of Special Design. Paper presented at the 39th International Astronautical Congress of the IAF, Bangalore, India, October 8-15, 1988. (IAF Paper No. 88-486)

[22] Ioseliani, K.K. and Khisambeyev, Sh.R. Predicting Mental Performance of Cosmonauts on Long-Term Flights. In: Kosmicheskaya Biologiya i Aviakosmicheskaya Meditsina. Tezisy Dokladov IX Vsesoyuznoy Konferentsii. Kaluga, June 19-21, 1990, pp. 236-237. (In Russian; English Abstract: USSR Space Life Sciences Digest 29: 52-53, 1991.)

[23] Alexandrov, A. and Radkovski, G. Psycho-Physiological Studies during the Flight of the Second Bulgarian Astronaut. Paper presented at the 40th International Astronautical Congress of the IAF, Malaga, Spain, October 7-13, 1989, 3 p. (IAF Paper 89-586)

[24] Adam, S., Gosbee, J., Callaghan, T.F., Diaz, M., and Koros, A. 90 Day Report for DSO-904 Assessment of Human Factors for the SLS-1 Mission. September 25, 1991.

[25] Compton, W.D. and Benson, C.D. Living and Working in Space: A History of Skylab. Washington, DC: NASA Headquarters, 1983.

[26] National Aeronautics and Space Administration. NASA/ESA CV-990 Spacelab Simulation: Airborne Science/Spacelab Experiments System Simulation (ASSESS) II. Washington, DC: NASA Headquarters, 1977.

[27] National Aeronautics and Space Administration Behavior and Performance Laboratory Multicultural Shuttle Crew Study, 1989.

[28] Garshnek, V. and Brown, J.W. Human Capabilities in Space Exploration and Utilization. In: Space Physiology and Medicine, Second Edition (Nicogossian, A.E., Huntoon, C.L., and Pool, S.L., eds.). Philadelphia, PA: Lea & Febiger, pp. 364-380, 1989.

[29] Adam, S., Holden, K., Gillan, D., and Rudisill, M. Microgravity Cursor Control Device Evaluation for Space Station Freedom Workstations. In: Fourth Annual Workshop on Space Operations Applications and Research (SOAR '90) (R.T. Savely, ed.). Houston TX: NASA Johnson Space Center, pp. 582-587, 1991. (NASA CP-3103)

[30] Gillan, D.J., Holden, K.L., Adam, S., Rudisill, M. and Magee, L. How Does Fitts' Law Fit Pointing and Dragging? In: Proceedings of the Association for Computing Machinery's Computer Human Interaction (CHI) Conference. New York, NY: Association for Computing Machinery, 1990.

[31] Adolf, J. and Beberness, B. Payload and General Support Computer (PGSC) Detailed Test Objective (DTO) #795 Postflight Report -- STS-41. Houston, TX: NASA Johnson Space Center, 21 pp., 1991. (JSC-25459)

[32] Bergeron, H.P. Single Pilot AFR Autopilot Complexity/Benefit Tradeoff Study. Journal of Aircraft 18(9): 705-706, 1981.

[33] National Aeronautics and Space Adminstration. Skylab Experiment M487, Habitability/Crew Quarters. Houston, TX: NASA Johnson Space Center, 1975. (NASA TM-X-59163)

[34] National Aeronautics and Space Administration. Skylab Experience Bulletin No. 26. The Methods and Importance of Man-Machine Engineering Evaluations in Zero-G. Houston, TX: NASA Johnson Space Center, 1976. (JSC-09560)

[35] Homick, J.L. Cabin Acoustical Noise. In: STS-1 Medical Report (Pool, S.L., Johnson, P.C., and Mason, J.A., eds.). Houston, TX: NASA Johnson Space Center, p. 79080, 1981. (NASA TM-58240)

[36] Homick, J.L. Cabin Acoustical Noise. In: STS-2 Medical Report (Pool, S.L., Johnson, P.C., and Mason, J.A., eds.). Houston, TX: NASA Johnson Space Center, p. 22, 1982. (NASA TM-58245)

[37] Bluth, B.J. and Helppie, M. Soviet Space Stations as Analogs, 2nd Edition. Washington, DC: NASA Headquarters, 567 pp., 1986. (NASA CR-180920)

[38] Petrov, Yu.A. Habitability of Spacecraft. In: Foundations of Space Biology and Medicine, Volume 3: Space Medicine and Biotechnology (Calvin, M. and Gazenko, O.G., eds.). Washington, DC: NASA Headquarters, pp. 157-192, 1975.

[39] Sevast'yanov, V.I. The Appearance of Certain Psychophysiological Characteristics of Man Under Conditions of Space Flight. In: Psikhologicheskiye Problemy Kosmicheskikh Poletov (Petrov, B.N., Lomov, B.F., and Samsonov, N.D., eds.). Moscow: Nauka Press, pp. 29-37, 1979. (In Russian; English translation: Psychological Problems of Space Flights, pp. 41-54, 1979. (NASA TM-75659))

[40] Oberg, J.E. Red Star in Orbit. New York, NY: Random House, 1981.

[41] Bluth, B.J. Soviet Space Stress. Science 81 2(7): 30-35, 1981.

[42] Popovich, P.R. and Artyukhin, Yu.I. Certain Characteristics of Crew Activity Under Space Flight Conditions. In: Psikhologicheskiye Problemy Kosmicheskikh Poletov (Petrov, B.N., Lomov, B.F., and Samsonov, N.D., eds.). Moscow: Nauka Press, pp. 38-41, 1979. (In Russian; English translation: Psychological Problems of Space Flights, pp. 54-59, 1979. (NASA TM-75659))

[43] Lomov, B.F. Psychological Problems of Space Flight. In: Psikhologicheskiye Problemy Kosmicheskikh Poletov (Petrov, B.N., Lomov, B.F., and Samsonov, N.D., eds.). Moscow: Nauka Press, pp. 5-16, 1979. (In Russian; English translation: Psychological Problems of Space Flights, pp. 4-23, 1979. (NASA TM-75659))

[44] Bluth, B.J. The Benefits and Dilemmas of an International Space Station. Acta Astronautica 11: 149-153, 1984.

[45] Myasnikov, V.I. Mental Status and Work Capacity of Salyut-6 Station Crew Members. Kosmicheskaya Biologiya i Aviakosmicheskaya Meditsina 17(6): 22-25, 1983. (In Russian; English translation: Space Biology and Aerospace Medicine 17(6): 30-33, 1983.)

[46] Gazenko, O.G., Myasnikov, V.I., and Uskov, F.N. Behavioral Control as a Tool in Evaluating the Functional State of Cosmonauts in Flight. Aviation, Space, and Environmental Medicine 47(11): 1226-1227, 1976.

[47] Nicogossian, A.E. Human Factors for Mars Mission. In: The NASA Mars Conference. Proceedings of the NASA Mars Conference held July 21-23, 1986 (Reiber, D.B., ed.). Volume 71: Science and Technology Series. San Diego, CA: American Astronautical Society, pp. 475-485, 1988.

[48] Earls, J.H. Human Adjustment to an Exotic Environment. Archives of General Psychiatry 20: 117-123, 1969.

[49] Serxner, J.L. An Experience in Submarine Psychiatry. American Journal of Psychiatry 125: 25-30, 1968.

[50] Tansey, W.A., Wilson, J.M., and Schaefer, K.E. Analysis of Health Data from 10 Years of Polaris Submarine Patrols. Undersea Biomedical Research (Submarine Suppl.), pp. S217-S246, 1979.

[51] Weybrew, B.B. Psychological and Psychophysiological Effects of Long Periods of Submergence. 1. Analysis of Data Collected during a 265-Hour, Completely Submerged, Habitability Cruise Made by the U.S.S. Nautilus (SSN571). New London, CT: U.S. Naval Medical Research Laboratory, 1957. (Report No. 28188)

[52] Deutsch, S. A Man-Systems Intregration Study of the Behavior of Crews and Habitability in Small Places. In: Project Tektite 2: Scientists in the Sea (Miller, J.W., Van Derwalker, J.G. and Waller, R.A., eds.). Washington, DC: U.S. Department of the Interior, pp. VIII 1-14, 1971.

[53] Helmreich, R. Behavioral Observations in an Undersea Habitat. Office of Naval Research Technical Report No. 16, 18 pp., 1971.

[54] Helmreich, R.L. Psychological Research in Tektite 2. Man-Environment Systems 3(2): 125-127, 1973.

[55] Radloff, R. and Helmreich, R. Groups under Stress: Psychological Research in Sealab II. New York, NY: Appleton-Cherry-Crofts, 259 pp., 1968.

[56] Gunderson, E.K.E. Mental Health Problems in Antarctica. Archives of Environmental Health 17: 558-564, 1968.

[57] Gunderson, E.K.E. Individual Behavior in Isolated or Confined Groups. In: Man in Isolation and Confinement (Rasmussen, J.E., ed.). Chicago, IL: Aldine Publishing Company, pp. 144-164, 1973.

[58] Strange, R.E. and Klein, W.F. Emotional and Social Adjustment of Recent U.S. Winter-Over Parties in Isolated Antarctic Stations. In: Polar Human Biology (Edholm, O.G. and Gunderson, E.K.E., eds.). London: Heinemann Medical Books, pp. 410-429, 1973.

[59] Angiboust, R. Psycho-Sociological Problems of Small, Isolated Groups Working under Extreme Conditions. In: Life Sciences Research and Lunar Medicine (Malina, F., ed.). New York, NY: Pergamon Press, pp. 11-20, 1967.

[60] Chambers, R.M. Isolation and Disorientation. In: Physiological Problems of Space Exploration (Hardy, J.D., ed.). Springfield, IL: Charles C. Thomas Publisher, pp. 231-292, 1964.

[61] Chambers, R.M. and Fried, R. Physiological Aspects of Space Flight. In: Physiology of Man in Space (Brown, J.H., ed.). New York, NY: Academic Press, pp. 173-256, 1963.

[62] Kubis, J.F. Isolation, Confinement, Group Dynamics in Long Duration Spaceflight. Astronautica Acta 17: 45-72, 1972.

[63] National Aeronautics and Space Administration. The Effects of Confinement on Long Duration Manned Space Flights. Proceedings of Symposium, November 17, 1966. Washington, DC: NASA Headquarters, 53 pp., 1966.

[64] Smith, S. Studies of Small Groups in Confinement. In: Sensory Deprivation: Fifteen Years of Research (Zubek, J.P., ed.). New York, NY: Appleton-Century-Crofts, pp. 374-485, 1969.

[65] Trego, R.E. and Sells, S.B. Toward Selection for Long-Duration Isolated Missions: A Literature Review. Fort Worth, TX: Institute for Behavioral Research, Texas Christian University, 49 pp., 1970.

[66] Del Vecchio, R.J., Goldman, A., Phillips, C.J., and Seitz, C.P. Life Science Investigations during the Thirty-Day Gulf Stream Drift Mission of the Grumman PX-15 (Ben Franklin) Submersible. In: Aerospace Medical Association, Preprints of Scientific Program. Annual Scientific Meeting, St. Louis, April 27-30, 1970, pp. 222-223.

[67] Grumman Aerospace Corporation. Use of the Ben Franklin Submersible as a Space Station Analog. Bethpage, NY: Grumman Aerospace Corporation, 1970.

[68] Helmreich, R.L. Prolonged Stress in Sealab II: A Field Study of Individual and Group Reactions. New Haven: Yale University Press, 96 pp., 1967.

[69] Pauli, D.C. and Clapper, G.P. (eds.) Project Sealab Report: An Experimental 45-Day Saturation Dive at 205 Feet. Sealab II Project Group. Washington, DC: Office of Naval Research, Department of the Navy, 434 pp., 1967. (ONR Report DR-153)

[70] Helmreich, R.L. Patterns of Aquanaut Behavior. In: Tektite 2: Scientists in the Sea (Miller, J.W., Van Derwalker, J.G. and Waller, R.A., eds.). Washington, DC: U.S. Department of the Interior, pp. VIII 30-49, 1971.

[71] Helmreich, R.L. The Tektite 2 Human Behavior Program. JSAS: Catalog of Selected Documents in Psychology, 2 (13) ms. 70, 1972.

[72] Miller, J.W., Van Derwalker, J.G., and Waller, R.A. (eds.) Tektite 2: Scientists in the Sea. Washington, DC: U.S. Department of the Interior, 615 pp., 1971.

[73] Christensen, J.M. and Talbot, J.M. Research Opportunities in Human Behavior and Performance. Washington, DC: NASA Headquarters, 82 pp., 1985. (NASA CR-3886)

[74] George Washington University. Studies of Social Group Dynamics under Isolated Conditions: Objective Summary of the Literature as It Relates to Potential Problems of Long Duration Space Flight. Washington, DC: NASA Headquarters, 295 pp., 1974. (NASA CR-2496)

[75] Barabasz, A.F. A Review of Antarctic Behavioral Research. In: From Antarctica to Outer Space: Life in Isolation and Confinement (Harrison, A.A., Clearwater, Y.A., and McKay, C.P., eds.). New York, NY: Springer-Verlag, pp. 21-30, 1990.

[76] Orlady, H.W. and Foushee, H.C. (eds.). Cockpit Resource Management Training. Washington, DC: NASA Headquarters, 308 pp., 1987. (NASA CP-2455)

[77] Foushee, H.C. and Helmreich, R.L. Group Interaction and Flight Crew Performance. In: Human Factors in Aviation (Weiner, E.L. and Nagel, D.C., eds.). San Diego, CA: Academic Press, pp. 189-227, 1988.

[78] Carroll, J.E. and Taggart, W.R. Cockpit Resource Management: A Tool for Improved Flight Safety (UAL CRM Training). In: Cockpit Resource Management Training (Orlady, H.W and Foushee, H.C., eds.). Washington, DC: NASA Headquarters, pp. 40-46, 1987. (NASA CP-2455)

[79] Garshnek, V. Exploration of Mars: The Human Aspect. Journal of the British Interplanetary Society 43: 475-488, 1990.

[80] Gazenko, O.G., Genin, A.M., and Egorov, A.D. Major Medical Results of the Salyut-6-Soyuz 185-Day Space Flight. Vol. II. Session D-5 of the 32nd Congress of the International Astronautical Federation, Rome, September 6-12, 1981.

[81] Johnston, R.S., Dietlein, L.F., and Berry, C.A. (eds.). Biomedical Results of Apollo. Washington, DC: NASA Headquarters, 592 pp., 1975. (NASA SP-368)

[82] Johnston, R.S. and Dietlein, L.F. (eds.). Biomedical Results from Skylab. Washington, DC: National Aeronautics and Space Administration, 481 pp., 1977. (NASA SP-377)

[83] Homick, J.L. and Miller, E.F., II. Apollo Flight Crew Vestibular Assessment. In: Biomedical Results of Apollo (Johnston, R.S., Dietlein, L.F., and Berry, C.A., eds.). Washington, DC: NASA Headquarters, pp. 323-340, 1975. (NASA SP-368)

[84] Homick, J.L. Space Motion Sickness. Acta Astronautica 6: 1259-1272, 1979.

[85] Yakovleva, I.Ya., Kornilova, L.N., Tarasov, I.K., and Alekseyev, V.N. Results of Studies of Cosmonauts' Vestibular Function and Spatial Perception. Kosmicheskaya Biologiya i Aviakosmicheskaya Meditsina 16(1): 20-26, 1982 (In Russian; English translation: Space Biology and Aerospace Medicine 16(1): 26-33, 1982.)

[86] Yuganov, Ye.M. and Kopanev, V.I. Physiology of the Sensory Sphere under Spaceflight Conditions. In: Foundations of Space Biology and Medicine, Volume II, Book 2. (Calvin, M. and Gazenko, O.G., eds.). Washington, DC: NASA Headquarters, pp. 571-599, 1975.

[87] Yuganov, E.M., Gorshkov, A.I., Kasyan, I.I., Bryanov, I.I., Kolosov, I.A., Kopanev, V.I., Lebedev, V.I., Popov, N.I., and Solodovnik, F.A. Vestibular Reactions of Cosmonauts During the Flight in the "Voskhod" Spaceship. Aerospace Medicine (July): 691-694, 1966.

[88] Beregovoy, G.T. The Role of the Human Factor in Space Flights. In: Psikhologicheskiye Problemy Kosmicheskikh Poletov (Petrov, B.N., Lomov, B.F., and Samsonov, N.D., eds.). Moscow: Nauka Press, pp. 17-24, 1979. (In Russian; English translation: Psychological Problems of Space Flights, pp. 23-34, 1979. (NASA TM-75659))

[89] Kubis, J.F., McLauchlin, E.J., Jackson, J.M., Rusnak, R., McBride, G.H., and Saxon, S.V. Task and Work Performance on Skylab Missions 2, 3, and 4: Time and Motion Study - Experiment M151. In: Biomedical Results of Skylab (Johnston, R.S. and Dietlein, L.F., eds.). Washington, DC: NASA Headquarters, pp. 136-154, 1977. (NASA SP-377)

[90] Watt, D.G.D., Money, K.E., Bondar, R.L., Thirsk, R.B., Garneau, M., and Scully-Power, P. Canadian Medical Experiments on Shuttle Flight 41G. Canadian Aeronautics and Space Journal 31(3): 215-226, 1985.

[91] Ross, H.E., Brodie, E.E., and Benson, A.J. Mass-Discrimination in Weightlessness and Readaptation to Earth's Gravity. Experimental Brain Research 64: 358-366, 1986.

[92] Young, L.R., Oman, C.M., Watt, D.G.D., Money, K.E., Lichtenberg, B.K., Kenyon, R.V., and Arrott, A.P. M.I.T./Canadian Vestibular Experiments on the Spacelab-1 Mission: 1. Sensory Adaptation to Weightlessness and Readaptation to One-g: An Overview. Experimental Brain Research 64: 291-298, 1986.

[93] Young, L.R., Oman, C.M., Watt, D.G.D., Money, K.E., and Lichtenberg, B.K. Spatial Orientation in Weightlessness and Readaptation to Earth's Gravity. Science 225: 205-208, 1984.

[94] Reschke, M.F., Anderson, D.J., and Homick, J.L. Vestibulo-Spinal Response Modification as Determined with the H-Reflex during the Spacelab-1 Flight. Experimental Brain Research 64: 367-379, 1986.

[95] Gill, J.R. and Foster, W.B. Science Experiments Summary. In: Gemini Summary Conference. Washington, DC: NASA Headquarters, pp. 291-306, 1967.

[96] Hawkins, W.R. and Zieglschmid, J.F. Clinical Aspects of Crew Health. In: Biomedical Results of Apollo (Johnston, R.S., Dietlein, L.F., and Berry, C.A., eds.). Washington, DC: NASA Headquarters, pp. 43-81, 1975. (NASA SP-368)

[97] Osborne, W.Z., Pinsky, L.S., and Bailey, J.V. Apollo Light Flash Investigations. In: Biomedical Results of Apollo (Johnston, R.S., Dietlein, L.F., and Berry, C.A., eds.). Washington, DC: NASA Headquarters, pp. 355-365, 1975. (NASA SP-368)

[98] Hoffman, R.A., Pinsky, L.S., Osborne, W.Z., and Bailey, J.V. Visual Light Flash Observations on Skylab 4. In: Biomedical Results from Skylab (Johnston, R.S. and Dietlein, L.F., eds.). Washington, DC: NASA Headquarters, pp. 127-130, 1977. (NASA SP-377)

[99] Task, H.L. and Genco, L.V. Effects of Short-Term Space Flight on Several Visual Functions. In: Results of the Life Sciences DSOs Conducted Aboard the Space Shuttle, 1981-1986 (Bungo, M.W., Bagian, T.M., Bowman, M.A., and Levitan, B.M., eds.). Houston, TX: NASA Johnson Space Center, pp. 173-178, 1987.

[100] Ginsburg, A.P. and Vanderploeg, J. Vision in Space: Near Vision Acuity and Contrast Sensitivity. In: Results of the Life Sciences DSOs Conducted Aboard the Space Shuttle, 1981-1986 (Bungo, M.W., Bagian, T.M., Bowman, M.A., and Levitan, B.M., eds.). Houston, TX: NASA, Johnson Space Center, pp. 179-182, 1987.

[101] Draeger, J., Wirt, H., and Schwartz, R. Tonometry under Microgravity Conditions. In: Proceedings of the Norderney Symposium on Scientific Results of the German Spacelab Mission D-1 (Sahm, P.R., Jansen, R., and Keller, M.H., eds.). Köln, Germany: DFVLR, pp. 503-509, 1987.

[102] Gorgiladze, G.I. and Maveyev, A.D. The Effect of Weightlessness on Eye Movement Responses. In: Kosmicheskaya Biologiya i Aviakosmicheskaya Meditsina. Tezisy Dokladov IX Vsesoyuznoy Konferentsii. Kaluga, June 19-21, 1990, pp. 51-52. (In Russian; English Abstract: USSR Space Life Sciences Digest 29: 91, 1991.)

[103] Dickson, K.J. Space Human Factors Publications: 1980-1990, Washington, DC: NASA Headquarters, 16 p., 1991. (NASA CR-4351)

[104] Czeisler, C.A., Johnson, M.P., Duffy, J.F., Brown, E.N., Ronda, J.M., and Kronauer, R.E. Exposure to Bright Light and Darkness to Treat Physiological Maladaptation to Night Work. The New England Journal of Medicine 322(18): 1253-1259, 1990.

[105] Monk, T.H. Sleep and Circadian Rhythms. Experimental Gerontology 26: 233-243, 1991.

[106] Helmreich, R.L., Holland, A.W., McFadden, T.J., Rose, R.M., and Santy, P.A. Strategies for Crew Selection for Long Duration Missions. Paper presented at AIAA Space Programs and Technologies Conference, Huntsville, AL, September 25-28, 1990, 5 pp., 1990a. (AIAA Paper 90-3762)

[107] Gugerty, L., Halgren, S., Gosbee, J., and Rudisill, M. Using GOMS Models and Hypertext to Create Representations of Medical Procedures for Online Display. Paper presented at Human Factors Society 35th Annual Meeting, Santa Monica, CA, 1991.

[108] Spampinato, P., Cadogan, D., McKee, T., and Kosmo, J.J. Advanced Technology Application in the Production of Spacesuit Gloves. Paper presented at the 20th Intersociety Conference on Environmental Systems, Williamsburg, VA, July 9-12, 1990. (SAE Paper 901322)

[109] Wiener, E.L. Future Research in Automation and Related Areas. Paper presented at the Society for Automative Engineers Conference Managing the Modern Cockpit: A Symposium for Line Pilots, Dallas, TX, December 4, 1990.

[110] Stuster, J.W. Space Station Habitability Recommendations Based on a Systematic Comparative Analysis of Analogous Conditions. Washington, DC: NASA Headquarters, 217 pp., 1986.

[111] Kanki, B.G. Teamwork in High-Risk Environments Analogous to Space. Paper presented at the AIAA Space Programs and Technologies Conference, Huntsville, AL, September 25-28, 1990, 7 pp., 1990. (AIAA Paper 90-3764)

[112] Helmreich, R.L., Wilhelm, J.A., Grogorich, S.E., and Chidester, T.R. Preliminary Results From the Evaluation of Cockpit Resource Management Training: Performance Ratings of Flightcrews. Aviation, Space, and Environmental Medicine 61(6): 576-579, 1990b.

[113] Hackman, J.R. (ed.) Groups That Work (And Those That Don't). San Francisco, CA: Jossey-Bass Publishers, 512 pp., 1990.

[114] Hackman, J.R. and Helmreich, R. Assessing the Behavior and Performance of Teams in Organizations: The Case of Air Transport Crews. In: Assessment for Decision (Peterson, D.R. and Fisherman, D.B., eds.). New Brunswick, NJ: Rutgers University Press, 43 pp., 1987.

[115] Klein, G., Orasanu, J., and Calderwood, R. (eds.). Decision Making in Action. Ablex Publishers, 1991, in press.

[116] Klute, G.K., Probe, J., and Greenisin, M.C. Exercise Induced Disturbances to the Microgravity Environment. Washington, DC: NASA Headquarters, 1991. (NASA TM-104736)

[117] Woolford, B., Pandya, A., and Maida, J. Development of Biomechanical Models for Human Factors Evaluations. In: Fourth Annual Workshop on Space Operations Applications and Research (SOAR '90) (Savely, R.T., ed.). Houston, TX: NASA Johnson Space Center, pp. 552-556, 1991. (NASA CP-3103)

[118] Perrone, J.A. Model for the Computation of Self-Motion in Biological Systems. Journal of the Optical Society of America A9: 177-194, 1992.

[119] Rumbaugh, D.M., Richardson, W.K., Washburn, D.A., Savage-Rumbaugh, E.S., and Hopkins, W.D. Rhesus Monkeys (Macaca mulatta), Video Tasks, and Implications for Stimulus-Response Spatial Contiguity. Journal of Comparative Psychology 103(1): 32-38, 1989.

THE ARCING RATE FOR A HIGH VOLTAGE SOLAR ARRAY: THEORY AND EXPERIMENTS

D. E. HASTINGS
Dept of Aeronautics and Astronautics, MIT
Cambridge, MA 02139

ABSTRACT. All solar arrays have biased surfaces which can be exposed to the space environment. It has been observed that when the array bias is less than a few hundred volts negative, then the exposed conductive surfaces may undergo arcing in the space plasma. Ground based and flight experiments are reviewed, and common results are extracted. A theory for arcing has been developed on these high voltage solar arrays which ascribes the arcing to electric field runaway at the interface of the plasma, conductor and solar cell dielectric. The theory is compared in detail to the flight experiment and shown to give a reasonable explanation for the data. Upcoming flight experiments to expand the arcing data base are described.

1. Introduction

For missions in space there are a variety of possible choices for power generation. For short-time, low-energy uses, the best power systems are electrochemical systems such as fuel cells. For long-term, low-energy uses radioisotope systems provide reliable power. For missions within the orbit of Mars, which lie in the range of 0.1 kW to 100 kW and have mission times from tens of days to many years, the power system of choice is a solar array. Historically, solar array systems in space have operated at low voltage. For example, most satellites have a solar array bus voltage of 28 Volts. The required power is generated by adding low voltage solar cell panels in parallel. However, future systems will require much higher operating voltages from the solar arrays. This is for two reasons. The first is that the resistive loss for a given power in the power distribution system is smaller for high voltage than for high current. The second related reason is that the mass of the power cables can be reduced by distributing the power at high voltage rather than high current. Proposed systems range from 160 Volts for the Space Station Freedom to thousands of volts for orbit transfer vehicles using solar electric propulsion.

The solar cells on an array usually have some part of the metallic interconnector between cells or part of the semiconductor exposed to the space environment. This is often by design since this allows the arrays

R. N. DeWitt et al. (eds.), The Behavior of Systems in the Space Environment, 779–794.

to be flexible and to deal with the problem of thermal expansion when the array enters and leaves eclipse. Even if all parts of the array are initially covered with an insulator, the possibility of particulate strikes in orbit will leave some parts of the interconnector or semiconductor exposed to the space environment after a period of time. All surfaces of any array which are exposed to the space environment and which can allow the passage of some current will collect current from the space plasma. In steady state the net current from the space plasma to the array and to any part of the spacecraft surface that the array is electrically connected with must be zero. This is the condition of no net charging and comes directly from the Poisson equation and Ampere's law. Since the electron mobility allows the electron random current flux to a spacecraft to be much larger than the ion current, most of the area of the array and attached conductive structure must collect ions so that the net current to the array and spacecraft structure will be zero. Hence the normal situation for a solar array electrically connected to a spacecraft will be for most of the current collecting surfaces on the array to be negatively biased with respect to space and therefore collect ions while the rest of the array will be positively biased and collect electrons.

In this paper we shall concentrate on the interaction of the negatively biased parts of a high voltage solar array with the space environment. The term "high voltage" used here is an operational term taken to mean that the array voltage is above a hundred volts so that some part of the array will expose a bias less than -100 Volts to the space environment.

High voltage solar arrays have been found to undergo two distinct sets of interactions with the space environment above a given threshold [Refs. 1,2,3]. It is found for the positively biased parts of the array that the current collection from the space environment can be anomalously large. For large voltages this can be a significant leakage power loss. For the negatively biased parts of the array it is observed that below a critical voltage, arc discharges occur on the solar array. These arc discharges give rise to electromagnetic noise and may also damage the solar cells [3]. Laboratory experiments where large arcs have been induced have shown substantial damage around the arc location.

2. Review of Experimental Work

A review of the laboratory experiments with respect to arcing has been given in Ref. [4]. Arcing in the experiments has been defined as a sudden current pulse much larger than the ambient current collection and typically lasting a few microseconds. The arcs are highly localized phenomena which have been observed to occur at the edge of the coverglass. They are sometimes, but not always, associated with the interconnector [Ref. 5].

Experimental work has been undertaken in ground based plasma chambers [Refs. 6,7,8,9] as well as in two flight experiments, the plasma interactions experiments I and II (PIX I and II) [Refs. 10,11]. The plasma and neutral gas environments in the plasma chambers are typically Argon or Nitrogen with a pressure range of 10^{-7} to 10^{-5} torr. The plasma density range is from 10^3 cm^{-3} to 10^6 cm^{-3}. The ram energy of the ions is 5 eV while the electron thermal energy (T_e) is in the range 0.05 to 0.1 eV.

The experimental work suggests the following observations: firstly, the key elements involved in the discharge process are the solar cell coverglass, the metallic interconnect and the plasma environment. This can be deduced from a set of experiments by Fujii et al [Ref. 9]. In their experiments a metallic plate biased to highly negative voltages was exposed to the plasma in a plasma tank. No arcing was observed except for very large negative voltages where the arc took place to the substrate. The plate was then partially covered with silica coverglasses. When actual solar cells were used with coverglasses, the arcing results were qualitatively similar. Secondly, there is a prebreakdown electron current that flows away from the interconnect prior to a discharge. This electron current was observed in the experiments of Fujii et al [Ref. 9] as well as the experiments of Synder [Ref. 7]. Calculation of the electron trajectories in the electric field of the interconnects indicate that the electrons must be coming from the interconnect and not from the plasma [Ref. 12].

The sequence of events associated with the arcing is the following: if a conductor with an insulating coverglass attached is put into the plasma and the conductor is initially not biased, then both the conductor and coverglass accumulate a negative charge. This occurs since the electrons are more easily collected than the ions. The final negative potential achieved ($V_{surface} \simeq -4T_e$) is such as to make the net electron flux equal to the net ion flux. If the conductor is now biased to a large negative potential, then the cover glass initially takes the same potential as the conductor and then slowly returns to a slightly negative potential with respect to the plasma. This is because ions are initially attracted to the system and accumulate on the surface of the coverglass. The potential drop between the conductor and the surface of the coverglass is almost equal to the bias voltage ($|V|$) where C is the capacitance per unit area of the coverglass. Below a certain voltage an electron current is observed to flow from the conductor. Some of the electrons leaving the interconnect strike the coverglass. There are two pieces of evidence which support this observation. First, there is observed to be a potential barrier over the interconnect [Ref. 3] which tends to keep the electrons from escaping to space. Secondly, the surface potential of the coverglass is observed to undergo frequent fluctuations towards negative potential [Ref. 13]. As the bias on the conductor becomes more negative, the magnitude of the voltage fluctuations on the coverglass increases, and below a critical voltage a discharge occurs from the interconnect. The discharge time is typically a few microseconds [Ref. 7], and once the discharge occurs the surface charge on the coverglass is neutralized, and the coverglass potential becomes the same as the conductor potential [Ref. 7]. The sequence now repeats itself with the coverglass slowly reaccumulating positive charge from the plasma.

The flight data from the PIX I and II experiments were taken at 900 km and show similar results to the ground based data. The current collection to the interconnects was measured to scale linearly with the voltage. There was substantial difference in the arc rate as compared to the ground based data. The arc rate R was measured [Ref. 14] to scale as

$$\dot{R} \sim n_i (T_i^{1/2}/m_i^{1/2}) V^a \qquad (1)$$

where $a \simeq 3$ for flight data and $a \simeq 5$ for ground data. In Eq. 1 all quantities with subscript i refer to ions. The ambient plasma density is n, the ambient temperature is T and the mass of the ambient plasma particles is m. The dependence of the arc rate on these quantities can be explained as due to the recharging of the coverglass surface by the thermal flux of ions. The thermal ion current scales as $n_i (T_i^{1/2}/m_i^{1/2}) r_s^2$ with a sheath radius r_s.

The voltage threshold appears to be of the order of -200 to -250 Volts in some ground based tests [Ref. 13] and -400 to -500 Volts in other tests [Refs. 3,9]. It has been suggested that the voltage threshold depends on the plasma density [Refs. 9,12], although the data range is not large and the evidence for a density dependence is ambiguous. If it exists it is very weak.

The results are shown in Fig. 1 along with the PIX-II ground experiments as well as data from two other ground based experiments. The results show that there does appear to be a threshold for arcing in the range of -200 Volts. For the PIX-II flight data it was found to be independent of the number of biased segments. The results also show the clear difference between the ground and flight data. This is due to facility effects such as secondary electron emission from the tank walls as well as the fact that the tank tests do not exactly simulate the orbital conditions. Typically the ground based tests are done at higher pressure and plasma density as well as in Argon or Nitrogen plasma as opposed to Oxygen. In addition, in orbit the ions ram into the solar cell and can therefore recharge the cells more rapidly than in the ground based experiments where thermal ions are used.

Recently, well diagnosed experiments have been carried out which have identified key factors contributing to the arcing of negatively biased high voltage solar cells [Ref. 15]. These experiments have led to a reduction of greater than a factor of 100 in the arc frequency of a single cell following proper remediation procedures. The experiments naturally led to and focussed upon the adhesive/encapsulant that is used to bond the protective coverglass to the solar cell. An image intensified CCD camera system recorded UV emission from the arc events which occurred exclusively along the interfacial region, and showed a bead of encapsulant along this entire edge. Elimination of this encapsulant bead reduced the arc frequency by two orders of magnitude. Water contamination was identified as a key contributor which enhances arcing of the encapsulated bead along the solar cell edge. Spectrally resolved measurements of the observable UV light showed a feature assignable to the OH (A-X) electronic emission which is common for water contaminated discharges. Experiments in which the solar cell temperature was raised to 85° showed a reduced arcing frequency, suggesting desorption of water. When the solar cell was exposed to water vapor the arcing frequency increased with all other parameters kept constant. Clean dry gases such as O_2, N_2 and Ar showed no enhancement of the arcing rate. Elimination of the exposed encapsulant eliminated any measurable sensitivity to the water vapor.

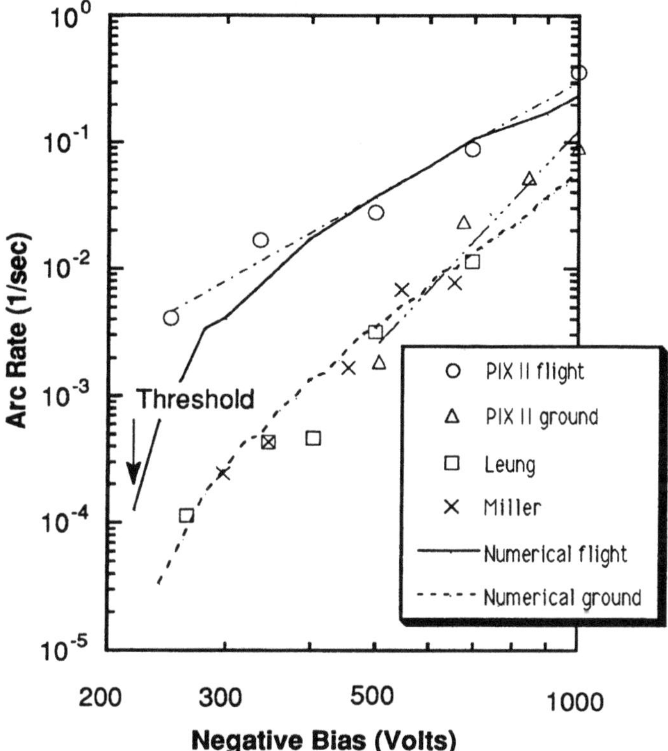

Fig 1: Experimental data for ground and flight experiments.

3. Theory

Two theoretical ideas have been published to explain the arcing on high voltage solar arrays. The first is in Ref. [12]. In this work it was proposed that there is a thin layer of insulating contaminant over each of the interconnects. Such contaminant could arise by exposure to air or be created in the manufacturing process. Ions from the space plasma are attracted by the negative potential on the interconnects. These ions accumulate in the surface layer resulting in a buildup of electric field in the layer. As the layer continues to charge, the internal field becomes large enough to cause electron emission into the space plasma. This electron current leads to subsequent heating and ionization in the layer. This is what is seen as the discharge.

The second hypothesis was advanced in Ref. [4]. There it was proposed that the prebreakdown current observed experimentally causes neutral gas molecules to be desorbed from the sides of the coverglass over the solar cells. These molecules build up over the interconnects, and arcing occurs inside this surface gas layer as a flashover discharge. An expression for the voltage threshold was derived, and the scalings with the gas and geometric properties were examined. The voltage threshold was independent of the plasma density and depended strongly on the geometric structure of the solar cell-interconnect connection.

In recent work, Cho and Hastings [Ref. 16] combined some of the ideas from Refs. [12] and [4] and studied the charging of the region near the plasma, dielectric and conductor interface (the triple junction). The model system that they studied is shown in Fig. 2.

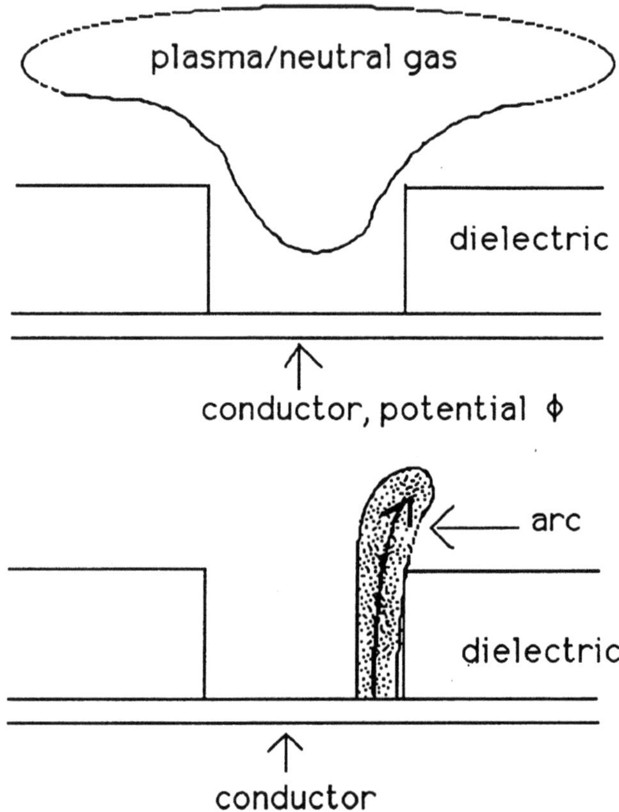

Fig. 2: Model system of the high voltage solar array - plasma interaction

It consists of dielectric material placed on a negatively biased conductor in a plasma environment. For a solar array, the dielectric material corresponds to the coverglass and adhesive and the conductor corresponds to the interconnect. The solar cell itself is neglected because its thickness is small compared to the coverglass and adhesive and because it is a semiconductor with a voltage drop across it of at most a volt or two.

Two types of charging mechanisms were considered. One was the charging due to ions from outside the electric sheath surrounding the solar arrays. Another was the charging due to electrons emitted from the conductor surface by enchanced field electron emission (EFEE). These charging processes were calculated by numerically integrating the particle orbits around the solar cell under the electric field which was consistent with the charging state of the dielectric material.

The numerical integration was done by a particle in cell simulation code. A typical grid is shown in Fig. 3. It concentrates cells in the vicinity of the dielectric-conductor-plasma interface (triple junction) so it can be studied with high fidelity.

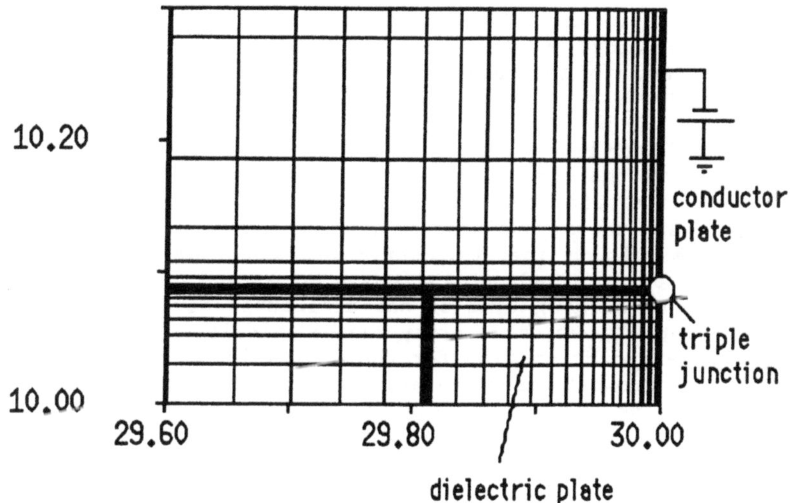

Fig. 3: Small scale typical grid structure for calculations.

They found three major results:

A. The ambient ions charge the dielectric front surface to the steady state potential ≃ 5 V, where 5 eV is the kinetic energy of the incoming ions with the orbital velocity in LEO. Therefore, at the steady state given by the ion charging, a strong electric field of $E \simeq V/d$ is created at the triple junction where V is the bias potential on the conductor, and d is the thickness of the dielectric.

B. After the electric field reaches the value of $E = V/d$ at the triple junction, if there is an emission site with a high field enhancement factor β on the conductor surface near the triple junction, then electrons can be emitted profusely and can charge the side surface. The functional dependence of the current density on the electric field at the emission site (E_e) is

$$j_{ec} = A(\beta^2 E_e^2)\exp(-\frac{B}{\beta E_e}), \quad (2)$$

where A *and* B are constants determined by work function of the surface ϕ_w. This charging due to enhanced field electron emission (EFEE) charges the side of the dielectric positive and can therefore enhance the electric field at the triple junction. It can develop very rapidly because of the strong exponential dependence of the current on the electric field. When the electric field doubles, the emission current increases by ten orders of magnitude. This incident current can desorb a significant amount of neutral gas from the surface and create a dense neutral cloud as high as 10^{21} m^{-3} over the surface. The electron current flowing through the neutral cloud can lead to a discharge like a surface flashover. Even if there is not a dense neutral cloud created over the dielectric, the electric field just over the surface may increase to the point where dielectric breakdown occurs in a thin layer along the side of the dielectric. For an initial conductor voltage of 500 Volts across a coverglass of 150 microns, the initial electric field imposed along the edge of the coverglass is 3.3 X 10^6 V/m. Typically, the coverglasses are held on with an adhesive which can have a dielectric breakdown strength as low as 2 X 10^7 V/m. Hence the electric field along the side of the coverglass only needs to be enhanced by an order of magnitude for dielectric breakdown of the adhesive or surface flashover of desorbed neutral gases to occur.

C. Once the charging time of the dielectric coverglass is known, the arcing rate for a given solar array can be calculated assuming that it takes a negligible amount of time for the ionization of neutral gases or for dielectric breakdown to occur relative to the field buildup time. The arcing rate is then defined as the inverse of the time which is necessary to build up the electric field and charge the surface with ions. The time between arcs is therefore given by

$$\tau_{arc} = \tau_{efee} + \tau_{ion}, \quad (3)$$

where τ_{efee} is the charging time due to enchanced field emission electrons, and τ_{ion} is the charging time due to ions. In Ref. [16], the charging times were calculated numerically using the particle in cell simulation.

4. Discussion

We now calculate the arc rate for the flight and ground based PIX II data [Ref. 11]. Ferguson [Ref. 14] calculated the arc rate for the PIX II flight experiment by taking the probability that no arc occurs during the experiment time of τ_{ex} = 16 *sec*. For a Poisson process, the probability that no arc occurs during time τ is $P_o = \exp(-R\tau)$ where R is the rate at which the random process is occurring. Therefore, calculating the probability P_o and taking $R = -ln(P_o)/\tau$, we can determine the true arc rate R, and this should be independent of the experiment time. We use this procedure to calculate the arc rate numerically for the PIX II flight and show the results in Fig. 1. The PIX II experiment was conducted with 456 2-cm x 2-cm solar cells. We divide the PIX II array surface into sections of area 0.01 m^2 and assume that arcs in the same section are all correlated but uncorrelated between different sections. Since the power supply was often shut down for more than 2 seconds in the flight experiment, it can be assumed that ions charge up the front surface while the power supply recovers. Therefore, the parameter ΔQ is between 10^{-12} and $10^{-14} C$. For the ground data, the experiment time of 1500 sec is chosen. Since there was no problem of power supply recovery reported in the ground experiment, ΔQ is assumed to be from 10^{-9} to 10^{-11} C. The emission site density is chosen as $n_{es} = 1 \times 10^5$ m^{-2} with an exponential distribution of field enhancement factors with average value $\beta_o = 150$ for the flight data and $n_{es} = 1 \times 10^6$ m^{-2}, $\beta_o = 120$ for the ground data. Since the samples used in the ground experiment and the flight experiment were different, it is possible that at the microscopic level the number of emission sites might be different. The plasma conditions were chosen as $n_e = 6.3 \times 10^9$ m^{-3} and $V_{ion} = 7.7 \times 10^3$ m/s of oxygen for the flight data and as $n_e = 2 \times 10^{10}$ m^{-3} and $T_i = 1$ eV of argon for the ground data. The ground data and ground numerical curve shown in Fig. 1 have been normalized to the values at $n_e = 6.3$ $10^9 m^{-3}$ and $V_{ion} = 7.7 \times 10^3 m/s$, assuming that the arc rate is proportional to the ion flux $n_e V_{ion}$.

The results are shown in Fig. 1 as the numerical curves. They show very good agreement with the data over the range the data exists. They predict a threshold when the charging process is exponentially slow and also predict a saturation for high voltages. The lower parts of the curves occur when the enhanced field electron emission charging is the slowest charging process in the system. The arc rate dependence on voltage here is exponential and enables a threshold voltage to be defined with a small uncertainty. The threshold voltage therefore can be defined as the voltage at which the arc rate is decaying very rapidly. The upper parts of the arc rate curves are dominated by the ion recharging time. This leads to the decrease in the rate of change of the arc frequency as can be clearly seen in the data. The fact that the arc rate scales with the density for the higher voltages can also be explained from the dominance of the ion recharging time since this scales directly with the density. Finally, the two theory curves show clearly the difference between the ground and flight data and provide an explanation of the difference. The difference between the two experiments occurred because the charge lost per event was different as well as the microscopic emission parameters. The difference in the charge lost affects the ion recharging time. The difference in the microscopic emission parameters affects the time to build up the electric field along the side surface.

From this good agreement between the theory and the experiment we can deduce that the important elements of the arc formation are the presence of a triple junction, the geometry and secondary electron emission properties of the side surface, the preparation and work functions of imperfections on the interconnector or other biased surfaces, the physical size of the solar cell module and the plasma density.

4.1 SEMI-VACUUM GAS BREAKDOWN

The arc rate in the previous section is the rate for the electric field and associated current to build up along the side of the dielectric. Once the field and current become large enough, then several different possibilities may occur. First, the current may just stop rising due to space charge limitations. In this case, the current and field will eventually decay, and all that would be observed is a nanosecond scale current pulse. This has been observed in some experiments. Secondly, the electric field may rise to the point that dielectric breakdown occurs of the adhesive near the triple junction. Thirdly, if there is desorbed neutral gas near the triple junction or enough ambient gas then a semi-vacuum gas breakdown may occur over the surface. In this section we estimate the threshold voltage that is necessary as a function of the neutral pressure, right over the surface, in order for a semi-vacuum gas breakdown to occur. This is motivated by the clear experimental observation [Ref. 15] that neutral gas plays a role in the arc process.

The neutral gas density over the surface is roughly estimated by dividing the neutral desorption flux by the desorption speed of the neutral particles. The desorption flux Γ_n is given by $\Gamma_n = \gamma j_e/e$ if the neutrals are desorbed due to electron bombardment, where γ is the ESD yield which is defined by the number of neutral particle desorbed by one incident electron, and j_e is the incident electron current. The ESD yields, γ, are seen between 0.01 and 0.1 for most materials, and the desorption speed of the neutrals is 10^2 - 10^3 m/sec, depending on its mass and the surface temperature. Since the electron current varies widely from 0 to 10^7 A/m^2 or even higher [Ref. 16], the resulting neutral density also varies from atmospheric (10^{25} m^{-3}) to vacuum.

The length scale of the solar array discharge is $d \leq 0.3mm$ if we take it from the thickness of the coverglass. Most of our values of the product pd (where p is the neutral gas pressure) fall into the region between Townsend gas breakdown and the vacuum breakdown. Little is known about this region compared to the other two regions. One possible explanation for the gas breakdown in this region is that, if the gap distance d is small enough, a prebreakdown electron current is emitted from the cathode surface by field emission and the feedback effect at the cathode is provided by field enhancement due to the space charge of the incoming ion current. We consider gas breakdown at low pd based on this hypothesis. For simplicity we consider a one-dimensional system filled with neutral gas of a certain density n_n with two biased electrodes but without any dielectric material. We assume that the field is already enhanced to a certain extent due to a whisker or dielectric impurity on the cathode surface.

The breakdown voltage is calculated analytically by solving one-dimensional equations with a simplified collision model. The results are plotted in the form of a Paschen curve with the Paschen discharge data in Fig. 4. The semi-vacuum breakdown voltage intersects with the Paschen curve at $P = 40 Torr$ and $V = 160 volt$. We also show the threshold voltage determined by the charging time of the dielectric material.

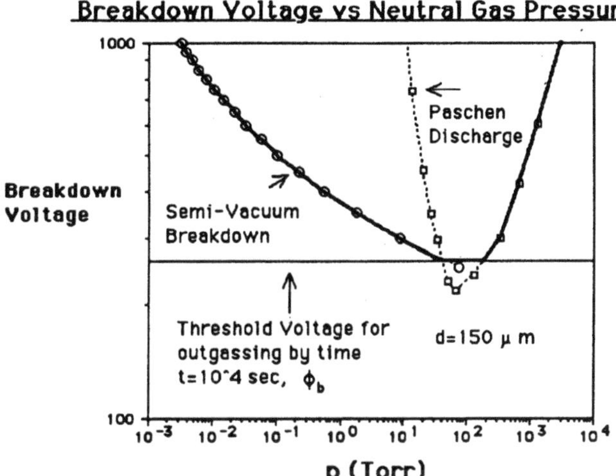

Fig. 4: Breakdown Voltage vs Neutral Gas Pressure.

A bias voltage below the threshold voltage cannot create a neutral cloud of enough density to undergo arcing because the electron current incident on the surface is too low. Therefore, when the bias voltage is lower than this voltage, we will not see arcing. When the bias voltage is higher than this threshold then arcing due to gas breakdown may occur depending on the amount of gas absorbed in the surface of the dielectric. This model clearly shows that the minimum voltage necessary for any gas breakdown over the surface to occur is of the order of 200 Volts. That is, high voltage systems biased with a smaller absolute value of the voltage than this minimum will not arc by this mechanism under any circumstances. This coincides with the observed experimental results.

4.2 MITIGATION STRATEGIES

The theory and experiments suggest that the key elements in the arc discharges on high voltage solar arrays are the existence of a triple junction, secondary electron emission from the dielectric, field enhancement sites on the conductor and the presence of neutral gas.

This understanding suggests several possible mitigation strategies. One obvious one is to hide the triple junction from the plasma. This could be achieved by causing the coverglass to overhang substantially or by completely encapsulating the conductor or semiconductor in a dielectric coat. This strategy has the possible advantage that as the solar array ages, it may develop cracks or suffer pinholes as a result of micrometeorite strikes. The major effect is to increase the threshold voltage by making the electric field charging time the dominant term over a larger range of voltage. Another strategy is to change the secondary emission properties of the coverglass and adhesive. This could be done by a simple doping procedure. If the secondary emission yield is reduced below one, then a dramatic reduction in the arc rate will be achieved. Finally, the reduction in the number of emission sites by cleaning up the number of whiskers or adhesive filaments will cause the arc rate to drop for no other changes in the plasma or geometry of the system [Ref. 15]. This can be realized by changes in the manufacturing process for making the solar cells. The effect of this can be seen in Fig. 5. A dramatic drop in the arc rate can be obtained by reducing the number of emission sites by an order of magnitude.

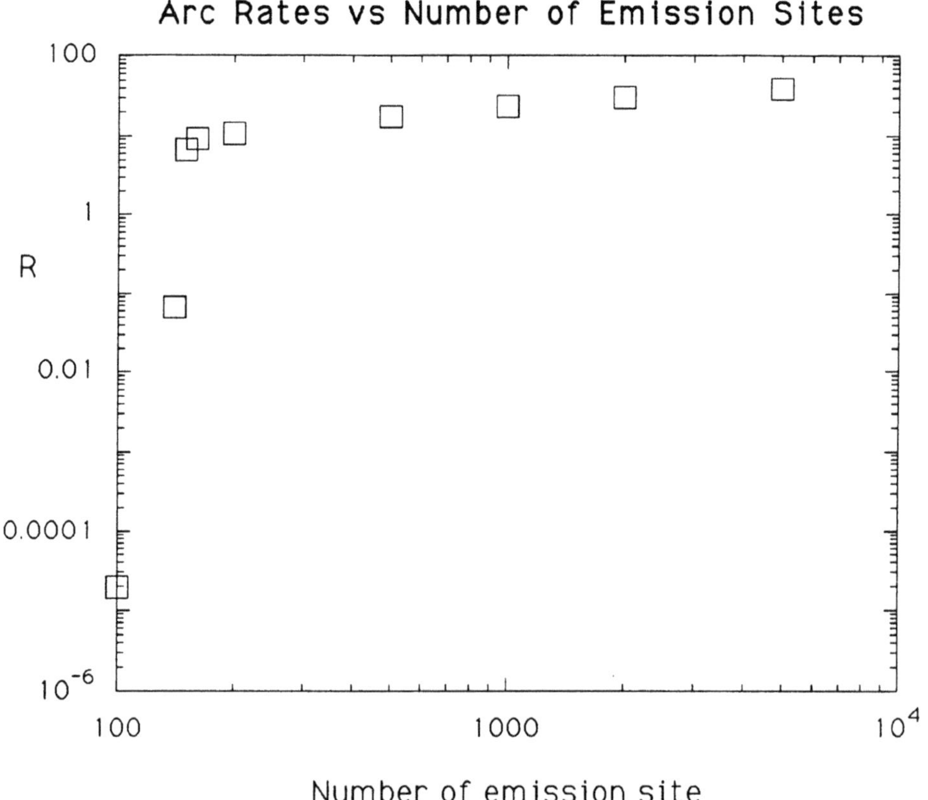

Fig. 5: Arc rate against number of emission sites for $|V| = 500$ Volts, $\beta = 480$ and $n_e = 1.1 \times 10^{12}$

5. Flight Experiments

A number of new flight experiments are currently under development. Japan is developing the Space Flyer Unit (SFU) [Ref. 17], which is a reusable free-flying space platform with eight integrated standardized payload Units. The first flight of the SFU will be launched by the Japanese rocket H-II in 1994 and retrieved by the U.S. Space Shuttle after six months of operation. The High Voltage Solar Array (HVSA) experiment is one of the payloads for the first flight of the SFU. A retractable membrane, based on the concept of the "Miuraori" folding, will be deployed in space. Four Solar Cell Modules (SCM) and an Electron Collector (EC) will be mounted on the 2D array along the main mast. Each SCM will have 135 series-connected solar cells, which will be 2 cm x 4 cm Silicon Back Surface Field and Reflection (BSFR) types, so that the single SCM will output 65 V at open voltage and 0.3 A of short circuit current under nominal conditions. The HVSA experiment system will be able to generate 260-280 V with all four modules connected in series. The maximum voltage will depend on the cell temperature. The EC will electrically ground the power circuit of HVSA to the ionosphere by passive collection of electrons. Table 1 lists the specifications of the HVSA Experiment. The plasma density, temperature, neutral pressure and the electromagnetic emission from the arcs will be monitored on the SFU by a plasma diagnostics package.

Table 1: Description of HVSA Flight Experiment.

2D Array configuration natural frequency	Miura-ori folding triangle 3.4-m x 3.4-m 2 masts 0.427 Hz
solar cell coverglass electron collector output power, voltage and current	2-cm x 4-cm Si cell Back Surface Field and Reflection type 0.1 mm thick antireflection coating and conductive coating gold-plated Al 260V max. 1.2A max. 80W max. series-parallel connection of 4 solar cell modules

Two other flight experiments that are planned for the near future are the Air Force Photovoltaic Array Space Power Plus Diagnostics experiment (PASP PLUS) and the NASA Solar Array Module Plasma Interactions Experiment (SAMPIE). PASP PLUS will be launched in 1993 on a Pegasus Launcher and will go into a highly elliptical, highly inclined (70°) orbit. The on-orbit lifetime is planned to be one year. The objectives of the experiment are to study solar array high voltage arcing and leakage currents, determine long term radiation degradation effects and to flight qualify advanced array designs. The experiment will carry eleven different array technologies and sixteen electrically isolated segments. The technologies that will be tested include different sizes, materials, mechanical mounting structure, energy conversion efficiencies and resistance to radiation damage. In contrast to the HVSA experiment, where the bias will be generated by the solar cells themselves, in PASP PLUS the cells will be biased by an on-board power supply from -500 Volts to + 500 Volts. The highly inclined orbit will allow the study of the high voltage solar cells interacting with the aurora.

SAMPIE is manifested for 1994 and will specifically study the arcing characteristics of standard silicon cells, Space-Station-design large silicon cells and very thin flexible silicon cells. Several arc mitigation strategies will be tested including the use of dielectric coatings and geometric modifications to the coverglasses. The ambient plasma parameters will be monitored. The arrays will be biased from -600 Volts to +600 Volts. An extensive program of ground tests is planned before the experiment in order to calibrate the data.

Each of the three flight experiments will add new information to the high voltage solar array arcing database. The HVSA and PASP PLUS experiments will study the arcing phenomena over an extended period of on orbit operation. The HVSA experiment will be the first to use the array voltage itself rather than a power supply (which has to be protected). The PASP PLUS experiment will essentially test all the currently planned solar array technologies for the next twenty years. The SAMPIE experiment will explore specific arc mitigation strategies.

6. Conclusions

The need for increasing power on board space systems implies that high voltage solar arrays will be considered as future sources of power. High voltage solar arrays are known to have several interactions with the space plasma environment. One that has received a great deal of attention has been the issue of arcing on high voltage solar arrays.

Detailed experiments indicate the importance of the dielectric, plasma interface with particular emphasis on the adhesive. A theoretical analysis has been developed based on the idea of the arcs being driven by electric field runaway at this interface. From this field runaway time and the ion charging time of the coverglass surface, it is possible to derive an arc rate. This expression depends on the surface preparation, the dielectric conductor geometry, the plasma density and other parameters and explains the data well. The experiments also show the importance of neutral gas, particularly water, in the discharge process. A theory based on the idea of semi-vacuum gas breakdown of this neutral gas gives a threshold voltage which coincides well with the measurements. Based on the theory and experiments, it is possible to construct specific arc mitigation strategies.

Future flight missions are currently being planned to explore the long term arcing problem as well as examine a wide range of solar cells and mitigation strategies.

Acknowledgments

This work was supported by the Air Force Office of Scientific Research under grant AFOSR-87-0340. The author would like to acknowledge help from Mengu Cho of MIT, Hitoshi Kuninaka of ISAS, Japan, Dave Hardy of AFGL and Bernie Upshulte of PSI.

References

[1] B.G. Herron, J.R. Bayless and J.D. Worden, *High Voltage Solar Array Technology*, Journal of Spacecraft and Rockets, 10, 457, 1973.

[2] N.J. Stevens, *Review of Interactions of Large Space Structures with the Environment, Space Systems and their Interactions with Earth's Space Environment*, Progress in Aeornautics and Astronautics, 71, AIAA, Washington, DC, p 437-454, 1980.

[3] H. Thiemann and K. Bogus, *Anomalous Current Collection and Arcing of Solar-Cell Modules in a Simulated High-Density Low-Earth-Orbit Plasma*. ESA Journal, 10, 43-57, 1986.

[4] D.E. Hastings, G. Weyl and D. Kaufman, *The Threshold Voltage for Arcing on Negatively Biased Solar Arrays*, Journal of Spacecraft and Rockets 27, 539-544, 1990.

[5] H. Thieman, R. W. Schunk and K. Bogus. *Where do negatively biased solar arrays arc?* Journal of Spacecraft and Rockets, 27, 563-565, 1990.

[6] K. L. Kennerud, *High Voltage Solar Array Experiments*, NASA CR-121280, 1974.

[7] D. B. Snyder, *Discharges on a Negatively Biased Solar Cell Array in a Charged Particle Environment*, Spacecraft Environment Interactions Technology Conference, Colorado Springs, CO, Oct 4-6, 1983, NASA CP-2359, p 379-388.

[8] D. B. Snyder and E. Tyree, *The Effect of Plasma on Solar Cell Array Arc Characteristics*, NASA TM-86887, 1985.

[9] H. Fujii, Y. Shibuya, T. Abe, K. Ijichi, R. Kasai and K. Duriki, *Laboratory Simulation of Plasma Interactions with High Voltage Solar Arrays*, Proceedings of the 15th International Symposium on Space Technology and Science, Tokyo, 1986.

[10] N.T. Gier and J.J. Stevens, *Plasma Interaction Experiment (PIX) Flight Results*, Spacecraft Charging Technology-1978, NASA CR-2071, p 295-314.

[11] N.T. Grier, *Plasma Interaction Experiment II: Laboratory and Flight Results*, Spacecraft Environment Interactions Technology Conference, Colorado Springs, CO, Oct 4-6, 1983, NASA CP-2359, p 333-348.

[12] D.E. Parks, G. Joneward, I. Katz and V. A. Davis, *Threshold-Determining Mechanisms for Discharges in High-Voltage Solar Arrays*, Journal of Spacecraft and Rockets, 24, 367-371, 1987.

[13] D.B. Synder, *Discharges on a Negatively Biased Solar Array*, NASA TM-83644, 1984.

[14] D.C. Ferguson, *The Voltage Threshold for Arcing for Solar Cells in LEO-Flight and Ground Test Results*, NASA TM-87259, March 1986.

[15] B.L. Upshulte, G.M. Weyl, W.J. Marinelli, E. Aifer, D. Hastings and D. Synder, *Significant Reduction in Arc Frequency of Negatively Biased Solar Cells: Observations, Diagnostics and Mitigation Techniques*, SPRAT Conference, Cleveland, OH, May 1991.

[16] M. Cho and D.E. Hastings, *Dielectric Charging Processes and Arcing Rates of High Voltage Solar Arrays*, Journal of Spacecraft and Rockets, 28, 698-706, 1991.

[17] K. Kuriki, M. Nagatoma and K. Ijichi, *Energetic Experiments on Space Stations*, Acta Astronautica, 14, 445, 1986.

ADVANCED SPACE PHOTOVOLTAIC TECHNOLOGY

D. J. FLOOD
National Aeronautics and
Space Administration
Lewis Research Center
Cleveland, Ohio 44135

1. Introduction

Solar arrays have been the predominate power sources in space for over thirty years, beginning with the launch of the first U.S. solar powered satellite in 1958 [Ref. 1]. Since that time, hundreds of kilowatts of solar power have been placed in orbit on commercial, civilian and military satellites, and small arrays have been transported and left behind on the surfaces of the Moon and Mars. At least one photovoltaic array has been in orbit for nearly two decades, and is still functional [Ref. 2]. Solar cell efficiencies on the earliest arrays were low, typically around 10%, and much had to be learned about the survivability of solar cells in the space environment. Enormous progress has been made since 1958, both in our understanding of the fundamental mechanisms which limit solar cell efficiency and lifetime, and in our ability to turn that understanding into real system improvements. In the pages that follow we will describe the technologies presently being pursued by the National Aeronautics and Space Administration to develop advanced solar cells and arrays, and discuss briefly the improved system capabilities which should result.

1.1. SPACE PHOTOVOLTAIC POWER SYSTEM DESCRIPTION

Figure 1 contains a block diagram of a complete photovoltaic power system. As can be seen, it is comprised of a number of subsystem elements, one of which is the solar array. As shown in Fig. 2, the array itself is a set of subsystems, with the result that designing and building just the solar array, let alone the entire satellite power system, requires an interdisciplinary, well-coordinated effort. We shall touch briefly on some of the important aspects of advanced array technology in section 4. The primary focus of the rest of the paper will be on advanced space solar cell research and development.

There are two figures of merit that are used to measure the performance of a space solar array, as well as the entire power system: power per unit mass in watts/kilogram (W/kg), and power per unit area in watts/sq. meter (W/Sq.M). These are referred to simply as specific power and area power density, respectively. The inverses of these quantities are

R. N. DeWitt et al. (eds.), The Behavior of Systems in the Space Environment, 795–823.

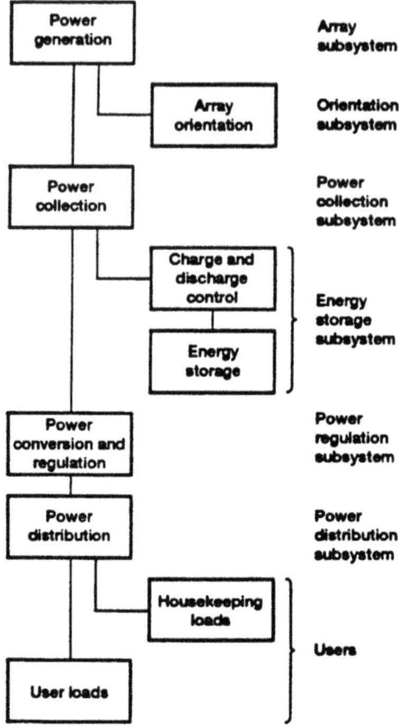

Fig. 1: Block diagram of a space photovoltaic power system.

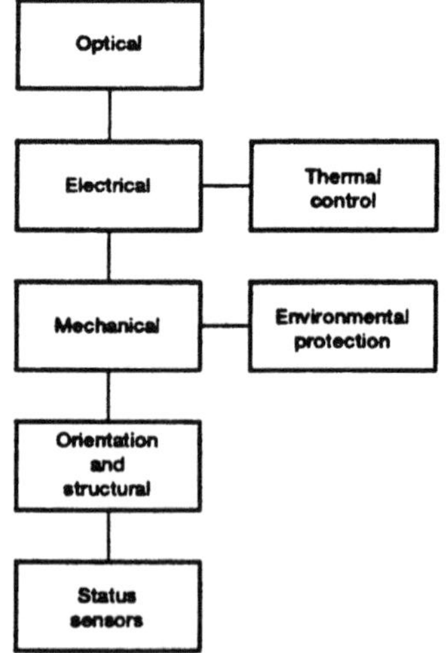

Fig. 2: Block diagram of a space photovoltaic array.

also often used, and are known as specific mass (kg/Kw), and specific area (Sq.M/Kw). Typical values for state-of-the-art (SOA) space solar arrays, using silicon solar cells mounted on rigid panels, are 30 to 40 W/kg and 90 to 110 W/Sq.M at the start of the mission, or beginning-of-life (BOL). The end-of-life (EOL) values for any given array are dependent on mission time and location. Chief among the factors affecting the ratio of EOL to BOL are radiation damage to the solar cells, followed by mechanical and electrical degradation of the cells, interconnections and array components from other environmental effects, such as plasma interactions and thermal cycling. Elimination, or at least substantial mitigation, of such effects is at the heart of all space photovoltaic device and system research and development efforts.

1.2. SPACE POWER SYSTEM APPLICATIONS AND REQUIREMENTS

Table 1 lists the broad mission categories into which the NASA program is roughly divided, some qualitative estimates of the power levels required for each, and the primary attributes any sort of power system must have if it is to be considered for use on such missions. The desired attributes are listed in relative priority order for each mission class, with the caveat that detailed trade studies are required to establish the proper ordering of priorities for any given mission. It is clear, however, that low mass and long lifetime are important power system technology drivers in virtually all potential space missions. Power system cost and size have greater or lesser importance depending on the mission objectives and operational environment (orbital, planetary surface, interplanetary, deep space, etc). For example, low total cross-sectional area is a critical attribute for the space station, because of the drag produced by the residual atmospheric density in the low altitude orbits in which it will fly [Ref. 3]. In this case, the array has an important effect on the total life cycle cost of the mission, because it directly affects the amount, and hence the cost, of constantly providing fuel to maintain the space station

Table 1. NASA Space Mission Categories

Mission subset	Power level	System attributes
Unmanned near earth (LEO, HEO, GEO) and planetary applications	Low to intermediate	Low mass, long life
Space station	High	Minimum area, low mass, low cost
GEO platform	Intermediate	Long life, low mass
Lunar base, manned planetary	Intermediate to high	Low mass, portability, long life
Electric propulsion orbit transfer (OTV)	High	Reusability, minimum area, low mass

at its proper altitude. Such considerations do not apply to a mission to the lunar surface, however. In that case, although total area may be important because of other factors such as ease of construction and deployment, it will not be a primary driver in selecting a particular technology for the mission. Specific power and resistance to proton radiation damage from solar flares will certainly be among the more important factors for selecting a lunar surface solar array technology.

1.3. SPACE SOLAR CELL AND ARRAY TECHNOLOGY DRIVERS

The most important specific technology drivers for advanced space solar cells which derive from the attributes described above are high efficiency and lifetime, with mass and cost of secondary importance. The principle reason is that the solar cells constitute a relatively lesser fraction of the total mass and cost of a system, while their efficiency and usable lifetime are major determinants of the balance-of-system (BOS) mass and cost. Cell efficiency determines array area, which in turn determines array mass. Array lifetime is mission-specific, and is loosely defined to be length of time the array operates before its output power has fallen to a level below that needed to operate the satellite (or surface system) reliably. In general, array lifetimes are determined by the rates at which solar cell electrical output degrades, assuming that the mechanical aspects of the cell and array have been properly engineered to withstand thermal cycling, vibration, and other operational and environmental effects. The chief cause of electrical degradation is bombardment of the cell by constituents of the natural charged particle radiation environment. The ability of a solar cell to operate while, or having been, subjected to such bombardment is a measure of the radiation resistance of the cell. It is measured by determining the ratio of the output power (P) remaining after absorbing a given dose of radiation to the initial output power (P_o) determined prior to such exposure (P/P_o). The extent to which a solar cell is radiation resistant depends on many factors: the material from which it is made (i.e., silicon, gallium arsenide, indium phosphide, etc.), its actual device structure, and its ability to anneal (or be annealed) as the damage occurs. We shall discuss the mechanisms involved in creating radiation damage in solar cells, and attempts to reduce it or eliminate it altogether, in section 3 when we discuss the various solar cell materials and structures now in use or under development.

2. Space Solar Cell Performance Verification

Proper design of a space photovoltaic power systems depends critically on the ability to accurately predict the performance of its component solar cells in the space environment. Given the inconvenience and cost of actually measuring solar cell performance in space, the usual procedure is to perform such measurements in the laboratory using a xenon arc lamp as a solar simulator. Figure 3 contains a plot of the solar spectrum outside the earth's atmosphere, known as the Air Mass Zero (AM0) spectrum, and that of a typical xenon arc lamp. The figure also contains a plot of the terrestrial solar spectrum taken at solar noon (Air Mass 1, or AM1), under

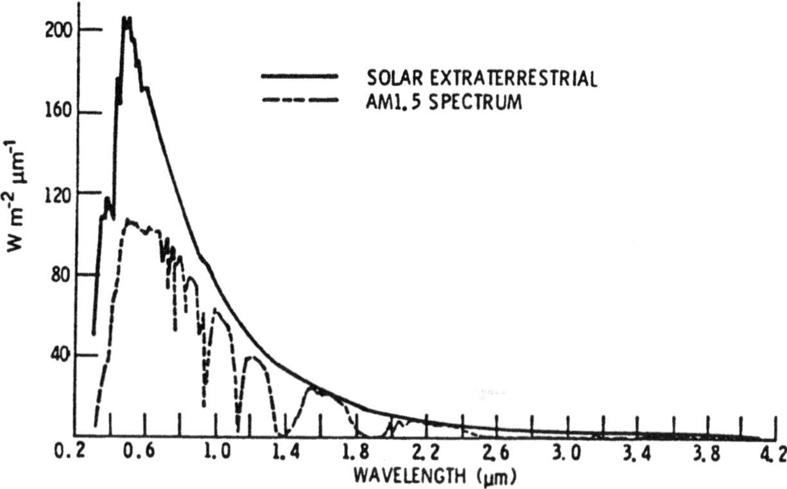

Fig. 3. Solar spectral irradience.

prescribed conditions of atmospheric clarity and moisture content. The differences between the terrestrial spectrum and the AMO spectrum are too great, and too variable because of uncontrollable meteorological conditions, to allow use of the former in predicting space solar cell performance with confidence. The match between the xenon arc lamp spectral intensity distribution and that of the AMO spectrum is much better, and the differences are repeatable and easily accounted for. Nonetheless, as will be shown in the next section, considerable care must still be exercised when using laboratory solar simulator measurement results to predict AMO performance values.

2.1. LABORATORY MEASUREMENT TECHNIQUES

The first step in the procedure is to obtain a "Standard cell," the total output of which has actually been measured in the AMO spectrum. The output of this "standard cell" is then used as a reference against which to adjust the intensity of the solar simulator. The proper simulator intensity is reached when the reference cell has the same output as obtained in the AMO spectrum. Before the measurement of the "unknown" cell can proceed, however, it is also necessary to determine the spectral response of both the standard cell and the unknown cell to make certain that they are closely matched. The spectral response of a solar cell is simply its absolute output under uniform illumination by a series of monochromatic light sources with precisely known intensities, and with wavelengths that span the AMO spectrum. This latter step is required because, although relatively well matched, the xenon arc lamp spectrum and the AMO spectrum do have important differences, most notably in the longer wavelength region where the typical solar simulator spectrum has several large spikes; differences also exist in the short wavelength region, where the AM0 spectrum is somewhat more intense than that of the simulator. If a reference cell which has a very strong spectral response at the shorter wavelengths and a relatively weaker response at the longer wavelengths is used to set the simulator output, the resulting spectrum will be "red-rich", with an artificially high intensity in the long wavelength region. The cell to be tested has the opposite trend in its spectral response, i.e., strong in the red and weaker in the blue; when it is placed in the same beam it will appear to have a higher output than it should. The difference can amount to several percent, depending on the degree of spectra response mismatch between the cells. The opposite condition can also occur, giving lower values for the unknown cell as well.

The net result of the above is that it becomes necessary to maintain a large "library" of AM0 calibrated solar cells of all sorts, with as many different types of spectral response as can be obtained. The problem is compounded because the radiation damage caused by charged particle bombardment affects the spectral response-matched standard cells in the library for each different material and structure as are likely to be encountered.

2.2. AM0 CALIBRATION TECHNIQUES

The need for large numbers of standard cells cannot be met by actual space flight calibration measurements, both because of the limited access to, and the cost of, even the simplest space missions. An alternative is to make the required measurements at very high altitudes, where the atmospheric attenuation is a very small correction, and the results are unaffected by water vapor or other meteorological conditions. The measurements are made using either a high altitude balloon or a high altitude aircraft [Refs. 4, 5]. Because the aircraft measurements are made at altitudes below 50,000 feet, a small correction must be made for the light absorbed by the ozone in the residual atmosphere above that altitude. The full I-V characteristics of the cells can, however, be measured under precisely controlled temperature conditions. On the other hand, while altitudes approaching 100,000 feet can be reached for the balloon measurements, essentially eliminating the need for atmospheric corrections, measurements are restricted only to a determination of the cell's output near its short circuit operating condition. In addition, the measurements are not made under controlled temperature conditions. The temperature at which the measurements are made is determined, however, and a correction made to the data to provide cell output current at the standard reporting condition of 25C. Both techniques give results that are in essential agreement with each other, and both have been used to build up the library of standard cells that NASA uses to make accurate laboratory measurements.

3. Advanced Space Solar Cells

Silicon solar cell arrays are still the primary sources of power for U.S., European and Japanese satellites. Although specialized improvements continue to be made in silicon solar cells for space applications, resulting in improved performance and reduced cost, they will not be discussed in this paper. The space silicon solar cell is considered a "mature technology", in which improvements are routinely accomplished by commercial vendors by implementing advanced fabrication techniques and cell designs into the manufacturing process. Our understanding of the fundamentals of high efficiency operation and radiation resistance in silicon solar cells is essentially complete [Refs. 6, 7, 8]. The same cannot be said for the advanced cell types and materials currently being investigated, however. Table 2 lists several advanced solar cell types that are of interest for use on planar arrays in high natural charged particle radiation environments, along with the data for commercial silicon cells. The laboratory efficiencies quoted for all cell types in the table are for 2 cm x 2 cm cells except where noted. The expected date of availability in each case is a subjective estimate of the time required to move the technology from the laboratory R&D phase through a successful demonstration on a pilot production line, with the very important caveat that R&D funding levels do not constrain any efforts to do so. Given sufficient funding for a subsequent manufacturing technology program, modest yields of cells with efficiencies at or near their laboratory values could be achieved in commercial production within a year or two of the dates listed.

Table 2. Advanced Planar Space Solar Cell Technology Status.

CELL TYPE	Cell STRUCTURE	PROJECTED EFFICIENCY	LABORATORY EFFICIENCY	COMMERCIAL EFFICIENCY	P/P_o (1E15 cm^{-2} 1MeV ELECTRONS)	P/P_o (E12 cm^{-2} 10MeV PROTONS)	ESTIMATED DATE AVAILABLE
THIN SILICON	62 μm SUBSTRATE n/p DIFFUSED BSF, BSR	14.5%	14.5%	13.7%	0.74	DAMAGE EQUIVALENCE TO 1MeV PROTONS KNOWN	NOW
GaAs	300 μm SUBSTRATE n/p, p/n OMCVD, LPE	23%	21.5%	20%	0.74	0.8	NOW
GaAs/Ge	75 μm SUBSTRATE 10 μm CELL, p/n OMCVD	23%	20.5%	20%	0.74	0.8	NOW
CLEFT GaAs/Si	62 μm SUBSTRATE, 6 μm CELL n/p, p/n OMCVD	23%	20%	-	0.74	0.8	> 5 yrs.
InP	300 μm SUBSTRATE n/p OMCVD	22%	19%	-	0.975	0.9	< 5 yrs.
InP/Ge	75 μm SUBSTRATE* 10 μm CELL, n/p OMCVD	20.5%	9%	-	0.975	-	5-7 yrs.
InP/Si	62 μm SUBSTRATE* 10 μm CELL, n/p OMCVD	18.5%	7%	-	-	-	7-10 yrs.
CuInSe2	5 μm FILM, 50 μm GLASS SUBSTRATE*	16%	11%	-	>0.98	0.9(1MeV)	< 5 yrs.

*PROJECTED THICKNESS
ALL RADIATION DAMAGE RESULTS WITHOUT COVERGLASSES

3.1. GALLIUM ARSENIDE SPACE SOLAR CELLS

Although our understanding of gallium arsenide space solar cells is not as complete as our understanding of silicon cells, it is good enough that a GaAs cell with a simple p/n structure for use on space arrays is now commercially available [Ref 9]. What is lacking is the same depth of knowledge, as now exists for silicon at the microscopic level, of the effects on cell performance from unwanted material imperfections and impurities, particularly with regard to radiation damage degradation. Theoretical analysis predicts small, but still significant, differences in the performance of p/n and n/p cells prior to and after radiation damage [Refs. 10, 11]. Such differences have not always been observed in practice, in part because material quality and fabrication techniques are not fully under control, and in part because theoretical models still need better data on key electronic and material properties.

Despite the present limits in our understanding of state-of-the-art GaAs solar cells, the fact remains that they are more efficient than silicon cells. Figure 4 contains a plot of the ideal AMO conversion efficiencies of solar cells made from various materials with differing bandgaps [Ref. 12]. The calculation assumes the cells to be operating

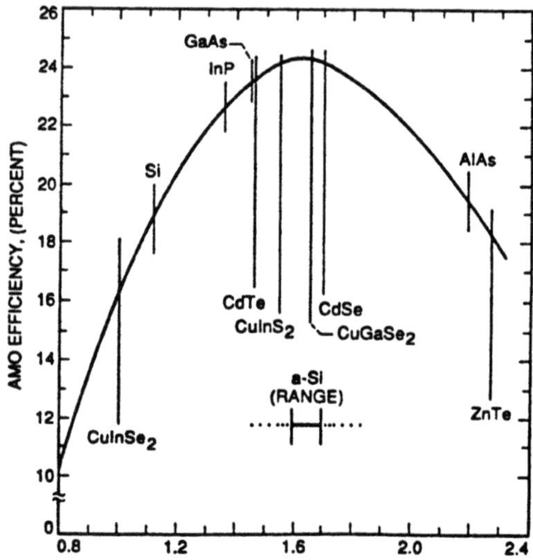

Fig. 4: Ideal efficiency for space solar cells in the air mass zero spectrum.

under ideal conditions, with no losses from material defects or unpassivated surface states taken into account, and therefore represents a simplified upper limit estimate of efficiency. The presence of a maximum in the curve is easily understood. For very low bandgap cells, most of the incoming solar energy is at energies well above that of the cell bandgap. Hence large numbers of electrons are excited from the valence band into unfilled states well above the conduction band minimum. They quickly interact with the semiconductor lattice, yield their excess energy as heat, and settle into the unfilled states at the bottom of the conduction band. The amount of energy lost to heat compared to that available to do work in an external circuit may be easily exceed 75% - 80% of the total incident energy for the low bandgap cells of technological interest for space applications, typically 0.5 eV or higher. (Clearly a cell with a vanishing, but not zero, bandgap will lose most of the energy from the absorbed photons to heat.) At the other end of the spectrum, cells with a bandgap energy higher than that of most of the incoming photons will simply not absorb any of them, again resulting in no net energy conversion. The exact position of the efficiency maximum depends on the shape of the spectral intensity distribution and the cell's operating temperature. For the ideal cell structure (i.e., with no other loss mechanisms) the maximum in the AM0 spectrum occurs for a bandgap of about 1.55 eV.

Laboratory efficiencies of 22.5%, which are very close to the values predicted in Fig. 4, have been attained in GaAs cells to date [Ref. 13]. It is important to note that efficiencies can be higher than the "ideal" values predicted by Fig. 4 because the calculation does not include any refinements to the basic solar cell structure which can enhance its performance. Refinements include such things as back surface fields, minority carrier mirrors, textured surfaces, light trapping geometries, etc., [Ref. 14]. Efficiencies exceeding 20% have been observed in carefully fabricated silicon cells [Ref. 15], significantly above the value predicted by the curve in Fig. 4. Similar results can be expected for GaAs cells with appropriate enhancements. An issue that arises, however, is the radiation resistance of cells that include efficiency enhancing features. Results obtained to date consistently indicate that the enhancements are strongly affected by radiation damage, so that cell output quickly degrades to values typical of "ordinary" cells at the higher accumulated fluences of charged particle radiation [Ref. 16].

In general, GaAs cells have four major advantages for space application when compared to silicon solar cells. They are (1) higher efficiency, as explained above; (2) higher temperature operation because of its higher bandgap; (3) higher radiation resistance; and (4) the potential for lower mass. Figures 5a and 5b illustrate items 1-3 above. Figure 5a shows the relative radiation damage degradation of silicon and

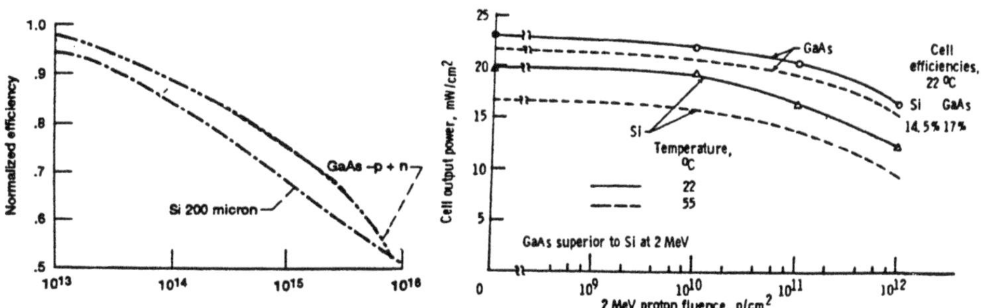

Fig. 5a: Normalized efficiencies of Si and GaAs solar cells after 1 MeV electron irradiation.

Fig. 5b: Comparison of Si and GaAs under 2 MeV proton irradiation.

GaAs cells under 1 MeV electron irradiation under standard laboratory measurement conditions. Figure 5b shows similar behavior under proton irradiation for two different temperatures, 22C and 60C. The lower temperature is close to the laboratory standard temperature of 25C, while the higher temperature is typical of actual operating temperatures on orbit. There is no longer debate about the relative superiority of GaAs over Si. The issue is now primarily the cost of GaAs cells, which, although higher at the cell level, has been shown to be comparable at the array level when the array will operate in a significant radiation environment [Ref. 9], such as is encountered in a geosynchronous or polar orbit.

The last, somewhat surprising advantage, given the much higher density of GaAs compared to Si, derives from the fact that 100% optical absorption is achieved in less than 5 microns of GaAs, while a nearly 300 micron thickness is required in silicon. High efficiency GaAs cells less than 6 microns thick, mounted on the underside of a standard space solar cell glass cover, have been demonstrated in planar and concentrator cell configurations [Ref. 17]. The potential for lower mass exists at both the cell level, as just described, and at the array level, as discussed in section 1. State-of-the-art rigid panel array masses are lower for the "Standard" GaAs cell configuration, i.e, a 6 micron active layer on a 200 micron thick GaAs substrate, than for silicon cells at the same array output power. The recently developed, and now commercially available, GaAs cell on a 75 micron thick germanium substrate [Ref. 18] makes GaAs all the more advantageous at the array level. Costs are lower than the "standard" GaAs cell, and the device is substantially more rugged [Ref. 9]. It is expected that GaAs cells will be used with increasing frequency as performance requirements on space power systems become more and more demanding.

3.2. INDIUM PHOSPHIDE SPACE SOLAR CELLS

While the radiation resistance of GaAs is significantly better than that in standard (i.e, 300 micron thick) Si cells, the advantage disappears when GaAs is compared to thin silicon cells. The data are shown in Fig. 6,

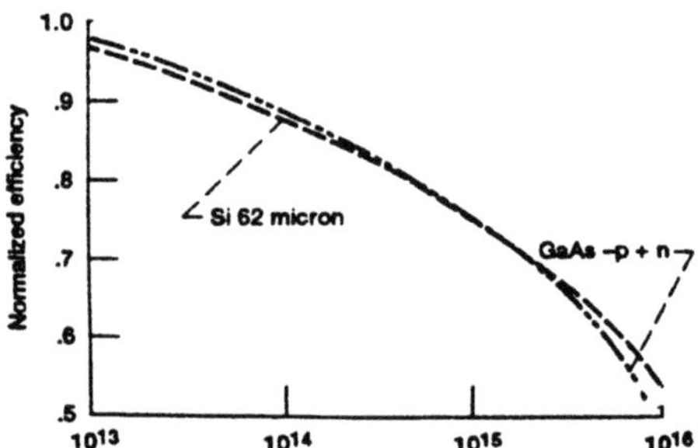

Fig. 6: Normalized output of GaAs and thin silicon solar cells after 1 MeV electron irradiation.

where the relative performances of both cell types are shown. The silicon cells in this case are 62 microns thick, and are designed for use on the newer, lightweight solar arrays that are currently under development by NASA [Ref. 19]. Advanced array technology will be discussed in more detail in the next section. The primary difference is that the lightweight arrays do not have rigid panels made from aluminum honeycomb, but instead have panels made from a thin polyimide sheet, such as Kapton. The aluminum honeycomb panels provide a significant amount of protection from the charged particle radiation that is incident on the backside of the array, while the thin polyimide panels do not. The thin silicon cells provide a trade-off between absolute efficiency and radiation resistance by providing less power than standard Si cells at BOL, but degrading less [Ref. 20]. Hence the absolute EOL efficiencies of the two silicon cell types are comparable. Both are substantially lower than GaAs cells, but the impact is far different. Unlike the situation with rigid panels, the cells account for a larger fraction of the total lightweight array mass than they do on the former. Because of the similar radiation resistance behavior, the absolute EOL specific power of a flexible array is higher with the thin silicon cells than with any of the GaAs cell types, except the stand-alone 6 micron cell. (The latter is still not yet a commercial device, and needs a substantial manufacturing technology development program to bring it to a viable commercial readiness.)

Improvement of the EOL specific power of such arrays requires the development of cells with significantly better radiation resistance than either GaAs or thin silicon, provided they also have one of two other characteristics: extreme lightweight, or efficiencies comparable to GaAs (i.e, at least 20% AMO). Candidates in the former category are the thin film cells such as amorphous silicon and copper indium diselenide, which shall be discussed in a later section. In the latter category, only single crystal solar cells have exhibited the requisite high efficiencies. What is required is the demonstration of high radiation resistance. Figure 7 is a plot of the relative efficiency degradation of InP solar cells as a

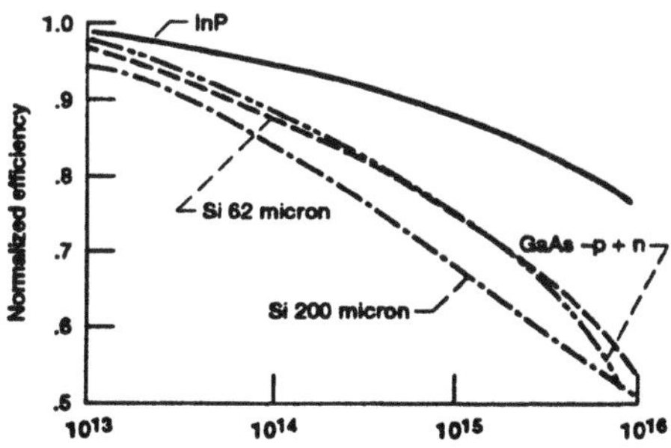

Fig. 7: Effect of 1 MeV electron fluence on InP, GaAs and Si solar cell normalized efficiencies.

function of 1 MeV electron irradiation. The normalized curves for GaAs and thin (62 micron) single crystal silicon are shown as well. Figure 8 contains the same sort of data for 10 MeV proton irradiations. The superior radiation resistance of InP is clearly evident. The room

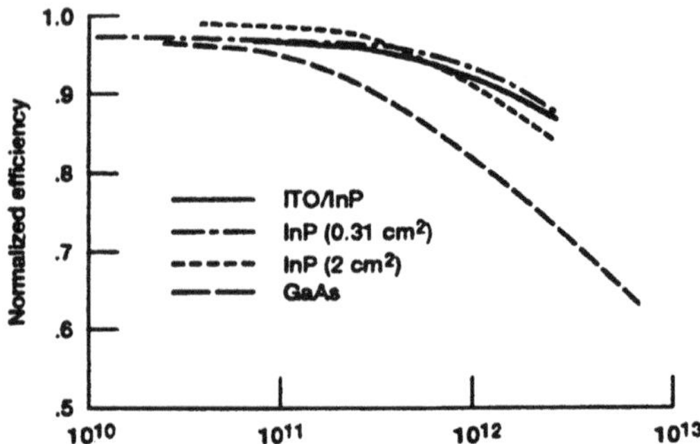

Fig. 8: 10 MeV proton radiation damage for GaAs and InP cells.

temperature bandgap of InP is 1.35 eV. From the plot in Fig. 4, it would appear that InP cells can be expected to have an AM0 efficiency comparable to that of GaAs cells. Efficiencies in excess of 19% AM0 have been demonstrated in laboratory devices [Ref. 21]. On this basis, the absolute EOL efficiency of InP will significantly exceed that of both Si and GaAs.

Figure 9 contains a plot of the normalized maximum power of heavily irradiated InP solar cells as a function of annealing temperature. The cells were produced by a closed tube diffusion process [Ref. 22]. Cells

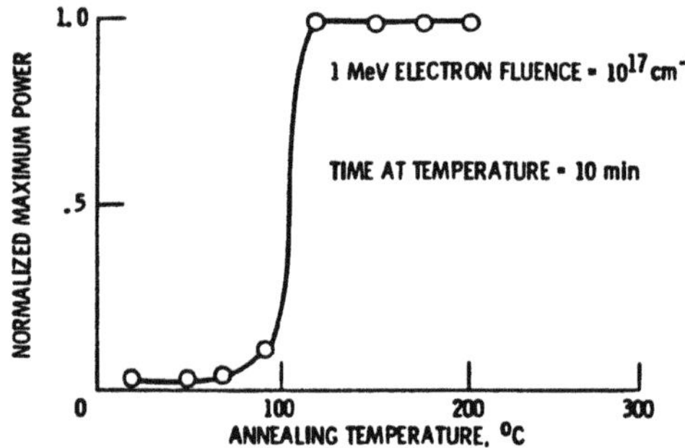

Fig. 9: Thermal annealing of 1 MeV electron damage in InP solar cells.

of this sort have already flow in space [Refs. 22, 23] and have exhibited outstanding radiation resistance [Ref. 24], as shown in Fig. 10. Although InP is somewhat less dense than GaAs, and needs only a 4 micron thick active layer, the cells are still significantly heavier than the 62 micron Si cell. To realize their full potential, InP cells need to be fabricated either on a silicon substrate, or as a stand alone thin cell, as has been described above for GaAs. Several studies have attempted to determine the optimum structure for InP cells for space application [Refs. 24-29]. Historically, interest developed in the n/p structure first, and the highest efficiency achieved thus far has been achieved in that structure. Some of the studies indicate, however, that the p/n cell may have a

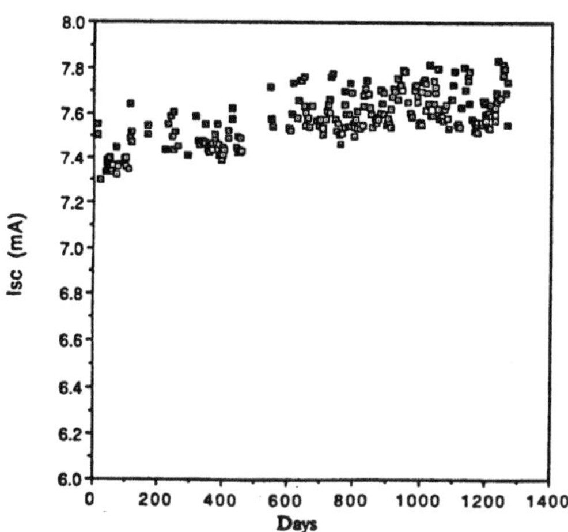

Fig. 10: On-orbit performance of InP solar cells.

slightly higher efficiency. That possibility, coupled with interest in growing InP directly on silicon substrates, makes the p/n structure the one of choice for future development. The p/n structure is favored for heteroepitaxial growth because n-type autodoping by silicon from the substrate occurs in the adjacent InP layer during the OMCVD growth. If the first layer is p-type, a p/n diode is formed which seriously degrades the performance of the cell [Ref. 30]. If the first layer is n-type, the autodoping may actually enhance the output of the cell in much the same manner as doping density gradients have been shown to work in silicon cells [Ref. 31].

Recent laboratory data has shown that the amount of degradation produced by 10 MeV protons is equivalent to that created by 1 MeV electrons at a fluence 650 times greater than the proton fluence [Ref. 32]. This compares to 3500 for silicon and 1000 for gallium arsenide [Ref. 33]. Moreover, 1 MeV electron radiation damaged InP solar cells appear to anneal significantly from minority carrier injection under forward bias (such as in normal operation in sunlight), as shown in Fig. 11 [Ref. 34]. This result may explain the lack of degradation shown in Fig. 10, and adds to the potential for little or no degradation of such cells on orbit.

An analysis of array performance under the orbital conditions anticipated for the Earth Observing Satellite (EOS) mission has been completed by Bailey, et al, [Ref. 35]. The study examined use of a heteroepitaxial InP solar cell, (5 micron thick InP active layer on 55 micron silicon substrates, with 50 micron thick coverglasses), on the baseline EOS array [Ref. 36]. The results show that significant array mass savings of more than 10% can accrue. Reducing the substrate and coverglass to a total of 20 microns or less is a next logical step for analysis and comparison. The calculation did not include any self-annealing effects, which would also add significantly to the weight saved, and would be

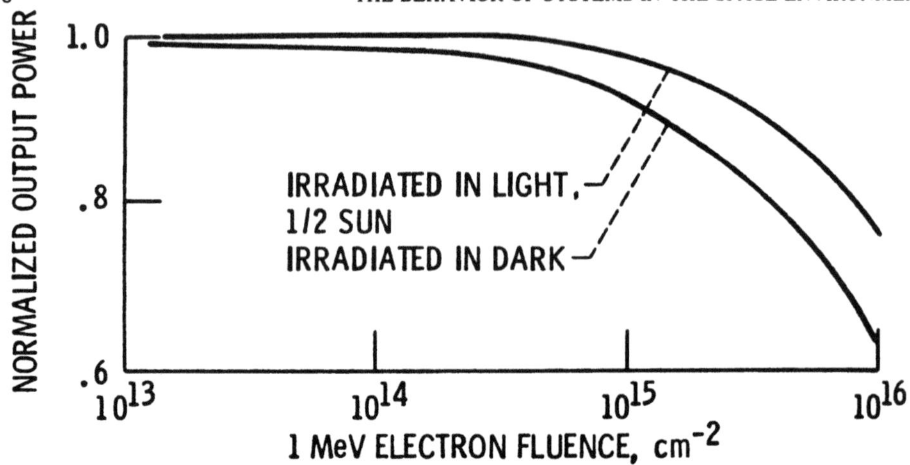

Fig. 11: Light-induced annealing in InP solar cells.

particularly applicable to electron-dominated orbits where the annealing appears to be most effective. Even so, use of the InP cell with its higher radiation resistance increases the EOS array specific power by more than 10%, and reduces array size by nearly 40%, compared to the baseline silicon cell presently planned for use.

Figure 12 contains a comparison of the projected lifetime of thin Si cells, GaAs cells, and InP cells in GEO. The comparison is made at the point where the accumulated fluence of 1 MeV electrons is projected to cause a 15% degradation in power for all three cells. The number of years in orbit was calculated by using the GEO equivalent fluence for silicon and must be taken as a caveat, since the true equivalent fluence is not yet known for InP. Nonetheless, the comparison is striking. InP cells offer the possibility of more than 30 year array lifetimes in GEO, with array specific powers well in excess of 300 W/kg.

Figure 13 provides an indication of what a future advanced InP based cell may include. The structure shown there is a monolithic, two junction cell which has the potential to be less than 10 microns thick. Analysis of two junction solar cells in the AM0 spectrum by Fan, et al [Ref. 37] indicate such as device could have an efficiency approaching 30% AM0. If the same sort of radiation damage resistance and annealing is found to be the case for such a device as is observed for InP cells, the possibility to achieve 30 year lifetimes in arrays with specific powers approaching 500 W/kg may become a reality. A device structure of the sort shown in Fig. 13 has been fabricated by Shen and co-workers [Ref. 38], but no AM0 efficiency data are available. Clearly, a great deal of work must be accomplished to determine the feasibility of achieving such performance, but the payoff in terms of radical advances in array specific power and satellite lifetime would be enormous.

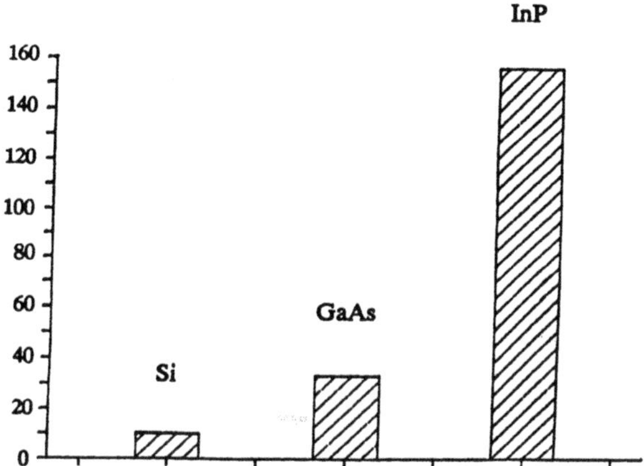

Fig. 12: Projected lifetime in GEO for Si, GaAs and InP solar cells.

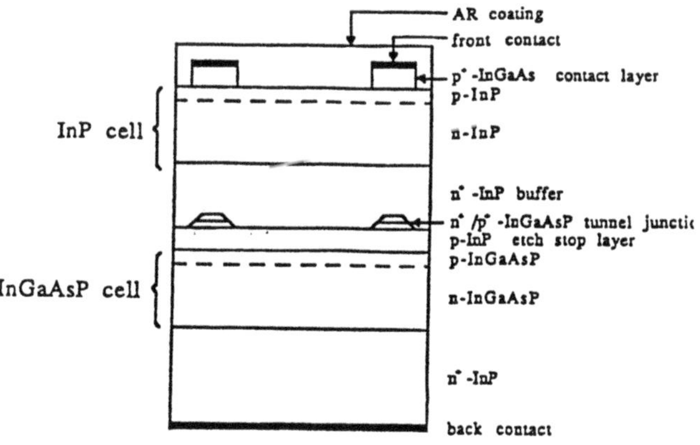

Fig. 13: InP/InGaAs monolithic, dual bandgap solar cell.

3.3. THIN FILM CELLS FOR SPACE APPLICATION

It has been axiomatic that high efficiency is the chief attribute sought in photovoltaic space power systems by mission planners, given that key reliability requirements are met first. The reasons for requiring high efficiency vary, and are dependent on both mission environment and array configuration. The usual result is that array size is restricted, placing a premium on cell efficiency. The area may be restricted because the array is body mounted, or, for deployable arrays, because onboard fuel storage is limited. The amount of fuel is important because of the need for momentum management and/or drag make-up (depending on the orbit), and strongly affects the on-orbit lifetime of the satellite. Other attributes such as radiation resistance, low mass and thermal cycle survivability have a greater or lessor degree of importance, depending on specific mission requirements.

Future mission requirements may well cause the traditional relative priorities of efficiency, radiation damage and mass to change, particularly as interest in establishing an infrastructure to support manned missions to the surface of the Moon and Mars grows. Power systems for surface operations place a new emphasis on low mass and transportability, or stowability, for the solar arrays. An additional element of the infrastructure required to support such missions is likely to be a system of cargo vehicles using electric propulsion. The contending power sources are solar arrays and nuclear reactors. Given the large power requirements expected, the end of life (EOL) specific power of the arrays becomes a dominant concern. Some early estimates are that EOL specific power must exceed 100 W/kg for at least one round trip from LEO to beyond the van Allen belts and back. The choices of LEO parking orbit and dwell time become critical in determining the most appropriate cell and array technologies to consider, and it now appears that system studies must trade ultrahigh EOL specific power carefully with array area when seeking to achieve optimum mission capability.

Included in Fig. 4 is an indication of the range of bandgaps for amorphous silicon cells, and their presently observed efficiencies. For the more well developed cells, such as Si and GaAs, observed efficiencies are very close to the predicted values. AM0 efficiencies obtained thus far for all the thin film cell types fall well below those values. The primary reason is that there are additional sources of efficiency loss because of the structural disorder of the material. It is not yet known if the efficiencies of these non-crystalline thin film cells will approach the values represented by the idealized calculation.

Although the thin film cells do not appear able to achieve efficiencies that compete with advanced, single crystal solar cells, they offer the potential for extremely high specific powers and low cost manufacturing techniques. The key technology issue is direct, monolithic fabrication of the cells and interconnects on space qualified, flexible substrates. There are at present three thin film cells of interest: amorphous silicon (a-Si), copper indium diselenide (CIS), and cadmium telluride (CdTe). Of these, only amorphous silicon has been fabricated appreciable quantity on flexible substrates of any sort. Materials used with various degrees of success include thin stainless steel and polyester sheet [Ref. 39], polyethylene terepthalate [Ref. 40] and polyimide (Kapton) [Refs. 41, 42]. Even though thin, the stainless steel substrate is still too heavy to yield high specific power arrays and is not of further interest in this discussion. Of the non metallic substrates, only the polyimide (Kapton) has been used in space solar arrays and has been shown to avoid degradation from the intense ultraviolet light in the AM0 spectrum. Efficiencies of comparable cell structures on stainless steel and polyethyene substrates are essentially the same, and comparable to those achieved on glass, the substrate most commonly used for terrestrial applications. Efficiencies of cells on Kapton are somewhat less than on glass or stainless steel, although efforts to develop structures on Kapton are most recent than on the other materials.

Fig. 14: Flexible amorphous silicon solar cell pilot production.

There are a large number of possible structures for a-Si solar cells, and not all of them have been fabricated on each of the flexible substrates. In fact, because of the large number of different cell types based on amorphous silicon, care must be exercised when comparing their efficiency and radiation resistance. Many reports of high radiation resistance came from measurements made on early, low efficiency cells, primarily single gap, single junction structures. Table 3 illustrates the situation. There is as yet little or no data on the radiation resistance of the more advanced, higher efficiency a-Si solar cell structures.

The inherent thinness of the active layers of single bandgap, single or tandem junction a-Si cells, coupled with the fact that minority carrier transport is dominated by electric field drift rather than diffusion, contributes in large measure to their observed high radiation resistance. However, higher efficiency structures will be thicker, since they are multiple bandgap structures that require more layers. The newer materials that will comprise the additional layers have not undergone extensive radiation damage testing, with the result that there is considerable

Table 3. Status of Thin Film Solar Celles for Space Applications.

[All efficiencies estimated from AM1 and AM1.5 measurements.]

Cell type	Cell structure	Projected efficiency, percent	Laboratory efficiency, percent	Commercial efficiency, percent	Radiation resistance, P/Po	
					1×10^{15} 1 MeV, e-	1×10^{13} 1 MeV, p+
a-Si	Single junction, single gap on rigid substrate	10	<9.0	<5.0	0.80	0.65
a-Si	Tandem junction, single gap on rigid substrate	12	9.9	<5.0	----	0.75
a-Si	Tandem junction, single gap on flexible substrate	10	5.5	----	----	----
a-Si	Tandem junction, dual gap on rigid substrate	15	8.6	----	----	----
a-Si	Monolithic, multiple bandgap on rigid substrate	18	10.9	----	----	----
CuInSe$_2$	3 μm cell, 1 μm window on glass substrate	>13	10.4	----	1.00	0.65
CuIn$_x$Ga$_{1-x}$Se	3 μm cell, 1 μm windown on glass substrate	>15	8.2	----	----	----
CdTe	Thin film on glass superstrate	>18	9.8	----	----	----
a-Si/CuInSe$_2$	Mechanically stacked tandem cell	>20	12.5	----	----	----

uncertainty about the degree of radiation resistance to expect from these cells.

For the sake of brevity, not all of the possible a-Si cell types are listed in the table, nor are all the individual substrates types differentiated (rigid substrates include both stainless steel and glass sheet, e.g), while only the results for cells on Kapton are included in the flexible substrate category. The very long term goal is to achieve a monolithic,multiple bandgap (triple junction, triple gap) 18% AM0 efficient cell on a flexible substrate with radiation resistance equal to or better than the best high efficiency single crystal cells. Clearly such a goal is ambitious. The payoff will be an ultralightweight solar array blanket with the potential for an EOL specific power (after an equivalent fluence of 1E17 1 MeV electrons/sq. cm) in excess of 1000 W/kg, and EOL area power densities above 150 W/sq. m.

We do not have space to discuss the problem in a-Si known as the Staebler-Wronski effect [Ref. 43], except to say that cell structures have recently been demonstrated in which it is limited to a 10% loss of power, compared to earlier values approaching 40% [Ref. 44]. Since it appears advantageous to have thin active layers in the cell to limit both Staebler-Wronski degradation and radiation damage, it is reasonable to expect that the multiple bandgap structure has a good chance of demonstrating excellent performance on both accounts.

Figure 14 shows an emerging section of a 50 micron thick, 33 cm wide Kapton ribbon that has been processed into a series of 30 cm x 30 cm amorphous silicon solar cell submodules in a roll-to-roll process. The cell structure produced is shown in Fig. 15. AM0 efficiencies approaching 5% have been achieved with the structure, although not as yet in the continuous processing technique [Ref. 45]. In a second approach, a Kapton film is laminated onto a thin stainless steel sheet and processed with the same techniques used to deposit and process a-Si directly on the stainless steel [Ref. 46]. After cell fabrication and interconnection are complete, the stainless steel is etched away, leaving a complete submodule in the form of a flexible, lightweight blanket. There are no radiation damage test results available for either submodule type. Although this work is very preliminary in nature, it does establish the feasibility of producing large area, monolithically fabricated, flexible amorphous silicon solar cells on Kapton. The next challenge is to fabricate tandem cells with efficiencies comparable to those achieved on glass substrates, followed by development of a suitable high efficiency cascade thin film solar cells.

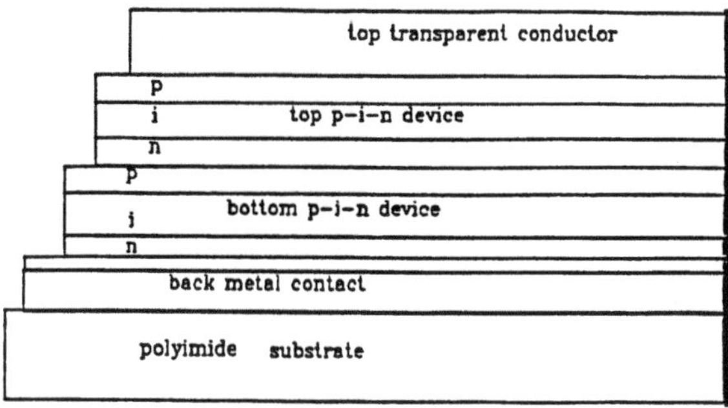

Fig. 15: Structure of flexible amorphous silicon solar cell.

CIS ($CuInSe_2$) cells on flexible, non metallic substrates have only recently begun to be investigated by NASA at the time of this writing, with the result that there is essentially no data to report. A related effort to deposit CIS cells on this metal foils has been reported [Ref. 47], but again, no efficiency or radiation damage data are available at this time. Radiation damage studies on CIS cells deposited on conventional glass substrates have shown superior resistance to 1 MeV electron radiation compared to the best single crystal cells of any type [Ref. 48], and good resistance to proton radiation damage [Ref. 49], as indicated Table 3. There is no reported work on deposition on flexible substrates for any of the remaining cells listed in Table 1. They are included here because they offer the potential for higher efficiency than CIS cells, with the possibility that they could be incorporated in monolithically integrated, flexible, thin film submodules.

3.4. MULTIPLE BANDGAP SPACE SOLAR CELLS

No attempt will be made to summarize all the multiple bandgap solar cell types that have potential for space application. Candidates range from two junction, mechanically stacked, two or four terminal devices to monolithically grown three junction structures with a variety of interconnect configurations. They can be combinations of thin film and single crystal devices, and can be designed for either planar or concentrator operation. The key point is that there have been rapid and significant advances in the technology for producing such cells [Refs. 50, 51, 52, 53].

The ability to accurately measure the performance of multiple bandgap cells, however, has only recently begun to be addressed [Ref. 54]. Work recently performed at NASA's Lewis Research Center has clearly shown the need for extreme care with MBG cell measurements. Results of experiments performed in the NASA high altitude aircraft to obtain the full I-V curve and the temperature dependence of MBG cells at low air mass show conclusively that conventional laboratory techniques will be misleading [Ref. 55]. Figure 16 illustrates the point. The variation with temperature of the bandgap of the upper cell has a major effect on overall device efficiency, as does the spectral content of the incident light used in the testing. Conventional laboratory simulators are often too rich in the red region of their output spectrum, and can give misleading (usually high) values for the MBG device efficiency. Measurement of the correct temperature dependence of such cells in the incorrect spectrum is virtually impossible. Until more advanced laboratory light sources are available, high altitude measurement of the full I-V characteristic and temperature dependence of the test device will be the only way to obtain correct results.

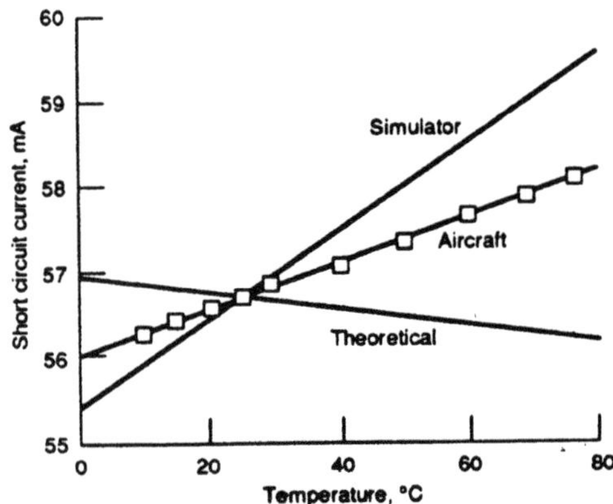

Fig. 16: AlGaAs/GaAs tandem solar cell short circuit current measurement.

Although multiple bandgap cells with up to three junctions are feasible, the gain in efficiency over two junction devices does not appear to justify the complexity in fabrication that appears necessary. For that reason, more attention has been paid to two junction devices. An issue not yet settled is which of two fundamentally different types of two junction cells is preferable for space application: monolithically grown cells or mechanically stacked cells. In the first case, the devices are most likely to have two terminals for interconnection to other cells on an array. Three, and even four terminal configurations are possible, with a great deal more complexity required in fabrication. In the second case, the devices are most likely to have four terminals, although again, two or three terminal configurations are possible, also with more difficulty in fabrication. From a practical standpoint, array designers have traditionally preferred two terminals; three or four terminals will increase the complexity of the interconnect design and wiring harness on the solar array. On the other hand, the monolithically grown cell will exhibit more radiation degradation than the four terminal device [Ref. 56], and has more stringent conditions on the selection of bandgaps to maximize its efficiency [Ref. 37]. As we shall see in the next section, the use of two junction, four terminal, mechanically stacked cells may actually be an advantage in concentrator arrays. The array configuration does not make it difficult to use new wiring harness designs, so that a unique voltage-matched interconnection scheme becomes possible.

4. Advanced Space Solar Array Technology

4.1. FLEXIBLE, LIGHTWEIGHT PLANAR ARRAYS

Currently flying solar arrays are essentially of two types: body mounted (i.e., the cells are attached directly to the body of the spacecraft); and a deployable rigid panel design, such as on the Skylab. Rigid panel arrays can be comprised of several panels hinged together, and can reach power levels of 10 to 20 kilowatts. The specific power is typically in the range from 20 to 30 W/kg at BOL. Such arrays grow rapidly in mass as they grow in power level, with the result that spacecraft momentum management becomes difficult. For that reason an alternative deployable array type has been developed which uses thin, flexible panels made of polyimide. The version under development by NASA is known as the Advanced Photovoltaic Solar Array, or APSA, as mentioned previously [Ref. 19]. Using the 62 micron silicon solar cell described above, the specific power at BOL exceeds 130 W/kg, about 4 times the best value for rigid arrays. The area power density is the same, since it is a function primarily of cell efficiency, although array design can affect it as well. Figure 17 shows a prototype 1 kilowatt wing undergoing tests for simulated zero gravity deployment.

A key feature of this sort of array design is that about half the total mass goes into the storage container, the mast, and the deployment system (motors, etc.). In general, these components do not change significantly as the array size changes from one or two kilowatts to over 10 kilowatts, with the result that array specific power actually grows by about 50% as the power grows to that level [Ref. 57]. As array power

Fig. 17: Advanced Photovoltaic Solar Array (APSA) prototype ground deployment test and demonstrations.

increases beyond that, the specific power remains essentially constant, as it does for rigid arrays. It is this feature that makes this array technology so attractive for all future missions requiring both increased power and reduced mass.

4.2. ADVANCED CONCENTRATOR ARRAY TECHNOLOGY

Figure 18 shows a 36 element submodule of an advanced space concentrator array concept now under development by NASA. A schematic of the basic conversion element is shown in Fig. 19. It consists of a unique, lightweight domed fresnel lens [Ref. 58] mounted over a high efficiency concentrator cell. The cell can be a single junction, two terminal cell, or a multiple bandgap cell with multiple terminals. This sort of technology transparency is one of the key features of the design; the second is the potential for low cost. The latter derives from the fact that concentrator arrays require a greatly reduced area to be covered by expensive semiconductor devices compared to planar arrays of the same output. An equivalent area must be covered by the lenses, but they are made inexpensively out of low cost materials.

The panel shown is designed for operation at a nominal 100X concentration ratio. Because the concentrator cell covers less than 5 percent of the panel area for the submodule shown, there is sufficient room on the panel to allow for innovative interconnection and wiring harness

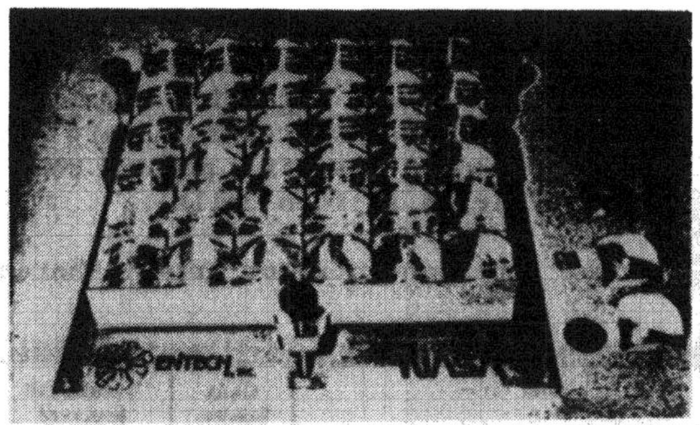

Fig. 18: Early prototype concentrator module with minidome lenses.

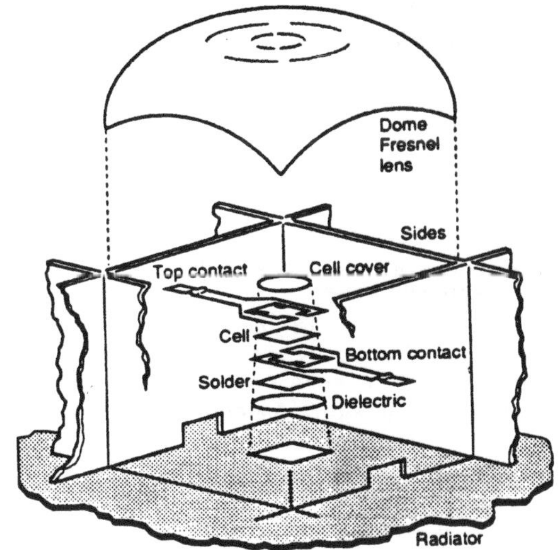

Fig. 19: Schematic diagram of concentrator lens/cell element.

designs to be used without any impact on the area power density, as would be the case for a typical planar array, such as the APSA. As a result, a mechanically stacked two junction cell, such as the 30% GaAs/GaSb device developed by Fraas, et al [Ref. 59] can be interconnected in a voltage-matched configuration [Ibid] rather than in a simple series (two terminal) or parallel (four terminal). Voltage matched interconnection is made possible because the operating voltage for GaAs is about three times higher than for GaSb. Hence, three GaSb cells can be connected in series with each other, and the series string can be placed in parallel with one GaAs cell. An immediate consequence is that the effects of higher operating temperature and radiation damage are not as severe as they would be for a two terminal device, and wiring design complexity is greatly reduced. This basic wiring scheme can be repeated until all the cells on the array are interconnected.

Table 4 provides a comparison of the estimated array parameters obtained when GaAs and GaAs/GaSb concentrator cells are used for conversion. The potential exists for significant gains in area power density with this technology compared to state-of-the-art rigid or flexible planar silicon solar arrays, with specific powers comparable to the baseline APSA (i.e., with 62 micron silicon cells).

Table 4. Advanced Minidome Lens Concentrator Arrayy Characteristics.

[Measured performance parameters for prototype cells and lenses are underlined.]

Item	GaAs baseline	Tandem cell improved
Cell type	GaAs	GaAs + GaSb
Cell efficiency (25 °C), percent	24	24 + 8 = [a]32
Cell operating temperature, °C	80	80 and 80
Cell efficiency (operating temperature), percent	22	29
Lens efficiency, percent	90	[b]95
Packing factor, percent	97	97
Mismatch/wiring, percent	93	93
Array efficiency, percent	18	25
Power density (W/m^2), percent	**245**	**340**
Panel mass (kg/m^2)	2.4	2.4
Structure mass (kg/m^2)	0.7	0.7
Array mass (kg/m^2)	3.1	3.1
Specific power (w/kg)	**79**	**110**

[a]Current performance measured at 30 percent AMO, 25 °C.
[b]With addition of anti-reflection coating.

The submodule shown in Fig. 18 has been designed to maintain essentially full output when the array is pointed off axis by up to ±1 degree. The pointing accuracy tolerance can be increased up to ±4 degrees with very little impact on array power. As shown in Fig. 20, on-axis array

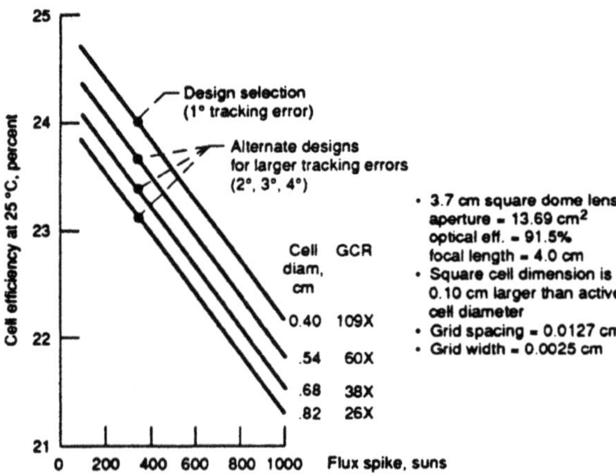

Fig. 20: Variation of concentrator cell efficiency with pointing error tolerance as a parameter.

maximum power will decrease by less than 4% when the pointing accuracy tolerance is increased to ±4 degrees. A 12 element submodule incorporating ± 2 degree pointing accuracy tolerance and the GaAs/GaSb tandem concentrator cell is currently under development for a flight test in early 1993, as part of a broader evaluation of space environmental effects on advanced solar cell and array technology [Ref. 60]. The flight test will determine the performance of a variety of advanced solar cell and array technologies in orbit. The data are expected to be useful not only for evaluating advanced array technology, but also for establishing acceptable levels for solar array operating voltages for a range of near earth orbits.

5. Conclusion

A wide variety of potential space solar cell technologies are beginning to emerge, ranging from ultra lightweight, flexible thin film submodules to advanced, high efficiency concentrator arrays using multiple bandgap solar cells. Blanket BOL specific powers exceeding 1000 W/Kg are now feasible, with only modest advances in thin film efficiency. The radiation resistance of future space solar cells may be such that array lifetimes may be extended by at least a factor of 3. Concentrator array BOL power densities exceeding 300 W/sq.m. are also possible, with specific powers competitive with current lightweight array technology.

References

[1] H.S. Rauschebach, Solar Cell Array Design Handbook, Van Nostrand Reinhold Company, (1980).

[2] M. Mirtich, Private Communication, Lewis Research Center, National Aeronautics and Space Administration, Cleveland, Ohio, (1991).

[3] D.L. Nored, D.T. Bernatowicz, Proceedings of the 21st Intersociety Energy Conversion Engineering Conference, San Diego, CA, Aug. 25-29, 1986; Chemical Society, Washington, DC, (1986) Vol. 3, pp. 1416-1422.

[4] B.E. Anspaugh, R.S. Weiss, "Results of the 1991 NASA/JPL Balloon Flight Solar Cell Calibration Program. JPL Publication 91-36, Jet Propulsion Laboratory, Pasadena, CA (1991).

[5] D.J. Brinker, Private communication, Lewis Research Center, National Aeronautics and Space Administration, Cleveland, Ohio, (1991).

[6] I. Weinberg and C.K. Swartz, Appl. Phys. Lett. 36, p. 693, (1980).

[7] I. Weinbert, S. Mehta and C.K. Swartz, Appl. Phys. Lett. 44, p. 1071, (1984).

[8] D.J. Flood and H.B. Brandhorst, Jr., "Current Topics in Photovoltaics," Vol. 2, Chapter 3, Academic Press, Inc., London (1987).

[9] G. Datum, Proceedings of the 22nd Photovoltaic Specialists Conference, Las Vegas, NV, Oct. 7-11, IEEE, New York, (1991).

[10] A.M. Barnett, R. H. Parekh, E.H. Mueller and C.C DeSalvo, Proceedings of the 17th Photovoltaic Specialists Conference, p. 50, IEEE, New York (1984).

[11] C. Goradia, M. Goradia and H. Curtis, Proceedings of the 17th Photovoltaic Specialists Conference, p. 56, IEEE, New York (1984).

[12] Solar Cells - "Outlook for Improved Efficiency Report by the AD Hoc Panel on Solar Cell Efficiency," National Research Council, National Academy of Sciences, Washington, DC, (1972).

[13] M. Ladle-Ristow, Proceedings of the ASME/JSES/KSES International Solar Energy Conference, Lahaina, Hawaii, April 5-11, 1992. To Be Published.

[14] M.A. Green, "High Efficiency Silicon Solar Cells," Trans Tech Publications, Ltd., Switzerland (1987).

[15] M.A. Green, et al, Proceedings of the 21st Photovoltaic Specialists Conference, p. 207, IEEE, New York, (1990).

[16] H.Y. Tada, J.R. Carter, B.E. Anspaugh and R.G. Downing, Solar Cell Radiation Handbook, Third Edition, Jet Propulsion Laboratory, Pasadena, CA (1982).

[17] Ron Gale, et al, Proceedings of the 20th Photovoltaic Specialists Conference, p. 446, IEEE, New York, (1988).

[18] C.L. Chu, P.A. Iles and W. Patterson, Proceedings of the Eleventh Conference on Space Photovoltaic Research and Technology, Lewis Research Center, Cleveland, OH, NASA CP-3121 (1991).

[19] P. Stella and R.M. Kurland, Proceedings of the 11th Conference on Space Photovoltaic Research and Technology, Lewis Research Center, Cleveland, Ohio, NASA CP-3107, (1989).

[20] D.J. Flood, Proceedings of the 4th International Photovoltaic Science and Engineering Conference, University of New South Wales, Sydney, Australia, (1989).

[21] C.J. Keavney, V.E. Haven and S.M. Vernon, Proceedings of the 2nd International Conference on Indium Phosphide and Related Materials, Denver, CO, (1990).

[22] M. Yamaguchi, Proceedings of the 20th Photovoltaic Specialists Conference, p. 880, IEEE, New York, (1988).

[23] M. Yamaguchi, et al, Proceedings of the 21st Photovoltaic Specialists Conference, p. 1198, IEEE New York, (1990).

[24] I. Weinberg, Solar Cells, Elsevier Press, The Netherlands, (1991). To Be Published.

[25] D.J. Brinker and I. Weinberg, Proceedings of the 21st Photovoltaic Specialists Conference, p. 1167, IEEE, New York, (1990).

[26] C. Goradia, J.V. Geir, and I. Weinberg, Solar Cells, Vol. 25, p. 235, (1988).

[27] S.L. Rhoades and A.M. Barnett, Proceedings of the 9th Conference on Space Photovoltaic Research and Technology, Lewis Research Center, NASA CP-3030, (1988).

[28] A.H. Yahio, et al, Proceedings of the 20th Photovoltaic Specialists Conference, p. 702, IEEE, New York, (1988).

[29] R.K. Jain and D.J. Flood, Proceedings of the 5th International Photovoltaic Science and Engineering Conference, Kyoto, Japan, (1990).

[30] Spire Corporation Final Report on NASA Contract NAS3-25283, (1989).

[31] M.P. Godlewski, C.R. Baraona and H.W. Brandhorst, Jr., Proceedings of the 10th Photovoltaic Specialists Conference, pg. 40, New York, (1973).

[32] T. Takamoto, et al, Proceedings of the 5th International Photovoltaic Science and Engineering Conference, p. 547, Kyoto, Japan, (1990).

[33] B.E. Anspaugh and R.G. Downing, Proceedings of the 19th Photovoltaic Specialists Conference, p. 23, IEEE, New York, (1984).

[34] K. Ando and M. Yamaguchi, Appl. Phys. Lett. 47, p. 846, (1984).

[35] S.G. Bailey, I. Weinberg, and D.J. Flood, NASA Technical Memorandum 105222, (1991).

[36] EOS Reference Handbook, Goddard Spaceflight Center, (1990).

[37] J.C.C. Fan and B.J. Palm, Proceedings of the 6th Conference on Space Photovoltaic Research and Technology, Lewis Research Center, NASA CP2314, (1983).

[38] C.C. Shen, et al, Proceedings of the 11th Conference on Space Photovoltaic Research and Technology, Lewis Research Center, NASA CP-3121, (1991).

[39] J.H. Hanak, et al, Proceedings of the 8th Conference on Space Photovoltaic Research and Technology, p. 121, Cleveland, Ohio, NASA CP-2475, (1986).

[40] N. Nakatani, et al, Proceedings of the 4th Photovoltaic Science and Engineering Conference, p. 639, Sydney, Australia, (1989).

[41] F.R. Jeffrey, et al, Proceedings of the 19th Photovoltaic Specialists Conference, p. 588, IEEE, New York, (1987).

[42] P. Nath, et al, Proceedings of the 5th Photovoltaic Science and Engineering Conference, p. 647, Kyoto, Japan, (1990).

[43] D.L. Staebler and C.R. Wronski, Applied Physics Letters, 31, p. 292, (1977).

[44] E.S. Sabisky and J.R. Stone, Proceedings of the 20th Photovoltaic Specialists Conference, p. 39, IEEE, New York, (1988).

[45] F.R. Jeffrey, et al, "Flexible, Lightweight Amorphous Silicon Solar Cells Tuned for the AMO Spectrum," Final NASA Contract NAS3-25825, (1990).

[46] S.B. Guha, "Development of Advanced High Energy Ultralight Photovoltaic Array for Non-Nuclear Space Power, "Final Contract Report, NASA CR-25458, (1990).

[47] J.A. Armstrong, "Flexible Copper-Indium-Disselenide Films and Devices for Space Applications," Proceedings of the 11th Conference on Space Photovoltaic Research and Technology, NASA CP-3121, (1991).

[48] B.J. Stanbery, et al, Proceedings of the 19th Photovoltaic Specialists Conference, p. 280, IEEE, New York, (1987).

[49] R.M. Burgess, et al, Proceedings of the 20th Photovoltaic Specialists Conference, p. 900, IEEE, New York, (1989).

[50] J.M. Olson, et al, Proceedings of the 11th Conference on Space Photovoltaic Research and Technology, Cleveland, Ohio, NASA CP-3121, (1991).

[51] N.P. Kim, et al, Ibid.

[52] K.R. Wickham, et al, Ibid.

[53] L.M. Fraas, et al, Proceedings of the 24th Intersociety Energy Conversion Engineering Conference, Washington, DC, (1989).

[54] G.E. Virshup, Proceedings of the 21st Photovoltaic Specialists Conference, p. 1249, IEEE, New York, (1990).

[55] D.J. Flood, NASA Technical Memorandum 104505, (1991).

[56] H.B. Curtis, C. K. Swartz and R. E. Hart, Jr., Proceedings of the 19th Photovoltaic Specialists Conference, p. 727, IEEE, New York, (1987).

[57] P.M. Stella and D.J. Flood, NASA Technical Memorandum 103284, (1990).

[58] M.J. O'Neill and M.P. Piszczor, Proceedings of the 9th Conference on Space Photovoltaic Research and Technology, p. 443, NASA CP-3107, (1989).

[59] L.M. Fraas, et al, Proceedings of the 24th Intersociety Energy Conversion Engineering Conference, pg. 815, Washington, DC (1989).

[60] D. Guidice, P. Severance and K. Reinhardt, Proceedings of the 11th Conference on Space Photovoltaic Research and Technology, p. 33-1, NASA CP-3121, (1991).

A BRIEF OVERVIEW OF ELECTRODYNAMIC TETHERS

DANIEL E. HASTINGS AND ROBIE I. SAMANTA ROY
Space Power and Propulsion Laboratory, Room 37-451
Department of Aeronautics and Astronautics
Massachusetts Institute of Technology
Cambridge, MA 02139 USA

1. Introduction

Tethers in space can be used for a wide variety of applications such as power generation, propulsion, remote atmospheric sensing, momentum transfer for orbital maneuvers, micro-gravity experimentation, and artificial gravity generation. These are only a few of the host of uses that have been envisioned and proposed for many years. In general, a tether is a long cable (even up to 100 km or more) that connects two or more spacecraft or scientific packages. Electrodynamic tethers are conducting wires that can be either insulated (in part or in whole) or bare, and that make use of an ambient magnetic field to induce a voltage drop across their length. The induced voltage is given by $|\mathbf{v}\times\mathbf{B}\bullet\mathbf{L}|$, where v is the relative velocity to the ambient plasma (the orbital velocity less the earth's rotational speed), B is the geomagnetic field strength ($2\text{-}6\times10^{-5}$T), and L is the length of the tether. For a 20 km tether in low earth orbit (LEO), this voltage would generally fluctuate between 1500 to 5300 volts open circuit, depending on the orbital inclination. If a current is allowed to circulate through the tether and a load, substantial power on the order of 15-30 kW can be generated. However, this power is generated at the cost of orbital energy. Hence an electrodynamic force $|ILxB|$, on the order of a few newtons, is exerted on the tether, lowering the orbit. On the other hand, with a sufficiently large power supply onboard the spacecraft, the direction of the current can be reversed and the tether becomes a thruster, raising its orbital height. Thus a spacecraft can use an electrodynamic tether system as a pure power generator (with a small rocket to periodically make-up for the drag), as a pure thruster, or in a combination of both roles.

R. N. DeWitt et al. (eds.), The Behavior of Systems in the Space Environment, 825–835.

2. History

The history of the concept of space tethers is traced out in an interesting summary by Grossi [Ref. 1], from which a few highlights will be drawn. The origin of tethers is placed during the last century when the Russian astronautics pioneer Kostantin Tsiolkovsky conceived of an "Orbital Tower" in 1895. His concept was a huge Eiffel Tower-like cable structure that would reach geosynchronous heights from the Earth's surface and could be used as a means for launching space flights. The next mention of tethers appeared in a Sunday edition of Pravda in 1960. Based on Tsiolkovsky's idea, Y.N. Artsutanov proposed a "Heavenly Funicular" which was a satellite in Geosynchronous Earth Orbit (GEO) with two long cables: one upwards, and one downwards towards the Earth's surface. These concepts captured the imagination of the famous science-fiction author Authur Clarke, who wrote an article in 1963 entitle, "Space Elevator". He also published the first novel containing long orbiting tethers, The Fountain of Paradise in 1979. On the more "academic" side, R.D. Moore proposed a "Geomagnetic Thruster" in 1966. This device was a conducting wire terminated at both ends by plasma contactors, the first true electrodynamic tether. Another innovative idea appeared in 1969. To reduce the transmitter power requirements of geostationary communications satellites, A. Collar and J. Flower proposed "A (Relatively) Low Altitude 24 Hour Satellite". This concept used a tether that extended down from a satellite in GEO to a transmitting subsatellite in LEO. Many other variations of these concepts followed, but a notable one was H. Alfvén's "Solar Wind Engine" in 1972. Alfvén proposed a 500 km superconducting cable, terminated with ion and electron emitters, that would utilize the solar magnetic field for propulsion. Meanwhile, in the experimental arena, the first tether experiments were conducted during the Gemini Program in the 1960's and were aimed at exploring the dynamics of tethered space vehicles.

The years 1972-3 saw the dawn of the Shuttle-borne tether era. M. Grossi and his colleagues at the Smithsonian Astrophysical Observatory (SAO), proposed using a tether as an orbiting ULF/ELF antenna to be carried on the Space Shuttle. Since then, much theoretical work has been done on tethers and the subject has grown substantially, culminating in the formation of a joint US-Italian space program with the goal of developing a Tethered Satellite System (TSS) to be operated from the Shuttle [Ref. 2]. The TSS program plans to launch the first tethered satellite TSS-1, a 20 km upwardly deployed conducting tether. TSS-1 will explore some basic dynamic, control, and electrodynamic issues and flew during 1992. A second program, TSS-2, will follow a couple of years later and will consist of a 100 km non-conducting tether deployed downward. This mission's objective is to conduct experiments mainly related to upper atmospheric physics.

3. General Principles of Operation

Consider a tether in LEO as in Fig. 1. For prograde low inclination orbits, the velocity vector (**v**) of the vertically orientated tether points eastwardly, and is almost perpendicular to the magnetic field lines (B)

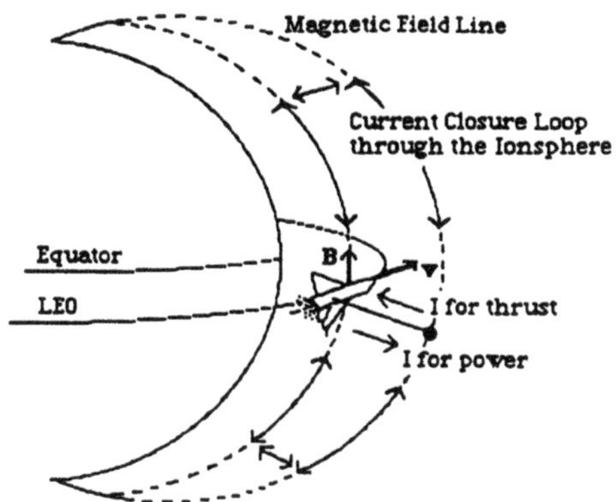

Fig. 1a: Schematic of tether in LEO.

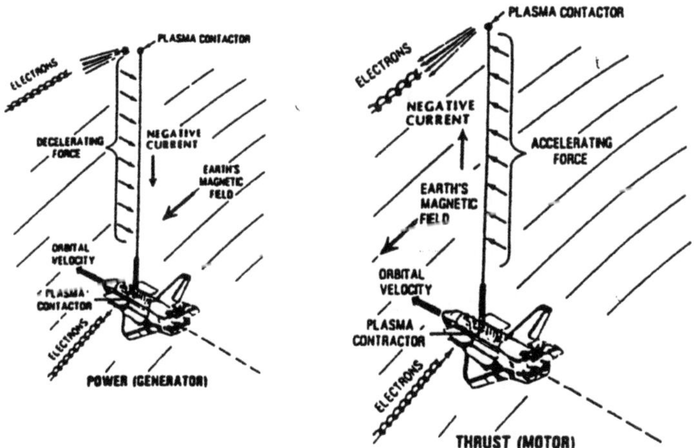

Fig. 1b Modes of electrodynamic tethers.

which run south to north. Charged particles in the tether will experience a force given by the Lorentz relation,

$$\mathbf{F} = q(\mathbf{E} + \mathbf{v} \times \mathbf{B}) \tag{1}$$

where **E** is the ambient electric field of the LEO plasma which is very small and will therefore be neglected. Hence the motional electric field **v** x **B** induces an EMF given by,

$$V_{\text{ind}} = \int \mathbf{v} \times B \cdot dl \tag{2a}$$

where dl is an element of length along the tether. Strictly speaking, the velocity **v** is the relative velocity of the tether to the ambient plasma that is co-rotating with the earth. However, the rotational speed of the earth is small compared to the orbital velocity of the tether, and usually **v** is taken to be the orbital velocity. In most cases, the tether is assumed to be straight so the induced voltage becomes,

$$V_{ind} = |\mathbf{v} \times \mathbf{B}L| \tag{2b}$$

where L is the length of the tether. If current is allowed to flow through the tether, the electromagnetic force on the tether will be,

$$F_{EM} = \int I \, dl \times B \tag{3a}$$

or when the tether is assumed rigid,

$$\mathbf{F}_{EM} = L\mathbf{I} \times \mathbf{B} \tag{3b}$$

The direction of the current will determine the direction of the force, whether it will be a thrusting force, or a drag force. For an upwards (or downwards) deployed tether, if the current flows up (away from the earth), power is generated at the cost of orbital energy since the electromagnetic force is anti-parallel to the direction of motion. On the other hand, if a power supply large enough to reverse the induced EMF drives the current down, the electromagnetic force vector acts parallel to the direction of motion. Of course, these directions are reversed for retro-grade orbits.

In an electrodynamic tether, the current that flows is actually the electrons that are collected from the ambient ionospheric plasma. Electrons are drawn in on one end of the tether, and are ejected out the other, the particular end depending on whether the tether is thrusting or generating power. The end of the tether that must collect electrons and/or eject ions is called the anode. The other end that ejects electrons and/or collects ions is called the cathode. Paramount to this process of current flow are two important issues. One is the ionospheric resistance to a current flowing through it. The other is the actual ability for the collecting end of the tether to collect the electrons with little voltage loss. The electron emitting end does not seem to pose a large problem since space tests have demonstrated that large currents can be ejected with little voltage drop.

The ionospheric impedance is actually due to a complicated electromagnetic phenomena. Analogous to a ship creating waves as it moves through the water, a moving conductor through a plasma generates electromagnetic waves. These waves dissipate energy with some effective resistance, called the radiation impedance, which has been examined in some detail [Refs. 3,4]. It has been found that radiation is emitted in three distinct bands, the Alfvén, lower hybrid, and upper hybrid bands. For a long tether, the lower hybrid band is of most significance and the impedance is highly dependent on the ambient ionospheric conditions, i.e., the electron density which varies considerably over an orbit. The impedance is also a strong function of the dimensions of the collecting ends of the tether system: the impedance is inversely proportional to the diameter. There is still some controversy over the impedances that will be encountered by long tether systems, but it is strongly believed that the order of magnitude will be around 10-20 ohms.

Research is still actively going on to investigate means of electron collection and the underlying physical processes. Several different options exist for electron collection such as a passive large surface (like a balloon), a passive grid, a plasma, or a light ion emitter. The most promising of these devices is the plasma contactor. The ambient electron density in low earth orbit (LEO) is rather low (10^{10-12} m^{-3}), so to collect the required current requires a very large surface area (~100-1000 m^2 for 1 amp). Instead of using a large physical area, plasma contactors create a plasma cloud that expands out and collects ambient electrons while emitting ions. It is important to note that these contactors operate by ejecting fully or partially ionized gas. For a device using argon, the mass flow rate is about 13 kg/yr/amp. A recently proposed method for electron collection [Ref. 5], is to leave part of the tether bare, i.e., to have only part or the whole of the conducting wire uninsulated. For a 20 km tether, up to ten kilometers, which represents a rather large area, will be positive to the plasma and can collect electrons. The inherent advantage of this scheme is the absence of the mass and complexity of a collecting contactor.

In the strict sense, the picture of electron emission and collection into the ionosphere as a DC phenomenon is not entirely correct. Electrons emitted and collected are constrained to travel along the magnetic field lines, or flux tubes which can be thought of as parallel transmission lines. These transmission lines are excited as the tether ends contact them, hence the phenomenon is fundamentally AC. However, since the magnetic field lines in reality form a continuous media, the current flow is DC.

A circuit equation can be written consisting of various voltage drops for a tether system. For a tether generating power we can write,

$$V_{IND} = \Delta V_A + \Delta V_C + IZ_I + IR_T + IR_L \tag{4a}$$

where ΔV_A and ΔV_C are the voltage drops across the anode and cathode respectively, Z_I is an effective ionospheric impedance, R_T is the tether ohmic resistance, and R_L is the load. If we define an efficiency, η, as

$$\eta = \frac{Power_{LOAD}}{Power_{TOTAL}} = \frac{IV_{LOAD}}{IV_{IND}} = \frac{V_{LOAD}}{V_{IND}} \tag{4b}$$

then, the circuit equation, Eq. (4a), can be rewritten as,

$$\Delta V_A + \Delta V_C + I(Z_I + R_T) = V_{IND}(1 - \eta) \tag{4c}$$

It can be shown that for any given operating conditions (i.e., V_{IND} and electron density), there exists a unique value of η where the power generated I^2R_L is maximized.

A similar equation can be written for a tether generating thrust, except now, an onboard power supply is required to reverse the current,

$$V_{IND} + \Delta V_A + \Delta V_C + I(Z_I + R_T) = V_{PS} \tag{4c}$$

where V_{PS} is the voltage of the power supply. Figures 2a,b show schematics of these tether circuits, and Fig. 3 shows a schematic of a bare tether.

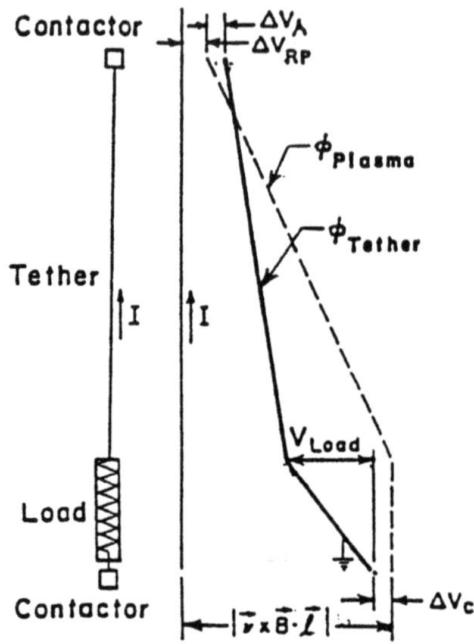

Fig. 2a: Diagram of tether as generator, deployed upwards [Ref. 15].

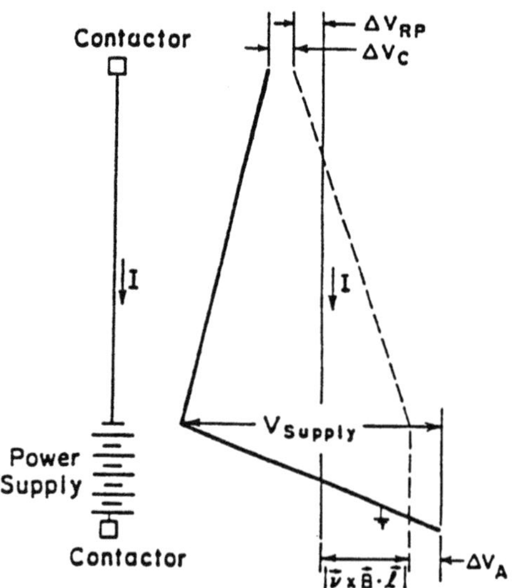

Fig. 2b: Diagram of tether as thruster, deployed upwards [Ref. 15].

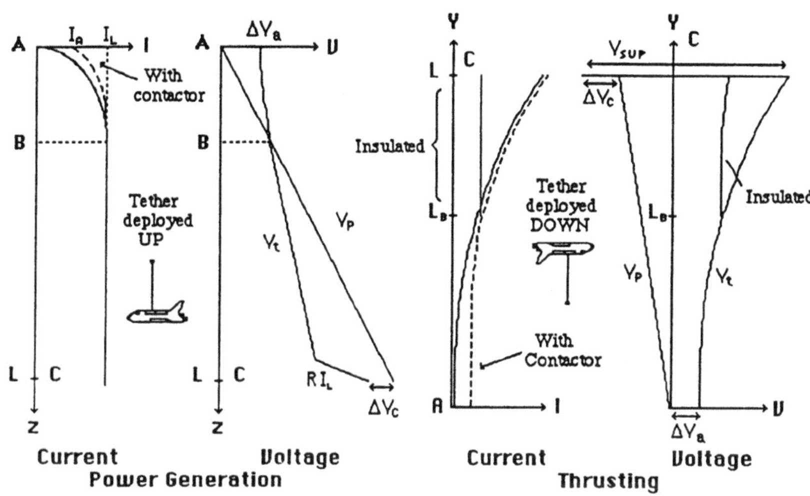

Fig. 3: Bare tether circuit schematics.

4. Past and Current Tether Research

Space experiments during the 1970's onboard the Space Electric Rocket Test (SERT) II showed that electrons could be ejected into the ionosphere with relatively little voltage loss. However, space experiments where multi-ampere currents of electrons are collected have never been performed. Hence, this area of electron collection has been examined theoretically at some length for some time. However, even today, the models are still far from being complete. Due to the lack of any in-space tests, the validity of models cannot be verified. The main complicating factor is the presence of the geomagnetic field which constrains electrons along the magnetic field lines, thus making the problem anisotropic. For typical values of $B \sim 2\text{-}6\times10^{-5}$T, the electron gyroradius ($m_e v/eB$) is about 2.5 cm. Due to the magnetized nature of the ambient plasma, spherical models describing the electron collection process and ground based experiments where magnetized effects are weak, are inadequate. Theoretical work has fallen into two categories: collisionless double layer theory describing space charge limited clouds, and collisional quasi-neutral theory [Ref. 6].

The various double layer models all have diverse hypotheses on the size of the inner core region of the plasma cloud where electrons are collected. Parks and Katz [Ref. 7] used the requirement of matching the cloud density with the ambient density. Dobrowolny and Iess [Ref. 8] presented another model where they required regularity of the self-consistent potential at the outer boundary of the core, $\partial\phi/\partial r(r_{core}) \approx 0$. Lastly, Wei and Wilbur [Ref. 9] applied consistent space charge limited flow in the core $m_i n_i U_i^2 \approx m_e n_e U_e^2$. More recently, Ahedo et al [Ref. 10] developed a rigorous theory for the steady-state collisionless interaction between a spherical anode and an unmagnetized plasma. It was found that the maximum current collected was in the inertial motion limit. Due to their neglect or weak inclusion of magnetized effects, these models are more applicable to ground based experiments, but not to space. Ground based experiments have not been able to fully simulate space conditions due to different plasma densities and magnetic fields.

The other contactor model category is collisional quasi-neutral theory. Hastings [Ref. 11] attempted to take magnetic effects into account by equating the electron collision frequency with the electron gyro frequency, $\nu_e(R_{core}) \approx \omega_{ce}$, at the outer region of the cloud. Hastings and Gatsonis [Ref. 12] examined a multi-region cloud that was anisotropic along the direction of orbital motion, not along the magnetic field. One drawback of this model was that it did not fully include anisotropic effects along the magnetic lines.

One of the latest models of contactor clouds, by Gerver, Hastings, and Oberhardt [Ref. 6], is an anisotropic model oriented along the magnetic field. This model has been recently extended to include neutral gas emissions and external ionization [Ref. 13].

Over the last several years, there have been a number of system or engineering studies conducted, concentrating on the electrodynamic aspects of tethers. These studies have examined the uses of tethers in propulsion and/or power generation applications. We will review three of such studies. One of the very first of these studies was by Gross and Arnold in 1984 [Ref. 14] who employed a computer simulation model that included a geomagnetic and ionospheric model. They found that the power generated varied by as much as ±20% due to magnetic field variations, and hence batteries were required for power leveling purposes. In addition, they included tether dynamics in their simulation, and found that the interaction between dynamics and electrodynamics (i.e., tether oscillations driven by the electromagnetic force) was only significant for very high current levels (>15 A). The concept of orbital energy storage, or mixed mode operation was also introduced. The idea was that a tether would thrust in the day drawing power from solar arrays, and generate power during the night; the power/thrust levels being adjusted so that the semimajor axis (i.e., the energy) of the orbit would remain constant. However, the major drawback of their approach was that they did not include very realistic models of the current collection process. Nevertheless, this study contributed greatly to the early understanding of the engineering issues involved with electrodynamic tethers.

In 1986, Martinez-Sanchez, Rivas, Prall, and Hastings [Refs. 15,16] conducted a very thorough systems study employing state of the art theories and components. A tether was studied as a stand-alone power generator, as a thruster for an orbital tug, and as a combination generator/thruster for orbital energy storage. This study concluded that electrodynamic tethers were marginally better in some respects and worse in others compared to other alternative space power and propulsion technologies. One of the key factors was the variability of the geomagnetic field which drove the system mass, as was first noted by Grossi and Arnold. The variability of the ionosphere was also cited as a factor, but not rigorously included in the calculations. Concern was expressed that the ionospheric density would drop too low during the night, so that tether operation would momentarily cease. The outstanding technical difficulties that were cited were ionospheric impedance, contactor performance, insulator fault behavior, and tether dynamics. Moreover, this work was based on several simplifying assumptions about the current closure loop and the ability of the tether to extract and emit electrons into the ionosphere. The ionospheric impedance was neglected, and the of voltage drops were taken to be independent of the current levels.

A recent study was conducted by Green, Wheelock, and Baginski [Ref. 17] on the proposed Getaway Tether Experiment (GATE), which consists of two small free-flying satellites launched from the shuttle connected by a one km conducting tether. Their objective was only to show crudely, the feasibility of the GATE to bilaterally transfer energy between stored electrical energy and orbital momentum. The major drawbacks of their model were the elementary magnetic model they used, the absence of an ionospheric model, and oversimplified contactor models.

Since the publication of the first two studies, the understanding of the current closure loop and its associated impedance has been considerably improved. In addition, physically based models have been developed for the current collection to a tether via a plasma cloud, although in many ways they are still far from being complete. A recent study [Ref. 18] incorporated these new models into an updated systems analysis to judge whether the previous conclusions are still valid or not. A detailed computer program was developed to simulate the performance of a tether in low earth orbit, and included a highly accurate model of the geomagnetic field, the latest International Reference Ionosphere model, realistic orbital dynamics, and temperature effects. In addition, the electron collection performance of a contactor and a bare wire tether, both separately and in combination, were compared and contrasted. The power and thrust generated by a bare wire tether was found to have a higher dependence on the geomagnetic and ionospheric fluctuations; however, depending on the performance of the contactor, the combination of a bare tether and contactor can substantially boost performance for power generation. As a pure thruster, the contactor tether was examined at constant current, voltage, thrust, and power, and it was found that the best mode of operation was with constant power, with resulting power/thrust ratios better than those for ion or MPD engines. For power generation, geomagnetic variability was still a major difficulty as observed in previous studies, but a control strategy was developed to greatly reduce the impact of this highly undesirable condition. In addition, operation at equatorial orbits was found to be much more beneficial for system performance. It is concluded that tethers offered greater potential than previously envisioned.

5. Concluding Remarks

Electrodynamic tethers in space offer much potential for a wide variety of missions. Although certain areas have been the subject of much research, there are still certain aspects where the level of understanding is inadequate. More work has yet to be done to provide a fuller understanding of all the detailed phenomena plasma of operation, but in the end, actual in-space experiments are imperative to determine the performance of not only contactors, but tether systems in general.

References

[1] M. Grossi, "Tether History and Historiography, " In *Second International Conference on tethers in Space*, Conference Proceedings, Bologna, Italy (1987).

[2] J. Laue, "Status of Tethered Satellite System (TSS) Development," In *Second International Conference on Tethers in Space*, Conference Proceedings, Bologna, Italy (1987).

[3] A. Barnett and S. Olbert, "Radiation and Waves by a Conducting Body moving through a Magnetized Plasma," *J. Geophys. Res.*, Vol. 91, pp. 117-135 (Sept. 1986).

[4] J. Wang and D.E. Hastings, "The Radiation Impedance of an Electrodynamic Tether with End Connectors," *Geophys. Res. Letters*, Vol. 14, May 1987, pp. 519-522.

[5] J. Sanmartin, M. Martinez-Sanchez, and E. Ahedo, "Bare Wire Anodes for Electrodynamic Tethers," Unpublished manuscript submitted to the *AIAA Journal* (1990).

[6] M. Gerver, D.E. Hastings, and M. Oberhardt, "Theory of Plasma Contactors in Ground-based Experiments and Low Earth Orbit," *J. of Spacecraft and Rockets*, Vol. 27, pp. 391-402 (1990).

[7] D. Parks and I. Katz, "Theory of Plasma Contactors for Electrodynamic Tethered Satellite Systems," *J. of Spacecraft and Rockets*, Vol. 24, pp. 245-249 (1987).

[8] M. Dobrowolny and L. Iess, "The Interaction of a Hollow Cathode with the Ionosphere," *Physics of Fluids* (B), Vol. 1, No. 9, pp. 1880-9 (1989).

[9] R. Wei and P. Wilbur, "Space charge limited Current Flow in a Spherical Double Sheath," *J. of Applied Physics*, Vol. 60, pp. 2280-2284 (1986).

[10] E. Ahedo, M. Martinez-Sanchez, and J. Sanmartin, "Current Collection by an Active Spherical Electrode in an Unmagnetized Plasma," Unpublished manuscript submitted to the *Physics of Fluids* (B) (1990).

[11] D.E. Hastings, "Theory of Plasma Contactors used in the Ionosphere," *J. of Spacecraft and Rockets*, Vol. 24, pp. 250-256 (1987).

[12] D.E. Hastings and N. Gatsonis, "Plasma Contactors for use with Electrodynamic Tethers for Power Generation," *Acta Astronautica*, Vol. 17, No. 8, pp. 827-836 (1988).

[13] R.I Samanta Roy and D.E. Hastings, "A Theory of Plasma Contactor Neutral Gas Emissions for Electrodynamics Tethers," To be published in the *J. of Spacecraft and Rockets* (1991).

[14] M. Grossi and D. Arnold, "Engineering Study of the Electrodynamic Tether as a Spaceborne Generator of Electric Power," Harvard-Smithsonian Astrophysical Observatory Report, NASA Contract NAS-8-35497 Final Report (1984).

[15] M. Martinez-Sanchez and D.E. Hastings, "A Systems Study of a 100kW Tether," *J. of Astronautical Sciences*, Vol. 35, pp. 75-96 (1987).

[16] M. Martinez-Sanchez, D. Rivas, J. Prall, D.E. Hastings, and R. Estes, "A Systems Study of a 100 kW Electrodynamic Tether, "NASA Contract NAS-3-24669 Final Report (1988).

[17] M. Green, D. Wheelock, and M. Baginski, "Electrodynamics of the Getaway Tether Experiment," *J. of Spacecraft and Rockets*, Vol. 26, pp. 452-459 (1989).

[18] R.I. Samanta Roy and D.E. Hastings, "A Systems Analysis of Electrodynamic Tethers," *Submitted to the J. of Spacecraft and Rockets* (1991).

BEYOND MARS

J. A. ANGELO, JR.
Science Applications International Corp.

ABSTRACT. The technical implications of an emerging Solar System Civilization on the design, performance requirements and operational features of future manned and unmanned space systems will be examined. Special emphasis will be placed on the following interesting properties/characteristics of a heliocentric civilization and how these particular properties/characteristics might influence the design and operation of advanced space systems in the 21st Century: (1) a true four dimensional environment (i.e. time and extended distances as measured by light-minutes to light hours); (2) independent biospheres which are physically isolated from the terrestrial biosphere; and (3) the availability of continuous multi-gravity level environments (ranging from microgravity to multi-g). Specific topics include: (1) the design and operational requirements for deep outer Solar System robotic space systems; (2) inter-biospheric transfer of human beings and cargo and the potential for extraterrestrial contamination (forward and backward); and (3) the potential consequences of multi-gravity level living.

1. Introduction

As we look out into the first portion of the next millennium, it is not unreasonable to speculate where the applications of space technology will take the human race in the next hundred or so years. Today, as a planetary people, we are becoming more aware each day that human activities in the 20th Century have enormously altered the nature of our home planet, the Earth. These activities often driven by population pressures and technology applications are producing observable changes in the composition of the our previously considered "infinite" natural sinks: the oceans and the atmosphere. There are also many noticeable (from space) alterations of the terrestrial landscape. [1,2] Consequently, in the context of the development of an emerging Solar System Civilization, it is also quite reasonable (and perhaps not too premature) to make the following two fundamental inquiries:

> First, what is the potential impact of the space environment on the manned and unmanned space systems that will be part of this Solar System Civilization?

and

R. N. DeWitt et al. (eds.), The Behavior of Systems in the Space Environment, 837–853.

Second, what is the potential impact of our Solar System Civilization (with its manned and unmanned space systems) on the overall space environment?

A companion series of deeply-philosophical questions also accompany these two fundamental inquires. These philosophical questions center around the basic question: How will access and use of the space environment alter the course and destiny of human development?

This lecture represents a brief, zeroth-order attempt at exploring the technical implications of humanity's Solar System Civilization with particular emphasis being placed on the design, performance requirements, and operational features of future manned and unmanned space systems. The space environment represents both a challenging obstacle and a unique opportunity. Fig. 1 presents some of the interesting properties or conditions found in space that have significant implications for our Solar System Civilization and the space systems necessary to support its development and growth. Of particular interest here are the following properties/conditions:

- access to continuous levels of both microgravity (in orbiting space systems) and other gravity levels (such as the lunar surface gravity level of 1.62 m/sec2 [about 1/6 g] and the Martian surface gravity level of 3.73 m/sec2 [about 1/3 g]);
- the existence of "independent" biospheres throughout heliocentric space which are physically isolated from the terrestrial biosphere; and
- space system operation and communication across great physical distances (measured in light-minutes to light-hours - see Fig. 2) [3].

It is assumed here that our Solar System Civilization will consist of at least the following elements by the year 2050:
an environmentally viable home planet, several permanent lunar settlements, several permanent space habitats in cislunar space, and at least one permanent Martian surface settlement. By the year 2100, it is further assumed that the Mars settlements have been expanded to include orbiting space bases (possibly using the Martian moons Deimos and Phobos), that semi-permanent to permanent habitats will have been established on the surface of Mercury, within the main asteroid belt and on selected moons of Jupiter and Saturn, and that advanced robotic space systems will have been sent on precursor interstellar missions [3-8]. Figs. 3 and 4 present possible development scenarios for the lunar and Martian portions of this Solar System civilization [3-5].

The three particular properties/conditions of a Solar System Civilization space environment mentioned above, might require us to expand our technical thinking horizons in many ways, including:

- multigravity level thinking (i.e., living systems would no longer be limited to just an Earth gravity environment and the field of gravitational biology will take on an extremely important and interesting role in 21st Century science);
- the plurality of biospheres (i.e., we have met the extraterrestrials and THEY ARE US!); and
- the challenges presented by the optic velocity limit in communication with and operation of advanced space systems.

INTERESTING PROPERTIES/CONDITIONS OF SOLAR SYSTEM SPACE ENVIRONMENT

- Microgravity
- Other Gravity Levels (e.g., Moon, Mars..)
- Extraterrestrial Resources
  - Energy (sunlight, 'fuels')
  - Materials (lunar, asteroidal, martian...)
- Physical Isolation From Terrestrial Biosphere
- Synoptic View of the Planet Earth (and other celestial bodies)
- Low Vibration Environment
- "Infinite" Heat Sink
- Physical Distances Not Available On Earth (e.g., light-minutes to light-hours)
- Hard Vacuum

Fig. 1.

SOLAR SYSTEM ENVIRONMENT (Distances)

- Astronomical Unit (AU)
  1 AU = 149.6 x 10*6 km = 8.32 light-min
- Earth-to-Mercury Distances
  Maximum: 1.35 AU (11.26 light-min)
  Minimum: 0.60 AU (4.98 light-min)
- Earth-to-Mars Distances
  Maximum: 2.67 AU (22.1 light-min)
  Minimum: 0.38 AU (3.17 light-min)
- Earth-to-Jupiter Distances
  Maximum: 6.40 AU (53.4 light-min)
  Minimum: 4.00 AU (33.3 light-min)
- Earth-to-Pluto Distances
  Maximum: 50.2 AU (418 light-min)
  Minimum: 28.5 AU (237 light-min)

Fig. 2.

STAGES OF LUNAR DEVELOPMENT
(2001 - 2101)

- Stage 1 - Automated Surface Exploration and Site Preparation
- Stage 2 - Initial Lunar Base (Human Population 7 - 15)
- Stage 3 - Early Lunar Settlements (Human Population 100 - 1000)
- Stage 4 - Mature Lunar Settlements (Human Population 1000 - 10,000)
- Stage 5 - Self-Sufficient Lunar Civilization (Human Population > 100,000)

Fig. 3.

21st CENTURY MARS DEVELOPMENT
(Possible Stages)

- Automated Exploration and Surface Investigations
- Initial Human Expeditions
  - Mars Sortie Mission
  - Initial Surface Base On Mars
  - Base on Phobos ('Natural' Space Station)
- Martian Civilization
  - Early Surface Settlements
  - Phobos ('Spaceport" to Outer Solar System)
  - Mature, Evolving Surface Settlements
  - Autonomous Planetary Civilization

Fig. 4.

It is interesting to further speculate here that as a result of expansion by men and smart machines into the space environment, creative minds in the 21st Century will be challenged by such issues as:

- How can we best use various gravity levels in conducting life sciences research?
- What happens when the living creatures or products from independent biospheres meet? (i.e. How do we avoid the problem of extraterrestrial contamination - either forward or backward?)
- What are suitable habitability design requirements for manned deep space exploration vessels and "routine" interplanetary spaceships?

and

- What levels of autonomy and artificial intelligence can we successfully build into advanced robotic exploration systems which will travel to the fringes of our Solar System and beyond?

Finally, it is not, perhaps, too speculative at this time, to suggest that frustrated by "luminal limits", we might even witness the rise of "superluminal" physics and its attendant space technology applications as part of our relentless expansion further and further into the space environment of our Solar System.

2. Characteristics of Extraterrestrial Civilizations

According to some scientists, intelligent life in the Galaxy might be thought of as experiencing three basic levels or types of technical civilization, when considered on a cosmic scale [3,9]. The Soviet astronomer, N.S. Kardashev (in examining the issue of information transmission by extraterrestrial civilizations in 1964), first postulated these three types of technologically developed civilizations on the basis of their overall energy use. A TYPE I civilization would represent a planetary civilization similar to the technology level found on Earth in the mid- to late 20th Century. This TYPE I civilization would command the use of somewhere between 10^{12} and 10^{16} watts of energy - the upper limit representing the amount of solar energy being intercepted by a "suitable" planet in to orbit around the parent star. For example, the solar energy flux at the Earth (i.e., the value of the solar constant outside the atmosphere), is approximately 1,370 watts per meter squared. A TYPE I civilization would also experience the development of nuclear energy and space technology. (See Fig. 5) As a TYPE I civilization begins to emerge from its home planet and expands out into the solar system around its parent star, this TYPE I civilization also transitions into the initial phases of a TYPE II civilization [3,9].

A TYPE II civilization would engage in feats of planetary engineering, expanding from its home planet through advances in space technology and extending its resource base throughout the local star system. (See Fig. 6) The eventual "energy use" upper limit of a TYPE II civilization might be taken as the creation of a Dyson Sphere. A Dyson Sphere is a postulated cluster of habitats and structures placed entirely around a star by an advanced civilization to intercept and use essentially all the radiant energy from the parent star. What physicist, Freeman J. Dyson, proposed in 1960 was that an advanced, intelligent civilization would eventually develop the space technologies necessary to rearrange the raw materials of all the planets in

TYPE I CIVILIZATION
(Characteristics)

- Planetary Society
- Developed Technology
  - Laws of Physics
  - Space Technology
  - Nuclear Technology
  - Electromagnetic Communications
  - Electronics
- Initiation of Spaceflight
  - Interplanetary
  - Permanent Space Bases

Fig. 5.

TYPE II CIVILIZATION
(Characteristics)

- Solar System Society
- Construction of Space Habitats
  (with DYSON SPHERE as upper limit)
- Long Societal Lifetimes
  (1,000 to 100,000 years)
- Ability to Perform Long-Term Planning
  (e.g., 1000-year time horizons)
- Possible Interstellar Contact/Communication
  Between TYPE II Civilizations

Fig. 6.

its solar system, creating a more efficient composite ecosphere around the parent star. Dyson further suggested that such advanced civilizations might be detected by the presence of thermal infrared emissions from such an "enclosed star system", in contrast to the normally anticipated visible radiation. Once this level of solar system civilization is achieved, the search for additional resources and the pressures of continued growth could encourage interstellar migrations. This would mark the start of a TYPE III civilization.

A TYPE III civilization would be an interstellar civilization consisting at first of several star systems (i.e. a "clustered" TYPE III civilization) and at maturity of a Galaxy-wide community of numerous intelligent species scattered over perhaps thousands or even millions of star systems. (See Fig. 7) In concept, a mature TYPE III civilization would be capable of harnessing the material and energy resources of an entire galaxy (typically containing some 10^{11} to 10^{12} stars) [3,9].

3. Multigravity Level Living

Throughout history visionaries have speculated over a future in which human beings understand the scientific truth about their origins, control the terrestrial environment and successfully live beyond the boundaries of Earth at various locations in our Solar System. Up until the Space Age, however, such speculations had often overlooked some fundamental facts - namely that the Universe is a complex and mostly inhospitable, and that life as we know it evolved in the protective shelter of the Earth's biosphere and its constant gravitational force. However, access to the space environment and various continuous multigravity levels (i.e., gravity levels ranging from essentially zero up to one-g) will enable 21st Century scientists to investigate the effects of gravity on numerous aspects of living systems. The knowledge obtained from space life sciences (including the subdisciplines: gravitational biology, exobiology, and biosphere engineering) will play a pivotal role in the development of our Solar System Civilization. To successfully conduct the advanced space missions now planned for the next century, information is needed concerning the existence of life beyond the Earth, the potential interactions between living organisms and planetary environments, and the possibilities for human beings to permanently inhabit the space environment (including the surfaces of other worlds) on a safe and productive basis [1,3].

For example, gravitational biology explores the scope and operating mechanisms of one of the strongest factors influencing life on our planet: gravity. This subdiscipline within space life sciences addresses fundamental questions concerning how living organisms perceive gravity, how gravity is involved in determining the developmental and physiological status of an organism, and how gravity has influenced the evolution of life on Earth. Although these questions are motivated mostly by scientific interest, space based research in this subdiscipline will also help determine whether living systems (including human beings) can function effectively for extended periods in microgravity or in the reduced gravity environments that will be encountered on the lunar or Martian surfaces.

TYPE III CIVILIZATION
(Characteristics)

- Interstellar Society
  - several star systems (Clustered TYPE III)
  - galactic in extent (Mature TYPE III)
- Interstellar Communication and Travel
- Very Long Societal Lifetimes
  - 1,000,000 years and longer
  - effectively "immortal" for planning purposes
- Magnificent Feats of Astroengineering
  - star system manipulation
  - "harvesting" black holes

Fig. 7.

The long-term space missions (associated with the development of a Solar System civilization) will require that crew members be able to safely adapt and readapt to varying gravity conditions. Unfortunately, scientific evidence is now lacking to respond to such important questions as:

- What changes are there in crew productivity and performance following prolonged exposure to microgravity or partial-gravity conditions?
- Can a human being live comfortably and work productively in a partial-gravity space facility that has a fixed- or variable rotation rate?
- What blend of micro-, partial-, and terrestrial (one-g) gravity levels will be needed to ensure crew safety and comfort on long-duration missions?
- What are the major engineering problems in developing rotating space facilities that provide various levels of partial gravity?
- After chronic exposure to a partial gravity environment (i.e., living permanently at a lunar or Martian base), can an "extraterrestrial" human being ever be productive and comfortable on the surface of Earth?

Therefore, the problems of adaption to various gravity environments in a Solar System civilization create a number of interesting engineering challenges for the space system designers of the 21st Century. One very important issue will be identification of the technical requirements for making a large, rotating spacecraft or space facility a habitable and productive environment. **Habitability** involves the design of space vehicle environments to support and enhance crew productivity, health, performance, safety and comfort. Contemporary studies of space system habitability emphasize the relationship between technological and human factors [1]. The extent to which engineered space system environments (including planetary surface bases) satisfy the basic needs and personal preferences of the individual determines the degree to which there is a "person-environment fit" or habitability. Some of the major space system factors related to habitability and crew well-being are [1]:

- volume
- lighting
- temperature and humidity
- vibroacoustics
- personal hygiene and waste management
- privacy
- food systems
- interior decors (both functional and aesthetic)
- leisure and recreation
- environmental monitoring and control (including buildup of pathogenic microorganisms in isolated habitats)

It is also interesting to note here that the multigravity level conditions potentially available in the space environment also offer some interesting ("nonterrestrial") options in the leisure and recreation area. It would not be unreasonable, therefore, for designers to explore and exploit these options in developing habitable space vehicles and facilities that have an inherent capability (by virtue of multigravity level access) to relieve crew boredom and stresses on long duration space missions.

A **variable-g facility** in low Earth orbit (LEO) would support key areas of gravitational biology research early in the development of our Solar System civilization and could also serve as a crew conditioning facility for return to Earth or entry into the multigravity environment of outer space.

4. Extraterrestrial Contamination Issues

In general, **extraterrestrial contamination** may be defined as the contamination of one world (or engineered-biosphere) by life forms, especially pathogenic microorganisms, from another world (or engineered-biosphere) [3,12-13]. Using the Earth and its biosphere as a reference, this planetary contamination process is called **forward contamination** if an extraterrestrial soil sample or the alien world itself is contaminated by contact with terrestrial organisms and **back contamination** if alien organisms are released into the Earth's biosphere.

The question of whether life exists elsewhere in our Solar System at present is still open [3]. But whether "natural" alien life forms exist and pose a threat to the terrestrial biosphere is only one part of the extraterrestrial contamination issue facing space system designers. It should be noted that a Solar System civilization will be accompanied by the creation of independent, engineered-biospheres throughout heliocentric space [3-8]. In time, variant microorganisms peculiar to one isolated biosphere may represent "alien", potentially pathogenic, microorganisms to other engineered-biospheres or even to the parent terrestrial biosphere. It is also quite possible that "hitchhiking microorganisms" from the biologically diverse terrestrial biosphere could represent a serious "forward contamination" threat to the carefully controlled microbiology environments of less biologically diverse engineered space habitats. Therefore, the designers of future space systems must pay special attention to potential extraterrestrial contamination issues, especially when the regular and routine transfer of people, animals, plants, and cargo takes place between the Earth's biosphere and other engineered-biospheres within the Solar System.

An alien species will usually not survive when introduced into a new ecological system, because it is unable to compete with native species that are better adapted to the particular environment. On occasion, however, alien species actually thrive, because the new environment is very suitable and native life-forms are unable to successfully defend themselves against these alien invaders. When this "war of the biological worlds" occurs, the result might be very well a permanent disruption of the host biosphere, with severe biological, environmental and economic consequences. For example, the bioregenerative life support systems now being considered for permanent space habitats could be especially sensitive to the disruptive effects of an "alien" microorganism invasion.

Consequently, in developing a Solar System civilization, it is obviously of extreme importance to recognize the potential hazard of extraterrestrial contamination (forward or back). Before any species is intentionally introduced into another planet's environment (or into another engineered-biosphere), we should carefully determine not only whether the organism is pathogenic to any native species but also whether the new organism will be able to force out native species - with destructive impact on the host biosphere [3].

At the start of the Space Age, exobiologists and planetary scientists were already aware of the potential extraterrestrial contamination problem - in either direction [3,12]. In fact, quarantine protocols were established to avoid the forward contamination of alien worlds by outbound unmanned spacecraft, as well as the back contamination of the terrestrial biosphere when lunar samples were returned to Earth as part of the U.S. Apollo program.

A quarantine is fundamentally a forced isolation to prevent the movement or spread of a contagious disease. Historically, quarantine was the period during which ships suspected of carrying persons or cargo with contagious diseases were detained at their port of arrival. The length of the quarantine (usually 40 days) was considered sufficient to cover the incubation period of most infectious terrestrial diseases. If no symptoms appeared at the end of the quarantine, then the travelers were permitted to disembark. In modern times, the term **quarantine** has obtained a new meaning: namely that of holding a suspect organism or infected person in strict isolation until the organism or the person is no longer capable of transmitting the disease. With the Apollo Program and the advent of a "lunar quarantine" program, the term has now taken on elements of both meanings [3].

Of special interest in future space missions to the planets and their moons is how we might avoid the potential hazard of back contamination of the terrestrial biosphere when robot spacecraft and human explorers bring back soil and rock samples for more detailed examination in laboratories on Earth. Similarly, with the emergence of a Solar System civilization in the mid- to late 21st, Century, we must also learn how to avoid the potential problems associated with inter-biosphere contamination.

At the beginning of the U.S. space program, a planetary quarantine protocol was established which required that outbound unmanned planetary missions be designed and configured to minimize the probability of forward (i.e., alien-world) contamination by terrestrial life-forms [3,12-13]. At the time it was reasoned that, if forward contamination did occur, it would compromise future attempts to search for and identify extraterrestrial life forms that might have arisen independently of the Earth's biosphere. As a design goal, these early spacecraft and probes had a probability of 1 in 1,000 or less that they could contaminate the target celestial body with terrestrial microorganisms. Decontamination, physical isolation (e.g., a prelaunch quarantine) and spacecraft design techniques have all been employed to support adherence to this planetary quarantine protocol [3, 12-13].

One simplified formula for describing the probability of planetary contamination is [3,12]:

$$P(c) = m\, P(r)\, P(g) \quad (1)$$

where

$P(c)$ is the probability of contamination of the target celestial body by terrestrial microorganisms

m is the microorganism burden

$P(r)$ is the probability of release of the terrestrial microorganisms from the spacecraft hardware

and

$P(g)$ is the probability of microorganism growth after release on a particular planet or celestial object.

As previously stated, P(c) had a spacecraft design goal value of less than or equal to 1 in 1,000. A value for the microorganism burden (m) was established by physically sampling an assembled spacecraft or probe. Then, through laboratory experiments, scientists empirically determined how much of this microorganism burden was reduced by subsequent sterilization and decontamination treatments. A value P(r) was obtained by placing duplicate spacecraft components in simulated planetary environments. Unfortunately, establishing a numerical value for P(g) was a bit more challenging. The technical intuition of exobiologists and some educated "guessing" were combined to create a best "guesstimate" for how well terrestrial microorganisms might thrive on alien worlds that had not yet been visited. As we keep expanding our knowledge and understanding of the environments on other worlds in our Solar System, we can also keep refining our estimates for P(g). In fact, just how well terrestrial life-forms grow on Mars, Venus, Titan, Europa, and a variety of other interesting celestial bodies will probably be the subject of actual in-situ experiments by 21st Century exobiologists [3].

The Lunar Receiving Laboratory (LRL) at the NASA Johnson Space Center in Houston provided Earth-based quarantine facilities for two years after the first lunar landing (July 20, 1969). What we learned during its operation can serve as a useful starting point for planning new quarantine facilities, Earth-based or space-based. In the future, such quarantine facilities will be needed: (1) to accept, handle and test extraterrestrial materials from Mars and other Solar System bodies of interest in our search for alien life forms (present or past); and (2) to inspect and (if necessary) isolate and disinfect the passengers, other living things or cargo that is being shipped between the independent biospheres that make up our Solar System civilization.

Again using the planet Earth as our reference frame, there are three fundamental approaches toward handling extraterrestrial samples to avoid back contamination. First, we could sterilize a sample while it is enroute to Earth from its native world. Second, we could place it in quarantine in a remotely located, maximum confinement facility on Earth while scientists examine the sample closely. Finally, we could also perform a preliminary hazard analysis (called the extraterrestrial protocol tests) on the alien sample in an orbiting quarantine facility before we allow the sample to enter the terrestrial biosphere [3,13]. Fig. 8 describes the event sequences and possible outcomes for extraterrestrial sample analyses performed in quarantine facilities on Earth and in space [13].

A space-based quarantine facility provides several distinct advantages (see Fig. 9) [13]:

(1) it eliminates the possibility of a sample-return spacecraft crashing and accidentally releasing its potentially hazardous cargo of alien microorganisms directly into the terrestrial biosphere;

(2) it guarantees that any alien microorganisms that might escape from confinement facilities within the orbiting complex cannot immediately enter the Earth's biosphere; and

(3) it ensures that all quarantine workers remain in total physical isolation during protocol testing.

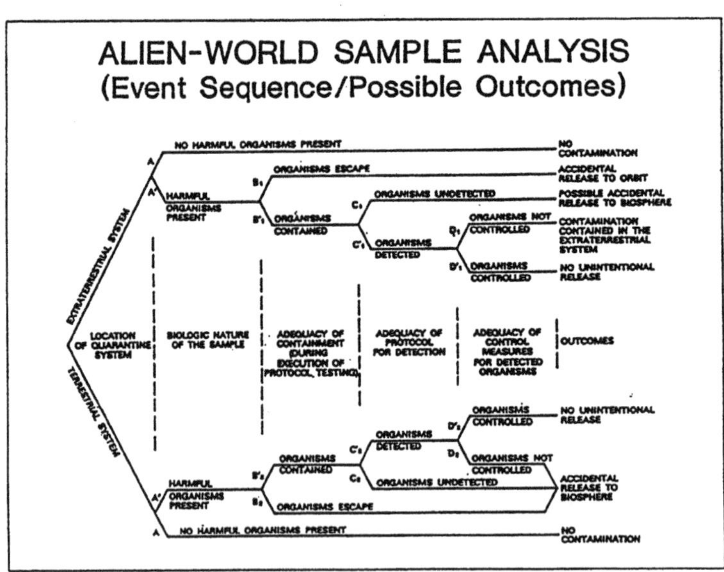

Fig. 8.

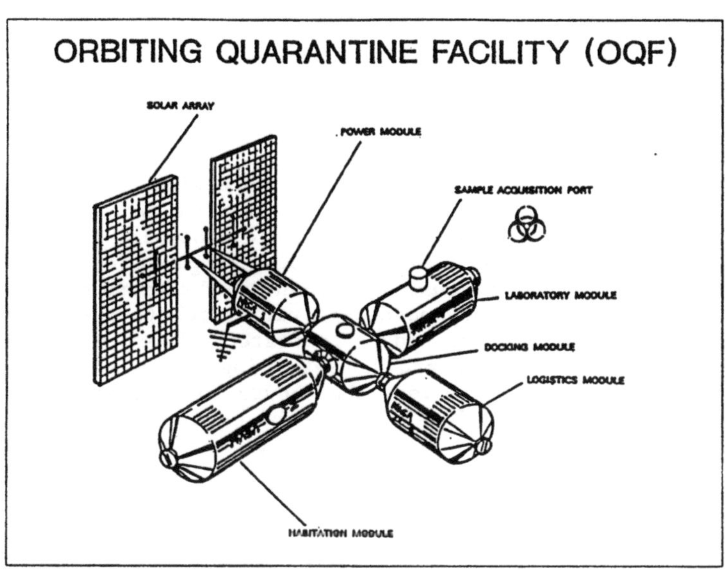

Fig. 9.

It is not too unreasonable, to speculate, therefore, that each independent biosphere in our Solar System civilization may be supported by an orbiting quarantine facility which conducts protocol testing on "alien world" scientific samples and "raw materials", and also on processed cargo, living things and human visitors from other independent biospheres. Interplanetary spacecraft will have to be designed to minimize the transport of potentially hazardous microorganisms from one world to another and to be capable of effectively interfacing with a variety of orbiting quarantine facilities which may serve as planetary system "ports of-entry". For example, the Jovian system might be served by one major orbiting quarantine facility, the planet Mars by a quarantine facility located on either of its two tiny moons, and the Earth Moon system by an orbiting quarantine complex at Lagrangian Point 4 or 5.

5. Deep Space Exploration Systems

As we push further into the outer regions of the Solar System with more ambitious exploration missions, future advanced robotic space systems will have to take over much of the data processing and information sorting activities that are now performed by human mission controllers on Earth. These robotic space explorers of tomorrow with advanced machine intelligence capable of making information gathering and processing decisions would have a large number of pattern classification templates or "world models" stored in their computer memories. These templates would represent the characteristics of objects or features of interest in a particular mission (e.g., models of the Titan's atmosphere). The advanced robot explorers would then compare the patterns or objects they "see" with those stored in their memories and discard any unnecessary or unusable data. As soon as something unusual was encountered, the smart machine explorer would examine this object or event even more closely. The advanced robotic explorer would then dutifully alert its human controllers on Earth (perhaps light hours away) and report the unusual findings. Through such automated selection and data filtering operations, these smart robots explorer would be able to capture interesting and perhaps rare transient phenomena, that would otherwise be totally lost to scientists on Earth (who could not exercise judgmental data acquisition decisions in a timely manner because of the many light hours distance) [3,14].

Quite similarly, these advanced exploration robots must be designed to continuously diagnose themselves and to initiate on board corrective actions and repairs for many problems and housekeeping adjustments that require "mission essential" responses in less than hours. Once again, because of the truly vast distances involved in exploring the outermost regions of our Solar System (see Fig. 2), mission controllers on Earth will only be able to make coarse temporal corrections (i.e. on the order of a day or so response time) to a particular mission plan, exploration protocol, or spacecraft repair sequence.

The advanced machine intelligence (or artificial intelligence) requirements for general purpose robotic exploration systems to the outer regions of the Solar System (and beyond) can be summarized mainly in terms of two fundamental tasks:

(1) the smart robot explorer must be capable of learning about new environments; and

(2) it must be able to formulate hypotheses about these new environments.

Hypothesis formation and learning represent the major problems in the successful development of machine intelligence. Deep interplanetary and interstellar robotic space systems will need a machine intelligence system capable of autonomously conducting intense studies of alien objects and alien worlds. The machine intelligence levels supporting these missions must be capable of producing scientific knowledge concerning previously unknown objects.

For a really autonomous, deep-space exploration system to undertake knowing and learning tasks, it must have the ability to "artificially" formulate hypotheses, using all three of the logical patterns of inference: analytic, inductive, and abductive [3,14]. Analytic inference is needed by the robot explorer system to process raw data and to identify, describe, predict and explain extraterrestrial events and processes in terms of existing knowledge structures. Inductive inference is needed so that the smart robotic explorer can formulate quantitative generalizations and abstract the common features of events and processes occurring on alien worlds. Such logic activities amount to the creation of new knowledge structures. Finally, abductive inference is needed by the smart robotic space system to formulate hypotheses about new scientific laws, theories, concepts, models, etc. The formulation of this type hypothesis is really the key to the ability to create a full range of new knowledge structures. An ability to create these new knowledge structures, in turn, is needed if human beings are to successfully explore and investigate alien worlds at the outermost regions of the Solar System and perhaps around neighboring star systems using smart robot systems as surrogates [3,14].

Although the three patterns of inference just described are distinct and independent, they can be ranked by order of difficulty or complexity. Analytic inference is at the low end of the new knowledge creation scale. An automated system that performs only this type of logic could probably successfully undertake only deep space reconnaissance missions. A machine explorer capable of performing both analytic and inductive inference could most likely successfully perform deep space missions that combine reconnaissance and exploration. This assumes, however, that the celestial object being visited is represented well enough by the world models with which the smart robotic explorer had been preprogrammed. However, if the target alien world cannot be well represented by such fundamental world models, then automated exploration missions will also require an ability to perform abductive inference. This logic pattern is the most difficult to perform and lies at the heart of knowledge creation. An advanced automated space system capable of abductive reasoning could successfully undertake missions combining reconnaissance, exploration and intensive study [3,14]. Fig. 10 summarizes the adaptive machine intelligence needed for very advanced deep space robotic exploration systems.

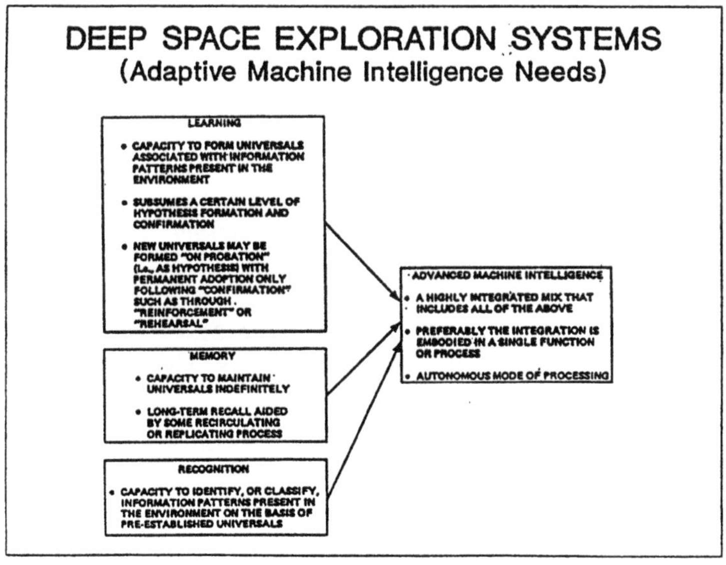

Fig. 10.

6. Summary

Taking a truly long term perspective, there appear to be two general pathways for mankind: either we are a very important biological entity in the overall evolutionary scheme of the Universe; or else we are an evolutionary dead end!

If we limit ourselves to just one fragile biosphere (our home planet Earth), then a major natural disaster or own technological foolhardiness could certainly terminate our existence - perhaps in just centuries or possibly in a few millennia. Excluding these unpleasant possibilities, without expansion into the space environment (and the subsequent "new thinking" and new technologies such an expansion stimulates), our one-world (planetary) society will simply stagnate and eventually reach some form of quasistable-equilibrium (much like the creatures in the global oceans).

On the other hand, if we respond to the challenges imposed upon us by the space environment as we attempt to become a Solar System civilization, then we (along with our very smart machine surrogates) have the opportunity of setting in motion a chain reaction that spreads life, organization, and consciousness across first our own Solar System and then out into the Galaxy - in an expansion wave limited perhaps only by the speed of light itself. This is the ultimate impact of the space environment!

References

[1] NASA, "Exploring the Living Universe: A Strategy for Space Life Sciences", A Report of the NASA Life Sciences Strategic Planning Study Committee, Washington, DC, June, 1988.

[2] Ginsberg, Irving W., and Angelo, Joseph A., Jr., (eds) Earth Observations and Global Change Decision Making, 1989: A National Partnership, Krieger Publishing Company, Malabar, Florida, 1990.

[3] Angelo, Joseph A., Jr., The Extraterrestrial Encyclopedia (revised edition), Facts On File Publishers, New York, 1991.

[4] Angelo, Joseph A., Jr., and Buden, David, "Nuclear Energy for Lunar Base Applications," 36th IAF Congress, IAF-85-180, Stockholm, Sweden, October 7-12, 1985.

[5] Angelo, Joseph A., Jr., and Buden, David, "A 21st Century Nuclear Power Strategy For Mars," 37th IAF Congress, IAF-86-322, Innsbruck, Austria, October 4-11, 1986.

[6] National Commission on Space (NCOS), Pioneering the Space Frontier, New York, Bantam, 1986.

[7] Ride, Sally, K., "Leadership and America's Future in Space," NASA, Washington, DC, 1987.

[8] Stafford, Thomas P. (Chairman), "America At the Threshold: Report of the Synthesis Group On America's Space Exploration Initiative," Washington, DC, May, 1991.

[9] Angelo, Joseph A., Jr., "The Characteristics of Extraterrestrial Civilizations and the Interstellar Imperative," 28th Space Congress, Cocoa Beach, Florida, April 1991.

[10] Stuster, Jack W., "Space Station Habitability Recommendations Based On A Systematic Comparative Analysis of Analogous Conditions," NASA CR 3943, 1986.

[11] Rockoff, L.A., et al, "Space Station Crew Safety Alternatives Study", Final Report, NASA CR 3856, 1985.

[12] Phillips, Charles R., "The Planetary Quarantine Program," NASA SP-4902, 1974.

[13] Vincenzi, Donald, and Bagby John R., (eds) "Orbiting Quarantine Facility," NASA SP-454, 1981.

[14] Freitas, Robert A., and Gilbreath, William P., "Advanced Automation for Space Missions," NASA Conf Pub 2255, 1982.

PLASMA IN THE ENVIRONS OF SPACECRAFT

A.D.R. PHELPS
Department of Physics and Applied Physics
University of Strathclyde
Glasgow G4 ONG
Scotland, UK

ABSTRACT. The state of matter surrounding spacecraft is usually plasma. The plasma parameters vary widely depending upon the location of the spacecraft. In the case of the earth it is possible to consider (a) a low earth orbit (b) a geostationary orbit and (c) an interplanetary flight. In contrast to laboratory plasmas, the electron number densities are typically much lower, the spacecraft is moving rapidly with respect to the plasma and there is the additional feature of photoelectron emission. Sheaths, wakes and spacecraft charging are phenomena that need a plasma physics treatment. The fact that external instrumentation, probes and antennae are immersed in a plasma medium, once the spacecraft has reached its orbit, results in plasma modifications to the behaviour of some of the spacecraft systems. Antennae intended for electromagnetic wave launching can also excite electrostatic plasma waves. The spacecraft potential can vary as a function of the photon flux. Charged particle collection and analysis can be grossly altered by plasma sheath effects. Spacecraft surfaces, including insulating components, are potentially immersed in an electrically conducting medium. It is concluded that, although there is now a reasonable awareness of many of these plasma effects, there is a need to improve laboratory plasma experimental simulations to complement in situ spacecraft plasma measurements and thereby to advance the understanding and control of plasma in the environs of spacecraft.

1. Introduction

A spacecraft in the ionosphere or magnetosphere may be regarded as very similar to a Langmuir probe in a laboratory plasma. Normally the laboratory probe is not isolated in the same way as a spacecraft. Examining the typical current-voltage characteristic of the Langmuir probe, however, there is a point where the probe draws zero net current. In this condition at "the floating potential" the Langmuir probe resembles the spacecraft in its isolated state. Normally the floating potential of a Langmuir probe is negative with respect to the plasma potential. The electrons moving faster than the ions usually charge the probe negative until eventually an equilibrium is reached where the negative potential of the probe repels

R. N. DeWitt et al. (eds.), The Behavior of Systems in the Space Environment, 855–859.

enough electrons from the plasma to balance the electron and ion fluxes. Additional factors experienced by the spacecraft that are not usually present in the laboratory situation are that the spacecraft is moving and it is subject to a solar photon flux on the dayside of the earth. For a satellite moving in the ionosphere the satellite velocity v_s is usually greater than the ion thermal velocity, but smaller than the electron thermal velocity i.e. $v_{Ti} < v_s < v_{Te}$. This implies that the satellite can sweep up ions in its path by a simple "ram" effect, whereas the electrons being more mobile tend to be collected as a thermal flux. Similarly in the wake immediately behind a spacecraft an absence of ions results. The effect of the photoemission is to create electrons that can take part in the electron/ion flux balance.

In (a) a low earth orbit the electron number density is typically in the range 10^{10} to 10^{12} m^{-3} whereas (b) further out in the magnetosphere the electron number density drops down to 10^{8} m^{-3} and eventually as low as 10^{6} m^{-3} where case (c) interplanetary flight becomes applicable. Considerable variations in the plasma density and temperature occur as a spacecraft moves from the solar wind region towards the earth and passes through the bow shock, the magnetosheath, the magnetopause [Ref. 1], the magnetosphere and down through the ionosphere. On such a trajectory the Debye length, which gives a measure of the thickness of an electrostatic sheath [Ref. 2] can vary from metres in the tenuous edges of the magnetosphere down to millimetres in the densest part of the ionosphere. A further complication is that in addition to the altitude scaling and to the strong variations of the ambient plasma with diurnal time, season, sunspot cycle and latitude [Ref. 3] there are local irregularities that have been measured with in situ probes on satellites [Ref. 4].

2. Plasma Wave Effects

Cold magnetoplasmas can typically support four electromagnetic wave branches. A warm magnetoplasma can also support propagating electrostatic waves such as electron plasma waves, ion acoustic waves and electron and ion Bernstein modes [Ref. 5]. Hence a spacecraft antenna that has been tested on the ground and shown to launch the anticipated electromagnetic waves has available to it many more modes into which to couple its energy once it is in orbit and it is immersed in the plasma. This phenomenon was observed in the 1960's when the first topside sounder satellites were launched and spectacular plasma wave effects were evident on the topside sounder echo data. J.O. Thomas et al [Refs. 6, 7] carried out a series of laboratory plasma experiments that helped to confirm that electron Bernstein modes were being launched, along with other plasma wave modes, in addition to the electromagnetic waves that the antennae were intended to launch. These laboratory experiments were extremely effective in that the plasma conditions could be controlled and the confirmatory details of the plasma wave dispersion could be measured. These relatively quick and inexpensive experiments complemented the longer time-scale and more expensive in situ satellite experiments.

Particle beams may also be used to excite waves and instabilities in the plasma surrounding the spacecraft. The electron cyclotron maser instability is a growing wave that is readily excited using a mildly relativistic electron beam [Ref. 8]. These experiments were carried out in the laboratory, but there is now a growing number of excellent experiments that have excited waves and controlled them directly from a spacecraft [Refs. 9-12].

Ion Bernstein modes, that have a similar structure to the electron Bernstein modes, but at lower frequencies linked to the ion cyclotron frequency have also been observed by in situ instrumentation on a satellite [Ref. 12]. In this case no deliberate attempt to excite them was made. It should be noted that there is a fundamental difference between the apparently similar electron and ion Bernstein mode dispersion relations. There is effectively only one type of electron Bernstein mode, whereas there are two types of ion Bernstein mode namely the "pure" ion Bernstein mode and the "neutralized" ion Bernstein mode. These modes are shown in Fig. 1.

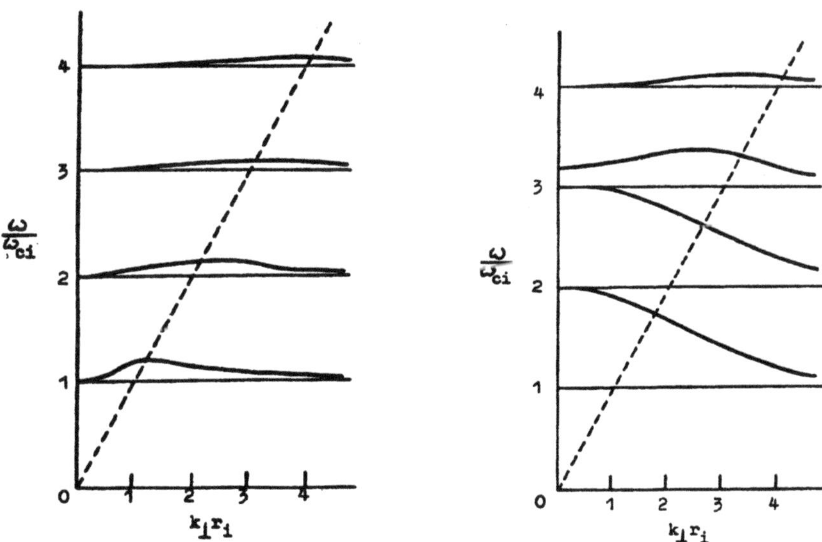

"neutralized" ion Bernstein mode "pure" ion Bernstein mode

Fig. 1.

This asymmetry with the electron case occurs because, whereas the massive ions can not move during the high frequency electron Bernstein mode oscillations, the light electrons can move during the neutralized ion Bernstein mode oscillations. It is usually the one that is observed since it occurs most readily. Only in rather special circumstances can one arrange the experimental conditions such as to generate the "pure" ion Bernstein mode.

3. Plasma Sheath Effects

The magnitude of the energy difference resulting from the electric potential difference between the spacecraft floating potential and the plasma potential is typically comparable with the thermal energy of an average electron in the plasma, i.e., the electric potential difference is comparable with kT_e/e. At low altitudes this is relatively small, but at high altitudes spacecraft surfaces can charge many kilovolts negative. Active experiments [Ref. 11], using charged particle beams to control the potential dropped across the spacecraft sheath, have been successful in establishing that effects such as arcing across insulators can be switched on and off. Tethered mother/daughter systems [Ref. 13], which correspond in laboratory plasma physics terms to a double Langmuir probe system, have been examined both experimentally and computationally. The wake effect results from the so-called "mesothermal" [Ref. 14] plasma flow around a moving spacecraft. This follows from the double velocity inequality mentioned in the Introduction above.

4. Conclusions

The important influence of the plasma surrounding a spacecraft on the operation of the spacecraft instrumentation, systems and communications has become increasingly clear. Several approaches are currently being used including direct in situ measurements on spacecraft, computational simulation using theoretical models and laboratory experiments designed to simulate the spacecraft environment. The latter two approaches have the benefits of being substantially less expensive and having short time-scale flexibility. The laboratory plasma experiments [Refs. 6, 7] have proven to be very successful at investigating and clarifying the mechanisms responsible for plasma wave excitation in the environs of spacecraft. A new laboratory plasma experiment is presently under construction at Strathclyde University. This experiment is designed to study the interaction of plasma particle fluxes with spacecraft surfaces.

5. Acknowledgements

This work is being supported by an SERC/MoD research grant. Helpful discussions with I.D. Chalmers are gratefully acknowledged.

6. References

[1] Phelps, A.D.R., "Interactions of plasmas with magnetic field boundaries", Planet. Space Sci., 21, 1497-1509, (1973).

[2] Phelps, A.D.R. and Allen, J.E., "A floating electrostatic sheath in a thermally produced plasma", Proc. R. Soc. Lond. A., 348, 221-233, (1976).

[3] Thomas, J.O. and Phelps, A.D.R., "The polar exospheric plasma", in Radar Propagation in the Arctic, AGARD-CP-97, 3.1-3.11, (1971).

[4] Phelps, A.D.R. and Sagalyn, R.C., "Plasma density irregularities in the high latitude topside ionosphere", J. Geophys. Res., 81, 515-523, (1976).

[5] Allen, J.E. and Phelps, A.D.R., "Waves and micro-instabilities in plasmas (linear effects)", Rep. Prog. Phys., 40, 1305-1368, (1977).

[6] Thomas, J.O., Andrews, M.K, Hall, T.A. and Phelps, A D.R., "Electron Bernstein mode propagation experiments", Proc. Xth Int. Conf. on Phen. in Ion. Gases, Oxford, (Oxford: Parsons), I, 314, (1971).

[7] Phelps, A D.R., "Apparatus appropriate for the excitation of electrostatic waves in space experiments and in related laboratory experiments", Imperial College Report No. SPT 101-71, University of London, (1971).

[8] Phelps, A.D.R. and Garvey, T., "Electron cyclotron maser emissions from pulsed electron beams", J.Phys.D:Appl.Phys., 19, 2051-2063, (1986).

[9] Winglee, R.M. and Kellogg, P.J., "Electron Beam injection during active experiments 1. Electromagnetic wave emissions", J. Geophys. Res., 95, 6167-6190, (1990).

[10] Cairns, I.H. and Gurnett, D.A., "Control of plasma waves associated with the space shuttle by the angle between the orbiter's velocity vector and the magnetic field", J. Geophys. Res., 96, 7591-7601, (1991).

[11] Olsen, R.C., Weddle, L.E. and Roeder, J.L., "Plasma wave observations during ion gun experiments", J. Geophys. Res., 95, 7759-7771, (1990).

[12] Phelps, A.D.R. and Sagalyn, R.C., "Observation of low frequency resonances in the topside ionosphere", Proc. 2nd European Conf. on Cosmic Plasma Physics, Culham, (London:Institute of Physics), 35, (1974).

[13] Neubert, T., Mandell, M.J., Sasaki, S., Gilchrist, B.E., Banks, P.M., Williamson, P.R., Raitt, W.J., Meyers, N.B., Oyama K.I., and Katz, I., "The sheath structure around a negatively charged rocket payload", J. Geophys. Res., 95, 6155-6165, (1990).

[14] Coggiola, E. and Soubeyran, A., "Mesothermal plasma flow around a negatively wake side biased cylinder", J. Geophys.Res., 96, 7613-7621, (1991).

RADIATION DAMAGE EFFECTS ON SPACECRAFT MATERIALS

W.S. WALTERS
A.E.A. Reactor Services
B 540.1, Harwell Laboratory
Oxfordshire OX11 ORA
United Kingdom

ABSTRACT. In geostationary orbit satellite materials are subject to radiation damage caused by trapped protons and electrons. This paper describes radiation damage testing carried out on novel imide - based conductive coatings applied to Kapton and glass. As background information, the radiation test facilities themselves are described, with a discussion of the relationship between the test conditions and the effects of the near earth orbit environment.

1. Proton Irradiation Facility

1.1 DESCRIPTION OF THE FACILITY

The proton irradiation facility consists of a Van de Graaff accelerator (High Voltage Engineering, Type AK) producing a proton beam of between 0.5 and 1.5 MeV with a beam current of between 1 n-amp and 1 μ-amp, which correspond to 10^{10} and 10^{13} protons cm^{-2} h^{-1}. The beam fluence allows the simulation of proton fluences of approximately 10^{14} p^{+}/cm^{2} (equivalent to a satellite operational life of 15 years at GEO) in a period of about 10 hours irradiation. The accelerator can produce a 10 cm diameter monoenergetic beam with a uniformity of ± 10% across the surface.

The facility is configured with a generator room immediately above a shielded target room. The proton beam is produced initially in a vertical flight tube, which is fed into a bending/analyzing magnet which provides a horizontal beam line in the target room, and with other ionic species removed from the beam. A gold foil is used to spread the beam and create some energy straggling enabling irradiations of larger components under more realistic conditions.

1.2 APPLICATIONS OF THE FACILITY

The 1.5 MeV Van de Graaff proton accelerator is dedicated to studies of spacecraft materials under conditions simulating exposure to protons in the solar wind and radiation belts. The proton flux forms an important part of the radiation spectrum to which satellites are exposed, both in

R. N. DeWitt et al. (eds.), The Behavior of Systems in the Space Environment, 861–872.

geostationary and low earth orbit. The trapped proton spectrum (Fig. 1) is heavily weighted towards the low energy end, consequently a 1.5 MeV particle beam is well suited to these studies.

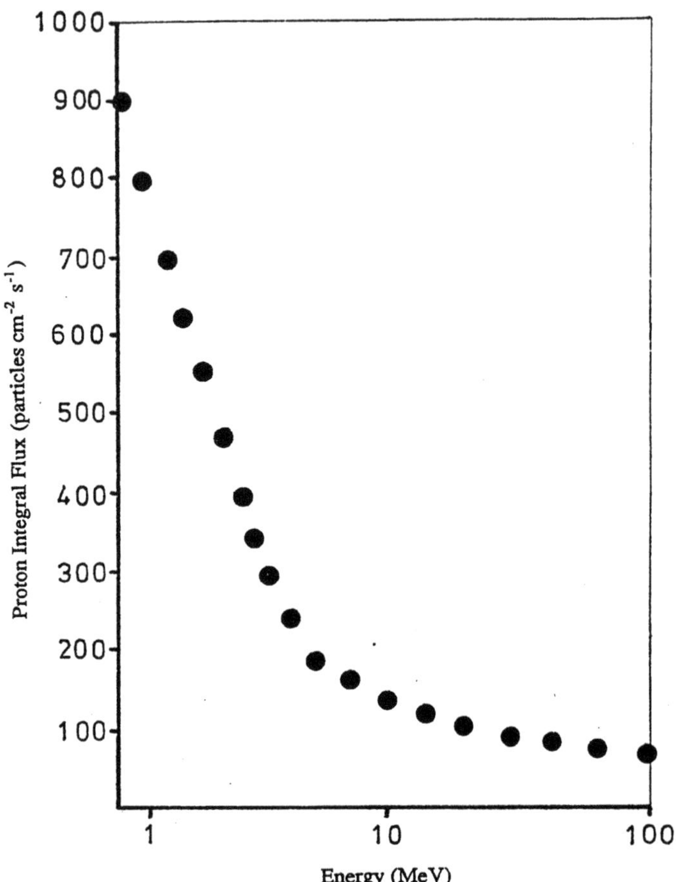

Fig. 1: Trapped proton energy spectrum for low earth (800 km) orbit.

The accelerator has been widely used for studies of proton irradiation effects on glass cover slips for solar cells. When irradiated, glasses in general are prone to discoloration, and this affects their ability to transmit light through to the solar cell. Special formulation glasses which are needed to overcome this problem have been extensively tested using this facility. When silica targets are irradiated using the facility, the targets may also fluoresce with a visible blue light emission.

2. Gamma Irradiation Studies

2.1 THE BASIS FOR GAMMA IRRADIATION TESTING

An economic and efficient way of screen testing the radiation stability of novel materials for use in space, especially with regard to electron irradiation, is to use a gamma irradiation facility. Gamma rays interact with the electrons in an irradiated material to produce energetic electrons which proceed to affect the material in a similar way to normal electron irradiation. In terms of radiation chemical effects, both processes cause ionization, secondary emission, free radical formation (in polymers, for example) and embrittlement. It is generally accepted that the predominant radiation chemical changes to polymer systems are caused by free radical processes, irrespective of how the radicals are formed [Ref. 1]. The difference is that most spacecraft materials will completely stop a flux of electrons, but will be relatively transparent to gamma radiation. However, it is possible to calculate the energy that is deposited in a material when irradiated with both electrons and gamma rays. The same order of effects may therefore be observed with both radiation types, by adjusting the exposure to gamma rays such that the total energy deposited is the same as for exposure to electrons.

An advantage of using a gamma irradiation facility is that the radiation dose rate may be increased significantly above the dose rate for electron irradiation for the satellite in orbit. This means that the material may be exposed to a "lifetime dose" (perhaps equivalent to several years in orbit) in a matter of days under gamma irradiation. This offers immediate benefits such as early identification of materials which are not radiation chemically stable.

2.2 APPLICATIONS OF THE GAMMA IRRADIATION FACILITY

Irradiation studies have been carried out on a range of candidate space materials (often polymers). These studies have been performed on samples of the material sealed in a container under vacuum (10^{-6} torr), so that any degradation followed by release of volatile products (outgassing) will be accurately reproduced.

Of particular interest for irradiation studies have been the range of conductive (metal oxide) coatings applied to Kapton [Ref. 2]. The preparation of the coatings is described more fully in an accompanying article by M.J. Duck. A number of test patches of Kapton (plain and coated) were prepared. The coatings included In_2O_3, SnO, ZnO, TiO_2, and carbon. A specimen of each coating type was irradiated in the gamma facility to a total exposure dose of 44 M rads, which is equivalent to 9 years total exposure at geostationary orbit [Ref. 3]. The test samples were then compared with the unirradiated control samples. The irradiated samples showed no visible sign of degradation and could be mechanically worked without any spalling or detachment of the applied coating. This demonstrates the very good radiation stability of the conductive coating materials applied to Kapton film.

By contrast, the irradiation study also included samples of Teflon sheet and expanded PTFE film, again as irradiated specimens and unirradiated controls. Both sheet and film samples of Teflon (PTFE) were seen to have degraded severely, and to have become embrittled, after a total exposure of 20 M rad (i.e. less than half the exposure given to the Kapton). The Teflon sheet (2 mm thick) had lost its flexural strength and had cracked in several places. The expanded PTFE film had lost all its tear strength, and was very easy to damage.

3. Comparative Studies of Radiation Damage Effects

3.1 THE BASIS FOR COMPARATIVE STUDIES

A number of radiation damage effects are possible when spacecraft materials are irradiated with either protons or electrons. As an illustration of comparative effects, the irradiation of glass with conductive coatings has been studied using both proton and gamma irradiation (using the secondary electron effect of gamma radiation to simulate primary electron irradiation). The required radiation dose from the proton accelerator was calculated to give the same absorbed dose to the samples as the total exposure dose to the samples from the gamma irradiation facility.

3.2 EXPERIMENTAL STUDIES

A number of borosilicate glass microscope slides were prepared with a range of conductive polyimide - based coatings, In_2O_3, ZnO, and TiO_2, with an uncoated control. These were irradiated with 1 MeV protons or ^{60}Co gamma radiation. Control samples of each material have been kept unirradiated. The gamma irradiation delivered an exposure dose of 2 M rad, equivalent to 5 months of exposure at geostationary orbit. The proton irradiation delivered a dose equivalent to 2 M rad, using 1 MeV protons.

It could be clearly seen that the gamma irradiation had produced a brown discoloration in the glass which is typical for borosilicate, caused by the dislocation of electrons and the creation of defect centers in the solid phase. This effect may be removed by thermal annealing. The proton irradiation had produced hardly any visible effect, all the damage being localized in the surface layer. The range of 1 MeV protons in borosilicate glass is approximately 13 μm, and in hydrocarbon polymers such as Kapton approximately 20 μm [Ref. 4], therefore protons passing through the conductive coating layer would only penetrate a few microns into the glass, limiting the radiation damage to the glass substrate.

3.3 CONCLUSIONS

It has been demonstrated that the radiation testing of satellite materials is necessary for any material which might be exposed to the space environment; materials which appear to be robust (e.g. teflon) may in fact degrade severely under irradiation.

The specialized coatings which have been developed for use on Kapton, to reduce surface charge build-up, have been shown to withstand radiation to an operational lifetime dose for GEO. This test result reinforces the claims of these materials to be suitable for use in space.

The radiation testing of each material should include both proton and electron (gamma) irradiation, as the two radiations may have different effects in the damage they cause to the bulk materials. This has been clearly seen in the experiments with glass slides coated with In_2O_3, ZnO, and TiO_2, and with an uncoated control. The proton irradiation has produced a scarcely visible amount of surface damage whilst the gamma irradiation has produced a clearly discernable amount of bulk material damage.

4. References

[1] Chapiro, A. (Ed.), "Radiation Chemistry of Polymeric Systems", Interscience, New York. (Volume XV in the High Polymers monograph series), 1962.

[2] Verdin, D. and Duck, J.J., "Surface Modifications to Minimise the Electrostatic Charging of Kapton in the Space Environment", J. of Electrostatics, 20, 123-129, 1987.

[3] Data taken from "The radiation design handbook" ESA PSS-01-609 European Space Agency, Noordwijk, The Netherlands, 1990 (draft).

[4] Janni, J.F., "Calculations of energy loss, range, pathlength, straggling, multiple scattering, and the probability of inelastic nuclear collisions for 0.1 to 1000 MeV protons", report AFWL-TR 65-150, 1966.

SURFACE CHARGING AND ITS PREVENTION

M.J. DUCK
A.E.A. Reactor Services
B 540.1, Harwell Laboratory
Oxfordshire OX11 ORA
United Kingdom

ABSTRACT. The charging effects which can occur in Kapton during geomagnetic substorms have been simulated using the Harwell electrostatic discharge test facility. The facility is described, and results are given for irradiations with electrons of up to 30 keV energy with films of differing thicknesses. The charging can by reduced by coating the Kapton with a dispersion of indium oxide in a soluble polyimide. Typical coatings with a 3:1 oxide/imide loading reduce the surface resistivity of the Kapton from 10^{18} to 10^{7} Ω/square. Some other preparations employing alternative fillers and possessing different thermo-optical and conductive properties are also described.

1. Introduction

A large fraction of the external surface of a satellite consists of dielectric materials which are used either for the passive temperature control of the satellite or as part of the power generation system. These materials include thermal control paints, polymer films, solar-cell cover glasses and deployable solar array substrates.

In the geosynchronous environment such materials are exposed to a number of hostile particles and radiations. In an accompanying article W.S. Walters discusses the damage caused by trapped proton and electron irradiations. In the present article the deleterious effects caused by incoming electron showers are investigated.

Since the exposed dielectrics can be of large area, are relatively thin and usually have a conductive backing, large capacitances exist which, under the conditions prevailing in geostationary orbit, can become electrostatically charged during geomagnetic substorms, resulting in arc discharges [Ref. 1]. These discharges can cause serious problems which arise from the current arc itself, from the accompanying electromagnetic pulse and from the surface potential of the satellite.

2. The Harwell ESD Test Facility

In order to investigate these problems the Harwell electrostatic discharge (ESD) facility was constructed; see Fig. 1. The irradiation chamber has a diameter of 750 mm, is 750 mm high, and operates at a pressure of approximately 2 x 10^{-7} torr. To avoid instantaneous differential charging of small areas of irradiated samples an electron gun giving a continuous and uniform beam over the whole area of the sample holder is employed. The gun is located at the top of a 1.2 m long flight-tube in which the electrons accelerate towards, and then pass through, an earthed mesh anode in the lid of the irradiation chamber. This source gives monoenergetic electrons at beam fluxes up to 10 nA.cm^{-2} and at electron energies from 0 to 30 keV. The beam current is automatically maintained constant at any preset level by means of a negative feedback circuit. The sample holder can accommodate specimens up to 35 cm in diameter and 2 cm thick, at temperatures from -180°C to +150°C. During irradiation the surface voltage of the sample is measured with a non-contacting electrostatic voltmeter probe mounted on an X-Y scanner, and the voltage profile over any part of the sample can be displayed on a potentiometric recorder. The scanner also carries a Fariday cup which is used to measure the beam distribution. The mean value for the coefficient of variation of beam uniformity was 17.5%, and the total beam flux remains constant to within 5% over periods of up to 10 hours.

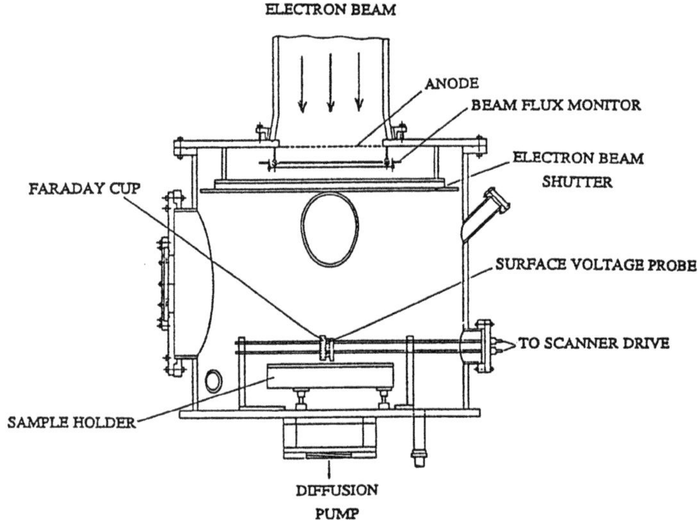

Fig. 1: The Harwell ESD test facility.

The facility is equipped with electrometers to monitor leakage currents from the irradiated samples, and the output signals are also displayed on potentiometric recorders. Inductively coupled probes are used to measure the charge flow associated with discharge pulses from the samples, the pulses being recorded either photographically on a 100 MHz storage oscilloscope, or with a 6 MHz transient recorder. Samples can be photographed during irradiation, and part of the sample can be exposed to simulated solar radiation, at an intensity of 0.9 solar constant, using the filtered output from a xenon lamp.

The ultimate surface potential attained by a spacecraft is partly governed by the secondary emission of low energy electrons. The Harwell test facility is fitted with a specially designed Faraday cup which can be directed towards the target material, and electrically biased to selectively measure the flux of low energy secondary electrons. These tests can be made on all types of materials, including dielectrics, metals, and mirrors.

3. Electrostatic Charging Measurements on Kapton.

The ESD facility has to date principally been employed to investigate the electrostatic behavior of Kapton - a popular dielectric material, since it exhibits suitable optical properties, shows little outgassing, and is generally resistant to radiation-induced changes. It is therefore widely used in thermal control blankets and as a substrate for flexible solar arrays.

Kapton films of various thicknesses were studied. The samples consisted of 140 mm diameter discs, the lower aluminized surfaces of which were earthed through an electrometer. The maximum surface voltages obtained at various incident energies for samples cooled to -180°C are shown in Fig. 2. In all cases these maximum values were only seen at or near the edges of the samples - the voltage profiles across the samples being saddle-shaped. [This phenomenon probably results from the mutual repulsion and resultant migration of the electrons.]

In all cases the surface voltage commenced to rise above zero at about 1.5 keV, this being the upper SEE (secondary electron emission) threshold energy for Kapton at -180°C. At first, the surface voltage rises linearly with respect to the incident energy. But, eventually the range of the incident electrons in the Kapton begins to approach the thickness of the film. At this point, the leakage current begins to increase, and the rate of rise of surface voltage begins to fall away, until it eventually reaches a maximum value and, finally, as the incident electrons penetrate through to the aluminum backing, any further increase in incident energy leads to a decrease in the surface voltage.

4. Conductive Coatings.

The susceptibility of dielectric materials such as Kapton to electrostatic charging and discharging may be reduced by treating the exposed surface with a conductive coating [Ref. 2]. Such coatings can be prepared by finely dispersing indium oxide powder in a soluble polyimide, the chemical similarity of the latter to Kapton leading to good adhesion. Typically, a 3.5 μm thick 3:1 (In_2O_3/imide) loaded coating gave a surface resistivity of 10 MΩ/square and only charged up to a maximum of about 240 volts when irradiated with electrons over the range 5 - 30 KeV.

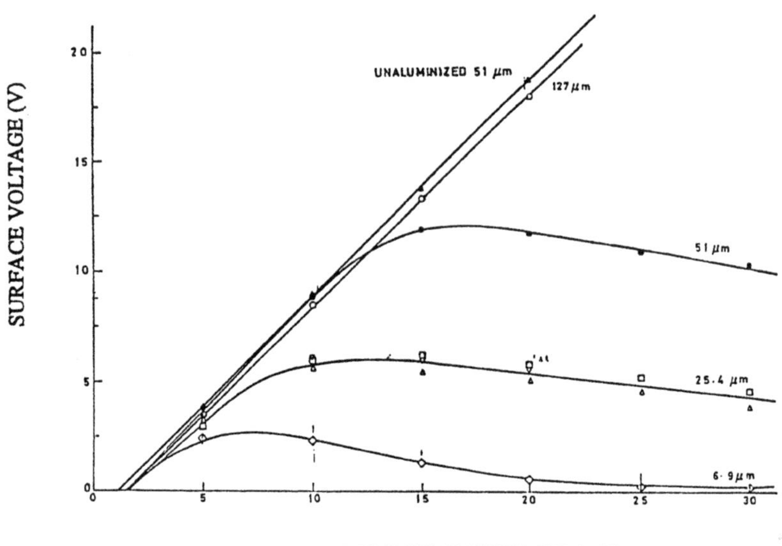

Fig. 2: Electrostatic performance of different thicknesses of Kapton irradiated at -180°C.

Figure 3 demonstrates that the surface voltage levels reached by coatings with a loading of 3:1 decreased as the thickness and/or loading of the coating was increased, and that raising the loading to 6:1 caused a further sharp drop in voltage. The surface voltages are only about 5% of those obtained by plain Kapton under comparable conditions.

The samples had been dried by vacuum outgassing at 125°C. Further heating at higher temperatures led to significant improvements in performance.

The coatings exhibited excellent adhesion to Kapton and this did not deteriorate or show any changes in surface resistivity during 1100 thermal cycles between +150°C and -150°C in a nitrogen atmosphere. The solar absorptance/emissivity ratio was always higher than that for plain Kapton (0.565) usually being about 0.8. The flexibility of the coatings was good.

Spacecraft have a requirement for coatings with a wide range of properties. These properties can be obtained by loading the polyimide with alternative fillers such as:

1. *Titanium dioxide*. This material is not conductive, but is highly reflective and gives coatings which are brilliant-white in appearance. Coatings prepared with a mixed titania/indium oxide filler retain this characteristic yet are still moderately conductive (10^2 – 10^6 MΩ/square, depending upon the ratio of the metal oxides).

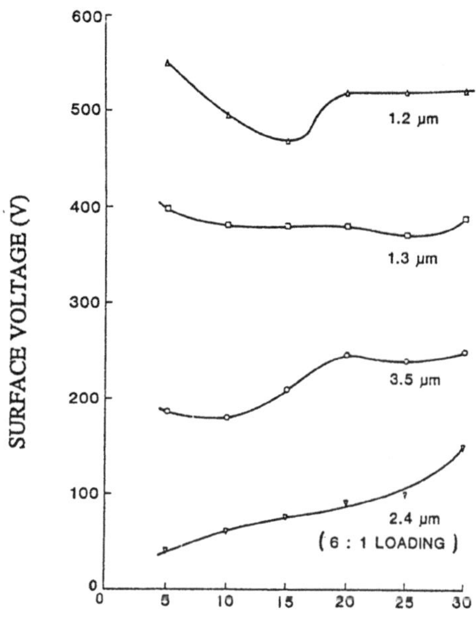

Fig. 3: Electrostatic performance of 3:1 In_2O_3 - loaded coatings of different thicknesses on Kapton at -180°C.

2. Stannic oxide. Coatings are whitish in appearance and somewhat less conductive than indium oxide coatings, giving surface conductivities in the range $10^2 - 10^5$ MΩ/square. They are RF-tuning friendly and are suitable for preparing tapes for special applications.
3. Zinc oxide. This material gives a product similar in appearance to that obtained with indium oxide. Coatings are initially non-conducting, but when irradiated with electrons, become conductive (and fluorescent) after a manner previously found with γ - radiation [Ref. 3].
4. Graphite. Highly conductive (1 - 10 kΩ/square) and matt-black, with a appearance very similar to that of commercial carbon-loaded Kapton. Suitable for applications where high thermo - optical absorptivity is required.

The coatings are easily prepared, and mixed loadings of any combination will further extend the range of properties obtainable. The major advantages of these coatings are their inherent stability, ease of application, and very low production costs. The production method is patented in the U.K. by the U.K. Atomic Energy Authority.

5. References

[1] Verdin, D., "Electrostatic Charging of SpacecraK Materials", J. of Electrostatics, 11, 249 - 263 (1982).
[2] Verdin, D. and Duck, M.J., "Surface Modifications to Minimise the Electrostatic Charging of Kapton in the Space Environment", J. of Electrostatics, 20, 123 - 139 (1987).
[3] Duck, M.J. and Nelson, R.L., "Surface Activity of Zinc Oxide", J. Chem. Soc. Faraday Trans. 1, 70, 436 - 449 (1974).

IMAGING OF X-RAYS AND ENERGETIC NEUTRAL ATOMS WITH ROTATING MODULATION COLLIMATORS

T.R. FISHER
Lockheed Space Payloads Dept., 91-21
3251 Hanover St. Palo Alto, CA 94304

ABSTRACT. Imaging of x-rays and particles is a valuable tool for understanding the structure and dynamics of planetary magnetospheres. Up to the present time, space-based imagers have usually been scanning collimated sensors or pinhole cameras. Future imagers to perform global imaging at large distances (out to 200 R_E) or to study fine structure in auroral arcs may require the increased efficiency of a coded aperture system. The technique of Rotating Modulation Collimators (RMC) is a coded aperture technique which can readily provide effective apertures >30 cm^2 and resolutions $<$ 1 deg across a 20 deg field-of-view for either x-rays or particles. A brief description of the technique is given, and simulated examples are presented showing an x-ray image of the Earth's aurora and an Energetic Neutral Atom (ENA) image of the Saturnian magnetosphere.

1. Introduction

Imaging of x-rays and particles has proven to be a valuable tool for understanding the structure of planetary magnetospheres and the dynamics of their interaction with the solar wind. From x-ray measurements, complemented by measurements at visible and UV wavelengths, we obtain information on the precipitation of particles at high latitudes in the Earth's auroral zones [Ref. 1]. Imaging of ENA's has recently been proposed as a technique for studying the ring current region of planetary magnetospheres [Ref. 2], including those of Earth, Jupiter, and Saturn. ENA's are produced by charge exchange between trapped ions and background gas, and after neutralization the particles travel on straight trajectories unaffected by electric and magnetic fields. Thus an ENA image provides a "snapshot" of the instantaneous ion distribution which can be used to study phenomena such as the evolution of substorms or plasma sheet dynamics.

In the case of both x-ray and ENA imaging, primary reliance up to the present has been on scanning collimated sensors or pinhole cameras for space-based instrumentation. Future applications, such as the resolution of fine structure in auroral arcs or global imaging of the Earth's ring current from high altitudes (e.g., on the Inner Magnetospheric Imager mission) may require the increased efficiency of coded aperture systems. This paper describes briefly the coded aperture imaging technique based on

R. N. DeWitt et al. (eds.), The Behavior of Systems in the Space Environment, 873–878.

Rotating Modulating Collimators (RMC) [Ref. 3] and presents examples of simulated x-ray and ENA images which demonstrate the capabilities of the technique. The RMC technique described here has been successfully implemented on WINKLER [Ref. 4], an imaging gamma-ray spectrometer for high-altitude balloon flights and terrestrial radiation monitoring, and a version of the technique is baselined for the High Energy Solar Physics mission (HESP) [Ref. 5].

2. Principles of RMC Imaging

The essential features of the RMC technique as applied here were defined by Mertz, Nakano, and Kilner [Ref. 3] in 1986. They showed that a configuration of modulation collimators could be employed to measure directly the amplitudes and phases of the low-order Fourier components of the source intensity distribution, and that these could then be used in an efficient algorithm to reconstruct the image of the source. The method by which this is accomplished is illustrated in Fig. 1. A typical modulation collimator is shown in Fig. 1(c). The upper grid casts a shadow of the source on the lower grid, and the intensity transmitted to the detector depends on the source position and the grid rotation angle. Let θ_o define the boundaries of the FOV. As a point in the field moves from $-\theta_o$ to θ_o in a plane perpendicular to the grid bars, the transmitted intensity displays a series of peaks and valleys as indicated in Fig. 1(a). The position of the peaks depends on the phasing between upper and lower grids. Fourier-like responses are obtained by taking differences between patterns with appropriate phases, as indicated in Fig. 1(b). Four independent measurements determine the amplitude and phase of a particular Fourier order, and these can be obtained from two collimator pairs.

If N_1 . . N_4 denote the four measurements, a one-dimensional reconstruction of the source intensity is given by:

$$S(\theta) = A_o + \Sigma\ A_n \sin(\pi n\theta/\theta_o) + B_n \cos(\pi n\theta/\theta_o);\ n = 1..M \tag{1}$$

$$A_n = N_{1n} - N_{2n} + N_{3n} - N_{4n};\ B_n = N_{1n} + N_{2n} - N_{3n} - N_{4n}$$

We obtain the two-dimensional source intensity by superimposing a set of one-dimensional reconstructions obtained as the grids rotate through 180 degrees:

$$I(x,y) = [I_o + \Sigma\ S_i(\Phi_i,\theta)]\alpha(x,y) \tag{2}$$

$$S_i(\Phi_i,\theta) = \Sigma\ [A_n(\Phi_i)\sin(\pi n\theta/\theta_o) + B_n(\Phi_i)\cos(\pi n\theta/\theta_o)]f_{in}$$

where $I(x,y)$ is the reconstructed source intensity, Φ_i denotes the grid angle, f_{in} is a filtering term, and $\alpha(x,y)$ corrects for vignetting by the collimator. Formally, this type of reconstruction is known as filtered back projection. The optimal choice of the filtering term is a subject of considerable importance which will be dealt with in detail elsewhere. In this paper, we will use the form $f_{in} = n/(1 + \lambda_i n^4)$ with λ_i adjusted to give

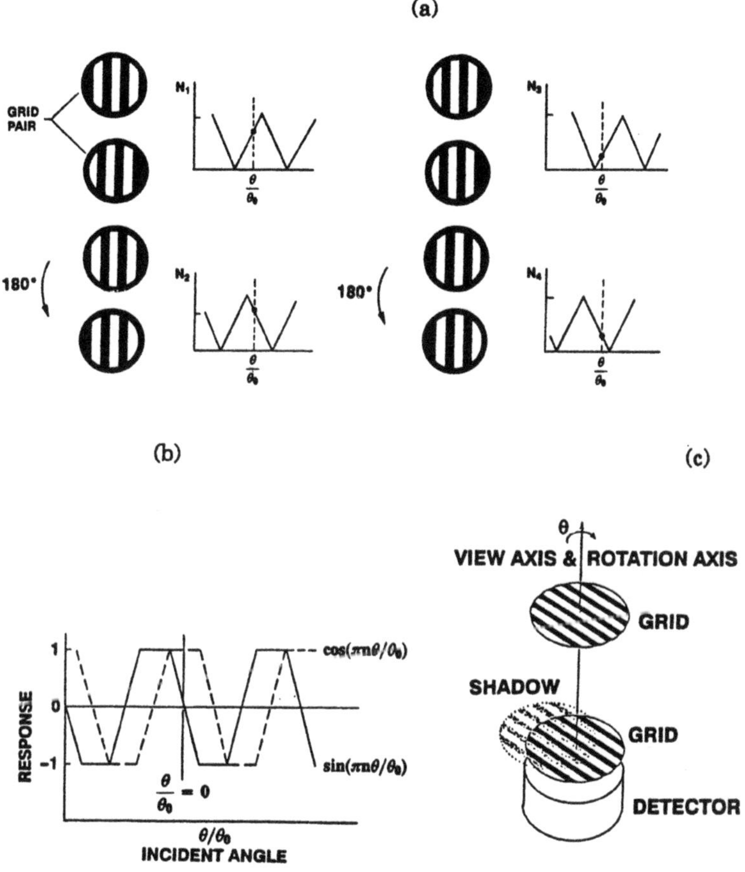

Fig. 1: Principle of RMC reconstruction: (a) phase relationship between upper and lower grids; (b) approximate Fourier-like responses obtained by combining $N_1..N_4$ as in equation (1); (c) example of a bi-grid modulation collimator.

an acceptable χ^2 fit to the entire data set. The "n" in the numerator removes the 1/r blurring introduced by the backprojection and the denominator corresponds to smoothing with a Laplacian or "thin plate spline" penalty functional. Also of considerable importance is the tradeoff between imaging resolution and FOV given approximately by $\Delta = 4.4\theta_0/\pi(M + .5)$, where Δ is the FWHM of the point spread function and M is the number of orders. This tradeoff must be carefully chosen for a particular application. For an 18 order RMC system with a 20 deg FOV, Δ is about 0.75 deg.

3. Simulations of X-ray and ENA Images

Figure 2 shows a simulated x-ray image of an auroral event as seen by viewing the Earth's north polar region from an altitude of $4R_E$. Figure 2(a) has been reproduced from the phenomenological model of Miller and Vondrak [Ref. 1], and has been chosen to show typical auroral features following initiation of a substorm: a western travelling surge, bright arcs, a "drizzle zone" and the residual diffuse aurora, all in decreasing order of luminosity. Figure 2(b) is a simulated RMC reconstruction corresponding to 9 orders, a 20 deg FOV, and 2×10^5 total counts (based

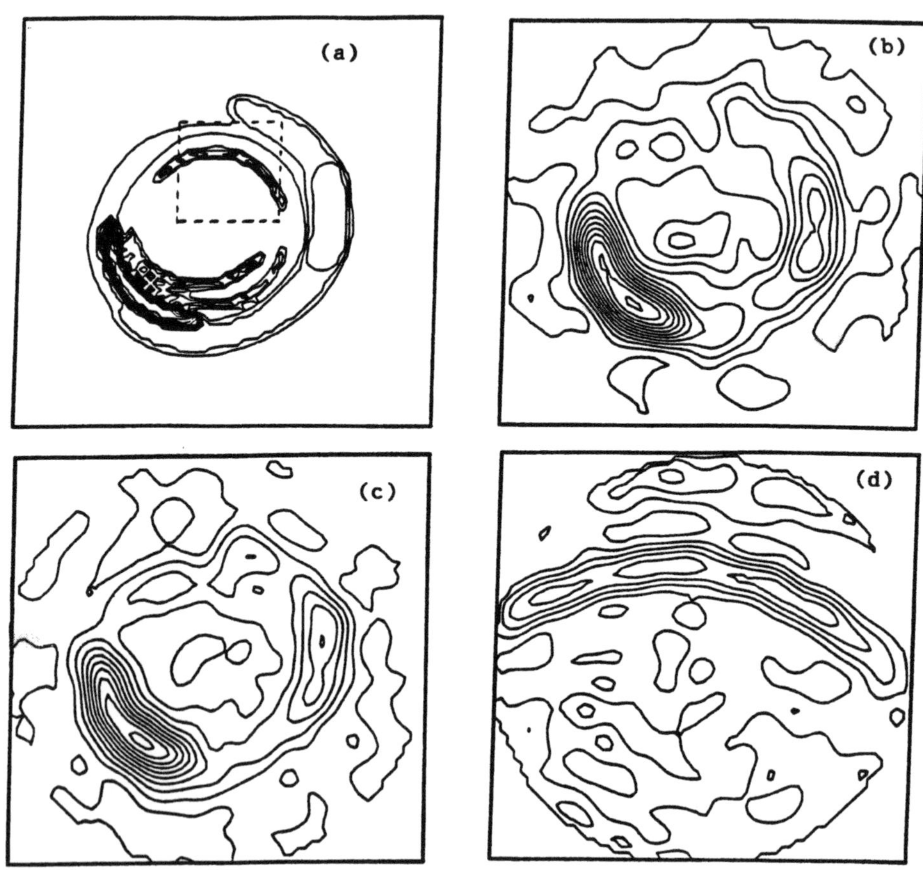

Fig. 2: Simulated x-ray images of the Earth's aurora: (a) phenomenological model from Ref. 1; (b) RMC reconstruction, 9 orders, 2×10^5 counts; (c) 18 orders, 4×10^5 counts, Laplacian smoothing; (d) reconstruction of area within dotted rectangle of (a) with 9 orders, 2×10^5 counts.

on a 5 minute recording time, 30 cm^2 effective area, 2-10 keV energy window). Figure 2(c) shows a similar reconstruction with 18 orders and 4×10^5 total counts which looks remarkably similar to Fig. 2(b). In fact, at this counting level, the higher orders > 9 contribute mainly noise and are removed by the Laplacian filter. In order to resolve finer structure, such as the discrete arc within the dotted rectangle of Fig. 2(a), it is preferable to reduce the FOV while keeping the same number of orders. This is illustrated in Fig. 2(d).

Figure 3 shows a simulated ENA image of the Saturnian magnetosphere taken from a distance of 66 R_S. The model is that of Curtis and Hsieh [Ref. 6], who estimate a total ENA flux in the energy range 10-100 keV of 6 $cm^{-2}s^{-1}$ at this distance. The reconstruction is for a 9-order RMC system with an effective area of 28 cm^2, giving about 2×10^5 total counts in a 20-minute interval. The simulation shows that the two lobes of the ring current distribution are resolved at this level. Longer exposures or larger detector areas would give improved definition of these features.

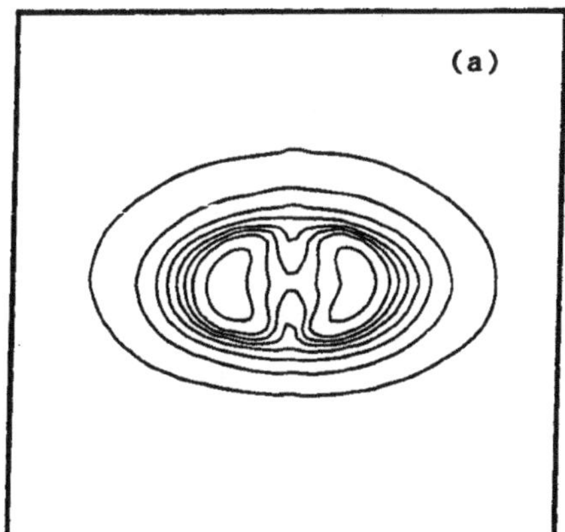

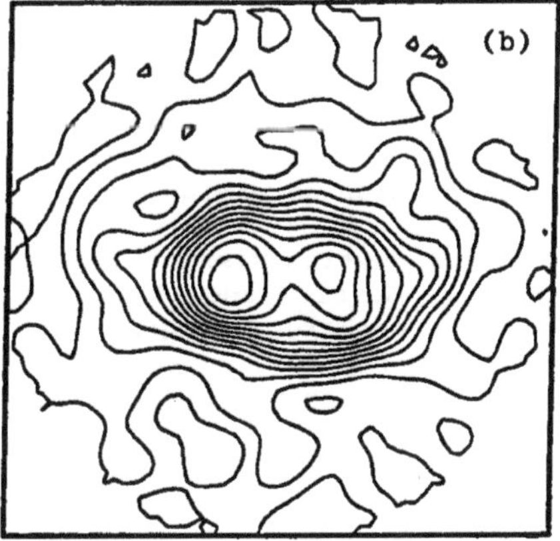

Fig. 3: Simulated ion ENA image of the Saturnian magnetosphere: (a) simulation from Ref. 6: (b) RMC reconstruction, 9 orders, 2×10^5 total counts, with Laplacian smoothing.

4. Conclusions

The simulations presented here show that the technique of RMC imaging merits consideration for the next generation of space-based imagers of x-rays and ENA's. In general, such RMC systems can be expected to combine effective areas > 30 cm^2 with fields-of-view of the order 20 x 20 deg and resolutions < 1 deg. The system parameters must be carefully chosen for a particular application, and simulations including the effects of realistic statistical noise should be performed to establish that features of interest can be resolved.

Acknowledgements The author acknowledges several useful conversations with members of the Lockheed Palo Alto Research Laboratory on the subject of RMC imaging and auroral phenomena. Preparation of this paper was supported by the Lockheed Independent Research Fund.

5. References

[1] See, for example, K.L. Miller and R. R. Vondrak, "A high-latitude phenomenological model of auroral precipitation and ionospheric effects," Radio Science 20, 431 (1985) for a good summary of references on this subject.

[2] H.D. Voss and L.G. Smith, "Global zones of energetic particle precipitation", Journal of Atm. and Terr. Phys. 42, 227 (1980); E.C. Roelof and D.J. Williams, "The terrestrial ring current: from in situ measurements to global images using energetic neutral atoms," Johns Hopkins APL Technical Digest Vol 11, Numbers 1 and 2, 72 (1990).

[3] L.N. Mertz, G.H. Nakano, and J.R. Kilner, "Rotational aperture synthesis for X-rays," J. Opt. Soc. Amer. 3, 2167 (1986).

[4] T.R. Fisher, J.W. Hamilton, J.D. Hawley, J.R. Kilner, M.J. Murphy, and G.H. Nakano, "Imaging of gamma rays with the WINKLER high-resolution Germanium Spectrometer," IEEE Trans. Nucl. Sci. 37, 1483 (1990).

[5] Space Physics Missions Handbook, NASA, section 6.11 (1991).

[6] C.C. Curtis and K.C. Hsieh, "Remote sensing of planetary magnetospheres: imaging via energetic neutral atoms," Solar System Plasma Physics, edited by J.H. Waite et. al., Geophysical Monograph 54, American Geophysical Union, Washington, D.C., p. 247, (1989).

TEST AND VALIDATION OF MCT DETECTORS FOR PMIRR ON THE MARS OBSERVER SPACECRAFT

S. ASLAM, C.L. HEPPLEWHITE and S.B. CALCUTT
Department of Physics,
Clarendon Laboratory,
University of Oxford, Parks Road,
Oxford OX 1 3PU, U.K.

ABSTRACT. This paper gives a brief description of the Mars Observer mission, the PMIRR instrument, its scientific objectives and the sub-assemblies that were built and tested by Oxford University. The mid-infrared detector assemblies used in the Pressure Modulator Infra-Red Radiometer (PMIRR) are detailed together with the tests that were performed to validate them.

1. Introduction

The Pressure Modulator Infra-red Radiometer (PMIRR) is an instrument designed to investigate key questions of the climatology of the Mars atmosphere [Ref. 1]. In particular, it aims to determine the temporal and spatial distribution, the abundance, sources and sinks of volatile material and dust over the seasonal cycle. Further, it explores the structure and aspects of the circulation of the atmosphere of Mars [Ref. 2].

PMIRR is a nine channel infrared radiometer employing pressure modulation and filter radiometry. Channels 1 and 4 employ single Mercury Cadmium Telluride (MCT) photovoltaic (PV) detectors. The performance of these detectors will be limited by instrument background radiation. Channels 2, 3 and 5 employ single MCT photoconductive (PC) detectors and will be detector noise limited [Ref. 3]. These 5 detectors share a common Focal Plane Assembly (FPA) and are cooled to 80K by a passive radiative cooler. The remaining detectors are DTGS pyroelectric detectors maintained at the instrument ambient temperature of 300K. All channels share a common scan mirror and telescope, which are used to map the atmosphere in three dimensions. Two pressure modulator cells (PMC), developed at Oxford are used [Ref. 4], one containing isotope enriched CO_2 and the other H_2O. The use of pressure modulation eliminates the optical errors associated with the selective chopping (SCR) [Ref. 5], while maintaining the large energy grasp of gas correlation.

The Mars Observer spacecraft is planned for launch in September 1992 and will be inserted into a polar phasing orbit at Mars after an eleven month interplanetary transit. The orbit at Mars will be adjusted to low

R. N. DeWitt et al. (eds.), The Behavior of Systems in the Space Environment, 879–887.

altitude, 361 km, near circular, polar, sun-synchronous, and the mission duration is expected to be 687 days (one Martian year).

Mars is known to have a weak magnetic field [Ref. 6] and measurements have been made of the ionospheric plasma [Refs. 7, 8] and of energetic particles in the vicinity of Mars [Ref. 9]. Knowledge of the conditions at the orbit of the spacecraft (s/c) and of the interplanetary transit permits design of the s/c and its payloads to operate reliably throughout the life of the mission. The payloads or instruments will be designed to survive and operate as described in document PD642-41, Mars Observer Payload Environmental Design Requirements.

The complete assembly will be subject to a variety of environmental tests, including accelerations and vibrations, thermal vacuum simulations, and electromagnetic compatibility (EMC) tests. In addition, electronic parts and materials will be selected according to their reliability and radiation hardness. For the PMIRR instrument, Oxford University supplied the MCT detectors and the pressure modulators. As a part of verification, these two sub-assemblies were subjected to launch vibration simulation, and the detectors were aligned and performance tested.

2. Design and Performance Requirements

PMIRR will view Mars via a two-axis, azimuth/elevation scan mirror, and each spectral channel, which shares this mirror, has a 3.1 mrad Field Of View (FOV), giving a 5km view in the vertical on the limb of Mars. Nadir scanning will also be used. In order to be sure that all channels view the same part of the atmosphere at a given time, all channels must be optically aligned, and for the retrieval of atmospheric composition from the measured signal, the response of each detector must be mapped. If this alignment is not performed, then signals from different channels may be weighted to different parts of the field of view in some unknown way. This validation of the mid infrared detectors represents a crucial aspect of the validation of this type of instrument used in space for remote sensing.

3. Optics/Detector Assembly

One of the optic/detector assemblies is shown schematically in Fig.1, and consists of (i) the detector element on a sapphire substrate, (ii) a Ti alloy optics housing, containing anti-reflection coated Ge doublet condensing lenses and (iii) a Ti-alloy detector baseplate. The size of the detector element is 150μm x 750μm.

4. Test and Validation

Accurate validation of the optic/detector assemblies was obtained iteratively by scanning a chopped hot pin-hole source across the detector FOV and mapping the detector array response. Each time the optic/detector assembly was cooled down in the space simulation chamber, the scanning

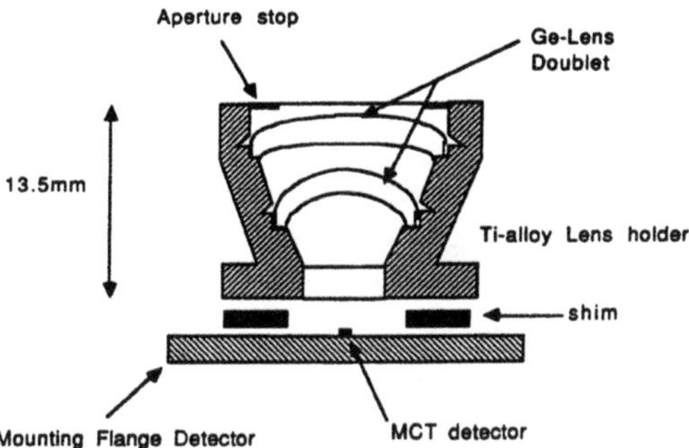

Fig. 1: Schematic of optic/detector assembly.

system had to be calibrated. The main purpose of this calibration was to determine the pinhole location relative to the origin of a reference plane; this was determined using three orthogonal position switches. The position switch locations were precisely known relative to the pinhole, through metrology. Knowledge of this and the detector plane to reference plane distance gave the accurate position of the pinhole centre to detector plane origin during any scan.

Adjustment of focus and alignment is achieved by changing the thicknesses of three shims under the optics housing. The best focus was established by determining the sharpest slope of the FOV response of the elements and was made to lie in the nominal focal plane within the required tolerances. The best focussed detector edge response provided a measure of abberated spot-size and hence a primary test of the accuracy of lens manufacture, alignment and focus.

The optic-axis of the optics mount is not generally perpendicular to the flat reference surface due to tolerance build up in the assembly of the optics mount. Using the above method, the alignment could be adjusted to better than ±0.3 mrad from the perpendicular of the reference surface.

5. Ground Support Equipment

A schematic of the Ground Support Equipment (GSE) is shown in Fig 2. The GSE consists of the vacuum system, calibration equipment, software, expendable materials and other equipment and fixtures required to operate, test and validate the detector elements in a contamination-free environment. Tests were performed to demonstrate that the GSE was functioning properly using sample PC and PV elements.

The vacuum system consists of a stainless steel chamber, a liquid nitrogen cooled baffle, a 2-stage rotary pump-backed 160/700 Diffstak, which can attain a total pressure better than 5 x 10^{-6} Torr. The system is fully automatic with programmed shut-down in case of leaks or power

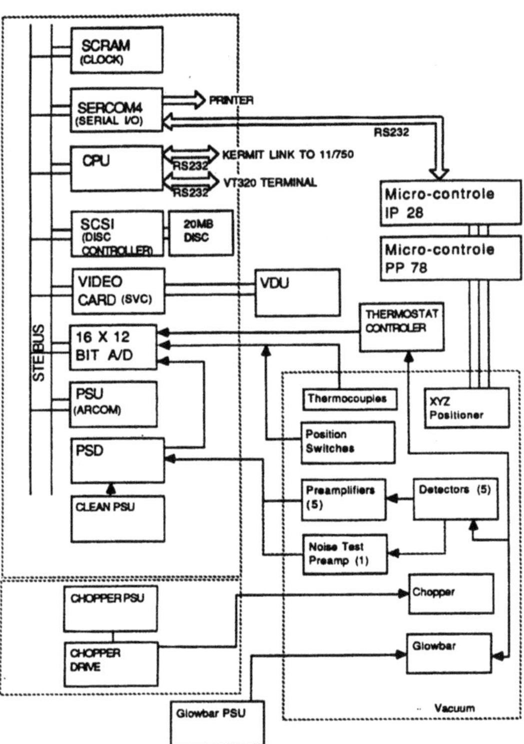

Fig. 2: PMIRR alignment rig control system.

failure. The cleanliness of the system was monitored by measuring the partial pressures of contaminants using a residual gas analyser, which during testing were less than 10^{-9} Torr. Contamination levels, during validation testing, were also monitored by analysing condensates formed during cool-down on calcium fluoride crystal windows and witness mirrors.

The detector FPA was bolted via a fixture onto a copper block which was mounted directly onto a liquid nitrogen dewar. Temperatures were monitored via calibrated platinum resistance thermometers fixed on the copper block on the reference plate and on the detector element baseplates.

A globar, pinhole and chopper assembly was used to provide a modulated blackbody radiation source. The pinhole source was scanned using a 1μm resolution commercial orthogonal micro-translation stage, driven from a control system running CPM/3. All temperatures and the detector signals were logged via a commercial analogue to digital card on an STE bus.

6. Some Results

Figures 3 and 4 show spectral response plots for flight channels 1 and 3 detectors. For the PV detector of channel 1 at a temperature of 85K the responsivity is R_i(7.75-8.02μm, 800Hz) = 4.81A/W, with a spectral detectivity of D*(7.75-8.02μm,800Hz,1) = 9 x 10^{10} cmHz$^{1/2}$/W. For the PC detector of channel 3, at the same temperature the responsivity is R_v(14.3-14.9μm, 800Hz) = 3.8 x 10^3V/W, with a detectivity of D*(14.3-14.9μm, 800Hz,1) = 5.12 x 10^{10} cmHz $^{1/2}$/W with an optimum bias current of 2mA.

Figure 5 shows a flight channel 5 PC detector responsivity image. The centroids are found independently for the Y and Z axes. For example, for each value of Y for which there is a recorded detector cross section, the 50% crossings are determined by linear interpolation. The 50% level is taken to be the half way between the minimum value recorded and an average of 9 points around the maximum value recorded. The 'moment of area' of the element at this Y value is taken as, M=Y * AREA, where AREA=FWHM *ΔY, FWHM is the distance between the 50% points, and ΔY is the sampling interval in the Y direction. The centroid is calculated from Ycentroid = Σ M/Σ AREA. The Z centroid is calculated similarly.

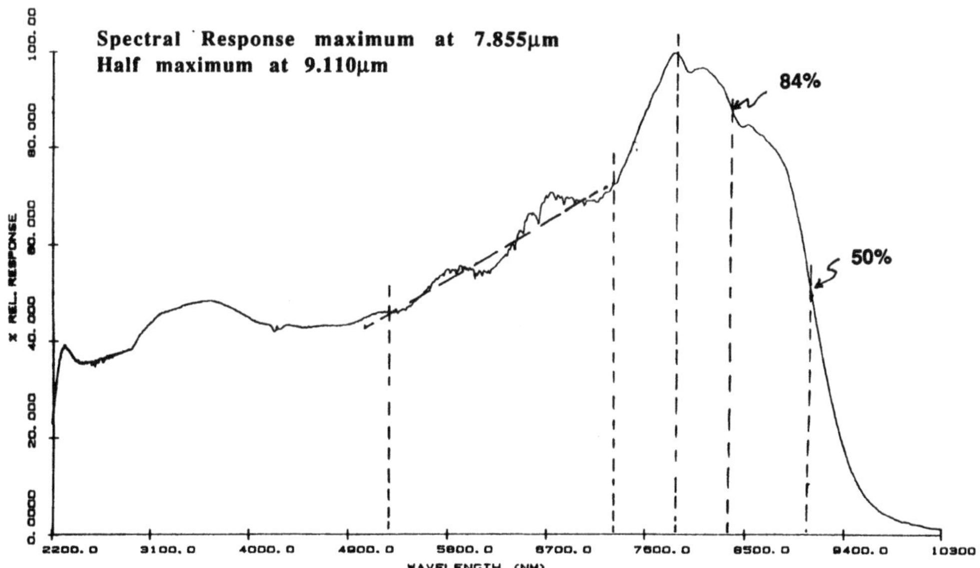

Fig. 3: PV detector response for channel 1.

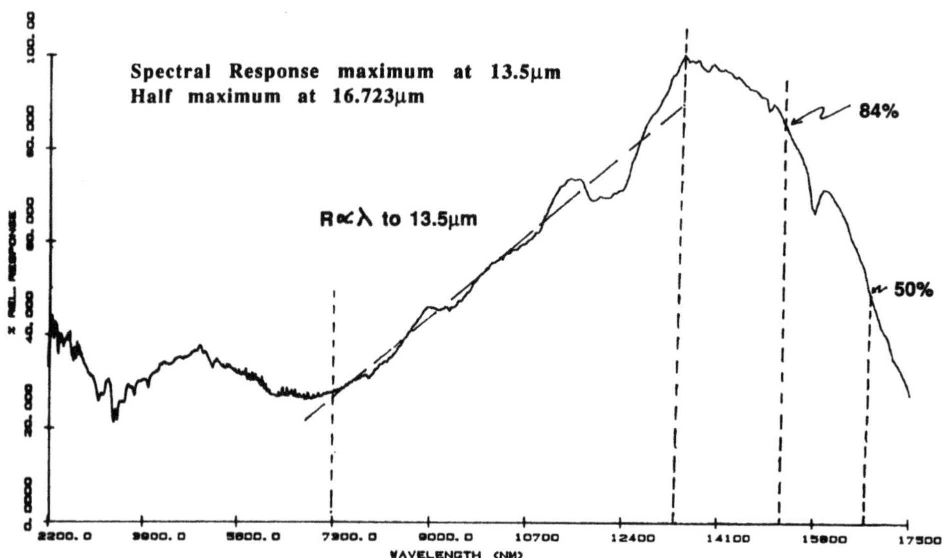

Fig. 4: PC detector response for channel 3.

final position of detector image.
Target box position is derived from rotation axis measured with nominal shims in place.
Angular error from target is 0.37 mrad (z direction), and 0.7 mrad (y direction)

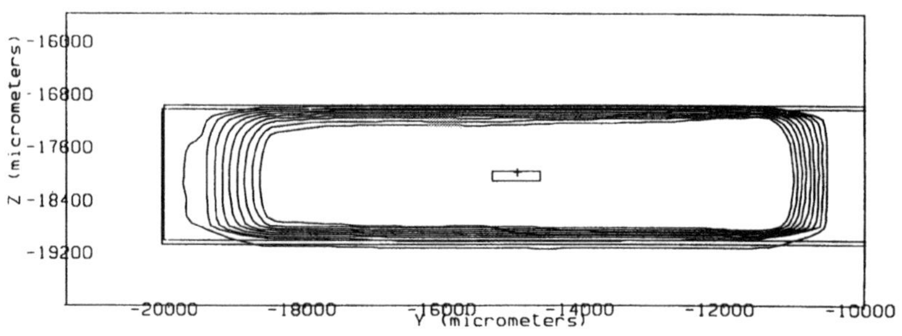

centroid at y=-14965., z=-18004. target at y=-14990., z=-18071.
file:ch5012.out X= 95000., scan dir:0

Fig. 5: PC detector response image.

Figure 6 shows a Y-line scan, through the centroid, for the channel 5 detector; the effect of minority carrier sweepout at the cathode is evident [Ref. 10]. Fig. 7 shows an aligned and focussed responsivity plot for channel 4.

7. Conclusion

The scientific objectives of the PMIRR instrument determine the design and performance requirements of the detector and optical assemblies. Many aspects of reliability and safety of the instrument also influence these requirements, for example use of approved parts and materials. The performance of the infra-red FPA is crucial to achieving the science objectives of PMIRR and the alignment and element mapping performed at Oxford fully characterised and verified the design and performance of these sub-assemblies.

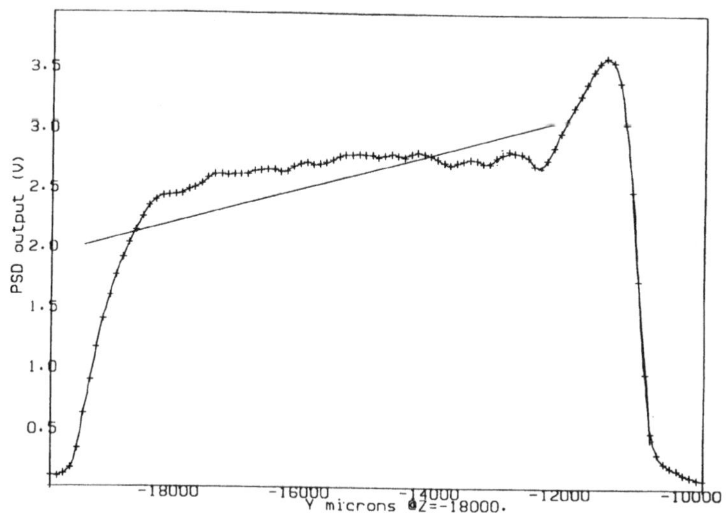

Fig. 6: Channel 5, Y-line scan.

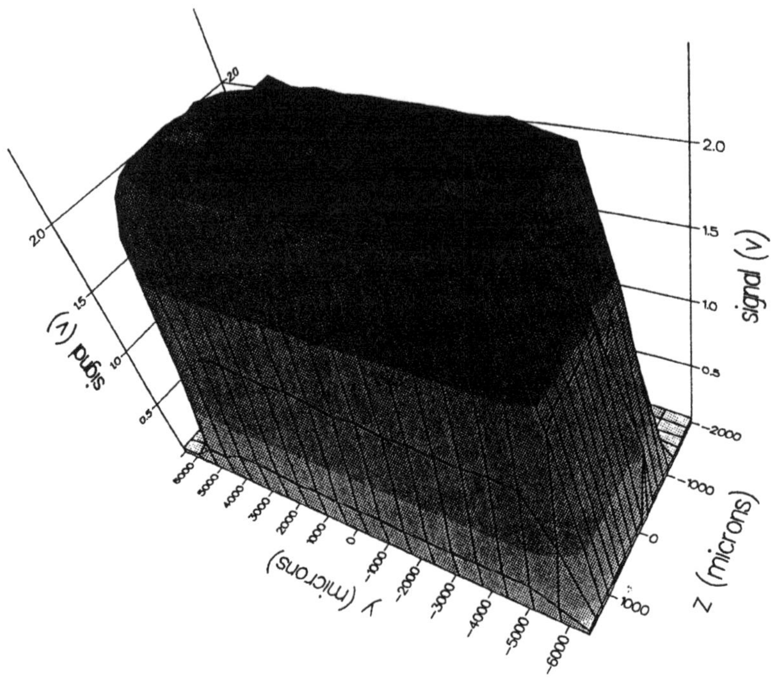

Fig. 7: PMIRR. Channel 4 PV detector response.

References

[1] McCleese, D.J., et. al., Remote sensing of the atmosphere of Mars using infrared pressure modulator and filter radiometry. *Appl. Optics*, 25, 23, 4232-4245, 1986.

[2] Taylor, F.W., Calcutt, S.B., Aslam, S., Hepplewhite, C. L., IR Sounder for Martian climate studies, *EGS, XVI,General Assembly*, Wiesbaden, Germany, 1991.

[3] Blouke, M.M, Burgett, C.B.,Williams, R.L., Sensitivity limits for extrinsic and intrinsic infrared detectors, *Infrared Physics*, 13, 61-71, 1973.

[4] Curtis, P.D., Houghton, J.T., Peskett, G.D., Rodgers, C.D., Remote sounding of atmospheric temperature from satellites. The pressure modulator radiometer for Nimbus F. *Proc. R. Soc. Lond.* A337, 135-150, 1974.

[5] Houghton, J.T., Smith, S.D. Remote sounding of atmospheric temperature from satellites. *Proc R. Soc.Lond.* A320, 23-33, 1970.

[6] Reidler, W. et al. Magnetic fields near Mars: first results. *Nature*. 341, 604-607, 1989.

[7] Kliore A.J., Fjeldbo G., Seidel B.L., Sykes M.J., Woiceshyn P.M. J *Geophys Res*. 78, 4331, 1973.

[8] Lundin, R. et al. First measurcments of the ionospheric plasma escape from Mars. *Nature*. 341, 609-612, 1989.

[9] Shutte, N.M., et al. Observation of electron and ion fluxes in the vicinity of Mars with the HARP spectrometer. *Nature*. 341, 614-616, 1989.

[10] Broudy, R.M., Mazurczyk, V.J. (HgCd)Te Photoconductive Detectors, *Semiconductors and Semimetals:* Mercury Cadmium Telluride, Editors, R.K. Willardson, A.C. Beer, 18, 157-199, 1981.

A COMBINED TEST FACILITY FOR SPACE RESEARCH AND TECHNOLOGY

G. STASEK, R. WIRTH and P. SEIDL
PTS Physikalisch-Technische Studien GmbH
Leinenweberstr. 16
D-78 Freiburg, Germany

ABSTRACT. The combined test facility IONAS (Ionospheric Atmospheric Simulator) is available at PTS, suitable for testing even large satellite subsystems in simulated LEO and GEO space environment. Numerous combinations with light-, plasma- and electron-sources as well as various optical systems cover a wide range of potential experiment-specific requirements. This paper summarizes the most important technical aspects of the facility and of the various add-on systems and presents typical applications.

1. Set-Up

1.1. GEOMETRY

Figure 1 shows schematically the set-up of IONAS and some major components.

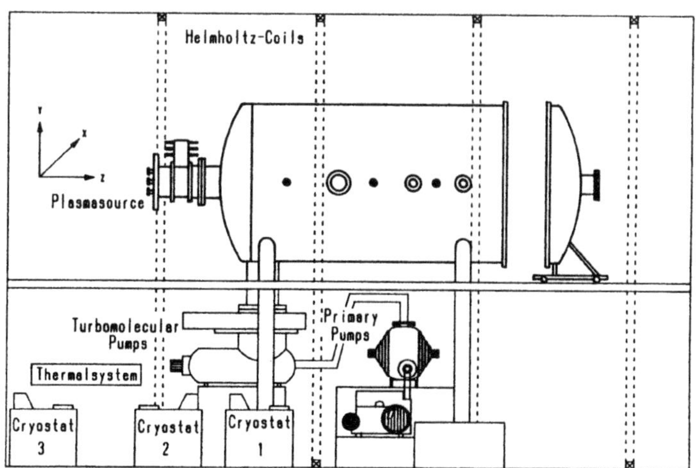

Fig. 1: Schematic Set-up of IONAS.

R. N. DeWitt et al. (eds.), The Behavior of Systems in the Space Environment, 889–895.

The facility is based on a horizontal stainless steel cylinder with a diameter of 2.5 m and a total length of 5.5 m. A sliding door provides easy access to the inner mounting platform consisting of removable segments with a total area of 1.5 x 4 m^2. The vacuum chamber has numerous flanges with inner diameters from 32 up to 500 mm, which fit to NW, ISO and U-Euronorm standards. They can be equipped with a variety of available feedthroughs for electrical signals, gases or liquids. Several computer-controlled mechanisms provide rotational and translational movements, e.g. for scanning of sensors around the test specimen. The facility access area at the vacuum door, as well as the adjacent test preparation and integration laboratory, are cleanrooms of class 100.

1.2. VACUUM SYSTEM

The pumping system of IONAS is based on turbomolecular pumps providing an oil-free vacuum and fast evacuation from atmospheric pressure down to 2 x 10^{-5} mb in about two hours. The typical background pressure of about 10^{-6} mb can be further improved by operating a cryogenic wall. A mass-spectrometer analyses the residual gas and monitors outgassing products from the test specimen, e.g., during TV-tests. Numerous valves, adaptors and small additional pumps allow a flexible realization of different vacuum-requirements such as differential pumping of subsystems.

1.3. HELMHOLTZ-COIL SYSTEM

IONAS is equipped with a three-axial Helmholtz-coil-system, which can generate a magnetic field in any direction with flux densities up to about 3.5 Gauss. A special application is the compensation of the earth magnetic field. The homogeneity is better than 1% in a volume of about 1 m^3. Field measurements are performed by means of a Foerster-probe. The Helmholtz-coil-system turns out to be particularly valuable for all applications involving low-energy electrons.

1.4. PLASMA SOURCE

Two plasma sources of the Kauffmann-type can be attached to the facility. Similar sources have been studied as propulsion systems for space vehicles. Usually one plasma-source is installed at the head side on the chamber axis. It can be separated by a large valve for maintainance operations without breaking the main vacuum.

1.5. PLASMA DIAGNOSTIC

The following diagnostic tools are available to analyze the plasma:

INSTRUMENT	MEASURED QUANTITY
Impedance Probe	N_e
Langmuir Probe	N_e, T_e, N_i
Retarding Potential Analyzer	T_e
Ion Time-of-Flight Measurement	v_i

Resonance Cone	N_e, T_e
Plasma Potential Probe	ϕ_p
Electric Field Mill	ϕ_p
Faraday Cup	j_e

Some of these techniques have been developed at PTS in the course of special studies [Refs. 1, 2].

1.6. ELECTRON-GUN

An electron gun designed for the energy range from 0.5 to 70 keV can be mounted at different flanges of IONAS. With the presently available high voltage supply, energies up to 35 keV and a total beam current of 1 mA are achievable.

The far-focussing-cathode (according to Steigerwald) of the electron gun operating together with a magnetic deflection system yields an electron beam of the small diameter (< 1 mm), which can be directed on different parts of the test specimen. Alternatively, the application of the thin-foil technique generates a widespread electron distribution of about 1 m diameter. This method was applied to homogeneously irradiate test specimens in the course of studies concerning spacecraft charging effects.

1.7. OPTICAL SYSTEMS

Many different sources are available for the generation of light covering a broad wavelength range from the X-rays through the infrared. Some of them are calibrated by the National Institute of Science and Technology (NIST) as listed in the summary below:

SOURCE	WAVELENGTH RANGE	REMARK
X-Ray	0.1 - 2 nm	max. 15 kV/80 mA
EUV-Lamp	20 - 200 nm	
Argon-Mini-Arc	160 - 335 nm	Calibrated
Deuterium Lamp	180 - 400 nm	
Quartz-Halogen-Lamp (FEL)	250 - 2500 nm	Calibrated

All sources can be completed with suitable monochromators or filters to provide irradiation at selected wavelengths. A further improvement of the facility in the near future will include a solar simulator.

A calibrated spectrometer can analyse optical emissions from the plasma as well as from test lamps. It operates computer-controlled in a broad wavelength range from 40 - 640 nm. Furthermore, calibrated detectors for the EUV as well as various channeltrons, multipliers and silicon detectors are available.

1.8. THERMAL SYSTEM

An essential extension of the applicability of IONAS was introduced by the thermal system. It is based on two completely independent systems supplied by very powerful cryostats. The first one performs cooling and heating of the test specimen, which can either be mounted on the PTS-coldplate (1 m x 1.5 m) or on its specific coldplate. Typical temperatures range from -50 to +95°C. The second system cools and heats the environment of the test specimen. It includes six hollow half-cylinders, which are mounted at the inner chamber wall and supplied with a special heat transfer fluid. They can be operated in the range from -45 to +95°C with maximum gradients of -0.9 and +1.1°C/min. Both systems are supplied with a special, oil-free heat transfer fluid, which is even used in space technology as a contacting temperature test medium. It evaporates without residues and eliminates any risk of damage and contamination for the test specimen in case of leakage. The temperatures are measured by numerous sensors (presently 48), which can be placed arbitrarily on the test specimen.

The data acquisition system, based on an AT-PC with appropriate interfaces, measures each temperature twice per minute, displays the result in digital and analog form and stores it. Any sensor can be chosen as a reference for each of the two systems to control the temperature. The cryostats are guided by the computer to either stabilize the reference temperature within typically 0.1°C or to change to another temperature level with a selected gradient (e.g. 0.5 °C/min). A special adaptive control algorithm yields stable conditions even in case of changing power dissipation as faced e.g. during different operation modes of the instrument under test. A second AT-PC ensures online access to all thermal data and flexible evaluation according to test-specific requirements (plots, listings, transfer to floppy disk).

Numerous precautions are implemented to ensure the safety of the test specimen. Inherent safety against leakages is provided by high-quality hoses and connectors and by the use of a special, non-contaminant and oil-free heat transfer fluid. An independent monitor system switches off the power for the test specimen as well as for the thermal system in case present pressure- or temperature-limits are exceeded. A third cryostat can be operated as additional support or serves as backup system.

2. Typical Applications

The study of electric double layers is one example for the activities performed in the field of fundamental research of plasma processes. An appreciable part of the past work was dedicated to the development and calibration of plasma sensors such as the resonance cone technique, the retarding potential analyzer, the impedance probe and the plasma-potential probe [Refs. 1, 2].

Studies regarding material-plasma interaction focussed on solar cells at elevated voltages in LEO-environment [Refs. 3-7]. The size of IONAS allows even for investigations of solar cell arrays as illustrated in Fig. 2. The research on spacecraft-charging effects in the course of past experiments required the simulation of GEO-conditions and the operation of the electron gun with the foil-technique [Refs. 8, 9].

Past activities concerning optical calibration include the calibration of EUV-photometers as well as studies on the emissions of

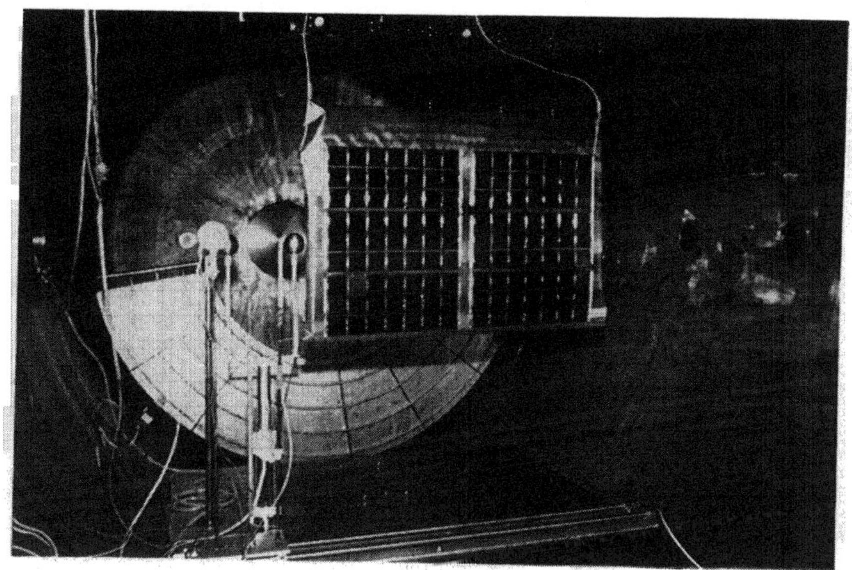

Fig. 2: Set-up for the interaction of simulated LEO with solar panel.

special discharge lamps. Thermal vacuum tests are a preferred application for IONAS. Large spacecraft subsystems for ERS-1, MAS (linear actuator, PEB), X-SAR and RMK were successfully tested in this facility (Fig. 3).

3. Summary

IONAS has been essentially improved since its first installation in 1974. The original intention, the study of plasma physics, has been extended with the development and calibration of plasma probes for the LEO and GEO environment. The completion with a powerful thermal system includes the potential of thermal vacuum tests on large spacecraft subsystems. Numerous combinations with additional electron- and plasma-sources as well as various optical systems yield a high degree of flexibility corresponding to a wide range of experiment specific requirements.

Fig. 3: Preparation of the MAS linear actuator for the thermal vacuum test in IONAS.

4. References

[1] A. Piel, G. Oelerich, H. Thiemann, Resonance Cones in nonthermal plasmas: Laboratory experiments; Proceedings of the 8th ESA Symposium on European Rocket and Balloon Programmes and Related Research, Sunne, Sweden, ESA SP-270, 1987.

[2] H. Thiemann, A. Hoffmann, P. Seidl, G. Stasek, P.V. Dharma Rao, Design and development of plasma probes for the Indian rocket program SPREAD-F; Final Report according to BMFT-Contract No. 01 TL 021/2, 1987.

[3] G. Stasek, Loss currents of solar cells under LEO-conditions; Proceedings of the 3rd European Symposium: Photovoltaic Generators in Space; Bath, England, ESA SP-173, 1982

[4] H. Thiemann, K. Bogus, Anomalous current collection and arcing of solar cell modules in a simulated high-density low-earth-orbit plasma; ESA-Journal, Vol. 10, 1986.

[5] H. Thiemann, P.V. Dharma Rao, Anomalies observed in plasma-module-interaction experiments; Proceedings of the 5th European Symposium: Photovoltaic Generators in Space; Scheveningen, The Netherlands, ESA SP-267, 1986.

[6] H. Thiemann, K. Bogus, High voltage solar cell modules in simulated low-earth-orbit plasma; Journal of Spacecraft and Rockets, Vol.25, No. 4, 1988.

[7] H. Thiemann, R.W. Schunk, L. Gerlach, Solar arrays in the LEO-plasma environment: A model for leakage current phenomena deduced from experimental and theoretical studies; Proceedings of the European Space Power Conference held in Madrid, Spain, ESA SP-294, 1989.

[8] E. Coggiola, L. Levy, D. Sarrail, H. Thiemann, Spacecraft Charging Study in Polar Orbit; Final Report according to ESA-Contract No. 7141.87, 1988.

[9] H. Thiemann, W. Noack, R. Wirth, ISO-Sunshield: ESD-Test Results; ESA Contract Report according to Purchase Order No. 192267, 1989.

IRRADIATION DAMAGE ASSESSMENT OF ELECTRONICS

J.S. BENNION, G.M. SANDQUIST, P.S. SHEEHAN
University of Utah
Nuclear Engineering Laboratory
Salt Lake City, UT 84112 U.S.A.

V.C. ROGERS
Rogers and Associates Engineering
P.O. Box 330
Salt Lake City, UT 84110

ABSTRACT. A university research reactor is employed to irradiate electronic components to assess radiation induced damage. The reactor facility, irradiation procedure and dosimetry are described. Semiconductor devices are irradiated following guidance set forth in military standards (MIL-STD-883C). Electrical properties of irradiated components are characterized and compared with pre-irradiation performance parameters to assess the degradation of critical parameters as a function of neutron fluence. Future experiments will investigate real-time irradiation damage effects as functions of dose rate and fluence.

1. Introduction

Radiation hazards existing in space include galactic cosmic rays, particles emanating from solar flares within the heliosphere, and charged particles confined to the magnetospheres of earth and other planets. Extraterrestrial radiation consists of approximately 87% protons, 12% alpha particles, and 1% heavy nuclei ranging in energy from 1 to 10^{13} GeV and forms a highly isotropic flux of relativistic particles [Refs. 1, 2]. Since the energies of these primary particles can greatly exceed the nuclear binding energy of all known elements, interaction with matter results in nuclear fragmentation and spallation. Secondary cascades of energetic protons, neutrons and other nuclear debris resulting from these primary interactions may produce significant localized areas of radiation damage in materials, primarily through the introduction of dislocations and various other defects in the crystalline lattice and the generation of thermal spikes. As a result, irradiated materials such as electronic components can suffer serious transient or permanent physical changes in their electrical properties. This resulting damage may alter or destroy the properties or performance anticipated for the materials in a space or other application where materials are exposed to radiation.

R. N. DeWitt et al. (eds.), The Behavior of Systems in the Space Environment, 897–903.

The U.S. Air Force is concerned about the potential malfunction of electrical systems of their defense missile systems subjected to intense radiation fields. Consequently, they have contracted with the University of Utah Nuclear Engineering Laboratory (UUNEL) to provide irradiation services to assess the nuclear hardness of sensitive individual electronic components [Ref. 3]. Although the radiation environment of space is, of course, not nearly as intense nor of the same composition as that found in a small research nuclear reactor, practical testing of components may be accomplished in a short time which can be reasonably extrapolated to systems operating in space and, other radiation environments. While this testing is crucial for assessing weapons effects and survivability, it will also become increasingly important for assessing the reliability of control and auxiliary subsystems of space nuclear propulsion systems as this technology matures and becomes more practicable.

2. Nuclear Reactor and Irradiation Facility

UUNEL operates a 100 kilowatt (thermal) TRIGA nuclear research reactor. The reactor core is located at the bottom of a 24-foot-deep pool of demineralized water which serves as heat sink, neutron moderator and radiation shield. A schematic diagram of the reactor core is presented in Fig. 1. The core is composed of a hexagonal configuration of standard TRIGA fuel elements containing 8.5% uranium enriched to 20% ^{235}U homogeneously dispersed as a uranium-zirconium hydride alloy, deuterium oxide (D_2O) reflector elements, and three control rods containing boron carbide. The core is surrounded by various radiation detectors to monitor the reactor power level. In addition, there are two external irradiation test stands. A thermal irradiator (TI) is filled with D_2O to provide an isotropic thermal neutron source. On the opposite side of the core is the fast irradiator (FI) used for the irradiation of electronic components. This fixture is provided with adjustable lead shielding for gamma radiation attenuation of the operating environment. The FI is situated directly adjacent to a face of the hexagonal core grid filled with fissile fuel to minimize moderation of fast neutrons emitted through fission. An aluminum sample holder is manually placed in the FI to expose contained samples to a known fluence of neutrons. The sample holder can accommodate additional gamma shielding or neutron filtering materials to simulate various composite materials.

The radiation field characteristics generated by the TRIGA reactor are summarized in Table 1. A continuous energy spectrum of neutrons is produced during sustained fission. Figure 2 illustrates typical fission spectra for a TRIGA water-moderated system, a bare-core reactor and a pure, unmoderated fission spectrum. Neutron energies extend more than eight orders of magnitude from the thermal region at 0.025 eV to beyond 10 MeV. The neutrons produced in the fission process have an average energy of about 2 MeV and a most probable energy of about 0.7 MeV as shown in Fig. 2.

FI - Fast Irradiator Test Stand
TI - Thermal Irradiator Test Stand
D - Deuterium Oxide Reflector Elements
F - Standard TRIGA Fuel Elements
FC - Fission Chamber
UI - Uncompensated Ion Chamber
CI - Compensated Ion Chamber
C - Control Rod

Fig. 1: Photograph of University of Utah TRIGA Mk. I reactor showing relative positions of core components, power monitoring detectors and external irradiation facilities.

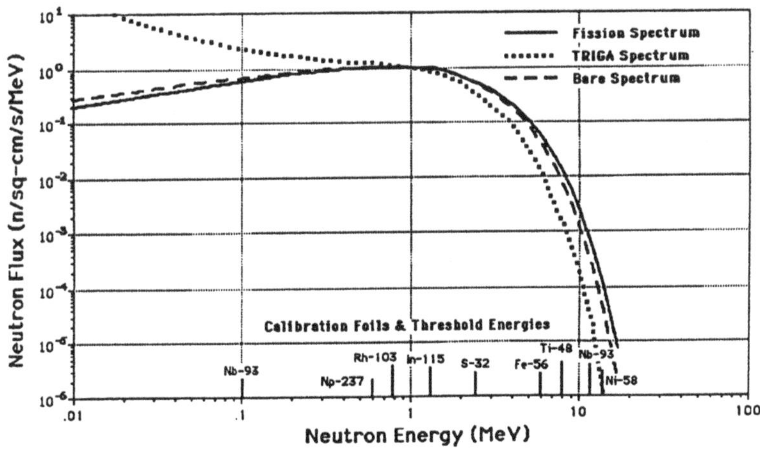

Fig. 2: Normalized neutron spectra of 235U-fuelled reactors. Several activation foils used to characterize fission spectra are indicated along the abscissa at their corresponding reaction threshold energies.

Table 1. Radiation characteristics of the UUNEL TRIGA nuclear reactor operating at a steady-state power output of 100 kilowatts.

Average thermal neutron flux	3×10^{12} n/cm^2-s
Average fast neutron flux (> 1 MeV)	5×10^{11} n/cm^2-s
Gamma radiation exposure rate	1×10^{7} R/hr

A theoretical description of the fission spectrum from thermal-neutron-induced fission in U-235 is given by Equation (1):

$$\phi(E) = 0.43\exp(-1.036\, E)\, \sinh\sqrt{2.29\, E} \tag{1}$$

where $\phi(E)$ is the energy-dependent neutron flux and E *is* the neutron energy in MeV [Ref. 4]. Because of the presence of nuclei which scatter neutrons the resulting neutron distribution is perturbed from a pure fission spectrum to a quasi-fast distribution for the bare-core Godiva system to a thermalized neutron distribution typical of the TRIGA water-moderated reactors all of which are depicted in Fig. 2.

For standardization, the damage equivalent silicon (DES) is specified which expresses the spectral fluence in terms of a 1-MeV-equivalent monoenergetic neutron fluence. The 1-MeV monoenergetic fluence is determined by various standard methodologies [Refs. 5-10]. The equivalent monoenergetic neutron fluence, ϕ_{eq} is described by Equation (2):

$$\phi_{eq}(E_0) = \frac{\int_{E_{min}}^{E_{max}} \phi(E)\, K_D(E)\, dE}{K_D(E_0)} \tag{2}$$

In this equation, $\phi(E)$ is the incident neutron energy-fluence spectral distribution, $K_D(E)$ is the energy-dependent neutron displacement kerma factor and $K_D(E_0)$ is the displacement kerma factor at the specified equivalent energy (i.e., 1 MeV) [10]. Activation foils are used to measure fluences above specific neutron threshold energies. Several of these foils are indicated at the corresponding threshold energies in Fig. 2. Sulfur foil dosimeters are used to measure the delivered neutron fluence during irradiations. Sulfur undergoes neutron activation via an (n,p) transmutation reaction to produce radioactive phosphorus-32, a pure beta emitter with a half-life of 14.7 days. The threshold energy for this reaction is 2.9 MeV. The activity of ^{32}P is used to calculate the delivered neutron fluence. The 1-MeV-equivalent fluence is obtained as the product of the fluence determined from the activation of the sulfur foils and the DES factor. The UUNEL is currently performing experiments in conjunction with the National Institute of Standards and Technology (NIST) to standardize their fission spectrum and determine the absolute DES factor for the UUNEL TRIGA reactor.

3. Irradiation Procedure

Electronic components are randomly sampled by lot number and date of manufacture for testing. A control sample is removed for reference and a pre-irradiation characterization is performed on all components to provide baseline data. The results of the pre-test characterization are compared with Department of Defense specifications. Components are irradiated under strictly controlled conditions, fixed orientation and geometry, and predetermined reactor power level and delivered energy. To avoid damaging susceptible components by electrostatic discharge (ESD), established ESD-prevention procedures are followed whenever components are handled directly. Several high purity sulfur foils are mounted on a conducting-foam target base with lithium fluoride thermoluminescent dosimeters to measure gamma radiation exposure. After the required dosimetry has been mounted, the electronic components are secured to the target within isodose regions of known fluence relative to the sulfur foil positions. The target is then hermetically-sealed within the sample holder which is lowered into the test stand adjacent to the reactor core. The reactor is then brought to low power and operated at steady state until the desired energy has been delivered. A computer monitors a TRIGA power channel, integrates the reactor power level and displays the energy delivered in units of kilowatt-hours. As soon as the desired energy is reached, the control rods are dropped as the sample holder is concurrently removed from the fast irradiator. The sample holder is kept submerged in the reactor tank for a minimum of 20 minutes to allow for decay of short-lived activation products to reduce exposure of laboratory personnel to radiation.

4. Post-Irradiation Electrical Characterization and Damage Assessment

The irradiated components are then removed from the sample holder and subsequently subjected to a post-irradiation electrical characterization which is compared to the pre-irradiation test to assess radiation-induced damage as a function of the delivered 1-MeV-equivalent neutron fluence. Radiation damage and response in semiconductor devices is manifested by parameter degradation [Ref. 11]. Evaluated parameters may change depending on the device type, length of time over which it must function or the environment in which it will be expected to perform.

Determination of component acceptability is based on electrical end-point limits. These limits represent the maximum tolerance in electrical performance for the particular application. The setting of these limits demands a thorough knowledge of the component to be tested, its interface within an electrical circuit, its operational environment and the "downstream" effects of its degradation or failure. Although neutron damage is generally considered as a permanent effect, a significant portion of the damage anneals *in situ* and within a short time after the test. Such effects must be considered as part of the damage assessment for the components.

4.1. FUTURE EXPERIMENTS

Future experiments are planned to investigate real-time irradiation effects on the performance characteristics of biased components. These experiments are designed to observe performance degradation as functions of dose rate and fluence.

5. Conclusion

A moderate flux nuclear research reactor is an effective system for providing the radiation types and intensities required for assessing damage in materials that might be exposed to radiation fields such as space, military, medical and industrial applications. The radiation field produced in a reactor can be modified and enhanced to deliver the radiation component of interest. Thus an intense gamma or neutron flux with a particular energy distribution can generally be developed. The application of this capability has been demonstrated here for damage assessment from fast neutrons to electronic components used in military and other systems.

References

[1] Simpson, J.A., 'Introduction to the galactic cosmic radiation,' in M.M. Shapiro (ed.), Composition and Origin of Cosmic Rays, D. Reidel Publishing Co., Dordrecht, pp. 1-24 (1983).

[2] Lund, Niels, 'Cosmic ray abundances, elemental and isotopic,' in M.M. Shapiro (ed.), Cosmic Radiation in Contemporary Astrophysics, D. Reidel Publishing Co., Dordrecht, pp. 1-26 (1984).

[3] J.S. Bennion and G.M. Sandquist, 'Report to the U.S. Air Force Concerning the Neutron Irradiation of Electronics by the University of Utah Nuclear Engineering Laboratory,' UTEC 90-19 (1989).

[4] R.E. Dahl and H.H. Yoshikawa, 'Fast Neutrons,' in J. Moteff (ed.), Neutron Fluence Measurements, International Atomic Energy Agency, Vienna, Chapter V (1970).

[5] MIL-STD-883C, 'Military Standard Test Methods and Procedures for Microelectronics,' U.S. Department of Defense (1983).

[6] ASTM E 265-88, 'Standard Test Method for Measuring Reaction Rates and Fast Neutron Fluences by Radioactivation of Sulfur-32,' American Society for Testing and Materials (1988).

[7] ASTM E 668-78, 'Standard Practice for the Application of Thermoluminescence Dosimetry (TLD) Systems for Determining Absorbed Dose in Radiation-Hardness Testing of Electronic Devices,' American Society for Testing and Materials (Reapproved 1984) .

[8] ASTM E 720-86, 'Standard Guide for Selection of a Set of Neutron-Activation Foils for Determining Neutron Spectra Used in Radiation-Hardness Testing of Electronics,' American Society for Testing and Materials (1986).

[9] ASTM E 721-85, 'Standard Method for Determining Neutron Energy Spectra With Neutron-Activation Foils for Radiation-Hardness Testing of Electronics,' American Society for Testing and Materials (1985).

[10] ASTM E 722-85, 'Standard Practice for Characterizing Neutron Energy Fluence Spectra in Terms of an Equivalent Monoenergetic Neutron Fluence for Radiation-Hardness Testing of Electronics,' American Society for Testing and Materials (1985).
[11] L.W. Ricketts, _Fundamentals of Nuclear Hardening of Electronic Equipment_, R. Krieger Publishing Co., Malabar, Florida (1986).

METEOROID AND DEBRIS ENVIRONMENT AND EFFECTS

D.R. ATKINSON, C.R. COOMBS, A.J. WATTS,
L.B. CROWELL, AND M. K ALLBROOKS
POD Associates, Inc
2309 Renard Place S.E., Suite 201
Albuquerque, NM 87106

ABSTRACT. The Long Duration Exposure Facility (LDEF) provides the first well quantified data for evaluating the meteoroid and debris environment models, as well as for evaluation of the synergistic effects of the low-Earth orbit space environments. Preliminary analysis of the LDEF data indicates that the cumulative number of impacts received in the spacecraft velocity (ram) direction were well predicted by the models. However, the time-dependent and directional distributions were not correctly predicted. In addition, data from LDEF related to impacts in thermal control paints indicate previously undocumented impact zone morphologies. Analysis of the atomic oxygen erosion patterns on these new morphologies may provide a method for determining the age and time-dependence of the impact features in these paints.

1. Introduction

Prior to the retrieval of the Long Duration Exposure Facility (LDEF) the data used in developing models of the small diameter (less than 0.1 millimeters) meteoroid and debris environment came primarily from the Solar Maximum Mission (SMM) materials returned in 1984. These materials consist of approximately one square meter of Multiple Layer Insulation (MLI) blankets and aluminum louvers exposed for approximately four years. The SMM flux and impactor diameter data was subject to errors due to the possibility of secondary ejecta impacts from the SMM solar panels and uncertainties in the correlation of penetration diameter to impactor diameters for different materials and velocities. In addition, the SMM provided no data for time or directional dependence of the flux.

By contrast LDEF was a gravity-gradient stabilized satellite launched into a 28.5° inclination circular orbit at 476 km. Exposed for 5 and 3/4 years with one end always pointed towards Earth and one side facing in the spacecraft velocity (ram) direction, LDEF decayed to ~ 330 km altitude before retrieval. LDEF was a twelve-sided open grid structure (~ 130 m^2) with structural members facing in 26 stable directions. The structural members were 6061-T6 aluminum (~ 15 m^2 total area, ~ 0.6 m^2 in each direction). The uniform material simplifies the data analysis by removing

R. N. DeWitt et al. (eds.), The Behavior of Systems in the Space Environment, 905–911.

the effects of differing target materials on crater diameters. Also, there are no penetrations or back-surface spallations due to the thickness of the structural members. In addition, the Interplanetary Dust Experiment (IDE) provided active time and direction dependent data for impacts throughout LDEFs first year of exposure. LDEF thus provides: 1) the first data for determining the time-dependent and directional distribution of meteoroids and debris, 2) uniform structural materials to reduce cratering versus impactor diameter conversion errors, and 3) more than 10,000 samples of various materials for determining hypervelocity impact effects in these materials.

During the LDEF deintegration at Kennedy Space Center, it was noticed that the thermal control paints and blankets possessed rings surrounding impact features. These rings covered a very large area and could not be explained. Examination of these rings to determine damage areas mechanisms, as well as correlations with time and impactor energies, has become the focus of numerous investigations including this one.

2. Method

Using data collected from the LDEF structure, the number of impacts per square meter versus crater diameter was calculated for each LDEF direction. Drs. Mulholland and Simon provided data from IDE for the one year flux of small particles (0.2 to 100 microns diameter) in each LDEF direction. Data was also collected from aluminum plates which were painted with Chemglaze A278 white thermal control paint. These plates were exposed to atomic oxygen (ram direction) and to ultraviolet light (all directions).

Impact features in the painted surfaces were examined to determine the damage areas and the damage mechanisms. In addition, analysis was performed to determine whether the damage areas provide information which allows the determination of the approximate age of craters. All data was collected using a Wild Leitz M8 stereo microscope.

3. Results

Figure 1 shows the number of impacts per square meter versus crater diameter for the structural members exposed in the ram direction. The figure also shows the data for the IDE experiment adjusted from one year exposure to 5 3/4 years exposure using the assumption that the flux remained constant throughout the exposure. For comparison, the figure shows the number of impacts per square meter versus crater diameter as predicted by: 1) the Cour Palais, et al., meteoroid environment model (NASA SP-8013, 1969) using the Erickson Kessler velocity distribution model; 2) the Kessler, et al., debris environment model (NASA TM-100471, 1987) including the improvements presented by Kessler in 1990; and 3) the composite of these meteoroid and debris environment models. As shown in the figure, the LDEF data agree with the models to within a factor of 2 to 3 for crater diameters greater than 30 microns. For smaller sizes, the LDEF data shows a much lower flux than predicted. Sources of uncertainty in the comparison include: cratering mechanics equations used to convert model predictions

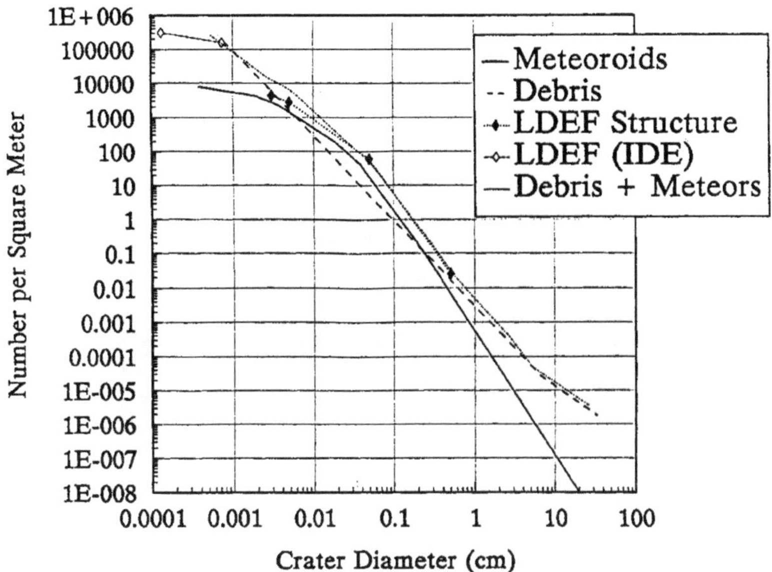

Fig. 1: Comparison of LDEF data to model predictions in the RAM direction. All curves are based on a 5.75 year exposure.

of impacts diameters to crater diameters; low statistics for large craters; lack of a beta meteoroid environment definition; and velocity, direction and density distribution and debris growth rate models. For this comparison, the model predictions assumed a constant altitude of 450 km for 5 3/4 years. In reality LDEF descended to ~ 330 km by the end of the mission.

For impacts into the Chemglaze A278 white thermal control painted surfaces, the impactor creates a crater which is often surrounded by a "spallation" region (paint removed around the crater) and a "dome" region where the paint delaminated and lifted from the substrate but was not completely removed. This is shown in Fig. 2. These front surface foliation and dome regions extend up to five times the greater diameter in the substrate.

Outside the spall region are "rings" (surface deposits on the paint) which can extend to greater than 20 times the crater diameter as shown in Fig. 3. The analysis of these rings has shown that some of the rings are caused by the top layer of paint rolling up and folding back across the top surface, and these rolls can be detected in the more degraded rings. The rings are apparently created through surface delaminations of the paint, followed by atomic oxygen erosion of these materials. Thus, it should be possible to use the various erosion states of the rings, as shown in Figs. 4 through 6, to determine the age of the impacts. The very young (Fig. 4) show very little erosion of the rolled-up layer of paint, whereas the very old (Fig. 6) show extensive erosion, removing most evidence of these layers.

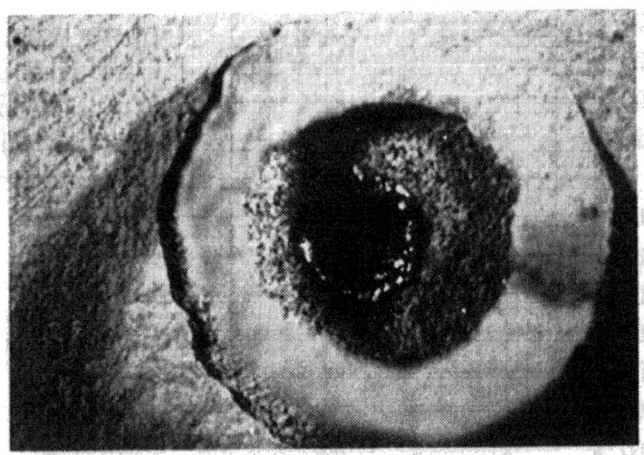

Fig. 2: Crater in leading edge grapple plate of LDEF. Note raised domical section. Field of view (FOV) is 1 mm.

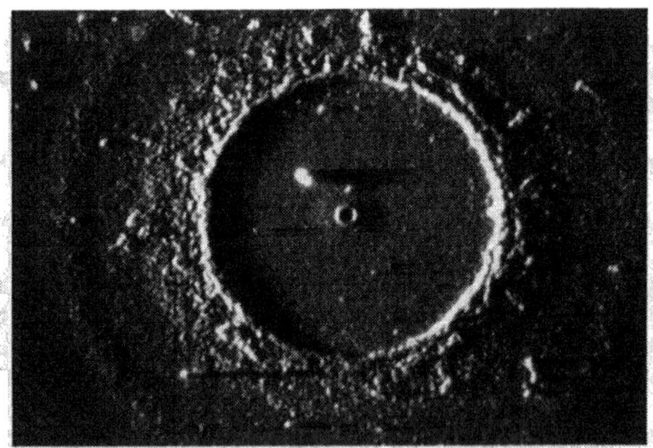

Fig. 3: Photograph illustrating ring structure commonly found around craters on leading edge surfaces. FOV=1.8 mm.

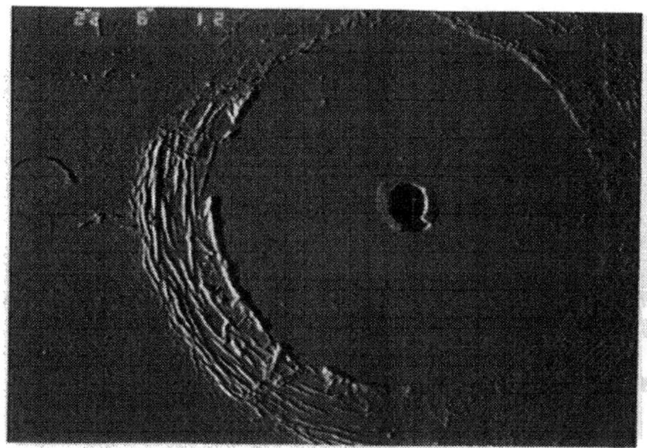

Fig. 4: Photograph of "early" erosion stage of ring structure. This impact is considered to be very young as its ring structure is still intact. FOV=5mm.

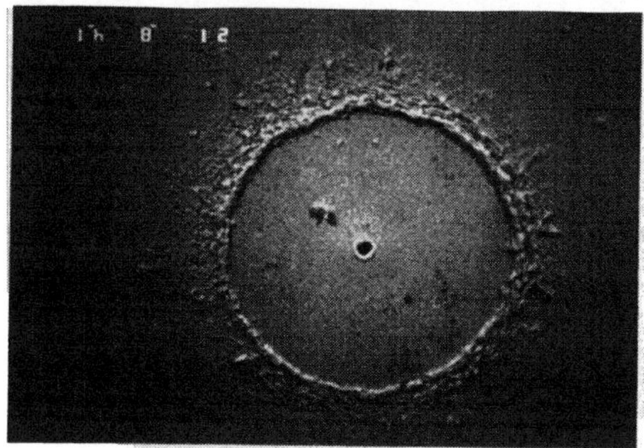

Fig. 5a: Photograph of "middle" stages of erosion. Rings are still somewhat intact. FOV=1.0 mm and 5mm respectively.

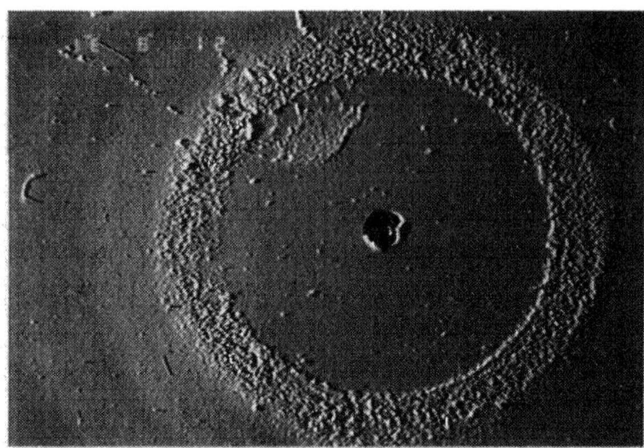

Fig. 5b: Photograph of "middle" stages of erosion. Rings are still somewhat intact. FOV=1.0 mm and 5mm respectively.

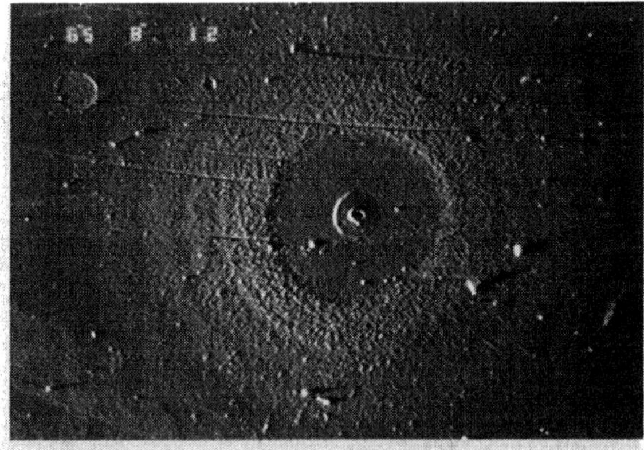

Fig. 6: Very "old" stage of erosion. Rings are degraded to invisible. FOV=4.5 mm.

4. Discussion

The meteoroid and debris model predictions match relatively well to the LDEF data for large craters. Improvements in data statistics, along with improved cratering mechanics equations, will probably improve this match.

For the small craters, the LDEF data do not match the model predictions very well, falling far below the predictions. With the current understanding of the distribution of meteoroids and debris, and the assumptions made in this comparison (constant altitude for 5 3/4 years), the model should over-predict the LDEF flux for all sizes. This is not the case, showing the need for improvement in the models, especially for the beta meteoroids and small orbital particles.

The newly reported morphology of "rings" surrounding impact in thermal control paint have been preliminarily attributed to top surface delaminations in the paint. Atomic oxygen erosion of the ring structures may be usable for identifying the actual age of the impact feature. In the future a test series will be completed to determine the validity of the proposed age determination methodology. This test series will include impacts into painted aluminum plates with varying levels of atomic oxygen exposure (from unexposed to the full LDEF exposure), and will expose impact features to these levels of atomic oxygen in an attempt to determine aging characteristics. In addition, on-going work is also examining the rings surrounding the penetration holes in thermal blanket materials.

The ring structures have a very large diameter (greater than 20 times the crater diameter), thus covering over 400 times the area of the crater. These areas of damage, outside of the impact crater areas, may be more detrimental to the operation of thermal control materials than the actual crater area itself. In order to determine the operational effects caused by meteoroid and debris impacts, future efforts will look at the effect of the rings on the operational performance of the thermal control materials.

ATOMIC OXYGEN PROTECTIVE COATINGS

B.A. BANKS, S.K. RUTLEDGE, K.K. de GROH, B.M. AUER
NASA Lewis Research Center
21000 Brookpark Road
Cleveland, Ohio 44135

C.M. HILL
Ohio Aerospace Institute
Brookpark, Ohio 44135

ABSTRACT. Atomic oxygen resident in low Earth orbit (LEO) impinges upon orbiting spacecraft such as Space Station Freedom (SSF) with sufficient flux to cause rapid oxidation and premature failure of organic spacecraft materials. Protective coatings consisting of metal oxides, fluoropolymer-filled metal oxides, and silicones can be used to minimize the reaction of atomic oxygen with organic materials. Such protective coatings are necessary for the long-term durability of polymeric films such as polyimide Kapton solar array blankets and other oxidizable materials. Defects in atomic oxygen protective coatings can enable atomic oxygen to react and oxidize the underlying polymeric material. The number and area of atomic oxygen defects is dependent upon surface irregularities, contamination during protective coating deposition, flexure or abrasion during materials processing, and micrometeoroid or debris impact in space. A combination of ground-based LEO simulation testing, in-space experiments, and Monte Carlo modeling have been utilized to forecast degradation modes of atomic oxygen protected materials exposed to sweeping atomic oxygen arrival conditions such as will occur on SSF.

1. Introduction

As a result of numerous in-space exposure tests of a wide variety of materials, it is now known that all organic materials are subject to varying degrees of oxidation by atomic oxygen in LEO. A list of atomic oxygen erosion yields for a wide variety of materials can be found in Reference 1. Efforts to identify alternative materials which possess atomic oxygen durability and meet the mechanical, optical, thermal, electrical, and other functional requirements demanded by their mission application has met with only minimal success. As a result, atomic oxygen durability has been accomplished through the application of protective coatings typically consisting of metal oxides, metal oxides with the addition of fluoropolymers, metals, and silicones [Ref. 2]. Such protective coatings

R. N. DeWitt et al. (eds.), The Behavior of Systems in the Space Environment, 913–919.

must themselves be atomic oxygen durable and be applied sufficiently thin to not adversely affect the required functional properties of the underlying oxidizable materials. Coating materials have been identified and demonstrated in space that meet these requirements, however their application as thin films presents a further complication [Refs. 2, 3]. Thin film deposition on most polymeric surfaces results in defect sites which allow atomic oxygen to attack the underlying polymeric material. Such defect sites are a result of deposition on surfaces that contain scratches, rills, voids, and contaminant particles. In addition, handling and further processing of materials can result in defects caused by abrasion. Thus, the effectiveness of atomic oxygen protective coatings and the ultimate durability of underlying vulnerable materials are primarily dependent upon the number and size of defects, as well as the nature of the atomic oxygen interaction processes which occur at the defect sites.

2. Ground Laboratory Evaluation of Atomic Oxygen Protective Coatings

Ground laboratory simulation of LEO atomic oxygen has been performed by a wide variety of neutral, plasma, and ion beam sources [Ref. 4]. Evaluations of protective coating effectiveness have frequently been performed in RF plasma ashers where typically, a 15.56 MHz discharge occurs in air or oxygen at 50-100 mtorr. Such facilities allow atomic oxygen, ionic oxygen, radicals and ions, to bombard materials. Thermal energies of a fraction of an electron volt are associated with arrival from all directions on the samples being evaluated. Atomic oxygen exposure fluences are measured as an effective fluence, where the oxidation of a polyimide Kapton control sample is used to calculate an effective fluence based on comparison with measured in-space atomic oxygen erosion of Kapton. More energetic directed neutral or ion beam systems have been useful for comparison of directionality effects and hold the potential for comparison of relative atomic oxygen erosion yields of various materials, as well as evaluation of synergistic effects such as ultraviolet radiation on materials.

The atomic oxygen protection of Kapton H photovoltaic blanket materials for SSF has been addressed by the use of sputter deposited SiO_x (where x = 1.9-2.0) films which are 1300 Å thick [Refs. 5, 6]. As previously mentioned, atomic oxygen defects can occur during fabrication or as a result of micrometeoroid or debris impact, however the number of defects occurring in space due to micrometeoroid and debris impact will probably be small compared to those resulting from deposition and handling [Ref. 6]. Typically, 10^3 defects/cm^2 greater in size than 5 μm diameter are found in these SiO_x thin film coatings on Kapton H [Ref. 6]. The results of atomic oxygen exposure of such protected Kapton in RF plasma ashers are shown in Fig. 1. Figure 1a is a scanning electron micrograph of SiO_x-protected Kapton after plasma asher exposure to a fluence of 4.45 x 10^{20} atoms/cm^2. Figure 1b shows the same location after adhesive tape removal of the SiO_x protective coating at defect sites. As can be seen comparing the figures, the atomic oxygen undercutting extends well beyond the diameter or width of the initial defects in the SiO_x-protective coating. Single direction atomic oxygen exposure in a directed oxygen beam produces far less undercutting provided the Kapton is not eroded down to

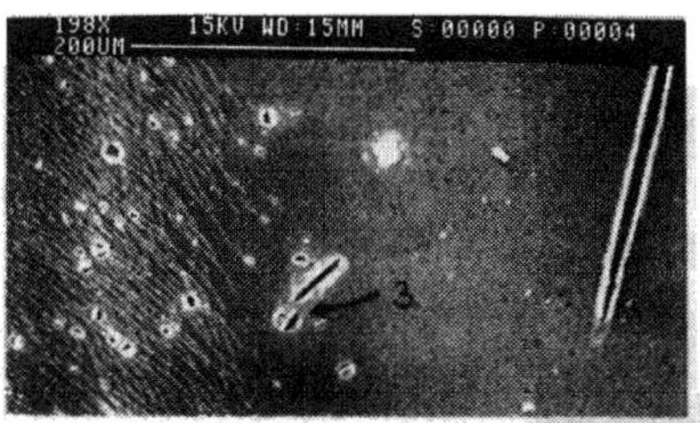

(A) Prior to adhesive tape removal of atomic oxygen undercut SiO_x.

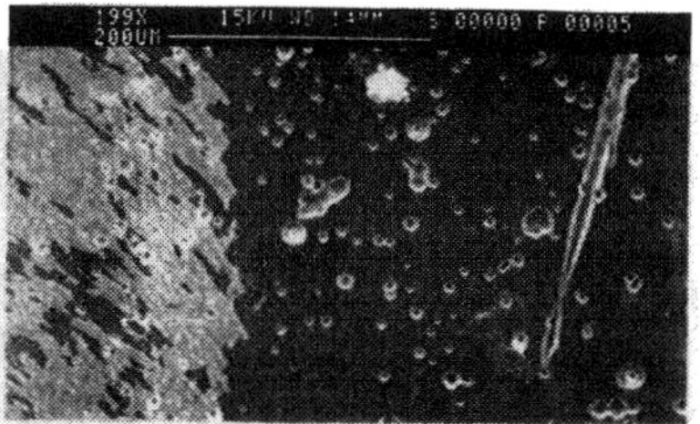

(B) After adhesive tape removal of the SiO_x-protective coating.

Fig. 1: Scanning electron photomicrographs of SiO_x-protected Kapton after plasma asher exposure to a fluence of 4.45 x 10^{20} atoms/cm^2.

a protective coating on the backside of the Kapton. Scattering off the back-side protective coating would cause wider undercutting by the trapped atomic oxygen. Two undercut defect sites indicated with an arrow and #3 on Fig. 1a are shown at higher magnification in Fig. 2. As can be seen by comparing the two undercut sites, a wide defect tends to produce a double indented undercut site, whereas a small defect tends to produce a smaller, singularly contoured cavity.

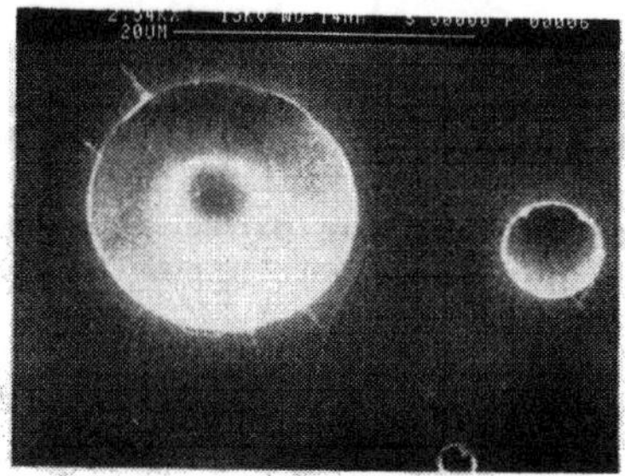

Fig. 2: Scanning electron photomicrograph of a large and small undercut defect site after plasma ashing and subsequent adhesive tape peeling of the protective coating.

Figure 3 shows a scanning electron micrograph of a sheet of Kapton H that was sputter-deposited on both sides with 1300 Å of SiO_x and plasma ashed to an atomic oxygen fluence of 4.4 x 10^{22} atoms/cm^2. Depending on which side of the protected Kapton had a defect, the undercut Kapton was chamfered such that the wider diameter was on the surface where the defect occurred. As can be seen in Fig. 4, high fluence exposure similar to that which would be experienced by photovoltaic blankets on SSF after 15 years in LEO results in extensive undercutting at defect sites, thus greatly reducing the structural integrity of the sheet. Atomic oxygen protective

Fig. 3: Photomicrograph of 0.045 mm (2 mil.) Kapton H which was protected on both sides with 1300Å-thick SiO_x after plasma asher exposure to an atomic oxygen fluence of 4.4×10^{22} atoms/cm^2.

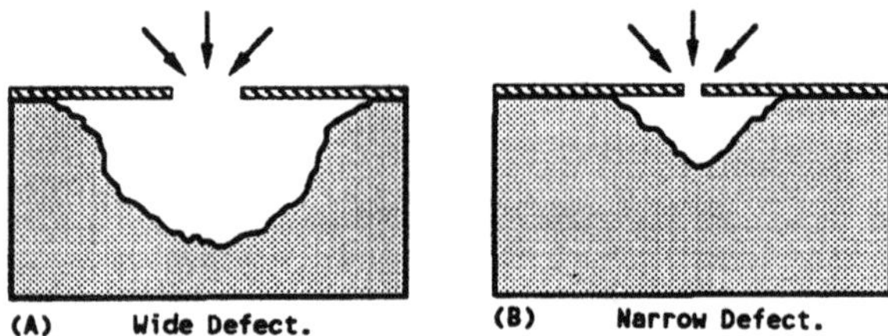

Fig. 4: Monte Carlo predictions for protective Kapton exposed to atomic oxygen in an RF plasma asher at defect sites in the protected coating.

coatings on solar concentrator reflector surfaces display some undercutting processes similar to SiO_x-protective coatings [Ref. 7].

3. Monte Carlo Simulation of Atomic Oxygen Interaction with Protected Materials at Defect Sites in Protective Coatings

A Monte Carlo modeling technique has been developed to simulate atomic oxygen erosion undercutting processes at defect sites in protective coatings on polymeric materials [Refs. 8, 9]. The Monte Carlo models utilize a ray tracing technique and assign probabilities of interaction with the unprotected polymers at defect sites, as well as allowing ejection and additional chances for reaction of atomic oxygen which does not react upon first impact. Figure 4 compares the Monte Carlo predictions for wide and narrow defects in protective coatings on Kapton exposed to an RF plasma asher environment. As can be seen by comparison of Fig. 2 and Fig. 4, the Monte Carlo predictions agree favorably with experimentally observed results for wide and narrow defect sites in protective coatings on polyimide Kapton. Although the Monte Carlo model simulates scratch or crack defects in protective coatings, and the experimental results shown in Fig. 2 are for circular pin window defects, the undercutting profiles show that scratches and pin windows appear to have the same general profile. Monte Carlo predictions for fixed direction space ram and sweeping space ram atomic oxygen exposure conditions are shown in Fig. 5. As can be seen, significant differences in the shape of the undercut pattern are anticipated to depend on the nature of the atomic oxygen attack. The models shown in Fig. 5 do not include transverse atomic oxygen velocities associated with thermal or orbital inclination contributions. Such transverse velocity contributions should widen the atomic oxygen undercutting significantly for the normal incidence space ram. Monte Carlo undercutting predictions for protective coatings over fiberglass epoxy for a beta cloth (glass fiber matte in an FEP Teflon matrix) indicate that fibers will not self-shield their polymeric matrix material from atomic oxygen attack, even if line-of-sight exposure to the underlying matrix material does not exist. This is the result of the scattering of unreacted atomic oxygen. Experimental evidence has shown the deep loss of matrix material and fiberglass epoxy in RF plasma ashers to be in agreement with Monte Carlo model predictions.

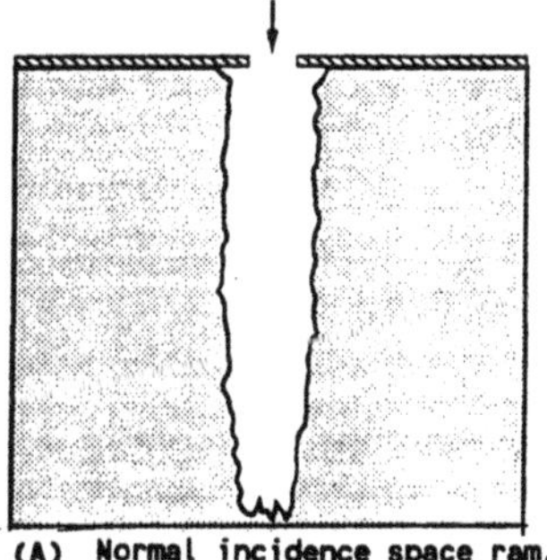
(A) Normal incidence space ram.

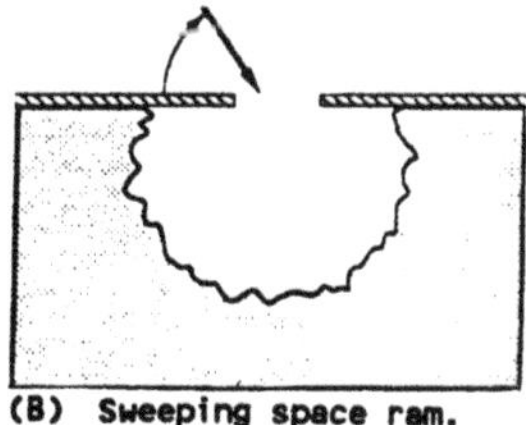
(B) Sweeping space ram.

Fig. 5: Monte Carlo predictions for space ram atomic oxygen attack at protective coating defect sites on polyimide Kapton.

4. Space Test Results of Atomic Oxygen Protective Coatings

The results of in-space exposure of SiO_2, SiO_2-4% PTFE, and Al_2O_3 protective coatings on Kapton H are given in References 1 and 3. Such tests indicate that the application of metal oxide protective films result in significant reductions in atomic oxygen erosion of materials. The addition of small percentages of fluoropolymer in SiO_2-protective coatings greatly increases their strain-to-failure, thus allowing reduced probability of tensile failure of the protective coating as a result of flexure or tension in the protected polymeric material. Micrometeoroid or debris impact of SiO_2-protective coatings causes local cracking but does not cause large area catastrophic failure of the coating [Ref. 10]. Figure 6 shows the effect of atomic oxygen undercutting at crack defect sites in aluminized

Kapton multilayer insulation, which was exposed to an atomic oxygen fluence of 5.77×10^{21} atoms/cm^2 on Long Duration Exposure Facility (LDEF). Figure 6a shows the aluminized Kapton with the aluminum film present after retrieval from LDEF. Figure 6b shows the exact same location with the 1000Å-thick aluminization chemically removed. As can be seen, significantly wider atomic oxygen undercutting occurs than the width of the initial defect. The wide undercutting may be a result of transverse atomic oxygen velocity components associated with the 1227 K thermal velocity of atomic oxygen that LDEF was exposed to, and transverse fluxes associated with scattered atomic oxygen as a result of impingement through openings of an incompletely removed surface sheet of aluminized Kapton [Ref. 11].

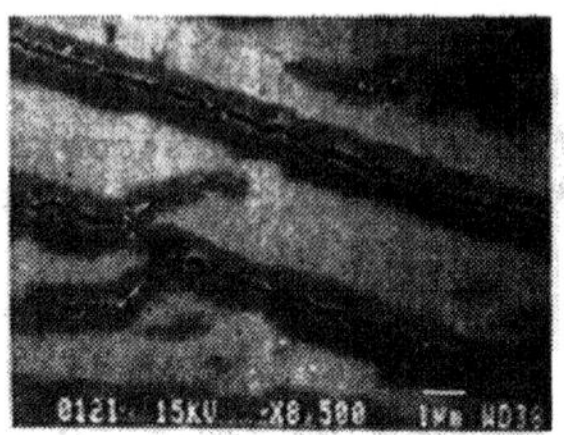

(A) With aluminum.

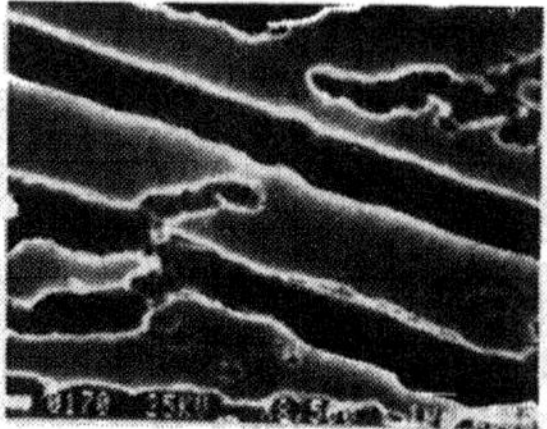

Aluminum chemically removed.

Fig. 6: Atomic oxygen undercutting at crack defect sites in aluminized Kapton multi-Layer insulation exposed on LDEF to an atomic oxygen fluence of 5.77×10^{21} atoms/cm^2.

5. Summary

Atomic oxygen protective coatings of SiO_x ($x = 1.9$-2.0), SiO_2 with 4% PTFE and Al_2O_3 are effective in increasing the durability of underlying polymer materials from atomic oxygen attack. Monte Carlo models have been developed which simulate undercutting phenomena observed in ground laboratory as well as in-space exposure of materials at defect sites in protective coatings. Defect sites in thin film protective coatings are the prime cause of atomic oxygen degradation of protected materials.

6. References

[1] Banks, B.A., et.al., 'Simulation of the Low-Earth-Orbital Atomic Oxygen Interaction with Materials by Means of an Oxygen Ion Beam,' NASA TM 101971 (1989).

[2] Banks, B.A., et.al., 'Ion Beam Sputter-Deposited Thin Film Coatings for Protection of Spacecraft Polymers in Low Earth Orbit,' NASA TM 87051 (1985).

[3] Banks, B.A., Mirtich, M.J., Rutledge, S.K., and Nahra, H.K., 'Protection of Solar Array Blankets from Attack by Low-Earth-Orbital Atomic Oxygen,' Proceedings of the 18th IEEE Photovoltaic Specialists Conference (1985).

[4] Banks, B.A., Rutledge, S.K., and Beatty, J.A., 'The NASA Atomic Oxygen Effects Test Program,' Presented at the 15th Space Simulation Conference (1988).

[5] Rutledge, S.K. and Mihelcic, J.A., 'Undercutting of Defects in Thin Film Protective Coatings on Polymer Surfaces Exposed to Atomic Oxygen,' NASA TM 101986 (1989).

[6] Banks, B.A., Rutledge, S.K., de Groh, K.K., 'Low Earth Orbital Atomic Oxygen, Micrometeoroid, and Debris Interactions with Photovoltaic Arrays,' Presented at the 11th Space Photovoltaic Research and Technology Conference (SPRAT XI) (1991).

[7] Banks, B.A., de Groh, K.K., Rutledge, S.K., Dever, T.M., Stueber, T.J., and Hotes, D., 'Performance and Durability of Space Solar Dynamic Power System Optical Surfaces,' Proceedings of the American Society of Mechanical Engineers (1990).

[8] Banks, B.A., Auer, B.M., Rutledge, S.K., and Hill, C.M., 'Atomic Oxygen Interactions with Solar Array Blankets at Protective Coating Defect Sites,' Proceedings of the Fourth Annual Workshop on Space Operations, Automations, and Robotics (SOAR '90) (1990).

[9] Banks, B.A., Rutledge, S.K., Auer, B.M., and DiFilippo, F., 'Atomic Oxygen Undercutting of Defects on SiO_2-Protected Polyimide Solar Array Blankets,' Published in Materials Degradation in Low Earth Orbit (LEO)," edited by V. Srinivasan, Banks, B.A., Minerals, Metals, and Materials Society (1990).

[10] Rutledge, S.K. and Olle, R.M., 'Durability Evaluation of Photovoltaic Blanket Materials Exposed on LDEF Tray S1003,' Presented at the 1st LDEF Data Conference (1991).

[11] De Groh, K.K. and Banks, B.A., 'Atomic Oxygen Undercutting of LDEF Aluminized Kapton Multilayer Insulation, Presented at the 1st LDEF Data Conference (1991).

EFFECTS OF ENVIRONMENTAL CONDITIONS ON THE BEHAVIOR OF GRAPHITE EPOXY COMPOSITE COUPONS

W.E. WOLFE and S.C. KIM
The Ohio State University
Department of Civil Engineering
2070 Neil Ave.
Columbus, Ohio USA 43210

ABSTRACT. A laboratory test apparatus, designed to subject a composite specimen to a low temperature and pressure environment and then load the specimen to failure in either tension or compression, is described. To demonstrate this apparatus, a static test program was conducted on 16-ply T300 graphite/epoxy coupons. Thirty-five coupons, consisting of five different ply configurations, were loaded in tension under either ambient or low temperature and pressure conditions. It was observed that environmental conditions do affect the elastic parameters as well as tensile strength and strain to failure. The effect was seen to be a function of ply orientation, with the greatest effects being measured on $(\pm 45)_{4s}$ coupons where the response to load is dominated by the epoxy matrix, and least for unidirectional 0_{16} specimens in which the fibers dominate the structural response. Elastic parameters obtained from the tests on the 0_{16}, $(0/90)_{4s}$ and $(\pm 45)_{4s}$ specimens were input into a finite element program used to calculate stress-strain response of the two coupons with different ply configurations. Laboratory tests were conducted on these specimens and the experimentally measured results for stiffness were compared with predictions from the finite element analyses. Good agreement was obtained for both environmental conditions studied.

1. Introduction

In the past several years there have been many cases of composite materials being incorporated into structural components. It is likely that in the future a substantial portion of the metallic materials presently used for structural purposes will be replaced by non-metallic composites. When compared with metals, composites are attractive to the structural designer because they have higher ratios of stiffness and strength to mass and possess good fatigue properties. Since composites have high specific strength and stiffness, incorporating them into the design of orbiting satellites and components allows engineers to reduce the structural mass of the system while maintaining mission requirements.

The space environment subjects structures to conditions that can be significantly different from those experienced on earth. The effect of low temperatures and pressures may result in performance much different from

R. N. DeWitt et al. (eds.), The Behavior of Systems in the Space Environment, 921–927.

what would be considered typical for earth-based systems. In this study, uniaxial tensile tests were conducted on graphite fiber/epoxy matrix composites at room temperature and pressure and at low temperatures and pressures. The response of each coupon was measured as it was loaded, and mechanical properties were determined for the two different environmental conditions. The primary objective of this study was to investigate the effect of changes in environmental conditions on the behavior of laminated coupons made in a variety of ply configurations. Stress-strain curves obtained for samples tested at room temperature were compared with those measured at low temperature and pressure, and the effects of these environmental conditions on strength and elastic modulus were investigated.

2. Test Equipment

A commercially available T300 graphite prepreg tape was used in the experimental program described in this paper. Composite plates were made by stacking 30.5cm by 30.5cm sheets of the tape in one of five different ply configurations. The five lay-ups, all symmetric about the mid-plane, were:

1) Unidirectional laminates with all fibers oriented in the direction of the load:	$(0)_{ms}$	m=8
2) Cross-ply laminates:	$(0/90)_{ms}$	m=4
3) Angle-ply laminate I:	$(+45/-45)_{ms}$	m=4
4) Angle-ply laminate II:	$(0/+45)_{ms}$	m=4
5) Quasi-isotropic laminates:	$(0/90/+45/-45)_{ms}$	m=2

The subscript m defines the number of sublaminate groups within the laminate, and s refers to symmetry about the specimen mid-plane. In all tests reported in this paper, m was chosen to make the total number of plies equal to 16. Each panel was cut into 2.5cm wide 30.5cm long coupons. All coupons were inspected ultrasonically and found to be free of fabrication flaws. The test equipment consisted of an environmental chamber, a high vacuum cryopump system, load frame and temperature and strain measurement systems. Figure 1 is an overall view of the equipment used in this study.

The function of the environmental chamber shown in Figure 1 was to maintain the low temperature and pressure generated by cryopump using liquid nitrogen as a coolant. A shroud was installed in the chamber around the specimen to reduce the time required to reach the desired test conditions. Limits of the chamber as configured in this program were 5×10^{-8} torr and 77°K. Actual test conditions were 9×10^{-8} torr and 150°K [Ref. 1].

The principal components of the loading system were a 44.6 KN load frame and load cell and a pair of wedge action grips. To avoid temperature effects on the load measurements, the load cell was positioned on the pull rod of load frame outside the environmental chamber. All tests were conducted at a constant displacement rate.

Fig. 1: OSU environmental chamber.

Four thermocouples were installed to measure temperatures in the chamber and on the specimen. Bonded foil resistance strain gage rosettes with a useable temperature range of 4° to 560°K and maximum strain ±2% were used during the tests to record specimen strain. Figure 2 shows a coupon, with strain gages and thermocouples attached, mounted in the grips just prior to the start of a test.

3. Experimental Results

Elastic properties for an orthotropic material system can be obtained from the results of tests conducted on 0, 0/90, and ±45 laminate coupons. The properties of interest are E_1, E_2, μ_{12}, and G_{12}. E_1 and E_2 are the moduli of elasticity in the fiber and transverse directions, respectively, μ_{12}, is the ratio of transverse strain to strain in the fiber direction, and G_{12} is the shear modulus. Table 1 lists the elastic parameters obtained from the test data for the 0, 0/90, and ±45 coupons for both the ambient and the reduced environment conditions. Also listed in the table are the failure stress and strain for each ply configuration tested. Plots of the measured stresses and strains for two ply configurations, 0/90 and ±45 are presented in Figure 3 for both environmental conditions. Note that the modulus is higher but that tensile strength and strain to failure are typically lower

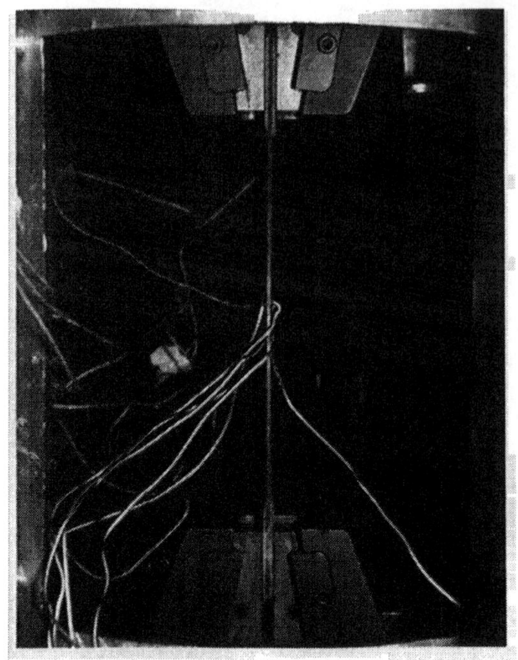

Fig. 2: Test specimen in chamber.

Table 1. Composite laminate properties.

Environmental Condition	Elastic Property			
	E_1 kPa	E_2 kPa	μ_{12}	G_{12} kPa
Ambient (295°K)	$138.75 x 10^3$	$3.31 x 10^3$	0.341	$4.90 x 10^3$
Low Temperature Low Pressure (150°K, $9 x 10^{-8}$ torr)	$142.3 x 10^3$	$8.83 x 10^3$	0.378	$6.55 x 10^3$
	Ambient (295°K)		Low Temperature (150°K) Low Pressure ($9 x 10^{-8}$ torr)	
Coupon	Tensile Strength (kPa)	Failure Strain %	Tensile Strength (kPa)	Failure Strain %
$(0)_{16}$	1358	0.91	1048	0.70
$(0/19)_{4s}$	669	0.90	579	0.75
$(+45/-45)_{4s}$	103*	1.0*	172*	1.2*
$(0/+45)_{4s}$	882	1.08	627	0.76
$(0/90/+45/-45)_{2s}$	524	1.03	469	0.81

* Strain gages failed

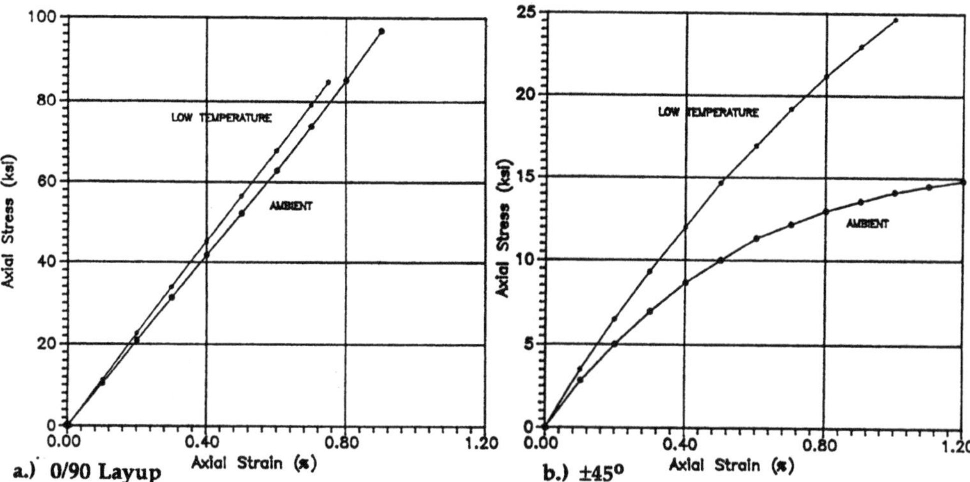

Fig. 3: Stress strain relationship for composite coupons for both environmental conditions.

for coupons tested in the low temperature and pressure conditions. This effect on properties is most pronounced in the ±45 laminates where behavior is dominated by the matrix.

4. Discussion

With the material parameters obtained from the testing program described in the previous section, a finite element program developed to analyze composite laminate structures [Ref. 2] was used to duplicate analytically the loading conditions the coupons had experienced in the laboratory. The elastic material constants obtained from the tests on the 0, 0/90, and ±45 coupons were used as inputs, and the response of the 0/45 and quasi-isotropic laminates to load was calculated. Figure 4 compares the predicted response of the two configurations with the experimental data for both environmental conditions. It can be seen that the finite element analysis predicts slightly smaller loads than were observed in the experiment. This is most likely the result of some slight stiffening of the coupon caused by a tendency toward reorientation of the fibers in the direction of loading as the specimen is tested. In general, however, good agreement between prediction and experiment was achieved in both cases.

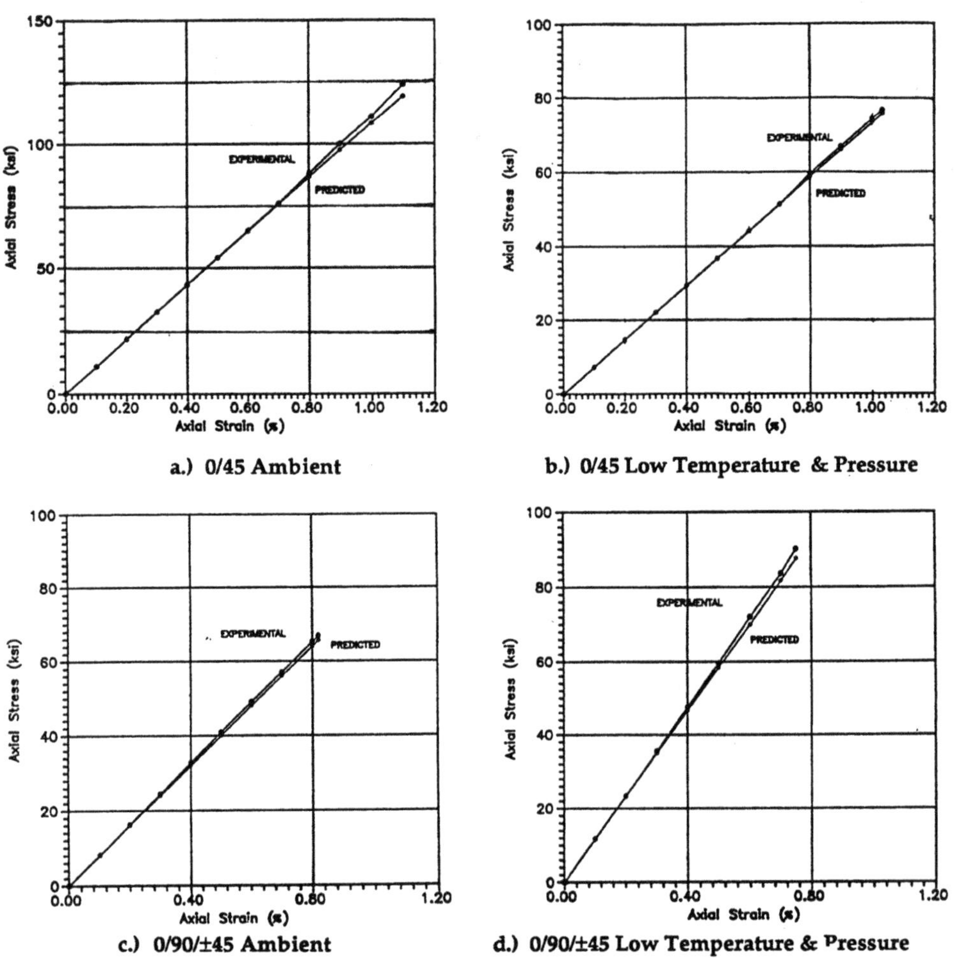

Fig. 4: Comparison of predicted response of the composite specimens with experiment data.

5. Acknowledgements

The prepreg tape used to make the coupons tested in the study was supplied by The Flight Dynamics Directorate, Wright Laboratory, Wright-Patterson AFB, Ohio. The authors are grateful for this support.

6. References

[1] Kim, S.C., "Effect of Environmental Conditions on the Behavior of Graphite/Epoxy Composites," M.S. Thesis, Ohio State University (1990).

[2] Sandhu, R.S., Wolfe, W.E., and Dandan, R.A., "A Computer Program for Finite Element Analysis of Laminated Composite Axisymmetric Solids," Report to the US Air Force, Aeronautical Systems Div., Wright-Patterson AFB, Structural Engineering Report No. 30 (1990).

ISEE-1 MEASUREMENTS OF THE NEAR EARTH ENERGETIC ELECTRON ENVIRONMENT

C. TRANQUILLE[1,2] and E.J. DALY[2]
[1]*Logica Space and Communications Ltd.,*
68 Newman Street, London W1A 4SE,
England
[2]*ESA/ESTEC,*
Postbus 299, 2200 AG Noordwijk,
The Netherlands

ABSTRACT. Omnidirectional electron fluxes measured by the Medium Energy Particle Instrument (D.J. Williams, APL/JHU) flown on the ISEE-1 satellite, are used to quantify the near earth electron environment. Data from a magnetometer (C.T. Russell, IGPP/UCLA) flown on the same spacecraft are also used to assess the validity of the magnetic field models employed to map the spacecraft orbit into geomagnetic *B* and *L* coordinates.

Average fluxes in *B* and *L* cells covered by the ISEE-1 orbit are compared with the AE8 data set, the standard NASA model for trapped electrons. At low altitudes, the two data sets agree qualitatively; at higher altitudes, the MEPI data imply that the AE8 fluxes fall off too quickly.

1. Analysis

The ISEE-1 orbit and an apogee of about 24 earth radii, a perigee height of 270 km and an inclination of 28°. Launch was in October 1977, and the MEPI instrument returned data for two years before a power supply unit failed. The instrument combined a motorized vertical scanning motion with the spin of the spacecraft to supply omnidirectional electron fluxes in 8 energy channels, ranging from 22 keV to 1.2 MeV. The data used in this analysis are 5-minute averages.

Using a combination of internal (IGRF 1980) and external (Tsyganenko, 1989) magnetic field models, the coverage of the ISEE-1 satellite can be mapped into geomagnetic *B* and *L* coordinates. The geomagnetic space can be divided into equally sized cells for binning flux measurements. The *B* and *L* coverage of ISEE-1 is plotted in Fig. 1. The geomagnetic equator is provided for reference. Baseline orbits of several forthcoming ESA astronomy missions (for which the trapped radiation environment can cause operational problems) are shown. The initial CRRES orbit is also plotted.

Assuming that flux measurements made in individual *B* and *L* cells follow a log-normal distribution, it is possible to obtain the mean logarithmic flux together with the standard deviation by plotting the cumulative frequency against the logarithm of the flux on probability graph

R. N. DeWitt et al. (eds.), The Behavior of Systems in the Space Environment, 929–933.

paper. If the distribution is indeed log-normal, a straight line should be obtained.

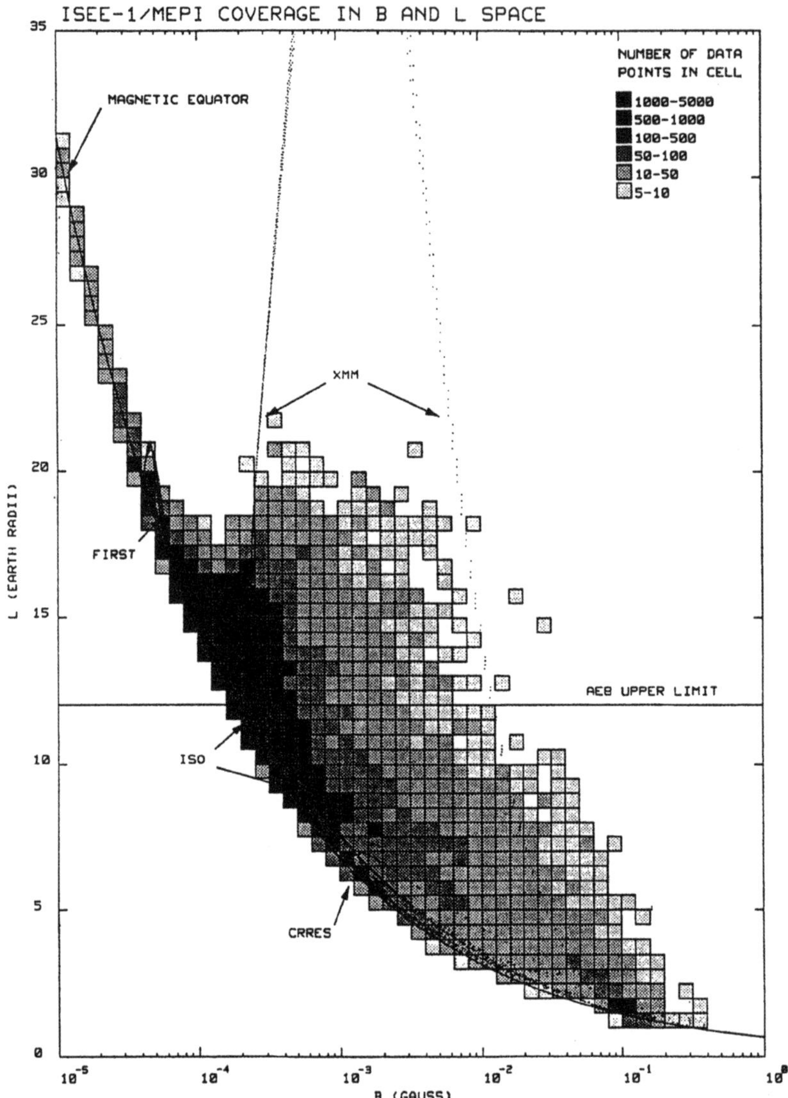

Fig. 1: ISEE-1/MEPI coverage in B and L space.

This method can be repeated for each cell covered by the ISEE-1 trajectory to provide the near earth electron flux distribution for each of the eight MEPI energy channels.

Figures 2 and 3 show an example of the cumulative probability flux distribution and electron fluxes as a function of geomagnetic coordinates, respectively, for the third MEPI energy channel. Direct comparison can be made with the AE8 model provided by NASA.

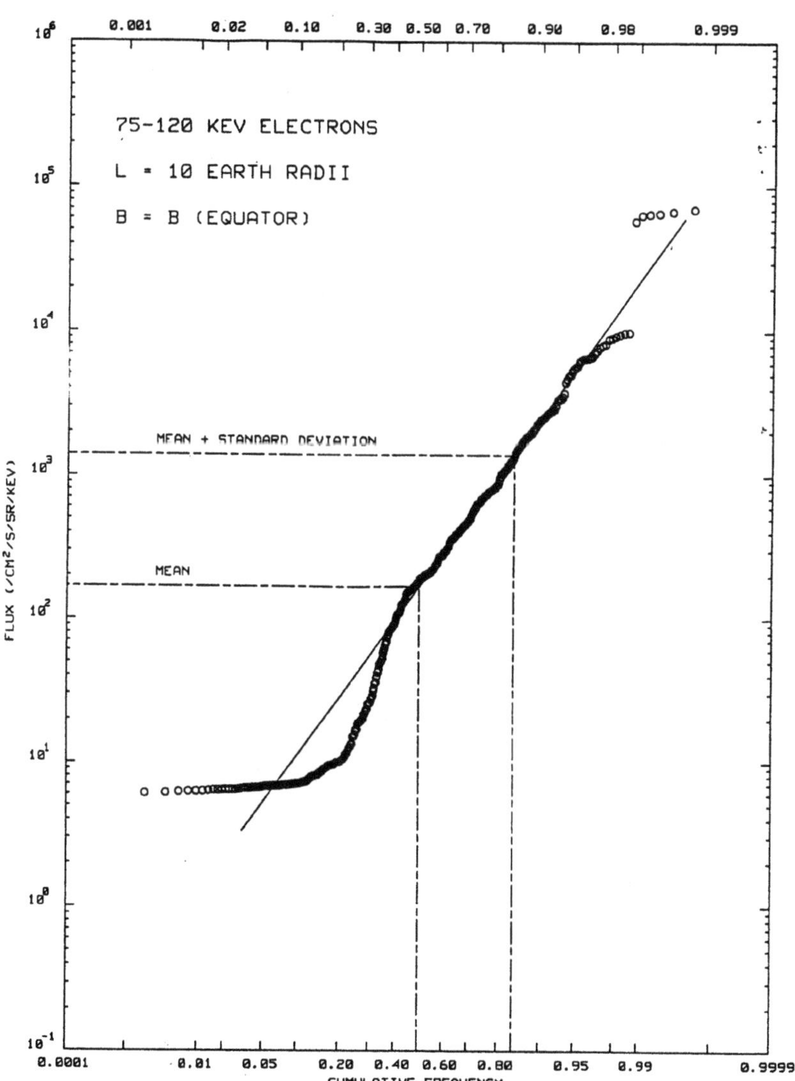

Fig. 2: ISEE-1/MEPI coverage in B and L space.

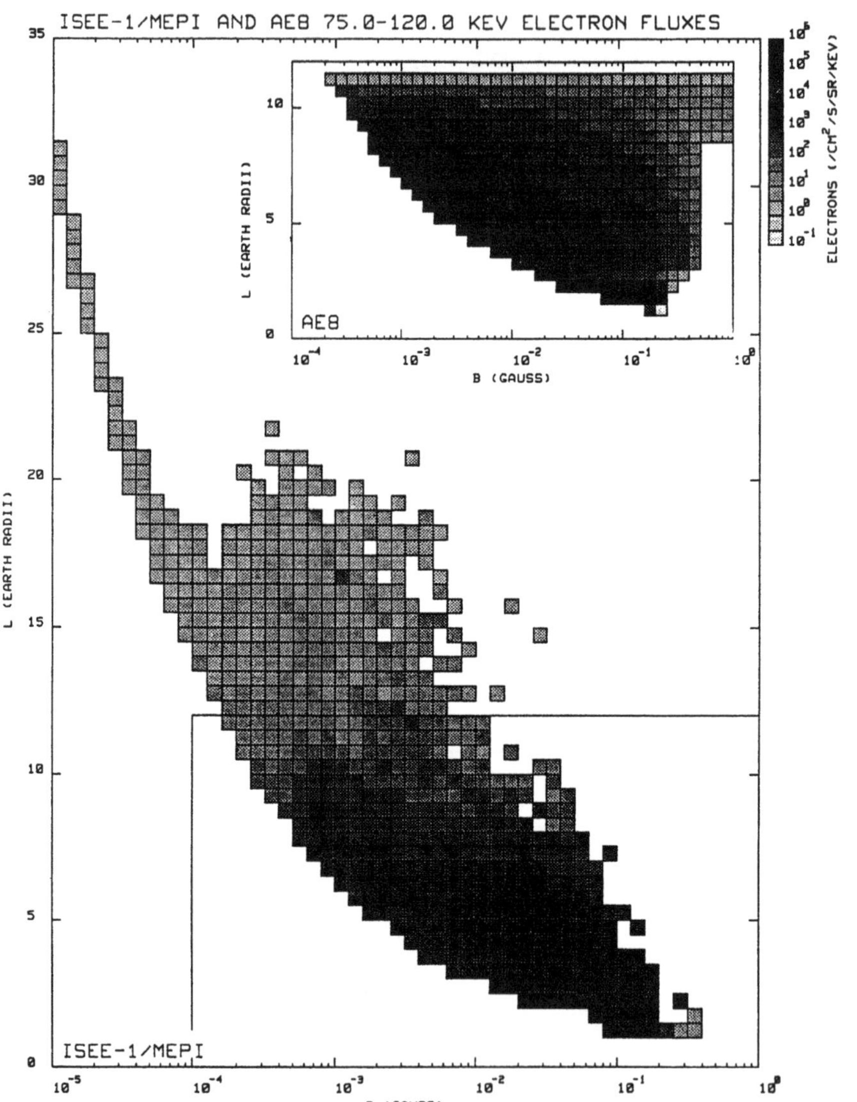

Fig. 3: ISEE-1/MEPI and AE8 75.0-120.0 keV electron fluxes.

A very large variation exists in electron fluxes as seen from the example of the distribution function. Electron fluxes are known to be very dynamic at high altitudes, but this feature is not reflected by the AE8 electron model, except for standard deviations which are provided for a subset of the AE8 database.

2. Acknowledgements

We thank D.J. Williams and C.T. Russell for providing us with ISEE-1 data.

LOW EARTH ORBIT SPACE PLASMA HIGH VOLTAGE SYSTEM INTERACTIONS

DALE C. FERGUSON
NASA Lewis Research Center
Cleveland, Ohio, USA 44135

ABSTRACT. Space power systems interact with the low Earth orbit (LEO) space plasma in two different ways. One way is the steady collection of current from the plasma onto exposed conductors. The relative currents collected by different parts of the system will determine the floating potential of the spacecraft. Also, steady state collected currents may lead to sputtering or heating of the system by ions or electrons, respectively. The second type of interaction is a short timescale arc into the space plasma, which may deplete the spacecraft of stored charge, damage surfaces, and produce EMI. Such arcs occur at high negative potentials relative to the space plasma potential, and depend on the steady state ion currents. New high voltage power systems incorporated into advanced spacecraft and space platforms may be endangered by these plasma interactions. Recent advances in laboratory testing and current collection modeling promise the capability of controlling, and perhaps even using, space plasma interactions to enable design of reliable high voltage space power systems.

1. Introduction

Previous space power systems have used low voltages, such as 28 V, and have had minimal interactions with the ionized plasma of the earth's upper atmosphere. With no exposed high voltages, such systems will come to an equilibrium close to the potential of the surrounding plasma. However, with the advent of large space power systems has come a desire for high efficiency power transmission. To power system designers, this implies high voltage systems, because the distribution losses go up as the square of the necessary current.

Modern designs for high power space applications typically use voltages of 160 to 200 V (eg. Space Station Freedom [SSF] and Advanced Photovoltaic Solar Array [APSA]). With such high distributed voltages, parts of the system must be at high potentials with respect to the ambient plasma. In the absence of a hard electrical ground in space, the power system will take on floating potentials such that electron currents collected by the more positive parts of the system will be balanced by the ions collected by the negative areas. In general, only connected electrical conductors must obey a "global" current balance condition. Insulators and

R. N. DeWitt et al. (eds.), *The Behavior of Systems in the Space Environment*, 935–942.

isolated conductors exposed to the plasma will locally balance electron and ion currents to their surfaces, resulting in a slightly negative surface potential.

Current balance requires that for unimpeded electron and ion current collection, the array will float with about 95% of its area negative and about 5% positive of the plasma potential. For a 160 V solar array hooked to an insulated structure, the most negative end of the array will be about 152 V negative of the plasma, and the positive end only 8 V positive. Sputtering and/or dielectric breakdown may result.

If parts of the solar array are forced to be at high positive potentials relative to the surrounding plasma, undesirable effects may occur. Above about 100 V positive, solar cells and arrays may collect anomalously high currents called "snapover", from the belief that they are caused by high surface potentials on conductors "snapping over", because of secondary electron effects, onto the surfaces of adjacent insulators. It may lead to high parasitic current power losses or localized heating on small exposed areas, which could lead to pyrolysis of Kapton adjacent to the conductor.

Ion current collection by conducting surfaces at high negative potentials is implicated in the second type of interactions - arcing to the space plasma. Arcs may discharge the entire electrically connected surface of the spacecraft or array, and are therefore potentially quite destructive. Solar array arcs into the plasma occur where conductors or semiconductors collecting ion current from the plasma are adjacent to insulators, such as coverslides or Kapton.

2. Steady State Current Collection

Figures 1 and 2 show the current collection behavior of a solar array in a plasma, with its conductors at a potential V relative to the plasma [Ref. 1]. Electron current collection is depressed for potentials less than about 100 V. This is because insulators surrounding the exposed conductors are slightly negative, to repel fast-moving electrons and allow the slower ions to balance current locally. These potentials, typically three to five times the plasma electron temperature, extend into space above the conductor, and may partially choke off the electron current. Above about 100-200 V, there is a transition to anomalously high electron current; the snapover phenomenon. Though there is disagreement about the mechanism of snapover [Ref. 2, 3], it may be due to secondary electron emission, where emitted electrons hop across the insulator until reaching the conductor surface.

By contrast, notice the extremely small ion collection currents. Ion collection currents are approximately linear with voltage up to where arcing occurs. Spacecraft speeds are much less than electron thermal speeds, so electrons are collected from all directions, at the thermal flux as modified by local potentials. However, the positive ions move much slower than the spacecraft, so their flux is the ram flux, modified by local potentials. Electron and ion current densities may be orders of magnitude lower in spacecraft wakes than in the undisturbed plasma.

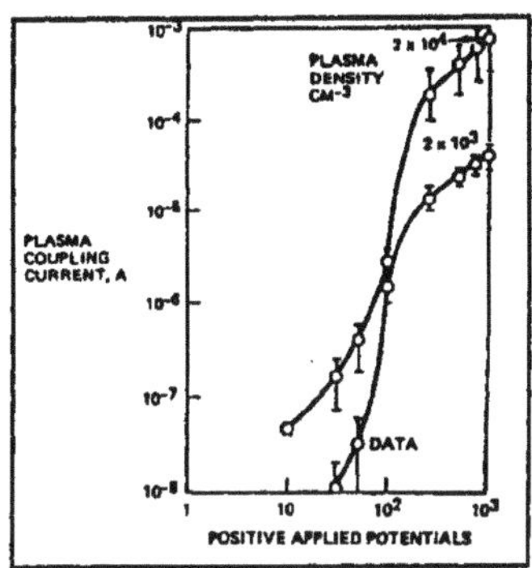

Fig. 1: Electron collection.

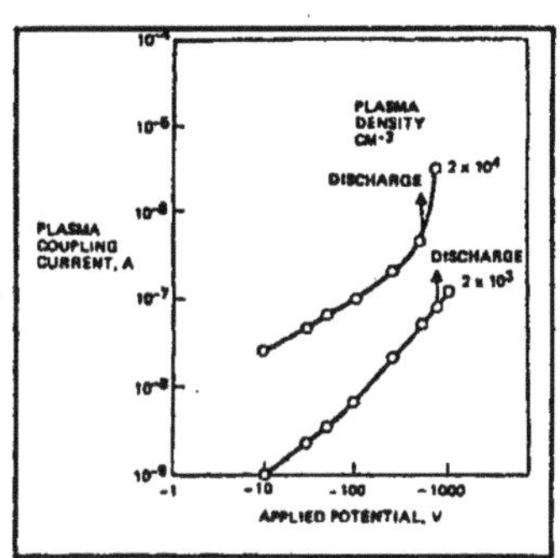

Fig. 2: Ion collection.

Simple models of array floating potentials yield about a 95% negative floating fraction. Thus it is reasonable, in the absence of very large ion collecting areas on the spacecraft, to assume that the array floats wholly negative; its most positive part at 0 volts relative to the plasma. This may lead to sputtering.

Sputtering is the physical removal of material from a surface by impact of incoming atoms or ions. Sputtered material may contaminate adjacent surfaces. Sputtering starts when exposed conductors are at the sputtering threshold below the plasma potential, typically between -10 and -30 volts. Yields are small for energies less than about 100 eV, so sputtering is only serious for negative potentials greater than this. Near holes in insulating coatings, sputtering ions will be focused to many times their undisturbed flux, exacerbating the problem, see Fig. 3 (courtesy of Joel Herr, Sverdrup Technology, Inc.). All previous space power systems have generated voltages less than about 100 volts, so sputtering was not important. However, for Space Station Freedom, sputtering may produce a loss of about 0.4 mils of material per year [Ref. 4]. Atomic oxygen protective coatings are usually much thinner than this, and if sputtered (ie near the edges of solar cells), their lifetime will be less than one year. Even 5 mil coatings may be lost during the lifetime of SSF. Sputtering problems are especially severe on rapidly switched components, because all of their insulating surfaces directly above conductors will usually be at very negative potentials, as the plasma ions cannot react quickly enough to neutralize the surfaces. Sputter coating may be a problem for solar cells, for their anti-reflective coatings may lose efficiency if covered with transparent material, or become opaque if coated with an opaque material.

Electron collection problems are important if the most negative end of the array becomes elevated to near the plasma potential. This may occur on negatively grounded arrays through thruster firings or effluent dumps, during arcs, or through purposeful increase of ion collection or decrease of electron current collection. During thruster firings on the negatively grounded SSF, the arrays may collect up to 10 amps of current [Ref. 4]. This will produce a 1.6 kW parasitic loss in the power system. More

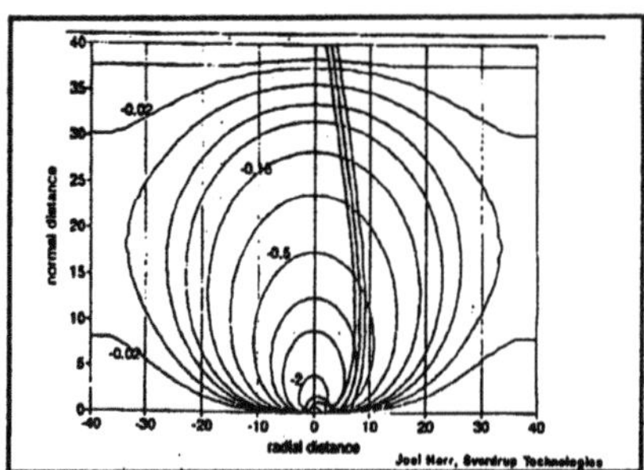

Fig. 3: Ion paths and equipotentials near an insulation pinhole.

importantly, temperature increases may occur in thin power system traces, leading to pyrolysis (charring) of Kapton or melting of copper or aluminum (as recently found by T. Morton, Sverdrup Technology, Inc.). This is only likely to occur under the following conditions:

1. The current-carrying trace is thin and covered with a poor heat conductor.
2. A hole large enough to prevent current chokeoff (about 60 mils, [Ref. 5]) but small enough to collect high snapover currents exists in the insulator covering.
3. The conductive trace is exposed to a high density LEO plasma in the ram direction.
4. The trace is above + 100 V with respect to the LEO plasma.
5. All above conditions hold for several seconds (perhaps 10 seconds).

Kapton pyrolysis occurred on a test panel-pair of SSF arrays in a vacuum chamber at + 450 volts [Ref. 6]. The charred area did not spread from the vicinity of the trace, but did significantly increase the effective electron current collecting area of the array until it was repaired with a Kapton patch. In space, holes for pyrolysis might be created by sputter or atomic oxygen enlargement of debris impact holes, or other array blanket defects.

It is possible to change the current-collection characteristics of solar cells and arrays through design practices. R. Chock [Ref. 5] has found by modeling SSF array current collection with 3-D computer codes (NASCAP/LEO) that the narrow spacing of the cells acts very well to choke off electron current collection in ground tests, and will do so somewhat under space conditions. Chock (Fig. 4) has shown that it may be possible to increase the overhangs of solar cell coverslides beyond the cell edges, and/or to decrease the gaps between solar cells, to produce an array which collects electrons no more efficiently than ions, and thereby to significantly influence the floating potential behavior of large space arrays. The manufacturing feasibility of these solutions is now being evaluated by major solar cell manufacturers. Caulking the gaps between cells on the most positive segments of solar arrays may be an alternative method of decreasing array electron collection and producing a more balanced distribution of array potentials.

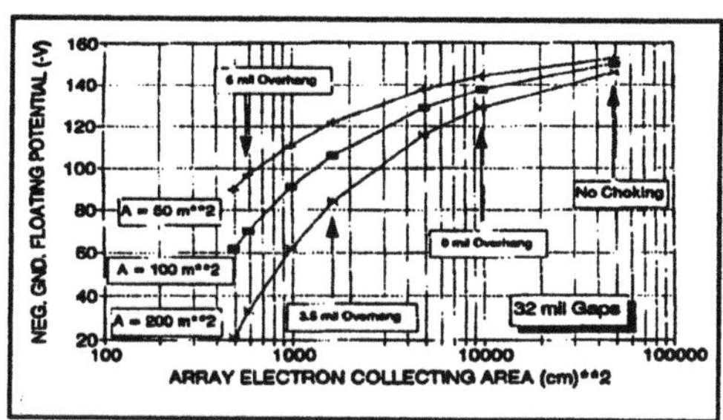

Fig. 4: SSF floating potentials with ion collecting area A vs electron collecting area.

3. Transient Events

Classical solar array arcing is well documented in both ground test and space flight conditions [Refs. 7, 8]. Figure 5 [Ref. 9] shows the voltage dependence of the sporadic arc rate for 2x2 and 2x4 cm standard silicon solar cells on the ground and in space. The same threshold seems to apply to all these data, about -230 V. SSF solar arrays arced into the plasma during tank tests at voltages of -205 V [Ref. 10]. It is not known whether this is the threshold voltage for them. Theories predict that the threshold voltage should depend on the conducting material exposed to the plasma, but the predictive ability of the existing theories is just now being explored [Ref. 11].

The arc rate for 2x2 standard cells depends linearly on the ion current collected and is a steep power-law of the voltage (at voltages above threshold). The arcs usually occur directly into the plasma, rather than to adjacent conductors. There seems to be no strong dependence of arc rate on number of possible arc sites. This may be due to a reset phenomenon occurring after each arc. In both ground and space testing, the arc rate has decreased to a constant level on a timescale of hours after immersion into the vacuum. Marinelli *et al* [Ref. 12] have found that this is most likely due to outgasing of adhesives, and a significant reduction in arc rate has been achieved by modifying solar cell coverslide adhesion and cleaning techniques. Increasing the solar cell coverslide overhangs may decrease the arc rate by decreasing local ion current collection. Chock [Ref. 5] has found that arc sites on the SSF array in ground tests preferentially occurred where coverslide overhangs were small. It is doubtful that such techniques will change the arcing threshold, however.

Arcs similar to classical solar cell arcs may occur on spacecraft surfaces with a dielectric coating of insufficient strength. Anodized aluminum surfaces have been seen in ground tests to arc into the plasma at potentials as small as -80 V. Large negative potentials on spacecraft may be the result of the electrical power grounding scheme, the voltage on the arrays, and the relative electron and ion current collection

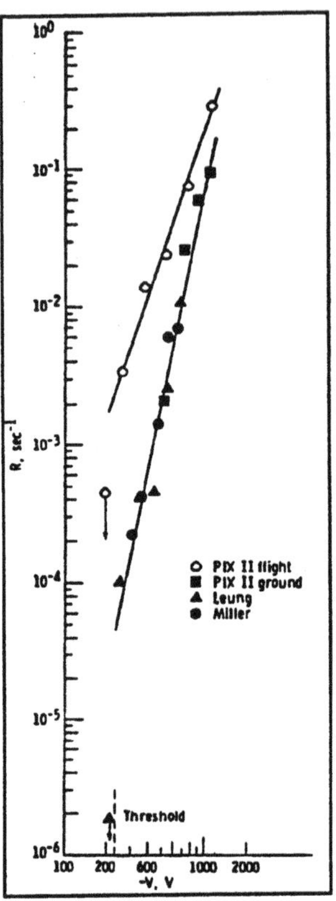

Fig. 5: Arc rate vs. voltage for LEO ram conditions.

characteristics of the solar arrays. They may therefore be controlled by changing the array coating potentials through coverslide and gap specifications, as well as by a proper grounding scheme and proper coatings. Arcs of all types seem to discharge the entire connected capacitance of the power system where they occur, and are therefore powerful current transfer events.

Figure 6 shows new laboratory results of arc strength versus connected capacitance in the system [Ref. 13]. For large capacitances, as on very large solar arrays or on large anodized spacecraft structure panels, peak arc currents may extend to thousands of amps. The limiting mechanism for peak arc currents has not yet been found. It is believed that large arcs produce a local plasma of such density that sufficient charge carriers exist for 1000 amp arcs. Large arcs may locally disrupt the surface, interrupt power for a short time, produce prompt contamination, and generate copious amounts of electromagnetic interference as shown in Figure 7 [Ref. 14]. It is desirable to limit the potential of spacecraft systems and arrays with respect to the plasma in order to prevent arcs, or to at least limit the amount of capacitance available to potential arc sites.

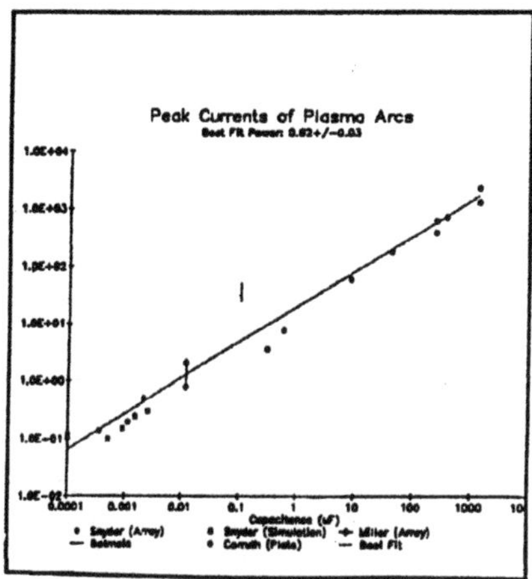

Fig. 6: Peak arc currents vs. connected capacitance.

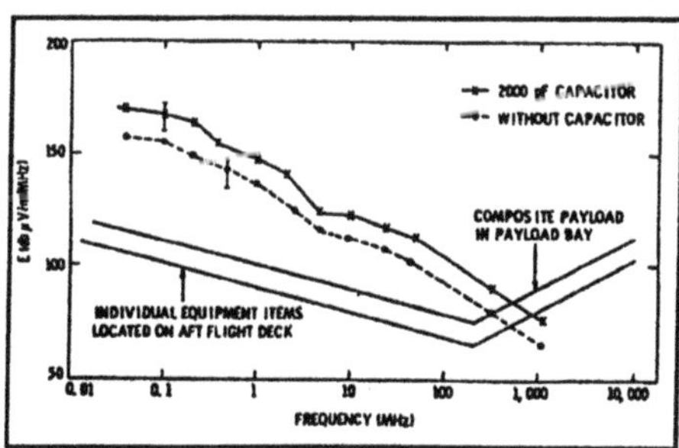

Fig. 7: EMI from solar array arcs.

4. References

[1] Stevens, N.J., and Stillwell, R.P., "Discharge Transient Coupling in Large Space Power Systems," in NASA CP-3059, p. 383 (1989).

[2] Gabriel, S.B., Garner, C.E., and Kitamura, S., "Experimental Measurements of the Plasma Sheath Around Pinhole Defects in a Simulated High-Voltage Solar Array," AIM Paper 83-0311 (1983).

[3] Thiemann, H., and Schunk, R.W., "Particle in Cell Simulations of Sheath Formation Around Biased Interconnectors in a Low-Earth-Orbit Plasma," J. Spacecraft and Rockets, 27, p. 554 (1990).

[4] Ferguson, D.C., Snyder, D.B., and Carruth, R., Final Report of the Joint Workshop on Evaluation of Impacts of Space Station Freedom Grounding Configurations, NASA LeRC, August 21, 1990 (1990).

[5] Chock, R. To be published (1991).

[6] Felder, M.C., "PV Plasma Interaction Test: Preliminary Results and Status," Final Report of the Joint Workshop on Evaluation of Impacts of Space Station Freedom Grounding Configurations, Ferguson, Snyder, and Carruth, eds., NASA LeRC, August 21, 1990 (1990).

[7] Snyder, D.B., "Characteristics of Arc Currents on a Negatively Biased Solar Cell Array in a Plasma," NASA TM-86887 (1984).

[8] Grier, N.T., "Plasma Interaction Experiment II: Laboratory and Flight Results," in NASA CP-2359 p 333, 1985 (1983).

[9] Ferguson, D.C., "The Voltage Threshold for Arcing for Solar Cells in LEO - Flight and Ground Test Results," NASA TM-87259 (1986).

[10] Nahra, H.K., Felder, M.C., Sater, B.L., and Staskus, J.V., "The Space Station Photovoltaic Panels Plasma Interaction Test Program: Test Plan and Results," NASA TM-102474 (1990).

[11] Hastings, D.E., Weyl, G., and Kaufman, D., "Threshold Voltage for Arcing on Negatively Biased Solar Arrays," J. Spacecraft and Rockets, 27, p. 539 (1990).

[12] Marinelli, W.J., Green, B.D. Upschulte, B.L., Weyl, G., Hastings, D., and Aifer, E., "Solar Cell Arcing: The Role of Outgassing and Contamination," proc. of the SOAR91 conf., Houston, Texas, July 9-11, 1991 (1991).

[13] Snyder, D.B. To be published (1991).

[14] Leung, P., "Characterization of EMI Generated by the Discharge of a 'VOLT' Solar Array Final Report," JPL Document D-2644, September, 1985 (1985).

[15] Jongeward, G.A., Kuharski, R.A., Kennedy, E., Wilcox, K.G., Stevens, N.J., Putnam, R.M., and Roche, J.C. (1990). "The Environment Power System Analysis Tool Development Program," in NASA CP-3089, p. 352.

[16] Mandell, M.J., Katz, I., Davis, V.A., and Kuharski, R.A. (1990). "NASCAP/LEO Calculations of Current Collection," in NASA CP-3089, p. 334.

PHOTOTHERMAL INVESTIGATION OF GOLD COATED KAPTON SAMPLES EXPOSED TO ATOMIC OXYGEN

A.W. WILLIAMS and N.J. WOOD
Department of Physics
University College of Swansea
University of Wales
Singleton Park
Swansea SA2 8PP UK

Abstract. The damage caused by atomic oxygen (ATOX) to coated materials has for the first time, been investigated using photothermal scanning techniques. Samples of gold coated polyimide Kapton were exposed to atomic oxygen fluences of 3 x 10^{19} and 3 x 10^{20} cm^{-2} in laboratory sources and subsequently scanned photothermally. The photothermal signal and phase correlated with optically visible damage and also showed invisible subsurface damage around defects.

1. Introduction

A major problem encountered by Low Earth Orbit (LEO) vehicles is bombardment by atomic oxygen. Materials such as Kapton when subjected to fluxes of 10^{15} atoms/cm/sec with energies of ≈ 5eV, are found to have erosion yields of 2.5 x 10^{-24} cm^3/atom which could mean the total loss of material during a 10 year flight. Such materials may be protected with a gold coating which has an insignificant erosion rate. The efficiency of such a coating may then depend on the initial defects present in the coating and on the subsequent handling of the coated material. There is strong evidence that atomic oxygen will attack polyimide substrates through defects in gold coatings and cause undercutting and subsurface damage [Refs. 1, 2].

Photothermal nondestructive evaluation (NDE) and characterization of solids is a growing and powerful technique that has been applied to numerous problems involving surface and subsurface cracks, inclusions and delaminations [Ref. 3]. The present paper describes the photothermal subsurface scanning of gold coated Kapton subjected to atomic oxygen bombardment.

R. N. DeWitt et al. (eds.), The Behavior of Systems in the Space Environment, 943–948.

2. Apparatus and Experimental Method

The apparatus consisted of an argon ion laser, optical modulator, piezoelectric detector element or photoacoustic cell, a dual phase lock-in amplifier, and X-Y translation stages. A computer controlled the translation stage driver, the lock-in amplifier and data acquisition system.

The principle of photothermal depth profiling and scanning is that an intensity modulated laser beam is optically absorbed producing a thermal wave in the sample. This thermal wave causes a layer of gas above the sample to be periodically heated giving an acoustic wave which is detected by a microphone. Alternatively the thermal waves may be detected by a piezoelectric detector coupled to the rear of the sample. The probe depth, μ, may be varied by changing the modulation frequency f, as μ is given by $(\alpha/\pi f)^{1/2}$, where α is the thermal diffusivity of the sample. Thus subsurface layers, inclusions or delaminations may be detected due to the scattering of the thermal waves, which modify both the photothermal signal and its phase lag. By moving the sample beneath the laser beam an X-Y raster scan at a particular depth within the sample can be achieved.

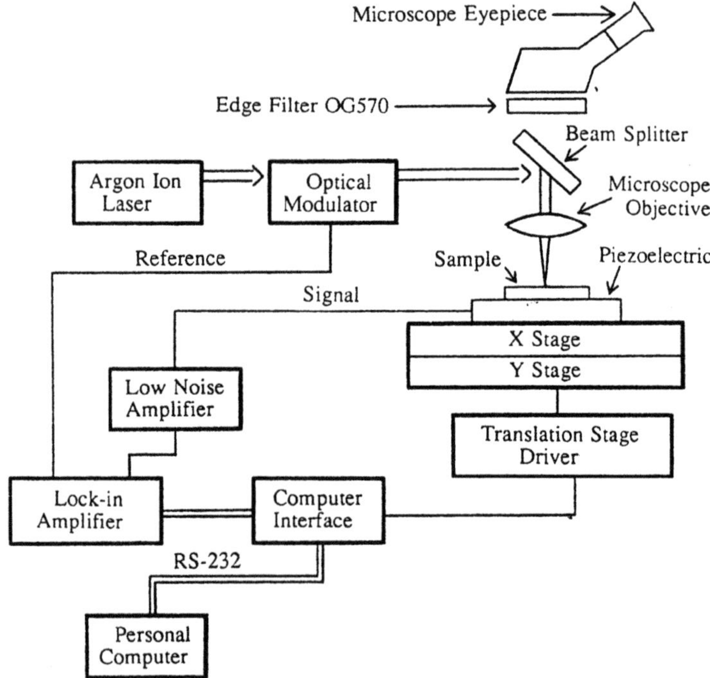

Fig. 1: Block diagram of the experimental apparatus.

For scans probed from the gold side of the sample a microscope objective was used to focus the light onto its surface, and the eye pieces, together with a suitable optical edge filter, enabled visible defects to be monitored as the scan progressed. Photothermal signal detection was achieved by bonding the Kapton side of the sample to a piezoelectric element. Scans from the Kapton side of the sample were performed using a photoacoustic cell, and the light was focused by means of a lens.

3. Sample Treatment

A 50μm Kapton sample coated with a 100nm of gold was exposed to an atomic oxygen fluence of 3 x 10^{19} atoms/cm^2 in a laboratory plasma asher. This sample showed a mass loss of 2.95mg from 25.50mg to 22.55mg. A second 12μm Kapton sample coated with 50nm of gold was exposed to an atomic oxygen fluence of 3 x 10^{20} atoms/cm^2. For both samples the gold was exposed to the above oxygen flux.

4. Results

Photothermal line scans were performed on the thinner 12μm Kapton sample. The sample was placed gold side down in a photoacoustic cell and the Kapton coated with a very thin carbon layer to act as an absorbing medium for the light. A Kapton thickness of 12μm corresponded to a modulation frequency of 220Hz, since the thermal diffusivity of Kapton (α_{kapton}) is 1 x 10^{-3} cm^2/s. Figures 2 and 3 show the phase lag of the photoacoustic signal over a distance of 3mm across the sample, at modulation frequencies ranging from 218Hz to 380Hz. Figure 2 shows a phase scan for an unexposed sample, while Fig. 3 shows a phase scan in which part of the sample has been exposed to ATOX, and part has been masked. Damage to the exposed part of the sample is clearly detectable.

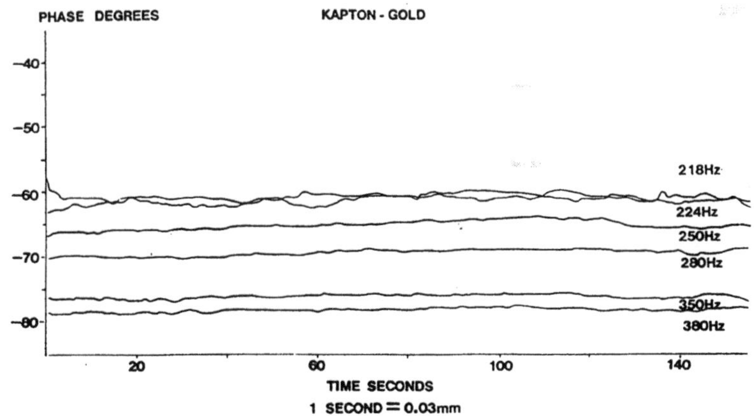

Fig. 2: Phase scan of an unexposed sample.

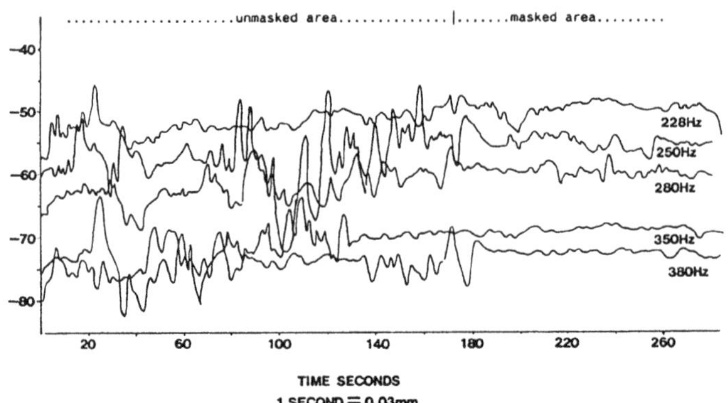

Fig. 3: Phase scan of an sample with part of the sample exposed to ATOX.

The 50μm Kapton sample was bonded with its gold side up to a piezoelectric detector element. Before scanning the sample was photographed, and a visible area of damage was selected, shown in Fig. 4. An X-Y photothermal scan was performed of this region. A higher modulation frequency, 1.8kHz, was chosen corresponding to a probe depth of $\approx$ 4μm into the Kapton. Thermal wave images of the area were produced comprising of 100 x 100 points at 10μm intervals across the surface, plotted for both the signal magnitude and the phase lag, over a total scan area of 1mm^2. A tightly focused laser spot $\approx$ 10μm was produced by using a microscope objective, and the power of the laser was kept down to prevent damage to the sample.

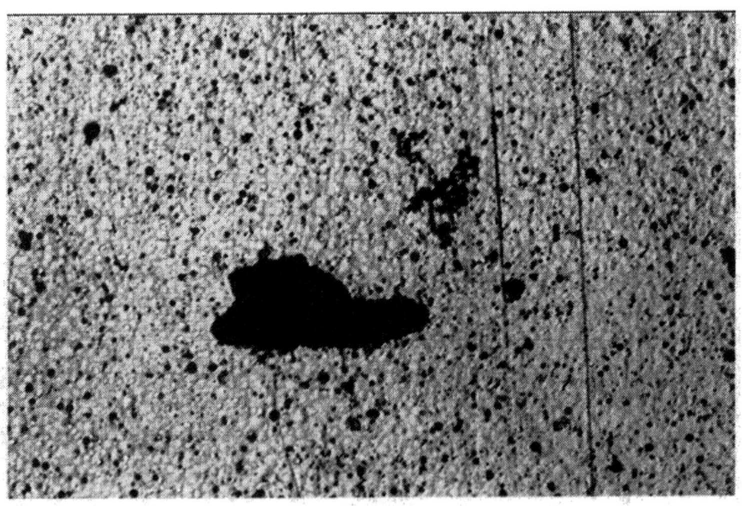

Fig. 4: Optical photograph of sample damage.

Figure 5 shows the photothermal signal magnitude response across the $1mm^2$ area. This response correlates with the optical image of the sample, since the magnitude of the thermal wave is dependent upon the optical absorption of the material. An increase in the signal is seen in the area where the gold layer is missing totally, due to an increase in the optical absorption of this area. At the edge of this damaged area the signal decreases. This coincides with the areas in which the gold is being undercut by ATOX damage. Thermal waves generated in the gold layer are attenuated by the additional air layer, preventing their transmission into the Kapton substrate and hence reducing the detected photothermal signal. A second region of lower signal can be seen in the top right of the area, which also implies delamination.

Figure 6 shows the phase lag of the photothermal signal across the sample. The variations in the phase lag are caused by differences in thermal wave velocities at the different points on the sample surface. Delamination causes an increase in this lag and hence predicts large scale damage.

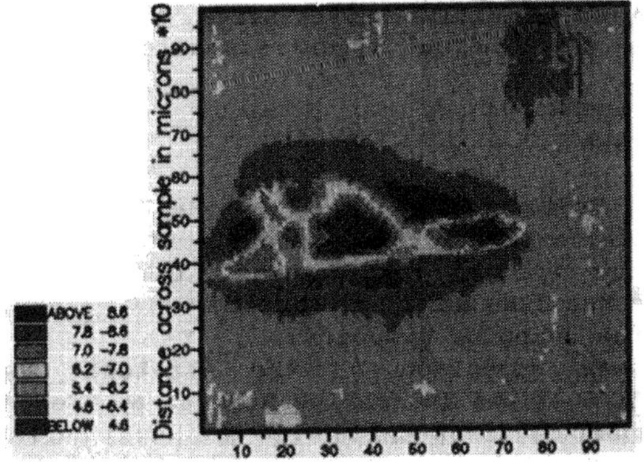

Fig. 5: Photothermal signal scan of damage in millivolts at 1.8 kHz

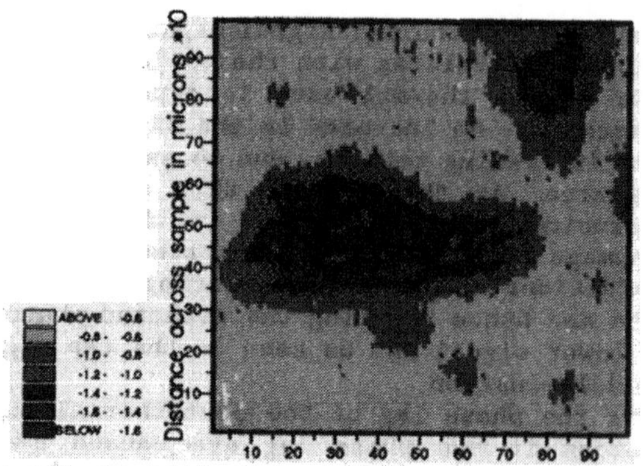

Fig. 6: Photothermal phase scan of damage in degrees at 1.8 kHz

5. Conclusion

Photothermal NDE methods have been used to characterize the damage caused to gold coated Kapton by atomic oxygen bombardment. Comparison of the photothermal scans with the optical images, indicate regions of delamination undetectable optically, as well as the undercutting of the gold by atomic oxygen at initial defect sites. The phase lag scans are more sensitive than the photothermal signal scans to delamination sites.

6. Acknowledgements

The authors thank Prof. Charles Neely and Prof. Bruce Tatarchuk, Auburn University, USA, for exposing the samples to atomic oxygen, Prof. C. Grey Morgan, University College of Swansea, U.K., for his advice and encouragement and Mr. C.A. Hall, University College of Swansea, U.K., for assistance with the computer graphics.

The work was supported by the Strategic Defense Initiative Organizations Office of the Innovative Science and Technology (SDIO/TNI), under contract number 60921-86-CA226 with the Naval Surface Warfare Center.

7. References

[1] 4th International Symposium on Spacecraft Materials in Space Environment, Toulouse, France, September 1988. ESA.

[2] Banks B.A., Rutledge S.K., Paulsen P.E., and Steuber T.J.,(1989) "Simulation of Low Earth Orbit Atomic Oxygen Interaction With Materials by Means of an Oxygen Ion Beam", NASA Technical Memorandum 101971.

[3] *Photoacoustic and Photothermal Phenomena*, Eds., P. Hess and J. Pelzl, Proceedings of the 5th International Topical Meeting, Heidelberg, Germany July 27-30, 1987.

APPENDIX B: NATO ASI ORGANIZING COMMITTEE

Dr. Robert N. DeWitt
U.S. Department of Energy
Defense Programs
Washington, DC 20585 U.S.A.
Tel: 301-903-3311
FAX: 301-903-5604

Dr. Dwight Duston
Director, Innovative Science & Technology Office
Strategic Defense Initiative Organization
Washington, DC 20301-7100 U.S.A
Tel: 703-693-1673
FAX: 703-693-1695

Dr. Anthony K. Hyder
Associate Vice President for Research
Professor of Aerospace Engineering
312 Main Building
University of Notre Dame
Notre Dame, IN 46556-5602
U.S.A.
Tel: 219-631-8591
FAX: 219-631-6630

Dr. Ranty H. Liang
Head, Materials Sciences Group
Applied Sciences and Microgravity Experiments
Jet Propulsion Laboratory
California Institute of Technology
4800 Oak Grove Dr.
Pasadena, CA 91109 U.S.A.
Tel: 818-354-6314
FAX: 818-393-6869

Dr. Jean Claude Mandeville
Department de Technologie Spatiale
CERT-ONERA
BP 4025
31055 Toulouse CEDEX France
Tel: 33-61-55-7160
FAX: 33-61-55-7172

Professor J.A.M. McDonnell
Head, Unit for Space Sciences
University of Kent
Canterbury, United Kingdom
CT27NR
Tel: 44-227-764000 EXT. 3788
FAX: 44-227-762-616

Dr. Carolyn K. Purvis
NASA/Lewis Research Center
21000 Brookpark Rd.
Cleveland, OH 44135 U.S.A.
Tel: 216-433-2307

Professor R. C. Tennyson
Institute for Aerospace Studies
University of Toronto
4925 Dufferin Street
Downs Views, Ontario CD M3H 5T6
Canada
Tel: 416-667-7710
FAX: 416-667-7799

Dr. Gordon L. Wrenn
Deputy Leader,
Spacecraft Environmental & Protective Section
Space Department
Royal Aerospace Establishment
Farnborough, Hampshire GU14 6TD
United Kingdom
Tel: 44-252-24461 EXT. 2258
FAX: 44-252-377121

APPENDIX C: NATO ASI LECTURERS

Dr. Joseph Angelo
Science Applications
International Corp.
700 South Babcock Street Ste. 300
Melbourne, FL 32901 U.S.A.

Dr. Raymond Askew
Space Power Institute
231 Leach Center
Auburn University, AL 36849
U.S.A.

Dr. Jacques Borde
Matra Espace--31 Rue des
Cosmonautes
F-31077 Toulouse Cedex
France

Dr. Jacques Bourrieau
Department de Technologie
Spatiale
CERT-ONERA
BP 4025
31055 Toulouse Cedex France

Prof. Stanley W.H. Cowley
Blackett Laboratory
Imperial College
London SW7 2BZ
United Kingdom

Dr. Eamonn J. Daly
European Space Agency, ESTEC/WMA
Postbus 299, 2200 AG
Noordwijk, The Netherlands

Dr. Vicente Domingo
European Space Agency, ESA/ESTEC
Space Science Department
Postbus 299, 2200 AG
Noordwijk, The Netherlands

Dr. Clive S. Dyer
Ministry of Defence (PE)
Space Systems Division
Q134 Building
Royal Aerospace Establishment
Farnborough, Hants, GU14 6TD
United Kingdom

Dr. Jean-Pierre Estienne
Matra Espace
31 Rue des Cosmonautes
F-31077 Toulouse Cedex France

Dr. Herbert Friedman
Naval Research Laboratory
Code 4190
4555 Overlook Avenue S.W.
Washington, DC 20375-5000
U.S.A.

Dr. M. Susan Gussenhoven-Shea
Phillips Laboratory, PL/PHP
Hanscom AFB, MA 01731 U.S.A.

Dr. Daniel Hastings
Dept. of Aeronautics &
Astronautics, 37-451
Massachusetts Institute of
Technology
Cambridge, MA 02139 U.S.A.

Dr. H. Gordon James
Communications Research Centre
P.O. Box 11490 Station H
Ottawa, Ontario K2H 8S2
Canada

Dr. Ira Katz
S-Cubed Division
Maxwell Laboratories, Inc.
P.O. Box 1620
La Jolla, CA 92038-1620 U.S.A.

Dr. Franz R. Krueger
Max Planck Institute fur Kernphysic
Post Fack 103980 D-6900
Heidelberg-1, Germany

Dr. Erich E. Kunhardt
Weber Research Institute
Polytechnic University
Route 110
Farmingdale, NY 11735 U.S.A.

Prof. Rodney V. Latham
Dept. Electrical Engineering & Applied Physics
Aston University
Aston Triangle, Birmingham B4 7ET
United Kingdom

Dr. Titus Mathews
Associate Vice-President (Academic)
The University of Calgary
Calgary, Alberta T2N 1N4
Canada

Prof. Ralph W. Nicholls
Centre for Research in Earth & Space Science
York University
4700 Keele Street
North York, Ontario
Canada

Dr. Alain Paillous
Department de Technologie Spatiale
CERT-ONERA
BP 4025
31055 Toulouse Cedex France

Dr. Manola Romero
Department de Technologie Spatiale
CERT-ONERA
BP 4025
31055 Toulouse Cedex France

Dr. J. Stephen
35 Appleford Drive
Abingdon OX142BZ
United Kingdom

Dr. Timothy J. Stevenson
Unit for Space Sciences
University of Kent at Canterbury
Canterbury CT2 7NR
United Kingdom

Dr. Janis H. Stoklosa
Space Human Factors Program, Code SBM
NASA Headquarters
600 Maryland Ave. S.W.
Washington, DC 20546 U.S.A.

APPENDIX D: NATO ASI ATTENDEES

Russ Alexander
Canadian Space Agency
1254 Major St.
Ottawa
Ontario, Canada K2C 2S2

Alfred Anderman
Flight Systems Design & Performance Group
Rockwell Space Systems Div.
MS/FB96, 12214 Lakewood Blvd.
Downey, CA 90241 U.S.A.

John Antoniades
Naval Research Lab
NRL Code 4751
Washington, DC 20375-5000
U.S.A.

Shahid Aslam
Oxford University
Atmospheric, Oceanic, and Planetary Physics
Clarendon Laboratory
Parks Road, Oxford OX1 3PU
United Kingdom

Brian Atkinson
Royal Aerospace Est.
Ministry of Defence
Farnborough
Hampshire, GU14 6TD
United Kingdom

Dale Atkinson
POD Associates, Inc.
2309 Renard Pl. S.E.
Suite 201
Albuquerque, NM 87106 U.S.A.

Virginia Ayres
Naval Surface Warfare Ctr.
Code R42
Silver Spring, MD 20903 U.S.A.

Richard Lee Balthazor
Univ. of Birmingham
Dept. of Physics and Space Research
Birmingham B15 2TT
United Kingdom

H.M. Banford
Scottish Univ. Research Ctr.
Birmiehill, East Kilbride
Gloscow, G75 OQU, Scotland

Bruce A. Banks
NASA-Lewis Research Ctr.
Mailstop 302-1
21000 Brookpark Road
Cleveland, OH 44138 U.S.A.

J. Douglas Beason
Office of Science and Technology Policy
Executive Office of the President
Washington, DC 20500 U.S.A.

John S. Bennion
Univ. of Utah
Nuclear Engineering Lab
1205 MEB
Salt Lake City, UT 84112 U.S.A.

Binay Bilgin
Univ. of Istanbul
Chemistry Division
Muhendislik Fakultesi
Kimya Bolumu, Avcilar
Istanbul, Turkey

Bernard Borg
Aerospatiale Society
Aircraft Division
Hermes Operational Group
B.P. n B 0411
31060 Toulouse Cedex
France

Jean-Luc Bouguereau
Alcatel Espace
26, avenue J-F Champollion
BP 1187-31037
Toulouse Cedex France

David E. Brinza
Jet Propulsion Laboratory
California Insitute of Tech.
Mail Stop 67-201
4800 Oak Grove Drive
Pasadena, CA 91109 U.S.A.

Carmine A. Carosella
Naval Research Lab
Code 4671
Washington, DC 20375 U.S.A.

Constantinos Cartalis
Univ. of Athens
6 Schina St.
Athens 11473 Greece

Ian D. Chalmers
Univ. of Strathclyde
204 George Street
Glasgow G11XW Scotland

Michel Collin
Commissariat a L'Energie
Atomique--Service PTN
BP 12, 91680
Bruyeres-le-Chatel,
France

David L. Cooke
Phillips Lab/WSSI
Hanscom AFB, MA 01731 U.S.A.

Gary R. Coulter
NASA, Life Sciences Division
Code SBR
600 Maryland Ave, SW; #600
Washington, DC 20024 U.S.A.

Sunil Deshpande
Space Sciences Unit
Univ. of Kent at Canterbury
Canterbury, Kent, CT2 7NR
United Kingdom

M. J. Duck
U.K. Atomic Energy Authority
Bldg. 540.1
Harwell Laboratory, Didcot,
Oxfordshire, OX11 ORA
United Kingdom

Dale C. Ferguson
NASA-Lewis Research Center
MS 302-1
21000 Brookpark Rd.
Cleveland, OH 44135 U.S.A.

Thornton R. Fisher
Lockheed Missiles and Space
Dept. 91-21 B/255
3251 Hanover St.
Palo Alto, CA 94304 U.S.A.

Dennis J. Flood
NASA-Lewis Research Center
MS 302-1 21000 Brookpark Rd.
Cleveland, OH 44135 U.S.A.

R. A. Fouracre
Univ. of Strathclyde
Dept. of Electronic &
Electrical Engineering
204 George Street
Glasgow G11XW, United Kingdom

Stephen B. Gabriel
Univ. of Southampton
Dept. of Aeronautics &
Astronomics
Highfield
Southampton S09 5NH
United Kingdom

Martin J. Given
Univ. of Strathclyde
Dept. of Electronic & Electrical
Engineering
204 George Street
Glasgow G11XW, United Kingdom

Kirk E. Hackett
EOARD
Box 14
FPO, NY 09510-0200 U.S.A.

M.R. Harris
Defence Research Agency
St. Andrews Road
Malvern, Worcs
WR143PS
United Kingdom

Christopher Hepplewhite
Oxford University
Dept. of Physics
Atmospheric Physics
Clarendon Lab, Parks Road
Oxford, OX1 3PU
United Kingdom

Richard Holdaway
Science & Engineering
Research Council
Rutherford Appleton Lab
Chilton, Didcot
Oxfordshire, United Kingdom

Yen Hsu
Space Program Planning Committee
Executive Yuan
106 Ho-Ping E. Rd.
Taipei, Taiwan 10636 R.O.C.

Calvin R. Johnson
Space Power Institute
Auburn University
231 Leach Center
Auburn University, AL 36849
U.S.A.

Stephen F. Knowlton
Auburn University
Dept. of Physics
206 Allison Laboratory
Auburn University, AL 36849
U.S.A.

Javaid R. Laghari
St. Univ. of New York/Buffalo
Dept. of Elec. & Comp. Engr.
316 Bonner Hall
Buffalo, NY 14260 U.S.A.

Philippe Mairet
Aerospatiale Society
Aircraft Division
Hermes Operational Group
B.P. n B 0411
31060 Toulouse Cedex
France

Diane J. Martin
3134 Babashaw Court
Fairfax, VA 22031-2001 U.S.A.

Darren McKnight
Kaman Sciences Corporation
2560 Huntington Ave, # 200
Alexandria, VA 22303 U.S.A.

William D. Morison
Univ. of Toronto Inst.-Aerospace
4925 Dufferin St., Downsview
Ontario, Canada M3H5T6

Steven Mullen
Space Sciences Unit
Univ. of Kent at Canterbury
Physics Laboratory
Kent CT2 7NR, United Kingdom

Michael J. Neish
Space Sciences Unit
Univ. of Kent at Canterbury
Canterbury, Kent CT2 7NR
United Kingdom

Annita Nerses
Polytechnic University
RTE 110
Farmingdale, NY 11735 U.S.A.

Norbert Nikolaizig
ESA/ESTEC
Kepler Laan 1
2200 AG Noordwijk
The Netherlands

Tugrul Ozel
Dokuz Eylul University
Department of Mech. Engin.
Faculty of Eng. & Architecture
35101 Bornova
Izmir, Turkey

J. D. Perez
Auburn University
Physics Dept.
206 Allison lab
Auburn University, AL 36849
U.S.A.

Alan D.R. Phelps
Strathclyde University
Dept. of Physics and Applied
Physics, John Anderson Bldg.
Glasgow G4 ONG SCOTLAND
United Kingdom

William A. Ranken
Los Alamos National Lab N-DP
PO Box 1663, MS-E552
Los Alamos, NM 87545 U.S.A.

Arto Salminen
Technical Research Centre of
Finland, Telecommunications
Otakaari 7 B
02150 Espoo
Finland

Ralph F. Schneider
Naval Surface Warfare Center
Code R42
10910 New Hampshire Ave.
Silver Spring, MD 20903-5000
U.S.A.

Prakash C. Sharma
Tuskegee University
Physics Department
Tuskegee University
Tuskegee, AL 36088 U.S.A.

Anthony R. Skinner
Ministry of Defence
Atomic Weapons Establishment
Building SB.43
Aldermaston
Reading, Berkshire RG7 4PR
United Kingdom

Gerhard Stasek
Physikalisch--Technische
Studien GMBH
Leinenweberstrasse 16
7800 Freiburg
Germany

Charles Stein
Phillips Laboratory
AFSC
Kirtland AFB, NM 87117 U.S.A.

Kenneth Sullivan
Space Sciences Unit
Univ. of Kent at Canterbury
Physics Laboratory
Kent CT2 7NR, United Kingdom

Anne C. Syljuasen
Polytechnic University
Weber Research Institute
Rte. 110
Farmingdale, NY 11735 U.S.A.

Cecil Tranquille
ESA/ESTEC
ESTEC (WMA) Postbus 299
2200 AG Noordwijk
The Netherlands

Yonhua Tzeng
Auburn University
200 Brown Hall
Dept. of Electrical Eng.
Auburn University, AL 36849
U.S.A.

John S. Wagner
Sandia National Laboratory
Plasma Theory Div. 1241
Albuquerque, NM 87185-5800
U.S.A.

W.S. Walters
Harwell Lab-UK Atomic Energy
Bldg. 540-1
Harwell Laboratory, Didcot
Oxfordshire OX11 ORA
United Kingdom

R.J. Will, MD
9 Elmhurst Place
Cincinnati, OH 45208 U.S.A.

Aled W. Williams
Dept. of Physics
Univ. College of Swansea
Singleton Park
Swansea SA2 8PP, United Kingdom

William E. Wolfe
Ohio State University
Department of Civil Engineering
470 Hitchcock Hall
2070 Neil Avenue
Columbus, OH 43210 U.S.A.

Gulgun Yalcinkaya
Tech. Univ. of Istanbul
Makina Fakultesi Gumussuyu
Gumussuyu 80191
Istanbul, Turkey

Hang Hong Yang
Industrial Technology Research Institute
195, Chung Hsing Rd.
Chutung, 31015 Hsinchu Taiwan
R.O.C.

INDEX

www.ingramcontent.com/pod-product-compliance
Ingram Content Group UK Ltd.
Pitfield, Milton Keynes, MK11 3LW, UK
UKHW021904190726
13853UKWH00003B/1398

* 9 7 8 9 4 0 1 1 2 0 4 9 4 *